PERIMETER AND AREA FORMULAS

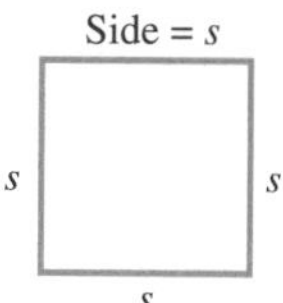

Square
$P = 4s$
$A = s^2$

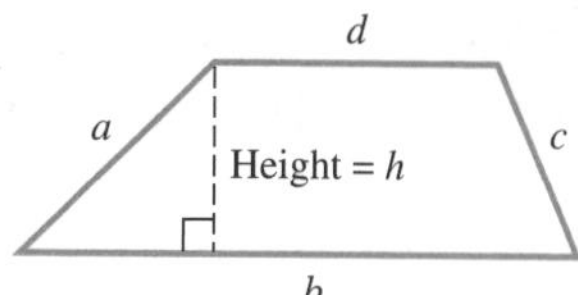

Trapezoid
$P = a + b + c + d$
$A = \frac{1}{2}h(b + d)$

Rectangle
$P = 2l + 2w$
$A = lw$

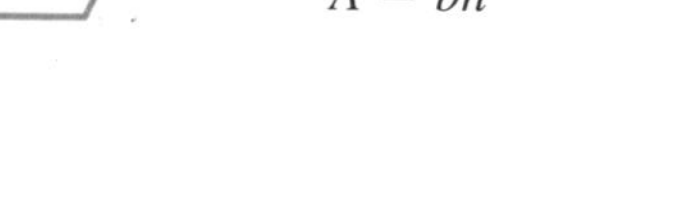

Parallelogram
$P = a + b + c + d$
$A = bh$

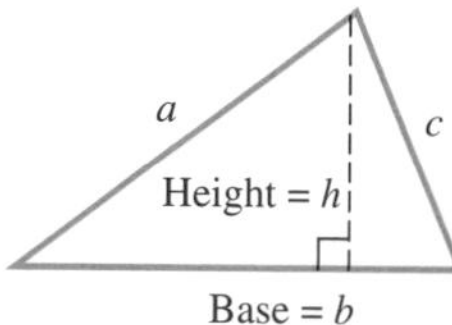

Triangle
$P = a + b + c$
$A = \frac{1}{2}bh$

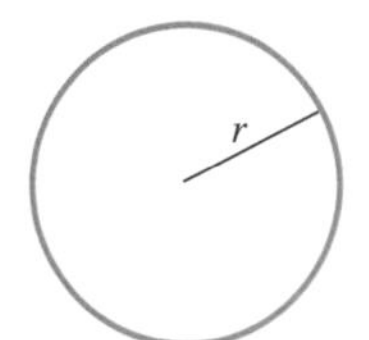

Circle
$C = 2\pi r$ or $C = \pi D$
where $\pi = 3.14$
$A = \pi r^2$

VOLUME FORMULAS

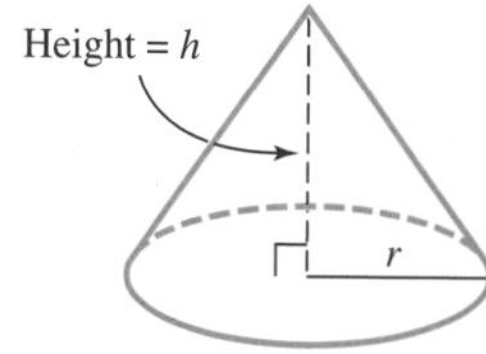

Cone
$V = \frac{1}{3}\pi r^2 h$

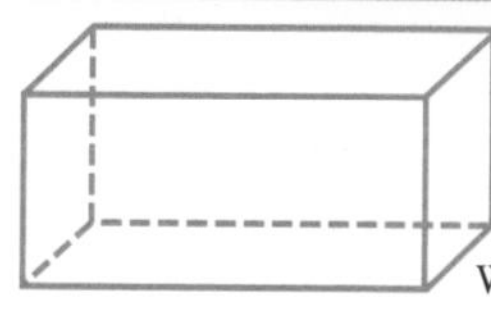

Rectangular solid
$V = lwh$

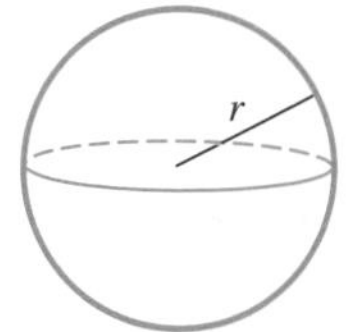

Sphere
$V = \frac{4}{3}\pi r^3$

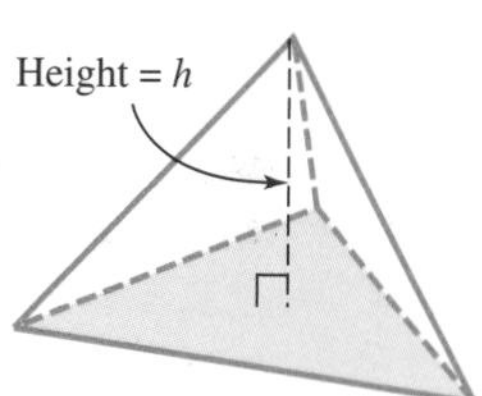

Pyramid
$V = \frac{1}{3}Bh$*
*B represents the area of the base.

Cylinder
$V = \pi r^2 h$

Make the most of your study time with ELEMENTARY Algebra f(x) Now™

WHAT DO YOU NEED TO LEARN *NOW*?

This powerful online learning companion helps you gauge your unique study needs and provides you with a *Personalized Learning Plan* so you can focus your study time where you need it most. Access to the program is included using the **1pass**™ access code packaged with this text.

TOTALLY INTEGRATED WITH THIS TEXT!

As you read the text, watch for references like the one at right. They direct you to corresponding media-enhanced activities found in **Elementary AlgebraNow**™. This page-by-page integration lets you fully explore key course concepts through the interactive learning environment of **Elementary AlgebraNow**.

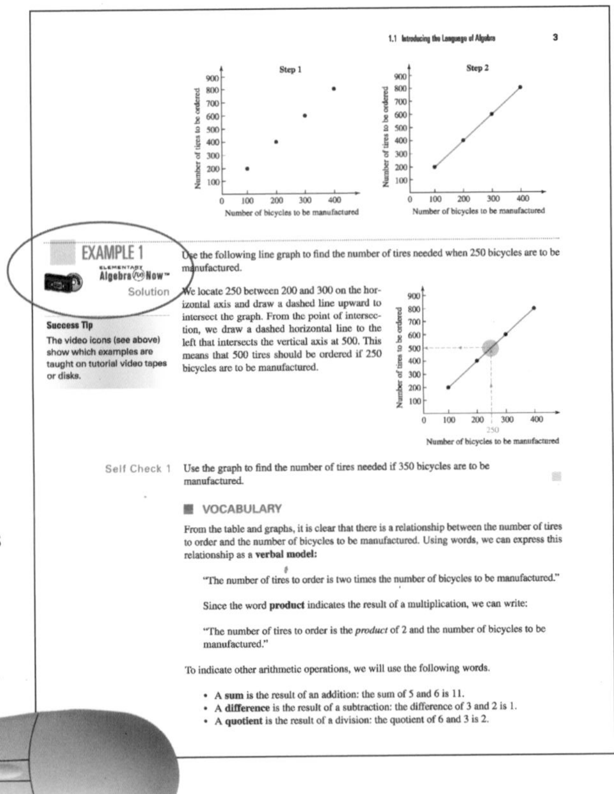

HOW DOES IT WORK?

Elementary AlgebraNow is made up of three powerful and easy-to-use components:

1) What Do I Know?
A diagnostic *Pre-Test* recognizes the key chapter concepts that you already know and highlights the areas that you need to study.

2) What Do I Need to Learn?
A *Personalized Learning Plan* outlines key elements for review and provides tutorial activities.

3) What Have I Learned?
A *Post-Test* assesses your mastery of core chapter concepts; results can be e-mailed to your instructor.

With a click of the mouse, the unique resources you'll access through **Elementary AlgebraNow** allow you to:

- Create a *Personalized Learning Plan* for each chapter that helps you focus on—and master—essential concepts
- Review for an exam by using the web quizzes and skillbuilder videos within the online tutorials
- Explore and reinforce mathematical concepts with **TLE Online Labs**
- Assess your understanding of core concepts by completing a *Post-Test* after you work through your *Personalized Learning Plan*

Study smarter—and make every minute count!

If an access card came with this book, you can start using **Elementary AlgebraNow** right away by following the directions on the card. If a card was not included, you may visit **http://1pass.thomson.com** to purchase electronic access.

Books in the Tussy and Gustafson Series

In hardcover:

Elementary Algebra, Third Edition
Intermediate Algebra, Third Edition
Elementary and Intermediate Algebra, Third Edition

In paperback:

Basic Mathematics for College Students, Second Edition
Basic Geometry for College Students
Prealgebra, Second Edition
Introductory Algebra, Second Edition
Intermediate Algebra, Second Edition
Developmental Mathematics

For more information, please visit www.brookscole.com

Edition

3

Elementary Algebra

Alan S. Tussy
Citrus College

R. David Gustafson
Rock Valley College

Australia • Canada • Mexico • Singapore • Spain
United Kingdom • United States

Executive Editor: Jennifer Huber
Publisher: Robert Pirtle
Assistant Editor: Rebecca Subity
Editorial Assistant: Sarah Woicicki
Technology Project Manager: Earl Perry
Marketing Manager: Greta Kleinert
Marketing Assistant: Jessica Bothwell
Advertising Project Manager: Bryan Vann
Project Manager, Editorial Production: Hal Humphrey
Senior Art Director: Vernon T. Boes
Print/Media Buyer: Barbara Britton
Permissions Editor: Joohee Lee
Production Service: Helen Walden
Text Designer: Kim Rokusek
Photo Researcher: Helen Walden
Illustrator: Lori Heckelman Illustration
Cover Designer: Patrick Devine
Cover Image: Kevin Tolman
Compositor: Graphic World, Inc.
Text and Cover Printer: Quebecor World/Taunton

Printed in the United States of America
3 4 5 6 7 08 07 06 05

Library of Congress Control Number: 2004106339

Student Edition: ISBN 0-495-18876-X

Annotated Instructor's Edition: ISBN 0-534-41915-1

Thomson Brooks/Cole
10 Davis Drive
Belmont, CA 94002
USA

Asia
Thomson Learning
5 Shenton Way #01-01
UIC Building
Singapore 068808

Australia/New Zealand
Thomson Learning
102 Dodds Street
Southbank, Victoria 3006
Australia

Canada
Nelson
1120 Birchmount Road
Toronto, Ontario M1K 5G4
Canada

Europe/Middle East/Africa
Thomson Learning
High Holborn House
50/51 Bedford Row
London WC1R 4LR
United Kingdom

Latin America
Thomson Learning
Seneca, 53
Colonia Polanco
11560 Mexico D.F.
Mexico

Spain/Portugal
Paraninfo
Calle Magallanes, 25
28015 Madrid, Spain

To **Margaret Kavelaar**
and
Gordon Guillaume,
two fine mathematics instructors
and outstanding
department chairpersons.

AST

To **Dr. John A. Schumaker,**
my teacher, mentor, colleague, and friend.

RDG

Contents

Chapter 1

An Introduction to Algebra 1

TLE Lesson 1: The Real Numbers
TLE Lesson 2: Order of Operations

1.1 Introducing the Language of Algebra 2
1.2 Fractions 9
1.3 The Real Numbers 21
1.4 Adding Real Numbers 32
1.5 Subtracting Real Numbers 41
1.6 Multiplying and Dividing Real Numbers 47
1.7 Exponents and Order of Operations 58
1.8 Algebraic Expressions 69

Accent on Teamwork 80
Key Concept: Variables 80
Chapter Review 82
Chapter Test 90

Chapter 2

Equations, Inequalities, and Problem Solving 92

TLE Lesson 3: Translating Written Phrases
TLE Lesson 4: Solving Equations Part 1
TLE Lesson 5: Solving Equations Part 2

2.1 Solving Equations 93
2.2 Problem Solving 103
2.3 Simplifying Algebraic Expressions 115
2.4 More about Solving Equations 124
2.5 Formulas 133
2.6 More about Problem Solving 145
2.7 Solving Inequalities 159

Accent on Teamwork 171
Key Concept: Simplify and Solve 171
Chapter Review 172
Chapter Test 178
Cumulative Review Exercises 180

Chapter 3 Linear Equations and Inequalities in Two Variables 183

TLE Lesson 6: Equations Containing Two Variables
TLE Lesson 7: Rate of Change and the Slope of a Line

3.1 Graphing Using the Rectangular Coordinate System 184
3.2 Graphing Linear Equations 195
3.3 More about Graphing Linear Equations 206
3.4 The Slope of a Line 217
3.5 Slope–Intercept Form 230
3.6 Point–Slope Form 241
3.7 Graphing Linear Inequalities 252

Accent on Teamwork 262
Key Concept: Describing Linear Relationships 263
Chapter Review 264
Chapter Test 269
Cumulative Review Exercises 271

Chapter 4 Exponents and Polynomials 274

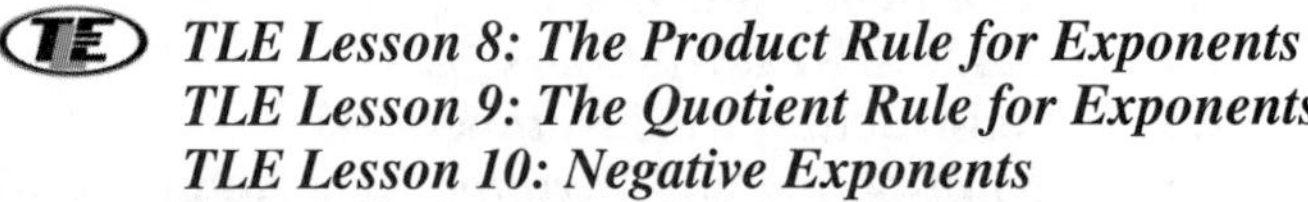
TLE Lesson 8: The Product Rule for Exponents
TLE Lesson 9: The Quotient Rule for Exponents
TLE Lesson 10: Negative Exponents

4.1 Rules for Exponents 275
4.2 Zero and Negative Exponents 284
4.3 Scientific Notation 293
4.4 Polynomials 300
4.5 Adding and Subtracting Polynomials 309
4.6 Multiplying Polynomials 318
4.7 Special Products 326
4.8 Division of Polynomials 333

Accent on Teamwork 342
Key Concept: Polynomials 343
Chapter Review 344
Chapter Test 349
Cumulative Review Exercises 351

Chapter 5 Factoring and Quadratic Equations 353

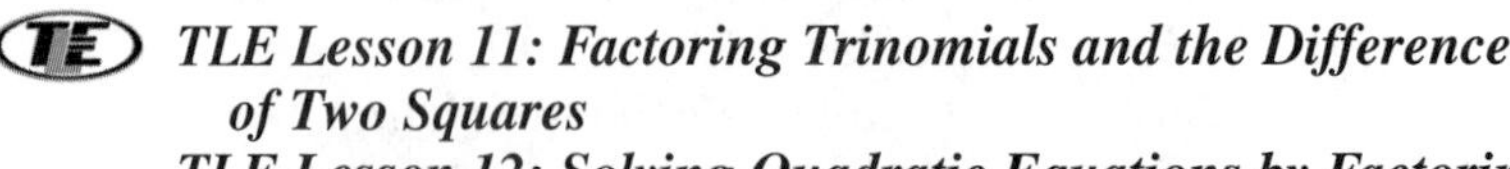
TLE Lesson 11: Factoring Trinomials and the Difference of Two Squares
TLE Lesson 12: Solving Quadratic Equations by Factoring

5.1 The Greatest Common Factor; Factoring by Grouping 354
5.2 Factoring Trinomials of the Form $x^2 + bx + c$ 363
5.3 Factoring Trinomials of the Form $ax^2 + bx + c$ 374
5.4 Factoring Perfect Square Trinomials and the Difference of Two Squares 384

5.5 Factoring the Sum and Difference of Two Cubes 391
5.6 A Factoring Strategy 395
5.7 Solving Quadratic Equations by Factoring 400

Accent on Teamwork 413

Key Concept: Factoring 414

Chapter Review 415

Chapter Test 419

Cumulative Review Exercises 420

Chapter 6 Rational Expressions and Equations 422

TLE Lesson 13: Adding and Subtracting Rational Expressions

6.1 Simplifying Rational Expressions 423
6.2 Multiplying and Dividing Rational Expressions 432
6.3 Addition and Subtraction with Like Denominators; Least Common Denominators 441
6.4 Addition and Subtraction with Unlike Denominators 450
6.5 Simplifying Complex Fractions 459
6.6 Solving Rational Equations 468
6.7 Problem Solving Using Rational Equations 476
6.8 Proportions and Similar Triangles 486

Accent on Teamwork 498

Key Concept: Expressions and Equations 499

Chapter Review 500

Chapter Test 506

Cumulative Review Exercises 508

Chapter 7 Solving Systems of Equations and Inequalities 510

 *TLE Lesson 14: Solving Systems of Equations by Graphing*

7.1 Solving Systems of Equations by Graphing 511
7.2 Solving Systems of Equations by Substitution 521
7.3 Solving Systems of Equations by Elimination 530
7.4 Problem Solving Using Systems of Equations 539
7.5 Solving Systems of Linear Inequalities 554

Accent on Teamwork 563

Key Concept: Three Methods for Solving Systems of Equations 564

Chapter Review 565

Chapter Test 569

Cumulative Review Exercises 571

Chapter 8

Radical Expressions and Equations 573

TLE Lesson 15: Using Square Roots

8.1 An Introduction to Square Roots 574
8.2 Simplifying Square Roots 584
8.3 Adding and Subtracting Radical Expressions 592
8.4 Multiplying and Dividing Radical Expressions 598
8.5 Solving Radical Equations 607
8.6 Higher-Order Roots and Rational Exponents 615

Accent on Teamwork 623

Key Concept: The Rules for Radicals 624

Chapter Review 625

Chapter Test 629

Cumulative Review Exercises 630

Chapter 9

Solving Quadratic Equations; Functions 633

TLE Lesson 16: The Quadratic Formula

9.1 The Square Root Property; Completing the Square 634
9.2 The Quadratic Formula 643
9.3 Complex Numbers 654
9.4 Graphing Quadratic Equations 661
9.5 An Introduction to Functions 671
9.6 Variation 682

Accent on Teamwork 692

Key Concept: Solving Quadratic Equations 693

Chapter Review 694

Chapter Test 701

Cumulative Review Exercises 703

Appendix I Statistics A-1
Appendix II Roots and Powers A-6
Appendix III Answers to Selected Exercises A-7

Index I-1

Preface

Algebra is a language in its own right. The purpose of this textbook is to teach students how to read, write, and think mathematically using the language of algebra. It is written for students studying algebra for the first time and for those who need a review of the basics. *Elementary Algebra,* Third Edition, employs a variety of instructional methods that reflect the recommendations of NCTM and AMATYC. You will find the vocabulary, practice, and well-defined pedagogy of a traditional approach. You will also find that we emphasize the reasoning, modeling, and communicating skills that are part of today's reform movement.

The third edition retains the basic philosophy of the second edition. However, we have made several improvements as a direct result of the comments and suggestions we received from instructors and students. Our goal has been to make the book more enjoyable to read, easier to understand, and more relevant.

NEW TO THIS MEDIA EDITION

- This new Media Edition is integrated with Elementary AlgebraNow™, a powerful online learning companion. Icons throughout the text indicate opportunities to reinforce concepts with online tutorials.
- Nine new chapter openers reference the *TLE* computer lessons that accompany each chapter.
- The new Language of Algebra features, along with Success Tips, Notation, Calculator Boxes, and Cautions, are presented in the margins to promote understanding and increased clarity.
- Many additional applications involving real-life data have been added.
- Answers to the popular *Self Check* feature have been relocated to the end of each section, right before the *Study Set.*
- Several high-level Challenge Problems have been added to each *Study Set.*
- The Accent on Teamwork feature has been redesigned to offer the instructor two or three collaborative activities per chapter that can be assigned as group work.
- More illustrations, diagrams, and color have been added for the visual learner.

REVISED TABLE OF CONTENTS

Chapter 1: *An Introduction to Algebra* Introductory concepts from Chapters 1 and 2 of the previous edition have been consolidated to create a less-repetitive, quicker-paced beginning for the course. More emphasis has been placed on simplifying and building fractions to ready students for Chapter 6, Rational Expressions and Equations. Addition, subtraction, multiplication, and division of signed numbers are now discussed in this chapter.

Chapter 2: *Equations, Inequalities, and Problem Solving* This chapter now contains two sections that are exclusively devoted to problem solving using the five-step problem-solving strategy. Circle graphs have been added to the study of percent.

Chapter 3: *Linear Equations and Inequalities in Two Variables* The nonlinear graphs previously in Section 3.2 were moved to Chapter 4. The slope section was rewritten so that slope of a line is introduced first, followed by a discussion of rates of change. Sections 3.5 and 3.6 have new titles: Slope–Intercept Form and Point–Slope Form. Graphing Linear Inequalities, previously a part of the Systems of Equations chapter, is now Section 3.7. With its placement here, students graph inequalities at two well-separated times during the course: in Chapter 3 and Chapter 7. New graphs and linear models of real-world data have been added as examples and as exercises in the Study Sets.

Chapter 4: *Exponents and Polynomials* Section 4.4, Polynomials, now includes graphs of nonlinear equations, such as $y = x^2$ and $y = x^3$. A separate section has been devoted to special products. Sections 4.7 and 4.8 from the second edition were combined so that division of polynomials is now covered in one section.

Chapter 5: *Factoring and Quadratic Equations* Previously Chapter 6, this chapter has been extensively rewritten. Added attention has been given to the grouping method (key number method) for factoring trinomials.

Chapter 6: *Rational Expressions and Equations* Unit conversion is now introduced in Section 6.2. Addition and subtraction of rational expressions is now presented in two parts: In Section 6.3, the denominators are like, and in Section 6.4, the denominators are unlike. The concepts of unit price and best buy have been added to Section 6.8, Proportions and Similar Triangles.

Chapter 7: *Solving Systems of Equations and Inequalities* Previously Chapter 8, many new real-life applications have been added.

Chapter 8: *Radical Expressions and Equations* This chapter, which was previously Chapter 5, has been rewritten so that only square roots are discussed in the first six sections. Solving radical equations now follows the sections that discuss simplifying radical expressions. In Section 8.6, higher-order roots and rational exponents are introduced.

Chapter 9: *Solving Quadratic Equations; Functions* This chapter is the result of some reorganization of some topics previously in Chapter 3, Chapter 6, and Chapter 7. A section on Complex Numbers has been added. Section 9.5, Introduction to Functions, now introduces this important concept using a real-world example. Section 9.6, Variation, is a consolidation of material that had previously appeared in two different sections of two different chapters.

ACKNOWLEDGMENTS

We are grateful to the following people who reviewed the manuscript at various stages of its development. They all had valuable suggestions that have been incorporated into the text.

The first and second editions:

Julia Brown
Atlantic Community College

Lynn B. Cade
Pensacola Junior College

Tim Caldwell
Meridian Community College

John Coburn
Saint Louis Community College–Florissant Valley

Sally Copeland
Johnson County Community College

Ben Cornelius
Oregon Institute of Technology

James Edmondson
Santa Barbara Community College

Hector L. Esteban
Valencia Community College

Judith Jones
Valencia Community College

Therese Jones
Amarillo College

Mauricio Marroquin
Los Angeles Valley College

Marilyn Mays
North Lake College

Elizabeth Morrison
Valencia Community College

Angelo Segalla
Orange Coast College

June Strohm
Pennsylvania State Community College–DuBois

Rita Sturgeon
San Bernardino Valley College

Jo Anne Temple
Texas Technical University

Sharon Testone
Onondaga Community College

Marilyn Treder
Rochester Community College

The third edition:

Cynthia J. Broughton
Arizona Western College

Don K. Brown
Macon State College

Light Bryant
Arizona Western College

John Scott Collins
Pima Community College

Lee Gibbs
Arizona Western College

Haile K. Haile
Minneapolis Community and Technical College

Suzanne Harris-Smith
Albuquerque Technical Vocational Institute

Kamal Hennayake
Chesapeake College

Lynn Marecek
Santa Ana College

Jamie McGill
East Tennessee State University

Margaret Michener
University of Nebraska, Kearney

Bernard J. Pina
Dona Ana Branch Community College

Carol Purcell
Century Community College

Daniel Russow
Arizona Western College

Donald W. Solomon
University of Wisconsin, Milwaukee

John Thoo
Yuba College

Susan M. Twigg
Wor-Wic Community College

Gizelle Worley
California State University, Stanislaus

We want to express our gratitude to Karl Hunsicker, Cathy Gong, Dave Ryba, Terry Damron, Marion Hammond, Lin Humphrey, Doug Keebaugh, Robin Carter, Tanja Rinkel, Bob Billups, Liz Tussy, Alexander Lee, Steve Odrich, and the Citrus College Library staff (including Barbara Rugeley) for their help with some of the application problems in the textbook.

Without the talents and dedication of the editorial, marketing, and production staff of Brooks/Cole, this revision of *Elementary Algebra* could not have been so well accomplished. We express our sincere appreciation for the hard work of Bob Pirtle, Jennifer Huber, Helen Walden, Lori Heckleman, Vernon Boes, Kim Rokusek, Sarah Woicicki, Greta Kleinert, Jessica Bothwell, Bryan Vann, Kirsten Markson, Rebecca Subity, Hal Humphrey, Tammy Fisher-Vasta, Christine Davis, Ellen Brownstein, Diane Koenig, and Graphic World for their help in creating the book.

Alan S. Tussy
R. David Gustafson

For the Student

SUCCESS IN ALGEBRA

To be successful in mathematics, you need to know how to study it. The following checklist will help you develop your own personal strategy to study and learn the material. The suggestions below require some time and self-discipline on your part, but it will be worth the effort. This will help you get the most out of the course.

As you read each of the following statements, place a check mark in the box if you can truthfully answer Yes. If you can't answer Yes, think of what you might do to make the suggestion part of your personal study plan. You should go over this checklist several times during the semester to be sure you are following it.

Preparing for the Class

- ❑ I have made a commitment to myself to give this course my best effort.
- ❑ I have the proper materials: a pencil with an eraser, paper, a notebook, a ruler, a calculator, and a calendar or day planner.
- ❑ I am willing to spend a minimum of two hours doing homework for every hour of class.
- ❑ I will try to work on this subject every day.
- ❑ I have a copy of the class syllabus. I understand the requirements of the course and how I will be graded.
- ❑ I have scheduled a free hour after the class to give me time to review my notes and begin the homework assignment.

Class Participation

- ❑ I know my instructor's name.
- ❑ I will regularly attend the class sessions and be on time.
- ❑ When I am absent, I will find out what the class studied, get a copy of any notes or handouts, and make up the work that was assigned when I was gone.
- ❑ I will sit where I can hear the instructor and see the chalkboard.
- ❑ I will pay attention in class and take careful notes.
- ❑ I will ask the instructor questions when I don't understand the material.
- ❑ When tests, quizzes, or homework papers are passed back and discussed in class, I will write down the correct solutions for the problems I missed so that I can learn from my mistakes.

Study Sessions

- ❑ I will find a comfortable and quiet place to study.
- ❑ I realize that reading a math book is different from reading a newspaper or a novel. Quite often, it will take more than one reading to understand the material.
- ❑ After studying an example in the textbook, I will work the accompanying Self Check.
- ❑ I will begin the homework assignment only after reading the assigned section.
- ❑ I will try to use the mathematical vocabulary mentioned in the book and used by my instructor when I am writing or talking about the topics studied in this course.
- ❑ I will look for opportunities to explain the material to others.
- ❑ I will check all my answers to the problems with those provided in the back of the book (or with the *Student Solutions Manual*) and resolve any differences.
- ❑ My homework will be organized and neat. My solutions will show all the necessary steps.
- ❑ I will work some review problems every day.

- ❑ After completing the homework assignment, I will read the next section to prepare for the coming class session.
- ❑ I will keep a notebook containing my class notes, homework papers, quizzes, tests, and any handouts—all in order by date.

Special Help

- ❑ I know my instructor's office hours and am willing to go in to ask for help.
- ❑ I have formed a study group with classmates that meets regularly to discuss the material and work on problems.
- ❑ When I need additional explanation of a topic, I use the tutorial videos and the interactive CD, as well as the website.
- ❑ I make use of extra tutorial assistance that my school offers for mathematics courses.
- ❑ I have purchased the *Students Solutions Manual* that accompanies this text, and I use it.

To follow each of these suggestions will take time. It takes a lot of practice to learn mathematics, just as with any other skill.

No doubt, you will sometimes become frustrated along the way. This is natural. When it occurs, take a break and come back to the material after you have had time to clear your thoughts. Keep in mind that the skills and discipline you learn in this course will help make for a brighter future. Good luck!

Elementary AlgebraNow Can Help You Succeed in Math

ELEMENTARY Algebra $f(x)$ Now™

Elementary AlgebraNow is a powerful online learning companion that helps you gauge your unique study needs and provides you with a personalized learning plan. The resources in Elementary AlgebraNow give you all the learning tools you need to master core concepts. Elementary AlgebraNow and this new edition of Tussy and Gustafson's *Elementary Algebra* enhance each other, providing you with a seamless learning system. Completely integrated with the textbook, icons in the text direct you to the online tutorials and TLE Labs found in Elementary AlgebraNow.

The Elementary AlgebraNow system consists of three powerful, easy-to-use assessment components:

1. What Do I Know?
   Take a diagnostic pre-test to find out.
2. What Do I Need to Learn?
   Your personalized learning plan helps you focus on the areas you need to study.
3. What Have I Learned?
   Take a post-test to see how your understanding of key concepts has improved; results can be e-mailed to your instructor.

WHAT YOU'LL FIND IN YOUR PERSONALIZED LEARNING PLAN

Text-specific tutorials, live online tutoring, and TLE online labs provide you with multiple tools to help you get a better grade.

Text-Specific Tutorials

The view of a tutorial looks like this. To navigate between chapters and sections, use the drop-down menu below the top navigation bar. This will give you access to the study activities available for each section.

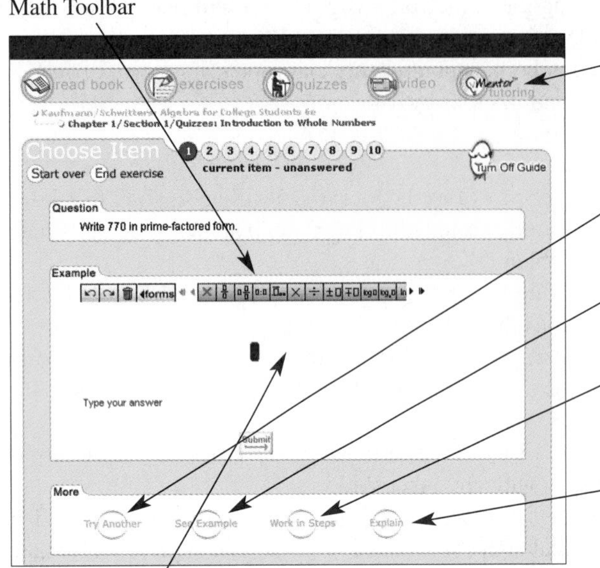

vMentor: Live online tutoring is only a click away. Tutors can take screen shots of your book and lead you through a problem with voice-over and visual aids.

Try Another: Click here to generate a new question or a new set of problems.

See Examples: Preworked examples provide you with additional help.

Work in Steps: Elementary AlgebraNow can guide you through a problem step-by-step.

Explain: Additional explanation from your book can help you with a problem.

Type your answer here.

Online Tutoring with *vMentor*

Access live online tutors and support through vMentor™. To access vMentor while you are working in the Exercises or "Tutorial" areas in iLrn, click on the **vMentor Tutoring** button at the top right of the navigation bar above the problem or exercise.

Next, click on the **vMentor** button; you will be taken to a Web page that lists the steps for entering a vMentor classroom. If you are a first-time user of vMentor, you might need to download Java software before entering the class for the first class. You can either take an Orientation Session or log in to a vClass from the links at the bottom of the opening screen.

All vMentor Tutoring is done through a vClass, an Internet-based virtual classroom that features two-way audio, a shared whiteboard, chat, messaging, and experienced tutors.

You can access vMentor during the following times:

Monday through Thursday:
5 p.m. to 12 a.m. Pacific Time
6 p.m. to 1 a.m. Mountain Time
7 p.m. to 2 a.m. Central Time
8 p.m. to 3 a.m. Eastern Time

Sunday:
12 p.m. to 12 a.m. Pacific Time
1 p.m. to 1 a.m. Mountain Time
2 p.m. to 2 a.m. Central Time
3 p.m. to 3 a.m. Eastern Time

If you need additional help using vMentor, you can access the Participant Guide at this Website: **http://www.elluminate.com/support/guide/pdf.**

TLE Online Labs

Use TLE Online Labs to explore and reinforce key concepts introduced in this text. These electronic labs give you access to additional instruction and practice problems, so you can explore each concept interactively, at your own pace. Not only will you be better prepared, but you will perform better in the class overall.

ACCESSING ELEMENTARY ALGEBRANOW THROUGH 1PASS

Registering with the PIN Code on the 1pass Card

Situation: Your instructor has not given you a PIN code for an online course, but you have a textbook with a 1pass PIN code. With 1pass, you have one simple PIN code access to all media resources associated with your textbook. Please refer to your 1pass card for a complete list of those resources.

Initial Log-in

To access your web gateway though 1pass:

1. Check the outside of your textbook to see if there is an additional 1pass card.
2. Take this card (and the additional 1pass card if appropriate) and go to http://1pass.thomson.com.
3. Type in your 1pass access code (or codes).
4. Follow the directions on the screen to set up your personal username and password.
5. Click through to launch your personal portal.
6. Access the media resources associated with your text . . . all the resources are just one click away.
7. Record your username and password for future visits and be sure to use the same username for all Thomson Learning resources.

For tech support, contact us at 1 (800) 423-0563.

You will be asked to enter a valid e-mail address and password. Save your password in a safe place. You will need them to log in the next time you use 1pass. Only your e-mail address and password will allow you to reenter 1pass.

Subsequent Log-in

1. Go to **http://1pass.thomson.com.**
2. Type your e-mail address and password (see boxed information above) in the "Existing Users" box; then click on **Login.**

Applications Index

Examples that are applications are shown with boldface page numbers.
Exercises that are applications are shown with lightface page numbers.

Business and Industry

Advertising, 174, 702
Automation, 249
Auto mechanics, 317
Blending coffee, 159, 631
Blueprints, 497, 598
Building construction, 692
Buying boats, 562
Buying furniture, 562
Candy, 351
Candy outlet stores, 568
Candy sales, 571
Carpentry, 142, 155, 267, 630
Celebrity endorsements, 568
Coffee blends, 682
Computer drafting, 240
Computing depreciation, 692
Concrete, 497
Construction, 191, 418, 692
Copiers, 268
Corporate earnings, **34**
Cost analysis, 669
Customer service, **64,** 410
Cutting steel, 508
Depreciation, 250, 266, 271, 622
Dog kennels, 485
Drafting, 30, 180
Economics, 519
Employment services, 240
Engine output, 251
Entrepreneurs, 141, 703
Exports, 114
Furniture, 383
Gear ratios, 496
Graphic design, 431
Greek architecture, 316
Income property, 240
Inventories, 261
Labor statistics, **106**
Landscaping, 216
Machine shops, 83
Making cheese, 158
Making tires, 552
Manufacturing, 309, **544**
Metal fabrication, 653
Milk production, 229
Mixing candy, 159, 273, 705
Mixing nuts, 553
Mixing perfume, 496
Mixtures, 178
Office work, 485
Oil reserves, 299
On-line sales, 632
Ordering furnace equipment, 562
Packaging fruit, 485
Parts list, 681
Payroll checks, **479**
Photographic chemicals, 682
Photography, **543**
Picture frames, **576**
Pitch, **223**
Power outage, 112
Power output, 706
Power usages, 614
Production costs, 239
Production planning, 251
Profit, 700
Profits and losses, 40
Quality control, 496, 583
Reading blueprints, 316
Rentals, 157
Retailing, 653
Retail stores, 332
Robotics, 155
Salaries, A-5
Sales, 486, A-4
Selling ice cream, 551
Sewing costs, 240
Snack foods, **152,** 494
Sports cars, 351
Supermarket displays, **302**
Supermarkets, 308
Supply and demand, 552
Suspension bridges, 308
Systems analysis, **103**
Studio tour, 112
Trade, 31
Tree trimming, 550
Trucking, 440
U.S. jobs, 46
Wal-mart, 229
Warehousing, 157
Wedding pictures, 551

Education

American college students, 156
Art history, 177, 283
Calculating grades, 362
Classroom space, 440
College boards, 519
College enrollments, **72,** 264, 565
College entrance exams, 114
College fees, 240
Counseling, 300
Crafts, 410
Crayons, **357**
The decimal numeration system, 292
Education, 539
English artists, 551
Faculty-student ratios, 496
Grades, **166,** 169

Grading papers, 485
History, 46, 411, 420, **542**
IQ tests, 293
Memory, 506
Natural light, 440
Paying tuition, **149**
Presidents, 411
Printers, 485
Saving for college, **289**
Science history, 308
Student loans, 552
Studying learning, **246**
U.S. presidents, 112
U.S. school construction, 506
Unit comparisons, 292

Electronics
Electricity, 661, 691, 700
Electronics, 40, 467, 476, 629, 661, 692

Entertainment
Academy awards, 418
Amusement park rides, 79, 592
Autographs, 177
Ballroom dancing, 551
Battleships, 192
Best-selling songs, 108
Birthday parties, 265
Buying compact discs, 562
Cable TV, 241
Camcorders, **212**
Camping, 440
Card games, 46
Carousels, 642
Checkers, 419
Children's stickers, 631
Choreography, 410
Collectibles, 174
Comics, 652
Crossword puzzles, 591, 669
Dancing, 591
Darts, 390
Dining, 529
Entertaining, 416
Entertainment, 112
Ferris wheels, 113, 628
Fireworks shows, 313
Five-card poker, 704
Gambling, 56
Game shows, **35**
Hollywood, 141
Miniatures, **491,** 494
Model railroads, 497
Motion pictures, **540**
Motown records, **405**
Movie losses, 39
Movie stunts, **648**
Number puzzles, 170
Orchestras, 112
Playground equipment, 597
Playing a flute, 102
Pogo sticks, 702
Raffles, 205
Rocketry, **74,** 78
Roller coasters, 652
Rose parade, 141
Scrabble, 68
Selling tickets, 551
Snacks, 412
Stamps, 324
Sweeps week, 90
Synthesizers, 102
Television, 583
Televisions, **649**
Theater, 112
Theater screens, 551
Theater seating, 626
Tightrope walkers, 418
Tipping, 113
Touring, 155
Toys, 325
TV news, 112
TV taping, **188**
TV trivia, 440
Vacationing, 570
Videotapes, 240
Yo-yos, 141

Farm Management
Cost of a well, 692
Crop damage, 497
Draining a tank, 509
Farming, 553, 691
Filling a tank, **480**
Pesticides, 562
Raising turkeys, 432
Water levels, **44**

Finance
Accounting, 56, 155
ATM receipt, 112
Banking, 141, **A-2**
Checking accounts, **224**
Comparing interest rates, 486
Comparing investments, **482,** 485, 486
Computing interest, 692
Computing a paycheck, 496
Credit cards, 39, 142
Currency exchange rate, 507
Financial planning, 158, 553, 570, 632
Interest, 299
Investing, 182, 569, 653
Investment income, 177
Investment plans, 157
Investment problem, 179
Investments, 158, 241, 505
Retirement, 157, 292
Retirement income, **134**
Saving for the future, 292
Savings bonds, 349
Service charges, 169
Stock exchange, 39
Stock market, 571

Geography
Alaska, 550
Bermuda triangle, 113
California coastline, **145**
Elevations, 567
Geography, 86, 143, 299
The gulf stream, 552
Hawaiian islands, 264
The jet stream, 552
Land elevations, 46
Latitude and longitude, 519
Maps, 192
Sahara Desert, 40
Statehood, 112
Temperature records, 46
U. S. temperatures, **44**
West Africa, 494

Geometry
Air conditioning, 341
The American Red Cross, 123
Angles, **541**
Appliances, 348
Areas, 114
Art, 156
Awards platforms, 607
Barbecuing, 144
Campers, 142
Camping, 176
Cargo space, 632
Checkerboards, 340
Chrome wheels, 173
Complementary angles, 156, 550
Concentric circles, 509
Counter space, 169
Designing a tent, 411
Drafting, 582
Fencing, 113, 598
Finding volumes, **137**
Firewood, 143
Flags, 652, 701
Framing pictures, 251
Furnace filters, 341
Gardening, 325, 347, 582
Geometry, 169, 179, 192, 264, 449, 551, 614, 691

Geometry tools, 420
Graphic arts, 325, 348
Growing sod, 79
Halloween, 176
Height of a building, 497
Height of a tree, **492,** 497, 509
Height of a triangle, 652, 704
Igloos, 143
Isosceles triangles, **147**
Ladders, 591
Land of the rising sun, 181
Landscaping, 143
Lawnmowers, 606
Luggage, 325
Mini-blinds, 341
Native American dwelling, 143
Packaging, 283, 308
Painting, 332
Photo enlargements, 496
Playpens, 332
Plotting points, 411
Pool, 340
Pool borders, 412
Pool design, 227
Porcelain figurines, 506
Projector screens, 606
Pyramids, 144
Right triangles, **407**
Roof design, **579**
Rubber meets the road, 143
Security gates, 696
Shortcuts, 584
Signage, 143
Signal flags, 332
Signs, 178
Solar heating, 155
Speakers, 623
Stop signs, 113
Storage, 332, 383
Storage cubes, 705
Supplementary angles, 156, 550
Surface area, 631
Surveying, 498
Swimming pools, 156
Tools, 79
TV towers, 156, 507
Vacuum cleaners, 627
Video cameras, 503

Home Management
Air conditioning, 571
Blending lawn seed, 158
Buying painting supplies, 551
Carpeting, **437**
Choosing a furnace, 552
Cider, 691
Cleaning windows, **201**
Clothes shopping, 571
Comparison shopping, **492,** 506, 507
Cooking, 180, 362, 484, 496, 509
Deck design, 583
Decking, 653
Decorating, 21
Depreciation, **53,** 193
Dining area improvements, **148**
Doing laundry, 449
Down payments, 178
Electric bills, 341
Energy usage, 68
Family budgets, 174
Filling a pool, 484, 632
Frames, 21
Furnace repairs, 486
Gardening, **146,** 627
Gardening tools, 412, 467
Gift shopping, 569
Grocery shopping, **490,** 494
Hiring baby-sitters, 251
Holiday decorating, 484
House cleaning, 505
Housekeeping, 205
House painting, 505
Insurance, 703
Insurance costs, 115
Interior decorating, 362
Landscaping, 419, **558**
Lawn mowers, **99**
Lawn sprinklers, 681
Mailing breakables, 394
Mixing fuel, 496
Mixing paint, 158
Painting, **71**
Painting equipment, 568
Playground equipment, 695
Playpens, 421
Real estate, 56
Real estate investments, 317
Roofing, 431, 704
Roofing houses, 484
Sewing, 123, 629
Shopping for clothes, 496
Shopping, 498, 568, 631
Sound systems, 177
Telephone calls, 702
Telephone rates, 552
Wrapping gifts, 68

Medicine and Health
Aging population, **108**
Anatomy, 249, 597
Antiseptics, 113
Backpacks, 630
Body temperatures, 179
Causes of death, 550
Child care, 114, 570
Clinical trials, 351
Counting calories, 155
CPR, 496
Dental records, 114
Dentistry, 192, 505, 629
Determining a child's dosage, **472**
Dosage, 691, 501
Exercise, 412, 505
Eyesight, 46
Eye surgery, **105**
First aid, 156
Fitness equipment, 626
Forensic medicine, 441
Handicap accessibility, 267
Health, 38
Health department, 670
Health risk, 507
Heart rates, 420
Losing weight, 704
Medical dosages, 432
Medical technology, **547**
Medical tests, 623
Medication, 691
Medicine, **48,** 475
Multiple births in the United States, 178
Nutrition, 114, 496, 572, 642
Orthopedics, 594
Physical fitness, 485, 507
Physical therapy, 551
Physiology, **A-2**
Safety codes, 170
Sick days, 181
Sunglasses, 325
Transplants, 519
Treadmills, 228
U.S. automobile accidents, 205
U.S. life expectancy, 115
U.S health care, 263
Viruses, A-5
Weight chart, 170
X-rays, **406**

Miscellaneous
The abacus, 653
Angle of elevation, 571
Antifreeze, 569
Auctions, 115
Avon products, 142
Bears, 440
Beverages, 421
Birds in flight, 485
Blending tea, 508
Bouncing balls, 284
Bridges, 670
Cash awards. 68

Cats, 205
Chain letters, 67
Charitable giving, **133**
Coast Guard rescues, **150**
Commemorative coins, 553
Communications, 341
Comparing temperatures, 170
Computers, 57
Converting temperatures, 251
Crime statistics, 9
Customer guarantees, 114
Daredevils, 652
Dinner plates, 331
Dolphins, 308, 410
Drainage, 227
Earthquakes, 652
Electricity, 349
Finding the rate, **135**
Flooding, 112
Gears, 706
Genealogy, 114
Goldsmith, 182
Got milk?, 251
Grand king size beds, 508
Hamster habitats, 143
Hardware, 20, 441, 597
Holiday decorating, 622
Identity theft, **109**
International alphabet, 440
Irrigation, 228
Lie detector tests, 46
Locks, 155
Mathematical formulas, 476
Melting ice, 395
Memorials, 143
Men's shoe sizes, **245**
Military science, 38
Mixing fuels, 158
Monarchy, 112
Musical instruments, 352
Niagara falls, 614
Number problem, **477,** 484, 505
Occupational testing, 169
Organ pipes, 431
Pennies, 271
Pets, 179, 373
Probability, 284
Quilts, 705
Rolling dice, 260
Salad bars, 240
Salt solutions, 158, 179
Search and rescue, 158, 631
Service clubs, 141
Sharpening pencils, 205
Signal flags, 390
Snack foods, 159
Solutions, 178
Submarines, 39
Surveys, 68
Talking, 440
Tape measures, 484
Target practice, 30
The *Titanic,* 497
Tubing, 411
TV coverage, 520
Valentine's Day, 141
Water heaters, 271
Water pressure, 192
Working two jobs, **257**
Work schedules, 269

Politics, Government, and the Military
Aircraft carriers, 362
Census, 194
Cleaning highways, 507
Cost of living, 174
Crash investigations, 568
Currency, 299
Daily tracking polls, 519
Driver's licenses, 497
Executive branch, 550
Federal outlays, 113
Freeway design, 250
Government debt, 31
Income tax, 114
Job testing, A-5
Marine corps, 551
The military, 193, 299, 696
The Motor City, 216
The national debt, 300
National parks, 155
NATO, 260
Naval operations, 317
Politics, 39, 40
Redevelopment, 562
Sewage treatment, 484
Taxes, 157
Toxic cleanup, 250
White-collar crime, **545**
World population, 345

Recreation and Hobbies
Aviation, 552
Comparing travel, 485
Flight path, 498
Fuel economy, 440
Jewelry making, 176
Photographic chemicals, 420
Photography, 373
River tours, 486
Speed of a plane, 508
Traffic signs, 502
Travel time, 486
Wind speed, 485, 505

Sports
Air hockey, 606
Archery, 143
Badminton, 642
Ballooning, 418
Baseball, 325, 582
Bicycle safety, 642
Billiards, 123, 204
Boating, 193, 412, 485, **546,** 552, 568
Camping, 598
Cycling, 158
Distance runners, **478**
Diving, **74,** 293, 346
Earned run averages, 467
Exhibition diving, 410
Football, 583, A-4
Golf, 39, 192
Grounds keeping, 484
High school sports, 529
Horses, 141
Indy 500, 176
Kayaking, 570
Making sporting goods, 251
Monuments, 614
Mountain bicycles, 156
NFL tickets, 205
Officiating, 409
Ping-Pong, 123, 205
Pole vaulting, 250
Professional hockey, 240
Projectiles, 670
Racing, 46, 351
Racing programs, 114
Sailing, **647**
Skateboarding, 144
Softball, 309, **405**
Sport fishing, A-5
Sports, 179
Sports equipment, 178
Swimming, 141, 193
Tennis, 112
Tether ball, 571
Track and field, 670
Trampoline, 250, 670
Walk-a-thons, 88
Walking and bicycling, 177
Water skiing, 316
Women's soccer, 173
Wrestling, 583

Science and Engineering
Anatomy, 112
Angstrom, 299
Antifreeze, 79
Archaeology, 629
Astronomy, 91, **296,** 298, 631

Atoms, 85, **296,** 299, 345
Bacterial growth, 56
Biology, 292, 669
Birds of prey, 561
Botany, 20
Bridges, **610**
Chemistry, 216, 240, 476
Comparing weights, **684**
Computing forces, 691
Computing pressures, 691
Data analysis, 467
Earth science, 562
Electronics, 293
Engineering, 156, 228, 504
Escape velocity, 642
Exploration, 299
Free fall, 350, 582
Gas law, **686**
Gas pressure, **688**
Gauges, 46
Genetics, 390
Gravity, 111, 692
Length of a meter, 299
Light, 57, 67
Lightning, 681
Light years, 299
Low temperatures, 141
Lunar gravity, 691
Marine biology, 553
Mars, 79
Metallurgy, 141
Mixing solutions, **151**
Pendulums, **576,** 614
Photographic chemicals, 158
Photography, **472**
Physics, 57, 102, 216, 390
Physics experiment, 440
Planets, 56, 113
Properties of water, 142
Protons, 299
Pulleys, 144
Reflections, 681
Space travel, 113
Speed of light, **437**
Speed of sound, 298, 299
Structural engineering, 496, 681
Temperature change, 56
Thermodynamics, 144
Time of flight, 410
Tornadoes, 158
Wavelengths, 299
Windmills, 622
Work, 182

Travel

Airlines, 57
Airline travel, **104**
Air traffic control, 158, 520
Auto insurance, 67
Aviation, 102, 705
Car radiators, 56
Car registration, 268
Car repairs, 412
Commercial jets, 228
Commuting distance, 691
Commuting time, 691
Driving safety, 419, 703
Fleet averages, 169
Fuel efficiency, A-5
Gas mileage, A-5
Grade of a road, 227
Group discounts, **236**
Highway design, 614
Highway safety, 582
Historic tours, 173
Insurance costs, 362
Navigation, 238
Parking, 324
The *Queen Mary,* 57
Rearview mirrors, 361
Road safety, 614
Road signs, 625
Road trips, 158, 272
Speed limits, 142
Speed of an airplane, 251
Speed of trains, 158
Sticker prices, 141
Stopping distance, 308
Tourism, 267
Traffic safety, 8
Travel times, 179
United airlines, 271
Vacationing, **72,** 681
Visibility, 622

Chapter

1 An Introduction to Algebra

ELEMENTARY
Algebra f(x) Now™

Throughout the chapter, this icon introduces resources on the Elementary AlgebraNow Web site, accessed through http://1pass.thomson.com, that will

- Help you test your knowledge of the material with a pre-test and a post-test
- Provide a personalized learning plan targeting areas you should study

1.1 Introducing the Language of Algebra
1.2 Fractions
1.3 The Real Numbers
1.4 Adding Real Numbers
1.5 Subtracting Real Numbers
1.6 Multiplying and Dividing Real Numbers
1.7 Exponents and Order of Operations
1.8 Algebraic Expressions
Accent on Teamwork
Key Concept
Chapter Review
Chapter Test

TLE Most banks are beehives of activity. Tellers and bank officials help customers make deposits and withdrawals, arrange loans, and set up credit card accounts. To describe these financial transactions, positive and negative numbers, fractions, and decimals are used. These numbers belong to a set that we call the *real numbers.*

To learn more about real numbers and how they are used in banking, visit *The Learning Equation* on the Internet at http://tle.brookscole.com. (The log-in instructions are in the Preface.) For Chapter 1, the online lessons are:

- *TLE* Lesson 1: The Real Numbers
- *TLE* Lesson 2: Order of Operations

Algebra is a mathematical language that can be used to solve many types of problems.

1.1 Introducing the Language of Algebra

- Tables and Graphs • Vocabulary • Notation
- Variables, Expressions, and Equations • Constructing Tables

Algebra is the result of contributions from many cultures over thousands of years. The word *algebra* comes from the title of the book *Ihm Al-jabr wa'l muqābalah,* written by the Arabian mathematician al-Khwarizmi around A.D. 800. Using the vocabulary and notation of algebra, we can mathematically **model** many situations in the real world. In this section, we begin to explore the language of algebra by introducing some of its basic components.

TABLES AND GRAPHS

In algebra, we use tables to show relationships between quantities. For example, the following table lists the number of bicycle tires a production planner must order when a given number of bicycles are to be manufactured. For a production run of, say, 300 bikes, we locate 300 in the left-hand column and then scan across the table to see that the company must order 600 tires.

Bicycles to be manufactured	Tires to order
100	200
200	400
300	600
400	800

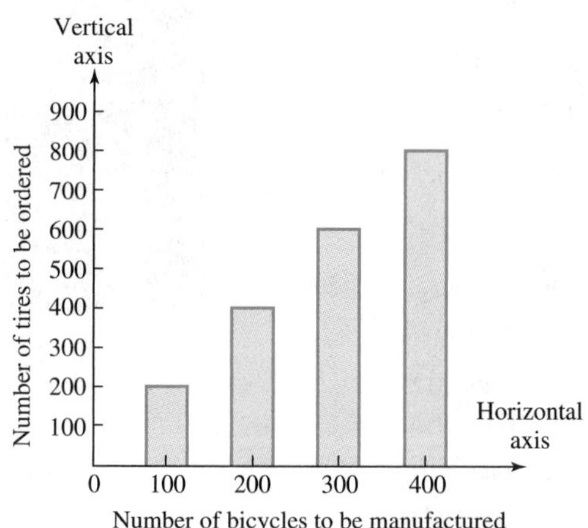

This information can also be presented using a graph. The **bar graph** (shown above) has a **horizontal axis** labeled "Number of bicycles to be manufactured." The labels are in units of 100 bicycles. The **vertical axis,** labeled "Number of tires to be ordered," is scaled in units of 100 tires. The height of a bar indicates the number of tires to order. For example, if 200 bikes are to be manufactured, we see that the bar extends to 400, meaning 400 tires are needed.

The Language of Algebra

Horizontal is a form of the word *horizon.* Think of the sun setting over the *horizon. Vertical* means in an upright position, like *vertical* blinds in a window.

Another way to present this information is with a **line graph.** Instead of using a bar to denote the number of tires to order, we use a dot drawn at the correct height. After drawing the data points for 100, 200, 300, and 400 bicycles, we connect them with line segments to create the following graph, on the right.

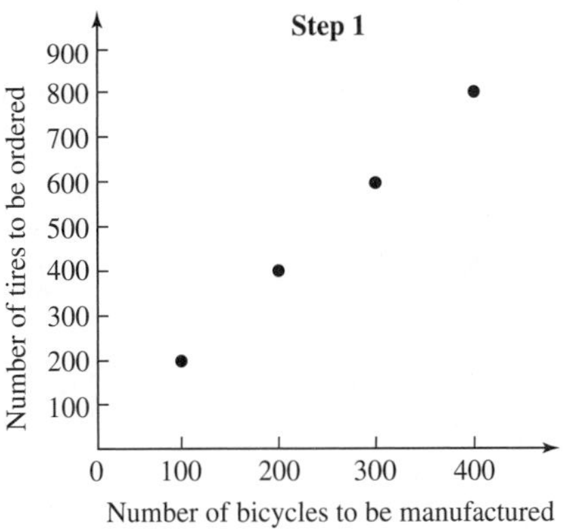

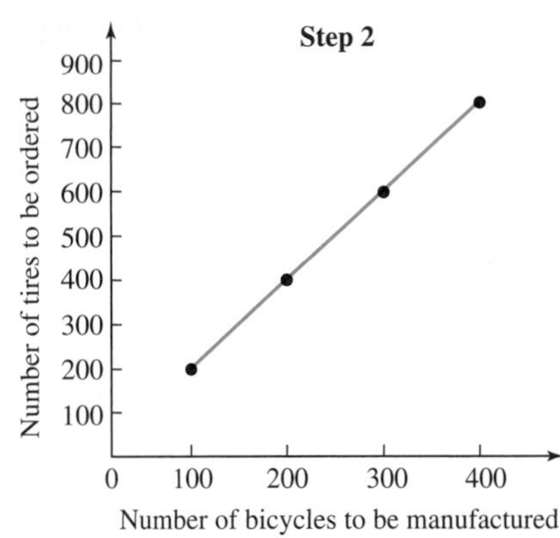

EXAMPLE 1

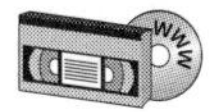

ELEMENTARY Algebra f(x) Now™

Use the following line graph to find the number of tires needed when 250 bicycles are to be manufactured.

Solution We locate 250 between 200 and 300 on the horizontal axis and draw a dashed line upward to intersect the graph. From the point of intersection, we draw a dashed horizontal line to the left that intersects the vertical axis at 500. This means that 500 tires should be ordered if 250 bicycles are to be manufactured.

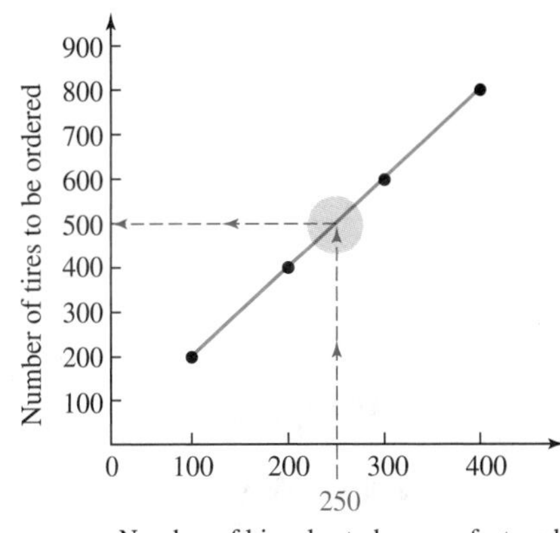

Success Tip

The video icons (see above) show which examples are taught on tutorial video tapes or disks.

Self Check 1 Use the graph to find the number of tires needed if 350 bicycles are to be manufactured.

VOCABULARY

From the table and graphs, it is clear that there is a relationship between the number of tires to order and the number of bicycles to be manufactured. Using words, we can express this relationship as a **verbal model:**

"The number of tires to order is two times the number of bicycles to be manufactured."

Since the word **product** indicates the result of a multiplication, we can write:

"The number of tires to order is the *product* of 2 and the number of bicycles to be manufactured."

To indicate other arithmetic operations, we will use the following words.

- A **sum** is the result of an addition: the sum of 5 and 6 is 11.
- A **difference** is the result of a subtraction: the difference of 3 and 2 is 1.
- A **quotient** is the result of a division: the quotient of 6 and 3 is 2.

NOTATION

Many symbols used in arithmetic are also used in algebra. For example, a $+$ symbol is used to indicate addition, a $-$ symbol is used to indicate subtraction, and an $=$ symbol means *is equal to.*

Since the letter x is often used in algebra and could be confused with the multiplication symbol $\times$, we normally write multiplication in other ways.

Symbols for Multiplication

Symbol	Meaning	Example
$\times$	Times sign	$6 \times 4 = 24$
$\cdot$	Raised dot	$6 \cdot 4 = 24$
( )	Parentheses	$(6)4 = 24$ or $6(4) = 24$ or $(6)(4) = 24$

In algebra, the symbol most often used to indicate division is the fraction bar.

Symbols for Division

Symbol	Meaning	Example
$\div$	Division sign	$24 \div 4 = 6$
$\overline{)}$	Long division	$4\overline{)24}$ with quotient 6
—	Fraction bar	$\frac{24}{4} = 6$

EXAMPLE 2

ELEMENTARY Algebra f(x) Now™

Express each statement in words, using one of the words *sum, product, difference,* or *quotient:* **a.** $\frac{22}{11} = 2$ and **b.** $22 + 11 = 33$.

Solution

a. We can represent the equal symbol $=$ with the word *is.* Since the fraction bar indicates division, we have: the quotient of 22 and 11 is 2.

b. The $+$ symbol indicates addition: the sum of 22 and 11 is 33.

Self Check 2 Express the following statement in words: $22 - 11 = 11$

VARIABLES, EXPRESSIONS, AND EQUATIONS

Another way to describe the tires–to–bicycles relationship uses *variables.* **Variables** are letters that stand for numbers. If we let the letter t stand for the number of tires to be ordered and b for the number of bicycles to be manufactured, we can translate the *verbal model* to mathematical symbols.

The Language of Algebra

The equal symbol = can be represented by words such as:

is gives yields equals

The symbol ≠ is read as *"is not equal to."*

The number of tires to order	is	two	times	the number of bicycles to be manufactured.
t	$=$	2	$\cdot$	b

The statement $t = 2 \cdot b$ is called an *equation.* An **equation** is a mathematical sentence that contains an $=$ symbol. Some examples are

$$3 + 5 = 8 \qquad x + 5 = 20 \qquad 17 - t = 14 - t \qquad p = 100 - d$$

When we multiply a variable by a number or multiply a variable by another variable, we can omit the symbol for multiplication.

$2b$ means $2 \cdot b$ $\quad$ xy means $x \cdot y$ $\quad$ abc means $a \cdot b \cdot c$

Using this form, we can write the equation $t = 2 \cdot b$ as $t = 2b$. The notation $2b$ on the right-hand side is called an **algebraic expression,** or more simply, an **expression.**

Algebraic Expressions Variables and/or numbers can be combined with the operations of addition, subtraction, multiplication, and division to create **algebraic expressions.**

Here are some examples of algebraic expressions.

$4a + 7$ — This expression is a combination of the numbers 4 and 7, the variable a, and the operations of multiplication and addition.

$\frac{10 - y}{3}$ — This expression is a combination of the numbers 10 and 3, the variable y, and the operations of subtraction and division.

$15mn(2m)$ — This expression is a combination of the numbers 15 and 2, the variables m and n, and the operation of multiplication.

In the bicycle manufacturing example, using the equation $t = 2b$ to describe the relationship has one major advantage over the other methods. It can be used to determine the number of tires needed for a production run of any size.

EXAMPLE 3

ELEMENTARY Algebra $f(x)$ Now™

Use the equation $t = 2b$ to find the number of tires needed for a production run of 178 bicycles.

Solution

$t = 2b$ — This is the describing equation.

$t = 2(178)$ — Replace b, which stands for the number of bicycles, with 178. Use parentheses to show the multiplication.

$t = 356$ — Do the multiplication.

If 178 bicycles are manufactured, 356 tires will be needed.

Self Check 3 Use the equation $t = 2b$ to find the number of tires needed if 604 bicycles are to be manufactured.

CONSTRUCTING TABLES

Equations such as $t = 2b$, which express a relationship between two or more variables, are called **formulas.** Some applications require the repeated use of a formula.

EXAMPLE 4

ELEMENTARY Algebra $f(x)$ Now™

Find the number of tires to order for production runs of 233 and 852 bicycles. Present the results in a table.

Bicycles to be manufactured b	Tires needed t
233	466
852	1,704

Solution **Step 1:** We construct a two-column table. Since b represents the number of bicycles to be manufactured, we use it as the heading of the first column. Since t represents the number of tires needed, we use it as the heading of the second column. Then we enter the size of each production run in the first column, as shown.

Step 2: In $t = 2b$, we replace b with 233 and with 852 and find each corresponding value of t.

$$t = 2b \qquad t = 2b$$
$$t = 2(233) \qquad t = 2(852)$$
$$t = 466 \qquad t = 1{,}704$$

Step 3: These results are entered in the second column.

Self Check 4 Find the number of tires needed for production runs of 87 and 487 bicycles. Present the results in a table.

Answers to Self Checks **1.** 700 **2.** The difference of 22 and 11 is 11. **3.** 1,208 **4.**

b	t
87	174
487	974

1.1 STUDY SET  ELEMENTARY Algebra f(x) Now™

VOCABULARY Fill in the blanks.

1. The answer to an addition problem is called the ______. The answer to a subtraction problem is called the ______.
2. The answer to a multiplication problem is called the ______. The answer to a division problem is called the ______.
3. ______ are letters that stand for numbers.
4. Variables and numbers can be combined with the operations of addition, subtraction, multiplication, and division to create algebraic ______.
5. An ______ is a mathematical sentence that contains an = symbol.
6. An equation such as $t = 2b$, which expresses a relationship between two or more variables, is called a ______.
7. The ______ axis of a graph extends left and right and the ______ axis extends up and down.
8. The word ______ comes from the title of a book written by an Arabian mathematician around A.D. 800.

CONCEPTS Classify each item as an algebraic expression or an equation.

9. $18 + m = 23$
10. $18 + m$
11. $y - 1$
12. $y - 1 = 2$
13. $30x$
14. $t = 16b$
15. $r = \frac{2}{3}$
16. $\frac{c - 7}{5}$
17. **a.** What operations does the expression $5x - 16$ contain?
    b. What variable does the expression contain?

18. a. What operations does the expression $\dfrac{12 + t}{25}$ contain?

b. What variable does the expression contain?

19. a. What operations does the equation $4 + 1 = 20 - m$ contain?

b. What variable does the equation contain?

20. a. What operations does the equation $y + 14 = 5(6)$ contain?

b. What variable does the equation contain?

21. Construct a line graph using the data in the table.

Hours worked	Pay (dollars)
1	20
2	40
3	60
4	80
5	100

22. Use the data in the graph to complete the table.

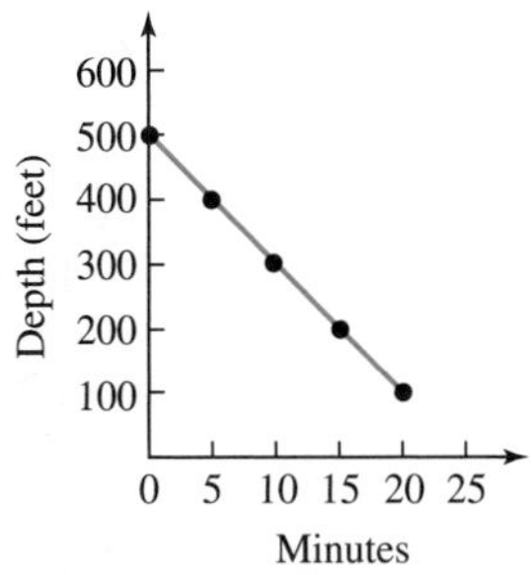

Minutes	Depth (feet)
0	
5	
10	
15	
20	

23. Explain what the dashed lines help us find in the graph.

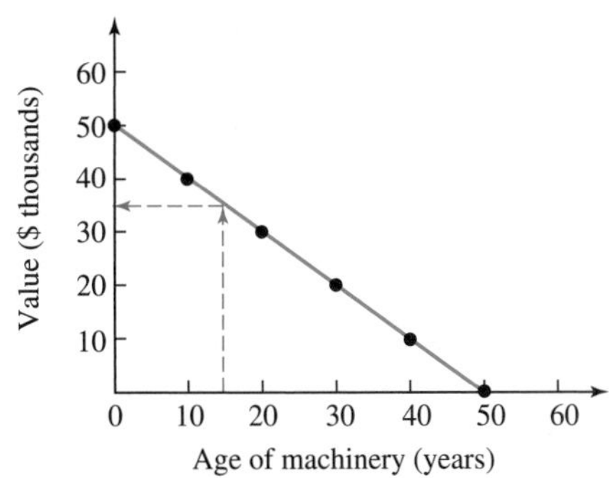

24. Use the line graph to find the income received from 30, 50, and 70 customers.

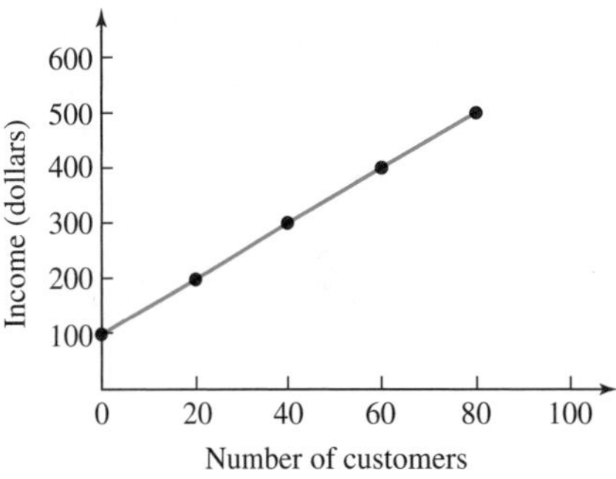

NOTATION **Write each multiplication using a raised dot and then parentheses.**

25. 5×6

26. 4×7

27. 34×75

28. 90×12

Write each expression without using a multiplication symbol or parentheses.

29. $4 \cdot x$

30. $5 \cdot y$

31. $3 \cdot r \cdot t$

32. $22 \cdot q \cdot s$

33. $l \cdot w$

34. $b \cdot h$

35. $P \cdot r \cdot t$

36. $l \cdot w \cdot h$

37. $2(w)$

38. $2(l)$

39. $(x)(y)$

40. $(r)(t)$

Write each division using a fraction bar.

41. $32 \div x$

42. $y \div 15$

43. $30\overline{)90}$

44. $20\overline{)80}$

PRACTICE **Express each statement using one of the words *sum, difference, product,* or *quotient.***

45. $8(2) = 16$

46. $45 \cdot 12 = 540$

47. $11 - 9 = 2$

48. $65 + 89 = 154$

49. $2x = 10$

50. $16t = 4$

51. $\dfrac{66}{11} = 6$

52. $12 \div 3 = 4$

Translate each verbal model into an equation. (Answers may vary, depending on the variables chosen.)

53. The sale price | is | $100 | minus | the discount.

54. The cost of dining out | equals | the cost of the meal | plus | $7 for parking.

55. 7 | times | the age of a dog in years | gives | the dog's equivalent human age.

56. The number of centuries | is | the number of years | divided by | 100.

57. The amount of sand that should be used is the product of 3 and the amount of cement used.
58. The number of waiters needed is the quotient of the number of customers and 10.
59. The weight of the truck is the sum of the weight of the engine and 1,200.
60. The number of classes still open is the difference of 150 and the number of classes that are closed.

61. The profit is the difference of the revenue and 600.

62. The distance is the product of the rate and 3.
63. The quotient of the number of laps run and 4 is the number of miles run.
64. The sum of the tax and 35 is the total cost.

Use the formula to complete each table.

65. $d = 360 + L$

Lunch time (minutes) L	School day (minutes) d
30	
40	
45	

66. $b = 1{,}024k$

Kilobytes k	Bytes b
1	
5	
10	

67. $t = 1{,}500 - d$

Deductions d	Take-home pay t
200	
300	
400	

68. $w = \frac{s}{12}$

Inches of snow s	Inches of water w
12	
24	
72	

Use the data to find a formula that describes the relationship between the two quantities.

69.

Eggs e	Dozens d
24	2
36	3
48	4

70.

Couples c	Individuals I
20	40
100	200
200	400

APPLICATIONS

71. TRAFFIC SAFETY As the railroad crossing guard drops, the measure of angle 1 ($\angle 1$) increases while the measure of $\angle 2$ decreases. At any instant the *sum* of the measures of the two angles is 90 degrees (denoted 90°). Complete the table. Then use the data to construct a line graph.

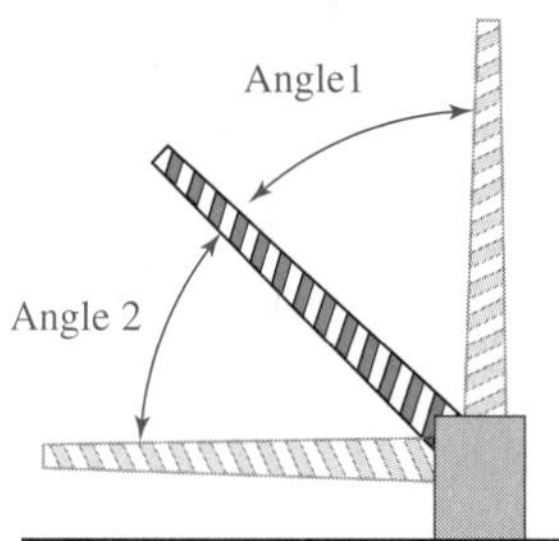

Angle 1 (degrees)	Angle 2 (degrees)
0	
30	
45	
60	
90	

72. U.S. CRIME STATISTICS Property crimes include burglary, theft, and motor vehicle theft. Graph the following property crime rate data using a bar graph. Is an overall trend apparent?

Year	Crimes per 1,000 households
1991	354
1992	325
1993	319
1994	310
1995	291
1996	266

Year	Crimes per 1,000 households
1997	248
1998	217
1999	198
2000	178
2001	167

Source: Bureau of Justice Statistics

WRITING

73. Many people misuse the word *equation* when discussing mathematics. What is an equation? Give an example.

74. Explain the difference between an algebraic expression and an equation. Give an example of each.

75. In this section, four methods for describing numerical relationships were discussed: tables, words, graphs, and equations. Which method do you think is the most useful? Explain why.

76. In your own words, define *horizontal* and *vertical.*

CHALLENGE PROBLEMS

77. Complete the table and the formula.

s	t
10	
18	19
	34
47	48

$t =$

78. Suppose $h = 4n$ and $n = 2g$. Complete the following formula: $h = \quad g$.

1.2 Fractions

- Factors and Prime Factorizations
- The Meaning of Fractions
- Multiplying and Dividing Fractions
- Building Equivalent Fractions
- Simplifying Fractions
- Adding and Subtracting Fractions
- Simplifying Answers
- Mixed Numbers

In arithmetic, you added, subtracted, multiplied, and divided **whole numbers:** 0, 1, 2, 3, 4, 5, and so on. Assuming that you have mastered those skills, we will now review the arithmetic of fractions.

FACTORS AND PRIME FACTORIZATIONS

To compute with fractions, we need to know how to *factor* whole numbers. To **factor** a number means to express it as a product of two or more numbers. For example, some ways to factor 8 are

$$1 \cdot 8, \qquad 4 \cdot 2, \qquad \text{and} \qquad 2 \cdot 2 \cdot 2$$

The numbers 1, 2, 4, and 8 that were used to write the products are called **factors** of 8.

The Language of Algebra

When we say "factor 8," we are using the word *factor* as a verb. When we say "2 is a *factor* of 8," we are using the word factor as a noun.

Sometimes a number has only two factors, itself and 1. We call these numbers *prime numbers.*

Prime Numbers and Composite Numbers

A **prime number** is a whole number greater than 1 that has only itself and 1 as factors. The first ten prime numbers are 2, 3, 5, 7, 11, 13, 17, 19, 23, and 29.

A **composite number** is a whole number, greater than 1, that is not prime. The first ten composite numbers are 4, 6, 8, 9, 10, 12, 14, 15, 16, and 18.

Every composite number can be factored into the product of two or more prime numbers. This product of these prime numbers is called its **prime factorization.**

EXAMPLE 1

Find the prime factorization of 210.

Solution First, write 210 as the product of two whole numbers other than 1.

The Language of Algebra

In Example 1, the prime factors are written in *ascending* order. To *ascend* means to move upward.

$210 = 10 \cdot 21$ The resulting prime factorization will be the same no matter which two factors of 210 you begin with.

Neither 10 nor 21 are prime numbers, so we factor each of them.

$210 = 2 \cdot 5 \cdot 3 \cdot 7$ Factor 10 as $2 \cdot 5$ and factor 21 as $3 \cdot 7$.

Writing the factors in ascending order, the **prime-factored form** of 210 is $2 \cdot 3 \cdot 5 \cdot 7$. Two other methods for prime factoring 210 are shown as follows.

Success Tip

When finding a prime factorization, remember that a whole number is divisible by

- 2 if it ends in 0, 2, 4, 6, or 8
- 3 if the sum of the digits is divisible by 3
- 5 if it ends in 0 or 5
- 10 if it ends in 0

Factor tree

Work downward. Factor each number as a product of two numbers other than 1 and itself until all factors are prime. ↓

```
  210
(7)   30
    6   (5)
 (3) (2)
```

Division ladder

```
   7
5)35
3)105
2)210
```

↑ Work upward. Perform repeated division by prime numbers until the final quotient is a prime number.

Either way, the factorization is $2 \cdot 3 \cdot 5 \cdot 7$.

Self Check 1 Find the prime factorization of 189.

THE MEANING OF FRACTIONS

In a fraction, the number above the **fraction bar** is called the **numerator,** and the number below is called the **denominator.**

fraction bar ⟶ $\frac{1}{2}$ ⟵ numerator / ⟵ denominator

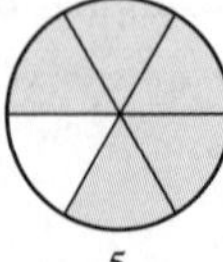

$\frac{5}{6}$

Fractions can describe the number of equal parts of a whole. To illustrate, we consider the circle with 5 of its 6 equal parts colored red. We say that $\frac{5}{6}$ (five-sixths) of the circle is shaded.

Fractions are also used to indicate division. For example, $\frac{8}{2}$ indicates that the numerator, 8, is to be divided by the denominator, 2:

$$\frac{8}{2} = 8 \div 2 = 4$$ We know that $\frac{8}{2} = 4$ because of its related multiplication statement: $4 \cdot 2 = 8$.

A numerator can be any number, including 0. If it is 0 and the denominator is not, the fraction is equal to 0. For example, $\frac{0}{6} = 0$ because $0 \cdot 6 = 0$.

A denominator can be any number except 0. To see why, consider $\frac{6}{0}$. If there were an answer to the division, then $\frac{6}{0}$ = answer. The related multiplication statement (answer $\cdot\ 0 = 6$) is impossible because no number, when multiplied by 0, gives 6. Fractions such as $\frac{6}{0}$, that indicate division of a nonzero number by 0, are undefined.

The Language of Algebra

The word *undefined* is used in mathematics to describe an expression that has no meaning.

If the numerator and denominator of a fraction are the same nonzero number, the fraction indicates division of a number by itself, and the result is 1. For example, $\frac{9}{9} = 1$.

If a denominator is 1, the fraction indicates division by 1, and the result is simply the numerator. For example, $\frac{5}{1} = 5$.

The facts about special fraction forms are summarized as follows.

Special Fraction Forms For any nonzero number a,

$$\frac{a}{a} = 1 \qquad \frac{a}{1} = a \qquad \frac{0}{a} = 0 \qquad \frac{a}{0} \text{ is undefined}$$

MULTIPLYING AND DIVIDING FRACTIONS

We now discuss how to add, subtract, multiply, and divide fractions. We begin with multiplication.

Multiplying Fractions To multiply two fractions, multiply the numerators and multiply the denominators.

Let a, b, c, and d represent numbers, where b and d are not 0,

$$\frac{a}{b} \cdot \frac{c}{d} = \frac{a \cdot c}{b \cdot d}$$

EXAMPLE 2 Multiply: $\frac{7}{8} \cdot \frac{3}{5}$.

ELEMENTARY Algebra $f(x)$ Now™

Solution

$$\frac{7}{8} \cdot \frac{3}{5} = \frac{7 \cdot 3}{8 \cdot 5}$$ Multiply the numerators and multiply the denominators.

$$= \frac{21}{40}$$ Do the multiplications.

Self Check 2 Multiply: $\frac{5}{9} \cdot \frac{2}{3}$.

One number is called the **reciprocal** of another if their product is 1. To find the reciprocal of a fraction, we invert its numerator and denominator.

Success Tip

Every number, except 0, has a reciprocal. Zero has no reciprocal, because the product of 0 and a number cannot be 1.

$\frac{3}{4}$ is the reciprocal of $\frac{4}{3}$, because $\frac{3}{4} \cdot \frac{4}{3} = \frac{12}{12} = 1$.

$\frac{1}{10}$ is the reciprocal of 10, because $\frac{1}{10} \cdot 10 = \frac{10}{10} = 1$.

We use reciprocals to divide fractions.

Dividing Fractions To divide fractions, multiply the first fraction by the reciprocal of the second.

Let a, b, c, and d represent numbers, where b, c, and d are not 0,

$$\frac{a}{b} \div \frac{c}{d} = \frac{a}{b} \cdot \frac{d}{c}$$

EXAMPLE 3 Divide: $\frac{1}{3} \div \frac{4}{5}$.

ELEMENTARY Algebra f(x) Now™

Solution

$\frac{1}{3} \div \frac{4}{5} = \frac{1}{3} \cdot \frac{5}{4}$ Multiply the first fraction by the reciprocal of the second. The reciprocal of $\frac{4}{5}$ is $\frac{5}{4}$.

$= \frac{1 \cdot 5}{3 \cdot 4}$ Multiply the numerators and multiply the denominators.

$= \frac{5}{12}$ Do the multiplications.

Self Check 3 Divide: $\frac{6}{25} \div \frac{1}{2}$.

BUILDING EQUIVALENT FRACTIONS

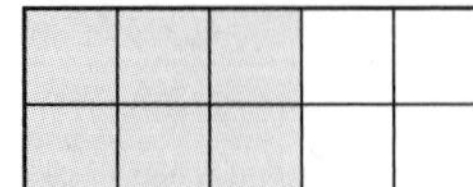

In the figure on the left, the rectangle is divided into 10 equal parts. Since 6 of those parts are red, $\frac{6}{10}$ of the figure is shaded.

Now consider the 5 equal parts of the rectangle created by the vertical lines. Since 3 of those parts are red, $\frac{3}{5}$ of the figure is shaded. We can conclude that $\frac{6}{10} = \frac{3}{5}$ because $\frac{6}{10}$ and $\frac{3}{5}$ represent the same shaded part of the rectangle. We say that $\frac{6}{10}$ and $\frac{3}{5}$ are *equivalent fractions.*

Equivalent Fractions Two fractions are **equivalent** if they represent the same number.

Writing a fraction as an equivalent fraction with a larger denominator is called **building** the fraction. To build a fraction, we multiply it by a form of the number 1. Since any number multiplied by 1 remains the same (identical), 1 is called the **multiplicative identity.**

Multiplication Property of 1

The product of 1 and any number is that number.
For any number a,

$$1 \cdot a = a \quad \text{and} \quad a \cdot 1 = a$$

EXAMPLE 4

Write $\frac{3}{5}$ as an equivalent fraction with a denominator of 35.

Solution We need to multiply the denominator of $\frac{3}{5}$ by 7 to obtain a denominator of 35. It follows that $\frac{7}{7}$ should be the form of 1 that is used to build $\frac{3}{5}$. Multiplying $\frac{3}{5}$ by $\frac{7}{7}$ does not change its value, because we are multiplying $\frac{3}{5}$ by 1.

$$\frac{3}{5} = \frac{3}{5} \cdot \frac{7}{7} \quad \frac{7}{7} = 1$$

$$= \frac{3 \cdot 7}{5 \cdot 7} \quad \text{Multiply the numerators and multiply the denominators.}$$

$$= \frac{21}{35}$$

Self Check 4 Write $\frac{5}{8}$ as an equivalent fraction with a denominator of 24.

SIMPLIFYING FRACTIONS

The Language of Algebra

The word *infinitely* is a form of the word *infinite,* which means endless.

Every fraction can be written in infinitely many equivalent forms. For example, some equivalent forms of $\frac{30}{36}$ are:

$$\frac{5}{6} = \frac{10}{12} = \frac{15}{18} = \frac{20}{24} = \frac{25}{30} = \frac{30}{36} = \frac{35}{42} = \frac{40}{48} = \frac{45}{54} = \frac{50}{60} = \cdots$$

Of all of the equivalent forms in which we can write a fraction, we often need to determine the one that is in *simplest form.*

Simplest Form of a Fraction

A fraction is in **simplest form,** or **lowest terms,** when the numerator and denominator have no common factors other than 1.

To **simplify a fraction,** we write it in simplest form by removing a factor equal to 1. For example, to simplify $\frac{30}{36}$, we note that the greatest factor common to the numerator and denominator is 6 and proceed as follows:

$$\frac{30}{36} = \frac{5 \cdot 6}{6 \cdot 6} \quad \text{Factor 30 and 36, using their greatest common factor, 6.}$$

$$= \frac{5}{6} \cdot \frac{6}{6} \quad \text{Use the rule for multiplying fractions in reverse: write } \frac{5 \cdot 6}{6 \cdot 6} \text{ as the product of two fractions, } \frac{5}{6} \text{ and } \frac{6}{6}.$$

$$= \frac{5}{6} \cdot 1 \quad \text{A number divided by itself is equal to 1: } \frac{6}{6} = 1.$$

$$= \frac{5}{6}$$ Use the multiplication property of 1: any number multiplied by 1 remains the same.

To simplify $\frac{30}{36}$, we removed a factor equal to 1 in the form of $\frac{6}{6}$. The result, $\frac{5}{6}$, is equivalent to $\frac{30}{36}$.

We can easily identify the greatest common factor of the numerator and the denominator of a fraction if we write them in prime-factored form.

EXAMPLE 5

Simplify each fraction, if possible: **a.** $\frac{63}{42}$ and **b.** $\frac{33}{40}$.

ELEMENTARY Algebra $f(x)$ Now™

Solution **a.** After prime factoring 63 and 42, we see that the greatest common factor of the numerator and the denominator is $3 \cdot 7 = 21$.

Language of Algebra

What do Calvin Klein, Sheryl Swoopes, and Dustin Hoffman have in common? They all attended a community college. The word *common* means shared by two or more. In this section, we will work with *common* factors and *common* denominators.

$$\frac{63}{42} = \frac{3 \cdot 3 \cdot 7}{2 \cdot 3 \cdot 7}$$ Write 63 and 42 in prime-factored form.

$$= \frac{3}{2} \cdot \frac{3 \cdot 7}{3 \cdot 7}$$ Write $\frac{3 \cdot 3 \cdot 7}{2 \cdot 3 \cdot 7}$ as the product of two fractions, $\frac{3}{2}$ and $\frac{3 \cdot 7}{3 \cdot 7}$.

$$= \frac{3}{2} \cdot 1$$ A nonzero number divided by itself is equal to 1: $\frac{3 \cdot 7}{3 \cdot 7} = 1$.

$$= \frac{3}{2}$$ Any number multiplied by 1 remains the same.

b. Prime factor 33 and 40.

$$\frac{33}{40} = \frac{3 \cdot 11}{2 \cdot 2 \cdot 2 \cdot 5}$$

Since the numerator and the denominator have no common factors other than 1, $\frac{33}{40}$ is in simplest form (lowest terms).

Self Check 5 Simplify each fraction, if possible: **a.** $\frac{24}{56}$ and **b.** $\frac{16}{125}$.

To streamline the simplifying process, we can replace pairs of factors common to the numerator and denominator with the equivalent fraction $\frac{1}{1}$.

EXAMPLE 6

Simplify: $\frac{90}{105}$.

Solution

$$\frac{90}{105} = \frac{2 \cdot 3 \cdot 3 \cdot 5}{3 \cdot 5 \cdot 7}$$ Write 90 and 105 in prime-factored form.

$$= \frac{2 \cdot \overset{1}{\cancel{3}} \cdot 3 \cdot \overset{1}{\cancel{5}}}{\underset{1}{\cancel{3}} \cdot \underset{1}{\cancel{5}} \cdot 7}$$ Slashes and 1's are used to show that $\frac{3}{3}$ and $\frac{5}{5}$ are replaced by the equivalent fraction $\frac{1}{1}$.

$$= \frac{6}{7}$$ Multiply to find the numerator: $2 \cdot 1 \cdot 3 \cdot 1 = 6$. Multiply to find the denominator: $1 \cdot 1 \cdot 7 = 7$.

Self Check 6 Simplify: $\frac{126}{70}$.

Simplifying a Fraction

1. Factor (or prime factor) the numerator and denominator to determine all the factors common to both.
2. Replace each pair of factors common to the numerator and denominator with the equivalent fraction $\frac{1}{1}$.
3. Multiply the remaining factors in the numerator and in the denominator.

Caution When all common factors of the numerator and the denominator of a fraction are removed, forgetting to write 1's above the slashes leads to a common mistake.

Correct

$$\frac{15}{45} = \frac{\overset{1}{\cancel{3}} \cdot \overset{1}{\cancel{5}}}{\underset{1}{\cancel{3}} \cdot 3 \cdot \underset{1}{\cancel{5}}} = \frac{1}{3}$$

Incorrect

$$\frac{15}{45} = \frac{\cancel{3} \cdot \cancel{5}}{\cancel{3} \cdot 3 \cdot \cancel{5}} = \frac{0}{3} = 0$$

ADDING AND SUBTRACTING FRACTIONS

To add or subtract fractions, they must have the same denominator.

Adding and Subtracting Fractions

To add (or subtract) two fractions with the same denominator, add (or subtract) their numerators and write the sum (or difference) over the common denominator.

Let a, b, and d represent numbers, where d is not 0,

$$\frac{a}{d} + \frac{b}{d} = \frac{a+b}{d} \qquad \frac{a}{d} - \frac{b}{d} = \frac{a-b}{d}$$

For example,

$$\frac{3}{7} + \frac{1}{7} = \frac{3+1}{7} = \frac{4}{7} \quad \text{and} \quad \frac{18}{25} - \frac{9}{25} = \frac{18-9}{25} = \frac{9}{25}$$

Caution Only factors common to the numerator and the denominator of a fraction can be removed. It is incorrect to remove the 5's in $\frac{5+8}{5}$, because 5 is not used as a factor in the expression $5 + 8$. This error leads to an incorrect answer of 9.

Correct

$$\frac{5+8}{5} = \frac{13}{5}$$

Incorrect

$$\frac{5+8}{5} = \frac{\overset{1}{\cancel{5}} + 8}{\underset{1}{\cancel{5}}} = \frac{9}{1} = 9$$

To add (or subtract) fractions with different denominators, we express them as equivalent fractions that have a common denominator. The smallest common denominator, called the **least** or **lowest common denominator,** is usually the easiest common denominator to use.

Least Common Denominator (LCD) The **least** or **lowest common denominator (LCD)** for a set of fractions is the smallest number each denominator will divide exactly (divide with no remainder).

Success Tip

To determine the LCD of two fractions, list the multiples of one of the denominators. The first number in the list that is exactly divisible by the other denominator is their LCD. For $\frac{2}{5}$ and $\frac{1}{3}$, the multiples of 5 are 5, 10, 15, 20, 25, ... Since 15 is the first number in the list that is exactly divisible by 3, the LCD is 15.

The denominators of $\frac{2}{5}$ and $\frac{1}{3}$ are 5 and 3. The numbers 5 and 3 divide many numbers exactly (30, 45, and 60, to name a few), but the smallest number that they divide exactly is 15. Thus, 15 is the LCD for $\frac{2}{5}$ and $\frac{1}{3}$.

To find $\frac{2}{5} + \frac{1}{3}$, we find equivalent fractions that have denominators of 15 and we use the rule for adding fractions.

$\frac{2}{5} + \frac{1}{3} = \frac{2}{5} \cdot \frac{3}{3} + \frac{1}{3} \cdot \frac{5}{5}$ Multiply $\frac{2}{5}$ by 1 in the form of $\frac{3}{3}$. Multiply $\frac{1}{3}$ by 1 in the form of $\frac{5}{5}$.

$= \frac{6}{15} + \frac{5}{15}$ Multiply the numerators and multiply the denominators. Note that the denominators are now the same.

$= \frac{6 + 5}{15}$ Add the numerators. Write the sum over the common denominator.

$= \frac{11}{15}$ Do the addition.

When adding (or subtracting) fractions with unlike denominators, the least common denominator is not always obvious. Prime factorization is helpful in determining the LCD.

EXAMPLE 7

Subtract: $\frac{3}{10} - \frac{5}{28}$.

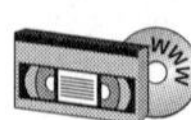

ELEMENTARY Algebra $f(x)$ Now™

Solution To find the LCD, we find the prime factorization of both denominators and use each prime factor the *greatest* number of times it appears in any one factorization:

$$\left.\begin{aligned} 10 &= 2 \cdot 5 \\ 28 &= 2 \cdot 2 \cdot 7 \end{aligned}\right\} \text{LCD} = 2 \cdot 2 \cdot 5 \cdot 7 = 140$$

2 appears twice in the factorization of 28.
5 appears once in the factorization of 10.
7 appears once in the factorization of 28.

Since 140 is the smallest number that 10 and 28 divide exactly, we write both fractions as fractions with the LCD of 140.

$\frac{3}{10} - \frac{5}{28} = \frac{3}{10} \cdot \frac{14}{14} - \frac{5}{28} \cdot \frac{5}{5}$ We must multiply 10 by 14 to obtain 140. We must multiply 28 by 5 to obtain 140.

$= \frac{42}{140} - \frac{25}{140}$ Do the multiplications.

$= \frac{42 - 25}{140}$ Subtract the numerators. Write the difference over the common denominator.

$= \frac{17}{140}$ Do the subtraction.

Self Check 7 Subtract: $\frac{11}{48} - \frac{7}{40}$.

SIMPLIFYING ANSWERS

When adding, subtracting, multiplying, or dividing fractions, remember to express the answer in simplest form.

EXAMPLE 8 ELEMENTARY Algebra f(x) Now™

Perform each operation: **a.** $45\left(\frac{4}{9}\right)$ and **b.** $\frac{5}{12} + \frac{3}{4}$.

Solution **a.**

$$45\left(\frac{4}{9}\right) = \frac{45}{1}\left(\frac{4}{9}\right)$$ Write 45 as a fraction: $45 = \frac{45}{1}$.

$$= \frac{45 \cdot 4}{1 \cdot 9}$$ Multiply the numerators. Multiply the denominators.

$$= \frac{\overset{1}{\cancel{3}} \cdot \overset{1}{\cancel{3}} \cdot 5 \cdot 2 \cdot 2}{1 \cdot \underset{1}{\cancel{3}} \cdot \underset{1}{\cancel{3}}}$$ To simplify the fraction, prime factor 45, 4, and 9. Then replace each $\frac{3}{3}$ with $\frac{1}{1}$.

$$= 20$$

b. Since the smallest number that 12 and 4 divide exactly is 12, the LCD is 12.

$$\frac{5}{12} + \frac{3}{4} = \frac{5}{12} + \frac{3}{4} \cdot \frac{\mathbf{3}}{\mathbf{3}}$$ $\frac{5}{12}$ already has a denominator of 12. Build $\frac{3}{4}$ so that its denominator is 12.

$$= \frac{5}{12} + \frac{9}{12}$$ Multiply the numerators and denominators in the second term. The denominators are now the same.

$$= \frac{14}{12}$$ Add the numerators, 5 and 9, to get 14. Write that sum over the common denominator.

$$= \frac{\overset{1}{\cancel{2}} \cdot 7}{\underset{1}{\cancel{2}} \cdot 6}$$ To simplify $\frac{14}{12}$, factor 14 and 12, using their greatest common factor, 2. Then remove $\frac{2}{2} = 1$.

$$= \frac{7}{6}$$

Self Check 8 Perform each operation: **a.** $24\left(\frac{8}{6}\right)$, **b.** $\frac{1}{15} + \frac{31}{30}$

MIXED NUMBERS

A **mixed number** represents the sum of a whole number and a fraction. For example, $5\frac{3}{4}$ means $5 + \frac{3}{4}$. To perform calculations involving mixed numbers, we often express them as improper fractions.

EXAMPLE 9

ELEMENTARY Algebra $f(x)$ Now™

Divide: $5\frac{3}{4} \div 2$.

Solution To write the mixed number $5\frac{3}{4}$ as a fraction, we use a two-step process.

$$5\frac{3}{4} = \frac{23}{4}$$

Step 1. Multiply the whole number by the denominator: $5 \cdot 4 = 20$. Then add that product to the numerator: $20 + 3 = 23$.
Step 2. Write the result from Step 1 over the denominator, 4.

The Language of Algebra

Fractions such as $\frac{23}{4}$, with a numerator greater than or equal to the denominator, are called **improper fractions.** This term is misleading. In algebra, such fractions are often preferable to their equivalent mixed number form.

Now we replace $5\frac{3}{4}$ with $\frac{23}{4}$ and divide.

$$5\frac{3}{4} \div 2 = \frac{23}{4} \div \frac{2}{1}$$ Write $5\frac{3}{4}$ as $\frac{23}{4}$. Write 2 as a fraction: $2 = \frac{2}{1}$.

$$= \frac{23}{4} \cdot \frac{1}{2}$$ Multiply by the reciprocal of $\frac{2}{1}$, which is $\frac{1}{2}$.

$$= \frac{23}{8}$$ Multiply the numerators. Multiply the denominators.

To write the answer, $\frac{23}{8}$, as a mixed number, we use a two-step process.

$$\frac{23}{8} = 2\frac{7}{8}$$

Step 1. Divide the numerator, 23, by the denominator, 8.
Step 2. The quotient, 2, is the whole-number part of the mixed number. Its fractional part is the remainder, 7, over the original denominator, 8.

$$\begin{array}{r} 2 \\ 8\overline{)23} \\ \underline{16} \\ 7 \end{array}$$

Self Check 9 Multiply: $1\frac{1}{8} \cdot 9$.

Answers to Self Checks **1.** $189 = 3 \cdot 3 \cdot 3 \cdot 7$ **2.** $\frac{10}{27}$ **3.** $\frac{12}{25}$ **4.** $\frac{15}{24}$ **5. a.** $\frac{3}{7}$, **b.** in simplest form **6.** $\frac{9}{5}$ **7.** $\frac{13}{240}$ **8. a.** 32, **b.** $\frac{11}{10}$ **9.** $\frac{81}{8} = 10\frac{1}{8}$

1.2 STUDY SET

ELEMENTARY Algebra $f(x)$ Now™

VOCABULARY **Fill in the blanks.**

1. Numbers that have only 1 and themselves as factors, such as 23, 37, and 41, are called _______ numbers.
2. When we write 60 as $20 \cdot 3$, we say that we have _________ 60. When we write 60 as $5 \cdot 3 \cdot 2 \cdot 2$, we say that we have written 60 in ____________ form.
3. The _________ of the fraction $\frac{3}{4}$ is 3, and the ___________ is 4.
4. A fraction is in ________ form, or _______ terms, when the numerator and denominator have no common factors other than 1.
5. Two fractions that represent the same number, such as $\frac{1}{2}$ and $\frac{2}{4}$, are called _________ fractions.
6. The number $\frac{2}{3}$ is the _________ of the number $\frac{3}{2}$, because their product is 1.
7. The ____________ common denominator for a set of fractions is the smallest number each denominator will divide exactly.
8. The _______ number $7\frac{1}{3}$ represents the sum of a whole number and a fraction: $7 + \frac{1}{3}$.

CONCEPTS

9. The prime factorization of a number is $2 \cdot 2 \cdot 3 \cdot 5$. What is the number?

10. Complete each fact about fractions. Assume $a \neq 0$.

$$\frac{a}{a} = \square \quad \frac{a}{1} = \square \quad \frac{0}{a} = \square \quad \frac{a}{0} \text{ is } \square$$

11. What equivalent fractions are shown in the illustration?

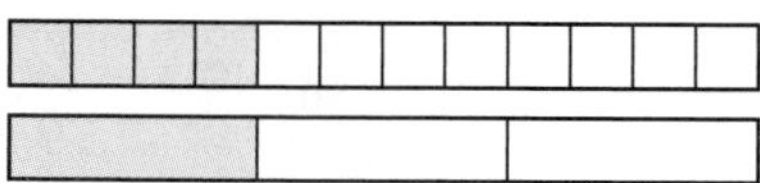

12. Complete each rule.

a. $\frac{a}{b} \cdot \frac{c}{d} = \square$ b. $\frac{a}{b} \div \frac{c}{d} = \square$

c. $\frac{a}{d} + \frac{b}{d} = \square$ d. $\frac{a}{d} - \frac{b}{d} = \square$

13. Simplify $\frac{2 \cdot 2 \cdot 3}{2 \cdot 3 \cdot 5}$.

14. To express $\frac{3}{8}$ as an equivalent fraction with a denominator of 40, by what number must we multiply the numerator and the denominator?

15. Complete each statement.

a. To build a fraction, we multiply it by $\square$ in the form of $\frac{2}{2}, \frac{3}{3}$, or $\frac{4}{4}$, and so on.

b. To simplify a fraction, we remove factors equal to $\square$ in the form of $\frac{2}{2}, \frac{3}{3}$, or $\frac{4}{4}$, and so on.

16. a. Give three numbers that 4 and 6 divide exactly.

b. What is the smallest number that 4 and 6 divide exactly?

17. The prime factorizations of 24 and 36 are

$$24 = 2 \cdot 2 \cdot 2 \cdot 3$$
$$36 = 2 \cdot 2 \cdot 3 \cdot 3$$

a. What is the greatest number of times 2 appears in any one factorization?

b. What is the greatest number of times 3 appears in any one factorization?

18. a. Write $2\frac{15}{16}$ as an improper fraction.

b. Write $\frac{49}{12}$ as a mixed number.

NOTATION

19. Consider $\frac{5}{16} = \frac{5}{16} \cdot \frac{3}{3}$.

a. What fraction is being built up?

b. Fill in the blank: $\frac{3}{3} = \square$.

c. What equivalent fraction is the result after building $\frac{5}{16}$?

20. Consider $\frac{70}{175} = \frac{\overset{1}{\cancel{7}} \cdot \overset{1}{\cancel{5}} \cdot 2}{\underset{1}{\cancel{7}} \cdot \underset{1}{\cancel{5}} \cdot 5}$.

a. What fraction is being simplified?

b. What are $\frac{7}{7}$ and $\frac{5}{5}$ replaced with?

c. What equivalent fraction is the result after simplifying?

PRACTICE

List the factors of each number.

21. 20
22. 50
23. 28
24. 36

Give the prime factorization of each number.

25. 75
26. 20
27. 28
28. 54
29. 117
30. 147
31. 220
32. 270

Build each fraction or whole number to an equivalent fraction having the indicated denominator.

33. $\frac{1}{3}$, denominator 9
34. $\frac{3}{8}$, denominator 24
35. $\frac{4}{9}$, denominator 54
36. $\frac{9}{16}$, denominator 64
37. 7, denominator 5
38. 12, denominator 3

Write each fraction in lowest terms. If the fraction is in lowest terms, so indicate.

39. $\frac{6}{12}$
40. $\frac{3}{9}$
41. $\frac{24}{18}$
42. $\frac{35}{14}$

43. $\frac{15}{20}$

44. $\frac{22}{77}$

45. $\frac{72}{64}$

46. $\frac{26}{21}$

47. $\frac{33}{56}$

48. $\frac{26}{39}$

49. $\frac{36}{225}$

50. $\frac{175}{490}$

Perform each operation and simplify the result when possible.

51. $\frac{1}{2} \cdot \frac{3}{5}$

52. $\frac{3}{4} \cdot \frac{5}{7}$

53. $\frac{4}{3}\left(\frac{6}{5}\right)$

54. $\frac{7}{8}\left(\frac{6}{15}\right)$

55. $\frac{5}{12} \cdot \frac{18}{5}$

56. $\frac{5}{4} \cdot \frac{12}{10}$

57. $21\left(\frac{10}{3}\right)$

58. $28\left(\frac{4}{7}\right)$

59. $7\frac{1}{2} \cdot 1\frac{2}{5}$

60. $3\frac{1}{4}\left(1\frac{1}{5}\right)$

61. $6 \cdot 2\frac{7}{24}$

62. $7 \cdot 1\frac{3}{28}$

63. $\frac{3}{5} \div \frac{2}{3}$

64. $\frac{4}{5} \div \frac{3}{7}$

65. $\frac{3}{4} \div \frac{6}{5}$

66. $\frac{3}{8} \div \frac{15}{28}$

67. $\frac{21}{35} \div \frac{3}{14}$

68. $\frac{23}{25} \div \frac{46}{5}$

69. $6 \div \frac{3}{14}$

70. $23 \div \frac{46}{5}$

71. $3\frac{1}{3} \div 1\frac{5}{6}$

72. $2\frac{1}{2} \div 1\frac{5}{8}$

73. $8 \div 3\frac{1}{5}$

74. $15 \div 3\frac{1}{3}$

75. $\frac{3}{5} + \frac{3}{5}$

76. $\frac{4}{13} - \frac{3}{13}$

77. $\frac{1}{6} + \frac{1}{24}$

78. $\frac{17}{25} - \frac{2}{5}$

79. $\frac{3}{5} + \frac{2}{3}$

80. $\frac{4}{3} + \frac{7}{2}$

81. $\frac{5}{12} + \frac{1}{3}$

82. $\frac{7}{15} + \frac{1}{5}$

83. $\frac{9}{4} - \frac{5}{6}$

84. $\frac{2}{15} + \frac{7}{9}$

85. $\frac{7}{10} - \frac{1}{14}$

86. $\frac{7}{25} + \frac{3}{10}$

87. $\frac{5}{14} - \frac{4}{21}$

88. $\frac{2}{33} + \frac{3}{22}$

89. $3 - \frac{3}{4}$

90. $\frac{17}{3} + 4$

91. $3\frac{3}{4} - 2\frac{1}{2}$

92. $15\frac{5}{6} + 11\frac{5}{8}$

93. $8\frac{2}{9} - 7\frac{2}{3}$

94. $3\frac{4}{5} - 3\frac{1}{10}$

APPLICATIONS

95. BOTANY To assess the effects of smog, botanists cut down a pine tree and measured the width of the growth rings for the last two years.

a. What was the growth over this two-year period?

b. What is the difference in the widths of the rings?

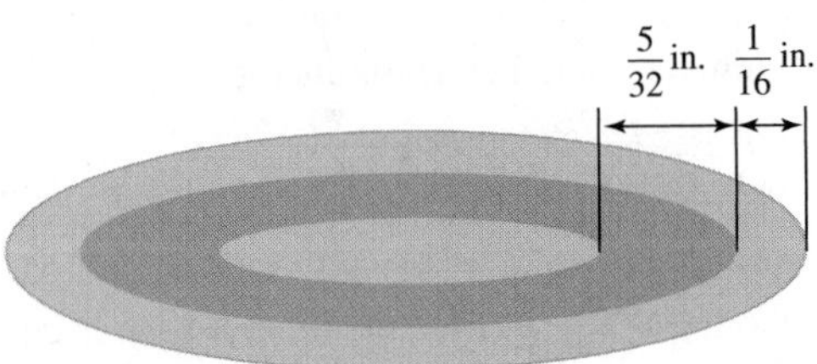

96. HARDWARE To secure the bracket to the stock, a bolt and a nut are used. How long should the threaded part of the bolt be?

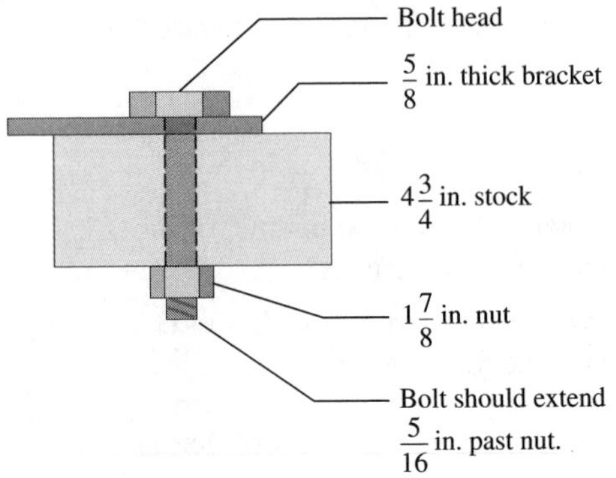

97. FRAMES How much molding is needed to produce the square picture frame shown?

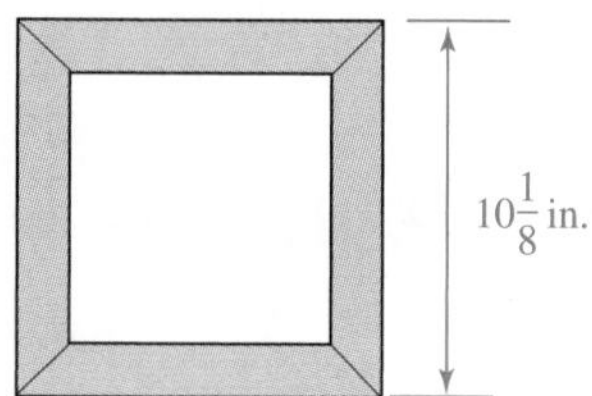

98. DECORATING The materials used to make a pillow are shown here. Examine the inventory list to decide how many pillows can be manufactured in one production run with the materials in stock.

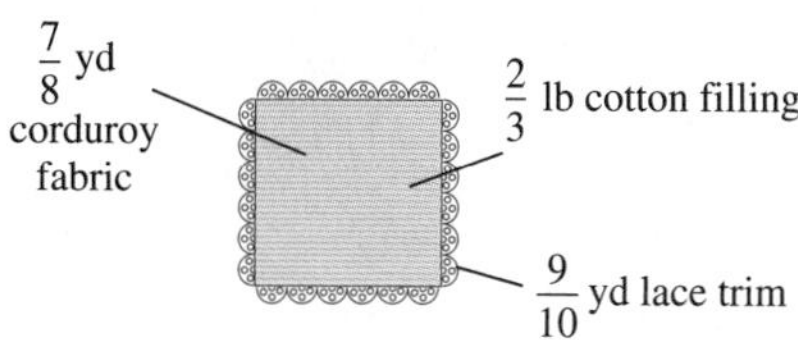

Factory Inventory List

Materials	Amount in stock
Lace trim	135 yd
Corduroy fabric	154 yd
Cotton filling	98 lb

WRITING

99. Explain how to add two fractions having unlike denominators.

100. To multiply two fractions, must they have like denominators? Explain.

101. What are equivalent fractions?

102. Explain the error in the following work.

$$\text{Add: } \frac{4}{3} + \frac{3}{2} = \frac{4}{\cancel{3}_1} + \frac{\cancel{3}^1}{5}$$

$$= \frac{4}{1} + \frac{1}{5}$$

$$= 4 + \frac{1}{5}$$

$$= 4\frac{1}{5}$$

REVIEW

Express each statement using one of the words sum, difference, product, or quotient.

103. $7 - 5 = 2$

104. $5(6) = 30$

105. $30 \div 15 = 2$

106. $12 + 12 = 24$

Use the formula to complete each table.

107. $T = 15g$

Number of gears g	Number of teeth T
10	
12	

108. $p = r - 200$

Revenue r	Profit p
1,000	
5,000	

CHALLENGE PROBLEMS

109. Which is larger: $\frac{11}{12}$ or $\frac{8}{9}$?

110. If the circle represents a whole, find the missing value.

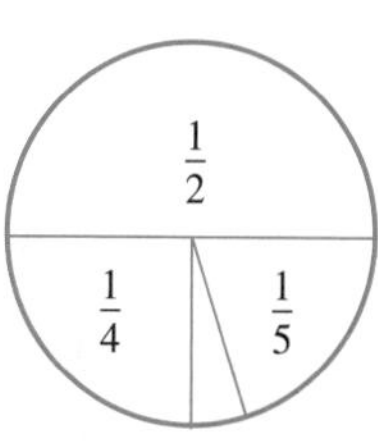

1.3 The Real Numbers

- The Integers
- Rational Numbers: Fractions and Mixed Numbers
- Rational Numbers: Decimals
- Irrational Numbers
- The Real Numbers
- Graphing on the Number Line
- Absolute Value

In this section, we will define many types of numbers. Then we will show that together, they form a collection or **set** of numbers called the *real numbers*.

THE INTEGERS

Natural numbers are the numbers that we use for counting. To write this set, we list its **members** (or **elements**) within **braces** { }.

Natural Numbers The set of **natural numbers** is {1, 2, 3, 4, 5, . . .}. Read as "the set containing one, two, three, four, five, and so on."

The three dots . . . in this definition mean that the list continues on forever.

The natural numbers, together with 0, form the set of **whole numbers.**

Whole Numbers The set of **whole numbers** is {0, 1, 2, 3, 4, 5, . . .}.

Although whole numbers are used in a wide variety of settings, they are not adequate for describing many real-life situations. For example, if you write a check for more than what's in your account, the account balance will be less than zero.

We can use the **number line** below to visualize numbers less than zero. A number line is straight and has uniform markings. The arrowheads indicate that it extends forever in both directions. For each natural number on the number line, there is a corresponding number, called its *opposite,* to the left of 0. In the illustration, we see that 3 and -3 (negative three) are opposites, as are -5 (negative five) and 5. Note that 0 is its own opposite.

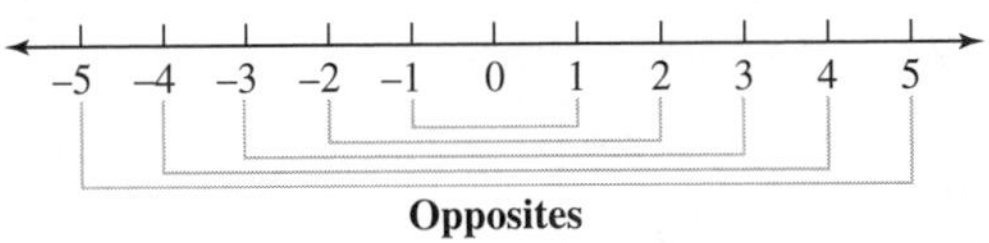

Opposites Two numbers that are the same distance from 0 on the number line, but on opposite sides of it, are called **opposites.**

The whole numbers, together with their opposites, form the set of **integers.**

Integers The set of **integers** is $\{. . . , -4, -3, -2, -1, 0, 1, 2, 3, 4, . . .\}$.

The Language of Algebra

The *positive integers* are: 1, 2, 3, 4, 5, ...

The *negative integers* are: $-1, -2, -3, -4, -5, ...$

On the number line, numbers greater than 0 are to the right of 0. They are called **positive numbers.** Positive numbers can be written with or without a **positive sign** +. For example, $2 = +2$ (positive two). They can be used to describe such quantities as an elevation above sea level (+3,000 ft) or a stock market gain (25 points).

Numbers less than 0 are to the left of 0 on the number line. They are called **negative numbers.** Negative numbers are always written with a **negative sign** $-$. They can be used to describe such quantities as an overdrawn checking account ($-$\$75) or a below-zero temperature ($-12°$).

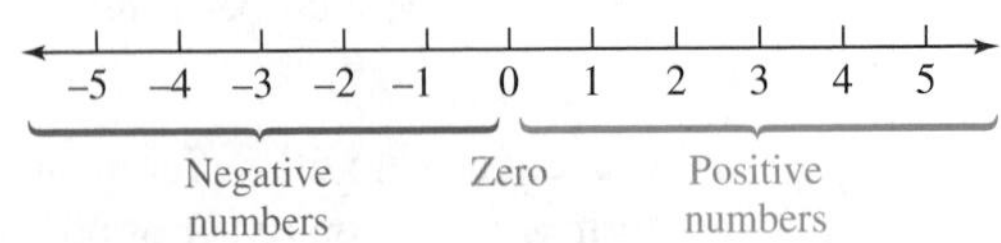

RATIONAL NUMBERS: FRACTIONS AND MIXED NUMBERS

Many situations cannot be described using integers. For example, a commute to work might take $\frac{1}{4}$ hour, or a hat might be size $7\frac{5}{8}$. To describe these situations, we need fractions, often called *rational numbers.*

Rational Numbers

A **rational number** is any number that can be expressed as a fraction with an integer numerator and a nonzero integer denominator.

Some examples of rational numbers are

$$\frac{1}{4}, \quad \frac{7}{8}, \quad \frac{25}{25}, \quad \text{and} \quad \frac{19}{12}$$

To show that negative fractions are rational numbers, we use the following fact.

Negative Fractions

Let a and b represent numbers, where b is not 0.

$$-\frac{a}{b} = \frac{-a}{b} = \frac{a}{-b}$$

To illustrate this rule, we consider $-\frac{11}{16}$. It is a rational number because it can be written as $\frac{-11}{16}$ or as $\frac{11}{-16}$.

Positive and negative mixed numbers are also rational numbers because they can be expressed as fractions. For example,

$$7\frac{5}{8} = \frac{61}{8} \qquad \text{and} \qquad -6\frac{1}{2} = -\frac{13}{2} = \frac{-13}{2}$$

The Language of Algebra

Rational numbers are so named because they can be expressed as the ratio (quotient) of two integers.

Any natural number, whole number, or integer can be expressed as a fraction with a denominator of 1. For example, $5 = \frac{5}{1}$, $0 = \frac{0}{1}$, and $-3 = \frac{-3}{1}$. Therefore, every natural number, whole number, and integer is also a rational number.

RATIONAL NUMBERS: DECIMALS

Many numerical quantities are written in decimal notation. For instance, a candy bar might cost \$0.89, a dragster might travel at 203.156 mph, or a business loss might be −\$4.7 million. These decimals are called **terminating decimals** because their representations terminate. As shown below, terminating decimals can be expressed as fractions. Therefore, terminating decimals are rational numbers.

The Language of Algebra

To *terminate* means to bring to an end. In the movie *The Terminator,* actor Arnold Schwarzenegger plays a heartless machine sent to Earth to bring an end to his enemies.

$$0.89 = \frac{89}{100} \qquad 203.156 = 203\frac{156}{1{,}000} = \frac{203{,}156}{1{,}000} \qquad -4.7 = -4\frac{7}{10} = \frac{-47}{10}$$

Decimals such as 0.3333 . . . and 2.8167167167 . . . , which have a digit (or block of digits) that repeats, are called **repeating decimals.** Since any repeating decimal can be expressed as a fraction, repeating decimals are rational numbers.

The set of rational numbers cannot be listed as we listed other sets in this section. Instead, we use **set-builder** notation.

Rational Numbers The set of rational numbers is

$$\left\{\frac{a}{b} \middle| a \text{ and } b \text{ are integers, with } b \neq 0.\right\}$$

Read as "the set of all numbers of the form $\frac{a}{b}$, such that a and b are integers, with $b \neq 0$."

To find the decimal equivalent for a fraction, we divide its numerator by its denominator. For example, to write $\frac{1}{4}$ and $\frac{5}{22}$ as decimals, we proceed as follows:

The Language of Algebra

Since every natural number belongs to the set of whole numbers, we say the set of natural numbers is a *subset* of the set of whole numbers. Similarly, the set of whole numbers is a subset of the set of integers, and the set of integers is a subset of the set of rational numbers.

$$\begin{array}{r} 0.25 \\ 4\overline{)1.00} \\ \underline{8} \\ 20 \\ \underline{20} \\ 0 \end{array}$$

Write a decimal point and additional zeros to the right of 1.

The remainder is 0.

$$\begin{array}{r} 0.22727\ldots \\ 22\overline{)5.00000} \\ \underline{4\,4} \\ 60 \\ \underline{44} \\ 160 \\ \underline{154} \\ 60 \\ \underline{44} \\ 160 \end{array}$$

Write a decimal point and additional zeros to the right of 5.

60 and 160 continually appear as remainders. Therefore, 2 and 7 will continually appear in the quotient.

The decimal equivalent of $\frac{1}{4}$ is 0.25 and the decimal equivalent of $\frac{5}{22}$ is 0.2272727 . . . . We can use an **overbar** to write repeating decimals in more compact form: $0.2272727\ldots = 0.2\overline{27}$. Here are more fractions and their decimal equivalents.

Terminating decimals	**Repeating decimals**
$\frac{1}{2} = 0.5$	$\frac{1}{6} = 0.16666\ldots$ or $0.1\overline{6}$
$\frac{5}{8} = 0.625$	$\frac{1}{3} = 0.3333\ldots$ or $0.\overline{3}$
$\frac{3}{4} = 0.75$	$\frac{5}{11} = 0.454545\ldots$ or $0.\overline{45}$

IRRATIONAL NUMBERS

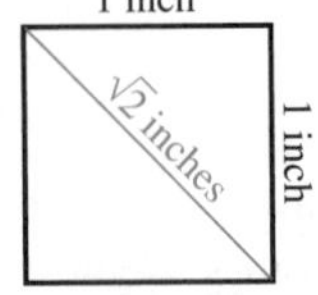

Numbers that cannot be expressed as a fraction with an integer numerator and an integer denominator are called **irrational numbers.** One example is the square root of 2, denoted $\sqrt{2}$. It is the number that, when multiplied by itself, gives 2: $\sqrt{2} \cdot \sqrt{2} = 2$. It can be shown that a square with sides of length 1 inch has a diagonal that is $\sqrt{2}$ inches long.

The number represented by the Greek letter π (pi) is another example of an irrational number. A circle, with a 1-inch diameter, has a circumference of π inches.

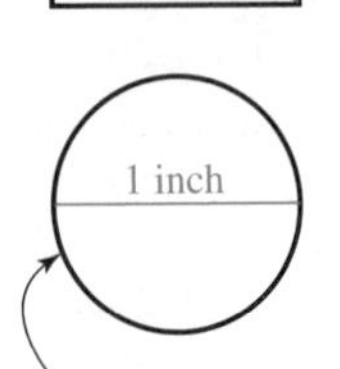

The distance around the circle is π inches.

Expressed in decimal form,

$$\sqrt{2} = 1.414213562\ldots \quad \text{and} \quad \pi = 3.141592654\ldots$$

These decimals neither terminate nor repeat.

Irrational Numbers An **irrational number** is a nonterminating, nonrepeating decimal. An irrational number cannot be expressed as a fraction with an integer numerator and an integer denominator.

Other examples of irrational numbers are:

$\sqrt{3} = 1.732050808\ldots$ $\quad -\sqrt{5} = -2.236067977\ldots$

$-\pi = -3.141592654\ldots$ $\quad 3\pi = 9.424777961\ldots$ $\quad$ 3π means $3 \cdot \pi$.

We can use a calculator to approximate the decimal value of an irrational number. To approximate $\sqrt{2}$ using a scientific calculator, we use the square root key $\sqrt{\ }$. To approximate π, we use the *pi* key π.

$\sqrt{2} \approx 1.414213562$ and $\pi \approx 3.141592654$ $\quad$ Read $\approx$ as "is approximately equal to."

Rounded to the nearest thousandth, $\sqrt{2} \approx 1.414$ and $\pi \approx 3.142$.

THE REAL NUMBERS

The set of **real numbers** is formed by combining the set of rational numbers and the set of irrational numbers. Every real number has a decimal representation. If it is rational, its corresponding decimal terminates or repeats. If it is irrational, its decimal representation is nonterminating and nonrepeating.

The Real Numbers A **real number** is any number that is a rational number or an irrational number.

The following diagram shows how various sets of numbers are related. Note that a number can belong to more than one set. For example, -6 is an integer, a rational number, and a real number.

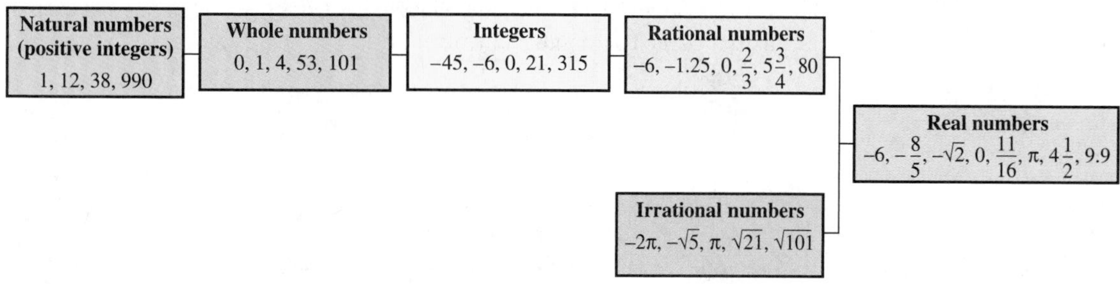

EXAMPLE 1

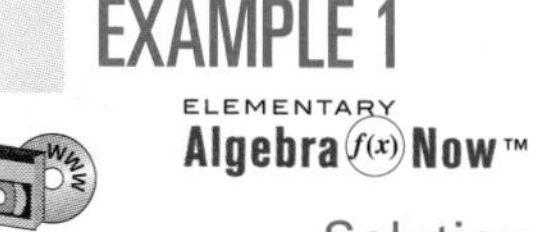

Which numbers in the following set are natural numbers, whole numbers, integers, rational numbers, irrational numbers, real numbers? $\left\{-3.4, \frac{2}{5}, 0, -6, 1\frac{3}{4}, \pi, 16\right\}$

Solution

Natural numbers: 16 $\quad$ 16 is a member of $\{1, 2, 3, 4, 5, \ldots\}$.

Whole numbers: 0, 16 $\quad$ 0 and 16 are members of $\{0, 1, 2, 3, 4, 5, \ldots\}$.

Integers: $0, -6, 16$ $\quad$ $0, -6$, and 16 are members of $\{\ldots, -3, -2, -1, 0, 1, 2, 3, \ldots\}$.

Rational numbers:

$$-3.4, \frac{2}{5}, 0, -6, 1\frac{3}{4}, 16$$

A rational number can be expressed as a fraction: $-3.4 = \frac{-34}{10}$, $0 = \frac{0}{1}$, $-6 = \frac{-6}{1}$, $1\frac{3}{4} = \frac{7}{4}$, and $16 = \frac{16}{1}$.

Irrational numbers: π $\quad$ $\pi = 3.1415\ldots$ is a nonterminating, nonrepeating decimal.

Real numbers:

$$-3.4, \frac{2}{5}, 0, -6, 1\frac{3}{4}, \pi, 16$$

Every natural number, whole number, integer, rational number, and irrational number is a real number.

Self Check 1 Use the instructions for Example 1 with the set $\left\{0.1, \sqrt{2}, -\frac{2}{7}, 45, -2, \frac{13}{4}, -6\frac{7}{8}\right\}$.

GRAPHING ON THE NUMBER LINE

Every real number corresponds to a point on the number line, and every point on the number line corresponds to exactly one real number. As we move right on the number line, the values of the numbers increase. As we move left, the values decrease. On the number line, we see that 5 is greater than -3, because 5 lies to the right of -3. Similarly, -3 is less than 5, because it lies to the left of 5.

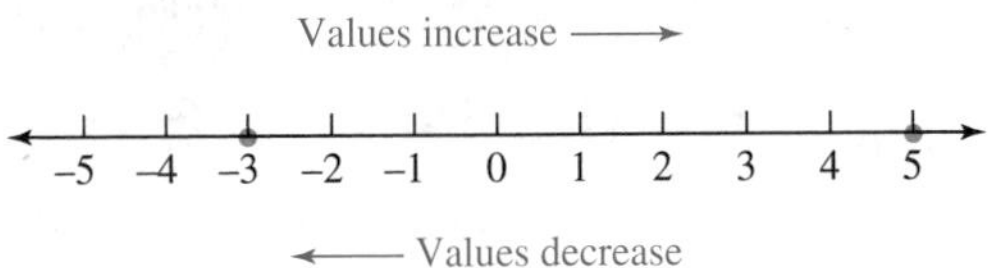

The Language of Algebra

The prefix *in* means *not.* For example:

inaccurate ↔ not accurate
inexpensive ↔ not expensive
inequality ↔ not equal

The **inequality symbol** $>$ means "is greater than." It is used to show that one number is greater than another. The inequality symbol $<$ means "is less than." It is used to show that one number is less than another. For example,

$5 > -3$ Read as "5 is greater than -3."

$-3 < 5$ Read as "-3 is less than 5."

To distinguish between these inequality symbols, remember that each one points to the smaller of the two numbers involved.

$5 > -3$ $-3 < 5$

Points to the smaller number.

EXAMPLE 2

ELEMENTARY Algebra f(x) Now™

Use one of the symbols $>$ or $<$ to make each statement true: **a.** $-4 \quad 4$, **b.** $-2 \quad -3$, **c.** $4.47 \quad 12.5$, and **d.** $\frac{3}{4} \quad \frac{5}{8}$.

Solution

a. Since -4 is to the left of 4 on the number line, we have $-4 < 4$.

b. Since -2 is to the right of -3 on the number line, we have $-2 > -3$.

c. Since 4.47 is to the left of 12.5 on the number line, we have $4.47 < 12.5$.

d. To compare fractions, express them in terms of the same denominator, preferably the LCD. If we write $\frac{3}{4}$ as an equivalent fraction with denominator 8, we see that $\frac{3}{4} = \frac{3 \cdot 2}{4 \cdot 2} = \frac{6}{8}$. Therefore, $\frac{3}{4} > \frac{5}{8}$.

The Language of Algebra

To state that a number x is positive, we can write $x > 0$. To state that a number x is negative, we can write $x < 0$.

To compare the fractions, we could also convert each to its decimal equivalent. Since $\frac{3}{4} = 0.75$ and $\frac{5}{8} = 0.625$, we know that $\frac{3}{4} > \frac{5}{8}$.

Self Check 2 Use one of the symbols $>$ or $<$ to make each statement true: **a.** $1 \quad -1$, **b.** $-5 \quad -4$, **c.** $6.7 \quad 4.999$, **d.** $\frac{3}{5} \quad \frac{2}{3}$.

To **graph a number** means to mark its position on the number line.

EXAMPLE 3

Graph each number in the following set: $\left\{-2.43, \sqrt{2}, 1, -0.\overline{3}, 2\frac{5}{6}, -\frac{3}{2}\right\}$.

ELEMENTARY Algebra $f(x)$ Now™

Solution We locate the position of each number on the number line, draw a bold dot, and label it. It is often helpful to approximate the value of a number or to write the number in an equivalent form to determine its location on the number line.

- To locate -2.43, we round it to the nearest tenth: $-2.43 \approx -2.4$.
- Use a calculator to find that $\sqrt{2} \approx 1.4$.
- Recall that $0.\overline{3} = 0.333\ldots = \frac{1}{3}$. Therefore, $-0.\overline{3} = -\frac{1}{3}$.
- In mixed-number form, $-\frac{3}{2} = -1\frac{1}{2}$. This is midway between -1 and -2.

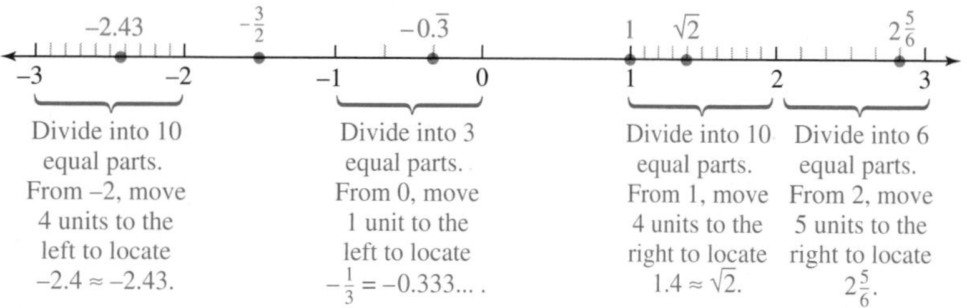

Self Check 3 Graph each number in the set: $\left\{1.7, \pi, -1\frac{3}{4}, 0.\overline{6}, \frac{5}{2}, -3\right\}$.

ABSOLUTE VALUE

A number line can be used to measure the distance from one number to another. For example, in the following figure we see that the distance from 0 to -4 is 4 units and the distance from 0 to 3 is 3 units.

4 units 3 units

−5 −4 −3 −2 −1 0 1 2 3 4 5

To express the distance that a number is from 0 on a number line, we often use absolute values.

Absolute Value The **absolute value** of a number is the distance from 0 to the number on the number line.

Success Tip

Since absolute value expresses distance, the absolute value of a number is always positive or zero.

To indicate the absolute value of a number, we write the number between two vertical bars. From the figure above, we see that $|-4| = 4$. This is read as "the absolute value of negative 4 is 4" and it tells us that the distance from 0 to -4 is 4 units. It also follows from the figure that $|3| = 3$.

EXAMPLE 4

Find each absolute value: **a.** $|18|$, **b.** $\left|-\frac{7}{8}\right|$, **c.** $|98.6|$, and **d.** $|0|$.

ELEMENTARY Algebra f(x) Now™

Solution

a. Since 18 is a distance of 18 from 0 on the number line, $|18| = 18$.
b. Since $-\frac{7}{8}$ is a distance of $\frac{7}{8}$ from 0 on the number line, $\left|-\frac{7}{8}\right| = \frac{7}{8}$.
c. Since 98.6 is a distance of 98.6 from 0 on the number line, $|98.6| = 98.6$.
d. Since 0 is a distance of 0 from 0 on the number line, $|0| = 0$.

Self Check 4 Find each absolute value: **a.** $|100|$, **b.** $|-4.7|$, **c.** $|-\sqrt{2}|$.

Answers to Self Checks **1.** natural numbers: 45; whole numbers: 45; integers: 45, -2; rational numbers: 0.1, $-\frac{2}{7}$, 45, -2, $\frac{13}{4}$, $-6\frac{7}{8}$; irrational numbers: $\sqrt{2}$; real numbers: all **2. a.** >, **b.** <, **c.** >, **d.** < **3.**

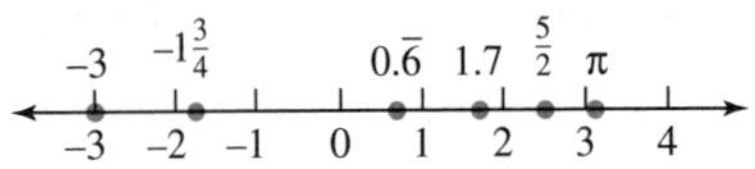

4. a. 100, **b.** 4.7, **c.** $\sqrt{2}$

1.3 STUDY SET

ELEMENTARY Algebra f(x) Now™

VOCABULARY Fill in the blanks.

1. The set of ________ numbers is {0, 1, 2, 3, 4, 5, . . .}.
2. The set of ________ numbers is {1, 2, 3, 4, 5, . . .}.
3. The set of ________ is {. . . , −2, −1, 0, 1, 2, . . .}.
4. Two numbers represented by points on the number line that are the same distance away from 0, but on opposite sides of it, are called ________.
5. Numbers less than zero are ________, and numbers greater than zero are ________.
6. The symbols < and > are ________ symbols.
7. A ________ number is any number that can be expressed as a fraction with an integer numerator and a nonzero integer denominator.
8. A decimal such as 0.25 is called a ________ decimal, and 0.333. . . is called a ________ decimal.
9. An ________ number is a nonterminating, nonrepeating decimal.
10. An ________ number cannot be expressed as a fraction.
11. Every point on the number line corresponds to exactly one ______ number.
12. The ________ of a number is the distance on the number line between the number and 0.

CONCEPTS

13. What concept is illustrated here?

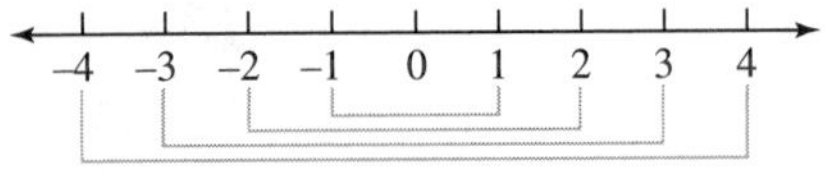

14. Fill in the blanks on the illustration.

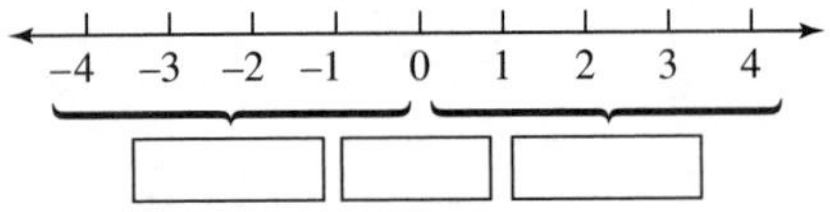

15. Show that each of the following numbers is a rational number by expressing it as a fraction with an integer numerator and a nonzero integer denominator: 6, -9, $-\frac{7}{8}$, $3\frac{1}{2}$, -0.3, 2.83.
16. Represent each situation using a signed number.
    a. A loss of \$15 million
    b. A rainfall total 0.75 inch below average
    c. A score $12\frac{1}{2}$ points under the standard
    d. A building foundation $\frac{5}{16}$ inch above grade

17. What numbers are a distance of 8 away from 5 on the number line?

18. Suppose m stands for a negative number. Use m, an inequality symbol, and 0 to express this fact.

19. Refer to the graph. Use an inequality symbol, $<$ or $>$, to make each statement true.

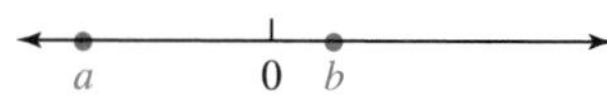

a. a ▢ b **b.** b ▢ a

c. b ▢ 0 and a ▢ 0 **d.** $|a|$ ▢ $|b|$

20. What is the length of the diagonal of the square shown below?

21. What is the circumference of the circle?

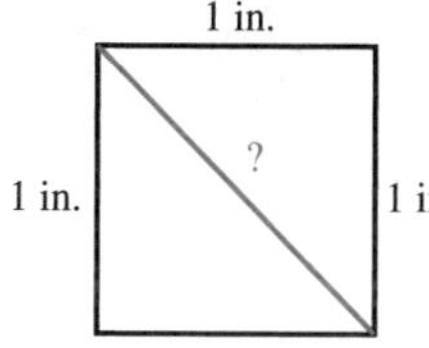

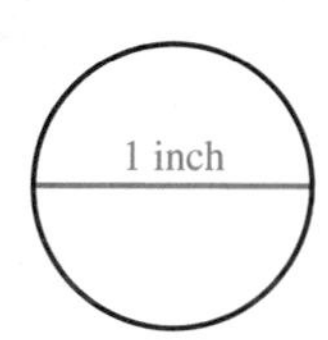

22. Place check marks in the table to show the set or sets to which each number belongs. For example, the checks show that $\sqrt{2}$ is an irrational and a real number.

	5	0	-3	$\frac{7}{8}$	0.17	$-9\frac{1}{4}$	$\sqrt{2}$	π
Real							✓	
Irrational							✓	
Rational								
Integer								
Whole								
Natural								

NOTATION **Fill in the blanks.**

23. $\sqrt{2}$ is read "the ________ of 2."

24. $|-15|$ is read "the ________ of -15."

25. The symbol $\neq$ means ________.

26. The symbols { }, called ________, are used when writing a set.

27. The symbol π is a letter from the ________ alphabet.

28. To find the decimal equivalent for the fraction $\frac{2}{3}$ we divide: $\square\overline{)\square}$

29. Write $-\frac{4}{5}$ in two other equivalent fractional forms.

30. Write each repeating decimal using an overbar.

a. 0.666. . . **b.** 0.2444. . .

c. 0.717171. . . **d.** 0.456456456. . .

PRACTICE **Write each fraction as a decimal. If the result is a repeating decimal, use an overbar.**

31. $\frac{5}{8}$ **32.** $\frac{3}{32}$

33. $\frac{1}{30}$ **34.** $\frac{7}{9}$

35. $\frac{21}{50}$ **36.** $\frac{2}{125}$

37. $\frac{5}{11}$ **38.** $\frac{1}{60}$

Insert one of the symbols $>$, $<$, or $=$ in the blank.

39. 5 ▢ 4 **40.** -5 ▢ -4

41. -2 ▢ -3 **42.** 0 ▢ 32

43. $|3.4|$ ▢ $\sqrt{2}$ **44.** 0.08 ▢ 0.079

45. $|-1.1|$ ▢ 1.2 **46.** -5.5 ▢ $-5\frac{1}{2}$

47. $-\frac{5}{8}$ ▢ $-\frac{3}{8}$ **48.** $-19\frac{2}{3}$ ▢ $-19\frac{1}{3}$

49. $\left|-\frac{15}{2}\right|$ ▢ 7.5 **50.** $\sqrt{2}$ ▢ π

51. $\frac{99}{100}$ ▢ 0.99 **52.** $|2|$ ▢ $|-2|$

53. 0.333. . . ▢ 0.3 **54.** $\left|-2\frac{2}{3}\right|$ ▢ $\frac{7}{3}$

55. 1 ▢ $\left|-\frac{15}{16}\right|$ **56.** $-0.666. . .$ ▢ 0

Decide whether each statement is true or false.

57. a. Every whole number is an integer.

b. Every integer is a natural number.

c. Every integer is a whole number.

d. Irrational numbers are nonterminating, nonrepeating decimals.

58. a. Irrational numbers are real numbers.

b. Every whole number is a rational number.

c. Every rational number can be written as a fraction.

d. Every rational number is a whole number.

59. **a.** Write the statement $-6 < -5$ using an inequality symbol that points in the other direction.

b. Write the statement $16 > -25$ using an inequality symbol that points in the other direction.

60. If we begin with the number -4 and find its opposite, and then find the opposite of that result, what number do we obtain?

Graph each set of numbers on the number line.

61. $\left\{-\pi, 4.25, -1\frac{1}{2}, -0.333\ldots, \sqrt{2}, -\frac{35}{8}, 3\right\}$

−5 −4 −3 −2 −1 0 1 2 3 4 5

62. $\left\{\pi, -2\frac{1}{8}, 2.75, -\sqrt{2}, \frac{17}{4}, 0.666\ldots, -3\right\}$

−5 −4 −3 −2 −1 0 1 2 3 4 5

APPLICATIONS

63. DRAFTING The drawing shows the dimensions of an aluminum bracket. Which numbers shown are natural numbers, whole numbers, integers, rational numbers, irrational numbers, and real numbers?

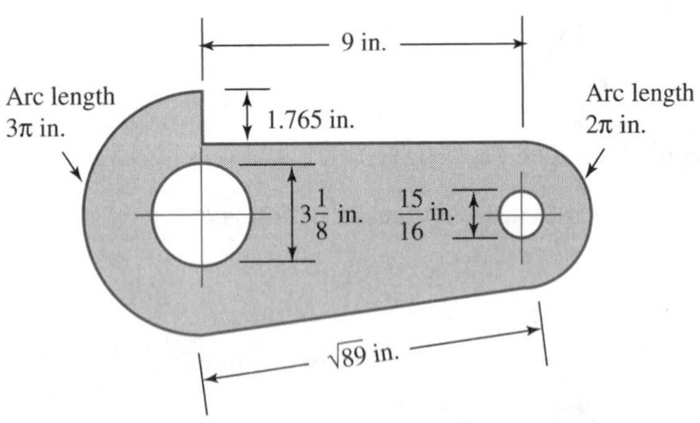

64. HISTORY Refer to the time line.

a. What basic unit was used to scale the time line?

b. On the time line, what symbolism is used to represent zero?

c. On the time line, which numbers could be thought of as positive and which as negative?

d. Express the dates for the Maya civilization using positive and negative numbers.

MAYA CIVILIZATION

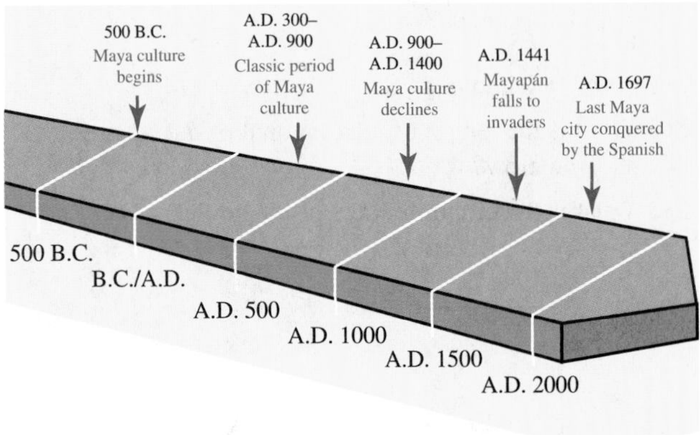

Based on data from *People in Time and Place, Western Hemisphere* (Silver Burdett & Ginn, 1991), p. 129

65. TARGET PRACTICE Which artillery shell landed farther from the target? How can the concept of absolute value be applied to answer this question?

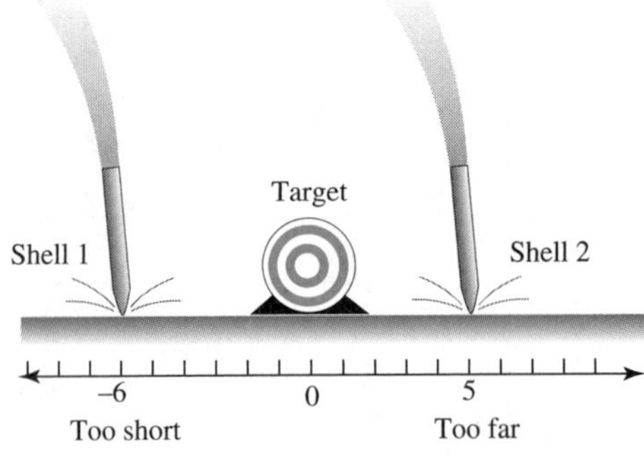

66. DRAFTING An architect's scale has several measuring edges. The edge marked 16 divides each inch into 16 equal parts. Find the decimal form for each fractional part of one inch that is highlighted on the scale.

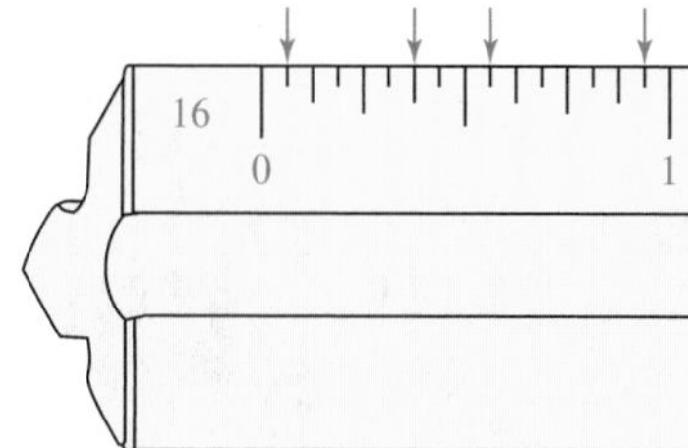

67. TRADE Each year from 1990 through 2002, the United States imported more goods and services from Japan than it exported to Japan. This caused trade deficits, which are represented by negative numbers on the graph.

a. In which three years was the deficit the worst? Estimate each of them.

b. In which year was the deficit the smallest? Estimate the deficit then.

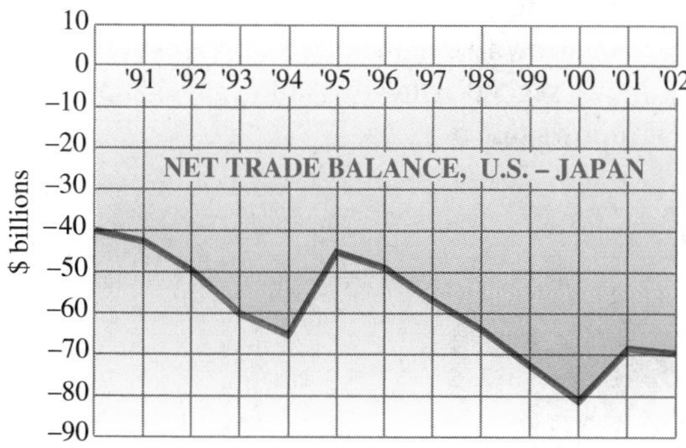

Source: U.S. Bureau of the Census

68. GOVERNMENT DEBT A budget deficit indicates that the government's expenditures were more than the revenue it took in that year. Deficits are represented by negative numbers on the graph.

a. For the years 1980–2001, when was the federal budget deficit the worst? Estimate the size of the deficit.

b. For the years 1980–2001, when did the first budget surplus occur? Estimate it. Explain what it means to have a budget surplus.

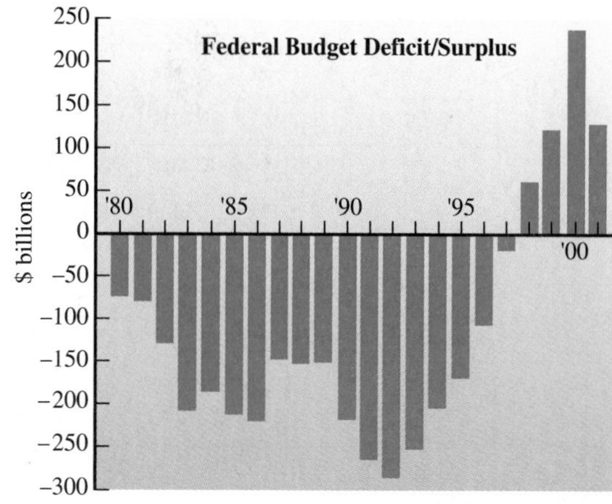

Source: U.S. Bureau of the Census

WRITING

69. Explain the difference between a rational and an irrational number.

70. Can two different numbers have the same absolute value? Explain.

71. Explain how to find the decimal equivalent of a fraction.

72. What is a real number?

REVIEW

73. Simplify: $\frac{24}{54}$.

74. Prime factor 60.

75. Find: $\frac{3}{4}\left(\frac{8}{5}\right)$.

76. Find: $5\frac{2}{3} \div 2\frac{5}{9}$.

77. Find: $\frac{3}{10} + \frac{2}{15}$.

78. Classify each of the following as an *expression* or an *equation.*

a. $2x$ **b.** $x = 2$

CHALLENGE PROBLEMS

79. Find the set of nonnegative integers.

80. Is 0.10100100010000 . . . a repeating decimal? Explain.

1.4 Adding Real Numbers

- Adding Two Numbers with the Same Sign
- Adding Two Numbers with Different Signs
- Opposites or Additive Inverses
- Properties of Addition

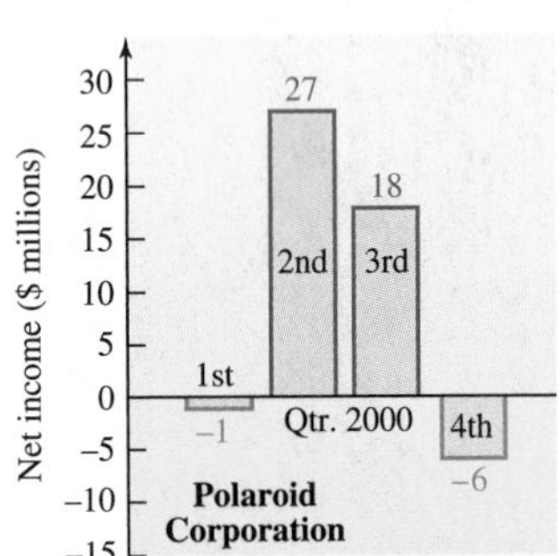

Source: Hoover's Online

Positive and negative numbers are called **signed numbers.** In the graph on the left, signed numbers are used to denote the financial performance of the Polaroid Corporation for the year 2000. The positive numbers indicate *profits* and the negative numbers indicate *losses.* To find Polaroid's net earnings (in millions of dollars), we need to calculate the following sum:

$$\text{Net earnings} = -1 + 27 + 18 + (-6)$$

In this section, we discuss how to perform this addition and others involving signed numbers.

ADDING TWO NUMBERS WITH THE SAME SIGN

A number line can be used to explain the addition of signed numbers. For example, to compute $5 + 2$, we begin at 0 and draw an arrow five units long that points right. It represents 5. From the tip of that arrow, we draw a second arrow two units long that points right. It represents 2. Since we end up at 7, it follows that $5 + 2 = 7$.

The Language of Algebra

The names of the parts of an addition fact are:

Addend *Addend* *Sum*

5 + 2 = 7

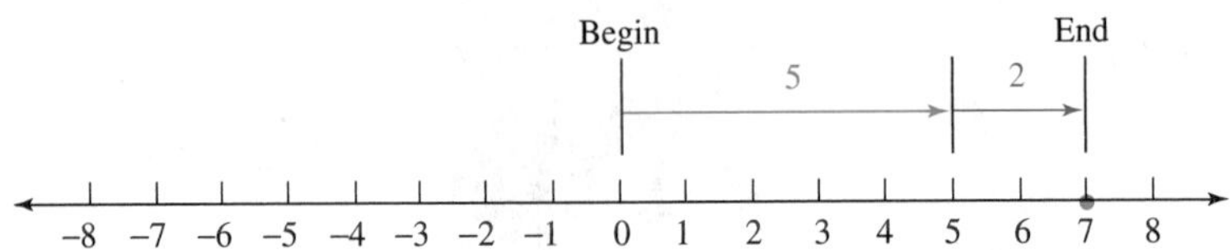

To compute $-5 + (-2)$, we begin at 0 and draw an arrow five units long that points left. It represents -5. From the tip of that arrow, we draw a second arrow two units long that points left. It represents -2. Since we end up at -7, it follows that $-5 + (-2) = -7$.

Notation

To avoid confusion, we write negative numbers within parentheses to separate the negative sign − from the addition symbol +.

$-5 + (-2)$

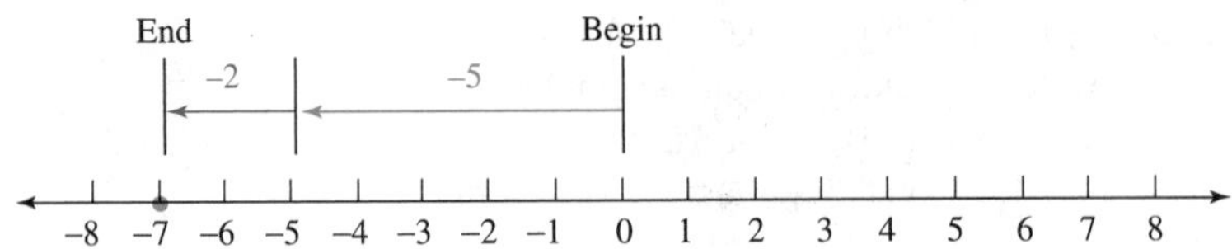

To check this result, think of the problem in terms of money. If you lost \$5 (-5) and then lost another \$2 (-2), you would have lost a total of \$7 (-7).

When we use a number line to add numbers with the same sign, the arrows point in the same direction and they build upon each other. Furthermore, the answer has the same sign as the numbers that we added. We can conclude that *the sum of two positive numbers is positive* and *the sum of two negative numbers is negative.*

Adding Two Numbers with the Same Sign

1. To add two positive numbers, add them. The answer is positive.
2. To add two negative numbers, add their absolute values and make the answer negative.

EXAMPLE 1 Find each sum: **a.** $-20 + (-15)$, **b.** $-7.89 + (-0.6)$, and **c.** $-\frac{1}{3} + \left(-\frac{1}{2}\right)$.

ELEMENTARY Algebra $f(x)$ Now™

Solution **a.** $-20 + (-15) = -35$ Add their absolute values, 20 and 15, to get 35. Make the answer negative.

b. Add their absolute values and make the answer negative: $-7.89 + (-0.6) = -8.49$.

c. Add their absolute values

$$\frac{1}{3} + \frac{1}{2} = \frac{2}{6} + \frac{3}{6}$$ The LCD is 6. Build each fraction: $\frac{1}{3} \cdot \frac{2}{2} = \frac{2}{6}$ and $\frac{1}{2} \cdot \frac{3}{3} = \frac{3}{6}$.

$$= \frac{5}{6}$$ Add the numerators and write the sum over the LCD.

and make the answer negative: $-\frac{1}{3} + \left(-\frac{1}{2}\right) = -\frac{5}{6}$.

Self Check 1 Find the sum: **a.** $-51 + (-9)$, **b.** $-12.3 + (-0.88)$, **c.** $-\frac{1}{4} + \left(-\frac{2}{3}\right)$.

ADDING TWO NUMBERS WITH DIFFERENT SIGNS

To compute $5 + (-2)$, we begin at 0 and draw an arrow five units long that points right. From the tip of that arrow, we draw a second arrow two units long that points left. Since we end up at 3, it follows that $5 + (-2) = 3$. In terms of money, if you won \$5 and then lost \$2, you would have \$3 left.

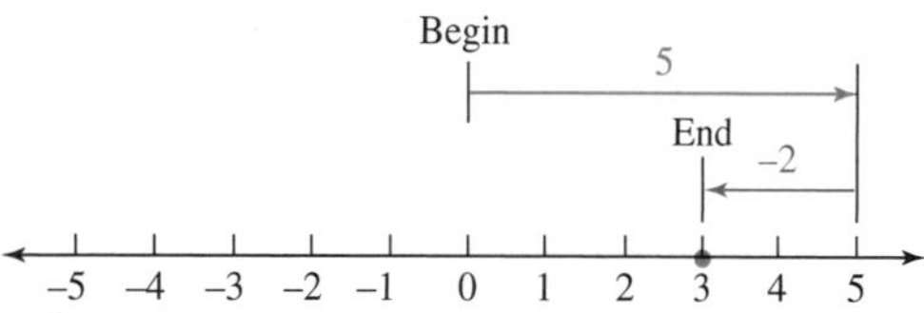

To compute $-5 + 2$, we begin at 0 and draw an arrow five units long that points left. From the tip of that arrow, we draw a second arrow two units long that points right. Since we end up at -3, it follows that $-5 + 2 = -3$. In terms of money, if you lost \$5 and then won \$2, you have lost \$3.

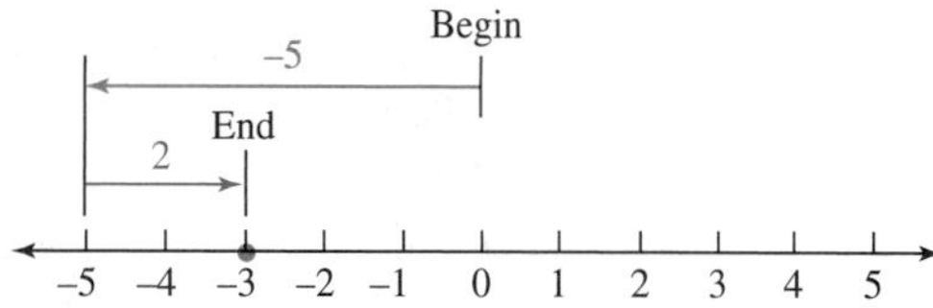

When we use a number line to add numbers with different signs, the arrows point in opposite directions and the longer arrow determines the sign of the answer. If the longer arrow represents a positive number, the sum is positive. If it represents a negative number, the sum is negative.

Adding Two Numbers with Different Signs To add a positive number and a negative number, subtract the smaller absolute value from the larger.

1. If the positive number has the larger absolute value, the answer is positive.
2. If the negative number has the larger absolute value, make the answer negative.

EXAMPLE 2

Find each sum: **a.** $-20 + 32$, **b.** $5.7 + (-7.4)$, and **c.** $-\frac{19}{25} + \frac{2}{5}$.

ELEMENTARY Algebra $f(x)$ Now™

Solution **a.** $-20 + 32 = 12$ Subtract the smaller absolute value from the larger: $32 - 20 = 12$. The positive number, 32, has the larger absolute value, so the answer is positive.

b. Subtract the smaller absolute value, 5.7, from the larger, 7.4, and determine the sign of the answer.

$5.7 + (-7.4) = -1.7$ The negative number, -7.4, has the larger absolute value, so we make the answer negative.

c. Since $\frac{2}{5} = \frac{10}{25}$, $-\frac{19}{25}$ has the larger absolute value. We subtract the smaller absolute value from the larger:

$$\frac{19}{25} - \frac{2}{5} = \frac{19}{25} - \frac{10}{25}$$ Replace $\frac{2}{5}$ with the equivalent fraction $\frac{10}{25}$.

$$= \frac{9}{25}$$ Subtract the numerators and write the difference over the LCD.

and determine the sign of the answer.

$$-\frac{19}{25} + \frac{10}{25} = -\frac{9}{25}$$ Since $-\frac{19}{25}$ has the larger absolute value, make the answer negative.

Calculators
Entering negative numbers

When using a calculator to add positive and negative numbers, we don't do anything special to enter positive numbers. To enter a negative number, say -1, on a scientific calculator, we press the *sign change* key +/− after entering 1. If we use a graphing calculator, we press the *negation* key (−) and then enter 1.

Self Check 2 Find each sum: **a.** $63 + (-87)$, **b.** $-6.27 + 8$, **c.** $-\frac{1}{10} + \frac{1}{2}$

EXAMPLE 3

ELEMENTARY Algebra $f(x)$ Now™

Corporate earnings. Find the net earnings of Polaroid Corporation for the year 2000 using the data in the graph on page 32.

Solution To find the net earnings, we add the quarterly profits and losses (in millions of dollars), performing the additions as they occur from left to right.

$$-1 + 27 + 18 + (-6) = 26 + 18 + (-6)$$ Add: $-1 + 27 = 26$.

$$= 44 + (-6)$$ Add: $26 + 18 = 44$.

$$= 38$$

In 2000, Polaroid's net earnings were $38 million.

Self Check 3 Add: $7 + (-13) + 8 + (-10)$.

PROPERTIES OF ADDITION

The addition of two numbers can be done in any order and the result is the same. For example, $8 + (-1) = 7$ and $-1 + 8 = 7$. This example illustrates that addition is **commutative.**

The Commutative Property of Addition

Changing the order when adding does not affect the answer.
Let a and b represent real numbers,

$$a + b = b + a$$

The Language of Algebra

Commutative is a form of the word *commute,* meaning to go back and forth.

In the following example, we add $-3 + 7 + 5$ in two ways. We will use grouping symbols ( ), called **parentheses,** to show this. Standard practice requires that the operation within the parentheses be performed first.

Method 1: Group −3 and 7

$$(-3 + 7) + 5 = 4 + 5$$
$$= 9$$

Method 2: Group 7 and 5

$$-3 + (7 + 5) = -3 + 12$$
$$= 9$$

It doesn't matter how we group the numbers in this addition; the result is 9. This example illustrates that addition is **associative.**

The Associative Property of Addition

Changing the grouping when adding does not affect the answer.
Let a, b, and c represent real numbers,

$$(a + b) + c = a + (b + c)$$

Sometimes, an application of the associative property can simplify a computation.

EXAMPLE 4

Find the sum: $98 + (2 + 17)$.

ELEMENTARY Algebra f(x) Now™

Solution If we use the associative property of addition to regroup, we have a convenient pair of numbers to add: $98 + 2 = 100$.

The Language of Algebra

Associative is a form of the word *associate,* meaning to join a group.

$$98 + (2 + 17) = (98 + 2) + 17$$
$$= 100 + 17$$
$$= 117$$

Self Check 4 Find the sum: $(39 + 25) + 75$.

EXAMPLE 5

ELEMENTARY Algebra f(x) Now™

Game shows. A contestant on *Jeopardy!* correctly answered the first question to win \$100, missed the second to lose \$200, correctly answered the third to win \$300, and missed the fourth to lose \$400. What is her score after answering four questions?

WHO'S WHO	WHODUNIT	WHO'S ON FIRST	HOUDINI	HOOVER	HOULIGANS
100	100	100	100	100	100
200	200	200	200	200	200
300	300	300	300	300	300
400	400	400	400	400	400
500	500	500	500	500	500

Solution We can represent money won by a positive number and money lost by a negative number. Her score is the sum of 100, −200, 300, and −400. Instead of doing the additions from left to right, we will use another approach. Applying the commutative and associative properties, we add the positives, add the negatives, and then add those results.

$100 + (-200) + 300 + (-400)$

$= (100 + 300) + [(-200) + (-400)]$ Reorder the numbers. Group the positives together. Group the negatives together using brackets [ ].

$= 400 + (-600)$ Add the positives. Add the negatives.

$= -200$ Add the results.

After four questions, her score was −$200, a loss of $200.

Self Check 5 Find $-6 + 1 + (-4) + (-5) + 9$.

The Language of Algebra

Identity is a form of the word *identical,* meaning the same. You have probably seen *identical* twins.

Whenever we add 0 to a number, the number remains the same:

$$8 + 0 = 8, \qquad 2.3 + 0 = 2.3, \qquad \text{and} \qquad 0 + (-16) = -16$$

These examples illustrate the **addition property of 0.** Since any number added to 0 remains the same, 0 is called the **identity element** for addition.

Addition Property of 0

When 0 is added to any real number, the result is the same real number.

For any real number a,

$$a + 0 = a \qquad \text{and} \qquad 0 + a = a$$

OPPOSITES OR ADDITIVE INVERSES

The Language of Algebra

Don't confuse the words *opposite* and *reciprocal.* The opposite of 4 is −4. The reciprocal of 4 is $\frac{1}{4}$.

Recall that two numbers that are the same distance from 0 on a number line, but on opposite sides of it, are called **opposites.** To develop a property of opposites, we will find $-4 + 4$ using a number line. We begin at 0 and draw an arrow four units long that points left, to represent −4. From the tip of that arrow, we draw a second arrow, four units long that points right, to represent 4. We end up at 0; therefore, $-4 + 4 = 0$.

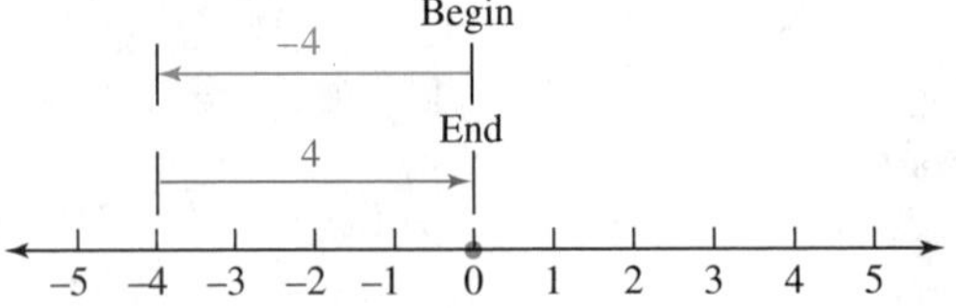

This example illustrates that when we add opposites, the result is 0. It also follows that whenever the sum of two numbers is 0, those numbers are opposites. For these reasons, opposites are also called **additive inverses.**

Addition Property of Opposites (Inverse Property of Addition)

The sum of a number and its opposite (additive inverse) is 0.

For any real number a and its opposite or additive inverse $-a$,

$$a + (-a) = 0$$ Read $-a$ as "the opposite of a."

EXAMPLE 6

Find the sum: $12 + (-5) + 6 + 5 + (-12)$.

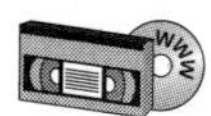

ELEMENTARY Algebra f(x) Now™

Solution The commutative and associative properties of addition enable us to add pairs of opposites: $12 + (-12) = 0$ and $-5 + 5 = 0$.

$$\underbrace{12 + \underbrace{(-5) + 6 + 5}_{\text{opposites}} + (-12)}_{\text{opposites}} = 0 + 0 + 6$$
$$= 6$$

Self Check 6 Find the sum: $8 + (-1) + 6 + 5 + (-8) + 1$.

Answers to Self Checks **1. a.** -60, **b.** -13.18, **c.** $-\frac{11}{12}$ **2. a.** -24, **b.** 1.73, **c.** $\frac{2}{5}$ **3.** -8 **4.** 139 **5.** -5 **6.** 11

1.4 STUDY SET

ELEMENTARY Algebra f(x) Now™

VOCABULARY **Fill in the blanks.**

1. Positive and negative numbers are called ________ numbers.
2. In the addition statement $-2 + 5 = 3$, the result, 3, is called the ______.
3. Two numbers that are the same distance from 0 on a number line, but on opposite sides of it, are called __________.
4. The _____________ property of addition states that changing the order when adding does not affect the answer. The ___________ property of addition states that changing the grouping when adding does not affect the answer.
5. Since any number added to 0 remains the same (is identical), the number 0 is called the _________ element for addition.
6. The sum of a number and its opposite or additive ________ is 0.

CONCEPTS **What addition fact is represented by each illustration?**

7.

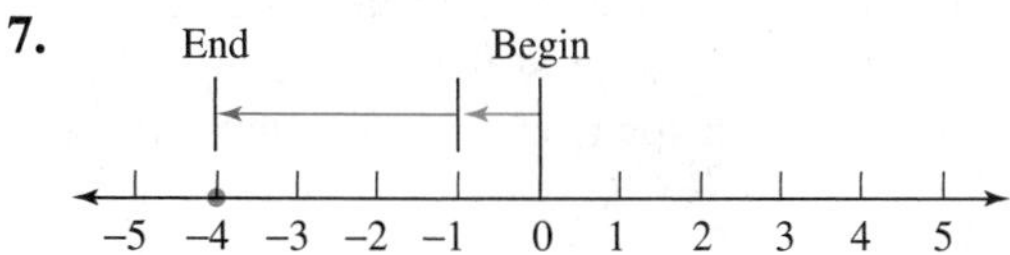

8.

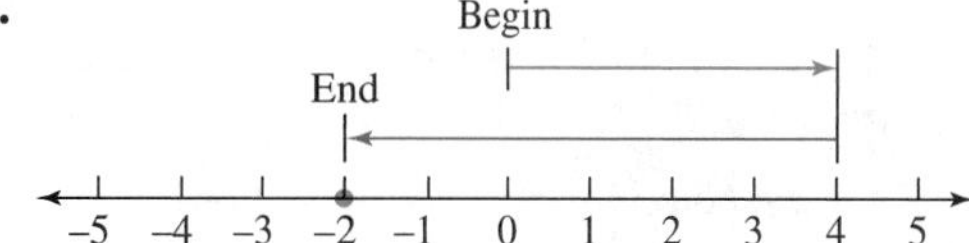

Add using a number line.

9. $-3 + (-2)$
10. $4 + (-6)$
11. $1 + (-4)$
12. $-3 + (-1)$
13. Complete each property of addition.
    a. $a + (-a) =$
    b. $a + 0 =$
    c. $a + b = b +$
    d. $(a + b) + c = a +$

For each addition, determine the sign of the answer.

14. $59 + (-64)$
15. $-87 + 98$
16. Circle any opposites in the following expression.

$$12 + (-3) + (-6) + 3 + 1$$

17. Use the commutative property of addition to complete each statement.

a. $-5 + 1 =$ ________

b. $15 + (-80.5) =$ ________

c. $-20 + (4 + 20) = -20 +$ ________

18. Use the associative property of addition to complete each statement.

a. $(-6 + 2) + 8 =$ ________

b. $-7 + (7 + 3) =$ ________

19. Find each sum.

a. $5 + (-5)$

b. $-2.2 + 2.2$

c. $0 + (-6)$

d. $-\frac{15}{16} + 0$

e. $-\frac{3}{4} + \frac{3}{4}$

f. $19 + (-19)$

20. Consider $-3 + 6 + (-9) + 8 + (-4)$.

a. Add all the positives in the expression.

b. Add all of the negatives.

c. Add the results from parts **a** and **b.**

NOTATION

21. Express the commutative property of addition using the variables x and y.

22. Express the associative property of addition using the variables x, y, and z.

23. In $7 + (8 + 9)$, which addition should be done first?

24. Insert parentheses where they are needed.

$$6 + -8 + -10$$

PRACTICE Add.

25. $6 + (-8)$

26. $4 + (-3)$

27. $-6 + 8$

28. $-21 + (-12)$

29. $-4 + (-4)$

30. $-5 + (-5)$

31. $9 + (-1)$

32. $11 + (-2)$

33. $-16 + 16$

34. $-25 + 25$

35. $-65 + (-12)$

36. $75 + (-13)$

37. $15 + (-11)$

38. $27 + (-30)$

39. $300 + (-335)$

40. $240 + (-340)$

41. $-10.5 + 2.3$

42. $-2.1 + 0.4$

43. $-9.1 + (-11)$

44. $-6.7 + (-7.1)$

45. $0.7 + (-0.5)$

46. $0.9 + (-0.2)$

47. $-\frac{9}{16} + \frac{7}{16}$

48. $-\frac{3}{4} + \frac{1}{4}$

49. $-\frac{1}{4} + \frac{2}{3}$

50. $\frac{3}{16} + \left(-\frac{1}{2}\right)$

51. $-\frac{4}{5} + \left(-\frac{1}{10}\right)$

52. $-\frac{3}{8} + \left(-\frac{1}{3}\right)$

53. $8 + (-5) + 13$

54. $17 + (-12) + (-23)$

55. $21 + (-27) + (-9)$

56. $-32 + 12 + 17$

57. $-27 + (-3) + (-13) + 22$

58. $53 + (-27) + (-32) + (-7)$

59. $-20 + (-16 + 10)$

60. $-13 + (-16 + 4)$

61. $19 + (-20 + 1)$

62. $33 + (-35 + 2)$

63. $(-7 + 8) + 2 + (-12 + 13)$

64. $(-9 + 5) + 10 + (-8 + 1)$

65. $-7 + 5 + (-10) + 7$

66. $-3 + 6 + (-9) + (-6)$

67. $-8 + 11 + (-11) + 8 + 1$

68. $2 + 15 + (-15) + 8 + (-2)$

69. $-2.1 + 6.5 + (-8.2) + 0.6$

70. $0.9 + 0.5 + (-0.2) + (-0.9)$

71. $-60 + 70 + (-10) + (-10) + 20$

72. $-100 + 200 + (-300) + (-100) + 200$

Apply the associative property of addition, and find the sum.

73. $-99 + (99 + 215)$

74. $67 + (-67 + 127)$

75. $(-112 + 56) + (-56)$

76. $(-67 + 5) + (-5)$

APPLICATIONS

77. MILITARY SCIENCE During a battle, an army retreated 1,500 meters, regrouped, and advanced 2,400 meters. The next day, it advanced another 1,250 meters. Find the army's net gain.

78. HEALTH Find the point total for the six risk factors (in blue) on the medical questionnaire. Then use the table to determine the patient's risk of contracting heart disease in the next 10 years.

Age		Total Cholesterol	
Age	Points	Reading	Points
34	−1	150	−3
Cholesterol		**Blood Pressure**	
HDL	Points	Systolic/Diastolic	Points
62	−2	124/100	3
Diabetic		**Smoker**	
	Points		Points
Yes	2	Yes	2

10-Year Heart Disease Risk

Total Points	**Risk**	Total Points	**Risk**
−2 or less	1%	5	4%
−1 to 1	2%	6	6%
2 to 3	3%	7	6%
4	4%	8	7%

79. GOLF The leaderboard shows the top finishers from the 1997 Masters Golf Tournament. Scores for each round are compared to *par,* the standard number of strokes necessary to complete the course. A score of −2, for example, indicates that the golfer used two strokes less than par to complete the course. A score of +5 indicates five strokes more than par. Determine the tournament total for each golfer.

Leaderboard

Round

	1	**2**	**3**	**4**	**Total**
Tiger Woods	−2	−6	−7	−3	
Tom Kite	+5	−3	−6	−2	
Tommy Tolles	0	0	0	−5	
Tom Watson	+3	−4	−3	0	

80. SUBMARINES A submarine was cruising at a depth of 1,250 feet. The captain gave the order to climb 550 feet. Relative to sea level, find the new depth of the sub.

81. CREDIT CARDS

a. What amounts in the monthly credit card statement could be represented by negative numbers?

b. Express the new balance as a negative number.

Previous Balance	New Purchases, Fees, Advances & Debts	Payments & Credits	New Balance
3,660.66	**1,408.78**	**3,826.58**	

04/21/03 Billing Date	**05/16/03** Date Payment Due	**9,100** Credit Line

Periodic rates may vary.
See reverse for explanation and important information.
Please allow sufficient time for mail to reach us.

82. POLITICS The following proposal to limit campaign contributions was on the ballot in a state election, and it passed. What will be the net fiscal impact on the state government?

212 Campaign Spending Limits YES ☐ NO ☐

Limits contributions to $200 in state campaigns. Fiscal impact: Costs of $4.5 million for implementation and enforcement. Increases state revenue by $6.7 million by eliminating tax deductions for lobbying.

83. MOVIE LOSSES According to the *Guinness Book of World Records 2000,* MGM's *Cutthroat Island* (1995), starring Geena Davis, cost about $100 million to produce, promote, and distribute. It has reportedly earned back just $11 million since being released. What dollar loss did the studio suffer on this film?

84. STOCK EXCHANGE Many newspapers publish daily summaries of the stock market's activity. The last entry on the line for June 12 indicates that one share of Walt Disney Co. stock lost $0.81 in value that day. How much did the value of a share of Disney stock rise or fall over the five-day period shown?

June 12	43.88	23.38	Disney	.21	0.5	87	−43	40.75	−.81
June 13	43.88	23.38	Disney	.21	0.5	86	−15	40.19	−.56
June 14	43.88	23.38	Disney	.21	0.5	87	−50	41.00	+.81
June 15	43.88	23.38	Disney	.21	0.5	89	−28	41.81	+.81
June 16	43.88	23.38	Disney				−15	41.19	−.63

Based on data from the *Los Angeles Times*

85. SAHARA DESERT From 1980 to 1990, a satellite was used to trace the expansion and contraction of the southern boundary of the Sahara Desert in Africa. If movement southward is represented with a negative number and movement northward with a positive number, use the data in the table to determine the net movement of the Sahara Desert boundary over the 10-year period.

Years	Distance/Direction
1980–1984	240 km/South
1984–1985	110 km/North
1985–1986	30 km/North
1986–1987	55 km/South
1987–1988	100 km/North
1988–1990	77 km/South

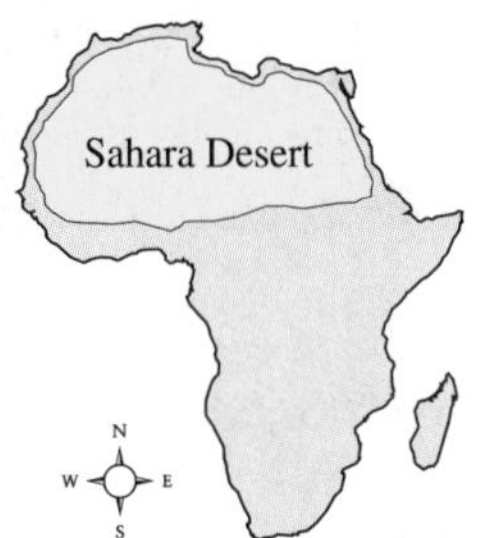

Based on data from A. Dolgoff, *Physical Geology* (D.C. Heath, 1996), p. 496

86. ELECTRONICS A closed circuit contains two batteries and three resistors. The sum of the voltages in the loop must be 0. Is it?

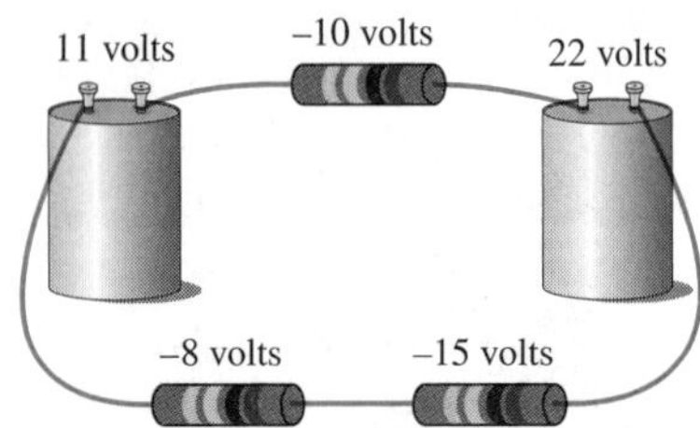

87. PROFITS AND LOSSES The 2001 quarterly profits and losses of Greyhound Bus Lines are shown in the table. Losses are denoted using parentheses. Use the data to construct a line graph. Then calculate the company's total net income for 2001.

Quarter	Net income ($ million)
1st	(10.7)
2nd	4.0
3rd	12.2
4th	(3.4)

Source: Edgar Online

88. POLITICS Six months before an election, the incumbent trailed the challenger by 18 points. To overtake her opponent, the campaign staff decided to use a four-part strategy. Each part of this plan is shown, with the anticipated point gain. With these gains, will the incumbent overtake the challenger on election day?

1. Intense TV ad blitz +10
2. Ask for union endorsement +2
3. Voter mailing +3
4. Get-out-the-vote campaign +1

REVIEW

89. True or false: Every real number can be expressed as a decimal.

90. Multiply: $\frac{1}{3} \cdot \frac{1}{3}$.

91. What two numbers are a distance of 6 away from -3 on the number line?

92. Graph $\left\{-2.5, \sqrt{2}, \frac{11}{3}, -0.333\ldots, 0.75\right\}$.

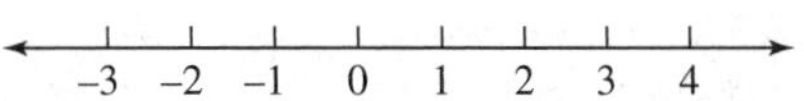

WRITING

93. Explain why the sum of two positive numbers is always positive and the sum of two negative numbers is always negative.

94. Explain why the sum of a negative number and a positive number is sometimes positive, sometimes negative, and sometimes zero.

CHALLENGE PROBLEMS

95. A set is said to be *closed under addition* if the sum of any two of its members is also a member of the set. Is the set $\{-1, 0, 1\}$ a closed set under addition? Explain.

96. Think of two numbers. First, add the absolute value of the two numbers, and write your answer. Second, add the two numbers, take the absolute value of that sum, and write that answer. Do the two answers agree? Can you find two numbers that produce different answers? When do you get answers that agree, and when don't you?

1.5 Subtracting Real Numbers

• Subtraction • Applications Involving Subtraction

In this section, we discuss a rule to use when subtracting signed numbers.

SUBTRACTION

A minus symbol $-$ is used to indicate subtraction. However, this symbol is also used in two other ways, depending on where it appears in an expression.

$5 - 18$	This is read as "five minus eighteen."
-5	This is usually read as "negative five." It could also be read as "the additive inverse of five" or "the opposite of five."
$-(-5)$	This is usually read as "the opposite of negative five." It could also be read as "the additive inverse of negative five."

In $-(-5)$, parentheses are used to write the opposite of a negative number. When such expressions are encountered in computations, we simplify them by finding the opposite of the number within the parentheses.

$-(-5) = 5$ The opposite of negative five is five.

This observation illustrates the following rule.

Opposite of an Opposite

The opposite of the opposite of a number is that number.

For any real number a,

$-(-a) = a$ Read as "the opposite of the opposite of a is a."

EXAMPLE 1

Simplify each expression: **a.** $-(-45)$, **b.** $-(-h)$, and **c.** $-|-10|$.

ELEMENTARY Algebra f(x) Now™

Solution

a. The number within the parentheses is -45. Its opposite is 45. Therefore, $-(-45) = 45$.

b. The opposite of the opposite of h is h. Therefore, $-(-h) = h$.

c. The notation $-|-10|$ means "the opposite of the absolute value of negative ten." Since $|-10| = 10$, we have:

$$-|-10| = -10$$

Self Check 1 Simplify each expression: **a.** $-(-1)$, **b.** $-(-y)$, and **c.** $-|-500|$.

The subtraction $5 - 2$ can be thought of as taking away 2 from 5. To use a number line to illustrate this, we begin at 0 and draw an arrow 5 units long that points to the right. From the tip of that arrow, we move back two units to the left. Since we end up at 3, it follows that $5 - 2 = 3$.

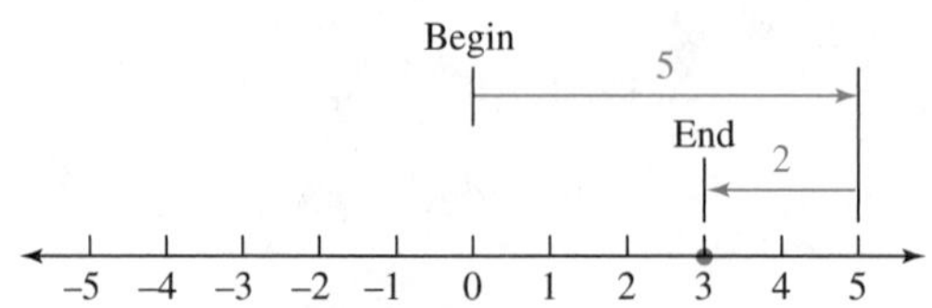

The Language of Algebra

The names of the parts of a subtraction fact are:

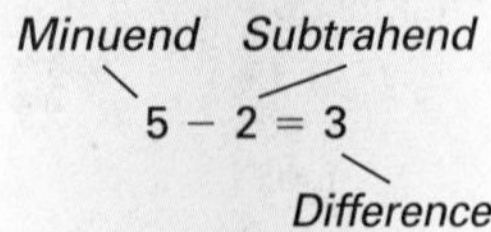

The previous illustration could also serve as a representation of the addition problem $5 + (-2)$. In the problem $5 - 2$, we subtracted 2 from 5. In the problem $5 + (-2)$, we added -2 to 5. In each case, the result is 3.

Subtracting 2. → $5 - 2 = 3$

Adding the opposite of 2. → $5 + (-2) = 3$

The results are the same.

These observations suggest the following definition.

Subtraction of Real Numbers

To subtract two real numbers, add the first number to the opposite (additive inverse) of the number to be subtracted.

Let a and b represent real numbers,

$$a - b = a + (-b)$$

EXAMPLE 2

Subtract: **a.** $-13 - 8$, **b.** $-7 - (-45)$, and **c.** $\frac{1}{4} - \left(-\frac{1}{8}\right)$.

ELEMENTARY Algebra f(x) Now™

Solution

a. We read $-13 - 8$ as "negative thirteen *minus* eight."

Change the subtraction to addition.

$$-13 - 8 = -13 + (-8) = -21$$ To subtract, add the opposite.

Change the number being subtracted to its opposite.

To check, we add the difference, -21, and the subtrahend, 8, to obtain the minuend, -13.

$$-21 + 8 = -13$$

The Language of Algebra

The rule for subtracting real numbers is often summarized as: *Subtraction is the same as adding the opposite.*

b. We read $-7 - (-45)$ as "negative seven *minus* negative forty-five."

Add

$$-7 - (-45) = -7 + 45 = 38$$ To subtract, add the opposite.

the opposite.

Check the result: $38 + (-45) = -7$.

Calculators
The subtraction key

When using a calculator to subtract signed numbers, be careful to distinguish between the *subtraction* key $-$ and the keys that are used to enter negative values: +/− on a scientific calculator and (−) on a graphing calculator.

c. $$\frac{1}{4} - \left(-\frac{1}{8}\right) = \frac{2}{8} - \left(-\frac{1}{8}\right) \quad \text{Build } \frac{1}{4}: \frac{1}{4} \cdot \frac{2}{2} = \frac{2}{8}.$$
$$= \frac{2}{8} + \frac{1}{8} \quad \text{To subtract, add the opposite.}$$
$$= \frac{3}{8}$$

Check the result: $\frac{3}{8} + \left(-\frac{1}{8}\right) = \frac{2}{8} = \frac{1}{4}$.

Self Check 2 Subtract: **a.** $-32 - 25$, **b.** $17 - (-12)$, and **c.** $-\frac{1}{3} - \left(-\frac{3}{4}\right)$.

EXAMPLE 3

ELEMENTARY Algebra f(x) Now™

Translate from words to symbols: **a.** Subtract 0.5 from 4.6, and **b.** subtract 4.6 from 0.5.

Solution **a.** The number to be subtracted is 0.5. When we translate to mathematical symbols, we must reverse the order in which 0.5 and 4.6 appear in the sentence.

Caution

When subtracting two numbers, it is important that we write them in the correct order, because, in general, $a - b \neq b - a$.

Subtract 0.5 from 4.6.

$$4.6 - 0.5 = 4.1$$

b. The number to be subtracted is 4.6. When we translate to mathematical symbols, we must reverse the order in which 4.6 and 0.5 appear in the sentence.

Subtract 4.6 from 0.5.

$$0.5 - 4.6 = 0.5 + (-4.6) \quad \text{Add the opposite of 4.6.}$$
$$= -4.1$$

Self Check 3 **a.** Subtract 2.2 from 4.9, and **b.** subtract 4.9 from 2.2.

EXAMPLE 4

Find: $-9 - 15 + 20 - (-6)$.

ELEMENTARY Algebra f(x) Now™

Solution We write each subtraction as addition of the opposite and add.

$$-9 - 15 + 20 - (-6) = -9 + (-15) + 20 + 6$$
$$= -24 + 26$$
$$= 2$$

Self Check 4 Find: $-40 - (-10) + 7 - (-15)$.

APPLICATIONS INVOLVING SUBTRACTION

Subtraction finds the difference between two numbers. When we find the difference between the maximum value and the minimum value of a collection of measurements, we are finding the **range** of the values.

EXAMPLE 5

ELEMENTARY Algebra $f(x)$ Now™

U.S. temperatures. The record high temperature in the United States was 134°F in Death Valley, California, on July 10, 1913. The record low was −80°F at Prospeck Creek, Alaska, on January 23, 1971. Find the temperature range for these extremes.

Solution To find the temperature range, we subtract the lowest temperature from the highest temperature.

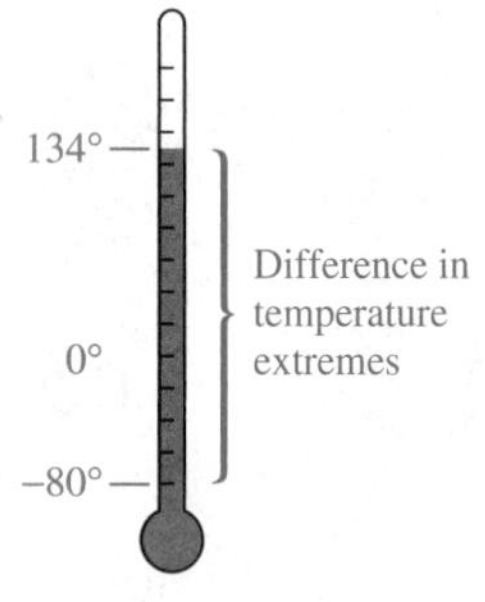

$$134 - (-80) = 134 + 80$$
$$= 214$$

The temperature range for these extremes is 214°F.

Many things change in our lives. The price of a gallon of gasoline, the amount of money we have in the bank, and our ages are just a few examples. In general, to find the change in a quantity, we subtract the earlier value from the later value.

EXAMPLE 6

ELEMENTARY Algebra $f(x)$ Now™

Water levels. In one week, the water level in a storage tank went from 16 feet above normal to 14 feet below normal. Find the change in the water level.

Solution We can represent a water level above normal using a positive number and a water level below normal using a negative number. To find the change in the water level, we subtract the previous measurement, 16, from the most recent measurement, −14.

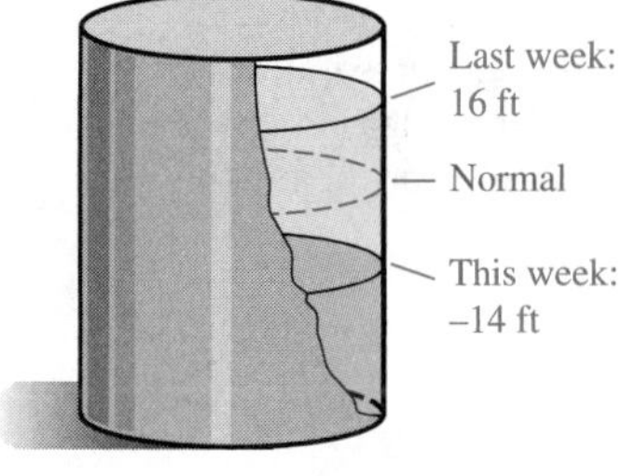

$$-14 - 16 = -14 + (-16)$$
$$= -30$$

The negative result indicates that the water level fell 30 feet that week.

Answers to Self Checks **1. a.** 1, **b.** y, **c.** −500 **2. a.** −57, **b.** 29, **c.** $\frac{5}{12}$ **3. a.** 2.7, **b.** −2.7 **4.** −8

1.5 STUDY SET

ELEMENTARY Algebra $f(x)$ Now™

VOCABULARY Fill in the blanks.

1. Two numbers that are the same distance from 0 on a number line, but on opposite sides of it, are called ________, or additive ________.
2. In the subtraction $-2 - 5 = -7$, the result of −7 is called the ________.
3. The difference between the maximum and the minimum value of a collection of measurements is called the ______ of the values.
4. To find the ________ in a quantity, subtract the earlier value from the later value.

CONCEPTS

5. Find the opposite, or additive inverse, of each number.

   a. 12 **b.** $-\frac{1}{5}$

   c. 2.71 **d.** 0

6. Complete each statement.
   a. $a - b = a +$ ____ b. $-(-a) =$ ____
   c. In general, $a - b$ ____ $b - a$.
7. Fill in the blanks.
   a. The opposite of the opposite of a number is that ________.
   b. To subtract two numbers, add the first number to the ________ of the number to be subtracted.
8. In each case, determine what number is being subtracted.
   a. $5 - 8$ b. $5 - (-8)$
   c. $-5 - 8$ d. $-5 - (-8)$

Apply the rule for subtraction and fill in the blanks.

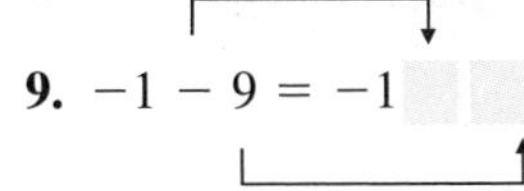

9. $-1 - 9 = -1$ ____ ____

10. $1 - (-9) = 1$ ____ ____

11. Use addition to check this subtraction: $15 - (-8) = 7$. Is the result correct?
12. Which expression below represents the phrase *subtract 6 from 2*?

$2 - 6$ or $6 - 2$

13. Write each subtraction in the following expression as addition of the opposite.

$$-10 - 8 + (-23) + 5 - (-34)$$

14. Simplify each expression.
   a. $-(-1)$ b. $-(-0.5)$
15. Simplify each expression.
   a. $-|-500|$ b. $-(-y)$

NOTATION

16. Write each phrase using symbols.
   a. Negative four
   b. One minus negative seven
   c. The opposite of negative two
   d. The opposite of the absolute value of negative three
   e. The opposite of the opposite of m

PRACTICE Subtract.

17. $4 - 7$
18. $1 - 6$
19. $2 - 15$
20. $3 - 14$
21. $8 - (-3)$
22. $17 - (-21)$
23. $-12 - 9$
24. $-25 - 17$
25. $0 - 6$
26. $0 - 9$
27. $0 - (-1)$
28. $0 - (-8)$
29. $10 - (-2)$
30. $11 - (-3)$
31. $-1 - (-3)$
32. $-1 - (-7)$
33. $20 - (-20)$
34. $30 - (-30)$
35. $-3 - (-3)$
36. $-6 - (-6)$
37. $-2 - (-7)$
38. $-9 - (-1)$
39. $-4 - 5$
40. $-3 - 4$
41. $-44 - 44$
42. $-33 - 33$
43. $0 - (-12)$
44. $0 - 12$
45. $-25 - (-25)$
46. $13 - (-13)$
47. $0 - 4$
48. $0 - (-3)$
49. $-19 - (-17)$
50. $-30 - (-11)$
51. $-\frac{1}{8} - \frac{3}{8}$
52. $-\frac{3}{4} - \frac{1}{4}$
53. $-\frac{9}{16} - \left(-\frac{1}{4}\right)$
54. $-\frac{1}{2} - \left(-\frac{1}{4}\right)$
55. $\frac{1}{3} - \frac{3}{4}$
56. $\frac{1}{6} - \frac{5}{8}$
57. $-0.9 - 0.2$
58. $-0.3 - 0.2$
59. $6.3 - 9.8$
60. $2.1 - 9.4$
61. $-1.5 - 0.8$
62. $-1.5 - (-0.8)$
63. $2.8 - (-1.8)$
64. $4.7 - (-1.9)$

65. Subtract -5 from 17.
66. Subtract 45 from -50.
67. Subtract 12 from -13.
68. Subtract -11 from -20.

Perform the operations.

69. $8 - 9 - 10$
70. $1 - 2 - 3$
71. $-25 - (-50) - 75$
72. $-33 - (-22) - 44$
73. $-6 + 8 - (-1) - 10$
74. $-4 + 5 - (-3) - 13$
75. $61 - (-62) + (-64) - 60$

76. $93 - (-92) + (-94) - 95$

77. $-6 - 7 - (-3) + 9$

78. $-1 - 3 - (-8) + 5$

79. $-20 - (-30) - 50 + 40$

80. $-24 - (-28) - 48 + 44$

APPLICATIONS

81. TEMPERATURE RECORDS Find the difference between the record high temperature of 108°F set in 1926 and the record low of −52°F set in 1979 for New York State.

82. LIE DETECTOR TESTS A burglar scored −18 on a lie detector test, a score that indicates deception. However, on a second test, he scored +3, a score that is inconclusive. Find the change in the scores.

83. LAND ELEVATIONS The elevation of Death Valley, California, is 282 feet below sea level. The elevation of the Dead Sea in Israel is 1,312 feet below sea level. Find the change in their elevations.

84. CARD GAMES Gonzalo won the second round of a card game and earned 50 points. Matt and Hydecki had to deduct the value of each of the cards left in their hands from their score on the first round. Use this information to update the score sheet. (Face cards are counted as 10 points and aces as 1 point.)

Matt

Hydecki

Running point total	Round 1	Round 2
Matt	+50	
Gonzalo	−15	
Hydecki	−2	

85. RACING To improve handling, drivers often adjust the angle of the wheels of their car. When the wheel leans out, the degree measure is considered positive. When the wheel leans in, the degree measure is considered negative. Find the change in the position of the wheel shown.

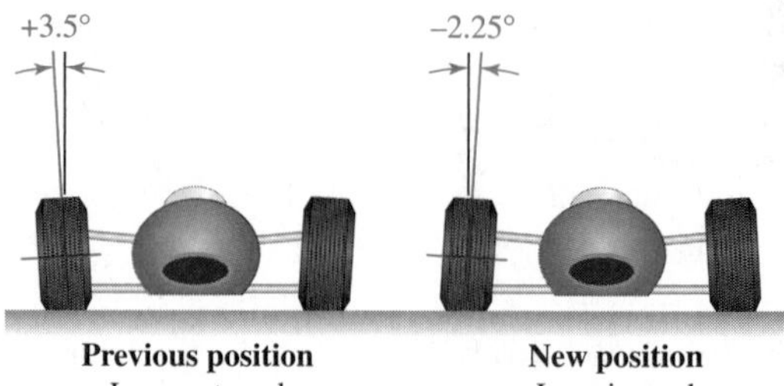

Previous position
Lean outward

New position
Lean inward

86. EYESIGHT Nearsightedness, the condition where near objects are clear and far objects are blurry, is measured using negative numbers. Farsightedness, the condition where far objects are clear and near objects are blurry, is measured using positive numbers. Find the range in the measurements shown.

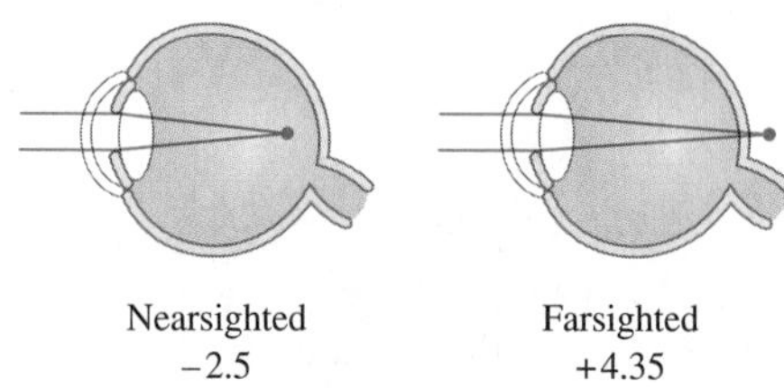

Nearsighted
−2.5

Farsighted
+4.35

87. HISTORY Plato, a famous Greek philosopher, died in 347 B.C. (−347) at the age of 81. When was he born?

88. HISTORY Julius Caesar, a famous Roman leader, died in 44 B.C. (−44) at the age of 56. When was he born?

89. GAUGES Many automobiles have an ammeter like that shown. If the headlights, which draw a current of 7 amps, and the radio, which draws a current of 6 amps, are both turned on, which way will the arrow move? What will be the new reading?

90. U.S. JOBS The graph shows the number of jobs gained or lost each month during the year 2002.

a. In what month were the most jobs gained? Estimate the number.

b. In what month were the most jobs lost? Estimate the number.

c. Find the range for these two extremes.

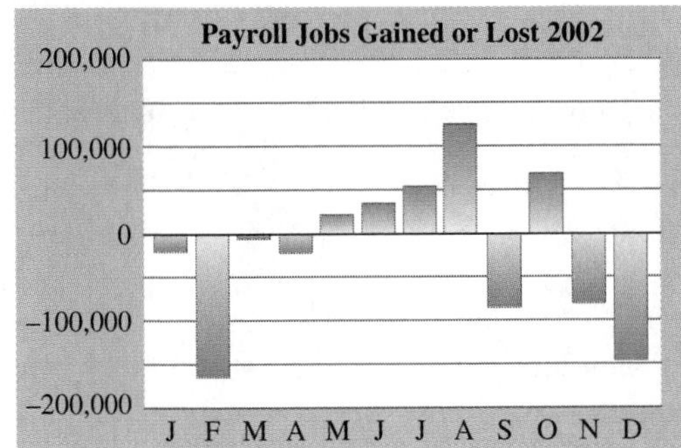

Source: Bureau of Labor Statistics

WRITING

91. Is subtracting 2 from 10 the same as subtracting 10 from 2? Explain.

92. How is $-1 - (-8)$ read?

93. Why is addition of signed numbers taught before subtraction of signed numbers?

94. Explain why we know that the answer to $4 - 10$ is negative without having to do any computation.

REVIEW

95. Find the prime factorization of 30.

96. What factor do the numerator and denominator of the fraction $\frac{15}{18}$ have in common? Simplify the fraction.

97. Build the fraction $\frac{3}{8}$ to an equivalent fraction with a denominator of 56.

98. Write the set of integers.

99. True or false: $-4 > -5$?

100. Use the associative property of addition to simplify the calculation: $-18 + (18 + 89)$.

CHALLENGE PROBLEMS

101. Suppose x is positive and y is negative. Determine whether each statement is true or false.

a. $x - y > 0$ **b.** $y - x < 0$

c. $-x < 0$ **d.** $-y < 0$

102. Find:

$$1 - 2 + 3 - 4 + 5 - 6 + \cdot \cdot \cdot + 99 - 100.$$

1.6 Multiplying and Dividing Real Numbers

- Multiplying Signed Numbers
- Properties of Multiplication
- Dividing Signed Numbers
- Properties of Division

In this section, we will develop rules for multiplying and dividing positive and negative numbers.

MULTIPLYING SIGNED NUMBERS

The Language of Algebra

The names of the parts of a multiplication fact are:

Factor *Factor* *Product*

4(3) = 12

Multiplication represents repeated addition. For example, 4(3) equals the sum of four 3's.

$$\begin{aligned} 4(3) &= 3 + 3 + 3 + 3 \\ &= 12 \end{aligned}$$

This example illustrates that *the product of two positive numbers is positive.*

To develop a rule for multiplying a positive number and a negative number, we will find $4(-3)$, which equals the sum of four -3's.

$$\begin{aligned} 4(-3) &= -3 + (-3) + (-3) + (-3) \\ &= -12 \end{aligned}$$

We see that the result is negative. As a check, think of the problem in terms of money. If you lose \$3 four times, you have lost a total of \$12, which is denoted −\$12. This example illustrates that *the product of a positive number and a negative number is negative.*

Multiplying Two Numbers with Unlike Signs To multiply a positive number and a negative number, multiply their absolute values. Then make the product negative.

EXAMPLE 1

Multiply: **a.** $8(-12)$, **b.** $-15 \cdot 5$, and **c.** $\frac{3}{4}\left(-\frac{4}{15}\right)$.

ELEMENTARY Algebra f(x) Now™

Solution

a. $8(-12) = -96$ Multiply the absolute values, 8 and 12, to get 96. The signs are unlike. Make the product negative.

b. $-15 \cdot 5 = -75$ Multiply the absolute values, 15 and 5, to get 75. The signs are unlike. Make the product negative.

The Language of Algebra

A positive number and a negative number are said to have *unlike* signs.

c. $\frac{3}{4}\left(-\frac{4}{15}\right) = -\frac{3 \cdot 4}{4 \cdot 15}$ Multiply the absolute values $\frac{3}{4}$ and $\frac{4}{15}$. The signs are unlike. Make the product negative.

$= -\frac{\overset{1}{\cancel{3}} \cdot \overset{1}{\cancel{4}}}{\underset{1}{\cancel{4}} \cdot \underset{1}{\cancel{3}} \cdot 5}$ To simplify the fraction, factor 15 as $3 \cdot 5$. Remove the common factors 3 and 4 in the numerator and denominator.

$= -\frac{1}{5}$

Self Check 1 Multiply: **a.** $20(-3)$, **b.** $-3 \cdot 5$, **c.** $-\frac{5}{8} \cdot \frac{16}{25}$

EXAMPLE 2

Medicine. A doctor changes the setting on a heart monitor so that the screen display is magnified by a factor of 1.5. If the current low reading is −14, what will it be after the setting is changed?

Solution To *magnify by a factor of 1.5* means to multiply by 1.5. Therefore, the new low will be 1.5(−14). To find this product, we multiply the absolute values, 1.5 and 14 and make the product negative.

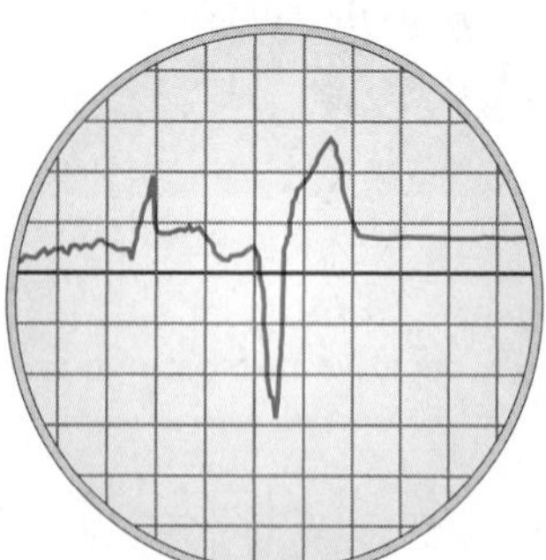

$1.5(-14) = -21$ Multiply absolute values, 1.5 and 14, to get 21. Since 1.5 and −14 have unlike signs, the product is negative.

When the setting is changed, the low reading will be −21.

To develop a rule for multiplying negative numbers, we will find −4(−1), −4(−2), and −4(−3). In the following list, we multiply −4 and a series of factors that decrease by 1. We know how to find the first four products. Graphing those results on a number line is helpful in determining the last three products.

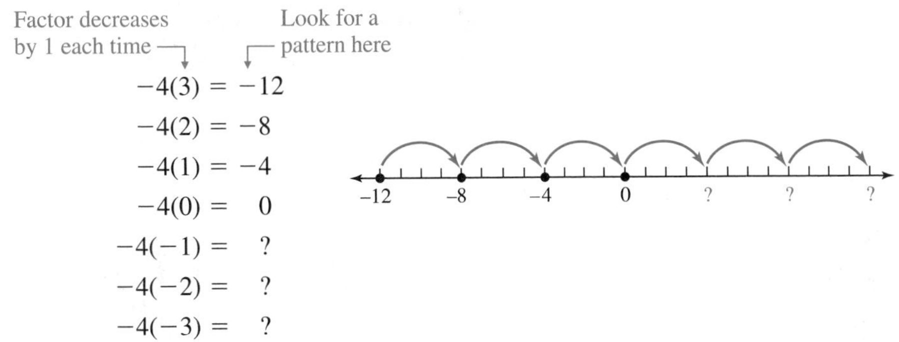

The Language of Algebra

Two negative numbers, as well as two positive numbers, are said to have *like* signs.

From the pattern, we see that the product increases by 4 each time. Thus,

$$-4(-1) = 4, \qquad -4(-2) = 8, \qquad \text{and} \qquad -4(-3) = 12$$

These results illustrate that *the product of two negative numbers is positive.* As a check, think of $-4(-3)$ as losing four debts of \$3. This is equivalent to gaining \$12. Therefore, $-4(-\$3) = \$12$.

Since the product of two positive numbers is positive, and the product of two negative numbers is also positive, we can summarize the multiplication rule as follows.

Multiplying Two Numbers with Like Signs

To multiply two real numbers with the same sign, multiply their absolute values. The product is positive.

EXAMPLE 3

Multiply: **a.** $-5(-6)$ and **b.** $\left(-\frac{1}{2}\right)\left(-\frac{5}{8}\right)$.

ELEMENTARY Algebra f(x) Now™

Solution

a. $-5(-6) = 30$ Multiply the absolute values, 5 and 6, to get 30. Since both factors are negative, the product is positive.

b. $\left(-\frac{1}{2}\right)\left(-\frac{5}{8}\right) = \frac{5}{16}$ Multiply the absolute values, $\frac{1}{2}$ and $\frac{5}{8}$, to get $\frac{5}{16}$. Since the factors have like signs, the product is positive.

Self Check 3 Multiply: **a.** $-15(-8)$, **b.** $-\frac{1}{4}\left(-\frac{1}{3}\right)$.

PROPERTIES OF MULTIPLICATION

The multiplication of two numbers can be done in any order; the result is the same. For example, $-6(5) = -30$ and $5(-6) = -30$. This shows that multiplication is **commutative.**

The Commutative Property of Multiplication

Changing the order when multiplying does not affect the answer.

Let a and b represent real numbers,

$$ab = ba$$

In the following example, we multiply $-3 \cdot 7 \cdot 5$ in two ways. Recall that the operation within the parentheses should be performed first.

Method 1: Group −3 and 7

$$(-3 \cdot 7)5 = (-21)5$$
$$= -105$$

Method 2: Group 7 and 5

$$-3(7 \cdot 5) = -3(35)$$
$$= -105$$

It doesn't matter how we group the numbers in this multiplication; the result is −105. This example illustrates that multiplication is **associative.**

The Associative Property of Multiplication

Changing the grouping when multiplying does not affect the answer.
Let a, b, and c represent real numbers,

$$(ab)c = a(bc)$$

EXAMPLE 4

Multiply: **a.** $-5(-37)(-2)$ and **b.** $-4(-3)(-2)(-1)$.

ELEMENTARY Algebra $f(x)$ Now™

Solution Using the commutative and associative properties of multiplication, we can reorder and regroup the factors to simplify computations.

a. Since it is easy to multiply by 10, we will find $-5(-2)$ first.

$$-5(-37)(-2) = 10(-37)$$
$$= -370$$

b. $-4(-3)(-2)(-1) = 12(2)$ Multiply the first two factors and multiply the last two factors.
$$= 24$$

Self Check 4 Multiply: **a.** $-25(-3)(-4)$, **b.** $-1(-2)(-3)(-3)$.

In Example 4a, we multiplied three negative numbers. In Example 4b, we multiplied four negative numbers. The results illustrate the following fact.

Multiplying Negative Numbers

The product of an even number of negative numbers is positive. The product of an odd number of negative numbers is negative.

Success Tip

If 0 is a factor in a multiplication, the product is 0. For example,

$$25(-6)(0)(-17) = 0$$

Whenever we multiply 0 and a number, the product is 0:

$$0 \cdot 8 = 0, \quad 6.5(0) = 0, \quad \text{and} \quad 0(-12) = 0$$

These examples illustrate the **multiplication property of 0.**

Multiplication Property of 0

The product of 0 and any real number is 0.
For any real number a,

$$0 \cdot a = 0 \quad \text{and} \quad a \cdot 0 = 0$$

Whenever we multiply a number by 1, the number remains the same:

$$1 \cdot 6 = 6, \quad 4.57 \cdot 1 = 4.57, \quad \text{and} \quad 1(-9) = -9$$

These examples illustrate the **multiplication property of 1.** Since any number multiplied by 1 remains the same (is identical), the number 1 is called the **identity element** for multiplication.

Multiplication Property of 1 (Identity Property of Multiplication)

The product of 1 and any number is that number.

For any real number a,

$$1 \cdot a = a \qquad \text{and} \qquad a \cdot 1 = a$$

Two numbers whose product is 1 are **reciprocals** or **multiplicative inverses** of each other. For example, 8 is the multiplicative inverse of $\frac{1}{8}$, and $\frac{1}{8}$ is the multiplicative inverse of 8, because $8 \cdot \frac{1}{8} = 1$. Likewise, $-\frac{3}{4}$ and $-\frac{4}{3}$ are multiplicative inverses because $-\frac{3}{4}\left(-\frac{4}{3}\right) = 1$. All real numbers, except 0, have a multiplicative inverse.

Multiplicative Inverses (Inverse Property of Multiplication)

The product of any number and its multiplicative inverse (reciprocal) is 1.

For any nonzero real number a,

$$a\left(\frac{1}{a}\right) = 1$$

EXAMPLE 5

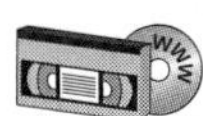

Find the reciprocal of each number. **a.** $\frac{2}{3}$, **b.** $-\frac{2}{3}$, and **c.** -11.

Solution

a. The reciprocal of $\frac{2}{3}$ is $\frac{3}{2}$ because $\frac{2}{3}\left(\frac{3}{2}\right) = 1$.

b. The reciprocal of $-\frac{2}{3}$ is $-\frac{3}{2}$ because $-\frac{2}{3}\left(-\frac{3}{2}\right) = 1$.

c. The reciprocal of -11 is $-\frac{1}{11}$ because $-11\left(-\frac{1}{11}\right) = 1$.

Self Check 5 Find the reciprocal of each number: **a.** $-\frac{15}{16}$, **b.** $\frac{15}{16}$, **c.** -27.

DIVIDING SIGNED NUMBERS

Every division fact can be written as an equivalent multiplication fact.

Division

Let a, b, and c represent real numbers, where $b \neq 0$,

$$\frac{a}{b} = c \qquad \text{provided that} \qquad c \cdot b = a$$

We can use this relationship between multiplication and division to develop rules for dividing signed numbers. For example,

$$\frac{15}{5} = 3 \qquad \text{because} \qquad 3(5) = 15$$

From this example, we see that *the quotient of two positive numbers is positive.*

To determine the quotient of two negative numbers, we consider $\frac{-15}{-5}$.

The Language of Algebra

The names of the parts of a division fact are:

Dividend *Quotient*

$\frac{15}{5} = 3$

Divisor

$$\frac{-15}{-5} = 3 \quad \text{because} \quad 3(-5) = -15$$

From this example, we see that the *quotient of two negative numbers is positive.*

To determine the quotient of a positive number and a negative number, we consider $\frac{15}{-5}$.

$$\frac{15}{-5} = -3 \quad \text{because} \quad -3(-5) = 15$$

From this example, we see that *the quotient of a positive number and a negative number is negative.*

To determine the quotient of a negative number and a positive number, we consider $\frac{-15}{5}$.

$$\frac{-15}{5} = -3 \quad \text{because} \quad -3(5) = -15$$

From this example, we see that *the quotient of a negative number and a positive number is negative.*

We summarize the rules from the previous examples and note that they are similar to the rules for multiplication.

Dividing Two Real Numbers

To divide two real numbers, divide their absolute values.

1. The quotient of two numbers with *like* signs is positive.
2. The quotient of two numbers with *unlike* signs is negative.

EXAMPLE 6

ELEMENTARY Algebra f(x) Now™

Find each quotient: **a.** $\frac{-81}{-9}$, **b.** $\frac{45}{-9}$, and **c.** $-2.87 \div 0.7$.

Solution

a. $\frac{-81}{-9} = 9$ Divide the absolute values, 81 by 9, to get 9. Since the signs are like, the quotient is positive.

Multiply to check the result: $9(-9) = -81$.

b. $\frac{45}{-9} = -5$ Divide the absolute values, 45 by 9, to get 5. Since the signs are unlike, make the quotient negative.

Multiply to check the result: $-5(-9) = 45$.

c. $-2.87 \div 0.7 = -4.1$ Since the signs are unlike, make the quotient negative.

Multiply to check the result: $-4.1(0.7) = -2.87$.

Self Check 6 Find each quotient: **a.** $\frac{-28}{-4}$, **b.** $\frac{75}{-25}$, **c.** $0.32 \div (-1.6)$.

EXAMPLE 7

ELEMENTARY Algebra f(x) Now™

Divide: $-\frac{5}{16} \div \left(-\frac{1}{2}\right)$.

Solution

$$-\frac{5}{16} \div \left(-\frac{1}{2}\right) = -\frac{5}{16}\left(-\frac{2}{1}\right)$$ Multiply the first fraction by the reciprocal of the second fraction. The reciprocal of $-\frac{1}{2}$ is $-\frac{2}{1}$.

$$= \frac{5 \cdot 2}{16 \cdot 1}$$ Multiply the absolute values $\frac{5}{16}$ and $\frac{2}{1}$. Since the signs are like, the product is positive.

$$= \frac{5 \cdot \overset{1}{\cancel{2}}}{\underset{1}{\cancel{2}} \cdot 8 \cdot 1}$$ To simplify the fraction, factor 16 as $2 \cdot 8$. Replace $\frac{2}{2}$ with $\frac{1}{1}$.

$$= \frac{5}{8}$$

Self Check 7 Divide: $\frac{3}{4} \div \left(-\frac{5}{8}\right)$.

EXAMPLE 8

ELEMENTARY Algebra f(x) Now™

Depreciation. Over an 8-year period, the value of a \$150,000 house fell at a uniform rate to \$110,000. Find the amount of depreciation per year.

Solution First, we find the change in the value of the house.

$$110{,}000 - 150{,}000 = -40{,}000$$ Subtract the previous value from the current value.

The Language of Algebra

Depreciation is a form of the word *depreciate,* meaning to lose value. You've probably heard that the minute you drive a new car off the lot, it has depreciated.

The result represents a drop in value of \$40,000. Since the depreciation occurred over 8 years, we divide −40,000 by 8.

$$\frac{-40{,}000}{8} = -5{,}000$$ Divide the absolute values, 40,000 by 8, to get 5,000, and make the quotient negative.

The house depreciated \$5,000 per year.

PROPERTIES OF DIVISION

Whenever we divide a number by 1, the quotient is that number:

$$\frac{12}{1} = 12, \qquad \frac{-80}{1} = -80, \qquad \text{and} \qquad 7.75 \div 1 = 7.75$$

Furthermore, whenever we divide a nonzero number by itself, the quotient is 1:

$$\frac{35}{35} = 1, \qquad \frac{-4}{-4} = 1, \qquad \text{and} \qquad 0.9 \div 0.9 = 1$$

These observations suggest the following properties of division.

Division Properties

Any number divided by 1 is the number itself. Any number (except 0) divided by itself is 1.

For any real number a,

$$\frac{a}{1} = a \qquad \text{and} \qquad \frac{a}{a} = 1 \qquad (\text{where } a \neq 0)$$

We will now consider division that involves zero. First, we examine division of zero. As an example, let's consider $\frac{0}{2}$. We know that

$$\frac{0}{2} = 0 \quad \text{because} \quad 0(2) = 0$$

The Language of Algebra

When we say a division by 0, such as $\frac{2}{0}$, is *undefined,* we mean that $\frac{2}{0}$ does not represent a real number.

From this example, we see that *0 divided by a nonzero number is 0.*

Next, we consider division by zero by considering $\frac{2}{0}$. We know that $\frac{2}{0}$ has no answer because there is no number we can multiply 0 by to get 2. We say that such a division is **undefined.**

These results suggest the following division facts.

Division Involving 0 For any nonzero real number a,

$$\frac{0}{a} = 0 \quad \text{and} \quad \frac{a}{0} \text{ is undefined.}$$

EXAMPLE 9

Find each quotient, if possible: **a.** $\frac{0}{13}$ and **b.** $\frac{-13}{0}$.

Solution

a. $\frac{0}{13} = 0$ because $0(13) = 0$.

b. Since $\frac{-13}{0}$ involves division by zero, the division is undefined.

Self Check 9 Find each quotient, if possible: **a.** $\frac{4}{0}$, **b.** $\frac{0}{17}$.

Answers to Self Checks

1. a. -60, **b.** -15, **c.** $-\frac{2}{5}$ **3. a.** 120, **b.** $\frac{1}{12}$ **4. a.** -300, **b.** 18
5. a. $-\frac{16}{15}$, **b.** $\frac{16}{15}$, **c.** $-\frac{1}{27}$ **6. a.** 7, **b.** -3, **c.** -0.2 **7.** $-\frac{6}{5}$
9. a. undefined, **b.** 0

1.6 STUDY SET

ELEMENTARY Algebra f(x) Now™

VOCABULARY Fill in the blanks.

1. The answer to a multiplication problem is called a ________. The answer to a division problem is called a ________.
2. The numbers -4 and -6 are said to have ______ signs. The numbers -10 and 12 are said to have _______ signs.
3. The __________ property of multiplication states that changing the order when multiplying does not affect the answer.
4. The __________ property of multiplication states that changing the grouping when multiplying does not affect the answer.
5. Division of a nonzero number by zero is __________.
6. If the product of two numbers is 1, the numbers are called __________ or ___________ inverses.

CONCEPTS Fill in the blanks.

7. The expression $-5 + (-5) + (-5) + (-5)$ can be represented by the multiplication $4(\quad)$.

8. The quotient of two numbers with ________ signs is negative.

9. The product of two negative numbers is ________.

10. The product of zero and any number is ____.

11. The product of ____ and any number is that number.

12. We know that $\frac{20}{-2} = -10$ because $-10(\quad) = 20$.

13. a. If we multiply two different numbers and the answer is 0, what must be true about one of the numbers?

b. If we multiply two different numbers and the answer is 1, what must be true about the numbers?

14. a. If we divide two numbers and the answer is 1, what must be true about the numbers?

b. If we divide two numbers and the answer is 0, what must be true about the numbers?

Let POS stand for a positive number and NEG stand for a negative number. Determine the sign of each result, if possible.

15. a. POS · NEG   b. POS + NEG

c. POS − NEG   d. $\frac{\text{POS}}{\text{NEG}}$

16. a. NEG · NEG   b. NEG + NEG

c. NEG − NEG   d. $\frac{\text{NEG}}{\text{NEG}}$

17. Complete each property of multiplication.

a. $a \cdot b = b \cdot$ b. $(ab)c =$

c. $0 \cdot a =$ d. $1 \cdot a =$

e. $a\left(\frac{1}{a}\right) = \quad (a \neq 0)$

18. Complete each property of division.

a. $\frac{a}{1} =$ b. $\frac{a}{a} = \quad (a \neq 0)$

c. $\frac{0}{a} = \quad (a \neq 0)$ d. $\frac{a}{0}$ is

19. Which property justifies each statement?

a. $-5(2 \cdot 17) = (-5 \cdot 2)17$

b. $-5\left(-\frac{1}{5}\right) = 1$

c. $-5 \cdot 2 = 2(-5)$

d. $-5(1) = -5$

20. Complete the table.

Number	Opposite (additive inverse)	Reciprocal (multiplicative inverse)
2		
$-\frac{4}{5}$		
−55		
1.75		

NOTATION

21. Write each sentence using symbols.

a. The product of negative four and negative five is twenty.

b. The quotient of sixteen and negative eight is negative two.

22. Write each expression without − signs.

a. $\frac{-1}{-2}$ b. $\frac{-7}{-8}$

PRACTICE Perform each operation.

23. $-2 \cdot 8$

24. $-3 \cdot 4$

25. $(-6)(-9)$

26. $(-8)(-7)$

27. $12(-5)$

28. $(-9)(11)$

29. $-6 \cdot 4$

30. $-8 \cdot 9$

31. $-20(40)$

32. $-10(10)$

33. $(-6)(-6)$

34. $(-1)(-1)$

35. $-0.6(-4)$

36. $-0.7(-8)$

37. $1.2(-0.4)$

38. $0(-0.2)$

39. $-1.1(-0.9)$

40. $-2.3(-3.1)$

41. $7.2(-2.1)$

42. $4.6(-5.4)$

43. $\frac{1}{2}\left(-\frac{3}{4}\right)$

44. $\frac{1}{3}\left(-\frac{5}{16}\right)$

45. $\left(-\frac{7}{8}\right)\left(-\frac{2}{21}\right)$

46. $\left(-\frac{5}{6}\right)\left(-\frac{2}{15}\right)$

47. $-\frac{16}{25} \cdot \frac{15}{64}$

48. $-\frac{15}{16} \cdot \frac{8}{25}$

49. $-1\frac{1}{4}\left(-\frac{3}{4}\right)$

50. $-1\frac{1}{8}\left(-\frac{3}{8}\right)$

51. $-5.2 \cdot 100$

52. $-1.17 \cdot 1{,}000$

53. $0(-22)$

54. $-8 \cdot 0$

55. $-3(-4)(0)$

56. $15(0)(-22)$

57. $3(-4)(-5)$

58. $(-2)(-4)(-5)$

59. $(-4)(3)(-7)$

60. $5(-3)(-4)$

61. $(-2)(-3)(-4)(-5)$

62. $(-3)(-4)(5)(-6)$

63. $\frac{1}{2}\left(-\frac{1}{3}\right)\left(-\frac{1}{4}\right)$

64. $\frac{1}{3}\left(-\frac{1}{5}\right)\left(-\frac{1}{7}\right)$

65. $-2(-3)(-4)(-5)(-6)$

66. $-9(-7)(-5)(-3)(-1)$

67. $-30 \div (-3)$

68. $-12 \div (-2)$

69. $\frac{-6}{-2}$

70. $\frac{-36}{9}$

71. $\frac{4}{-2}$

72. $\frac{-9}{3}$

73. $\frac{80}{-20}$

74. $\frac{-66}{33}$

75. $\frac{17}{-17}$

76. $\frac{-24}{24}$

77. $\frac{-110}{-110}$

78. $\frac{-200}{-200}$

79. $\frac{-160}{40}$

80. $\frac{-250}{-50}$

81. $\frac{320}{-16}$

82. $\frac{-180}{36}$

83. $\frac{0.5}{-100}$

84. $\frac{-1.7}{10}$

85. $\frac{0}{150}$

86. $\frac{225}{0}$

87. $\frac{-17}{0}$

88. $\frac{0}{-12}$

89. $-\frac{1}{3} \div \frac{4}{5}$

90. $-\frac{2}{3} \div \frac{7}{8}$

91. $-\frac{9}{16} \div \left(-\frac{3}{20}\right)$

92. $-\frac{4}{5} \div \left(-\frac{8}{25}\right)$

93. $-3\frac{3}{8} \div \left(-2\frac{1}{4}\right)$

94. $-3\frac{4}{15} \div \left(-2\frac{1}{10}\right)$

95. $\frac{-23.5}{5}$

96. $\frac{-337.8}{6}$

97. $\frac{-24.24}{-0.8}$

98. $\frac{-55.02}{-0.7}$

Use the associative property of multiplication to find each product.

99. $-\frac{1}{2}(2 \cdot 67)$

100. $\left(-\frac{5}{16} \cdot \frac{1}{7}\right)7$

101. $-0.2(10 \cdot 3)$

102. $-1.5(100 \cdot 4)$

APPLICATIONS

103. TEMPERATURE CHANGE In a lab, the temperature of a fluid was decreased 6° per hour for 12 hours. What signed number indicates the change in temperature?

104. BACTERIAL GROWTH To slowly warm a bacterial culture, biologists programmed a heating pad under the culture to increase the temperature 4° every hour for 6 hours. What signed number indicates the change in the temperature of the pad?

105. GAMBLING A gambler places a $40 bet and loses. He then decides to go "double or nothing" and loses again. Feeling that his luck has to change, he goes "double or nothing" once more and, for the third time, loses. What signed number indicates his gambling losses?

106. REAL ESTATE A house has depreciated $1,250 each year for 8 years. What signed number indicates its change in value over that time period?

107. PLANETS The temperature on Pluto gets as low as −386° F. This is twice as low as the lowest temperature reached on Jupiter. What is the lowest temperature on Jupiter?

108. CAR RADIATORS The instructions on the back of a container of antifreeze state, "A 50/50 mixture of antifreeze and water protects against freeze-ups down to −34° F, while a 60/40 mix protects against freeze-ups down to one and one-half times that temperature." To what temperature does the 60/40 mixture protect?

109. ACCOUNTING For 1999, the total net income for Converse, the sports shoe company, was about −$22.8 million. The company's losses for 2000 were even worse, by a factor of about 1.9. What signed number indicates the company's total net income that year?

110. AIRLINES In the income statement for Trans World Airlines, numbers within parentheses represent a loss. Complete the statement given these facts. The second and fourth quarter losses were approximately the same and totaled $-\$1{,}000$ million. The third quarter loss was about $\frac{3}{5}$ of the first quarter loss.

TWA INCOME STATEMENT				2002
All amounts in millions of US dollars	1st Qtr	2nd Qtr	3rd Qtr	4th Qtr
	(1,550)	(?)	(?)	(?)

111. THE QUEEN MARY The ocean liner Queen Mary was commissioned in 1936 and cost \$22,500,000 to build. In 1967, the ship was purchased by the city of Long Beach, California for \$3,450,000 and now serves as a hotel and convention center. What signed number indicates the annual average depreciation of the Queen Mary over the 31-year period from 1936 to 1967? Round to the nearest dollar.

112. COMPUTERS The formula = A1∗B1∗C1 in cell D1 of the following spreadsheet instructs the computer to multiply the values in cells A1, B1, and C1 and to print the result *in place of the formula* in cell D1. (The symbol ∗ represents multiplication.) What value will the computer print in the cell D1? What values will be printed in cells D2 and D3?

Microsoft Excel-Book 1

File Edit View Insert Format Tools

	A	B	C	D
1	4	–5	–17	= A1∗B1∗C1
2	22	–30	14	= A2∗B2∗C2
3	–60	–20	–34	= A3∗B3∗C3
4				

Sheet 1 Sheet 2 Sheet 3 Sheet 4 Sheet 5

113. PHYSICS An oscilloscope displays electrical signals which appear as wavy lines on a screen. By switching the magnification setting to × 2, for example, the height of the "peak" and the depth of the "valley" of a graph will be doubled. Use signed numbers to indicate the height and depth of the display for each setting of the magnification dial.

a. normal **b.** × 0.5

c. × 1.5 **d.** × 2

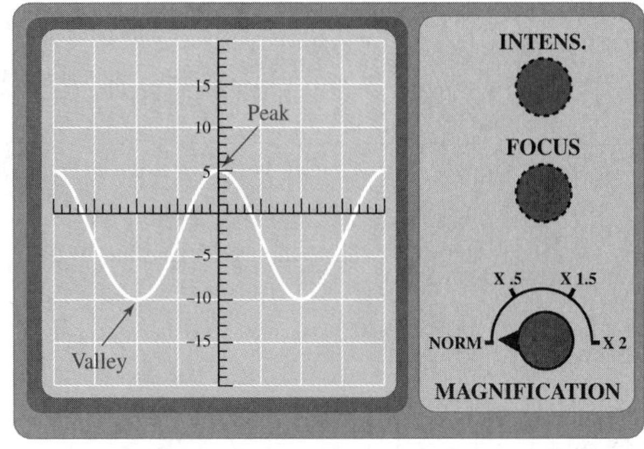

114. LIGHT Water acts as a selective filter of light. In the illustration, we see that red light waves penetrate water only to a depth of about 5 meters. How many times deeper does

a. yellow light penetrate than red light?

b. green light penetrate than orange light?

c. blue light penetrate than yellow light?

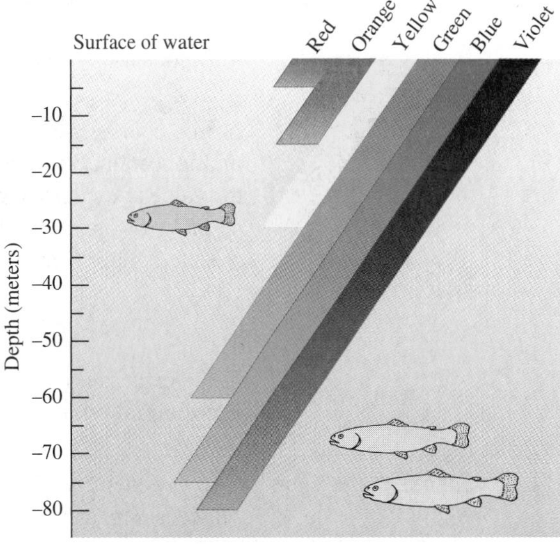

WRITING

115. When a calculator was used to compute 16 ÷ 0, the message shown appeared on the display screen. Explain what the message means.

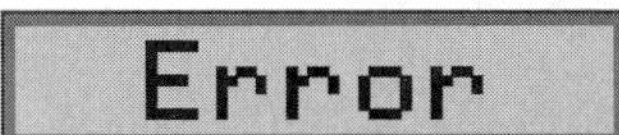

116. **a.** Find $-1(8)$. In general, what is the result when a number is multiplied by -1?

b. Find $\frac{8}{-1}$. In general, what is the result when a number is divided by -1?

117. What is wrong with the following statement?

A negative and a positive is a negative.

118. Is 80 divided by -5 the same as -5 divided by 80? Explain.

REVIEW

119. Add: $-3 + (-4) + (-5) + 4 + 3$.

120. Write the subtraction statement $-3 - (-5)$ as addition of the opposite.

121. Find $\frac{1}{2} + \frac{1}{4} + \frac{1}{3}$ and express the result as a decimal.

122. Describe the balance in a checking account that is overdrawn \$65 using a signed number.

123. Remove the common factors of the numerator and denominator to simplify the fraction: $\frac{2 \cdot 3 \cdot 5 \cdot 5}{2 \cdot 5 \cdot 5 \cdot 7}$.

124. Find $|-2{,}345|$.

CHALLENGE PROBLEMS

125. If the product of five numbers is negative, how many of them could be negative? Explain.

126. Suppose a is a positive number and b is a negative number. Determine whether the given expression is positive or negative.

a. $-a(-b)$ **b.** $\frac{-a}{b}$

c. $\frac{-a}{a}$ **d.** $\frac{1}{b}$

1.7 Exponents and Order of Operations

• Exponents • Order of Operations • Grouping Symbols • The Mean (Average)

In this course, we will perform six operations with real numbers: addition, subtraction, multiplication, division, raising to a power, and finding a root. Often, we will have to find the value of expressions that involve more than one operation. In this section, we introduce an order-of-operations rule to follow in such cases. But first, we discuss a way to write repeated multiplication using *exponents.*

EXPONENTS

In the expression $3 \cdot 3 \cdot 3 \cdot 3 \cdot 3$, the number 3 repeats as a factor five times. We can use **exponential notation** to write this product more concisely.

Exponent and Base An **exponent** is used to indicate repeated multiplication. It tells how many times the **base** is used as a factor.

$$\underbrace{3 \cdot 3 \cdot 3 \cdot 3 \cdot 3}_{\text{Five repeated factors of 3.}} = 3^5$$

The exponent is 5. The base is 3.

The Language of Algebra

5^2 represents the area of a square with sides 5 units long. 4^3 represents the volume of a cube with sides 4 units long.

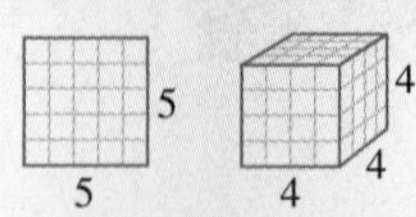

In the **exponential expression** 3^5, 3 is the base, and 5 is the exponent. The expression 3^5 is called a power of 3. Some other examples of exponential expressions are:

5^2 Read as "5 to the second power" or "5 squared."

4^3 Read as "4 to the third power" or "4 cubed."

$(-2)^5$ Read as "-2 to the fifth power."

EXAMPLE 1

Write each expression using exponents: **a.** $7 \cdot 7 \cdot 7$, **b.** $(-5)(-5)(-5)(-5)(-5)$, **c.** sixteen cubed, and **d.** $8 \cdot 8 \cdot 15 \cdot 15 \cdot 15 \cdot 15$.

Solution

a. We can represent this repeated multiplication with an exponential expression having a base of 7 and an exponent of 3: $7 \cdot 7 \cdot 7 = 7^3$.

b. The factor -5 is repeated five times: $(-5)(-5)(-5)(-5)(-5) = (-5)^5$.

c. Sixteen cubed can be written as 16^3.

d. $8 \cdot 8 \cdot 15 \cdot 15 \cdot 15 \cdot 15 = 8^2 \cdot 15^4$

Self Check 1

Write each expression using exponents: **a.** $(12)(12)(12)(12)(12)(12)$, **b.** $2 \cdot 9 \cdot 9 \cdot 9$, **c.** fifty squared, **d.** $(-30)(-30)(-30)$.

EXAMPLE 2

Write each product using exponents: **a.** $a \cdot a \cdot a \cdot a \cdot a \cdot a$ and **b.** $4 \cdot \pi \cdot r \cdot r$.

Solution

a. $a \cdot a \cdot a \cdot a \cdot a \cdot a = a^6$ $\quad$ a is repeated as a factor 6 times.

b. $4 \cdot \pi \cdot r \cdot r = 4\pi r^2$ $\quad$ r is repeated as a factor 2 times.

Self Check 2

Write each product using exponents: **a.** $y \cdot y \cdot y \cdot y$, **b.** $12 \cdot b \cdot b \cdot b \cdot c$.

EXAMPLE 3

Find the value of each expression: **a.** 5^3, **b.** 10^1, **c.** $(-3)^4$, and **d.** $(-3)^5$.

ELEMENTARY Algebra *f(x)* Now™

Solution

a. $5^3 = 5 \cdot 5 \cdot 5 = 125$ $\quad$ The base is 5, the exponent is 3.

b. $10^1 = 10$ $\quad$ The base is 10. Since the exponent is 1, we write the base once.

c.
$$\begin{aligned} (-3)^4 &= (-3)(-3)(-3)(-3) && \text{Write } -3 \text{ as a factor four times.} \\ &= 9(-3)(-3) && \text{Work from left to right.} \\ &= -27(-3) \\ &= 81 \end{aligned}$$

d.
$$\begin{aligned} (-3)^5 &= (-3)(-3)(-3)(-3)(-3) && \text{Write } -3 \text{ as a factor five times.} \\ &= 9(-3)(-3)(-3) && \text{Work from left to right.} \\ &= -27(-3)(-3) \\ &= 81(-3) \\ &= -243 \end{aligned}$$

Notation

If a number or a variable is written without an exponent, we assume the number or variable has an understood exponent of 1. For example,

$$x = x^1 \quad \text{and} \quad 8 = 8^1$$

Self Check 3

Evaluate: **a.** 2^5, **b.** 9^1, **c.** $(-6)^2$, **d.** $(-5)^3$.

Caution A common mistake when evaluating an exponential expression is to multiply the base and the exponent. For example, $5^3 \neq 15$. As we saw in Example 3a, $5^3 = 5 \cdot 5 \cdot 5 = 125$.

In part c of Example 3, we raised -3 to an even power, and the result was positive. In part d, we raised -3 to an odd power, and the result was negative. These results illustrate the following rule.

Even and Odd Powers of a Negative Number

When a negative number is raised to an even power, the result is positive.

When a negative number is raised to an odd power, the result is negative.

Calculators

Finding a power

The squaring key x^2 can be used to find the square of a number. To raise a number to a power, we use the y^x key on a scientific calculator and the ^ key on a graphing calculator.

Although the expressions $(-4)^2$ and -4^2 look alike, they are not. When we find the value of each expression, it becomes clear that they are not equivalent.

$$(-4)^2 = (-4)(-4) \quad \text{The base is } -4\text{, the exponent is 2.}$$
$$= 16$$

$$-4^2 = -(4 \cdot 4) \quad \text{The base is 4, the exponent is 2.}$$
$$= -16$$

Different results

EXAMPLE 4

Evaluate: -2^4.

Solution

$$-2^4 = -(2 \cdot 2 \cdot 2 \cdot 2)$$ Since the base is 2, and the exponent is 4, write 2 as a factor four times within parentheses.

$$= -16$$ Do the multiplication within the parentheses.

Self Check 4 Evaluate: -5^4.

EXAMPLE 5

Evaluate: **a.** $\left(-\frac{2}{3}\right)^3$ and **b.** $(0.6)^2$.

Solution

a. $\left(-\frac{2}{3}\right)^3 = \left(-\frac{2}{3}\right)\left(-\frac{2}{3}\right)\left(-\frac{2}{3}\right)$ Since $-\frac{2}{3}$ is the base and 3 is the exponent, we write $-\frac{2}{3}$ as a factor three times.

$= \frac{4}{9}\left(-\frac{2}{3}\right)$ Work from left to right. $\left(-\frac{2}{3}\right)\left(-\frac{2}{3}\right) = \frac{4}{9}$

$= -\frac{8}{27}$ We say $-\frac{8}{27}$ is the *cube* of $-\frac{2}{3}$.

b. $(0.6)^2 = (0.6)(0.6)$ Since 0.6 is the base and 2 is the exponent, we write 0.6 as a factor two times.

$= 0.36$ We say 0.36 is the *square* of 0.6.

Self Check 5 Evaluate: **a.** $\left(-\frac{3}{4}\right)^3$, **b.** $(-0.3)^2$.

ORDER OF OPERATIONS

Suppose you have been asked to contact a friend if you see a Rolex watch for sale when you are traveling in Europe. While in Switzerland, you find the watch and send the e-mail message shown on the left. The next day, you get the response shown on the right.

E-Mail
FOUND WATCH. $5,000. SHOULD I BUY IT FOR YOU?

E-Mail
NO PRICE TOO HIGH! REPEAT...NO! PRICE TOO HIGH.

Something is wrong. The first part of the response (No price too high!) says to buy the watch at any price. The second part (No! Price too high.) says not to buy it, because it's too expensive. The placement of the exclamation point makes us read the two parts of the response differently, resulting in different meanings. When reading a mathematical statement, the same kind of confusion is possible. For example, consider the expression

$$2 + 3 \cdot 6$$

We can evaluate this expression in two ways. We can add first, and then multiply. Or we can multiply first, and then add. However, the results are different.

$2 + 3 \cdot 6 = 5 \cdot 6$ Add 2 and 3 first.
$= 30$ Multiply 5 and 6.

$2 + 3 \cdot 6 = 2 + 18$ Multiply 3 and 6 first.
$= 20$ Add 2 and 18.

Different results

If we don't establish a uniform order of operations, the expression has two different values. To avoid this possibility, we will always use the following set of priority rules.

Order of Operations

1. Perform all calculations within parentheses and other grouping symbols following the order listed in Steps 2–4 below, working from the innermost pair of grouping symbols to the outermost pair.
2. Evaluate all exponential expressions.
3. Perform all multiplications and divisions as they occur from left to right.
4. Perform all additions and subtractions as they occur from left to right.

When grouping symbols have been removed, repeat Steps 2–4 to complete the calculation.

If a fraction is present, evaluate the expression above and the expression below the bar separately. Then do the division indicated by the fraction bar, if possible.

It isn't necessary to apply all of these steps in every problem. For example, the expression $2 + 3 \cdot 6$ does not contain any parentheses, and there are no exponential expressions. So we look for multiplications and divisions to perform and proceed as follows:

$2 + 3 \cdot 6 = 2 + 18$ Do the multiplication first: $3 \cdot 6 = 18$.
$= 20$ Do the addition.

EXAMPLE 6

Evaluate: $3 \cdot 2^3 - 4$.

Solution Three operations need to be performed to find the value of this expression. By the rules for the order of operations, we evaluate 2^3 first.

$$\begin{aligned} 3 \cdot 2^3 - 4 &= 3 \cdot 8 - 4 && \text{Evaluate the exponential expression: } 2^3 = 8. \\ &= 24 - 4 && \text{Do the multiplication: } 3 \cdot 8 = 24. \\ &= 20 && \text{Do the subtraction.} \end{aligned}$$

Self Check 6 Evaluate: $2 \cdot 3^2 + 17$.

EXAMPLE 7

Evaluate: $-30 - 4 \cdot 5 + 9$.

Solution This expression involves subtraction, multiplication, and addition. The rules for the order of operations tell us to multiply first.

The Language of Algebra

Sometimes, the word *simplify* is used in place of the word *evaluate.*

$$\begin{aligned} -30 - 4 \cdot 5 + 9 &= -30 - 20 + 9 && \text{Do the multiplication: } 4 \cdot 5 = 20. \\ &= -50 + 9 && \text{Working from left to right, do the subtraction: } -30 - 20 = -30 + (-20) = -50. \\ &= -41 && \text{Do the addition.} \end{aligned}$$

Self Check 7 Evaluate: $-40 - 9 \cdot 4 + 10$.

EXAMPLE 8

Evaluate: $160 \div (-4) - 6(-2)3$.

Solution Although this expression contains parentheses, there are no operations to perform within them. Since there are no exponents, we do multiplications and divisions as they are encountered from left to right.

$$\begin{aligned} 160 \div (-4) - 6(-2)3 &= -40 - 6(-2)3 && \text{Do the division: } 160 \div (-4) = -40. \\ &= -40 - (-12)3 && \text{Do the multiplication: } 6(-2) = -12. \\ &= -40 - (-36) && \text{Do the multiplication: } (-12)3 = -36. \\ &= -40 + 36 && \text{Write the subtraction as addition of the opposite.} \\ &= -4 && \text{Do the addition.} \end{aligned}$$

Self Check 8 Evaluate: $240 \div (-8) - 3(-2)4$.

GROUPING SYMBOLS

Grouping symbols serve as mathematical punctuation marks. They help determine the order in which an expression is to be evaluated. Examples of grouping symbols are parentheses ( ), brackets [ ], absolute value symbols | |, and the fraction bar —.

EXAMPLE 9

Evaluate: $(6 - 3)^2$.

Solution This expression contains parentheses. By the rules for the order of operations, we must perform the operation within the parentheses first.

$$(6 - 3)^2 = 3^2 \quad \text{Do the subtraction: } 6 - 3 = 3.$$
$$= 9 \quad \text{Evaluate the exponential expression.}$$

Self Check 9 Evaluate: $(12 - 6)^3$.

EXAMPLE 10

Evaluate: $5^3 + 2(-8 - 3 \cdot 2)$.

ELEMENTARY Algebra f(x) Now™

Solution We begin by performing the operations within the parentheses in the proper order: multiplication first, and then subtraction.

Notation

Multiplication is indicated when a number is next to a grouping symbol.

↓
$5^3 + 2(-8 - 3 \cdot 2)$

$$5^3 + 2(-8 - 3 \cdot 2) = 5^3 + 2(-8 - 6) \quad \text{Do the multiplication: } 3 \cdot 2 = 6.$$
$$= 5^3 + 2(-14) \quad \text{Do the subtraction: } -8 - 6 = -14.$$
$$= 125 + 2(-14) \quad \text{Evaluate } 5^3.$$
$$= 125 + (-28) \quad \text{Do the multiplication.}$$
$$= 97 \quad \text{Do the addition.}$$

Self Check 10 Evaluate: $1^3 + 6(-6 - 3 \cdot 0)$.

Expressions can contain two or more pairs of grouping symbols. To evaluate the following expression, we begin by working within the innermost pair of grouping symbols. Then we work within the outermost pair.

Innermost pair
↓ ↓
$-4[-2 - 3(4 - 8^2)] - 2$
↑ ↑
Outermost pair

EXAMPLE 11

Evaluate: $-4[-2 - 3(4 - 8^2)] - 2$.

ELEMENTARY Algebra f(x) Now™

Solution We do the work within the innermost grouping symbols (the parentheses) first.

$$-4[-2 - 3(4 - 8^2)] - 2$$
$$= -4[-2 - 3(4 - 64)] - 2 \quad \text{Evaluate the exponential expression within the parentheses: } 8^2 = 64.$$
$$= -4[-2 - 3(-60)] - 2 \quad \text{Do the subtraction within the parentheses: } 4 - 64 = 4 + (-64) = -60.$$
$$= -4[-2 - (-180)] - 2 \quad \text{Do the multiplication within the brackets: } 3(-60) = -180.$$
$$= -4[178] - 2 \quad \text{Do the subtraction within the brackets: } -2 - (-180) = -2 + 180 = 178.$$
$$= -712 - 2 \quad \text{Do the multiplication: } -4[178] = -712.$$
$$= -714 \quad \text{Do the subtraction.}$$

Self Check 11 Evaluate: $-5[2(5^2 - 15) + 4] - 10$

EXAMPLE 12

Evaluate: $\dfrac{-3(3 + 2) + 5}{17 - 3(-4)}$.

Solution We simplify the numerator and the denominator separately.

> **Calculators**
> ***Order of operations***
> Calculators have the order of operations built in. A left parenthesis key (and a right parenthesis key) should be used when grouping symbols, including a fraction bar, are needed.

$$\frac{-3(3 + 2) + 5}{17 - 3(-4)} = \frac{-3(5) + 5}{17 - (-12)}$$ In the numerator, do the addition within the parentheses. In the denominator, do the multiplication.

$$= \frac{-15 + 5}{17 + 12}$$ In the numerator, do the multiplication. In the denominator, write the subtraction as addition of the opposite of -12, which is 12.

$$= \frac{-10}{29}$$ Do the additions.

$$= -\frac{10}{29}$$ Write the $-$ sign in front of the fraction: $\frac{-10}{29} = -\frac{10}{29}$.

Self Check 12 Evaluate: $\dfrac{-4(-2 + 8) + 6}{8 - 5(-2)}$.

EXAMPLE 13

Evaluate: $10|9 - 15| - 2^5$.

ELEMENTARY Algebra f(x) Now™

Solution The absolute value bars are grouping symbols. We do the calculation within them first.

$$10|9 - 15| - 2^5 = 10|-6| - 2^5$$ Subtract: $9 - 15 = 9 + (-15) = -6$.

$$= 10(6) - 2^5$$ Find the absolute value: $|-6| = 6$.

$$= 10(6) - 32$$ Evaluate the exponential expression: $2^5 = 32$.

$$= 60 - 32$$ Do the multiplication: $10(6) = 60$.

$$= 28$$ Do the subtraction.

Self Check 13 Evaluate: $10^3 + 3|24 - 25|$.

THE MEAN (AVERAGE)

The **arithmetic mean** (or **average**) of a set of numbers is a value around which the values of the numbers are grouped.

Finding an Arithmetic Mean To find the **mean** of a set of values, divide the sum of the values by the number of values.

EXAMPLE 14

ELEMENTARY Algebra f(x) Now™

Customer service. To measure its effectiveness in serving customers, a store had the telephone company electronically record the number of times the telephone rang before an employee answered it. The results of the week-long survey are shown in the table. We see that for 11 calls, the phone was answered after it rang 1 time. For 46 calls, the phone was answered after it rang 2 times, and so on. Find the *average* number of times the phone rang before an employee answered it that week.

Number of rings	Number of calls
1	11
2	46
3	45
4	28
5	20

Solution To find the total number of rings, we multiply each *number of rings* (1, 2, 3, 4, and 5 rings) by the respective number of occurrences and add those subtotals.

$$\text{Total number of rings} = 11(1) + 46(2) + 45(3) + 28(4) + 20(5)$$

The total number of calls received was $11 + 46 + 45 + 28 + 20$. To find the average, we divide the total number of rings by the total number of calls.

$$\text{Average} = \frac{11(1) + 46(2) + 45(3) + 28(4) + 20(5)}{11 + 46 + 45 + 28 + 20}$$

$$= \frac{11 + 92 + 135 + 112 + 100}{150}$$ In the numerator, do the multiplications. In the denominator, do the additions.

$$= \frac{450}{150}$$ Do the addition.

$$= 3$$

The average number of times the phone rang before it was answered was 3.

Answers to Self Checks **1. a.** 12^6, **b.** $2 \cdot 9^3$, **c.** 50^2, **d.** $(-30)^3$ **2. a.** y^4, **b.** $12b^3c$ **3. a.** 32, **b.** 9, **c.** 36, **d.** -125 **4.** -625 **5. a.** $-\frac{27}{64}$, **b.** 0.09 **6.** 35 **7.** -66 **8.** -6 **9.** 216 **10.** -35 **11.** -130 **12.** -1 **13.** 1,003

1.7 STUDY SET

ELEMENTARY Algebra f(x) Now™

VOCABULARY Fill in the blanks.

1. In the exponential expression 3^2, 3 is the ______, and 2 is the ______.

2. 10^2 can be read as ten ______, and 10^3 can be read as ten ______.

3. 7^5 is the fifth ______ of seven.

4. An ______ is used to represent repeated multiplication.

5. The rules for the ______ of operations guarantee that an evaluation of a numerical expression will result in a single answer.

6. The arithmetic ______ or ______ of a set of numbers is a value around which the values of the numbers are grouped.

CONCEPTS

7. Given: $4 + 5 \cdot 6$.

a. Evaluate the expression in two different ways and state the two possible results.

b. Which result from part a is correct, and why?

8. a. What repeated multiplication does 5^3 represent?

b. Write a multiplication statement in which the factor x is repeated 4 times. Then write the expression in simpler form using an exponent.

9. In the expression $-8 + 2[15 - (-6 + 1)]$, which grouping symbols are innermost, and which are outermost?

10. a. What operations does the expression $12 + 5^2(-3)$ contain?

b. In what order should they be performed?

11. a. What operations does the expression $20 - (-2)^2 + 3(-1)$ contain?

b. In what order should they be performed?

12. Consider the expression $\frac{36 - 4(7)}{2(10 - 8)}$. In the numerator, what operation should be done first? In the denominator, what operation should be done first?

13. To evaluate each expression, what operation should be done first?
a. $24 - 4 + 2$ b. $24 \div 4 \cdot 2$

14. To evaluate each expression, what operation should be done first?
a. $-80 - 3 + 5 - 2^2$
b. $-80 - (3 + 5) - 2^2$
c. $-80 - 3 + (5 - 2)^2$

15. To evaluate each expression, what operation should be done first?
a. $(65 - 3)^3$
b. $65 - 3^3$
c. $6(5) - (3)^3$ d. $65 \cdot 3^3$

16. a. How is the mean (or average) of a set of scores found?
b. Find the average of 75, 81, 47, and 53.

NOTATION

17. Fill in the blanks.
a. $3^1 = \square$ b. $x^1 = \square$
c. $9 = 9^{\square}$ d. $y = y^{\square}$

18. Tell the name of each grouping symbol: ( ), [ ], | |, and —.

19. a. In the expression $(-5)^2$, what is the base?
b. In the expression -5^2, what is the base?

20. Write each expression using symbols.
a. Negative two squared.
b. The opposite of two squared.

Complete the evaluation of each expression.

21. $-19 - 2[(1 + 2) \cdot 3] = -19 - 2[\square \cdot 3]$
$= -19 - 2[\square]$
$= -19 - \square$
$= -37$

22. $\frac{46 - 2^3}{-3(5) - 4} = \frac{46 - \square}{\square - 4}$
$= \frac{\square}{\square}$
$= -2$

PRACTICE **Write each product using exponents.**

23. $3 \cdot 3 \cdot 3 \cdot 3$
24. $m \cdot m \cdot m \cdot m \cdot m$
25. $10 \cdot 10 \cdot k \cdot k \cdot k$
26. $5(5)(5)(i)(i)$
27. $8 \cdot \pi \cdot r \cdot r \cdot r$
28. $4 \cdot \pi \cdot r \cdot r$
29. $6(x)(x)(y)(y)(y)$
30. $76 \cdot s \cdot s \cdot s \cdot s \cdot t$

Evaluate each expression.

31. $(-6)^2$
32. -6^2
33. -4^4
34. $(-4)^4$
35. $(-5)^3$
36. -5^3
37. $-(-6)^4$
38. $-(-7)^2$
39. $(-0.4)^2$
40. $(-0.5)^2$
41. $\left(-\frac{2}{5}\right)^3$
42. $\left(-\frac{1}{4}\right)^3$
43. $3 - 5 \cdot 4$
44. $-4 \cdot 6 + 5$
45. $3 \cdot 8^2$
46. $(3 \cdot 4)^2$
47. $8 \cdot 5^1 - 4 \div 2$
48. $9 \cdot 5^1 - 6 \div 3$
49. $100 - 8(10) + 60$
50. $50 - 2(5) - 7$
51. $-22 - (15 - 3)$
52. $-(33 - 8) - 10$
53. $-2(9) - 2(5)$
54. $-75 - 7^2$
55. $5^2 + 13^2$
56. $3^3 - 2^3$
57. $-4(6 + 5)$
58. $-3(5 - 4)$
59. $(9 - 3)(9 - 9)$
60. $-(-8 - 6)(6 - 6)$
61. $(-1 - 18)2$
62. $-5(128 - 5^3)^2$
63. $-2(-1)^2 + 3(-1) - 3$
64. $-4(-3)^2 + 3(-3) - 1$
65. $4^2 - (-2)^2$
66. $(-5 - 2)^2$
67. $12 + 2\left(-\frac{9}{3}\right) - (-2)$
68. $2 + 3\left(-\frac{25}{5}\right) - (-4)$
69. $\frac{-2 - 5}{-7 - (-7)}$
70. $\frac{-3 - (-1)}{-2 - (-2)}$

71. $200 - (-6 + 5)^3$

72. $19 - (-45 + 41)^3$

73. $|5 \cdot 2^2 \cdot 4| - 30$

74. $2 + |3 \cdot 2^2 \cdot 4|$

75. $[6(5) - 5(5)]4$

76. $175 - 2 \cdot 3^4$

77. $-6(130 - 4^3)$

78. $-5(150 - 3^3)$

79. $(17 - 5 \cdot 2)^3$

80. $(4 + 2 \cdot 3)^4$

81. $-5(-2)^3(3)^2$

82. $-3(-2)^5(2)^2$

83. $-2\left(\frac{15}{-5}\right) - \frac{6}{2} + 9$

84. $-6\left(\frac{25}{-5}\right) - \frac{36}{9} + 1$

85. $\frac{5 \cdot 50 - 160}{-9}$

86. $\frac{5(68 - 32)}{-9}$

87. $\frac{2(6 - 1)}{16 - (-4)^2}$

88. $\frac{6 - (-1)}{4 - 2^2}$

89. $5(10 + 2) - 1$

90. $14 + 3(7 - 5)$

91. $64 - 6[15 + (-3)3]$

92. $4 + 2[26 + 5(-3)]$

93. $(12 - 2)^3$

94. $(-2)^3\left(\frac{-6}{2}\right)(-1)$

95. $(-3)^3\left(\frac{-4}{2}\right)(-1)$

96. $\frac{-5 - 3^3}{2^3}$

97. $\frac{1}{2}\left(\frac{1}{8}\right) + \left(-\frac{1}{4}\right)^2$

98. $-\frac{1}{9}\left(\frac{1}{4}\right) + \left(-\frac{1}{6}\right)^2$

99. $-2|4 - 8|$

100. $-5|1 - 8|$

101. $|7 - 8(4 - 7)|$

102. $|9 - 5(1 - 8)|$

103. $3 + 2[-1 - 4(5)]$

104. $4 + 2[-7 - 3(9)]$

105. $-3[5^2 - (7 - 3)^2]$

106. $3 - [3^3 + (3 - 1)^3]$

107. $-(2 \cdot 3 - 4)^3$

108. $-(3 \cdot 5 - 2 \cdot 6)^2$

109. $\frac{(3 + 5)^2 + |-2|}{-2(5 - 8)}$

110. $\frac{|-25| - 8(-5)}{2^4 - 29}$

111. $\frac{2[-4 - 2(3 - 1)]}{3(-3)(-2)}$

112. $\frac{3[-9 + 2(7 - 3)]}{(5 - 8)(7 - 9)}$

113. $\frac{|6 - 4| + 2|-4|}{26 - 2^4}$

114. $\frac{4|9 - 7| + |-7|}{3^2 - 2^2}$

115. $\frac{(4^3 - 10) + (-4)}{5^2 - (-4)(-5)}$

116. $\frac{(6 - 5)^4 - (-21)}{(-9)(-3) - 4^2}$

117. $\frac{72 - (2 - 2 \cdot 1)}{10^2 - (90 + 2^2)}$

118. $\frac{13^2 - 5^2}{-3(5 - 9)}$

119. $-\left(\frac{40 - 1^3 - 2^4}{3(2 + 5) + 2}\right)$

120. $-\left(\frac{8^2 - 10}{2(3)(4) - 5(3)}\right)$

APPLICATIONS

121. LIGHT The illustration shows that the light energy that passes through the first unit of area, 1 yard away from the bulb, spreads out as it travels away from the source. How much area does that light energy cover 2 yards, 3 yards, and 4 yards from the bulb? Express each answer using exponents.

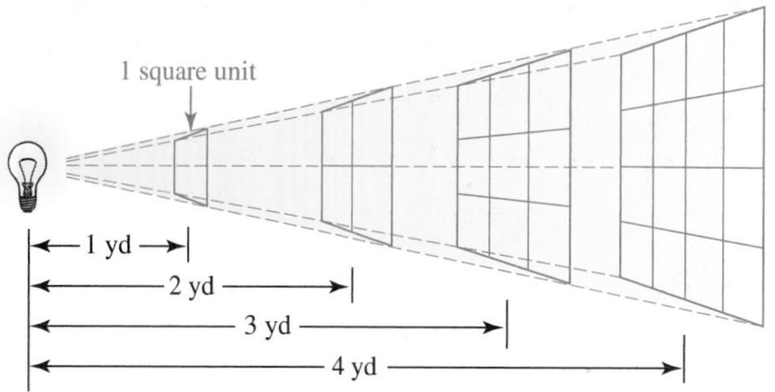

122. CHAIN LETTERS A store owner sent two friends a letter advertising her store's low prices. The ad closed with the following request: "Please send a copy of this letter to two of your friends."

a. Assume that all those receiving letters respond and that everyone in the chain receives just one letter. Complete the table.

b. How many letters will be circulated in the 10th level of the mailing?

Level	Numbers of letters circulated
1st	$2 = 2^1$
2nd	$\square = 2^{\square}$
3rd	$\square = 2^{\square}$
4th	$\square = 2^{\square}$

123. AUTO INSURANCE See the following premium comparison. What is the average six-month insurance premium?

Allstate	\$2,672	Mercury	\$1,370
Auto Club	\$1,680	State Farm	\$2,737
Farmers	\$2,485	20th Century	\$1,692

Criteria: Six-month premium. Husband, 45, drives a 1995 Explorer, 12,000 annual miles. Wife, 43, drives a 1996 Dodge Caravan, 12,000 annual miles. Son, 17, is an occasional operator. All have clean driving records.

124. ENERGY USAGE Find the average number of therms of natural gas used per month.

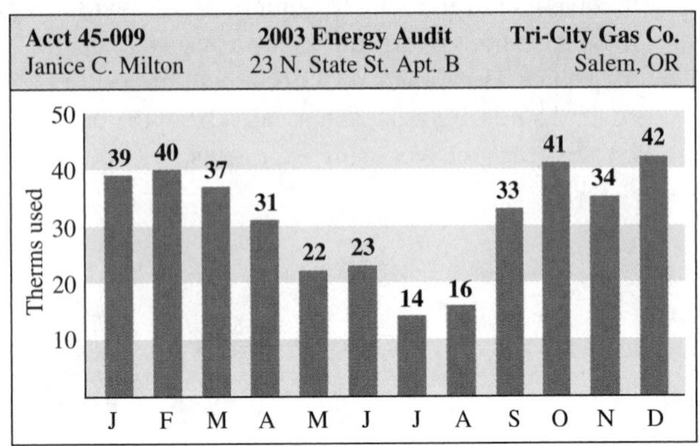

125. CASH AWARDS A contest is to be part of a promotional kickoff for a new children's cereal. The prizes to be awarded are shown.

a. How much money will be awarded in the promotion?

b. What is the average cash prize?

Coloring Contest

Grand prize: Disney World vacation plus $2,500
Four 1st place prizes of $500
Thirty-five 2nd place prizes of $150
Eighty-five 3rd place prizes of $25

126. SURVEYS Some students were asked to rate the food at their college cafeteria on a scale from 1 to 5. The responses are shown on the tally sheet. Find the average rating.

Poor		Fair		Excellent
1	2	3	4	5
	|||	|||	~~||||~~	~~||||~~ ||||

127. WRAPPING GIFTS How much ribbon is needed to wrap the package if 15 inches of ribbon are needed to make the bow?

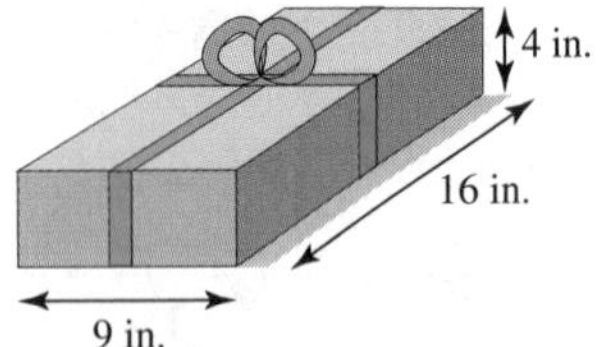

128. SCRABBLE Illustration (a) shows a portion of the game board before and Illustration (b) shows it after the word *QUARTZY* is played. Determine the score. (The number on each tile gives the point value of the letter.)

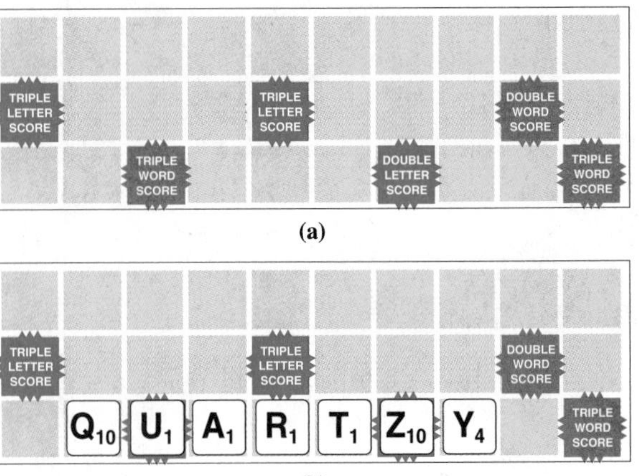

WRITING

129. Explain the difference between 2^3 and 3^2.

130. Explain why rules for the order of operations are necessary.

131. What does it mean when we say, do all additions and subtractions as they occur from left to right?

132. In what settings do you encounter or use the concept of arithmetic mean (average) in your everyday life?

REVIEW

133. Match each term with the proper operation.

a. sum	**i.** division
b. difference	**ii.** addition
c. product	**iii.** subtraction
d. quotient	**iv.** multiplication

134. **a.** What is the opposite of -8?

b. What is the reciprocal of -8?

CHALLENGE PROBLEMS

135. Using each of the numbers 2, 3, and 4 only once, what is the greatest value that the following expression can have?

$$(\square^{\square})^{\square}$$

136. Insert a pair of parentheses into the expression so that it has a value of 40.

$$4 \cdot 3^2 - 4 \cdot 2$$

1.8 Algebraic Expressions

- Algebraic Expressions
- Translating from Words to Symbols
- Writing Algebraic Expressions
- Analyzing Problems
- Number and Value Problems
- Evaluating Algebraic Expressions

Since problems in algebra are often presented in words, the ability to interpret what you read is important. In this section, we will introduce several strategies that will help you translate words into algebraic expressions.

ALGEBRAIC EXPRESSIONS

Recall that variables and/or numbers can be combined with the operations of arithmetic to create **algebraic expressions.** Addition symbols separate these expressions into parts called *terms.* For example, the expression $x + 8$ has two terms.

The Language of Algebra

Note the difference between *terms* and *factors.* In the expression $x + 8$, x and 8 are *terms.* In the expression $8x$, x and 8 are *factors.*

$$\underset{\text{First term}}{x} \quad + \quad \underset{\text{Second term}}{8}$$

Since subtraction can be written as addition of the opposite, the expression $a^2 - 3a - 9$ has three terms.

$$a^2 - 3a - 9 = \underset{\text{First term}}{a^2} + \underset{\text{Second term}}{(-3a)} + \underset{\text{Third term}}{(-9)}$$

Notation

By the commutative property of multiplication, $r6 = 6r$ and $-15b^2a = -15ab^2$. However, we usually write the numerical factor first and the variable factors in alphabetical order.

In general, a **term** is a product or quotient of numbers and/or variables. A single number or variable is also a term. Examples of terms include:

$$8, \quad y, \quad 6r, \quad -w^3, \quad 3.7x^5, \quad \frac{3}{n}, \quad -15ab^2$$

The numerical factor of a term is called the **coefficient** of the term. For example, the term $6r$ has a coefficient of 6 because $6r = 6 \cdot r$. The coefficient of $-15ab^2$ is -15 because $-15ab^2 = -15 \cdot ab^2$. More examples are shown in the following table.

Term	Coefficient	
$8y^2$	8	
$-0.9pq$	-0.9	
$\frac{3}{4}b$	$\frac{3}{4}$	This term could be written $\frac{3b}{4}$.
$-\frac{x}{6}$	$-\frac{1}{6}$	This term could be written $-\frac{1}{6}x$.
x	1	$x = 1x$.
$-t$	-1	$-t = -1t$.
15	15	

The Language of Algebra

Terms such as x and yz^3 have *implied* coefficients of 1. *Implied* means suggested without being precisely expressed.

A term, such as 15, that consists of a single number is called a **constant term.**

EXAMPLE 1

Identify the coefficient of each term in the expression $7x^2 - x + 6$.

ELEMENTARY Algebra $f(x)$ Now™

Solution If we write $7x^2 - x + 6$ as $7x^2 + (-x) + 6$, we see that it has three terms: $7x^2$, $-x$, and 6.

The coefficient of $7x^2$ is 7.
The coefficient of $-x$ is -1.
The coefficient of 6 is 6.

Self Check 1 Identify the coefficient of each term in the expression $p^3 - 12p^2 + 3p - 4$.

TRANSLATING FROM WORDS TO SYMBOLS

In the following tables, we list words and phrases and show how they can be translated into algebraic expressions.

Addition	
the sum of a and 8	$a + 8$
4 plus c	$4 + c$
16 added to m	$m + 16$
4 more than t	$t + 4$
20 greater than F	$F + 20$
T increased by r	$T + r$
exceeds y by 35	$y + 35$

Subtraction	
the difference of 23 and P	$23 - P$
550 minus h	$550 - h$
18 less than w	$w - 18$
7 decreased by j	$7 - j$
M reduced by x	$M - x$
12 subtracted from L	$L - 12$
5 less f	$5 - f$

The Language of Algebra

When a translation involves the phrase *less than,* note how the terms are reversed.

18 less than w

$w - 18$

Multiplication	
the product of 4 and x	$4x$
20 times B	$20B$
twice r	$2r$
triple the profit P	$3P$
$\frac{3}{4}$ of m	$\frac{3}{4}m$

Division	
the quotient of R and 19	$\frac{R}{19}$
s divided by d	$\frac{s}{d}$
the ratio of c to d	$\frac{c}{d}$
k split into 4 equal parts	$\frac{k}{4}$

EXAMPLE 2

Write each phrase as an algebraic expression: **a.** one-half of the profit P, **b.** 5 less than the capacity c, and **c.** the product of the weight w and 2,000, increased by 300.

Solution

a. Key phrase: *One-half of* **Translation:** multiply by $\frac{1}{2}$

The translation is: $\frac{1}{2}P$.

b. Key phrase: *less than* **Translation:** subtract

Sometimes, thinking in terms of specific numbers makes translating easier. Suppose the capacity was 100. Then 5 *less than* 100 would be $100 - 5$. If the capacity is c, then we need to make it 5 less. The translation is: $c - 5$.

c. We are given: The product of the weight w and 2,000, increased by 300.

Key word: *product* **Translation:** multiply
Key phrase: *increased by* **Translation:** add

The comma after 2,000 means w is first multiplied by 2,000 and then 300 is added to that product. The translation is: $2{,}000w + 300$.

Self Check 2 Write each phrase as an algebraic expression: **a.** 80 less than the total t, **b.** $\frac{2}{3}$ of the time T, **c.** the difference of twice a and 15.

WRITING ALGEBRAIC EXPRESSIONS

When solving problems, we usually begin by letting a variable stand for an unknown quantity.

EXAMPLE 3

ELEMENTARY Algebra $f(x)$ Now™

Swimming. The pool shown is x feet wide. If it is to be sectioned into 8 equally wide swimming lanes, write an algebraic expression that represents the width of each lane.

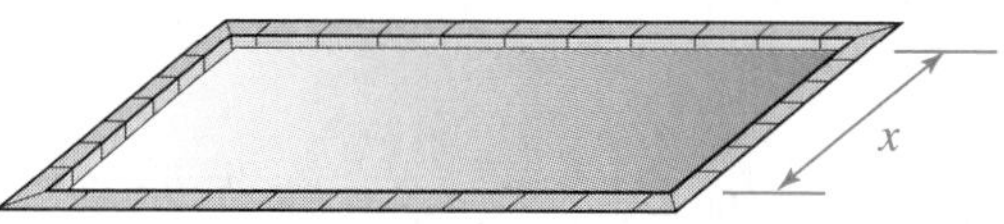

Solution We let x = the width of the swimming pool (in feet).

Key phrase: *sectioned into 8 equally wide lanes* **Translation:** divide

The width of each lane is $\frac{x}{8}$ feet.

Self Check 3 When a secretary rides the bus to work, it takes her m minutes. If she drives her own car, her travel time exceeds this by 15 minutes. How long does it take her to get to work by car?

EXAMPLE 4

Painting. A 10-inch-long paintbrush has two parts: a handle and bristles. Choose a variable to represent the length of one of the parts. Then write an expression to represent the length of the other part.

Solution There are two approaches. First, refer to the following drawing on the left. If we let h = the length of the handle (in inches), the length of the bristles is $10 - h$.

Now refer to the drawing on the right. If we let b = the length of the bristles (in inches), the length of the handle is $10 - b$.

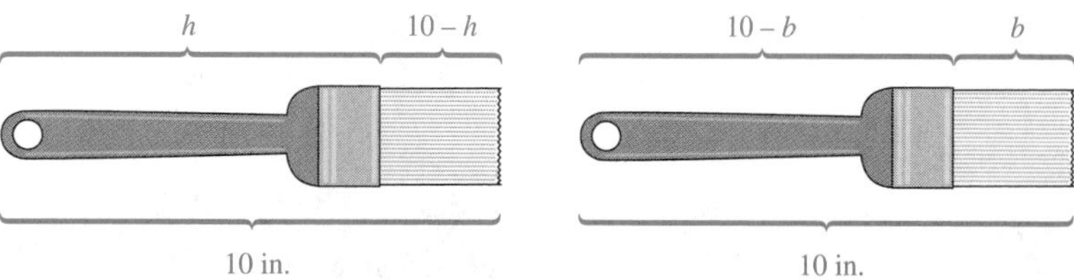

Self Check 4 Part of a \$900 donation to a preschool was designated to go to the scholarship fund, the remainder to the building fund. Choose a variable to represent the amount donated to one of the funds. Write an expression for the amount donated to the other fund.

EXAMPLE 5

College enrollments. In the second semester, enrollment in a retraining program at a college was 32 more than twice that of the first semester. Choose a variable to represent the enrollment in one of the semesters. Then write an expression for the enrollment in the other semester.

Solution Since the second-semester enrollment is expressed in terms of the first-semester enrollment, we let $x =$ the enrollment in the first semester.

Key phrase: *more than* **Translation:** add

Key word: *twice* **Translation:** multiply by 2

The enrollment for the second semester is $2x + 32$.

Self Check 5 In an election, the incumbent received 55 fewer votes than three times the challenger's votes. Choose a variable to represent the number of votes received by one candidate. Write an expression for the number of votes received by the other.

ANALYZING PROBLEMS

When solving problems, we aren't always given key words or key phrases to help establish what mathematical operation to use. Sometimes a careful reading of the problem is needed to determine any hidden operations.

EXAMPLE 6

Vacationing. Disneyland, in California, was in operation 16 years before the opening of Disney World in Florida. Euro Disney, in France, was constructed 21 years after Disney World. Write an algebraic expression to represent the age (in years) of each Disney attraction.

Solution The ages of Disneyland and Euro Disney are both related to the age of Disney World. Therefore, we will let $x =$ the age of Disney World.

In carefully reading the problem, we see that Disneyland was built 16 years *before* Disney World. That makes its age 16 years more than that of Disney World.

Attraction	Age
Disneyland	$x + 16$
Disney World	x
Euro Disney	$x - 21$

$x + 16 =$ the age of Disneyland

Euro Disney was built 21 years *after* Disney World. That makes its age 21 years less than that of Disney World.

$x - 21 =$ the age of Euro Disney

EXAMPLE 7

How many months are in x years?

Solution Since there are no key words, we must carefully analyze the problem. It is often helpful to consider some specific cases. For example, let's calculate the number of months in 1 year, 2 years, and 3 years. When we write the results in a table, a pattern is apparent.

Number of years	Number of months
1	12
2	24
3	36
x	$12x$

We multiply the number of years by 12 to find the number of months.

The number of months in x years is $12 \cdot x$ or $12x$.

Self Check 7 How many days is h hours?

NUMBER AND VALUE PROBLEMS

In some problems, we must distinguish between *the number of* and *the value of* the unknown quantity. For example, to find the value of 3 quarters, we multiply the number of quarters by the value (in cents) of one quarter. Therefore, the value of 3 quarters is $3 \cdot 25$ cents $= 75$ cents.

The same distinction must be made if the number is unknown. For example, the value of n nickels is not n cents. The value of n nickels is $n \cdot 5$ cents $= 5n$ cents. For problems of this type, we will use the relationship

Number · value = total value

EXAMPLE 8 Find the total value of **a.** five dimes, **b.** q quarters, and **c.** $x + 1$ half-dollars.

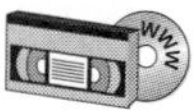

Solution To find the total value (in cents) of each collection of coins, we multiply the number of coins by the value (in cents) of one coin, as shown in the table.

Type of coin	Number	· Value	= Total value
Dime	5	10	50
Quarter	q	25	$25q$
Half-dollar	$x + 1$	50	$50(x + 1)$

← $q \cdot 25$ is written $25q$.

Self Check 8 Find the value of **a.** six fifty-dollar savings bonds, **b.** t one-hundred-dollar savings bonds, **c.** $x - 4$ one-thousand-dollar savings bonds.

EVALUATING ALGEBRAIC EXPRESSIONS

To evaluate an algebraic expression, we substitute given numbers for each variable and do the necessary calculations.

EXAMPLE 9 Evaluate each expression if $x = 3$ and $y = -4$: **a.** $y^3 + y^2$, **b.** $-y - x$, **c.** $|5xy - 7|$, and **d.** $\dfrac{y - 0}{x - (-1)}$.

Solution **a.** $y^3 + y^2 = (-4)^3 + (-4)^2$ Substitute -4 for each y. We must write -4 within parentheses so that it is the base of each exponential expression.

$= -64 + 16$ Evaluate each exponential expression.

$= -48$

Caution

When replacing a variable with its numerical value, we must often write the replacement number within parentheses to convey the proper meaning.

b. $-y - x = -(-4) - 3$ Substitute 3 for x and -4 for y.

$= 4 - 3$ Simplify: $-(-4) = 4$.

$= 1$

c. $|5xy - 7| = |5(3)(-4) - 7|$ Substitute 3 for x and -4 for y.

$= |-60 - 7|$ Do the multiplication within the absolute value symbols, working left to right: $5(3)(-4) = 15(-4) = -60$.

$= |-67|$ Do the subtraction: $-60 - 7 = -60 + (-7) = -67$.

$= 67$ Find the absolute value of -67.

d. $\frac{y - 0}{x - (-1)} = \frac{-4 - 0}{3 - (-1)}$ Substitute 3 for x and -4 for y.

$= \frac{-4}{4}$ In the denominator, do the subtraction: $3 - (-1) = 3 + 1 = 4$.

$= -1$

Self Check 9 Evaluate each expression if $a = -2$ and $b = 5$: **a.** $|a^3 + b^2|$, **b.** $-a + 2ab$, and **c.** $\frac{a + 2}{b - 3}$.

EXAMPLE 10

ELEMENTARY Algebra f(x) Now™

Diving. Fins are used by divers to provide a larger area to push against the water. The fin shown on the right is in the shape of a trapezoid. The expression $\frac{1}{2}h(b + d)$ gives the area of a trapezoid, where h is the height and b and d are the lengths of the lower and upper bases, respectively. Find the area of the fin.

$d = 8.5$ in.

$h = 14$ in.

$b = 3.5$ in.

Solution $\frac{1}{2}h(b + d) = \frac{1}{2}(14)(3.5 + 8.5)$ Substitute 14 for h, 3.5 for b, and 8.5 for d.

$= \frac{1}{2}(14)(12)$ Do the addition within the parentheses.

$= 84$ $\frac{1}{2}(14) = 7$ and $7(12) = 84$.

The area of the fin is 84 square inches.

EXAMPLE 11

ELEMENTARY Algebra f(x) Now™

Rocketry. If a toy rocket is shot into the air with an initial velocity of 80 feet per second, its height (in feet) after t seconds in flight is given by

$$80t - 16t^2$$

How many seconds after the launch will it hit the ground?

Solution We can substitute positive values for t, the time in flight, until we find the one that gives a height of 0. At that time, the rocket will be on the ground. We begin by finding the height after the rocket has been in flight for 1 second ($t = 1$).

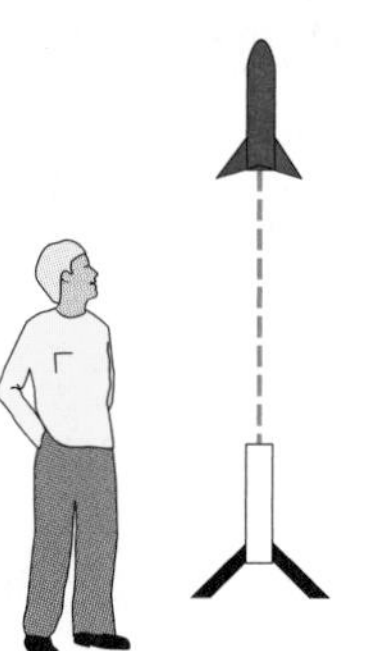

$$80t - 16t^2 = 80(1) - 16(1)^2 \quad \text{Substitute 1 for } t.$$
$$= 64$$

As we evaluate $80t - 16t^2$ for several more values of t, we record each result in a table.

t	$80t - 16t^2$
1	64
2	96
3	96
4	64
5	0

Evaluate for $t = 2$: $80t - 16t^2 = 80(2) - 16(2)^2 = 96$

Evaluate for $t = 3$: $80t - 16t^2 = 80(3) - 16(3)^2 = 96$

Evaluate for $t = 4$: $80t - 16t^2 = 80(4) - 16(4)^2 = 64$

Evaluate for $t = 5$: $80t - 16t^2 = 80(5) - 16(5)^2 = 0$

The height of the rocket is 0 when $t = 5$. The rocket will hit the ground 5 seconds after being launched.

Self Check 11 In Example 11, suppose the height of the rocket is given by $112t - 16t^2$. In how many seconds after launch would the rocket hit the ground?

The two columns of the table in Example 11 can be headed with the terms **input** and **output.** The values of t are the inputs into the expression $80t - 16t^2$, and the resulting values are the outputs.

Input	Output
1	64
2	96
3	96
4	64
5	0

Answers to Self Checks **1.** 1, −12, 3, −4 **2. a.** $t - 80$, **b.** $\frac{2}{3}T$, **c.** $2a - 15$ **3.** $(m + 15)$ minutes **4.** s = amount donated to scholarship fund in dollars; $900 - s$ = amount donated to building fund **5.** x = the number of votes received by the challenger; $3x - 55$ = the number of votes received by the incumbent **7.** $\frac{h}{24}$ **8. a.** \$300, **b.** \$100t, **c.** \$1,000$(x - 4)$ **9. a.** 17, **b.** −18, **c.** 0 **11.** 7 sec

1.8 STUDY SET

ELEMENTARY Algebra$f(x)$Now™

VOCABULARY **Fill in the blanks.**

1. Variables and/or numbers can be combined with the operations of arithmetic to create algebraic __________.
2. Addition symbols separate algebraic expressions into parts called ________.
3. A term, such as 27, that consists of a single number is called a _________ term.

4. The ________ of $10x$ is 10.

5. To ________ an algebraic expression, we substitute the values for the variables and simplify.

6. Consider the expression $2a + 8$. When we replace a with 4, we say we are ________ a value for the variable.

7. $2x + 5$ is an example of an algebraic ________, whereas $2x + 5 = 7$ is an example of an ________.

8. When we evaluate an expression for several values of x, we can keep track of the results in an input/output ______.

CONCEPTS

9. Consider the expression $11x^2 - 6x - 9$.
   a. How many terms does this expression have?
   b. What is the coefficient of the first term?
   c. What is the coefficient of the second term?
   d. What is the constant term?

10. Complete the table.

Term	Coefficient
$6m$	
$-75t$	
w	
$\frac{1}{2}bh$	
$\frac{x}{5}$	

11. In each expression, tell whether c is used as a *factor* or as a *term*.
   a. $c + 32$
   b. $5c$
   c. $-18bc$
   d. $a + b + c$
   e. $24c + 6$
   f. $c - 9$

12. **a.** Write a term that has an implied coefficient of 1.
   b. Write a term that has an implied coefficient of -1.

13. **a.** Complete the table on the left to determine how many days are in w weeks.
   b. Complete the table on the right.

Number of weeks	Number of days
1	
2	
3	
w	

Number of seconds	Number of minutes
60	
120	
180	
s	

14. **a.** If the knife is 12 inches long, how long is the blade?

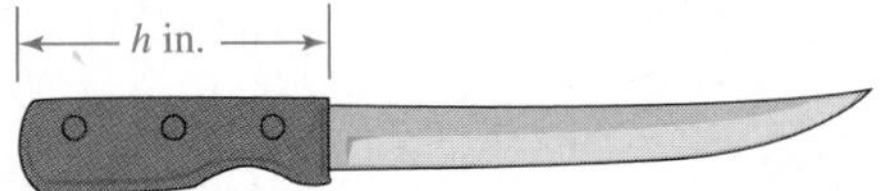

   b. A student inherited \$5,000 and deposits x dollars in American Savings. Write an expression to show how much she has left to deposit in a City Mutual account.

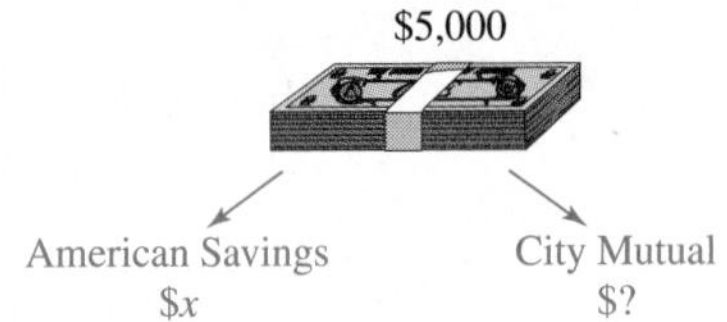

   c. Suppose solution 1 is poured into solution 2. Write an expression to show how many ounces of the mixture there will be.

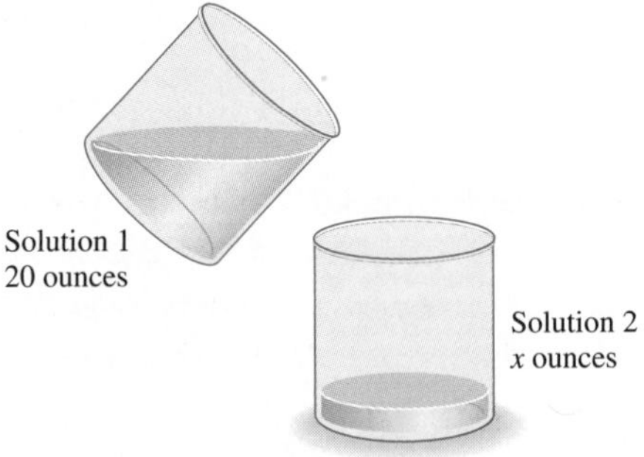

15. **a.** The weight of the van in the illustration is 500 pounds less than twice the weight of the car. Express the weight of the van and the car using the variable x.

b. If the actual weight of the car is 2,000 pounds, what is the weight of the van?

16. a. If we let b represent the length of the beam, write an algebraic expression for the length of the pipe.

b. If we let p represent the length of the pipe, write an algebraic expression for the length of the beam.

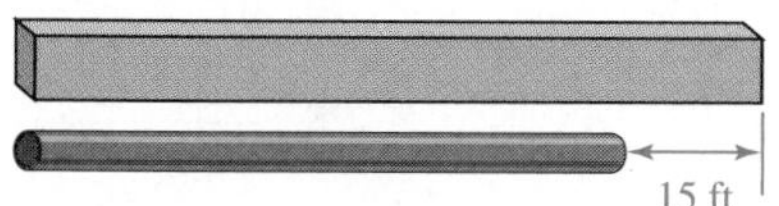

17. Complete the table.

Type of coin	Number ·	Value (in cents) =	Total value (in cents)
Nickel	6		
Dime	d		
Half-dollar	$x + 5$		

18. If $x = -9$, find the value of $-x$.

NOTATION Complete each solution.

19. Evaluate the expression $9a - a^2$ for $a = 5$.

$$\begin{aligned} 9a - a^2 &= 9(\square) - (5)^2 \\ &= 9(5) - \square \\ &= \square - 25 \\ &= 20 \end{aligned}$$

20. Evaluate $-x + 6y$ for $x = -2$ and $y = 4$.

$$\begin{aligned} -x + 6y &= -(\square) + 6(\square) \\ &= \square + 24 \\ &= 26 \end{aligned}$$

21. Write each term in standard form.

a. $y8$ **b.** $c2d$

c. $15xs$ **d.** $b^2(-9)a^3$

Fill in the blanks.

22. a. $\frac{2}{3}m = \frac{\square}{3}$ **b.** $\frac{t}{3} = \square\, t$

c. $-\frac{w}{2} = \square\, w$ **d.** $-\frac{5}{3}s = -\frac{\square}{3}$

e. $d = \square\, d$ **f.** $-h = \square\, h$

PRACTICE Translate each phrase to an algebraic expression. If no variable is given, use x as the variable.

23. The sum of the length l and 15

24. The difference of a number and 10

25. The product of a number and 50

26. Three-fourths of the population p

27. The ratio of the amount won w and lost l

28. The tax t added to c

29. P increased by p

30. 21 less than the total height h

31. The square of k minus 2,005

32. s subtracted from S

33. J reduced by 500

34. Twice the attendance a

35. 1,000 split n equal ways

36. Exceeds the cost c by 25,000

37. 90 more than the current price p

38. 64 divided by the cube of y

39. The total of 35, h, and 300

40. x decreased by 17

41. 680 fewer than the entire population p

42. Triple the number of expected participants

43. The product of d and 4, decreased by 15

44. Forty-five more than the quotient of y and 6

45. Twice the sum of 200 and t

46. The square of the quantity 14 less than x

47. The absolute value of the difference of a and 2

48. The absolute value of a, decreased by 2

49. How many minutes are there in **a.** 5 hours and **b.** h hours?

50. A woman watches television x hours a day. Express the number of hours she watches TV **a.** in a week and **b.** in a year.

51. a. How many feet are in y yards?

b. How many yards are in f feet?

52. A sales clerk earns \$$x$ an hour. How much does he earn in **a.** an 8-hour day and **b.** a 40-hour week?

53. If a car rental agency charges 29¢ a mile, express the rental fee if a car is driven x miles.

54. A model's skirt is x inches long. The designer then lets the hem down 2 inches. How can we express the length (in inches) of the altered skirt?

55. A soft drink manufacturer produced c cans of cola during the morning shift. Write an expression for how many six-packs of cola can be assembled from the morning shift's production.

56. The tag on a new pair of 36-inch-long jeans warns that after washing, they will shrink x inches in length. Express the length (in inches) of the jeans after they are washed.

57. A caravan of b cars, each carrying 5 people, traveled to the state capital for a political rally. Express how many people were in the caravan.

58. A caterer always prepares food for 10 more people than the order specifies. If p people are to attend a reception, write an expression for the number of people she should prepare for.

59. Tickets to a circus cost \$5 each. Express how much tickets will cost for a family of x people if they also pay for two of their neighbors.

60. If each egg is worth e¢, express the value (in cents) of a dozen eggs.

Complete each table.

61.

x	$x^3 - 1$
0	
−1	
−3	

62.

g	$g^2 - 7g + 1$
0	
7	
−10	

63.

s	$\frac{5s+36}{s}$
1	
6	
−12	

64.

a	$2{,}500a + a^3$
2	
4	
−5	

65.

Input x	Output $2x - \frac{x}{2}$
100	
−300	

66.

Input x	Output $\frac{x}{3} + \frac{x}{4}$
12	
−36	

67.

x	$(x + 1)(x + 5)$
−1	
−5	
−6	

68.

x	$\frac{1}{x+8}$
−7	
−9	
−8	

Evaluate each expression, given that $x = 3$, $y = -2$, and $z = -4$.

69. $3y^2 - 6y - 4$

70. $-z^2 - z - 12$

71. $(3 + x)y$

72. $(4 + z)y$

73. $(x + y)^2 - |z + y|$

74. $[(z - 1)(z + 1)]^2$

75. $(4x)^2 + 3y^2$

76. $4x^2 + (3y)^2$

77. $-\frac{2x + y^3}{y + 2z}$

78. $-\frac{2z^2 - y}{2x - y^2}$

Evaluate each expression for the given values of the variables.

79. $b^2 - 4ac$ for $a = -1$, $b = 5$, and $c = -2$

80. $(x - a)^2 + (y - b)^2$ for $x = -2$, $y = 1$, $a = 5$, and $b = -3$

81. $a^2 + 2ab + b^2$ for $a = -5$ and $b = -1$

82. $\frac{x - a}{y - b}$ for $x = -2$, $y = 1$, $a = 5$, and $b = 2$

83. $\frac{n}{2}[2a + (n - 1)d]$ for $n = 10$, $a = -4$, and $d = 6$

84. $\frac{a(1 - r^n)}{1 - r}$ for $a = -5$, $r = 2$, and $n = 3$

85. $\frac{a^2 + b^2}{2}$ for $a = 0$ and $b = -10$

86. $(y^3 - 52y^2)^2$ for $y = 0$

APPLICATIONS

87. ROCKETRY The expression $64t - 16t^2$ gives the height of a toy rocket (in feet) t seconds after being launched. Find the height of the rocket for each of the times shown.

t	h
1	
2	
3	
4	

88. AMUSEMENT PARK RIDES The distance in feet that an object will fall in t seconds is given by the expression $16t^2$. Find the distance that riders on "Drop Zone" will fall during the times listed in the table.

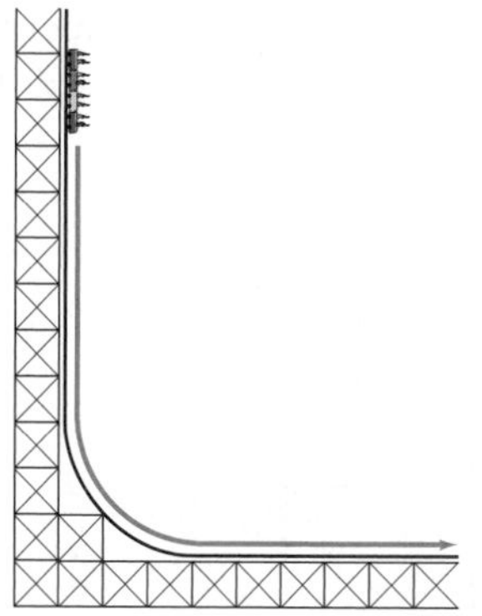

Time (seconds)	Distance (feet)
1	
2	
3	
4	

89. ANTIFREEZE The expression $\frac{5(F - 32)}{9}$ converts a temperature in degrees Fahrenheit (given as F) to degrees Celsius. Convert the temperatures listed on the container of antifreeze shown to degrees Celsius. Round to the nearest degree.

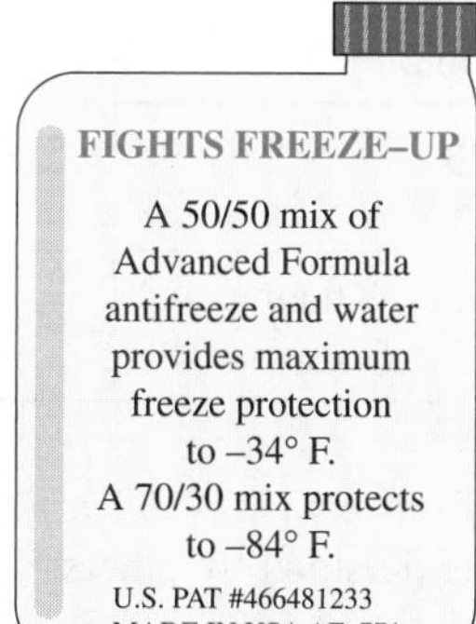

90. MARS The expression $\frac{9C + 160}{5}$ converts a temperature in degrees Celsius (represented by C) to a temperature in degrees Fahrenheit. On Mars, daily temperatures average $-33°$ C. Convert this to degrees Fahrenheit. Round to the nearest degree.

91. TOOLS The utility knife blade shown is in the shape of a trapezoid. Find the area of the front face of the blade.

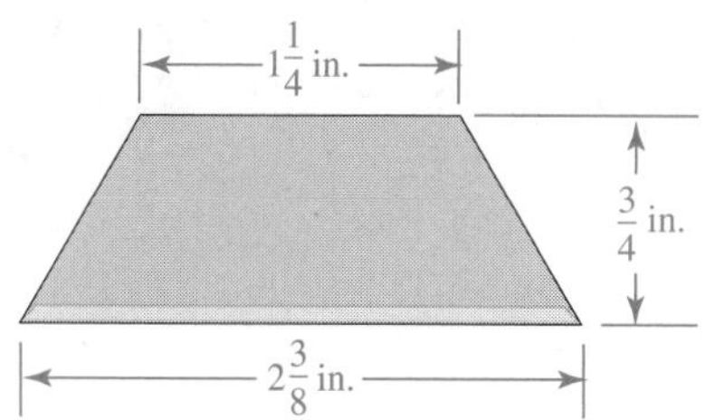

92. GROWING SOD To determine the number of square feet of sod remaining in a field after filling an order, the manager of a sod farm uses the expression $20{,}000 - 3s$ (where s is the number of strips the customer has ordered). To sod a soccer field, a city orders 7,000 strips of sod. Evaluate the expression for this value of s and explain the result.

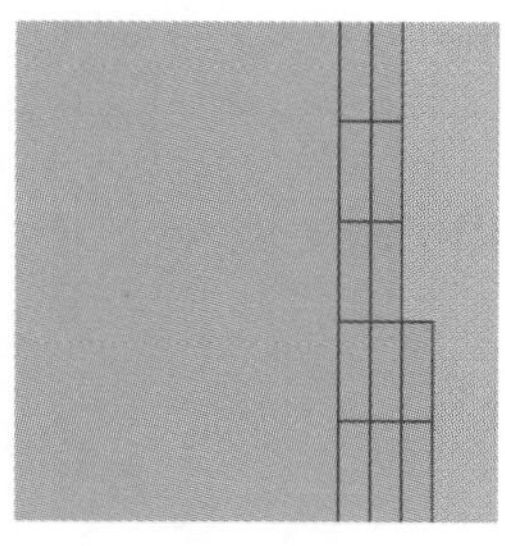

Strips of sod, cut and ready to be loaded on a truck for delivery

WRITING

93. What is an algebraic expression? Give some examples.

94. What is a variable? How are variables used in this section?

95. In this section, we substituted a number for a variable. List some other uses of the word *substitute* that you encounter in everyday life.

96. Explain why d dimes are not worth d¢.

REVIEW

97. Find the LCD for $\frac{5}{12}$ and $\frac{1}{15}$.

98. Remove the common factors of the numerator and denominator to simplify: $\frac{3 \cdot 5 \cdot 5}{3 \cdot 5 \cdot 5 \cdot 11}$.

99. Evaluate: $(-2 \cdot 3)^3 - 10 + 1$.

100. Find the result when $\frac{7}{8}$ is multiplied by its reciprocal.

CHALLENGE PROBLEMS

101. Evaluate:
$(8 - 1)(8 - 2)(8 - 3) \cdot \cdots \cdot (8 - 49)(8 - 50)$

102. If the values of x and y were doubled, what would happen to the value of $17xy$?

ACCENT ON TEAMWORK

WRITING FRACTIONS AS DECIMALS

Overview: This is a good activity to try at the beginning of the course. You can become acquainted with other students in your class while you review the process for finding decimal equivalents of fractions.

Instructions: Form groups of 6 students. Select one person from your group to record the group's responses on the questionnaire. Express the results in fraction form and in decimal form.

What fraction (decimal) of the students in your group . . .	Fraction	Decimal
have the letter *a* in their first names?		
have a birthday in January or February?		
work full-time or part-time?		
have ever been on television?		
live more than 10 miles from the campus?		
say that summer is their favorite season of the year?		

WRITING MATHEMATICAL SOLUTIONS

Overview: A major objective of this course is to learn how to put your thinking on paper in a form that can be read and understood by others. This activity will help you develop that ability.

Instructions: Form groups of 2 or 3 students. Have someone read the following problem out loud. Discuss it in your group. Then work together to present a written solution. Exchange your solution with another group and see whether you understand their explanation.

> When three professors attending a convention in Las Vegas registered at the hotel, they were told that the room rate was \$120. Each professor paid his \$40 share. Later the desk clerk realized that the cost of the room should have been \$115. To fix the mistake, she sent a bellhop to the room to refund the \$5 overcharge. Realizing that \$5 could not be evenly divided among the three professors and not wanting to start a quarrel, the bellhop refunded only \$3 and kept the other \$2. Since each professor received a \$1 refund, each paid \$39 for the room, and the bellhop kept \$2. This gives \$39 + \$39 + \$39 + \$2, or \$119. What happened to the other \$1?

KEY CONCEPT: VARIABLES

One of the major objectives of this course is for you to become comfortable working with **variables.** In Chapter 1, we have used the concept of variables in several ways.

STATING MATHEMATICAL PROPERTIES

Variables have been used to state properties of mathematics. Match each statement in words with its proper description expressed with a variable (or variables). Assume that a, b, and c are real numbers and that there are no divisions by zero.

1. A nonzero number divided by itself is 1.

2. When we add opposites, the result is 0.

3. Two numbers can be multiplied in either order to get the same result.

4. It doesn't matter how we group numbers in multiplication.

5. When we multiply a number and its reciprocal, the result is 1.

6. When we multiply a number and 0, the result is 0.

7. It doesn't matter how we group numbers in addition.

8. Any number divided by 1 is the number itself.

9. Two numbers can be added in either order to get the same result.

10. When we add a number and 0, the number remains the same.

a. $(ab)c = a(bc)$

b. $a\left(\frac{1}{a}\right) = 1$

c. $(a + b) + c = a + (b + c)$

d. $a + b = b + a$

e. $a \cdot 0 = 0$

f. $\frac{a}{a} = 1$

g. $a + 0 = a$

h. $ab = ba$

i. $\frac{a}{1} = a$

j. $a + (-a) = 0$

STATING RELATIONSHIPS BETWEEN QUANTITIES

Variables are letters that stand for numbers. We have used **formulas** to express known relationships between two or more variables.

11. Translate the word model to an equation (formula) that mathematically describes the situation: The total cost is the sum of the purchase price of the item and the sales tax.

12. Use the data in the table to state the relationship between the quantities using a formula.

Picnic tables	Benches needed
2	4
3	6
4	8

WRITING ALGEBRAIC EXPRESSIONS

We have combined variables and numbers with the operations of arithmetic to create **algebraic expressions.**

13. One year, a cruise company did x million dollars' worth of business. After a celebrity was signed as a spokeswoman for the company, its business increased by \$4 million the next year. Write an expression that represents the amount of business the cruise company had in the year the celebrity was the spokeswoman.

14. Evaluate the expression for the given values of the variable, and enter the results in the table.

x	$3x^2 - 2x + 1$
0	
4	
6	

CHAPTER REVIEW

ELEMENTARY Algebra $f(x)$ Now™

SECTION 1.1 Introducing the Language of Algebra

CONCEPTS

Tables, bar graphs, and *line graphs* are used to describe numerical relationships.

REVIEW EXERCISES

The line graph shows the number of cars parked in a parking structure from 6 P.M. to 12 midnight on a Saturday.

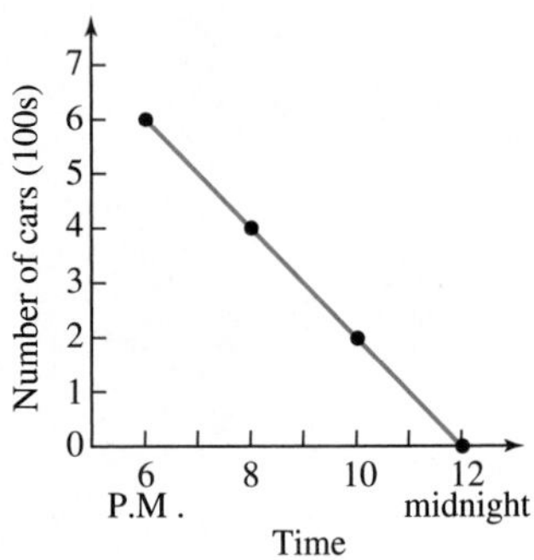

1. What units are used to scale the horizontal and vertical axes?

2. How many cars were in the parking structure at 11 P.M.?

3. At what time did the parking structure have 500 cars in it?

The result of an addition is called the *sum;* of a subtraction, the *difference;* of a multiplication, the *product;* and of a division, the *quotient.*

Express each statement in words.

4. $15 - 3 = 12$

5. $15 + 3 = 18$

6. $15 \div 3 = 5$

7. $15 \cdot 3 = 45$

8. Write the multiplication 4×9 with a raised dot and then with parentheses.

9. Write the division $9 \div 3$ using a fraction bar.

Variables are letters used to stand for numbers.

An *equation* is a mathematical sentence that contains an = symbol. Variables and/or numbers can be combined with the operations of arithmetic to create *algebraic expressions.*

Equations that express a known relationship between two or more variables are called *formulas.*

Write each multiplication without a multiplication symbol.

10. $8 \cdot b$

11. $P \cdot r \cdot t$

Classify each item as either an expression or an equation.

12. $5 = 2x + 3$

13. $2x + 3$

14. Use the formula $n = b + 5$ to complete the table.

Number of brackets (b)	Number of nails (n)
5	
10	
20	

SECTION 1.2 Fractions

Whole numbers can be written as the product of two or more whole-number *factors.*

A *prime number* is a whole number greater than 1 that has only 1 and itself as factors.

Division by 0 is *undefined.*

15. a. Write 24 as the product of two factors.

b. Write 24 as the product of three factors.

c. List the factors of 24.

Give the prime factorization of each number, if possible.

16. 54 **17.** 147 **18.** 385 **19.** 41

Perform each division, if possible.

20. $\frac{12}{12}$ **21.** $\frac{0}{10}$

Simplify each fraction.

22. $\frac{20}{35}$ **23.** $\frac{24}{18}$

Build each number to an equivalent fraction with the indicated denominator.

24. $\frac{5}{8}$, denominator 64 **25.** 12, denominator 3

To multiply two fractions, multiply their numerators and multiply their denominators.

To divide two fractions, multiply the first fraction by the reciprocal of the second fraction.

To add (or subtract) fractions with the same denominator, add (or subtract) the numerators and keep the common denominator.

The *least common denominator (LCD)* for a set of fractions is the smallest number each denominator will divide exactly.

Perform each operation.

26. $\frac{1}{8} \cdot \frac{7}{8}$ **27.** $\frac{16}{35} \cdot \frac{25}{48}$ **28.** $\frac{1}{3} \div \frac{15}{16}$ **29.** $16\frac{1}{4} \div 5$

30. $\frac{17}{25} - \frac{7}{25}$ **31.** $\frac{8}{11} - \frac{1}{2}$ **32.** $\frac{1}{4} + \frac{2}{3}$ **33.** $4\frac{1}{9} - 3\frac{5}{6}$

34. MACHINE SHOPS How much must be milled off the $\frac{17}{24}$-inch-thick steel rod so that the collar will slip over it?

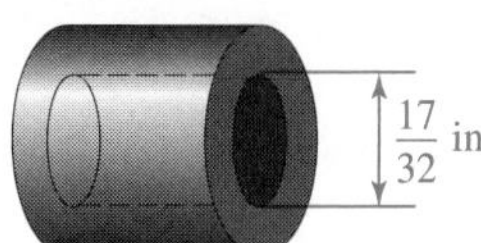

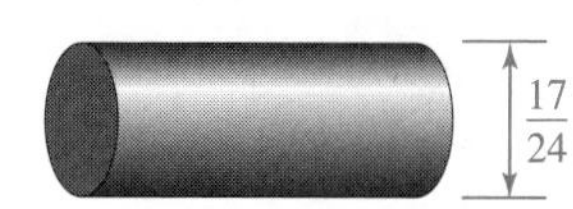

Steel rod

SECTION 1.3 The Real Numbers

The *natural numbers:* $\{1, 2, 3, 4, 5, 6, \ldots\}$
The *whole numbers:* $\{0, 1, 2, 3, 4, 5, 6, \ldots\}$
The *integers:* $\{\ldots, -2, -1, 0, 1, 2, \ldots\}$

Two *inequality symbols* are
$>$ "is greater than"
$<$ "is less than"

A *rational number* is any number that can be expressed as a fraction with an integer numerator and a nonzero integer denominator.

To write a fraction as a decimal, divide the numerator by the denominator.

An *irrational number* is a nonterminating, nonrepeating decimal. Irrational numbers cannot be written as the ratio of two integers.

A *real number* is any number that is either a rational or an irrational number.

The natural numbers are a *subset* of the whole numbers. The whole numbers are a subset of the integers. The integers are a subset of the rational numbers. The rational numbers are a subset of the real numbers.

35. Which number is a whole number but not a natural number?

Represent each of these situations with a signed number.

36. A budget deficit of \$65 billion

37. 206 feet below sea level

Use one of the symbols $>$ or $<$ to make each statement true.

38. $0 \quad 5$

39. $-12 \quad -13$

Show that each of the following numbers is a rational number by expressing it as a fraction.

40. 0.7

41. $4\frac{2}{3}$

Write each fraction as a decimal. Use an overbar if the result is a repeating decimal.

42. $\frac{1}{250}$

43. $\frac{17}{22}$

44. Graph each number on a number line: $\{\pi, 0.333\ldots, 3.75, \sqrt{2}, -\frac{17}{4}, \frac{7}{8}, -2\}$.

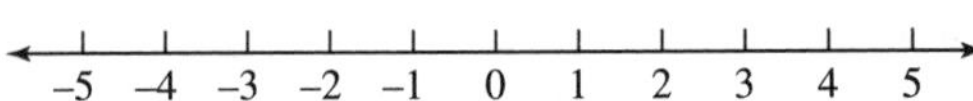

Decide whether each statement is true or false.

45. All integers are whole numbers.

46. π is a rational number.

47. The set of real numbers corresponds to all points on the number line.

48. A real number is either rational or irrational.

49. Tell which numbers in the given set are natural numbers, whole numbers, integers, rational numbers, irrational numbers, and real numbers.

$$\{-\tfrac{4}{5}, 99.99, 0, \sqrt{2}, -12, 4\tfrac{1}{2}, 0.666\ldots, 8\}$$

The *absolute value* of a number is the distance on the number line between the number and 0.

Insert one of the symbols >, <, or = in the blank to make each statement true.

50. $|-6| \quad |5|$

51. $-9 \quad |-10|$

SECTION 1.4 Adding Real Numbers

To add two real numbers with *like signs,* add their absolute values and attach their common sign to the sum.

To add two real numbers with *unlike signs,* subtract their absolute values, the smaller from the larger. To that result, attach the sign of the number with the larger absolute value.

The *commutative* and *associative* properties of addition:

$a + b = b + a$

$(a + b) + c = a + (b + c)$

Add the numbers.

52. $-45 + (-37)$

53. $25 + (-13)$

54. $0 + (-7)$

55. $-7 + 7$

56. $12 + (-8) + (-15)$

57. $-9.9 + (-2.4)$

58. $\frac{5}{16} + \left(-\frac{1}{2}\right)$

59. $35 + (-13) + (-17) + 6$

Tell what property of addition guarantees that the quantities are equal.

60. $-2 + 5 = 5 + (-2)$

61. $(-2 + 5) + 1 = -2 + (5 + 1)$

62. $80 + (-80) = 0$

63. $-5.75 + 0 = -5.75$

64. ATOMS An atom is composed of protons, neutrons, and electrons. A proton has a positive charge (represented by +1), a neutron has no charge, and an electron has a negative charge (−1). A simple model of an atom is shown here. What is its net charge?

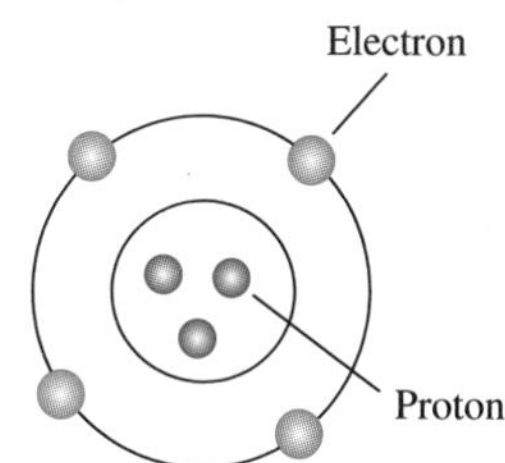

SECTION 1.5 Subtracting Real Numbers

Two numbers represented by points on a number line that are the same distance away from 0, but on opposite sides of it, are called *opposites.*

To *subtract* real numbers, add the opposite:

$$a - b = a + (-b)$$

Write the expression in simpler form.

65. The opposite of 10

66. The opposite of -3

67. $-\left(-\frac{9}{16}\right)$

68. $-|-4|$

Perform the operations.

69. $45 - 64$

70. $-17 - 32$

71. $-27 - (-12)$

72. $3.6 - (-2.1)$

73. $0 - 10$

74. $-33 + 7 - 5 - (-2)$

75. GEOGRAPHY The tallest peak on Earth is Mount Everest, at 29,028 feet, and the greatest ocean depth is the Mariana Trench, at $-36{,}205$ feet. Find the difference in the two elevations.

SECTION 1.6 Multiplying and Dividing Real Numbers

To multiply two real numbers, multiply their absolute values.

1. The product of two real numbers with *like signs* is positive.
2. The product of two real numbers with *unlike signs* is negative.

The *commutative* and *associative* properties of multiplication:

$$ab = ba$$
$$(ab)c = a(bc)$$

Multiply.

76. $-8 \cdot 7$

77. $(-9)(-6)$

78. $2(-3)(-2)$

79. $(-4)(-1)(-3)(-3)$

80. $-1.2(-5.3)$

81. $0.002(-1{,}000)$

82. $-\frac{2}{3}\left(\frac{1}{5}\right)$

83. $-6(-3)(0)(-1)$

Tell what property of multiplication guarantees that the quantities are equal.

84. $(2 \cdot 3)5 = 2(3 \cdot 5)$

85. $(-5)(-6) = (-6)(-5)$

86. $-6 \cdot 1 = -6$

87. $\frac{1}{2}(2) = 1$

88. What is the additive inverse (opposite) of -3?

89. What is the multiplicative inverse (reciprocal) of -3?

To divide two real numbers, divide their absolute values.

1. The quotient of two real numbers with *like signs* is positive.
2. The quotient of two real numbers with *unlike signs* is negative.

Division *of zero* by a nonzero number is 0. Division *by zero* is undefined.

Perform each division.

90. $\frac{44}{-44}$

91. $\frac{-100}{25}$

92. $\frac{-81}{-27}$

93. $-\frac{3}{5} \div \frac{1}{2}$

94. $\frac{-60}{0}$

95. $\frac{-4.5}{1}$

96. Find the high and low reading that is displayed on the screen of the emissions-testing device.

97. The picture on the screen can be magnified by switching a setting on the monitor. What would be the new high and low if every value were to be doubled?

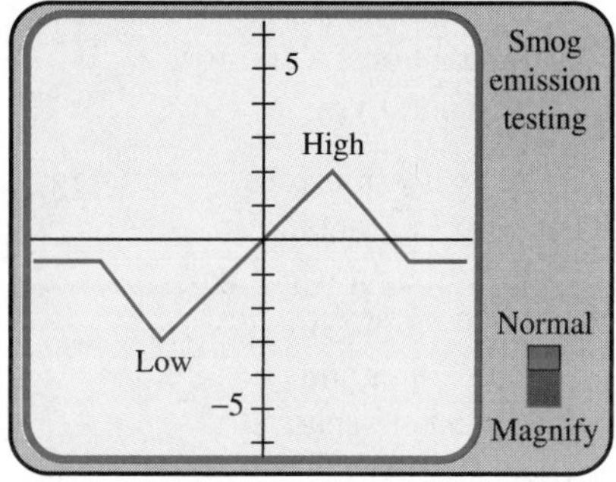

SECTION 1.7 Exponents and Order of Operations

An *exponent* is used to represent repeated multiplication. In the *exponential expression* a^n, a is the *base,* and n is the *exponent.*

Write each expression using exponents.

98. $8 \cdot 8 \cdot 8 \cdot 8 \cdot 8$

99. $a(a)(a)(a)$

100. $9 \cdot \pi \cdot r \cdot r$

101. $x \cdot x \cdot x \cdot y \cdot y \cdot y \cdot y$

Order of operations:

1. Do all calculations within grouping symbols, working from the innermost pair to the outermost pair, in the following order:
2. Evaluate all exponential expressions.
3. Do all multiplications and divisions, working from left to right.
4. Do all additions and subtractions, working from left to right.

If the expression does not contain grouping symbols, begin with Step 2. In a fraction, simplify the numerator and denominator separately. Then simplify the fraction, if possible.

Evaluate each expression.

102. 9^2 **103.** $\left(-\frac{2}{3}\right)^3$ **104.** 2^5 **105.** 50^1

106. How many operations does the expression $5 \cdot 4 - 3^2 + 1$ contain, and in what order should they be performed?

Evaluate each expression.

107. $2 + 5 \cdot 3$

108. $24 - 3(6)(4)$

109. $-(6 - 3)^2$

110. $4^3 + 2(-6 - 2 \cdot 2)$

111. $10 - 5[-3 - 2(5 - 7^2)] - 5$

112. $\frac{-4(4 + 2) - 4}{|-18 - 4(5)|}$

113. $(-3)^3\left(\frac{-8}{2}\right) + 5$

114. $-9^2 + (-9)^2$

The *arithmetic mean* (or *average*) is a value around which number values are grouped.

$$\text{Mean} = \frac{\text{sum of values}}{\text{number of values}}$$

115. WALK-A-THONS Use the data in the table to find the average (mean) donation to a charity walk-a-thon.

Donation	Number received
\$5	20
\$10	65
\$20	25
\$50	5
\$100	10

SECTION 1.8 Algebraic Expressions

Addition signs separate algebraic expressions into terms.

How many terms are in each expression?

116. $3x^2 + 2x - 5$

117. $-12xyz$

In a term, the numerical factor is called the *coefficient.*

Identify the coefficient of each term.

118. $2x - 5$

119. $16x^2 - 5x + 25$

120. $\frac{x}{2} + y$

121. $9.6t^2 - t$

In order to describe numerical relationships, we need to translate the words of a problem into mathematical symbols.

Write each phrase as an algebraic expression.

122. 25 more than the height h

123. 15 less than the cutoff score s

124. $\frac{1}{2}$ of the time t

125. If we let n represent the length of the nail, write an algebraic expression for the length of the bolt (in inches).

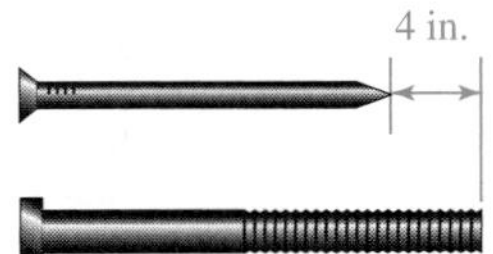

126. If we let b represent the length of the bolt, write an algebraic expression for the length of the nail (in inches).

Sometimes we must rely on common sense and insight to find *hidden operations.*

127. How many years are in d decades?

128. Five years after a house was constructed, a patio was added. How old, in years, is the patio if the house is x years old?

Number · value = total value

129. Complete the table.

Type of coin	Number	Value (¢)	Total value (¢)
Nickel	6	5	
Dime	d	10	

When we replace the variable, or variables, in an algebraic expression with specific numbers and then apply the rules for the order of operations, we are *evaluating* the algebraic expression.

130. Complete the table.

x	$20x - x^3$
0	
1	
−4	

Evaluate each algebraic expression for the given value(s) of the variable(s).

131. $b^2 - 4ac$ for $b = -10$, $a = 3$, and $c = 5$

132. $\frac{x + y}{-x - z}$ for $x = 19$, $y = 17$, and $z = -18$

CHAPTER 1 TEST

The line graph shows the cost to hire a security guard. Use the graph to answer Problems 1 and 2.

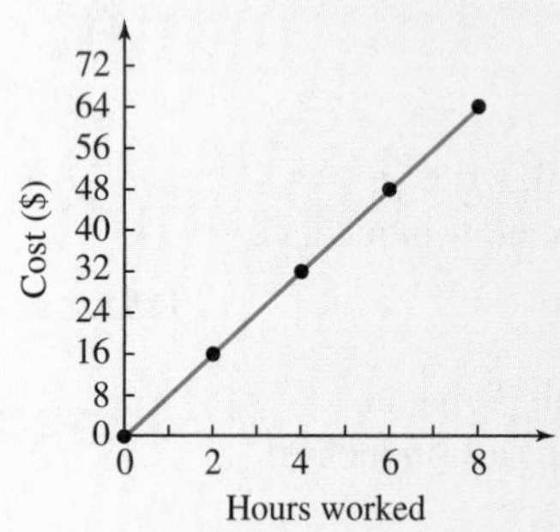

1. What will it cost to hire a security guard for 3 hours?

2. If a school was billed \$40 for hiring a security guard for a dance, for how long did the guard work?

3. Use the formula $f = \frac{a}{5}$ to complete the table.

Area in square miles (a)	Number of fire stations (f)
15	
100	
350	

4. Give the prime factorization of 180.

5. Simplify: $\frac{42}{105}$.

6. Divide: $\frac{15}{16} \div \frac{5}{8}$.

7. Subtract: $\frac{11}{12} - \frac{2}{9}$.

8. Add: $1\frac{2}{3} + 8\frac{2}{5}$.

9. SHOPPING Find the cost of the fruit on the scale.

10. Write $\frac{5}{6}$ as a decimal.

11. Graph each member of the set on the number line.

$\left\{-1\frac{1}{4}, \sqrt{2}, -3.75, \frac{7}{2}, 0.5, -3\right\}$

−5 −4 −3 −2 −1 0 1 2 3 4 5

12. Decide whether each statement is true or false.

a. Every integer is a rational number.
b. Every rational number is an integer.
c. π is an irrational number.
d. 0 is a whole number.

13. Describe the set of real numbers.

14. Insert the proper symbol, > or <, in the blank to make each statement true.

a. $-2 \;\square\; -3$ **b.** $-|-7| \;\square\; 8$
c. $|-4| \;\square\; -(-5)$ **d.** $\left|-\frac{7}{8}\right| \;\square\; 0.5$

15. SWEEPS WEEK During "sweeps week," television networks make a special effort to gain viewers by showing unusually flashy programming. Use the information in the graph to determine the average daily gain (or loss) of ratings points by a network for the 7-day "sweeps period."

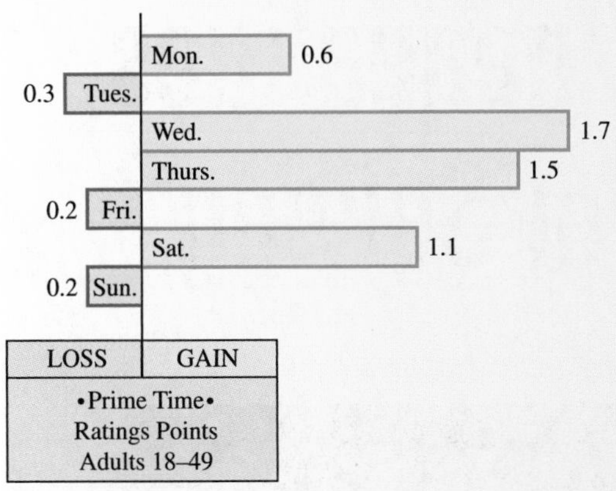

Perform the operations.

16. $(-6) + 8 + (-4)$

17. $-\frac{1}{2} + \frac{7}{8}$

18. $-10 - (-4)$

19. $(-2)(-3)(-5)$

20. $\frac{-22}{-11}$

21. $-6.1(0.4)$

22. $\frac{0}{-3}$

23. $0 - 3$

24. $3 + (-3)$

25. $-30 + 50 - 10 - (-40)$

26. $\left(-\frac{3}{5}\right)^3$

27. ASTRONOMY *Magnitude* is a term used in astronomy to designate the brightness of celestial objects as viewed from Earth. Smaller magnitudes are associated with brighter objects, and larger magnitudes refer to fainter objects. By how many magnitudes do a full moon and the sun differ?

Object	Magnitude
Sun	-26.5
Full moon	-12.5

28. What property of real numbers is illustrated below? $(-12 + 97) + 3 = -12 + (97 + 3)$.

29. Write each product using exponents:

a. $9(9)(9)(9)(9)$ **b.** $3 \cdot x \cdot x \cdot z \cdot z \cdot z$.

30. Evaluate: $8 + 2 \cdot 3^4$.

31. Evaluate: $9^2 - 3[45 - 3(6 + 4)]$.

32. Evaluate: $\frac{3(40 - 2^3)}{-2(6 - 4)^2}$.

33. Evaluate: -10^2.

34. Evaluate $3(x - y) - 5(x + y)$ for $x = 2$ and $y = -5$.

35. Complete the table.

x	$2x - \frac{30}{x}$
5	
10	
-30	

36. Translate to an algebraic expression: seven more than twice the width w.

37. A rock band recorded x songs for a CD. Technicians had to delete two songs from the album because of poor sound quality. Express the number of songs on the CD using an algebraic expression.

38. Find the value of q quarters in cents.

39. Explain the difference between an expression and an equation.

40. How many terms are in the expression $4x^2 + 5x - 7$? What is the coefficient of the second term?

Chapter

2 Equations, Inequalities, and Problem Solving

ELEMENTARY
Algebra f(x) Now™

Throughout the chapter, this icon introduces resources on the Elementary AlgebraNow Web site, accessed through http://1pass.thomson.com, that will

- Help you test your knowledge of the material with a pre-test and a post-test
- Provide a personalized learning plan targeting areas you should study

2.1 Solving Equations

2.2 Problem Solving

2.3 Simplifying Algebraic Expressions

2.4 More about Solving Equations

2.5 Formulas

2.6 More about Problem Solving

2.7 Solving Inequalities

Accent on Teamwork

Key Concept

Chapter Review

Chapter Test

Cumulative Review Exercises

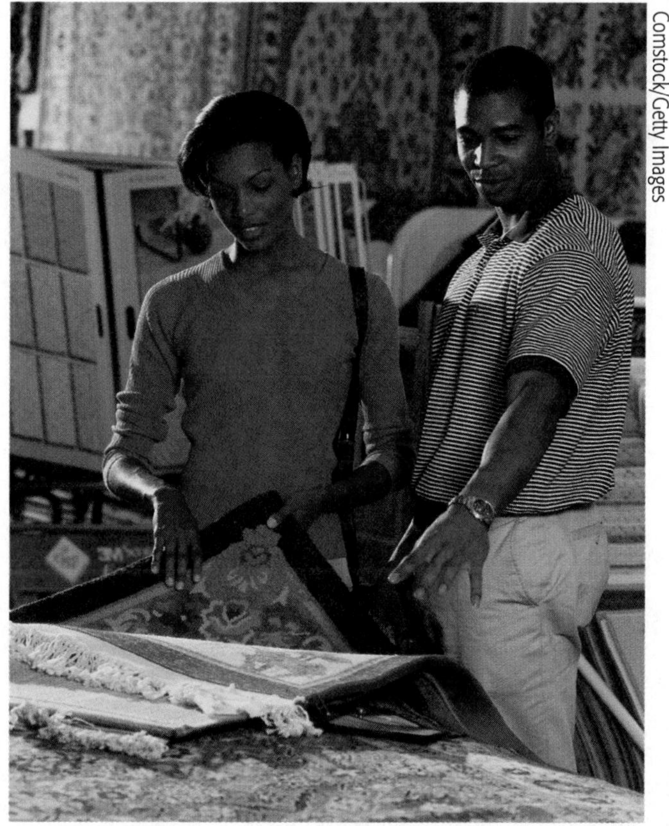

Comstock/Getty Images

When shopping for big-ticket items such as appliances, automobiles, or home furnishings, we often have to stay within a budget. Besides price, the quality and dependability of an item must also be considered. The problem-solving skills that we will discuss can help you make wise decisions when making large purchases such as these.

To learn more about the use of algebra in the marketplace, visit *The Learning Equation* on the Internet at http://tle.brookscole.com. (Log-in instructions are in the Preface.) For Chapter 2, the online lessons are:

- *TLE* Lesson 3: Translating Written Phrases
- *TLE* Lesson 4: Solving Equations Part 1
- *TLE* Lesson 5: Solving Equations Part 2

One of the most useful concepts in algebra is the equation. Writing and then solving an equation is a powerful problem-solving strategy.

2.1 Solving Equations

- Equations and Solutions
- The Addition Property of Equality
- The Subtraction Property of Equality
- The Multiplication Property of Equality
- The Division Property of Equality
- Geometry

In this section, we introduce basic types of equations and discuss four fundamental properties that are used to solve them.

EQUATIONS AND SOLUTIONS

An **equation** is a statement indicating that two expressions are equal. An example is $x + 5 = 15$. The equal symbol separates the equation into two parts: The expression $x + 5$ is the **left-hand side** and 15 is the **right-hand side.** The letter x is the **variable** (or the **unknown**). The sides of an equation can be reversed, so we can write $x + 5 = 15$ or $15 = x + 5$.

- An equation can be true: $6 + 3 = 9$.
- An equation can be false: $2 + 4 = 7$.
- An equation can be neither true nor false. For example, $x + 5 = 15$ is neither true nor false because we don't know what number x represents.

An equation that contains a variable is made true or false by substituting a number for the variable. If we substitute 10 for x in $x + 5 = 15$, the resulting equation is true: $10 + 5 \doteq 15$. If we substitute 1 for x, the resulting equation is false: $1 + 5 = 15$. A number that makes an equation true is called a **solution** and it is said to *satisfy* the equation. Therefore, 10 is a solution of $x + 5 = 15$, and 1 is not.

EXAMPLE 1

Is 9 a solution of $3y - 1 = 2y + 7$?

ELEMENTARY Algebra f(x) Now™

Solution We begin by substituting 9 for each y in the equation. Then we evaluate each side separately. If 9 is a solution, we will obtain a true statement.

The Language of Algebra
Read $\stackrel{?}{=}$ as "is possibly equal to."

Evaluate the expression on the left-hand side.

$$3y - 1 = 2y + 7$$
$$3(9) - 1 \stackrel{?}{=} 2(9) + 7$$
$$27 - 1 \stackrel{?}{=} 18 + 7$$
$$26 = 25$$

Evaluate the expression on the right-hand side.

Since $26 = 25$ is false, 9 is not a solution.

Self Check 1 Is 25 a solution of $10 - x = 35 - 2x$?

THE ADDITION PROPERTY OF EQUALITY

To **solve an equation** means to find all values of the variable that make the equation true. We can develop an understanding of how to solve equations by referring to the scales shown on the right.

The first scale represents the equation $x - 2 = 3$. The scale is in balance because the weight on the left-hand side, $(x - 2)$ grams, and the weight on the right-hand side, 3 grams, are equal. To find x, we must add 2 grams to the left-hand side. To keep the scale in balance, we must also add 2 grams to the right-hand side. After doing this, we see in the second illustration that x grams is balanced by 5 grams. Therefore, x must be 5. We say that we have solved the equation $x - 2 = 3$ and that the solution is 5.

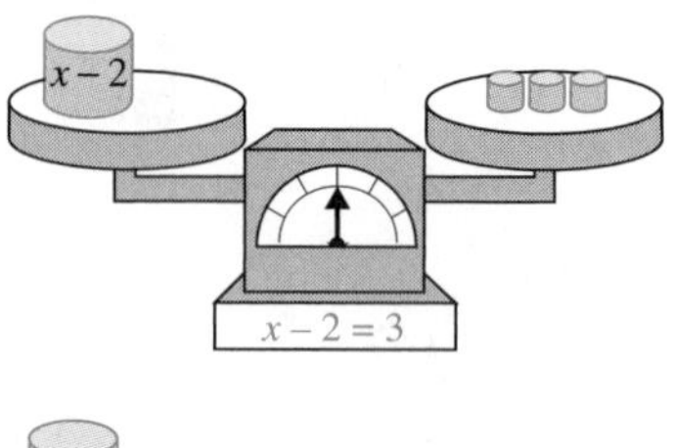

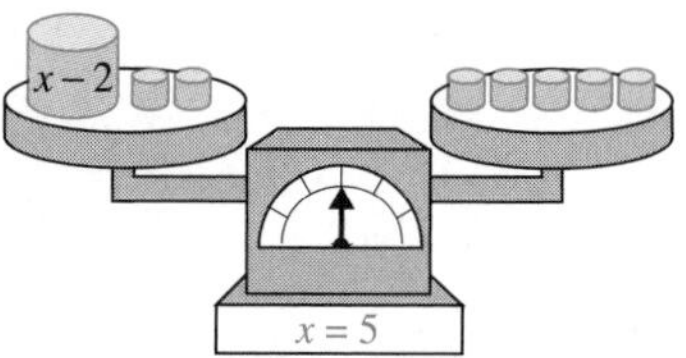

In this example, we solved $x - 2 = 3$ by transforming it to a simpler *equivalent equation,* $x = 5$.

Equivalent Equations Equations with the same solutions are called **equivalent equations.**

The procedure that we used suggests the following property of equality.

Addition Property of Equality Adding the same number to both sides of an equation does not change its solution.

For any real numbers a, b, and c,

if $a = b$, then $a + c = b + c$

When we use this property, the resulting equation is equivalent to the original one. We will now show how it is used to solve $x - 2 = 3$ algebraically.

EXAMPLE 2

Solve: $x - 2 = 3$.

Solution To isolate x on the left-hand side of the equation, we use the addition property of equality. We can undo the subtraction of 2 by adding 2 to both sides.

$$x - 2 = 3$$
$$x - 2 + 2 = 3 + 2 \quad \text{Add 2 to both sides.}$$
$$x + 0 = 5 \quad \text{Do the addition: } -2 + 2 = 0.$$
$$x = 5 \quad \text{When 0 is added to a number, the result is the same number.}$$

The Language of Algebra

When solving an equation, we want to *isolate* the variable on one side of the equation. The word *isolate* means to be alone or by itself.

To check, we substitute 5 for x in the original equation and simplify. If 5 is a solution, we will obtain a true statement.

$$x - 2 = 3 \quad \text{This is the original equation.}$$
$$5 - 2 \stackrel{?}{=} 3 \quad \text{Substitute 5 for } x.$$
$$3 = 3 \quad \text{True.}$$

Since the statement is true, 5 is the solution.

Self Check 2 Solve: $n - 16 = 33$.

EXAMPLE 3

Solve: $-19 = y - 7$.

Solution To isolate y on the right-hand side, we use the addition property of equality. We can undo the subtraction of 7 by adding 7 to both sides.

Notation

We may solve an equation so that the variable is isolated on either side of the equation. In Example 3, note that $-12 = y$ is equivalent to $y = -12$.

$$-19 = y - 7$$
$$-19 + 7 = y - 7 + 7 \quad \text{Add 7 to both sides.}$$
$$-12 = y \quad \text{Do the addition: } -7 + 7 = 0.$$

To check, we substitute -12 for y in the original equation and simplify.

$$-19 = y - 7 \quad \text{This is the original equation.}$$
$$-19 \stackrel{?}{=} -12 - 7 \quad \text{Substitute } -12 \text{ for } y.$$
$$-19 = -19 \quad \text{True.}$$

Since the statement is true, the solution is -12.

Self Check 3 Solve: $-5 = b - 38$.

EXAMPLE 4

Solve: $-27 + g = -3$.

ELEMENTARY Algebra f(x) Now™

Solution To isolate g, we use the addition property of equality. We can eliminate -27 on the left-hand side by adding its opposite (additive inverse) to both sides.

Caution

After checking a result, be careful when stating your conclusion. For Example 4, it would be incorrect to say:

The solution is -3.

The number we were checking was 24, not -3.

$$-27 + g = -3$$
$$-27 + g + 27 = -3 + 27 \quad \text{Add 27 to both sides.}$$
$$g = 24 \quad \text{Do the addition: } -27 + 27 = 0.$$

Check:

$$-27 + g = -3 \quad \text{This is the original equation.}$$
$$-27 + 24 \stackrel{?}{=} -3 \quad \text{Substitute 24 for } g.$$
$$-3 = -3 \quad \text{True.}$$

The solution is 24.

Self Check 4 Solve: $-20 + n = 29$.

THE SUBTRACTION PROPERTY OF EQUALITY

The first scale shown on the next page represents the equation $x + 2 = 5$. The scale is in balance because the weight on the left-hand side, $(x + 2)$ grams, and the weight on the right-hand side, 5 grams, are equal. To find x, we isolate it by subtracting 2 grams from the left-hand side. To keep the scale in balance, we must also subtract 2 grams from the right-hand side. After doing this, we see in the second illustration that x grams is balanced by 3 grams. Therefore, x must be 3. We say that we have solved the equation $x + 2 = 5$ and that the solution is 3.

In this example, we solved $x + 2 = 5$ by transforming it to a simpler equivalent equation, $x = 3$. The process that we used to isolate x on the left-hand side of the scale suggests the following property of equality.

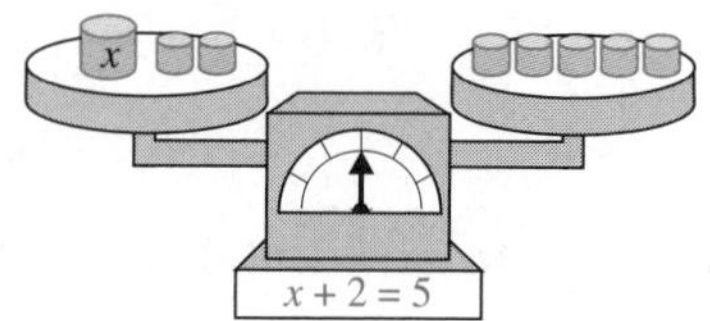

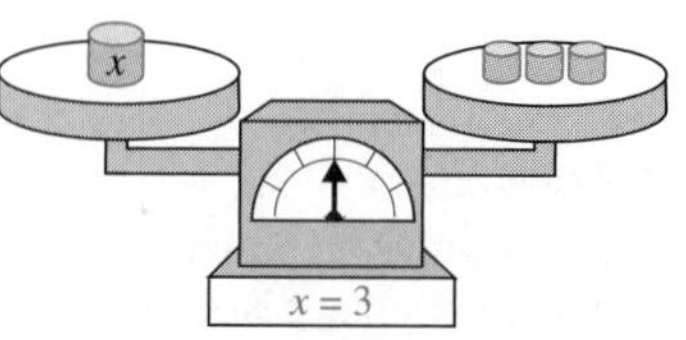

Subtraction Property of Equality Subtracting the same number from both sides of an equation does not change its solution. For any real numbers a, b, and c,

$$\text{if } a = b, \text{ then } a - c = b - c$$

When we use this property, the resulting equation is equivalent to the original one. We will now show how it is used to solve $x + 2 = 5$ algebraically.

EXAMPLE 5

Solve: $x + 2 = 5$.

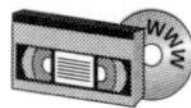

Solution To isolate x on the left-hand side, we use the subtraction property of equality. We can undo the addition of 2 by subtracting 2 from both sides.

$$x + 2 = 5$$
$$x + 2 - \mathbf{2} = 5 - \mathbf{2} \quad \text{Subtract 2 from both sides.}$$
$$x + 0 = 3$$
$$x = 3$$

Check:

$$x + 2 = 5 \quad \text{This is the original equation.}$$
$$\mathbf{3} + 2 \stackrel{?}{=} 5 \quad \text{Substitute 3 for } x.$$
$$5 = 5 \quad \text{True.}$$

The solution is 3.

Self Check 5 Solve: $x + 24 = 50$.

EXAMPLE 6

Solve: $54.9 + m = 45.2$.

ELEMENTARY Algebra $f(x)$ Now™

Solution To isolate m, we can undo the addition of 54.9 by subtracting 54.9 from both sides.

$$54.9 + m = 45.2$$
$$54.9 + m - \mathbf{54.9} = 45.2 - \mathbf{54.9} \quad \text{Subtract 54.9 from both sides.}$$
$$m = -9.7$$

Check:

$$54.9 + m = 45.2 \quad \text{This is the original equation.}$$
$$54.9 + (\mathbf{-9.7}) \stackrel{?}{=} 45.2 \quad \text{Substitute } -9.7 \text{ for } m.$$
$$45.2 = 45.2 \quad \text{True.}$$

The solution is -9.7.

Self Check 6 Solve: $0.7 + a = 0.2$

THE MULTIPLICATION PROPERTY OF EQUALITY

We can think of the first scale shown below as representing the equation $\frac{x}{3} = 25$. The weight on the left-hand side is $\frac{x}{3}$ grams, and the weight on the right-hand side is 25 grams. Because the weights are equal, the scale is in balance. To find x, we can triple (multiply by 3) the weight on each side. When we do this, the scale will remain in balance. We see that x grams will be balanced by 75 grams. Therefore, x is 75.

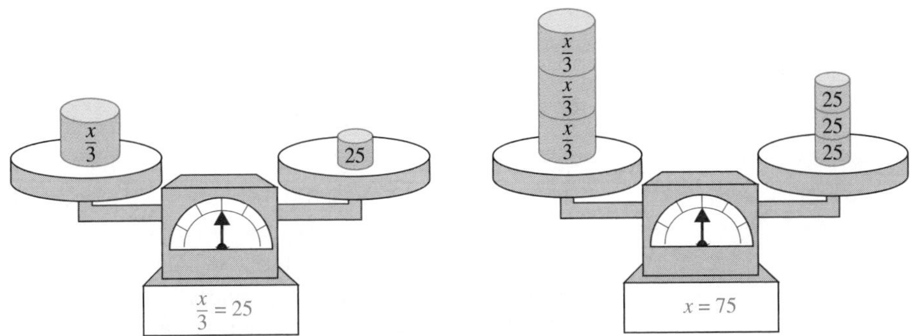

The process used to isolate x on the left-hand side of the scale suggests the following property of equality.

Multiplication Property of Equality

Multiplying both sides of an equation by the same nonzero number does not change its solution.

For any real numbers a, b, and c, where c is not 0,

if $a = b$, then $ca = cb$

When we use this property, the resulting equation is equivalent to the original one. We will now show how it is used to solve $\frac{x}{3} = 25$ algebraically.

EXAMPLE 7 Solve: $\frac{x}{3} = 25$.

ELEMENTARY Algebra $f(x)$ Now™

Solution To isolate x on the left-hand side, we use the multiplication property of equality. We can undo the division by 3 by multiplying both sides by 3.

$$\frac{x}{3} = 25$$

$$\mathbf{3} \cdot \frac{x}{3} = \mathbf{3} \cdot 25 \quad \text{Multiply both sides by 3.}$$

$$\frac{3x}{3} = 75 \quad \text{Do the multiplications.}$$

$$1x = 75 \quad \text{Simplify the fraction by removing the common factor of 3 in the numerator and denominator: } \tfrac{3}{3} = 1.$$

$$x = 75 \quad \text{The product of 1 and any number is that number.}$$

Check: $\frac{x}{3} = 25$ This is the original equation.

$\frac{75}{3} \stackrel{?}{=} 25$ Substitute 75 for x.

$25 = 25$ True.

The solution is 75.

Self Check 7 Solve: $\frac{b}{24} = 3$.

THE DIVISION PROPERTY OF EQUALITY

We will now consider how to solve the equation $2x = 8$. Since $2x$ means $2 \cdot x$, the equation can be written as $2 \cdot x = 8$. The first scale represents this equation.

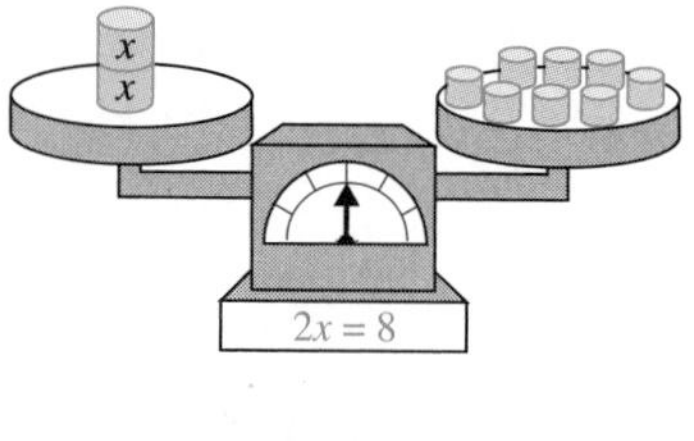

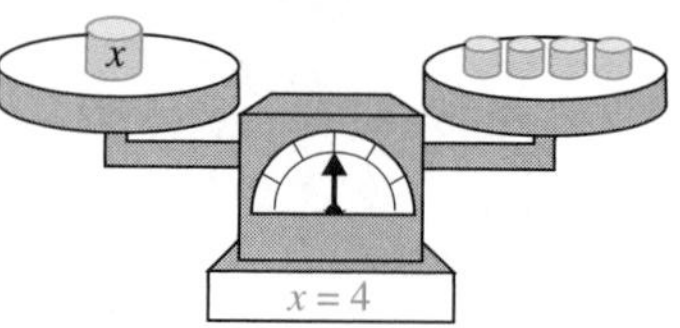

The weight on the left-hand side is $2 \cdot x$ grams and the weight on the right-hand side is 8 grams. Because these weights are equal, the scale is in balance. To find x, we remove half of the weight from each side. This is equivalent to dividing the weight on both sides by 2. When we do this, the scale remains in balance. We see that x grams is balanced by 4 grams. Therefore, x is 4. We say that we have solved the equation $2x = 8$ and that the solution is 4.

In this example, we solved $2x = 8$ by transforming it to a simpler equivalent equation, $x = 4$. The procedure that we used suggests the following property of equality.

Division Property of Equality

Dividing both sides of an equation by the same nonzero number does not change its solution.

For any real numbers a, b, and c, where c is not 0,

$$\text{if } a = b, \text{ then } \frac{a}{c} = \frac{b}{c}$$

When we use this property, the resulting equation is equivalent to the original one. We will now show how it is used to solve $2x = 8$ algebraically.

EXAMPLE 8 Solve: $2x = 8$.

Solution To isolate x on the left-hand side, we use the division property of equality to undo the multiplication by 2 by dividing both sides of the equation by 2.

$$2x = 8$$

$$\frac{2x}{2} = \frac{8}{2}$$ Divide both sides by 2.

$$1x = 4$$ Simplify the fraction by removing the common factor of 2 in the numerator and denominator: $\frac{2}{2} = 1$.

$$x = 4$$ The product of 1 and any number is that number: $1x = x$.

Check: $2x = 8$ This is the original equation.

$2(4) \stackrel{?}{=} 8$ Substitute 4 for x.

$8 = 8$ True.

The solution is 4.

Self Check 8 Solve: $16x = 176$.

EXAMPLE 9

ELEMENTARY Algebra *f(x)* Now™

Solve: $-6.02 = -8.6t$.

Solution To isolate t on the right-hand side, we can undo the multiplication by -8.6 by dividing both sides by -8.6.

$$-6.02 = -8.6t$$

$$\frac{-6.02}{-8.6} = \frac{-8.6t}{-8.6}$$ Divide both sides by -8.6.

$$0.7 = t$$ Do the divisions.

The check is left to the student. The solution is 0.7.

Notation

In Example 9, if you prefer to isolate the variable on the left-hand side, you can solve $-6.02 = -8.6t$ by reversing both sides and solving $-8.6t = -6.02$.

Self Check 9 Solve: $10.04 = -0.4r$.

GEOMETRY

The following three figures are called **angles.** Angles are measured in **degrees,** denoted by the symbol °. If an angle measures 90° (ninety degrees), it is called a **right angle.** If an angle measures 180°, it is called a **straight angle.**

47°

An angle

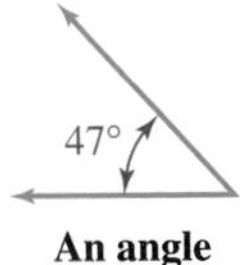

A right angle

A ⌐ symbol can be used to denote a right angle.

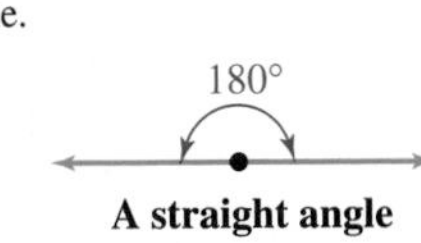

A straight angle

EXAMPLE 10

ELEMENTARY Algebra f(x) Now™

Lawn mowers. Find x, the angle that the mower's handle makes with the ground.

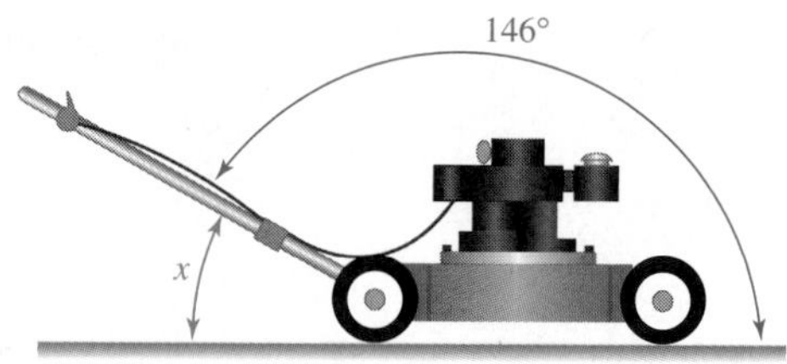

Solution We can use algebra to find the unknown angle labeled x. Since the sum of the measures of the angles is 180°, we have

$$x + 146 = 180$$
$$x + 146 - \mathbf{146} = 180 - \mathbf{146} \quad \text{Subtract 146 from both sides.}$$
$$x = 34$$

The angle that the handle makes with the ground is 34°.

Answers to Self Checks **1.** yes **2.** 49 **3.** 33 **4.** 49 **5.** 26 **6.** −0.5 **7.** 72 **8.** 11 **9.** −25.1

2.1 STUDY SET

ELEMENTARY Algebra f(x) Now™

VOCABULARY Fill in the blanks.

1. An ________ is a statement indicating that two expressions are equal.
2. Any number that makes an equation true when substituted for its variable is said to ________ the equation. Such numbers are called ________.
3. To ________ the solution of an equation, we substitute the value for the variable in the original equation and see whether the result is a true statement.
4. In $30 = t - 12$, the ________ side of the equation is $t - 12$.
5. Equations with the same solutions are called ________ equations.
6. To ________ an equation means to find all values of the variable that make the equation true.
7. To solve an equation, we ________ the variable on one side of the equal symbol.
8. When solving an equation, the objective is to find all values of the ________ that will make the equation true.

CONCEPTS

9. **a.** What equation does the balanced scale represent?

   b. What must be done to isolate x on the left-hand side?

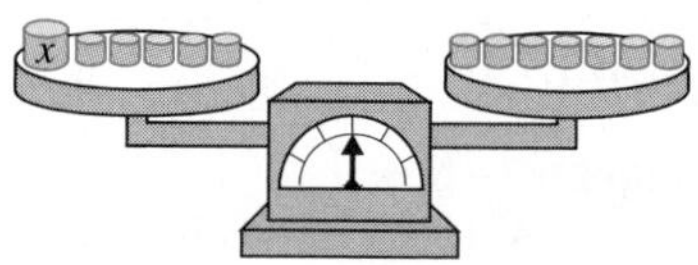

10. **a.** What equation does the balanced scale represent?

    b. What must be done to isolate x on the left-hand side?

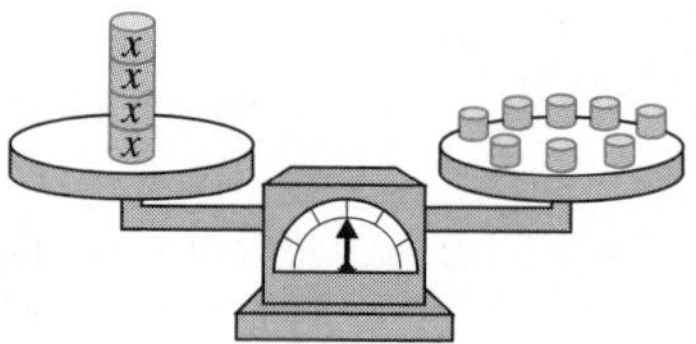

11. Given $x + 6 = 12$,
    a. What forms the left-hand side of the equation?
    b. Is this equation true or false?
    c. Is 5 a solution?
    d. Does 6 satisfy the equation?
12. For each equation, tell what operation is performed on the variable. Then tell how to undo that operation to isolate the variable.
    a. $x - 8 = 24$
    b. $x + 8 = 24$

c. $\frac{x}{8} = 24$

d. $8x = 24$

Complete each flow chart.

13.

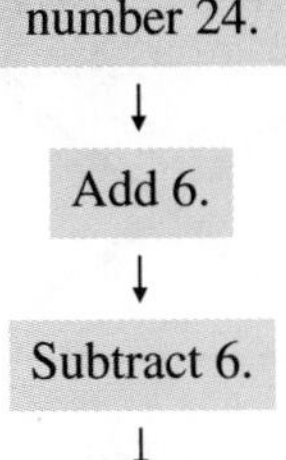

The result is ___.

14.

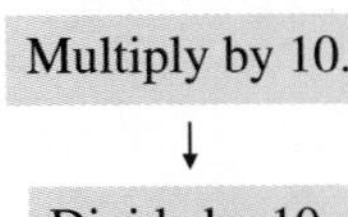

The result is ___.

15.

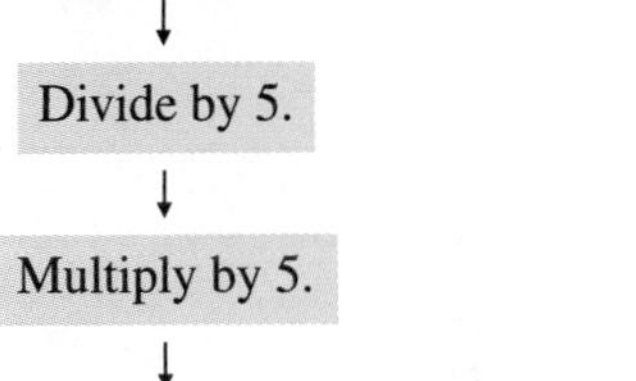

The result is ___.

16.

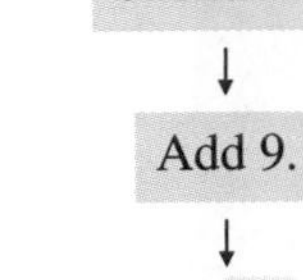

The result is ___.

17. Complete the following properties of equality.

a. If $x = y$, then $x + c = y +$ ___ and $x - c = y -$ ___.

b. If $x = y$, then $cx =$ ___ y and $\frac{x}{c} = \frac{y}{\_\_\_}$ $(c \neq 0)$.

18. a. When solving $\frac{h}{10} = 20$, do we multiply both sides of the equation by 10 or 20?

b. When solving $4k = 16$, do we subtract 4 from both sides of the equation or divide both sides by 4?

19. Simplify each expression.

a. $x + 7 - 7$

b. $y - 2 + 2$

c. $\frac{5t}{5}$

d. $6 \cdot \frac{h}{6}$

20. Complete each equation.

a. $x + 20 =$ ___

b. $x + 20 =$ ___

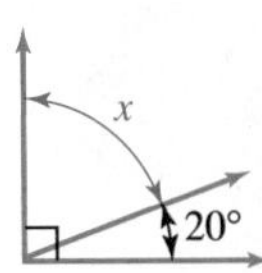

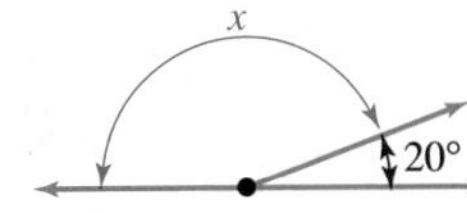

NOTATION **Complete each solution.**

21. Solve: $x + 15 = 45$

$x + 15 - \_\_\_ = 45 - \_\_\_$

$x = 30$

Check: $x + 15 = 45$

$\_\_\_ + 15 \stackrel{?}{=} 45$

$45 = 45$

___ is a solution.

22. Solve: $8x = 40$

$\frac{8x}{\_\_\_} = \frac{40}{\_\_\_}$

$x = 5$

Check: $8x = 40$

$8(\_\_\_) \stackrel{?}{=} 40$

$40 = 40$

___ is a solution.

23. a. What does the symbol $\stackrel{?}{=}$ mean?

b. Write twenty-seven degrees using symbols.

24. If you solve an equation and obtain $50 = x$, can you write $x = 50$?

PRACTICE **Check to see whether the given number is a solution of the equation.**

25. $6, x + 12 = 18$

26. $110, x - 50 = 60$

27. $-8, 2b + 3 = -15$

28. $-2, 5t - 4 = -16$

29. $5, 0.5x = 2.9$

30. $3.5, 1.2 + x = 4.7$

31. $-6, 33 - \frac{x}{2} = 30$

32. $-8, \frac{x}{4} + 98 = 100$

33. $20, |c - 8| = 10$

34. $20, |30 - r| = 15$

35. $12, 3x - 2 = 4x - 5$

36. $5, 5y + 8 = 3y - 2$

37. $-3, x^2 - x - 6 = 0$

38. $-2, y^2 + 5y - 3 = 0$

39. $1, \frac{2}{a+1} + 5 = \frac{12}{a+1}$

40. $4, \frac{2t}{t-2} - \frac{4}{t-2} = 1$

41. $-3, (x-4)(x+3) = 0$

42. $5, (2x+1)(x-5) = 0$

Use a property of equality to solve each equation. Then check the result.

43. $x + 7 = 10$

44. $y + 15 = 24$

45. $a - 5 = 66$

46. $x - 34 = 19$

47. $0 = n - 9$

48. $3 = m - 20$

49. $9 + p = 9$

50. $88 + j = 88$

51. $x - 16 = -25$

52. $y - 12 = -13$

53. $a + 3 = 0$

54. $m + 1 = 0$

55. $f + 3.5 = 1.2$

56. $h + 9.4 = 8.1$

57. $-8 + p = -44$

58. $-2 + k = -41$

59. $8.9 = -4.1 + t$

60. $7.7 = -3.2 + s$

61. $d - \frac{1}{9} = \frac{7}{9}$

62. $\frac{7}{15} = b - \frac{1}{15}$

63. $s + \frac{4}{25} = \frac{11}{25}$

64. $\frac{8}{3} = h + \frac{1}{3}$

65. $4x = 16$

66. $5y = 45$

67. $369 = 9c$

68. $840 = 105t$

69. $4f = 0$

70. $0 = 60k$

71. $23b = 23$

72. $16 = 16h$

73. $-8h = 48$

74. $-9a = 72$

75. $-100 = -5g$

76. $-80 = -5w$

77. $-3.4y = -1.7$

78. $-2.1x = -1.26$

79. $\frac{x}{15} = 3$

80. $\frac{y}{7} = 12$

81. $0 = \frac{v}{11}$

82. $\frac{d}{49} = 0$

83. $\frac{w}{-7} = 15$

84. $\frac{h}{-2} = 3$

85. $\frac{d}{-7} = -3$

86. $\frac{c}{-2} = -11$

87. $\frac{y}{0.6} = -4.4$

88. $\frac{y}{0.8} = -2.9$

89. $a + 456{,}932 = 1{,}708{,}921$

90. $229{,}989 = x - 84{,}863$

91. $-1{,}563x = 43{,}764$

92. $999 = \frac{y}{-5{,}565}$

APPLICATIONS

93. SYNTHESIZERS Find the unknown angle measure.

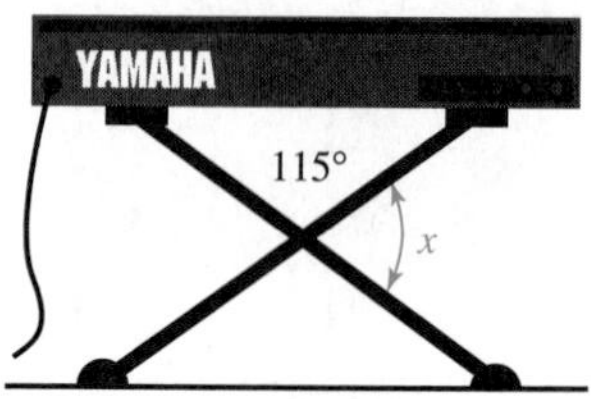

94. PHYSICS A 15-pound block is suspended with two ropes, one of which is horizontal. Find the unknown angle measure.

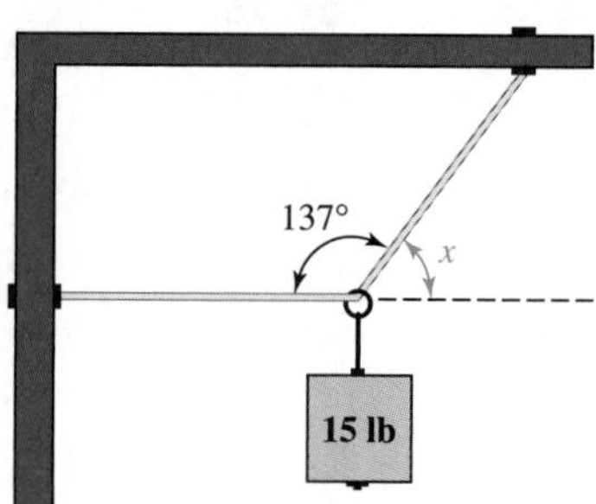

95. AVIATION How many degrees from the horizontal position are the wings of the airplane?

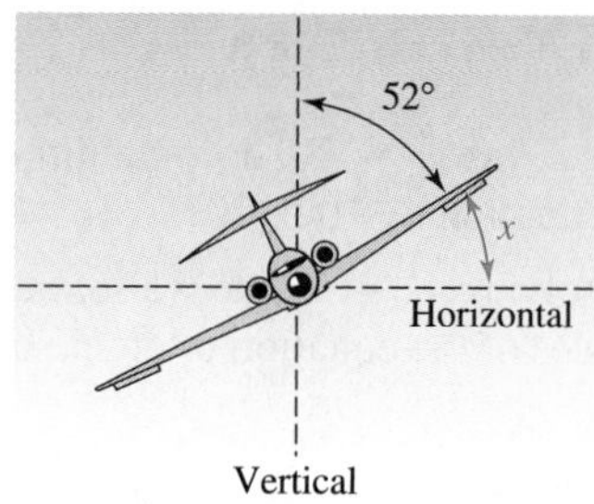

96. PLAYING A FLUTE How many degrees from the horizontal is the position of the flute?

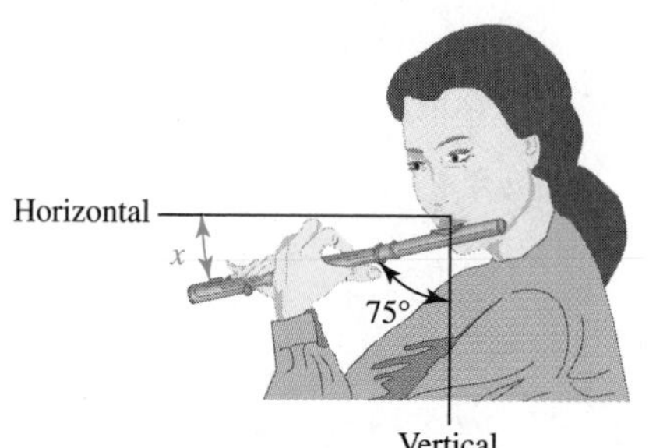

WRITING

97. What does it mean to solve an equation?

98. When solving an equation, we *isolate* the variable on one side of the equation. Write a sentence in which the word *isolate* is used in a different context.

99. Explain the error in the following work.

$$\begin{aligned} \text{Solve:}\quad x + 2 &= 40 \\ x + 2 - 2 &= 40 \\ x &= 40 \end{aligned}$$

100. After solving an equation, how do we check the result?

REVIEW

101. Evaluate $-9 - 3x$ for $x = -3$.

102. Write a formula that would give the number of eggs in d dozen.

103. Translate to symbols: Subtract x from 45.

104. Evaluate: $\dfrac{2^3 + 3(5 - 3)}{15 - 4 \cdot 2}$.

CHALLENGE PROBLEMS

105. If $a + 80 = 50$, what is $a - 80$?

106. Find two solutions of $|x + 1| = 100$.

2.2 Problem Solving

- A Problem-Solving Strategy
- Drawing Diagrams
- Constructing Tables
- Solving Percent Problems

The Language of Algebra

A *strategy* is a plan for achieving a goal. Businesses often hire firms to develop an advertising *strategy* that will increase the sales of their products.

In this section, we combine the translating skills discussed in Chapter 1 and the equation-solving skills discussed in Section 2.1 to solve many applied problems.

A PROBLEM-SOLVING STRATEGY

To become a good problem solver, you need a plan to follow, such as the following five-step strategy.

Strategy for Problem Solving

1. **Analyze the problem** by reading it carefully to understand the given facts. What information is given? What are you asked to find? What vocabulary is given? Often, a diagram or table will help you visualize the facts of the problem.
2. **Form an equation** by picking a variable to represent the numerical value to be found. Then express all other unknown quantities as expressions involving that variable. Key words or phrases can be helpful. Finally, translate the words of the problem into an equation.
3. **Solve the equation.**
4. **State the conclusion.**
5. **Check the result** in the words of the problem.

EXAMPLE 1

ELEMENTARY Algebra$f(x)$Now™

Systems analysis. A company's telephone use would have to increase by 350 calls per hour before the system would reach the maximum capacity of 1,500 calls per hour. Currently, how many calls are being made each hour on the system?

Analyze the Problem

- The maximum capacity of the system is 1,500 calls per hour.
- If the number of calls increases by 350, the system will reach capacity.
- We are to find the number of calls currently being made each hour.

Form an Equation Let $n =$ the number of calls currently being made each hour. To form an equation, we look for a key word or phrase in the problem.

Key phrase: *increase by 350* **Translation:** addition

The key phrase tells us to add 350 to the current number of calls to obtain an expression for the maximum capacity of the system. Now we translate the words of the problem into an equation.

The current number of calls per hour	increased by	350	equals	the maximum capacity of the system.
n	$+$	350	$=$	1,500

Solve the Equation

$$n + 350 = 1{,}500$$

$$n + 350 - 350 = 1{,}500 - 350$$ To undo the addition of 350, subtract 350 from both sides.

$$n = 1{,}150$$ Do the subtractions.

State the Conclusion Currently, 1,150 calls are being made per hour.

Check the Result If 1,150 calls are currently being made each hour and an increase of 350 calls per hour occurs, then $1{,}150 + 350 = 1{,}500$ calls will be made each hour. This is the capacity of the system. The answer, 1,150, checks.

DRAWING DIAGRAMS

Diagrams are often helpful because they enable us to visualize the given facts of a problem.

EXAMPLE 2

Airline travel. On a book tour that took her from New York City to Chicago to Los Angeles and back to New York City, an author flew a total of 4,910 miles. The flight from New York to Chicago was 714 miles, and the flight from Chicago to L.A. was 1,745 miles. How long was the direct flight back to New York City?

Analyze the Problem

- The total miles flown on the tour was 4,910.
- The flight from New York to Chicago was 714 miles.
- The flight from Chicago to L.A. was 1,745 miles.
- We are to find the length of the flight from L.A. to New York City.

In the diagram, we see that the three parts of the tour form a triangle. We know the lengths of two of the sides of the triangle, 714 and 1,745, and the perimeter of the triangle, 4,910.

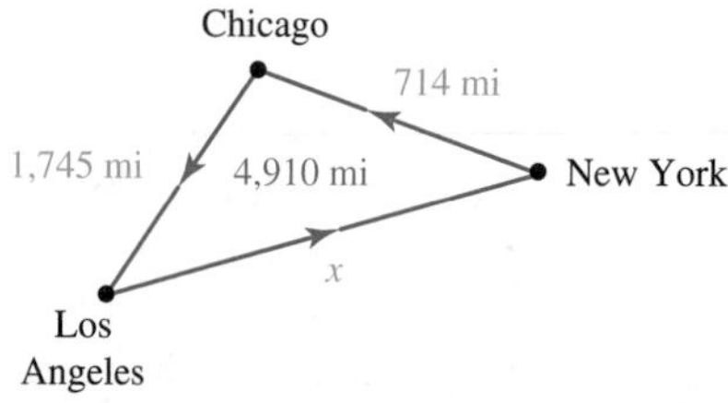

Form an Equation We will let x = the length (in miles) of the flight from L.A. to New York and label the appropriate side of the triangle in the diagram. We then have:

The miles from New York to Chicago	plus	the miles from Chicago to L.A.	plus	the miles from L.A. to New York	is	4,910.
714	+	1,745	+	x	=	4,910

Notation

When forming an equation, we usually don't include the units.

$714 \cancel{mi} + 1{,}745 \cancel{mi} + x \cancel{mi} = 4{,}910 \cancel{mi}$

Solve the Equation

$$714 + 1{,}745 + x = 4{,}910$$

$$2{,}459 + x = 4{,}910$$ Simplify the left-hand side of the equation: $714 + 1{,}745 = 2{,}459$.

$$2{,}459 + x - 2{,}459 = 4{,}910 - 2{,}459$$ Subtract 2,459 from both sides to isolate x.

$$x = 2{,}451$$ Do the subtractions.

State the Conclusion The flight from L.A. to New York was 2,451 miles.

Check the Result If we add the three flight lengths, we get $714 + 1{,}745 + 2{,}451 = 4{,}910$. This was the total number of miles flown on the book tour. The answer, 2,451, checks.

EXAMPLE 3

Eye surgery. A technique called **radial keratotomy** is sometimes used to correct nearsightedness. This procedure involves equally spaced incisions in the cornea, as shown in the figure. Find the angle between each incision. (Round to the nearest tenth of a degree.)

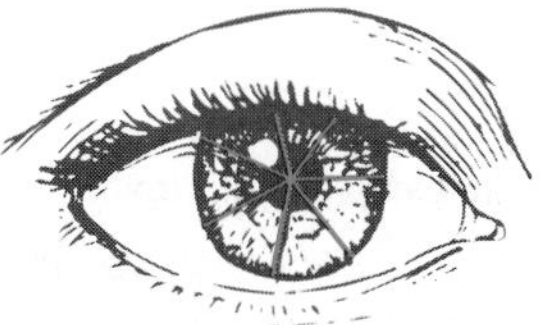

Analyze the Problem A diagram labeled with the known and unknown information is shown.

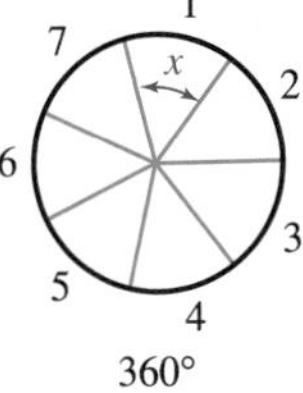

- There are 7 angles of equal measure.
- One complete revolution is 360°.
- Find the measure of one angle.

Form an Equation We will let x = the measure of one of the angles. We then have:

7	times	the measure of one of the angles	is	360°
7	·	x	=	360

Solve the Equation

$$7x = 360$$

$$\frac{7x}{7} = \frac{360}{7}$$ Divide both sides by 7.

$$x = 51.42\ldots$$ Do the division.

$$x \approx 51.4$$ Round to the nearest tenth of a degree.

State the Conclusion The incisions are approximately 51.4° apart.

Check the Result The angle measure is approximately 51°. Since $51 \cdot 7 = 357$, and this is close to 360, the result of 51.4° seems reasonable.

CONSTRUCTING TABLES

Sometimes it is helpful to organize the given facts of the problem in a table.

EXAMPLE 4

ELEMENTARY Algebra $f(x)$ Now™

Labor statistics. The number of women (16 years or older) in the U.S. labor force has grown steadily over the past 40 years. From 1960 to 1970, the number grew by 8 million. By 1980, it had increased an additional 12 million. By 2000, the number rose another 21 million; by the end of that year, 63 million women were in the labor force. How many women were in the labor force in 1960?

Analyze the Problem

- The number grew by 8 million, increased by 12 million, and rose 21 million.
- The number of women in the labor force in 2000 was 63 million.
- We are to find the number of women in the U.S. labor force in 1960.

Form an Equation We will let x = the number of women (in millions) in the labor force in 1960. We can write algebraic expressions to represent the number of women in the work force in 1970, 1980, and 2000 by translating key words.

The Language of Algebra

The word *cumulative* means to increase by successive additions. You've probably heard of a *cumulative* final exam—one that covers all of the material that was studied in the course.

Year	Women in the labor force (millions)
1960	x
1970	$x + 8$
1980	$x + 8 + 12$
2000	$x + 8 + 12 + 21$

This table helps us keep track of the cumulative total as the number of women in the work force increased over the years.

Key word: *grew* **Translation:** addition

Key word: *increased* **Translation:** addition

Key word: *rose* **Translation:** addition

There are two ways to represent the number of women (in millions) in the 2000 labor force: $x + 8 + 12 + 21$ and 63. Therefore,

$$x + 8 + 12 + 21 = 63$$

Solve the Equation

$$x + 8 + 12 + 21 = 63$$

$$x + 41 = 63 \quad \text{Simplify: } 8 + 12 + 21 = 41.$$

$$x + 41 - \mathbf{41} = 63 - \mathbf{41} \quad \text{To undo the addition of 41, subtract 41 from both sides.}$$

$$x = 22 \quad \text{Do the subtractions.}$$

State the Conclusion There were 22 million women in the U.S. labor force in 1960.

Check the Result Adding the number of women (in millions) in the labor force in 1960 and the increases, we get $22 + 8 + 12 + 21 = 63$. In 2000 there were 63 million, so the answer, 22, checks.

SOLVING PERCENT PROBLEMS

Percents are often used to present numeric information. Stores use them to advertise discounts, manufacturers use them to describe the contents of their products, and banks use them to list interest rates for loans and savings accounts.

Percent means parts per one hundred. For example, 93% means 93 out of 100 or $\frac{93}{100}$. There are three types of percent problems. Examples of these are as follows:

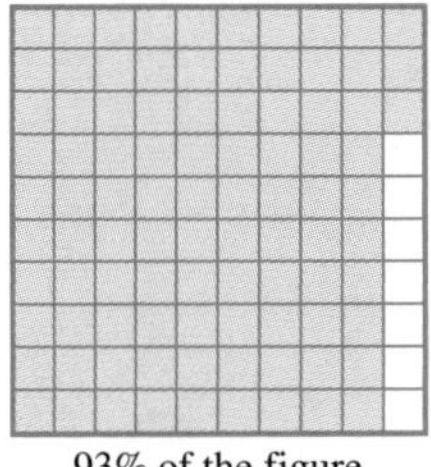

93% of the figure is shaded.

- What number is 8% of 215?
- 14 is what percent of 52?
- 82 is 20.3% of what number?

EXAMPLE 5

What number is 8% of 215?

ELEMENTARY Algebra*f(x)* Now™

Solution First, we translate the words into an equation. Here the word *of* indicates multiplication, and the word *is* means equals.

What number	is	8%	of	215?	
↓	↓	↓	↓	↓	
x	$=$	8%	$\cdot$	215	Translate to mathematical symbols.

To do the multiplication on the right-hand side of the equation, we must change the percent to a decimal (or a fraction). To change 8% to a decimal, we proceed as follows.

$8\% = 8.0\%$ The number 8 has an understood decimal point to the right of the 8.

$= .08\,0$ Drop the % symbol and divide 8.0 by 100 by moving the decimal point 2 places to the left.

To complete the solution, we replace 8% with its decimal equivalent, 0.08, and do the multiplication.

$x = 8\% \cdot 215$ This is the original equation.

$x = 0.08 \cdot 215$ $8\% = 0.08$

$x = 17.2$ Do the multiplication.

We have found that 17.2 is 8% of 215.

The Language of Algebra

The names of the parts of a percent sentence are:

17.2	is	8%	of	215.
amount		percent		base

They are related by the formula:

Amount = percent · base

Self Check 5 What number is 5.6% of 40?

One method for solving applied percent problems is to read the problem carefully, and use the given facts to write a **percent sentence** of the form:

We enter the appropriate numbers in the first two blanks, and the word "what" in the remaining blank. Then we translate the sentence to mathematical symbols and solve the resulting equation.

EXAMPLE 6

Best-selling songs. In 1993, Whitney Houston's "I Will Always Love You" led the *Billboard*'s music charts for 14 weeks. What percent of the year did she have the #1 song? (Round to the nearest one percent.)

Analyze the Problem

- For 14 out of 52 weeks in a year, she had the #1 song.
- We are to find what percent of the year she had the #1 song.

Form an Equation Let $x =$ the unknown percent and translate the words of the problem into an equation.

14	is	what percent	of	52?	
14	=	x	$\cdot$	52	14 is the amount, x is the percent, and 52 is the base.

Solve the Equation

$$14 = x \cdot 52$$

$$14 = 52x \quad \text{Write } x \cdot 52 \text{ as } 52x.$$

$$\frac{14}{52} = \frac{52x}{52} \quad \text{To isolate } x \text{, undo the multiplication by 52 by dividing both sides by 52.}$$

$$0.2692308 \approx x \quad \text{Use a calculator to do the division.}$$

We were asked to find what *percent* of the year she had the #1 song. To change the decimal 0.2692307 to a percent, we proceed as follows.

$$0\,26.92308\% \approx x \quad \text{Multiply 0.2692308 by 100 by moving the decimal point 2 places to the right, and then insert a \% symbol.}$$

$$26.92308\% \approx x$$

$$27\% \approx x \quad \text{Round 26.92308\% to the nearest one percent.}$$

State the Conclusion To the nearest one percent, Whitney Houston had the #1 song for 27% of the year.

Check the Result We can check this result using estimation. Fourteen out of 52 weeks is approximately $\frac{14}{50}$ or $\frac{28}{100}$, which is 28%. The answer, 27%, seems reasonable.

EXAMPLE 7

ELEMENTARY Algebra $f(x)$ Now™

Aging population. By the year 2050, the U.S. Bureau of the Census predicts that about 82 million residents will be 65 years of age or older. The **circle graph** (or **pie chart**) indicates that age group will make up 20.3% of the population. If the prediction is correct, what will the population of the United States be in 2050? (Round to the nearest million.)

Projection of the 2050 U.S. Population by Age

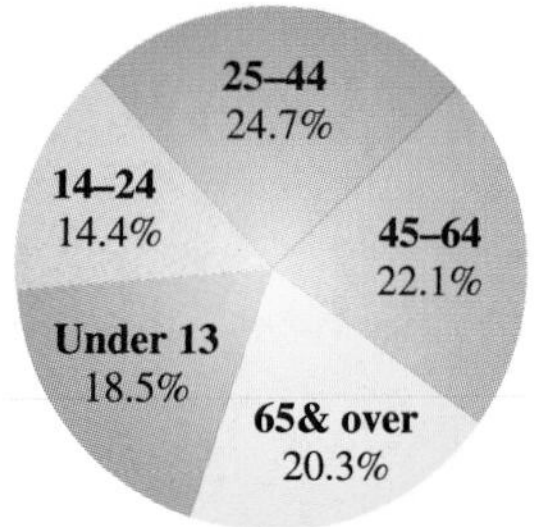

Source: U.S. Bureau of the Census (2002).

Analyze the Problem

- 82 million people will be 65 years of age or older in 2050.
- 20.3% of the population will be 65 years of age or older in 2050.
- We are to find the predicted U.S. population for the year 2050.

Form an Equation Let x = the predicted population in 2050. Translate to form an equation.

82	is	20.3%	of	what number?	
82	=	20.3%	·	x	82 is the amount, 20.3% is the percent, and x is the base.

Solve the Equation

$$82 = 20.3\% \cdot x$$

$$82 = 0.203 \cdot x \quad \text{Change 20.3\% to a decimal: 20.3\% = 0.203.}$$

$$82 = 0.203x \quad \text{Write } 0.203 \cdot x \text{ as } 0.203x.$$

$$\frac{82}{\mathbf{0.203}} = \frac{0.203x}{\mathbf{0.203}} \quad \text{To undo the multiplication by 0.203, divide both sides by 0.203.}$$

$$403.94 \approx x \quad \text{Use a calculator to do the division.}$$

$$404 \approx x \quad \text{Round to the nearest one million.}$$

State the Conclusion The census bureau is predicting a population of about 404 million in the year 2050.

Check the Result 82 million out of a population of 404 million is about $\frac{80}{400} = \frac{40}{200} = \frac{20}{100}$, or 20%. The answer of 404 million seems reasonable.

Percents are often used to describe how a quantity has changed. For example, a health care provider might increase the cost of medical insurance by 3%, or a police department might decrease the number of officers assigned to street patrols by 10%. To describe such changes, we use **percent of increase** or **percent of decrease.**

EXAMPLE 8

ELEMENTARY Algebra $f(x)$ Now™

Identity theft. The Federal Trade Commission receives complaints involving the theft of someone's identity information, such as a credit card/Social Security number or cell phone account. Refer to the data in the table. What was the percent of increase in the number of complaints from 2001 to 2002? (Round to the nearest percent.)

Number of Complaints

Year	2001	2002
	86,000	162,000

Caution

Always find the percent of increase (or decrease) with respect to the *original* amount.

Analyze the Problem To find the *amount of increase,* we subtract the number of complaints in 2001 from the number of complaints in 2002.

$$162{,}000 - 86{,}000 = 76{,}000 \quad \text{Subtract the earlier number from the later number.}$$

- The number of complaints increased by 76,000.
- Find what percent of the previous number of complaints a 76,000 increase is.

Form an Equation Let x = the unknown percent and translate the words to an equation.

76,000	is	what percent	of	86,000?	
76,000	=	x	·	86,000	76,000 is the amount, x is the percent, and 86,000 is the base.

Solve the Equation

$$76{,}000 = x \cdot 86{,}000$$

$$76{,}000 = 86{,}000x \qquad \text{Write } x \cdot 86{,}000 \text{ as } 86{,}000x.$$

$$\frac{76{,}000}{86{,}000} = \frac{86{,}000x}{86{,}000} \qquad \text{To undo the multiplication by 86,000, divide both sides by 86,000.}$$

$$0.88372093 \approx x \qquad \text{Use a calculator to do the division.}$$

$$0\,88.372093\% \approx x$$ To write the decimal as a percent, multiply by 100 by moving the decimal point two places to the right and insert a % symbol.

$$88\% \approx x \qquad \text{Round to the nearest percent.}$$

State the Conclusion In 2002, the number of complaints increased by about 88%.

Check the Result A 100% increase in complaints would be 86,000 more complaints. Therefore, it seems reasonable that 76,000 complaints is an 88% increase.

Answer to Self Check **5.** 2.24

2.2 STUDY SET

ELEMENTARY Algebra $f(x)$ Now™

VOCABULARY **Fill in the blanks.**

1. A letter that is used to represent a number is called a ________.

2. An ________ is a mathematical statement that two quantities are equal. To ______ an equation means to find all the values of the variable that make the equation true.

3. To solve an applied problem, we let a ________ represent the unknown quantity. Then we write an ________ that models the situation. Finally, we ______ the equation for the variable to find the unknown.

4. ________ means parts per one hundred.

5. In the statement "10 is 50% of 20," 10 is called the ________, 50% is the ________, and 20 is the ______.

6. In mathematics, the word *of* often indicates ____________, and ____ means equals.

Write an equation to describe each situation.

7. A college choir's tour of three cities covers 1,240 miles.

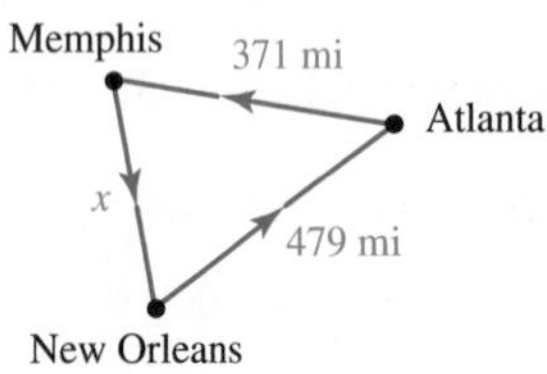

8. A woman participated in a triathlon.

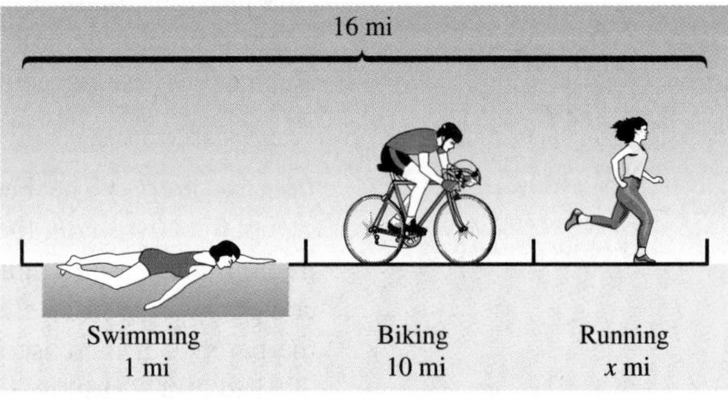

9. An airliner changed its altitude to avoid storm clouds.

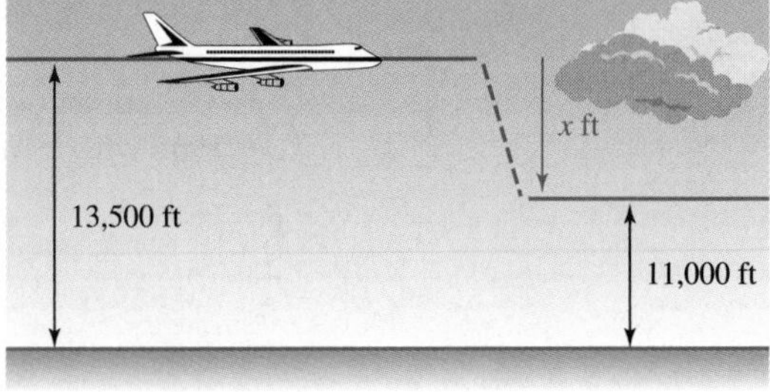

10. The sections of a 430-page book were assembled.

Section	Number of pages
Table of Contents	4
Preface	x
Text	400
Index	12

11. A hamburger chain sold a total of 31 million hamburgers in its first 4 years of business.

Year in business	Running total of hamburgers sold (millions)
1	x
2	$x + 5$
3	$x + 5 + 8$
4	$x + 5 + 8 + 16$

12. A pie was cut into equal-sized slices.

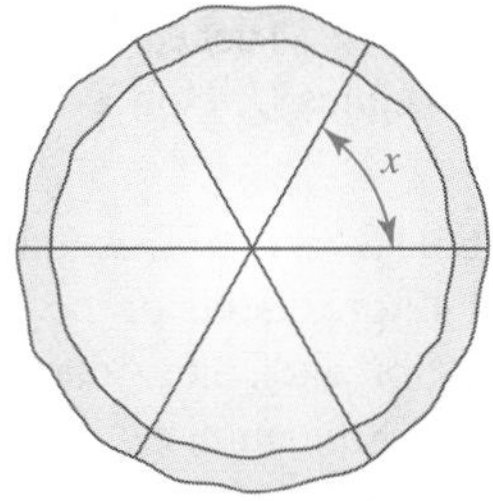

13. One method for solving percent problems is to read the problem carefully and use the given facts to write a percent sentence. What is the basic form of a percent sentence?

14.

High School Sports Programs	
Girl's Water Polo—Number of Participants	
1981	**2001**
282	14,792

Source: National Federation of State High School Associations

a. Find the *amount* of increase in participation.

b. Fill in blanks to find the percent of increase in participation: ____ is ____ % of ____

NOTATION **Translate each sentence into an equation.**

15. 12 is 40% of what number?
16. 99 is what percent of 200?
17. When computing with percents, the percent must be changed to a decimal or a fraction. Change each percent to a decimal.
    a. 35% **b.** 3.5%
    c. 350% **d.** $\frac{1}{2}\%$
18. Change each decimal to a percent.
    a. 0.9 **b.** 0.09
    c. 9 **d.** 0.999

PRACTICE **Translate each sentence into an equation, and then solve it.**

19. What number is 48% of 650?
20. What number is 60% of 200?
21. What percent of 300 is 78?
22. What percent of 325 is 143?
23. 75 is 25% of what number?
24. 78 is 6% of what number?
25. What number is 92.4% of 50?
26. What number is 2.8% of 220?
27. What percent of 16.8 is 0.42?
28. What percent of 2,352 is 199.92?
29. 128.1 is 8.75% of what number?
30. 1.12 is 140% of what number?

APPLICATIONS

31. GRAVITY Suppose an astronaut, in full gear, was weighed on Earth. (See the illustration.) The weight of an object on Earth is 6 times greater than what it is on the moon. If we let $x =$ the weight the scale would read on the moon, which equation is true?

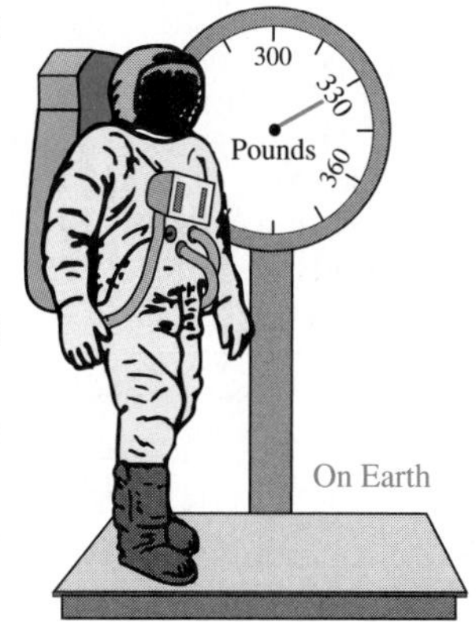

$330x = 6$ $\quad x + 6 = 330$ $\quad \frac{x}{6} = 330$ $\quad 6x = 330$

32. POWER OUTAGE The electrical system in a building automatically shuts down when the meter shown reads 85. Suppose we let $x =$ the amount the reading must increase to cause the system to shut down. Which equation is true?

$85 + x = 60 \quad 60 + x = 85 \quad 60x = 85 \quad 60 - 85 = x$

You can probably solve Problems 33–38 without algebra. Nevertheless, use the methods discussed in this section so that you can gain experience with writing and then solving an equation to find the unknown.

33. MONARCHY George III reigned as king of Great Britain for 59 years. This is 4 years less than the longest-reigning British monarch, Queen Victoria. For how many years did Queen Victoria rule?

34. TENNIS Billie Jean King won 40 Grand Slam tennis titles in her career. This is 14 less than the all-time leader, Martina Navratilova. How many Grand Slam titles did Navratilova win?

35. ATM RECEIPT Use the information on the automatic-teller receipt to find the balance in the account before the withdrawal.

HOME SAVINGS OF AMERICA			
TRAN.	DAT	TIM	T RM
0286.	1/16/03	11:46 AM	HSOA822
CARD NO.			61258
WITHDRAWAL OF			\$35.00
FROM CHECKING ACCT.			3325256-612
CHECKING BAL.			\$287.00

36. ENTERTAINMENT According to *Forbes* magazine, Oprah Winfrey made an estimated \$150 million in 2002. This was \$92 million more than Mariah Carey's estimated earnings for that year. How much did Mariah Carey make in 2002?

37. TV NEWS An interview with a world leader was edited into equally long segments and broadcast in parts over a 3-day period on a TV news program. If each daily segment of the interview lasted 9 minutes, how long was the original interview?

38. FLOODING Torrential rains caused the width of a river to swell to 84 feet. If this was twice its normal size, how wide was the river before the flooding?

Use a table to help organize the facts of the problem, then find the solution.

39. STATEHOOD From 1800 to 1850, 15 states joined the Union. From 1851 to 1900, an additional 14 states entered. Three states joined from 1901 to 1950. Since then, Alaska and Hawaii are the only others to enter the Union. How many states were part of the Union prior to 1800?

40. STUDIO TOUR Over a 4-year span, improvements in a Hollywood movie studio tour caused it to take longer. The first year, 10 minutes were added to the tour length. In the second, third, and fourth years, 5 minutes were added each year. If the tour now lasts 135 minutes, how long was it originally?

41. THEATER The play *Romeo and Juliet,* by William Shakespeare, has 5 acts and a total of 24 scenes. The second act has the most scenes, 6. The third and fourth acts each have 5 scenes. The last act has the least number of scenes, 3. How many scenes are in the first act?

42. U.S. PRESIDENTS As of December 31, 1999, there had been 42 presidents of the United States. George Washington and John Adams were the only presidents in the 18th century (1700–1799). During the 19th century (1800–1899), there were 23 presidents. How many presidents were there during the 20th century (1900–1999)?

43. ORCHESTRAS A 98-member orchestra is made up of a woodwind section with 19 musicians, a brass section with 23 players, a two-person percussion section, and a large string section. How many musicians make up the string section of the orchestra?

44. ANATOMY A premed student has to know the names of all 206 bones that make up the human skeleton. So far, she has memorized the names of the 60 bones in the feet and legs, the 31 bones in the torso, and the 55 bones in the neck and head. How many more names does she have to memorize?

Draw a diagram to help organize the facts of the problem, and then find the solution.

45. BERMUDA TRIANGLE The Bermuda Triangle is a triangular region in the Atlantic Ocean where many ships and airplanes have disappeared. The perimeter of the triangle is about 3,075 miles. It is formed by three imaginary lines. The first, 1,100 miles long, is from Melbourne, Florida, to Puerto Rico. The second, 1,000 miles long, stretches from Puerto Rico to Bermuda. The third extends from Bermuda back to Florida. Find its length.

46. FENCING To cut down on vandalism, a lot on which a house was to be constructed was completely fenced. The north side of the lot was 205 feet in length. The west and east sides were 275 and 210 feet long, respectively. If 945 feet of fencing was used, how long is the south side of the lot?

47. SPACE TRAVEL The 364-foot-tall *Saturn V* rocket carried the first astronauts to the moon. Its first, second, and third stages were 138, 98, and 46 feet tall, respectively. Atop the third stage was the lunar module, and from it extended a 28-foot escape tower. How tall was the lunar module?

48. PLANETS Mercury, Venus, and Earth have approximately circular orbits around the sun. Earth is the farthest from the sun, at 93 million miles, and Mercury is the closest, at 36 million miles. The orbit of Venus is about 31 million miles from that of Mercury. How far is Earth's orbit from that of Venus?

49. STOP SIGNS Find the measure of one angle of the octagonal stop sign. (*Hint:* The sum of the measures of the angles of an octagon is 1,080°.)

50. FERRIS WHEELS What is the measure of the angle between each of the "spokes" of the Ferris wheel?

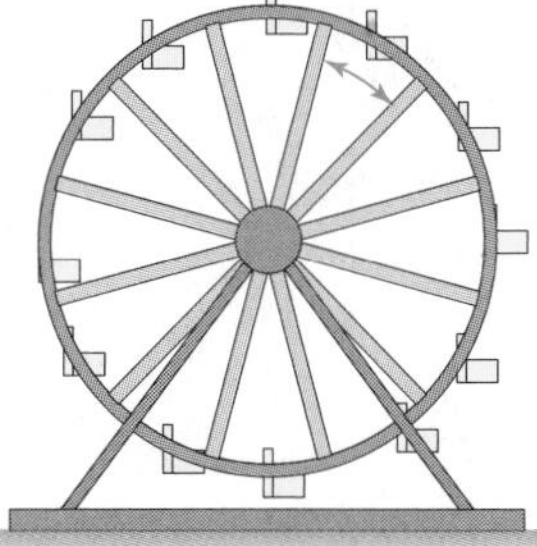

PERCENT PROBLEMS

51. ANTISEPTICS Use the facts on the label to determine the amount of pure hydrogen peroxide in the bottle.

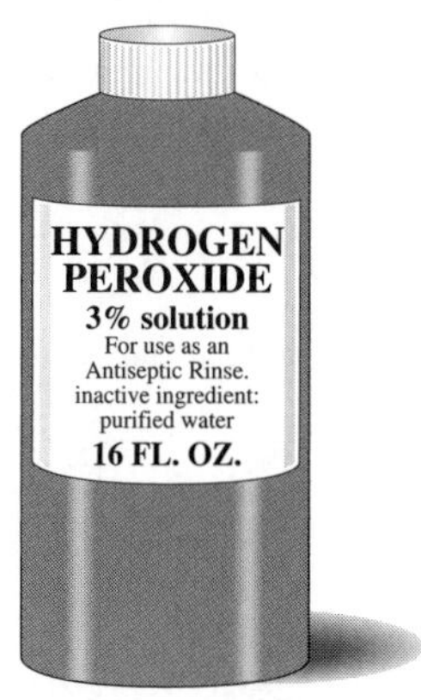

52. TIPPING When paying with a Visa card, the user must fill in the amount of the gratuity (tip) and then compute the total. Complete the sales receipt if a 15% tip, rounded up to the nearest dollar, is to be left for the waiter.

53. FEDERAL OUTLAYS The **circle graph** shows the breakdown of the U.S. federal budget for fiscal year 2001. If total spending was approximately $1,900 billion, how much was paid for Social Security, Medicare, and other retirement programs?

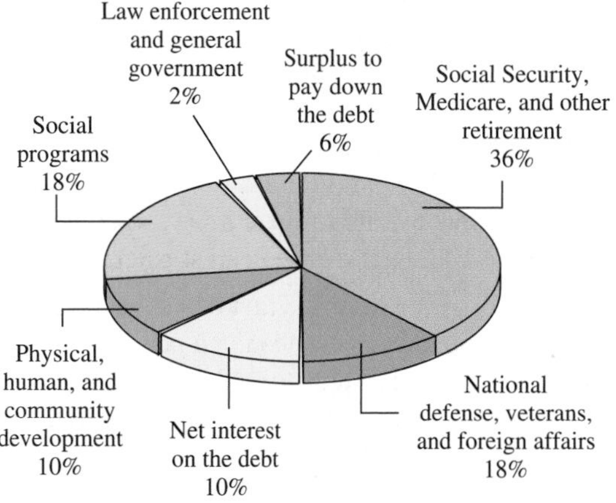

Based on 2002 Federal Income Tax Form 1040

54. INCOME TAX Use the tax table to compute the amount of federal income tax if the amount of taxable income entered on Form 1040, line 40, is $39,909.

If the amount on Form 1040, line 40, is: *Over—*	*But not over—*	Enter on Form 1040, line 41	*of the amount over—*
$0	$7,000	.... 10%	$0
7,000	28,400	$700.00 + 15%	7,000
28,400	68,800	3,910.00 + 25%	28,400
68,800	143,500	14,010.00 + 28%	68,800
143,500	311,950	34,926.00 + 33%	143,500
311,950	....	90,514.50 + 35%	311,950

55. COLLEGE ENTRANCE EXAMS On the Scholastic Aptitude Test, or SAT, a high school senior scored 550 on the mathematics portion and 700 on the verbal portion. What percent of the maximum 1,600 points did this student receive?

56. GENEALOGY Through an extensive computer search, a genealogist determined that worldwide, 180 out of every 10 million people had his last name. What percent is this?

57. DENTAL RECORDS On the dental chart for an adult patient, the dentist marks each tooth that has had a filling. To the nearest percent, what percent of this patient's teeth have fillings?

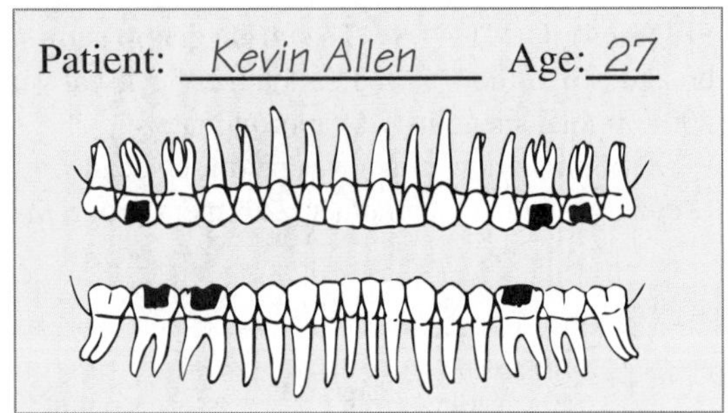

58. AREAS The total area of the 50 states and the District of Columbia is 3,618,770 square miles. If Alaska covers 591,004 square miles, what percent is this of the U.S. total (to the nearest percent)?

59. CHILD CARE After the first day of registration, 84 children had been enrolled in a new day care center. That represented 70% of the available slots. What was the maximum number of children the center could enroll?

60. RACING PROGRAMS One month before a stock car race, the sale of ads for the official race program was slow. Only 12 pages, or just 60% of the available pages, had been sold. What was the total number of pages devoted to advertising in the program?

61. NUTRITION The Nutrition Facts label from a can of clam chowder is shown.

a. Find the number of grams of saturated fat in one serving. What percent of a person's recommended daily intake is this?

Nutrition Facts

Serving Size 1 cup (240mL)
Servings Per Container about 2

Amount per serving	
Calories 240	Calories from Fat 140
	% Daily Value*
Total Fat 15 g	**23%**
Saturated Fat 5 g	**25%**
Cholesterol 10 mg	**3%**
Sodium 980 mg	**41%**
Total Carbohydrate 21 g	**7%**
Dietary Fiber 2 g	**8%**
Sugars 1 g	
Protein 7 g	

b. Determine the recommended number of grams of saturated fat that a person should consume daily.

62. CUSTOMER GUARANTEES To assure its customers of low prices, the Home Club offers a "10% Plus" guarantee. If the customer finds the same item selling for less somewhere else, he or she receives the difference in price plus 10% of the difference. A woman bought miniblinds at the Home Club for $120 but later saw the same blinds on sale for $98 at another store. How much can she expect to be reimbursed?

63. EXPORTS The bar graph shows United States exports to Mexico for the years 1992 through 2001. Between what two years was there the greatest percent of decrease in exports?

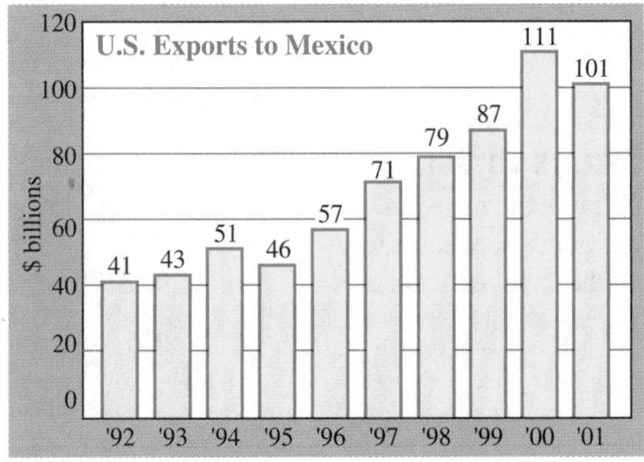

Based on data from www.census.gov/foreign-trade

64. AUCTIONS A pearl necklace of former First Lady Jacqueline Kennedy Onassis, originally valued at \$700, was sold at auction in 1996 for \$211,500. What was the percent of increase in the value of the necklace? (Round to the nearest percent.)

65. INSURANCE COSTS A college student's good grades earned her a student discount on her car insurance premium. What was the percent of decrease, to the nearest percent, if her annual premium was lowered from \$1,050 to \$925?

66. U.S. LIFE EXPECTANCY Use the following life expectancy data for 1900 and 2000 to determine the percent of increase for males and for females. Round to the nearest percent.

Years of life expected at birth		
	Male	**Female**
1900	46.3 yr	48.3 yr
2000	74.1 yr	79.5 yr

WRITING

67. Explain the relationship in a percent problem between the amount, the percent, and the base.

68. Write a real-life situation that could be described by "9 is what percent of 20?"

69. Explain why 150% of a number is more than the number.

70. Explain why "Find 9% of 100" is an easy problem to solve.

REVIEW

71. Divide: $-\frac{16}{25} \div \left(-\frac{4}{15}\right)$.

72. What two numbers are a distance of 8 away from 4 on the number line?

73. Is -34 a solution of $x + 15 = -49$?

74. Evaluate: $2 + 3[24 - 2(2 - 5)]$.

CHALLENGE PROBLEMS

75. SOAP A soap advertises itself as $99\frac{44}{100}\%$ pure. First, tell what percent of the soap is impurities. Then express your answer as a decimal.

76. Express $\frac{1}{20}$ of 1% as a percent using decimal notation.

2.3 Simplifying Algebraic Expressions

- Simplifying Products
- The Distributive Property
- Like Terms
- Combining Like Terms

In algebra, we frequently replace one algebraic expression with another that is equivalent and simpler in form. That process, called *simplifying an algebraic expression,* often involves the use of one or more properties of real numbers.

SIMPLIFYING PRODUCTS

Recall that the associative property of multiplication enables us to change the grouping of factors involved in a multiplication. The commutative property of multiplication enables us to change the order of the factors. These properties can be used to simplify certain products. For example, let's simplify $8(4x)$.

$$\begin{aligned} 8(4x) &= 8 \cdot (4 \cdot x) && 4x = 4 \cdot x. \\ &= (8 \cdot 4) \cdot x && \text{Use the associative property of multiplication to group 4 with 8.} \\ &= 32x && \text{Do the multiplication within the parentheses.} \end{aligned}$$

We have found that $8(4x) = 32x$. We say that $8(4x)$ and $32x$ are **equivalent expressions** because for each value of x, they represent the same number.

If $x = 10$		*If* $x = -3$	
$8(4x) = 8[4(10)]$	$32x = 32(10)$	$8(4x) = 8[4(-3)]$	$32x = 32(-3)$
$= 8(40)$	$= 320$	$= 8(-12)$	$= -96$
$= 320$		$= -96$	

EXAMPLE 1

Simplify each expression: **a.** $-9(3b)$, **b.** $15a(6)$, **c.** $35\left(\frac{4}{5}x\right)$, **d.** $\frac{8}{3} \cdot \frac{3}{8}r$, and **e.** $3(7p)(-5)$.

Solution

a. $-9(3b) = -27b$

b. $15a(6) = 90a$ Use the commutative property of multiplication to reorder the factors.

c. $35\left(\frac{4}{5}x\right) = \left(35 \cdot \frac{4}{5}\right)x$ Use the associative property of multiplication to regroup the factors.

$= 28x$ Factor 35 as $5 \cdot 7$ to do the multiplication.

d. $\frac{8}{3} \cdot \frac{3}{8}r = \left(\frac{8}{3} \cdot \frac{3}{8}\right)r$ Use the associative property of multiplication to regroup the factors.

$= 1r$ Multiply.

$= r$ Simplify.

e. $3(7p)(-5) = -105p$ Use the commutative property of multiplication to reorder the factors.

The Language of Algebra

Two of the most often encountered instructions in algebra are *simplify* and *solve.* Remember that we *simplify* expressions and we *solve* equations.

Self Check 1 Multiply: **a.** $9 \cdot 6s$, **b.** $36\left(\frac{2}{9}y\right)$, **c.** $\frac{2}{3} \cdot \frac{3}{2}m$, and **d.** $-4(6u)(-2)$.

THE DISTRIBUTIVE PROPERTY

Another property that is used to simplify algebraic expressions is the **distributive property.** To introduce it, we will evaluate $4(5 + 3)$ in two ways.

Method 1 ***Order of operations:*** Compute the sum within the parentheses first.

$4(5 + 3) = 4(8)$

$= 32$ Do the multiplication.

Method 2 ***The distributive property:*** Multiply 5 and 3 by 4 and then add the results.

$4(5 + 3) = 4(5) + 4(3)$ Distribute the multiplication by 4.

$= 20 + 12$ Do the multiplications.

$= 32$

Each method gives a result of 32. This observation suggests the following property.

The Distributive Property For any real numbers a, b, and c,

$$a(b + c) = ab + ac$$

The Language of Algebra

Formally, it is called the distributive property of multiplication over addition. When we use it to write a product, such as $5(x + 2)$ as a sum, $5x + 10$, we say that we have *removed* or *cleared* the parentheses.

To illustrate one use of the distributive property, let's consider the expression $5(x + 2)$. Since we are not given the value of x, we cannot add x and 2 within the parentheses. However, we can distribute the multiplication by the factor of 5 that is outside the parentheses to x and to 2 and add those products.

$$\begin{aligned} 5(x + 2) &= 5(x) + 5(2) && \text{Distribute the multiplication by 5.} \\ &= 5x + 10 \end{aligned}$$

Since subtraction is the same as adding the opposite, the distributive property also holds for subtraction.

$$a(b - c) = ab - ac$$

EXAMPLE 2

ELEMENTARY Algebra f(x) Now™

Use the distributive property to remove parentheses: **a.** $6(a + 9)$, **b.** $3(3b - 8)$, **c.** $-12(a + 1)$, **d.** $-6(-3y - 8)$, and **e.** $15\left(\frac{x}{3} + \frac{2}{5}\right)$.

Solution

a. $6(a + 9) = 6 \cdot a + 6 \cdot 9$ Distribute the multiplication by 6.

$= 6a + 54$

The Language of Algebra

We read $6(a + 9)$ as "six times the *quantity* of a plus nine." The word *quantity* alerts us to the grouping symbols in the expression.

b. $3(3b - 8) = 3(3b) - 3(8)$ Distribute the multiplication by 3.

$= 9b - 24$ Do the multiplications.

c. $-12(a + 1) = -12(a) + (-12)(1)$ Distribute the multiplication by -12.

$= -12a - 12$

d. $-6(-3y - 8) = -6(-3y) - (-6)(8)$ Distribute the multiplication by -6.

$= 18y + 48$

e. $15\left(\frac{x}{3} + \frac{2}{5}\right) = 15 \cdot \frac{x}{3} + 15 \cdot \frac{2}{5}$ Distribute the multiplication by 15.

$= 5x + 6$

Self Check 2 Use the distributive property to remove parentheses: **a.** $5(p + 2)$, **b.** $4(2x - 1)$, **c.** $-8(2x - 4)$, and **d.** $24\left(\frac{y}{6} + \frac{3}{8}\right)$.

Caution The distributive property does not apply to every expression that contains parentheses—only those where multiplication is distributed over addition (or subtraction). For example, to simplify $6(5x)$, we do not use the distributive property.

Correct	*Incorrect*
$6(5x) = (6 \cdot 5)x = 30x$	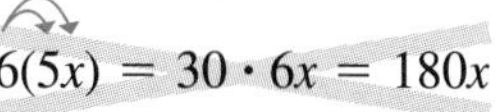 $6(5x) = 30 \cdot 6x = 180x$

The distributive property can be extended to several other useful forms. Since multiplication is commutative:

$$(b + c)a = ba + ca \qquad (b - c)a = ba - ca$$

For situations in which there are more than two terms within parentheses:

$$a(b + c + d) = ab + ac + ad \qquad a(b - c - d) = ab - ac - ad$$

EXAMPLE 3

ELEMENTARY Algebra f(x) Now™

Multiply: **a.** $(6x + 4y)\frac{1}{2}$, **b.** $2(a - 3b)8$, and **c.** $-0.3(3a - 4b + 7)$.

Solution

a. $(6x + 4y)\frac{1}{2} = (6x)\frac{1}{2} + (4y)\frac{1}{2}$ Distribute the multiplication by $\frac{1}{2}$.

$= 3x + 2y$ Do the multiplications.

b. $2(a - 3b)8 = 2 \cdot 8(a - 3b)$ Use the commutative property of multiplication to reorder the factors.

$= 16(a - 3b)$ Do the multiplication.

$= 16a - 48b$ Distribute the multiplication by 16.

c. $-0.3(3a - 4b + 7) = -0.3(3a) - (-0.3)(4b) + (-0.3)(7)$ Distribute the multiplication by -0.3.

$= -0.9a + 1.2b - 2.1$ Do the three multiplications.

Self Check 3 Multiply: **a.** $(-6x - 24y)\frac{1}{3}$, **b.** $6(c - 2d)9$, **c.** $-0.7(2r + 5s - 8)$

Success Tip

Note that distributing the multiplication by -1 changes the sign of each term within the parentheses.

We can use the distributive property to find the opposite of a sum. For example, to find $-(x + 10)$, we interpret the $-$ symbol as a factor of -1, and proceed as follows:

$-(x + 10) = -1(x + 10)$ Replace the $-$ symbol with -1.

$= -1(x) + (-1)(10)$ Distribute the multiplication by -1.

$= -x - 10$ Multiply.

In general, we have the following property of real numbers.

The Opposite of a Sum The opposite of a sum is the sum of the opposites. For any real numbers a and b,

$$-(a + b) = -a + (-b)$$

EXAMPLE 4

Simplify: $-(-9s - 3)$.

ELEMENTARY Algebra f(x) Now™

Solution

$-(-9s - 3) = -1(-9s - 3)$ Replace the $-$ symbol in front of the parentheses with -1.

$= -1(-9s) - (-1)(3)$ Distribute the multiplication by -1.

$= 9s + 3$

Self Check 4 Simplify: $-(-5x + 18)$.

LIKE TERMS

The distributive property can be used to simplify certain sums and differences. But before we can discuss this, we need to introduce some new vocabulary.

Like Terms **Like terms** are terms with exactly the same variables raised to exactly the same powers. Any constant terms in an expression are considered to be like terms. Terms that are not like terms are called **unlike terms.**

Here are several examples.

Like terms	*Unlike terms*	
$4x$ and $7x$	$4x$ and $7y$	Different variables
$-10p^2$ and $25p^2$	$-10p$ and $25p^2$	Same variable, different powers
$\frac{1}{3}c^3d$ and c^3d	$\frac{1}{3}c^3d$ and c^3	Different variables

EXAMPLE 5

List the like terms in each expression: **a.** $7r + 5 + 3r$, **b.** $6x^4 - 6x^2 - 6x$, and **c.** $-17m^3 + 3 - 2 + m^3$.

Solution

a. $7r + 5 + 3r$ contains the like terms $7r$ and $3r$.

b. $6x^4 - 6x^2 - 6x$ contains no like terms.

c. $-17m^3 + 3 - 2 + m^3$ contains two pairs of like terms: $-17m^3$ and m^3 are like terms, and the constant terms, 3 and -2, are like terms.

Self Check 5 List the like terms: **a.** $5x - 2y + 7y$ and **b.** $-5p^2 - 12 + 17p^2 + 2$.

COMBINING LIKE TERMS

To add or subtract objects, they must have the same units. For example, we can add dollars to dollars and inches to inches, but we cannot add dollars to inches. When simplifying algebraic expressions, we can only add or subtract like terms.

Success Tip

When looking for like terms, don't look at the coefficients of the terms. Consider only the variable factors of each term.

This expression can be simplified, because it contains like terms.	This expression cannot be simplified, because its terms are not like terms.
$3x + 4x$	$3x + 4y$

Recall that the distributive property can be written in the following forms:

$$(b + c)a = ba + ca \qquad (b - c)a = ba - ca$$

We can use these forms of the distributive property in reverse to simplify a sum or difference of like terms. For example, we can simplify $3x + 4x$ as follows:

$$\begin{aligned} 3x + 4x &= (3 + 4)x \\ &= 7x \end{aligned}$$

We can simplify $15m^2 - 9m^2$ in a similar way:

$$15m^2 - 9m^2 = (15 - 9)m^2$$
$$= 6m^2$$

In each case, we say that we *combined like terms.* These examples suggest the following general rule.

Combining Like Terms To add or subtract like terms, combine their coefficients and keep the same variables with the same exponents.

EXAMPLE 6

Simplify each expression, if possible: **a.** $-2x + 11x$, **b.** $-8p + (-2p) + 4p$, **c.** $0.5s^2 - 0.3s^2$, and **d.** $4w + 6$.

Solution
a. $-2x + 11x = 9x$ Think: $(-2 + 11)x = 9x$.
b. $-8p + (-2p) + 4p = -6p$ Think: $[-8 + (-2) + 4]p = -6p$.
c. $0.5s^2 - 0.3s^2 = 0.2s^2$ Think: $(0.5 - 0.3)s^2 = 0.2s^2$.
d. Since $4w$ and 6 are not like terms, they cannot be combined.

Self Check 6 Simplify, if possible: **a.** $-3x + 5x$, **b.** $-6y + (-6y) + 9y$, **c.** $4.4s^4 - 3.9s^4$, and **d.** $4a - 2$.

EXAMPLE 7

ELEMENTARY Algebra f(x) Now™

Simplify by combining like terms: **a.** $16t - 15t$, **b.** $16t - t$, **c.** $15t - 16t$, and **d.** $16t + t$.

Solution
a. $16t - 15t = t$ Think: $(16 - 15)t = 1t = t$.
b. $16t - t = 15t$ Think: $16t - 1t = (16 - 1)t = 15t$.
c. $15t - 16t = -t$ Think: $(15 - 16)t = -1t = -t$.
d. $16t + t = 17t$ Think: $16t + 1t = (16 + 1)t = 17t$.

Self Check 7 Simplify: **a.** $9h - h$, **b.** $9h + h$, **c.** $9h - 8h$, and **d.** $8h - 9h$.

EXAMPLE 8

Simplify: $6t - 8 - 4t + 1$.

Solution We combine the like terms that involve the variable t and we combine the constant terms.

ELEMENTARY Algebra f(x) Now™

Think: $(6 - 4)t = 2t$

$$6t - 8 - 4t + 1 = 2t - 7$$

Think: $-8 + 1 = -7$

Self Check 8 Simplify: $50 + 70a - 60 - 10a$.

EXAMPLE 9

Simplify: $4(x + 5) - 5 - (2x - 4)$.

ELEMENTARY Algebra f(x) Now™

Solution To simplify, we use the distributive property and combine like terms.

$$4(x + 5) - 5 - (2x - 4) = 4(x + 5) - 5 - 1(2x - 4)$$

$$= 4x + 20 - 5 - 2x + 4 \quad \text{Distribute the multiplication by 4 and } -1.$$

$$= 2x + 19 \quad \text{Simplify.}$$

Self Check 9 Simplify: $6(3y - 1) + 2 - (-3y + 4)$.

Answers to Self Checks **1. a.** $54s$, **b.** $8y$, **c.** m, **d.** $48u$ **2. a.** $5p + 10$, **b.** $8x + 4$, **c.** $-16x + 32$, **d.** $4y + 9$ **3. a.** $-2x - 8y$, **b.** $54c - 108d$, **c.** $-1.4r - 3.5s + 5.6$ **4.** $5x - 18$ **5. a.** $-2y$ and $7y$ **b.** $-5p^2$ and $17p^2$; -12 and 2 **6. a.** $2x$, **b.** $-3y$, **c.** $0.5s^4$, **d.** $4a - 2$ **7. a.** $8h$, **b.** $10h$, **c.** h, **d.** $-h$ **8.** $60a - 10$ **9.** $21y - 8$

2.3 STUDY SET

ELEMENTARY Algebra f(x) Now™

VOCABULARY Fill in the blanks.

1. We can use the associative property of multiplication to ________ the expression $5(6x)$.
2. The ____________ property of multiplication allows us to reorder factors. The ____________ property of multiplication allows us to regroup factors.
3. We simplify ____________ and we solve ____________.
4. We can use the ____________ property to remove or clear parentheses in the expression $2(x + 8)$.
5. We call $-(c + 9)$ the __________ of a sum.
6. Terms such as $7x^2$ and $5x^2$, which have the same variables raised to exactly the same exponents, are called ______ terms.
7. The ___________ of the term $-23y$ is -23.
8. When we write $9x + x$ as $10x$, we say we have __________ like terms.

CONCEPTS

9. Fill in the blanks to simplify each product.
   a. $5 \cdot 6t = (\_\_\_ \cdot \_\_\_)t = \_\_\_\, t$
   b. $-8(2x)(4) = (\_\_\_ \cdot \_\_\_ \cdot \_\_\_)x = \_\_\_\, x$
10. Fill in the blanks.
   a. $a(b + c) = ab + \_\_\_$ **b.** $a(b - c) = \_\_\_ - ac$
   c. $(b + c)a = ba + \_\_\_$ **d.** $(b - c)a = \_\_\_ - ca$
   e. $a(b + c + d) = \_\_\_ + ac + \_\_\_$
   f. $-(a + b) = -a + \_\_\_$
11. Consider $3(x + 6)$. Why can't we add x and 6 within the parentheses?
12. Fill in the blanks.
   a. $2(x + 4) = 2x \_\_\_ 8$
   b. $2(x - 4) = 2x \_\_\_ 8$
   c. $-2(x + 4) = -2x \_\_\_ 8$
   d. $-2(-x - 4) = 2x \_\_\_ 8$
13. Fill in the blanks.

$$-(x + 10) = \_\_\_(x + 10)$$

   Distributing the multiplication by -1 changes the ______ of each term within the parentheses.
14. Consider $33x - 8x^2 - 21x + 6$. Identify the terms of the expression and the coefficient of each term.
15. For each expression, identify any like terms.
   a. $3a + 8 + 2a$ **b.** $10 - 13h + 12$
   c. $3x^2 + 3x + 3$ **d.** $9y^2 - 9m - 8y^2$
16. Complete this statement: To add like terms, add their ____________ and keep the same __________ and exponents.
17. Fill in the blanks to combine like terms.
   a. $4m + 6m = (\_\_\_)m = \_\_\_\, m$
   b. $30n - 50n = (\_\_\_)n = \_\_\_\, n$
   c. $12 - 32d + 15 = -32d + \_\_\_$

18. Simplify each expression, if possible.
a. $5(2x)$ and $5 + 2x$
b. $6(-7x)$ and $6 - 7x$
c. $2(3x)(3)$ and $2 + 3x + 3$

NOTATION Complete each solution.

19. Translate to symbols.
a. Six times the quantity of h minus four.
b. The opposite of the sum of z and sixteen.

20. Write an equivalent expression for the given expression using fewer symbols.
a. $1x$ **b.** $-1d$ **c.** $0m$
d. $5x - (-1)$ **e.** $16t + (-6)$

21. A student compared her answers to six homework problems to the answers in the back of the book. Are her answers equivalent? Write *yes* or *no*.

Student's answer	Book's answer	Equivalent?
$10x$	$10 + x$	
$3 + y$	$y + 3$	
$5 - 8a$	$8a - 5$	
$3x + 4$	$3(x + 4)$	
$3 - 2x$	$-2x + 3$	
$h^2 + (-16)$	$h^2 - 16$	

22. Draw arrows to illustrate how we distribute the factor outside the parentheses over the terms within the parentheses.
a. $8(6g + 7)$ **b.** $(4x^2 - x + 6)2$

PRACTICE Simplify each expression.

23. $9(7m)$
24. $12n(8)$
25. $5(-7q)$
26. $-7(5t)$
27. $5t \cdot 60$
28. $70a \cdot 10$
29. $(-5.6x)(-2)$
30. $(-4.4x)(-3)$
31. $\frac{5}{3} \cdot \frac{3}{5}g$
32. $\frac{9}{7} \cdot \frac{7}{9}k$
33. $12\left(\frac{5}{12}x\right)$
34. $15\left(\frac{4}{15}w\right)$
35. $8\left(\frac{3}{4}y\right)$
36. $27\left(\frac{2}{3}x\right)$
37. $-\frac{15}{4}\left(-\frac{4s}{15}\right)$
38. $-\frac{50}{3}\left(-\frac{3h}{50}\right)$
39. $24\left(-\frac{5}{6}r\right)$
40. $\frac{3}{4} \cdot \frac{1}{2}g$
41. $5(4c)(3)$
42. $9(2h)(2)$
43. $-4(-6)(-4m)$
44. $-5(-9)(-4n)$

Remove parentheses, and simplify.

45. $5(x + 3)$
46. $4(x + 2)$
47. $6(6c - 7)$
48. $9(9d - 3)$
49. $(3t + 2)8$
50. $(2q + 1)9$
51. $0.4(x - 4)$
52. $-2.2(2q + 1)$
53. $-5(-t - 1)$
54. $-8(-r - 1)$
55. $-4(3x + 5)$
56. $-4(6r + 4)$
57. $(13c - 3)(-6)$
58. $(10s - 11)(-2)$
59. $-\frac{2}{3}(3w - 6)$
60. $\frac{1}{2}(2y - 8)$
61. $45\left(\frac{x}{5} + \frac{2}{9}\right)$
62. $35\left(\frac{y}{5} + \frac{8}{7}\right)$
63. $60\left(\frac{3}{20}r - \frac{4}{15}\right)$
64. $72\left(\frac{7}{8}f - \frac{8}{9}\right)$
65. $-(x - 7)$
66. $-(y + 1)$
67. $-(-5.6y + 7)$
68. $-(-4.8a - 3)$
69. $2(4d + 5)5$
70. $4(2w + 3)5$
71. $-6(r + 5)2$
72. $-7(b + 3)3$
73. $-(-x - y + 5)$
74. $-(-14 + 3p - t)$
75. $5(1.2x - 4.2y - 3.2z)$
76. $5(2.4a + 5.4b - 6.4c)$

Simplify each expression.

77. $3x + 17x$
78. $12y - 15y$
79. $-4x + 4x$
80. $-16y + 16y$
81. $-7b^2 + 7b^2$
82. $-2c^3 + 2c^3$
83. $13r - 12r$
84. $25s + s$

85. $36y + y$

86. $32a - a$

87. $43s^3 - 44s^3$

88. $8j^3 - 9j^3$

89. $23w + 5 - 23w$

90. $19x + 3 - 19x$

91. $-4r - 7r + 2r - r$

92. $-v - 3v + 6v + 2v$

93. $a + a + a$

94. $t - t - t - t$

95. $0 - 3x$

96. $0 - 4a$

97. $3x + 5x - 7x$

98. $-5.7m + 4.3m$

99. $\frac{3}{5}t + \frac{1}{5}t$

100. $\frac{3}{16}x - \frac{5}{16}x$

101. $-0.2r - (-0.6r)$

102. $-1.1m - (-2.4m)$

103. $2z + 5(z - 3)$

104. $12(m + 11) - 11$

105. $3x + 4 - 5x + 1$

106. $4b + 9 - 9b + 9$

107. $10(2d - 7) + 4$

108. $5(3x - 2) + 5$

109. $-(c + 7) - 2(c - 3)$

110. $-(z + 2) + 5(3 - z)$

111. $2(s - 7) - (s - 2)$

112. $4(d - 3) - (d - 1)$

113. $6 - 4(-3c - 7)$

114. $10 - 5(-5g - 1)$

115. $36\left(\frac{2}{9}x - \frac{3}{4}\right) + 36\left(\frac{1}{2}\right)$

116. $40\left(\frac{3}{8}y - \frac{1}{4}\right) + 40\left(\frac{4}{5}\right)$

APPLICATIONS

117. THE AMERICAN RED CROSS In 1891, Clara Barton founded the Red Cross. Its symbol is a white flag bearing a red cross. If each side of the cross has length x, write an algebraic expression for the perimeter of the cross.

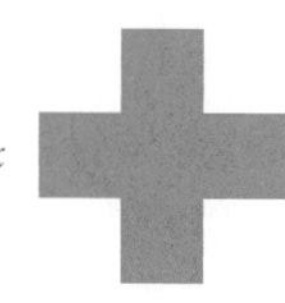

118. BILLIARDS Billiard tables vary in size, but all tables are twice as long as they are wide.

a. If the billiard table is x feet wide, write an expression involving x that represents its length.

b. Write an expression for the perimeter of the table.

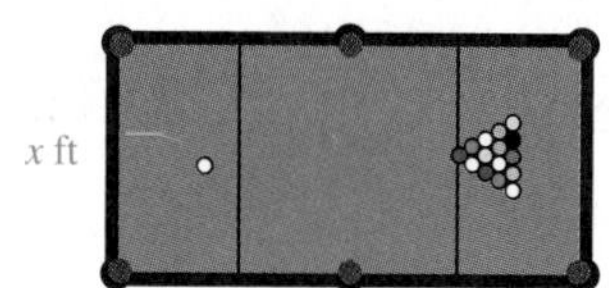

119. PING-PONG Write an expression for the perimeter of the Ping-Pong table.

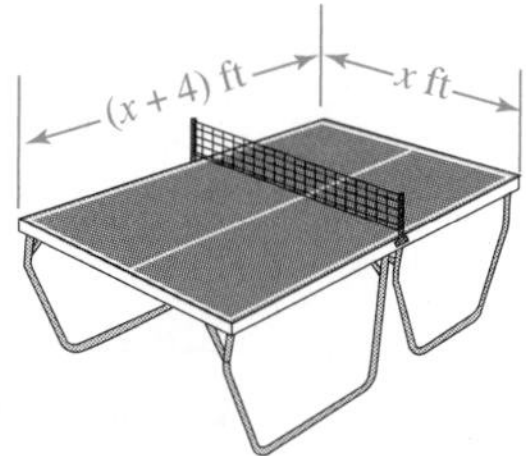

120. SEWING Write an expression for the length of the yellow trim needed to outline a pennant with the given side lengths.

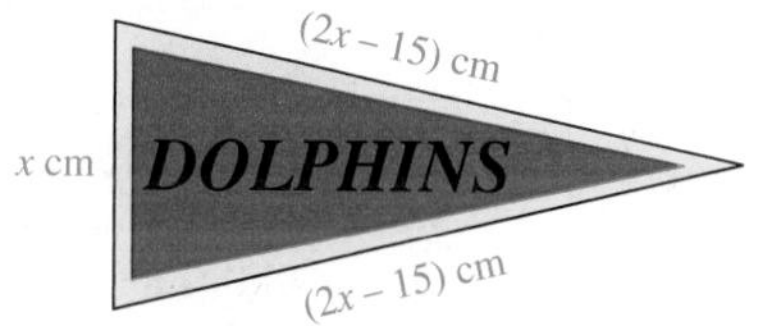

WRITING

121. Explain why the distributive property applies to $2(3 + x)$ but not to $2(3x)$.

122. Tell how to combine like terms.

REVIEW

Evaluate each expression for $x = -3$, $y = -5$, and $z = 0$.

123. $x^2z(y^3 - z)$

124. $|y^3 - z|$

125. $\frac{x - y^2}{2y - 1 + x}$

126. $\frac{2y + 1}{x} - x$

CHALLENGE PROBLEMS

127. The quantities below do not have the same units and cannot be added as shown. Is there a way to find the sum? If so, what is it?

2 inches + 2 feet + 2 yards

128. Fill in the blanks: $\square(\square - \square) = -75x + 40$.

Simplify.

129. $-2[x + 4(2x + 1)]$

130. $-5[y + 2(3y + 4)]$

2.4 More about Solving Equations

- Using More Than One Property of Equality
- Simplifying Expressions to Solve Equations
- Identities and Contradictions

We have solved simple equations by using properties of equality. We will now expand our equation-solving skills by considering more complicated equations. The objective is to develop a general strategy that we can use to solve any kind of *linear equation in one variable.*

Linear Equation in One Variable

A **linear equation in one variable** can be written in the form

$$ax + b = c$$

where a, b, and c are real numbers and $a \neq 0$.

USING MORE THAN ONE PROPERTY OF EQUALITY

Sometimes we must use several properties of equality to solve an equation. For example, on the left-hand side of $2x + 6 = 10$, the variable is multiplied by 2, and then 6 is added to that product. To isolate x, we use the order of operations rules in reverse. First, we undo the addition of 6, and then we undo the multiplication by 2.

$$2x + 6 = 10$$

$$2x + 6 - \mathbf{6} = 10 - \mathbf{6}$$ To undo the addition of 6, subtract 6 from both sides.

$$2x = 4$$ Do the subtractions.

$$\frac{2x}{\mathbf{2}} = \frac{4}{\mathbf{2}}$$ To undo the multiplication by 2, divide both sides by 2.

$$x = 2$$

The solution is 2.

EXAMPLE 1 Solve: $-12x + 5 = 17$.

ELEMENTARY Algebra $f(x)$ Now™

Solution On the left-hand side of the equation, x is multiplied by -12, and then 5 is added to that product. To isolate x, we undo the operations in the opposite order.

- To undo the addition of 5, we subtract 5 from both sides.
- To undo the multiplication by -12, we divide both sides by -12.

$$-12x + 5 = 17$$

$$-12x + 5 - \mathbf{5} = 17 - \mathbf{5}$$ Subtract 5 from both sides.

$$-12x = 12$$ Do the subtractions: $5 - 5 = 0$ and $17 - 5 = 12$.

$$\frac{-12x}{\mathbf{-12}} = \frac{12}{\mathbf{-12}}$$ Divide both sides by -12.

$$x = -1$$ Do the divisions.

Caution

When checking solutions, always use the original equation.

Check:

$$-12x + 5 = 17$$

$$-12(-1) + 5 \stackrel{?}{=} 17 \quad \text{Substitute } -1 \text{ for } x.$$

$$12 + 5 \stackrel{?}{=} 17 \quad \text{Do the multiplication.}$$

$$17 = 17 \quad \text{True.}$$

The solution is -1.

Self Check 1 Solve: $8x - 13 = 43$.

EXAMPLE 2

Solve: $\frac{2x}{3} = -6$.

Solution On the left-hand side, x is multiplied by 2, and then that product is divided by 3. To solve this equation, we must undo these operations in the opposite order.

- To undo the division of 3, we multiply both sides by 3.
- To undo the multiplication by 2, we divide both sides by 2.

$$\frac{2x}{3} = -6$$

$$3\left(\frac{2x}{3}\right) = 3(-6) \quad \text{Multiply both sides by 3.}$$

$$2x = -18 \quad \text{On the left-hand side: } 3\left(\frac{2x}{3}\right) = \frac{\overset{1}{\cancel{3}} \cdot 2 \cdot x}{\underset{1}{\cancel{3}}} = 2x.$$

$$\frac{2x}{2} = \frac{-18}{2} \quad \text{Divide both sides by 2.}$$

$$x = -9 \quad \text{Do the divisions.}$$

Check:

$$\frac{2x}{3} = -6$$

$$\frac{2(-9)}{3} \stackrel{?}{=} -6 \quad \text{Substitute } -9 \text{ for } x.$$

$$\frac{-18}{3} \stackrel{?}{=} -6 \quad \text{Do the multiplication.}$$

$$-6 = -6 \quad \text{True.}$$

The solution is -9.

Self Check 2 Solve: $\frac{7h}{16} = -14$.

Recall that the product of a number and its **reciprocal,** or **multiplicative inverse,** is 1. We can use this fact to solve the equation from Example 2 in a different way. Since $\frac{2x}{3}$ is the same as $\frac{2}{3}x$, the equation can be written

$$\frac{2}{3}x = -6 \quad \text{Note that the coefficient of } x \text{ is } \frac{2}{3}.$$

When the coefficient of the variable is a fraction, we can isolate the variable by multiplying both sides by the reciprocal.

$$\frac{3}{2}\left(\frac{2}{3}x\right) = \frac{3}{2}(-6) \quad \text{Multiply both sides by the reciprocal of } \frac{2}{3}\text{, which is } \frac{3}{2}.$$

$$\left(\frac{3}{2} \cdot \frac{2}{3}\right)x = \frac{3}{2}(-6) \quad \text{On the left-hand side, regroup the factors.}$$

$$1x = -9 \quad \text{Do the multiplications: } \tfrac{3}{2} \cdot \tfrac{2}{3} = 1 \text{ and } \tfrac{3}{2}(-6) = -9.$$

$$x = -9 \quad \text{Simplify.}$$

EXAMPLE 3

ELEMENTARY Algebra $f(x)$ Now™

Solve: $-\frac{5}{8}m - 2 = -12$.

Solution The coefficient of m is the fraction $-\frac{5}{8}$. We proceed as follows.

- To undo the subtraction of 2, we add 2 to both sides.
- To undo the multiplication by $-\frac{5}{8}$, we multiply both sides by its reciprocal, $-\frac{8}{5}$.

$$-\frac{5}{8}m - 2 = -12$$

$$-\frac{5}{8}m - 2 + 2 = -12 + 2 \quad \text{Add 2 to both sides.}$$

$$-\frac{5}{8}m = -10 \quad \text{Do the additions: } -2 + 2 = 0 \text{ and } -12 + 2 = -10.$$

$$-\frac{8}{5}\left(-\frac{5}{8}m\right) = -\frac{8}{5}(-10) \quad \text{Multiply both sides by } -\frac{8}{5}.$$

$$m = 16 \quad \text{On the left-hand side: } -\tfrac{8}{5}(-\tfrac{5}{8})m = 1m = m.$$

$$\text{On the right-hand side: } -\tfrac{8}{5}(-10) = \frac{8 \cdot 2 \cdot \cancel{5}^{1}}{\cancel{5}_{1}} = 16.$$

Check that 16 is the solution.

Self Check 3 Solve: $\frac{7}{12}a - 6 = -27$.

EXAMPLE 4

Solve: $-0.2 = -0.8 - y$.

Solution We begin by eliminating -0.8 from the right-hand side. We can do this by adding 0.8 to both sides.

$$-0.2 = -0.8 - y$$

$$-0.2 + 0.8 = -0.8 - y + 0.8 \quad \text{Add 0.8 to both sides.}$$

$$0.6 = -y \quad \text{Do the additions.}$$

Since the term $-y$ has an understood coefficient of -1, the equation can be rewritten as $0.6 = -1y$. To isolate y, either multiply both sides or divide both sides by -1.

Method 1

$$\begin{aligned} 0.6 &= -1y \\ -1(0.6) &= -1(-1y) \quad \text{Multiply both sides by } -1. \\ -0.6 &= y \end{aligned}$$

Method 2

$$\begin{aligned} 0.6 &= -1y \\ \frac{0.6}{-1} &= \frac{-1y}{-1} \quad \text{To undo the multiplication by } -1\text{, divide both sides by } -1. \\ -0.6 &= y \end{aligned}$$

Check that -0.6 is the solution.

Self Check 4 Solve: $-6.6 - m = -2.7$.

SIMPLIFYING EXPRESSIONS TO SOLVE EQUATIONS

When solving equations, we should simplify the expressions that make up the left- and right-hand sides before applying any properties of equality. Often, that involves removing parentheses and/or combining like terms.

EXAMPLE 5

Solve: $3(k + 1) - 5k = 0$.

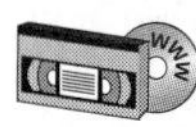

Solution

$$\begin{aligned} 3(k + 1) - 5k &= 0 \\ 3k + 3 - 5k &= 0 && \text{Distribute the multiplication by 3.} \\ -2k + 3 &= 0 && \text{Combine like terms: } 3k - 5k = -2k. \\ -2k + 3 - 3 &= 0 - 3 && \text{To undo the addition of 3, subtract 3 from both sides.} \\ -2k &= -3 && \text{Do the subtractions: } 3 - 3 = 0 \text{ and } 0 - 3 = -3. \\ \frac{-2k}{-2} &= \frac{-3}{-2} && \text{To undo the multiplication by } -2\text{, divide both sides by } -2. \\ k &= \frac{3}{2} && \text{Simplify: } \frac{-3}{-2} = \frac{3}{2}. \end{aligned}$$

Check:

$$\begin{aligned} 3(k + 1) - 5k &= 0 \\ 3\left(\frac{3}{2} + 1\right) - 5\left(\frac{3}{2}\right) &\stackrel{?}{=} 0 && \text{Substitute } \frac{3}{2} \text{ for } k. \\ 3\left(\frac{5}{2}\right) - 5\left(\frac{3}{2}\right) &\stackrel{?}{=} 0 && \text{Do the addition within the parentheses. Think of 1 as } \tfrac{2}{2} \text{ and then add: } \tfrac{3}{2} + \tfrac{2}{2} = \tfrac{5}{2}. \\ \frac{15}{2} - \frac{15}{2} &\stackrel{?}{=} 0 && \text{Do the multiplications.} \\ 0 &= 0 && \text{True.} \end{aligned}$$

The solution is $\frac{3}{2}$.

Self Check 5 Solve: $-5(x - 3) + 3x = 11$.

EXAMPLE 6

Solve: $3x - 15 = 4x + 36$.

ELEMENTARY Algebra f(x) Now™

Solution To solve for x, all the terms containing x must be on the same side of the equation. We can eliminate $3x$ from the left-hand side by subtracting $3x$ from both sides.

$$3x - 15 = 4x + 36$$

$$3x - 15 - 3x = 4x + 36 - 3x \quad \text{Subtract } 3x \text{ from both sides.}$$

$$-15 = x + 36 \quad \text{Combine like terms: } 3x - 3x = 0 \text{ and } 4x - 3x = x.$$

$$-15 - 36 = x + 36 - 36 \quad \text{To undo the addition of 36, subtract 36 from both sides.}$$

$$-51 = x \quad \text{Do the subtractions.}$$

Success Tip

We could have eliminated $4x$ from the right-hand side by subtracting $4x$ from both sides:

$$3x - 15 - 4x = 4x + 36 - 4x$$
$$-x - 15 = 36$$

However, it is usually easier to isolate the variable term on the side that will result in a *positive* coefficient.

Check:

$$3x - 15 = 4x + 36$$

$$3(-51) - 15 \stackrel{?}{=} 4(-51) + 36 \quad \text{Substitute } -51 \text{ for } x.$$

$$-153 - 15 \stackrel{?}{=} -204 + 36 \quad \text{Do the multiplications.}$$

$$-168 = -168 \quad \text{True.}$$

The solution is -51.

Self Check 6 Solve: $30 + 6n = 4n - 2$.

Equations are usually easier to solve if they don't involve fractions. We can use the multiplication property of equality to *clear* an equation of fractions by multiplying both sides of the equation by the least common denominator.

EXAMPLE 7

Solve: $\frac{x}{6} + \frac{5}{2} = \frac{1}{3}$.

ELEMENTARY Algebra f(x) Now™

Solution To clear the equation of fractions, we multiply both sides by the LCD, 6.

$$\frac{x}{6} + \frac{5}{2} = \frac{1}{3}$$

$$6\left(\frac{x}{6} + \frac{5}{2}\right) = 6\left(\frac{1}{3}\right)$$

Multiply both sides by the LCD of $\frac{x}{6}$, $\frac{5}{2}$, and $\frac{1}{3}$, which is 6. The parentheses are used to show that both $\frac{x}{6}$ and $\frac{5}{2}$ must be multiplied by 6.

$$6\left(\frac{x}{6}\right) + 6\left(\frac{5}{2}\right) = 6\left(\frac{1}{3}\right) \quad \text{On the left-hand side, distribute the multiplication by 6.}$$

$$x + 15 = 2 \quad \text{Simplify.}$$

$$x + 15 - 15 = 2 - 15 \quad \text{To undo the addition of 15, subtract 15 from both sides.}$$

$$x = -13$$

Success Tip

Before multiplying both sides of an equation by the LCD, enclose the left-hand side and enclose the right-hand side with parentheses:

$$\left(\frac{x}{6} + \frac{5}{2}\right) = \left(\frac{1}{3}\right)$$

Check that -13 is the solution.

Self Check 7 Solve: $\frac{x}{4} + \frac{1}{2} = -\frac{1}{8}$.

The preceding examples suggest the following strategy for solving equations.

Strategy for Solving Equations

1. Clear the equation of fractions.
2. Use the distributive property to remove parentheses, if necessary.
3. Combine like terms, if necessary.
4. Undo the operations of addition and subtraction to get the variables on one side and the constant terms on the other.
5. Undo the operations of multiplication and division to isolate the variable.
6. Check the result.

The next example illustrates an important point: not all of these steps are necessary to solve every equation.

EXAMPLE 8 Solve: $\dfrac{7m + 5}{5} = -4m + 1$.

ELEMENTARY Algebra $f(x)$ Now™

Solution

$$\frac{7m + 5}{5} = -4m + 1$$

$$5\left(\frac{7m + 5}{5}\right) = 5(-4m + 1)$$ Clear the equation of the fraction by multiplying both sides by 5.

$$7m + 5 = -20m + 5$$ On the left-hand side, divide out the common factor of 5 in the numerator and denominator. On the right-hand side, distribute the multiplication by 5.

$$7m + 5 + \mathbf{20m} = -20m + 5 + \mathbf{20m}$$ To eliminate the term $-20m$ on the right-hand side, add $20m$ to both sides.

$$27m + 5 = 5$$ Combine like terms: $7m + 20m = 27m$ and $-20m + 20m = 0$.

$$27m + 5 - \mathbf{5} = 5 - \mathbf{5}$$ To undo the addition of 5 on the left-hand side, subtract 5 from both sides.

$$27m = 0$$ Do the subtractions.

$$\frac{27m}{\mathbf{27}} = \frac{0}{\mathbf{27}}$$ To undo the multiplication by 27, divide both sides by 27.

$$m = 0$$ 0 divided by any nonzero number is 0.

Substitute 0 for m in $\dfrac{7m + 5}{5} = -4m + 1$ to check that the solution is 0.

Caution

Remember that when you multiply one side of an equation by a nonzero number, you must multiply the other side of the equation by the same number.

Self Check 8 Solve: $6c + 2 = \dfrac{-c + 18}{9}$.

IDENTITIES AND CONTRADICTIONS

Each of the equations in Examples 1 through 8 had exactly one solution. However, some equations have no solutions while others have infinitely many solutions.

An equation that is true for all values of its variable is called an **identity.** An example is

$x + x = 2x$ If we substitute -10 for x, we get the true statement $-20 = -20$. If we substitute 0 for x, we get $0 = 0$. If we substitute 7 for x, we get $14 = 14$, and so on.

It is apparent that in an identity, we can replace x with any number and the equation will be true. We say that $x + x = 2x$ has infinitely many solutions.

An equation that is not true for any values of its variable is called a **contradiction.** An example is

$x = x + 1$ No number is 1 greater than itself.

We say that $x = x + 1$ has no solutions.

EXAMPLE 9

Solve: $3(x + 8) + 5x = 2(12 + 4x)$.

ELEMENTARY Algebra f(x) Now™

Solution

$3(x + 8) + 5x = 2(12 + 4x)$

$3x + 24 + 5x = 24 + 8x$ Distribute the multiplication by 3 and by 2.

$8x + 24 = 24 + 8x$ Combine like terms: $3x + 5x = 8x$. Note that the sides of the equation are identical.

$8x + 24 - \mathbf{8x} = 24 + 8x - \mathbf{8x}$ To eliminate the term $8x$ on the right-hand side, subtract $8x$ from both sides.

$24 = 24$ Combine like terms: $8x - 8x = 0$.

The terms involving x drop out and the result, $24 = 24$, is true. This means that any number substituted for x in the original equation will yield a true statement. Therefore, every real number is a solution and this equation is an identity.

Self Check 9 Solve: $3(x + 5) - 4(x + 4) = -x - 1$.

EXAMPLE 10

Solve: $3(d + 7) - d = 2(d + 10)$.

Solution

$3(d + 7) - d = 2(d + 10)$

$3d + 21 - d = 2d + 20$ Distribute the multiplication by 3 and by 2.

$2d + 21 = 2d + 20$ Combine like terms: $3d - d = 2d$.

$2d + 21 - \mathbf{2d} = 2d + 20 - \mathbf{2d}$ To eliminate the term $2d$ on the right-hand side, subtract $2d$ from both sides.

$21 = 20$ Combine like terms: $2d - 2d = 0$.

The Language of Algebra

Contradiction is a form of the word *contradict,* meaning conflicting ideas. During a trial, evidence might be introduced that *contradicts* the testimony of a witness.

The terms involving d drop out and the result, $21 = 20$, is false. This means that any number that is substituted for x in the original equation will yield a false statement. This equation has no solution and it is a contradiction.

Self Check 10 Solve: $-4(c - 3) + 2c = 2(10 - c)$.

Answers to Self Checks **1.** 7 **2.** −32 **3.** −36 **4.** −3.9 **5.** 2 **6.** −16 **7.** $-\frac{5}{2}$ **8.** 0 **9.** all real numbers; this equation is an identity **10.** no solution; this equation is a contradiction

2.4 STUDY SET  ELEMENTARY Algebra f(x) Now™

VOCABULARY **Fill in the blanks.**

1. An equation is a statement indicating that two expressions are ______.
2. To ______ an equation means to find all of the values of the variable that make the equation a true statement.
3. After solving an equation, we can check our result by substituting that value for the variable in the ______ equation.
4. The product of a number and its ________ is 1.
5. An equation that is true for all values of its variable is called an ________.
6. An equation that is not true for any values of its variable is called a __________.

CONCEPTS **Fill in the blanks.**

7. To solve the equation $2x - 7 = 21$, we first undo the ________ of 7 by adding 7 to both sides. Then we undo the __________ by 2 by dividing both sides by 2.
8. To solve the equation $\frac{x}{2} + 3 = 5$, we first undo the ________ of 3 by subtracting 3 from both sides. Then we undo the ________ by 2 by multiplying both sides by 2.
9. To solve $\frac{s}{3} + \frac{1}{4} = -\frac{1}{2}$, we can clear the equation of the fractions by ________ both sides by 12.
10. To solve $15d = -2(3d + 7) + 2$, we begin by using the ________ property to remove parentheses.
11. One method of solving $-\frac{4}{5}x = 8$ is to multiply both sides of the equation by the reciprocal of $-\frac{4}{5}$. What is the reciprocal of $-\frac{4}{5}$?
12. a. Combine like terms on the left-hand side of $6x - 8 - 8x = -24$.
    b. Combine like terms on the right-hand side of $5a + 1 = 9a + 16 + a$.
    c. Combine like terms on both sides of $12 - 3r + 5r = -8 - r - 2$.
13. Find the LCD of the fractions in the equation $\frac{x}{3} - \frac{4}{5} = \frac{1}{2}$.
14. Simplify: $20\left(\frac{3}{5}x\right)$.
15. What must you multiply both sides of $\frac{2}{3} - \frac{b}{2} = -\frac{4}{3}$ by to clear the equation of fractions?
16. a. Simplify: $3x + 5 - x$.
    b. Solve: $3x + 5 - x = 9$.
    c. Evaluate $3x + 5 - x$ for $x = 9$.
    d. Check: Is -1 a solution of $3x + 5 - x = 9$?

NOTATION **Complete the solution.**

17.
$$\begin{aligned} 2x - 7 &= 21 \\ 2x - 7 + \square &= 21 + \square \\ 2x &= \square \\ \frac{2x}{\square} &= \frac{28}{\square} \\ x &= 14 \end{aligned}$$

18. Identify the like terms on the right-hand side of $5(9a + 1) = a + 6 - 7a$.
19. Fill in the blanks.
    a. $-x = \square x$. b. $\frac{3x}{5} = \square x$.
20. What does the symbol $\stackrel{?}{=}$ mean?

PRACTICE **Solve each equation and check all solutions.**

21. $2x + 5 = 17$
22. $3x - 5 = 13$
23. $5q - 2 = 23$
24. $4p + 3 = 43$
25. $-33 = 5t + 2$
26. $-55 = 3w + 5$
27. $20 = -x$
28. $10 = -a$
29. $-g = -4$
30. $-u = -20$
31. $1.2 - x = -1.7$
31. $0.6 = 4.1 - x$
33. $-3p + 7 = -3$
34. $-2r + 8 = -1$
35. $0 - 2y = 8$
36. $0 - 7x = -21$
37. $-8 - 3c = 0$
38. $-5 - 2d = 0$
39. $\frac{5}{6}k = 10$
40. $\frac{2c}{5} = 2$
41. $-\frac{7}{16}h = 21$
42. $-\frac{5}{8}h = 15$
43. $-\frac{t}{3} + 2 = 6$
44. $\frac{x}{5} - 5 = -12$
45. $2(-3) + 4y = 14$
46. $4(-1) + 3y = 8$

47. $7(0) - 4y = 17$

48. $3x - 4(0) = 2$

49. $10.08 = 4(0.5x + 2.5)$

50. $-3.28 = 8(1.5y - 0.5)$

51. $-(4 - m) = -10$

52. $-(6 - t) = -12$

53. $15s + 8 - s = 7 + 1$

54. $-7t - 9 + t = -10 + 1$

55. $-3(2y - 2) - y = 5$

56. $-(3a + 1) + a = 2$

57. $3x - 8 - 4x - 7x = -2 - 8$

58. $-6t - 7t - 5t - 1 = 12 - 3$

59. $4(5b) + 2(6b - 1) = -34$

60. $2(3x) + 5(3x - 1) = 58$

61. $9(x + 11) + 5(13 - x) = 0$

62. $-(19 - 3s) - (8s + 1) = 35$

63. $60r - 50 = 15r - 5$

64. $100f - 75 = 50f + 75$

65. $8y - 3 = 4y + 15$

66. $7 + 3w = 4 + 9w$

67. $5x + 7.2 = 4x$

68. $3x + 2.5 = 2x$

69. $8y + 328 = 4y$

70. $9y + 369 = 6y$

71. $15x = x$

72. $7y = 8y$

73. $3(a + 2) = 2(a - 7)$

74. $9(t - 1) = 6(t + 2) - t$

75. $2 - 3(x - 5) = 4(x - 1)$

76. $2 - (4x + 7) = 3 + 2(x + 2)$

77. $\frac{x + 5}{3} = 11$

78. $\frac{x + 2}{13} = 3$

79. $\frac{y}{6} + \frac{y}{4} = -1$

80. $\frac{x}{3} + \frac{x}{4} = -2$

81. $-\frac{2}{9} = \frac{5x}{6} - \frac{1}{3}$

82. $\frac{2}{3} = -\frac{2x}{3} + \frac{3}{4}$

83. $\frac{2}{3}y + 2 = \frac{1}{5} + y$

84. $\frac{2}{5}x + 1 = \frac{1}{3} + x$

85. $-\frac{3}{4}n + 2n = \frac{1}{2}n + \frac{13}{3}$

86. $-\frac{5}{6}n - 3n = \frac{1}{3}n + \frac{11}{9}$

87. $\frac{10 - 5s}{3} = s$

88. $\frac{40 - 8s}{5} = -2s$

89. $\frac{5(1 - x)}{6} = -x$

90. $\frac{3(14 - u)}{8} = -3u$

91. $\frac{3(d - 8)}{4} = \frac{2(d + 1)}{3}$

92. $\frac{3(c - 2)}{2} = \frac{2(2c + 3)}{5}$

93. $\frac{1}{2}(x + 3) + \frac{3}{4}(x - 2) = x + 1$

94. $\frac{3}{2}(t + 2) + \frac{1}{6}(t + 2) = 2 + t$

95. $8x + 3(2 - x) = 5(x + 2) - 4$

96. $5(x + 2) = 5x - 2$

97. $-3(s + 2) = -2(s + 4) - s$

98. $21(b - 1) + 3 = 3(7b - 6)$

99. $2(3z + 4) = 2(3z - 2) + 13$

100. $x + 7 = \frac{2x + 6}{2} + 4$

101. $4(y - 3) - y = 3(y - 4)$

102. $5(x + 3) - 3x = 2(x + 8)$

Solve each equation.

103. $\frac{h}{709} - 23{,}898 = -19{,}678$

104. $9.35 - 1.4y = 7.32 + 1.5y$

WRITING

105. To solve $3x - 4 = 5x + 1$, one student began by subtracting $3x$ from both sides. Another student solved the same equation by first subtracting $5x$ from both sides. Will the students get the same solution? Explain why or why not.

106. What does it mean to clear an equation such as $\frac{1}{4} + \frac{x}{2} = \frac{3}{8}$ of the fractions?

107. Explain the error in the following solution.

Solve: $2x + 4 = 30$.

$$2x + 4 = 30$$
$$\frac{2x}{2} + 4 = \frac{30}{2}$$
$$x + 4 = 15$$
$$x + 4 - 4 = 15 - 4$$
$$x = 11$$

108. Write an equation that is an identity. Explain why every number is a solution.

REVIEW

109. Subtract: $-8 - (-8)$.

110. Add: $\frac{1}{8} + \frac{1}{8}$.

111. Multiply: $\frac{1}{8} \cdot \frac{1}{8}$.

112. Divide: $\frac{0.8}{8}$.

113. Simplify: $8x + 8 + 8x - 8$.

114. Evaluate: -1^8.

CHALLENGE PROBLEMS

115. In this section, we discussed equations that have no solution, one solution, and an infinite number of solutions. Do you think an equation could have exactly two solutions? If so, give an example.

116. The equation $4x - 3y = 5$ contains two different variables. Solve the equation by determining a value of x and a value for y that make the equation true.

2.5 Formulas

- Formulas from Business
- Formulas from Science
- Formulas from Geometry
- Solving for a Specified Variable

A **formula** is an equation that states a known relationship between two or more variables. Formulas are used in fields such as economics, physical education, biology, automotive repair, and nursing. In this section, we will consider formulas from business, science, and geometry.

FORMULAS FROM BUSINESS

A formula for retail price: To make a profit, a merchant must sell an item for more than he or she paid for it. The price at which the merchant sells the product, called the **retail price,** is the sum of what the item cost the merchant plus the **markup.**

Retail price = cost + markup

Using r to represent the retail price, c the cost, and m the markup, we can write this formula as

$$r = c + m$$

A formula for profit: The **profit** a business makes is the difference between the **revenue** (the money it takes in) and the costs.

Profit = revenue − costs

Using p to represent the profit, r the revenue, and c the costs, we can write this formula as

$$p = r - c$$

EXAMPLE 1

Charitable giving. In 2001, the Salvation Army received \$2.31 billion in revenue. Of that amount, \$1.92 billion went directly toward program services. Find the 2001 administrative costs of the organization.

Solution The charity collected \$2.31 billion. We can think of the \$1.92 billion that was spent on programs as profit. We need to find the administrative costs, c.

$$p = r - c$$ This is the formula for profit.

$$\mathbf{1.92} = \mathbf{2.31} - c$$ Substitute 1.92 for p and 2.31 for r.

$$1.92 - \mathbf{2.31} = 2.31 - c - \mathbf{2.31}$$ To eliminate 2.31, subtract 2.31 from both sides.

$$-0.39 = -c$$ Do the subtractions.

$$\frac{-0.39}{\mathbf{-1}} = \frac{-c}{\mathbf{-1}}$$ Since $-c = -1c$, divide (or multiply) both sides by -1.

$$0.39 = c$$

In 2001, the Salvation Army had administrative costs of \$0.39 billion.

Self Check 1 A PTA spaghetti dinner made a profit of \$275.50. If the cost to host the dinner was \$1,235, how much revenue did it generate?

The Language of Algebra

The word *annual* means occurring once a year. An *annual* interest rate is the interest rate paid per year.

A formula for simple interest: When money is borrowed, the lender expects to be paid back the amount of the loan plus an additional charge for the use of the money. The additional charge is called **interest.** When money is deposited in a bank, the depositor is paid for the use of the money. The money the deposit earns is also called interest.

Interest is computed in two ways: either as **simple interest** or as **compound interest.** Simple interest is the product of the principal (the amount of money that is invested, deposited, or borrowed), the annual interest rate, and the length of time in years.

Interest = principal · rate · time

Using I to represent the simple interest, P the principal, r the annual interest rate, and t the time in years, we can write this formula as

$$I = Prt$$

EXAMPLE 2

ELEMENTARY Algebra $f(x)$ Now™

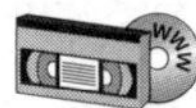

Retirement income. One year after investing \$15,000, a retired couple received a check for \$1,125 in interest. Find the interest rate their money earned that year.

Solution The couple invested \$15,000 (the principal) for 1 year (the time) and made \$1,125 (the interest). We need to find the annual interest rate, r.

Caution

When using the formula $I = Prt$, always write the interest rate r (which is given as a percent) as a decimal or fraction before performing any calculations.

$$I = Prt$$ This is the formula for simple interest.

$$\mathbf{1{,}125} = \mathbf{15{,}000}r(\mathbf{1})$$ Substitute 1,125 for I, 15,000 for P, and 1 for t.

$$1{,}125 = 15{,}000r$$ Simplify.

$$\frac{1{,}125}{\mathbf{15{,}000}} = \frac{15{,}000r}{\mathbf{15{,}000}}$$ To solve for r, undo the multiplication by 15,000 by dividing both sides by 15,000.

$$0.075 = r$$ Do the divisions.

$$7.5\% = r$$ To write 0.075 as a percent, multiply 0.075 by 100 by moving the decimal point two places to the right and inserting a % symbol.

The couple received an annual rate of 7.5% that year on their investment.

	P	· r	· t =	I
Investment	15,000	0.075	1	1,125

Self Check 2 A father loaned his daughter \$12,200 at a 2% annual simple interest rate for a down payment on a house. If the interest on the loan amounted to \$610, for how long was the loan?

FORMULAS FROM SCIENCE

A formula for distance traveled: If we know the average rate (of speed) at which we will be traveling and the time we will be traveling at that rate, we can find the distance traveled.

$$\text{Distance} = \text{rate} \cdot \text{time}$$

Using d to represent the distance, r the average rate, and t the time, we can write this formula as

$$d = rt$$

EXAMPLE 3

ELEMENTARY Algebra f(x) Now™

Finding the rate. As they migrate from the Bering Sea to Baja California, gray whales swim for about 20 hours each day, covering a distance of approximately 70 miles. Estimate their average swimming rate in miles per hour (mph).

Solution Since the distance d is 70 miles and the time t is 20 hours, we substitute 70 for d and 20 for t in the formula $d = rt$ and solve for r.

Caution

When using the formula $d = rt$, make sure the units are consistent. For example, if the rate is given in miles per hour, the time must be expressed in hours.

$$d = rt$$

$$70 = r(20) \quad \text{Substitute 70 for } d \text{ and 20 for } t.$$

$$\frac{70}{20} = \frac{20r}{20} \quad \text{To undo the multiplication by 20, divide both sides by 20.}$$

$$3.5 = r$$

	r	$\cdot\ t$	$= d$
Gray whale	3.5	20	70

The whales' average swimming rate is 3.5 mph.

Self Check 3 An elevator travels at an average rate of 288 feet per minute. How long will it take the elevator to climb 30 stories, a distance of 360 feet?

A formula for converting temperatures: A message board flashes two temperature readings, one in degrees Fahrenheit and one in degrees Celsius. The Fahrenheit scale is used in the American system of measurement. The Celsius scale is used in the metric system. The formula that relates a Fahrenheit temperature F to a Celsius temperature C is:

$$C = \frac{5}{9}(F - 32)$$

EXAMPLE 4

Convert the temperature shown on the City Savings sign to degrees Fahrenheit.

ELEMENTARY Algebra f(x) Now™

Solution Since the temperature C in degrees Celsius is 30°, we substitute 30 for C in the formula and solve for F.

$$C = \frac{5}{9}(F - 32)$$ This is the temperature conversion formula.

$$30 = \frac{5}{9}(F - 32)$$ Substitute 30 for C.

$$\frac{9}{5} \cdot 30 = \frac{9}{5} \cdot \frac{5}{9}(F - 32)$$ To undo the multiplication by $\frac{5}{9}$, multiply both sides by the reciprocal of $\frac{5}{9}$.

$$54 = F - 32$$ Do the multiplications.

$$54 + 32 = F - 32 + 32$$ To undo the subtraction of 32, add 32 to both sides.

$$86 = F$$

30°C is equivalent to 86°F.

Self Check 4 Change −175°C, the temperature on Saturn, to degrees Fahrenheit.

FORMULAS FROM GEOMETRY

To find the **perimeter** of a geometric figure, we find the distance around the figure by computing the sum of the lengths of its sides. Perimeter is measured in linear units, such as inches, feet, yards, and meters. The **area** of a figure is the amount of surface that it encloses. Area is measured in square units, such as square inches, square feet, square yards, and square meters (denoted as in.^{2}, ft^2, yd^2, and m^2, respectively). Many formulas for perimeter and area are shown inside the front cover of the book.

EXAMPLE 5

ELEMENTARY Algebra $f(x)$ Now™

The flag of Eritrea, a country in east Africa, is shown below. **a.** Find the perimeter of the flag. **b.** Find the area of the red triangular region of the flag.

Solution

a. The perimeter of the flag is given by the formula $P = 2l + 2w$, where l is the length and w is the width of the rectangle.

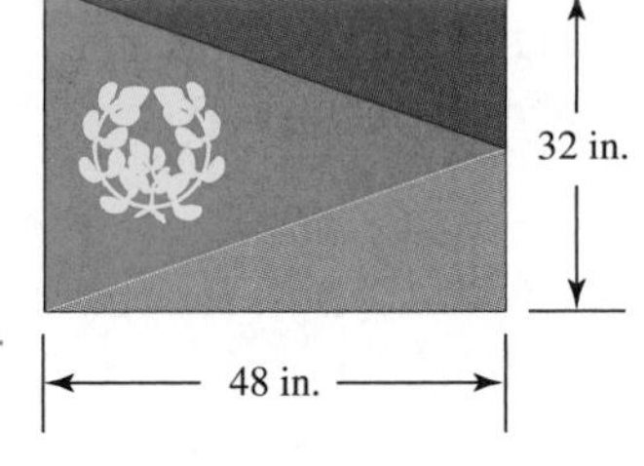

$$P = 2l + 2w$$

$$P = 2(48) + 2(32)$$ Substitute 48 for l and 32 for w.

$$= 96 + 64$$

$$= 160$$

The perimeter of the flag is 160 inches.

b. The area of a triangle is given by the formula $A = \frac{1}{2}bh$, where b is the length of the base and h is the height. With the triangle positioned as it is, the base is 32 inches and the height is 48 inches.

$$A = \frac{1}{2}bh$$

$$A = \frac{1}{2}(32)(48)$$ Substitute 32 for b and 48 for h.

$$= 16(48)$$ Multiply.

$$= 768$$

The area of the red triangular region of the flag is 768 in.^{2}.

Self Check 5 **a.** Find the perimeter of a square with sides 6 inches long. **b.** Find the area of a triangle with base of 8 meters and height of 13 meters.

Formulas involving circles: A **circle** is the set of all points on a flat surface that are a fixed distance from a point called its **center.** A segment drawn from the center to a point on the circle is called a **radius.** Since a **diameter** of a circle is a segment passing through the center that joins two points on the circle, the diameter D of a circle is twice as long as its radius r. The perimeter of a circle is called its **circumference** C.

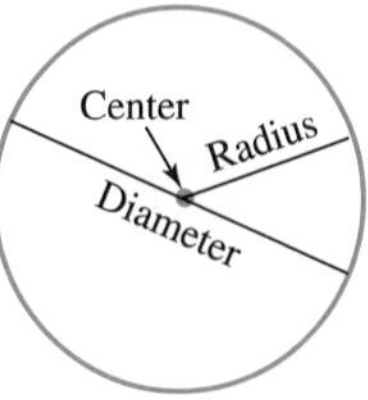

EXAMPLE 6

To the nearest tenth, find the area of a circle with a diameter of 14 feet.

Solution The radius is one-half the diameter, or 7 feet. To find the area we substitute 7 for r in the formula for the area of a circle and proceed as follows.

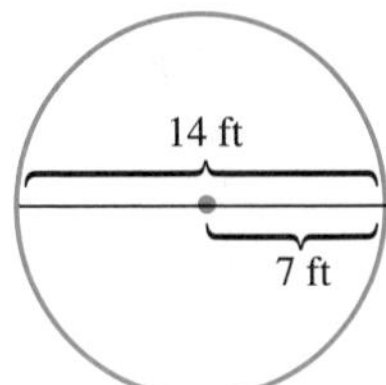

$A = \pi r^2$ πr^2 means $\pi \cdot r^2$.

$A = \pi(7)^2$ Substitute 7 for r.

$= 49\pi$ Evaluate the exponential expression: $7^2 = 49$. The exact area is 49π ft^2.

≈ 153.93804 Using a scientific calculator, enter these numbers and press these keys: 49 $\times$ π $=$. If you do not have a calculator, use 3.14 as an approximation of π.

To the nearest tenth, the area is 153.9 ft^2.

Notation

We found the area of the circle to be *exactly* 49π ft^2. This form of the answer is convenient, but not very informative. To get a better feel for the area, we computed $49 \cdot \pi$ and rounded to the nearest tenth.

Self Check 6 To the nearest hundredth, find the circumference of the circle.

The **volume** of a three-dimensional geometric solid is the amount of space it encloses. Volume is measured in cubic units, such as cubic inches, cubic feet, and cubic meters (denoted as in.^{3}, ft^3, and m^3, respectively). Many formulas for volume are shown inside the front cover of the book.

EXAMPLE 7

Finding volumes. To the nearest tenth, find the volume of the cylinder.

ELEMENTARY Algebra f(x) Now™

Solution Since the radius of a circle is one-half its diameter, the radius of the cylinder is $\frac{1}{2}(6 \text{ cm}) = 3$ cm. The height of the cylinder is 12 cm. We substitute 3 for r and 12 for h in the formula for volume and proceed as follows.

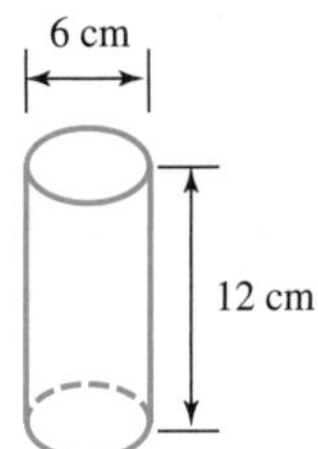

$V = \pi r^2 h$ $\pi r^2 h$ means $\pi \cdot r^2 \cdot h$.

$V = \pi(3)^2(12)$ Substitute 3 for r and 12 for h.

$= \pi(9)(12)$ Evaluate the exponential expression.

$= 108\pi$ Multiply. The exact volume is 108π cm^3.

≈ 339.2920066 Use a calculator.

To the nearest tenth, the volume is 339.3 cubic centimeters. This can be written as 339.3 cm^3.

Self Check 7 Find the volume of each figure: **a.** a rectangular solid with length 7 inches, width 12 inches, and height 15 inches; and **b.** a cone whose base has radius 12 meters and whose height is 9 meters. Give the answer to the nearest tenth.

SOLVING FOR A SPECIFIED VARIABLE

Suppose we wish to find the bases of several triangles whose areas and heights are known. It could be tedious to substitute values for A and h into the formula and then repeatedly solve the formula for b. A better way is to solve the formula $A = \frac{1}{2}bh$ for b first, and then substitute values for A and h and compute b directly.

To **solve an equation for a variable** means to isolate that variable on one side of the equation, with all other terms on the opposite side.

EXAMPLE 8

Solve $A = \frac{1}{2}bh$ for b.

Solution

$A = \frac{1}{2}bh$ — We must isolate b on one side of the equation.

$2 \cdot A = 2 \cdot \frac{1}{2}bh$ — To clear the equation of the fraction, multiply both sides by 2.

$2A = bh$ — Simplify.

$\frac{2A}{h} = \frac{bh}{h}$ — To undo the multiplication by h, divide both sides by h.

$\frac{2A}{h} = b$ — On the right-hand side, remove the common factor of h: $\frac{b\not{h}^{1}}{\not{h}_{1}} = b$.

$b = \frac{2A}{h}$ — Reverse the sides to write b on the left.

Self Check 8 Solve $V = lwh$ for w.

EXAMPLE 9

Solve $P = 2l + 2w$ for l.

Solution

$P = 2l + 2w$ — We must isolate l on one side of the equation.

$P - 2w = 2l + 2w - 2w$ — To undo the addition of $2w$, subtract $2w$ from both sides.

$P - 2w = 2l$ — Combine like terms: $2w - 2w = 0$.

$\frac{P - 2w}{2} = \frac{2l}{2}$ — To undo the multiplication by 2, divide both sides by 2.

$\frac{P - 2w}{2} = l$ — Simplify the right-hand side.

We can write the result as $l = \frac{P - 2w}{2}$.

Caution

In Example 9, do not try to simplify the result this way:

$$l = \frac{P - \overset{1}{\not{2}}w}{\underset{1}{\not{2}}}$$

This step is incorrect because 2 is not a factor of the entire numerator.

Self Check 9 Solve $P = 2l + 2w$ for w.

EXAMPLE 10

In Chapter 3, we will work with equations that involve the variables x and y, such as $2y - 4 = 3x$. Solve this equation for y.

Solution

We must isolate y on one side of the equation.

$$2y - 4 = 3x$$

$$2y - 4 + 4 = 3x + 4$$ To undo the subtraction of 4, add 4 to both sides.

$$2y = 3x + 4$$ Do the addition.

$$\frac{2y}{2} = \frac{3x + 4}{2}$$ To undo the multiplication by 2, divide both sides by 2.

$$y = \frac{3x}{2} + \frac{4}{2}$$ On the right-hand side, write $\frac{3x+4}{2}$ as the sum of two fractions with like denominators, $\frac{3x}{2}$ and $\frac{4}{2}$.

$$y = \frac{3}{2}x + 2$$ Write $\frac{3x}{2}$ as $\frac{3}{2}x$. Simplify: $\frac{4}{2} = 2$.

Self Check 10 Solve $3y + 12 = x$ for y.

EXAMPLE 11

Solve $V = \pi r^2 h$ for r^2.

Solution

We must isolate r^2 on one side of the equation.

$$V = \pi r^2 h$$

$$\frac{V}{\pi h} = \frac{\pi r^2 h}{\pi h}$$ To undo the multiplication by π and h on the right-hand side, divide both sides by πh.

$$\frac{V}{\pi h} = r^2$$ Remove the common factors of π and h: $\frac{\overset{1}{\cancel{\pi}} r^2 \overset{1}{\cancel{h}}}{\underset{1}{\cancel{\pi}} \underset{1}{\cancel{h}}} = r^2$.

$$r^2 = \frac{V}{\pi h}$$ Reverse the sides of the equation so that r^2 is on the left.

Caution

When solving for a variable, that variable must be isolated on one side of the equation.

Self Check 11 Solve $a^2 + b^2 = c^2$ for b^2.

Answers to Self Checks **1.** \$1,510.50 **2.** 2.5 years **3.** 1.25 minutes **4.** −283°F **5. a.** 24 in., **b.** 52 m^2 **6.** 43.98 ft **7. a.** 1,260 in.^{3}, **b.** 1,357.2 m^3 **8.** $w = \frac{V}{lh}$ **9.** $w = \frac{P - 2l}{2}$ **10.** $y = \frac{1}{3}x - 4$ **11.** $b^2 = c^2 - a^2$

2.5 STUDY SET

ELEMENTARY Algebra f(x) Now™

VOCABULARY **Fill in the blanks.**

1. A ________ is an equation that is used to state a known relationship between two or more variables.
2. The ________ of a three-dimensional geometric solid is the amount of space it encloses.
3. The distance around a geometric figure is called its ________.
4. A ________ is the set of all points on a flat surface that are a fixed distance from a point called its center.
5. A line segment drawn from the center of a circle to a point on the circle is called a ________.
6. The amount of surface that is enclosed by a geometric figure is called its ________.
7. The perimeter of a circle is called its ________.
8. A line segment passing through the center of a circle and connecting two points on the circle is called a ________.

CONCEPTS

9. Use variables to write the formula relating the following:

a. Time, distance, rate

b. Markup, retail price, cost

c. Costs, revenue, profit

d. Interest rate, time, interest, principal

e. Circumference, radius

10. Complete the table.

Principal ·	rate ·	time =	interest
\$2,500	5%	2 yr	
\$15,000	4.8%	1 yr	

11. Complete the table to find how far light and sound travel in 60 seconds. (*Hint:* mi/sec means miles per second.)

	Rate ·	time =	distance
Light	186,282 mi/sec	60 sec	
Sound	1,088 ft/sec	60 sec	

12. Give the name of each figure.

a.

b.

c.

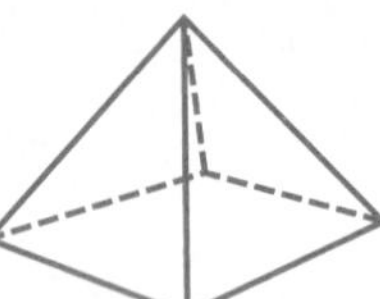

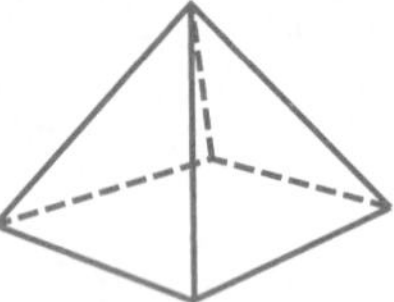

d.

e.

f.

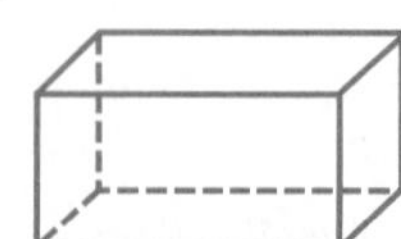

13. Tell which concept, perimeter, circumference, area, or volume, should be used to find the following:

a. The amount of storage in a freezer

b. How far a bicycle tire rolls in one revolution

c. The amount of land making up the Sahara Desert

d. The distance around a Monopoly game board

14. Tell which unit of measurement, ft, ft^2, or ft^3, would be appropriate when finding the following:

a. The amount of storage inside a safe

b. The ground covered by a sleeping bag lying on the floor

c. The distance the tip of an airplane propeller travels in one revolution

d. The size of the trunk of a car

15. a. Write an expression for the perimeter of the figure.

b. Write an expression for the area of the figure.

16. WHEELCHAIRS

a. Find the diameter of the rear wheel.

b. Find the radius of the front wheel.

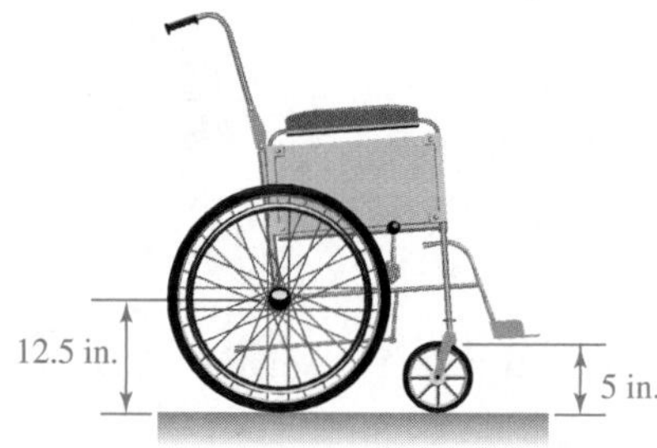

NOTATION **Complete the solution.**

17. Solve $Ax + By = C$ for y.

$$Ax + By = C$$
$$Ax + By - \square = C - \square$$
$$\square = C - Ax$$
$$\frac{By}{\square} = \frac{C - Ax}{\square}$$
$$y = \frac{C - Ax}{B}$$

18. Enter the missing formulas in each table.

a. The table contains information about an investment earning simple interest.

	? · ? · ? = ?			
Certificate of deposit	\$3,500	0.04	1 yr	\$140

b. The table contains information about a trip made by a cross-country skier.

	? · ? = ?		
Skier	3 mph	2 hr	6 mi

19. a. Approximate π to the nearest hundredth.

b. What does 98π mean?

c. In the formula for the volume of a cylinder, $V = \pi r^2 h$, what does r represent? What does h represent?

20. a. What does ft^2 mean?

b. What does in.^3 mean?

PRACTICE Use a formula to solve each problem.

21. SWIMMING In 1930, a man swam down the Mississippi River from Minneapolis to New Orleans, a total of 1,826 miles. He was in the water for 742 hours. To the nearest tenth, what was his average swimming rate?

22. ROSE PARADE Rose Parade floats travel down the 5.5-mile-long parade route at a rate of 2.5 mph. How long will it take a float to complete the route if there are no delays?

23. HOLLYWOOD Figures for the summer of 1998 showed that the movie *Saving Private Ryan* had U.S. box-office receipts of \$190 million. What were the production costs to make the movie if, at that time, the studio had made a \$125 million profit?

24. SERVICE CLUBS After expenses of \$55.15 were paid, a Rotary Club donated \$875.85 in proceeds from a pancake breakfast to a local health clinic. How much did the pancake breakfast gross?

25. ENTREPRENEURS To start a mobile dog-grooming service, a woman borrowed \$2,500. If the loan was for 2 years and the amount of interest was \$175, what simple interest rate was she charged?

26. BANKING Three years after opening an account that paid 6.45% annually, a depositor withdrew the \$3,483 in interest earned. How much money was left in the account?

27. METALLURGY Change 2,212°C, the temperature at which silver boils, to degrees Fahrenheit. Round to the nearest degree.

28. LOW TEMPERATURES Cryobiologists freeze living matter to preserve it for future use. They can work with temperatures as low as −270°C. Change this to degrees Fahrenheit.

29. VALENTINE'S DAY Find the markup on a dozen roses if a florist buys them wholesale for \$12.95 and sells them for \$37.50.

30. STICKER PRICES The factory invoice for a minivan shows that the dealer paid \$16,264.55 for the vehicle. If the sticker price of the van is \$18,202, how much over factory invoice is the sticker price?

31. YO-YOS How far does a yo-yo travel during one revolution of the "around the world" trick if the length of the string is 21 inches?

32. HORSES A horse trots in a perfect circle around its trainer at the end of a 28-foot-long rope. How far does the horse travel as it circles the trainer once?

Solve each formula for the given variable.

33. $E = IR$; for R

34. $d = rt$; for t

35. $V = lwh$; for w

36. $I = Prt$; for r

37. $C = 2\pi r$; for r

38. $V = \pi r^2 h$; for h

39. $A = \dfrac{Bh}{3}$; for h

40. $C = \dfrac{Rt}{7}$; for R

41. $w = \dfrac{s}{f}$; for f

42. $P = \dfrac{ab}{c}$; for c

43. $P = a + b + c$; for b

44. $a + b + c = 180$; for a

45. $T = 2r + 2t$; for r

46. $y = mx + b$; for x

47. $Ax + By = C$; for x

48. $A = P + Prt$; for t

49. $K = \dfrac{1}{2}mv^2$; for m

50. $V = \dfrac{1}{3}\pi r^2 h$; for h

51. $A = \frac{a + b + c}{3}$; for c

52. $x = \frac{a + b}{2}$; for b

53. $2E = \frac{T - t}{9}$; for t

54. $D = \frac{C - s}{n}$; for s

55. $s = 4\pi r^2$; for r^2

56. $E = mc^2$; for c^2

57. $Kg = \frac{wv^2}{2}$; for v^2

58. $c^2 = a^2 + b^2$; for a^2

59. $V = \frac{4}{3}\pi r^3$; for r^3

60. $A = \frac{\pi r^2 S}{360}$; for r^2

61. $\frac{M}{2} - 9.9 = 2.1B$; for M

62. $\frac{G}{0.5} + 16r = -8t$; for G

63. $S = 2\pi rh + 2\pi r^2$; for h

64. $c = bn + 16t^2$; for t^2

65. $3x + y = 9$; for y

66. $-5x + y = 4$

67. $3y - 9 = x$; for y

68. $5y - 25 = x$; for y

69. $4y + 16 = -3x$; for y

70. $6y + 12 = -5x$; for y

71. $A = \frac{1}{2}h(b + d)$; for b

72. $C = \frac{1}{4}s(t - d)$; for t

73. $\frac{7}{8}c + w = 9$; for c

74. $\frac{3}{4}m - t = 5b$; for m

APPLICATIONS

75. PROPERTIES OF WATER The boiling point and the freezing point of water are to be given in both degrees Celsius and degrees Fahrenheit on the thermometer. Find the missing degree measures.

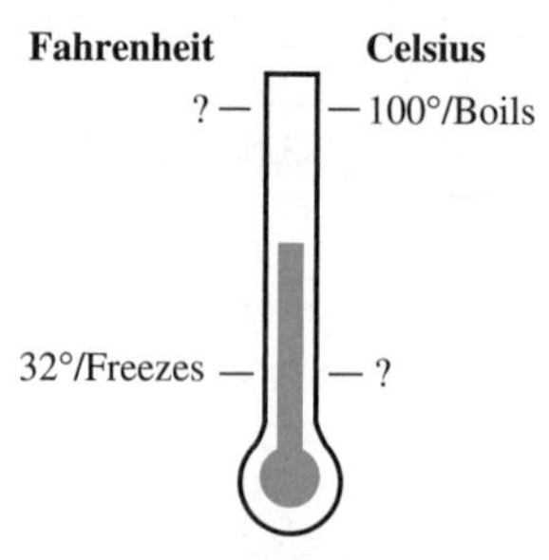

76. SPEED LIMITS Several state speed limits for trucks are shown. At each of these speeds, how far would a truck travel in $2\frac{1}{2}$ hours?

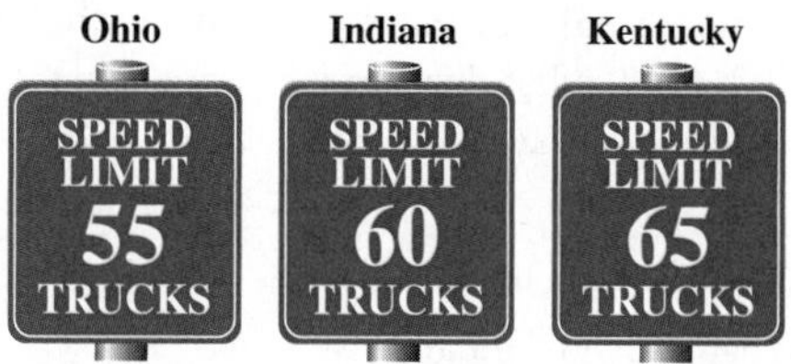

77. AVON PRODUCTS Complete the financial statement from the Hoover's Online business Web site.

Quarterly financials income statement (dollar amounts in millions except per share amounts)	**Quarter ending Dec 02**	**Quarter ending Sep 02**
Revenue	1,854.1	1,463.4
Cost of goods sold	679.5	506.5
Gross profit		

78. CREDIT CARDS The finance charge section of a person's credit card statement says, "annual percentage rate (APR) is 19.8%." Determine how much finance charges (interest) the card owner would have to pay if the account's average balance for the year was \$2,500.

79. CARPENTRY Find the perimeter and area of the truss.

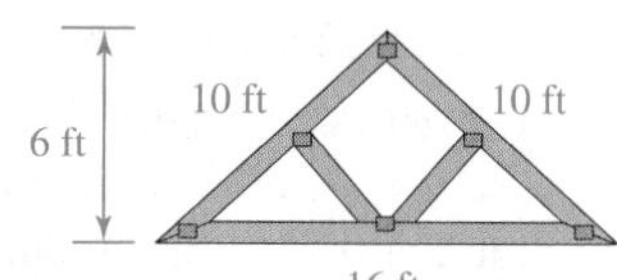

80. CAMPERS Find the area of the window of the camper shell on the next page.

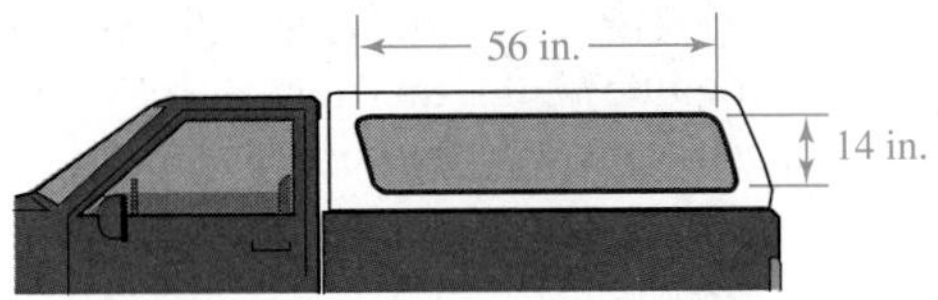

81. ARCHERY To the nearest tenth, find the circumference and area of the target.

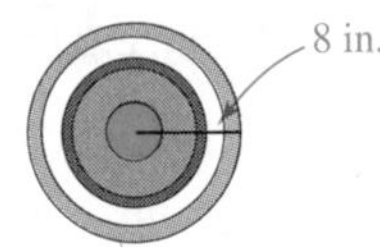

82. GEOGRAPHY The circumference of the Earth is about 25,000 miles. Find its diameter to the nearest mile.

83. LANDSCAPING Find the perimeter and the area of the redwood trellis.

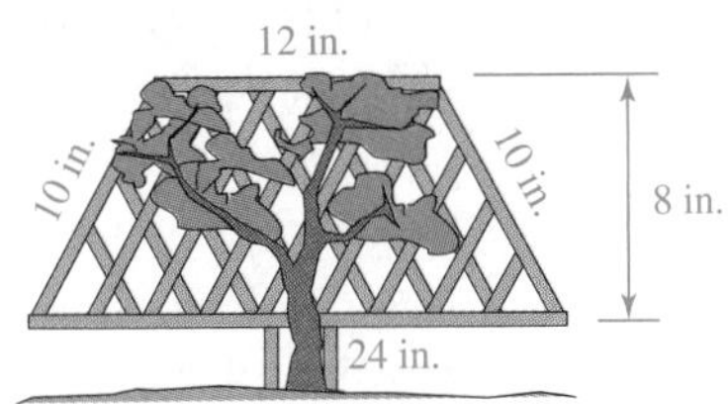

84. HAMSTER HABITATS Find the amount of space in the tube.

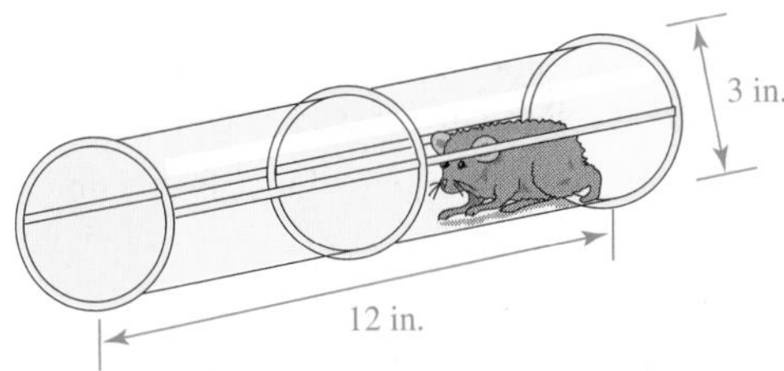

85. MEMORIALS The Vietnam Veterans Memorial is a black granite wall recognizing the more than 58,000 Americans who lost their lives or remain missing. Find the total area of the two triangular-shaped surfaces on which the names are inscribed.

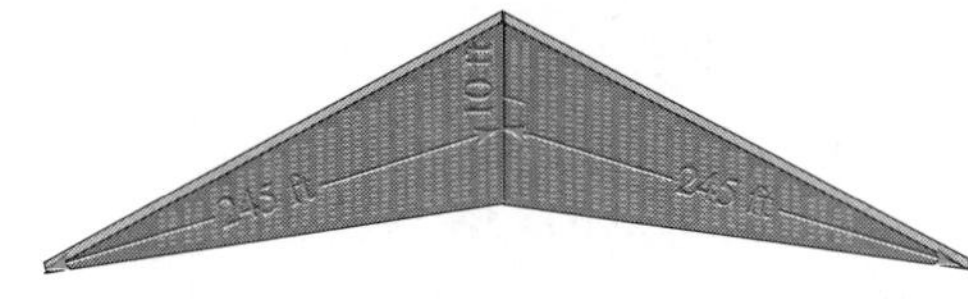

86. SIGNAGE Find the perimeter and area of the service station sign.

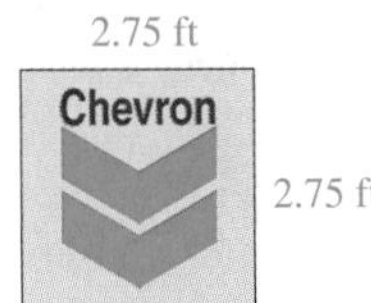

87. RUBBER MEETS THE ROAD A sport truck tire has the road surface footprint shown here. Estimate the perimeter and area of the tire's footprint.

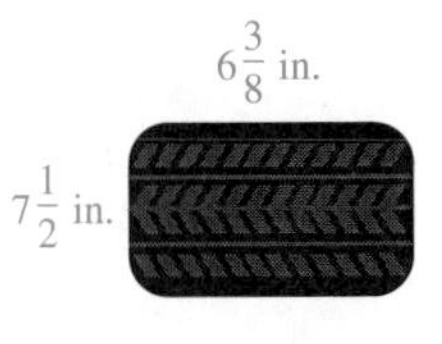

88. SOFTBALL The strike zone in fast-pitch softball is between the batter's armpit and the top of her knees, as shown. Find the area of the strike zone.

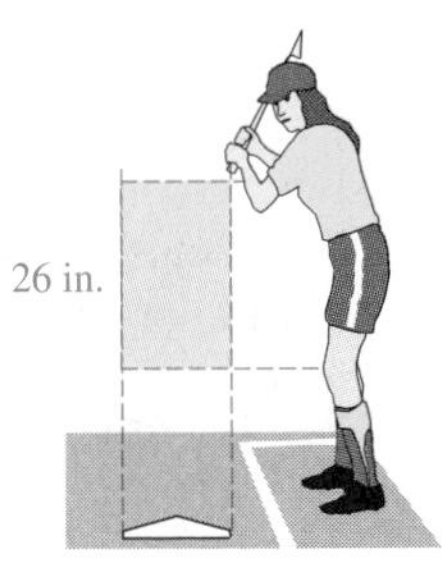

89. FIREWOOD Find the area on which the wood is stacked and the volume the cord of firewood occupies.

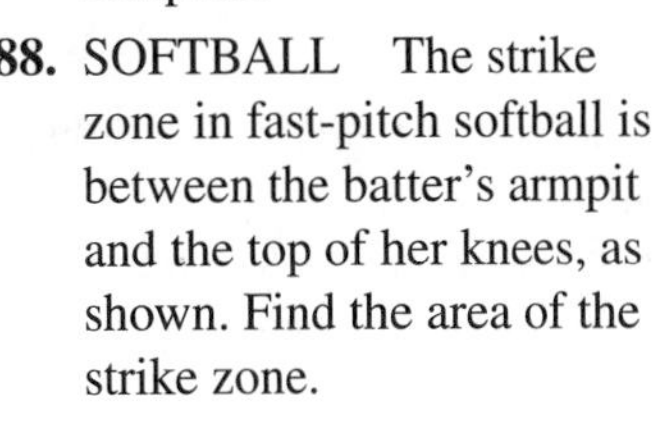

90. NATIVE AMERICAN DWELLINGS The teepees constructed by the Blackfoot Indians were cone-shaped tents made of long poles and animal hide, about 10 feet high and about 15 feet across at the ground. Estimate the volume of a teepee with these dimensions, to the nearest cubic foot.

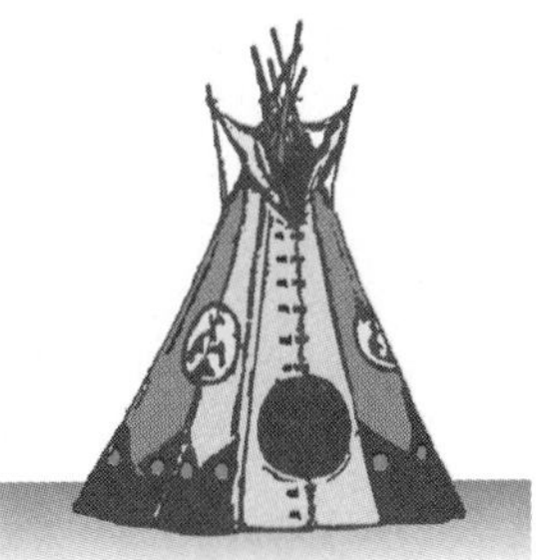

91. IGLOOS During long journeys, some Canadian Inuit (Eskimos) built winter houses of snow blocks stacked in the dome shape shown on the next page. Estimate the volume of an igloo having an interior height of 5.5 feet to the nearest cubic foot.

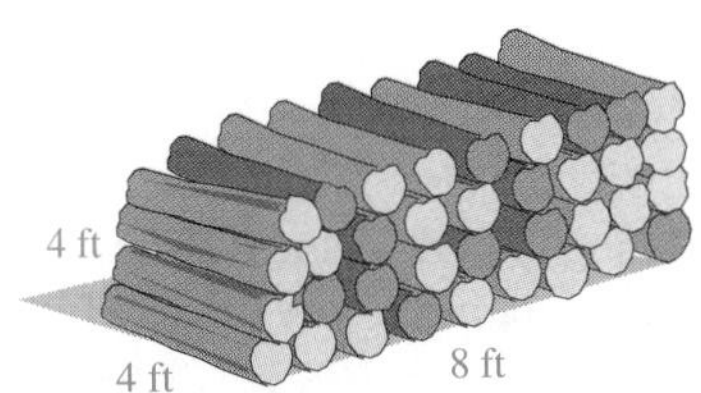

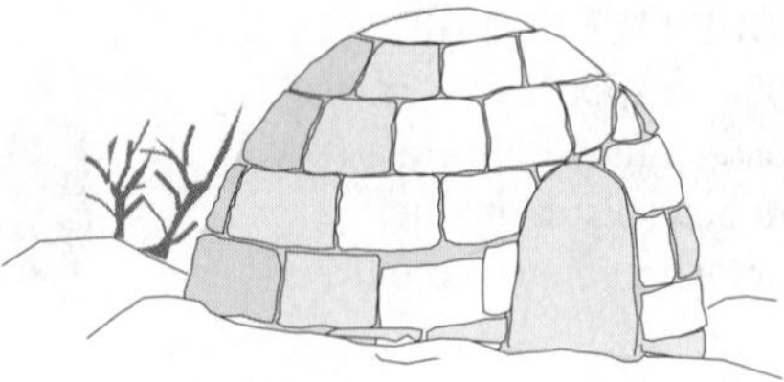

92. PYRAMIDS The Great Pyramid at Giza in northern Egypt is one of the most famous works of architecture in the world. Use the information in the illustration to find the volume to the nearest cubic foot.

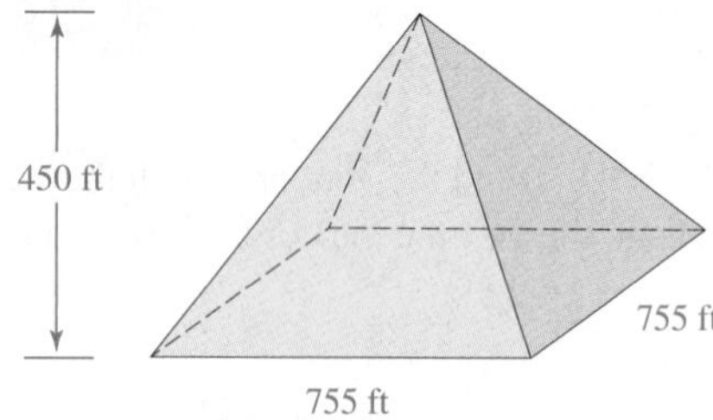

93. BARBECUING Use the fact that the fish is 18 inches long to find the area of the barbecue grill to the nearest square inch.

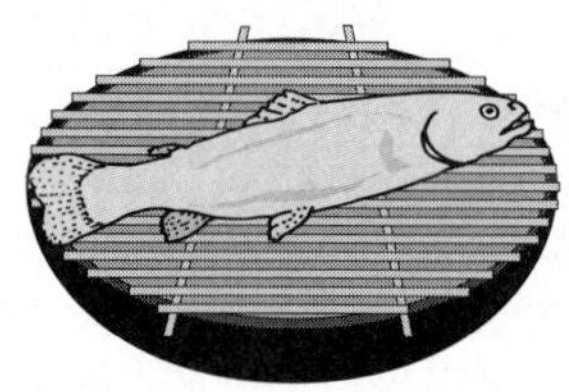

94. SKATEBOARDING A half-pipe ramp used for skateboarding is in the shape of a semicircle with a radius of 8 feet. To the nearest tenth of a foot, what is the length of the arc that the skateboarder travels on the ramp?

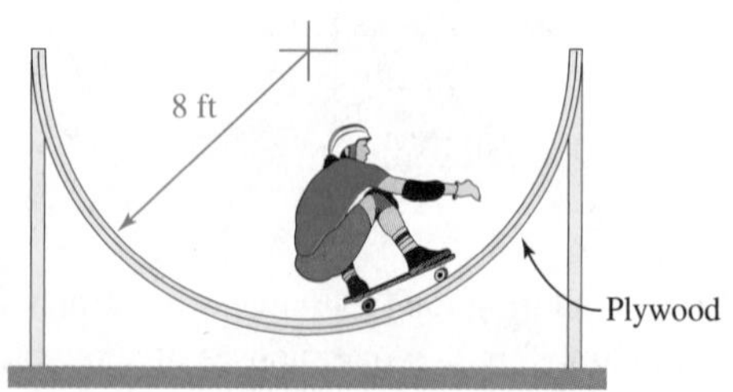

95. PULLEYS The approximate length L of a belt joining two pulleys of radii r and R feet with centers D feet apart is given by the formula

$$L = 2D + 3.25(r + R)$$

Solve the formula for D.

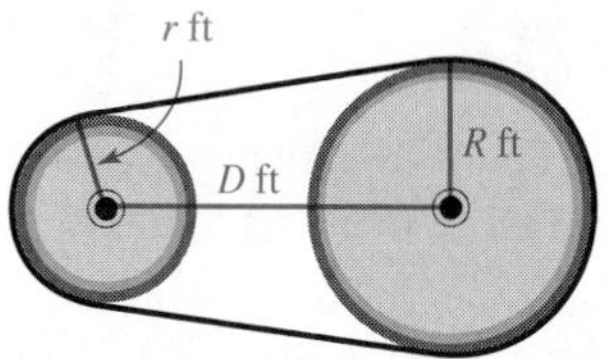

96. THERMODYNAMICS The Gibbs free-energy function is given by $G = U - TS + pV$. Solve this formula for the pressure p.

WRITING

97. After solving $A = B + C + D$ for B, a student compared her answer with that at the back of the textbook.

Student's answer: $B = A - C - D$
Book's answer: $B = A - D - C$

Could this problem have two different-looking answers? Explain why or why not.

98. Suppose the volume of a cylinder is 28 cubic feet. Explain why it is incorrect to express the volume as 28^3 ft.

99. Explain the difference between what perimeter measures and what area measures.

100. Explain the error made below.

$$y = \frac{3x + \overset{1}{\cancel{2}}}{\underset{1}{\cancel{2}}}$$

REVIEW

101. Find 82% of 168.

102. 29.05 is what percent of 415?

103. What percent of 200 is 30?

104. A woman bought a coat for \$98.95 and some gloves for \$7.95. If the sales tax was 6%, how much did the purchase cost her?

CHALLENGE PROBLEMS

105. In mathematics, letters from the Greek alphabet are often used as variables. Solve the following equation for α (read as "alpha"), the first letter of the Greek alphabet.

$$-7(\alpha - \beta) - (4\alpha - \theta) = \frac{\alpha}{2}$$

106. When a car of mass m collides with a wall, the energy of the collision is given by the formula $E = \frac{1}{2}mv^2$. Compare the energy of two collisions: a car striking a wall at 30 mph, and at 60 mph.

2.6 More about Problem Solving

- Finding More than One Unknown
- Solving Number–Value Problems
- Solving Uniform Motion Problems
- Solving Geometric Problems
- Solving Investment Problems
- Solving Mixture Problems

In this section, we will solve several types of problems using the five-step problem-solving strategy.

FINDING MORE THAN ONE UNKNOWN

EXAMPLE 1

California coastline. The first part of California's magnificent 17-Mile Drive begins at the Pacific Grove entrance and continues to Seal Rock. It is 1 mile longer than the second part of the drive, which extends from Seal Rock to the Lone Cypress. The final part of the tour winds through the Monterey Peninsula, eventually returning to the entrance. This part of the drive is 1 mile longer than four times the length of the second part. How long is each part of 17-Mile Drive?

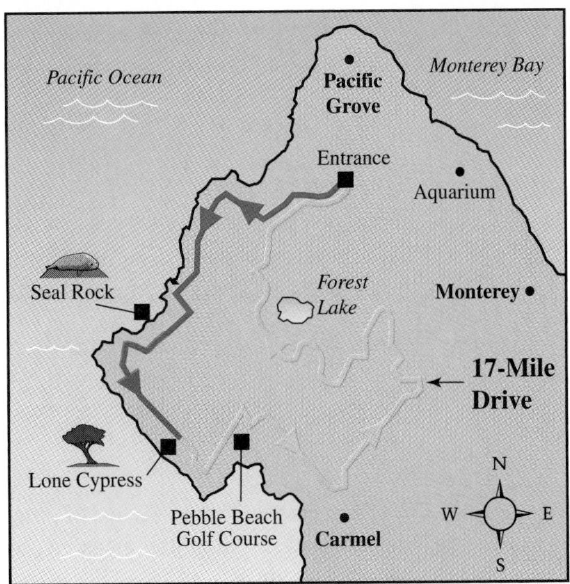

Analyze the Problem The drive is composed of three parts. We need to find the length of each part. We can straighten out the winding 17-Mile Drive and model it with a line segment.

Form an Equation Since the lengths of the first part and of the third part of the drive are related to the length of the second part, we will let x represent the length of that part. We then express the other lengths in terms of x. Let

$$x = \text{the length of the second part of the drive}$$
$$x + 1 = \text{the length of the first part of the drive}$$
$$4x + 1 = \text{the length of the third part of the drive}$$

Caution

For this problem, one common mistake is to let

x = the length of each part of the drive

The three parts of the drive have different lengths; x cannot represent three different distances.

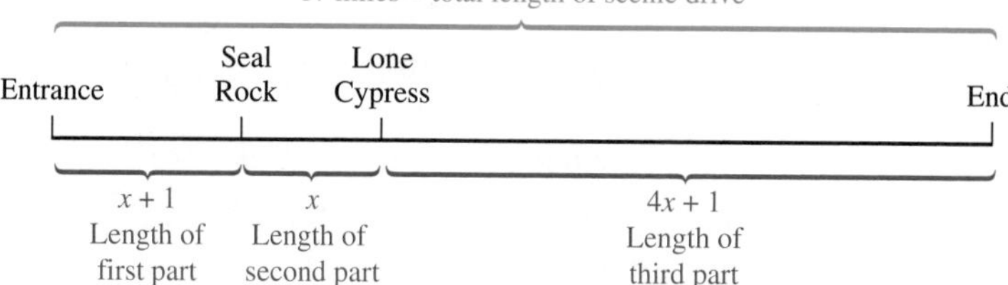

The sum of the lengths of the three parts must be 17 miles.

The length of part 1	plus	the length of part 2	plus	the length of part 3	equals	the total length.
$x + 1$	$+$	x	$+$	$4x + 1$	$=$	17

Solve the Equation

$$x + 1 + x + 4x + 1 = 17$$

$$6x + 2 = 17 \quad \text{Combine like terms: } x + x + 4x = 6x \text{ and } 1 + 1 = 2.$$

$$6x = 15 \quad \text{To undo the addition of 2, subtract 2 from both sides.}$$

$$\frac{6x}{6} = \frac{15}{6} \quad \text{To undo the multiplication by 6, divide both sides by 6.}$$

$$x = 2.5$$

Recall that x represents the length of the second part of the drive. To find the lengths of the first and third parts, we evaluate $x + 1$ and $4x + 1$ for $x = 2.5$.

First part of drive	***Third part of drive***	
$x + 1 = 2.5 + 1$	$4x + 1 = 4(2.5) + 1$	Substitute 2.5 for x.
$= 3.5$	$= 11$	

State the Conclusion The first part of the drive is 3.5 miles long, the second part is 2.5 miles long, and the third part is 11 miles long.

Check the Result Since 3.5 mi + 2.5 mi + 11 mi = 17 mi, the answers check.

SOLVING GEOMETRIC PROBLEMS

EXAMPLE 2

A gardener wants to use 62 feet of fencing bought at a garage sale to enclose a rectangular-shaped garden. Find the dimensions of the garden if its length is to be 4 feet longer than twice its width.

The Language of Algebra
Dimensions are measurements of length, width, and thickness. Perhaps you have viewed a 3-D (dimensional) movie wearing the special glasses.

Analyze the Problem A sketch is often helpful when solving problems about geometric figures. We know that the length of the garden is to be 4 feet longer than twice the width. We also know that its perimeter is to be 62 feet. Recall that the perimeter of a rectangle is given by the formula $P = 2l + 2w$.

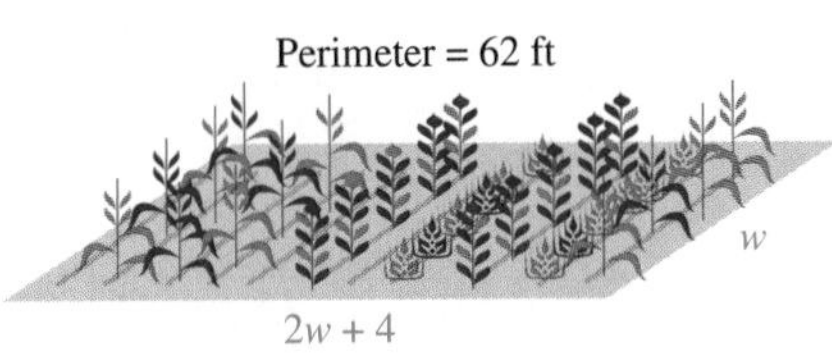

Form an Equation Since the length of the garden is given in terms of the width, we will let $w =$ the width of the garden. Then the length $= 2w + 4$.

2	times	the length	plus	2	times	the width	is	the perimeter.
2	$\cdot$	$(2w + 4)$	$+$	2	$\cdot$	w	$=$	62

Solve the Equation

$2(2w + 4) + 2w = 62$ Be sure to write the parentheses so that the entire expression $2w + 4$ is multiplied by 2.

$4w + 8 + 2w = 62$ Use the distributive property to remove parentheses.

$6w + 8 = 62$ Combine like terms: $4w + 2w = 6w$.

$6w = 54$ To undo the addition of 8, subtract 8 from both sides.

$w = 9$ To undo the multiplication by 6, divide both sides by 6.

State the Conclusion The width of the garden is 9 feet. Since $2w + 4 = 2(9) + 4 = 22$, the length is 22 feet.

Check the Result If the garden has a width of 9 feet and a length of 22 feet, its length is 4 feet longer than twice the width $(2 \cdot 9 + 4 = 22)$. Since its perimeter is $2 \cdot 22 \text{ ft} + 2 \cdot 9 \text{ ft} = 62 \text{ ft}$, the answers check.

EXAMPLE 3

Isosceles triangles. If the vertex angle of an isosceles triangle is 56°, find the measure of each base angle.

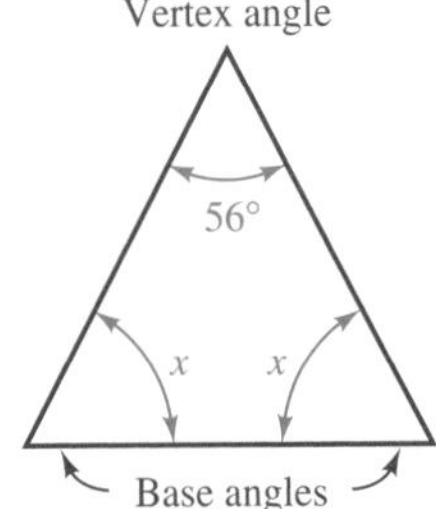

Analyze the Problem An **isosceles triangle** has two sides of equal length, which meet to form the **vertex angle.** In this case, the measurement of the vertex angle is 56°. We can sketch the triangle as shown. The **base angles** opposite the equal sides are also equal. We need to find their measure.

Form an Equation If we let $x =$ the measure of one base angle, the measure of the other base angle is also x. Since the sum of the angles of any triangle is 180°, the sum of the base angles and the vertex angle is 180°. We can use this fact to form the equation.

One base angle	plus	the other base angle	plus	the vertex angle	is	180°.
x	$+$	x	$+$	56	$=$	180

Solve the Equation

$$x + x + 56 = 180$$
$$2x + 56 = 180 \quad \text{Combine like terms: } x + x = 2x.$$
$$2x = 124 \quad \text{To undo the addition of 56, subtract 56 from both sides.}$$
$$x = 62 \quad \text{To undo the multiplication by 2, divide both sides by 2.}$$

State the Conclusion The measure of each base angle is 62°.

Check the Result Since 62° + 62° + 56° = 180°, the answer checks.

SOLVING NUMBER–VALUE PROBLEMS

Some problems deal with items that have monetary value. In these problems, we must distinguish between the *number of* and the *value of* the items. For problems of this type, we will use the fact that

Number · value = total value

EXAMPLE 4

Dining area improvements. A restaurant owner needs to purchase some tables, chairs, and dinner plates for the dining area of her establishment. She plans to buy four chairs and four plates for each new table. She also plans to buy 20 additional plates in case of breakage. If a table costs \$100, a chair \$50, and a plate \$5, how many of each can she buy if she takes out a loan for \$6,500 to pay for the new items?

Analyze the Problem We know the *value* of each item: Tables cost \$100, chairs cost \$50, and plates cost \$5 each. We need to find the *number* of tables, chairs, and plates she can purchase for \$6,500.

Form an Equation The number of chairs and plates she needs depends on the number of tables she buys. So we let t = the number of tables to be purchased. Since every table requires four chairs and four plates, she needs to order $4t$ chairs. Because 20 additional plates are needed, she should order $(4t + 20)$ plates. We can organize the facts of the problem in a table.

	Number ·	Value =	Total value
Tables	t	100	$100t$
Chairs	$4t$	50	$50(4t)$
Plates	$4t + 20$	5	$5(4t + 20)$

Enter this information first. Multiply to get each of these entries.

We can use the information in the last column of the table to form an equation.

The value of the tables	plus	the value of the chairs	plus	the value of the plates	equals	the total value of the purchase.
$100t$	$+$	$50(4t)$	$+$	$5(4t + 20)$	$=$	6,500

Solve the Equation

$$100t + 50(4t) + 5(4t + 20) = 6{,}500$$
$$100t + 200t + 20t + 100 = 6{,}500 \quad \text{Do the multiplications.}$$
$$320t + 100 = 6{,}500 \quad \text{Combine like terms.}$$
$$320t = 6{,}400 \quad \text{Subtract 100 from both sides.}$$
$$t = 20 \quad \text{Divide both sides by 320.}$$

To find the number of chairs and plates to buy, we evaluate $4t$ and $4t + 20$ for $t = 20$.

Chairs: $4t = 4(\mathbf{20}) = 80$ ***Plates:*** $4t + 20 = 4(\mathbf{20}) + 20 = 100$ Substitute 20 for t.

State the Conclusion The owner needs to buy 20 tables, 80 chairs, and 100 plates.

Check the Result The total value of 20 tables is 20($100) = $2,000, the total value of 80 chairs is 80($50) = $4,000, and the total value of 100 plates is 100($5) = $500. Because the total purchase is $2,000 + $4,000 + $500 = $6,500, the answers check.

SOLVING INVESTMENT PROBLEMS

To find the amount of simple interest I an investment earns, we use the formula

$$I = Prt$$

where P is the principal, r is the annual rate, and t is the time in years.

EXAMPLE 5

Paying tuition. A college student invested the $12,000 inheritance he received and decided to use the annual interest earned to pay his tuition costs of $945. The highest rate offered by a bank at that time was 6% annual simple interest. At this rate, he could not earn the needed $945, so he invested some of the money in a riskier, but more profitable, investment offering a 9% return. How much did he invest at each rate?

Analyze the Problem We know that $12,000 was invested for 1 year at two rates: 6% and 9%. We are asked to find the amount invested at each rate so that the total return would be $945.

Form an Equation Let x = the amount invested at 6%. Then $12{,}000 - x$ = the amount invested at 9%. To organize the facts of the problem, we enter the principal, rate, time, and interest earned in a table.

	P ·	r ·	t =	I
Bank	x	0.06	1	$0.06x$
Riskier investment	$12{,}000 - x$	0.09	1	$0.09(12{,}000 - x)$

Enter this information first. Multiply to get each of these entries.

We can use the information in the last column of the table to form an equation.

The interest earned at 6%	plus	the interest earned at 9%	equals	the total interest.
$0.06x$	$+$	$0.09(12{,}000 - x)$	$=$	945

Solve the Equation

$$0.06x + 0.09(12{,}000 - x) = 945$$

$$100[0.06x + 0.09(12{,}000 - x)] = 100(945)$$ Multiply both sides by 100 to clear the equation of decimals.

$$100(0.06x) + 100(0.09)(12{,}000 - x) = 100(945)$$ Distribute the multiplication by 100.

$$6x + 9(12{,}000 - x) = 94{,}500$$ Do the multiplications by 100.

$$6x + 108{,}000 - 9x = 94{,}500$$ Use the distributive property.

$$-3x + 108{,}000 = 94{,}500$$ Combine like terms.

$$-3x = -13{,}500$$ Subtract 108,000 from both sides.

$$x = 4{,}500$$ Divide both sides by -3.

Success Tip

We can *clear an equation of decimals* by multiplying both sides by a power of 10. In Example 5, we multiply the hundredths by 100 to move each decimal point two places to the right:

$100(0.06) = 6 \quad 100(0.09) = 9$

State the Conclusion The student invested \$4,500 at 6% and \$12,000 − \$4,500 = \$7,500 at 9%.

Check the Result The first investment earned 0.06(\$4,500), or \$270. The second earned 0.09(\$7,500), or \$675. The total return was \$270 + \$675 = \$945. The answers check.

SOLVING UNIFORM MOTION PROBLEMS

If we know the rate r at which we will be traveling and the time t we will be traveling at that rate, we can find the distance d traveled by using the formula

$$d = rt$$

EXAMPLE 6

Coast Guard rescues. A cargo ship, heading into port, radios the Coast Guard that it is experiencing engine trouble and that its speed has dropped to 3 knots (3 nautical miles per hour). Immediately, a Coast Guard cutter leaves port and speeds at a rate of 25 knots directly toward the disabled ship, which is 21 nautical miles away. How long will it take the Coast Guard to reach the cargo ship?

Analyze the Problem We know the *rate* of each ship (25 knots and 3 knots), and we know that they must close a *distance* of 21 nautical miles between them. We don't know the *time* it will take to do this.

Form an Equation Let t = the time it takes for the ships to meet. Using $d = rt$, we find that $25t$ represents the distance traveled by the Coast Guard cutter and $3t$ represents the distance traveled by the cargo ship. We can organize the facts of the problem in a table.

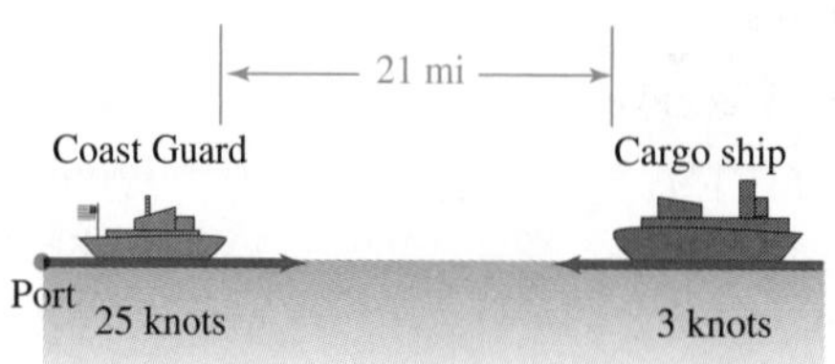

	r	$\cdot\ t$	$= d$
Coast Guard cutter	25	t	$25t$
Cargo ship	3	t	$3t$

Enter this information first.

We can use the information in the last column of the table to form an equation.

The distance the Coast Guard cutter travels	plus	the distance the cargo ship travels	equals	the initial distance between the two ships.
$25t$	$+$	$3t$	$=$	21

Solve the Equation

$$25t + 3t = 21$$

$$28t = 21 \quad \text{Combine like terms.}$$

$$t = \frac{21}{28} \quad \text{Divide both sides by 28.}$$

$$t = \frac{3}{4} \quad \text{Simplify the fraction.}$$

State the Conclusion The ships will meet in $\frac{3}{4}$ hr, or 45 minutes.

Check the Result In $\frac{3}{4}$ hr, the Coast Guard cutter travels $25 \cdot \frac{3}{4} = \frac{75}{4}$ nautical miles, and the cargo ship travels $3 \cdot \frac{3}{4} = \frac{9}{4}$ nautical miles. Together, they travel $\frac{75}{4} + \frac{9}{4} = \frac{84}{4} = 21$ nautical miles. This is the initial distance between the ships; the answer checks.

SOLVING MIXTURE PROBLEMS

We now discuss how to solve mixture problems. In the first type, a *liquid mixture* of a desired strength is made from two solutions with different concentrations. In the second type, a *dry mixture* of a specified value is created from two differently priced components.

EXAMPLE 7

ELEMENTARY Algebra f(x) Now™

Mixing solutions. A chemistry experiment calls for a 30% sulfuric acid solution. If the lab supply room has only 50% and 20% sulfuric acid solutions, how much of each should be mixed to obtain 12 liters of a 30% acid solution?

Analyze the Problem The 50% solution is too strong and the 20% solution is too weak. We must find how much of each should be combined to obtain 12 liters of a 30% solution.

Form an Equation If x = the number of liters of the 50% solution used in the mixture, the remaining $(12 - x)$ liters must be the 20% solution.

The amount of pure acid in each solution is given by

Amount of solution · strength of solution = amount of pure acid

A table is helpful in organizing the facts of the problem.

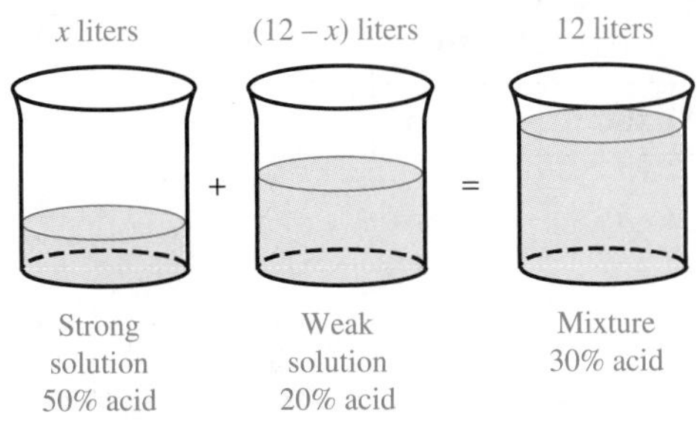

	Amount	· Strength	= Amount of acid
Strong	x	0.50	$0.50x$
Weak	$12 - x$	0.20	$0.20(12 - x)$
Mixture	12	0.30	$12(0.30)$

Enter this information first. Multiply to get each of these entries.

We can use the information in the last column of the table to form an equation.

The acid in the 50% solution	plus	the acid in the 20% solution	equals	the acid in the final mixture.
$0.50x$	$+$	$0.20(12 - x)$	$=$	$12(0.30)$

Solve the Equation

$$0.50x + 0.20(12 - x) = 12(0.30) \quad 50\% = 0.50,\ 20\% = 0.20, \text{ and } 30\% = 0.30.$$

$$0.5x + 2.4 - 0.2x = 3.6 \quad \text{Distribute the multiplication by 0.20.}$$

$$0.3x + 2.4 = 3.6 \quad \text{Combine like terms.}$$

$$0.3x = 1.2 \quad \text{Subtract 2.4 from both sides.}$$

$$\frac{0.3x}{0.3} = \frac{1.2}{0.3} \quad \text{To undo the multiplication by 0.3, divide both sides by 0.3.}$$

$$x = 4$$

Success Tip

We could begin by multiplying both sides of the equation by 10 to clear it of the decimals.

State the Conclusion 4 liters of 50% solution and $12 - 4 = 8$ liters of 20% solution should be used.

Check the Result
Acid in 4 liters of the 50% solution: $0.50(4) = 2.0$ liters.
Acid in 8 liters of the 20% solution: $0.20(8) = 1.6$ liters.
(3.6 liters)
Acid in 12 liters of the 30% mixture: $0.30(12) = 3.6$ liters. The answers check.

EXAMPLE 8

Snack foods. Because cashews priced at \$9 per pound were not selling, a produce clerk decided to combine them with less expensive peanuts and sell the mixture for \$7 per pound. How many pounds of peanuts, selling at \$6 per pound, should be mixed with 50 pounds of cashews to obtain such a mixture?

Analyze the Problem We need to determine how many pounds of peanuts to mix with 50 pounds of cashews to obtain a mixture worth \$7 per pound.

Form an Equation Let $x =$ the number of pounds of peanuts to use in the mixture. Since 50 pounds of cashews will be combined with the peanuts, the mixture will weigh $50 + x$ pounds. The value of the mixture and of each of the components of the mixture is given by

$$\text{Amount} \cdot \text{price} = \text{total value}$$

We can organize the facts of the problem in a table.

	Amount ·	Price =	Total value
Peanuts	x	6	$6x$
Cashews	50	9	450
Mixture	$50 + x$	7	$7(50 + x)$

Enter this information first. (Amount and Price columns)
Multiply to get each of these entries. (Total value column)

We can use the information in the last column of the table to form an equation.

The value of the peanuts	plus	the value of the cashews	equals	the value of the mixture.
$6x$	$+$	450	$=$	$7(50 + x)$

Solve the Equation

$6x + 450 = 7(50 + x)$

$6x + 450 = 350 + 7x$ Distribute the multiplication by 7.

$450 = 350 + x$ Subtract $6x$ from both sides.

$100 = x$ Subtract 350 from both sides.

State the Conclusion 100 pounds of peanuts should be used in the mixture.

Check the Result

Value of 100 pounds of peanuts, at \$6 per pound: 100(6) = \$600.
Value of 50 pounds of cashews, at \$9 per pound: 50(9) = \$450. } \$1,050
Value of 150 pounds of the mixture, at \$7 per pound: \$1,050. The answer checks.

2.6 STUDY SET

ELEMENTARY Algebra f(x) Now™

VOCABULARY Fill in the blanks.

1. The ________ of a triangle or a rectangle is the distance around it.
2. An ________ triangle is a triangle with two sides of the same length.
3. The equal sides of an isosceles triangle meet to form the ________ angle. The angles opposite the equal sides are called ______ angles, and they have equal measures.
4. When asked to find the dimensions of a rectangle, we are to find its ________ and ________.

CONCEPTS

5. A plumber wants to cut a 17-foot pipe into three sections. The longest section is to be three times as long as the shortest, and the middle-sized section is to be 2 feet longer than the shortest.
   a. Complete the diagram.

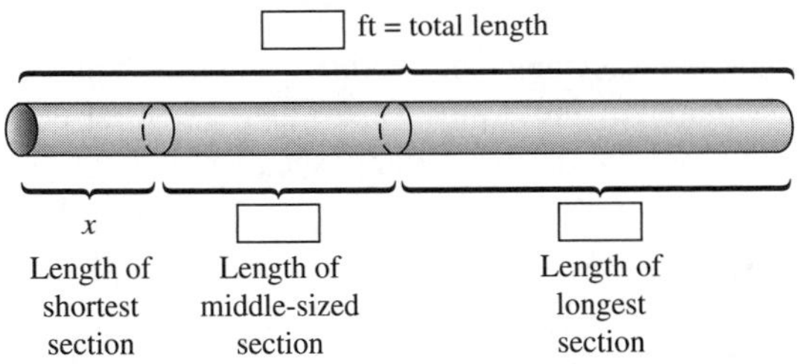

   b. To solve this problem, an equation is formed, it is solved, and it is found that $x = 3$. How long is each section of pipe?

6. Complete the expression, which represents the perimeter of the rectangle shown.

   $2(\quad) + \quad x$

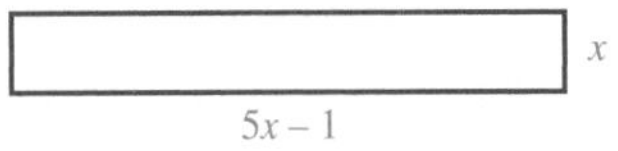

7. What is the sum of the measures of the angles of any triangle?
8. **a.** Complete the table, which shows the inventory of nylon brushes that a paint store carries.

	Number	· Value	= Total value
1 inch	$\frac{x}{2}$	4	
2 inch	x	5	
3 inch	$x + 10$	7	

   b. Which type of brush does the store have the greatest number of?
   c. What is the least expensive brush?
   d. What is the total value of the inventory of nylon brushes?

9. In the advertisement, what are the principal, the rate, and the time for the investment opportunity shown?

> **Invest in Mini Malls!**
> Builder seeks daring people who want to earn big $$$$$$. In just 1 year, you will earn a gigantic 14% on an investment of only $30,000! Call now.

10. a. Complete the table, which gives the details about two investments made by a retired couple.

	P	$\cdot\ r$	$\cdot\ t =$	I
Certificate of deposit	x	0.04	1	
Brother-in-law's business	$2x$	0.06	1	

b. How much more money was invested in the brother-in-law's business than in the certificate of deposit?

c. What is the total amount of interest the couple will make from these investments?

11. When a husband and wife leave for work, they drive in opposite directions. Their average speeds are different; however, their drives last the same amount of time. Complete the table, which gives the details of each person's commute.

	r	$\cdot\ t =$	d
Husband	35	t	
Wife	45		

12. a. How many gallons of acid are there in the second barrel?

b. Suppose the contents of the two barrels are poured into an empty third barrel. How many gallons of liquid will the third barrel contain?

c. What would be a *reasonable* estimate of the concentration of the solution in the third barrel: 15%, 35%, or 60% acid?

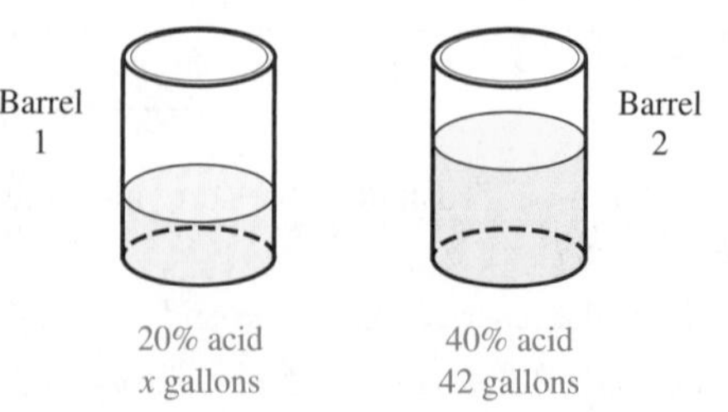

13. a. Two oil and vinegar salad dressings are combined to make a new mixture. Complete the table.

	Amount	· Strength	= Pure vinegar
Strong	x	0.06	
Weak		0.03	
Mixture	10	0.05	

b. Two antifreeze solutions are combined to form a mixture. Complete the table.

	Amount	· Strength	= Pure antifreeze
Strong	6	0.50	
Weak	x	0.25	
Mixture		0.30	

14. Use the information in the table to fill in the blanks. How many pounds of a ________ diet supplement worth $15.95 per pound should be mixed with _____ pounds of a vitamin diet supplement worth ________ per pound to obtain a ________ that is worth ________ per pound?

	Amount	· Value =	Total value
Protein	x	15.95	$15.95x$
Vitamin	20	7.99	20(7.99)
Mixture	$x + 20$	12.50	$12.50(x + 20)$

NOTATION

15. What concept about decimal multiplication is shown?

$$100(0.08) = 8$$

16. True or false: $x(0.09) = 0.09x$?

17. Suppose a rectangle has width w and length $2w - 3$. Explain why the expression $2 \cdot 2w - 3 + 2w$ does not represent the perimeter of the rectangle.

18. Write 5.5% as a decimal.

PRACTICE **Solve each equation by first clearing it of decimals.**

19. $0.08x + 0.07(15{,}000 - x) = 1{,}110$

20. $0.108x + 0.07(16{,}000 - x) = 1{,}500$

APPLICATIONS

21. CARPENTRY A 12-foot board has been cut into two sections, one twice as long as the other. How long is each section?

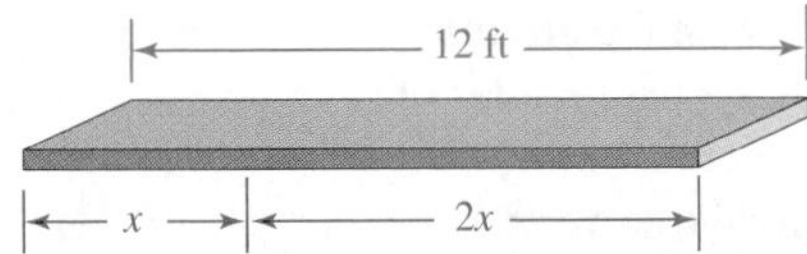

22. ROBOTICS The robotic arm will extend a total distance of 18 feet. Find the length of each section.

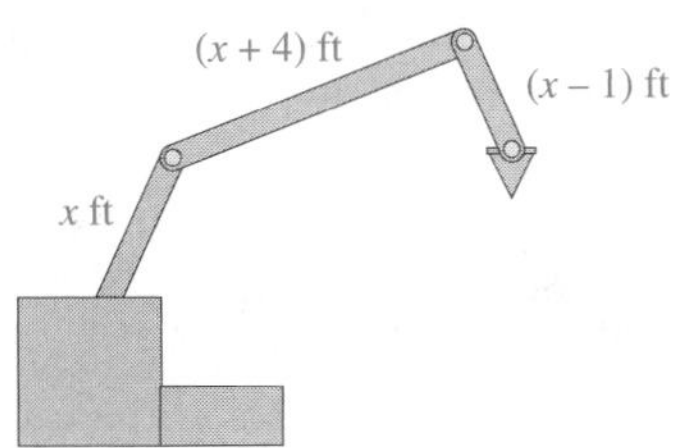

23. NATIONAL PARKS The Natchez Trace Parkway is a historical 444-mile route from Natchez, Mississippi, to Nashville, Tennessee. A couple drove the Trace in four days. Each day they drove 6 miles more than the previous day. How many miles did they drive each day?

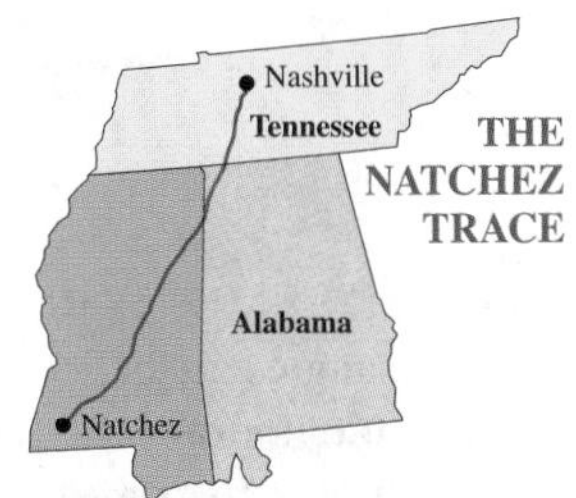

24. SOLAR HEATING One solar panel is 3.4 feet wider than the other. Find the width of each panel.

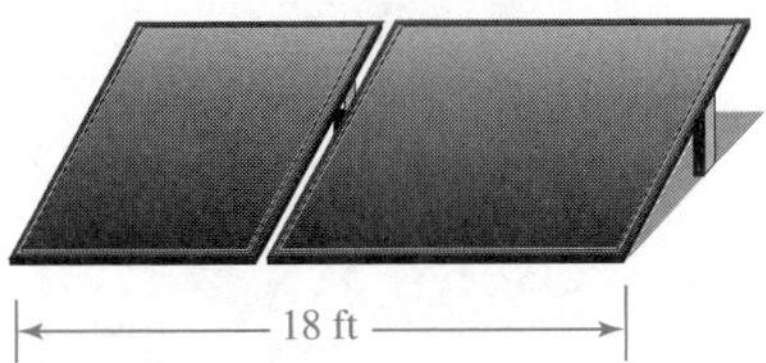

25. TOURING A rock group plans to travel for a total of 38 weeks, making three concert stops. They will be in Japan for 4 more weeks than they will be in Australia. Their stay in Sweden will be 2 weeks shorter than that in Australia. How many weeks will they be in each country?

26. LOCKS The three numbers of the combination for a lock are **consecutive integers,** and their sum is 81. (Consecutive integers follow each other, like 7, 8, 9.) Find the combination. (*Hint:* If x represents the smallest integer, $x + 1$ represents the next integer, and $x + 2$ represents the largest integer.)

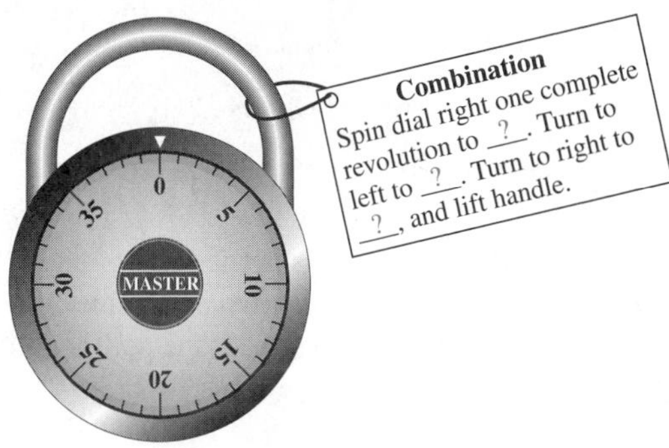

27. COUNTING CALORIES A slice of pie with a scoop of ice cream has 850 calories. The calories in the pie alone are 100 more than twice the calories in the ice cream alone. How many calories are in each food?

28. WASTE DISPOSAL Two tanks hold a total of 45 gallons of a toxic solvent. One tank holds 6 gallons more than twice the amount in the other. How many gallons does each tank hold?

29. ACCOUNTING Determine the 2002 income of Sears, Roebuck and Co. for each quarter from the data in the graph. (Source: Hoover's Online Internet service.)

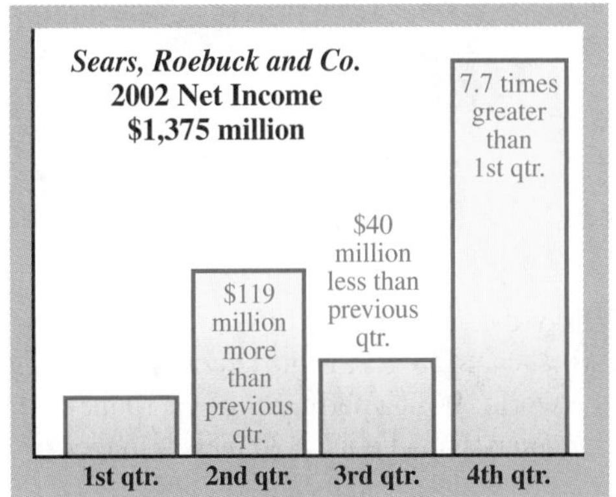

30. AMERICAN COLLEGE STUDENTS A 1999–2000 survey found that the percent of college students that worked part time was 1% more than twice the percent that did not work. The percent that worked full time was 1% less than twice the percent that did not work. Find the missing percents in the graph. (*Hint:* The sum of the percents is 100%.)

Most students work

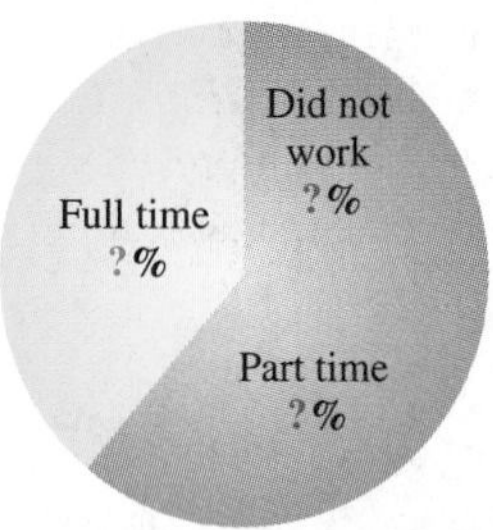

Source: National Center for Education Statistics

31. ENGINEERING A truss is in the form of an isosceles triangle. Each of the two equal sides is 4 feet shorter than the third side. If the perimeter is 25 feet, find the lengths of the sides.

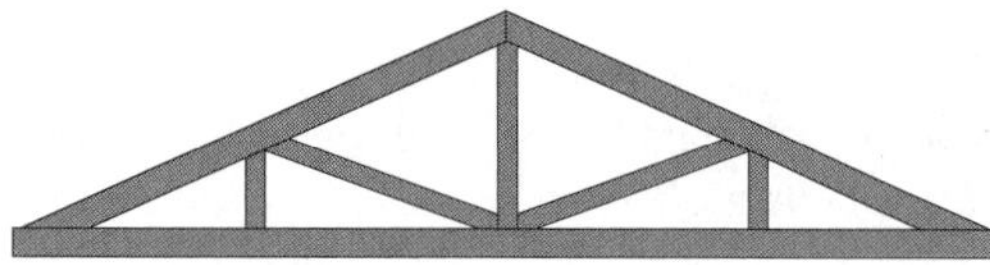

32. FIRST AID A sling is in the shape of an isosceles triangle with a perimeter of 144 inches. The longest side of the sling is 18 inches longer than either of the other two sides. Find the lengths of each side.

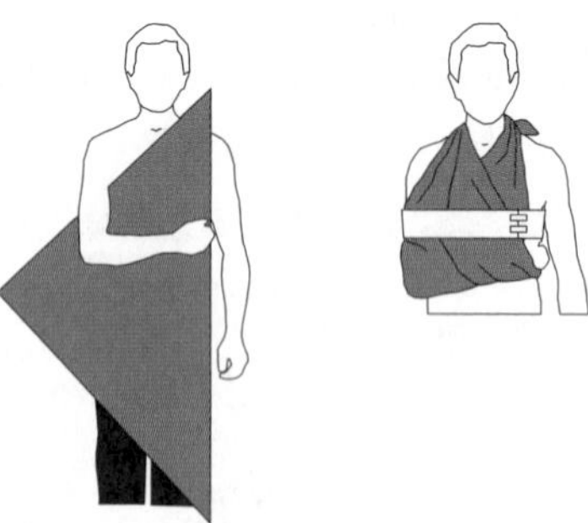

33. SWIMMING POOLS The seawater Orthlieb Pool in Casablanca, Morocco, is the largest swimming pool in the world. With a perimeter of 1,110 meters, this rectangular-shaped pool is 30 meters longer than 6 times its width. Find its dimensions.

34. ART The *Mona Lisa* was completed by Leonardo da Vinci in 1506. The length of the picture is 11.75 inches shorter than twice the width. If the perimeter of the picture is 102.5 inches, find its dimensions.

35. TV TOWERS The two guy wires supporting a tower form an isosceles triangle with the ground. Each of the base angles of the triangle is 4 times the third angle (the vertex angle). Find the measure of the vertex angle.

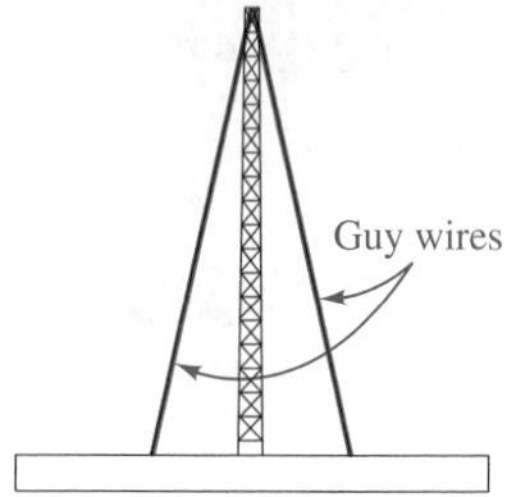

36. MOUNTAIN BICYCLES For the bicycle frame shown, the angle that the horizontal crossbar makes with the seat support is 15° less than twice the angle at the steering column. The angle at the pedal gear is 25° more than the angle at the steering column. Find these three angle measures.

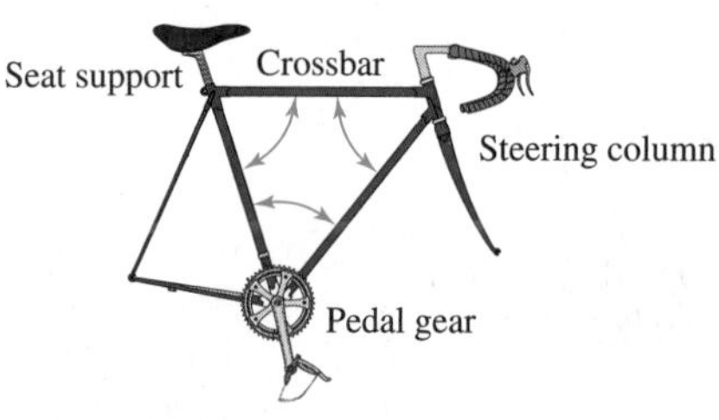

37. COMPLEMENTARY ANGLES Two angles are called **complementary angles** when the sum of their measures is 90°. Find the measures of the complementary angles shown here.

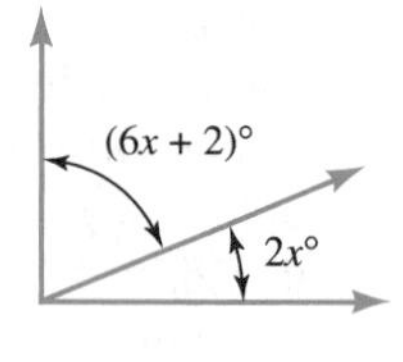

38. SUPPLEMENTARY ANGLES Two angles are called **supplementary angles** when the sum of their measures is 180°. Find the measures of the supplementary angles shown here.

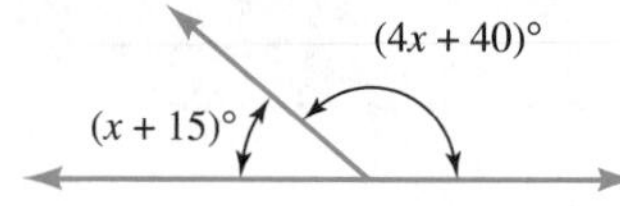

39. RENTALS The owners of an apartment building rent 1-, 2-, and 3-bedroom units. They rent equal numbers of each, with the monthly rents given in the table. If the total monthly income is $36,550, how many of each type of unit are there?

Unit	Rent
One-bedroom	$550
Two-bedroom	$700
Three-bedroom	$900

40. WAREHOUSING A store warehouses 40 more portables than big-screen TV sets, and 15 more consoles than big-screen sets. Storage costs for the different TV sets are shown in the table. If storage costs $276 per month, how many big-screen sets are in stock?

Type of TV	Monthly cost
Portable	$1.50
Console	$4.00
Big-screen	$7.50

41. SOFTWARE Three software applications are priced as shown. Spreadsheet and database programs sold in equal numbers, but 15 more word processing applications were sold than the other two combined. If the three applications generated sales of $72,000, how many spreadsheets were sold?

Software	Price
Spreadsheet	$150
Database	$195
Word processing	$210

42. INVENTORIES With summer approaching, the number of air conditioners sold is expected to be double that of stoves and refrigerators combined. Stoves sell for $350, refrigerators for $450, and air conditioners for $500, and sales of $56,000 are expected. If stoves and refrigerators sell in equal numbers, how many of each appliance should be stocked?

43. INVESTMENTS Equal amounts are invested in each of three accounts paying 7%, 8%, and 10.5% annually. If one year's combined interest income is $1,249.50, how much is invested in each account?

44. RETIREMENT A professor wants to supplement her pension with investment interest. If she invests $28,000 at 6% interest, how much more would she have to invest at 7% to achieve a goal of $3,500 per year in supplemental income?

45. INVESTMENT PLANS A financial planner recommends a plan for a client who has $65,000 to invest. (See the chart.) At the end of the presentation, the client asks, "How much will be invested at each rate?" Answer this question using the given information.

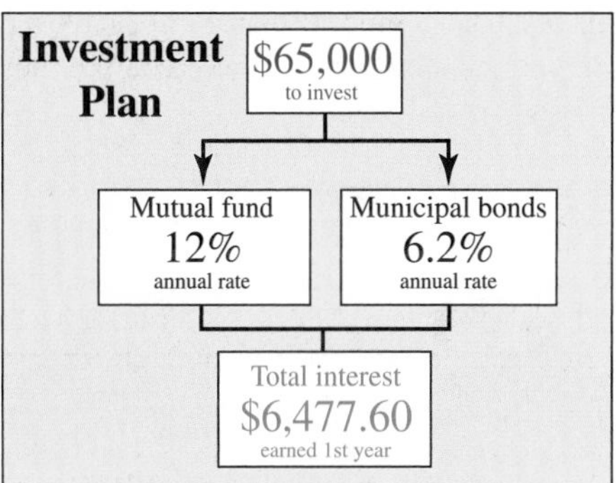

46. TAXES On January 2, 2003, Terrell Washington opened two savings accounts. At the end of the year his bank mailed him the form shown below for income tax purposes. If a total of $15,000 was initially deposited and if no further deposits or withdrawals were made, how much money was originally deposited in account number 721?

USA HOME SAVINGS

Copy B
For Recipient
Interest Income

This is important tax information and is being furnished to the Internal Revenue Service.

OMB No. 1545-0112
2003
Form 1099–INT

RECIPIENT'S name
TERRELL WASHINGTON

Acct. Number	Annual Percent Yield	Early Withdrawal Penalty
822	5%	.00
721	4.5%	.00

Total Interest Income 720.00

47. FINANCIAL PLANNING A plumber has a choice of two investment plans:

- An insured fund that pays 11% interest
- A risky investment that pays a 13% return

If the same amount invested at the higher rate would generate an extra $150 per year, how much does the plumber have to invest?

48. INVESTMENTS The amount of annual interest earned by $8,000 invested at a certain rate is $200 less than $12,000 would earn at a rate 1% lower. At what rate is the $8,000 invested?

49. TORNADOES During a storm, two teams of scientists leave a university at the same time in specially designed vans to search for tornadoes. The first team travels east at 20 mph and the second travels west at 25 mph. If their radios have a range of up to 90 miles, how long will it be before they lose radio contact?

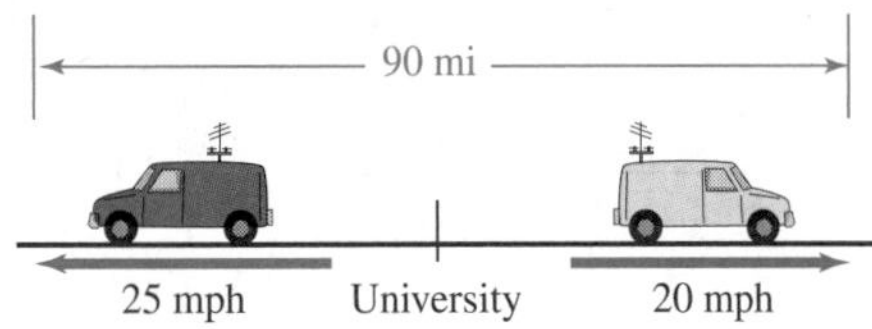

50. SEARCH AND RESCUE Two search-and-rescue teams leave base at the same time looking for a lost boy. The first team, on foot, heads north at 2 mph and the other, on horseback, south at 4 mph. How long will it take them to search a combined distance of 21 miles between them?

51. AIR TRAFFIC CONTROL An airliner leaves Berlin, Germany, headed for Montreal, Canada, flying at an average speed of 450 mph. At the same time, an airliner leaves Montreal headed for Berlin, averaging 500 mph. If the airports are 3,800 miles apart, when will the air traffic controllers have to make the pilots aware that the planes are passing each other?

52. SPEED OF TRAINS Two trains are 330 miles apart, and their speeds differ by 20 mph. Find the speed of each train if they are traveling toward each other and will meet in 3 hours.

53. ROAD TRIPS A car averaged 40 mph for part of a trip and 50 mph for the remainder. If the 5-hour trip covered 210 miles, for how long did the car average 40 mph?

54. CYCLING A cyclist leaves his training base for a morning workout, riding at the rate of 18 mph. One hour later, his support staff leaves the base in a car going 45 mph in the same direction. How long will it take the support staff to catch up with the cyclist?

55. PHOTOGRAPHIC CHEMICALS A photographer wishes to mix 2 liters of a 5% acetic acid solution with a 10% solution to get a 7% solution. How many liters of 10% solution must be added?

56. SALT SOLUTIONS How many gallons of a 3% salt solution must be mixed with 50 gallons of a 7% solution to obtain a 5% solution?

57. ANTISEPTIC SOLUTIONS A nurse wants to add water to 30 ounces of a 10% solution of benzalkonium chloride to dilute it to an 8% solution. How much water must she add? (*Hint:* Water is 0% benzalkonium chloride.)

58. MAKING CHEESE To make low-fat cottage cheese, milk containing 4% butterfat is mixed with milk containing 1% butterfat to obtain 15 gallons of a mixture containing 2% butterfat. How many gallons of the richer milk must be used?

59. MIXING FUELS How many gallons of fuel costing $1.15 per gallon must be mixed with 20 gallons of a fuel costing $0.85 per gallon to obtain a mixture costing $1 per gallon?

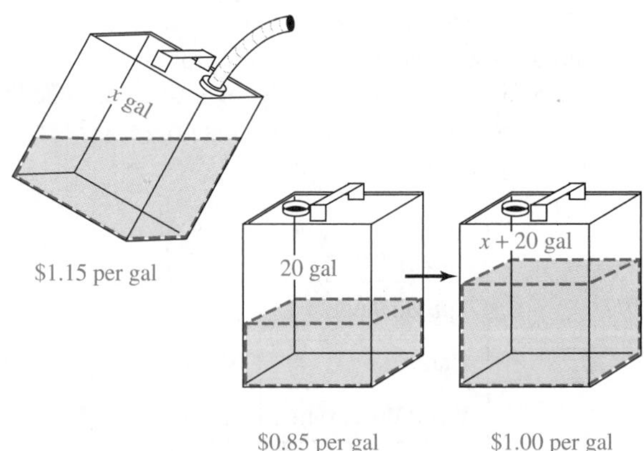

60. MIXING PAINT Paint costing $19 per gallon is to be mixed with $3-per-gallon thinner to make 16 gallons of a paint that can be sold for $14 per gallon. How much paint thinner should be used?

61. BLENDING LAWN SEED A store sells bluegrass seed for $6 per pound and ryegrass seed for $3 per pound. How much ryegrass must be mixed with 100 pounds of bluegrass to obtain a blend that will sell for $5 per pound?

62. BLENDING COFFEE A store sells regular coffee for \$4 a pound and gourmet coffee for \$7 a pound. To get rid of 40 pounds of the gourmet coffee, a shopkeeper makes a blend to put on sale for \$5 a pound. How many pounds of regular coffee should he use?

63. MIXING CANDY Lemon drops worth \$1.90 per pound are to be mixed with jelly beans that cost \$1.20 per pound to make 100 pounds of a mixture worth \$1.48 per pound. How many pounds of each candy should be used?

64. SNACK FOODS A bag of peanuts is worth \$.30 less than a bag of cashews. Equal amounts of peanuts and cashews are used to make 40 bags of a mixture that sells for \$1.05 per bag. How much is a bag of cashews worth?

WRITING

65. Create a mixture problem of your own, and solve it.

66. Is it possible to mix a 10% sugar solution with a 20% sugar solution to get a 30% sugar solution?

67. A car travels at 60 mph for 15 minutes. Why can't we multiply the rate, 60, and the time, 15, to find the distance traveled by the car?

68. Create a geometry problem that could be answered by solving the equation $2w + 2(w + 5) = 26$.

REVIEW **Remove parentheses.**

69. $-25(2x - 5)$

70. $-12(3a + 4b - 32)$

71. $-(-3x - 3)$

72. $\frac{1}{2}(4b - 8)$

73. $(4y - 4)4$

74. $3(5t + 1)2$

CHALLENGE PROBLEMS

75. EVAPORATION How much water must be boiled away to increase the concentration of 300 milliliters of a 2% salt solution to a 3% salt solution?

76. TESTING A teacher awarded 4 points for each correct answer and deducted 2 points for each incorrect answer when grading a 50-question true-false test. A student scored 56 points on the test, and did not leave any questions unanswered. How many questions did the student answer correctly?

2.7 Solving Inequalities

- Inequalities and Solutions
- Graphing Inequalities and Interval Notation
- Solving Inequalities
- Graphing Compound Inequalities
- Solving Compound Inequalities
- Applications

In our daily lives, we often speak of one value being greater than or less than another value.

- To melt ice, the temperature must be *greater than* 32° F.
- An airplane is rated to fly at altitudes that are *less than* 36,000 feet.
- To earn a B, a student needed a final exam score of *at least* 80%.

In mathematics, we use *inequalities* to show that one expression is greater than or less than another expression.

INEQUALITIES AND SOLUTIONS

An **inequality** is a statement that contains one of the following symbols.

Inequality Symbols	$<$ is less than	$>$ is greater than
	$\leq$ is less than or equal to	$\geq$ is greater than or equal to

Some examples of inequalities are:

$$2 < 3, \quad 1.1 \geq -2.7, \quad x + 1 > 5, \quad \text{and} \quad 5(x + 1) \leq 2(x - 3)$$

The Language of Algebra

Because $<$ requires one number to be strictly less than another number and $>$ requires one number to be strictly greater than another number, $<$ and $>$ are called *strict inequalities.*

An inequality may be true, false, or neither true nor false.

- $9 \geq 9$ is true because $9 = 9$.
- $37 < 24$ is false.
- $x + 1 > 5$ is neither true nor false because we don't know what number x represents.

An inequality that contains a variable can be made true or false, depending on the number that we substitute for the variable. If we substitute 10 for x in $x + 1 > 5$, the resulting inequality is true: $10 + 1 > 5$. If we substitute 1 for x, the result is false: $1 + 1 > 5$. A number that makes an inequality true is called a **solution** of the inequality. Therefore, 10 is a solution of $x + 1 > 5$ and 1 is not.

EXAMPLE 1

Is 9 a solution of $2x + 4 \leq 21$?

ELEMENTARY Algebra f(x) Now™

Solution We substitute 9 for x in the inequality and evaluate the left-hand side. If 9 is a solution, we will obtain a true statement.

$$2x + 4 \leq 21$$
$$2(9) + 4 \stackrel{?}{\leq} 21$$
$$18 + 4 \stackrel{?}{\leq} 21$$
$$22 \leq 21$$

The statement $22 \leq 21$ is false because neither $22 < 21$ nor $22 = 21$ is true. Therefore, 9 is not a solution.

Self Check 1 Is 2 a solution of $3x - 1 \geq 0$?

GRAPHING INEQUALITIES AND INTERVAL NOTATION

The **solution set** of an inequality is the set of all of its solutions. Some solution sets are easy to determine. For example, if we replace the variable in $x > -3$ with a number greater than -3, the resulting inequality will be true. Because there are infinitely many real numbers greater than -3, it follows that $x > -3$ has infinitely many solutions. We can illustrate the solution set on a number line by **graphing the inequality.**

To graph $x > -3$, we shade all the points on the number line that are to the right of -3. We use a shaded arrowhead to show that the solutions continue indefinitely to the right. A **parenthesis** or an **open circle** is drawn at the endpoint -3 to indicate that -3 is not part of the graph.

The Language of Algebra

The *infinity* symbol ∞ does not represent a number. It indicates that an interval extends to the right without end.

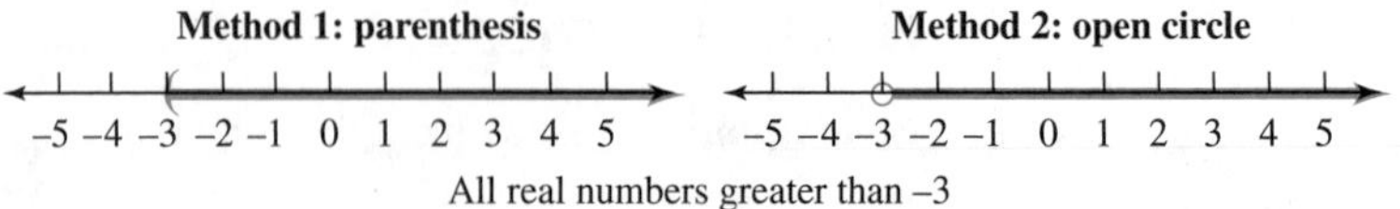

Graphs of inequalities are **intervals** on the number line. The graph of $x > -3$ can be expressed in **interval notation** as $(-3, \infty)$. Again, the left parenthesis indicates that -3 is not included in the interval. The **infinity symbol** ∞ indicates that the interval continues without end to the right.

EXAMPLE 2

Graph: $x \leq 2$.

Solution If we replace x with a number less than or equal to 2, the resulting inequality will be true. To graph the solution set, we shade the point 2 and all points to the left of 2 on the number line. A **bracket** or a **closed circle** is drawn at the endpoint 2 to indicate that 2 is part of the graph.

Notation

Since we use parentheses and brackets in interval notation, we will use them to graph inequalities. Note that parentheses, not brackets, are written next to infinity symbols:

$(-3, \infty)$ $(-\infty, 2]$

Method 1: bracket

−5 −4 −3 −2 −1 0 1 2 3 4 5

Method 2: closed circle

−5 −4 −3 −2 −1 0 1 2 3 4 5

All real numbers less than or equal to 2

The interval is written as $(-\infty, 2]$. The bracket indicates that 2 is included in the interval. The **negative infinity symbol** $-\infty$ shows that the interval continues indefinitely to the left.

Self Check 2 Graph: $x \geq 0$.

SOLVING INEQUALITIES

To **solve an inequality** means to find all values of the variable that make the inequality true. Inequalities are solved by isolating the variable on one side. We will use the following properties of inequality to do this.

Addition and Subtraction Properties of Inequality

Adding the same number to, or subtracting the same number from, both sides of an inequality does not change the solutions.

For any real numbers a, b, and c,

If $a < b$, then $a + c < b + c$.

If $a < b$, then $a - c < b - c$.

Similar statements can be made for the symbols $\leq$, $>$, or $\geq$.

After applying one of these properties, the resulting inequality is equivalent to the original one. Like equivalent equations, **equivalent inequalities** have the same solution set.

EXAMPLE 3

Solve: $x + 3 > 2$.

ELEMENTARY Algebra $f(x)$ Now™

Solution We can use the subtraction property of inequality to isolate x on the left-hand side.

$$x + 3 > 2$$

$$x + 3 - 3 > 2 - 3 \quad \text{To undo the addition of 3, subtract 3 from both sides.}$$

$$x > -1$$

All real numbers greater than -1 are solutions of $x + 3 > 2$. The solution set can be written as $(-1, \infty)$. The graph of the solution set is shown below.

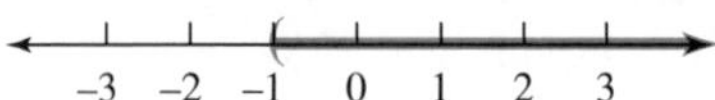

Since there are infinitely many solutions, we cannot check each of them. As an informal check, we can pick two numbers in the graph, say 1 and 30, substitute each for x in the original inequality, and see whether true statements result.

Check:

$x + 3 > 2$		$x + 3 > 2$	
$1 + 3 \overset{?}{>} 2$	Substitute 1 for x.	$30 + 3 \overset{?}{>} 2$	Substitute 30 for x.
$4 > 2$	This is a true inequality.	$33 > 2$	This is a true inequality.

The solution set appears to be correct.

Self Check 3 Solve: $x - 3 < -2$. Write the solution set in interval notation and graph it.

As with equations, there are properties for multiplying and dividing both sides of an inequality by the same number. To develop what is called *the multiplication property of inequality,* consider the true statement $2 < 5$. If both sides are multiplied by a positive number, such as 3, another true inequality results.

$$2 < 5$$
$$3 \cdot 2 < 3 \cdot 5 \quad \text{Multiply both sides by 3.}$$
$$6 < 15 \quad \text{This is a true inequality.}$$

However, if we multiply both sides of $2 < 5$ by a negative number, such as -3, the direction of the inequality symbol is reversed to produce another true inequality.

$$2 < 5$$
$$-3 \cdot 2 > -3 \cdot 5 \quad \text{Multiply both sides by the negative number } -3 \text{ and reverse the direction of the inequality.}$$
$$-6 > -15 \quad \text{This is a true inequality.}$$

The inequality $-6 > -15$ is true because -6 is to the right of -15 on the number line.

Dividing both sides of an inequality by the same negative number also requires that the direction of the inequality symbol be reversed.

$$-4 < 6 \quad \text{A true inequality.}$$
$$\frac{-4}{-2} > \frac{6}{-2} \quad \text{Divide both sides by } -2 \text{ and change} < \text{to} >.$$
$$2 > -3 \quad \text{This is a true inequality.}$$

These examples illustrate the multiplication and division properties of inequality.

Multiplication and Division Properties of Inequality

Multiplying or dividing both sides of an inequality by the same positive number does not change the solutions.

For any real numbers a, b, and c, where c is positive,

$$\text{If } a < b, \text{ then } ac < bc. \qquad \text{If } a < b, \text{ then } \frac{a}{c} < \frac{b}{c}.$$

If we multiply or divide both sides of an inequality by a negative number, the direction of the inequality symbol must be reversed for the inequalities to have the same solutions.

For any real numbers a, b, and c, where c is negative,

$$\text{If } a < b, \text{ then } ac > bc. \qquad \text{If } a < b, \text{ then } \frac{a}{c} > \frac{b}{c}.$$

Similar statements can be made for the symbols $\leq$, $>$, or $\geq$.

EXAMPLE 4

Solve: $-\frac{3}{2}t \geq -12$.

Solution To undo the multiplication by $-\frac{3}{2}$, we multiply both sides by the reciprocal, which is $-\frac{2}{3}$.

$$-\frac{3}{2}t \geq -12$$

$$-\frac{2}{3}\left(-\frac{3}{2}t\right) \leq -\frac{2}{3}(-12) \quad \text{Multiply both sides by } -\frac{2}{3}. \text{ Change } \geq \text{ to } \leq.$$

$$t \leq 8 \quad \text{Do the multiplications.}$$

The solution set is $(-\infty, 8]$ and it is graphed as shown.

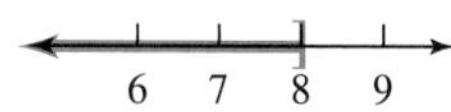

Self Check 4 Solve: $-\frac{h}{20} \leq 10$. Write the solution set in interval notation and graph it.

EXAMPLE 5

Solve: $-5 > 3x + 7$.

ELEMENTARY Algebra f(x) Now™

Solution

$$-5 > 3x + 7$$

$$-5 - 7 > 3x + 7 - 7 \quad \text{To undo the addition of 7, subtract 7 from both sides.}$$

$$-12 > 3x \quad \text{Do the subtractions.}$$

$$\frac{-12}{3} > \frac{3x}{3} \quad \text{To undo the multiplication by 3, divide both sides by 3.}$$

$$-4 > x \quad \text{Do the divisions.}$$

Caution

In Example 5, don't be confused by the negative number on the left-hand side. We didn't reverse the $>$ symbol because we divided both sides by *positive* 3.

$$\frac{-12}{3} > \frac{3x}{3}$$

To find the solution set, it is useful to write $-4 > x$ with the variable on the left-hand side. If -4 is greater than x, then x must be less than -4.

$$x < -4$$

The solution set is $(-\infty, -4)$ whose graph is shown.

−6 −5 −4 −3

Self Check 5 Solve: $-13 < 2r - 7$. Write the solution set in interval notation and graph it.

EXAMPLE 6

Solve: $5.1 - 3a < 19.5$.

ELEMENTARY Algebra f(x) Now™

Solution

$5.1 - 3a < 19.5$	
$5.1 - 3a - \mathbf{5.1} < 19.5 - \mathbf{5.1}$	To isolate $-3a$ on the left-hand side, subtract 5.1 from both sides.
$-3a < 14.4$	Do the subtractions.
$\frac{-3a}{\mathbf{-3}} > \frac{14.4}{\mathbf{-3}}$	To undo the multiplication by -3, divide both sides by -3. Since we are dividing by a negative number, we reverse the direction of the $<$ symbol.
$a > -4.8$	Do the divisions.

The solution set is $(-4.8, \infty)$ whose graph is shown.

Self Check 6 Solve: $-9n + 1.8 > -17.1$. Write the solution set in interval notation and graph it.

EXAMPLE 7

Solve: $8(y + 1) \geq 2(y - 4) + y$.

ELEMENTARY Algebra f(x) Now™

Solution To solve this inequality, we follow the same strategy used for solving equations.

$8(y + 1) \geq 2(y - 4) + y$	
$8y + 8 \geq 2y - 8 + y$	Distribute the multiplication by 8 and by 2.
$8y + 8 \geq 3y - 8$	Combine like terms: $2y + y = 3y$.
$8y + 8 - \mathbf{3y} \geq 3y - 8 - \mathbf{3y}$	To eliminate $3y$ from the right-hand side, subtract $3y$ from both sides.
$5y + 8 \geq -8$	Combine like terms on both sides.
$5y + 8 - \mathbf{8} \geq -8 - \mathbf{8}$	To undo the addition of 8, subtract 8 from both sides.
$5y \geq -16$	Do the subtractions.
$\frac{5y}{\mathbf{5}} \geq \frac{-16}{\mathbf{5}}$	To undo the multiplication by 5, divide both sides by 5.
$y \geq -\frac{16}{5}$	

The solution set is $\left[-\frac{16}{5}, \infty\right)$. To graph it, we note that $-\frac{16}{5} = -3\frac{1}{5}$.

Self Check 7 Solve: $5(b - 2) \geq -(b - 3) + 2b$. Write the solution set in interval notation and graph it.

GRAPHING COMPOUND INEQUALITIES

The Language of Algebra

The word *compound* means made up of individual parts. For example, a *compound* inequality has three parts. In writing classes, students learn about *compound* sentences.

Two inequalities can be combined into a **compound inequality** to show that an expression lies between two fixed values. For example, $-2 < x < 3$ is a combination of

$$-2 < x \qquad \text{and} \qquad x < 3$$

It indicates that x is greater than -2 and that x is also less than 3. The solution set of $-2 < x < 3$ consists of all numbers that lie between -2 and 3, and we write it as $(-2, 3)$. The graph of the compound inequality is shown below.

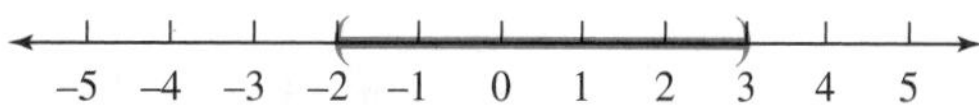

EXAMPLE 8

Graph: $-4 \le x < 0$.

Solution If we replace the variable in $-4 \le x < 0$ with a number between -4 and 0, including -4, the resulting compound inequality will be true. Therefore, the solution set is $[-4, 0)$. To graph the interval, we draw a bracket at -4, a parenthesis at 0, and shade in between.

To check, we pick a number in the graph, such as -2, and see whether it satisfies the inequality. Since $-4 \le -2 < 0$ is true, the answer appears to be correct.

Self Check 8 Graph $-2 \le x < 1$ and write the solution set in interval notation.

SOLVING COMPOUND INEQUALITIES

To solve compound inequalities, we use the same methods used for solving equations. However, we will apply the properties of inequality to all *three* parts of the inequality.

EXAMPLE 9

Solve: $-4 < 2(x - 1) \le 4$.

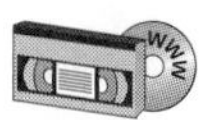

ELEMENTARY Algebra $f(x)$ Now™

Solution

$$-4 < 2(x - 1) \le 4$$

$$-4 < 2x - 2 \le 4 \qquad \text{Distribute the multiplication by 2.}$$

$$-4 + 2 < 2x - 2 + 2 \le 4 + 2 \qquad \text{To undo the subtraction of 2, add 2 to all three parts.}$$

$$-2 < 2x \le 6 \qquad \text{Do the additions.}$$

$$\frac{-2}{2} < \frac{2x}{2} \le \frac{6}{2} \qquad \text{To isolate } x\text{, we undo the multiplication by 2 by dividing all three parts by 2.}$$

$$-1 < x \le 3$$

The solution set is $(-1, 3]$ and its graph is shown.

Self Check 9 Solve: $-6 \le 3(t + 2) \le 6$. Write the solution set in interval notation and graph it.

APPLICATIONS

When solving problems, phrases such as "not more than," or "should exceed" suggest that an *inequality* should be written instead of an *equation.*

EXAMPLE 10

Grades. A student has scores of 72%, 74%, and 78% on three exams. What percent score does he need on the last exam to earn no less than a grade of B (80%)?

Analyze the Problem We know three scores. We are to find what the student must score on the last exam to earn at least a B grade.

Form an Inequality We can let $x =$ the score on the fourth (and last) exam. To find the average grade, we add the four scores and divide by 4. To earn no less than a grade of B, the student's average must be greater than or equal to 80%.

The average of the four grades	must be greater than or equal to	80.
$\frac{72 + 74 + 78 + x}{4}$	$\geq$	80

Solve the Inequality We can solve this inequality for x.

$$\frac{224 + x}{4} \geq 80 \quad \text{Combine like terms in the numerator: } 72 + 74 + 78 = 224.$$

$$224 + x \geq 320 \quad \text{To clear the inequality of the fraction, multiply both sides by 4.}$$

$$x \geq 96 \quad \text{To undo the addition of 224, subtract 224 from both sides.}$$

State the Conclusion To earn a B, the student must score 96% or better on the last exam. Assuming the student cannot score higher than 100% on the exam, the solution set is written as [96, 100]. The graph is shown below.

92 93 94 95 96 97 98 99 100

Check the Result Pick some numbers in the interval, and verify that the average of the four scores will be 80% or greater.

Answers to Self Checks

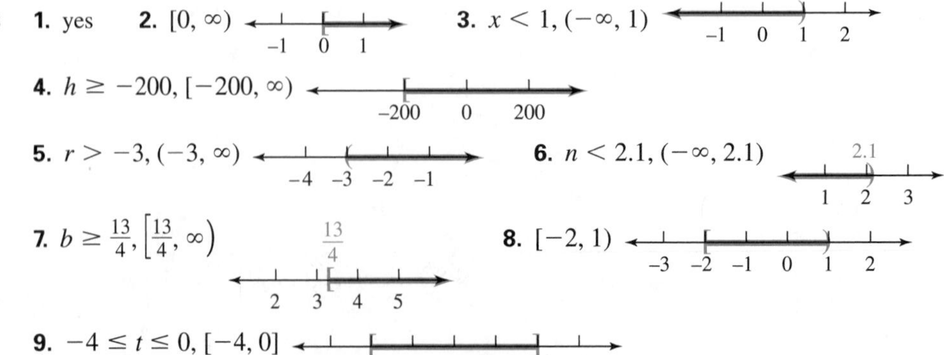

2.7 STUDY SET  ELEMENTARY Algebra $f(x)$ Now™

VOCABULARY **Fill in the blanks.**

1. An __________ is a statement that contains one of the following symbols: $>$, $\geq$, $<$, or $\leq$.
2. A number that makes an inequality true is called a ________ of the inequality. The solution _____ of an inequality is the set of all solutions.
3. To _______ an inequality means to find all the values of the variable that make the inequality true.
4. Graphs of inequalities are _________ on the number line.
5. The solution set of $x > 2$ can be expressed in ________ notation as $(2, \infty)$.
6. The inequality $-4 < x \leq 10$ is an example of a __________ inequality.

CONCEPTS

7. Decide whether each statement is true or false.
   a. $35 \geq 34$ **b.** $-16 \leq -17$
   c. $\frac{3}{4} \leq 0.75$ **d.** $-0.6 \geq -0.5$
8. Decide whether each number is a solution of $3x + 7 < 4x - 2$.
   a. 12 **b.** -6
   c. 0 **d.** 9
9. Write each inequality so that the inequality symbol points in the opposite direction.
   a. $17 \geq -2$ **b.** $32 < x$
10. The solution set of an inequality is graphed as shown.

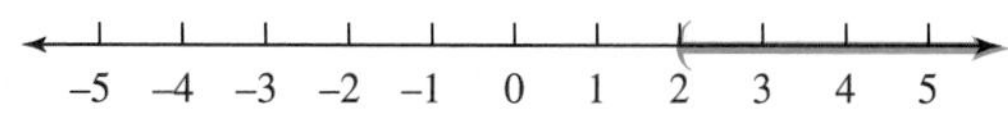

Which of the following numbers, when substituted for the variable in that inequality, would make it true?

3 -3 2 4.5

Fill in the blanks.

11. **a.** Adding the same number to, or subtracting the ______ number from, both sides of an inequality does not change the solutions.
    b. Multiplying or dividing both sides of an inequality by the same ________ number does not change the solutions.
12. If we multiply or divide both sides of an inequality by a ________ number, the direction of the inequality symbol must be reversed for the inequalities to have the same solutions.
13. Solve $x + 2 > 10$ and give the solution set:
    a. using a graph
    b. using interval notation
    c. in words
14. The solution set of a compound inequality is graphed as shown.

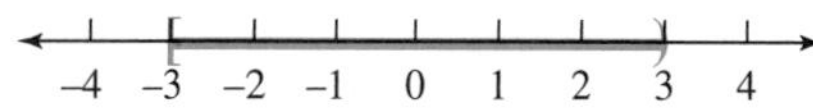

Which of the following numbers, when substituted for the variable in that compound inequality, would make it true?

3 -3 2.75 -3.5

15. To solve compound inequalities, the properties of inequalities are applied to all ______ parts of the inequality.
16. Solve $-4 < 2x < 12$ and give the solution set:
    a. using a graph
    b. using interval notation
    c. in words

NOTATION **Fill in the blanks.**

17. **a.** The symbol $<$ means "___________," and the symbol $>$ means "____________."
    b. The symbol $\geq$ means "____________ or equal to," and the symbol $\leq$ means "is less than __________."
18. In the interval $[4, 8)$, the endpoint 4 is ________, but the endpoint 8 is not included.
19. Give an example of each symbol: bracket, parenthesis, infinity, negative infinity.
20. Tell what is wrong with the notation that is used.
    a. The graph of the solution set for $x > 1$:

b. The solution set for $x > 5$ is $(5, \infty]$.

c. The solution set for $x < 7$ is $(7, -\infty)$.

d. $5 > 2 < 10$ is a true compound inequality.

Complete the solution to solve each inequality.

21.
$$4x - 5 \geq 7$$
$$4x - 5 + \underline{\quad} \geq 7 + \underline{\quad}$$
$$4x \geq \underline{\quad}$$
$$\frac{4x}{\underline{\quad}} \geq \frac{12}{\underline{\quad}}$$
$$x \geq 3$$

Solution set: $[\underline{\quad}, \infty)$

22. $-6x > 12$
$$\frac{-6x}{\underline{\quad}} \ \underline{\quad} \ \frac{12}{-6}$$
$$x < \underline{\quad}$$

Solution set: $(\underline{\quad}, -2)$

PRACTICE **Graph each inequality and describe the graph using interval notation.**

23. $x < 5$

24. $x \geq -2$

25. $-3 < x \leq 1$

26. $-1 \leq x \leq 3$

Write the inequality that is represented by each graph. Then describe the graph using interval notation.

27. 

28. 

29. 

30. 

Solve each inequality. Write the solution set in interval notation and graph it.

31. $x + 2 > 5$

32. $x + 5 \geq 2$

33. $3 + x < 2$

34. $5 + x > 3$

35. $g - 30 \geq -20$

36. $h - 18 \leq -3$

37. $\frac{2}{3}x \geq 2$

38. $\frac{3}{4}x < 3$

39. $\frac{y}{4} + 1 \leq -9$

40. $\frac{r}{8} - 7 \geq -8$

41. $7x - 1 > 5$

42. $3x + 10 \leq 5$

43. $0.5 \geq 2x - 0.3$

44. $0.8 > 7x - 0.04$

45. $-30y \leq -600$

46. $-6y \geq -600$

47. $-\frac{7}{8}x \leq 21$

48. $-\frac{3}{16}x \geq -9$

49. $-1 \leq -\frac{1}{2}n$

50. $-3 \geq -\frac{1}{3}t$

51. $\frac{m}{-42} - 1 > -1$

52. $\frac{a}{-25} + 3 < 3$

53. $-x - 3 \leq 7$

54. $-x - 9 > 3$

55. $-3x - 7 > -1$

56. $-5x + 7 \leq 12$

57. $-4x + 6 > 17$

58. $-3x - 0.5 < 0.4$

59. $2x + 9 \leq x + 8$

60. $3x + 7 \leq 4x - 2$

61. $9x + 13 \geq 8x$

62. $7x - 16 < 6x$

63. $8x + 4 > -(3x - 4)$

64. $7x + 6 \geq -(x - 6)$

65. $0.4x + 0.4 \leq 0.1x + 0.85$

66. $0.05 - 0.5x \leq -0.7 - 0.8x$

67. $7 < \frac{5}{3}a - 3$

68. $5 > \frac{7}{2}a - 9$

69. $7 - x \leq 3x - 2$

70. $9 - 3x \geq 6 + x$

71. $8(5 - x) \leq 10(8 - x)$

72. $17(3 - x) \geq 3 - 13x$

73. $\frac{1}{2} + \frac{n}{5} > \frac{3}{4}$

74. $\frac{1}{3} + \frac{c}{5} > -\frac{3}{2}$

75. $-\frac{2}{3} \geq \frac{2y}{3} - \frac{3}{4}$

76. $-\frac{2}{9} \geq \frac{5x}{6} - \frac{1}{3}$

77. $\frac{6x + 1}{4} \leq x + 1$

78. $\frac{3x - 10}{5} \leq x + 4$

79. $\frac{5}{2}(7x - 15) + x \geq \frac{13}{2}x - \frac{3}{2}$

80. $\frac{5}{3}(x + 1) \leq -x + \frac{2}{3}$

Solve each compound inequality. Write the solution set in interval notation and graph it.

81. $2 < x - 5 < 5$

82. $-8 < t - 8 < 8$

83. $0 \leq x + 10 \leq 10$

84. $-9 \leq x + 8 < 1$

85. $-3 \leq \frac{c}{2} \leq 5$

86. $-12 < \frac{b}{3} < 0$

87. $3 \leq 2x - 1 < 5$

88. $4 < 3x - 5 \leq 7$

89. $4 < -2x < 10$

90. $-4 \leq -4x < 12$

91. $0 < 10 - 5x \leq 15$

92. $-18 \leq 9(x - 5) < 27$

Solve each inequality. Write the solution set in interval notation and graph it.

93. $9(0.05 - 0.3x) + 0.162 \leq 0.081 + 15x$

94. $-1{,}630 \leq \frac{b + 312{,}451}{47} < 42{,}616$

APPLICATIONS

95. GRADES A student has test scores of 68%, 75%, and 79% in a government class. What must she score on the last exam to earn a B (80% or better) in the course?

96. OCCUPATIONAL TESTING Before taking on a client, an employment agency requires the applicant to average at least 70% on a battery of four job skills tests. If an applicant scored 70%, 74%, and 84% on the first three exams, what must he score on the fourth test to maintain a 70% or better average?

97. FLEET AVERAGES A car manufacturer produces three models in equal quantities. One model has an economy rating of 17 miles per gallon, and the second model is rated for 19 mpg. If government regulations require the manufacturer to have a fleet average of at least 21 mpg, what economy rating is required for the third model?

98. SERVICE CHARGES When the average daily balance of a customer's checking account falls below \$500 in any week, the bank assesses a \$5 service charge. The table shows the daily balances of one customer. What must Friday's balance be to avoid the service charge?

Day	Balance
Monday	\$540.00
Tuesday	\$435.50
Wednesday	\$345.30
Thursday	\$310.00

99. GEOMETRY The perimeter of an equilateral triangle is at most 57 feet. What could the length of a side be? (*Hint:* All three sides of an equilateral triangle are equal.)

100. GEOMETRY The perimeter of a square is no less than 68 centimeters. How long can a side be?

101. COUNTER SPACE In a large discount store, a rectangular counter is being built for the customer service department. If designers have determined that the outside perimeter of the counter (shown in red) needs to be at least 150 feet, determine the acceptable values for x.

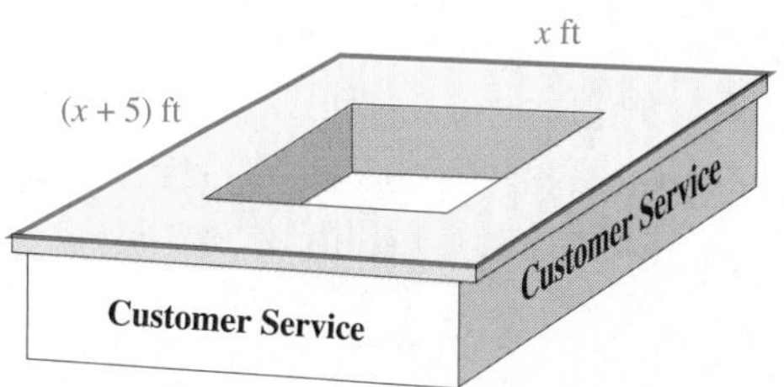

102. NUMBER PUZZLES What numbers satisfy the condition: Four more than three times the number is at most 10?

103. SAFETY CODES The illustration shows the acceptable and preferred angles of "pitch" or slope for ladders, stairs, and ramps. Use a compound inequality to describe each safe-angle range.

a. ramps or inclines

b. stairs

c. preferred range for stairs

d. ladders with cleats

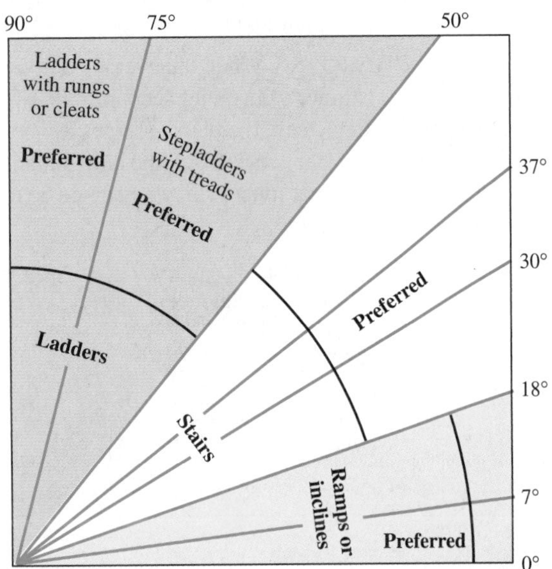

104. COMPARING TEMPERATURES To hold the temperature of a room between 19° and 22° Celsius, what Fahrenheit temperatures must be maintained? (*Hint:* Fahrenheit temperature (F) and Celsius temperature (C) are related by the formula $C = \frac{5}{9}(F - 32)$.)

105. WEIGHT CHART The graph is used to classify the weight of a baby boy from birth to 1 year. Estimate the weight range w for boys in the following classifications, using a compound inequality:

a. 10 months old, "heavy"

b. 5 months old, "light"

c. 8 months old, "average"

d. 3 months old, "moderately light"

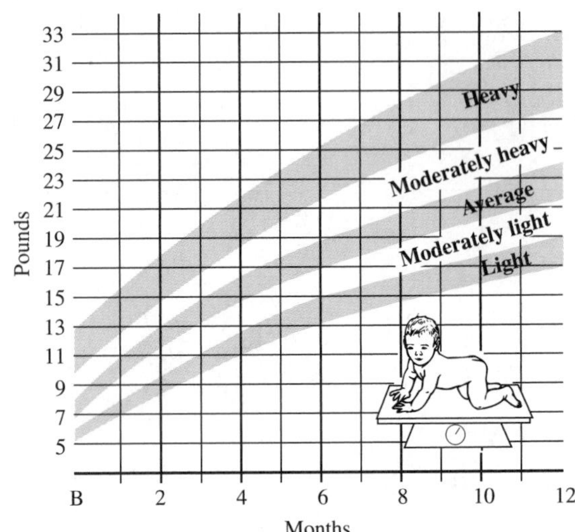

Based on data from *Better Homes and Gardens Baby Book* (Meredith Corp., 1969)

106. NUMBER PUZZLE What *whole* numbers satisfy the condition: Twice the number decreased by 1 is between 50 and 60?

WRITING

107. Explain why multiplying both sides of an inequality by a negative number reverses the direction of the inequality.

108. Explain the use of parentheses and brackets for graphing intervals.

REVIEW **Evaluate each expression.**

109. -5^3

110. $(-3)^4$

Complete each table.

111.

x	$x^2 - 3$
−2	
0	
3	

112.

x	$\frac{x}{3} + 2$
−6	
0	
12	

CHALLENGE PROBLEMS

113. Solve the inequality. Write the solution set in interval notation and graph it.

$$3 - x < 5 < 7 - x$$

114. Use a guess-and-check approach to solve $\frac{1}{x} > 1$. Write the solution set in interval notation and graph it.

ACCENT ON TEAMWORK

TRANSLATING KEY WORDS AND PHRASES

Overview: Students often say that the most challenging step of the five-step problem-solving strategy is forming an equation. This activity is designed to make that step easier by improving your translating skills.

Instructions: Form groups of 3 or 4 students. Select one person from your group to record the group's responses. Determine whether addition, subtraction, multiplication, or division is suggested by each of the following words or phrases. Then use the word or phrase in a sentence to illustrate its meaning.

deflate	recede	partition	evaporate	amplify
bisect	augment	hike	erode	boost
annexed	diminish	plummet	upsurge	wane
quadruple	corrode	taper off	trisect	broaden

GEOMETRIC SNACKS

Overview: This activity is designed to improve your ability to identify geometric figures and recall their perimeter, area, and volume formulas.

Instructions: Review the geometric figures and formulas inside the front cover. Then find some snack foods that have the shapes of the figures. For example, tortilla chips can be triangular in shape, and malted milk balls are spheres. If you are unable to find a particular shape already available, make a snack in that shape. Bring your collection of snacks to the next class. Form groups of 3 or 4 students. In your group, discuss the various shapes of the snacks, as well as their respective perimeter, area, and volume formulas.

COMPUTER SPREADSHEETS

Overview: In this activity, you will get some experience working with a spreadsheet.

Instructions: Form groups of 3 or 4 students. Examine the following spreadsheet, which consists of cells named by column and row. For example, 7 is entered in cell B3. In any cell you may enter data or a formula. For each formula in cells D1–D4 and E1–E4, the computer performs a calculation using values entered in other cells and prints the result in place of the formula. Find the value that will be printed in each formula cell. The symbol * means multiply, / means divide, and ∧ means raise to a power.

	A	B	C	D	E
1	−8	20	−6	$= 2*B1 - 3*C1 + 4$	$= B1 - 3*A1\wedge 2$
2	39	2	−1	$= A2/(B2 - C2)$	$= B3*B2*C2*2$
3	50	7	3	$= A3/5 + C3\wedge 3$	$= 65 - 2*(B3 - 5)\wedge 5$
4	6.8	−2.8	−0.5	$= 100*A4 + B4*C4$	$= A4/10 + A3/2*5$

KEY CONCEPT: SIMPLIFY AND SOLVE

Two of the most often used instructions in this book are **simplify** and **solve.** In algebra, we *simplify expressions* and we *solve equations and inequalities.*

To simplify an expression, we write it in a less complicated form. To do so, we apply the rules of arithmetic as well as algebraic concepts such as combining like terms, the distributive property, and the properties of 0 and 1.

To solve an equation or an inequality means to find the numbers that make the equation or inequality true when substituted for its variable. We use the addition, subtraction, multiplication, and division properties of equality or inequality to solve equations and inequalities. Quite often, we must simplify expressions on the left- or right-hand sides of an equation or inequality when solving it.

Use the procedures and the properties that we have studied to simplify the expression in part a and to solve the equation or inequality in part b.

Simplify	**Solve**
1. a. $-3x + 2 + 5x - 10$	**b.** $-3x + 2 + 5x - 10 = 4$
2. a. $4(y + 2) - 3(y + 1)$	**b.** $4(y + 2) = 3(y + 1)$
3. a. $\frac{1}{3}a + \frac{1}{3}a$	**b.** $\frac{1}{3}a + \frac{1}{3} = \frac{1}{2}$
4. a. $-(2x + 10)$	**b.** $-2x \geq -10$
5. a. $\frac{2}{3}(x - 2) - \frac{1}{6}(4x - 8)$	**b.** $\frac{2}{3}(x - 2) - \frac{1}{6}(4x - 8) = 0$

6. In the student's work on the right, where was the mistake made? Explain what the student did wrong.

Simplify: $2(x + 3) - x - 12$.

$$\begin{aligned} 2(x + 3) - x - 12 &= 2x + 6 - x - 12 \\ &= x - 6 \\ 0 &= x - 6 \\ 0 + 6 &= x - 6 + 6 \\ 6 &= x \end{aligned}$$

CHAPTER REVIEW

ELEMENTARY Algebra $f(x)$ Now™

SECTION 2.1 Solving Equations

CONCEPTS

A number that makes an equation a true statement when substituted for the variable is called a *solution* of the equation.

REVIEW EXERCISES

Decide whether the given number is a solution of the equation.

1. $84, x - 34 = 50$

2. $3, 5y + 2 = 12$

3. $-30, \frac{x}{5} = 6$

4. $2, a^2 - a - 1 = 0$

5. $-3, 5b - 2 = 3b - 8$

6. $1, \frac{2}{y + 1} = \frac{12}{y + 1} - 5$

To *solve an equation,* isolate the variable on one side of the equation by undoing the operations performed on it.

Equations with the same solutions are called *equivalent equations.*

If the same number is added to, or subtracted from, both sides of an equation, an equivalent equation results.

If both sides of an equation are multiplied, or divided, by the same nonzero number, an equivalent equation results.

7. Fill in the blanks: To solve $x + 8 = 10$ means to find all the values of the ________ that make the equation a ______ statement.

Solve each equation. Check each result.

8. $x - 9 = 12$

9. $y + 15 = -32$

10. $a + 3.7 = 16.9$

11. $100 = -7 + r$

12. $120 = 15c$

13. $t - \frac{1}{2} = \frac{1}{2}$

14. $\frac{t}{8} = -12$

15. $3 = \frac{q}{2.6}$

16. $6b = 0$

17. $\frac{x}{14} = 0$

18. GEOMETRY Find the unknown angle measure.

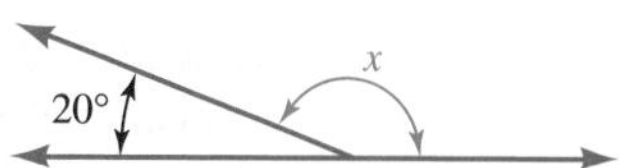

SECTION 2.2 Problem Solving

To solve a problem, follow these steps:

1. Analyze the problem.
2. Form an equation.
3. Solve the equation.
4. State the conclusion.
5. Check the result.

Drawing a diagram or creating a table is often helpful in problem solving.

19. WOMEN'S SOCCER The U.S. National Women's team won the 1999 World Cup. On the 20-player roster were 3 goalkeepers, 6 defenders, 6 midfielders, and a group of forwards. How many forwards were on the team?

20. HISTORIC TOURS A driving tour of three historic cities is an 858-mile round trip. Beginning in Boston, the drive to Philadelphia is 296 miles. From Philadelphia to Washington, D.C., is another 133 miles. How long will the return trip to Boston be?

21. CHROME WHEELS Find the measure of the angle between each of the spokes on the wheel shown.

To solve applied percent problems, use the facts of the problem to write a percent sentence of the form:

is % of .

Translate the sentence to mathematical symbols: *is* translates to an = symbol and *of* means multiply. Then, solve the resulting equation.

22. ADVERTISING In 2001, \$231 billion was spent on advertising in the United States. The circle graph gives the individual expenditures in percents. Find the amount of money spent on television advertising. Round to the nearest billion dollars.

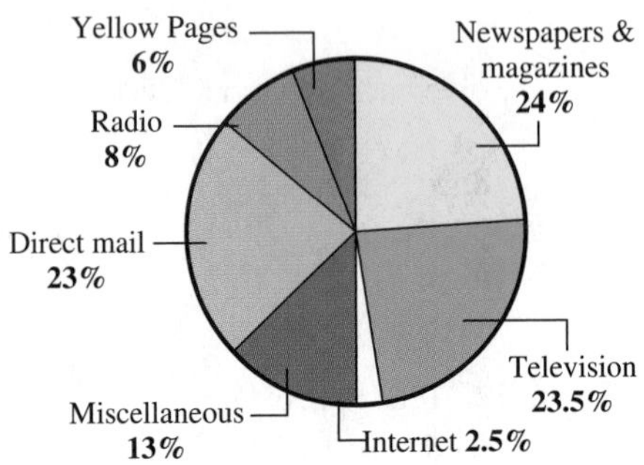

Based on data from *Advertising Age*

23. COST OF LIVING A retired trucker receives a monthly Social Security check of \$764. If she is to receive a 3.5% cost-of-living increase soon, how much larger will her check be?

24. 4.81 is 2.5% of what number?

25. FAMILY BUDGETS It is recommended that a family pay no more than 30% of its monthly income (after taxes) on housing. If a family has an after-tax income of \$1,890 per month and pays \$625 in housing costs each month, are they within the recommended range?

To find *percent of increase or decrease,* find what percent the increase or decrease is of the original amount.

26. COLLECTIBLES A collector of football trading cards paid \$6 for a 1984 Dan Marino rookie card several years ago. If the card is now worth \$100, what is the percent of increase in the card's value? (Round to the nearest one percent.)

SECTION 2.3 Simplifying Algebraic Expressions

To *simplify* an algebraic expression means to write it in less complicated form.

The *distributive property:*

$a(b + c) = ab + ac$

$a(b - c) = ab - ac$

$a(b + c + d) = ab + ac + ad$

Simplify each expression.

27. $-4(7w)$

28. $-3r(-5)$

29. $3(-2x)(-4)$

30. $0.4(5.2f)$

31. $15\left(\frac{3}{5}a\right)$

32. $\frac{7}{2} \cdot \frac{2}{7}r$

Remove parentheses.

33. $5(x + 3)$

34. $-2(2x + 3 - y)$

35. $-(a - 4)$

36. $\frac{3}{4}(4c - 8)$

37. $40\left(\frac{x}{2} + \frac{4}{5}\right)$

38. $2(-3c - 7)(2.1)$

Like terms are terms with exactly the same variables raised to exactly the same powers.

To add or subtract like terms, combine their coefficients and keep the same variables with the same exponents.

Simplify each expression by combining like terms.

39. $8p + 5p - 4p$

40. $-5m + 2 - 2m - 2$

41. $n + n + n + n$

42. $5(p - 2) - 2(3p + 4)$

43. $55.7k - 55.6k$

44. $8a^3 + 4a^3 - 20a^3$

45. $\frac{3}{5}w - \left(-\frac{2}{5}w\right)$

46. $36\left(\frac{1}{9}h - \frac{3}{4}\right) + 36\left(\frac{1}{3}\right)$

47. Write an algebraic expression in simplified form for the perimeter of the triangle.

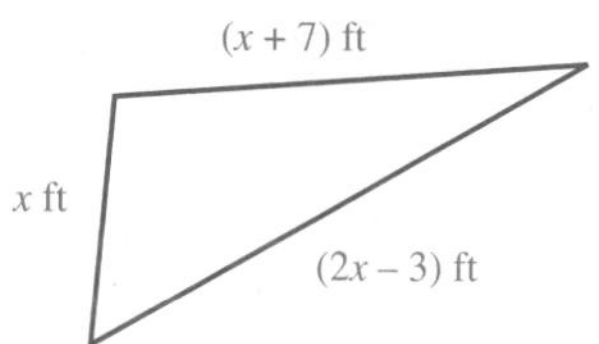

SECTION 2.4 More about Solving Equations

To solve an equation means to find all the values of the variable that, when substituted for the variable, make a true statement.

An equation that is true for all values of its variable is called an *identity*.

An equation that is not true for any values of its variable is called a *contradiction*.

Solve each equation. Check the result.

48. $5x + 4 = 14$

49. $98.6 - t = 129.2$

50. $\frac{n}{5} - 2 = 4$

51. $\frac{b - 5}{4} = -6$

52. $5(2x - 4) - 5x = 0$

53. $-2(x - 5) = 5(-3x + 4) + 3$

54. $\frac{3}{4} = \frac{1}{2} + \frac{d}{5}$

55. $-\frac{2}{3}f = 4$

56. $\frac{3(2 - c)}{2} = \frac{-2(2c + 3)}{5}$

57. $\frac{b}{3} + \frac{11}{9} + 3b = -\frac{5}{6}b$

58. $3(a + 8) = 6(a + 4) - 3a$

59. $2(y + 10) + y = 3(y + 8)$

SECTION 2.5 Formulas

A *formula* is an equation that is used to state a known relationship between two or more variables.

Retail price: $r = c + m$

Profit: $p = r - c$

Distance: $d = rt$

Temperature: $C = \frac{5}{9}(F - 32)$

Formulas from geometry:

Square: $P = 4s, A = s^2$

Rectangle: $P = 2l + 2w,$ $A = lw$

Triangle: $P = a + b + c$ $A = \frac{1}{2}bh$

Trapezoid: $P = a + b + c + d$ $A = \frac{1}{2}h(b + d)$

Circle: $D = 2r$ $C = 2\pi r$ $A = \pi r^2$

Rectangular solid: $V = lwh$

Cylinder: $V = \pi r^2 h$

Pyramid: $V = \frac{1}{3}Bh$ (B is the area of the base)

Cone: $V = \frac{1}{3}\pi r^2 h$

Sphere: $V = \frac{4}{3}\pi r^3$

60. Find the markup on a CD player whose wholesale cost is \$219 and whose retail price is \$395.

61. One month, a restaurant had sales of \$13,500 and made a profit of \$1,700. Find the expenses for the month.

62. INDY 500 In 2002, the winner of the Indianapolis 500-mile automobile race averaged 166.499 mph. To the nearest hundredth of an hour, how long did it take him to complete the race?

63. JEWELRY MAKING Gold melts at about 1,065°C. Change this to degrees Fahrenheit.

64. CAMPING Find the perimeter of the air mattress.

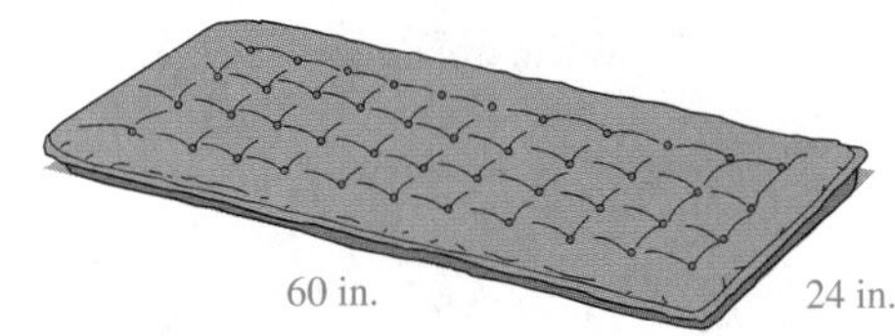

65. CAMPING Find the amount of sleeping area on the top surface of the air mattress.

66. Find the area of a triangle with a base 17 meters long and a height of 9 meters.

67. Find the area of a trapezoid with bases 11 inches and 13 inches long and a height of 12 inches.

68. To the nearest hundredth, find the circumference of a circle with a radius of 8 centimeters.

69. To the nearest hundredth, find the area of the circle in Problem 68.

70. CAMPING Find the approximate volume of the air mattress in Problem 64 if it is 3 inches thick.

71. Find the volume of a 12-foot cylinder whose circular base has a radius of 0.5 feet. Give the result to the nearest tenth.

72. Find the volume of a pyramid that has a square base, measuring 6 feet on a side, and a height of 10 feet.

73. HALLOWEEN After being cleaned out, a spherical-shaped pumpkin has an inside diameter of 9 inches. To the nearest hundredth, what is its volume?

Solve each formula for the required variable.

74. $A = 2\pi rh$ for h

75. $A - BC = \frac{G - K}{3}$ for G

76. $a^2 + b^2 = c^2$ for b^2

77. $4y - 16 = 3x$ for y

SECTION 2.6 More about Problem Solving

To solve problems, use the five-step problem-solving strategy.

1. Analyze the problem.
2. Form an equation.
3. Solve the equation.
4. State the conclusion.
5. Check the result.

78. SOUND SYSTEMS A 45-foot-long speaker wire is to be cut into three pieces. One piece is to be 15 feet long. Of the remaining pieces, one must be 2 feet less than 3 times the length of the other. Find the length of the shorter piece.

79. AUTOGRAPHS Kesha collected the autographs of 8 more television celebrities than she has of movie stars. Each TV celebrity autograph is worth \$75 and each movie star autograph is worth \$250. If her collection is valued at \$1,900, how many of each type of autograph does she have?

80. ART HISTORY *American Gothic* was painted in 1930 by Grant Wood. The length of the rectangular painting is 5 inches more than the width. Find the dimensions of the painting if it has a perimeter of $109\frac{1}{2}$ inches.

The sum of the measures of the angles of a triangle is 180°.

81. Find the missing angle measures of the triangle.

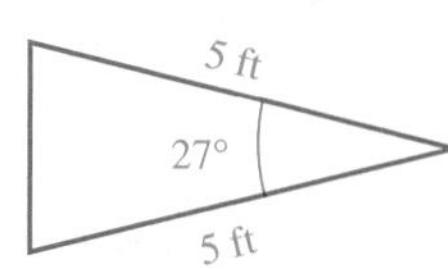

Total value = number · value

Interest = principal · rate · time

$I = Prt$

82. Write an expression to represent the value of x video games each costing \$45.

83. INVESTMENT INCOME A woman has \$27,000. Part is invested for 1 year in a certificate of deposit paying 7% interest, and the remaining amount in a cash management fund paying 9%. After 1 year, the total interest on the two investments is \$2,110. How much is invested at each rate?

Distance = rate · time

$d = rt$

84. WALKING AND BICYCLING A bicycle path is 5 miles long. A man walks from one end at the rate of 3 mph. At the same time, a friend bicycles from the other end, traveling at 12 mph. In how many minutes will they meet?

The value v of a commodity is its price per pound p times the number of pounds n:

$$v = pn$$

85. MIXTURES A store manager mixes candy worth 90¢ per pound with gumdrops worth \$1.50 per pound to make 20 pounds of a mixture worth \$1.20 per pound. How many pounds of each kind of candy does he use?

86. SOLUTIONS How much acetic acid is in x gallons of a solution that is 12% acetic acid?

SECTION 2.7 Solving Inequalities

An *inequality* is a mathematical expression that contains a $>$, $<$, $\geq$, or $\leq$ symbol.

A *solution of an inequality* is any number that makes the inequality true.

A *parenthesis* indicates that a number is not on the graph. A *bracket* indicates that a number is included in the graph.

Interval notation can be used to describe a set of real numbers.

Solve each inequality. Write the solution set in interval notation and graph it.

87. $3x + 2 < 5$

88. $-\frac{3}{4}x \geq -9$

89. $\frac{3}{4} < \frac{d}{5} + \frac{1}{2}$

90. $5(3 - x) \leq 3(x - 3)$

91. $8 < x + 2 < 13$

92. $0 \leq 3 - 2x < 10$

93. SPORTS EQUIPMENT The acceptable weight of Ping-Pong balls used in competition can range from 2.40 to 2.53 grams. Express this range using a compound inequality.

94. SIGNS A large office complex has a strict policy about signs. Any sign to be posted in the building must meet three requirements:

- It must be rectangular in shape.
- Its width must be 18 inches.
- Its perimeter is not to exceed 132 inches.

What possible sign lengths meet these specifications?

CHAPTER 2 TEST

ELEMENTARY Algebra $f(x)$ Now™

1. Is 3 a solution of $2x + 3 = 4x - 6$?

2. EXERCISING Find x.

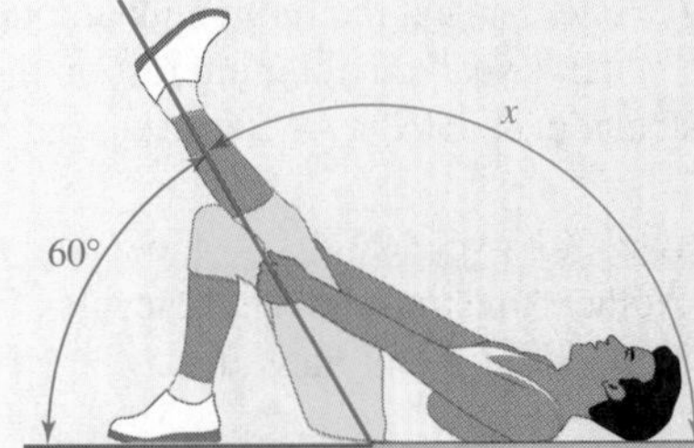

3. MULTIPLE BIRTHS IN THE UNITED STATES In 2000, about 7,322 women gave birth to three or more babies at one time. This is about seven times the number of such births in 1971, 29 years earlier. How many multiple births occurred in 1971?

4. DOWN PAYMENTS To buy a house, a woman was required to make a down payment of \$11,400. What did the house sell for if this was 15% of the purchase price?

5. SPORTS In sports, percentages are most often expressed as three-place decimals instead of percents. For example, if a basketball player makes 75.8% of his free throws, the sports page will list this as .758. Use this format to complete the table.

All-time best regular-season winning percentages		
Team	**Won–lost record**	**Winning percentage**
1996 Chicago Bulls Basketball	72–10	
1972 Miami Dolphins Football	14–0	

6. BODY TEMPERATURES Suppose a person's body temperature rises from 98.6°F to 101.6°F. What is the percent increase? Round to the nearest one percent.

7. Find the expression that represents the perimeter of the rectangle.

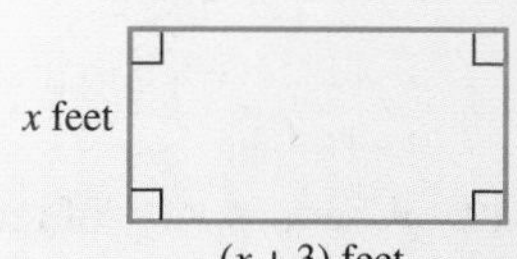

8. What property is illustrated below?

$$2(x + 7) = 2x + 2(7)$$

Simplify each expression.

9. $5(-4x)$

10. $-8(-7t)(4)$

11. $\frac{4}{5}(15a + 5) - 16a$

12. $-1.1d^2 - 3.8d^2 - d^2$

Solve each equation.

13. $5h + 8 - 3h + h = 8$

14. $\frac{4}{5}t = -4$

15. $\frac{11(b - 1)}{5} = 3b - 2$

16. $0.8x + 1.4 = 2.9 + 0.2x$

17. $\frac{m}{2} - \frac{1}{3} = \frac{1}{4}$

18. $23 - 5(x + 10) = -12$

19. Solve the equation for the variable indicated.
$A = P + Prt$; for r

20. On its first night of business, a pizza parlor brought in \$445. The owner estimated his costs that night to be \$295. What was the profit?

21. Find the Celsius temperature reading if the Fahrenheit reading is 14°.

22. PETS The spherical fishbowl is three-quarters full of water. To the nearest cubic inch, find the volume of water in the bowl.

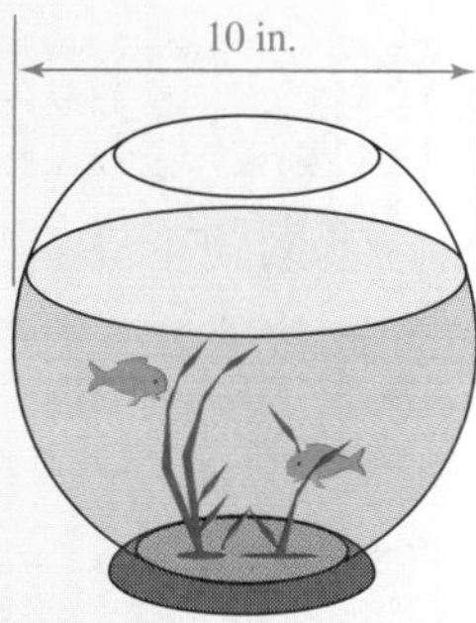

23. TRAVEL TIMES A car leaves Rockford, Illinois, at the rate of 65 mph, bound for Madison, Wisconsin. At the same time, a truck leaves Madison at the rate of 55 mph, bound for Rockford. If the cities are 72 miles apart, how long will it take for the car and the truck to meet?

24. SALT SOLUTIONS How many liters of a 2% brine solution must be added to 30 liters of a 10% brine solution to dilute it to an 8% solution?

25. GEOMETRY If the vertex angle of an isosceles triangle is 44°, find the measure of each base angle.

26. INVESTMENT PROBLEM Part of \$13,750 is invested at 9% annual interest, and the rest is invested at 8%. After one year, the accounts paid \$1,185 in interest. How much was invested at the lower rate?

Solve each inequality. Write the solution set in interval notation and graph it.

27. $-8x - 20 \le 4$

28. $-4 \le 2(x + 1) < 10$

29. DRAFTING In the illustration, the ± (read "plus or minus") symbol means that the width of a plug a manufacturer produces can range from 1.497 − 0.001 inches to 1.497 + 0.001 inches. Write the range of acceptable widths w for the plug using a compound inequality.

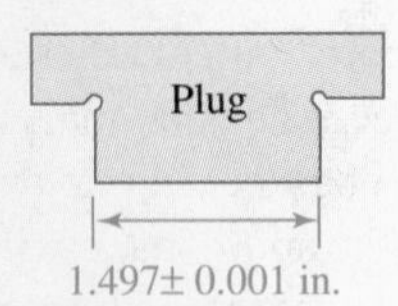

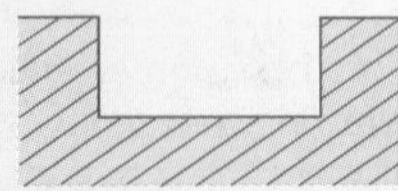

30. Solve: $2(y - 7) - 3y = -(y - 3) - 17$. Explain why the solution set is all real numbers.

CHAPTERS 1–2 CUMULATIVE REVIEW EXERCISES

1. Classify each of the following as an equation or an expression.
a. $4m - 3 + 2m$ **b.** $4m = 3 + 2m$

2. Use the formula $t = \frac{w}{5}$ to complete the table.

Weight (lb)	Cooking time (hr)
15	
20	
25	

3. Give the prime factorization of 200.

4. Simplify: $\frac{24}{36}$.

5. Multiply: $\frac{11}{21}\left(-\frac{14}{33}\right)$.

6. COOKING A recipe calls for $\frac{3}{4}$ cup of flour, and the only measuring container you have holds $\frac{1}{8}$ of a cup. How many $\frac{1}{8}$ cups of flour would you need to add to follow the recipe?

7. Add: $\frac{4}{5} + \frac{2}{3}$.

8. Subtract: $42\frac{1}{8} - 29\frac{2}{3}$.

9. Write $\frac{15}{16}$ as a decimal.

10. Multiply: 0.45(100).

11. Evaluate each expression.
a. $|-65|$ **b.** $-|-12|$

12. What property of real numbers is illustrated below?

$x \cdot 5 = 5x$

Classify each number as a natural number, a whole number, an integer, a rational number, an irrational number, and a real number. Each number may have several classifications.

13. 3

14. -1.95

15. $\frac{17}{20}$

16. π

17. Write each product using exponents.
a. $4 \cdot 4 \cdot 4$ **b.** $\pi \cdot r \cdot r \cdot h$

18. Perform each operation.
a. $-6 + (-12) + 8$
b. $-15 - (-1)$
c. $2(-32)$
d. $\frac{0}{35}$ **e.** $\frac{-11}{11}$

19. Write each phrase as an algebraic expression.
a. The sum of the width w and 12.
b. Four less than a number n.

20. SICK DAYS Use the data in the table to find the average (mean) number of sick days used by this group of employees this year.

Name	Sick days	Name	Sick days
Chung	4	Ryba	0
Cruz	8	Nguyen	5
Damron	3	Tomaka	4
Hammond	2	Young	6

21. Complete the table.

x	$x^2 - 3$
-2	
0	
3	

22. Translate to mathematical symbols.

The loudness of a stereo speaker	is	2,000	divided by	the square of the distance of the listener from the speaker.

23. LAND OF THE RISING SUN The flag of Japan is a red disc (representing sincerity and passion) on a white background (representing honesty and purity).
a. What is the area of the rectangular-shaped flag?
b. To the nearest tenth of a square foot, what is the area of the red disc?
c. Use the results from parts a and b to find what percent of the area of the Japanese flag is occupied by the red disc.

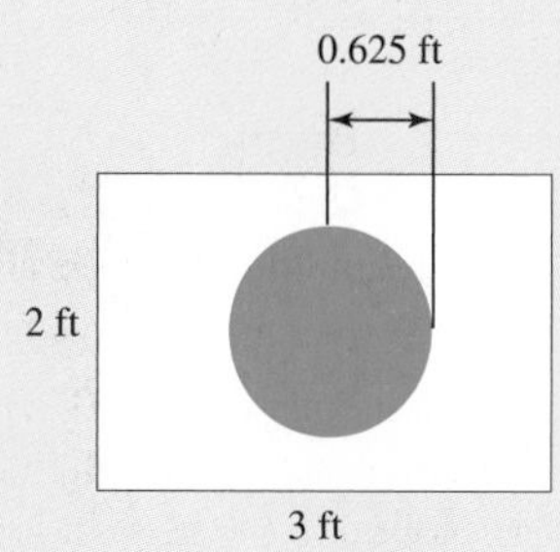

24. 45 is 15% of what number?

Let $x = -5$, $y = 3$, and $z = 0$. Evaluate each expression.

25. $(3x - 2y)z$

26. $\dfrac{x - 3y + |z|}{2 - x}$

27. $x^2 - y^2 + z^2$

28. $\dfrac{x}{y} + \dfrac{y + 2}{3 - z}$

Simplify each expression.

29. $-8(4d)$

30. $5(2x - 3y + 1)$

31. $2x + 3x - x$

32. $3a^2 + 6a^2 - 17a^2$

33. $\dfrac{2}{3}(15t - 30) + t - 30$

34. $5(t - 4) + 3t$

35. What is the length of the longest side of the triangle shown below?

36. Write an algebraic expression in simplest form for the perimeter of the triangle.

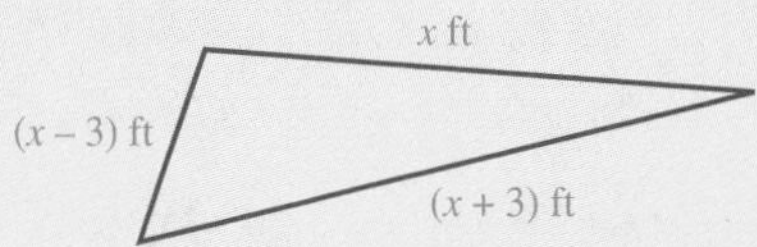

Solve each equation.

37. $3x - 4 = 23$

38. $\dfrac{x}{5} + 3 = 7$

39. $-5p + 0.7 = 3.7$

40. $\dfrac{y - 4}{5} = 3 - y$

41. $-\dfrac{4}{5}x = 16$

42. $\dfrac{1}{2} + \dfrac{x}{5} = \dfrac{3}{4}$

43. $-9(n + 2) - 2(n - 3) = 10$

44. $\frac{2}{3}(r - 2) = \frac{1}{6}(4r - 1) + 1$

45. Find the area of a rectangle with sides of 5 meters and 13 meters.

46. Find the volume of a cone that is 10 centimeters tall and has a circular base whose diameter is 12 centimeters. Round to the nearest hundredth.

47. Solve $V = \frac{1}{3}\pi r^2 h$ for r^2.

48. WORK Physicists say that work is done when an object is moved a distance d by a force F. To find the work done, we can use the formula $W = Fd$. Find the work done in lifting the bundle of newspapers onto the workbench. (*Hint:* The force that must be applied to lift the newspapers is equal to the weight of the newspapers.)

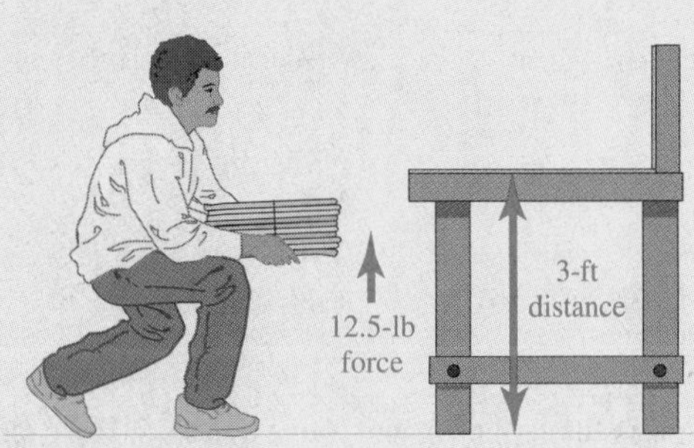

49. WORK See Exercise 48. Find the weight of a 1-gallon can of paint if the amount of work done to lift it onto the workbench is 28.35 foot-pounds.

50. Find the unknown angle measures.

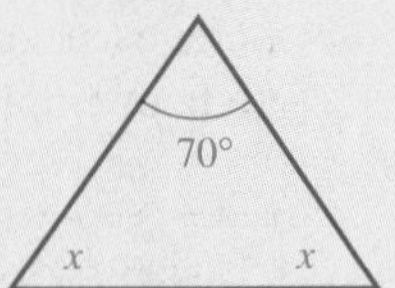

51. INVESTING An investment club invested part of \$10,000 at 9% annual interest and the rest at 8%. If the annual income from these investments was \$860, how much was invested at 8%?

52. GOLDSMITH How many ounces of a 40% gold alloy must be mixed with 10 ounces of a 10% gold alloy to obtain an alloy that is 25% gold?

Solve each inequality. Write the solution set in interval notation and graph it.

53. $x - 4 > -6$

54. $-6x \geq -12$

55. $8x + 4 \geq 5x + 1$

56. $-1 \leq 2x + 1 < 5$

Chapter

3 Linear Equations and Inequalities in Two Variables

ELEMENTARY **Algebra $f(x)$ Now™**

Throughout the chapter, this icon introduces resources on the Elementary AlgebraNow Web site, accessed through **http://1pass.thomson.com**, that will

- Help you test your knowledge of the material with a pre-test and a post-test
- Provide a personalized learning plan targeting areas you should study

3.1 Graphing Using the Rectangular Coordinate System

3.2 Graphing Linear Equations

3.3 More about Graphing Linear Equations

3.4 The Slope of a Line

3.5 Slope–Intercept Form

3.6 Point–Slope Form

3.7 Graphing Linear Inequalities

Accent on Teamwork

Key Concept

Chapter Review

Chapter Test

Cumulative Review Exercises

Getty/Stone/Glen Allison

Snow skiing is one of our country's most popular recreational activities. Whether on the gentle incline of a cross-country trip or racing down a near-vertical mountainside, a skier constantly adapts to the steepness of the course. In this chapter, we discuss lines and a means of measuring their steepness, called *slope.* The concept of slope has a wide variety of applications, including roofing, road design, and handicap accessible ramps.

To learn more about the slope of a line, visit *The Learning Equation* on the Internet at http://tle.brookscole.com. (The log-in instructions are in the Preface.) For Chapter 3, the following online lessons are:

- *TLE* Lesson 6: Equations Containing Two Variables
- *TLE* Lesson 7: Rate of Change and the Slope of a Line

Relationships between two quantities can be described by a table, a graph, or an equation.

3.1 Graphing Using the Rectangular Coordinate System

- The Rectangular Coordinate System
- Graphing Mathematical Relationships
- Reading Graphs

It is often said, "A picture is worth a thousand words." That is certainly true when it comes to graphs. Graphs present data in an attractive and informative way. **Bar graphs** enable us to make quick comparisons, **line graphs** let us notice trends, and **circle graphs** show the relationship of the parts to the whole.

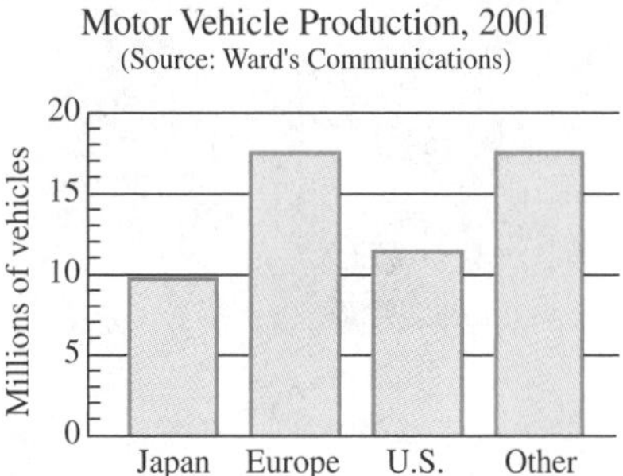

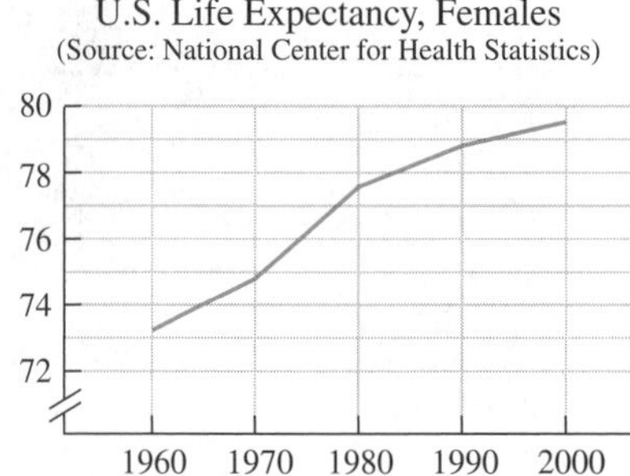

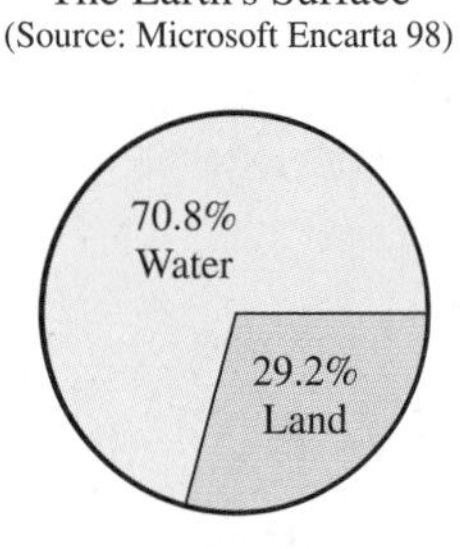

We will now introduce another type of graph that is widely used in mathematics called a *rectangular coordinate graph.*

THE RECTANGULAR COORDINATE SYSTEM

When designing the Gateway Arch in St. Louis, architects created a mathematical model called a **rectangular coordinate graph.** This graph, shown below on the right, is drawn on a grid called a **rectangular coordinate system.** This coordinate system is also called a **Cartesian coordinate system,** after the 17th-century French mathematician René Descartes.

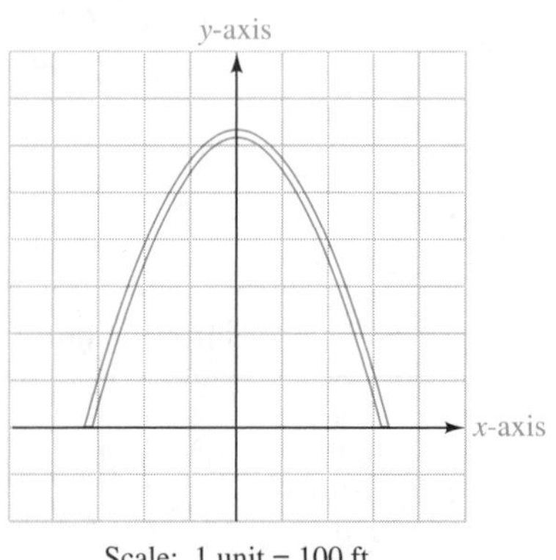

Scale: 1 unit = 100 ft

A rectangular coordinate system is formed by two perpendicular number lines. The horizontal number line is usually called the ***x*-axis,** and the vertical number line is usually called the ***y*-axis.** On the x-axis, the positive direction is to the right. On the y-axis, the positive direction is upward. Each axis should be scaled to fit the data. For example, the axes of the graph of the arch are scaled in units of 100 feet.

The point where the axes intersect is called the **origin.** This is the zero point on each axis. The axes form a **coordinate plane,** and they divide it into four regions called **quadrants,** which are numbered using Roman numerals.

The Language of Algebra

The word *axis* is used in mathematics and science. For example, Earth rotates on its *axis* once every 24 hours.

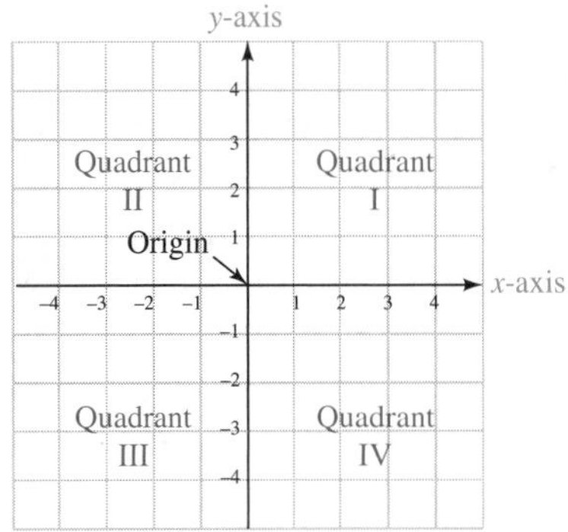

Each point in a coordinate plane can be identified by an **ordered pair** of real numbers x and y written in the form (x, y). The first number x in the pair is called the ***x*-coordinate,** and the second number y is called the ***y*-coordinate.** Some examples of such pairs are $(3, -4)$, $\left(-1, -\frac{3}{2}\right)$, and $(0, 2.5)$.

$(3, -4)$

↑ The x-coordinate ↑ The y-coordinate

Notation

Don't be confused by this new use of parentheses. The notation (3, −4) represents a point on the coordinate plane, whereas 3(−4) indicates multiplication. Also, don't confuse the ordered pair with interval notation.

The process of locating a point in the coordinate plane is called **graphing** or **plotting** the point. Below, we use blue arrows to show how to graph the point with coordinates $(3, -4)$. Since the x-coordinate, 3, is positive, we start at the origin and move 3 units to the *right* along the x-axis. Since the y-coordinate, -4, is negative, we then move *down* 4 units, and draw a dot. This locates the point $(3, -4)$.

In the figure, red arrows are used to show how to plot the point $(-4, 3)$. We start at the origin, move 4 units to the *left* along the x-axis, then move *up* 3 units and draw a dot. This locates the point $(-4, 3)$.

The Language of Algebra

Note that the point (3, −4) has a different location than the point (−4, 3). Since the order of the coordinates of a point is important, we call them *ordered pairs.*

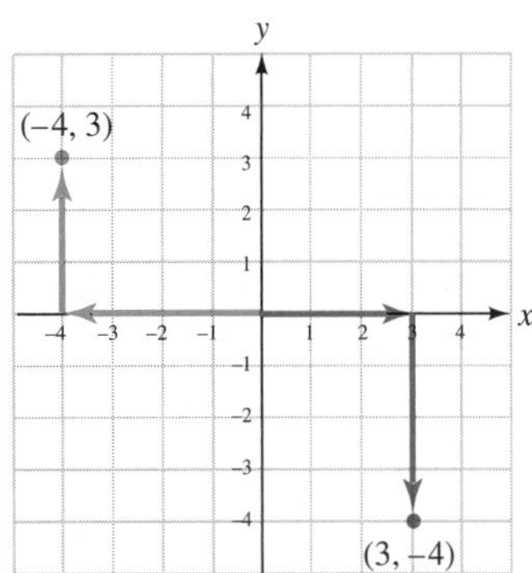

EXAMPLE 1

ELEMENTARY Algebra f(x) Now™

Plot each point and state the quadrant in which it lies.

a. $(4, 4)$ **b.** $\left(-1, -\frac{7}{2}\right)$ **c.** $(0, 2.5)$ **d.** $(-3, 0)$ **e.** $(0, 0)$

Solution

a. To plot the point $(4, 4)$, we begin at the origin, move 4 units to the *right* on the x-axis, and then move 4 units *up*. The point lies in quadrant I.

b. To plot the point $\left(-1, -\frac{7}{2}\right)$, we begin at the origin, move 1 unit to the *left*, and then move $\frac{7}{2}$ units, or $3\frac{1}{2}$ units, *down*. The point lies in quadrant III.

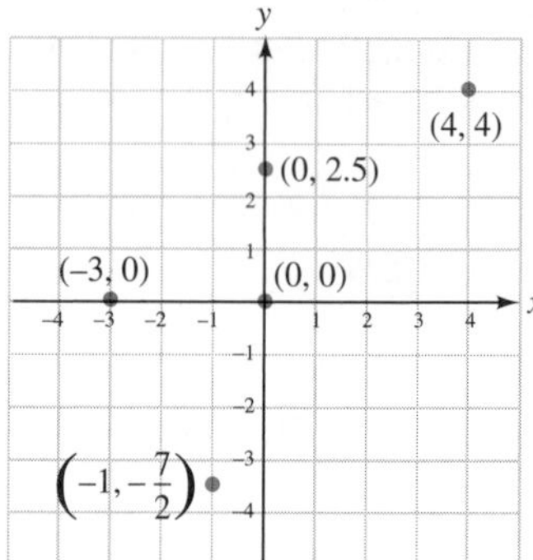

c. To plot the point $(0, 2.5)$, we begin at the origin and do not move right or left, because the x-coordinate is 0. Since the y-coordinate is positive, we move 2.5 units *up*. The point lies on the y-axis.

d. To plot the point $(-3, 0)$, we begin at the origin and move 3 units to the *left*. Since the y-coordinate is 0, we do not move up or down. The point lies on the x-axis.

e. To plot the point $(0, 0)$, we begin at the origin, and we remain there because both coordinates are 0. The point with coordinates $(0, 0)$ is the origin.

Success Tip

Points with an x-coordinate that is 0 lie on the y-axis. Points with a y-coordinate that is 0 lie on the x-axis. Points that lie on an axis are not considered to be in any quadrant.

Self Check 1 Plot the points:

a. $(2, -2)$ **b.** $(-4, 0)$ **c.** $\left(1.5, \frac{5}{2}\right)$ **d.** $(0, 5)$

EXAMPLE 2

ELEMENTARY Algebra f(x) Now™

Find the coordinates of points A, B, C, D, E, and F plotted below.

Notation

Points are labeled with capital letters. The notation $A(2, 3)$ indicates that point A has coordinates $(2, 3)$.

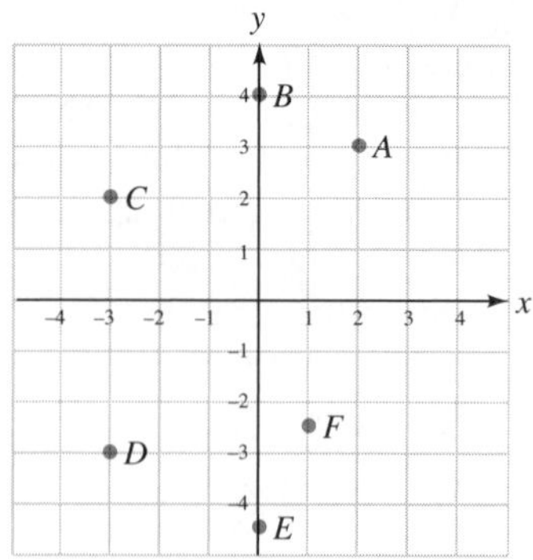

Solution To locate point A, we start at the origin, move 2 units to the right, and then 3 units up. Its coordinates are $(2, 3)$. The coordinates of the other points are found in the same manner.

$B(0, 4)$ $C(-3, 2)$ $D(-3, -3)$ $E(0, -4.5)$ $F(1, -2.5)$

Self Check 2 Find the coordinates of each point.

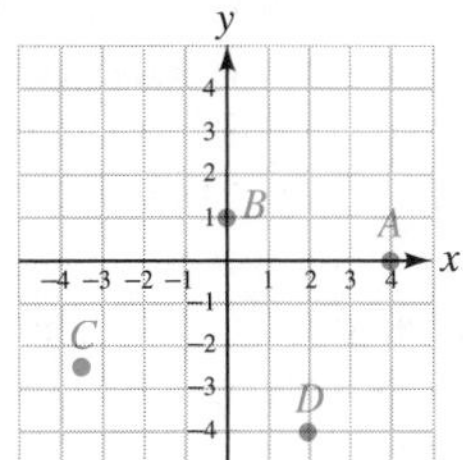

GRAPHING MATHEMATICAL RELATIONSHIPS

Every day, we deal with quantities that are related:

- The time it takes to cook a roast depends on the weight of the roast.
- The money we earn depends on the number of hours we work.
- The sales tax that we pay depends on the price of the item purchased.

We can use graphs to visualize such relationships. For example, suppose a tub is filling with water, as shown below. Obviously, the amount of water in the tub depends on how long the water has been running. To graph this relationship, we can use the measurements that were taken as the tub began to fill.

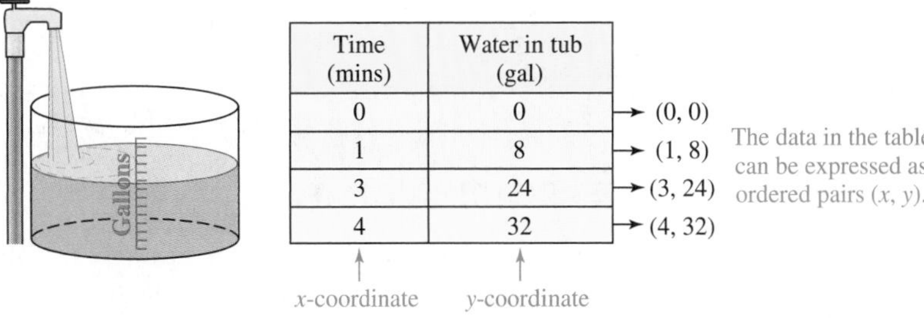

The data in each row of the table can be written as an ordered pair and plotted on a rectangular coordinate system. Since the first coordinate of each ordered pair is a time, we label the x-axis *Time (min).* The second coordinate is an amount of water, so we label the y-axis *Amount of water (gal).* The y-axis is scaled in larger units (multiples of 4 gallons) because the size of the data ranges from 0 to 32 gallons.

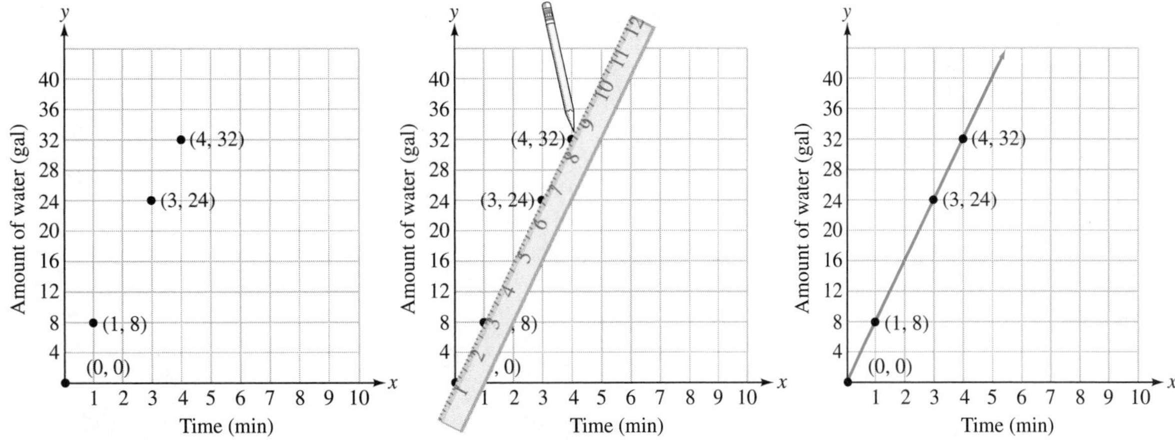

After plotting the ordered pairs, we use a straightedge to draw a line through the points. As we would expect, the completed graph shows that the amount of water in the tub increases steadily as the water is allowed to run.

We can use the graph to determine the amount of water in the tub at various times. For example, the green dashed line on the graph below shows that in 2 minutes, the tub will contain 16 gallons of water. This process, called **interpolation,** uses known information to predict values that are not known but are *within* the range of the data. The blue dashed line on the graph shows that in 5 minutes, the tub will contain 40 gallons of water. This process, called **extrapolation,** uses known information to predict values that are not known and are *outside* the range of the data.

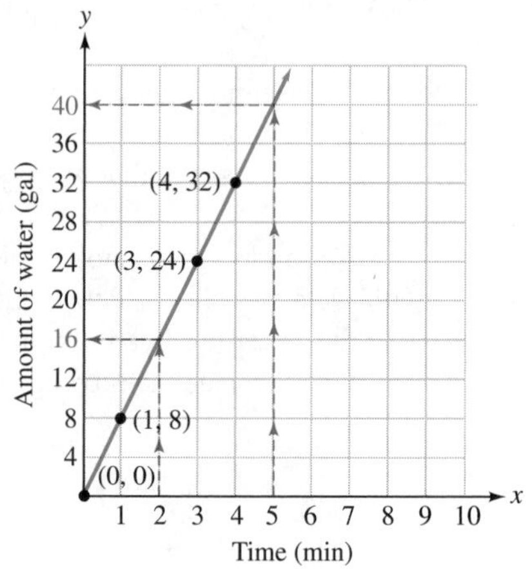

READING GRAPHS

Since graphs are becoming an increasingly popular way to present information, the ability to read and interpret them is becoming ever more important.

EXAMPLE 3

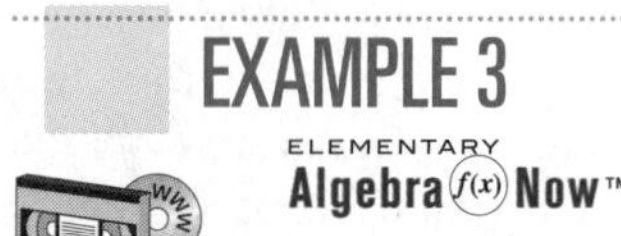

TV taping. The graph below shows the number of people in an audience before, during, and after the taping of a television show. Use the graph to answer the following questions.

a. How many people were in the audience when the taping began?
b. At what times were there exactly 100 people in the audience?
c. How long did it take the audience to leave after the taping ended?

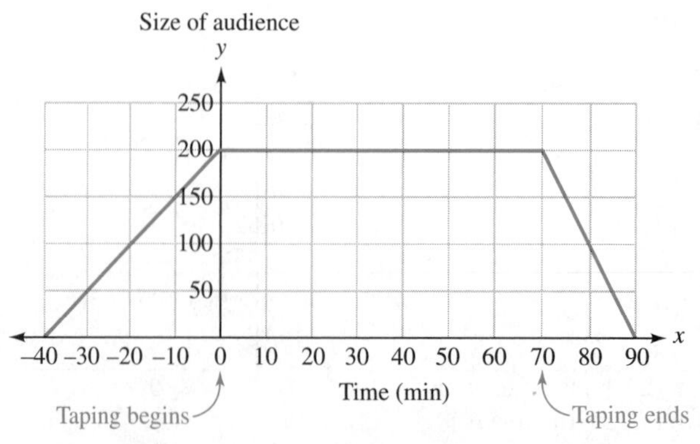

The Language of Algebra

A rectangular coordinate system is a *grid*—a network of uniformly spaced perpendicular lines. At times, some large U.S. cities have such horrible traffic congestion that vehicles can barely move, if at all. The condition is called *gridlock.*

Solution For each part of the solution, refer to the graph below. We can use the coordinates of specific points on the graph to answer these questions.

a. The time when the taping began is represented by 0 on the x-axis. The point on the graph directly above 0 is (0, 200). The y-coordinate indicates that 200 people were in the audience when the taping began.

b. We can draw a horizontal line passing through 100 on the y-axis. Since the line intersects the graph twice, at (−20, 100) and at (80, 100), there are two times when 100 people were in the audience. The x-coordinates of the points tell us those times: 20 minutes before the taping began, and 80 minutes after.

c. The x-coordinate of the point (70, 200) tells us when the audience began to leave. The x-coordinate of (90, 0) tells when the exiting was completed. Subtracting the x-coordinates, we see that it took $90 - 70 = 20$ minutes for the audience to leave.

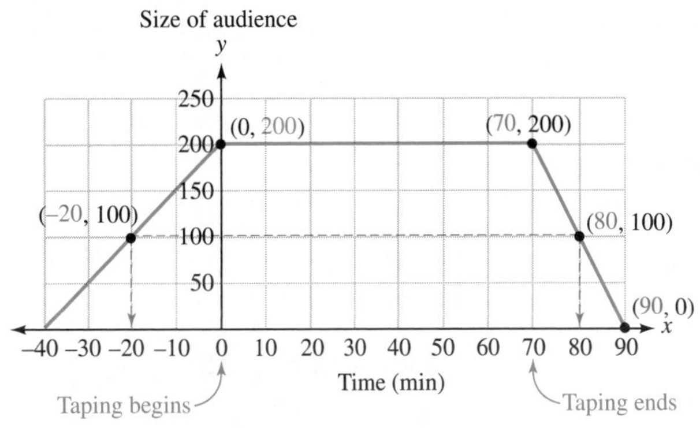

Self Check 3 Use the graph to answer the following questions.

a. At what times were there exactly 50 people in the audience?
b. How many people were in the audience when the taping took place?
c. When were the first audience members allowed into the taping session?

Answers to Self Checks

1.

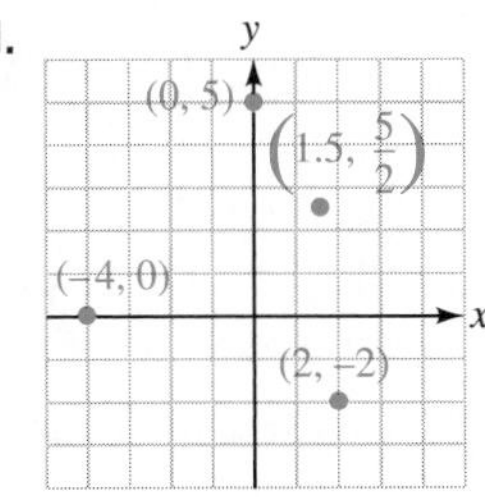

2. $A(4, 0)$; $B(0, 1)$; $C(-3.5, -2.5)$; $D(2, -4)$

3. a. 30 min before and 85 min after taping began, **b.** 200, **c.** 40 min before taping began

3.1 STUDY SET  ELEMENTARY Algebra $f(x)$ Now™

VOCABULARY **Fill in the blanks.**

1. $(-1, -5)$ is called an ________ pair.

2. In the ordered pair $\left(-\frac{3}{2}, -5\right)$, the ____________ is -5.

3. A rectangular coordinate system is formed by two perpendicular number lines called the ________ and the ________. The point where the axes cross is called the ________.

4. The x- and y-axes divide the coordinate plane into four regions called ________.
5. The point with coordinates (4, 2) can be graphed on a ________ coordinate system.
6. The process of locating the position of a point on a coordinate plane is called ____________ the point.

CONCEPTS

7. Fill in the blanks.
   a. To plot the point with coordinates $(-5, 4)$, we start at the ______ and move 5 units to the ______ and then move 4 units ______.
   b. To plot the point with coordinates $\left(6, -\frac{3}{2}\right)$, we start at the ______ and move 6 units to the ______ and then move $\frac{3}{2}$ units ______.
8. In which quadrant is each point located?
   a. $(-2, 7)$ b. $(3, 16)$
   c. $(-1, -2.75)$ d. $(50, -16)$
   e. $\left(\frac{1}{2}, \frac{15}{16}\right)$ f. $(-6, \pi)$
9. a. In which quadrants are the second coordinates of points positive?
   b. In which quadrants are the first coordinates of points negative?
   c. In which quadrant do points with a negative x-coordinate and a positive y-coordinate lie?
   d. In which quadrant do points with a positive x-coordinate and a negative y-coordinate lie?
10. FARMING
    a. Write each row of data in the table as an ordered pair.

Rain (inches)	Bushels produced	
2	10	→
4	15	→
8	25	→

    b. Plot the ordered pairs on the following graph. Then draw a straight line through the points.
    c. Use the graph to determine how many bushels will be produced if 6 inches of rain fall.

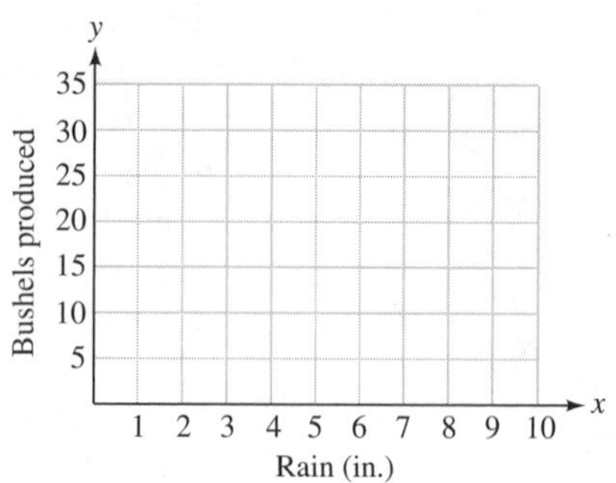

The graph that follows gives the heart rate of a woman before, during, and after an aerobic workout. In Problems 11–18, use the graph to answer the questions.

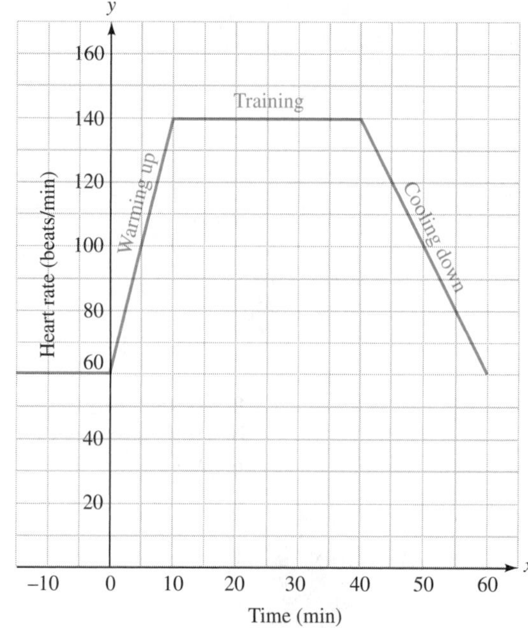

11. What was her heart rate before beginning the workout?
12. After beginning her workout, how long did it take the woman to reach her training-zone heart rate?
13. What was the woman's heart rate half an hour after beginning the workout?
14. For how long did the woman work out at her training zone?
15. At what time was her heart rate 100 beats per minute?

16. How long was her cool-down period?

17. What was the difference in the woman's heart rate before the workout and after the cool-down period?

18. What was her approximate heart rate 8 minutes after beginning?

19. BAR GRAPHS Use the graph on page 184 to estimate the difference in the number of motor vehicles produced by Europe and the United States in 2001.

20. CIRCLE GRAPHS The surface area of Earth is about 197,000,000 square miles. Use the graph on page 184 to find how many square miles are covered with water.

NOTATION

21. Explain the difference between (3, 5), 3(5), and 5(3 + 5).

22. In the table, which column contains values associated with the vertical axis of a graph?

x	y
2	0
5	3
−1	−3

23. Do these ordered pairs name the same point? $\left(2.5, -\frac{7}{2}\right), \left(2\frac{1}{2}, -3.5\right), \left(2.5, -3\frac{1}{2}\right)$

24. Do (3, 2) and (2, 3) represent the same point?

25. In the ordered pair (4, 5), is the number 4 associated with the horizontal or the vertical axis?

26. Fill in the blank: In the notation $A(4, 5)$, the capital letter A is used to name a ______.

PRACTICE

27. Complete the coordinates for each point.

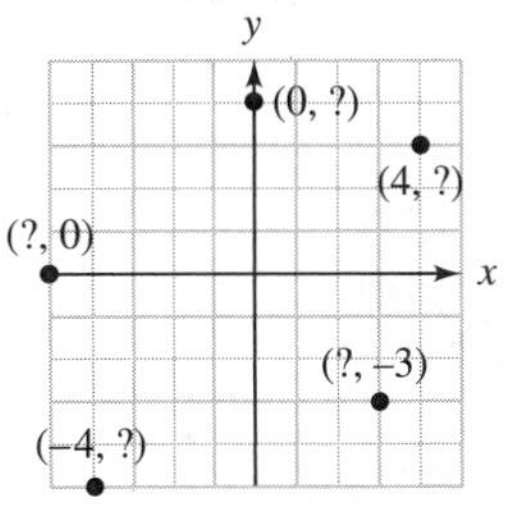

28. Find the coordinates of points A, B, C, D, E, and F.

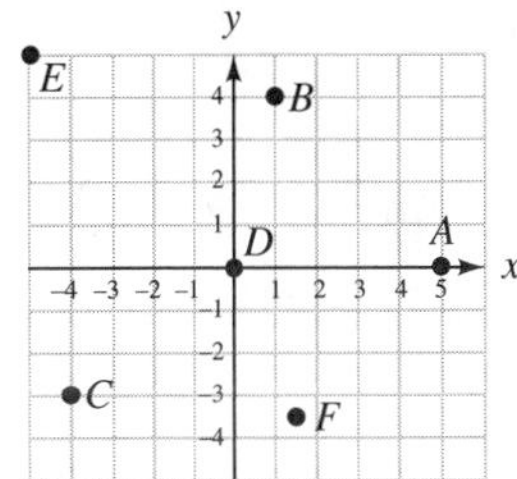

29. Graph each point: $(-3, 4)$, $(4, 3.5)$, $\left(-2, -\frac{5}{2}\right)$, $(0, -4)$, $\left(\frac{3}{2}, 0\right)$, $(3, -4)$.

30. Graph each point: $(4, 4)$, $(0.5, -3)$, $(-4, -4)$, $(0, -1)$, $(0, 0)$, $(0, 3)$, $(-2, 0)$.

APPLICATIONS

31. CONSTRUCTION The following graph shows a side view of a bridge design. Find the coordinates of each rivet, weld, and anchor.

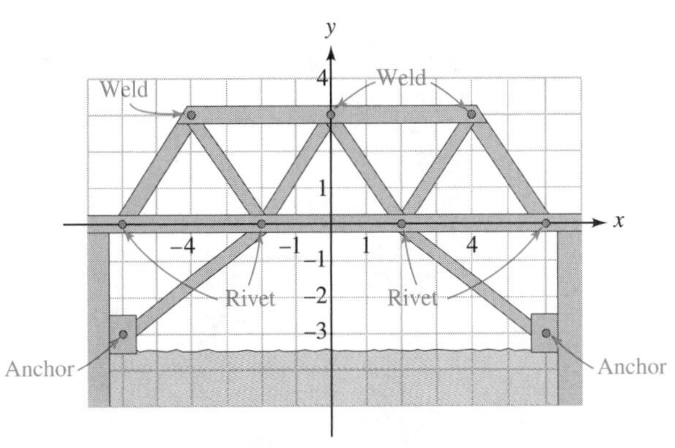

Scale: 1 unit = 8 ft

32. GOLF To correct her swing, the golfer is videotaped and then has her image displayed on a computer monitor so that it can be analyzed by a golf pro. Give the coordinates of the points that are highlighted on the arc of her swing.

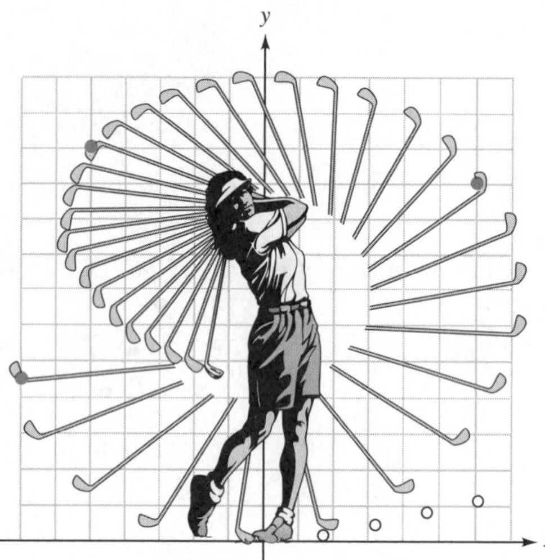

33. BATTLESHIP In the game Battleship, the player uses coordinates to drop depth charges from a battleship to hit a hidden submarine. What coordinates should be used to make three hits on the exposed submarine shown? Express each answer in the form (letter, number).

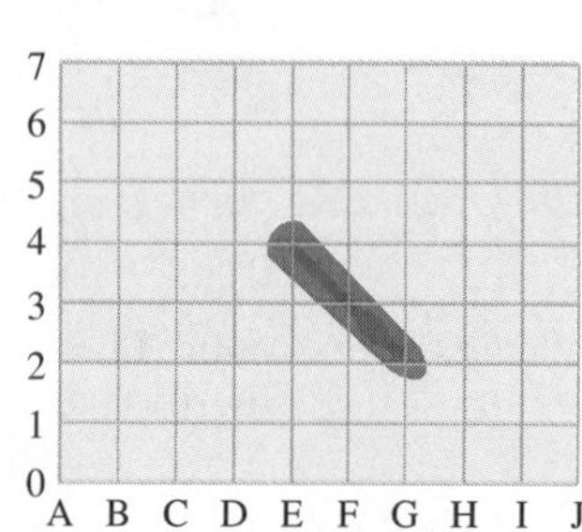

34. DENTISTRY Orthodontists describe teeth as being located in one of four *quadrants* as shown below.

a. How many teeth are in the *upper left quadrant?*

b. Why would the upper left quadrant appear on the right in the illustration?

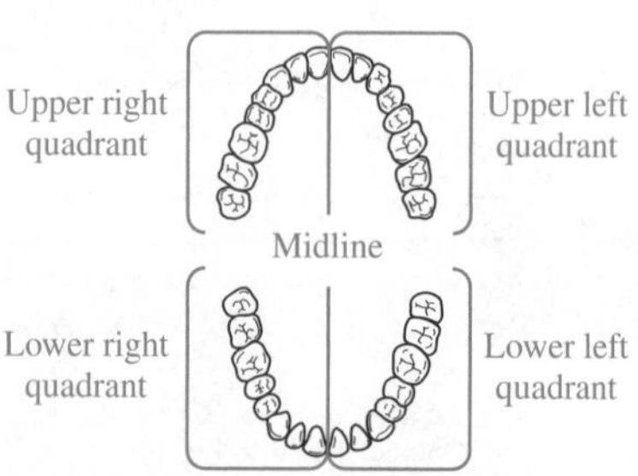

35. MAPS Use coordinates that have the form (number, letter) to locate each of the following on the map: Rockford, Mount Carroll, Harvard, and the intersection of state Highway 251 and U.S. Highway 30.

36. WATER PRESSURE The graphs show how the path of a stream of water changes when the hose is held at two different angles.

a. At which angle does the stream of water shoot up higher? How much higher?

b. At which angle does the stream of water shoot out farther? How much farther?

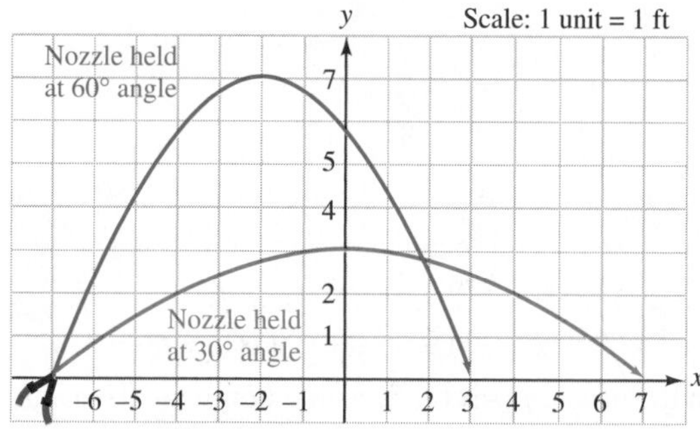

37. GEOMETRY Three vertices (corners) of a rectangle are the points (2, 1), (6, 1), and (6, 4). Find the coordinates of the fourth vertex. Then find the area of the rectangle.

38. GEOMETRY Three vertices (corners) of a right triangle are the points $(-1, -7)$, $(-5, -7)$, and $(-5, -2)$. Find the area of the triangle.

39. THE MILITARY The table below shows the number of miles that a tank can be driven on a given number of gallons of diesel fuel. Write the data in the table as ordered pairs and plot them. Then draw a straight line through the points.

a. How far can the tank go on 7 gallons of fuel?

b. How many gallons of fuel are needed to travel a distance of 20 miles?

c. How far can the tank go on 6.5 gallons of fuel?

Fuel (gal)	Distance (mi)
2	10
3	15
5	25

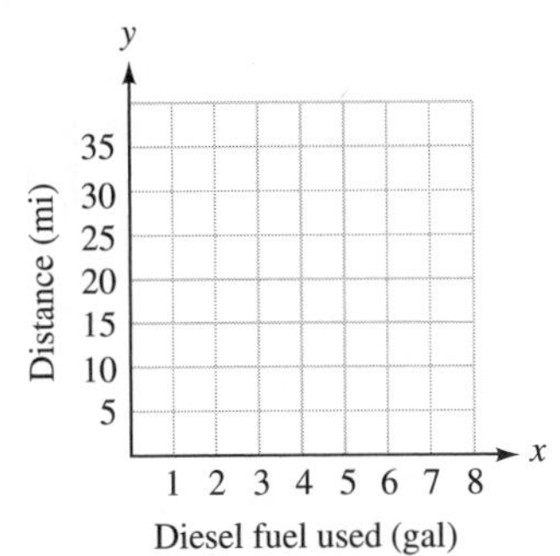

40. BOATING The table below shows the cost to rent a sailboat for a given number of hours. Write the data in the table as ordered pairs and plot them. Then draw a straight line through the points.

a. What does it cost to rent the boat for 3 hours?

b. For how long can the boat be rented for $60?

Rental time (hr)	Cost ($)
2	20
4	30
9	55

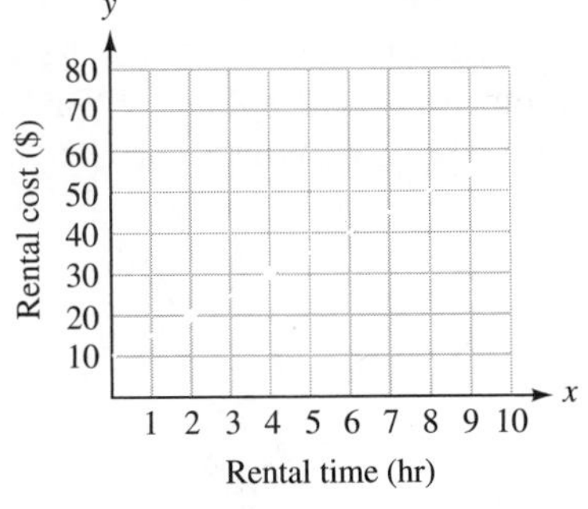

41. DEPRECIATION The table below shows the value (in thousands of dollars) of a car at various lengths of time after its purchase. Write the data in the table as ordered pairs and plot them. Then draw a straight line passing through the points.

a. What does the point (3, 7) on the graph tell you?

b. Find the value of the car when it is 7 years old.

c. After how many years will the car be worth $2,500?

Age (yr)	Value ($1,000)
3	7
4	5.5
5	4

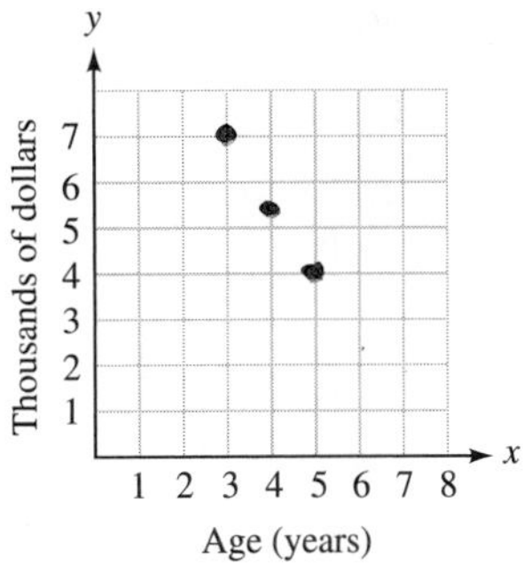

42. SWIMMING The table below shows the number of people at a public swimming pool at various times during the day. Write the data in the table as ordered pairs and plot them. (On the x-axis, 0 represents noon, 1 represents 1 PM, and so on.) Then draw a straight line passing through the points.

a. How many people will be at the pool at 6 PM?

b. At what time will there be 250 people at the pool?

c. When will the number of people at the pool be half of what it was at noon?

Time	Number of people
0	350
3	200
5	100

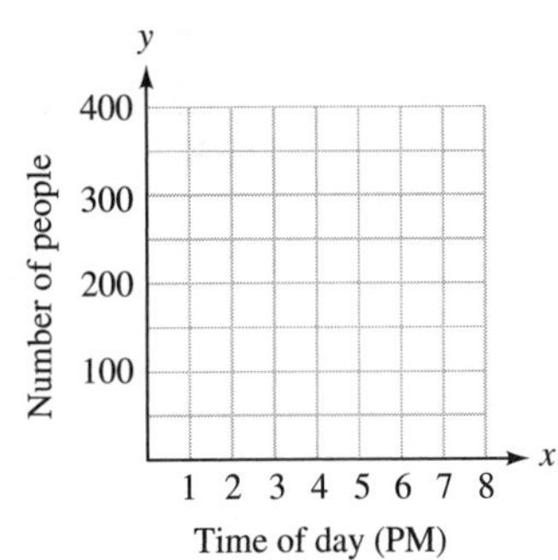

In Problems 43–44, refer to the following information. The approximate population of the United States for the years 1950–2000 is given by the straight line graph shown below.

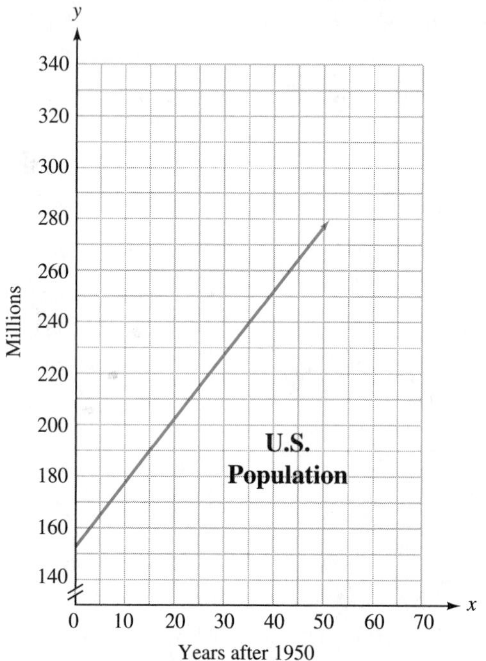

43. CENSUS An official count of the population, known as the Census, is conducted every 10 years in the United States, as required by the Constitution. Use the graph to estimate the Census numbers.

Census year	U.S. Population (millions)
1950	
1960	
1970	
1980	
1990	
2000	

44. EXTRAPOLATION On the graph, use a straightedge to predict the Census numbers for 2010 and 2020.

Census year	U.S. Population (millions)
2010	
2020	

WRITING

45. Explain why the point $(-3, 3)$ is not the same as the point $(3, -3)$.

46. Explain how to plot the point $(-2, 5)$.

47. Explain why the coordinates of the origin are $(0, 0)$.

48. Explain this diagram.

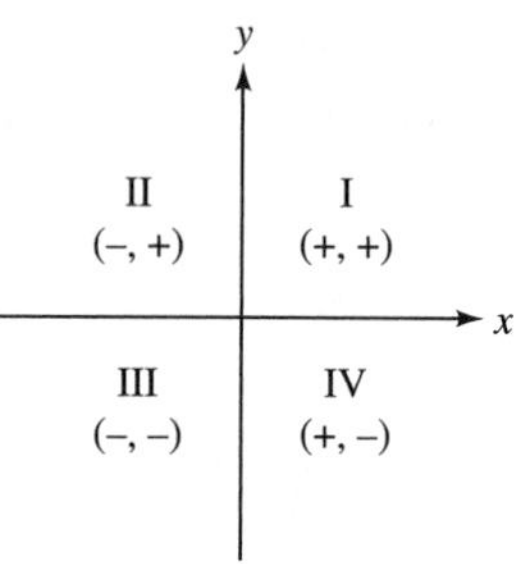

REVIEW

49. Solve $AC = \frac{2}{3}h - T$ for h.

50. Solve: $5(x + 1) \leq 2(x - 3)$. Write the solution set in interval notation and graph it.

51. Evaluate: $\dfrac{-4(4 + 2) - 2^3}{|-12 - 4(5)|}$.

52. Simplify: $\dfrac{24}{54}$.

CHALLENGE PROBLEMS

53. In what quadrant does a point lie if the *sum* of its coordinates is negative and the *product* of its coordinates is positive?

54. Draw a line segment $\overline{AB}$ with endpoints $A(6, 5)$ and $B(-4, 5)$. Suppose that the x-coordinate of a point C is the average of the x-coordinates of points A and B, and the y-coordinate of point C is the average of the y-coordinates of points A and B. Find the coordinates of point C. Why is C called the midpoint of $\overline{AB}$?

3.2 Graphing Linear Equations

- Solutions of Equations in Two Variables
- Constructing Tables of Solutions
- Graphing Linear Equations
- Applications

In this section, we discuss equations that contain two variables. Such equations are often used to describe algebraic relationships between two quantities. To see a mathematical picture of these relationships, we will construct graphs of their equations.

SOLUTIONS OF EQUATIONS IN TWO VARIABLES

We have previously solved **equations in one variable.** For example, $x + 3 = 9$ is an equation in x. If we subtract 3 from both sides, we see that 6 is the solution. To verify this, we replace x with 6 and note that the result is a true statement: $9 = 9$.

In this chapter, we extend our equation-solving skills to find solutions of **equations in two variables.** To begin, let's consider $y = x - 1$, an equation in x and y.

Notation

Equations in two variables often involve the variables x and y. However, other letters can be used. For example, $a - 3b = 5$ and $n = 4m + 6$ are equations in two variables.

A solution of $y = x - 1$ is a *pair* of values, one for x and one for y, that make the equation true. To illustrate, suppose x is 5 and y is 4. Then we have:

$$y = x - 1$$
$$4 \stackrel{?}{=} 5 - 1 \quad \text{Substitute 5 for } x \text{ and 4 for } y.$$
$$4 = 4 \quad \text{True.}$$

Since the result is a true statement, $x = 5$ and $y = 4$ is a solution of $y = x - 1$. We write the solution as the ordered pair (5, 4), with the value of x listed first. We say that (5, 4) *satisfies* the equation.

In general, a **solution of an equation in two variables** is an ordered pair of numbers that makes the equation a true statement.

EXAMPLE 1

Is $(-1, -3)$ a solution of $y = x - 1$?

ELEMENTARY Algebra f(x) Now™

Solution We substitute -1 for x and -3 for y and see whether the resulting equation is true.

$$y = x - 1$$
$$-3 \stackrel{?}{=} -1 - 1 \quad \text{Substitute } -1 \text{ for } x \text{ and } -3 \text{ for } y.$$
$$-3 = -2$$

Since $-3 = -2$ is false, $(-1, -3)$ is *not* a solution.

Self Check 1 Is (9, 8) a solution of $y = x - 1$?

We have seen that solutions of equations in two variables are written as ordered pairs. If only one of the values of an ordered-pair solution is known, we can substitute it into the equation to determine the other value.

EXAMPLE 2

Complete the solution $(-5, \quad)$ of the equation $y = -2x + 3$.

Solution In the ordered pair, the x-value is -5; the y-value is not known. To find y, we substitute -5 for x in the equation and evaluate the right-hand side.

$$y = -2x + 3$$
$$y = -2(-5) + 3 \quad \text{Substitute } -5 \text{ for } x.$$
$$y = 10 + 3 \quad \text{Do the multiplication.}$$
$$y = 13$$

The completed ordered pair is $(-5, 13)$.

Self Check 2 Complete the solution $(-2, \ )$ of the equation $y = 4x - 2$.

Solutions of equations in two variables are often listed in a **table of solutions** (or **table of values**).

EXAMPLE 3

ELEMENTARY Algebra f(x) Now™

Complete the table of solutions for $3x + 2y = 5$.

x	y	(x, y)
7		(7,)
	4	(, 4)

Solution In the first row, we are given an x-value of 7. To find the corresponding y-value, we substitute 7 for x and solve for y.

$$3x + 2y = 5$$
$$3(7) + 2y = 5 \quad \text{Substitute 7 for } x.$$
$$21 + 2y = 5 \quad \text{Do the multiplication.}$$
$$2y = -16 \quad \text{Subtract 21 from both sides.}$$
$$y = -8 \quad \text{Divide both sides by 2.}$$

A solution of $3x + 2y = 5$ is $(7, -8)$.

In the second row, we are given a y-value of 4. To find the corresponding x-value, we substitute 4 for y and solve for x.

$$3x + 2y = 5$$
$$3x + 2(4) = 5 \quad \text{Substitute 4 for } y.$$
$$3x + 8 = 5 \quad \text{Do the multiplication.}$$
$$3x = -3 \quad \text{Subtract 8 from both sides.}$$
$$x = -1 \quad \text{Divide both sides by 3.}$$

Another solution is $(-1, 4)$. The completed table is as follows:

x	y	(x, y)
7	−8	(7, −8)
−1	4	(−1, 4)

Self Check 3 Complete the table of solutions for $3x + 2y = 5$.

x	y	(x, y)
	-2	$(\;\; , -2)$
5		$(5, \;\;)$

CONSTRUCTING TABLES OF SOLUTIONS

To find a solution of an equation in two variables, we can select a number, substitute it for one of the variables, and find the corresponding value of the other variable. For example, to find a solution of $y = x - 1$, we can select a value for x, say, -4, substitute -4 for x in the equation, and find y.

x	y	(x, y)
-4	-5	$(-4, -5)$

$$y = x - 1$$
$$y = -4 - 1 \quad \text{Substitute } -4 \text{ for } x.$$
$$y = -5$$

The ordered pair $(-4, -5)$ is a solution. We list it in the table on the left.

To find another solution of $y = x - 1$, we select another value for x, say, -2, and find the corresponding y-value.

x	y	(x, y)
-4	-5	$(-4, -5)$
-2	-3	$(-2, -3)$

$$y = x - 1$$
$$y = -2 - 1 \quad \text{Substitute } -2 \text{ for } x.$$
$$y = -3$$

A second solution is $(-2, -3)$, and we list it in the table of solutions.

If we let $x = 0$, we can find a third ordered pair that satisfies $y = x - 1$.

x	y	(x, y)
-4	-5	$(-4, -5)$
-2	-3	$(-2, -3)$
0	-1	$(0, -1)$

$$y = x - 1$$
$$y = 0 - 1 \quad \text{Substitute 0 for } x.$$
$$y = -1$$

A third solution is $(0, -1)$, which we also add to the table of solutions.

We can find a fourth solution by letting $x = 2$, and a fifth solution by letting $x = 4$.

x	y	(x, y)
-4	-5	$(-4, -5)$
-2	-3	$(-2, -3)$
0	-1	$(0, -1)$
2	1	$(2, 1)$
4	3	$(4, 3)$

$$y = x - 1$$
$$y = 2 - 1 \quad \text{Substitute 2 for } x.$$
$$y = 1$$

$$y = x - 1$$
$$y = 4 - 1 \quad \text{Substitute 4 for } x.$$
$$y = 3$$

A fourth solution is $(2, 1)$ and a fifth solution is $(4, 3)$. We add them to the table.

Since we can choose any real number for x, and since any choice of x will give a corresponding value of y, it is apparent that the equation $y = x - 1$ has *infinitely many solutions.* We have found five of them: $(-4, -5)$, $(-2, -3)$, $(0, -1)$, $(2, 1)$, and $(4, 3)$.

GRAPHING LINEAR EQUATIONS

It would be impossible to list the infinitely many solutions of the equation $y = x - 1$. To show all of its solutions, we draw a mathematical "picture" of them. We call this picture the *graph of the equation.*

To graph $y = x - 1$, we plot the ordered pairs shown in the table on a rectangular coordinate system. Then we draw a straight line through the points, because *the graph of any solution of* $y = x - 1$ *will lie on this line.* Furthermore, every point on this line represents a solution. We call the line the **graph of the equation;** it represents all of the solutions of $y = x - 1$.

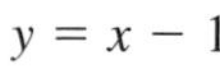

x	y	(x, y)
−4	−5	(−4, −5)
−2	−3	(−2, −3)
0	−1	(0, −1)
2	1	(2, 1)
4	3	(4, 3)

Construct a table of solutions.

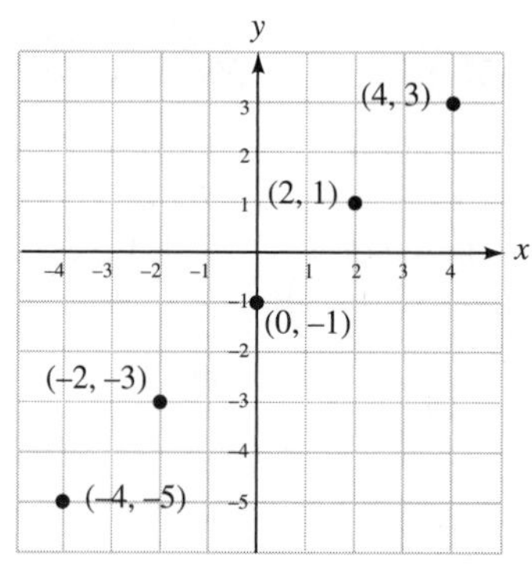

Plot the ordered pairs.

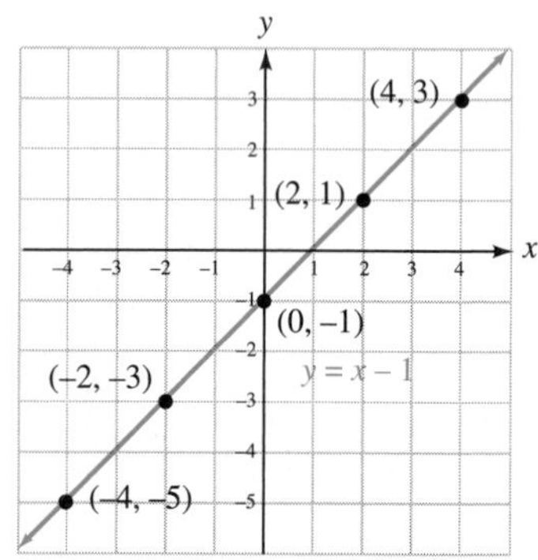

Draw a straight line through the points. This is the *graph of the equation*.

The equation $y = x - 1$ is said to be *linear* because its graph is a line. By definition, a linear equation in two variables is any equation that can be written in the following **general** (or **standard**) **form.**

Linear Equations

A **linear equation in two variables** is an equation that can be written in the form

$$Ax + By = C$$

where A, B, and C are real numbers and A and B are not both 0.

Success Tip

The exponent on each variable of a linear equation is an understood 1. For example, $y = 2x + 4$ can be thought of as $y^1 = 2x^1 + 4$.

Some more examples of linear equations are

$$5x - 6y = 10, \qquad 3y = -2x - 12, \qquad \text{and} \qquad y = 2x + 4$$

Linear equations can be graphed in several ways. Generally, the form in which an equation is written determines the method that we use to graph it. To graph linear equations solved for y, such as $y = 2x + 4$, we can use the following method.

Graphing Linear Equations Solved for y

1. Find three solutions of the equation by selecting three values for x and calculating the corresponding values of y.
2. Plot the solutions on a rectangular coordinate system.
3. Draw a straight line passing through the points. If the points do not lie on a line, check your computations.

EXAMPLE 4

ELEMENTARY Algebra $f(x)$ Now™

Graph: $y = 2x + 4$.

Solution To find three solutions of this linear equation, we select three values of x that will make the computations easy. Then we find each corresponding value of y.

Success Tip

When selecting x-values for a table of solutions, a rule-of-thumb is to choose a negative number, a positive number, and 0. When $x = 0$, the computations to find y are usually quite simple.

***If* $x = -2$**	***If* $x = 0$**	***If* $x = 2$**
$y = 2x + 4$	$y = 2x + 4$	$y = 2x + 4$
$y = 2(-2) + 4$	$y = 2(0) + 4$	$y = 2(2) + 4$
$y = -4 + 4$	$y = 0 + 4$	$y = 4 + 4$
$y = 0$	$y = 4$	$y = 8$
$(-2, 0)$ is a solution.	$(0, 4)$ is a solution.	$(2, 8)$ is a solution.

We enter the results in a table of solutions and plot the points. Then we draw a straight line through the points and label it $y = 2x + 4$.

Success Tip

Since two points determine a line, only two points are needed to graph a linear equation. However, we should plot a third point as a check. If the three points do not lie on a straight line, then at least one of them is in error.

$y = 2x + 4$

x	y	(x, y)
-2	0	$(-2, 0)$
0	4	$(0, 4)$
2	8	$(2, 8)$

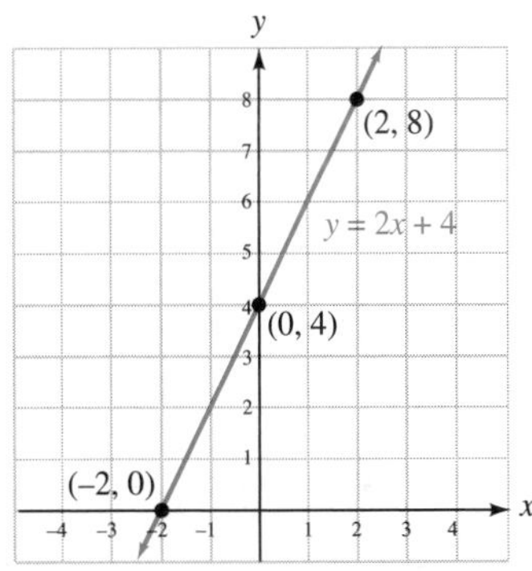

Self Check 4 Graph: $y = 2x - 3$.

EXAMPLE 5

ELEMENTARY Algebra $f(x)$ Now™

Graph: $y = -3x$.

Solution To find three solutions, we begin by choosing three x-values: -1, 0, and 1. Then we find the corresponding values of y. If $x = -1$, we have

$$y = -3x$$
$$y = -3(-1) \quad \text{Substitute } -1 \text{ for } x.$$
$$y = 3$$

$(-1, 3)$ is a solution.

In a similar manner, we find the y-values for x-values of 0 and 1, and record the results in a table of solutions. After plotting the ordered pairs, we draw a straight line through the points and label it $y = -3x$.

$y = -3x$

x	y	(x, y)
−1	3	(−1, 3)
0	0	(0, 0)
1	−3	(1, −3)

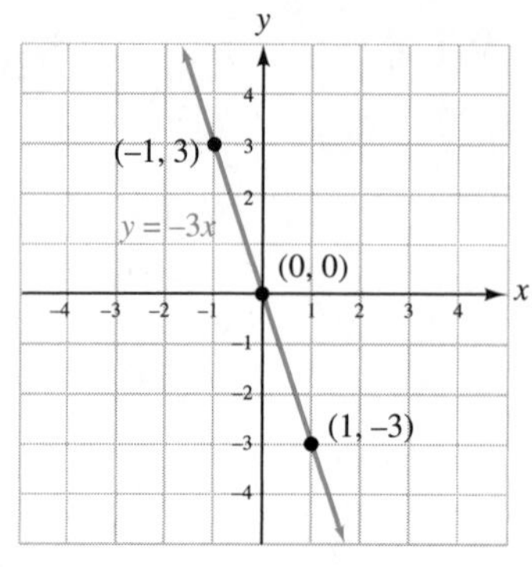

Self Check 5 Graph: $y = -4x$.

EXAMPLE 6

Graph: $3y = -2x - 12$.

ELEMENTARY Algebra $f(x)$ Now™

Solution The graph of $3y = -2x - 12$ can be found by solving for y and graphing the *equivalent equation* that results. To isolate y, we proceed as follows.

$$3y = -2x - 12$$

$$\frac{3y}{3} = \frac{-2x}{3} - \frac{12}{3}$$ To undo the multiplication by 3, divide both sides by 3.

$$y = -\frac{2}{3}x - 4$$ Write $\frac{-2x}{3}$ as $-\frac{2}{3}x$. Simplify: $\frac{12}{3} = 4$.

To find solutions of $y = -\frac{2}{3}x - 4$, each value of x must be multiplied by $-\frac{2}{3}$. This computation is made easier if we select x-values that are *multiples of 3,* such as −3, 0, and 6. For example, if $x = -3$, we have

$$y = -\frac{2}{3}x - 4$$

$$y = -\frac{2}{3}(-3) - 4$$ Substitute −3 for x.

$$y = 2 - 4$$ Do the multiplication: $-\frac{2}{3}(-3) = 2$. This step is simpler if we choose x-values that are multiples of 3.

$$y = -2$$

(−3, −2) is a solution.

Two more solutions, one for $x = 0$ and one for $x = 6$, are found in a similar way, and entered in a table. We plot the ordered pairs, draw a straight line through the points, and label the line as $y = -\frac{2}{3}x - 4$ or as $3y = -2x - 12$.

$y = -\frac{2}{3}x - 4$

x	y	(x, y)
−3	−2	(−3, −2)
0	−4	(0, −4)
6	−8	(6, −8)

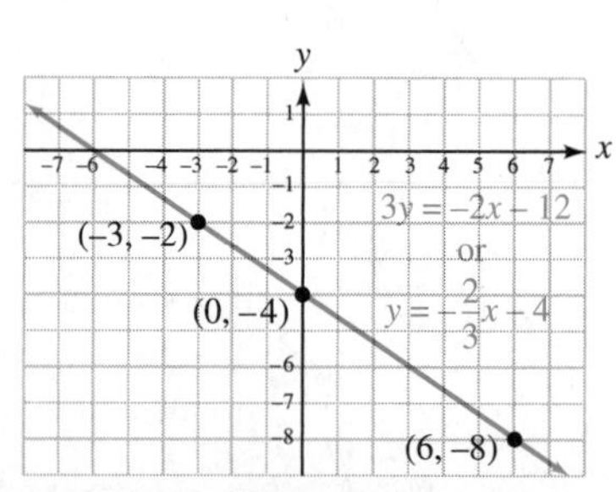

Self Check 6 Solve $2y = 5x + 2$ for y. Then graph the equation.

APPLICATIONS

When linear equations are used to model real-life situations, they are often written in variables other than x and y. In such cases, we must make the appropriate changes when labeling the table of solutions and the graph of the equation.

EXAMPLE 7

Cleaning windows. The linear equation $A = -0.03n + 32$ estimates the amount A of glass cleaner (in ounces) that is left in the bottle after the sprayer trigger has been pulled a total of n times. Graph the equation and use the graph to estimate the amount of cleaner that is left after 500 sprays.

Solution Since A depends on n in the equation $A = -0.03n + 32$, solutions will have the form (n, A). To find three solutions, we begin by selecting three values of n. Because the number of trials cannot be negative, and the computations to find A involve decimal multiplication, we select 0, 100, and 1,000. For example, if $n = 100$, we have

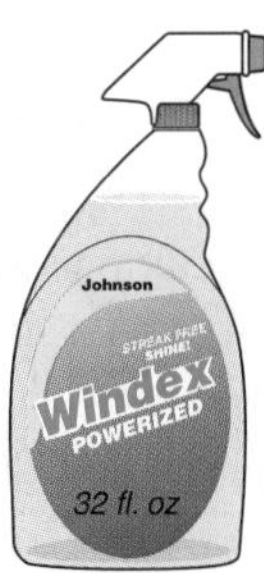

$A = -0.03n + 32$

$A = -0.03(100) + 32$

$A = -3 + 32$ Do the multiplication: $-0.03(100) = -3$.

$A = 29$

(100, 29) is a solution. After 100 sprays, 29 ounces of cleaner are left in the bottle.

In a similar manner, solutions are found for $n = 0$ and $n = 1{,}000$, and listed in the following table. Then the ordered pairs are plotted and a straight line is drawn through the points.

To graphically estimate the amount of solution that is left after 500 sprays, we draw the dashed blue lines, as shown. Reading on the vertical A-axis, we see that after 500 sprays, about 17 ounces of glass cleaner would be left.

$A = -0.03n + 32$

n	A	(n, A)
0	32	(0, 32)
100	29	(100, 29)
1,000	2	(1,000, 2)

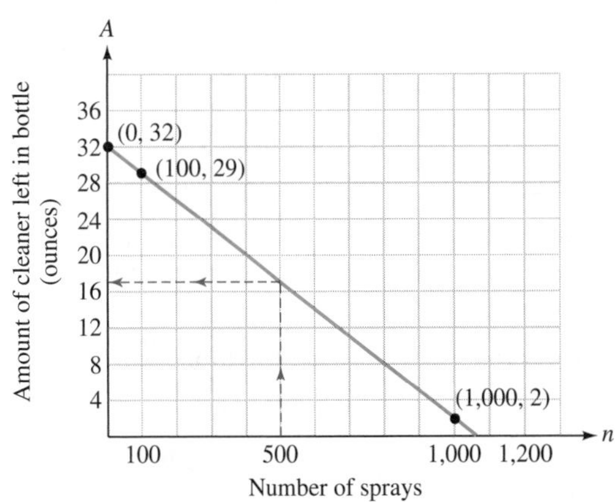

Answers to Self Checks 1. yes 2. $(-2, -10)$ 3.

x	y	(x, y)
3	-2	$(3, -2)$
5	-5	$(5, -5)$

4. 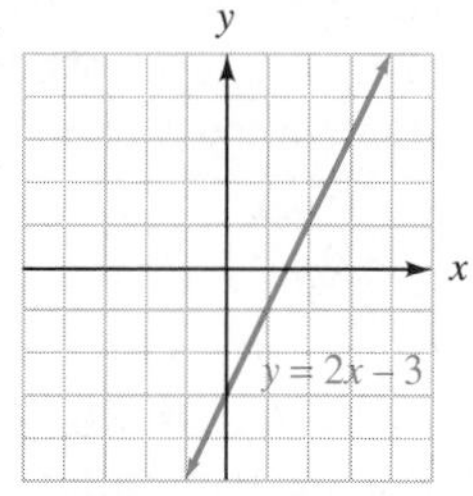

5. 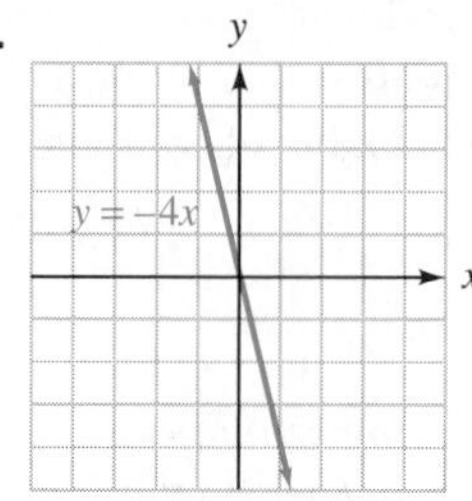

6. 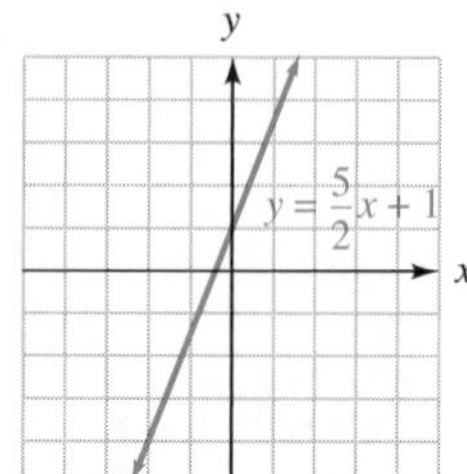

3.2 STUDY SET

ELEMENTARY Algebra f(x) Now™

VOCABULARY Fill in the blanks.

1. We say $y = 2x + 5$ is an equation in ______ variables, x and y.
2. A ________ of an equation in two variables is an ordered pair of numbers that makes the equation a true statement.
3. Solutions of equations in two variables are often listed in a ______ of solutions.
4. The line that represents all of the solutions of a linear equation is called the ______ of the equation.
5. The equation $y = 3x + 8$ is said to be ______ because its graph is a line.
6. The ______________ form of a linear equation in two variables is $Ax + By = C$.
7. A linear equation in two variables has ________ many solutions.
8. Two points ________ a line.

CONCEPTS

9. Consider the equation $y = -2x + 6$.
   a. How many variables does the equation contain?
   b. Does $(4, -2)$ satisfy the equation?
   c. Is $(-3, 0)$ a solution?
   d. How many solutions does this equation have?
10. To graph a linear equation, three solutions were found, they were plotted (in black), and a straight line was drawn through them, as shown below.
    a. Construct the table of solutions for this graph.
    b. From the graph, determine three other solutions of the equation.

x	y	(x, y)

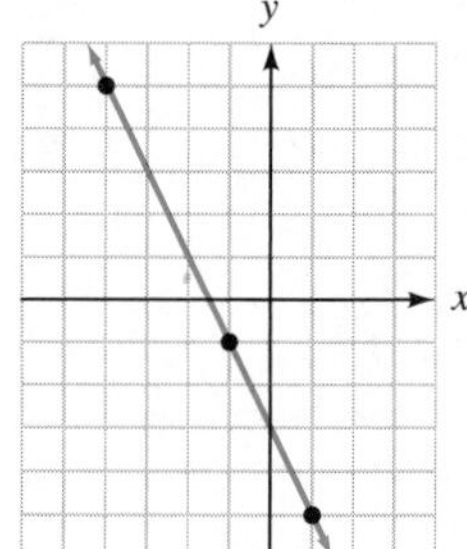

11. The graph of $y = -2x - 3$ is shown in Problem 10. Fill in the blanks: Every point on the graph represents an ordered-pair ________ of $y = -2x - 3$ and every ordered-pair solution is a ______ on the graph.
12. The graph of a linear equation is shown on the next page.
    a. If the coordinates of point M are substituted into the equation, will the result be true or false?
    b. If the coordinates of point N are substituted into the equation, will the result be true or false?

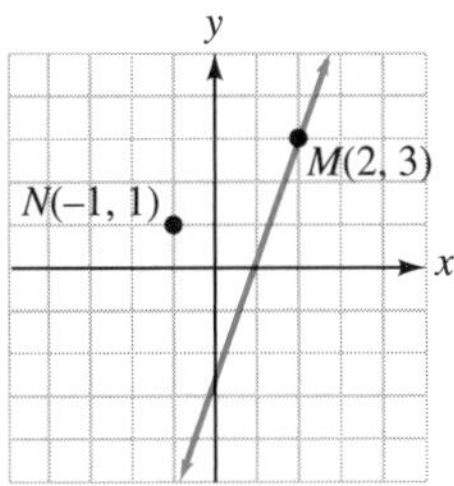

13. A student found three solutions of a linear equation and plotted them as shown in part (a) of the illustration.

a. What conclusion can be made?

b. What is wrong with the graph of $y = x - 3$, shown in illustration (b)?

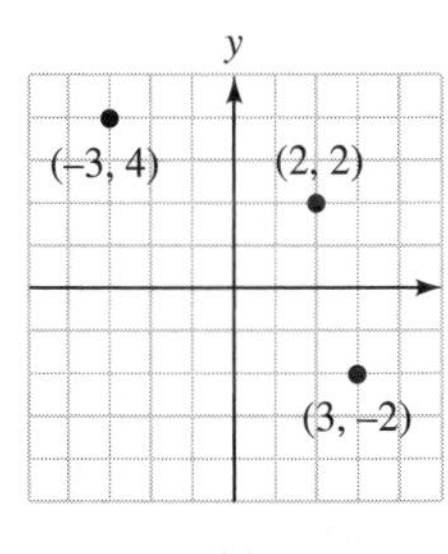

(a)

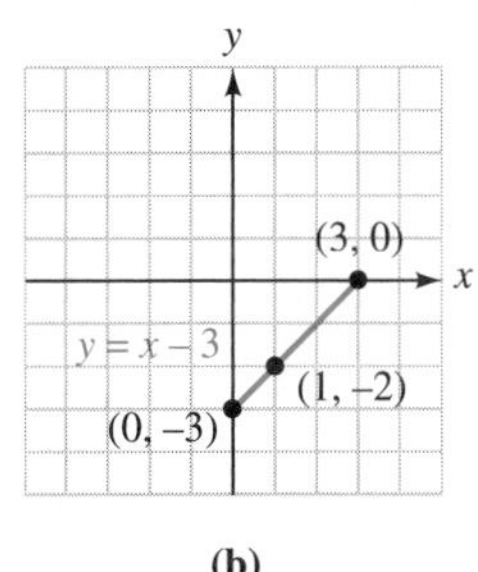

(b)

14. a. Rewrite the linear equation $y = \frac{1}{2}x + 7$ showing the understood exponents on the variables.

$$y^{\square} = \tfrac{1}{2}x^{\square} + 7$$

b. Explain why $y = x^2 + 2$ and $y = x^3 - 4$ are not linear equations.

15. Solve each equation for y.

a. $5y = 10x + 5$

b. $3y = -5x - 6$

c. $-7y = -x + 21$

16. a. Complete three lines of the table of solutions for $y = \frac{4}{5}x + 2$. Use the three values of x that make the computations the easiest.

x	y	(x, y)
−5		
−3		
0		
4		
10		

b. Complete three lines of the table of solutions for $T = 0.6r + 560$. Use the three values of r that make the computations the easiest.

r	T	(r, T)
−12		
−10		
0		
8		
100		

NOTATION Complete the solution.

17. Verify that $(-2, 6)$ is a solution of $y = -x + 4$.

$$y = -x + 4$$
$$\stackrel{?}{=} -(\quad) + 4$$
$$6 \stackrel{?}{=} \quad + 4$$
$$6 =$$

18. Determine whether each statement is true or false.

a. $\frac{9x}{8} = \frac{9}{8}x$ **b.** $-\frac{1}{6}x = -\frac{x}{6}$

c. When we divide both sides of $2y = 3x + 10$ by 2, we obtain $y = 3x + 5$.

19. Complete the labeling of the table of solutions and graph of $c = -a + 4$.

		(,)
−1	5	(−1, 5)
0	4	(0, 4)
2	2	(2, 2)

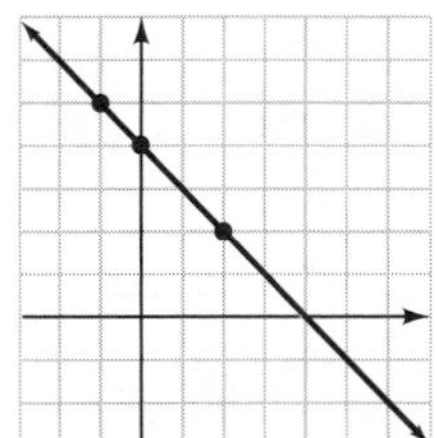

20. A table of solutions for a linear equation is shown below. When constructing the graph of the equation, how would you scale the x-axis and the y-axis?

x	y	(x, y)
−20	600	(−20, 600)
5	100	(5, 100)
35	−500	(35, −500)

PRACTICE **Determine whether each equation has the given ordered pair as a solution.**

21. $y = 5x - 4; (1, 1)$

22. $y = -2x + 3; (2, -1)$

23. $y = -\frac{3}{4}x + 8; (-8, 12)$

24. $y = \frac{1}{6}x - 2; (-12, 4)$

25. $7x - 2y = 3; (2, 6)$

26. $10x - y = 10; (0, 0)$

27. $x + 12y = -12; (0, -1)$

28. $-2x + 3y = 0; (-3, -2)$

For each equation, complete the solution.

29. $y = -5x - 4; (-3, \quad)$

30. $y = 8x + 30; (-6, \quad)$

31. $4x - 5y = -4; (\quad, 4)$

32. $7x + y = -12; (\quad, 2)$

Complete each table of solutions.

33. $y = 2x - 4$

x	y	(x, y)
8		
	8	

34. $y = 3x + 1$

x	y	(x, y)
−3		
	−2	

35. $3x - y = -2$

x	y	(x, y)
−5		
	−1	

36. $5x - 2y = -15$

x	y	(x, y)
5		
	0	

Construct a table of solutions and then graph each equation.

37. $y = 2x - 3$

38. $y = 3x + 1$

39. $y = 5x - 4$

40. $y = 6x - 3$

41. $y = x$

42. $y = 4x$

43. $y = -3x + 2$

44. $y = -2x + 1$

45. $y = -x - 1$

46. $y = -x + 2$

47. $y = \frac{x}{3}$

48. $y = -\frac{x}{3} - 1$

49. $y = -\frac{1}{2}x$

50. $y = \frac{3}{4}x$

51. $y = \frac{3}{8}x - 6$

52. $y = \frac{5}{6}x - 5$

53. $y = \frac{2}{3}x - 2$

54. $y = -\frac{3}{2}x + 2$

Solve each equation for *y* and then graph it.

55. $7y = -2x$

56. $6y = -4x$

57. $3y = 12x + 15$

58. $5y = 20x - 30$

59. $5y = x + 20$

60. $4y = x - 16$

61. $y + 1 = 7x$

62. $y - 3 = 2x$

APPLICATIONS

63. BILLIARDS The path traveled by the black eight ball on a game-winning shot is described by two linear equations. Complete the table of solutions for each equation and then graph the path of the ball.

$y = 2x - 4$

x	y	(x, y)
1		
2		
4		

$y = -2x + 12$

x	y	(x, y)
4		
6		
8		

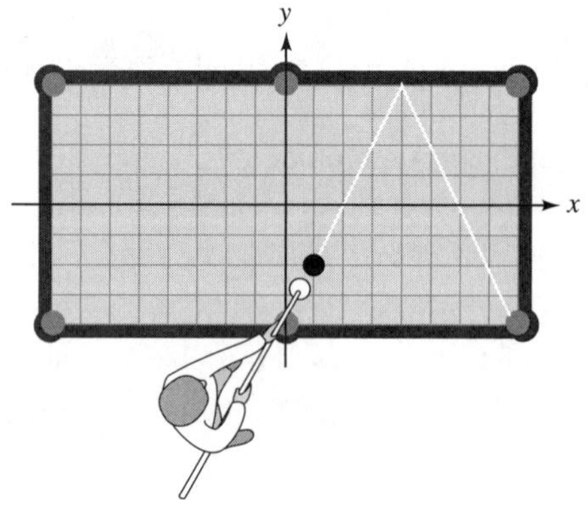

64. PING-PONG The path traveled by a Ping-Pong ball is described by two linear equations. Complete the table of solutions for each equation and then graph the path of the ball.

$y = \frac{1}{2}x + \frac{3}{2}$

x	y	(x, y)
7		
3		
−3		

$y = -\frac{1}{2}x - \frac{3}{2}$

x	y	(x, y)
−3		
−5		
−7		

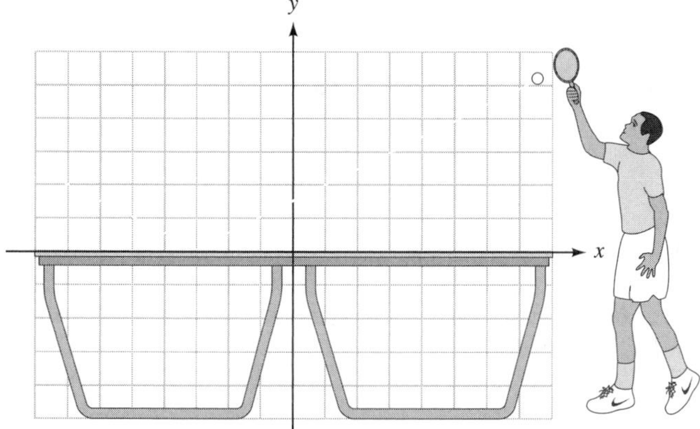

65. HOUSEKEEPING The linear equation $A = -0.02n + 16$ estimates the amount A of furniture polish (in ounces) that is left in the bottle after the sprayer trigger has been pulled a total of n times. Graph the equation and use the graph to estimate the amount of polish that is left after 650 sprays.

66. SHARPENING PENCILS The linear equation $L = -0.04t + 8$ estimates the length L (in inches) of a pencil after it has been inserted into a sharpener and the handle turned a total of t times. Graph the equation and use the graph to estimate the length of the pencil after 75 turns of the handle.

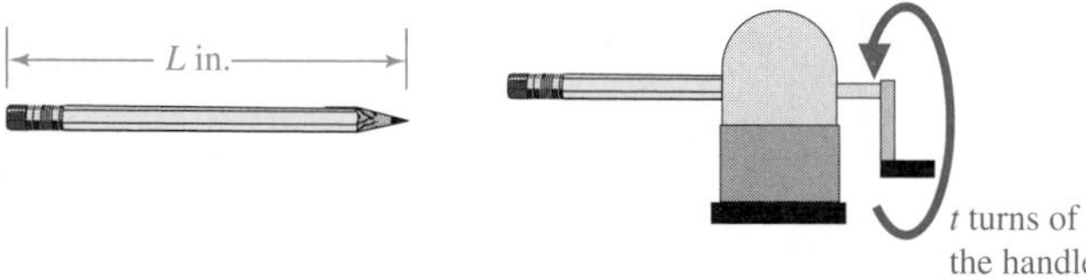

67. NFL TICKETS The average ticket price p to a National Football League game during the years 1990–2002 is approximated by $p = \frac{9}{4}t + 23$, where t is the number of years after 1990. Graph this equation and use the graph to predict the average ticket price in 2010. (Source: Team Marketing Report, NFL.)

68. U.S. AUTOMOBILE ACCIDENTS The number n of lives saved by seat belts during the years 1995–2001 is estimated by $n = 392t + 9{,}970$, where t is the number of years after 1995. Graph this equation and use the graph to predict the number of lives that will be saved by seat belts in 2020. (Source: Bureau of Transportation Statistics.)

69. RAFFLES A private school is going to sell raffle tickets as a fund raiser. Suppose the number n of raffle tickets that will be sold is predicted by the equation $n = -20p + 300$, where p is the price of a raffle ticket in dollars. Graph the equation and use the graph to predict the number of raffle tickets that will be sold at a price of $6.

70. CATS The number n of cat owners (in millions) in the U.S. during the years 1994–2002 is estimated by $n = \frac{3}{5}t + 31.5$, where t is the number of years after 1994. Graph this equation and use the graph to predict the number of cat owners in the U.S. in 2014. (Source: Pet Food Institute.)

WRITING

71. When we say that $(-2, -6)$ is a solution of $y = x - 4$, what do we mean?

72. What is a table of solutions?

73. What does it mean when we say that a linear equation in two variables has infinitely many solutions?

74. A linear equation and a graph are two ways of mathematically describing a relationship between two quantities. Which do you think is more informative and why?

75. From geometry, we know that two points determine a line. Explain why it is a good practice when graphing linear equations to find and plot three solutions instead of just two.

76. On a quiz, students were asked to graph $y = 3x - 1$. One student made the table of solutions on the left. Another student made the table on the right. The tables are completely different. Which table is incorrect? Or could they both be correct? Explain.

x	y	(x, y)
0	−1	(0, −1)
2	5	(2, 5)
3	8	(3, 8)

x	y	(x, y)
−2	−7	(−2, −7)
−1	−4	(−1, −4)
1	2	(1, 2)

REVIEW

77. Simplify: $-(-5 - 4c)$.

78. Denote the set of integers.

79. Evaluate: $-2^2 + 2^2$.

80. Find the volume, to the nearest tenth, of a sphere with radius 6 feet.

81. Evaluate: $1 + 2[-3 - 4(2 - 8^2)]$.

82. Solve: $-2(a + 3) = 3(a - 5)$.

CHALLENGE PROBLEMS **Graph each of the following *nonlinear* equations in two variables by constructing a table of solutions consisting of seven ordered pairs. These equations are called nonlinear, because their graphs are not straight lines.**

83. $y = x^2 + 1$

84. $y = x^3 - 2$

85. $y = |x| - 2$

86. $y = (x + 2)^2$

3.3 More about Graphing Linear Equations

- Intercepts
- The Intercept Method
- Graphing Horizontal and Vertical Lines
- Information from Intercepts
- Graphing Calculators

In this section, we graph a linear equation by determining the points where the graph intersects the x-axis and the y-axis. These points are called the *intercepts* of the graph.

INTERCEPTS

The graph of $y = 2x - 4$ is shown below. We see that the graph crosses the y-axis at the point $(0, -4)$; this point is called the **y-intercept** of the graph. The graph crosses the x-axis at the point $(2, 0)$; this point is called the **x-intercept** of the graph.

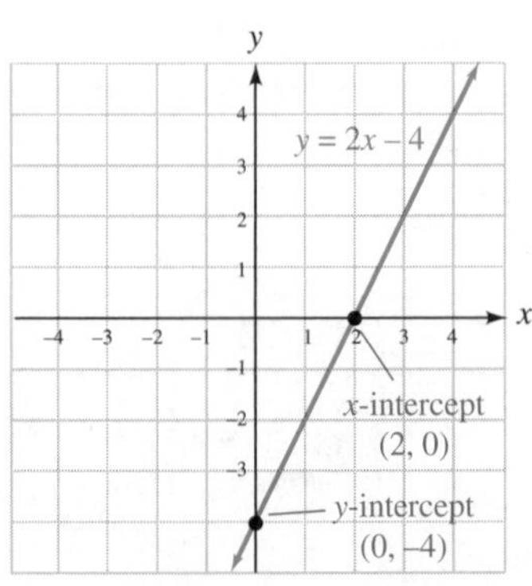

EXAMPLE 1

ELEMENTARY Algebra $f(x)$ Now™

For each graph, identify the x-intercept and the y-intercept.

a.

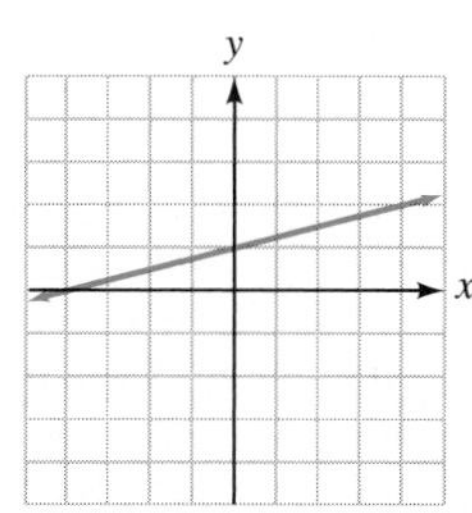

b.

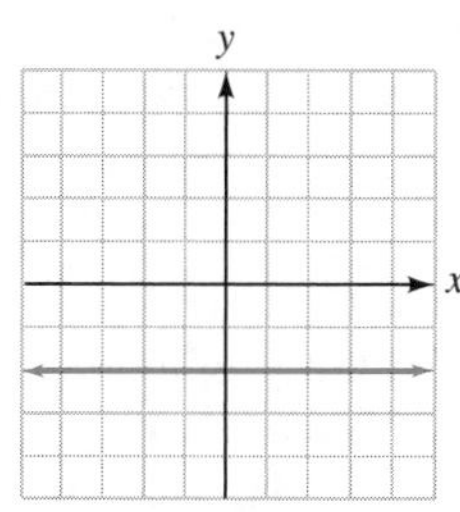

c.

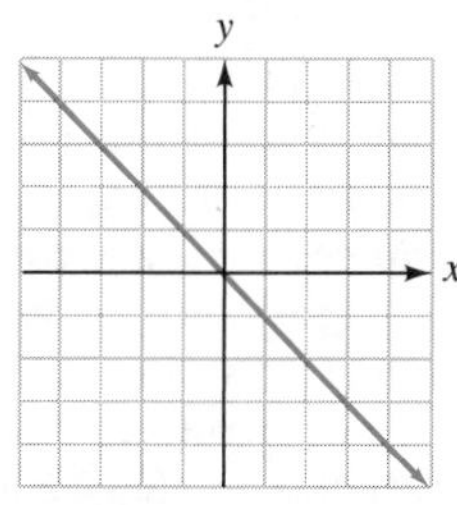

Solution

a. The graph crosses the x-axis at $(-4, 0)$. This is the x-intercept. The graph crosses the y-axis at $(0, 1)$. This is the y-intercept.

b. Since the horizontal line does not cross the x-axis, there is no x-intercept. The graph crosses the y-axis at $(0, -2)$. This is the y-intercept.

c. The graph crosses the x-axis and the y-axis at the same point, the origin. Therefore the x-intercept is $(0, 0)$ and the y-intercept is $(0, 0)$.

Self Check 1 Identify the x-intercept and the y-intercept of the graph.

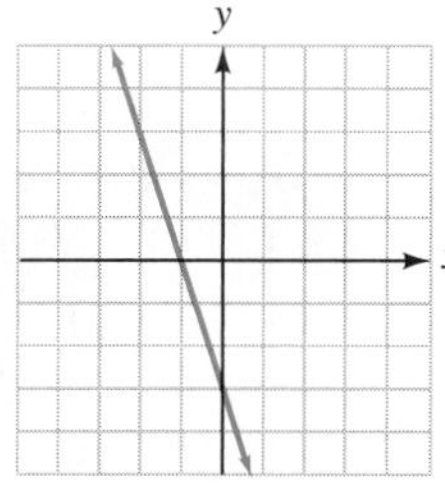

From the previous examples, we see that a y-intercept has an x-coordinate of 0, and an x-intercept has a y-coordinate of 0. These observations suggest the following procedures for finding the intercepts of a graph from its equation.

Finding Intercepts

To find the y-intercept, substitute 0 for x in the given equation and solve for y.
To find the x-intercept, substitute 0 for y in the given equation and solve for x.

THE INTERCEPT METHOD

Plotting the x- and y-intercepts of a graph and drawing a straight line through them is called the **intercept method of graphing a line.** This method is useful when graphing linear equations written in the general form $Ax + By = C$.

EXAMPLE 2

Graph $x - 3y = 6$ by finding the intercepts.

ELEMENTARY Algebra $f(x)$ Now™

Solution To find the y-intercept, let $x = 0$ and solve for y. To find the x-intercept, let $y = 0$ and solve for x.

The Language of Algebra

The point where a line *intersects* the *x*- or *y*-axis is called an intercept. The word *intersect* means to cut through or cross. A famous tourist attraction in Southern California is the *intersection* of Hollywood Blvd. & Vine St.

***y*-intercept**		***x*-intercept**	
$x - 3y = 6$		$x - 3y = 6$	
$0 - 3y = 6$	Substitute 0 for x.	$x - 3(0) = 6$	Substitute 0 for y.
$-3y = 6$		$x - 0 = 6$	
$y = -2$	Divide both sides by -3.	$x = 6$	

The y-intercept is $(0, -2)$ and the x-intercept is $(6, 0)$. Since each intercept of the graph is a solution of the equation, we enter the intercepts in a table of solutions.

As a check, we find one more point on the line. We select a convenient value for x, say, 3, and find the corresponding value of y.

$$
\begin{aligned}
x - 3y &= 6 \\
3 - 3y &= 6 && \text{Substitute 3 for } x. \\
-3y &= 3 && \text{Subtract 3 from both sides.} \\
y &= -1 && \text{Divide both sides by } -3.
\end{aligned}
$$

Therefore, $(3, -1)$ is a solution. It is also entered in the table.

The intercepts and the check point are plotted, a straight line is drawn through them, and the line is labeled $x - 3y = 6$.

$x - 3y = 6$

x	y	(x, y)	
0	-2	$(0, -2)$	← y-intercept
6	0	$(6, 0)$	← x-intercept
3	-1	$(3, -1)$	← check point

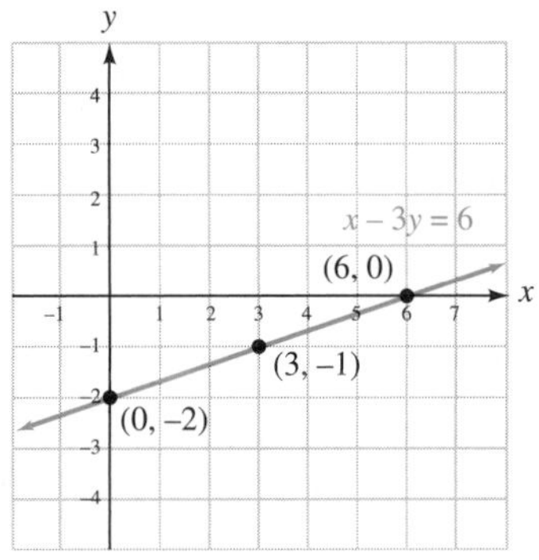

Self Check 2 Graph $x - 2y = 2$ by finding the intercepts.

The computations for finding intercepts can be simplified if we realize what logically follows when we substitute 0 for y or 0 for x in an equation written in the form $Ax + By = C$.

EXAMPLE 3

Graph $4x + 3y = -12$ by finding the intercepts.

ELEMENTARY Algebra f(x) Now™

Solution When we substitute 0 for x, it follows that the term $4x$ will be equal to 0. Therefore, to find the y-intercept, we can cover $4x$ and solve the equation that remains for y.

$$
\begin{aligned}
\square + 3y &= -12 && \text{If } x = 0, \text{ then } 4x = 4(0) = 0. \text{ Cover this term.} \\
y &= -4 && \text{To solve } 3y = -12, \text{ divide both sides by 3.}
\end{aligned}
$$

The y-intercept is $(0, -4)$.

When we substitute 0 for y, it follows that the term $3y$ will be equal to 0. Therefore, to find the x-intercept, we can cover $3y$ and solve the equation that remains for x.

$$4x \qquad = -12 \qquad \text{If } y = 0\text{, then } 3y = 3(0) = 0\text{. Cover this term.}$$
$$x = -3 \qquad \text{To solve } 4x = -12\text{, divide both sides by 4.}$$

The x-intercept is $(-3, 0)$.

A third solution can be found by selecting a convenient value for x and finding the corresponding value for y. If we choose $x = -6$, we find that $y = 4$. The solution $(-6, 4)$ is entered in the table, and the equation is graphed as shown.

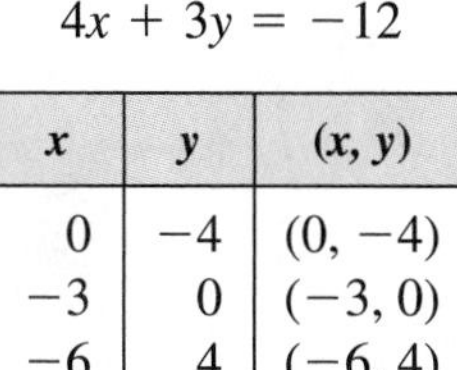

$4x + 3y = -12$

x	y	(x, y)
0	−4	(0, −4)
−3	0	(−3, 0)
−6	4	(−6, 4)

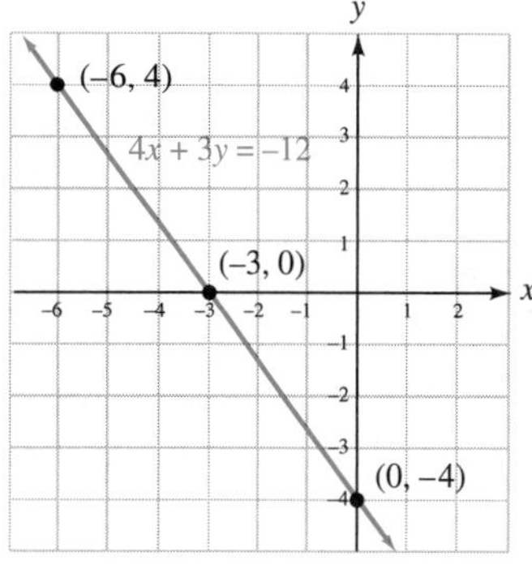

Self Check 3 Graph $2x + 5y = -10$ by finding the intercepts.

EXAMPLE 4

Graph $3x = -5y + 8$ by finding the intercepts.

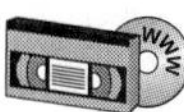

Solution We find the intercepts and select $x = 1$ to find a check point.

***y*-intercept: $x = 0$**

$$3x = -5y + 8$$
$$3(0) = -5y + 8$$
$$0 = -5y + 8$$
$$-8 = -5y$$
$$\frac{8}{5} = y$$
$$1\tfrac{3}{5} = y$$

***x*-intercept: $y = 0$**

$$3x = -5y + 8$$
$$3x = -5(0) + 8$$
$$3x = 8$$
$$x = \frac{8}{3}$$
$$x = 2\tfrac{2}{3}$$

***Check point:* $x = 1$**

$$3x = -5y + 8$$
$$3(1) = -5y + 8$$
$$3 = -5y + 8$$
$$-5 = -5y$$
$$1 = y$$

The y-intercept is $\left(0, 1\tfrac{3}{5}\right)$, the x-intercept is $\left(2\tfrac{2}{3}, 0\right)$, and the check point is $(1, 1)$. The ordered pairs are plotted as shown.

Success Tip

When graphing, it is often helpful to write any coordinates that are improper fractions as mixed numbers. For example: $\left(\frac{8}{3}, 0\right) = \left(2\tfrac{2}{3}, 0\right)$.

$3x = -5y + 8$

x	y	(x, y)
0	$1\tfrac{3}{5}$	$\left(0, 1\tfrac{3}{5}\right)$
$\frac{8}{3} = 2\tfrac{2}{3}$	0	$\left(2\tfrac{2}{3}, 0\right)$
1	1	(1, 1)

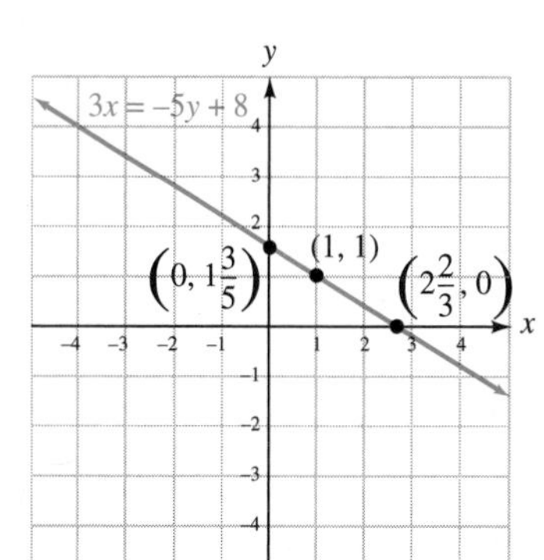

Self Check 4 Graph $8x = -4y + 15$ by finding the intercepts.

GRAPHING HORIZONTAL AND VERTICAL LINES

Equations such as $y = 4$ and $x = -3$ are linear equations, because they can be written in the general form $Ax + By = C$.

$y = 4$	is equivalent to	$0x + 1y = 4$
$x = -3$	is equivalent to	$1x + 0y = -3$

We now discuss how to graph these types of linear equations.

EXAMPLE 5

Graph: $y = 4$.

ELEMENTARY Algebra $f(x)$ Now™

Solution We can write the equation in general form as $0x + y = 4$. Since the coefficient of x is 0, the numbers chosen for x have no effect on y. The value of y is always 4. For example, if $x = 2$, we have

$$0x + y = 4 \quad \text{This is the original equation written in general form.}$$
$$0(2) + y = 4 \quad \text{Substitute 2 for } x.$$
$$y = 4 \quad \text{Simplify the left-hand side.}$$

One solution is (2, 4). To find two more solutions, we choose $x = 0$ and $x = -3$. For any x-value, the y-value is always 4, so we enter (0, 4) and $(-3, 4)$ in the table. If we plot the ordered pairs and draw a straight line through the points, the result is a horizontal line. The y-intercept is (0, 4) and there is no x-intercept.

$y = 4$

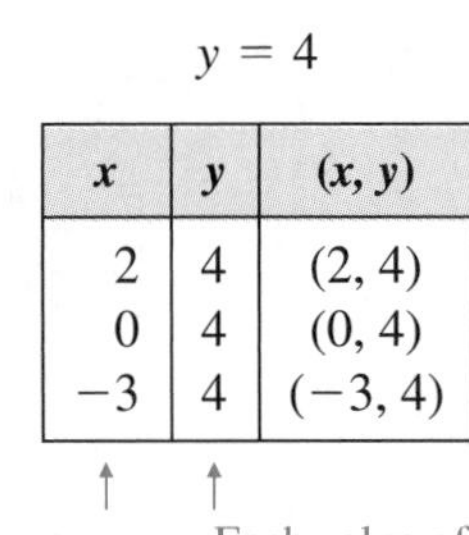

x	y	(x, y)
2	4	(2, 4)
0	4	(0, 4)
-3	4	$(-3, 4)$

↑ Choose any number for x. ↑ Each value of y must be 4.

Self Check 5 Graph: $y = -2$.

EXAMPLE 6

Graph: $x = -3$.

ELEMENTARY Algebra $f(x)$ Now™

Solution We can write the equation in general form as $x + 0y = -3$. Since the coefficient of y is 0, the numbers chosen for y have no effect on x. The value of x is always -3. For example, if $y = -2$, we have

$$x + 0y = -3 \quad \text{This is the original equation written in general form.}$$
$$x + 0(-2) = -3 \quad \text{Substitute } -2 \text{ for } y.$$
$$x = -3 \quad \text{Simplify the left-hand side.}$$

One solution is $(-3, -2)$. To find two more solutions, we choose $y = 0$ and $y = 3$. For any y-value, the x-value is always -3, so we enter $(-3, 0)$ and $(-3, 3)$ in the table. If we plot the ordered pairs and draw a straight line through the points, the result is a vertical line. The x-intercept is $(-3, 0)$ and there is no y-intercept.

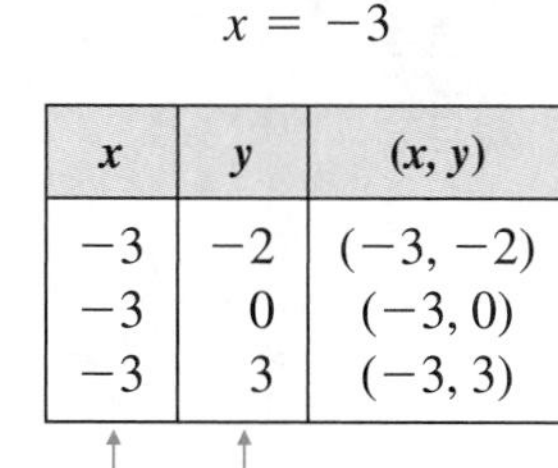

$x = -3$

x	y	(x, y)
-3	-2	$(-3, -2)$
-3	0	$(-3, 0)$
-3	3	$(-3, 3)$

Each value of x must be -3. Choose any number for y.

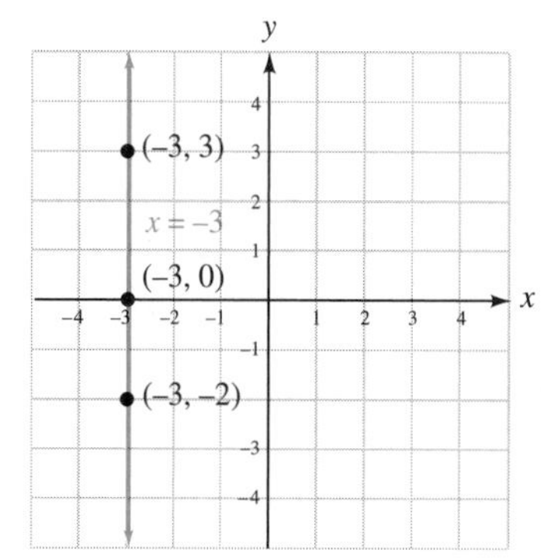

Self Check 6 Graph: $x = 4$.

From the results of Examples 5 and 6, we have the following facts.

Equations of Horizontal and Vertical Lines

The equation $y = b$ represents the horizontal line that intersects the y-axis at $(0, b)$.
The equation $x = a$ represents the vertical line that intersects the x-axis at $(a, 0)$.

The Language of Algebra

Two *parallel* lines are always the same distance apart. For example, think of the rails of train tracks. When graphing lines, remember that $y = b$ is *parallel* to the x-axis and $x = a$ is *parallel* to the y-axis.

The graph of the equation $y = 0$ has special significance; it is the x-axis. Similarly, the graph of the equation $x = 0$ is the y-axis.

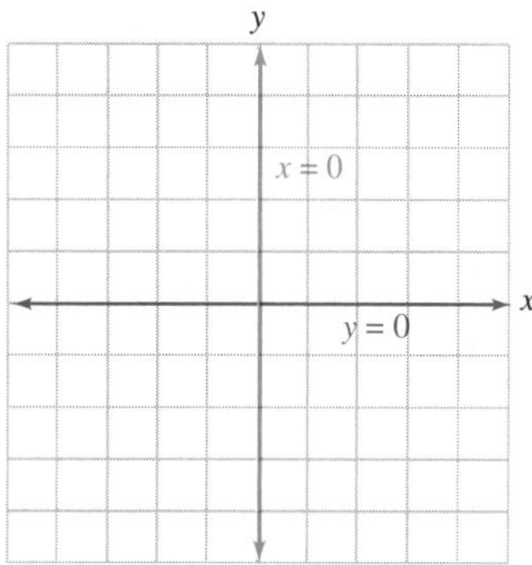

INFORMATION FROM INTERCEPTS

The ability to read and interpret graphs is a valuable skill. When analyzing a graph, we should locate and examine the intercepts. As the following example illustrates, the coordinates of the intercepts can yield useful information.

EXAMPLE 7

Camcorders. The number of feet of videotape that remain on a cassette depends on the number of minutes of videotaping that has already occurred. The following graph shows the relationship between these two quantities for a cassette in standard play mode. What information do the intercepts give about the cassette?

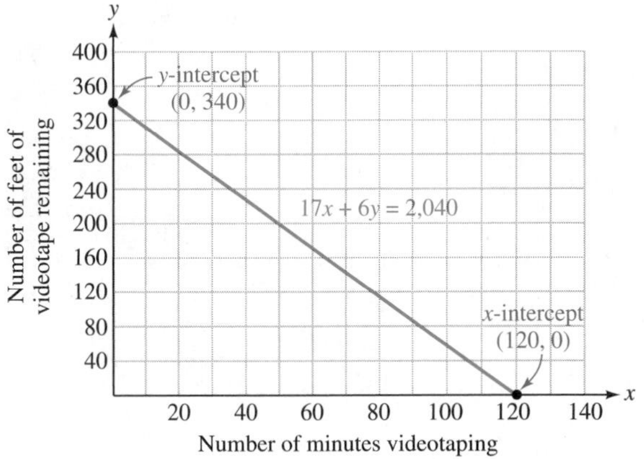

Solution The y-intercept of (0, 340) indicates that when 0 minutes of taping have occurred, 340 feet of videotape are available. That is, the cassette contains 340 feet of videotape.

The x-coordinate of (120, 0) indicates that when 120 minutes of videotaping have occurred, 0 feet of tape are available. In other words, the cassette can hold 120 minutes, or 2 hours, of videotaping.

Courtesy of Texas Instruments

GRAPHING CALCULATORS

So far, we have graphed linear equations by making tables of solutions and plotting points. A graphing calculator can make the task of graphing much easier.

However, a graphing calculator does not take the place of a working knowledge of the topics discussed in this chapter. It should serve as an aid to enhance your study of algebra.

The Viewing Window The screen on which a graph is displayed is called the **viewing window.** The **standard window** has settings of

$$\text{Xmin} = -10, \quad \text{Xmax} = 10, \quad \text{Ymin} = -10, \quad \text{and} \quad \text{Ymax} = 10$$

which indicate that the minimum x- and y-coordinates used in the graph will be -10, and that the maximum x- and y-coordinates will be 10.

Graphing an Equation To graph $y = x - 1$ using a graphing calculator, we press the Y = key and enter $x - 1$ after the symbol Y_1. Then we press the GRAPH key to see the graph.

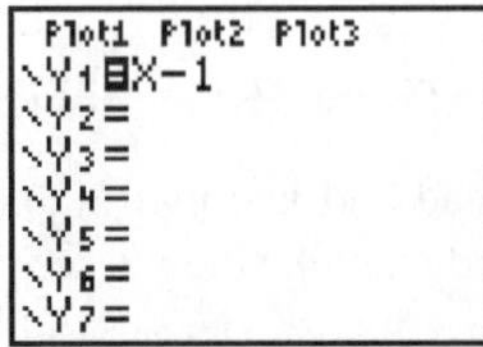

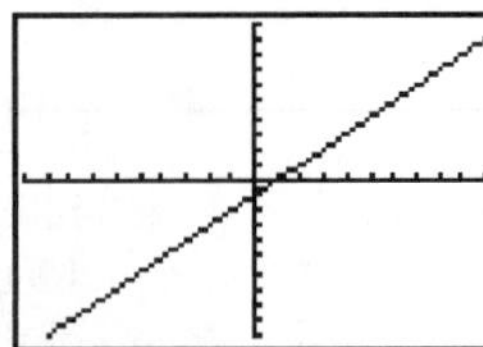

Changing the Viewing Window We can change the viewing window by pressing the WINDOW key and entering -4 for the minimum x- and y-coordinates and 4 for the maximum x- and y-coordinates. Then we press the GRAPH key to see the graph of $y = x - 1$ in more detail.

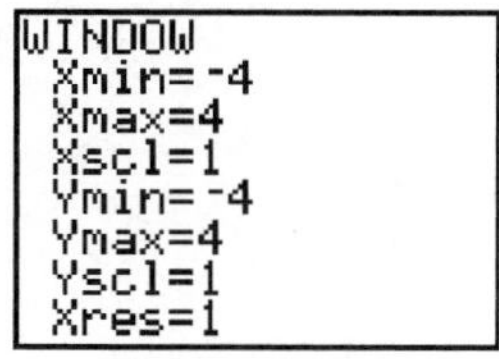

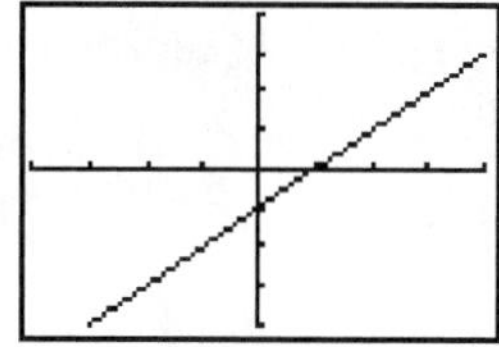

Solving an Equation for y To graph $3x + 2y = 12$, we must first solve the equation for y.

$$3x + 2y = 12$$

$$2y = -3x + 12 \quad \text{Subtract } 3x \text{ from both sides.}$$

$$y = -\frac{3}{2}x + 6 \quad \text{Divide both sides by 2.}$$

Next, we press the WINDOW key to reenter the standard window settings and press the Y = key to enter $y = -\frac{3}{2}x + 6$. Then we press the GRAPH key to see the graph.

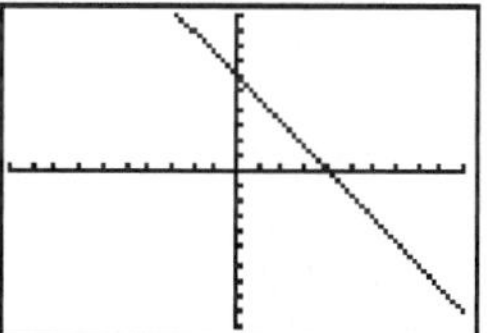

Answers to Self Checks

1. $(-1, 0)$; $(0, -3)$

2.
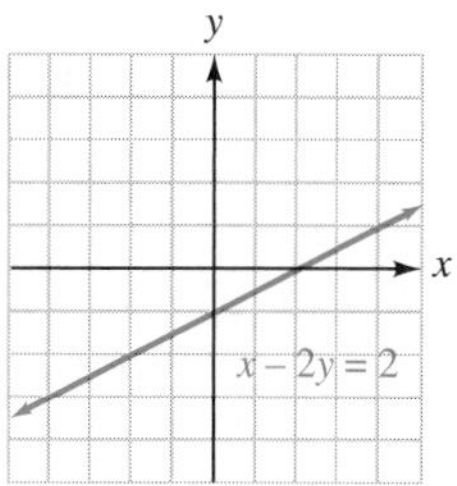

3.
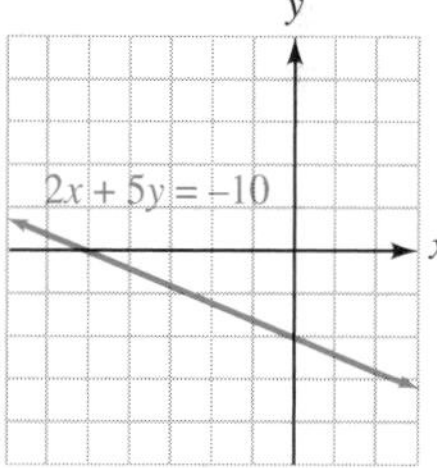

4.
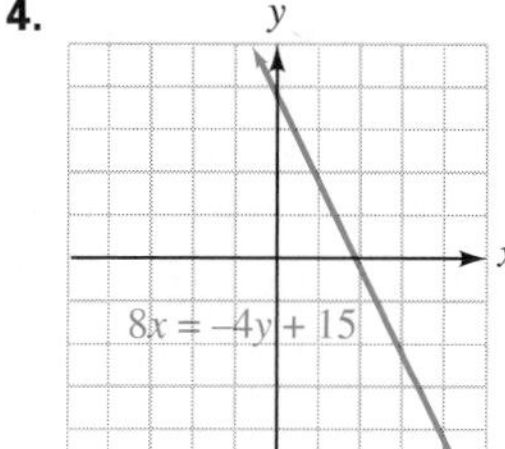

5.
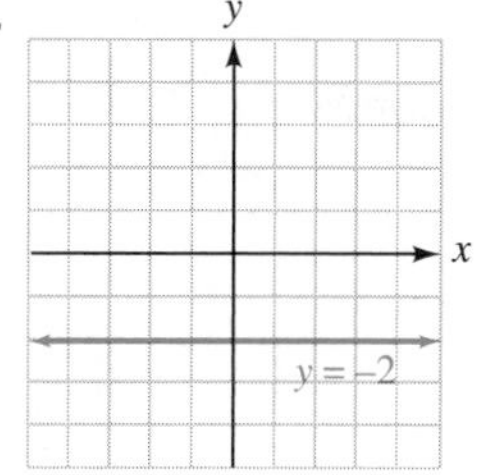

6.
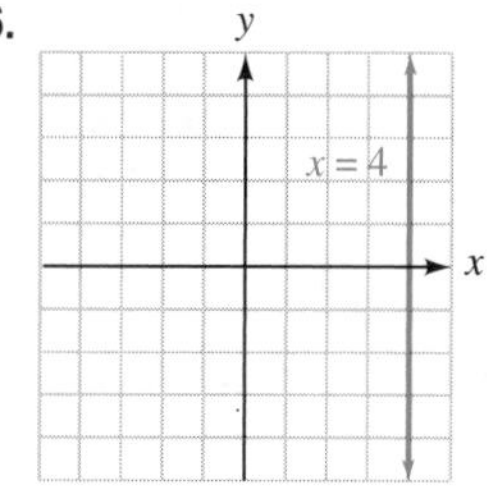

3.3 STUDY SET

ELEMENTARY Algebra $f(x)$ Now™

VOCABULARY **Fill in the blanks.**

1. We say $5x + 3y = 10$ is an equation in _____ variables, x and y.
2. $2x - 3y = 6$ is a _______ equation; its graph is a line.
3. The equation $2x - 3y = 7$ is written in _______________ form.
4. The __________ of a line is the point where the line intersects the x-axis.
5. The y-intercept of a line is the point where the line _______________ the y-axis.
6. The graph of $y = 4$ is a __________ line, with y-intercept (0, 4).

CONCEPTS **Identify the intercepts of each graph.**

7.

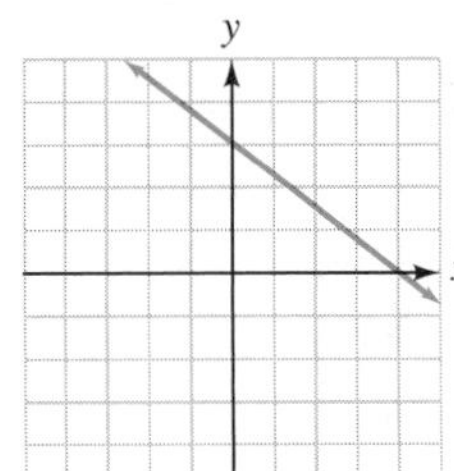

8.

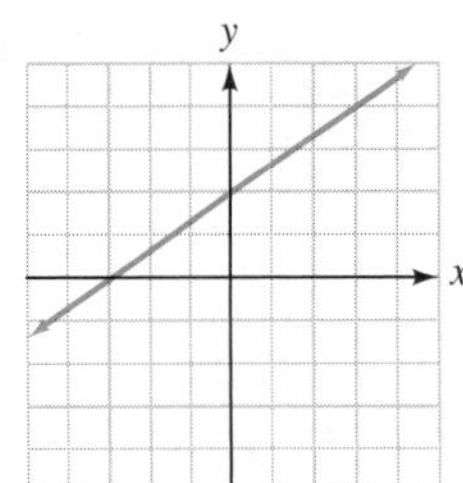

9.

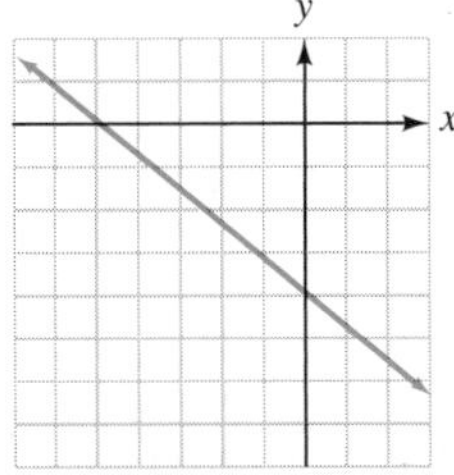

10.

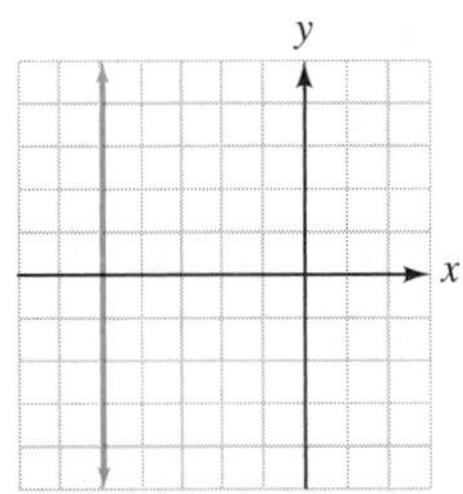

11.

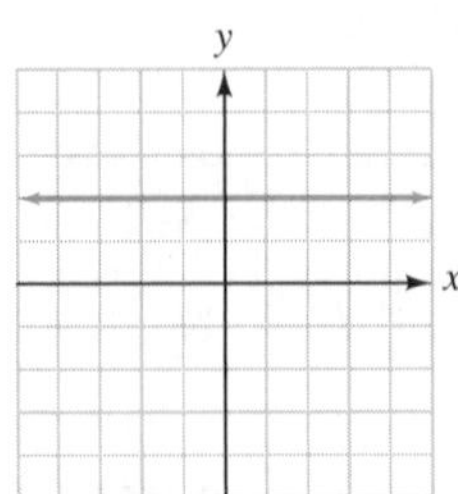

12.

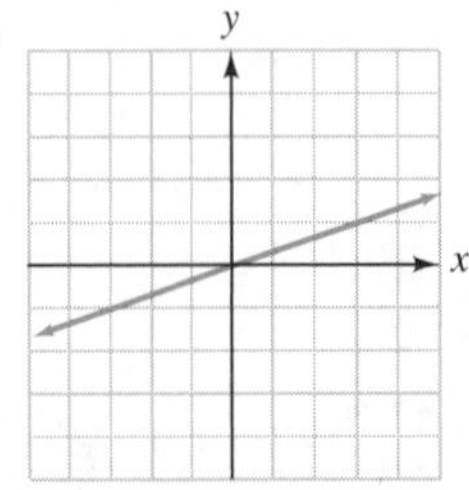

13. Estimate the intercepts of the line in the graph.

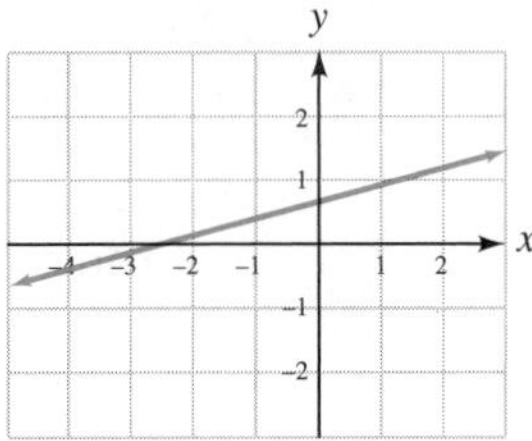

14. Fill in the blanks.
    a. To find the y-intercept of the graph of a line, substitute ☐ for x in the equation and solve for ☐.
    b. To find the x-intercept of the graph of a line, substitute ☐ for y in the equation and solve for ☐.

15. Consider the linear equation $3x + 2y = 6$.
    a. If we let $x = 0$, which term of the equation is equal to 0?
    b. Solve the equation that remains. What is the y-intercept of the graph of $3x + 2y = 6$?

16. In the table of solutions, which entry is the y-intercept of the graph and which entry is the x-intercept of the graph?

x	y	(x, y)
6	0	(6, 0)
0	−2	(0, −2)
−3	−3	(−3, −3)

17. What is the maximum number of intercepts that a line may have? What is the minimum number of intercepts that a line may have?

18. It is known that the value of a certain piece of farm machinery will steadily decrease after it is purchased.
  a. From the graph, which intercept tells the purchase price of the machinery? What was that price?
  b. Which intercept tells when the machinery will have lost all of its value? When is that?

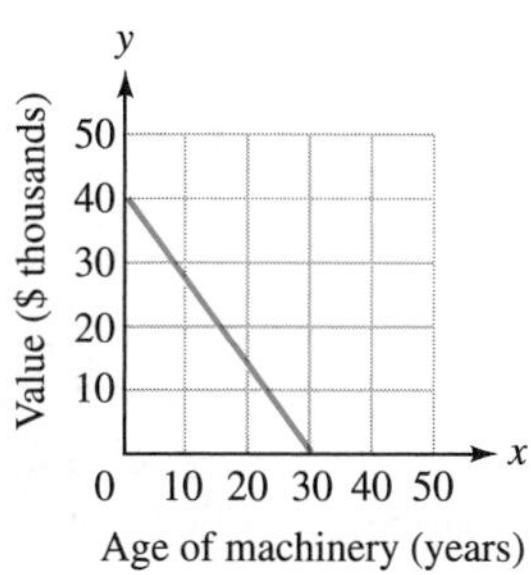

19. Match each graph with its equation

a. $x = 2$	b. $y = 2$	c. $y = 2x$
d. $2x - y = 2$	e. $y = 2x + 2$	f. $y = -2x$

i.
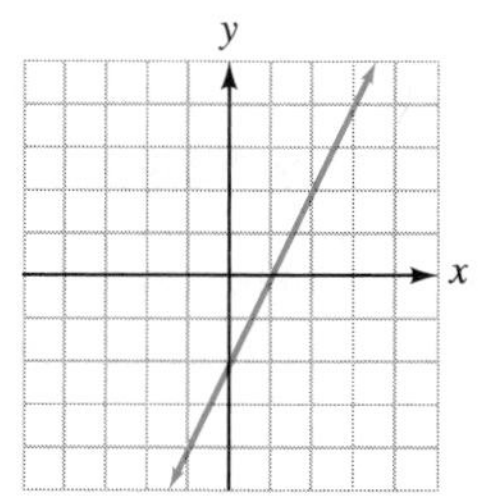

ii.
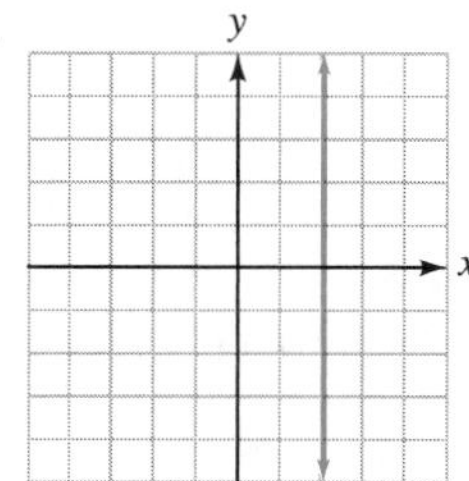

iii.
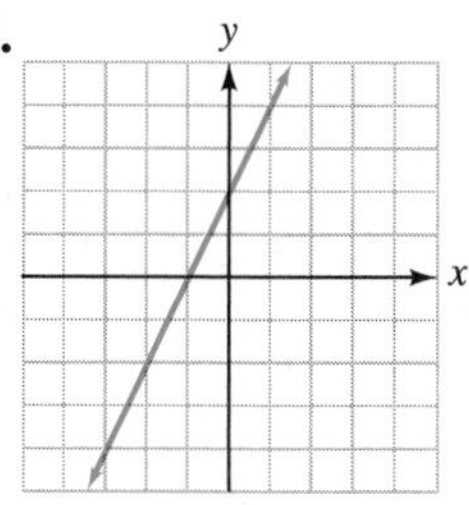

iv.
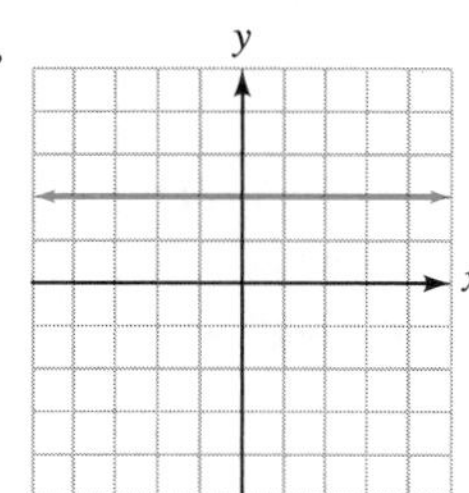

v.
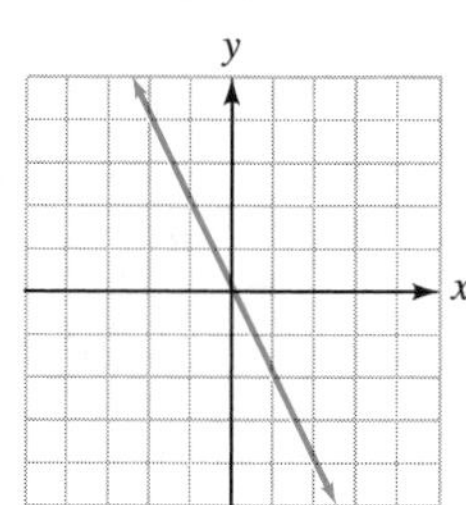

vi.
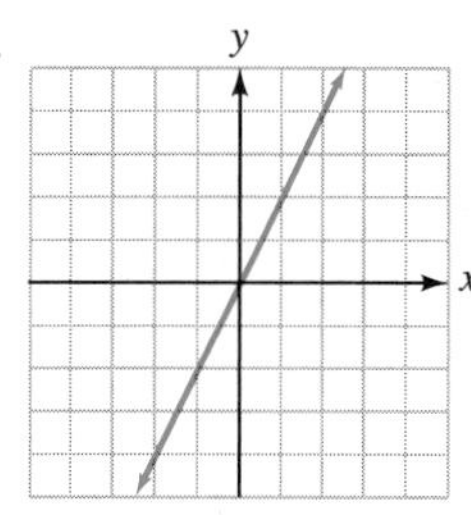

20. What linear equation could have the table of solutions shown below?

x	y	(x, y)
1	8	(1, 8)
0	8	(0, 8)
−1	8	(−1, 8)

NOTATION

21. a. Does the point (0, 6) lie on the x-axis or the y-axis?
  b. Is it correct to say that the point (0, 0) lies on the x-axis *and* on the y-axis?

22. True or false: $x = 5$ is equivalent to $1x + 0y = 5$.

23. What is the equation of the x-axis? What is the equation of the y-axis?

24. Write any coordinates that are improper fractions as mixed numbers.
  a. $\left(\frac{7}{2}, 0\right)$ b. $\left(0, -\frac{17}{3}\right)$

PRACTICE Use the intercept method to graph each equation.

25. $4x + 5y = 20$
26. $3x + 4y = 12$
27. $x - y = -3$
28. $x - y = 3$
29. $5x + 15y = -15$
30. $8x + 4y = -24$
31. $x + 2y = -2$
32. $x + 2y = -4$
33. $4x - 3y = 12$
34. $5x - 10y = 20$
35. $3x + y = -3$
36. $2x - y = -2$
37. $9x - 4y = -9$
38. $5x - 4y = -15$
39. $8 = 3x + 4y$
40. $9 = 2x + 3y$
41. $4x - 2y = 6$
42. $6x - 3y = 3$
43. $3x - 4y = 11$
44. $5x - 4y = 13$
45. $9x + 3y = 10$
46. $4x + 4y = 5$
47. $3x = -15 - 5y$
48. $x = 5 - 5y$
49. $-4x = 8 - 2y$
50. $-5x = 10 + 5y$
51. $7x = 4y - 12$
52. $7x = 5y - 15$
53. $y - 3x = -\frac{4}{3}$
54. $y - 2x = -\frac{9}{8}$

Graph each equation.

55. $y = 4$

56. $y = -3$

57. $x = -2$

58. $x = 5$

59. $y = -\frac{1}{2}$

60. $y = \frac{5}{2}$

61. $x = \frac{4}{3}$

62. $x = -\frac{5}{3}$

63. $y - 2 = 0$

64. $x + 1 = 0$

65. $-2x + 3 = 11$

66. $-3y + 2 = 5$

APPLICATIONS

67. CHEMISTRY The relationship between the temperature and volume of a gas at a constant pressure is graphed below. The T-intercept of this graph is a very important scientific fact. It represents the lowest possible temperature, called **absolute zero.**

a. Estimate absolute zero.

b. What is the volume of the gas when the temperature is absolute zero?

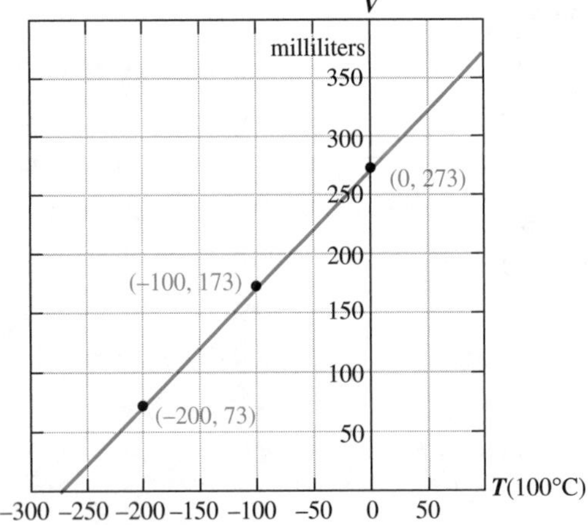

68. PHYSICS The graph shows the length L of a stretched spring (in inches) as different weights w (in pounds) are attached to it. What information about the spring does the L-intercept give us?

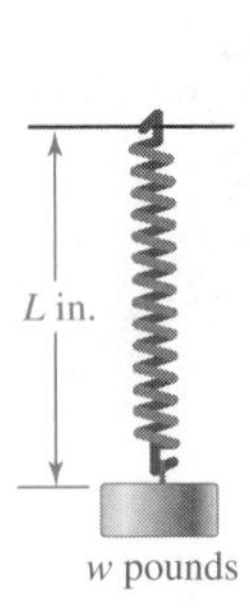

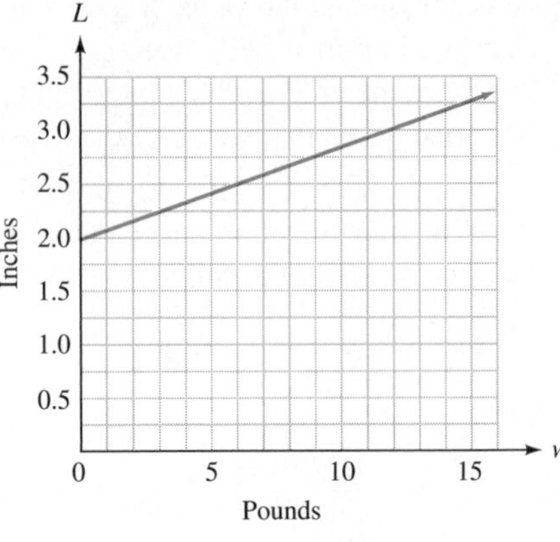

69. LANDSCAPING A developer is going to purchase x trees and y shrubs to landscape a new office complex. The trees cost \$50 each and the shrubs cost \$25 each. His budget is \$5,000. This situation is modeled by the equation $50x + 25y = 5{,}000$. Use the intercept method to graph it.

a. What information is given by the y-intercept?

b. What information is given by the x-intercept?

70. THE MOTOR CITY The linear equation $y = -192{,}000x + 1{,}850{,}000$ models the population y of Detroit, Michigan, where x is the number of decades since 1950. Without graphing, find the y-intercept of the graph. Then explain what it means.

WRITING

71. To graph $3x + 2y = 12$, a student found the intercepts and a check point, and graphed them, as shown on the left. Instead of drawing a crooked line through the points, what should he have done?

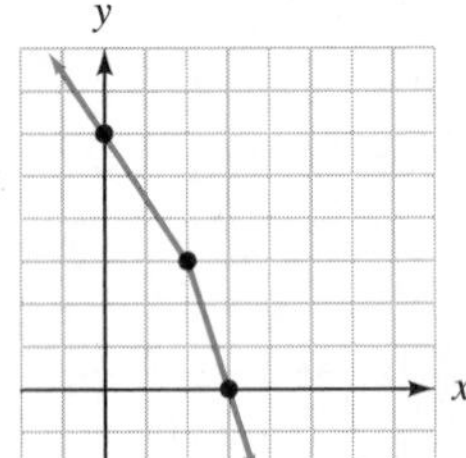

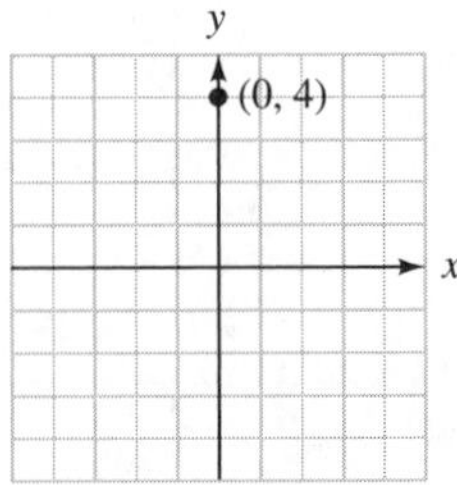

72. A student graphed the linear equation $y = 4$, as shown above on the right. Explain his error.

73. How do we find the intercepts of the graph of an equation without having to graph the equation?

74. In Section 3.2, we discussed a method to graph $y = 2x - 3$. In Section 3.3, we discussed a method to graph $2x + 3y = 6$. Briefly explain the steps involved in each method.

REVIEW

75. Simplify $\frac{3 \cdot 5 \cdot 5}{3 \cdot 5 \cdot 5 \cdot 5}$ by removing the common factors.

76. Simplify: $4\left(\frac{d}{2} - 3\right) - 5\left(\frac{2}{5}d - 1\right)$.

77. Translate: Six less than twice x.

78. Is -5 a solution of $2(3x + 10) = 5x + 6$?

CHALLENGE PROBLEMS

79. Where will the line $y = b$ intersect the line $x = a$?

80. Write an equation of the line that has an x-intercept of $(4, 0)$ and a y-intercept of $(0, 3)$.

3.4 The Slope of a Line

- Finding the Slope of a Line from Its Graph
- The Slope Formula
- Slopes of Horizontal and Vertical Lines
- Applications of Slope
- Rates of Change

In Sections 3.2 and 3.3, we graphed linear equations. All of the graphs were similar in one sense—they were lines. However, the lines slanted in different ways and had varying degrees of steepness. In this section, we introduce a means of measuring the steepness of a line. We call this measure the *slope of the line,* and it can be found in several ways.

FINDING THE SLOPE OF A LINE FROM ITS GRAPH

The **slope of a line** is a comparison of the vertical change to the corresponding horizontal change as we move along the line. The comparison is expressed as a **ratio** (a quotient of two numbers).

The Language of Algebra

Ratios are used in many settings. Mechanics speak of gear *ratios.* Colleges advertise their student-to-teacher *ratios.* Banks calculate debt-to-income *ratios* for loan applicants.

As an example, let's find the slope of the line graphed on the next page. To begin, we select two points on the line, $P(4, 2)$ and $Q(10, 7)$. As we move from point P to point Q, the y-coordinates change from 2 to 7. Therefore, the vertical change, called the **rise,** is $7 - 2$ or 5 units.

As we move from point P to point Q, the x-coordinates change from 4 to 10. Therefore, the horizontal change, called the **run,** is $10 - 4$ or 6 units.

The slope of a line is defined to be *the ratio of the vertical change to the horizontal change.* So we have

$$\text{slope} = \frac{\text{vertical change}}{\text{horizontal change}} = \frac{\text{change in } y}{\text{change in } x} = \frac{\text{rise}}{\text{run}} = \frac{5}{6}$$

The slope of the line is $\frac{5}{6}$. This indicates that there is a vertical change (rise) of 5 units for each horizontal change (run) of 6 units.

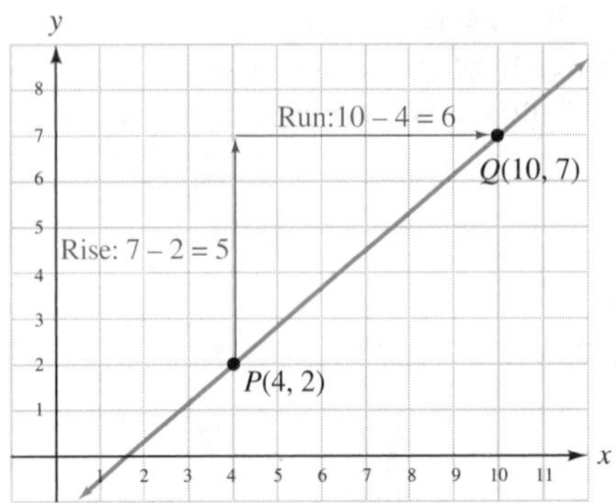

EXAMPLE 1

ELEMENTARY Algebra f(x) Now™

Find the slope of the line graphed below.

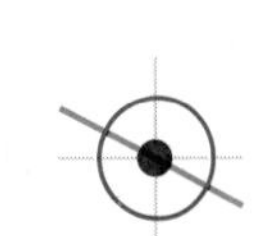

Pick a point on the line that also lies on the intersection of two grid lines.

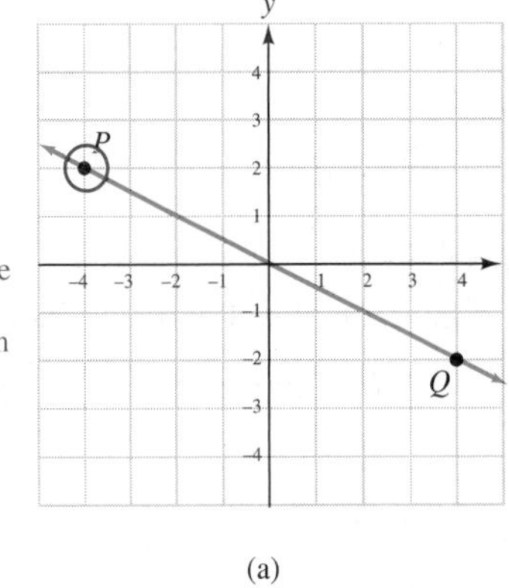

(a)

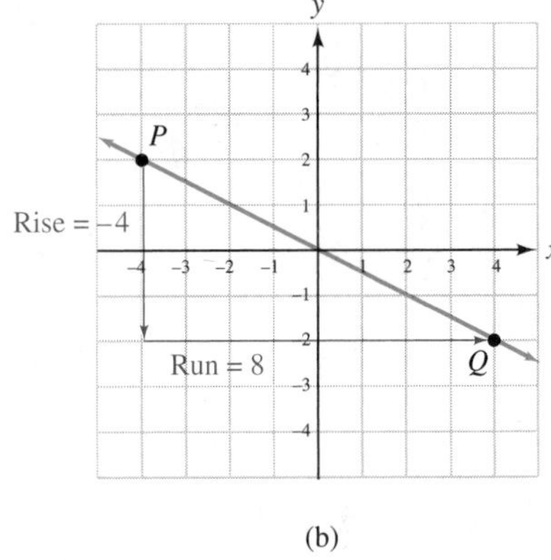

(b)

Solution We begin by choosing two points on the line, P and Q, as shown in illustration (a). One way to move from P to Q is shown in illustration (b). Starting at P, we move downward, a rise of -4, and then to the right, a run of 8, to reach Q. These steps create a right triangle called a **slope triangle.**

To find the slope of the line, we write a ratio of the rise to the run. By tradition, the letter m is used to denote slope, so we have

$$m = \frac{\text{rise}}{\text{run}} \quad \text{Slope is the ratio (quotient) of rise to run.}$$

$$m = \frac{-4}{8} \quad \text{Substitute } -4 \text{ for the rise and 8 for the run.}$$

$$m = -\frac{1}{2} \quad \text{Simplify the fraction.}$$

The slope of the line is $-\frac{1}{2}$.

The two-step process to move from P to Q can be reversed. Starting at P, we can move to the right, a run of 8; and then downward, a rise of -4, to reach Q. With this approach, the slope triangle is above the line. When we form the ratio to find the slope, we get the same result as before:

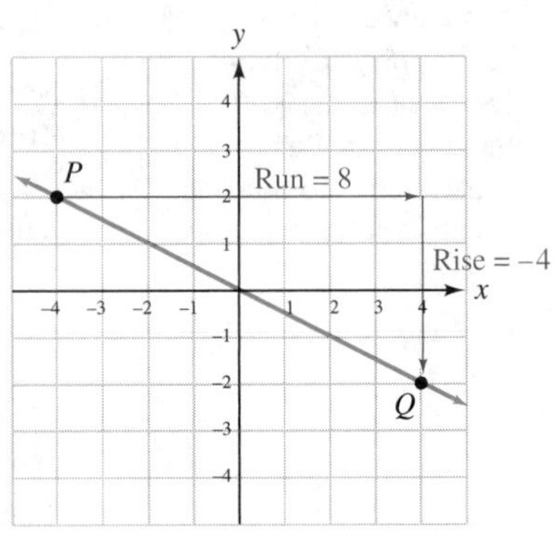

The Language of Algebra

The symbol m is used to denote the slope of a line. Many historians credit this to the fact that it is the first letter of the French word *monter,* meaning to ascend or to climb.

Success Tip

When drawing a slope triangle, remember that upward movements are positive, downward movements are negative, movements to the right are positive, and movements to the left are negative.

$$m = \frac{\text{rise}}{\text{run}} = \frac{-4}{8} = -\frac{1}{2}$$

Self Check 1 Find the slope of the line shown above using two points different from those used in the solution of Example 1.

The identical answers from Example 1 and its Self Check illustrate an important fact about slope: The same value will be obtained no matter which two points on a line are used to find the slope.

THE SLOPE FORMULA

We can generalize the graphic method for finding slope to develop a slope formula. To begin, we select two points on a line, as shown in the figure below. Call them P and Q. To distinguish between the coordinates of these points, we use **subscript notation.**

The Language of Algebra

The prefix *sub* means below or beneath, as in submarine or subway. In x_2, the *subscript* 2 is written lower than the variable.

- Point P is denoted as $P(x_1, y_1)$. Read as "point P with coordinates of x sub 1 and y sub 1."
- Point Q is denoted as $Q(x_2, y_2)$. Read as "point Q with coordinates of x sub 2 and y sub 2."

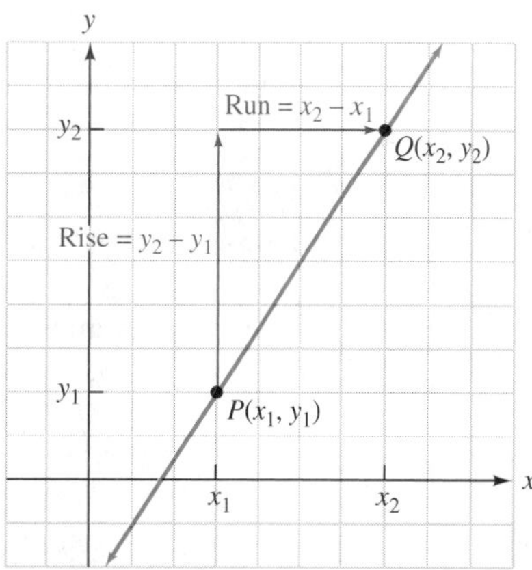

As we move from point P to point Q, the rise is the difference of the y-coordinates: $y_2 - y_1$. The run is the difference of the x-coordinates: $x_2 - x_1$. Since the slope is the ratio $\frac{\text{rise}}{\text{run}}$, we have the following formula for calculating slope.

Slope of a Line The **slope** of a line passing through points (x_1, y_1) and (x_2, y_2) is

$$m = \frac{\text{vertical change}}{\text{horizontal change}} = \frac{\text{change in } y}{\text{change in } x} = \frac{\text{rise}}{\text{run}} = \frac{y_2 - y_1}{x_2 - x_1} \quad \text{if } x_2 \neq x_1.$$

EXAMPLE 2

Find the slope of the line passing through (1, 2) and (3, 8). Then graph the line.

ELEMENTARY Algebra *f(x)* Now™

Solution When using the slope formula, it makes no difference which point you call (x_1, y_1) and which point you call (x_2, y_2). If we let (x_1, y_1) be (1, 2) and (x_2, y_2) be (3, 8), then

$m = \frac{y_2 - y_1}{x_2 - x_1}$ This is the slope formula.

Notation

We can write slopes that are integers in $\frac{\text{rise}}{\text{run}}$ form by writing them as fractions with a denominator of 1. For example, $m = 3 = \frac{3}{1}$ or $m = -5 = \frac{-5}{1}$.

$m = \frac{8 - 2}{3 - 1}$ Substitute 8 for y_2, 2 for y_1, 3 for x_2, and 1 for x_1.

$m = \frac{6}{2}$ Do the subtractions.

$m = 3$ Simplify. Think of this as a $\frac{3}{1}$ rise-to-run ratio.

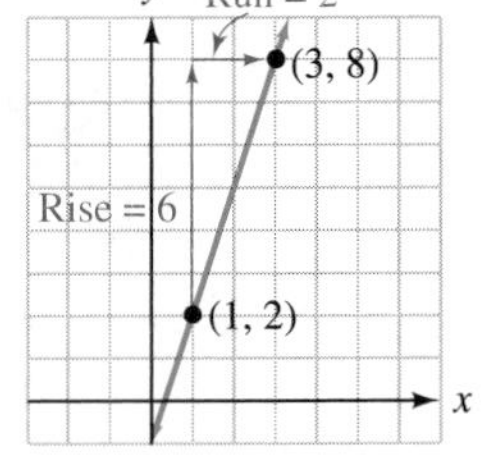

The slope of the line is 3. The graph of the line, including the slope triangle, is shown here. Note that we obtain the same value for the slope if we let $(x_1, y_1) = (3, 8)$ and $(x_2, y_2) = (1, 2)$.

$$m = \frac{y_2 - y_1}{x_2 - x_1} = \frac{2 - 8}{1 - 3} = \frac{-6}{-2} = 3$$

Self Check 2 Find the slope of the line passing through (2, 1) and (4, 11).

Caution When using the slope formula, be sure to subtract the y-coordinates and the x-coordinates in the same order. Otherwise, your answer will have the wrong sign.

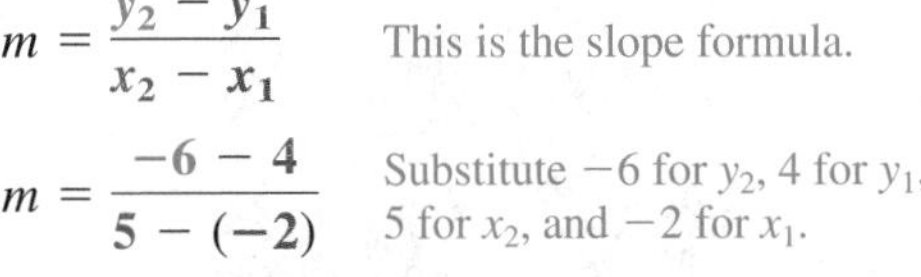

$$m \neq \frac{y_2 - y_1}{x_1 - x_2} \quad \text{and} \quad m \neq \frac{y_1 - y_2}{x_2 - x_1}$$

EXAMPLE 3

Find the slope of the line that passes through $(-2, 4)$ and $(5, -6)$ and graph the line.

ELEMENTARY Algebra f(x) Now™

Solution Since we know the coordinates of two points on the line, we can find its slope. If (x_1, y_1) is $(-2, 4)$ and (x_2, y_2) is $(5, -6)$, then

Notation

Slopes are normally written as fractions, sometimes as decimals, but never as mixed numbers.

As with any fractional answer, always express slope in lowest terms.

$m = \frac{y_2 - y_1}{x_2 - x_1}$ This is the slope formula.

$m = \frac{-6 - 4}{5 - (-2)}$ Substitute -6 for y_2, 4 for y_1, 5 for x_2, and -2 for x_1.

$m = -\frac{10}{7}$ Do the subtractions. We may write the result as $\frac{-10}{7}$ or $-\frac{10}{7}$.

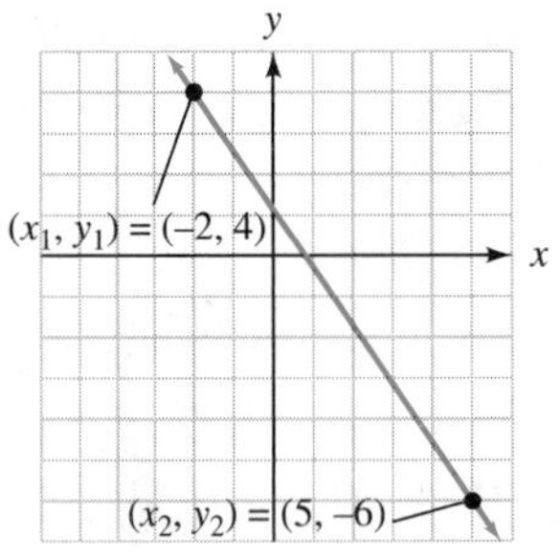

The slope of the line is $-\frac{10}{7}$. In the graph, we see that the line falls from left to right—a fact indicated by its negative slope.

Self Check 3 Find the slope of the line that passes through $(-1, -2)$ and $(1, -7)$.

From the previous examples, we see that the slope of the line in Example 2 was positive and the slopes of the lines in Examples 1 and 3 were negative. In general, lines that rise from left to right have a positive slope. Lines that fall from left to right have a negative slope.

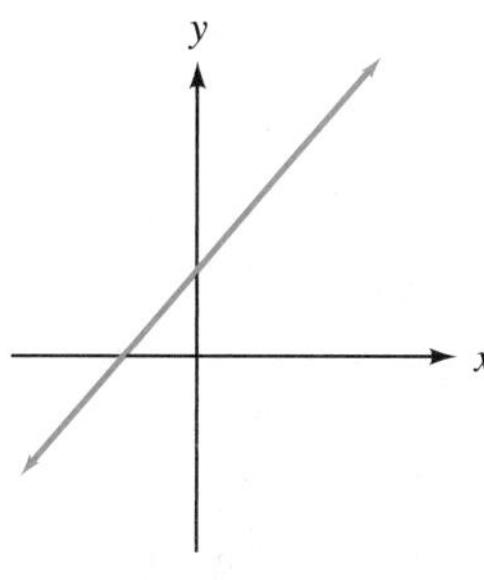

Positive slope

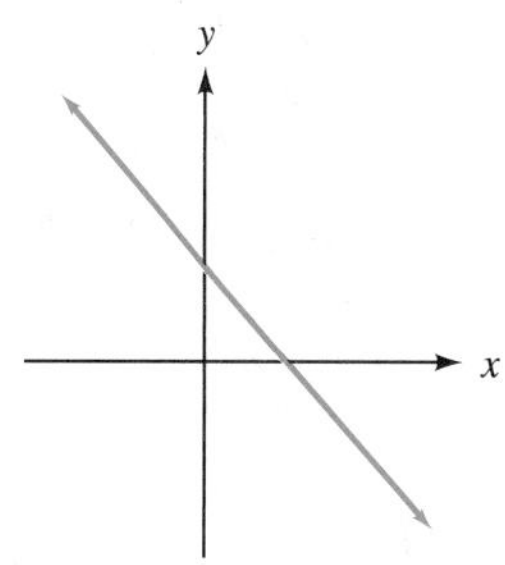

Negative slope

When we compare lines with positive slopes, we see that the larger the slope, the steeper the line. For example, a line with slope 3 is steeper than a line with slope $\frac{5}{6}$, and a line with slope $\frac{5}{6}$ is steeper than a line with slope $\frac{1}{4}$.

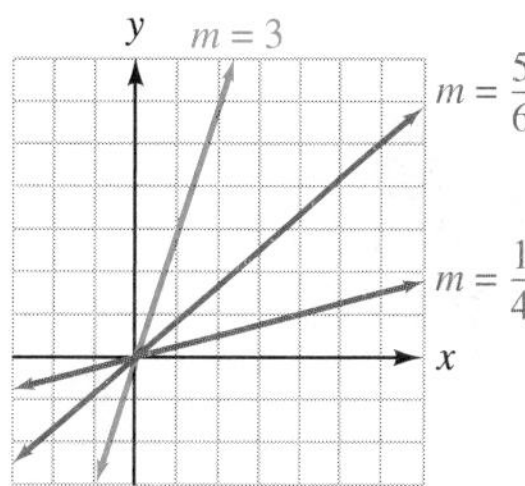

SLOPES OF HORIZONTAL AND VERTICAL LINES

In the next two examples, we calculate the slope of a horizontal line and we show that a vertical line has no defined slope.

EXAMPLE 4

Find the slope of the line $y = 3$.

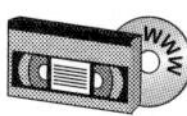

Solution The graph of $y = 3$ is a horizontal line. To find its slope, we need to know two points on the line. From the graph, we select $(-2, 3)$ and $(3, 3)$. If (x_1, y_1) is $(-2, 3)$ and (x_2, y_2) is $(3, 3)$, we have

$$m = \frac{y_2 - y_1}{x_2 - x_1}$$ This is the slope formula.

$$m = \frac{3 - 3}{3 - (-2)}$$ Substitute 3 for y_2, 3 for y_1, 3 for x_2, and -2 for x_1.

$$m = \frac{0}{5}$$ Simplify the numerator and the denominator.

$$m = 0$$

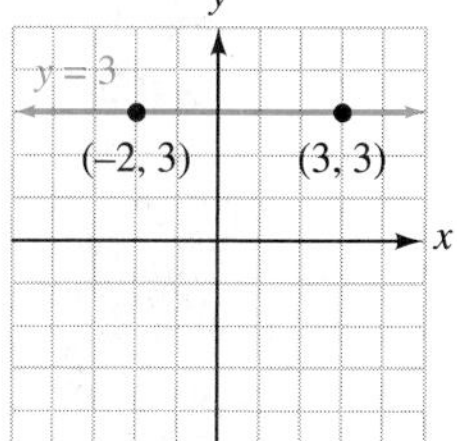

The slope of the line $y = 3$ is 0.

The y-coordinates of any two points on a horizontal line will be the same, and the x-coordinates will be different. Thus, the numerator of

$$\frac{y_2 - y_1}{x_2 - x_1}$$

will always be zero, and the denominator will always be nonzero. Therefore, the slope of a horizontal line is zero.

EXAMPLE 5

If possible, find the slope of the line $x = -2$.

ELEMENTARY Algebra $f(x)$ Now™

Solution The graph of $x = -2$ is a vertical line. To find its slope, we need to know two points on the line. From the graph, we select $(-2, 3)$ and $(-2, -1)$. If (x_2, y_2) is $(-2, 3)$ and (x_1, y_1) is $(-2, -1)$, we have

$$m = \frac{y_2 - y_1}{x_2 - x_1}$$ This is the slope formula.

$$m = \frac{3 - (-1)}{-2 - (-2)}$$ Substitute 3 for y_2, -1 for y_1, -2 for x_2, and -2 for x_1.

$$m = \frac{4}{0}$$ Simplify the numerator and the denominator.

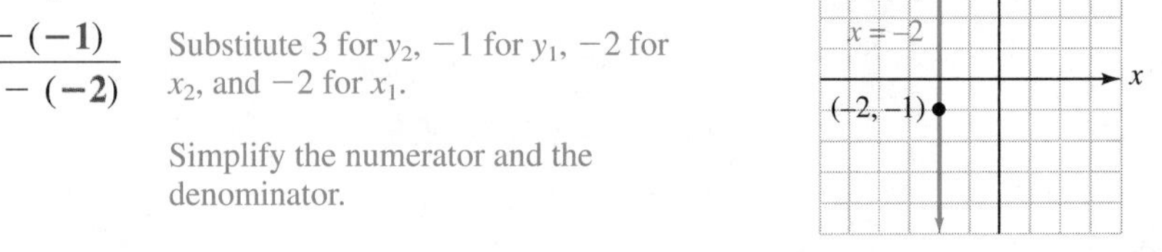

Since division by zero is undefined, $\frac{4}{0}$ has no meaning. The slope of the line $x = -2$ is undefined.

The y-coordinates of any two points on a vertical line will be different, and the x-coordinates will be the same. Thus, the numerator of

$$\frac{y_2 - y_1}{x_2 - x_1}$$

will always be nonzero, and the denominator will always be zero. Therefore, the slope of a vertical line is undefined.

We now summarize the results from Examples 4 and 5.

Slopes of Horizontal and Vertical Lines Horizontal lines (lines with equations of the form $y = b$) have a slope of 0. Vertical lines (lines with equations of the form $x = a$) have undefined slope.

The Language of Algebra

Undefined and *0* do not mean the same thing. A horizontal line has a defined slope; it is 0. A vertical line does not have a defined slope; we say its slope is *undefined.*

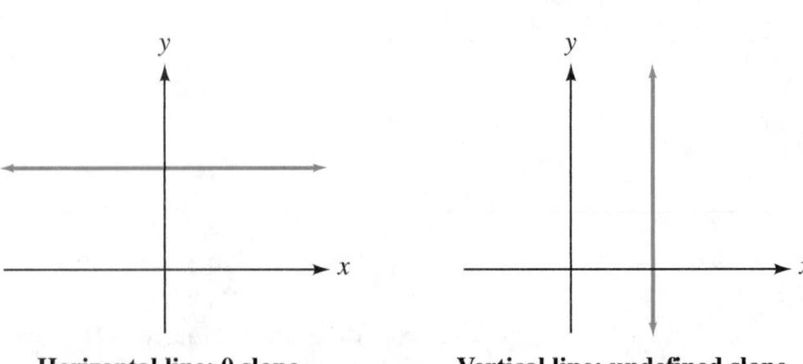

Horizontal line: 0 slope **Vertical line: undefined slope**

APPLICATIONS OF SLOPE

The concept of slope has many applications. For example, architects use slope when designing ramps and roofs. Truckers must be aware of the slope, or *grade,* of the roads they travel. Mountain bikers ride up rocky trails and snow skiers speed down steep slopes.

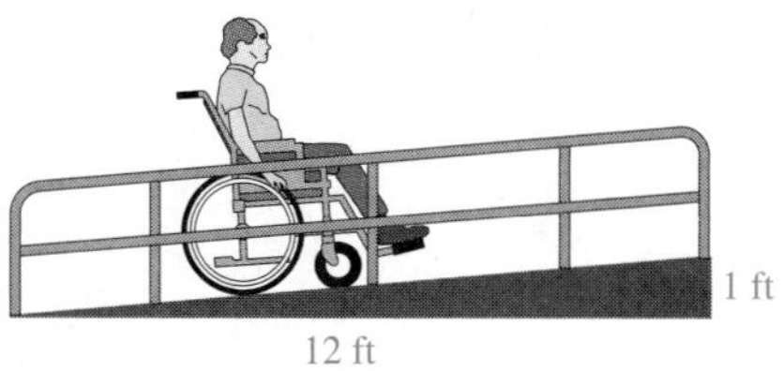

The Americans with Disabilities Act provides a guideline for the steepness of a ramp. The maximum slope for a wheelchair ramp is 1 foot of rise for every 12 feet of run: $m = \frac{1}{12}$.

The **grade** of an incline is its slope expressed as a percent: A 15% grade means a rise of 15 feet for every run of 100 feet: $m = \frac{15}{100}$.

EXAMPLE 6

ELEMENTARY Algebra f(x) Now™

Architecture. **Pitch** is the incline of a roof expressed as a ratio of the vertical rise to the horizontal run. Find the pitch of the roof shown in the illustration.

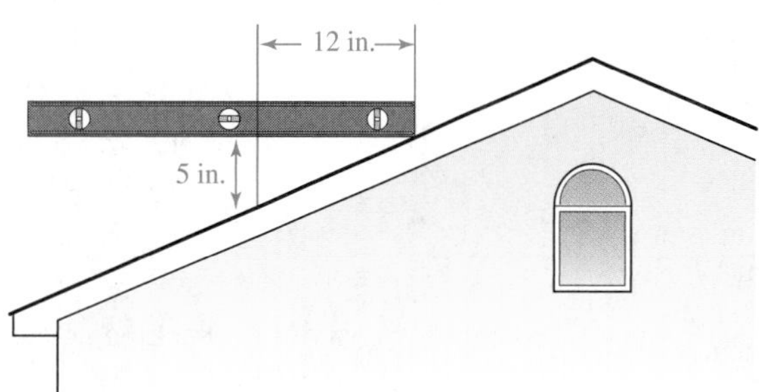

Solution From the definition, we recognize that the pitch of a roof is simply its slope. In the illustration, a level is used to create a slope triangle. The rise is 5 and the run is 12.

$$\begin{aligned} m &= \frac{\text{rise}}{\text{run}} \\ &= \frac{5}{12} \end{aligned}$$

The roof has a $\frac{5}{12}$ pitch.

RATES OF CHANGE

We have seen that the slope of a line is a *ratio* of two numbers. For many applications, however, we often attach units to a slope calculation. When we do so, we say that we have found a **rate of change.**

EXAMPLE 7

ELEMENTARY Algebra f(x) Now™

Checking accounts. A checking plan at a bank charges customers a fixed monthly fee plus a small service charge for each check written. The relationship between the monthly cost y and the number x of checks written is graphed below. At what rate does the monthly cost change?

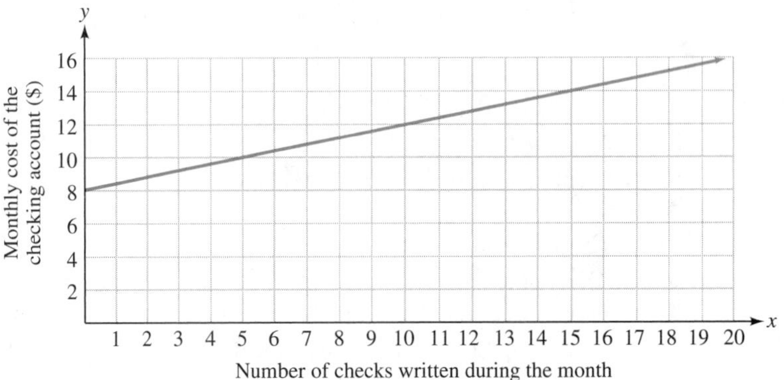

Solution To find the rate of change, we find the slope of the line and attach the proper units. Two points on the line are (5, 10) and (15, 14). If we let $(x_1, y_1) = (5, 10)$ and $(x_2, y_2) = (15, 14)$, we have

$$\text{Rate of change} = \frac{(y_2 - y_1)\text{ dollars}}{(x_2 - x_1)\text{ checks}}$$ This is the slope formula with the appropriate units attached.

$$= \frac{(14 - 10)\text{ dollars}}{(15 - 5)\text{ checks}}$$ Substitute 14 for y_2, 10 for y_1, 15 for x_2, and 5 for x_1.

$$= \frac{4\text{ dollars}}{10\text{ checks}}$$ Do the subtractions.

$$= \frac{2\text{ dollars}}{5\text{ checks}}$$ Simplify the fraction.

The Language of Algebra

The preposition *per* means for each, or for every. When we say the rate of change is 40¢ *per* check, we mean 40¢ for each check.

The monthly cost of the checking account increases $2 for every 5 checks written.

We can express $\frac{2}{5}$ in decimal form by dividing the numerator by the denominator. Then we can write the rate of change in two other ways, using the word *per,* which indicates division.

Rate of change = \$0.40 per check or Rate of change = 40¢ per check

Answers to Self Checks **1.** $-\frac{1}{2}$ **2.** 5 **3.** $-\frac{5}{2}$

3.4 STUDY SET

ELEMENTARY Algebra f(x) Now™

VOCABULARY **Fill in the blanks.**

1. A _______ is the quotient of two numbers.

2. The _______ of a line is a measure of the line's steepness.

3. The _______ of a line is defined to be the ratio of the change in y to the change in x.

4. $m = \frac{\_\_\_\_\_\_\_\_}{\text{horizontal change}} = \frac{\text{change in } y}{\_\_\_\_\_\_\_\_} = \frac{\text{rise}}{\_\_\_\_}$

5. The rate of _______ of a linear relationship can be found by finding the slope of the graph of the line and attaching the proper units.

6. __________ lines have a slope of 0. Vertical lines have __________ slope.

CONCEPTS

7. Fill in the blanks.
   a. A line with positive slope ______ from left to right.
   b. A line with negative slope ______ from left to right.
8. Which line graphed has
   a. a positive slope?
   b. a negative slope?
   c. zero slope?
   d. undefined slope?

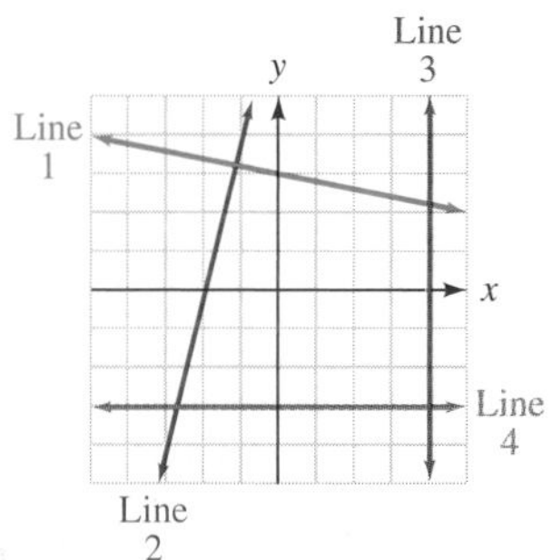

9. Suppose the rise of a line is 2 and the run is 15. Write the ratio of rise to run.
10. Consider the following graph of the line and the slope triangle.
    a. What is the rise?
    b. What is the run?
    c. What is the slope of the line?

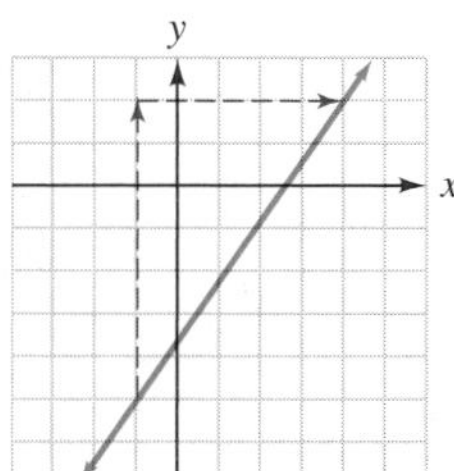

11. Consider the graph of the line and the slope triangle shown in the next column.
    a. What is the rise?
    b. What is the run?
    c. What is the slope of the line?

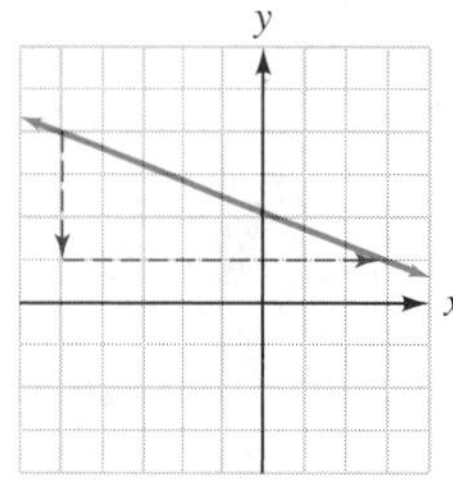

12. Which two labeled points should be used to find the slope of the line?

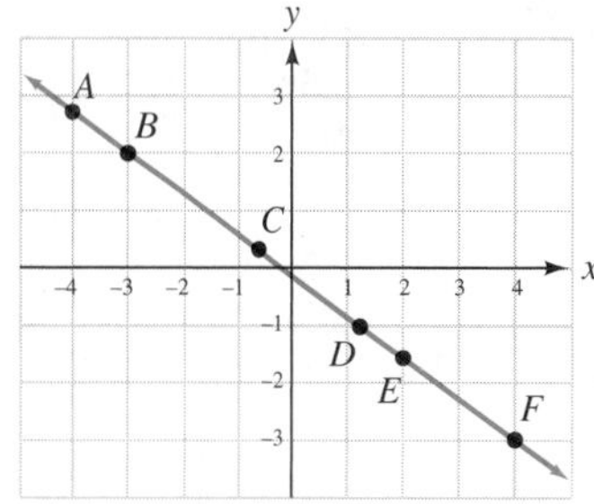

13. In this problem, you are to find the slope of the line graphed below by drawing a slope triangle.
    a. Find the slope using points A and B.
    b. Find the slope using points B and C.
    c. Find the slope using points A and C.
    d. What observation is suggested by your answers to parts a, b, and c?

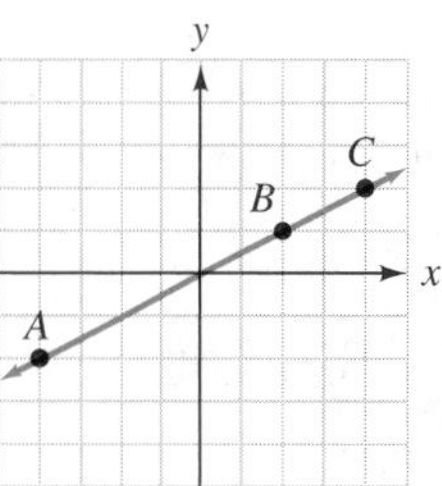

14. Evaluate each expression.
    a. $\dfrac{10 - 4}{6 - 5}$ b. $\dfrac{-1 - 1}{-2 - (-7)}$
15. Express each of the following slope calculations in a better way.
    a. $m = \dfrac{0}{6}$ b. $m = \dfrac{8}{0}$

16. Simplify each slope.

a. $m = \frac{3}{12}$ b. $m = -\frac{9}{6}$

c. $m = \frac{-4}{4}$ d. $m = \frac{-10}{-5}$

17. The *grade* of an incline is its slope expressed as a percent. What grade is represented by each of the following slopes?

a. $m = \frac{2}{5}$ b. $m = \frac{3}{20}$

18. GROWTH RATES Refer to the graph. The slope of the line is 3. Fill in the correct units: The rate of change of the boy's height is 3 ___ per ___.

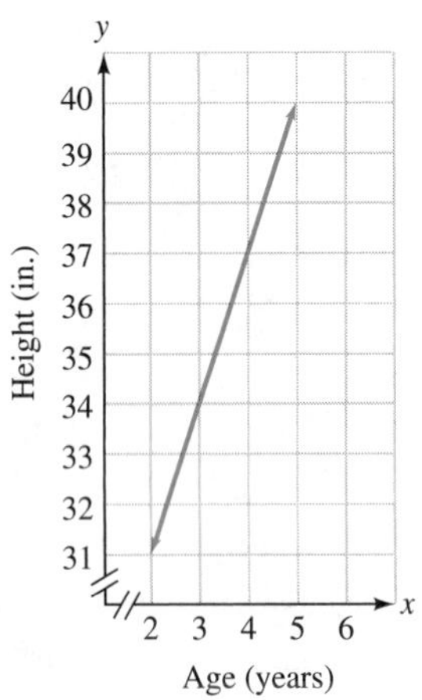

NOTATION

19. What is the formula used to find the slope of the line passing through (x_1, y_1) and (x_2, y_2)?

20. Fill in the blanks to state the slope formula in words: m equals y _____ two minus y _____ one _______________ x sub _____ minus x sub _____.

21. Explain the difference between y^2 and y_2.

22. Consider the points (7, 2) and (−4, 1). If we let $y_2 = 1$, then what is x_2?

PRACTICE Find the slope of each line, if possible.

23.

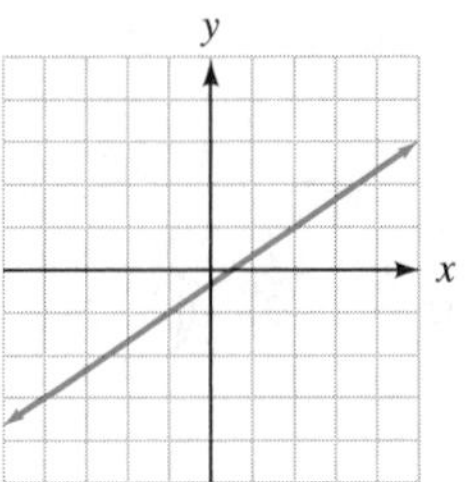

24.

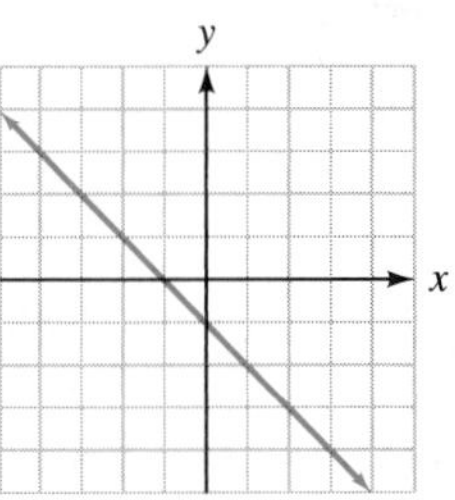

25.

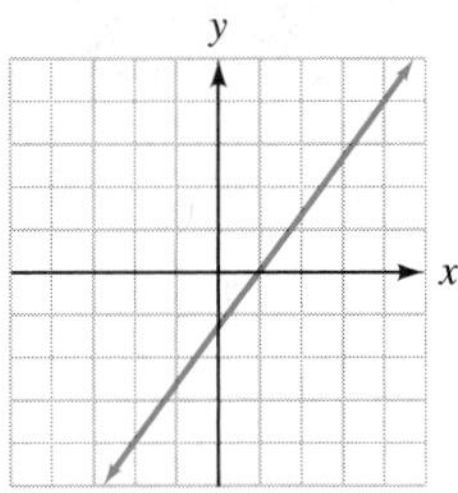

26.

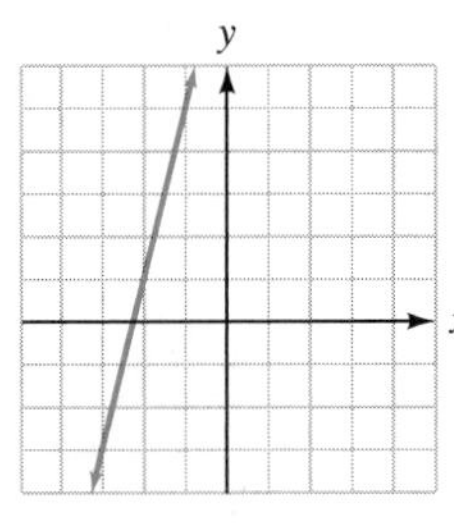

27.

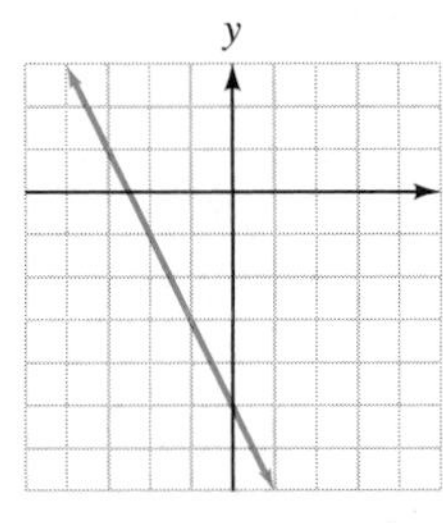

28.

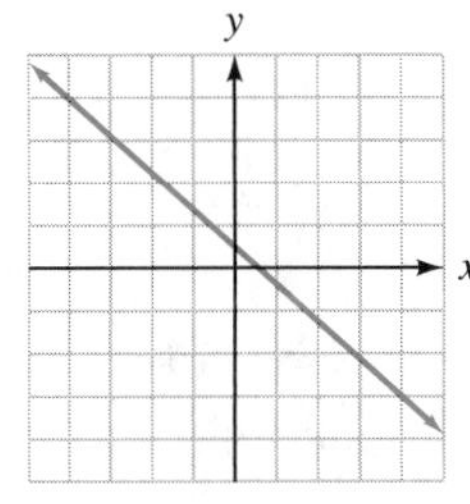

29.

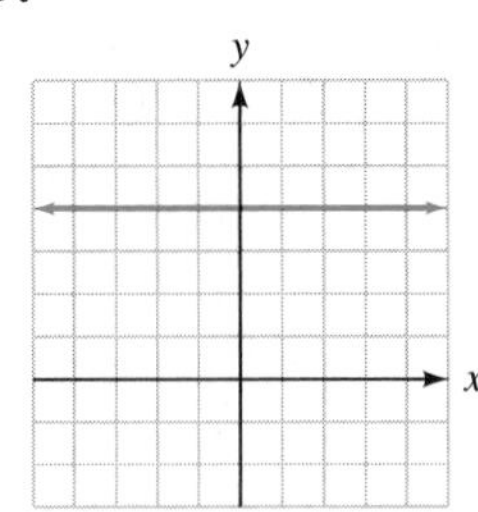

30.

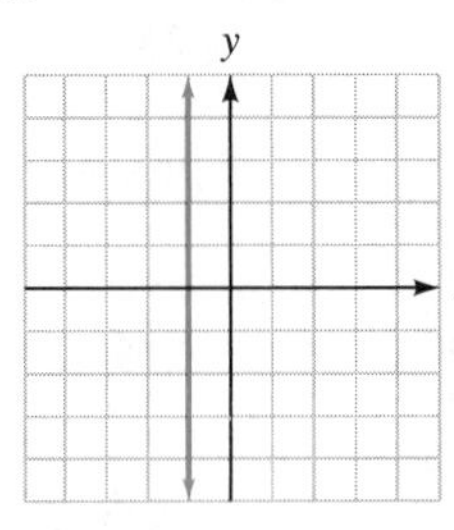

31.

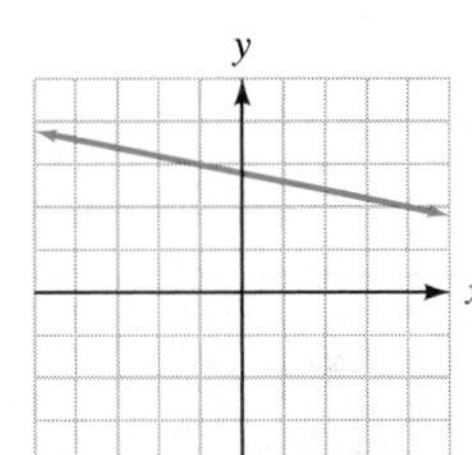

32.

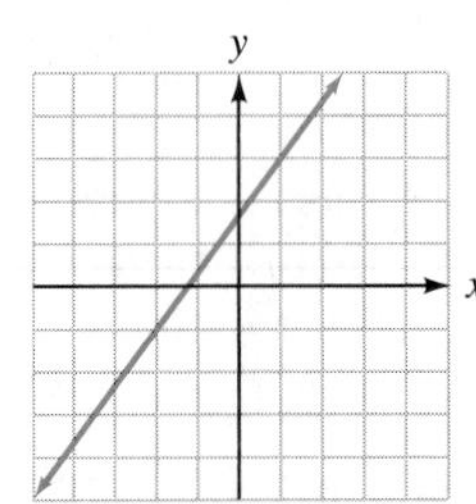

Find the slope of the line passing through the given points, when possible.

33. (2, 4) and (1, 3)
34. (1, 3) and (2, 5)
35. (3, 4) and (2, 7)
36. (3, 6) and (5, 2)
37. (0, 0) and (4, 5)
38. (4, 3) and (7, 8)
39. (−3, 5) and (−5, 6)
40. (6, −2) and (−3, 2)
41. (−2, −2) and (−12, −8)
42. (−1, −2) and (−10, −5)
43. (5, 7) and (−4, 7)
44. (−1, −12) and (6, −12)
45. (8, −4) and (8, −3)
46. (−2, 8) and (−2, 15)
47. (−6, 0) and (0, −4)
48. (0, −9) and (−6, 0)
49. (−2.5, 1.75) and (−0.5, −7.75)
50. (6.4, −7.2) and (−8.8, 4.2)

Determine the slope of the graph of the line that has the given table of solutions.

51.

x	y	(x, y)
−3	−1	(−3, −1)
1	2	(1, 2)
5	5	(5, 5)

52.

x	y	(x, y)
−3	6	(−3, 6)
0	2	(0, 2)
3	−2	(3, −2)

53.

x	y	(x, y)
−3	6	(−3, 6)
0	6	(0, 6)
3	6	(3, 6)

54.

x	y	(x, y)
4	−5	(4, −5)
4	0	(4, 0)
4	3	(4, 3)

Find the slope of each line, if possible.

55. $x = 6$
56. $y = -2$
57. $y = 0$
58. $x = 0$
59. $x = -2$
60. $y = 8$
61. $y = -3$
62. $x = 6$

APPLICATIONS

63. POOL DESIGN Find the slope of the bottom of the swimming pool as it drops off from the shallow end to the deep end.

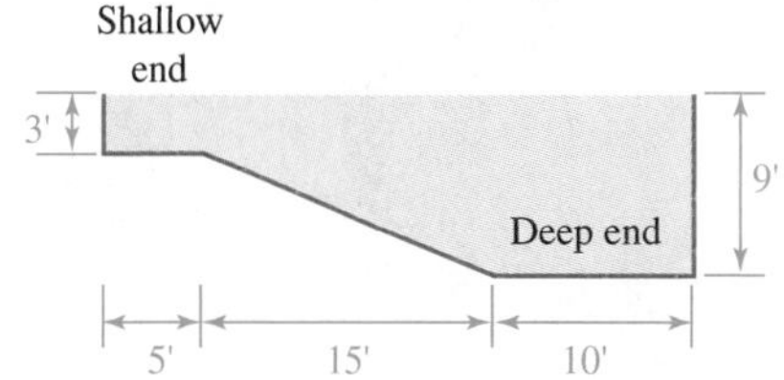

64. DRAINAGE To measure the amount of fall (slope) of a concrete patio slab, a 10-foot-long 2-by-4, a 1-foot ruler, and a level were used. Find the amount of fall in the slab.

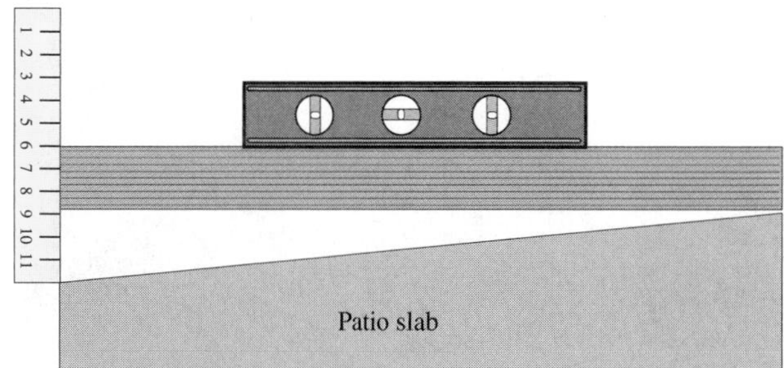

65. GRADE OF A ROAD The vertical fall of the road shown in the illustration is 264 feet for a horizontal run of 1 mile. Find the slope of the decline and use that information to complete the roadside warning sign for truckers.

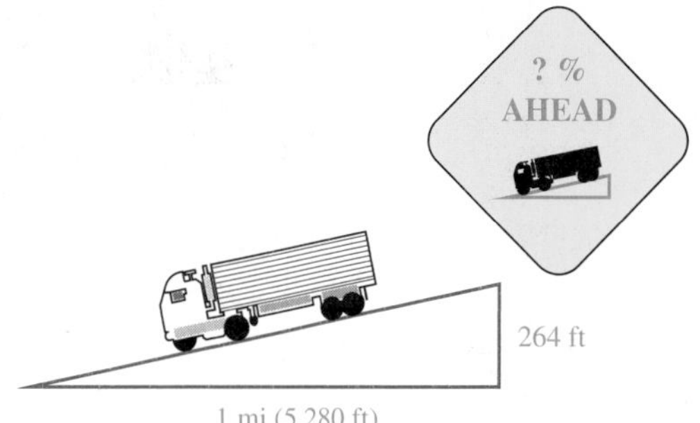

66. TREADMILLS For each height setting listed in the table, find the resulting slope of the jogging surface of the treadmill. Express each incline as a percent.

Height setting	% incline
2 inches	
4 inches	
6 inches	

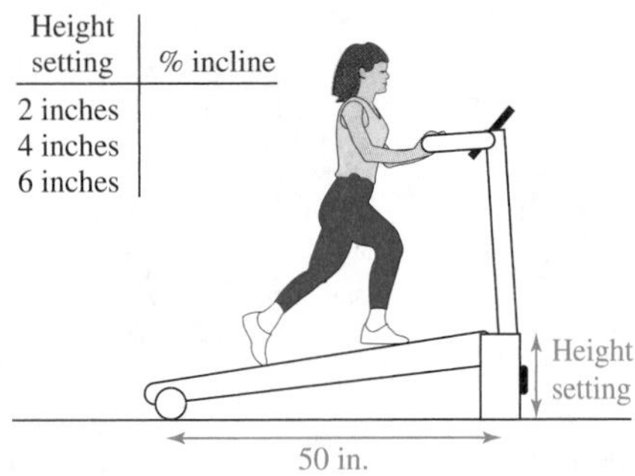

67. ENGINEERING The illustrations show two ramp designs.

a. Find the slope of the ramp in design 1.

b. Find the slopes of the ramps in design 2.

c. Give one advantage and one drawback of each design.

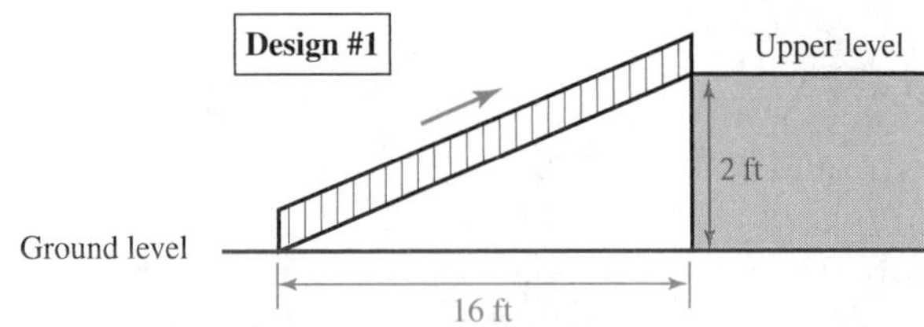

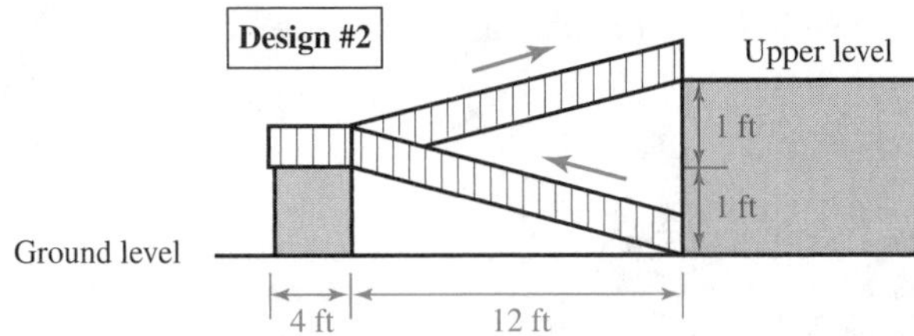

68. ARCHITECTURE Since the pitch of the roof of the house shown is to be $\frac{2}{5}$, there will be a 2-foot rise for every 5-foot run. Draw the roof line if it is to pass through the given black points. Find the coordinates of the peak of the roof.

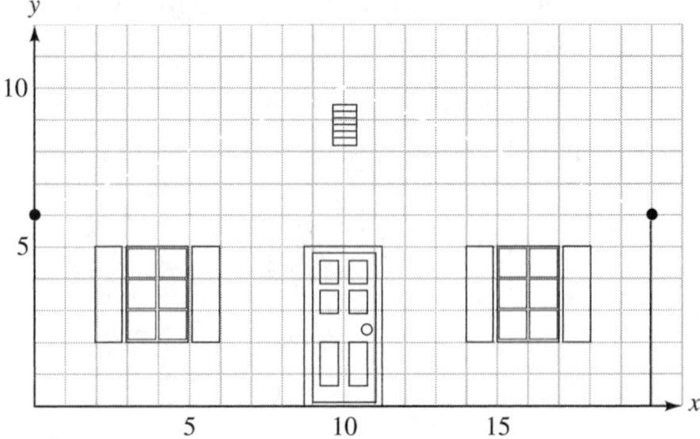

69. IRRIGATION The graph shows the number of gallons of water remaining in a reservoir as water is discharged from it to irrigate a field. Find the rate of change in the number of gallons of water in the reservoir.

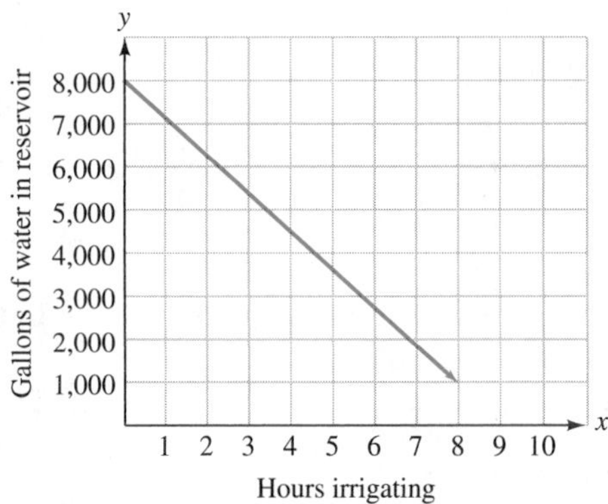

70. COMMERCIAL JETS Examine the graph and consider trips of more than 7,000 miles by a Boeing 777. Use a rate of change to estimate how the maximum payload decreases as the distance traveled increases.

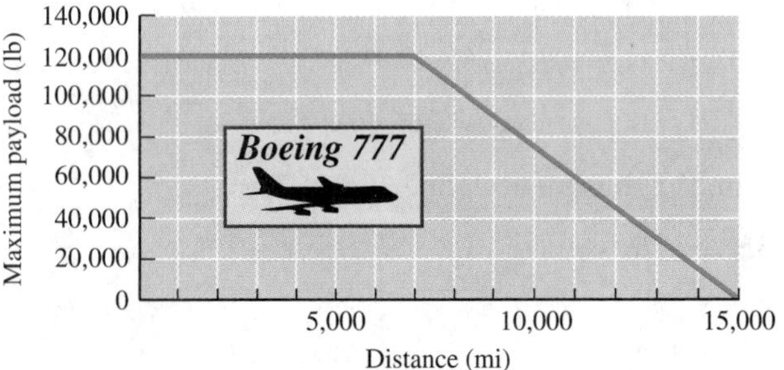

Based on data from Lawrence Livermore National Laboratory and *Los Angeles Times* (October 22, 1998).

71. MILK PRODUCTION The following graph approximates the amount of milk produced per cow in the U.S. for the years 1993–2002. Find the rate of change.

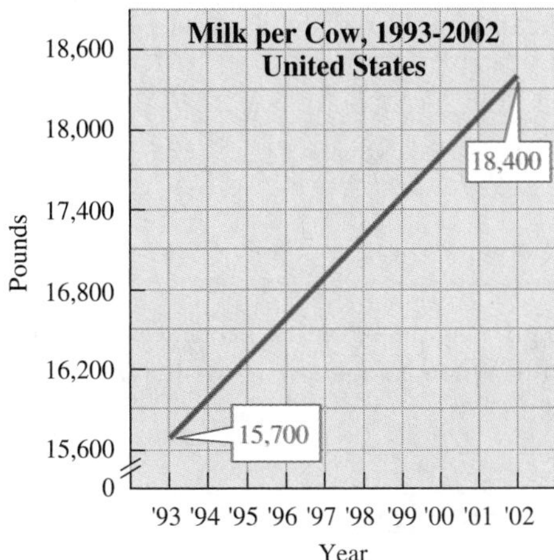

Source: United States Department of Agriculture

72. WAL-MART The graph below approximates the net sales of Wal-Mart for the years 1991–2002.
   a. Find the rate of change in sales for the years 1991–1998.
   b. Find the rate of change in sales for the years 1998–2002.

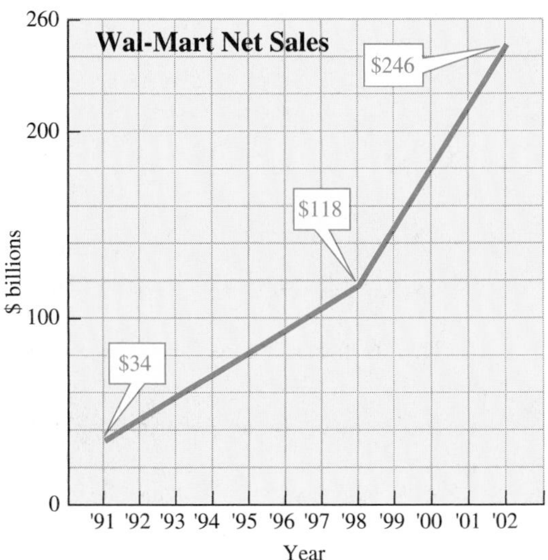

Based on data from Wal-Mart, *USA TODAY* (November 6, 1998), and Hoover's online.

WRITING

73. Explain why the slope of a vertical line is undefined.

74. How do we distinguish between a line with positive slope and a line with negative slope?

75. Give an example of a rate of change that government officials might be interested in knowing so they can plan for the future needs of our country.

76. Explain the difference between a rate of change that is positive and one that is negative. Give an example of each.

REVIEW

77. HALLOWEEN CANDY A candy maker wants to make a 60-pound mixture of two candies to sell for \$2 per pound. If black licorice bits sell for \$1.90 per pound and orange gumdrops sell for \$2.20 per pound, how many pounds of each should be used?

78. MEDICATIONS A doctor prescribes an ointment that is 2% hydrocortisone. A pharmacist has 1% and 5% concentrations in stock. How many ounces of each should the pharmacist use to make a 1-ounce tube?

CHALLENGE PROBLEMS

79. Use the concept of slope to determine whether $A(-50, -10)$, $B(20, 0)$, and $C(34, 2)$ all lie on the same straight line.

80. A line having slope $\frac{2}{3}$ passes through the point $(10, -12)$. What is the y-coordinate of another point on the line whose x-coordinate is 16?

3.5 Slope–Intercept Form

- Slope–Intercept Form of the Equation of a Line
- Using the Slope and y-Intercept to Graph a Line
- Writing the Equation of a Line
- Parallel and Perpendicular Lines
- Applications

Linear equations appear in many forms. Some examples are:

$$y = 2x + 1, \qquad 3x - 5y = 15, \qquad 8y = 6x + 7, \qquad \text{and} \qquad x = -4$$

Of all of the ways in which a linear equation can be written, one form, called *slope–intercept form,* is probably the most useful. When an equation is written in this form, two important features of its graph are evident.

SLOPE–INTERCEPT FORM OF THE EQUATION OF A LINE

To explore the relationship between a linear equation and its graph, let's consider $y = 2x + 1$ and its graph.

$y = 2x + 1$

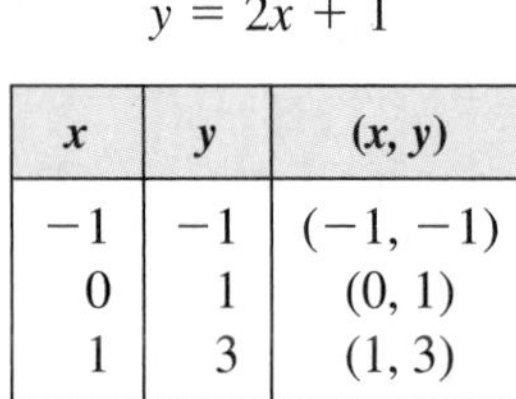

x	y	(x, y)
-1	-1	$(-1, -1)$
0	1	$(0, 1)$
1	3	$(1, 3)$

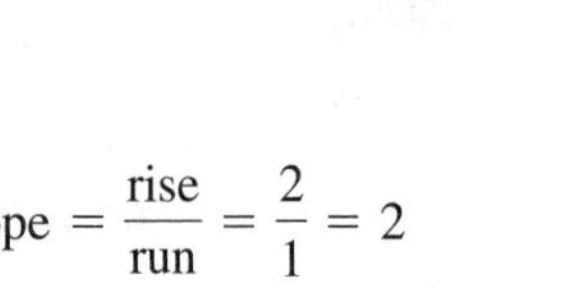

$$\text{Slope} = \frac{\text{rise}}{\text{run}} = \frac{2}{1} = 2$$

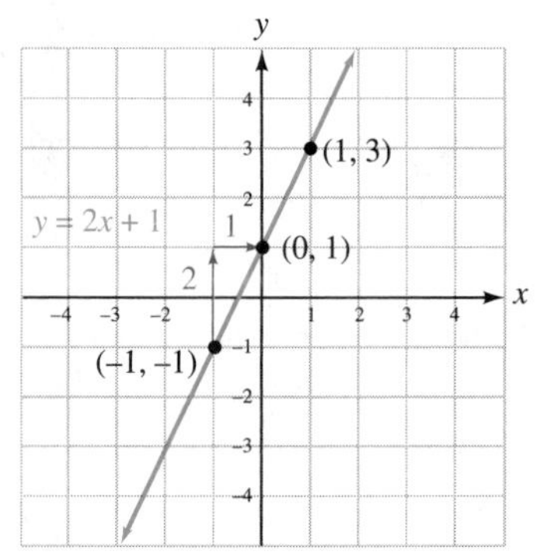

A close examination of the equation and the graph leads to two observations:

- The graph crosses the y-axis at 1. This is the same as the constant term in $y = 2x + 1$.
- The slope of the line is 2. This is the same as the coefficient of x in $y = 2x + 1$.

It appears that the slope and y-intercept of the graph of $y = 2x + 1$ can be determined from the equation.

$$y = 2x + 1$$

↑ The slope of the line is 2.

↑ The y-intercept is $(0, 1)$.

These observations suggest the following form of an equation of a line.

Slope–Intercept Form of the Equation of a Line

If a linear equation is written in the form

$$y = mx + b$$

the graph of the equation is a line with slope m and y-intercept $(0, b)$.

EXAMPLE 1

Find the slope and the y-intercept of the graph of each equation.

a. $y = 6x - 2$ **b.** $y = -\frac{5}{4}x$ **c.** $y = \frac{x}{2} + 3$ **d.** $y = 5 + 12x$

Solution

a. If we write the subtraction as the addition of the opposite, the equation will be in $y = mx + b$ form:

$$\begin{array}{l} y = 6x + (-2) \\ \quad\;\; \uparrow \qquad \uparrow \\ y = mx + \;\; b \end{array}$$

Since $m = 6$ and $b = -2$, the slope of the line is 6 and the y-intercept is $(0, -2)$.

Caution

For equations in $y = mx + b$ form, the slope of the line is the *coefficient* of x, not the x-term. For example, the graph of $y = 6x - 5$ has slope 6, *not* $6x$.

b. Writing $y = -\frac{5}{4}x$ in slope–intercept form, we have

$$y = -\frac{5}{4}x + \mathbf{0} \qquad \text{Add 0 to make the value of } b \text{ obvious.}$$

Since $m = -\frac{5}{4}$ and $b = 0$, the slope of the line is $-\frac{5}{4}$ and the y-intercept is $(0, 0)$.

c. Since $\frac{x}{2}$ means $\frac{1}{2}x$, we can rewrite $y = \frac{x}{2} + 3$ as

$$y = \frac{1}{2}x + 3$$

We see that $m = \frac{1}{2}$ and $b = 3$, so the slope of the line is $\frac{1}{2}$ and the y-intercept is $(0, 3)$.

d. We can use the commutative property of addition to reorder the terms on the right-hand side of the equation so that it is in $y = mx + b$ form.

$$y = 12x + 5$$

The slope is 12 and the y-intercept is $(0, 5)$.

Self Check 1 Find the slope and the y-intercept:

a. $y = -5x - 1$ **b.** $y = \frac{7}{8}x$ **c.** $y = 5 - \frac{x}{3}$

The equation of any nonvertical line can be written in slope–intercept form. To do so, we apply the properties of equality to solve for y.

EXAMPLE 2

Find the slope and y-intercept of the line whose equation is $8x + y = 9$.

ELEMENTARY Algebra $f(x)$ Now™

Solution The slope and y-intercept are not immediately apparent because the equation is not in slope–intercept form. To write it in $y = mx + b$ form, we isolate y on the left-hand side.

$$8x + y = 9$$

$$8x + y - \mathbf{8x} = \mathbf{-8x} + 9 \quad \text{To undo the addition of } 8x\text{, subtract } 8x \text{ from both sides.}$$

$$y = \mathbf{-8x} + \mathbf{9} \quad \text{On the left-hand side, combine like terms: } 8x - 8x = 0.$$

The slope is -8. The -intercept is $(0, 9)$.

Self Check 2 Find the slope and y-intercept of the line whose equation is $9x + y = -4$.

EXAMPLE 3

Find the slope and y-intercept of the line with the given equation.

ELEMENTARY Algebra $f(x)$ Now™

a. $x + 4y = 16$ **b.** $-9x - 3y = 11$

Solution **a.** To write the equation in slope–intercept form, we solve for y.

$$x + 4y = 16$$

$$x + 4y - \mathbf{x} = \mathbf{-x} + 16 \quad \text{To undo the addition of } x\text{, subtract } x \text{ from both sides.}$$

$$4y = -x + 16 \quad \text{Simplify the left-hand side.}$$

$$\frac{4y}{\mathbf{4}} = \frac{-x + 16}{\mathbf{4}} \quad \text{To undo the multiplication by 4, divide both sides by 4.}$$

$$y = \frac{-x}{4} + \frac{16}{4} \quad \text{On the right-hand side, rewrite } \tfrac{-x + 16}{4} \text{ as the sum of two fractions with like denominators, } \tfrac{-x}{4} \text{ and } \tfrac{16}{4}.$$

$$y = -\frac{1}{4}x + 4 \quad \text{Write } \tfrac{-x}{4} \text{ as } -\tfrac{1}{4}x\text{. Simplify: } \tfrac{16}{4} = 4.$$

Since $m = -\frac{1}{4}$ and $b = 4$, the slope is $-\frac{1}{4}$ and the y-intercept is $(0, 4)$.

b. To write the equation in $y = mx + b$ form, we isolate y on the left-hand side.

$$-9x - 3y = 11$$

$$-3y = 9x + 11 \quad \text{To eliminate the term } -9x \text{ on the left-hand side, add } 9x \text{ to both sides: } -9x + 9x = 0.$$

$$\frac{-3y}{\mathbf{-3}} = \frac{9x}{\mathbf{-3}} + \frac{11}{\mathbf{-3}} \quad \text{To undo the multiplication by } -3\text{, divide both sides by } -3.$$

$$y = -3x - \frac{11}{3} \quad \text{Simplify.}$$

Since $m = -3$ and $b = -\frac{11}{3}$, the slope is -3 and the y-intercept is $\left(0, -\frac{11}{3}\right)$.

Self Check 3 Find the slope and y-intercept of the line whose equation is $10x + 2y = 7$.

USING THE SLOPE AND y-INTERCEPT TO GRAPH A LINE

If we know the slope and y-intercept of a line, we can graph the line. To illustrate this, we graph $y = 5x - 4$, a line with slope 5 and y-intercept $(0, -4)$.

We begin by plotting the y-intercept. If we write the slope as $\frac{5}{1}$, we see that the rise is 5 and the run is 1. From $(0, -4)$, we move 5 units upward and then 1 unit to the right. This locates a second point on the line, $(1, 1)$. Then we draw a line through the two points. The result is a line with y-intercept $(0, -4)$ and slope 5.

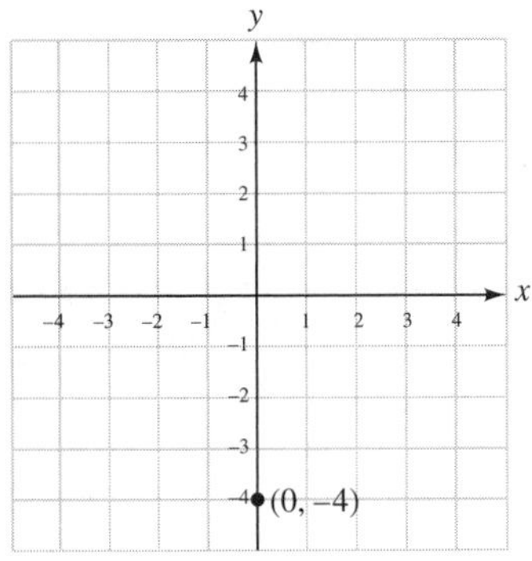

Plot the y-intercept, $(0, -4)$.

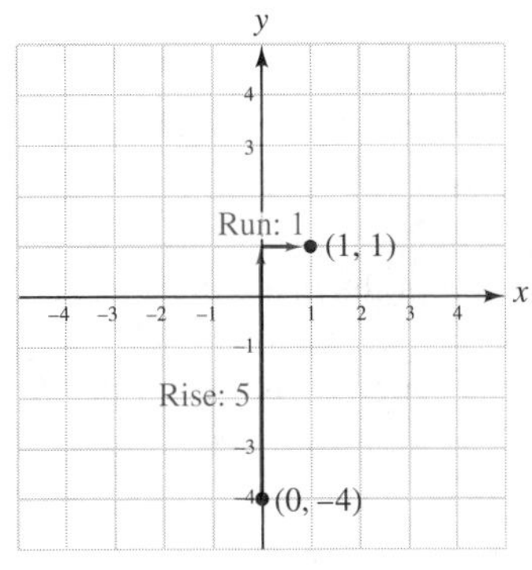

From the y-intercept, draw the rise and run components of the slope triangle.

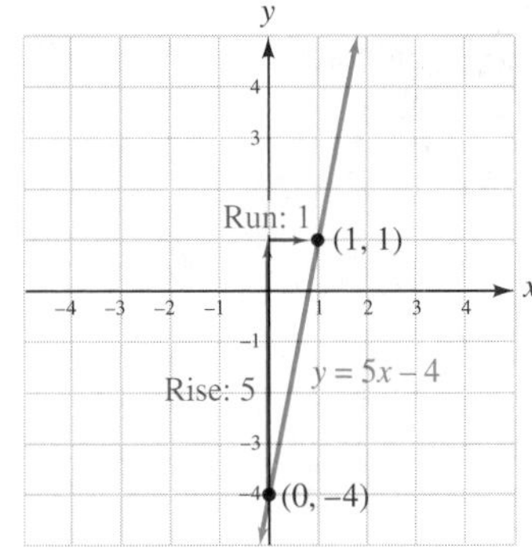

Use a straightedge to draw a line through the two points.

EXAMPLE 4

Graph the line whose equation is $y = -\frac{4}{3}x + 2$.

ELEMENTARY Algebra f(x) Now™

Solution The slope of the line is $-\frac{4}{3}$, which can be expressed as $\frac{-4}{3}$. After plotting the y-intercept, $(0, 2)$, we move 4 units downward and then 3 units to the right. This locates a second point on the line, $(3, -2)$. From this point, we can move another 4 units downward and 3 units to the right, to locate a *third point* on the line, $(6, -6)$. Then we draw a line through the three points to obtain a line with y-intercept $(0, 2)$ and slope $-\frac{4}{3}$.

Caution

When using the y-intercept and the slope to graph a line, remember to draw the slope triangle from the y-intercept, *not* from the origin.

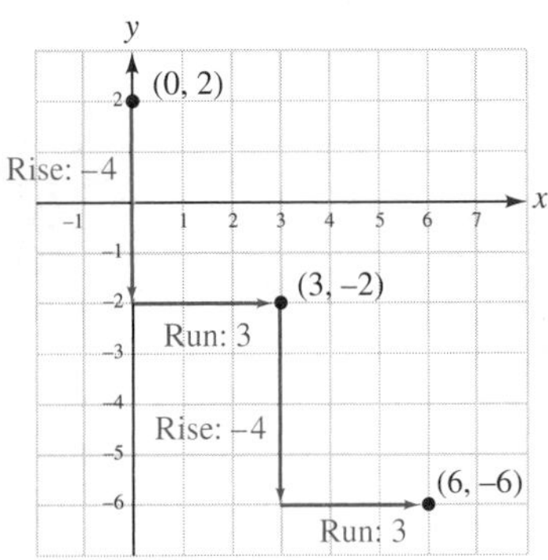

Plot the y-intercept. From $(0, 2)$, draw the rise and run components of a slope triangle. From $(3, -2)$, draw another slope triangle.

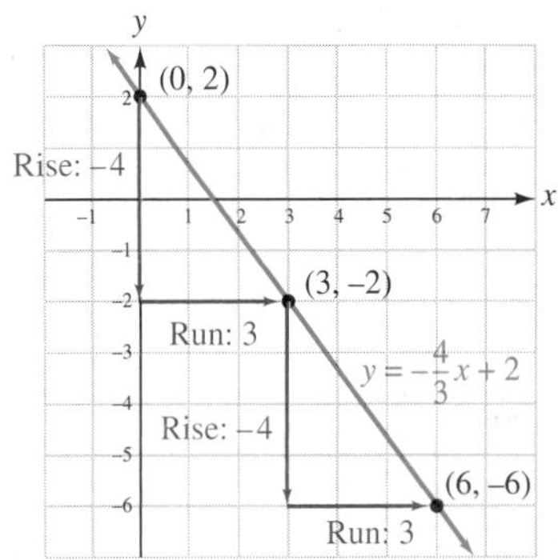

Use a straightedge to draw a line through the three points.

Self Check 4 Graph the line whose equation is $y = -\frac{5}{6}x + 2$.

WRITING THE EQUATION OF A LINE

If we are given the slope and y-intercept of a line, we can write an equation of the line by substituting for m and b in the slope–intercept form.

EXAMPLE 5

Write an equation of the line with slope -1 and y-intercept $(0, 9)$.

ELEMENTARY Algebra $f(x)$ Now™

Solution If the slope is -1 and the y-intercept is $(0, 9)$, then $m = -1$ and $b = 9$.

$y = mx + b$ This is the slope–intercept form.

$y = -1x + 9$ Substitute -1 for m and 9 for b.

$y = -x + 9$ Simplify: $-1x = -x$.

The equation of the line with slope -1 and y-intercept $(0, 9)$ is $y = -x + 9$.

Self Check 5 Write the equation of the line with slope 1 and y-intercept $(0, -12)$.

PARALLEL AND PERPENDICULAR LINES

Two lines that lie in the same plane are **parallel** if they do not intersect. When graphed, parallel lines have the same slope, but different y-intercepts.

Parallel Lines Two different lines with the same slope are parallel.

EXAMPLE 6

Graph $y = -\frac{2}{3}x$ and $y = -\frac{2}{3}x + 3$ on the same coordinate system.

Solution The graph of the first equation is a line with slope $-\frac{2}{3}$ and y-intercept $(0, 0)$. The graph of the second equation is a line with slope $-\frac{2}{3}$ and y-intercept $(0, 3)$. Since the lines have the same slope, they are parallel.

The Language of Algebra

The word *parallel* is used in many settings: drivers *parallel* park, and gymnasts perform on the *parallel* bars.

Self Check 6 Graph $y = \frac{5}{2}x - 2$ and $y = \frac{5}{2}x$ on the same coordinate system.

Unlike parallel lines, **perpendicular lines** intersect. And more important, they intersect to form four right angles (angles with measure 90°). The two lines graphed below are perpendicular. In the figure, the symbol ⌉ is used to denote a right angle.

The Language of Algebra

The word *perpendicular* is used in construction. For example, the monument shown is *perpendicular* to the ground.

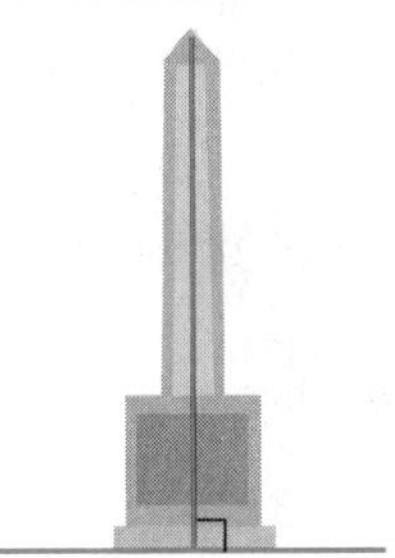

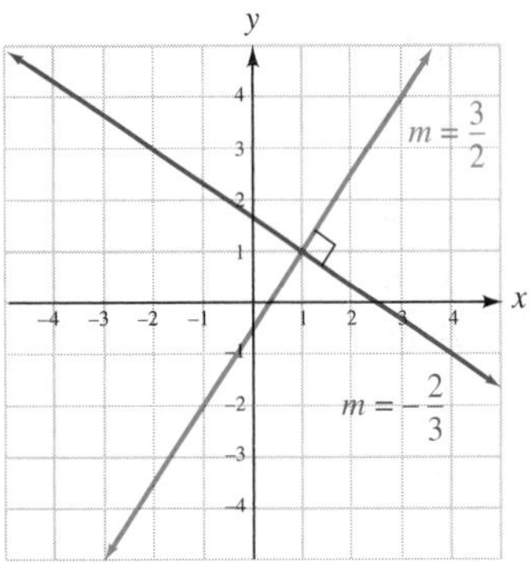

The product of the slopes of two (nonvertical) perpendicular lines is -1. To illustrate this, consider the lines in the illustration, with slopes $\frac{3}{2}$ and $-\frac{2}{3}$. If we find the product of their slopes, we have

$$\frac{3}{2}\left(-\frac{2}{3}\right) = -\frac{6}{6} = -1$$

Note that the slopes of the lines in this example, $\frac{3}{2}$ and $-\frac{2}{3}$, are **negative** (or **opposite**) **reciprocals.** We can use the term *negative reciprocal* to express the relationship between the slopes of perpendicular lines in an alternate way.

Slopes of Perpendicular Lines

1. Two nonvertical lines are perpendicular if the product of the slopes is -1, that is, if their slopes are negative reciprocals.
   In symbols, two lines with slopes m_1 and m_2 are perpendicular if

$$m_1 \cdot m_2 = -1 \quad \text{or} \quad m_1 = -\frac{1}{m_2}$$

2. A horizontal line with 0 slope is perpendicular to a vertical line with undefined slope.

EXAMPLE 7

ELEMENTARY Algebra f(x) Now™

Determine whether the graphs of $y = -5x + 6$ and $y = \frac{x}{5} - 2$ are parallel, perpendicular, or neither.

Solution The slope of the line $y = -5x + 6$ is -5. The slope of the line $y = \frac{x}{5} - 2$ is $\frac{1}{5}$. (Recall that $\frac{x}{5} = \frac{1}{5}x$.) Since the slopes are not equal, the lines are not parallel. If we find the product of their slopes, we have

$$-5\left(\frac{1}{5}\right) = -\frac{5}{5} = -1 \qquad -5 \text{ and } \frac{1}{5} \text{ are negative reciprocals.}$$

Since the product of their slopes is -1, the lines are perpendicular.

Self Check 7 Determine whether the graphs of $y = 4x + 6$ and $y = \frac{1}{4}x$ are parallel, perpendicular, or neither.

APPLICATIONS

In the next example, as a means of making the equation more descriptive, we replace x and y in $y = mx + b$ with two other variables.

EXAMPLE 8

ELEMENTARY Algebra $f(x)$ Now™

Group discounts. To promote group sales for an Alaskan cruise, a travel agency reduces the regular cost of \$4,500 by \$5 for each person traveling in the group.

a. Write a linear equation in slope–intercept form that finds the cost c of the cruise, if a group of p people travel together.

b. Use the equation to determine the cost if a group of 55 retired teachers travel together.

Cruise to Alaska
\$4,500 per person
Group discounts available*
*For groups of up to 100

Solution **a.** Since the cost c of the cruise depends on the number p of people traveling in the group, the equation will have the form $c = mp + b$. We need to determine m and b.

The cost of the cruise steadily *decreases* as the number of people in the group increases. This rate of change, -5 dollars per person, is the slope of the graph of the equation. Thus m is -5.

If a group of 0 people take the cruise, there will be no discount; the cruise will cost \$4,500. Written as an ordered pair of the form (p, c), we have $(0, 4{,}500)$. When graphed, this would be the c-intercept. Thus, b is 4,500.

Substituting for m and b, we obtain the linear equation that models this situation.

$$c = -5p + 4{,}500$$

b. To find the cost of the cruise for a group of 55, we proceed as follows:

$$c = -5p + 4{,}500$$
$$c = -5(55) + 4{,}500 \quad \text{Substitute 55 for } p\text{, the number of people in the group.}$$
$$c = -275 + 4{,}500$$
$$= 4{,}225$$

If a group of 55 people travel together, the Alaskan cruise will cost each person \$4,225.

Self Check 8 Write a linear equation in slope–intercept form that finds the cost c of the cruise if a \$10-per-person discount is offered for groups.

Answers to Self Checks

1. a. $m = -5, (0, -1)$ **b.** $m = \frac{7}{8}, (0, 0)$ **c.** $m = -\frac{1}{3}, (0, 5)$

2. $m = -9; (0, -4)$

3. $m = -5; \left(0, \frac{7}{2}\right)$

4.

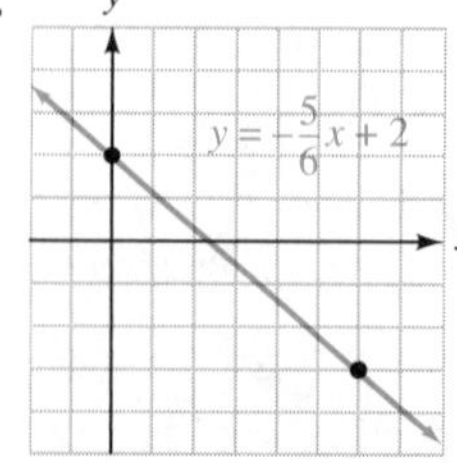

5. $y = x - 12$

6.

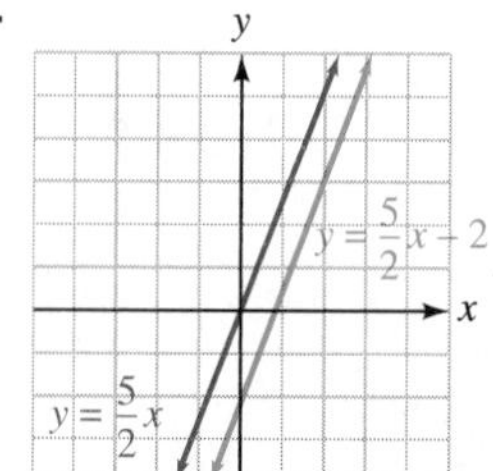

7. neither

8. $c = -10p + 4{,}500$

3.5 STUDY SET  ELEMENTARY Algebra f(x) Now™

VOCABULARY Fill in the blanks.

1. The equation $y = mx + b$ is called the ____________ form of the equation of a line.
2. The graph of the linear equation $y = mx + b$ has a __________ of $(0, b)$ and a ______ of m.
3. ________ lines do not intersect.
4. The slope of a line is a _____ of change.
5. The numbers $\frac{5}{6}$ and $-\frac{6}{5}$ are called negative __________. Their product is -1.
6. The product of the slopes of ____________ lines is -1.

CONCEPTS

7. Tell whether each equation is in slope–intercept form.
   a. $7x + 4y = 2$
   b. $5y = 2x - 3$
   c. $y = 6x + 1$
   d. $x = 4y - 8$
   e. $y = \frac{x}{5} - 3$
   f. $y = 2x$
8. a. How do we solve $4x + y = 9$ for y?

   b. How do we solve $-2x + y = 9$ for y?
9. a. To solve $5y = 10x + 20$ for y, both sides of the equation were divided by 5. Complete the solution.

$$\frac{5y}{5} = \frac{10x}{5} + \frac{20}{5}$$
$$\square = \square + \square$$

   b. To solve $-2y = 6x - 12$ for y, both sides of the equation were divided by -2. Complete the solution.

$$\frac{-2y}{-2} = \frac{6x}{-2} - \frac{12}{-2}$$
$$\square = \square \; \square \; 6$$

10. Examine the work shown in the following graph. An equation in slope–intercept form is in the process of being graphed.
    a. What is the y-intercept of the line?
    b. What is the slope of the line?
    c. What equation is being graphed?
    d. One more step needs to be completed. What is it?

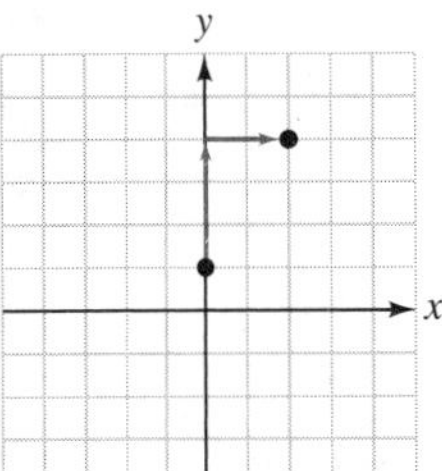

11. Use the graph of the line to determine m and b. Then write the equation of the line in slope–intercept form.

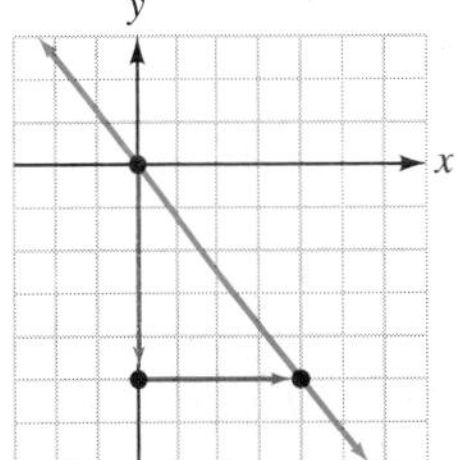

12. Find the negative reciprocal of each number.

Number	Negative reciprocal
6	
$\frac{7}{8}$	
$-\frac{1}{4}$	
1	

13. Fill in the blanks.
    a. Two different lines with the same slope are ________.
    b. If the slopes of two lines are negative reciprocals, the lines are ____________.
    c. The product of the slopes of perpendicular lines is $\square$.

14. a. What is the y-intercept of Line 1?

b. What do Line 1 and Line 2 have in common? How are they different?

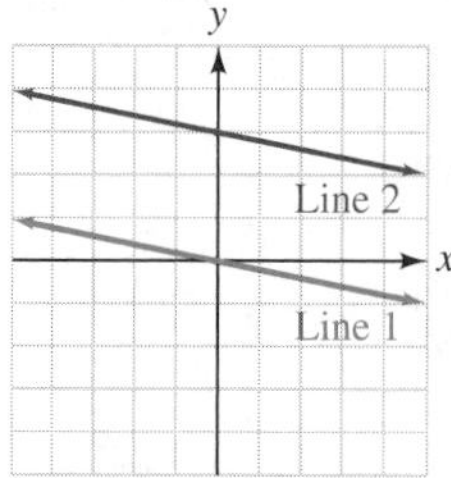

15. The slope of Line 1 shown in the following illustration is 2.

a. What is the slope of Line 2?

b. What is the slope of Line 3?

c. What is the slope of Line 4?

d. Which lines have the same y-intercept?

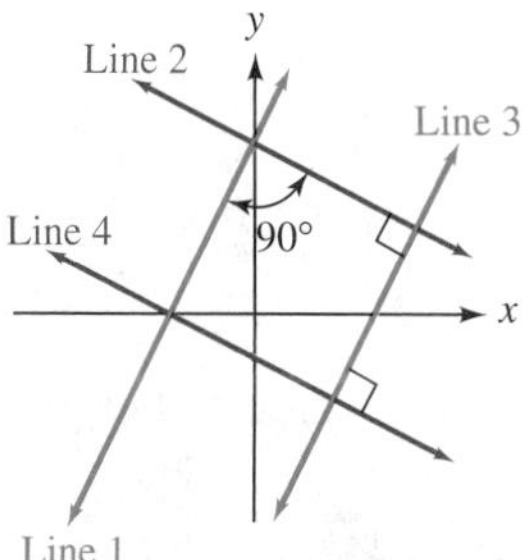

16. NAVIGATION The graph shows the recommended speed at which a ship should proceed into head waves of various heights.

a. What information does the y-intercept of the graph give?

b. What is the rate of change in the recommended speed of the ship as the wave height increases?

c. Write the equation of the graph.

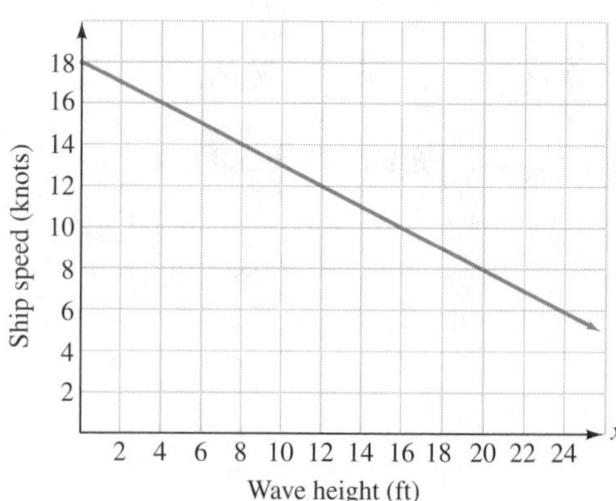

NOTATION **Complete the solution by solving the equation for y. Then find the slope and the y-intercept of its graph.**

17.

$$2x + 5y = 15$$
$$2x + 5y - \square = \square + 15$$
$$\square = -2x + 15$$
$$\frac{5y}{\square} = \frac{-2x}{\square} + \frac{15}{\square}$$
$$y = -\frac{2}{5}x + \square$$

The slope is $\square$ and the y-intercept is $\square$.

18. What is the slope–intercept form of the equation of a line?

19. Simplify each expression.

a. $\frac{8x}{2}$ **b.** $\frac{8x}{6}$

c. $\frac{-8x}{-8}$ **d.** $\frac{-16}{8}$

20. Tell whether each statement is true or false.

a. $\frac{x}{6} = \frac{1}{6}x$ **b.** $\frac{5}{3}x = \frac{5x}{3}$

21. What does the symbol ⌉ denote?

22. Write the phrase *ninety degrees* in symbols.

PRACTICE **Find the slope and the y-intercept of the graph of each equation.**

23. $y = 4x + 2$ **24.** $y = -4x - 2$

25. $y = -5x - 8$ **26.** $y = 7x + 3$

27. $4x - 2 = y$ **28.** $6x - 1 = y$

29. $y = \frac{x}{4} - \frac{1}{2}$

30. $y = \frac{x}{15} - \frac{3}{4}$

31. $y = \frac{1}{2}x + 6$

32. $y = \frac{4}{5}x - 9$

33. $y = 6 - x$

34. $y = 12 + 4x$

35. $x + y = 8$

36. $x - y = -30$

37. $6y = x - 6$

38. $2y = x + 20$

39. $7y = -14x + 49$

40. $9y = -27x + 36$

41. $-4y = 6x - 4$

42. $-6y = 8x + 6$

43. $2x + 3y = 6$

44. $4x + 5y = 25$

45. $3x - 5y = 15$

46. $x - 6y = 6$

47. $-6x + 6y = -11$

48. $-4x + 4y = -9$

49. $y = x$

50. $y = -x$

51. $y = -5x$

52. $y = 14x$

53. $y = -2$

54. $y = 30$

55. $-5y - 2 = 0$

56. $3y - 13 = 0$

Write an equation of the line with the given slope and *y*-intercept. Then graph it.

57. Slope 5, y-intercept $(0, -3)$

58. Slope -2, y-intercept $(0, 1)$

59. Slope $\frac{1}{4}$, y-intercept $(0, -2)$

60. Slope $\frac{1}{3}$, y-intercept $(0, -5)$

61. Slope -3, y-intercept $(0, 6)$

62. Slope 4, y-intercept $(0, -1)$

63. Slope $-\frac{8}{3}$, y-intercept $(0, 5)$

64. Slope $-\frac{7}{6}$, y-intercept $(0, 2)$

Find the slope and the *y*-intercept of the graph of each equation. Then graph the equation.

65. $y = 3x + 3$

66. $y = -3x + 5$

67. $y = -\frac{x}{2} + 2$

68. $y = \frac{x}{3}$

69. $y = -3x$

70. $y = -4x$

71. $4x + y = -4$

72. $2x + y = -6$

73. $3x + 4y = 16$

74. $2x + 3y = 9$

75. $10x - 5y = 5$

76. $4x - 2y = 6$

For each pair of equations, determine whether their graphs are parallel, perpendicular, or neither.

77. $y = 6x + 8$
$y = 6x$

78. $y = 3x - 15$
$y = -\frac{1}{3}x + 4$

79. $y = x$
$y = -x$

80. $y = \frac{1}{2}x - \frac{4}{5}$
$y = 0.5x + 3$

81. $y = -2x - 9$
$y = 2x - 9$

82. $y = \frac{3}{4}x + 1$
$y = \frac{4}{3}x - 5$

83. $x - y = 12$
$-2x + 2y = -23$

84. $y = -3x + 1$
$3y = x - 5$

85. $x = 9$
$y = 8$

86. $-x + 4y = 10$
$2y + 16 = -8x$

APPLICATIONS

87. PRODUCTION COSTS A television production company charges a basic fee of \$5,000 and then \$2,000 an hour when filming a commercial.

a. Write a linear equation that describes the relationship between the total production costs c and the hours h of filming.

b. Use your answer to part a to find the production costs if a commercial required 8 hours of filming.

88. COLLEGE FEES Each semester, students enrolling at a community college must pay tuition costs of \$20 per unit as well as a \$40 student services fee.

a. Write a linear equation that gives the total fees t to be paid by a student enrolling at the college and taking x units.

b. Use your answer to part a to find the enrollment cost for a student taking 12 units.

89. CHEMISTRY A portion of a student's chemistry lab manual is shown below. Use the information to write a linear equation relating the temperature F (in degrees Fahrenheit) of the compound to the time t (in minutes) elapsed during the lab procedure.

Chem. Lab #1 Aug. 13
Step 1: Removed compound from freezer @ −10°F.
Step 2: Used heating unit to raise temperature of compound 5° F every minute.

90. INCOME PROPERTY Use the information in the newspaper advertisement to write a linear equation that gives the amount of income A (in dollars) the apartment owner will receive when the unit is rented for m months.

APARTMENT FOR RENT
1 bedroom/1 bath, with garage
\$500 per month + \$250 nonrefundable security fee.

91. EMPLOYMENT SERVICE A policy statement of LIZCO, Inc., is shown below. Suppose a secretary had to pay an employment service \$500 to get placed in a new job at LIZCO. Write a linear equation that tells the secretary the actual cost c of the employment service to her m months after being hired.

Policy no. 23452–A new hire will be reimbursed by LIZCO for any employment service fees paid by the employee at the rate of \$20 per month.

92. VIDEOTAPES A VHS videocassette contains 800 feet of tape. In the long play mode (LP), it plays 10 feet of tape every 3 minutes. Write a linear equation that relates the number of feet f of tape yet to be played and the number of minutes m the tape has been playing.

93. SEWING COSTS A tailor charges a basic fee of \$20 plus \$5 per letter to sew an athlete's name on the back of a jacket.

a. Write a linear equation that will find the cost c to have a name containing x letters sewn on the back of a jacket.

b. Graph the equation.

c. Suppose the tailor raises the basic fee to \$30. On your graph from part b, draw the new graph showing the increased cost.

94. SALAD BARS For lunch, a delicatessen offers a "Salad and Soda" special where customers serve themselves at a well-stocked salad bar. The cost is \$1.00 for the drink and 20¢ an ounce for the salad.

a. Write a linear equation that will find the cost c of a "Salad and Soda" lunch when a salad weighing x ounces is purchased.

b. Graph the equation.

c. How would the graph from part b change if the delicatessen began charging \$2.00 for the drink?

d. How would the graph from part b change if the cost of the salad changed to 30¢ an ounce?

95. PROFESSIONAL HOCKEY Use the following facts to write a linear equation in slope–intercept form that approximates the average price of a National Hockey League ticket for the years 1995–2002.

- Let t represent the number of years since 1995 and c the average cost of a ticket.
- In 1995, the average ticket price was \$33.50.
- From 1995 to 2002, the average ticket price increased \$2 per year.

(Source: Team Marketing Report, NHL)

96. COMPUTER DRAFTING The illustration shows a computer-generated drawing of an automobile engine mount. When the designer clicks the mouse on a line of the drawing, the computer finds the equation of the line. Determine whether the two lines selected in the drawing are perpendicular.

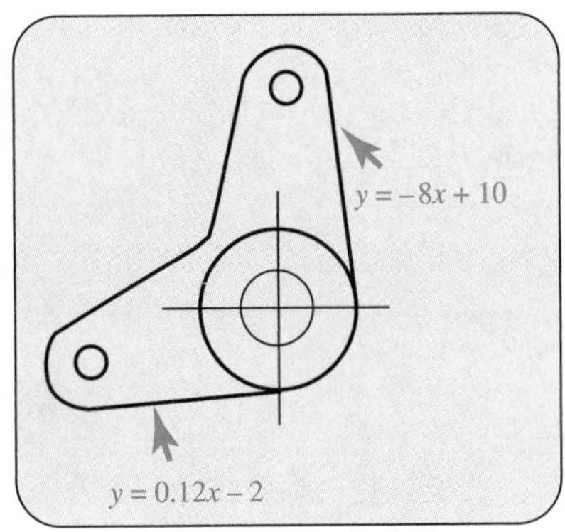

WRITING

97. Why is $y = mx + b$ called the slope–intercept form of the equation of a line?

98. On a quiz, a student was asked to find the slope of the graph of $y = 2x + 3$. She answered: $m = 2x$. Her instructor marked it wrong. Explain why the answer is incorrect.

REVIEW

99. CABLE TV A 186-foot television cable is to be cut into four pieces. Find the length of each piece if each successive piece is 3 feet longer than the previous one.

100. INVESTMENTS Joni received $25,000 as part of a settlement in a class action lawsuit. She invested some money at 10% and the rest at 9% simple interest rates. If her total annual income from these two investments was $2,430, how much did she invest at each rate?

CHALLENGE PROBLEMS

101. If the graph of $y = mx + b$ passes through quadrants I, II, and IV, what can be known about the constants m and b?

102. The equation $y = \frac{3}{4}x - 5$ is in slope–intercept form. Write it in general form, $Ax + By = C$, where $A > 0$.

3.6 Point–Slope Form

- Point–Slope Form of the Equation of a Line
- Writing the Equation of a Line
- Horizontal and Vertical Lines
- Using a Point and the Slope to Graph a Line
- Applications

If we know the slope of a line and its y-intercept, we can use the slope–intercept form to write the equation of the line. The question that now arises is, can *any* point on the line be used in combination with its slope to write its equation? In this section, we answer this question.

POINT–SLOPE FORM OF THE EQUATION OF A LINE

Refer to the line graphed on the left. Since the slope triangle has rise 3 and run 1, the slope of the line is $\frac{3}{1} = 3$. To develop a new form for the equation of a line, we will find the slope of this line in another way.

If we pick another point on the line and call it $Q(x, y)$, we can find the slope of the line by substituting the coordinates of points P and Q into the slope formula.

$$\frac{y_2 - y_1}{x_2 - x_1} = m$$

$$\frac{y - 1}{x - 2} = m \quad \text{Substitute } y \text{ for } y_2,\ 1 \text{ for } y_1,\ x \text{ for } x_2, \text{ and } 2 \text{ for } x_1.$$

Since the slope of the line is 3, we can substitute 3 for m in the previous equation.

$$\frac{y-1}{x-2} = m$$

$$\frac{y-1}{x-2} = 3$$

We then multiply both sides by $x - 2$ to get

$$\frac{y-1}{x-2}(x-2) = 3(x-2) \quad \text{Clear the equation of the fraction.}$$

$$y - 1 = 3(x-2) \quad \text{Simplify the left-hand side. Remove the common factor } x-2 \text{ in the numerator and denominator: } \frac{y-1}{\cancel{x-2}} \cdot \frac{\cancel{x-2}}{1}.$$

The resulting equation displays the slope of the line and the coordinates of one point on the line:

Slope of the line ↓

$$y - \mathbf{1} = \mathbf{3}(x - \mathbf{2})$$

↑ y-coordinate of the point ↑ x-coordinate of the point

In general, suppose we know that the slope of a line is m and that the line passes through the point (x_1, y_1). Then if (x, y) is any other point on the line, we can use the definition of slope to write

$$\frac{y - y_1}{x - x_1} = m$$

If we multiply both sides by $x - x_1$, we have

$$y - y_1 = m(x - x_1)$$

This form of a linear equation is called **point–slope form.** It can be used to write the equation of a line when the slope and one point on the line are known.

Point–Slope Form of the Equation of a Line

If a line with slope m passes through the point (x_1, y_1), the equation of the line is

$$y - y_1 = m(x - x_1)$$

WRITING THE EQUATION OF A LINE

If we are given the slope of a line and a point on the line, we can write an equation of the line by substituting for m, x_1, and y_1 in the point–slope form.

EXAMPLE 1

Write an equation of the line that has slope -8 and passes through $(-1, 5)$. Write the answer in slope–intercept form.

Solution We begin by writing an equation in point–slope form.

$$y - y_1 = m(x - x_1)$$
$$y - \mathbf{5} = \mathbf{-8}[x - (\mathbf{-1})] \quad \text{Substitute } -8 \text{ for } m, -1 \text{ for } x_1, \text{ and } 5 \text{ for } y_1.$$
$$y - 5 = -8(x + 1) \quad \text{Simplify within the brackets.}$$

To write this equation in slope–intercept form, solve for y.

$$y - 5 = -8(x + 1)$$
$$y - 5 = -8x - 8 \quad \text{Distribute the multiplication by } -8.$$
$$y = -8x - 3 \quad \text{To undo the subtraction of 5, add 5 to both sides.}$$

In slope–intercept form, the equation is $y = -8x - 3$.

To verify this result, we note that $m = -8$. Therefore, the slope of the line is -8, as required. To see whether the line passes through $(-1, 5)$, we substitute -1 for x and 5 for y in the equation. If this point is on the line, a true statement should result.

$$y = -8x - 3$$
$$\mathbf{5} \stackrel{?}{=} -8(\mathbf{-1}) - 3$$
$$5 \stackrel{?}{=} 8 - 3$$
$$5 = 5 \quad \text{True.}$$

Self Check 1 Write an equation of the line that has slope -2 and passes through $(4, -3)$. Write the answer in slope–intercept form.

In the next example, we show that it is possible to write the equation of a line when we know the coordinates of two points on the line.

EXAMPLE 2

ELEMENTARY **Algebra** $f(x)$ **Now**™

Write an equation of the line that passes through $(-2, 6)$ and $(4, 7)$. Write the answer in slope–intercept form.

Solution We begin by writing an equation in point–slope form. To find the slope of the line, we use the slope formula.

$$m = \frac{y_2 - y_1}{x_2 - x_1} = \frac{\mathbf{7 - 6}}{\mathbf{4 - (-2)}} = \frac{1}{6}$$

> **Success Tip**
> In Example 2, either of the given points can be used as (x_1, y_1) when writing the point–slope equation.
> Looking ahead, we usually choose the point whose coordinates will make the computations the easiest.

Either point on the line can serve as (x_1, y_1). If we choose $(4, 7)$, we have

$$y - y_1 = m(x - x_1) \quad \text{This is the point–slope form.}$$
$$y - \mathbf{7} = \frac{\mathbf{1}}{\mathbf{6}}(x - \mathbf{4}) \quad \text{Substitute } \frac{1}{6} \text{ for } m, 7 \text{ for } y_1, \text{ and } 4 \text{ for } x_1.$$

To write this equation in slope–intercept form, solve for y.

$$y - 7 = \frac{1}{6}x - \frac{2}{3}$$ Distribute the multiplication by $\frac{1}{6}$.

$$y = \frac{1}{6}x - \frac{4}{6} + \frac{42}{6}$$ To undo the subtraction of 7, add 7 to both sides. On the right-hand side, add 7 in the form $\frac{42}{6}$ so that it can be added to $-\frac{4}{6}$.

$$y = \frac{1}{6}x + \frac{19}{3}$$ This is slope–intercept form.

An equation of the line that passes through $(-2, 6)$ and $(4, 7)$ is $y = \frac{1}{6}x + \frac{19}{3}$.

Self Check 2 Write an equation of the line that passes through $(-5, 4)$ and $(8, -6)$. Write the answer in slope–intercept form.

HORIZONTAL AND VERTICAL LINES

We have previously graphed horizontal and vertical lines. We will now discuss how to write their equations.

EXAMPLE 3

ELEMENTARY Algebra $f(x)$ Now™

Write an equation of each line and graph it: **a.** a horizontal line passing through $(-2, -4)$ and **b.** a vertical line passing through $(1, 3)$.

Solution **a.** The equation of a horizontal line can be written in the form $y = b$. Since the y-coordinate of $(-2, -4)$ is -4, the equation of the line is $y = -4$. The graph is shown in the figure.

b. The equation of a vertical line can be written in the form $x = a$. Since the x-coordinate of $(1, 3)$ is 1, the equation of the line is $x = 1$. The graph is shown in the figure.

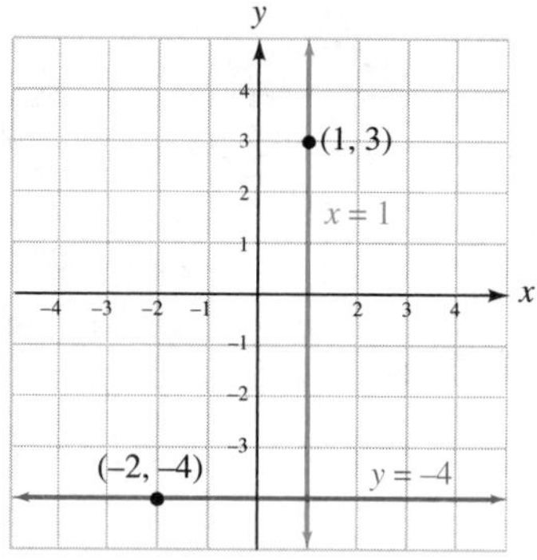

Self Check 3 Write an equation of each line: **a.** a horizontal line passing through $(3, 2)$ **b.** a vertical line passing through $(-1, -3)$

USING A POINT AND THE SLOPE TO GRAPH A LINE

If we know the coordinates of a point on a line, and if we know the slope of the line, we can use the slope to determine a second point on the line.

EXAMPLE 4

ELEMENTARY Algebra $f(x)$ Now™

Graph the line with slope $\frac{2}{5}$ that passes through $(-1, -3)$.

Solution We begin by plotting the point $(-1, -3)$. From there, we move 2 units up and then 5 units to the right. This puts us at a second point on the line, $(4, -1)$. We use a straightedge to draw a line through the two points.

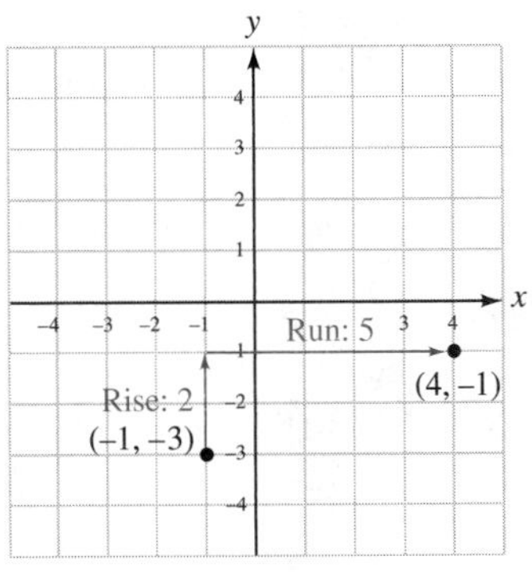

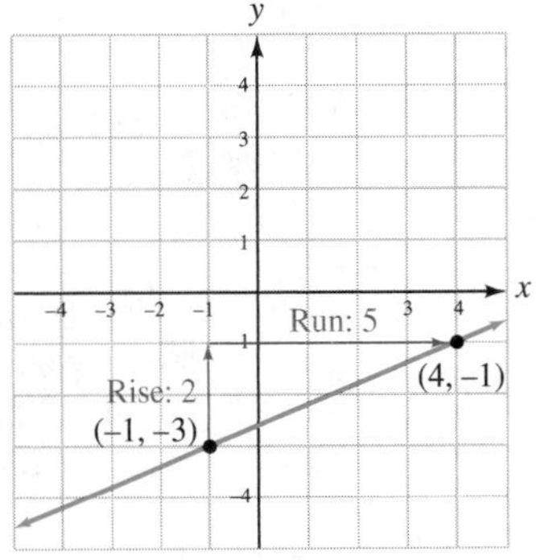

Self Check 4 Graph the line with slope -4 that passes through $(-4, 2)$.

APPLICATIONS

A variety of situations can be described by a linear equation. Quite often, these equations are written using variables other than x and y. In such cases, it is helpful to determine what an ordered-pair solution of the equation would look like.

EXAMPLE 5

ELEMENTARY Algebra f(x) Now™

Men's shoe sizes. The length (in inches) of a man's foot is *not* his shoe size. For example, the smallest adult men's shoe size is 5, and it fits a 9-inch-long foot. There is, however, a linear relationship between the two. It can be stated this way: Shoe size increases by 3 sizes for each 1-inch increase in foot length.

a. Write a linear equation that relates shoe size s to foot length L.

b. Shaquille O'Neal's foot is about 14.6 inches long. What size shoe does he wear?

Solution **a.** Since shoe size s depends on the length L of the foot, ordered pairs have the form (L, s). The relationship is linear, so the graph of the equation we are to write is a line.

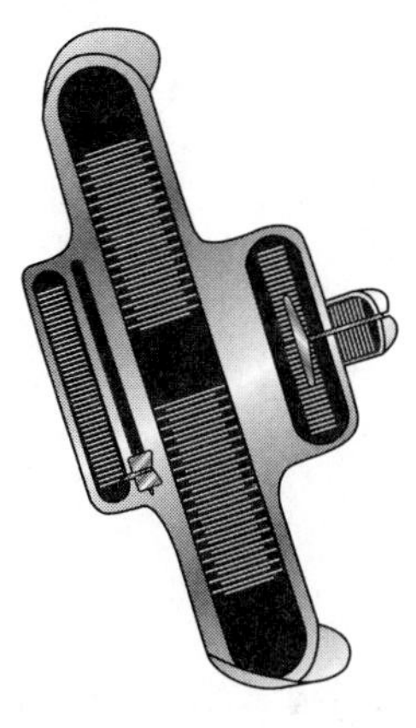

- The line's slope is the rate of change: $\frac{3 \text{ sizes}}{1 \text{ inch}}$. Therefore, $m = 3$.
- A 9-inch-long foot wears size 5, so the line passes through $(9, 5)$.

We substitute these facts into the point–slope form and solve for s.

$s - s_1 = m(L - L_1)$	This is the point–slope form using the variables L and s.
$s - 5 = 3(L - 9)$	Substitute 3 for m, 9 for L_1, and 5 for s_1.
$s - 5 = 3L - 27$	Distribute the multiplication by 3.
$s = 3L - 22$	Add 5 to both sides.

The equation relating men's shoe size and foot length is $s = 3L - 22$.

b. To find Shaquille's shoe size, we substitute 14.6 inches for L in the equation.

$$s = 3L - 22$$
$$s = 3(14.6) - 22$$
$$s = 43.8 - 22$$
$$s = 21.8$$

Since men's shoes only come in full- and half-sizes, we round 21.8 up to 22. Shaquille O'Neal wears size 22 shoes.

EXAMPLE 6

ELEMENTARY Algebra f(x) Now™

Studying learning. In a series of trials, a rat was released in a maze to search for food. Researchers recorded the time that it took the rat to complete the maze on a **scatter diagram.** After the 40th trial, they drew a straight line through the data to obtain a model of the rat's performance. Write an equation of the line in slope–intercept form.

The Language of Algebra

The term *scatter diagram* is somewhat misleading. Often, the data points are not scattered loosely about. In this case, they fall, more or less, along an imaginary straight line, indicating a linear relationship.

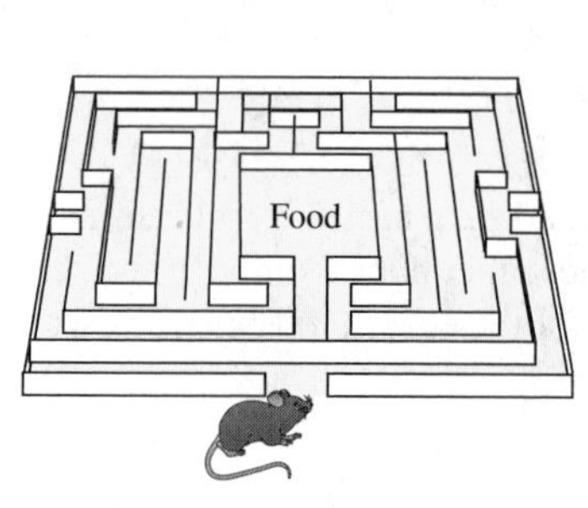

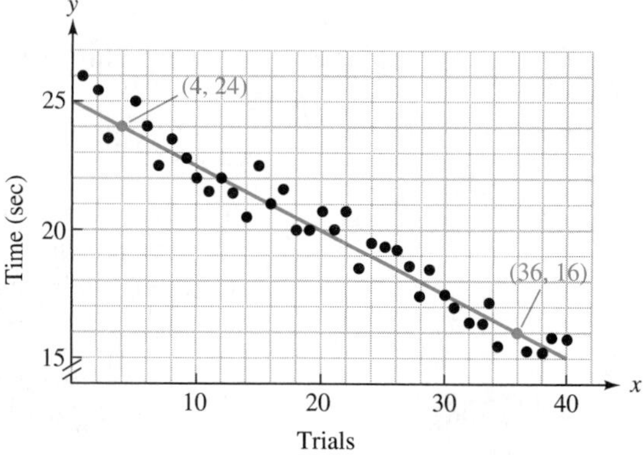

Solution We begin by writing a point–slope equation. The line passes through several points; we will use (4, 24) and (36, 16) to find the slope.

$$m = \frac{y_2 - y_1}{x_2 - x_1} = \frac{16 - 24}{36 - 4} = \frac{-8}{32} = -\frac{1}{4}$$

Any point on the line can serve as (x_1, y_1). We will use (4, 24).

$$y - y_1 = m(x - x_1) \quad \text{This is the point–slope form.}$$
$$y - 24 = -\frac{1}{4}(x - 4) \quad \text{Substitute } -\frac{1}{4} \text{ for } m, 4 \text{ for } x_1, \text{ and } 24 \text{ for } y_1.$$

To write this equation in slope–intercept form, solve for y.

$$y - 24 = -\frac{1}{4}x + 1 \quad \text{Distribute the multiplication by } -\frac{1}{4}: -\frac{1}{4}(-4) = 1.$$
$$y = -\frac{1}{4}x + 25 \quad \text{Add 24 to both sides.}$$

A linear equation that models the rat's performance on the maze is $y = -\frac{1}{4}x + 25$, where x is the number of the trial and y is the time it took, in seconds.

Answers to Self Checks

1. $y = -2x + 5$
2. $y = -\frac{10}{13}x + \frac{2}{13}$
3. a. $y = 2$, b. $x = -1$
4. 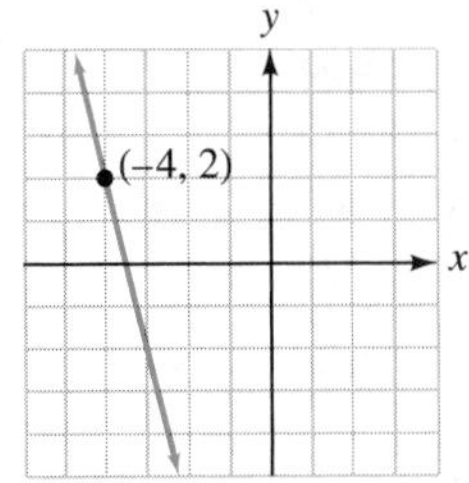

3.6 STUDY SET

ELEMENTARY Algebra $f(x)$ Now™

VOCABULARY Fill in the blanks.

1. $y - y_1 = m(x - x_1)$ is called the __________ form of the equation of a line.
2. $y = mx + b$ is called __________ form of the equation of a line.

CONCEPTS

3. Tell what form each equation is written in.
   a. $y - 4 = 2(x - 5)$
   b. $y = 2x + 15$
4. The following equations are written in point–slope form. What point does the graph of the equation pass through, and what is the line's slope?
   a. $y - 2 = 6(x - 7)$
   b. $y + 3 = -8(x + 1)$
5. Simplify each expression.
   a. $x - (-6)$ b. $y - (-9)$
6. Use the distributive property to remove parentheses.
   a. $-3(x - 2)$ b. $\frac{4}{5}(x + 10)$
   c. $-\frac{1}{3}(x + 4)$ d. $4[x - (-12)]$
7. Find the slope of the line that passes through $(2, -3)$ and $(-4, 12)$.
8. Simplify: $\frac{1}{8}x - \frac{5}{8} + 2$.
9. On a quiz, a student was asked to write the slope–intercept equation of a line with slope 4 that passes through $(-1, 3)$. His answer was $y = 4x + 7$.
   a. Does $y = 4x + 7$ have the required slope?
   b. Does $y = 4x + 7$ pass through the point $(-1, 3)$?
10. Fill in the blanks.
    a. The equation of a horizontal line has the form ___ $= b$.
    b. The equation of a vertical line has the form ___ $= a$.
11. Examine the work shown in the following illustration, where a line is being graphed.
    a. What point does the line pass through?
    b. What is the slope of the line?
    c. What is a second point on the line?

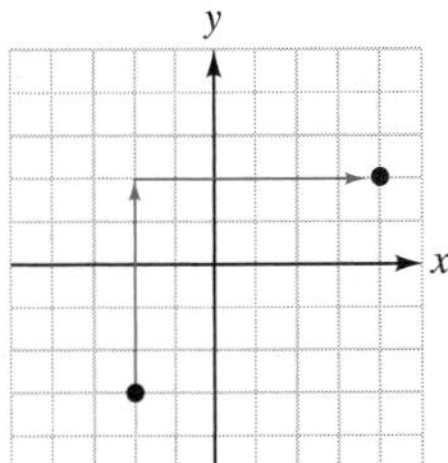

12. a. Find two points on the line graphed below whose coordinates are integers.
    b. What is the slope of the line?
    c. Use your answers to parts a and b to write the equation of the line. Give the answer in point–slope form.

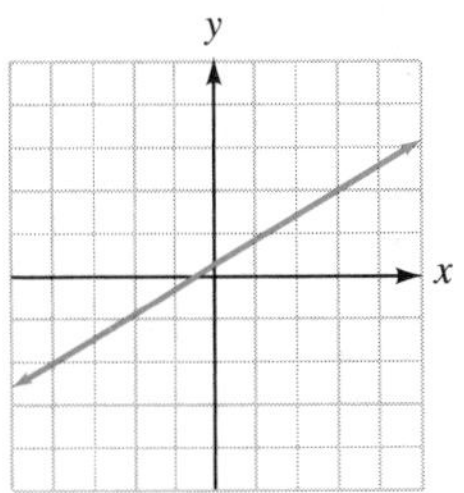

13. In each of the following cases, is the given information sufficient to write an equation of the line?
a. It passes through $(2, -7)$.
b. Its slope is $-\frac{3}{4}$.
c. Its y-intercept is $(0, -1)$ and its slope is 2.
d. It has the following table of solutions.

x	y
2	3
−3	−6

14. In each of the following cases, is the given information sufficient to write an equation of the line?
a. It is horizontal.
b. It is vertical and passes through $(-1, 1)$.
c. Its y-intercept is $(0, 5)$.
d. It has the following table of solutions.

x	y
4	5

15. Suppose you are asked to write an equation of the line in the scatter diagram below. What two points would you use to write the point–slope equation?

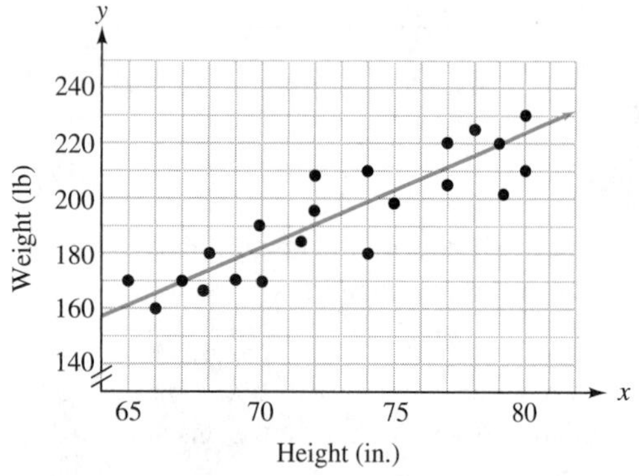

16. In each of the following cases, a linear relationship between two quantities is described. If the relationship were graphed, what would be the slope of the line?
a. The sales of new cars increased by 15 every 2 months.
b. There were 35 fewer robberies for each dozen police officers added to the force.
c. Withdrawals were occurring at the rate of $700 every 45 minutes.
d. One acre of forest is being destroyed every 30 seconds.

NOTATION

17. What is the point–slope form of the equation of a line?

18. Fill in the blanks to state the point–slope form in words.

y minus y ______ one equals m ______ the quantity of x ______ x sub ______.

19. Consider these steps:

$$y - 3 = 2(x + 1)$$
$$y - 3 = 2x + 2$$
$$y = 2x + 5$$

Now fill in the blanks. The original equation was in ______ form. After solving for ▢, we obtain an equation in ______ form.

Complete the solution.

20. Write the equation of the line with slope -2 that passes through the point $(-1, 5)$. Write the answer in slope–intercept form.

$$y - y_1 = m(x - x_1)$$
$$y - \square = -2[x - (\square)]$$
$$y - 5 = -2x - \square$$
$$y = -2x + \square$$

PRACTICE **Use the point–slope form to write an equation of the line with the given slope and point.**

21. Slope 3, passes through $(2, 1)$

22. Slope 2, passes through $(4, 3)$

23. Slope $-\frac{4}{5}$, passes through $(-5, -1)$

24. Slope $-\frac{7}{8}$, passes through $(-2, -9)$

Use the point–slope form to write an equation of the line with the given slope and point. Then write your result in slope–intercept form.

25. Slope $\frac{1}{5}$, passes through (10, 1)

26. Slope $\frac{1}{4}$, passes through (8, 1)

27. Slope −5, passes through (−9, 8)

28. Slope −4, passes through (−2, 10)

29. Slope $-\frac{4}{3}$,

x	y
6	−4

30. Slope $-\frac{3}{2}$,

x	y
−2	1

31. Slope $-\frac{11}{6}$, passes through (2, −6)

32. Slope $-\frac{5}{4}$, passes through (2, 0)

33. Slope $-\frac{2}{3}$, passes through (3, 0)

34. Slope $-\frac{2}{5}$, passes through (15, 0)

35. Slope 8, passes through (0, 4)

36. Slope 6, passes through (0, −4)

37. Slope −3, passes through the origin

38. Slope −1, passes through the origin

Write an equation of the line that passes through the two given points. Write your result in slope–intercept form.

39. Passes through (1, 7) and (−2, 1)

40. Passes through (−2, 2) and (2, −8)

41.

x	y
−4	3
2	0

42.

x	y
−1	−4
1	−2

43. Passes through (5, 5) and (7, 5)

44. Passes through (−2, 1) and (−2, 15)

45. Passes through (5, 1) and (−5, 0)

46. Passes through (−3, 0) and (3, 1)

47. Passes through (−8, 2) and (−8, 17)

48. Passes through $\left(\frac{2}{3}, 2\right)$ and (0, 2)

49. Passes through (5, 0) and (−11, −4)

50. Passes through (7, −3) and (−5, 1)

Write the equation of the line with the given characteristics.

51. Vertical, passes through (4, 5)

52. Vertical, passes through (−2, −5)

53. Horizontal, passes through (4, 5)

54. Horizontal, passes through (−2, −5)

Graph the line that passes through the given point and has the given slope.

55. (1, −2), slope −1

56. (−4, 1), slope −3

57. (5, −3), $m = \frac{3}{4}$

58. (2, −4), $m = \frac{2}{3}$

59. (−2, −3), slope 2

60. (−3, −3), slope 4

61. (4, −3), slope $-\frac{7}{8}$

62. (4, 2), slope $-\frac{1}{5}$

APPLICATIONS

63. ANATOMY There is a linear relationship between a woman's height and the length of her radius bone. It can be stated this way: Height increases by 3.9 inches for each 1-inch increase in the length of the radius. Suppose a 64-inch-tall woman has a 9-inch-long radius bone. Use this information to write a linear equation that relates height h to the length r of the radius. Write your answer in slope–intercept form.

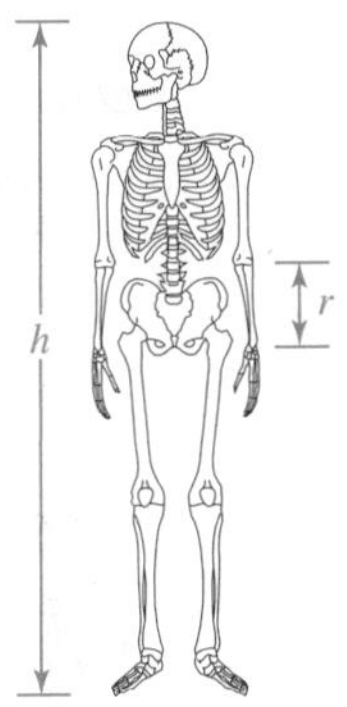

64. AUTOMATION An automated production line uses distilled water at a rate of 300 gallons every 2 hours to make shampoo. After the line had run for 7 hours, planners noted that 2,500 gallons of distilled water remained in the storage tank. Write a linear equation relating the time t in hours since the production line began and the number g of gallons of distilled water in the storage tank. Write the answer in slope–intercept form.

65. POLE VAULTING Write the equations of the lines that describe the positions of the pole for parts 1, 3, and 4 of the jump. Write the answers in slope–intercept form.

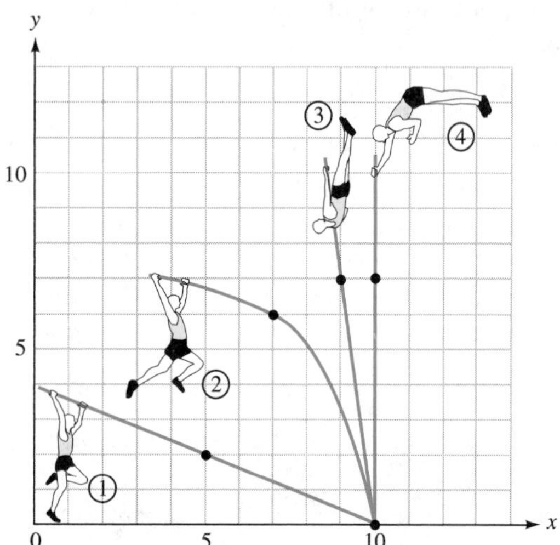

66. FREEWAY DESIGN The graph below shows the route of a proposed freeway.

a. Give the coordinates of the points where the proposed freeway will join Interstate 25 and Highway 40.

b. Write the equation of the line that describes the route of the proposed freeway. Give the answer in slope–intercept form.

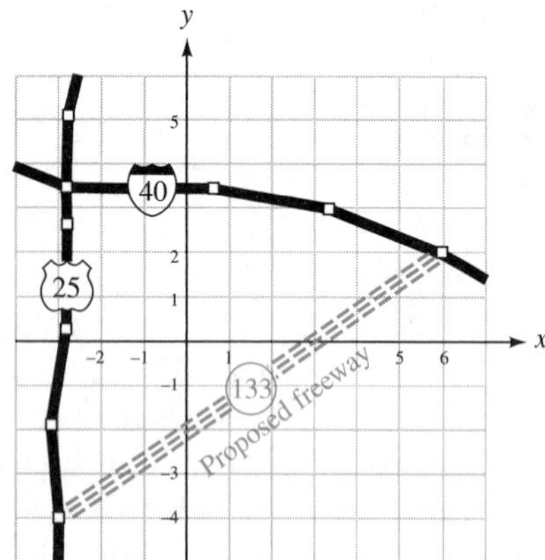

67. TOXIC CLEANUP Three months after cleanup began at a dump site, 800 cubic yards of toxic waste had yet to be removed. Two months later, that number had been lowered to 720 cubic yards.

a. Write an equation that describes the linear relationship between the length of time m (in months) the cleanup crew has been working and the number of cubic yards y of toxic waste remaining. Give the answer in slope–intercept form.

b. Use your answer to part a to predict the number of cubic yards of waste that will still be on the site one year after the cleanup project began.

68. DEPRECIATION To lower its corporate income tax, accountants of a large company depreciated a word processing system over several years using a linear model, as shown in the worksheet.

a. Write a linear equation relating the years since the system was purchased, x, and its value, y, in dollars. Give the answer in slope–intercept form.

b. Find the purchase price of the system.

Tax Worksheet

Method of depreciation: *Linear*

Property	Value	Years after purchase
Word processing system	*$60,000*	2
"	*$30,000*	4

69. TRAMPOLINES There is a linear relationship between the length of the protective pad that wraps around a trampoline and the radius of the trampoline. Use the data in the table to write an equation that gives the length l of pad needed for any trampoline with radius r. Write the answer in slope–intercept form. Use units of feet for both l and r.

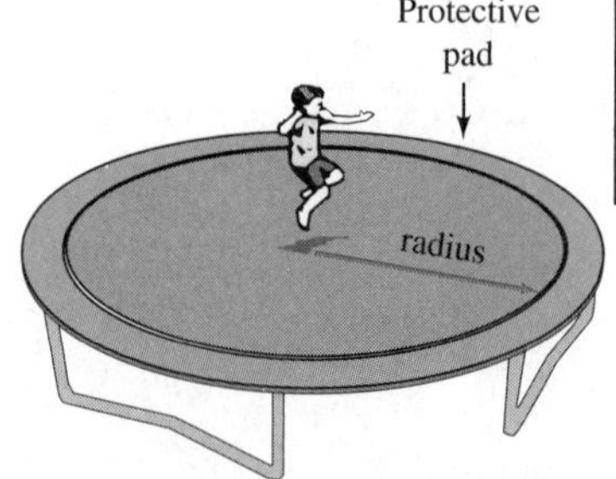

Radius	Pad length
3 ft	19 ft
7 ft	44 ft

70. CONVERTING TEMPERATURES The relationship between Fahrenheit temperature, F, and Celsius temperature, C, is linear.

a. Use the data in the illustration to write two ordered pairs of the form (C, F).

b. Use your answer to part a to write a linear equation relating the Fahrenheit and Celsius scales. Write the answer in slope–intercept form.

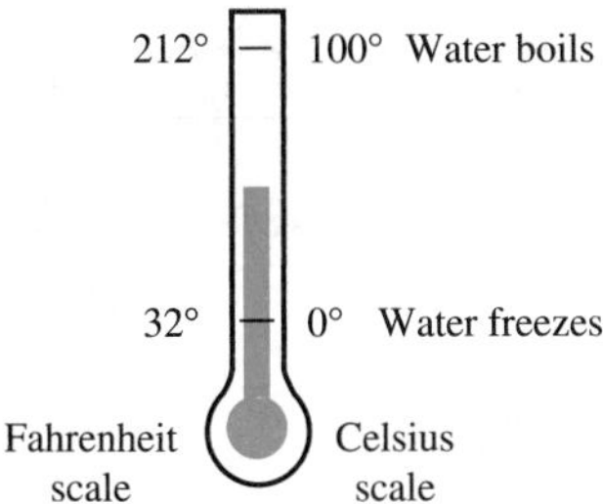

71. GOT MILK? The scatter diagram shows per capita milk consumption in the U.S. for the years 1975–2000. A straight line can be used to model the data.

a. Use two points on the line to write its equation. Write the answer in slope–intercept form.

b. Use your answer to part a to predict the per capita milk consumption in 2015.

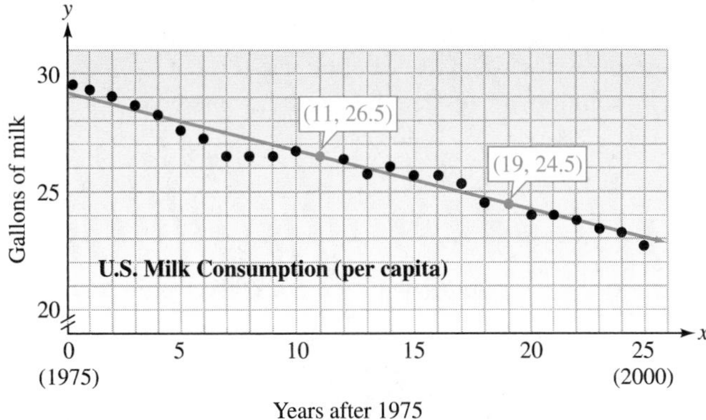

Source: United States Department of Agriculture

72. ENGINE OUTPUT The horsepower produced by an automobile engine was recorded for various engine speeds in the range of 2,400–4,800 revolutions per minute (rpm). The data were recorded on the following scatter diagram. Write an equation of the line that models the relationship between engine speed s and horsepower h. Give the answer in slope–intercept form.

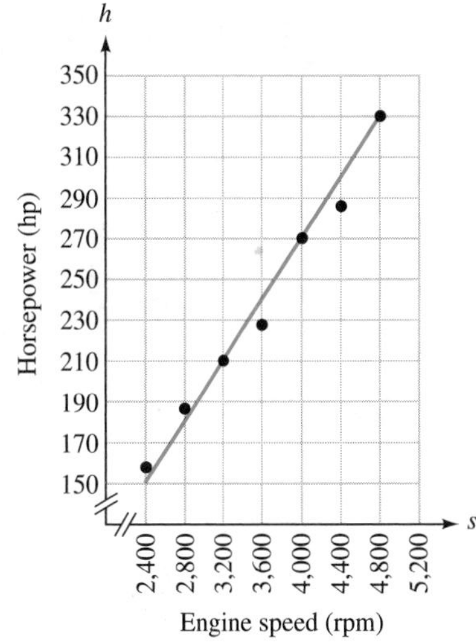

WRITING

73. Why is $y - y_1 = m(x - x_1)$ called the point–slope form of the equation of a line?

74. If we know two points that a line passes through, we can write its equation. Explain how this is done.

75. Explain the steps involved in writing $y - 6 = 4(x - 1)$ in slope–intercept form.

76. Think of several points on the graph of the horizontal line $y = 4$. What do the points have in common? How do they differ?

REVIEW

77. FRAMING PICTURES The length of a rectangular picture is 5 inches greater than twice the width. If the perimeter is 112 inches, find the dimensions of the frame.

78. SPEED OF AN AIRPLANE Two planes are 6,000 miles apart, and their speeds differ by 200 mph. They travel toward each other and meet in 5 hours. Find the speed of the slower plane.

CHALLENGE PROBLEMS

79. Write an equation of the line that passes through $(2, 5)$ and is parallel to the line $y = 4x - 7$. Give the answer in slope–intercept form.

80. Write an equation of the line that passes through $(-6, 3)$ and is perpendicular to the line $y = -3x - 12$. Give the answer in slope–intercept form.

3.7 Graphing Linear Inequalities

- Linear Inequalities and Solutions
- Graphing Linear Inequalities
- Applications

Recall that an **inequality** is a statement that contains one of the symbols $<$, $\leq$, $>$, or $\geq$. Inequalities *in one variable,* such as $x + 6 < 8$ and $5x + 3 \geq 4x$, were solved in Section 2.7. Because they have an infinite number of solutions, we represented their solution sets graphically, by shading intervals on a number line.

We now extend that concept to linear inequalities *in two variables,* as we introduce a procedure that is used to graph their solution sets.

LINEAR INEQUALITIES AND SOLUTIONS

If the $=$ symbol in a linear equation in two variables is replaced with an inequality symbol, we have a **linear inequality in two variables.** Some examples are

$$x - y \leq 5, \qquad 4x + 3y < -6, \qquad \text{and} \qquad y > 2x$$

As with linear equations, a **solution of a linear inequality** is an ordered pair of numbers that makes the inequality true.

EXAMPLE 1

ELEMENTARY Algebra f(x) Now™

Determine whether each ordered pair is a solution of $x - y \leq 5$. Then graph each solution: **a.** (4, 2), **b.** (0, −6), and **c.** (1, −4).

Solution In each case, we substitute the x-coordinate for x and the y-coordinate for y in the inequality $x - y \leq 5$. If the ordered pair is a solution, a true statement will be obtained.

a. For (4, 2):

$x - y \leq 5$ This is the original inequality.

$4 - 2 \overset{?}{\leq} 5$ Substitute 4 for x and 2 for y. The symbol $\overset{?}{\leq}$ is read as "is possibly less than or equal to."

$2 \leq 5$ True.

Because $2 \leq 5$ is true, (4, 2) is a solution of $x - y \leq 5$. We say that (4, 2) *satisfies* the inequality. The solution is graphed as shown on the right.

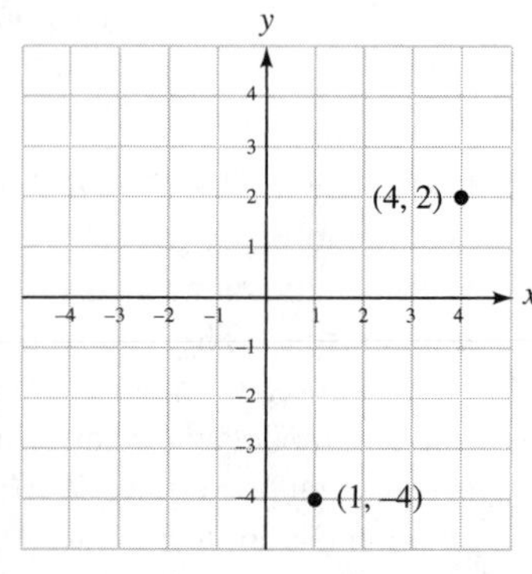

b. For (0, −6):

$x - y \leq 5$ This is the original inequality.

$0 - (-6) \overset{?}{\leq} 5$ Substitute 0 for x and −6 for y.

$6 \leq 5$ False.

Because $6 \leq 5$ is false, (0, −6) is not a solution.

c. For $(1, -4)$:

$$x - y \leq 5 \quad \text{This is the original inequality.}$$
$$1 - (-4) \stackrel{?}{\leq} 5 \quad \text{Substitute 1 for } x \text{ and } -4 \text{ for } y.$$
$$5 \leq 5 \quad \text{True.}$$

Because $5 \leq 5$ is true, $(1, -4)$ is a solution, and we graph it as shown.

Self Check 1 Using the inequality in Example 1, determine whether each ordered pair is a solution: **a.** $(8, 2)$, **b.** $(4, -1)$, **c.** $(-2, 4)$, and **d.** $(-3, -5)$.

In Example 1, we graphed several of the solutions of $x - y \leq 5$. There are many more ordered pairs (x, y) that make the inequality true. It would not be reasonable to plot each of them. Fortunately, there is an easier way to show all of the solutions.

GRAPHING LINEAR INEQUALITIES

The graph of a linear inequality is a picture that represents the set of all points whose coordinates satisfy the inequality. In general, such graphs are areas bounded by a line. We call those areas **half-planes,** and we use a two-step procedure to find them.

EXAMPLE 2

Graph: $x - y \leq 5$.

ELEMENTARY Algebra f(x) Now™

Solution Since the inequality symbol $\leq$ includes an equal symbol, the graph of $x - y \leq 5$ includes the graph of $x - y = 5$.

Step 1: To graph $x - y = 5$, we use the intercept method, as shown in part (a) of the illustration. The resulting line, called a **boundary line,** divides the coordinate plane into two half-planes. To show that the points on the boundary line are solutions of $x - y \leq 5$, we draw it as a solid line.

$x - y = 5$

x	y	(x, y)
0	-5	$(0, -5)$
5	0	$(5, 0)$
6	1	$(6, 1)$

Let $x = 0$, find y. Let $y = 0$, find x. As a check, let $x = 6$ and find y.

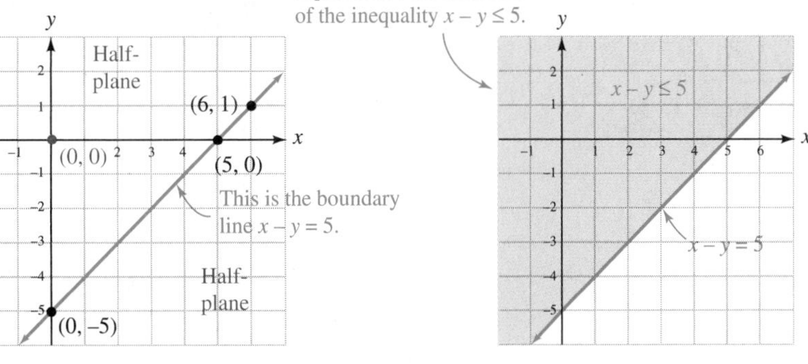

(a) (b)

Step 2: Since the inequality $x - y \leq 5$ also allows $x - y$ to be less than 5, other ordered pairs, besides those on the boundary, satisfy the inequality. For example, consider the origin, with coordinates (0, 0). If we substitute 0 for x and 0 for y in the given inequality, we have

$$x - y \leq 5$$
$$0 - 0 \overset{?}{\leq} 5$$
$$0 \leq 5 \quad \text{True.}$$

Because $0 \leq 5$, the coordinates of the origin satisfy $x - y \leq 5$. In fact, the coordinates of every point on the *same side* of the line as the origin satisfy the inequality. To indicate this, we shade the half-plane that contains (0, 0), as shown in part (b). Every point in the shaded half-plane and every point on the boundary line satisfies $x - y \leq 5$.

As an *informal* check, we can pick an ordered pair that lies in the shaded region and one that does not lie in the shaded region. When we substitute their coordinates into the inequality, we should obtain a true statement and then a false statement.

For* (3, 1), *in the shaded region:

$$x - y \leq 5$$
$$3 - 1 \overset{?}{\leq} 5$$
$$2 \leq 5 \quad \text{True.}$$

For* (5, −4), *not in the shaded region:

$$x - y \leq 5$$
$$5 - (-4) \overset{?}{\leq} 5$$
$$9 \leq 5 \quad \text{False.}$$

Self Check 2 Graph: $x - y \leq 2$.

EXAMPLE 3

Graph: $4x + 3y < -6$.

ELEMENTARY Algebra $f(x)$ Now™

Solution To find the boundary line, we graph $4x + 3y = -6$. Since the inequality symbol $<$ does *not* include an equal symbol, the points on the graph of $4x + 3y = -6$ are not part of the graph of $4x + 3y < -6$. We draw the boundary line as a dashed line to show this. See part (a) of the illustration.

$4x + 3y = -6$

x	y	(x, y)
0	−2	(0, −2)
$-\frac{3}{2}$	0	$(-\frac{3}{2}, 0)$
−3	2	(−3, 2)

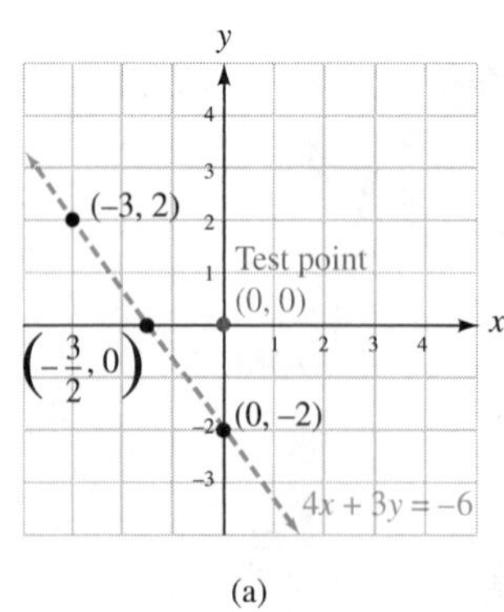

(a)

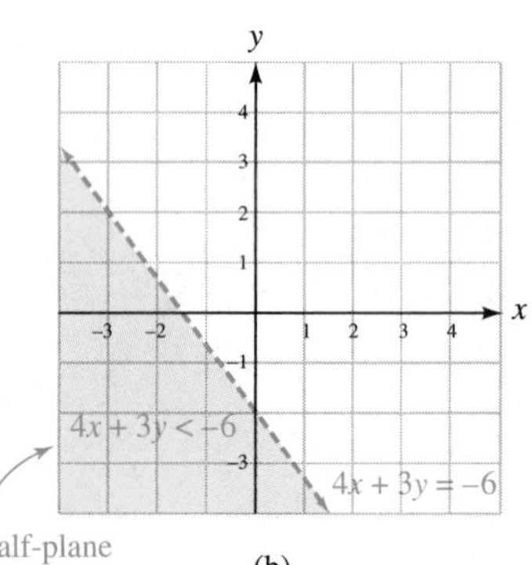

This shaded half-plane represents all the solutions of the inequality $4x + 3y < -6$.

(b)

Caution

When using a test point to determine which half-plane to shade, remember to substitute the coordinates into the given inequality, not the equation for the boundary.

To determine which half-plane to shade, we substitute the coordinates of a point that lies on one side of the boundary line into $4x + 3y < -6$. We choose the origin (0, 0) as the **test point** because the computations are usually easy when they involve 0. We substitute 0 for x and 0 for y in the inequality.

$$4x + 3y < -6$$
$$4(0) + 3(0) \overset{?}{<} -6 \quad \text{The symbol } \overset{?}{<} \text{ is read as "is possibly less than."}$$
$$0 + 0 \overset{?}{<} -6$$
$$0 < -6 \quad \text{False.}$$

Since $0 < -6$ is a false statement, the point (0, 0) does not satisfy the inequality. This indicates that it is *not* on the side of the dashed line we wish to shade. Instead, we shade the other side of the boundary line. The graph of the solution set of $4x + 3y < -6$ is the half-plane below the dashed line, as shown in part (b).

Self Check 3 Graph: $5x + 6y < -15$.

EXAMPLE 4

Graph: $y > 2x$.

ELEMENTARY Algebra f(x) Now™

Solution To find the boundary line, we graph $y = 2x$. Since the symbol $>$ does *not* include an equal symbol, the points on the graph of $y = 2x$ are not part of the graph of $y > 2x$. Therefore, the boundary line should be dashed, as shown in part (a) of the illustration.

Success Tip

Draw a solid boundary line if the inequality has $\leq$ or $\geq$. Draw a dashed line if the inequality has $<$ or $>$.

$y = 2x$

x	y	(x, y)
0	0	(0, 0)
−1	−2	(−1, −2)
1	2	(1, 2)

↑ Choose three values for x and find the corresponding values of y.

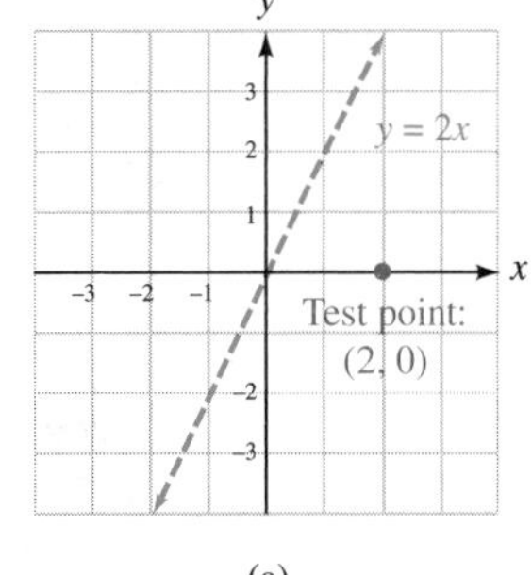

(a)

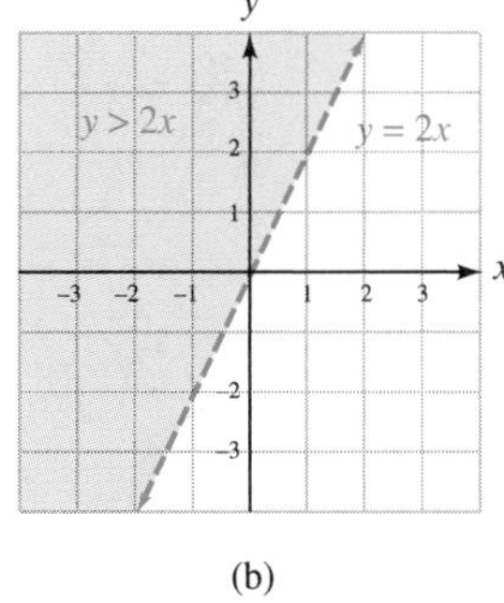

(b)

Success Tip

The origin (0, 0) is a smart choice for a test point because computations involving 0 are usually easy. If the origin is on the boundary, choose a test point not on the boundary that has one coordinate that is 0, such as (0, 1) or (2, 0).

To determine which half-plane to shade, we substitute the coordinates of a point that lies on one side of the boundary line into $y > 2x$. Since the origin is on the boundary, it cannot serve as a test point. One of the many possible choices for a test point is (2, 0), because it does not lie on the boundary line. To see whether it satisfies $y > 2x$, we substitute 2 for x and 0 for y in the inequality.

$$y > 2x$$
$$0 \overset{?}{>} 2(2) \quad \text{The symbol } \overset{?}{>} \text{ is read as "is possibly greater than."}$$
$$0 > 4 \quad \text{False.}$$

Since $0 > 4$ is a false statement, the point (2, 0) does not satisfy the inequality. We shade the half-plane that does not contain (2, 0), as shown in part (b).

Self Check 4 Graph: $y < 3x$.

EXAMPLE 5

Graph each linear inequality: **a.** $x < -3$ and **b.** $y \geq 0$.

ELEMENTARY Algebra f(x) Now™

Solution **a.** Because the inequality contains a < symbol, we draw the boundary, $x = -3$, as a dashed vertical line. We can use (0, 0) as the test point.

$x < -3$

$0 < -3$ Substitute 0 for x. The y-coordinate of (0, 0) is not used.

Since the result is false, we shade the half-plane that does not contain (0, 0), as shown in part (a) of the illustration below. Note that the solution consists of all points that have an x-coordinate that is less than -3.

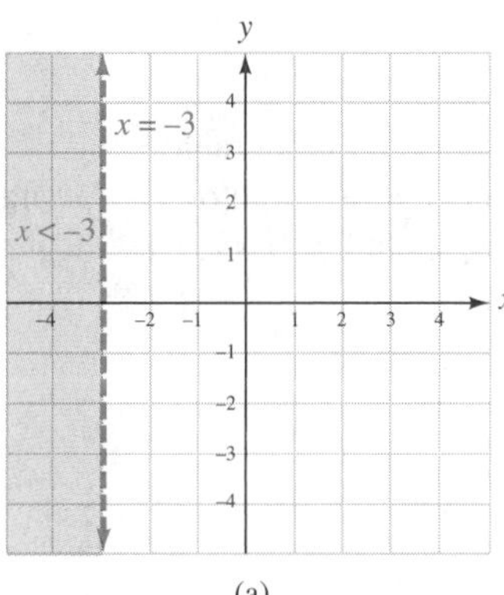

(a)

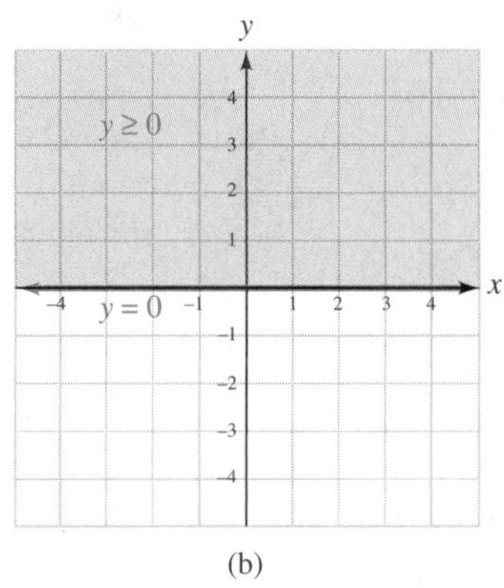

(b)

b. Because the inequality contains a ≥ symbol, we draw the boundary, $y = 0$, as a solid horizontal line. (Recall that the graph of $y = 0$ is the x-axis.) Next, we choose a test point not on the boundary. The point (0, 1) is a convenient choice.

$y \geq 0$

$1 \geq 0$ Substitute 1 for y. The x-coordinate of (0, 1) is not used.

Since the result is true, we shade the half-plane that contains (0, 1), as shown in part (b) above. Note that the solution consists of all points that have a y-coordinate that is greater than or equal to 0.

Self Check 5 Graph each linear inequality: **a.** $x \geq 2$ and **b.** $y < 4$.

The following is a summary of the procedure for graphing linear inequalities.

Graphing Linear Inequalities in Two Variables

1. Graph the boundary line of the region. If the inequality allows the possibility of equality (the symbol is either $\leq$ or $\geq$), draw the boundary line as a solid line. If equality is not allowed ($<$ or $>$), draw the boundary line as a dashed line.
2. Pick a test point that is on one side of the boundary line. (Use the origin if possible.) Replace x and y in the inequality with the coordinates of that point. If the inequality is satisfied, shade the side that contains that point. If the inequality is not satisfied, shade the other side of the boundary.

APPLICATIONS

When solving applied problems, phrases such as *at least, at most,* and *should not exceed* indicate that an inequality should be used.

EXAMPLE 6

Working two jobs. Carlos has two part-time jobs, one paying \$10 per hour and another paying \$12 per hour. If x represents the number of hours he works on the first job, and y represents the number of hours he works on the second, the graph of $10x + 12y \geq 240$ shows the possible ways he can schedule his time to earn at least \$240 per week to pay his college expenses. Find three possible combinations of hours he can work to achieve his financial goal.

Solution The graph of the inequality is shown below in part (a) of the illustration. Any point in the shaded region represents a possible way Carlos can schedule his time and earn \$240 or more per week. If each shift is a whole number of hours long, the highlighted points in part (b) represent the acceptable combinations. Three such combinations are

(6, 24): 6 hours on the first job, 24 hours on the second job
(12, 12): 12 hours on the first job, 12 hours on the second job
(22, 4): 22 hours on the first job, 4 hours on the second job

To verify one combination, suppose Carlos works 22 hours on the first job and 4 hours on the second job. He will earn

$$\$10(22) + \$12(4) = \$220 + \$48$$
$$= \$268$$

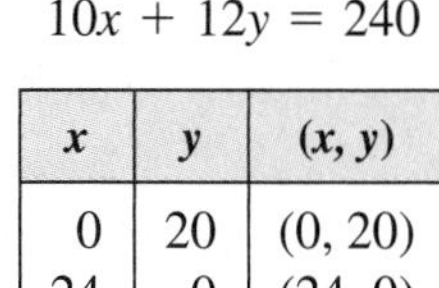

$10x + 12y = 240$

x	y	(x, y)
0	20	(0, 20)
24	0	(24, 0)

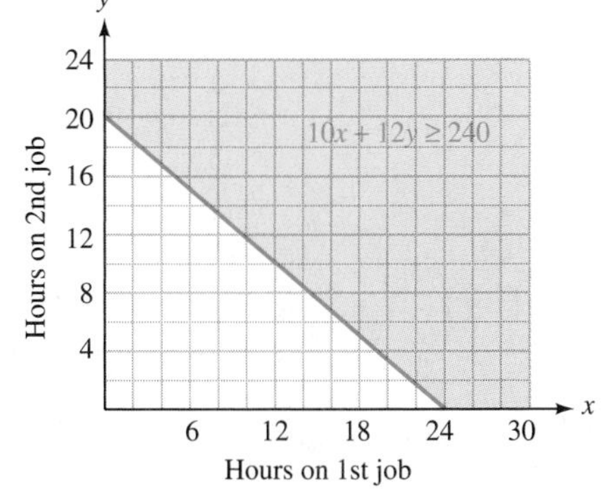

(a)

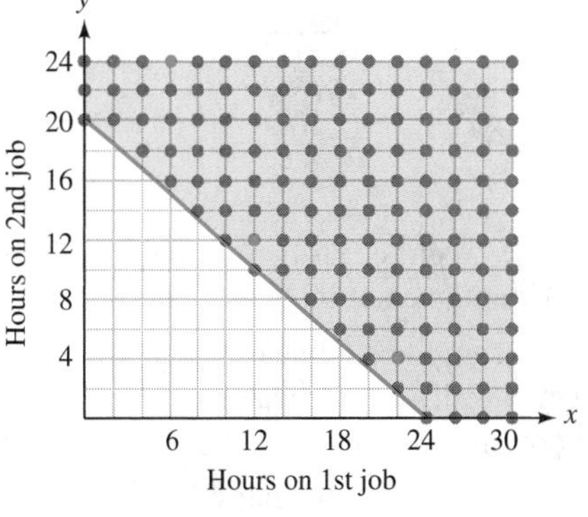

(b)

Answers to Self Checks **1. a.** not a solution, **b.** solution, **c.** solution, **d.** solution

2.

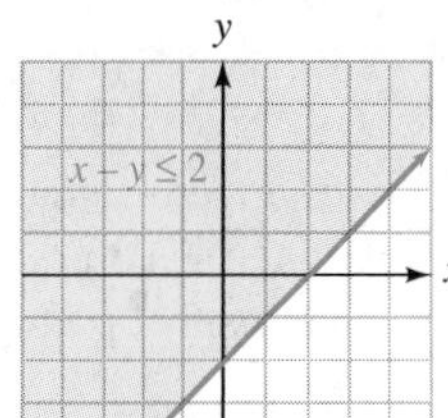

3.

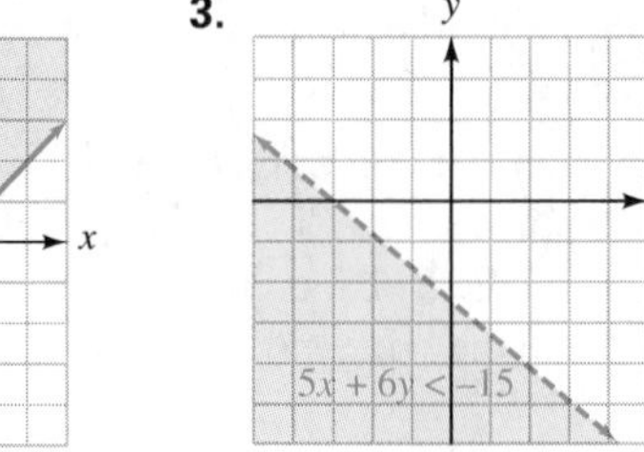

4.

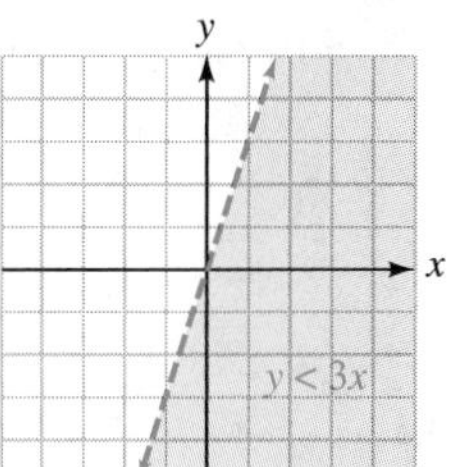

5.

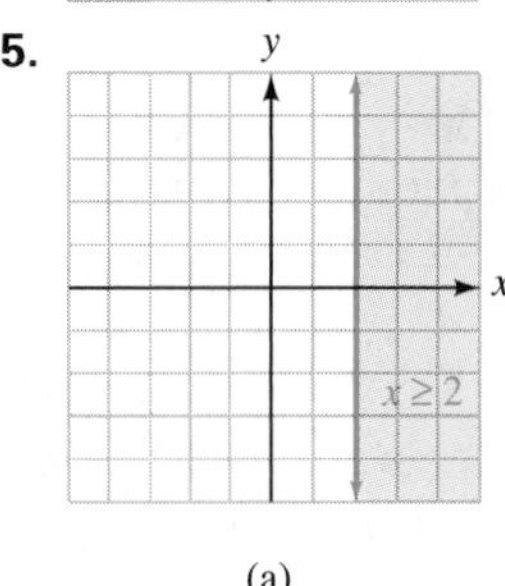

(a)

(b)

3.7 STUDY SET

ELEMENTARY Algebra f(x) Now™

VOCABULARY **Fill in the blanks.**

1. An __________ is a statement that contains one of the symbols $<$, $\leq$, $>$, or $\geq$.

2. $2x - y \leq 4$ is a _______ inequality in two variables.

3. A ________ of a linear inequality is an ordered pair of numbers that makes the inequality true.

4. (7, 2) is a solution of $x - y > 1$. We say that (7, 2) ________ the inequality.

5. In the graph, the line $2x - y = 4$ is the _________.

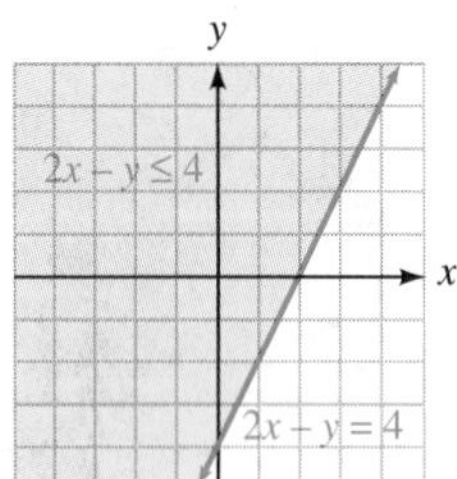

6. In the graph for Problem 5, the line $2x - y = 4$ divides the coordinate plane into two ___________.

7. When graphing a linear inequality, we determine which half-plane to shade by substituting the coordinates of a test ______ into the inequality.

8. A ________ line indicates that points on the boundary are not solutions. A ______ line indicates that points on the boundary are solutions.

CONCEPTS

9. Decide whether each statement is true or false.

a. $2 \leq 0$ **b.** $-5 > 0$

c. $0 \geq 0$ **d.** $0 < -6$

10. Decide whether each ordered pair is a solution of $5x - 3y \geq 0$.

a. (1, 1) **b.** $(-2, -3)$

c. (0, 0) **d.** $\left(\frac{1}{5}, \frac{4}{3}\right)$

11. Decide whether each ordered pair is a solution of $x + 4y < -1$.

a. (3, 1) **b.** $(-2, 0)$

c. $(-0.5, 0.2)$ **d.** $\left(-2, \frac{1}{4}\right)$

12. Fill in the blanks to explain the procedure for graphing a linear inequality.
Step 1: Graph the __________.
Step 2: Use a test point to determine which side to _______.

13. Decide whether the graph of each linear inequality includes the boundary line.

a. $y > -x$ **b.** $5x - 3y \leq -2$

14. The boundary for the graph of a linear inequality follows. Why can't the origin be used as a test point to decide which side to shade?

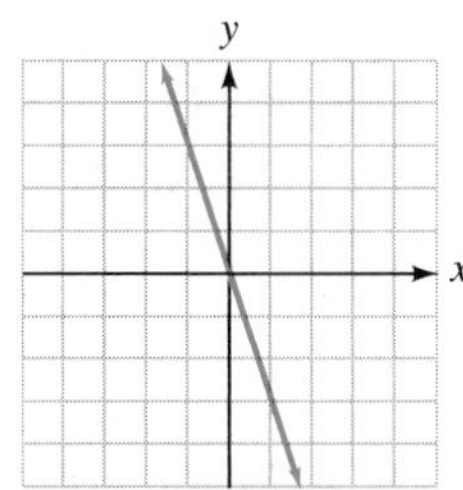

15. If a false statement results when the coordinates of a test point are substituted into a linear inequality, which half-plane should be shaded to represent the solution of the inequality?

16. A linear inequality has been graphed. Tell whether each point satisfies the inequality.
 a. $(1, -3)$
 b. $(-2, -1)$
 c. $(2, 3)$
 d. $(3, -4)$

17. A linear inequality has been graphed. Tell whether each point satisfies the inequality.
 a. $(2, 1)$
 b. $(-2, -4)$
 c. $(4, -2)$
 d. $(-3, 4)$

18. To graph linear inequalities, we must be able to graph boundary lines. Complete the table of solutions for each given boundary line.
 a. $5x - 3y = 15$

x	y	(x, y)
0		
	0	
1		

 b. $y = 3x - 2$

x	y	(x, y)
−1		
0		
	4	

19. **a.** Is the graph of $y = 5$ a vertical or horizontal line?

 b. Is the graph of $x = -6$ a vertical or horizontal line?

20. To decide how many pallets x and barrels y a delivery truck can hold, a dispatcher refers to the loading sheet below. Can a truck make a delivery of 4 pallets and 10 barrels in one trip?

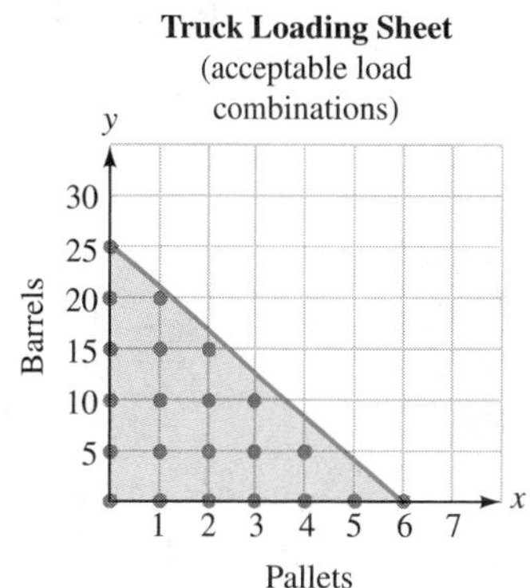

NOTATION

21. Write the meaning of each symbol in words.
 a. $<$ **b.** $>$
 c. $\leq$ **d.** $\geq$

22. When graphing linear inequalities, which inequality symbols are associated with a dashed boundary line?

23. When graphing linear inequalities, which inequality symbols are associated with a solid boundary line?

24. How do we read the symbol $\overset{?}{>}$?

PRACTICE **Complete the graph by shading the correct side of the boundary.**

25. $x - y \geq -2$

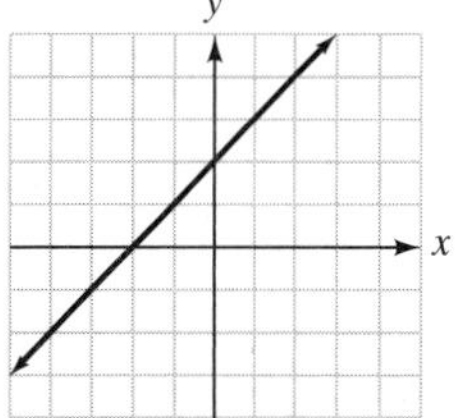

26. $x - y < 3$

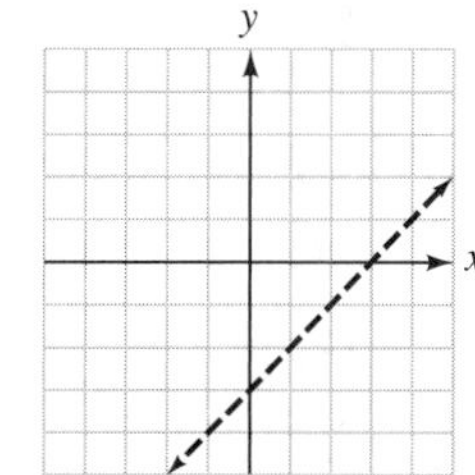

27. $y > 2x - 4$

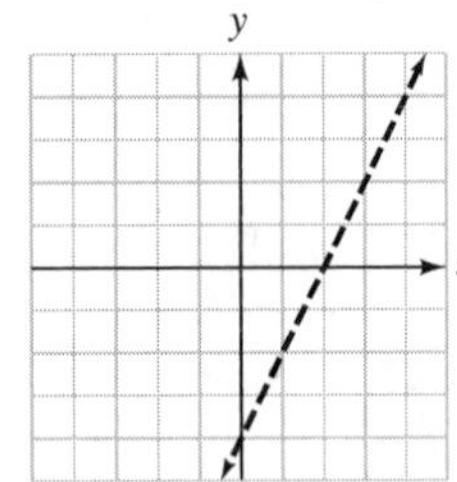

28. $y \leq -x + 1$

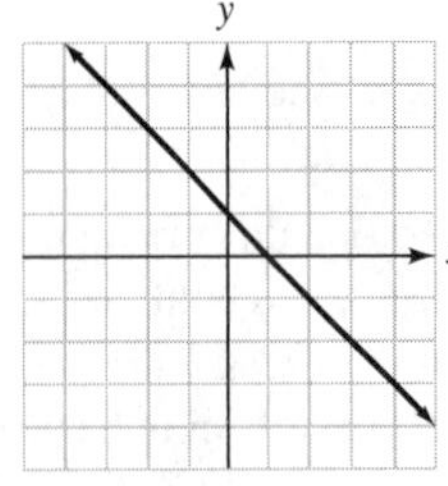

29. $x - 2y \geq 4$

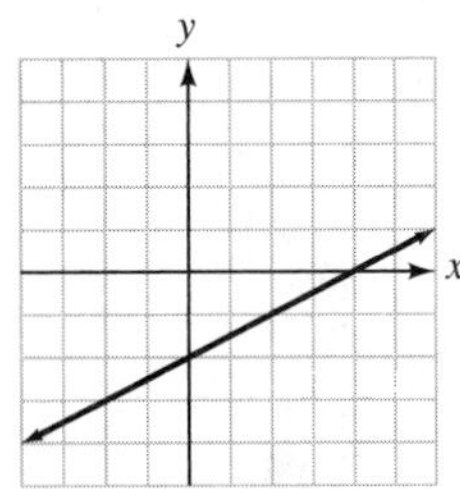

30. $3x + 2y > 12$

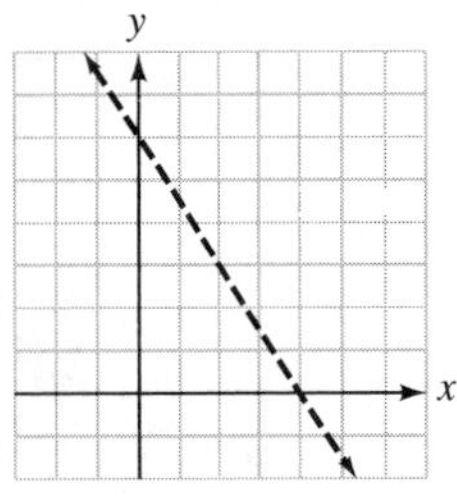

31. $y \leq 4x$

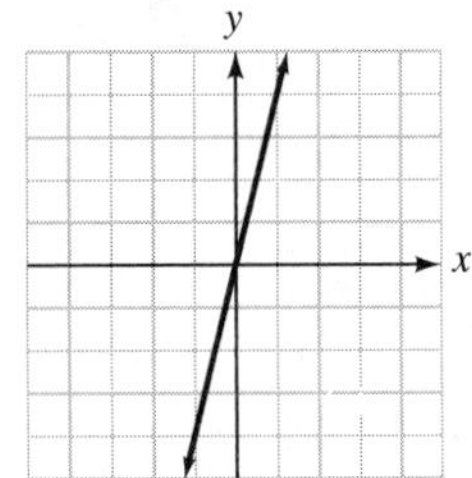

32. $y + 2x < 0$

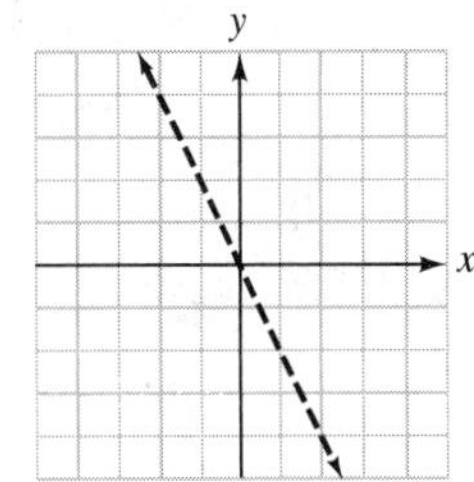

Graph each inequality.

33. $x + y \geq 3$

34. $x + y < 2$

35. $3x - 4y > 12$

36. $5x + 4y \geq 20$

37. $2x + 3y \leq -12$

38. $3x - 2y > 6$

39. $y < 2x - 1$

40. $y > x + 1$

41. $y < -3x + 2$

42. $y \geq -2x + 5$

43. $y \geq -\frac{3}{2}x + 1$

44. $y < \frac{x}{3} - 1$

45. $x - 2y \geq 4$

46. $4x + y \geq -4$

47. $2y - x < 8$

48. $y - x \geq 0$

49. $y + x < 0$

50. $y + 9x \geq 3$

51. $y \geq 2x$

52. $y < 3x$

53. $y < -\frac{x}{2}$

54. $y \geq x$

55. $x < 2$

56. $y > -3$

57. $y \leq 1$

58. $x \geq -4$

59. $x \leq 0$

60. $y < 0$

61. $7x - 2y < 21$

62. $3x - 3y \geq -10$

63. $2x - 3y \geq 4$

64. $4x + 3y < 6$

65. $5x + 3y < 0$

66. $2x + 5y > 0$

APPLICATIONS

67. ROLLING DICE The points on the graph represent all of the possible outcomes when two fair dice are rolled a single time. For example, (5, 2), shown in red, represents a 5 on the first die and a 2 on the second. Which of the following sentences best describes the outcomes that lie in the shaded area?

(i) Their sum is at most 6.
(ii) Their sum exceeds 6.
(iii) Their sum does not exceed 6.
(iv) Their sum is at least 6.

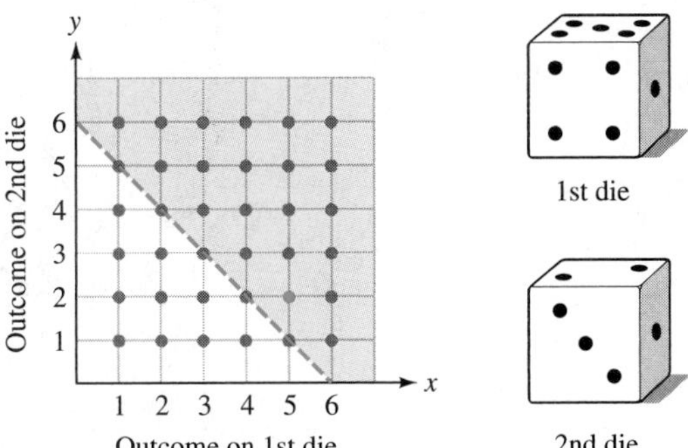

68. NATO In March 1999, NATO aircraft and cruise missiles targeted Serbian military forces that were south of the 44th parallel in Yugoslavia, Montenegro, and Kosovo. Shade the geographic area that NATO was trying to rid of Serbian forces.

Based on data from *Los Angeles Times* (March 24, 1999)

69. PRODUCTION PLANNING It costs a bakery \$3 to make a cake and \$4 to make a pie. If x represents the number of cakes made, and y represents the number of pies made, the graph of $3x + 4y \le 120$ shows the possible combinations of cakes and pies that can be produced so that costs do not exceed \$120 per day. Graph the inequality. Then find three possible combinations of pies and cakes that can be made so that the daily costs are not exceeded.

70. HIRING BABYSITTERS Mrs. Cansino has a choice of two babysitters. Sitter 1 charges \$6 per hour, and Sitter 2 charges \$7 per hour. If x represents the number of hours she uses Sitter 1 and y represents the number of hours she uses Sitter 2, the graph of $6x + 7y \le 42$ shows the possible ways she can hire the sitters and not spend more than \$42 per week. Graph the inequality. Then find three possible ways she can hire the babysitters so that her weekly budget for babysitting is not exceeded.

71. INVENTORIES A clothing store advertises that it maintains an inventory of at least \$4,400 worth of men's jackets at all times. At the store, leather jackets cost \$100 and nylon jackets cost \$88. If x represents the number of leather jackets in stock and y represents the number of nylon jackets in stock, the graph of $100x + 88y \ge 4{,}400$ shows the possible ways the jackets can be stocked. Graph the inequality. Then find three possible combinations of leather and nylon jackets so that the store lives up to its advertising claim.

72. MAKING SPORTING GOODS A sporting goods manufacturer allocates at least 2,400 units of production time per day to make baseballs and footballs. It takes 20 units of time to make a baseball and 30 units of time to make a football. If x represents the number of baseballs made and y represents the number of footballs made, the graph of $20x + 30y \ge 2{,}400$ shows the possible ways to schedule the production time. Graph the inequality. Then find three possible combinations of production time for the company to make baseballs and footballs.

WRITING

73. Explain how to decide which side of the boundary line to shade when graphing a linear inequality in two variables.

74. Why is the origin usually a good test point to choose when graphing a linear inequality?

75. Why is (0, 0) not an acceptable choice for a test point when graphing a linear inequality whose boundary passes through the origin?

76. Explain the difference between the graph of the solution set of $x + 1 > 8$, an inequality *in one variable,* and the graph of $x + y > 8$, an inequality *in two variables.*

REVIEW

77. Solve $A = P + Prt$ for t.

78. What is the sum of the measures of the three angles of any triangle?

79. Simplify: $40\left(\frac{3}{8}x - \frac{1}{4}\right) + 40\left(\frac{4}{5}\right)$.

80. Evaluate: $-4 + 5 - (-3) - 13$.

CHALLENGE PROBLEMS

81. Find a linear inequality that has the following graph.

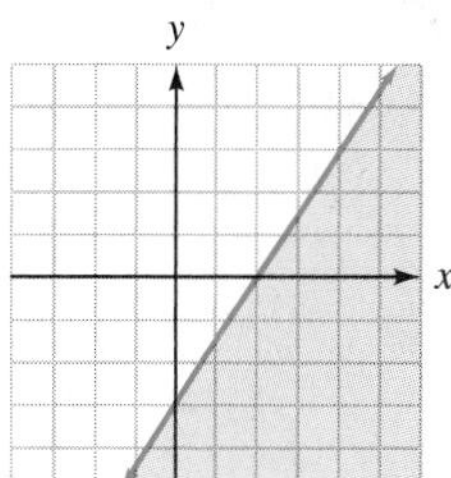

82. Graph the inequality: $4x - 3(x + 2y) \ge -6y$.

ACCENT ON TEAMWORK

MATCHING GAME

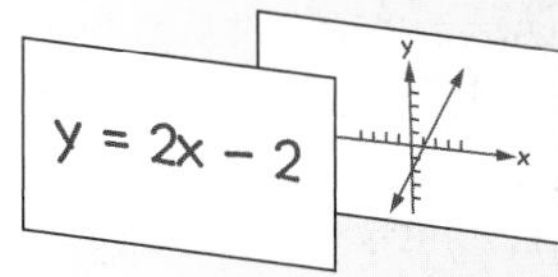

Overview: This activity is designed to improve your understanding of linear equations and their graphs.

Instructions: Form groups of 3 or 4 students. Then distribute ten 3 × 5 cards to every student in the class. Each student is to create a "deck" of equation–graph cards by following these steps.

1. On the first card, write a linear equation. (You may get ideas from the problems in the Study Sets in Sections 3.2, 3.3, and 3.5.) On the second card, draw the graph of that equation. Then, write "1" on the *backs* of those two cards to identify them as an equation–graph pair.
2. Create four more equation–graph pairs using the remaining cards. Label them pairs 2, 3, 4, and 5. When finished, shuffle your cards, and exchange decks with another member of your group.
3. Match each equation with its graph. To check your work, you can turn the cards over and compare the numbers on the back. If time allows, exchange decks of cards with another member of your group.

HEIGHT AND ARM SPAN

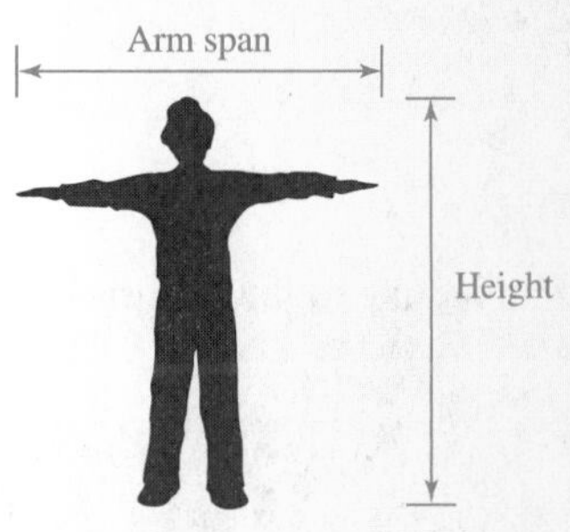

Overview: In this activity, you will explore the relationship between a person's height and arm span. Arm span is defined to be the distance between the tips of a person's fingers when their arms are held out to the side.

Instructions: Form groups of 5 or 6 students. Measure the height and arm span of each person in your group, and record the results in a table like the one shown below.

Name	Height (in.)	Arm span (in.)
1.		
2.		
3.		
4.		
5.		
6.		

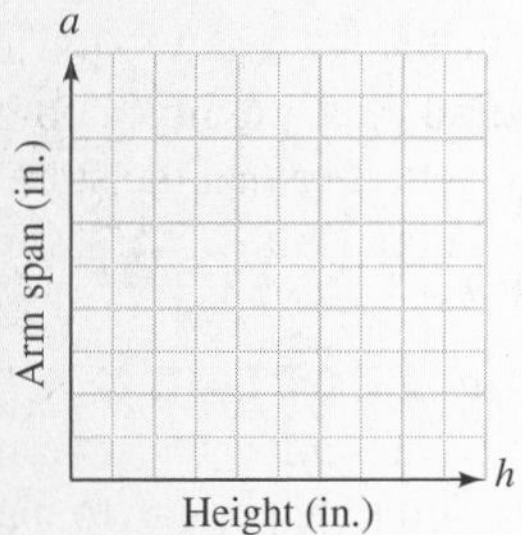

Plot the data in the table as ordered pairs of the form (height, arm span) on a graph like the one shown above. Then draw a straight-line model that best fits the data points.

Pick two convenient points on the line and find its slope. Use the point–slope form $a - a_1 = m(h - h_1)$ to find an equation of the line. Then, write the equation in slope–intercept form.

Ask a person from another group for his or her height measurement. Substitute that value into your linear model to predict that person's arm span. How close is your prediction to the person's actual arm span?

(From *Activities for Beginning and Intermediate Algebra* by Debbie Garrison, Judy Jones, and Jolene Rhodes)

KEY CONCEPT: DESCRIBING LINEAR RELATIONSHIPS

In Chapter 3, we discussed ways to mathematically describe linear relationships between two quantities using equations and graphs.

EQUATIONS IN TWO VARIABLES

The general form of the equation of a line is $Ax + By = C$. Two very useful forms of the equation of a line are the slope–intercept form and the point–slope form.

1. Write the equation of a line with a slope of -3 and a y-intercept of $(0, -4)$.
2. Write the equation of the line that passes through $(5, 2)$ and $(-5, 0)$. Give the answer in slope–intercept form.
3. CRICKETS The equation $T = \frac{1}{4}c + 40$ predicts the outdoor temperature T in degrees Fahrenheit using the number c of cricket chirps per minute. Find the temperature if a cricket chirps 160 times in one minute.
4. U.S. HEALTH CARE For the year 1990, the per capita health care expenditure was about \$2,660. Since then, the rate of increase has been about \$209 per year. Write a linear equation to model this. Let x represent the number of years since 1990 and let y represent the yearly per capita expenditure. Use your answer to predict the per capita expenditure in 2020. (Source: Centers for Medicare and Medicaid Services)

RECTANGULAR COORDINATE GRAPHS

The graph of an equation is a "picture" of all of its solutions. A thorough examination of a graph can yield a lot of useful information.

5. Complete the table of solutions for $2x - 4y = 8$. Then graph the equation.

$$2x - 4y = 8$$

x	y
0	
	0
-2	

6. a. What information does the y-intercept give?

   b. What is the slope of the line and what does it tell?

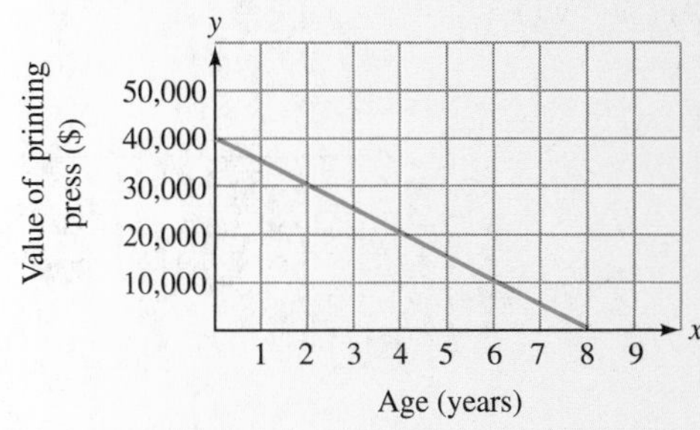

7. a. Find a point on the line.

   b. Determine the slope of the line.

   c. Write the equation of the line in slope–intercept form.

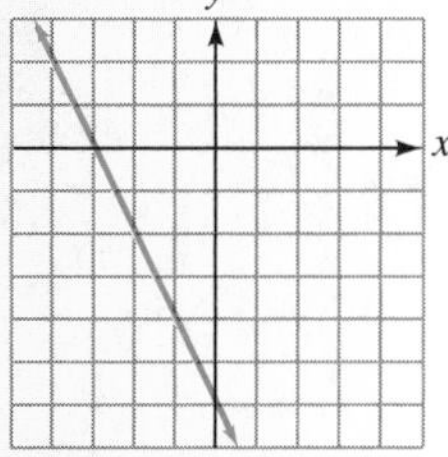

8. Write the equation of the line that passes through $(1, -1)$ and is parallel to the line graphed to the right.

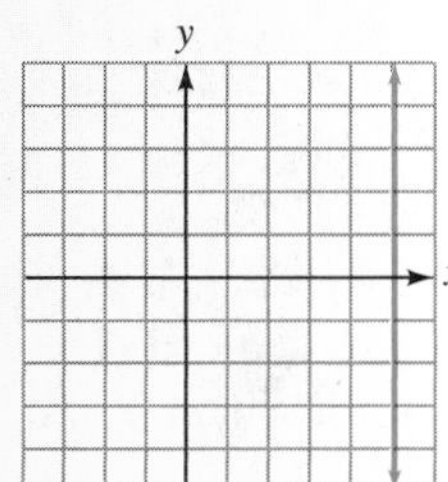

CHAPTER REVIEW

ELEMENTARY Algebra $f(x)$ Now™

SECTION 3.1 Graphing Using the Rectangular Coordinate System

CONCEPTS

A *rectangular coordinate system* is composed of a horizontal number line called the *x*-axis and a vertical number line called the *y*-axis.

To *graph* ordered pairs means to locate their position on a coordinate system.

The two axes divide the coordinate plane into four regions called *quadrants*.

REVIEW EXERCISES

1. Graph the points with coordinates $(-1, 3)$, $(0, 1.5)$, $(-4, -4)$, $\left(2, \frac{7}{2}\right)$, and $(4, 0)$.

2. HAWAIIAN ISLANDS Estimate the coordinates of Oahu using an ordered pair of the form (longitude, latitude).

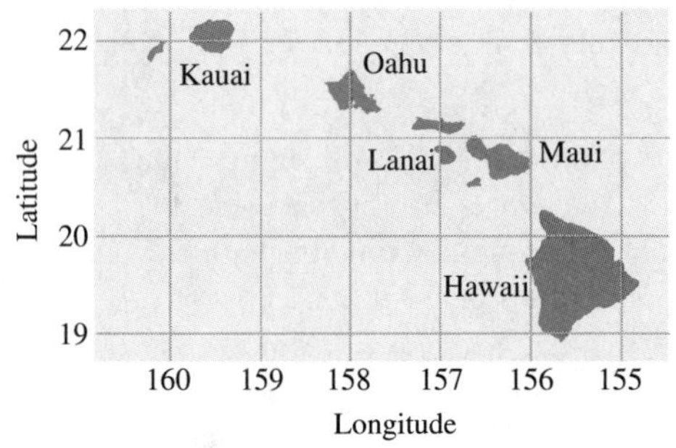

3. In what quadrant does the point $(-3, -4)$ lie?

4. What are the coordinates of the origin?

5. GEOMETRY Three vertices (corners) of a square are the points $(-5, 4)$, $(-5, -2)$, and $(1, -2)$. Find the coordinates of the fourth vertex and find the area of the square.

6. COLLEGE ENROLLMENTS The graph gives the number of students enrolled at a college for the period from 4 weeks before to 5 weeks after the semester began.

a. What was the maximum enrollment and when did it occur?

b. How many students had enrolled 2 weeks before the semester began?

c. When was enrollment 2,250?

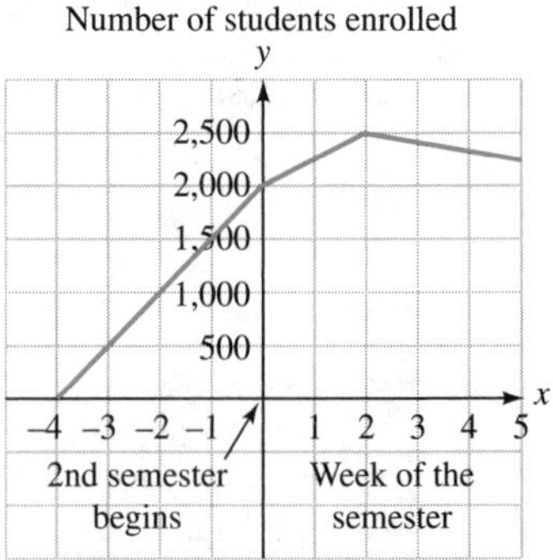

SECTION 3.2 Graphing Linear Equations

A *solution* of an equation in two variables is an ordered pair of numbers that makes the equation a true statement.

An equation whose graph is a straight line and whose variables are raised to the first power is called a *linear equation.*

The *general* or *standard form* of a linear equation is $Ax + By = C$, where A, B, and C are real numbers and A and B are not both zero.

To graph a linear equation solved for y:

1. Find three solutions by selecting three x-values and finding the corresponding y-values.
2. Plot each ordered-pair solution.
3. Draw a straight line through the points.

7. Is $(-3, -2)$ a solution of $y = 2x + 4$?

8. Complete the table of solutions for $3x + 2y = -18$.

x	y	(x, y)
-2		
	3	

9. Which of the following equations are not linear equations?

$8x - 2y = 6 \qquad y = x^2 + 1 \qquad y = x \qquad 3y = -x + 4 \qquad y - x^3 = 0$

Graph each equation by constructing a table of solutions.

10. $y = 4x - 2$

11. $5y = -5x + 15$ (Solve for y first.)

12. The graph of a linear equation is shown here.

a. When the coordinates of point A are substituted into the equation, will a true or false statement result?

b. When the coordinates of point B are substituted into the equation, will a true or false statement result?

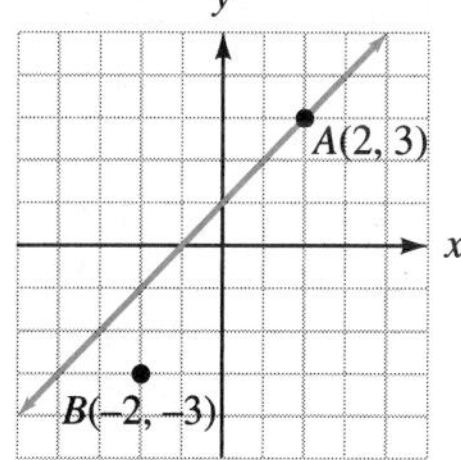

13. BIRTHDAY PARTIES A restaurant offers a party package for children that includes everything: food, drinks, cake, and party favors. The cost c, in dollars, is given by the linear equation $c = 8n + 50$, where n is the number of children attending the party. Graph the equation and use the graph to estimate the cost of a party if 18 children attend.

SECTION 3.3 More about Graphing Linear Equations

The point where a line intersects the x-axis is called the *x-intercept.* The point where a line intersects the y-axis is called the *y-intercept.*

14. Identify the x- and y-intercepts of the graph.

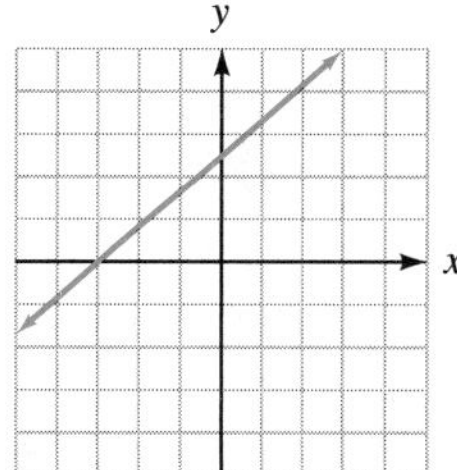

To find the y-intercept, substitute 0 for x in the given equation and solve for y. To find the x-intercept, substitute 0 for y in the given equation and solve for x.

The equation $y = b$ represents the horizontal line that intersects the y-axis at $(0, b)$. The equation $x = a$ represents the vertical line that intersects the x-axis at $(a, 0)$.

15. Graph $-4x + 2y = 8$ by finding its x- and y-intercepts.

16. Graph: $y = 4$.

17. Graph: $x = -1$.

18. DEPRECIATION The graph shows how the value of some sound equipment decreased over the years. Find the intercepts of the graph. What information do the intercepts give about the equipment?

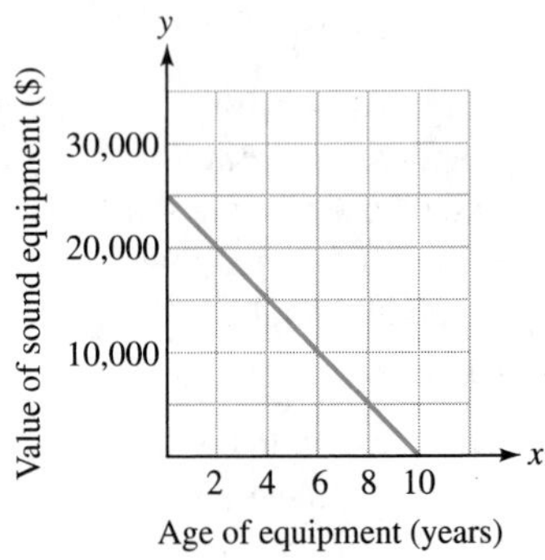

SECTION 3.4 The Slope of a Line

The *slope m* of a line is a number that measures "steepness" by finding the ratio $\frac{\text{rise}}{\text{run}}$.

$$m = \frac{\text{change in the } y\text{-values}}{\text{change in the } x\text{-values}}$$

Lines that rise from left to right have a *positive slope,* and lines that fall from left to right have a *negative slope.* Horizontal lines have a slope of zero. Vertical lines have *undefined* slope.

If (x_1, y_1) and (x_2, y_2) are two points on a nonvertical line, the slope m of the line is

$$m = \frac{y_2 - y_1}{x_2 - x_1}$$

In each case, find the slope of the line.

19.

20.

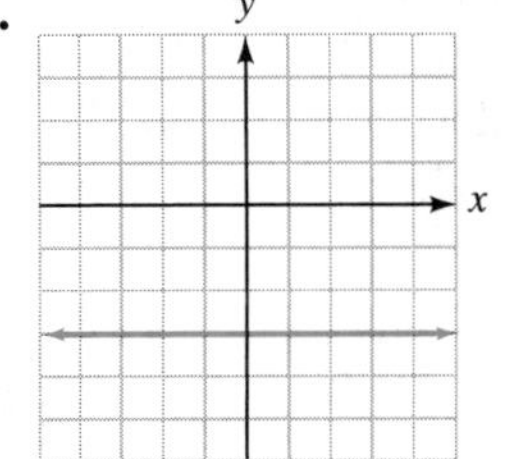

21. The line with the table of solutions shown below

x	y	(x, y)
2	-3	$(2, -3)$
4	-17	$(4, -17)$

22. The line passing through points $(1, -4)$ and $(3, -7)$

The *pitch* of a roof is its slope.

The *grade* of an incline or decline is its slope expressed as a percent.

When units are attached to a slope it becomes a *rate of change.*

23. CARPENTRY Trusses like the one shown here will be used to construct the roof of a shed. Find the pitch of the roof.

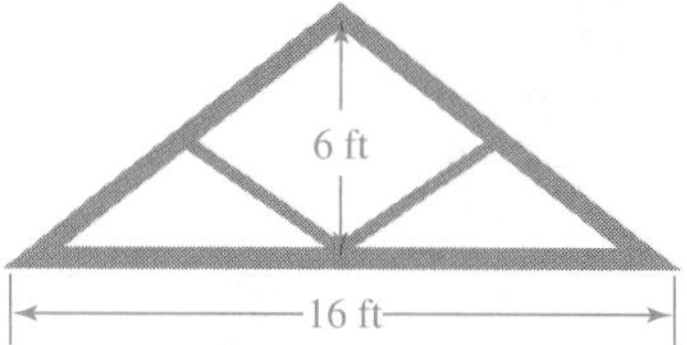

24. HANDICAP ACCESSIBILITY Find the grade of the ramp. Round to the nearest tenth of a percent.

25. TOURISM The graph shows the number of international travelers to the United States from 1986 to 2000, in two-year increments.

a. Between 1992 and 1994 the largest decline in the number of visitors occurred. What was the rate of change?

b. Between 1986 and 1988 the largest increase in the number of visitors occurred. What was the rate of change?

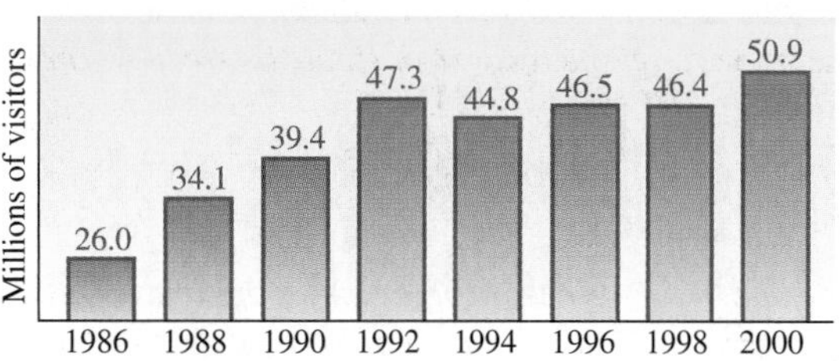

Based on data from *World Almanac* 2003.

SECTION 3.5 Slope–Intercept Form

If a linear equation is written in *slope–intercept* form,

$$y = mx + b$$

the graph of the equation is a line with slope m and y-intercept $(0, b)$.

Find the slope and the y-intercept of each line.

26. $y = \frac{3}{4}x - 2$

27. $y = -4x$

28. $y = \frac{x}{8} + 10$

29. $7x + 5y = -21$

30. Write an equation of the line with slope -6 and y-intercept $(0, 4)$.

31. Find the slope and the y-intercept of the line whose equation is $9x - 3y = 15$. Then graph it.

The *rate of change* is the slope of the graph of a linear equation.

32. COPIERS A business buys a used copy machine that, when purchased, has already produced 75,000 copies.

a. If the business plans to run 300 copies a week, write a linear equation that would find the number of copies c the machine has made in its lifetime after the business has used it for w weeks.

b. Use your result in part a to predict the total number of copies that will have been made on the machine 1 year, or 52 weeks, after being purchased by the business.

Two lines with the same slope are *parallel.*

The product of the slopes of *perpendicular* lines is -1.

33. Without graphing, determine whether graphs of the given pairs of lines would be parallel, perpendicular, or neither.

a. $y = -\frac{2}{3}x + 6$

$y = -\frac{2}{3}x - 6$

b. $x + 5y = -10$

$y = 5x$

SECTION 3.6 Point–Slope Form

If a line with slope m passes through the point (x_1, y_1), the equation of the line in *point–slope* form is

$$y - y_1 = m(x - x_1)$$

Write an equation of a line with the given slope that passes through the given point. Give the answer in slope–intercept form and graph the equation.

34. $m = 3, (1, 5)$

35. $m = -\frac{1}{2}, (-4, -1)$

Write an equation of the line with the following characteristics. Give the answer in slope–intercept form.

36. passing through $(3, 7)$ and $(-6, 1)$

37. horizontal, passing through $(6, -8)$

38. CAR REGISTRATION When it was 2 years old, the annual registration fee for a Dodge Caravan was \$380. When it was 4 years old, the registration fee dropped to \$310. If the relationship is linear, write an equation that gives the registration fee f in dollars for the van when it is x years old.

SECTION 3.7 Graphing Linear Inequalities

An ordered pair (x, y) is a *solution* of an inequality in x and y if a true statement results when the variables are replaced by the coordinates of the ordered pair.

39. Determine whether each ordered pair is a solution of $2x - y \leq -4$.

a. $(0, 5)$

b. $(2, 8)$

c. $(-3, -2)$

d. $\left(\frac{1}{2}, -5\right)$

To graph a linear inequality:

1. Graph the *boundary line.* Draw a solid line if the inequality contains ≤ or ≥ and a dashed line if it contains < or >.
2. Pick a *test point* on one side of the boundary. Use the origin if possible. Replace x and y with the coordinates of that point. If the inequality is satisfied, shade the side that contains the point. If the inequality is not satisfied, shade the other side.

Graph each inequality.

40. $x - y < 5$

41. $2x - 3y \geq 6$

42. $y \leq -2x$

43. $y < -4$

44. The graph of a linear inequality is shown. Would a true or a false statement result if the coordinates of

a. point A were substituted into the inequality?

b. point B were substituted into the inequality?

c. point C were substituted into the inequality?

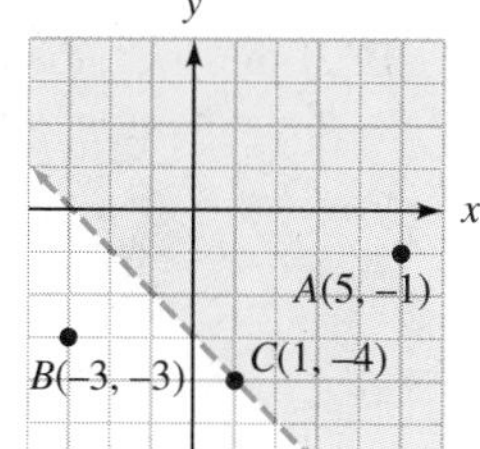

45. WORK SCHEDULES A student told her employer that during the school year, she would be available for up to 30 hours a week, working either 3- or 5-hour shifts. If x represents the number of 3-hour shifts she works and y represents the number of 5-hour shifts she works, the inequality $3x + 5y \leq 30$ shows the possible combinations of shifts she can work. Graph the inequality, then find three possible combinations.

46. Explain the difference between an equation and an inequality.

CHAPTER 3 TEST  ELEMENTARY Algebra *f(x)* Now™

The graph shows the number of dogs being boarded in a kennel over a 3-day holiday weekend.

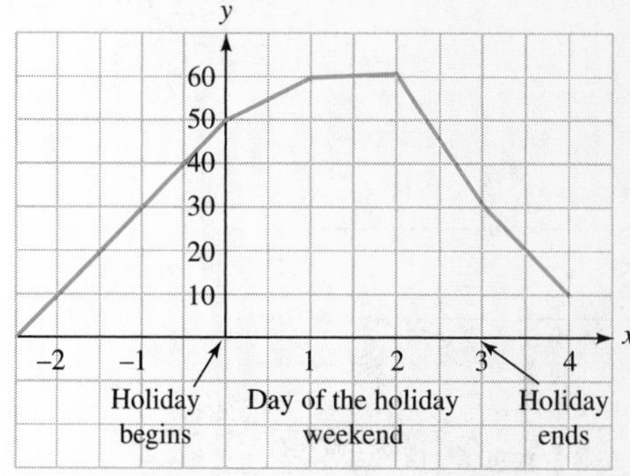

1. How many dogs were in the kennel 2 days before the holiday?

2. What is the maximum number of dogs that were boarded on the holiday weekend?

3. When were there 30 dogs in the kennel?

4. What information does the y-intercept of the graph give?

5. Draw a rectangular coordinate system and label each quadrant.

6. Complete the table of solutions for the linear equation.

$$x + 4y = 6$$

x	y	(x, y)
2		
	3	

7. Is $(-3, -4)$ a solution of $3x - 4y = 7$?

8. The graph of a linear equation is shown below.

a. If the coordinates of point C are substituted into the equation, will the result be true or false?

b. If the coordinates of point D are substituted into the equation, will the result be true or false?

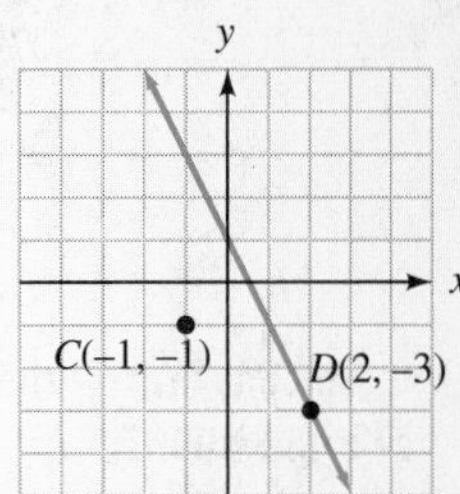

9. Graph: $y = \dfrac{x}{3}$.

10. Graph: $8x + 4y = -24$.

11. What are the x- and y-intercepts of the graph of $2x - 3y = 6$?

12. Find the slope of the line.

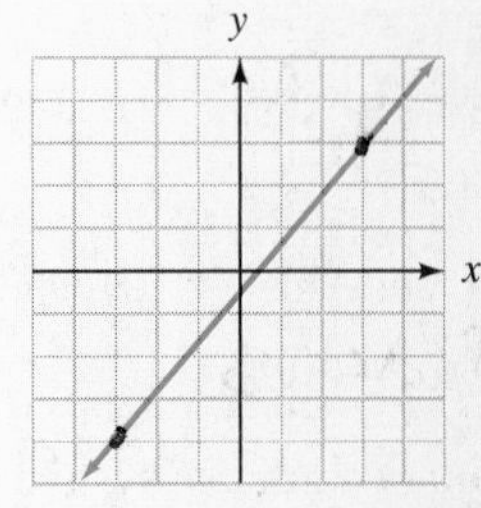

13. Find the slope of the line passing through $(-1, 3)$ and $(3, -1)$.

14. What is the slope of a vertical line?

15. What is the slope of a line that is perpendicular to a line with slope $-\frac{7}{8}$?

16. When graphed, are the lines $y = 2x + 6$ and $y = 2x$ parallel, perpendicular, or neither?

In Problems 17 and 18, refer to the illustration at the bottom of the page that shows the elevation changes in a 26-mile marathon course.

17. Find the rate of change of the decline on which the woman is running.

18. Find the rate of change of the incline on which the man is running.

19. Graph: $x = -4$.

20. Graph the line passing through $(-2, -4)$ having a slope of $\frac{2}{3}$.

21. Find the slope and the y-intercept of $x + 2y = 8$.

22. Write an equation of the line passing through $(-2, 5)$ with slope 7. Give the answer in slope–intercept form.

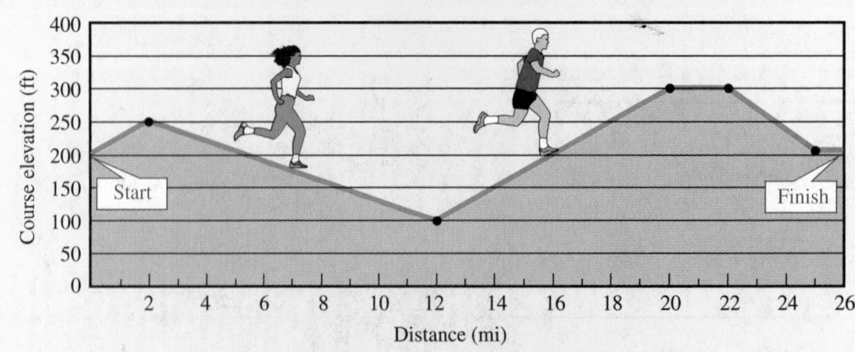

23. DEPRECIATION After it is purchased, a \$15,000 computer loses \$1,500 in resale value every year. Write a linear equation that gives the resale value v of the computer x years after being purchased.

24. Determine whether (6, 1) is a solution of $2x - 4y \geq 8$.

25. WATER HEATERS The scatter diagram shows how excessively high temperatures affect the life of a water heater. Write an equation of the line that models the data for water temperatures between 140° and 180°. Let T represent the temperature of the water in degrees Fahrenheit and y represent the expected life of the heater in years. Give the answer in slope–intercept form.

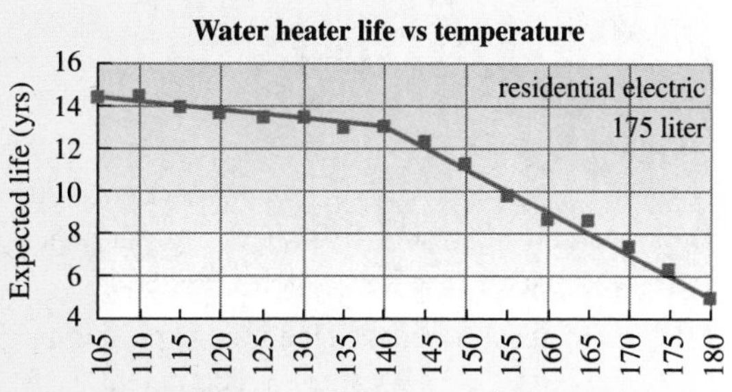

Source: www.uniongas.com/WaterHeating

26. Graph the inequality $x - y > -2$.

CHAPTERS 1–3 CUMULATIVE REVIEW EXERCISES

1. UNITED AIRLINES On May 2, 2003, UAL Corporation, the parent company of United Airlines, announced its 11th straight quarterly loss. (See the graph below.)

a. During this time span, in which quarter was the loss the greatest?

b. Estimate the corporation's losses for the year 2002.

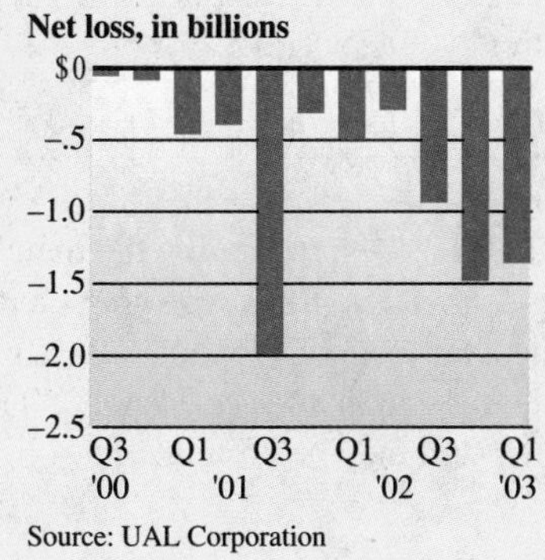

Source: UAL Corporation

2. Give the prime factorization of 108.

3. Write $\frac{1}{250}$ as a decimal.

4. Determine whether each statement is true or false.

a. Every whole number is an integer.

b. Every integer is a real number.

c. 0 is a whole number, an integer, and a rational number.

5. PENNIES A 2002 telephone survey of adults asked whether the penny should be discontinued from the national currency. The results are shown in the circle graph. If 781 people favored keeping the penny, how many took part in the survey?

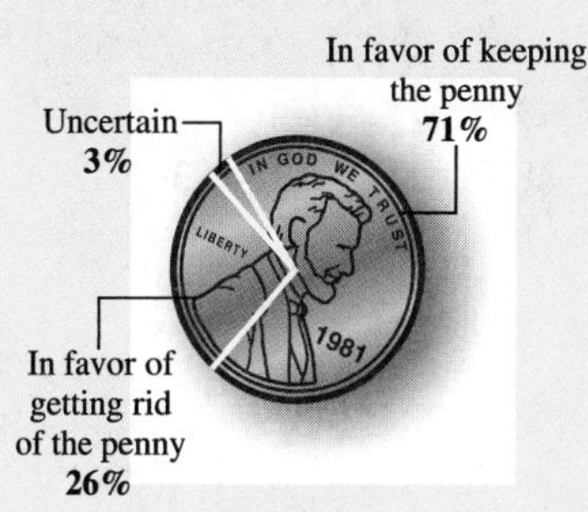

Based on data from Coinstar

6. Evaluate: $\left|\frac{(6-5)^4-(-21)}{-27+4^2}\right|$.

7. Evaluate $b^2 - 4ac$ for $a = 2$, $b = -8$, and $c = 4$.

8. Suppose x sheets from a 500-sheet ream of paper have been used. How many sheets are left?

9. How many terms does the algebraic expression $3x^2 - 2x + 1$ have? What is the coefficient of the second term?

10. Use the distributive property to remove parentheses.
    a. $2(x + 4)$ b. $2(x - 4)$
    c. $-2(x + 4)$ d. $-2(x - 4)$

Simplify each expression.

11. $5a + 10 - a$
12. $-7(9t)$
13. $-2b^2 + 6b^2$
14. $5(-17)(0)(2)$
15. $(a + 2) - (a - 2)$
16. $-4(-5)(-8a)$
17. $-y - y - y$
18. $\frac{3}{2}(4x - 8) + x$

Solve each equation.

19. $3x - 5 = 13$
20. $1.2 - x = -1.7$
21. $\frac{2x}{3} - 2 = 4$
22. $\frac{y-2}{7} = -3$
23. $-3(2y - 2) - y = 5$
24. $9y - 3 = 6y$
25. $\frac{1}{3} + \frac{c}{5} = -\frac{3}{2}$
26. $5(x + 2) = 5x - 2$
27. $-x = -99$
28. $3c - 2 = \frac{11(c-1)}{5}$

29. HIGH HEELS Find x.

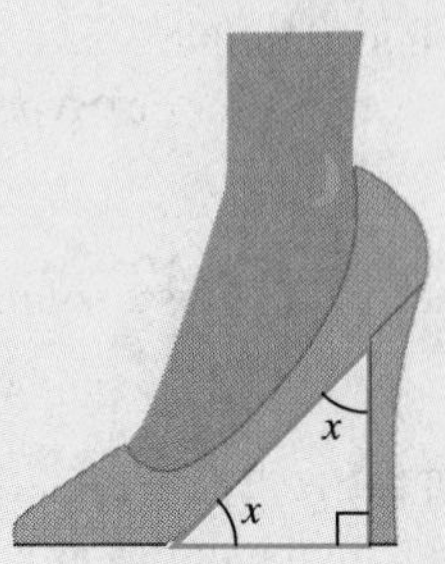

30. Solve for h:

$$S = 2\pi rh + 2\pi r^2$$

31. Find the perimeter and the area of the gauze pad of the bandage.

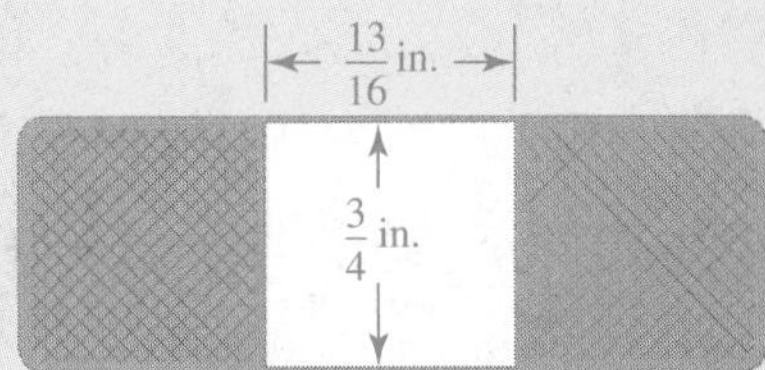

32. Two sides of an isosceles triangle are 3 feet and 4 feet long. What are the possible perimeters of the triangle?

33. Complete the table.

	% acid	Liters	Amount of acid
50% solution	0.50	x	
25% solution	0.25	$13 - x$	
30% mixture	0.30	13	

34. ROAD TRIPS A bus, carrying the members of a marching band, and a truck, carrying their instruments, leave a high school at the same time. The bus travels at 60 mph and the truck at 50 mph. In how many hours will they be 75 miles apart?

35. MIXING CANDY Candy corn worth \$1.90 per pound is to be mixed with black gumdrops that cost \$1.20 per pound to make 200 pounds of a mixture worth \$1.48 per pound. How many pounds of each candy should be used?

Solve each inequality. Write the solution set in interval notation and graph it.

36. $-\frac{3}{16}x \geq -9$

37. $8x + 4 > 3x + 4$

38. Is $(-2, 4)$ a solution of $y = 2x - 8$?

Graph each equation.

39. $y = x$

40. $4y + 2x = -8$

41. What is the slope of the graph of the line $y = 5$?

42. What is the slope of the line passing through $(-2, 4)$ and $(5, -6)$?

43. ROOFING What is the pitch of the roof?

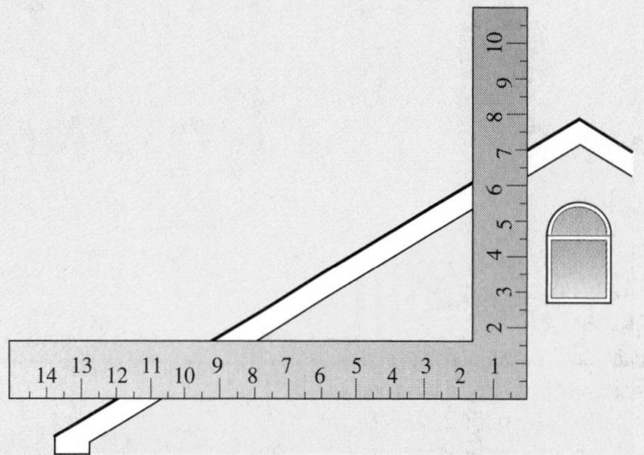

44. Find the slope and the y-intercept of the graph of the line described by $4x - 6y = -12$.

45. Write an equation of the line that has slope -2 and y-intercept $(0, 1)$.

46. Write an equation of the line that has slope $-\frac{7}{8}$ and passes through $(2, -9)$. Express the answer in point–slope form.

47. Is $(-6, 0)$ a solution of $y \geq -x - 6$?

48. Graph the inequality $x < 4$ on a rectangular coordinate system.

Chapter

4 Exponents and Polynomials

ELEMENTARY **Algebra $f(x)$ Now™**

Throughout the chapter, this icon introduces resources on the Elementary AlgebraNow Web site, accessed through **http://1pass.thomson.com**, that will

- Help you test your knowledge of the material with a pre-test and a post-test
- Provide a personalized learning plan targeting areas you should study

4.1 Rules for Exponents

4.2 Zero and Negative Exponents

4.3 Scientific Notation

4.4 Polynomials

4.5 Adding and Subtracting Polynomials

4.6 Multiplying Polynomials

4.7 Special Products

4.8 Division of Polynomials

Accent on Teamwork

Key Concept

Chapter Review

Chapter Test

Cumulative Review Exercises

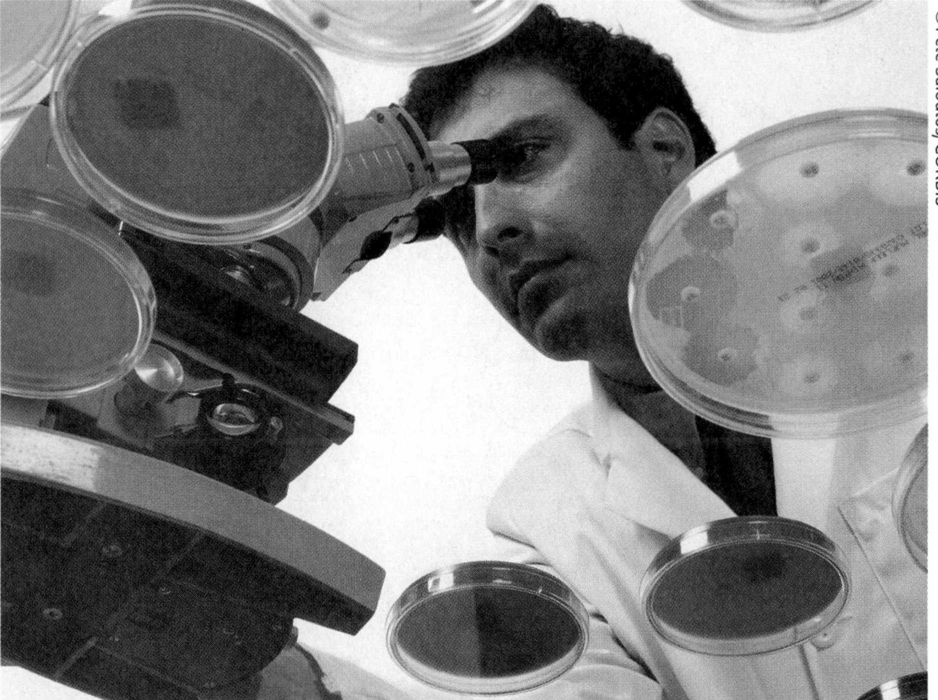

Under certain conditions, bacteria can grow at an incredible rate. To model the growth, scientists use exponents. In this chapter, we will discuss several rules for simplifying expressions involving exponents. We will also see the role exponents play as part of a compact notation that is used to represent extremely large numbers, such as the distance from the Earth to the sun, and extremely small numbers, such as the mass of a proton.

To learn more about exponents visit *The Learning Equation* on the Internet at http://tle.brookscole.com. (The log-in instructions are in the Preface.) For Chapter 4, the following online lessons are:

- *TLE* Lesson 8: The Product Rule for Exponents
- *TLE* Lesson 9: The Quotient Rule for Exponents
- *TLE* Lesson 10: Negative Exponents

In this chapter, we introduce the rules for exponents and use them to add, subtract, multiply, and divide polynomials.

4.1 Rules for Exponents

- Multiplying Exponential Expressions That Have Like Bases
- Dividing Exponential Expressions That Have Like Bases
- Raising an Exponential Expression to a Power
- Powers of Products and Quotients

Recall that an **exponent** indicates repeated multiplication. It tells how many times the **base** is used as a factor. For example, 3^5 represents the product of five 3's.

$$\underset{\text{base}}{3}^{\overset{\text{exponent}}{5}} = \overbrace{3 \cdot 3 \cdot 3 \cdot 3 \cdot 3}^{\text{5 factors of 3}}$$

In general, we have the following definition for exponents that are natural numbers: 1, 2, 3, 4, 5, and so on.

Natural-Number Exponents

A natural-number exponent tells how many times its base is to be used as a factor. For any number x and any natural number n,

$$x^n = \overbrace{x \cdot x \cdot x \cdot \cdots \cdot x}^{n \text{ factors of } x}$$

Notation

When an exponent is written outside parentheses, the expression within the parentheses is the base.

$(\underset{\text{base}}{-2s})^3 \leftarrow$ exponent

Expressions of the form x^n are called **exponential expressions.** The base of an exponential expression can be a number, a variable, or a combination of numbers and variables. Some examples are:

$10^5 = 10 \cdot 10 \cdot 10 \cdot 10 \cdot 10$ Read 10^5 as "10 to the fifth power."

$y^2 = y \cdot y$ Read y^2 as "y to the second power" or "y squared."

$(-2s)^3 = (-2s)(-2s)(-2s)$ Read $(-2s)^3$ as "the quantity of $-2s$ cubed."

$-8^4 = -(8 \cdot 8 \cdot 8 \cdot 8)$ Read -8^4 as "the opposite of 8 to the fourth power."

In this section, we continue our study of exponents as we discuss how to simplify exponential expressions that are multiplied, divided, and raised to powers.

MULTIPLYING EXPONENTIAL EXPRESSIONS THAT HAVE LIKE BASES

To develop a rule for multiplying exponential expressions that have the same base, we consider the product $6^2 \cdot 6^3$. Since 6^2 means that 6 is to be used as a factor two times, and 6^3 means that 6 is to be used as a factor three times, we have

$$6^2 \cdot 6^3 = \overbrace{6 \cdot 6}^{\text{2 factors of 6}} \cdot \overbrace{6 \cdot 6 \cdot 6}^{\text{3 factors of 6}}$$
$$= \overbrace{6 \cdot 6 \cdot 6 \cdot 6 \cdot 6}^{\text{5 factors of 6}}$$
$$= 6^5$$

We can quickly find this result if we keep the common base 6 and add the exponents on 6^2 and 6^3.

$$6^2 \cdot 6^3 = 6^{2+3} = 6^5$$

This example suggests the following rule for exponents.

Product Rule for Exponents

To multiply exponential expressions with the same base, keep the common base and add the exponents. For any number x and any natural numbers m and n,

$$x^m \cdot x^n = x^{m+n}$$

EXAMPLE 1

Simplify: **a.** $9^5(9^6)$, **b.** $x^3 \cdot x^4$, **c.** y^2y^4y, **d.** $(x + 2)^8(x + 2)^7$, and **e.** $(c^2d^3)(c^4d^5)$.

ELEMENTARY Algebra f(x) Now™

Solution **a.** To simplify $9^5(9^6)$ means to write it in an equivalent form using one base and one exponent.

$$9^5(9^6) = 9^{5+6} \quad \text{Keep the common base, which is 9, and add the exponents.}$$
$$= 9^{11}$$

Notation

An exponent of 1 means the base is to be used as a factor 1 time. For example, $y^1 = y$.

b. $x^3 \cdot x^4 = x^{3+4}$ Keep the common base x and add the exponents.
$= x^7$

c. $y^2y^4y = y^{2+4}y$ Working from left to right, keep the common base y and add the exponents.
$= y^6y$ Do the addition.
$= y^6y^1$ Write y as y^1.
$= y^{6+1}$ Keep the common base y and add the exponents.
$= y^7$

d. $(x + 2)^8(x + 2)^7 = (x + 2)^{8+7}$ Keep the common base $x + 2$ and add the exponents.
$= (x + 2)^{15}$

e. $(c^2d^3)(c^4d^5) = (c^2c^4)(d^3d^5)$ Group like bases together.
$= (c^{2+4})(d^{3+5})$ Keep the common base c and add the exponents. Keep the common base d and add the exponents.
$= c^6d^8$

Self Check 1 Simplify: **a.** $7^8(7^7)$, **b.** $z \cdot z^3$, **c.** $x^2x^3x^6$, **d.** $(y - 1)^5(y - 1)^5$, **e.** $(s^4t^3)(s^4t^4)$.

DIVIDING EXPONENTIAL EXPRESSIONS THAT HAVE LIKE BASES

To develop a rule for dividing exponential expressions that have the same base, we consider the quotient

$$\frac{4^5}{4^2}$$

where the exponent in the numerator is greater than the exponent in the denominator. We can simplify this fraction as follows:

$$\frac{4^5}{4^2} = \frac{4 \cdot 4 \cdot 4 \cdot 4 \cdot 4}{4 \cdot 4}$$

$$= \frac{\overset{1}{\cancel{4}} \cdot \overset{1}{\cancel{4}} \cdot 4 \cdot 4 \cdot 4}{\underset{1}{\cancel{4}} \cdot \underset{1}{\cancel{4}}}$$ Remove the common factors of 4 in the numerator and denominator.

$$= 4^3$$

We can quickly find this result if we keep the common base 4 and subtract the exponents on 4^5 and 4^2.

$$\frac{4^5}{4^2} = 4^{5-2} = 4^3$$

This example suggests another rule for exponents.

Quotient Rule for Exponents

To divide exponential expressions with the same base, keep the common base and subtract the exponents. For any nonzero number x and any natural numbers m and n, where $m > n$,

$$\frac{x^m}{x^n} = x^{m-n}$$

EXAMPLE 2

ELEMENTARY Algebra f(x) Now™

Simplify each expression. Assume that there are no divisions by 0. **a.** $\frac{20^{16}}{20^9}$, **b.** $\frac{x^9}{x^3}$, **c.** $\frac{(75n)^{12}}{(75n)^{11}}$, and **d.** $\frac{a^3b^8}{ab^5}$.

Solution

a. To simplify $\frac{20^{16}}{20^9}$ means to write it in an equivalent form using one base and one exponent.

$$\frac{20^{16}}{20^9} = 20^{16-9}$$ Keep the common base, which is 20, and subtract the exponents.

$$= 20^7$$

b. $\frac{x^9}{x^3} = x^{9-3}$ Keep the common base x and subtract the exponents.

$$= x^6$$

c. $\frac{(75n)^{12}}{(75n)^{11}} = (75n)^{12-11}$ Keep the common base $75n$ and subtract the exponents.

$$= (75n)^1$$

$$= 75n$$ Any number raised to the first power is simply that number.

d. $\frac{a^3b^8}{ab^5} = \frac{a^3}{a} \cdot \frac{b^8}{b^5}$ Group the common bases together.

$$= a^{3-1}b^{8-5}$$ Keep the common base a and subtract the exponents. Keep the common base b and subtract the exponents.

$$= a^2b^3$$ Do the subtractions.

Self Check 2 Simplify: **a.** $\frac{55^{30}}{55^5}$, **b.** $\frac{a^5}{a^3}$, **c.** $\frac{(8t)^8}{(8t)^7}$, **d.** $\frac{b^{15}c^4}{b^4c}$.

EXAMPLE 3

Simplify: $\frac{a^3a^5a^7}{a^4a}$.

ELEMENTARY Algebra f(x) Now™

Solution We simplify the numerator and denominator separately and proceed as follows.

$$\frac{a^3a^5a^7}{a^4a} = \frac{a^{15}}{a^5}$$ In the numerator, keep the common base a and add the exponents. In the denominator, keep the common base a and add the exponents.

$$= a^{15-5}$$ Keep the common base a and subtract the exponents.

$$= a^{10}$$

Self Check 3 Simplify: $\frac{b^2b^6b}{b^4b^4}$.

Caution Recall that like terms are terms with exactly the same variables raised to exactly the same powers. To add or subtract exponential expressions, they must be like terms. To multiply or divide exponential expressions, only the bases need to be the same.

$x^5 + x^2$ They are not like terms; the exponents are different. We cannot add.

$x^2 + x^2 = 2x^2$ They are like terms; we can add. Recall that $x^2 = 1x^2$.

$x^5 \cdot x^2 = x^7$ The bases are the same; we can multiply.

$\frac{x^5}{x^2} = x^3$ The bases are the same; we can divide.

RAISING AN EXPONENTIAL EXPRESSION TO A POWER

To develop another rule for exponents, we consider $(5^3)^4$. Here, an exponential expression, 5^3, is raised to a power. Since 5^3 is the base and 4 is the exponent, $(5^3)^4$ can be written as $5^3 \cdot 5^3 \cdot 5^3 \cdot 5^3$. Because each of the four factors of 5^3 contains three factors of 5, there are $4 \cdot 3$ or 12 factors of 5.

The Language of Algebra

An exponential expression raised to a power, such as $(5^3)^4$, is also called a *power of a power.*

$$(5^3)^4 = 5^3 \cdot 5^3 \cdot 5^3 \cdot 5^3$$

$$= \overbrace{\underbrace{5 \cdot 5 \cdot 5}_{5^3} \cdot \underbrace{5 \cdot 5 \cdot 5}_{5^3} \cdot \underbrace{5 \cdot 5 \cdot 5}_{5^3} \cdot \underbrace{5 \cdot 5 \cdot 5}_{5^3}}^{\text{12 factors of 5}}$$

$$= 5^{12}$$

We can quickly find this result if we keep the common base 5 and multiply the exponents.

$$(5^3)^4 = 5^{3 \cdot 4} = 5^{12}$$

This example suggests the following rule for exponents.

Power Rule for Exponents To raise an exponential expression to a power, keep the base and multiply the exponents. For any number x and any natural numbers m and n,

$$(x^m)^n = x^{mn}$$

EXAMPLE 4

Simplify: **a.** $(2^3)^7$ and **b.** $(z^8)^8$.

Solution **a.** To simplify $(2^3)^7$ means to write it in an equivalent form using one base and one exponent.

$$(2^3)^7 = 2^{3\cdot 7} \quad \text{Keep the base and multiply the exponents.}$$
$$= 2^{21}$$

b. $$(z^8)^8 = z^{8\cdot 8} \quad \text{Keep the base and multiply the exponents.}$$
$$= z^{64}$$

Self Check 4 Simplify: **a.** $(4^6)^5$, **b.** $(y^5)^2$.

EXAMPLE 5

Simplify: **a.** $(x^2x^5)^2$ and **b.** $(z^2)^4(z^3)^3$.

ELEMENTARY Algebra f(x) Now™

Solution **a.** We begin by using the product rule for exponents. Then we use the power rule.

$$(x^2x^5)^2 = (x^7)^2 \quad \text{Within the parentheses, keep the base and add the exponents.}$$
$$= x^{14} \quad \text{Keep the base and multiply the exponents.}$$

b. We begin by using the power rule for exponents twice. Then we use the product rule.

$$(z^2)^4(z^3)^3 = z^8z^9 \quad \text{For each power of } z \text{ raised to a power, keep the base and multiply the exponents.}$$
$$= z^{17} \quad \text{Keep the common base } z \text{ and add the exponents.}$$

Self Check 5 Simplify: **a.** $(a^4a^3)^3$, **b.** $(a^3)^3(a^4)^2$.

POWERS OF PRODUCTS AND QUOTIENTS

To develop more rules for exponents, we consider the expression $(2x)^3$, which is a *power of the product* of 2 and x, and the expression $\left(\frac{2}{x}\right)^3$, which is a *power of the quotient* of 2 and x.

$$(2x)^3 = (2x)(2x)(2x)$$
$$= (2\cdot 2\cdot 2)(x\cdot x\cdot x)$$
$$= 2^3x^3$$
$$= 8x^3$$

$$\left(\frac{2}{x}\right)^3 = \left(\frac{2}{x}\right)\left(\frac{2}{x}\right)\left(\frac{2}{x}\right) \quad \text{Assume } x \neq 0.$$
$$= \frac{2\cdot 2\cdot 2}{x\cdot x\cdot x} \quad \text{Multiply the numerators. Multiply the denominators.}$$
$$= \frac{2^3}{x^3}$$
$$= \frac{8}{x^3}$$

These examples suggest the following rules for exponents.

Powers of a Product and a Quotient

To raise a product to a power, raise each factor of the product to that power. To raise a quotient to a power, raise the numerator and the denominator to that power. For any numbers x and y, and any natural number n,

$$(xy)^n = x^ny^n \quad \text{and} \quad \left(\frac{x}{y}\right)^n = \frac{x^n}{y^n}, \quad \text{where } y \neq 0$$

EXAMPLE 6

Simplify: **a.** $(3c)^3$, **b.** $(x^2y^3)^5$, and **c.** $(-2a^3b)^2$.

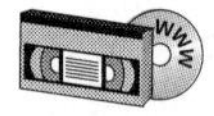

ELEMENTARY Algebra $f(x)$ Now™

Caution

There is no rule for the power of a sum or power of a difference. To show why, consider this example:

$$(3+2)^2 \stackrel{?}{=} 3^2 + 2^2$$
$$5^2 \stackrel{?}{=} 9 + 4$$
$$25 \neq 13$$

Solution **a.** Since $3c$ is the product of 3 and c, the expression $(3c)^3$ is a power of a product.

$$(3c)^3 = 3^3c^3 \quad \text{Raise each factor of the product } 3c \text{ to the 3rd power.}$$
$$= 27c^3 \quad \text{Evaluate } 3^3.$$

b. $(x^2y^3)^5 = (x^2)^5(y^3)^5$ Raise each factor of the product x^2y^3 to the 5th power.

$= x^{10}y^{15}$ For each power of a power, keep the base and multiply the exponents.

c. $(-2a^3b)^2 = (-2)^2(a^3)^2b^2$ Raise each of the three factors of the product $-2a^3b$ to the 2nd power.

$= 4a^6b^2$ Evaluate $(-2)^2$. Keep the base a and multiply the exponents.

Self Check 6 Simplify: **a.** $(2t)^4$, **b.** $(c^3d^4)^6$, **c.** $(-3ab^5)^3$.

EXAMPLE 7

ELEMENTARY Algebra $f(x)$ Now™

Simplify: $\dfrac{(a^3b^4)^2}{ab^5}$.

Solution

$\dfrac{(a^3b^4)^2}{ab^5} = \dfrac{(a^3)^2(b^4)^2}{ab^5}$ In the numerator, raise each factor within the parentheses to the 2nd power.

$= \dfrac{a^6b^8}{ab^5}$ In the numerator, for each power of a power, keep the base and multiply the exponents.

$= a^{6-1}b^{8-5}$ Keep each of the bases, a and b, and subtract the exponents.

$= a^5b^3$ Do the subtractions.

Self Check 7 Simplify: $\dfrac{(c^4d^5)^3}{c^2d^3}$.

EXAMPLE 8

Simplify: **a.** $\left(\dfrac{4}{k}\right)^3$ and **b.** $\left(\dfrac{3x^2}{2y^3}\right)^5$.

Solution **a.** Since $\frac{4}{k}$ is the quotient of 4 and k, the expression $\left(\frac{4}{k}\right)^3$ is a power of a quotient.

$\left(\dfrac{4}{k}\right)^3 = \dfrac{4^3}{k^3}$ Raise the numerator and denominator to the 3rd power.

$= \dfrac{64}{k^3}$ Evaluate 4^3.

b. $\left(\dfrac{3x^2}{2y^3}\right)^5 = \dfrac{(3x^2)^5}{(2y^3)^5}$ Raise the numerator and the denominator to the 5th power.

$= \dfrac{3^5(x^2)^5}{2^5(y^3)^5}$ In the numerator and denominator, raise each factor within the parentheses to the 5th power.

$= \dfrac{243x^{10}}{32y^{15}}$ Evaluate 3^5 and 2^5. For each power of a power, keep the base and multiply the exponents.

Self Check 8 Simplify: **a.** $\left(\frac{x}{7}\right)^3$, **b.** $\left(\frac{2x^3}{3y^2}\right)^4$.

EXAMPLE 9

Simplify: $\frac{(5b)^9}{(5b)^6}$.

Solution

$\frac{(5b)^9}{(5b)^6} = (5b)^{9-6}$ Keep the common base $5b$, and subtract the exponents.

$= (5b)^3$ Do the subtraction.

$= 5^3b^3$ Raise each factor within the parentheses to the 3rd power.

$= 125b^3$ Evaluate 5^3.

Self Check 9 Simplify: $\frac{(-2h)^{20}}{(-2h)^{14}}$.

The rules for natural-number exponents are summarized as follows.

Rules for Exponents If m and n represent natural numbers and there are no divisions by zero, then

1. $x^mx^n = x^{m+n}$ 2. $\frac{x^m}{x^n} = x^{m-n}$ 3. $(x^m)^n = x^{mn}$

4. $(xy)^n = x^ny^n$ 5. $\left(\frac{x}{y}\right)^n = \frac{x^n}{y^n}$

Answers to Self Checks **1. a.** 7^{15}, **b.** z^4, **c.** x^{11}, **d.** $(y-1)^{10}$, **e.** s^8t^7 **2. a.** 55^{25}, **b.** a^2, **c.** $8t$, **d.** $b^{11}c^3$ **3.** b **4. a.** 4^{30}, **b.** y^{10} **5. a.** a^{21}, **b.** a^{17} **6. a.** $16t^4$, **b.** $c^{18}d^{24}$, **c.** $-27a^3b^{15}$ **7.** $c^{10}d^{12}$ **8. a.** $\frac{x^3}{343}$, **b.** $\frac{16x^{12}}{81y^8}$ **9.** $64h^6$

4.1 STUDY SET

ELEMENTARY Algebra f(x) Now™

VOCABULARY Fill in the blanks.

1. Expressions such as x^4, 10^3, and $(5t)^2$ are called ________ expressions.
2. The _____ of the expression 5^3 is 5. The _______ is 3.
3. The expression x^4 represents a repeated multiplication where x is to be written as a ______ four times.
4. $3^4 \cdot 3^8$ is a ______ of exponential expressions with the same base and $\frac{x^4}{x^2}$ is a ______ of exponential expressions with the same base.
5. $(h^3)^7$ is a _____ of an exponential expression.
6. The expression $(2xy)^3$ is a power of a ______ and $\left(\frac{x}{c}\right)^5$ is a power of a ______.

CONCEPTS Fill in the blanks.

7. **a.** $(3x)^4$ means $\square \cdot \square \cdot \square \cdot \square$.
   b. Using an exponent, $(-5y)(-5y)(-5y)$ can be written as $\square$.
8. **a.** $x^mx^n = \square$ **b.** $(xy)^n = \square$
   c. $\left(\frac{a}{b}\right)^n = \square$ **d.** $(a^b)^c = \square$
   e. $\frac{x^m}{x^n} = \square$ **f.** $x = x^{\square}$

9. a. Write a power of a product that has two factors.

b. Write a power of a quotient.

10. a. To simplify $(2y^3z^2)^4$, how many factors within the parentheses must be raised to the fourth power?

b. To simplify $\left(\frac{y^3}{z^2}\right)^4$ what two expressions must be raised to the fourth power?

Simplify each expression, if possible.

11. a. $x^2 + x^2$ **b.** $x^2 - x^2$ **c.** $x^2 \cdot x^2$ **d.** $a^3 \cdot a^4$

12. a. $x^2 + x$ **b.** $x^2 - x$ **c.** $x^2 \cdot x$ **d.** $\frac{x^2}{x}$

13. a. $x^3 + x^2$ **b.** $x^3 - x^2$ **c.** $x^3 \cdot x^2$ **d.** $\frac{x^3}{x^2}$

14. a. $x^3 + y^3$ **b.** $x^3 - y^3$ **c.** x^3y^3 **d.** $\frac{x^3}{y^3}$

Evaluate each expression.

15. a. $(-4)^2$ **b.** -4^2

16. a. $(-5)^2$ **b.** -5^2

NOTATION **Complete each solution.**

17. $(x^4x^2)^3 = (\square)^3$

$= x^{\square}$

18. $\frac{a^3a^4}{a^2} = \frac{\square}{a^2}$

$= a^{\square - 2}$

$= a^{\square}$

Identify the base and the exponent in each expression.

19. 4^3

20. $(-8)^2$

21. x^5

22. $\left(\frac{5}{x}\right)^3$

23. $(-3x)^2$

24. $-x^4$

25. $-\frac{1}{3}y^6$

26. $3.14r^4$

27. $(y + 9)^4$

28. $(2xy)^{10}$

29. $(-3ab)^7$

30. $-(z - 2)^3$

Write each repeated multiplication without using exponents.

31. x^5

32. $(-7y)^4$

33. $\left(\frac{t}{2}\right)^3$

34. c^3d^2

35. $(x - 5)^2$

36. $(m + 4)^3$

Write each expression using an exponent.

37. $4t(4t)(4t)(4t)$

38. $-5u(-5u)$

39. $-4 \cdot t \cdot t \cdot t$

40. $-5 \cdot u \cdot u$

41. $(x - y)(x - y)(x - y)$

42. $\left(\frac{x}{c}\right)\left(\frac{x}{c}\right)\left(\frac{x}{c}\right)\left(\frac{x}{c}\right)$

PRACTICE **Write each expression as an expression involving one base and one exponent.**

43. $12^3 \cdot 12^4$

44. $3^4 \cdot 3^6$

45. $2(2^3)(2^2)$

46. $5(5^5)(5^3)$

47. $a^3 \cdot a^3$

48. $m^7 \cdot m^7$

49. x^4x^3

50. y^5y^2

51. a^3aa^5

52. b^2b^3b

53. $(-7)^2(-7)^3$

54. $(-10)^8(-10)^3$

55. $(8t)^{20}(8t)^{40}$

56. $(9p)^{80}(9p)^{10}$

57. $(n - 1)^2(n - 1)$

58. $(s - 14)(s - 14)^8$

59. $y^3(y^2y^4)$

60. $(y^4y)y^6$

61. $\frac{8^{12}}{8^4}$

62. $\frac{10^4}{10^2}$

63. $\frac{x^{15}}{x^3}$

64. $\frac{y^6}{y^3}$

65. $\frac{c^{10}}{c^9}$

66. $\frac{h^{20}}{h^{10}}$

67. $\frac{(k-2)^{15}}{(k-2)}$

68. $\frac{(m+8)^{20}}{(m+8)}$

69. $(3^2)^4$

70. $(4^3)^3$

71. $(y^5)^3$

72. $(b^3)^6$

73. $(m^{50})^{10}$

74. $(n^{25})^4$

Simplify. Assume there are no divisions by 0.

75. $(a^2b^3)(a^3b^3)$

76. $(u^3v^5)(u^4v^5)$

77. $(cd^4)(cd)$

78. $ab^3c^4 \cdot ab^4c^2$

79. $xy^2 \cdot x^2y$

80. $s^8t^2s^2t^7$

81. $\frac{y^3y^4}{yy^2}$

82. $\frac{b^4b^5}{b^2b^3}$

83. $\frac{c^3d^7}{cd}$

84. $\frac{r^8s^9}{rs}$

85. $(x^2x^3)^5$

86. $(y^3y^4)^4$

87. $(3zz^2z^3)^5$

88. $(4t^3t^6t^2)^2$

89. $(3n^8)^2$

90. $(y^3y)^2(y^2)^2$

91. $(uv)^4$

92. $(2m^4)^3$

93. $(a^3b^2)^3$

94. $(r^3s^2)^2$

95. $(-2r^2s^3)^3$

96. $(-3x^2y^4)^2$

97. $\left(\frac{a}{b}\right)^3$

98. $\left(\frac{r}{s}\right)^4$

99. $\left(\frac{x^2}{y^3}\right)^5$

100. $\left(\frac{u^4}{v^2}\right)^6$

101. $\left(\frac{-2a}{b}\right)^5$

102. $\left(\frac{-2t}{3}\right)^4$

103. $\frac{(6k)^7}{(6k)^4}$

104. $\frac{(-3a)^{12}}{(-3a)^{10}}$

105. $\frac{(a^2b)^{15}}{(a^2b)^9}$

106. $\frac{(s^2t^3)^4}{(s^2t^3)^2}$

107. $\frac{a^2a^3a^4}{(a^4)^2}$

108. $\frac{(aa^2)^3}{a^2a^3}$

109. $\frac{(ab^2)^3}{(ab)^2}$

110. $\frac{(m^3n^4)^3}{(mn^2)^3}$

111. $\frac{(r^4s^3)^4}{(rs^3)^3}$

112. $\frac{(x^2y^5)^5}{(x^3y)^2}$

113. $\left(\frac{y^3y}{2yy^2}\right)^3$

114. $\left(\frac{2y^3y}{yy^2}\right)^3$

115. $\left(\frac{3t^3t^4t^5}{4t^2t^6}\right)^3$

116. $\left(\frac{4t^3t^4t^5}{3t^2t^6}\right)^3$

APPLICATIONS **Write an expression for the area or volume of each figure. Leave π in your answer.**

117.

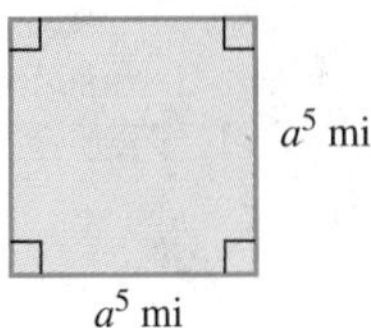

118.

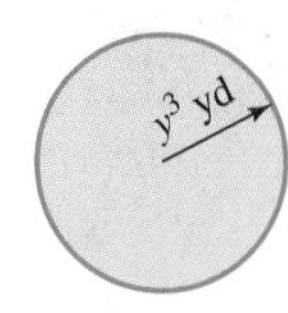

119.

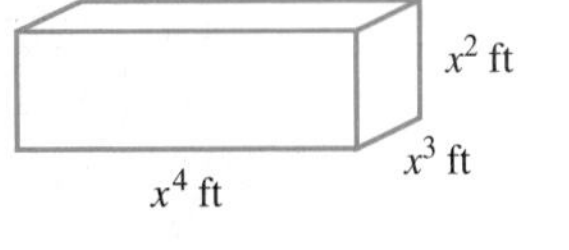

120.

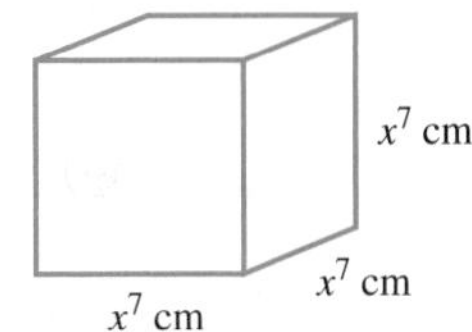

121. ART HISTORY Leonardo da Vinci's drawing relating a human figure to a square and a circle is shown.

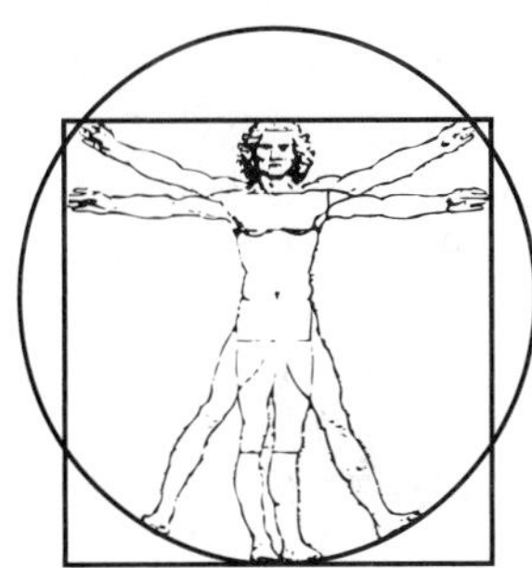

a. Find the area of the square if the man's height is $5x$ feet.

b. Find the area of the circle if the distance from his waist to his feet is $3a$ feet. Leave π in your answer.

122. PACKAGING Find the volume of the bowling ball and the cardboard box it is packaged in. Leave π in your answer.

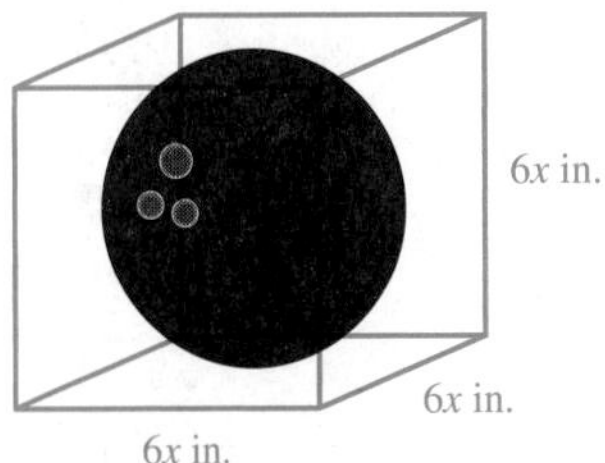

123. BOUNCING BALLS A ball is dropped from a height of 32 feet and always rebounds to one-half of its previous height. Draw a diagram of the path of the ball, showing four bounces. Then explain why the expressions $32\left(\frac{1}{2}\right)$, $32\left(\frac{1}{2}\right)^2$, $32\left(\frac{1}{2}\right)^3$, and $32\left(\frac{1}{2}\right)^4$ represent the height of the ball on the first, second, third, and fourth bounces, respectively. Find the heights of the first four bounces.

124. PROBABILITY The probability that a couple will have n baby boys in a row is given by the formula $\left(\frac{1}{2}\right)^n$. Find the probability that a couple will have four baby boys in a row.

WRITING

125. Explain the mistake in the following work.

$$2^3 \cdot 2^2 = 4^5$$
$$= 1{,}024$$

126. Are the expressions $2x^3$ and $(2x)^3$ equivalent? Explain.

127. Explain why we can simplify $x^4 \cdot x^5$, but cannot simplify $x^4 + x^5$.

128. Explain the power-of-a-product rule for exponents. Then give an example that shows why there is no power-of-a-sum rule for exponents.

REVIEW **Match each equation with its graph below.**

129. $y = 2x - 1$

130. $y = 3x - 1$

131. $y = 3$

132. $x = 3$

a.

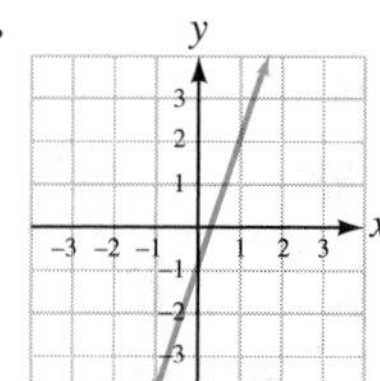

b.

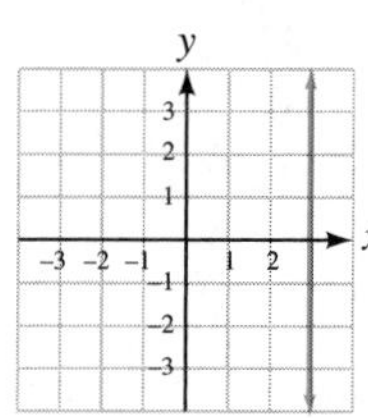

c.

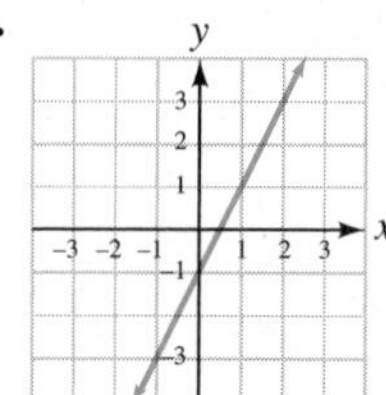

d.

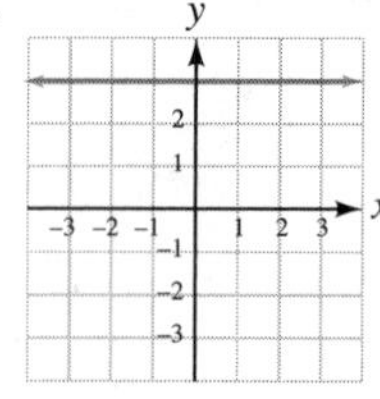

CHALLENGE PROBLEMS

133. Simplify each expression. The variables represent natural numbers.

a. $x^{2m}x^{3m}$

b. $(y^{5c})^4$

c. $\dfrac{m^{8x}}{m^{4x}}$

d. $(2a^{6y})^4$

134. Is the operation of raising to a power commutative? That is, is $a^b = b^a$? Explain.

4.2 Zero and Negative Exponents

- Zero Exponents
- Negative Integer Exponents
- Negative Exponents Appearing in Fractions
- Applications

We now extend the discussion of natural-number exponents to include exponents that are zero and exponents that are negative integers.

ZERO EXPONENTS

To develop the definition of a zero exponent, we will simplify the expression

$$\frac{5^3}{5^3}$$

in two ways and compare the results.

First, we apply the quotient rule for exponents, where we subtract the equal exponents in the numerator and denominator. The result is 5^0. In the second approach, we write 5^3 as $5 \cdot 5 \cdot 5$ and remove the common factors of 5 in the numerator and denominator. The result is 1.

$$\frac{5^3}{5^3} = 5^{3-3} = 5^0 \qquad \frac{5^3}{5^3} = \frac{\overset{1}{\cancel{5}} \cdot \overset{1}{\cancel{5}} \cdot \overset{1}{\cancel{5}}}{\underset{1}{\cancel{5}} \cdot \underset{1}{\cancel{5}} \cdot \underset{1}{\cancel{5}}} = 1$$

These must be equal.

Since $\frac{5^3}{5^3} = 5^0$ and $\frac{5^3}{5^3} = 1$, we conclude that $5^0 = 1$. This observation suggests the following definition.

Zero Exponents Any nonzero base raised to the 0 power is 1. For any nonzero real number x,

$$x^0 = 1$$

EXAMPLE 1

ELEMENTARY Algebra $f(x)$ Now™

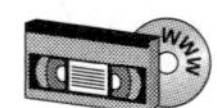

Simplify each expression, and assume $a \neq 0$: **a.** $(-8)^0$, **b.** $\left(\frac{14}{15}\right)^0$, **c.** $(3a)^0$, and **d.** $3a^0$.

Solution

a. $(-8)^0 = 1$ Any nonzero number raised to the 0 power is 1.

b. $\left(\frac{14}{15}\right)^0 = 1$

c. $(3a)^0 = 1$ Because of the parentheses, the base is $3a$. Any nonzero base raised to the 0 power is 1.

d. $3a^0 = 3 \cdot a^0$ Since there are no parentheses, the base is a.

$= 3 \cdot 1$

$= 3$

The Language of Algebra

Note that the zero exponent definition doesn't define 0^0. This expression is said to be an *indeterminate form,* which is beyond the scope of this book.

Self Check 1 Simplify each expression: **a.** $(0.75)^0$, **b.** $-5c^0d$, **c.** $(5c)^0$.

NEGATIVE INTEGER EXPONENTS

The Language of Algebra

The *negative integers* are: $-1, -2, -3, -4, -5$, and so on.

To develop the definition of a negative exponent, we will simplify

$$\frac{6^2}{6^5}$$

in two ways and compare the results.

If we apply the quotient rule for exponents, where we subtract the greater exponent in the denominator from the lesser exponent in the numerator, we obtain 6^{-3}. In the second approach, we remove the two common factors of 6 to obtain $\frac{1}{6^3}$.

$$\frac{6^2}{6^5} = 6^{2-5} = 6^{-3} \qquad \frac{6^2}{6^5} = \frac{\overset{1}{\cancel{6}} \cdot \overset{1}{\cancel{6}}}{\underset{1}{\cancel{6}} \cdot \underset{1}{\cancel{6}} \cdot 6 \cdot 6 \cdot 6} = \frac{1}{6^3}$$

These must be equal.

Since $\frac{6^2}{6^5} = 6^{-3}$ and $\frac{6^2}{6^5} = \frac{1}{6^3}$, we conclude that $6^{-3} = \frac{1}{6^3}$. Note that 6^{-3} is equal to the reciprocal of 6^3. This observation suggests the following definition.

Negative Exponents For any nonzero real number x and any integer n,

$$x^{-n} = \frac{1}{x^n}$$

In words, x^{-n} is the reciprocal of x^n.

From the definition, we see that another way to write x^{-n} is to write its reciprocal and change the sign of the exponent. For example,

$$5^{-4} = \frac{1}{5^4}$$ Think of the reciprocal of 5^{-4}, which is $\frac{1}{5^{-4}}$. Then change the sign of the exponent.

EXAMPLE 2

Simplify: **a.** 3^{-2}, **b.** y^{-1}, and **c.** $(-2)^{-3}$.

ELEMENTARY Algebra f(x) Now™

Solution

a. $3^{-2} = \frac{1}{3^2}$ Write the reciprocal of 3^{-2} and change the sign of the exponent.

$= \frac{1}{9}$ $3^2 = 9$.

Caution

A negative exponent does not indicate a negative number. It indicates a reciprocal.

$$4^{-2} = \frac{1}{4^2} = \frac{1}{16}$$

b. $y^{-1} = \frac{1}{y^1}$ Write the reciprocal of y^{-1} and change the sign of the exponent.

$= \frac{1}{y}$

c. $(-2)^{-3} = \frac{1}{(-2)^3}$ Because of the parentheses, the base is -2. Write the reciprocal of $(-2)^{-3}$ and change the exponent from -3 to 3.

$= -\frac{1}{8}$ $(-2)^3 = -8$.

Self Check 2 Simplify: **a.** 8^{-2}, **b.** x^{-5}, **c.** $(-3)^{-3}$.

EXAMPLE 3

Simplify: **a.** $9m^{-3}$ and **b.** -5^{-2}.

ELEMENTARY Algebra f(x) Now™

Solution

a. $9m^{-3} = 9 \cdot m^{-3}$ The base is m.

$= 9 \cdot \frac{1}{m^3}$ Write the reciprocal of m^{-3} and change the sign of the exponent.

$= \frac{9}{m^3}$ Multiply.

Caution

Don't confuse *negative numbers* with *negative exponents.* For example, the expressions -2 and 2^{-1} are not the same.

$$2^{-1} = \frac{1}{2^1} = \frac{1}{2}$$

b. $-5^{-2} = -1 \cdot 5^{-2}$ The base is 5.

$= -1 \cdot \frac{1}{5^2}$ Write the reciprocal of 5^{-2} and change the sign of the exponent.

$= -\frac{1}{25}$ Evaluate 5^2 and multiply.

Self Check 3 Simplify: **a.** $12h^{-9}$ and **b.** -2^{-4}.

NEGATIVE EXPONENTS APPEARING IN FRACTIONS

Negative exponents can appear in the numerator and/or the denominator of a fraction. To develop rules for such situations, we consider the following example.

$$\frac{a^{-4}}{b^{-3}} = \frac{\frac{1}{a^4}}{\frac{1}{b^3}} = \frac{1}{a^4} \cdot \frac{b^3}{1} = \frac{b^3}{a^4}$$

We can obtain this result in a simpler way. Beginning with $\frac{a^{-4}}{b^{-3}}$, move a^{-4} to the denominator and change the sign of the exponent. Then, move b^{-3} to the numerator and change the sign of the exponent.

$$\frac{a^{-4}}{b^{-3}} = \frac{b^3}{a^4}$$

This example suggests the following rules.

Changing from Negative to Positive Exponents

A factor can be moved from the denominator to the numerator or from the numerator to the denominator of a fraction if the sign of its exponent is changed. For any nonzero real numbers x and y, and any integers m and n,

$$\frac{1}{x^{-n}} = x^n \quad \text{and} \quad \frac{x^{-m}}{y^{-n}} = \frac{y^n}{x^m}$$

These rules streamline the process when simplifying expressions involving negative exponents.

EXAMPLE 4

ELEMENTARY Algebra f(x) Now™

Write each expression using positive exponents only: **a.** $\frac{1}{d^{-10}}$, **b.** $\frac{2^{-3}}{3^{-4}}$, and **c.** $\frac{s^{-2}}{5t^{-9}}$.

Solution

a. $\frac{1}{d^{-10}} = d^{10}$ Move d^{-10} to the numerator and change the sign of the exponent.

b. $\frac{2^{-3}}{3^{-4}} = \frac{3^4}{2^3}$ Move 2^{-3} to the denominator and change the sign of the exponent. Move 3^{-4} to the numerator and change the sign of the exponent.

$= \frac{81}{8}$ $3^4 = 81$ and $2^3 = 8$.

c. $\frac{s^{-2}}{5t^{-9}} = \frac{t^9}{5s^2}$ Move s^{-2} to the denominator and change the sign of the exponent. Since $5t^{-9}$ has no parentheses, t is the base. Move t^{-9} to the numerator and change the sign of the exponent.

Caution

This rule does not permit moving *terms* that have negative exponents. For example,

$$\frac{3^{-2} + 8}{5} \neq \frac{8}{3^2 \cdot 5}$$

Self Check 4 Write each expression using positive exponents only: **a.** $\frac{1}{w^{-5}}$, **b.** $\frac{5^{-2}}{4^{-3}}$, and **c.** $\frac{h^{-6}}{8a^{-7}}$.

The rules for exponents involving products, powers, and quotients are also true for zero and negative exponents.

Summary of Exponent Rules If m and n represent integers and there are no divisions by zero, then

Product rule	*Quotient rule*	*Power rule*
$x^m \cdot x^n = x^{m+n}$	$\dfrac{x^m}{x^n} = x^{m-n}$	$(x^m)^n = x^{mn}$
Power of a product	***Power of a quotient***	***Zero exponent***
$(xy)^n = x^n y^n$	$\left(\dfrac{x}{y}\right)^n = \dfrac{x^n}{y^n}$	$x^0 = 1$
Negative exponent	***Negative exponent***	***Negative exponent***
$x^{-n} = \dfrac{1}{x^n}$	$\dfrac{1}{x^{-n}} = x^n$	$\dfrac{x^{-m}}{y^{-n}} = \dfrac{y^n}{x^m}$

The rules for exponents are used to simplify expressions. In general, an expression involving exponents is simplified when

- Each base occurs only once
- There are no parentheses
- There are no negative or zero exponents

EXAMPLE 5

Simplify: **a.** $\left(\dfrac{5}{16}\right)^{-1}$, **b.** $\dfrac{x^3}{x^7}$, and **c.** $(x^3)^{-2}$.

ELEMENTARY Algebra f(x) Now™

Solution

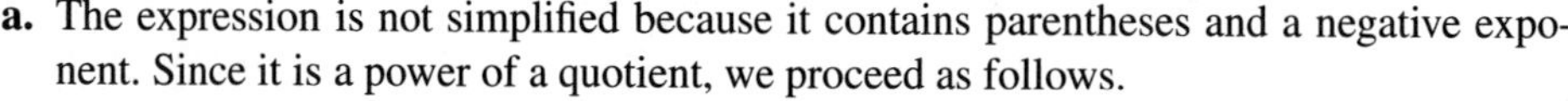

a. The expression is not simplified because it contains parentheses and a negative exponent. Since it is a power of a quotient, we proceed as follows.

$$\left(\frac{5}{16}\right)^{-1} = \frac{5^{-1}}{16^{-1}} \quad \text{Raise the numerator and the denominator to the } -1 \text{ power.}$$

$$= \frac{16^1}{5^1} \quad \begin{array}{l}\text{Move } 5^{-1} \text{ to the denominator. Change the sign of the exponent.}\\ \text{Move } 16^{-1} \text{ to the numerator. Change the sign of the exponent.}\end{array}$$

$$= \frac{16}{5}$$

b. The expression is not simplified because the base x occurs more than once. Since it is a quotient of like bases, we proceed as follows.

$$\frac{x^3}{x^7} = x^{3-7} \quad \text{Keep the base } x \text{ and subtract the exponents.}$$

$$= x^{-4} \quad \text{Do the subtraction: } 3 - 7 = -4.$$

$$= \frac{1}{x^4} \quad \text{Write the reciprocal of } x^{-4} \text{ and change the sign of the exponent.}$$

c. The expression is not simplified because it contains parentheses and a negative exponent. Since it is a power of a power, we proceed as follows.

$$(x^3)^{-2} = x^{-6} \quad \text{Multiply exponents.}$$

$$= \frac{1}{x^6} \quad \text{Write the reciprocal of } x^{-6} \text{ and change the sign of the exponent.}$$

Self Check 5 Simplify: **a.** $\left(\frac{3}{7}\right)^{-2}$, **b.** $\frac{a^3}{a^8}$, and **c.** $(n^4)^{-5}$.

EXAMPLE 6

Simplify each expression: **a.** $\frac{y^{-4}y^{-3}}{y^{-20}}$, **b.** $\frac{7^{-1}a^3b^4}{6^{-2}a^5b^2}$, and **c.** $\left(\frac{x^3y^2}{xy^{-3}}\right)^{-2}$.

Solution **a.**

$$\frac{y^{-4}y^{-3}}{y^{-20}} = \frac{y^{-7}}{y^{-20}}$$ In the numerator, add exponents: $-4 + (-3) = -7$.

$$= y^{-7-(-20)}$$ Keep the common base y and subtract exponents.

$$= y^{13}$$ $-7 - (-20) = -7 + 20 = 13$.

b.

$$\frac{7^{-1}a^3b^4}{6^{-2}a^5b^2} = \frac{6^2a^3b^4}{7^1a^5b^2}$$ Move 7^{-1} to the denominator. Change the sign of the exponent. Move 6^{-2} to the numerator. Change the sign of the exponent.

$$= \frac{36a^{3-5}b^{4-2}}{7}$$ Use the quotient rule twice and subtract exponents.

$$= \frac{36a^{-2}b^2}{7}$$ Do the subtractions.

$$= \frac{36b^2}{7a^2}$$ Move a^{-2} to the denominator and change the sign of the exponent.

c.

$$\left(\frac{x^3y^2}{xy^{-3}}\right)^{-2} = \left(x^{3-1}y^{2-(-3)}\right)^{-2}$$ Use the quotient rule twice and subtract exponents.

$$= (x^2y^5)^{-2}$$ Do the subtractions.

$$= \frac{1}{(x^2y^5)^2}$$ Write the reciprocal of $(x^2y^5)^{-2}$ and change the sign of the exponent.

$$= \frac{1}{x^4y^{10}}$$ Raise each factor within the parentheses to the 2nd power.

Self Check 6 Simplify: **a.** $\frac{a^{-4}a^{-5}}{a^{-3}}$, **b.** $\frac{1^{-4}x^5y^3}{9^{-2}x^3y^6}$, and **c.** $\left(\frac{c^2d^2}{c^4d^{-3}}\right)^{-3}$.

APPLICATIONS

When we determine what sum of money must be invested today to be worth a given amount in the future, we are computing **present value.** The formula for present value P involves a negative exponent.

$$P = A(1 + i)^{-n}$$

where $\$A$ is the amount of money that is needed in n years, and i is the annual interest rate (expressed as a decimal) that the investment earns.

EXAMPLE 7

ELEMENTARY Algebra $f(x)$ Now™

Saving for college. How much money should the grandparents of a newborn baby girl invest at a 6% annual rate so that on her 18th birthday, she will have a college fund of $20,000?

Solution We substitute 20,000 for A, 0.06 for i, and 18 for n in the formula to find P.

$P = A(1 + i)^{-n}$
$P = 20{,}000(1 + 0.06)^{-18}$
$P = 20{,}000(1.06)^{-18}$ Do the addition within the parentheses.

To evaluate $20{,}000(1.06)^{-18}$ with a scientific calculator, we use the exponential key y^x.

Enter 1.06 y^x 18 +/− × 20000 = .

Since the result is $P \approx 7{,}006.88$, the grandparents must invest $7,006.88 to have $20,000 in 18 years.

Answers to Self Checks **1. a.** 1, **b.** $-5d$, **c.** 1 **2. a.** $\frac{1}{64}$, **b.** $\frac{1}{x^5}$, **c.** $-\frac{1}{27}$ **3. a.** $\frac{12}{h^9}$, **b.** $-\frac{1}{16}$ **4. a.** w^5, **b.** $\frac{64}{25}$, **c.** $\frac{a^7}{8h^6}$ **5. a.** $\frac{49}{9}$, **b.** $\frac{1}{a^5}$, **c.** $\frac{1}{n^{20}}$ **6. a.** $\frac{1}{a^6}$, **b.** $\frac{81x^2}{y^3}$, **c.** $\frac{c^6}{d^{15}}$

4.2 STUDY SET

ELEMENTARY Algebra Now™

VOCABULARY Fill in the blanks.

1. In the expression 8^{-3}, 8 is the ______ and −3 is the ______.
2. In the expression 5^{-1}, the exponent is a ______ integer.
3. Another way to write 2^{-3} is to write its ______ and change the sign of the exponent:

$$2^{-3} = \frac{1}{2^{\square}}$$

4. To ______ $(y^{-4})^{-3}$, keep the base y and multiply the exponents to get y^{12}.

CONCEPTS

5. In part a, fill in the blanks to simplify the fraction in two ways. Then complete the sentence in part b.

a. $\frac{6^4}{6^4} = 6^{\square}$ $\qquad \frac{6^4}{6^4} = \frac{\square \cdot \square \cdot \square \cdot \square}{6 \cdot 6 \cdot 6 \cdot 6}$

$= 6^{\square}$ $\qquad = \square$

These must be equal.

b. So we define 6^0 to be ___, and in general, if x is any nonzero real number, then $x^0 =$ ___.

6. In part a, fill in the blanks to simplify the fraction in two ways. Then complete the sentence in part b.

a. $\frac{8^3}{8^5} = 8^{\square}$ $\qquad \frac{8^3}{8^5} = \frac{\square \cdot \square \cdot \square}{8 \cdot 8 \cdot 8 \cdot 8 \cdot 8}$

$= 8^{\square}$ $\qquad = \frac{1}{8^{\square}}$

These must be equal.

b. So we define 8^{-2} to be ___, and in general, if x is any nonzero real number, then $x^{-n} =$ ___.

Complete each table.

7.

x	3^x
2	
1	
0	
−1	
−2	

8.

x	$(-9)^x$
2	
1	
0	
−1	
−2	

9. Complete each rule for exponents.

a. $x^m \cdot x^n =$ ___ **b.** $\frac{x^m}{x^n} =$ ___

c. $(x^m)^n =$ ___ **d.** $(xy)^n =$ ___

e. $\left(\frac{x}{y}\right)^n =$ ___ **f.** $x^{-n} =$ ___

g. $\frac{1}{x^{-n}} =$ ___ **h.** $\frac{x^{-m}}{y^{-n}} =$ ___

i. $x^0 =$ ___

Fill in the blanks.

10. A factor can be moved from the denominator to the numerator or from the numerator to the denominator of a fraction if the sign of its exponent is ________.

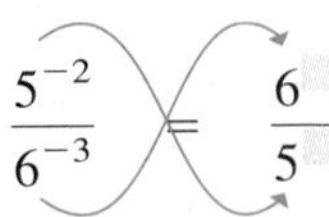

$$\frac{5^{-2}}{6^{-3}} = \frac{6^{\square}}{5^{\square}}$$

11. In general, an expression involving exponents is simplified when

- Each base occurs only ______
- There are _____ parentheses
- There are no negative or zero __________

12. Decide whether each statement is true or false.

a. $6^{-2} = -36$ **b.** $6^{-2} = -\frac{1}{36}$

c. $-6 = \frac{1}{6}$ **d.** $6^{-2} < 6^{-1}$

NOTATION **Complete each solution.**

13. $(y^5y^3)^{-5} = (\square)^{-5}$
$= y^{\square}$
$= \frac{1}{y^{\square}}$

14. $\left(\frac{a^2b^3}{a^{-3}b}\right)^{-3} = (a^{2-\square}b^{\square-1})^{-3}$
$= (a^{\square}b^{\square})^{-3}$
$= \frac{1}{(a^5b^2)^{\square}}$
$= \frac{1}{a^{15}b^{\square}}$

15. Complete the table.

Expression	Base	Exponent
4^{-2}		
$6x^{-5}$		
$\left(\frac{3}{y}\right)^{-8}$		
-7^{-1}		
$(-2)^{-3}$		
$10a^0b$		

16. The formula for present value is

$$P = A(1 + i)^{-n}$$

On the right-hand side, what is the base and what is the exponent?

PRACTICE **Simplify each expression. Write each answer without using parentheses or negative exponents.**

17. 7^0
18. 9^0
19. $\left(\frac{1}{4}\right)^0$
20. $\left(\frac{3}{8}\right)^0$
21. $2x^0$
22. $(2x)^0$
23. $(-x)^0$
24. $-x^0$
25. $\left(\frac{a^2b^3}{ab^4}\right)^0$
26. $\frac{2}{3}\left(\frac{xyz}{x^2y}\right)^0$
27. $\frac{5}{2x^0}$
28. $\frac{4}{3a^0}$
29. $-15x^0y$
30. $24g^0h^2$
31. 12^{-2}
32. 11^{-2}
33. $(-4)^{-1}$
34. $(-8)^{-1}$
35. $44g^{-6}$
36. $55g^{-3}$
37. $(-10)^{-3}$
38. $(-20)^{-2}$
39. -4^{-3}
40. -6^{-3}
41. $-(-4)^{-3}$
42. $-(-4)^{-2}$
43. x^{-2}
44. y^{-3}
45. $-b^{-5}$
46. $-c^{-4}$
47. $\left(\frac{1}{6}\right)^{-2}$
48. $\left(\frac{1}{7}\right)^{-2}$
49. $\left(-\frac{1}{2}\right)^{-3}$
50. $\left(-\frac{1}{5}\right)^{-3}$
51. $\left(\frac{7}{8}\right)^{-1}$
52. $\left(\frac{16}{5}\right)^{-1}$
53. $\frac{1}{5^{-3}}$
54. $\frac{1}{3^{-3}}$
55. $\frac{2^{-4}}{3^{-1}}$
56. $\frac{7^{-2}}{2^{-3}}$
57. $\frac{a^{-5}}{b^{-2}}$
58. $\frac{r^{-6}}{s^{-1}}$
59. $\frac{r^{-50}}{r^{-70}}$
60. $\frac{m^{-30}}{m^{-40}}$
61. $-\frac{1}{p^{-10}}$
62. $-\frac{1}{n^{-30}}$

63. $\dfrac{h^{-5}}{h^2}$ **64.** $\dfrac{y^{-3}}{y^4}$

65. $\dfrac{8}{s^{-1}}$ **66.** $\dfrac{6}{k^{-2}}$

67. $\dfrac{h}{h^{-6}}$ **68.** $\dfrac{w}{w^{-9}}$

69. $(2y)^{-4}$ **70.** $(-3x)^{-1}$

71. $(ab^2)^{-3}$ **72.** $(m^2n^3)^{-2}$

73. $2^5 \cdot 2^{-2}$ **74.** $10^2 \cdot 10^{-4}$

75. $\left(\dfrac{y^4}{3}\right)^{-2}$ **76.** $\left(\dfrac{p^3}{2}\right)^{-3}$

77. $\dfrac{y^4}{y^5}$ **78.** $\dfrac{t^7}{t^{10}}$

79. $\dfrac{(r^2)^3}{(r^3)^4}$ **80.** $\dfrac{(b^3)^4}{(b^5)^4}$

81. $\dfrac{4s^{-5}}{t^{-2}}$ **82.** $\dfrac{9k^{-8}}{m^{-2}}$

83. $(5d^{-2})^3$ **84.** $(9s^{-6})^2$

85. $\dfrac{-2a^{-4}}{a^{-8}}$ **86.** $\dfrac{-3b^{-3}}{b^{-9}}$

87. $x^{-3}x^{-3}x^{-3}$ **88.** $y^{-2}y^{-2}y^{-2}$

89. $\dfrac{t(t^{-2})^{-2}}{t^{-5}}$ **90.** $\dfrac{d(d^{-3})^{-3}}{d^{-7}}$

91. $\dfrac{y^4y^3}{y^4y^{-2}}$ **92.** $\dfrac{x^{12}x^{-7}}{x^3x^4}$

93. $\dfrac{2a^4a^{-2}}{a^2a^0}$ **94.** $\dfrac{3b^0b^3}{b^{-3}b^4}$

95. $(ab^2)^{-2}$ **96.** $(c^2d^3)^{-2}$

97. $(x^2y)^{-3}$ **98.** $(-xy^2)^{-4}$

99. $(x^{-4}x^3)^3$ **100.** $(y^{-2}y)^3$

101. $(-2x^3y^{-2})^{-5}$ **102.** $(-3u^{-2}v^3)^{-3}$

103. $\left(\dfrac{a^3}{a^{-4}}\right)^2$ **104.** $\left(\dfrac{a^4}{a^{-3}}\right)^3$

105. $\left(\dfrac{4x^2}{3x^{-5}}\right)^4$ **106.** $\left(\dfrac{-3r^4r^{-3}}{r^{-3}r^7}\right)^3$

107. $\left(\dfrac{y^3z^{-2}}{3y^{-4}z^3}\right)^2$ **108.** $\left(\dfrac{6xy^3}{x^{-1}y}\right)^3$

109. $\dfrac{2^{-2}g^{-2}h^{-3}}{9^{-1}h^{-3}}$ **110.** $\dfrac{5^{-1}x^{-2}y^{-3}}{8^{-2}x^{-11}}$

Evaluate each expression.

111. $5^0 + (-7)^0$ **112.** $-4^0 + 5^0$

113. $2^{-2} + 4^{-1}$ **114.** $-9^{-1} + 9^{-2}$

115. $9^0 - 9^{-1}$ **116.** $7^{-1} - 7^0$

APPLICATIONS

117. THE DECIMAL NUMERATION SYSTEM Decimal numbers are written by putting digits into place-value columns that are separated by a decimal point. Express the value of each of the columns using a power of 10.

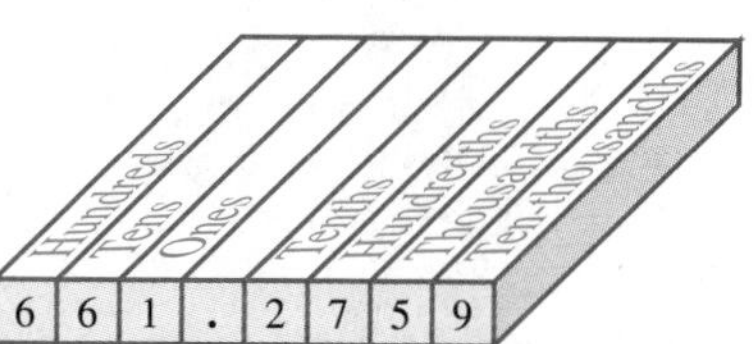

118. UNIT COMPARISONS Consider the relative sizes of the items listed in the table. In the column titled "measurement," write the most appropriate number from the following list. Each number is used only once.

10^0 meter 10^{-1} meter 10^{-2} meter
10^{-3} meter 10^{-4} meter 10^{-5} meter

Item	Measurement (m)
Thickness of a dime	
Height of a bathroom sink	
Length of a pencil eraser	
Thickness of soap bubble film	
Width of a video cassette	
Thickness of a piece of paper	

119. RETIREMENT How much money should a young married couple invest now at an 8% annual rate if they want to have \$100,000 in the bank when they reach retirement age in 40 years?

120. SAVING FOR THE FUTURE How much money should a nursery school owner invest now at a 4% annual rate if he wants to have \$20,000 in the bank in 5 years?

121. BIOLOGY During reproduction, the time required for a population to double is called the **generation time.** If b bacteria are introduced into a medium, then after the generation time has elapsed, there will be

$2b$ bacteria. After n generations, there will be $b \cdot 2^n$ bacteria. Explain what this expression represents when $n = 0$.

122. ELECTRONICS The total resistance R of a certain circuit is given by

$$R = \left(\frac{1}{R_1} + \frac{1}{R_2}\right)^{-1} + R_3$$

Find R if $R_1 = 4$, $R_2 = 2$, and $R_3 = 1$.

WRITING

123. Explain how you would help a friend understand that 2^{-3} is not equal to -8.

124. What does it mean to simplify $\left(\frac{xy^3}{2x^{-3}}\right)^{-2}$?

REVIEW

125. IQ TESTS An IQ (intelligence quotient) is a score derived from the formula

$$\text{IQ} = \frac{\text{mental age}}{\text{chronological age}} \cdot 100$$

Find the mental age of a 10-year-old girl if she has an IQ of 135.

126. DIVING When you are under water, the pressure in your ears is given by the formula

$$\text{Pressure} = \text{depth} \cdot \text{density of water}$$

Find the density of water (in lb/ft^3) if, at a depth of 9 feet, the pressure on your eardrum is $561.6\ \text{lb/ft}^2$.

127. Write the equation of the line having slope $\frac{3}{4}$ and y-intercept -5.

128. Write an equation of the line that passes through $(4, 4)$ and $(-6, -6)$. Express the answer in slope–intercept form.

CHALLENGE PROBLEMS

129. Simplify each expression. Assume there are no divisions by 0.

a. $r^{5m}r^{-6m}$ **b.** $\frac{x^{3n}}{x^{6n}}$

130. If a positive number x is raised to a negative power, is the result greater than, equal to, or less than x? Explore the possibilities.

4.3 Scientific Notation

- Converting from Scientific to Standard Notation
- Writing Numbers in Scientific Notation
- Computations with Scientific Notation

Scientists often deal with extremely large and small numbers. Two examples are shown below.

The distance from the Earth to the sun is approximately 150,000,000 kilometers.

The influenza virus, which causes "flu" symptoms of cough, sore throat, headache, and congestion, has a diameter of 0.00000256 inch.

Because 150,000,000 and 0.00000256 contain many zeros, they are difficult to read and cumbersome to work with in calculations. In this section, we will discuss a more convenient form in which we can write such numbers.

CONVERTING FROM SCIENTIFIC TO STANDARD NOTATION

Scientific notation provides a compact way of writing very large or very small numbers.

Scientific Notation A positive number is written in **scientific notation** when it is written in the form $N \times 10^n$, where $1 \le N < 10$ and n is an integer.

Some examples of numbers written in scientific notation are shown below. Each is the product of a decimal number that is at least 1, but less than 10, and a power of 10.

Notation

A raised dot · is sometimes used when writing scientific notation.

$3.67 \times 10^2 = 3.67 \cdot 10^2$

An integer exponent

$$3.67 \times 10^2 \qquad 2.15 \times 10^{-3} \qquad 9.875 \times 10^{22}$$

A decimal that is at least 1, but less than 10

A number written in scientific notation can be written in **standard notation** by performing the indicated multiplication. For example, to write 3.67×10^2 in standard notation, we recall that multiplying a decimal by 100 moves the decimal point 2 places to the right.

$$3.67 \times 10^2 = 3.67 \times 100 = 367.$$

To write 2.15×10^{-3} in standard notation, we recall from arithmetic that dividing a decimal by 1,000 moves the decimal point 3 places to the left.

$$2.15 \times 10^{-3} = 2.15 \times \frac{1}{10^3} = 2.15 \times \frac{1}{1{,}000} = \frac{2.15}{1{,}000} = 0.00215$$

In 3.67×10^2 and 2.15×10^{-3}, the exponent gives the number of decimal places that the decimal point moves, and the sign of the exponent indicates the direction in which it moves. Applying this observation to several other examples, we have

Success Tip

Since $10^0 = 1$, scientific notation involving 10^0 is easily simplified. For example,

$9.7 \times 10^0 = 9.7 \times 1 = 9.7$

$5.32 \times 10^6 = 5320000.$ Move the decimal point 6 places to the right.

$1.95 \times 10^{-5} = 0.0000195$ Move the decimal point 5 places to the left.

$9.7 \times 10^0 = 9.7$ There is no movement of the decimal point.

The following procedure summarizes our observations.

Converting from Scientific to Standard Notation

1. If the exponent is positive, move the decimal point the same number of places to the right as the exponent.
2. If the exponent is negative, move the decimal point the same number of places to the left as the absolute value of the exponent.

EXAMPLE 1

Write each number in standard notation: **a.** 3.467×10^5 and **b.** 8.9×10^{-4}.

ELEMENTARY Algebra $f(x)$ Now™

Solution **a.** Since the exponent is 5, the decimal point moves 5 places to the right.

346700. To move 5 places to the right, two placeholder zeros must be written.

Thus, $3.467 \times 10^5 = 346{,}700$

b. Since the exponent is -4, the decimal point moves 4 places to the left.

0.00089 To move 4 places to the left, three placeholder zeros must be written.

Thus, $8.9 \times 10^{-4} = 0.00089$

Self Check 1 Write each number in standard notation: **a.** 4.88×10^6 and **b.** 9.8×10^{-3}.

WRITING NUMBERS IN SCIENTIFIC NOTATION

The next example shows how to write a number in scientific notation.

EXAMPLE 2

ELEMENTARY Algebra $f(x)$ Now™

Write each number in scientific notation: **a.** 150,000,000, **b.** 0.00000256, and **c.** 432×10^5.

Solution **a.** We must write 150,000,000 as the product of a number between 1 and 10 and a power of 10. We note that 1.5 lies between 1 and 10. To obtain 150,000,000, the decimal point in 1.5 must be moved 8 places to the right.

150000000.

This will happen if we multiply 1.5 by 10^8. Therefore,

$$150{,}000{,}000 = 1.5 \times 10^8$$

b. We must write 0.00000256 as the product of a number between 1 and 10 and a power of 10. We note that 2.56 lies between 1 and 10. To obtain 0.00000256, the decimal point in 2.56 must be moved 6 places to the left.

0000002.56

This will happen if we multiply 2.56 by 10^{-6}. Therefore,

$$0.00000256 = 2.56 \times 10^{-6}$$

c. The number 432×10^5 is not written in scientific notation because 432 is not a number between 1 and 10. To write this number in scientific notation, we proceed as follows:

$$432 \times 10^5 = 4.32 \times 10^2 \times 10^5$$ Write 432 in scientific notation.

$$= 4.32 \times 10^7$$ Use the product rule to find $10^2 \times 10^5$. Keep the base of 10 and add the exponents.

Written in scientific notation, 432×10^5 is 4.32×10^7.

Notation

When writing numbers in scientific notation, keep the negative exponents. Don't apply the negative exponent rule.

2.56×10^{-6} ~~$2.56 \times \frac{1}{10^6}$~~

Calculators

Scientific notation

When displaying very large or very small answers, a scientific calculator uses scientific notation. The displays below show the results when two powers are computed.

$(453.46)^5 = 1.917321395 \times 10^{13}$

1.917321395 13

$(0.0005)^{12} = 2.44140625 \times 10^{-40}$

2.44140625 -40

Self Check 2 Write each number in scientific notation: **a.** 93,000,000 **b.** 0.00009055, and **c.** 85×10^{-3}.

The results from Example 2 illustrate the following forms to use when converting numbers from standard to scientific notation.

For real numbers between 0 and 1:	$\square \times 10^{\text{negative integer}}$
For real numbers at least 1, but less than 10:	$\square \times 10^{0}$
For real numbers greater than or equal to 10:	$\square \times 10^{\text{positive integer}}$

COMPUTATIONS WITH SCIENTIFIC NOTATION

Another advantage of scientific notation becomes apparent when we evaluate products or quotients that involve very large or small numbers. If we express those numbers in scientific notation, we can use rules for exponents to make the calculations easier.

EXAMPLE 3

ELEMENTARY Algebra f(x) Now™

Astronomy. Except for the sun, the nearest star visible to the naked eye from most parts of the United States is Sirius. Light from Sirius reaches Earth in about 70,000 hours. If light travels at approximately 670,000,000 mph, how far from Earth is Sirius?

Solution We are given the rate at which light travels (670,000,000 mph) and the time it takes the light to travel from Sirius to Earth (70,000 hr). We can find the distance the light travels using the formula $d = rt$.

$d = rt$	
$d = \mathbf{670{,}000{,}000}(\mathbf{70{,}000})$	Substitute 670,000,000 for r and 70,000 for t.
$= (6.7 \times 10^8)(7.0 \times 10^4)$	Write each number in scientific notation.
$= (6.7 \cdot 7.0) \times (10^8 \times 10^4)$	Group the numbers together and the powers of 10 together.
$= (6.7 \cdot 7.0) \times 10^{8+4}$	Use the product rule to find $10^8 \times 10^4$. Keep the base 10 and add exponents.
$= 46.9 \times 10^{12}$	Do the multiplication. Do the addition.

We note that 46.9 is not between 0 and 1, so 46.9×10^{12} is not written in scientific notation. To answer in scientific notation, we proceed as follows.

$= 4.69 \times 10^1 \times 10^{12}$	Write 46.9 in scientific notation as 4.69×10^1.
$= 4.69 \times 10^{13}$	Keep the base of 10 and add the exponents.

Sirius is approximately 4.69×10^{13} or 46,900,000,000,000 miles from Earth.

EXAMPLE 4

ELEMENTARY Algebra f(x) Now™

Atoms. Scientific notation is used in chemistry. As an example, we can approximate the weight (in grams) of one atom of the heaviest naturally occurring element, uranium, by evaluating the following expression.

$$\frac{2.4 \times 10^2}{6.0 \times 10^{23}}$$

Solution

$$\frac{2.4 \times 10^2}{6.0 \times 10^{23}} = \frac{2.4}{6.0} \times \frac{10^2}{10^{23}}$$ Divide the numbers and the powers of 10 separately.

$$= \frac{2.4}{6.0} \times 10^{2-23}$$ For the powers of 10, use the quotient rule. Keep the base 10 and subtract the exponents.

$$= 0.4 \times 10^{-21}$$ Do the division. Then subtract the exponents.

$$= 4.0 \times 10^{-1} \times 10^{-21}$$ Write 0.4 in scientific notation as 4.0×10^{-1}.

$$= 4.0 \times 10^{-22}$$ Keep the base of 10 and add the exponents.

Calculators
Entering scientific notation

We can evaluate the expression from Example 4 by entering the numbers in scientific notation using the EE key on a scientific calculator:

2.4 EE 2 ÷ 6 EE 23 =

One atom of uranium weighs 4.0×10^{-22} gram. Written in standard notation, this is 0.0000000000000000000004 g.

Self Check 4 Find the approximate weight (in grams) of one atom of gold by evaluating

$$\frac{1.98 \times 10^2}{6.0 \times 10^{23}}$$

Answers to Self Checks **1. a.** 4,880,000, **b.** 0.0098 **2. a.** 9.3×10^7, **b.** 9.055×10^{-5}, **c.** 8.5×10^{-2} **4.** 3.3×10^{-22} g

4.3 STUDY SET

ELEMENTARY Algebra $f(x)$ Now™

VOCABULARY Fill in the blanks.

1. 4.84×10^9 and 1.05×10^{-2} are written in ________ notation.
2. The number 125,000 is written in ________ notation.
3. Scientific ________ provides a compact way of writing very large or very small numbers.
4. A number written in scientific notation is the product of a ________ that is at least 1, but less than 10, and a power of ____.
5. 10^{34}, 10^{50}, and 10^{-14} are ________ of 10.
6. The numbers 2, 5, 8, and 15 are examples of ________ integers. The numbers -1, -6, -30, and -45 are examples of negative ________.

CONCEPTS Fill in the blanks.

7. When we multiply a decimal by 10^5, the decimal point moves 5 places to the ______.
   When we multiply a decimal by 10^{-7}, the decimal point moves 7 places to the _____.
8. Multiplying a decimal by 10^0 does not move the decimal point, because $10^0 = $ ____.
9. The arrows show the movement of a decimal point. By what power of 10 was each decimal multiplied?
   a. 0.0 0 0 0 0 0 5 56
   b. 8, 0 4 1, 0 0 0, 0 0 0.
10. Fill in the blanks to describe the procedure for converting a number from scientific notation to standard form.
    a. If the exponent is positive, move the decimal point the same number of places to the ______ as the exponent.
    b. If the exponent is negative, move the decimal point the same number of places to the _____ as the absolute value of the exponent.
11. **a.** When a real number greater than 1 is written in scientific notation, the exponent on 10 is a ________ integer.
    b. When a real number between 0 and 1 is written in scientific notation, the exponent on 10 is a ________ integer.
12. A portion of the scientific notation for a positive number is blocked out. Decide whether each statement is true or false.
    a. ■ $\times 10^{-8} > 0$ **b.** ■ $\times 10^{-6} > 1$
    c. ■ $\times 10^3 > 0$ **d.** ■ $\times 10^5 > 10$

Fill in the blanks to write each number in scientific notation.

13. **a.** $7{,}700 = $ ____ $\times 10^3$
    b. $500{,}000 = $ ____ $\times 10^5$
    c. $114{,}000{,}000 = 1.14 \times 10^{\square}$

14. **a.** $0.0082 = \square \times 10^{-3}$

b. $0.0000001 = \square \times 10^{-7}$

c. $0.00003457 = 3.457 \times 10^{\square}$

15. Write each expression so that the decimal numbers are grouped together and the powers of ten are grouped together.

a. $(5.1 \times 10^9)(1.5 \times 10^{22})$

b. $\dfrac{8.8 \times 10^{30}}{2.2 \times 10^{19}}$

16. Simplify each expression.

a. $10^{24} \times 10^{33}$ **b.** $\dfrac{10^{50}}{10^{36}}$

c. $(10^{18})^3$ **d.** $\dfrac{10^{15} \times 10^{27}}{10^{40}}$

NOTATION

17. Fill in the blanks. A positive number is written in scientific notation when it is written in the form $N \times 10^n$, where $\square \leq N < \square$ and n is an ________

18. In the basic form for scientific notation, what type of number can be written as the exponent?

$$\blacksquare \times 10^{\blacksquare}$$

What type of number can be written as the coefficient?

PRACTICE **Write each number in standard notation.**

19. 2.3×10^2 **20.** 3.75×10^4

21. 8.12×10^5 **22.** 1.2×10^3

23. 1.15×10^{-3} **24.** 4.9×10^{-2}

25. 9.76×10^{-4} **26.** 7.63×10^{-5}

27. 6.001×10^6 **28.** 9.998×10^5

29. 2.718×10^0 **30.** 3.14×10^0

31. 6.789×10^{-2} **32.** 4.321×10^{-1}

33. 2.0×10^{-5} **34.** 7.0×10^{-6}

35. 9.0×10^9 **36.** 8.0×10^8

Write each number in scientific notation.

37. 23,000 **38.** 4,750

39. 1,700,000 **40.** 290,000

41. 0.062 **42.** 0.00073

43. 0.0000051 **44.** 0.04

45. 5,000,000,000 **46.** 7,000,000

47. 0.0000003 **48.** 0.0001

49. 909,000,000 **50.** 7,007,000,000

51. 0.0345 **52.** 0.000000567

53. 9 **54.** 2

55. 1,718,000,000,000,000,000

56. 44,180,000,000,000,000,000

57. 0.0000000000000123

58. 0.00000000000000000555

Use scientific notation to perform the calculations. Give all answers in standard notation.

59. $(3.4 \times 10^2)(2.1 \times 10^3)$

60. $(4.1 \times 10^{-3})(3.4 \times 10^4)$

61. $(8.4 \times 10^{-13})(4.8 \times 10^9)$

62. $(5.5 \times 10^{-15})(2.2 \times 10^{13})$

63. $\dfrac{9.3 \times 10^2}{3.1 \times 10^{-2}}$ **64.** $\dfrac{7.2 \times 10^6}{1.2 \times 10^8}$

65. $\dfrac{0.00000129}{0.0003}$ **66.** $\dfrac{169{,}000{,}000{,}000}{26{,}000{,}000}$

67. (0.0000000056)(5,500,000)

68. (0.000000061)(3,500,000,000)

69. $\dfrac{96{,}000}{(12{,}000)(0.00004)}$ **70.** $\dfrac{(0.48)(14{,}400{,}000)}{96{,}000{,}000}$

Find each power.

71. $(456.4)^6$

72. $(0.009)^{-6}$

73. 225^{-5}

74. $\left(\dfrac{1}{3}\right)^{-55}$

APPLICATIONS

75. ASTRONOMY The distance from Earth to Alpha Centauri (the nearest star outside our solar system) is about 25,700,000,000,000 miles. Express this number in scientific notation.

76. SPEED OF SOUND The speed of sound in air is 33,100 centimeters per second. Express this number in scientific notation.

77. GEOGRAPHY The largest ocean in the world is the Pacific Ocean, which covers 6.38×10^7 square miles. Express this number in standard notation.

78. ATOMS The number of atoms in 1 gram of iron is approximately 1.08×10^{22}. Express this number in standard notation.

79. LENGTH OF A METER One meter is approximately 0.00622 mile. Use scientific notation to express this number.

80. ANGSTROMS One angstrom is 1.0×10^{-7} millimeter. Express this number in standard notation.

81. WAVELENGTHS Transmitters, vacuum tubes, and lights emit energy that can be modeled as a wave, as shown below. Examples of the most common types of electromagnetic waves are given in the table. List the wavelengths in order from shortest to longest.

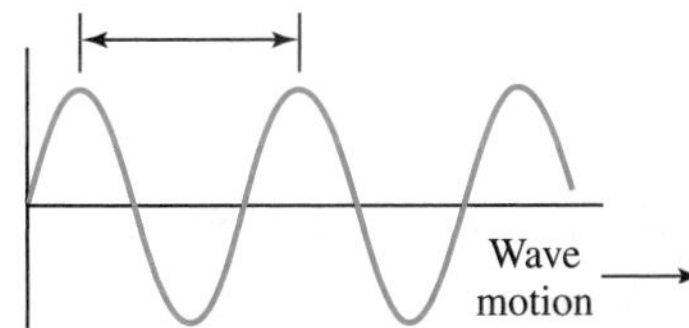

Type	Use	Wavelength (m)
visible light	lighting	9.3×10^{-6}
infrared	photography	3.7×10^{-5}
x-ray	medical	2.3×10^{-11}
radio wave	communication	3.0×10^{2}
gamma ray	treating cancer	8.9×10^{-14}
microwave	cooking	1.1×10^{-2}
ultraviolet	sun lamp	6.1×10^{-8}

82. EXPLORATION On July 4, 1997, the *Pathfinder,* carrying the rover vehicle called Sojourner, landed on Mars to perform a scientific investigation of the planet. The distance from Mars to Earth is approximately 3.5×10^7 miles. Use scientific notation to express this distance in feet. (*Hint:* 5,280 feet = 1 mile.)

83. PROTONS The mass of one proton is approximately 1.7×10^{-24} gram. Use scientific notation to express the mass of 1 million protons.

84. SPEED OF SOUND The speed of sound in air is approximately 3.3×10^4 centimeters per second. Use scientific notation to express this speed in kilometers per second. (*Hint:* 100 centimeters = 1 meter and 1,000 meters = 1 kilometer.)

85. LIGHT YEARS One light year is about 5.87×10^{12} miles. Use scientific notation to express this distance in feet. (*Hint:* 5,280 feet = 1 mile.)

86. OIL RESERVES As of January 1, 2001, Saudi Arabia was believed to have crude oil reserves of about 2.617×10^{11} barrels. A barrel contains 42 gallons of oil. Use scientific notation to express its oil reserves in gallons. (Source: *The World Almanac and Book of Facts 2003.*)

87. INTEREST As of December 2000, the Federal Deposit Insurance Corporation (FDIC) reported that the total insured deposits in U.S. banks and savings and loans was approximately 5.2×10^{12} dollars. If this money was invested at a rate of 4% simple annual interest, how much would it earn in 1 year? Use scientific notation to express the answer. (Source: *The World Almanac and Book of Facts 2003.*)

88. CURRENCY As of June 30, 2002, the U.S. Treasury reported that the number of $20 bills in circulation was approximately 4.84×10^9. What was the total value of the currency? Use scientific notation to express the answer. (Source: *The World Almanac and Book of Facts 2003.*)

89. THE MILITARY The graph shows the number of U.S. troops for 1983–2001. Estimate each of the following and express your answers in scientific and standard notation.

a. The number of troops in 1993

b. The largest number of troops during these years

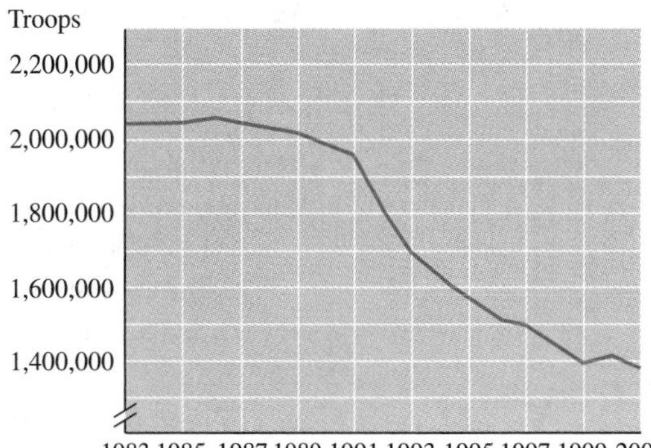

Source: The U.S. Department of Defense

90. THE NATIONAL DEBT The graph shows the growth of the national debt for the fiscal years 1993–2002.

a. Use scientific notation to express the debt as of 1994, 1995, and 2001.

b. In 2002, the population of the United States was about 2.88×10^8. Estimate the share of the debt for each man, woman, and child in the United States. Answer in standard notation.

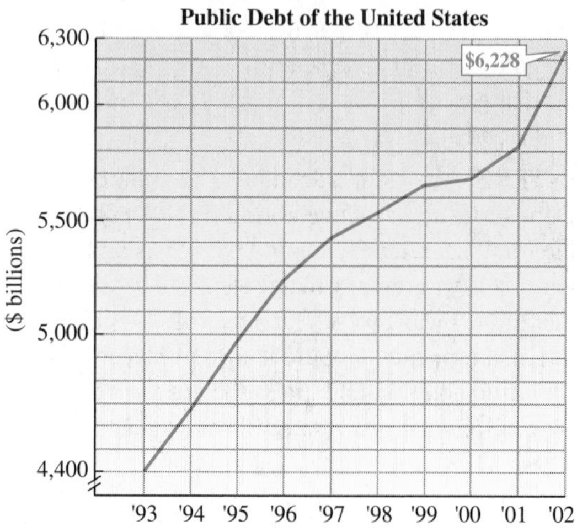

Source: The U.S. Department of the Treasury

WRITING

91. In what situations would scientific notation be more convenient than standard notation?

92. To multiply a number by a power of 10, we move the decimal point. Which way, and how far? Explain.

93. 2.3×10^{-3} contains a negative sign but represents a positive number. Explain.

94. Explain why 237.8×10^8 is not written in scientific notation.

REVIEW

95. If $y = -1$, find the value of $-5y^{55}$.

96. What is the y-intercept of the graph of $y = -3x - 5$?

97. COUNSELING In the first year of her practice, a family counselor saw 75 clients. In her second year, the number of clients grew to 105. If a linear trend continues, write an equation that gives the number of clients c the counselor will have t years after beginning her practice.

98. Is $(0, -5)$ a solution of $2x + 3y \geq -14$?

CHALLENGE PROBLEMS

99. Consider 2.5×10^{-4}.

a. What is its opposite?

b. What is its reciprocal?

100. a. Write one million and one millionth in scientific notation.

b. By what number must we multiply one millionth to get one million?

4.4 Polynomials

• Polynomials • Evaluating Polynomials • Graphing Nonlinear Equations

The simplest types of algebraic expressions are *polynomials.* In this section, we will define polynomials and introduce the vocabulary that is used to describe them.

POLYNOMIALS

The Language of Algebra

The prefix *poly* means many. Some other words that begin with this prefix are *poly*gon and *poly*unsaturated.

Recall that in the expression $x^4 + 6x^2 + 4x + 25$, x^4, $6x^2$, $4x$, and 25 are called **terms** and that the numerical factor of each term is called its **coefficient.** Therefore, x^4, $6x^2$, $4x$, and 25 have coefficients of 1, 6, 4, and 25, respectively.

A **polynomial** is an expression that consists of one or more terms. The exponents on the variables in a polynomial must be whole numbers.

Polynomials A **polynomial** is a single term or a sum of terms in which all variables have whole-number exponents. No variable appears in a denominator.

Here are some examples of polynomials:

$$3x + 2, \quad 4y^2 - 2y - 3, \quad -8xy^2, \quad \text{and} \quad a^3 + 3a^2b + 3ab^2 + b^3$$

Caution

The expression $6x^3 + 4x^{-2}$ is not a polynomial, because of the negative exponent on the variable in $4x^{-2}$. Similarly, $y^2 - \frac{7}{y} + 1$ is not a polynomial, because $\frac{7}{y}$ can be written $7y^{-1}$.

The polynomial $3x + 2$ is the sum of two terms, $3x$ and 2, and we say it is a **polynomial in *x*.** A single number is called a **constant,** and so its last term, 2, is called the **constant term.**

Since $4y^2 - 2y - 3$ can be written as $4y^2 + (-2y) + (-3)$, it is the sum of three terms, $4y^2$, $-2y$, and -3. It is written in **descending powers of *y*,** because the exponents on y decrease from left to right. When a polynomial is written in descending powers, the first term, in this case $4y^2$, is called the **lead term.** The coefficient of the lead term, in this case 4, is called the **lead coefficient.**

$-8xy^2$ is a polynomial with just one term. We say that it is a **polynomial in *x* and *y*.** The four-term polynomial $a^3 + 3a^2b + 3ab^2 + b^3$ is written in descending powers of a and **ascending powers** of b.

The Language of Algebra

The prefix *mono* means one; Jay Leno begins the *Tonight Show* with a monologue. The prefix *bi* means two, as in bicycle or binoculars. The prefix *tri* means three, as in triangle or the *Lord of the Rings Trilogy.*

Polynomials are classified according to the number of terms they have. A polynomial with exactly one term is called a **monomial;** exactly two terms, a **binomial;** and exactly three terms, a **trinomial.** Polynomials with four or more terms have no special names.

Polynomials

Monomials	Binomials	Trinomials
$-6x$	$3u^3 - 4u^2$	$-5t^2 + 4t + 3$
$5x^2y$	$18a^2b + 4ab$	$27x^3 - 6x - 2$
29	$-29z^{17} - 1$	$a^2 + 2ab + b^2$

Polynomials and their terms can be described according to the exponents on their variables.

Degree of a Term

The **degree of a term** of a polynomial in one variable is the value of the exponent on the variable. If a polynomial is in more than one variable, the **degree of a term** is the sum of the exponents on the variables. The **degree of a nonzero constant** is 0.

Caution

Since 0 could be written as $0x^0$, $0x^1$, $0x^2$, and so on, the constant 0 has no defined degree.

Here are some examples:

$7x^6$ has degree 6.
$-2a^4$ has degree 4.
$47x^2y$ has degree 3 because x^2y can be written x^2y^1 and $2 + 1 = 3$.
8 has degree 0 since it can be written as $8x^0$.

We can determine the *degree of a polynomial* by considering the degrees of each of its terms.

Degree of a Polynomial

The **degree of a polynomial** is the same as the highest degree of any term of the polynomial.

EXAMPLE 1

Describe each polynomial: **a.** $d^3 + 4d^2 - 16$, **b.** $\frac{1}{2}x^2 - x$, and **c.** $-6y^{14} - 15y^9z^9 + 25y^8z^{10} + 4yz^{11}$.

Solution

a. Since $d^3 + 4d^2 - 16$ has three terms, it is a trinomial. It is written in descending powers of d. The highest degree of any term is 3, so it is of degree 3.

Term	Coefficient	Degree
d^3	1	3
$4d^2$	4	2
-16	-16	0

b. Since $\frac{1}{2}x^2 - x$ has two terms, it is a binomial. It is written in descending powers of x. The highest degree of any term is 2, so it is of degree 2.

Term	Coefficient	Degree
$\frac{1}{2}x^2$	$\frac{1}{2}$	2
$-x$	-1	1

c. $-6y^{14} - 15y^9z^9 + 25y^8z^{10} + 4yz^{11}$ is a polynomial with 4 terms. It is written in descending powers of y and ascending powers of z. The highest degree of any term is 18, so it is of degree 18.

Term	Coefficient	Degree
$-6y^{14}$	-6	14
$-15y^9z^9$	-15	18
$25y^8z^{10}$	25	18
$4yz^{11}$	4	12

The Language of Algebra

To *descend* means to move from higher to lower. A classic scene in the movie "Gone with the Wind" is Scarlett O'Hara *descending* the grand staircase. In $x^3 + x^2 + x$, the powers of x are written in *descending* order, as we read from left to right.

Self Check 1 Describe each polynomial: **a.** $x^2 + 4x - 16$ and **b.** $-14s^5t + s^4t^3$.

EVALUATING POLYNOMIALS

A polynomial can have different values depending on the number that is substituted for its variable.

EXAMPLE 2

Evaluate $3x^2 + 4x - 5$ for **a.** $x = 0$ and **b.** $x = -2$.

ELEMENTARY Algebra f(x) Now™

Solution We substitute the given value for each x and follow the rules for the order of operations.

a.
$$\begin{aligned} 3x^2 + 4x - 5 &= 3(0)^2 + 4(0) - 5 && \text{Substitute 0 for } x. \\ &= 3(0) + 4(0) - 5 \\ &= 0 + 0 - 5 \\ &= -5 \end{aligned}$$

b.
$$\begin{aligned} 3x^2 + 4x - 5 &= 3(-2)^2 + 4(-2) - 5 && \text{Substitute } -2 \text{ for } x. \\ &= 3(4) + 4(-2) - 5 \\ &= 12 + (-8) - 5 \\ &= -1 \end{aligned}$$

Self Check 2 Evaluate $-x^3 + x - 2x + 3$ for **a.** $x = 0$ and **b.** $x = -3$.

EXAMPLE 3

Supermarket displays. The polynomial

$$\frac{1}{3}c^3 + \frac{1}{2}c^2 + \frac{1}{6}c$$

gives the number of cans used in a display shaped like a square pyramid, having a square base formed by c cans per side. Find the number of cans used in the display shown here.

Solution Since each side of the square base of the display is formed by 4 cans, c is 4.

$$\frac{1}{3}c^3 + \frac{1}{2}c^2 + \frac{1}{6}c = \frac{1}{3}(4)^3 + \frac{1}{2}(4)^2 + \frac{1}{6}(4) \quad \text{Substitute 4 for } c.$$

$$= \frac{1}{3}(64) + \frac{1}{2}(16) + \frac{1}{6}(4) \quad \text{Find the powers.}$$

$$= \frac{64}{3} + 8 + \frac{2}{3} \quad \text{Do the multiplication, and then simplify: } \tfrac{4}{6} = \tfrac{2}{3}.$$

$$= 30 \quad \text{Add the fractions: } \tfrac{64}{3} + \tfrac{2}{3} = \tfrac{66}{3} = 22.$$

30 cans of soup were used in the display.

GRAPHING NONLINEAR EQUATIONS

We have graphed equations such as $y = x$ and $y = 2x - 3$. These equations are said to be linear equations because their graphs are straight lines. Note that the right-hand side of each equation is a polynomial of degree 1.

$$y = x \qquad y = 2x - 3$$

The degree of each polynomial is 1.

We can also graph equations defined by polynomials with degrees greater than 1.

$y = x^2$ — The degree of this polynomial is 2.

$y = -x^2 + 2$ — The degree of this polynomial is 2.

$y = x^3 + 1$ — The degree of this polynomial is 3.

EXAMPLE 4

Graph: $y = x^2$.

Solution To make a table of solutions, we choose values for x and find the corresponding values of y. If $x = -3$, we have

$$y = x^2$$
$$y = (-3)^2 \quad \text{Substitute } -3 \text{ for } x.$$
$$y = 9$$

Thus, $(-3, 9)$ is a solution. In a similar manner, we find the corresponding y-values for x-values of $-2, -1, 0, 1, 2$, and 3. If we plot the ordered pairs listed in the table and join the points with a smooth curve, we get the graph shown below, which is called a **parabola.**

The Language of Algebra

The cup-like shape of a *parabola* has many real-life applications. Did you know that a satellite TV dish is more formally known in the electronics industry as a *parabolic* dish?

$y = x^2$

x	y	(x, y)
-3	9	$(-3, 9)$
-2	4	$(-2, 4)$
-1	1	$(-1, 1)$
0	0	$(0, 0)$
1	1	$(1, 1)$
2	4	$(2, 4)$
3	9	$(3, 9)$

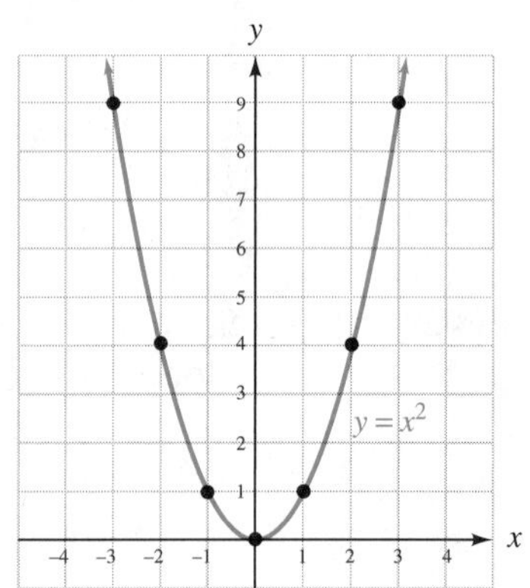

Self Check 4 Graph: $y = x^2 - 2$.

EXAMPLE 5

Graph: $y = -x^2 + 2$.

ELEMENTARY Algebra f(x) Now™

Solution To make a table of solutions, we select x-values of $-3, -2, -1, 0, 1, 2,$ and 3 and find each corresponding y-value. For example, if $x = -3$, we have

$$y = -x^2 + 2$$
$$y = -(-3)^2 + 2 \quad \text{Substitute } -3 \text{ for } x.$$
$$y = -(9) + 2$$
$$y = -7$$

The ordered pair $(-3, -7)$ is a solution. Six other solutions appear in the table. After plotting each pair, we join the points with a smooth curve to obtain the graph, a parabola opening downward.

x	y	(x, y)
-3	-7	$(-3, -7)$
-2	-2	$(-2, -2)$
-1	1	$(-1, 1)$
0	2	$(0, 2)$
1	1	$(1, 1)$
2	-2	$(2, -2)$
3	-7	$(3, -7)$

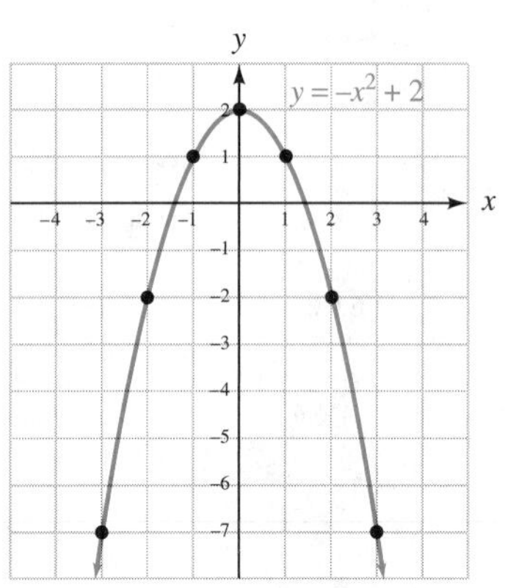

Self Check 5 Graph: $y = -x^2$.

EXAMPLE 6

Graph: $y = x^3 + 1$.

ELEMENTARY Algebra f(x) Now™

Solution If we let $x = -2$, we have

$$y = x^3 + 1$$
$$y = (-2)^3 + 1 \quad \text{Substitute } -2 \text{ for } x.$$
$$y = -8 + 1$$
$$y = -7$$

The ordered pair $(-2, -7)$ is a solution. This pair and others that satisfy the equation are listed in the table. Plotting the ordered pairs and joining the points with a smooth curve gives us the graph.

$y = x^3 + 1$

x	y	(x, y)
−2	−7	(−2, −7)
−1	0	(−1, 0)
0	1	(0, 1)
1	2	(1, 2)
2	9	(2, 9)

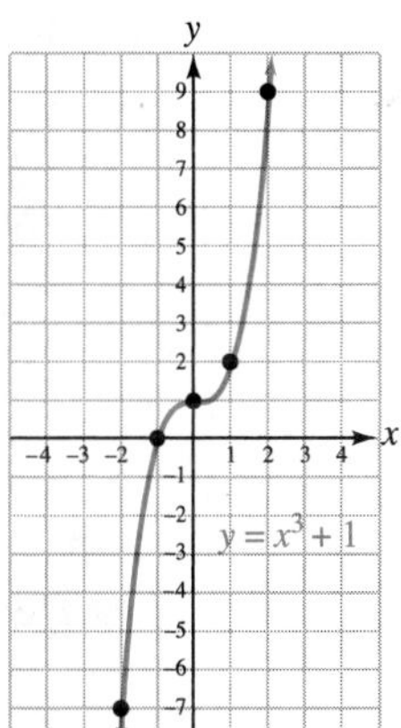

Self Check 6 Graph: $y = x^3 - 1$.

Since the graphs of $y = x^2$, $y = -x^2 + 2$, and $y = x^3 + 1$ are not lines, they are called **nonlinear equations.**

Answers to Self Checks **1. a.** trinomial of degree 2; terms: x^2, $4x$, -16; coefficients: 1, 4, −16; degree: 2, 1, 0. **b.** binomial of degree 7; terms: $-14s^5t$, s^4t^3; coefficients: −14, 1; degree: 6, 7 **2. a.** 3, **b.** 33 **4.** The graph has the same shape, but is 2 units lower.

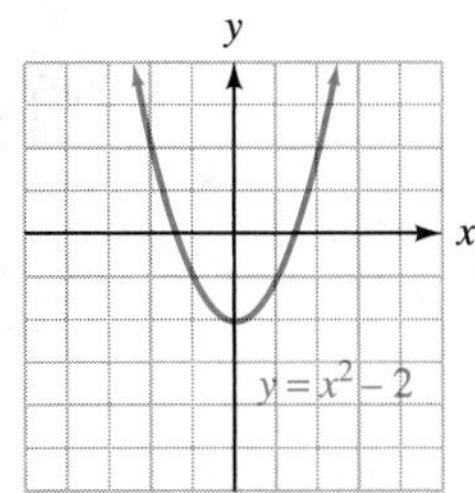

5.

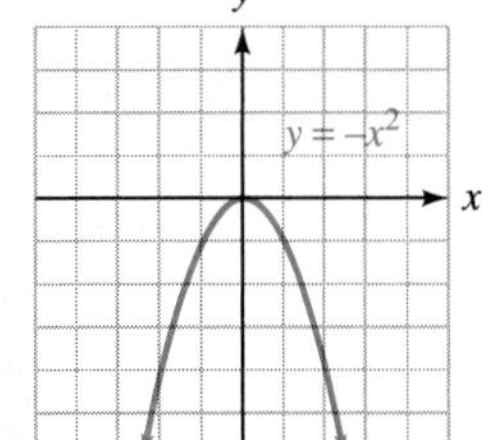

6.

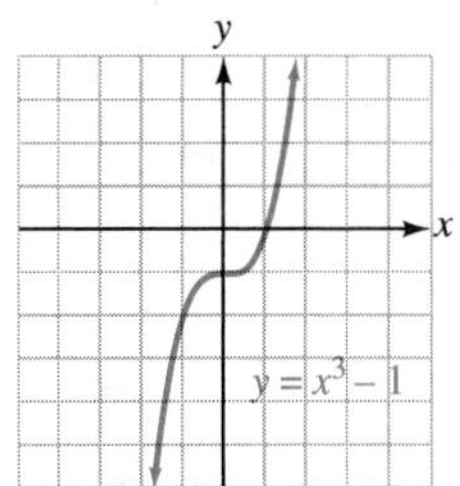

4.4 STUDY SET  ELEMENTARY Algebra f(x) Now™

VOCABULARY **Fill in the blanks.**

1. A __________ is a term or a sum of terms in which all variables have whole-number exponents.
2. The numerical __________ of the term $-25x^2y^3$ is -25.
3. The degree of a polynomial is the same as the degree of its ______ with the highest degree.
4. A __________ is a polynomial with one term. A __________ is a polynomial with two terms. A __________ is a polynomial with three terms.
5. The ________ of the monomial $3x^7$ is 7.
6. For the polynomial $6x^2 + 3x - 1$, the ______ term is $6x^2$, and the lead __________ is 6. The ________ term is -1.
7. $x^3 - 6x^2 + 9x - 2$ is a polynomial in ___ and is written in ____________________ powers of x.
8. To ________ the polynomial $x^2 - 2x + 1$ for $x = 6$, we substitute 6 for x and simplify.
9. Because the graph of $y = x^3$ is not a straight line, we call $y = x^3$ a __________ equation.
10. The graph of $y = x^2$ is a cup-shaped curve called a ________.

CONCEPTS **Decide whether each expression is a polynomial.**

11. **a.** $x^3 - 5x^2 - 2$ **b.** $x^{-4} - 5x$
c. $x^2 - \dfrac{1}{2x} + 3$ **d.** $x^3 - 1$
e. $x^2 - y^2$ **f.** $a^4 + a^3 + a^2 + a$

12. Complete the table for each polynomial.

a. $8x^2 + x - 7$

Term	Coefficient	Degree

b. $y^4 - y^3 + 16y^2 + 3y$

Term	Coefficient	Degree

c. $8a^6b^3 - 27ab$

Term	Coefficient	Degree

13. Is $(-1, 2)$ a solution of $y = x^2 + 3$?

14. Complete the table of solutions.

$y = x^3 + 5$

x	y	(x, y)
-2		
-1		
0		
1		
2		

15. To graph $y = x^2 - 4$, a table of solutions is constructed and a graph is drawn. Explain the error.

$y = x^2 - 4$

x	y	(x, y)
0	-4	$(0, -4)$
2	0	$(2, 0)$

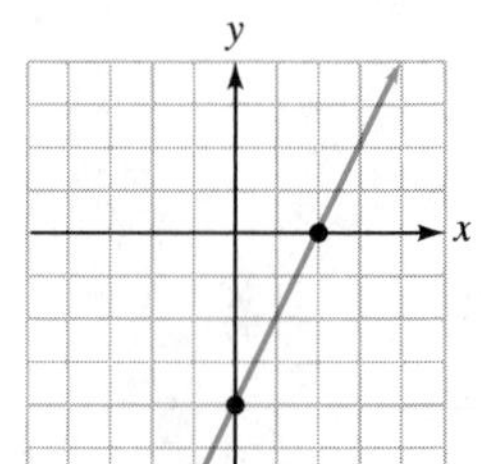

16. Explain the error in the graph of $y = x^2$.

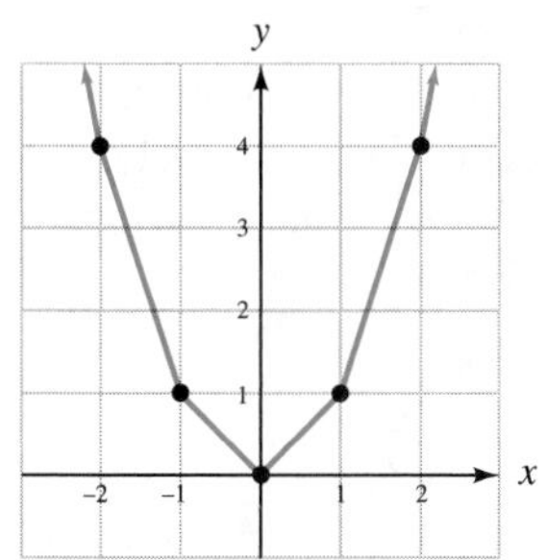

NOTATION **Complete the solution.**

17. Evaluate $-2x^2 + 3x - 1$ for $x = -2$.

$$\begin{aligned} -2x^2 + 3x - 1 &= -2(\quad)^2 + 3(\quad) - 1 \\ &= -2(\quad) + 3(\quad) - 1 \\ &= \quad + (-6) - 1 \\ &= \quad - 1 \\ &= \quad \end{aligned}$$

18. a. Write $x - 9 + 3x^2$ in descending powers of x.

b. Write $-2xy + y^2 + x^2$ in ascending powers of y.

19. Which of the following equations are defined by a polynomial of degree 2 or higher?

$y = x^2 - 1 \qquad y = x - 1 \qquad y = x^3 - 1$

20. The expression $x + y$ is a binomial. Is xy also a binomial? Explain why or why not.

PRACTICE **Classify each polynomial as a monomial, a binomial, a trinomial, or none of these.**

21. $3x + 7$ **22.** $3y - 5$
23. $y^2 + 4y + 3$ **24.** $3xy$
25. $3z^2$ **26.** $3x^4 - 2x^3 + 3x - 1$
27. $t - 32$ **28.** $9x^2y^3z^4$
29. $s^2 - 23s + 31$ **30.** $2x^3 - 5x^2 + 6x - 3$
31. $3x^5 - x^4 - 3x^3 + 7$ **32.** x^3
33. $2a^2 - 3ab + b^2$ **34.** $a^3 - b^3$

Find the degree of each polynomial.

35. $3x^4$ **36.** $3x^5$
37. $-2x^2 + 3x + 1$ **38.** $-5x^4 + 3x^2 - 3x$
39. $3x - 5$ **40.** $y^3 + 4y^2$
41. $-5r^2s^2 - r^3s + 3$ **42.** $4r^2s^3 - 5r^2s^8$
43. $x^{12} + 3x^2y^3$ **44.** $17ab^5 - 12a^3b$
45. 38 **46.** -24

Evaluate each expression.

47. $-9x + 1$ for
a. $x = 5$ **b.** $x = -4$

48. $-10x + 6$ for
a. $x = 8$ **b.** $x = -6$

49. $3x^2 - 2x + 8$ for
a. $x = 1$ **b.** $x = 0$

50. $4x^2 + 2x - 8$ for
a. $x = -1$ **b.** $x = -10$

51. $-x^2 - 6$ for
a. $x = -4$ **b.** $x = 20$

52. $-x^3 + 2$ for
a. $x = -3$ **b.** $x = 10$

53. $x^3 + 3x^2 + 2x + 4$ for
a. $x = 2$ **b.** $x = -2$

54. $x^3 - 3x^2 - x + 9$ for
a. $x = 3$ **b.** $x = -3$

55. $x^4 - x^3 + x^2 + 2x - 1$ for
a. $x = 1$ **b.** $x = -1$

56. $-x^4 + x^3 + x^2 + x + 1$ for
a. $x = 1$ **b.** $x = -1$

Complete the table of solutions and graph the equation.

57. $y = x^2$

x	y
−3	
−2	
−1	
0	
1	
2	
3	

58. $y = x^3$

x	y
−2	
−1	
0	
1	
2	

Construct a table of solutions. Then graph the equation.

59. $y = x^2 + 1$

60. $y = x^2 - 4$

61. $y = -x^2 - 2$

62. $y = -x^2 + 1$

63. $y = 2x^2 - 3$

64. $y = -2x^2 + 2$

65. $y = x^3 + 2$

66. $y = x^3 + 4$

67. $y = -x^3 - 1$

68. $y = -x^3$

APPLICATIONS

69. SUPERMARKETS A grocer plans to set up a pyramid-shaped display of cantaloupes like that shown in Example 3. If each side of the square base of the display is made of six cantaloupes, how many will be used in the display?

70. PACKAGING To make boxes, a manufacturer cuts equal-sized squares from each corner of a 10 in. × 12 in. piece of cardboard, and then folds up the sides. The polynomial $4x^3 - 44x^2 + 120x$ gives the volume (in cubic inches) of the resulting box when a square with sides x inches long is cut from each corner. Find the volume of a box if 3-inch squares are cut out.

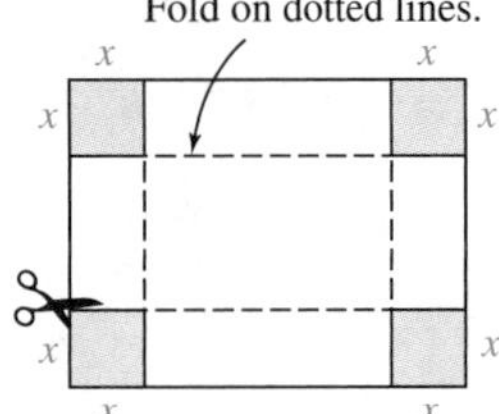

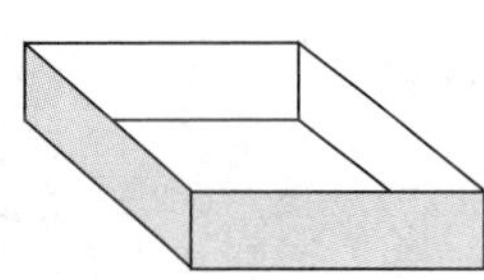

71. STOPPING DISTANCE The number of feet that a car travels before stopping depends on the driver's reaction time and the braking distance, as shown in the illustration. For one driver, the stopping distance is given by the polynomial

$$0.04v^2 + 0.9v$$

where v is the velocity of the car. Find the stopping distance when the driver is traveling at 30 mph.

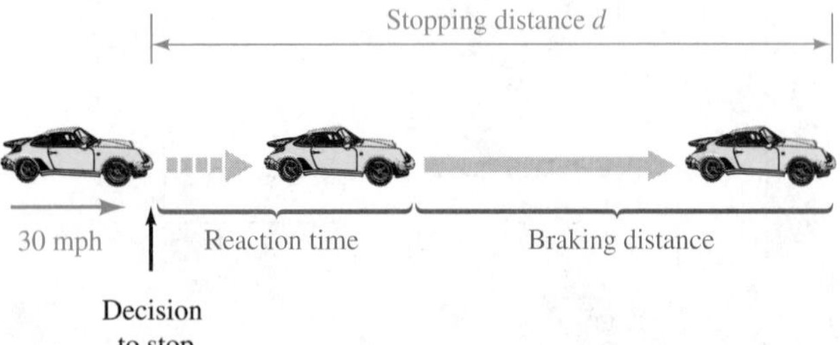

72. SUSPENSION BRIDGES The polynomial

$$-0.0000001s^4 + 0.0066667s^2 + 400$$

approximates the length of the cable between the two vertical towers of a suspension bridge, where s is the sag in the cable. Estimate the length of the cable if the sag is 24.6 feet.

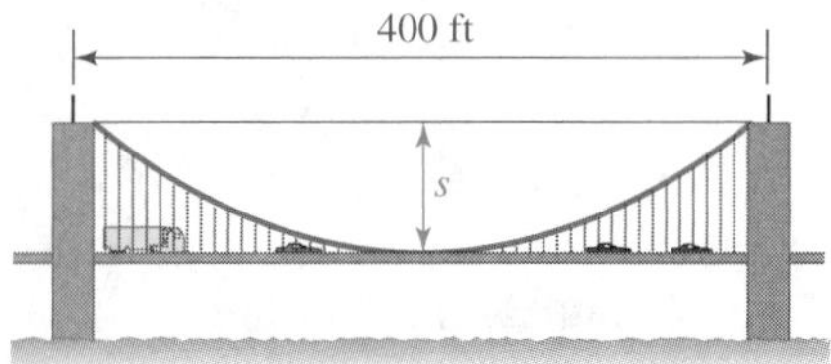

73. SCIENCE HISTORY The renowned Italian scientist Galileo Galilei (1564–1642) built an incline plane like that shown to study falling objects. As the ball rolled down, he measured the time it took the ball to travel different distances. Graph the data and then connect the points with a smooth curve.

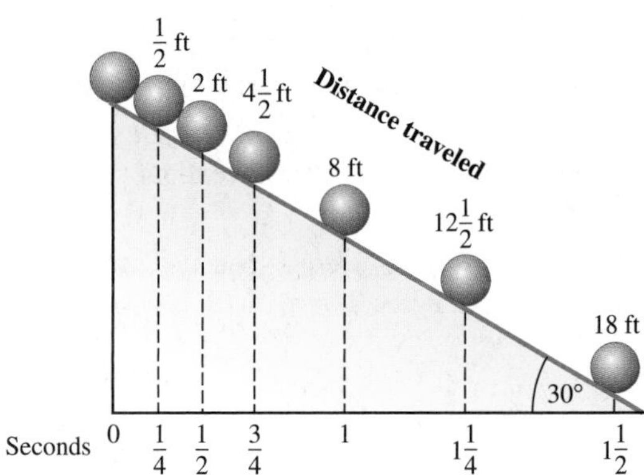

74. DOLPHINS At a marine park, three trained dolphins jump in unison over an arching stream of water whose path can be described by the equation

$$y = -0.05x^2 + 2x$$

Given the takeoff points for each dolphin, how high must each jump to clear the stream of water?

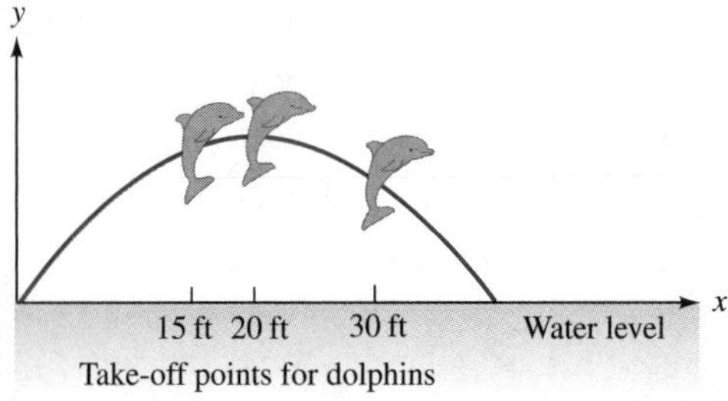

75. MANUFACTURING The graph shows the relationship between the length l (in inches) of a machine bolt and the cost C (in cents) to manufacture it.

a. What information does the point (2, 8) on the graph give us?

b. How much does it cost to make a 7-inch bolt?

c. What length bolt is the least expensive to make?

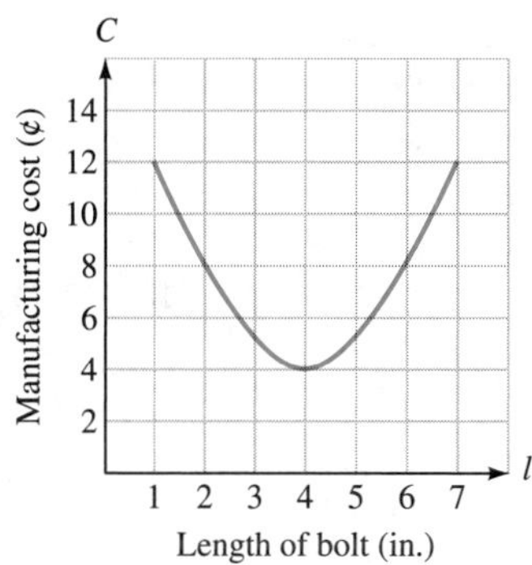

76. SOFTBALL The graph shows the relationship between the distance d (in feet) traveled by a batted softball and the height h (in feet) it attains.

a. What information does the point (40, 40) on the graph give us?

b. At what distance from home plate does the ball reach its maximum height?

c. Where will the ball land?

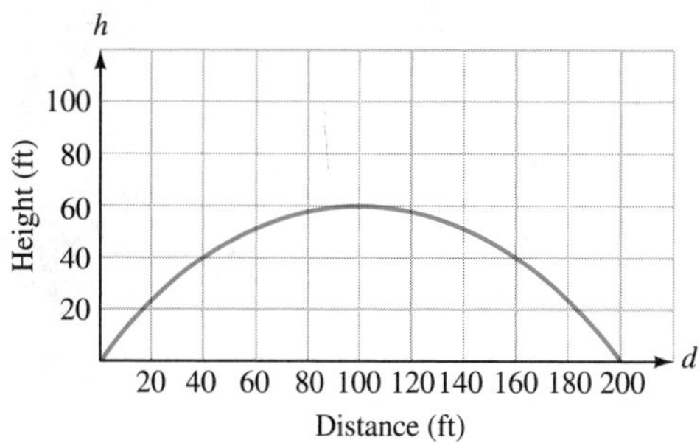

WRITING

77. Describe how to determine the degree of a polynomial.

78. List some words that contain the prefixes *mono, bi,* or *tri.*

REVIEW **Solve each inequality. Write the solution set in interval notation and graph it.**

79. $-4(3y + 2) \le 28$

80. $-5 < 3t + 4 \le 13$

Simplify each expression.

81. $(x^2x^4)^3$

82. $(a^2)^3(a^3)^2$

83. $\left(\frac{y^2y^5}{y^4}\right)^3$

84. $\left(\frac{2t^3}{t}\right)^{-4}$

CHALLENGE PROBLEMS

85. Find a three-term polynomial of degree 2 whose value will be 1 when it is evaluated for $x = -2$.

86. Graph: $y = 2x^3 - 3x^2 - 11x + 6$

4.5 Adding and Subtracting Polynomials

- Simplifying Polynomials by Combining Like Terms
- Adding Polynomials
- Subtracting Polynomials
- Adding and Subtracting Multiples of Polynomials
- Applications

The heights of the Seattle Space Needle and the Eiffel Tower are shown in the illustration. We can find the difference in their heights by subtracting 607 from 984.

Arithmetic

$984 - 607 = 377$

The difference in height is 377 feet.

Algebra

$(x^2 - 3x + 2) - (5x - 10) = ?$

The heights of the two Greek columns shown above are expressed as polynomials. To find the difference in their heights, we must subtract $5x - 10$ from $x^2 - 3x + 2$. In this section, we will learn how to perform this subtraction.

SIMPLIFYING POLYNOMIALS BY COMBINING LIKE TERMS

Recall that **like terms** have the same variables with the same exponents:

Like terms	***Unlike terms***
$-7x$ and $15x$	$-7x$ and $15a$
$4y^3$ and $16y^3$	$4y^3$ and $16y^2$
$\frac{1}{2}xy^2$ and $-\frac{1}{3}xy^2$	$\frac{1}{2}xy^2$ and $-\frac{1}{3}x^2y$

The Language of Algebra

Simplifying the sum or difference of like terms is called *combining like terms.*

Also recall that to **combine like terms,** we combine their coefficients and keep the same variables with the same exponents. For example,

$$4y + 5y = (4 + 5)y \qquad 8x^2 - x^2 = (8 - 1)x^2$$
$$= 9y \qquad\qquad = 7x^2$$

Polynomials with like terms can be simplified by combining like terms.

EXAMPLE 1

ELEMENTARY Algebra f(x) Now™

Simplify each polynomial: **a.** $4x^4 + 81x^4$, **b.** $17x^2y^2 + 2x^2y - 6x^2y^2$, **c.** $-3r - 4r + 6r$, and **d.** $ab + 8 - 15 + 4ab$.

Solution

a. $4x^4 + 81x^4 = 85x^4$ $\quad (4 + 81)x^4 = 85x^4$.

b. The first and third terms are like terms.

$$17x^2y^2 + 2x^2y - 6x^2y^2 = 11x^2y^2 + 2x^2y \qquad (17 - 6)x^2y^2 = 11x^2y^2.$$

c. $-3r - 4r + 6r = -r$ $\quad (-3 - 4 + 6)r = -1r = -r$.

d. The first and fourth terms are like terms, and the second and third terms are like terms.

$$ab + 8 - 15 + 4ab = 5ab - 7 \qquad (1 + 4)ab = 5ab \text{ and } 8 - 15 = -7.$$

Self Check 1

Simplify each polynomial: **a.** $6m^4 + 3m^4$, **b.** $17s^3t + 3s^2t - 6s^3t$, **c.** $-19x + 21x - x$, and **d.** $rs + 3r - 5rs + 4r$.

ADDING POLYNOMIALS

When adding polynomials horizontally, each polynomial is usually enclosed within parentheses. For example,

$$(3x^2 + 6x + 7) + (2x - 5)$$

is the sum of a trinomial and a binomial. To find the sum, we reorder and regroup the terms so that like terms are together.

$$\begin{aligned}(3x^2 + 6x + 7) + (2x - 5) &= 3x^2 + (6x + 2x) + (7 - 5)\\ &= 3x^2 + 8x + 2 \quad \text{Combine like terms.}\end{aligned}$$

This example suggests the following rule.

Adding Polynomials To add polynomials, combine their like terms.

EXAMPLE 2

ELEMENTARY Algebra f(x) Now™

Add the polynomials: **a.** $(-6a^3 + 5a^2 - 7a + 9) + (4a^3 - 5a^2 - a - 8)$ and **b.** $(16g^2 - h^2) + (4g^2 + 2gh + 10h^2)$.

Solution Reorder and regroup the terms to get like terms together and combine like terms.

Notation

When performing operations on polynomials, it is standard practice to write the terms of a result in descending powers of one variable.

a. $(-6a^3 + 5a^2 - 7a + 9) + (4a^3 - 5a^2 - a - 8)$

$$\begin{aligned}&= (-6a^3 + 4a^3) + (5a^2 - 5a^2) + (-7a - a) + (9 - 8)\\ &= -2a^3 + 0a^2 + (-8a) + 1 \quad \text{Combine like terms.}\\ &= -2a^3 - 8a + 1\end{aligned}$$

b. $(16g^2 - h^2) + (4g^2 + 2gh + 10h^2)$

$$\begin{aligned}&= (16g^2 + 4g^2) + 2gh + (-h^2 + 10h^2)\\ &= 20g^2 + 2gh + 9h^2 \quad \text{Combine like terms.}\end{aligned}$$

Self Check 2 Add the polynomials: **a.** $(2a^2 - a + 4) + (5a^2 + 6a - 5)$ and **b.** $(7x^2 - 2xy - y^2) + (4x^2 - y^2)$.

Polynomials can also be added vertically by aligning like terms in columns.

EXAMPLE 3

Add $4x^2 - 3$ and $3x^2 - 8x + 8$.

ELEMENTARY Algebra f(x) Now™

Solution We write one polynomial underneath the other. Since the first polynomial does not have an x-term, we leave a space so that the constant terms can be aligned. Next, we add like terms, column by column, writing each result under the horizontal bar.

Notation

In performing a vertical addition, any missing term may be written with a coefficient of 0:

$$\begin{array}{r}4x^2 + 0x - 3\\ \underline{3x^2 - 8x + 8}\end{array}$$

$$\begin{array}{rl}4x^2 - 3 & \text{In the } x^2\text{-column, find } 4x^2 + 3x^2.\\ +\ \underline{3x^2 - 8x + 8} & \text{In the } x\text{-column, find } 0x + (-8x).\\ 7x^2 - 8x + 5 & \text{In the constant column, find } -3 + 8.\end{array}$$

Self Check 3 Add $4q^2 - 7$ and $2q^2 - 8q + 9$ vertically.

SUBTRACTING POLYNOMIALS

Because of the distributive property, we can remove parentheses enclosing several terms when the sign preceding the parentheses is a $-$ sign. We simply drop the $-$ sign and the parentheses, and *change the sign of every term within the parentheses.*

$$\begin{aligned}-(3x^2 + 3x - 2) &= -1(3x^2 + 3x - 2)\\ &= -1(3x^2) + (-1)(3x) + (-1)(-2)\\ &= -3x^2 + (-3x) + 2\\ &= -3x^2 - 3x + 2\end{aligned}$$

This suggests a way to subtract polynomials.

Subtracting Polynomials To subtract two polynomials, change the signs of the terms of the polynomial being subtracted, drop the parentheses, and combine like terms.

EXAMPLE 4

ELEMENTARY Algebra f(x) Now™

Subtract the polynomials: **a.** $(3a^2 - 4a - 6) - (2a^2 - a + 9)$ and **b.** $(-t^3u + 2t^2u - u + 1) - (-3t^2u - u + 8)$

Solution **a.** $(3a^2 - 4a - 6) - (2a^2 - a + 9)$

$= 3a^2 - 4a - 6 - 2a^2 + a - 9$ Change the sign of each term of $2a^2 - a + 9$ and drop the parentheses.

$= a^2 - 3a - 15$ Combine like terms.

b. $(-t^3u + 2t^2u - u + 1) - (-3t^2u - u + 8)$

$= -t^3u + 2t^2u - u + 1 + 3t^2u + u - 8$ Change the sign of each term of $-3t^2u - u + 8$ and drop the parentheses.

$= -t^3u + 5t^2u - 7$ Combine like terms.

Caution

When combining like terms, the exponents on the variables stay the same. Don't incorrectly apply a rule for exponents and add the exponents.

Self Check 4 Subtract the polynomials: **a.** $(8a^3 - 5a^2 + 5) - (a^3 - a^2 - 7)$ and **b.** $(x^2y - 2x + y - 2) - (6x + 9y - 2)$.

EXAMPLE 5

Subtract $12a - 7$ from the sum of $6a + 5$ and $4a - 10$.

Solution We will use brackets to show that $(12a - 7)$ is to be subtracted from the *sum* of $(6a + 5)$ and $(4a - 10)$.

$$[(6a + 5) + (4a - 10)] - (12a - 7)$$

Next, we remove the grouping symbols to obtain

$= 6a + 5 + 4a - 10 - 12a + 7$ Change the sign of each term within $(12a - 7)$ and drop the parentheses.

$= -2a + 2$ Combine like terms.

Self Check 5 Subtract $-2q^2 - 2q$ from the sum of $q^2 - 6q$ and $3q^2 + q$.

Polynomials can also be subtracted vertically by aligning like terms in columns.

EXAMPLE 6

Subtract $3x^2 - 2x + 3$ from $2x^2 + 4x - 1$ using vertical form.

Solution Since $3x^2 - 2x + 3$ is to be subtracted from $2x^2 + 4x - 1$, we write $3x^2 - 2x + 3$ underneath $2x^2 + 4x - 1$. Then we add the opposite of $3x^2 - 2x + 3$ by changing the sign of each of its terms, and combining like terms, column-by-column.

$$\begin{array}{r} 2x^2 + 4x - 1 \\ -\ \underline{3x^2 - 2x + 3} \end{array} \longrightarrow \begin{array}{r} 2x^2 + 4x - 1 \\ +\ \underline{-3x^2 + 2x - 3} \\ -x^2 + 6x - 4 \end{array}$$

In the x^2-column, find $2x^2 + (-3x^2)$.
In the x-column, find $4x + 2x$.
In the constant-column, find $-1 + (-3)$.

Self Check 6 Subtract $2p^2 + 2p - 8$ from $5p^2 - 6p + 7$.

ADDING AND SUBTRACTING MULTIPLES OF POLYNOMIALS

Because of the distributive property, we can remove parentheses enclosing several terms when a monomial precedes the parentheses. We simply multiply every term within the parentheses by that monomial. For example, to add $3(2x + 5)$ and $2(4x - 3)$, we proceed as follows:

$$3(2x + 5) + 2(4x - 3) = 6x + 15 + 8x - 6$$ Use the distributive property to remove parentheses.

$$= 6x + 8x + 15 - 6$$ $15 + 8x = 8x + 15$.

$$= 14x + 9$$ Combine like terms.

EXAMPLE 7

ELEMENTARY Algebra f(x) Now™

Remove parentheses and simplify.

a. $3(x^2 + 4x) + 2(x^2 - 4) = 3x^2 + 12x + 2x^2 - 8$
$= 5x^2 + 12x - 8$

b. $-8(y^2 - 2y + 3) - 4(2y^2 + y - 6) = -8y^2 + 16y - 24 - 8y^2 - 4y + 24$
$= -16y^2 + 12y$

Self Check 7 Remove parentheses and simplify: $2(a^2 - 3a) + 5(a^2 + 2a)$

APPLICATIONS

EXAMPLE 8

ELEMENTARY Algebra f(x) Now™

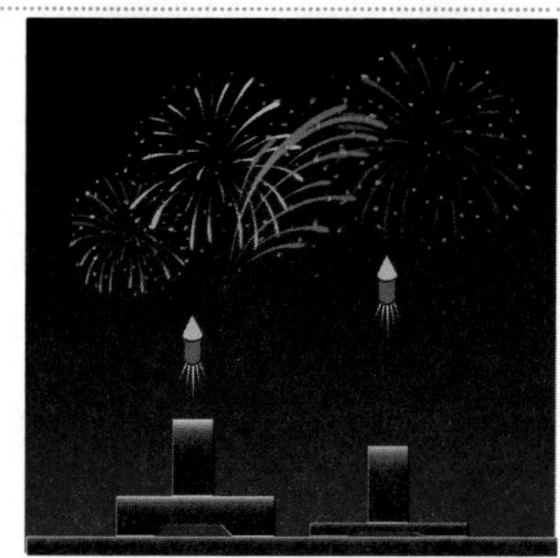

Fireworks shows. Two fireworks shells are simultaneously fired upward from different platforms. The height of the first shell is $(-16t^2 + 160t + 3)$ feet and the height of the higher-traveling second shell is $(-16t^2 + 200t + 1)$ feet, after t seconds.

a. Find a polynomial that represents the difference in the heights of the shells.

b. In 5 seconds, the first shell reaches its peak and explodes. How much higher is the second shell at that time?

Solution **a.** To find the difference in heights, we subtract the height of the first shell from the height of the second.

$$(-16t^2 + 200t + 1) - (-16t^2 + 160t + 3)$$
$$= -16t^2 + 200t + 1 + 16t^2 - 160t - 3$$ Change the sign of each term of $-16t^2 + 160t + 3$ and remove parentheses.
$$= 40t - 2$$ Combine like terms.

The difference in the heights of the shells t seconds after being fired is $(40t - 2)$ feet.

b. If we substitute 5 for t in the polynomial found in part a, we have

$$40t - 2 = 40(5) - 2 = 200 - 2 = 198$$

When the first shell explodes, the second shell will be 198 feet higher than the first shell.

Answers to Self Checks **1. a.** $9m^4$, **b.** $11s^3t + 3s^2t$, **c.** x, **d.** $-4rs + 7r$ **2. a.** $7a^2 + 5a - 1$, **b.** $11x^2 - 2xy - 2y^2$ **3.** $6q^2 - 8q + 2$ **4. a.** $7a^3 - 4a^2 + 12$ **b.** $x^2y - 8x - 8y$ **5.** $6q^2 - 3q$ **6.** $3p^2 - 8p + 15$ **7.** $7a^2 + 4a$

4.5 STUDY SET  ELEMENTARY Algebra f(x) Now™

VOCABULARY **Fill in the blanks.**

1. The expression $(b^3 - b^2 - 9b + 1) + (b^3 - b^2 - 9b + 1)$ is the sum of two ______.
2. The expression $(b^2 - 9b + 11) - (4b^2 - 14b)$ is the ______ of a trinomial and a binomial.
3. ______ terms have the same variables with the same exponents.
4. Simplifying the sum or difference of like terms is called ______ like terms.
5. The polynomial $2t^4 + 3t^3 - 4t^2 + 5t - 6$ is written in ______ powers of t.
6. When a $-$ symbol precedes a grouping symbol, as in $-(m^2 + 12m - 1)$, we interpret it as a factor of ______.

CONCEPTS **Fill in the blanks.**

7. To add polynomials, ______ like terms.
8. To subtract polynomials, ______ the signs of the terms of the polynomial being subtracted, drop parentheses, and combine like terms.
9. Subtract using vertical form:

$$\begin{array}{r} 8x^2 - 7x - 1 \\ -\ \underline{4x^2 + 6x - 9} \end{array} \longrightarrow \begin{array}{r} 8x^2 - 7x - 1 \\ +\ \underline{\quad\ \ - 6x \quad 9} \end{array}$$

10. $(y^2 - 6y + 7) - (2y^2 - 4) = y^2 - 6y + 7 \quad 2y^2 \quad 4$
11. Simplify.
    a. $2x^2 + 3x^2$ b. $15m^3 - m^3$
    c. $8a^3b - ab$ d. $6cd + 4c^2d$
12. Explain the error.

$$7x^2y + 6x^2y = 13x^4y^2$$

13. Write without parentheses.
    a. $-(5x^2 - 8x + 23)$ b. $-(-5y^4 + 3y^2 - 7)$
14. What is the result when the addition is done in the x-column?

$$\begin{array}{r} 4x^2 + \ \ x - 12 \\ \underline{+\ 5x^2 - 8x + 23} \end{array}$$

NOTATION **Complete the solution.**

15. $(6x^2 + 2x + 3) - (4x^2 - 7x + 1)$
    $= 6x^2 + 2x + 3 \quad 4x^2 \quad 7x \quad 1$
    $= \quad + 9x + \quad$

16. Fill in the blank:
$-(7x^2 - 6x + 12) = -\ \_\_\ (7x^2 - 6x + 12)$.

17. True or false: $5x^2 - 120 = 5x^2 + 0x - 120$.

18. Write $3x^2 - 9 + 6x - 12x^3$ in descending powers of x.

PRACTICE **Simplify each polynomial.**

19. $8t^2 + 4t^2$

20. $15x^2 + 10x^2$

21. $-32u^3 - 16u^3$

22. $-25x^3 - 7x^3$

23. $1.8x - 1.9x$

24. $1.7y - 2.2y$

25. $\frac{1}{2}st + \frac{3}{2}st$

26. $\frac{2}{5}at + \frac{1}{5}at$

27. $3r - 4r + 7r$

28. $-2b + 7b - 3b$

29. $-4ab + 4ab - ab$

30. $xy - 4xy - 2xy$

31. $10x^2 - 8x + 9x - 9x^2$

32. $-3y^2 - y - 6y^2 + 7y$

33. $6x^3 + 8x^4 + 7x^3 + (-8x^4)$

34. $-3rt - 7t^2 - 6rt + 7t^2 + (-6r^2)$

35. $4x^2y + 5 - 6x^3y - 3x^2y + 2x^3y$

36. $5b - 9ab^2 + 10a^3b - 8ab^2 - 9a^3b$

37. $\frac{2}{3}d^2 - \frac{1}{4}c^2 + \frac{5}{6}c^2 - \frac{1}{2}cd + \frac{1}{3}d^2$

38. $1 + \frac{3}{5}s^2 - \frac{2}{5}t^2 - \frac{1}{2}s^2 - \frac{7}{10}st - \frac{3}{10}st$

Perform the operations.

39. $(3x + 7) + (4x - 3)$

40. $(2y - 3) + (4y + 7)$

41. $(9a^2 + 3a) - (2a - 4a^2)$

42. $(4b^2 + 3b) - (7b - b^2)$

43. $(2x + 3y) + (5x - 10y)$

44. $(5x - 8y) - (-2x + 5y)$

45. $(-8x - 3y) - (-11x + y)$

46. $(-4a + b) + (5a - b)$

47. $(3x^2 - 3x - 2) + (3x^2 + 4x - 3)$

48. $(3a^2 - 2a + 4) - (a^2 - 3a + 7)$

49. $(2b^2 + 3b - 5) - (2b^2 - 4b - 9)$

50. $(4c^2 + 3c - 2) + (3c^2 + 4c + 2)$

51. $(2x^2 - 3x + 1) - (4x^2 - 3x + 2) + (2x^2 + 3x + 2)$

52. $(-3z^2 - 4z + 7) + (2z^2 + 2z - 1) - (2z^2 - 3z + 7)$

53. $(-4h^3 + 5h^2 + 15) - (h^3 - 15)$

54. $(-c^5 + 5c^4 - 12) - (2c^5 - c^4)$

55. $(1.04x^2 + 2.07x - 5.01) + (1.33x - 2.98x^2 + 5.02)$

56. $(0.03f^2 + 0.25g) - (0.17g - 0.23f^2)$

57. $\left(\frac{7}{8}r^4 + \frac{5}{3}r^2 - \frac{9}{4}\right) - \left(-\frac{3}{8}r^4 - \frac{2}{3}r^2 - \frac{1}{4}\right)$

58. $\left(\frac{1}{16}r^6 + \frac{1}{2}r^3 - \frac{11}{12}\right) + \left(\frac{9}{16}r^6 + \frac{9}{2}r^3 + \frac{1}{12}\right)$

59.
$$\begin{array}{r} 3x^2 + 4x + 5 \\ +\ \underline{2x^2 - 3x + 6} \end{array}$$

60.
$$\begin{array}{r} -6x^3 - 4x^2 + 7 \\ +\ \underline{-7x^3 + 9x^2 } \end{array}$$

61.
$$\begin{array}{r} 3x^2 + 4x - 5 \\ -\ \underline{-2x^2 - 2x + 3} \end{array}$$

62.
$$\begin{array}{r} 3y^2 - 4y + 7 \\ -\ \underline{6y^2 - 6y - 13} \end{array}$$

63.
$$\begin{array}{r} 4x^3 + 4x^2 - 3x + 10 \\ -\ \underline{5x^3 - 2x^2 - 4x - 4} \end{array}$$

64.
$$\begin{array}{r} 2x^3 + 2x^2 - 3x + 5 \\ +\ \underline{3x^3 - 4x^2 - x - 7} \end{array}$$

65.
$$\begin{array}{r} -3x^3 + 4x^2 - 4x + 9 \\ +\ \underline{2x^3 + 9x - 3} \end{array}$$

66.
$$\begin{array}{r} -3x^2 + 4x + 25 \\ +\ \underline{5x^2 - 12} \end{array}$$

67.
$$\begin{array}{r} 3x^3 + 4x^2 + 7x + 12 \\ -\ \underline{-4x^3 + 6x^2 + 9x - 3} \end{array}$$

68.
$$\begin{array}{r} -2x^2y^2 + 12y^2 \\ -\ \underline{10x^2y^2 + 9xy - 24y^2} \end{array}$$

69. $2(x + 3) + 4(x - 2)$

70. $3(y - 4) - 5(y + 3)$

71. $-2(x^2 + 7x - 1) - 3(x^2 - 2x + 2)$

72. $-5(y^2 - 2y - 6) + 6(2y^2 + 2y - 5)$

73. $2(2y^2 - 2y + 2) - 4(3y^2 - 4y - 1) + 4(y^3 - y^2 - y)$

74. $-4(z^2 - 5z) - 5(4z^2 - 1) + 6(2z - 3)$

75. Subtract $s^2 + 4s + 2$ from $5s^2 - s + 9$.

76. Subtract $4p^2 - 4p - 40$ from $10p^2 - p - 30$.

77. Subtract $-y^5 + 5y^4 - 1.2$ from $2y^5 - y^4$.

78. Subtract $-4w^3 + 5w^2 + 7.6$ from $w^3 - 15w^2$.

79. Find the difference when $t^3 - 2t^2 + 2$ is subtracted from the sum of $3t^3 + t^2$ and $-t^3 + 6t - 3$.

80. Find the difference when $-3z^3 - 4z + 7$ is subtracted from the sum of $2z^2 + 3z - 7$ and $-4z^3 - 2z - 3$.

81. Find the sum when $3x^2 + 4x - 7$ is added to the sum of $-2x^2 - 7x + 1$ and $-4x^2 + 8x - 1$.

82. Find the difference when $32x^2 - 17x + 45$ is subtracted from the sum of $23x^2 - 12x - 7$ and $-11x^2 + 12x + 7$.

APPLICATIONS

83. GREEK ARCHITECTURE

a. Find a polynomial that represents the difference in the heights of the columns shown at the beginning of this section.

b. If the columns were stacked one atop the other, to what height would they reach?

Find the polynomial that represents the perimeter of each figure.

84. a.

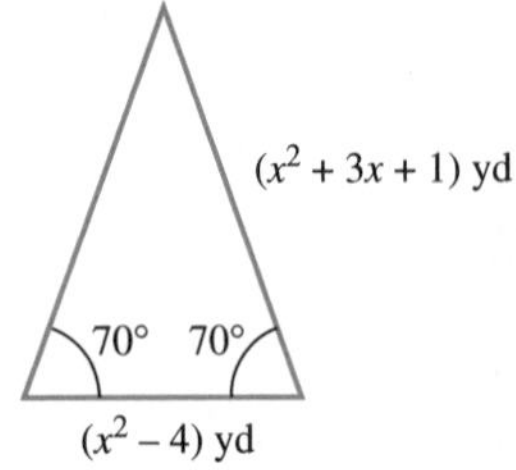

b.

85. JETS Find the polynomial representing the length of the passenger jet.

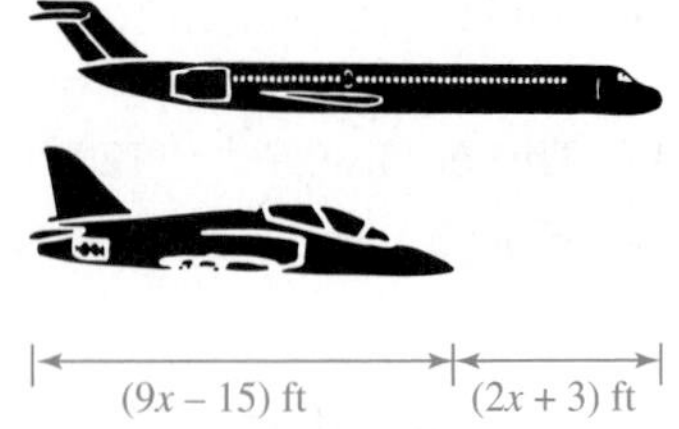

86. PIÑATAS Find the polynomial that represents the length of the rope used to hold up the piñata.

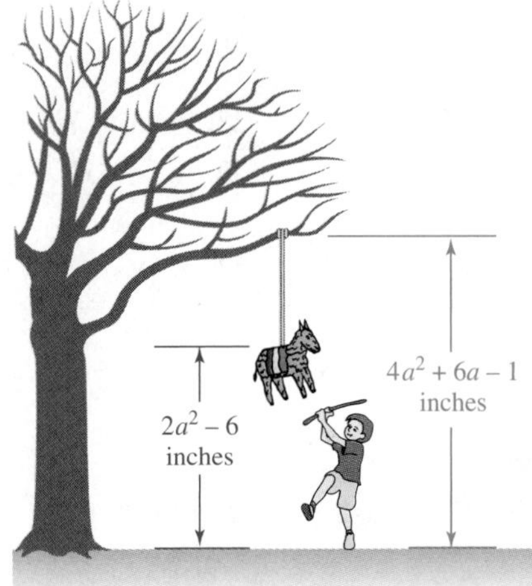

87. READING BLUEPRINTS

a. Find a polynomial that represents the difference in the length and width of the one-bedroom apartment shown in the illustration.

b. Find the perimeter of the apartment.

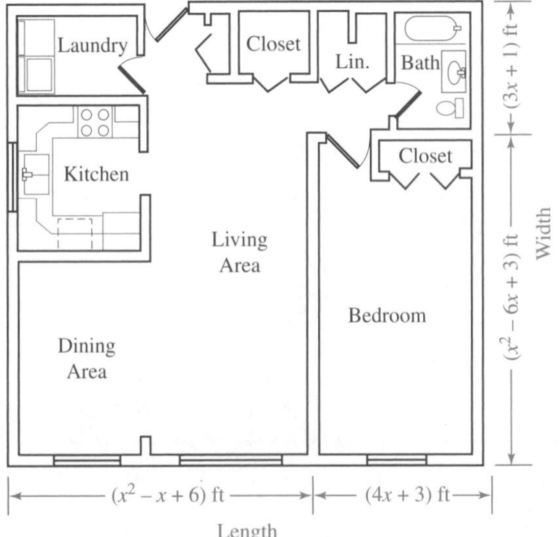

88. AUTO MECHANICS Find the polynomial representing the length of the fan belt. The dimensions are in inches. Leave π in your answer.

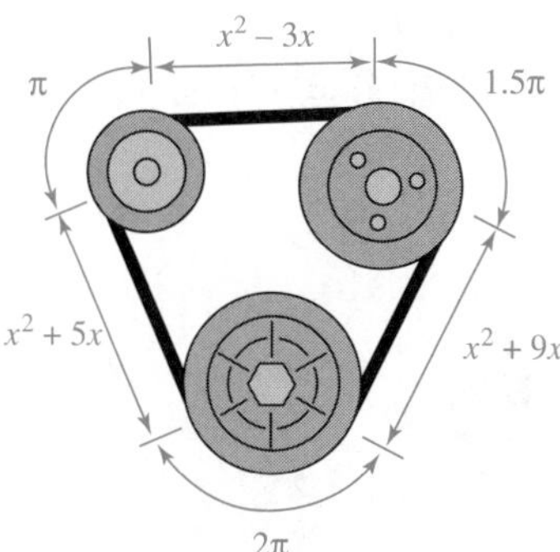

89. NAVAL OPERATIONS Two warning flares are simultaneously fired upward from different parts of a ship. The height of the first flare is $(-16t^2 + 128t + 20)$ feet and the height of the higher-traveling second flare is $(-16t^2 + 150t + 40)$ feet, after t seconds.

a. Find a polynomial that represents the difference in the heights of the flares.

b. In 4 seconds, the first flare reaches its peak, explodes, and lights up the sky. How much higher is the second flare at that time?

90. REAL ESTATE INVESTMENTS A house is purchased for \$105,000 and is expected to appreciate \$900 per year. Therefore, its value in x years is given by the polynomial $900x + 105{,}000$.

a. A second house is purchased at the same time for \$120,000 and is expected to appreciate \$1,000 per year. Find a polynomial that predicts the value of the second house in x years.

b. Find a polynomial that predicts the combined value of the houses x years after their purchase.

c. Use your answer to part b to find the predicted value of the two houses in 20 years.

WRITING

91. How do you recognize like terms?

92. How do you add like terms?

93. Explain the concept that is illustrated by the statement

$$-(x^2 + 3x - 1) = -1(x^2 + 3x - 1)$$

94. Explain the mistake made in the solution.
Simplify: $(12x^2 - 4) - (3x^2 - 1)$

$$\begin{aligned}(12x^2 - 4) - (3x^2 - 1) &= 12x^2 - 4 - 3x^2 - 1\\ &= 9x^2 - 5\end{aligned}$$

95. What can be concluded if

a. you add two polynomials and get 0?

b. you subtract two polynomials and get 0?

96. Explain the error.
Subtract $(2d^2 - d - 3)$ from $(d^2 - 9)$:

$$(2d^2 - d - 3) - (d^2 - 9)$$

REVIEW

97. What is the sum of the measures of the angles of a triangle?

98. Graph: $y = -\frac{x}{2} + 2$.

99. Graph: $2x + 3y = 9$.

100. CURLING IRONS A curling iron is plugged into a 110-volt electrical outlet and used for $\frac{1}{4}$ hour. If its resistance is 10 ohms, find the electrical power (in kilowatt hours, kwh) used by the curling iron by applying the formula

$$\text{kwh} = \frac{(\text{volts})^2}{1{,}000 \cdot \text{ohms}} \cdot \text{hours}$$

CHALLENGE PROBLEMS

101. What polynomial must be added to $2x^2 - x + 3$ so that the sum is $6x^2 - 7x - 8$?

102. Is the sum of two trinomials always a trinomial?

4.6 Multiplying Polynomials

- Multiplying Monomials
- Multiplying a Polynomial by a Monomial
- Multiplying Binomials
- The FOIL Method
- Multiplying Polynomials

The dimensions of a dollar bill are shown in the illustration. We can find its area by multiplying its length and width.

6.5 cm

15.6 cm

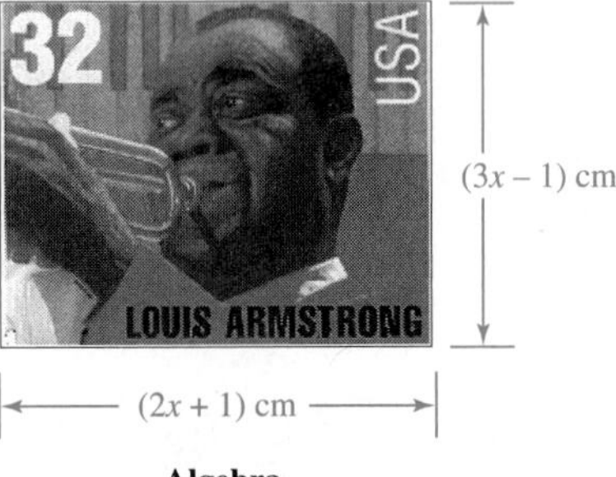

(3x – 1) cm

(2x + 1) cm

Arithmetic	Algebra
$15.6(6.5) = 101.4$	$(2x + 1)(3x - 1) = ?$
The area is 101.4 cm^2.	

The length and width of the postage stamp shown above are expressed as polynomials. To find the area of the stamp, we need to multiply $2x + 1$ by $3x - 1$. In this section, we will discuss the rules for multiplying polynomials. To begin, we consider the product of two monomials.

MULTIPLYING MONOMIALS

Success Tip

In this section, you will see that every polynomial multiplication is a series of monomial multiplications.

We have seen that to multiply the monomials $8x^2$ and $3x^4$, we proceed as follows.

$(8x^2)(3x^4) = (8 \cdot 3)(x^2 \cdot x^4)$ Group the coefficients together and the variables together.

$= 24x^6$ $x^2 \cdot x^4 = x^{2+4} = x^6$.

This example suggests the following rule.

Multiplying Monomials To multiply two monomials, multiply the numerical factors (the coefficients) and then multiply the variable factors.

EXAMPLE 1

ELEMENTARY Algebra f(x) Now™

Multiply: **a.** $(6r)(r)$, **b.** $3t^4(-2t^5)$, **c.** $(-2a^2b^3)(-5ab^2)$, and **d.** $-4y^5z^2(2y^3z^3)(3yz)$.

Solution

a. $(6r)(r) = 6r^2$ Recall that $r = 1r$. $6 \cdot 1 = 6$ and $r \cdot r = r^2$.

b. $(3t^4)(-2t^5) = -6t^9$ $3(-2) = -6$ and $t^4 \cdot t^5 = t^9$.

c. $-2a^2b^3(-5ab^2) = 10a^3b^5$ $-2(-5) = 10$, $a^2 \cdot a = a^3$, and $b^3 \cdot b^2 = b^5$.

d. $-4y^5z^2(2y^3z^3)(3yz) = -24y^9z^6$ $-4(2)(3) = -24$, $y^5 \cdot y^3 \cdot y = y^9$, and $z^2 \cdot z^3 \cdot z = z^6$.

Self Check 1 Multiply: **a.** $18t(t)$, **b.** $-10d^8(-6d^3)$, and **c.** $(5a^2b^3)(-6a^3b^4)(ab)$.

MULTIPLYING A POLYNOMIAL BY A MONOMIAL

To find the product of a polynomial and a monomial, we use the distributive property. To multiply $2x + 4$ by $5x$, for example, we proceed as follows.

$$5x(2x + 4) = 5x(2x) + 5x(4) \quad \text{Distribute the multiplication by } 5x.$$
$$= 10x^2 + 20x \quad \text{Multiply the monomials.}$$

This example suggests the following rule.

Multiplying Polynomials by Monomials To multiply a monomial and a polynomial, multiply each term of the polynomial by the monomial.

EXAMPLE 2

ELEMENTARY Algebra $f(x)$ Now™

Multiply: **a.** $3a^2(3a^2 - 5a + 2)$, **b.** $-2xz^3(6x^3z + x^2z^2 - xz^3 + 7z^4)$, and **c.** $(-m^4 - 25)(4m^3)$.

Solution **a.** Multiply each term of $3a^2 - 5a + 2$ by $3a^2$.

$$3a^2(3a^2 - 5a + 2)$$
$$= 3a^2(3a^2) + 3a^2(-5a) + 3a^2(2) \quad \text{Distribute the multiplication by } 3a^2.$$
$$= 9a^4 - 15a^3 + 6a^2 \quad \text{Multiply the monomials.}$$

b. Multiply each term of $6x^3z + x^2z^2 - xz^3 + 7z^4$ by $-2xz^3$.

$$-2xz^3(6x^3z + x^2z^2 - xz^3 + 7z^4)$$
$$= -2xz^3(6x^3z) - 2xz^3(x^2z^2) - 2xz^3(-xz^3) - 2xz^3(7z^4)$$
$$= -12x^4z^4 - 2x^3z^5 + 2x^2z^6 - 14xz^7 \quad \text{Multiply the monomials.}$$

c. Multiply each term of $-m^4 - 25$ by $4m^3$.

$$(-m^4 - 25)(4m^3) = -m^4(4m^3) - 25(4m^3) \quad \text{Distribute the multiplication by } 4m^3.$$
$$= -4m^7 - 100m^3 \quad \text{Multiply the monomials.}$$

Success Tip

The rectangle can be used to visualize polynomial multiplication. The total area is $x(x + 2)$ and the sum of the two smaller areas is $x^2 + 2x$. Thus,

$$x(x + 2) = x^2 + 2x$$

x	x^2	$2x$
	x	2

$x + 2$

Self Check 2 Multiply: **a.** $5c^2(4c^3 - 9c - 8)$ and **b.** $-s^2t^2(-s^4t^2 + s^3t^3 - st^4 + 7s)$.

MULTIPLYING BINOMIALS

The distributive property can also be used to multiply binomials. For example, to multiply $2a + 4$ and $3a + 5$, we think of $2a + 4$ as a single quantity and distribute it over each term of $3a + 5$.

$$(2a + 4)(3a + 5) = (2a + 4)3a + (2a + 4)5$$
$$= 3a(2a + 4) + 5(2a + 4) \quad \text{Use the commutative property of multiplication.}$$

$= 3a(2a) + 3a(4) + 5(2a) + 5(4)$ Distribute the multiplication by $3a$. Distribute the multiplication by 5.

$= 6a^2 + 12a + 10a + 20$ Multiply the monomials.

$= 6a^2 + 22a + 20$ Combine like terms.

This example suggests the following rule.

Multiplying Binomials To multiply two binomials, multiply each term of one binomial by each term of the other binomial, and then combine like terms.

EXAMPLE 3

Multiply: $(5x - 8)(x + 1)$.

ELEMENTARY Algebra $f(x)$ Now™

Solution To find the product, we multiply $x + 1$ by $5x$ and by -8. In that way, each term of $x + 1$ will be multiplied by each term of $5x - 8$.

$(5x - 8)(x + 1) = 5x(x + 1) - 8(x + 1)$

$= 5x^2 + 5x - 8x - 8$ Distribute the multiplication by $5x$. Distribute the multiplication by -8.

$= 5x^2 - 3x - 8$ Combine like terms.

Self Check 3 Multiply: $(9y + 3)(y - 4)$.

THE FOIL METHOD

We can use a shortcut method, called the **FOIL method,** to multiply binomials. FOIL is an acronym for **F**irst terms, **O**uter terms, **I**nner terms, **L**ast terms. To use the FOIL method to multiply $2a + 4$ by $3a + 5$, we

1. multiply the **F**irst terms $2a$ and $3a$ to obtain $6a^2$,
2. multiply the **O**uter terms $2a$ and 5 to obtain $10a$,
3. multiply the **I**nner terms 4 and $3a$ to obtain $12a$, and
4. multiply the **L**ast terms 4 and 5 to obtain 20.

Then we simplify the resulting polynomial, if possible.

First terms Last terms

$(2a + 4)(3a + 5) = 2a(3a) + 2a(5) + 4(3a) + 4(5)$

Inner terms

$= 6a^2 + 10a + 12a + 20$ Do the multiplications.

$= 6a^2 + 22a + 20$ Combine like terms.

Outer terms

The Language of Algebra

The acronym FOIL helps us remember the order to follow when multiplying two binomials. Another popular acronym is PEMDAS. It represents the order of operations rules: **P**arentheses, **E**xponents, **M**ultiply, **D**ivide, **A**dd, **S**ubtract.

EXAMPLE 4

ELEMENTARY Algebra $f(x)$ Now™

Find each product.

Solution **a.** $(x + 5)(x + 7) = x(x) + x(7) + 5(x) + 5(7)$
$= x^2 + 7x + 5x + 35$
$= x^2 + 12x + 35$ Combine like terms.

Success Tip

The area of the large rectangle is given by $(x + 5)(x + 7)$. The sum of the areas of the smaller rectangles is $x^2 + 7x + 5x + 35$ or $x^2 + 12x + 35$. Thus,

$(x + 5)(x + 7) = x^2 + 12x + 35.$

	x	7
5	$5x$	35
x	x^2	$7x$

(sides: $x + 5$ and $x + 7$)

b. $(3x + 4)(2x - 3) = 3x(2x) + 3x(-3) + 4(2x) + 4(-3)$
$= 6x^2 - 9x + 8x - 12$
$= 6x^2 - x - 12$ Combine like terms.

c. $(2r - 3s)(2r + t) = 2r(2r) + 2r(t) - 3s(2r) - 3s(t)$
$= 4r^2 + 2rt - 6rs - 3st$

d. $(3a^2 - 7b)(a^2 - b) = 3a^2(a^2) + 3a^2(-b) - 7b(a^2) - 7b(-b)$
$= 3a^4 - 3a^2b - 7a^2b + 7b^2$
$= 3a^4 - 10a^2b + 7b^2$ Combine like terms.

(F, O, I, L arrows mark First, Outer, Inner, Last products.)

Self Check 4 Find each product: **a.** $(y + 3)(y + 1)$, **b.** $(2a - 1)(3a + 2)$, **c.** $(5y^3 - 2z)(2y^3 - z)$

MULTIPLYING POLYNOMIALS

To develop a general rule for multiplying any two polynomials, we will find the product of $2x + 3$ and $3x^2 + 3x + 5$. In the solution, the distributive property is used four times.

$$(2x + 3)(3x^2 + 3x + 5) = (2x + 3)3x^2 + (2x + 3)3x + (2x + 3)5$$
$$= 2x(3x^2) + 3(3x^2) + 2x(3x) + 3(3x) + 2x(5) + 3(5)$$
$$= 6x^3 + 9x^2 + 6x^2 + 9x + 10x + 15 \quad \text{Multiply the monomials.}$$
$$= 6x^3 + 15x^2 + 19x + 15 \quad \text{Combine like terms.}$$

In the second line of the solution, note that each term of $3x^2 + 3x - 5$ has been multiplied by each term of $2x + 3$. This example suggests the following rule.

Multiplying Polynomials To multiply two polynomials, multiply each term of one polynomial by each term of the other polynomial, and then combine like terms.

EXAMPLE 5

Multiply: $(6y^3 + y^2 - 8y + 1)(7y + 3)$.

ELEMENTARY Algebra f(x) Now™

Solution Multiply each term of $7y + 3$ by each term of $6y^3 + y^2 - 8y + 1$.

$$\begin{aligned}(6y^3 + y^2 - 8y + 1)(7y + 3) &= 6y^3(7y) + 6y^3(3) + y^2(7y) + y^2(3) - 8y(7y) - 8y(3) + 1(7y) + 1(3)\\ &= 42y^4 + 18y^3 + 7y^3 + 3y^2 - 56y^2 - 24y + 7y + 3\\ &= 42y^4 + 25y^3 - 53y^2 - 17y + 3\end{aligned}$$

Self Check 5 Multiply: $(2a^4 - a^2 - a)(3a^2 - 1)$.

It is often convenient to multiply polynomials using a vertical format similar to that used to multiply whole numbers.

EXAMPLE 6

ELEMENTARY Algebra f(x) Now™

a. Multiply:

$$\begin{array}{rrrrl} & 3a^2 & -\ 4a & +\ 7 & \\ & & 2a & +\ 5 & \\ \hline & 15a^2 & -\ 20a & +\ 35 & \text{Multiply } 3a^2 - 4a + 7 \text{ by } 5.\\ 6a^3 & -\ 8a^2 & +\ 14a & & \text{Multiply } 3a^2 - 4a + 7 \text{ by } 2a.\\ \hline 6a^3 & +\ 7a^2 & -\ 6a & +\ 35 & \text{In each column, combine like terms.}\end{array}$$

b. Multiply:

$$\begin{array}{rrrrrl} & 6y^3 & & -\ 5y & +\ 4 & \\ & & -4y^2 & & -\ 3 & \\ \hline & -18y^3 & & +\ 15y & -\ 12 & \text{Multiply } 6y^3 - 5y + 4 \text{ by } -3.\\ -24y^5 & +\ 20y^3 & -\ 16y^2 & & & \text{Multiply } 6y^3 - 5y + 4 \text{ by } -4y^2.\\ \hline -24y^5 & +\ 2y^3 & -\ 16y^2 & +\ 15y & -\ 12 & \text{Leave a space for any missing powers of } y. \text{ In each column, combine like terms.}\end{array}$$

Self Check 6 Multiply: **a.** $(3x + 2)(2x^2 - 4x + 5)$ and **b.** $(-2x^2 + 3)(2x^2 - 4x - 1)$.

When finding the product of three polynomials, we begin by multiplying any two of them, and then we multiply that result by the third polynomial.

EXAMPLE 7

Find the product: $-3a(4a + 1)(a - 7)$.

Solution We find the product of $4a + 1$ and $a - 7$ and then multiply that result by $-3a$.

$$\begin{aligned}-3a(4a + 1)(a - 7) &= -3a(4a^2 - 28a + a - 7) && \text{Multiply the two binomials.}\\ &= -3a(4a^2 - 27a - 7) && \text{Combine like terms.}\\ &= -12a^3 + 81a^2 + 21a && \text{Distribute the multiplication by } -3a.\end{aligned}$$

Self Check 7 Find the product: $-2y(y + 3)(3y - 2)$

Answers to Self Checks

1. a. $18t^2$, **b.** $60d^{11}$, **c.** $-30a^6b^8$ **2. a.** $20c^5 - 45c^3 - 40c^2$, **b.** $s^6t^4 - s^5t^5 + s^3t^6 - 7s^3t^2$ **3.** $9y^2 - 33y - 12$ **4. a.** $y^2 + 4y + 3$, **b.** $6a^2 + a - 2$, **c.** $10y^6 - 9y^3z + 2z^2$ **5.** $6a^6 - 5a^4 - 3a^3 + a^2 + a$ **6. a.** $6x^3 - 8x^2 + 7x + 10$, **b.** $-4x^4 + 8x^3 + 8x^2 - 12x - 3$ **7.** $-6y^3 - 14y^2 + 12y$

4.6 STUDY SET

ELEMENTARY Algebra $f(x)$ Now™

VOCABULARY **Fill in the blanks.**

1. The expression $(2x^3)(3x^4)$ is the product of two ________.
2. To find the product $5a^3(a^2 - 2a + 8)$, we ________ the multiplication by $5a^3$.
3. The expression $(2a - 4)(3a + 5)$ is the product of two ________.
4. In the acronym FOIL, F stands for ______ terms, O for ______ terms, I for ______ terms, and L for ______ terms.
5. To simplify $y^2 + 3y + 2y + 6$, we combine the ______ terms $3y$ and $2y$ to get $y^2 + 5y + 6$.
6. The expression $(2a - 4)(3a^2 + 5a - 1)$ is the product of a ________ and a ________.

CONCEPTS **Fill in the blanks.**

7. **a.** To multiply two monomials, multiply the numerical factors (the ________) and then multiply the ________ factors.
   b. To multiply two polynomials, multiply ______ term of one polynomial by ______ term of the other polynomial, and then combine like terms.
   c. When multiplying three polynomials, we begin by multiplying ______ two of them, and then we multiply that result by the ______ polynomial.
   d. To find the area of a rectangle, multiply the ______ by the width.
8. To multiply two monomials, we use the commutative and associative properties of multiplication to reorder and regroup. Fill in the blanks.

$$(9n^3)(8n^2) = (9 \cdot \quad)(\quad \cdot n^2) = \quad$$

9. Label each arrow using one of the letters F, O, I, or L. Then fill in the blanks.

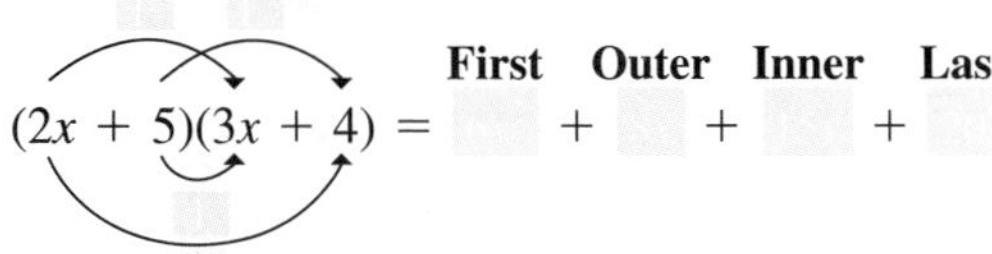

10. Simplify each polynomial by combining like terms.
    a. $6x^2 - 8x + 9x - 12$
    b. $5x^4 + 3ax^2 + 5ax^2 + 3a^2$

11. **a.** Add: $(x - 4) + (x + 8)$.
    b. Subtract: $(x - 4) - (x + 8)$.
    c. Multiply: $(x - 4)(x + 8)$.
12. Fill in the blanks in the multiplication.

$$\begin{array}{r} 3x^2 + 4x - 2 \\ 2x + 3 \\ \hline \quad + 12x - 6 \\ 6x^3 + 8x^2 - 4x \quad\quad \\ \hline \quad + 17x^2 + \quad - 6 \end{array}$$

NOTATION **Complete each solution.**

13. $7x(3x^2 - 2x + 5) = \quad(3x^2) - \quad(2x) + \quad(5)$
    $= \quad - 14x^2 + 35x$
14. $(2x + 5)(3x - 2) = 2x(3x) - \quad(2) + \quad(3x) - \quad(2)$
    $= 6x^2 - \quad + \quad - 10$
    $= 6x^2 + \quad - 10$

PRACTICE **Find each product.**

15. $15m(m)$
16. $4s^9(s)$
17. $(3x^2)(4x^3)$
18. $(-2a^3)(3a^2)$
19. $(3b^2)(-2b)(4b^3)$
20. $(3y)(2y^2)(-y^4)$
21. $(2x^2y^3)(3x^3y^2)$
22. $(-5x^3y^6)(x^2y^2)$
23. $(8a^5)\left(-\frac{1}{4}a^6\right)$
24. $\left(-\frac{2}{3}x^6\right)(9x^3)$
25. $(1.2c^3)(5c^3)$
26. $(2.5h^4)(2h^4)$
27. $\left(\frac{1}{2}a\right)\left(\frac{1}{8}a^4\right)(a^5)$
28. $\left(\frac{1}{3}b\right)\left(\frac{7}{6}b\right)(b^4)$
29. $(x^2y^5)(x^2z^5)(-3z^3)$
30. $(-r^4st^2)(2r^2st)(rst)$
31. $3(x + 4)$
32. $-3(a - 2)$
33. $-4(t^2 + 7)$
34. $6(s^2 - 3)$
35. $3x(x - 2)$
36. $4y(y + 5)$
37. $-2x^2(3x^2 - x)$
38. $4b^3(2b^2 - 2b)$
39. $(x^2 - 12x)(6x^{12})$
40. $(w^9 - 11w)(2w^7)$

41. $\frac{5}{8}t^2(t^6 + 8t^2)$

42. $\frac{4}{9}a^2(9a^3 + a^2)$

43. $0.3p^5(0.4p^4 - 6p^2)$

44. $0.5u^5(0.4u^6 - 0.5u^3)$

45. $-4x^2z(3x^2 - z)$

46. $3xy(x + y)$

47. $2x^2(3x^2 + 4x - 7)$

48. $3y^3(2y^2 - 7y - 8)$

49. $3a(4a^2 + 3a - 4)$

50. $-2x(3x^2 - 3x + 2)$

51. $(-2a^2)(-3a^3)(3a - 2)$

52. $(3x)(-2x^2)(x + 4)$

53. $(y - 3)(y + 5)$

54. $(a + 4)(a + 5)$

55. $(t + 4)(2t - 3)$

56. $(3x - 2)(x + 4)$

57. $(2y - 5)(3y + 7)$

58. $(3x - 5)(2x + 1)$

59. $(2x + 3)(2x - 5)$

60. $(x + 3)(2x - 3)$

61. $(a + b)(a + b)$

62. $(m - n)(m - n)$

63. $(3a - 2b)(4a + b)$

64. $(2t + 3s)(3t - s)$

65. $(t^2 - 3)(t^2 + 4)$

66. $(s^3 + 6)(s^3 - 8)$

67. $(-3t + 2s)(2t - 3s)$

68. $(4t - u)(-3t + u)$

69. $\left(4a - \frac{5}{9}r\right)\left(2a + \frac{3}{4}r\right)$

70. $\left(5c - \frac{2}{3}t\right)\left(2c + \frac{1}{5}t\right)$

71. $4(2x + 1)(x - 2)$

72. $-5(3a - 2)(2a + 3)$

73. $3a(a + b)(a - b)$

74. $-2r(r + s)(r + s)$

75. $(x + 2)(x^2 - 2x + 3)$

76. $(x - 5)(x^2 + 2x - 3)$

77. $(4t + 3)(t^2 + 2t + 3)$

78. $(3x + 1)(2x^2 - 3x + 1)$

79. $(x^2 + 6x + 7)(2x - 5)$

80. $(y^2 - 2y + 1)(4y + 8)$

81. $(-3x + y)(x^2 - 8xy + 16y^2)$

82. $(3x - y)(x^2 + 3xy - y^2)$

83. $(r^2 - r + 3)(r^2 - 4r - 5)$

84. $(w^2 + w - 9)(w^2 - w + 3)$

Perform each multiplication.

85. $\begin{array}{r} x^2 - 2x + 1 \\ \underline{x + 2} \end{array}$

86. $\begin{array}{r} 5r^2 + r + 6 \\ \underline{2r - 1} \end{array}$

87. $\begin{array}{r} 4x^2 + 3x - 4 \\ \underline{3x + 2} \end{array}$

88. $\begin{array}{r} x^2 - x + 1 \\ \underline{x + 1} \end{array}$

89. $\begin{array}{r} 2a^2 + 3a + 1 \\ \underline{3a^2 - 2a + 4} \end{array}$

90. $\begin{array}{r} 3y^2 + 2y - 4 \\ \underline{2y^2 - 4y + 3} \end{array}$

APPLICATIONS

91. STAMPS Find a polynomial that represents the area of the Louis Armstrong stamp shown at the beginning of the section.

92. PARKING Find a polynomial to represent the total area of the van-accessible parking space and its access aisle.

Find the area of each figure. Leave π in your answer.

93.

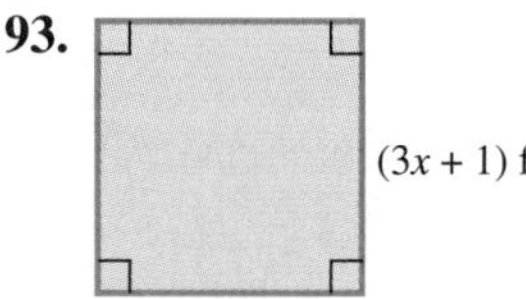

94.

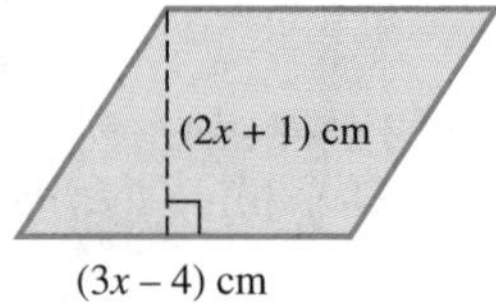

95.

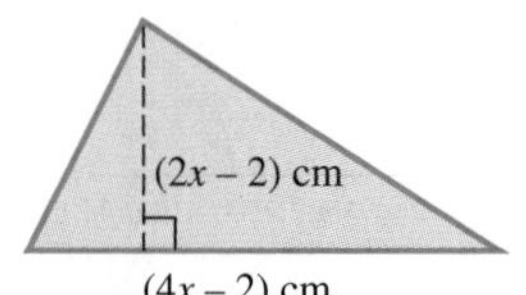

96.

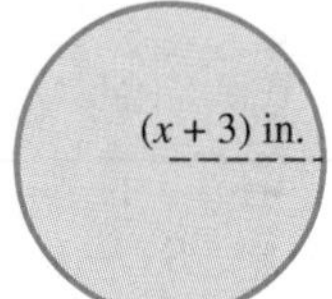

97. SUNGLASSES An ellipse is an oval-shaped closed curve. The area of an ellipse is approximately $3.14ab$, where a is its length and b is its width. Find a polynomial that approximates the total area of the elliptical-shaped lenses of the sunglasses.

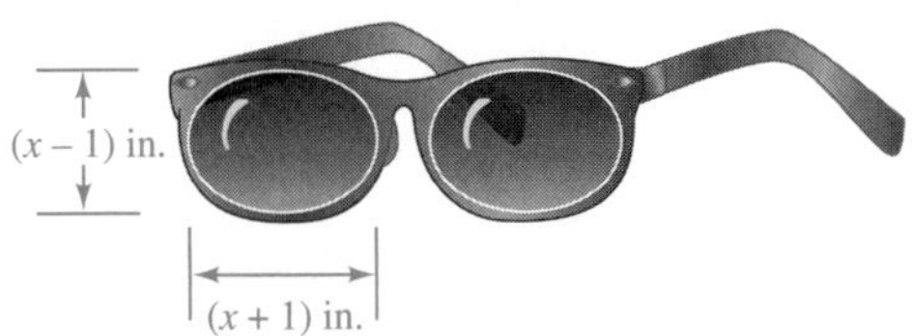

98. GARDENING

a. What is the area of the region planted with corn? tomatoes? beans? carrots? Use your answers to find the total area of the garden.

b. What is the length of the garden? What is its width? Use your answers to find its area.

c. How do the answers from parts a and b for the area of the garden compare?

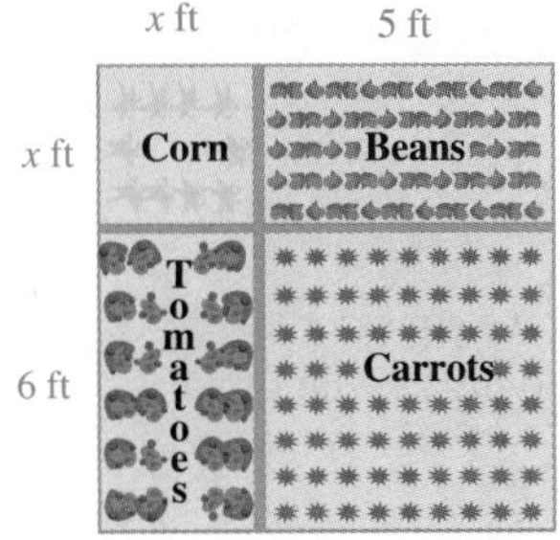

99. TOYS Find a polynomial that represents the area of the screen of the Etch A Sketch®

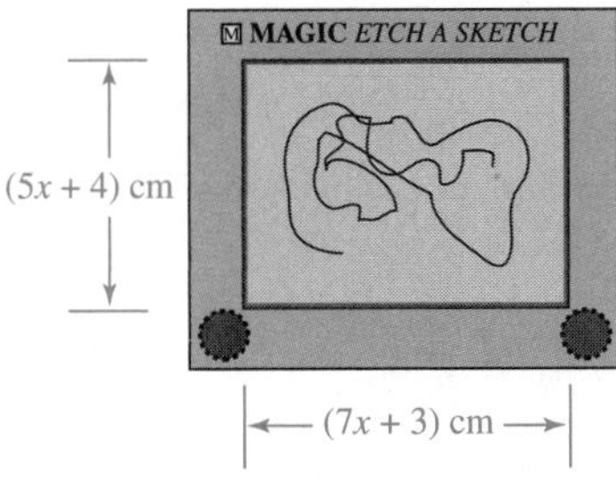

100. GRAPHIC ARTS A graduation announcement has a 1-inch-wide border around the written text. Find a polynomial that represents the area of the announcement.

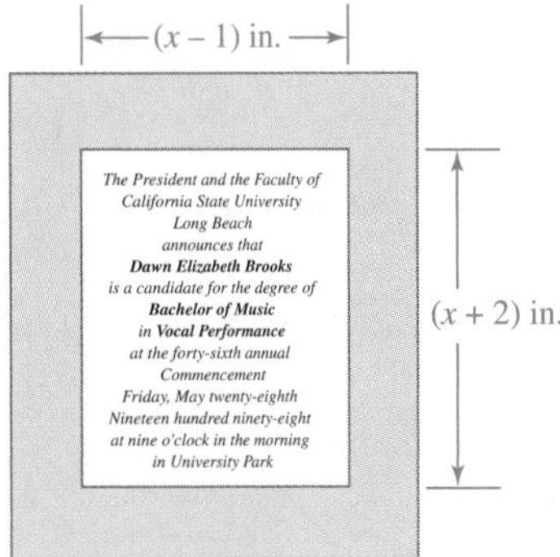

101. LUGGAGE Find a polynomial that represents the volume of the garment bag.

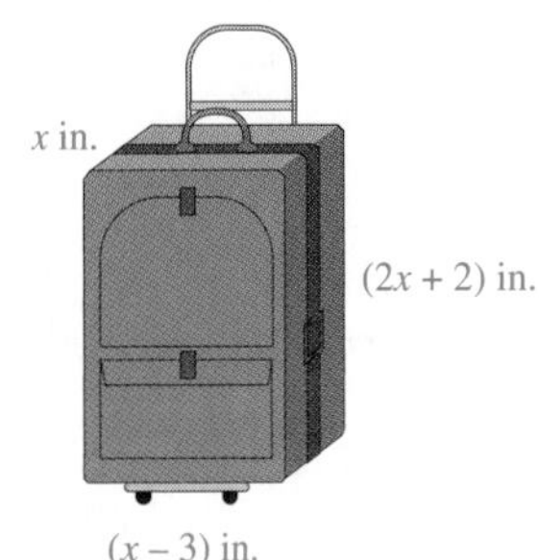

102. BASEBALL Find a polynomial that represents the amount of space within the batting cage.

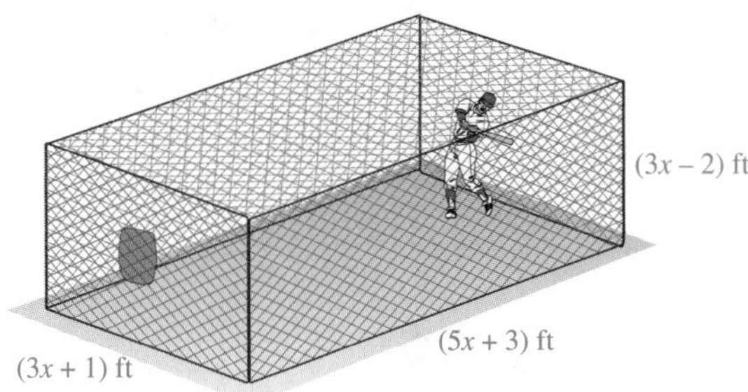

WRITING

103. Is the product of a monomial and a monomial always a monomial? Explain why or why not.

104. Explain this diagram.

$$(5x + 6)(7x - 1)$$

105. Explain why the FOIL method cannot be used to find $(3x + 2)(4x^2 - x + 10)$.

106. Explain the error:

$$(x + 3)(x - 2) = x^2 - 6$$

REVIEW

107. What is the slope of Line 1?

108. What is the slope of Line 2?

109. What is the slope of Line 3?

110. What is the slope of the x-axis?

111. What is the y-intercept of Line 1?

112. What is the x-intercept of Line 1?

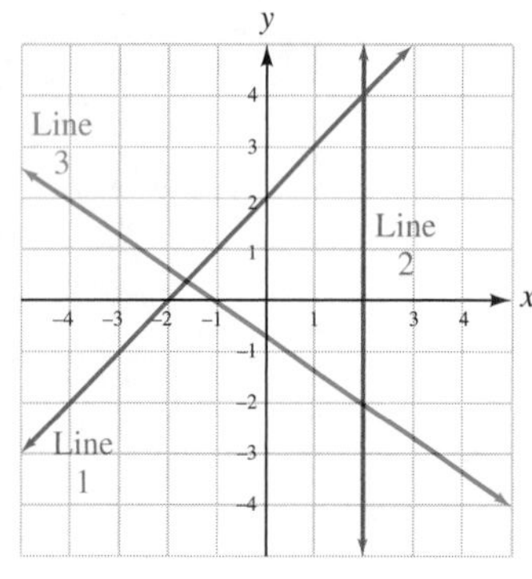

CHALLENGE PROBLEMS

113. Find each of the following.

a. $(x - 1)(x + 1)$

b. $(x - 1)(x^2 + x + 1)$

c. $(x - 1)(x^3 + x^2 + x + 1)$

114. Solve: $(y - 1)(y + 6) = (y - 3)(y - 2) + 8$.

4.7 Special Products

- Squaring a Binomial
- The Product of a Sum and Difference
- Higher Powers of Binomials
- Simplifying Expressions Containing Polynomials

Certain products of binomials, called **special products,** occur so frequently that it is worthwhile to learn their forms.

SQUARING A BINOMIAL

To develop a formula to find the *square of a sum,* we consider $(x + y)^2$. We can use the definition of exponent and the procedure for multiplying two binomials to find the product.

$$(x + y)^2 = (x + y)(x + y) \quad \text{In } (x + y)^2 \text{, the base is } (x + y) \text{ and the exponent is 2.}$$
$$= x^2 + xy + xy + y^2 \quad \text{Multiply the binomials.}$$
$$= x^2 + 2xy + y^2 \quad \text{Combine like terms: } xy + xy = 1xy + 1xy = 2xy.$$

Note that the terms of the result are related to the terms of the binomial that was squared.

$$(x + y)^2 = x^2 + 2xy + y^2$$

- y^2: The square of the second term, y.
- $2xy$: Twice the product of the first and second terms, x and y.
- x^2: The square of the first term, x.

Success Tip

The illustration can be used to visualize a special product. The area of the large square is $(x + y)(x + y) = (x + y)^2$. The sum of the four smaller areas is $x^2 + xy + xy + y^2$ or $x^2 + 2xy + y^2$. Thus,

$$(x + y)^2 = x^2 + 2xy + y^2$$

	x	y
y	xy	y^2
x	x^2	xy

(side lengths: $x + y$ by $x + y$)

To develop a formula to find the *square of a difference,* we consider $(x - y)^2$.

$$(x - y)^2 = (x - y)(x - y) \quad \text{In } (x - y)^2 \text{, the base is } (x - y) \text{ and the exponent is 2.}$$
$$= x^2 - xy - xy + y^2 \quad \text{Multiply the binomials.}$$
$$= x^2 - 2xy + y^2 \quad \text{Combine like terms: } -xy - xy = -2xy.$$

Again, the terms of the result are related to the terms of the binomial that was squared.

$$(x - y)^2 = x^2 - 2xy + y^2$$

- x^2: The square of the first term, x.
- $-2xy$: Twice the product of the first and second terms, x and $-y$.
- y^2: The square of the second term, $-y$.

The results of these two examples suggest the following rules.

Squaring a Binomial

The **square of a binomial** is the square of its first term, plus twice the product of both of its terms, plus the square of its second term.

$$(x + y)^2 = x^2 + 2xy + y^2 \qquad (x - y)^2 = x^2 - 2xy + y^2$$

EXAMPLE 1

Find each square: **a.** $(t + 9)^2$, **b.** $(8a - 5)^2$, **c.** $(-d + 0.5)^2$, and **d.** $(c^3 - 7d)^2$.

ELEMENTARY Algebra f(x) Now™

Solution

a. $(t + 9)^2$ is the square of a sum. The first term is t and the second term is 9.

$$(t + 9)^2 = \underbrace{t^2}_{\text{The square of the first term, } t.} + \underbrace{2(t)(9)}_{\text{Twice the product of both terms.}} + \underbrace{9^2}_{\text{The square of the second term, 9.}}$$

$$= t^2 + 18t + 81$$

The Language of Algebra

When squaring a binomial, the result is called a *perfect square trinomial.* For example,

$$(t + 9)^2 = \underbrace{t^2 + 18t + 81}_{\text{Perfect square trinomial}}$$

b. $(8a - 5)^2$ is the square of a difference. The first term is $8a$ and the second term is -5.

$$(8a - 5)^2 = \underbrace{(8a)^2}_{\text{The square of the first term, } 8a.} - \underbrace{2(8a)(5)}_{\text{Twice the product of both terms.}} + \underbrace{(-5)^2}_{\text{The square of the second term, } -5.}$$

$$= 64a^2 - 80a + 25$$ Use the power rule for products: $(8a)^2 = 8^2a^2 = 64a^2$.

Caution

Remember that the square of a binomial is a *trinomial.* A common error when squaring a binomial is to forget the middle term of the product. For example,

$$(x + 3)^2 \neq x^2 + 9$$

Missing $6x$

c. $(-d + 0.5)^2$ is the square of a sum. The first term is $-d$ and the second term is 0.5.

$$(-d + 0.5)^2 = (-d)^2 + 2(-d)(0.5) + (0.5)^2$$
$$= d^2 - d + 0.25$$

d. $(c^3 - 7d)^2$ is the square of a difference. The first term is c^3 and the second term is $-7d$.

$$(c^3 - 7d)^2 = (c^3)^2 - 2(c^3)(7d) + (-7d)^2$$
$$= c^6 - 14c^3d + 49d^2$$ Use rules for exponents to find $(c^3)^2$ and $(-7d)^2$.

Self Check 1 Find each square: **a.** $(r + 6)^2$, **b.** $(7g - 2)^2$, **c.** $(-v + 0.8)^2$, and **d.** $(w^4 - 3y)^2$

THE PRODUCT OF A SUM AND DIFFERENCE

The final special product is the product of two binomials that differ only in the signs of the second terms. To develop a rule to find the product of a *sum and a difference,* we consider $(x + y)(x - y)$.

$$\begin{aligned}(x + y)(x - y) &= x^2 - xy + xy - y^2 && \text{Multiply the binomials.}\\ &= x^2 - y^2 && \text{Combine like terms: } -xy + xy = 0.\end{aligned}$$

Success Tip

We can use the FOIL method to find each of the special products discussed in this section. However, these forms occur so often, it is worthwhile to learn the special product rules.

Note that when we combined like terms, we added opposites. This will always be the case for products of this type; the sum of the outer and inner products will be 0.

$$(x + y)(x - y) = x^2 - y^2$$

The square of the second term, y.

The square of the first term, x.

These observations suggest the following rule.

Multiplying the Sum and Difference of Two Terms

The product of the sum and difference of the two terms x and y is the square of x minus the square of y.

$$(x + y)(x - y) = x^2 - y^2$$

EXAMPLE 2

ELEMENTARY Algebra f(x) Now™

Multiply: **a.** $(m + 2)(m - 2)$, **b.** $(3y + 4)(3y - 4)$, **c.** $\left(b - \frac{2}{3}\right)\left(b + \frac{2}{3}\right)$, and **d.** $(t^4 - 6u)(t^4 + 6u)$.

Solution

a. In $m + 2$, the first term is m and the second term is 2.

$$(m + 2)(m - 2) = \underbrace{m^2}_{\text{The square of the first term, } m.} - \underbrace{2^2}_{\text{The square of the second term, 2.}}$$

$$= m^2 - 4$$

The Language of Algebra

When multiplying the sum and difference of two terms, the result is called a *difference of two squares.* For example:

$(m + 2)(m - 2) = m^2 - 4$

Difference of two squares

b. In $3y + 4$, the first term is $3y$ and the second term is 4.

$$\begin{aligned}(3y + 4)(3y - 4) &= (3y)^2 - 4^2\\ &= 9y^2 - 16 && (3y)^2 = 3^2y^2 = 9y^2.\end{aligned}$$

c. By the commutative property of multiplication, the special product rule can be written with the factor containing the $-$ symbol first. That is, $(x - y)(x + y) = x^2 - y^2$. So we have

$$\begin{aligned}\left(b - \frac{2}{3}\right)\left(b + \frac{2}{3}\right) &= b^2 - \left(\frac{2}{3}\right)^2\\ &= b^2 - \frac{4}{9}\end{aligned}$$

d. Apply the sum and difference special product rule and simplify.

$$\begin{aligned}(t^4 - 6u)(t^4 + 6u) &= (t^4)^2 - (6u)^2\\ &= t^8 - 36u^2 && (t^4)^2 = t^{4 \cdot 2} = t^8 \text{ and } (6u)^2 = 6^2u^2 = 36u^2.\end{aligned}$$

Self Check 2 Multiply: **a.** $(b + 4)(b - 4)$, **b.** $(5m + 9)(5m - 9)$, **c.** $\left(s - \frac{3}{4}\right)\left(s + \frac{3}{4}\right)$, and **d.** $(c^3 + 2d)(c^3 - 2d)$.

EXAMPLE 3

Paper towels. The amount of space (volume) occupied by the paper on a roll of paper towels is given by the expression $\pi h(R + r)(R - r)$, where R is the outer radius and r is the inner radius. Perform the indicated multiplication.

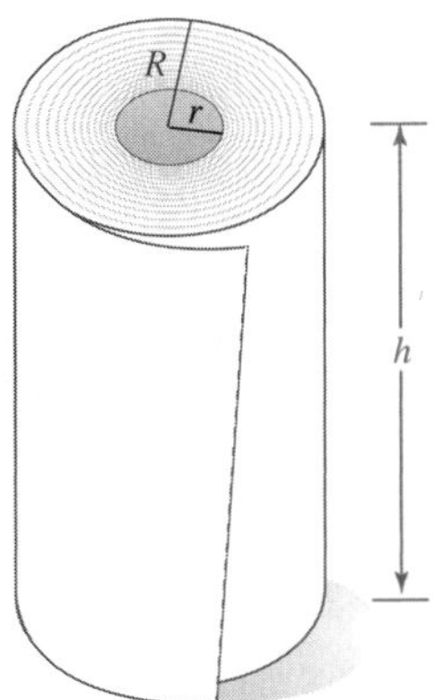

Solution We can multiply the polynomials πh, $(R + r)$, and $(R - r)$ in any order. Since $(R + r)$ and $(R - r)$ form a special product, we will do that multiplication first. Then we multiply the result by the monomial πh.

$$\pi h(R + r)(R - r) = \pi h(R^2 - r^2)$$ To find $(R + r)(R - r)$, use a special product rule. The first term squared is R^2. The second term squared is r^2.

$$= \pi hR^2 - \pi hr^2$$ Distribute the multiplication by πh.

HIGHER POWERS OF BINOMIALS

To find the third, fourth, or even higher powers of a binomial, we can use the special product rules.

EXAMPLE 4

Expand: $(x + 1)^3$.

ELEMENTARY Algebra f(x) Now™

Solution Write $(x + 1)^3$ as $(x + 1)^2(x + 1)$ so that a special product rule can be used.

$$(x + 1)^3 = (x + 1)^2(x + 1)$$

$$= (x^2 + 2x + 1)(x + 1)$$ Find $(x + 1)^2$ using the square of a sum rule.

$$= x^2(x) + x^2(1) + 2x(x) + 2x(1) + 1(x) + 1(1)$$ Multiply each term of $x + 1$ by each term of $x^2 + 2x + 1$.

$$= x^3 + x^2 + 2x^2 + 2x + x + 1$$ Multiply the monomials.

$$= x^3 + 3x^2 + 3x + 1$$ Combine like terms.

The Language of Algebra

To *expand* means to increase in size. When we *expand* a power of a binomial, the result is an expression that has more terms than the original binomial.

Self Check 4 Expand: $(n - 3)^3$.

SIMPLIFYING EXPRESSIONS CONTAINING POLYNOMIALS

We can use the methods discussed in Sections 4.5, 4.6, and 4.7 to simplify expressions that involve addition, subtraction, and multiplication of polynomials.

EXAMPLE 5

ELEMENTARY Algebra f(x) Now™

Simplify each expression: **a.** $(x + 1)(x - 2) + 3x(x + 3)$ and **b.** $(3y - 2)^2 - (y - 5)(y + 5)$.

Solution **a.** $(x + 1)(x - 2) + 3x(x + 3)$

$= x^2 - x - 2 + 3x^2 + 9x$ Use the FOIL method to find $(x + 1)(x - 2)$. Distribute the multiplication by $3x$.

$= 4x^2 + 8x - 2$ Combine like terms.

b. $(3y - 2)^2 - (y - 5)(y + 5)$

$= 9y^2 - 12y + 4 - (y^2 - 25)$ Find the special products: $(3y - 2)^2$ and $(y - 5)(y + 5)$. Write $y^2 - 25$ within parentheses so that both terms get subtracted.

$= 9y^2 - 12y + 4 - 1(y^2 - 25)$ Interpret the $-$ symbol preceding the parenthesis as -1.

$= 9y^2 - 12y + 4 - y^2 + 25$ Distribute the multiplication by -1.

$= 8y^2 - 12y + 29$ Combine like terms.

Self Check 5 Simplify each expression: **a.** $(x - 4)(x + 6) + 5x(2x - 1)$ and **b.** $(a + 9)(a - 9) - (2a - 4)^2$.

Answers to Self Checks **1. a.** $r^2 + 12r + 36$, **b.** $49g^2 - 28g + 4$, **c.** $v^2 - 1.6v + 0.64$, **d.** $w^8 - 6w^4y + 9y^2$ **2. a.** $b^2 - 16$, **b.** $25m^2 - 81$, **c.** $s^2 - \frac{9}{16}$, **d.** $c^6 - 4d^2$ **4.** $n^3 - 9n^2 + 27n - 27$ **5. a.** $11x^2 - 3x - 24$, **b.** $-3a^2 + 16a - 97$

4.7 STUDY SET

ELEMENTARY Algebra Now™

VOCABULARY Fill in the blanks.

1. The first ______ of $3x + 6$ is $3x$ and the second ______ is 6.
2. $(x + 4)^2$ is the ______ of a sum and $(m - 9)^2$ is the square of a ______.
3. $(b + 1)(b - 1)$ is the product of the ______ and difference of two terms.
4. Since $x^2 + 16x + 64$ is the square of $x + 8$, it is called a ______ square trinomial.
5. An expression of the form $x^2 - y^2$ is called a ______ of two squares.
6. Expressions of the form $(x + y)^2$, $(x - y)^2$, and $(x + y)(x - y)$ occur so frequently in algebra that they are called special ______.

CONCEPTS

7. Complete the rules for exponents.
   a. $(x^m)^n =$ ______ **b.** $(xy)^n =$ ______
8. Simplify each expression.
   a. $(5x)^2$ **b.** $2(x)(3)$
   c. $(d^3)^2$ **d.** $-2(6y)(5)$
9. Fill in the blanks.
   a. $(a + 5)(a + 5) = (a + 5)^{\square}$
   b. $(n - 12)(n - 12) = (n - 12)^{\square}$
10. Complete the special product.

$$(x + y)^2 = x^2 + 2xy + y^2$$

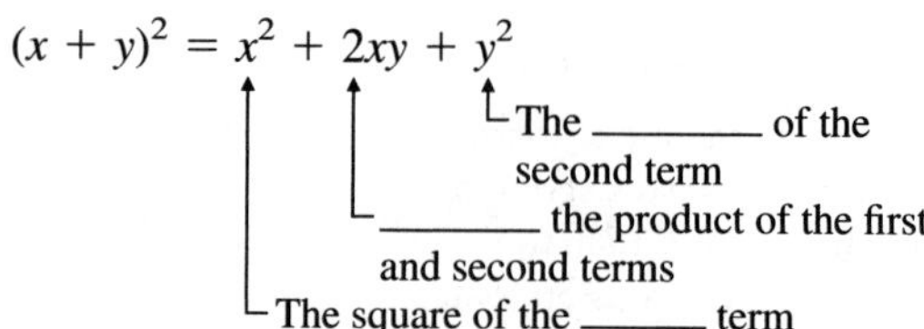

11. Complete the special product.

$$(x + y)(x - y) = x^2 - y^2$$

The square of the ______ term
The ______ of the first term

12. Fill in the blanks.
    a. $(x + 2)^3 = (x + 2)^{\square}(x + 2)$
    b. $(c - 1)^4 = (c - 1)^2(c - 1)^{\square}$

NOTATION Complete each solution.

13. $(x + 4)^2 = \square^2 + 2(x)(\square) + \square^2$
    $= x^2 + \square + 16$

14. $(6r - 1)^2 = (\square)^2 \square 2(6r)(1) + (-1)^2$
$= \square - \square + 1$

15. $(s + 5)(s - 5) = \square^2 \square \square^2$
$= s^2 - \square$

16. $(h - 3)(h + 3) = \square^2 \square \square^2$
$= \square - 9$

PRACTICE **Find each product.**

17. $(x + 1)^2$
18. $(y + 7)^2$
19. $(r + 2)^2$
20. $(n + 10)^2$
21. $(m - 6)^2$
22. $(b - 1)^2$
23. $(f - 8)^2$
24. $(w - 9)^2$
25. $(d + 7)(d - 7)$
26. $(t + 2)(t - 2)$
27. $(n + 6)(n - 6)$
28. $(a + 12)(a - 12)$
29. $(4x + 5)^2$
30. $(6y + 3)^2$
31. $(7m - 2)^2$
32. $(9b - 2)^2$
33. $(y^2 + 9)^2$
34. $(d^2 + 2)^2$
35. $(2v^3 - 8)^2$
36. $(8x^4 - 3)^2$
37. $(4f + 0.4)(4f - 0.4)$
38. $(4t + 0.6)(4t - 0.6)$
39. $(3n + 1)(3n - 1)$
40. $(5a + 4)(5a - 4)$
41. $(1 - 3y)^2$
42. $(1 - 4a)^2$
43. $(x - 2y)^2$
44. $(3a + 2b)^2$
45. $(2a - 3b)^2$
46. $(2x + 5y)^2$
47. $\left(s + \frac{3}{4}\right)^2$
48. $\left(y - \frac{5}{3}\right)^2$
49. $(a + b)^2$
50. $(c + d)^2$
51. $(r - s)^2$
52. $(t - u)^2$
53. $\left(6b + \frac{1}{2}\right)\left(6b - \frac{1}{2}\right)$
54. $\left(4h + \frac{2}{3}\right)\left(4h - \frac{2}{3}\right)$
55. $(r + 10s)^2$
56. $(m + 8n)^2$
57. $(6 - 2d^3)^2$
58. $(6 - 5p^2)^2$
59. $-(8x + 3)^2$
60. $-(4b - 8)^2$
61. $-(5 - 6g)(5 + 6g)$
62. $-(6 - c^2)(6 + c^2)$
63. $3x(2x + 3)^2$
64. $4y(3y + 4)^2$
65. $-5d(4d - 1)^2$
66. $-2h(7h - 2)^2$
67. $4d(d^2 + g^3)(d^2 - g^3)$
68. $8y(x^2 + y^2)(x^2 - y^2)$

Expand each binomial.

69. $(x + 4)^3$
70. $(y + 2)^3$
71. $(n - 6)^3$
72. $(m - 5)^3$
73. $(2g - 3)^3$
74. $(3x - 2)^3$
75. $(n - 2)^4$
76. $(c + d)^4$

Perform the operations.

77. $2t(t + 2) + (t - 1)(t + 9)$
78. $3y(y + 2) + (y + 1)(y - 1)$
79. $(x + y)(x - y) + x(x + y)$
80. $(3x + 4)(2x - 2) - (2x + 1)(x + 3)$
81. $(3x - 2)^2 + (2x + 1)^2$
82. $(4a - 3)^2 + (a + 6)^2$
83. $(m + 10)^2 - (m - 8)^2$
84. $(5y - 1)^2 - (y + 7)(y - 7)$

APPLICATIONS

85. DINNER PLATES The expression $\frac{\pi}{4}(D - d)(D + d)$ estimates the difference in area of two plates, the larger with diameter D and smaller with diameter d, as shown on the next page. Perform the indicated multiplication.

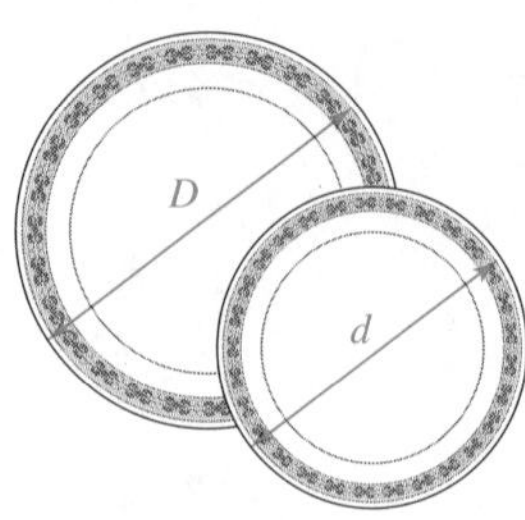

86. RETAIL STORES Refer to the illustration, which shows the floor space of a department store.

a. What area do the men's items occupy? the women's items? the children's items? the household items? Use these answers to find the total floor space of the department store.

b. What is the length of the floor space? the width? Use these answers and a special product rule to find the floor space area.

c. How do your final answers from parts a and b compare?

y ft | Children's items | Men's items
x ft | Household items | Women's items
x ft | y ft

87. PLAYPENS Find a polynomial that represent the area of the floor of the playpen.

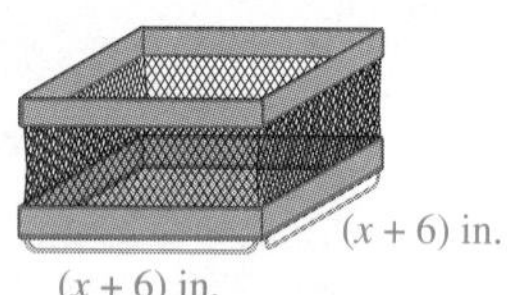

88. STORAGE Find a polynomial that represents the volume of the cubicle.

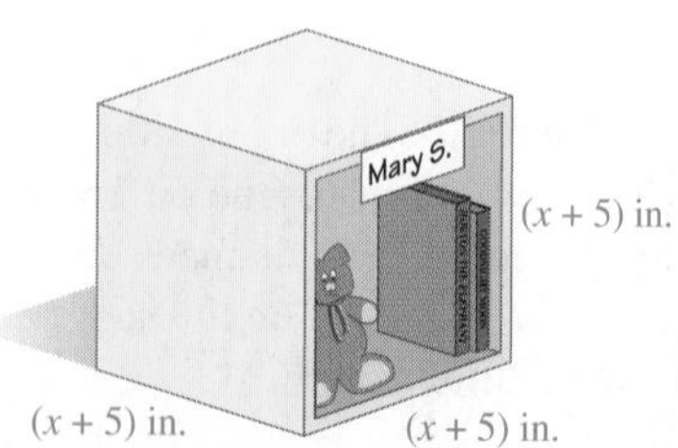

89. PAINTING To purchase the correct amount of enamel to paint the two garage doors, a painter must find their areas. Find a polynomial that gives the number of square feet to be painted. All dimensions are in feet, and the windows are squares with sides of x feet.

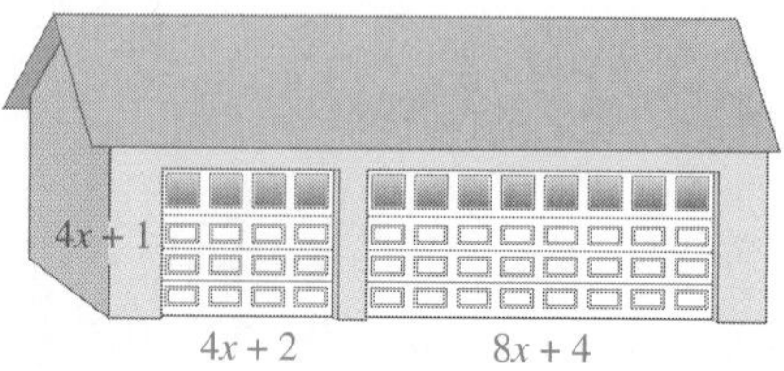

90. SIGNAL FLAGS Find a polynomial that represents the area in blue of the maritime signal flag for the letter p. The dimensions are given in centimeters.

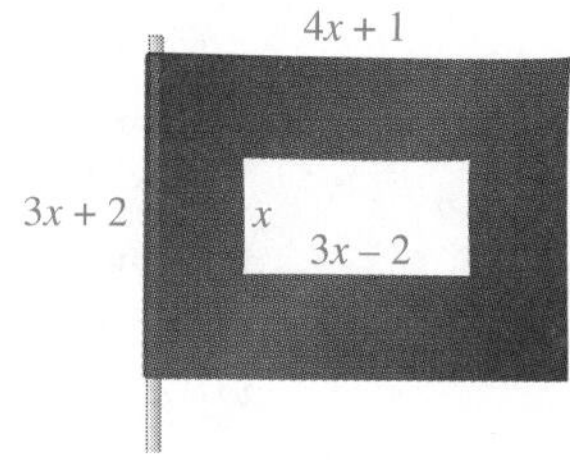

WRITING

91. What is a binomial? Explain how to square it.

92. Writing $(x + y)^2$ as $x^2 + y^2$ illustrates a common error. Explain.

93. We can find $(2x + 3)^2$ and $(5y - 6)^2$ using the FOIL method or using special product rules. Explain why the special product rules are faster.

94. Explain why $(x + 2)^3 \neq x^3 + 8$.

REVIEW

95. Find the prime factorization of 189.

96. Complete each statement. For any nonzero number a,

a. $\frac{a}{a} =$ 　　 **b.** $\frac{a}{1} =$

c. $\frac{0}{a} =$ 　　 **d.** $\frac{a}{0}$ is

97. Simplify: $\frac{30}{36}$. 　　 **98.** Add: $\frac{5}{12} + \frac{1}{4}$.

99. Multiply: $\frac{7}{8} \cdot \frac{3}{5}$. 　　 **100.** Divide: $\frac{1}{3} \div \frac{4}{5}$.

CHALLENGE PROBLEMS

101. **a.** Find two binomials whose product is a binomial.

b. Find two binomials whose product is a trinomial.

c. Find two binomials whose product is a four-term polynomial.

102. A special product rule can be used to find $31 \cdot 29$.

$$31 \cdot 29 = (30 + 1)(30 - 1) = 30^2 - 1^2 = 900 - 1 = 899$$

Use this method to find $52 \cdot 48$.

4.8 Division of Polynomials

- Dividing a Monomial by a Monomial
- Dividing a Polynomial by a Monomial
- Dividing a Polynomial by a Polynomial

In this section, we will conclude our study of operations with polynomials by discussing division of polynomials. To begin, we consider the simplest case, the quotient of two monomials.

DIVIDING A MONOMIAL BY A MONOMIAL

To divide monomials, we can use the method for simplifying fractions or the quotient rule for exponents.

EXAMPLE 1

Simplify: **a.** $\frac{21x^5}{7x^2}$ and **b.** $\frac{10r^6s}{6rs^3}$.

ELEMENTARY Algebra $f(x)$ Now™

Solution

By simplifying fractions

a. $\frac{21x^5}{7x^2} = \frac{3 \cdot \overset{1}{\cancel{7}} \cdot \overset{1}{\cancel{x}} \cdot \overset{1}{\cancel{x}} \cdot x \cdot x \cdot x}{\underset{1}{\cancel{7}} \cdot \underset{1}{\cancel{x}} \cdot \underset{1}{\cancel{x}}}$

$= 3x^3$

b. $\frac{10r^6s}{6rs^3} = \frac{\overset{1}{\cancel{2}} \cdot 5 \cdot \overset{1}{\cancel{r}} \cdot r \cdot r \cdot r \cdot r \cdot r \cdot \overset{1}{\cancel{s}}}{\underset{1}{\cancel{2}} \cdot 3 \cdot \underset{1}{\cancel{r}} \cdot \underset{1}{\cancel{s}} \cdot s \cdot s}$

$= \frac{5r^5}{3s^2}$

Using the rules for exponents

$\frac{21x^5}{7x^2} = 3x^{5-2}$ Divide the coefficients.

$= 3x^3$

$\frac{10r^6s}{6rs^3} = \frac{5}{3}r^{6-1}s^{1-3}$ Simplify $\frac{10}{6}$.

$= \frac{5}{3}r^5s^{-2}$

$= \frac{5r^5}{3s^2}$ Move s^{-2} to the denominator and change the sign of the exponent.

Success Tip

In this section, you will see that regardless of the number of terms involved, every polynomial division is a series of monomial divisions.

Self Check 1 Simplify: **a.** $\frac{30y^4}{5y^2}$ and **b.** $\frac{8c^2d^6}{32c^5d^2}$.

DIVIDING A POLYNOMIAL BY A MONOMIAL

Recall that to add two fractions with the same denominator, we add their numerators and keep the common denominator.

$$\frac{a}{d} + \frac{b}{d} = \frac{a + b}{d}$$

We can use this rule in reverse to divide polynomials by monomials.

Dividing a Polynomial by a Monomial

To divide a polynomial by a monomial, divide each term of the polynomial by the monomial. Let a, b, and d represent monomials, where d is not 0

$$\frac{a + b}{d} = \frac{a}{d} + \frac{b}{d}$$

EXAMPLE 2

Divide **a.** $\frac{9x^2 + 6x}{3x}$ and **b.** $\frac{12a^4b^3 - 18a^3b^2 + 2a^2}{6a^2b^2}$.

ELEMENTARY Algebra f(x) Now™

Solution **a.** Here, we have a binomial divided by a monomial.

$$\frac{9x^2 + 6x}{3x} = \frac{9x^2}{3x} + \frac{6x}{3x}$$ Divide each term of the numerator by the denominator, $3x$.

$$= 3x^{2-1} + 2x^{1-1}$$ Do each monomial division. Divide the coefficients. Subtract the exponents.

$$= 3x + 2$$ Recall that $x^0 = 1$.

The Language of Algebra

The names of the parts of a division statement are

Dividend $\frac{9x^2 + 6x}{3x} = 3x + 2$ *Divisor* *Quotient*

Check: We multiply the divisor, $3x$, and the quotient, $3x + 2$. The result should be the dividend, $9x^2 + 6x$.

$$3x(3x + 2) = 9x^2 + 6x$$ The answer checks.

b. Here, we have a trinomial divided by a monomial.

$$\frac{12a^4b^3 - 18a^3b^2 + 2a^2}{6a^2b^2} = \frac{12a^4b^3}{6a^2b^2} - \frac{18a^3b^2}{6a^2b^2} + \frac{2a^2}{6a^2b^2}$$ Divide each term of the numerator by the denominator, $6a^2b^2$.

$$= 2a^{4-2}b^{3-2} - 3a^{3-2}b^{2-2} + \frac{a^{2-2}}{3b^2}$$ Do each monomial division. Simplify: $\frac{2}{6} = \frac{1}{3}$.

$$= 2a^2b - 3a + \frac{1}{3b^2}$$

Success Tip

The sum, difference, and product of two polynomials are always polynomials. However, as seen in Example 2b, the quotient of two polynomials is not always a polynomial.

Recall that the variables in a polynomial must have whole-number exponents. Therefore, the result, $2a^2b - 3a + \frac{1}{3b^2}$, is not a polynomial because the last term can be written $\frac{1}{3}b^{-2}$.

Check: $$6a^2b^2\left(2a^2b - 3a + \frac{1}{3b^2}\right) = 12a^4b^3 - 18a^3b^2 + 2a^2$$ The answer checks.

Self Check 2 Divide: **a.** $\frac{50h^3 + 15h^2}{5h^2}$ and **b.** $\frac{22s^5t^2 - s^4t^3 + 44s^2t^4}{11s^2t^2}$.

EXAMPLE 3

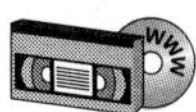

Formulas. The area of a trapezoid is given by the formula $A = \frac{1}{2}h(B + b)$, where B and b are its bases and h is its height. Solve the formula for b.

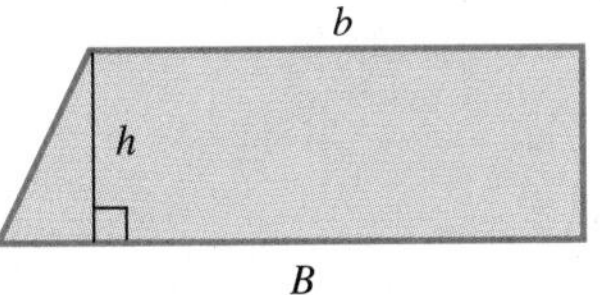

$$A = \frac{1}{2}h(B + b)$$

$$2A = h(B + b)$$ Multiply both sides by 2 to clear the equation of the fraction.

$$2A = hB + hb$$ Distribute the multiplication by h.

$$2A - \mathbf{hB} = hB + hb - \mathbf{hB}$$ Subtract hB from both sides.

$$2A - hB = hb$$ Combine like terms: $hB - hB = 0$.

$$\frac{2A - hB}{\mathbf{h}} = \frac{hb}{\mathbf{h}}$$ To undo the multiplication by h, divide both sides by h.

$$\frac{2A - hB}{h} = b$$

We can stop at this point, or, as shown below, we can perform the division on the left-hand side of the equation.

$$\frac{2A}{h} - \frac{hB}{h} = b$$ Divide each term of the numerator by the denominator, h.

$$\frac{2A}{h} - B = b$$ Simplify: $\frac{\overset{1}{\cancel{h}}B}{\underset{1}{\cancel{h}}} = B$.

DIVIDING A POLYNOMIAL BY A POLYNOMIAL

To divide one polynomial by another, we use a method similar to long division in arithmetic.

EXAMPLE 4

Divide $x^2 + 5x + 6$ by $x + 2$.

ELEMENTARY Algebra f(x) Now™

Solution Here the divisor is $x + 2$, and the dividend is $x^2 + 5x + 6$.

Success Tip

The long-division process is a series of four steps that are repeated: Divide, multiply, subtract, and bring down.

Step 1:
$$x + 2\overline{)x^2 + 5x + 6}\quad \text{quotient: } x$$
Divide x^2 by x: $\frac{x^2}{x} = x$. Write the result, x, above the division symbol.

Step 2:
$$\begin{array}{r} x \\ x + 2\overline{)x^2 + 5x + 6} \\ \underline{x^2 + 2x} \end{array}$$
Multiply each term of the divisor by x. Write the result, $x^2 + 2x$, under $x^2 + 5x$, and draw a line.

Step 3:
$$\begin{array}{r} x \\ x + 2\overline{)x^2 + 5x + 6} \\ \underline{x^2 + 2x} \\ 3x + 6 \end{array}$$
Subtract $x^2 + 2x$ from $x^2 + 5x$. Work vertically, column by column: $x^2 - x^2 = 0$ and $5x - 2x = 3x$.

Bring down the 6.

Step 4:
$$\begin{array}{r} x + 3 \\ x + 2\overline{)x^2 + 5x + 6} \\ \underline{x^2 + 2x} \\ 3x + 6 \end{array}$$
Divide $3x$ by x: $\frac{3x}{x} = 3$. Write $+ 3$ above the division symbol to form the second term of the quotient.

$$\text{Step 5: } x+2\overline{)x^2+5x+6}\quad \text{quotient } x+3;\quad \begin{array}{r} x^2+2x \\ \hline 3x+6 \\ 3x+6 \\ \hline \end{array}$$

Multiply each term of the divisor by 3. Write the result, $3x + 6$, under $3x + 6$, and draw a line.

$$\text{Step 6: } x+2\overline{)x^2+5x+6}\quad \text{quotient } x+3;\quad \begin{array}{r} x^2+2x \\ \hline 3x+6 \\ 3x+6 \\ \hline 0 \end{array}$$

Subtract $3x + 6$ from $3x + 6$. Work vertically: $3x - 3x = 0$ and $6 - 6 = 0$.

0 ← This is the remainder.

The quotient is $x + 3$ and the remainder is 0.

Step 7: Check the work by verifying that $(x + 2)(x + 3)$ is $x^2 + 5x + 6$.

$$\begin{aligned}(x+2)(x+3) &= x^2+3x+2x+6 \\ &= x^2+5x+6\end{aligned}$$

The answer checks.

Self Check 4 Divide $x^2 + 7x + 12$ by $x + 3$.

EXAMPLE 5

Divide: $\dfrac{6x^2 - 7x - 2}{2x - 1}$.

ELEMENTARY Algebra f(x) Now™

Solution Here the divisor is $2x - 1$ and the dividend is $6x^2 - 7x - 2$.

$$\text{Step 1: } 2x-1\overline{)6x^2-7x-2}\quad \text{quotient } 3x$$

Divide $6x^2$ by $2x$: $\frac{6x^2}{2x} = 3x$. Write the result, $3x$, above the division symbol.

$$\text{Step 2: } 2x-1\overline{)6x^2-7x-2}\quad \text{quotient } 3x;\quad \begin{array}{r} 6x^2-3x \\ \hline \end{array}$$

Multiply each term of the divisor by $3x$. Write the result, $6x^2 - 3x$, under $6x^2 - 7x$, and draw a line.

$$\text{Step 3: } 2x-1\overline{)6x^2-7x-2}\quad \text{quotient } 3x;\quad \begin{array}{r} 6x^2-3x \\ \hline -4x-2 \end{array}$$

Subtract $6x^2 - 3x$ from $6x^2 - 7x$. Work vertically: $6x^2 - 6x^2 = 0$ and $-7x - (-3x) = -7x + 3x = -4x$.

Bring down the -2.

$$\text{Step 4: } 2x-1\overline{)6x^2-7x-2}\quad \text{quotient } 3x-2;\quad \begin{array}{r} 6x^2-3x \\ \hline -4x-2 \end{array}$$

Divide $-4x$ by $2x$: $\frac{-4x}{2x} = -2$. Write -2 above the division symbol to form the second term of the quotient.

$$\text{Step 5: } 2x-1\overline{)6x^2-7x-2}\quad \text{quotient } 3x-2;\quad \begin{array}{r} 6x^2-3x \\ \hline -4x-2 \\ -4x+2 \\ \hline \end{array}$$

Multiply each term of the divisor by -2. Write the result, $-4x + 2$, under $-4x - 2$, and draw a line.

$$\begin{array}{r} 3x \;\; - 2 \\ \textit{Step 6: } 2x - 1\overline{)6x^2 - 7x - 2} \\ \underline{6x^2 - 3x} \\ -4x - 2 \\ \underline{-4x + 2} \\ -4 \end{array}$$

Subtract $-4x + 2$ from $-4x - 2$. Work vertically: $-4x - (-4x) = -4x + 4x = 0$ and $-2 - 2 = -4$.

Success Tip

The division process for polynomials continues until the degree of the remainder is less than the degree of the divisor. In Example 5, the remainder, −4, has degree 0. The divisor, $2x - 1$, has degree 1. Therefore, the division ends.

Here the quotient is $3x - 2$ and the remainder is -4. It is common to write the answer as either

$$3x - 2 + \frac{-4}{2x - 1} \quad \text{or} \quad 3x - 2 - \frac{4}{2x - 1} \qquad \text{Quotient} + \frac{\text{remainder}}{\text{divisor}}.$$

Step 7: We can check the answer using the fact that for any division:

divisor · quotient + remainder = dividend

$$\begin{aligned} (2x - 1)(3x - 2) + (-4) &= 6x^2 - 4x - 3x + 2 + (-4) \\ &= 6x^2 - 7x - 2 \qquad \text{The answer checks.} \end{aligned}$$

Self Check 5 Divide: $\dfrac{8x^2 + 6x - 3}{2x + 3}$.

The division method works best when the terms of the divisor and the dividend are written in descending powers of the variable. If the powers in the dividend or divisor are not in descending order, we use the commutative property of addition to write them that way.

EXAMPLE 6

Divide $4x^2 + 2x^3 + 12 - 2x$ by $x + 3$.

ELEMENTARY Algebra f(x) Now™

Solution If we write the dividend in descending powers of x, the division is routine.

$$\begin{array}{r} 2x^2 - 2x \;\; + 4 \\ x + 3\overline{)2x^3 + 4x^2 - 2x + 12} \\ \underline{2x^3 + 6x^2} \qquad\qquad\;\; \\ -2x^2 - 2x \qquad \\ \underline{-2x^2 - 6x} \qquad \\ 4x + 12 \\ \underline{4x + 12} \\ 0 \end{array}$$

$\frac{2x^3}{x} = 2x^2$.

$\frac{-2x^2}{x} = -2x$.

$\frac{4x}{x} = 4$.

$$\begin{aligned} \textbf{\textit{Check: }} (x + 3)(2x^2 - 2x + 4) &= 2x^3 - 2x^2 + 4x + 6x^2 - 6x + 12 \\ &= 2x^3 + 4x^2 - 2x + 12 \end{aligned}$$

Self Check 6 Divide $x^2 - 10x + 6x^3 + 4$ by $2x - 1$.

When we write the terms of a dividend in descending powers, we must determine whether some powers of the variable are missing. If any are missing, we should write such terms with a coefficient of 0 or leave blank spaces for them.

EXAMPLE 7

Divide: $\dfrac{27x^3 + 1}{3x + 1}$.

ELEMENTARY Algebra $f(x)$ Now™

Solution The dividend, $27x^3 + 1$, does not have an x^2-term or an x-term. We can either insert a $0x^2$ term and a $0x$ term as placeholders, or leave spaces for them.

$$\begin{array}{r l}
9x^2 - 3x + 1 & \\
3x + 1\overline{)27x^3 + 0x^2 + 0x + 1} & \frac{27x^3}{3x} = 9x^2. \\
\underline{27x^3 + 9x^2} \qquad\qquad & \\
-9x^2 + 0x \qquad\quad & \frac{-9x^2}{3x} = -3x. \\
\underline{-9x^2 - 3x} \qquad\quad & \\
3x + 1 & \frac{3x}{3x} = 1. \\
\underline{3x + 1} & \\
0 &
\end{array}$$

Check: $(3x + 1)(9x^2 - 3x + 1) = 27x^3 - 9x^2 + 3x + 9x^2 - 3x + 1$
$= 27x^3 + 1$

Self Check 7 Divide: $\dfrac{x^2 - 9}{x - 3}$.

Answers to Self Checks **1. a.** $6y^2$, **b.** $\frac{d^4}{4c^3}$ **2. a.** $10h + 3$, **b.** $2s^3 - \frac{s^2t}{11} + 4t^2$ **4.** $x + 4$ **5.** $4x - 3 + \frac{6}{2x + 3}$ **6.** $3x^2 + 2x - 4$ **7.** $x + 3$

4.8 STUDY SET

ELEMENTARY Algebra $f(x)$ Now™

VOCABULARY **Fill in the blanks.**

1. The __________ of $\frac{15x^2 - 25x}{5x}$ is $15x^2 - 25x$ and the __________ is $5x$.
2. The expression $\frac{18x^7}{9x^4}$ is a monomial divided by a __________.
3. The expression $\frac{6x^3y - 4x^2y^2 + 8xy^3 - 2y^4}{2x^4}$ is a __________ divided by a monomial.
4. The expression $\frac{x^2 - 8x + 12}{x - 6}$ is a trinomial divided by a __________.
5. The powers of x in $2x^4 + 3x^3 + 4x^2 - 7x - 8$ are written in __________ order.
6.

$$\begin{array}{r l}
\downarrow \quad x - 2 \leftarrow & \\
x - 6\overline{)x^2 - 8x - 4} \leftarrow & \\
\underline{x^2 - 6x} \qquad & \\
-2x - 4 & \\
\underline{-2x + 12} & \\
-16 \leftarrow &
\end{array}$$

7. The expression $5x^2 + 6$ is missing an x-term. We can insert a __________ $0x$ term and write it as $5x^2 + 0x + 6$.

CONCEPTS **Fill in the blanks.**

8. **a.** To divide a polynomial by a monomial, divide each ______ of the polynomial by the monomial.

   b. $\dfrac{18x + 9}{9} = \dfrac{18x}{\square} + \dfrac{9}{\square}$

   c. $\dfrac{30x^2 + 12x - 24}{6} = \dfrac{30x^2}{\square} + \dfrac{12x}{\square} - \dfrac{24}{\square}$

9. Complete each rule of exponents.

   a. $\dfrac{x^m}{x^n} = \square$ **b.** $x^{-n} = \square$

Simplify.

10. **a.** $\dfrac{x^8}{x^3}$ **b.** $\dfrac{y^5}{y^7}$

c. $\dfrac{a^4}{a^4}$

d. h^{1-5}

e. s^{10-4}

f. t^{5-5}

11. Write each polynomial with the powers in descending order.

a. $5x^2 + 7x^3 - 3x - 9$

b. $9x + 2x^2 - x^3 + 6x^4$

12. The long division process is a series of four steps that are repeated. Put them in the correct order:

subtract   multiply   bring down   divide

13. In the long division below, a subtraction must be performed. What is the answer to the subtraction?

$$\begin{array}{r} x \phantom{{}- 9x - 6} \\ x - 7\overline{)x^2 - 9x - 6} \\ \underline{x^2 - 7x} \phantom{{}- 6} \end{array}$$

14. Fill in the blanks: To check an answer of a long division, we use the fact that

divisor · ______ + remainder = ______

15. Using long division, a student found that

$$\frac{3x^2 + 8x + 4}{3x + 2} = x + 2$$

Check to see whether the result is correct.

16. Using long division, a student found that

$$\frac{x^2 + 4x - 20}{x - 3} = x + 7 + \frac{1}{x - 3}$$

Check to see whether the result is correct.

NOTATION Complete each solution.

17.
$$\frac{28x^5 - x^3 + 7x^2}{7x^2} = \frac{28x^5}{\square} - \frac{\square}{7x^2} + \frac{7x^2}{\square}$$
$$= 4x^{5-2} - \frac{x^{3-2}}{\square} + x^{\square}$$
$$= \square - \frac{x}{7} + \square$$

18.
$$\begin{array}{r} \square + 2 \phantom{{}+ 5} \\ x + 2\overline{)x^2 + 4x + 5} \\ \underline{x^2 + \square} \phantom{{}+ 5} \\ \square + 5 \\ \underline{2x + 4} \\ \square \end{array}$$

19. Insert placeholders for each missing term in the polynomial.

a. $5x^4 + 2x^2 - 1$

b. $-3x^5 - 2x^3 + 4x - 6$

20. True or false: $6x + 4 + \dfrac{-3}{x + 2} = 6x + 4 - \dfrac{3}{x + 2}$.

PRACTICE Perform each division by simplifying the fraction.

21. $\dfrac{8}{6}$

22. $\dfrac{15}{9}$

23. $\dfrac{x^5}{x^2}$

24. $\dfrac{a^{12}}{a^8}$

25. $\dfrac{45m^{10}}{9m^5}$

26. $\dfrac{24n^{12}}{8n^4}$

27. $\dfrac{12h^8}{9h^6}$

28. $\dfrac{22b^9}{6b^6}$

29. $\dfrac{-3d^4}{15d^8}$

30. $\dfrac{-4x^3}{16x^5}$

31. $\dfrac{r^3s^2}{rs^3}$

32. $\dfrac{y^4z^3}{y^2z^2}$

33. $\dfrac{8x^3y^2}{4xy^3}$

34. $\dfrac{-3y^3z}{6yz^2}$

35. $\dfrac{-16r^3y^2}{-4r^2y^4}$

36. $\dfrac{-35xyz^2}{-7x^2yz}$

37. $\dfrac{-65rs^2t}{15r^2s^3t}$

38. $\dfrac{112u^3z^6}{-42u^3z^6}$

Perform each division.

39. $\dfrac{6x + 9}{3}$

40. $\dfrac{8x + 12}{4}$

41. $\dfrac{8x^9 - 32x^6}{4x^4}$

42. $\dfrac{30y^8 + 40y^7}{10y^6}$

43. $\dfrac{6h^{12} + 48h^9}{24h^{10}}$

44. $\dfrac{4x^{14} - 36x^8}{36x^{12}}$

45. $\dfrac{-18w^6 - 9}{9w^4}$

46. $\dfrac{-40f^4 + 16}{8f^3}$

47. $\dfrac{9s^8 - 18s^5 + 12s^4}{3s^3}$

48. $\dfrac{16b^{10} + 4b^6 - 20b^4}{4b^2}$

49. $\dfrac{7c^5 + 21c^4 - 14c^3 - 35c}{7c^2}$

50. $\dfrac{12r^{15} - 48r^{12} + r^{10} - 18r^8}{6r^{10}}$

51. $\dfrac{5x - 10y}{25xy}$

52. $\dfrac{2x - 32}{16x}$

53. $\dfrac{15a^3b^2 - 10a^2b^3}{5a^2b^2}$

54. $\dfrac{9a^4b^3 - 16a^3b^4}{12a^2b}$

55. $\dfrac{12x^3y^2 - 8x^2y - 4x}{4xy}$

56. $\dfrac{12a^2b^2 - 8a^2b - 4ab}{4ab}$

57. $\dfrac{-25x^2y + 30xy^2 - 5xy}{-5xy}$

58. $\dfrac{-30a^2b^2 - 15a^2b - 10ab^2}{-10ab}$

59. Divide $x^2 + 8x + 12$ by $x + 2$.

60. Divide $x^2 + 5x + 6$ by $x + 2$.

61. Divide $y^2 + 13y + 12$ by $y + 1$.

62. Divide $z^2 + 7z + 12$ by $z + 3$.

63. $\dfrac{6a^2 + 5a - 6}{2a + 3}$

64. $\dfrac{3b^2 - 5b + 2}{3b - 2}$

65. $\dfrac{3b^2 + 11b + 6}{3b + 2}$

66. $\dfrac{8a^2 + 2a - 3}{2a - 1}$

67. $5x + 3\overline{)11x + 10x^2 + 3}$

68. $2x - 7\overline{)-x - 21 + 2x^2}$

69. $4 + 2x\overline{)-10x - 28 + 2x^2}$

70. $1 + 3x\overline{)9x^2 + 1 + 6x}$

71. $2x - 1\overline{)x - 2 + 6x^2}$

72. $2 + x\overline{)3x + 2x^2 - 2}$

73. $2x + 3\overline{)2x^3 + 7x^2 + 4x - 3}$

74. $2x - 1\overline{)2x^3 - 3x^2 + 5x - 2}$

75. $3x + 2\overline{)6x^3 + 10x^2 + 7x + 2}$

76. $4x + 3\overline{)4x^3 - 5x^2 - 2x + 3}$

77. $2x + 1\overline{)2x^3 + 3x^2 + 3x + 1}$

78. $3x - 2\overline{)6x^3 - x^2 + 4x - 4}$

79. $\dfrac{x^2 - 1}{x - 1}$

80. $\dfrac{x^2 - 9}{x + 3}$

81. $\dfrac{4x^2 - 9}{2x + 3}$

82. $\dfrac{25x^2 - 16}{5x - 4}$

83. $\dfrac{x^3 + 1}{x + 1}$

84. $\dfrac{x^3 - 8}{x - 2}$

85. $\dfrac{a^3 + a}{a + 3}$

86. $\dfrac{y^3 - 50}{y - 5}$

87. $\dfrac{2x^2 + 5x + 2}{2x + 3}$

88. $\dfrac{3x^2 - 8x + 3}{3x - 2}$

89. $\dfrac{4x^2 + 6x - 1}{2x + 1}$

90. $\dfrac{6x^2 - 11x + 2}{3x - 1}$

91. $\dfrac{2x^3 + 7x^2 + 4x + 3}{2x + 3}$

92. $\dfrac{2x^3 + 4x^2 - 2x + 3}{x - 2}$

93. $\dfrac{6x^3 + x^2 + 2x + 1}{3x - 1}$

94. $\dfrac{3y^3 - 4y^2 + 2y + 3}{y + 3}$

APPLICATIONS

95. POOL The rack shown in the illustration is used to set up the balls for a game of pool. If the perimeter of the rack, in inches, is given by the polynomial $6x^2 - 3x + 9$, what is the length of one side?

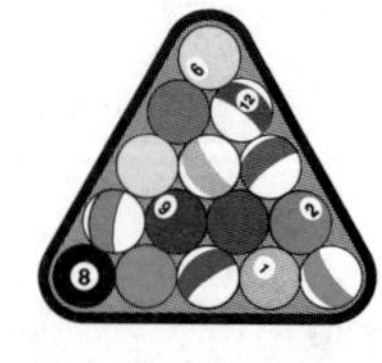

96. CHECKERBOARD If the perimeter (in inches) of a checkerboard is $12x^2 - 8x + 32$, what is the length of one side?

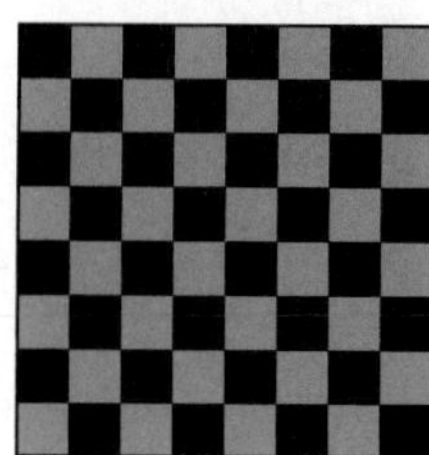

97. **a.** Solve the formula $d = rt$ for t.

b. Use your answer from part a to complete the table.

	r	$\cdot\ t$	$= d$
Motorcycle	$2x$		$6x^3$

c. Use your answer to part a to complete the table.

	r	$\cdot\ t$	$= d$
Subway	$x - 3$		$x^2 + x - 12$

98. FURNACE FILTERS The area of the furnace filter is $(x^2 - 2x - 24)$ square inches. Find its length.

99. AIR CONDITIONING If the volume occupied by the air conditioning unit is $(36x^3 - 24x^2)$ cubic feet, find its height.

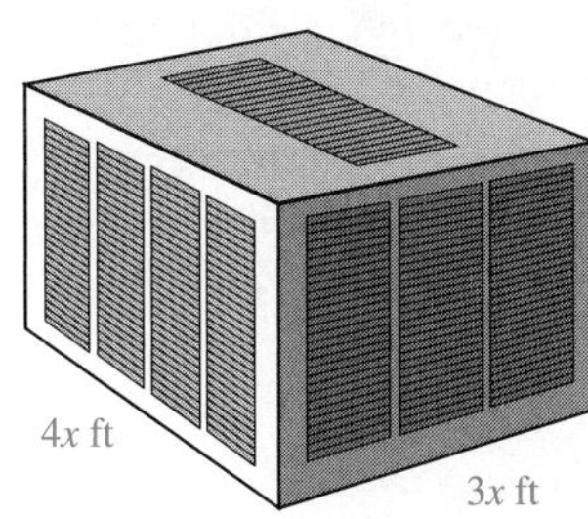

100. MINI-BLINDS The area covered by the mini-blinds is $(3x^3 - 6x)$ square feet. How long are the blinds?

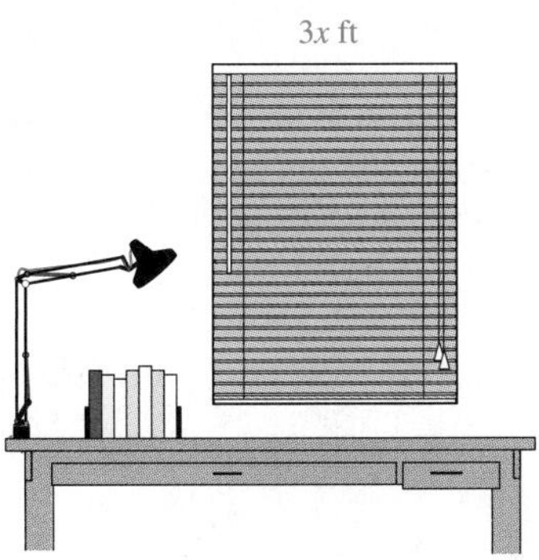

101. COMMUNICATIONS Telephone poles were installed every $(2x - 3)$ feet along a stretch of railroad track $(8x^3 - 6x^2 + 5x - 21)$ feet long. How many poles were used?

102. ELECTRIC BILLS On an electric bill, the formula

$$A = \frac{0.08x + 5}{x}$$

is used to compute the average cost of x kilowatt hours of electricity. Could the following formula be used instead?

$$A = 0.08 + \frac{5}{x}$$

WRITING

103. Explain how to check the following long division.

$$\begin{array}{r} x + 5 \\ 3x + 5\overline{)3x^2 + 20x - 5} \\ \underline{3x^2 + 5x} \qquad \\ 15x - 5 \\ \underline{15x + 25} \\ -30 \end{array}$$

104. Explain the difference in the methods used to divide $\frac{x^2 - 3x + 2}{x}$ as compared to $\frac{x^2 - 3x + 2}{x - 2}$.

105. How do you know when to stop the long division process when dividing polynomials?

106. When dividing $x^3 + 1$ by $x + 1$, why is it helpful to write $x^3 + 1$ as $x^3 + 0x^2 + 0x + 1$?

REVIEW

107. Write an equation of the line with slope $-\frac{11}{6}$ that passes through $(2, -6)$. Write the answer in slope–intercept form.

108. Solve $S = 2\pi rh + 2\pi r^2$ for h.

109. Evaluate: $-10(18 - 4^2)^3$.

110. Evaluate: -5^2.

CHALLENGE PROBLEMS

111. Divide: $\dfrac{6x^{6m}y^{6n} + 15x^{4m}y^{7n} - 24x^{2m}y^{8n}}{3x^{2m}y^n}$.

112. Divide: $\dfrac{6a^3 - 17a^2b + 14ab^2 - 3b^3}{2a - 3b}$.

ACCENT ON TEAMWORK

EVALUATING POLYNOMIALS USING LONG DIVISION

Overview: In this activity, you will learn an alternate way to evaluate polynomials.
Instructions: Form groups of 2 or 3 students. Have each student evaluate the polynomial $2x^2 - 3x - 5$ for $x = 1$ and for $x = 3$.

Next, divide the polynomial by $x - 1$ and by $x - 3$. What do you notice about the remainders of these divisions when compared to your answers to the evaluations?

Continue by evaluating the polynomial for $x = 2$ and dividing it by $x - 2$. Does the pattern still hold?

Finally, evaluate the polynomial for $x = -2$. By what must you divide to get a remainder that matches the evaluation?

Does the pattern hold for other polynomials? Try some polynomials of your own, experiment, and report your conclusions.

SHIFTING THE GRAPH OF $y = x^2$

Overview: In this activity, some minor changes are made to the equation $y = x^2$. You are to determine how these changes affect the position of the graph.
Instructions: Form groups of 2 or 3 students. Graph the following equations on the *same* coordinate system using a table of solutions as shown. Compare the graphs. How are they similar? How do they differ?

$$y = x^2$$
$$y = (x - 1)^2$$
$$y = (x + 1)^2$$

x	y
−4	
−3	
−2	
−1	
0	
1	
2	
3	
4	

On another coordinate system, graph $y = x^2$, $y = (x - 2)^2$, and $y = (x + 2)^2$. Compare these graphs. How are they similar? How do they differ?

From these examples, what conclusions can be made about shifting the graph of $y = x^2$ to the left and to the right?

BINOMIAL MULTIPLICATION AND THE AREA OF RECTANGLES

Overview: In this activity, rectangles are used to visualize binomial multiplication.
Instructions: Form groups of 2 or 3 students. Study the figure. The area of the large rectangle is given by $(x + 2)(x + 3)$. The area of the large rectangle is also the sum of the areas of the four smaller rectangles: $x^2 + 3x + 2x + 6$. Thus,

$$(x + 2)(x + 3) = x^2 + 3x + 2x + 6 = x^2 + 5x + 6.$$

		x	2
$x + 3$	3	$3x$	6
	x	x^2	$2x$

(width: $x + 2$)

Draw three similar models to represent the following products.

1. $(x + 4)(x + 5)$ **2.** $x(x + 6)$ **3.** $(x + 2)^2$

KEY CONCEPT: POLYNOMIALS

A **polynomial** is a single term or a sum of terms in which all variables have whole-number exponents. Some examples are

$$-16a^4b, \quad y + 8, \quad x^2 + 2xy - y^2, \quad \text{and} \quad x^3 - 2x^2 + 6x - 8$$

THE VOCABULARY OF POLYNOMIALS

1. Consider $x^3 - 2x^2 + 6x - 8$.
   a. Fill in: This is a polynomial in ___. It is written in __________ powers of ___.
   b. How many terms does the polynomial have?
   c. Give the degree of each term.
   d. What is the degree of the polynomial?
   e. Give the coefficient of each term.

2. Classify each polynomial as a monomial, binomial, trinomial, or none of these.
   a. $x^2 - y^2$
   b. $s^2t + st^2 - st + 1$
   c. $4y^2 - 10y + 16$
   d. $15h$

OPERATIONS WITH POLYNOMIALS

Just like numbers, polynomials can be added, subtracted, multiplied, divided, and raised to powers. The key to performing these operations with polynomials is knowing how to perform these operations with monomials.

Perform the operations.

3. $4x^3 + 3x^3$
4. $7m^{10} + (-6m^{10})$
5. $7a^2b - 9a^2b$
6. $(6y^5)(-7y^8)$
7. $\dfrac{16c^4d^5}{8c^2d^6}$
8. $(5f^3)^2$

Fill in the blanks.

9. To add polynomials, ________ like terms.
10. To subtract two polynomials, change the ______ of the terms of the polynomial being subtracted, drop parentheses, and combine like terms.
11. To multiply two polynomials, multiply ______ term of one polynomial by ______ term of the other polynomial and combine like terms.
12. To divide a polynomial by a monomial, divide each ______ of the polynomial by the monomial.
13. To divide two polynomials, use the ______ division method.

Perform the operations.

14. $(8x^3 + 4x^2 - 8x + 1) + (6x^3 - 5x^2 - 2x + 3)$
15. $(20s^3t + s^2t^2 - 6st^3) - (8s^3t - 9s^2t^2 + 12st^3)$
16. $(2x + 3)(x - 8)$
17. $(2x^2 + 3)^2$
18. $(4h^5 + 8t)(4h^5 - 8t)$
19. $(y^2 + y - 6)(y + 3)$
20. $\dfrac{9x^6 + 27x^7 - 18x^5}{3x^2}$
21. $\dfrac{x^3 + 3x^2 + 5x + 3}{x + 1}$

CHAPTER REVIEW

ELEMENTARY Algebra f(x) Now™

SECTION 4.1 Rules for Exponents

CONCEPTS

If n represents a natural number, then

$$x^n = \underbrace{x \cdot x \cdot x \cdot \cdots \cdot x}_{n \text{ factors of } x}$$

where x is called the *base* and n is called the *exponent.*

Rules for exponents:
If m and n represent integers, and there are no divisions by 0, then

$$x^m x^n = x^{m+n}$$

$$\frac{x^m}{x^n} = x^{m-n}$$

$$(x^m)^n = x^{mn}$$

$$(xy)^n = x^n y^n$$

$$\left(\frac{x}{y}\right)^n = \frac{x^n}{y^n}$$

REVIEW EXERCISES

1. Write the repeated multiplication that is indicated.

a. $-3x^4$ **b.** $\left(\frac{1}{2}pq\right)^3$

2. Identify the base and the exponent in each expression.

a. $2x^6$ **b.** $(2x)^6$

Evaluate each expression.

3. 5^3

4. $(-8)^2$

5. -8^2

6. $(5 - 3)^2$

Simplify each expression. Assume there are no divisions by 0.

7. $7^4 \cdot 7^8$

8. $mmnn$

9. $(y^7)^3$

10. $(3x)^4$

11. $b^3b^4b^5$

12. $-z^2(z^3y^2)$

13. $(-16s)^2s$

14. $(2x^2y)^2$

15. $(x^2x^3)^3$

16. $\left(\frac{x^2y}{xy^2}\right)^2$

17. $\frac{(m-25)^{16}}{(m-25)^4}$

18. $\frac{(5y^2z^3)^3}{(yz)^5}$

Find the area or the volume of each figure, whichever is appropriate.

19.

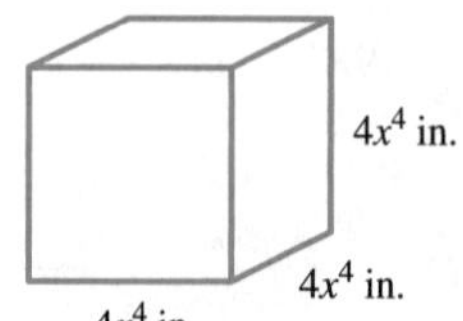

20.

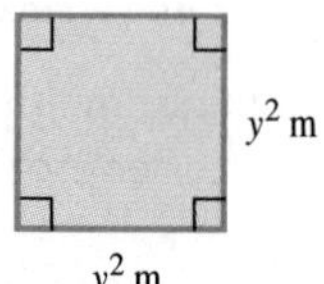

SECTION 4.2 Zero and Negative Exponents

For any nonzero real numbers x and y and any integers m and n,

$$x^0 = 1$$

$$x^{-n} = \frac{1}{x^n}$$

$$\frac{1}{x^{-n}} = x^n$$

$$\frac{x^{-m}}{y^{-n}} = \frac{y^n}{x^m}$$

Simplify each expression. Write each answer without using negative exponents or parentheses.

21. x^0

22. $(3x^2y^2)^0$

23. $(3x^0)^2$

24. 10^{-3}

25. $\left(\frac{3}{4}\right)^{-1}$

26. -5^{-2}

27. x^{-5}

28. $-6y^4y^{-5}$

29. $\frac{7^{-2}}{2^{-3}}$

30. $(x^{-3}x^{-4})^{-2}$

31. $\left(\frac{-3r^4r^{-3}}{r^{-3}r^7}\right)^3$

32. $\left(\frac{4z^4}{z^3}\right)^{-2}$

SECTION 4.3 Scientific Notation

A positive number is written in *scientific notation* when it is written in the form $N \times 10^n$, where $1 \le N < 10$ and n is an integer.

Write each number in scientific notation.

33. 728

34. 9,370,000,000,000,000

35. 0.0136

36. 0.00942

37. 0.018×10^{-2}

38. 753×10^3

Write each number in standard notation.

39. 7.26×10^5

40. 3.91×10^{-8}

41. 2.68×10^0

42. 5.76×10^1

Scientific notation provides an easier way to perform computations involving very large or very small numbers.

Evaluate each expression by first writing each number in scientific notation. Then do the arithmetic. Express the result in standard notation.

43. $\frac{(0.00012)(0.00004)}{0.00000016}$

44. $\frac{(4{,}800)(20{,}000)}{600{,}000}$

45. WORLD POPULATION As of 2003, the world's population was estimated to be 6.31 billion. Write this number in standard notation and in scientific notation.

46. ATOMS The illustration shows a cross section of an atom. How many nuclei, placed end to end, would it take to stretch across the atom?

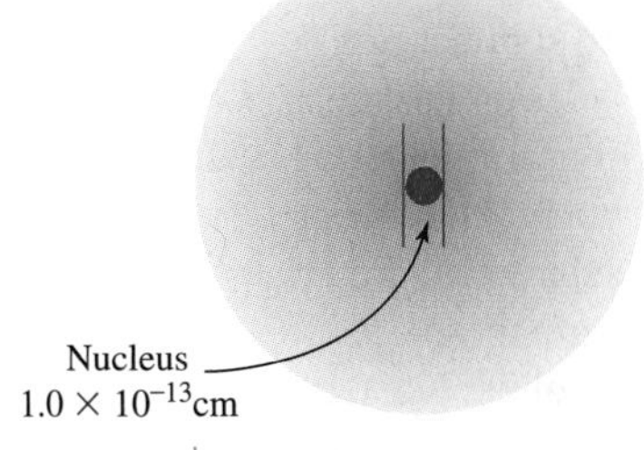

SECTION 4.4 Polynomials

A *polynomial* is a single term or a sum of terms in which all variables have whole-number exponents.

47. Consider the polynomial $3x^3 - x^2 + x + 10$.

a. How many terms does the polynomial have?

b. What is the lead term?

c. What is the coefficient of each term?

d. What is the constant term?

The *degree of a term* of a polynomial in one variable is the value of the exponent on the variable. If a polynomial is in more than one variable, the *degree of a term* is the sum of the exponents on the variables. The *degree of a nonzero constant* is 0.

The *degree of a polynomial* is the highest degree of any term of the polynomial.

48. Find the degree of each polynomial and classify it as a monomial, binomial, trinomial, or none of these.

a. $13x^7$ **b.** $-16a^2b$

c. $5^3x + x^2$ **d.** $-3x^5 + x - 1$

e. $9xy^2 + 21x^3y^3$ **f.** $4s^4 - 3s^2 + 5s + 4$

49. Evaluate $-x^5 - 3x^4 + 3$ for $x = 0$ and $x = -2$.

The graph of a *nonlinear equation* is not a straight line.

A *parabola* is a cup-shaped curve.

50. DIVING See the illustration. The number of inches that the woman deflects the diving board is given by the polynomial

$$0.1875x^2 - 0.0078125x^3$$

where x is the number of feet that she stands from the front anchor point of the board. Find the amount of deflection if she stands on the end of the diving board, 8 feet from the anchor point.

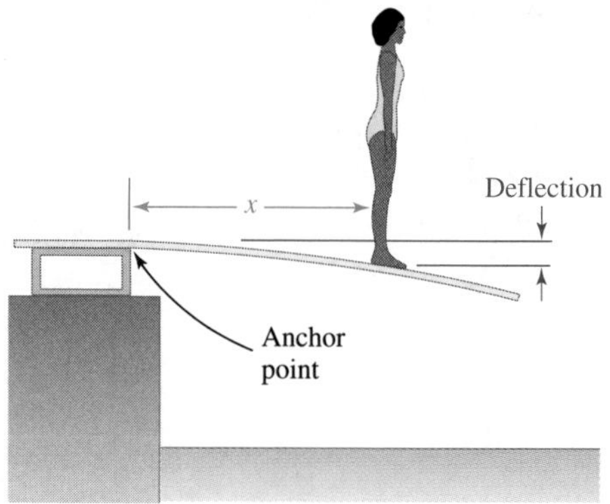

Construct a table of solutions and then graph the equation.

51. $y = x^2$ **52.** $y = x^3 + 1$

SECTION 4.5 Adding and Subtracting Polynomials

To add polynomials, combine like terms.

To subtract two polynomials, change the signs of the terms of the polynomial being subtracted, drop the parentheses, and combine like terms.

Simplify each polynomial.

53. $6y^3 + 8y^4 + 7y^3 + (-8y^4)$ **54.** $4a^2b + 5 - 6a^3b - 3a^2b + 2a^3b$

55. $\frac{5}{6}x^2 + \frac{1}{3}y^2 - \frac{1}{4}x^2 - \frac{3}{4}xy + \frac{2}{3}y^2$ **56.** $-(c^5 + 5c^4 - 12)$

Perform the operations.

57. $(2r^6 + 14r^3) + (23r^6 - 5r^3 + 5r)$

58. $(7a^2 + 2a - 5) - (3a^2 - 2a + 1)$

59. $(3r^3s + r^2s^2 - 3rs^3 - 3s^4) + (r^3s - 8r^2s^2 - 4rs^3 + s^4)$

60. $3(9x^2 + 3x + 7) - 2(11x^2 - 5x + 9)$

61. Find the difference when $-3z^3 - 4z + 7$ is subtracted from the sum of $2z^2 + 3z - 7$ and $-4z^3 - 2z - 3$.

62.
$$\begin{array}{r} 3x^2 + 5x + 2 \\ +\ \underline{x^2 - 3x + 6} \end{array}$$

63.
$$\begin{array}{r} 20x^3 \qquad\quad + 12x \\ -\ \underline{12x^3 + 7x^2 - 7x} \end{array}$$

64. GARDENING Find a polynomial that represents the length of the wooden handle of the shovel.

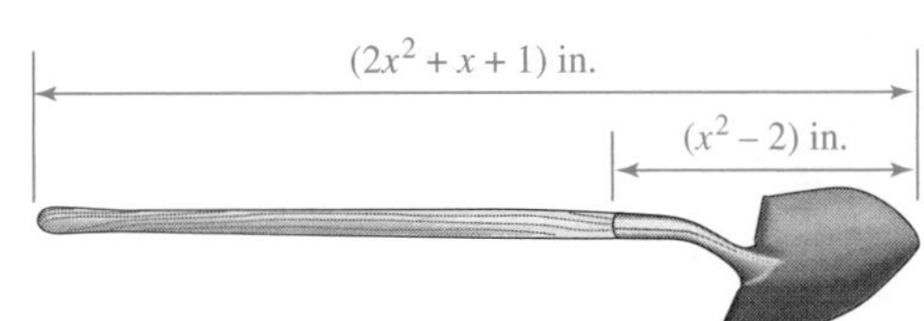

SECTION 4.6 Multiplying Polynomials

To multiply two monomials, multiply the numerical factors (the coefficients) and then multiply the variable factors.

To multiply a monomial and a polynomial, multiply each term of the polynomial by the monomial.

To multiply two binomials, use the *FOIL method:*

F: First
O: Outer
I: Inner
L: Last

To multiply two polynomials, multiply each term of one polynomial by each term of the other polynomial, and then combine like terms.

Find each product.

65. $(2x^2)(5x)$

66. $(-6x^4z^3)(x^6z^2)$

67. $5b^3 \cdot 6b^2 \cdot 4b^6$

68. $\frac{2}{3}h^5(3h^9 + 12h^6)$

69. $x^2y(y^2 - xy)$

70. $3n^2(3n^2 - 5n + 2)$

71. $2x(3x^4)(x + 2)$

72. $-a^2b^2(-a^4b^2 + a^3b^3 - ab^4 + 7a)$

73. $(x + 3)(x + 2)$

74. $(2x + 1)(x - 1)$

75. $(3a - 3)(2a + 2)$

76. $6(a - 1)(a + 1)$

77. $(a - b)(2a + b)$

78. $(3n^4 - 5n^2)(2n^4 - n^2)$

79. $(2a - 3)(4a^2 + 6a + 9)$

80. $(8x^2 + x - 2)(7x^2 + x - 1)$

81. Multiply:
$$\begin{array}{r} 4x^2 - 2x + 1 \\ \underline{2x + 1} \end{array}$$

82. APPLIANCES Find the perimeter of the base, the area of the base, and the volume occupied by the dishwasher.

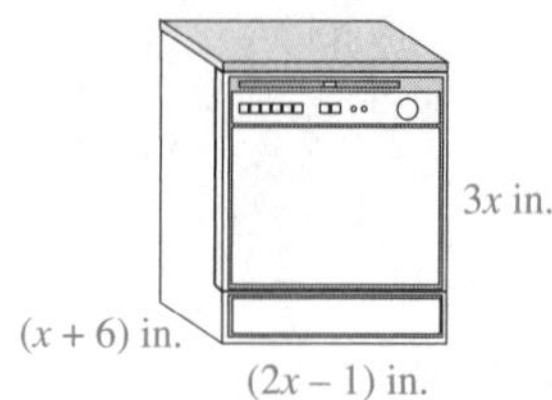

SECTION 4.7 Special Products

Special products:

$(x + y)^2 = x^2 + 2xy + y^2$

$(x - y)^2 = x^2 - 2xy + y^2$

$(x + y)(x - y) = x^2 - y^2$

Find each product.

83. $(x + 3)(x + 3)$

84. $(2x - 0.9)(2x + 0.9)$

85. $(a - 3)^2$

86. $(x + 4)^2$

87. $(-2y + 1)^2$

88. $(y^2 + 1)(y^2 - 1)$

89. $(6r^2 + 10s)^2$

90. $-(8a - 3)^2$

91. $80s(r^2 + s^2)(r^2 - s^2)$

92. $4b(3b - 4)^2$

93. $\left(t - \frac{3}{4}\right)^2$

94. $(m + 2)^3$

95. Perform the operations.

$$(5c - 1)^2 - (c + 6)(c - 6)$$

96. GRAPHIC ARTS A Dr. Martin Luther King poster has his picture with a $\frac{1}{2}$-inch wide border around it. The length of the poster is $(x + 3)$ inches and the width is $(x - 1)$ inches. Find a polynomial that represents the area of the picture of Dr. King.

SECTION 4.8 Division of Polynomials

To divide monomials, use the method for simplifying fractions or use the rules for exponents.

Perform each division.

97. $\dfrac{16n^8}{8n^5}$

98. $\dfrac{-14x^2y}{21xy^3}$

99. $\dfrac{a^{15} - 24a^8}{6a^{12}}$

100. $\dfrac{15a^2b + 20ab^2 - 25ab}{5ab}$

To divide a polynomial by a monomial, divide each term of the numerator by the denominator.

Long division is used to divide one polynomial by another. When a division has a remainder, write the answer in the form

$$\text{Quotient} + \frac{\text{remainder}}{\text{divisor}}$$

The division method works best when the exponents of the terms of the divisor and the dividend are written in descending order.

When the dividend is missing a term, write it with a coefficient of zero or leave a blank space.

101. $x - 1\overline{)x^2 - 6x + 5}$

102. $\dfrac{2x^2 + 3 + 7x}{x + 3}$

103. $2x - 1\overline{)6x^3 + x^2 + 1}$

104. $3x + 1\overline{)-13x - 4 + 9x^3}$

105. Use multiplication to show that the answer when dividing $3y^2 + 11y + 6$ by $y + 3$ is $3y + 2$.

106. SAVINGS BONDS How many \$50 savings bonds would have a total value of $(50x + 250)$ dollars?

CHAPTER 4 TEST

ELEMENTARY Algebra Now™

1. Use exponents to rewrite $2xxxyyyy$.

2. Evaluate: -6^2.

Simplify each expression. Write answers without using parentheses or negative exponents.

3. $y^2(yy^3)$

4. $(2x^3)^5(x^2)^3$

5. $3x^0$

6. $2y^{-5}y^2$

7. 5^{-3}

8. $\dfrac{(x + 1)^{15}}{(x + 1)^6}$

9. $\dfrac{y^2}{yy^{-2}}$

10. $\left(\dfrac{a^2b^{-1}}{4a^3b^{-2}}\right)^{-3}$

11. Find the volume of a cube that has sides of length $10y^4$ inches.

12. ELECTRICITY One ampere (amp) corresponds to the flow of 6,250,000,000,000,000,000 electrons per second past any point in a direct current (DC) circuit. Write this number in scientific notation.

13. Write 9.3×10^{-5} in standard notation.

14. Evaluate: $(2.3 \times 10^{18})(4.0 \times 10^{-15})$. Write the answer in standard notation.

15. Complete the table for the polynomial $x^4 + 8x^2 - 12$.

Term	Coefficient	Degree
Degree of the polynomial		

16. Identify $3x^2 + 2$ as a monomial, binomial, or trinomial.

17. Find the degree of the polynomial $3x^2y^3 + 2x^3y - 5x^2y$.

18. Graph: $y = x^2 + 2$.

19. FREE FALL A visitor standing on the rim of the Grand Canyon drops a rock over the side. The distance (in feet) that the rock is from the canyon floor t seconds after being dropped is given by the polynomial $-16t^2 + 5{,}184$. Find the position of the rock 18 seconds after being dropped. Explain your answer.

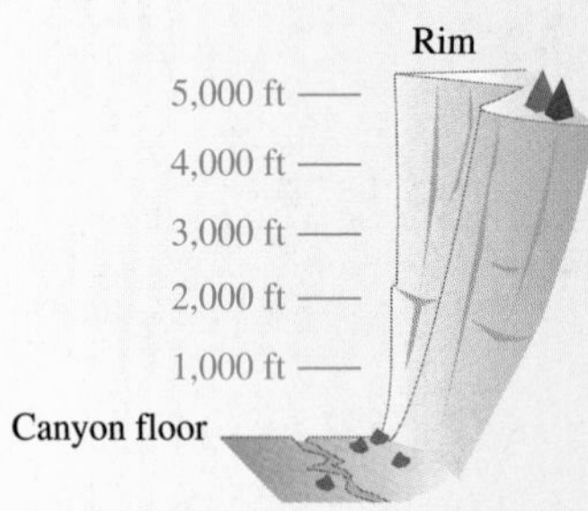

20. Simplify: $4a^2b + 5 - 6a^3b - 3a^2b + 2a^3b$.

Perform the operations:

21. $(3a^2 - 4a - 6) + (2a^2 - a + 9)$

22. Subtract $(b^3c - 3bc + 12)$ from $(7b^3c - 5bc)$.

23.
$$\begin{array}{r} -5y^3 + 4y^2 - 11y + 3 \\ -\underline{-2y^3 - 14y^2 + 17y - 32} \end{array}$$

24. Simplify: $-6(x - y) + 2(x + y)$.

Find each product.

25. $(-2x^3)(2x^2y)$

26. $3y^2(y^2 - 2y + 3)$

27. $(2x - 5)(3x + 4)$

28. $(2x - 3)(x^2 - 2x + 4)$

29. $(1 + 10c)(1 - 10c)$

30. $(7b^3 - 3)^2$

31. Perform the operations: $(x + y)(x - y) + x(x + y)$

Perform each division.

32. $\dfrac{6a^2 - 12b^2}{24ab}$

33. $\dfrac{2x^2 - x - 6}{2x + 3}$

34. $2x - 1\overline{)6x^3 + x^2 + 1}$

35. Find the width of the rectangle.

Area: $(x^2 - 6x + 5)$ ft^2

Length: $(x - 1)$ ft

CHAPTERS 1–4 CUMULATIVE REVIEW EXERCISES

1. SPORTS CARS The graph shows the Porsche vehicle sales in the United States for the years 1986–2001.
   a. In what year were sales the lowest?
   b. In what year were sales the greatest?
   c. Between what two years was there the greatest increase in sales?

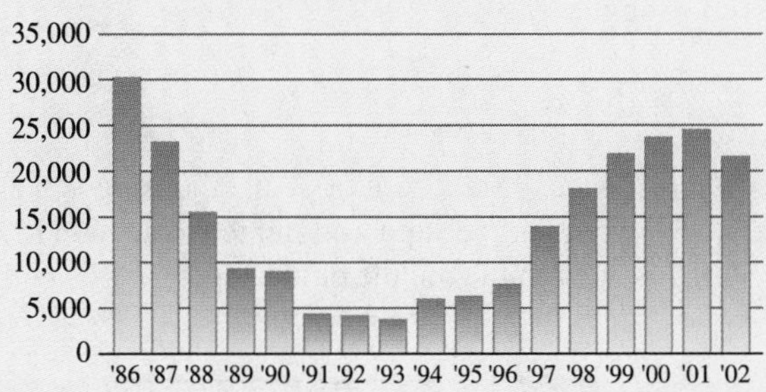

2. Divide: $\frac{3}{4} \div \frac{6}{5}$.

3. Subtract: $\frac{7}{10} - \frac{1}{14}$.

4. Is π a rational or irrational number?

5. RACING Suppose a driver has completed x laps of a 250-lap race. Write an expression for how many more laps he must make to finish the race.

6. CLINICAL TRIALS In a clinical test of Aricept, a drug to treat Alzheimer's disease, one group of patients took a placebo (a sugar pill) while another group took the actual medication. Use the data in the table to determine the number of patients in each group who experienced nausea.

Comparison of rates of adverse events in patients		
Adverse event	**Group 1—Placebo (number = 300)**	**Group 2—Aricept (number = 320)**
Nausea	6%	5%

7. Give the prime factorization of 100.

8. Graph each member of the set on the number line.

$$\left\{-2\frac{1}{4}, \sqrt{2}, -1.75, \frac{7}{2}, 0.5\right\}$$

9. Write $\frac{2}{3}$ as a decimal.

10. What property of real numbers is illustrated?

$$3(2x) = (3 \cdot 2)x$$

11. What is the value of d dimes?

Simplify each expression.

12. $13r - 12r$

13. $27\left(\frac{2}{3}x\right)$

14. $4(d - 3) - (d - 1)$

15. $(13c - 3)(-6)$

Evaluate each expression.

16. $-3^2 + |4^2 - 5^2|$

17. $(4 - 5)^{20}$

18. $\frac{-3 - (-7)}{2^2 - 3}$

19. $12 - 2[1 - (-8 + 2)]$

Solve each equation.

20. $3(x - 5) + 2 = 2x$

21. $\frac{x - 5}{3} - 5 = 7$

22. $\frac{2}{5}x + 1 = \frac{1}{3} + x$

23. $-\frac{5}{8}h = 15$

24. Solve: $8(4 + x) > 10(6 + x)$. Write the solution set in interval notation and graph it.

25. Solve: $A = \frac{1}{2}h(b + B)$ for h.

26. CANDY The owner of a candy store wants to make a 30-pound mixture of two candies to sell for \$2 per pound. If red licorice bits sell for \$1.90 per pound and lemon gumdrops sell for \$2.20 per pound, how many pounds of each should be used?

Graph each equation.

27. $4x - 3y = 12$

28. $x = 4$

Find the slope of the line with the given properties.

29. Passing through $(-2, 4)$ and $(6, 8)$

30. A line that is horizontal

31. An equation of $2x - 3y = 12$

Write an equation of the line with the following properties.

32. Slope $= \frac{2}{3}$, y-intercept $= (0, 5)$

33. Passing through $(-2, 4)$ and $(6, 10)$

34. A horizontal line passing through $(2, 4)$

35. Are the graphs of the lines parallel or perpendicular?

$$y = -\frac{3}{4}x + \frac{15}{4}$$

$$4x - 3y = 25$$

36. Subtract:
$$\begin{array}{r} 17x^4 - 3x^2 - 65x - 12 \\ -\ 23x^4 + 14x^2 + 3x - 23 \\ \hline \end{array}$$

Simplify each expression. Write each answer without using parentheses or negative exponents.

37. $(-3x^2y^4)^2$

38. $(2y)^{-4}$

39. $(x^3x^4)^2$

40. $ab^3c^4 \cdot ab^4c^2$

41. $\dfrac{a^4b^0}{a^{-3}}$

42. $\left(\dfrac{4t^3t^4t^5}{3t^2t^6}\right)^3$

Perform the operations.

43. $(4c^2 + 3c - 2) + (3c^2 + 4c + 2)$

44. $3x(2x + 3)^2$

45. $(2t + 3s)(3t - s)$

46. $5x + 3\overline{)11x + 10x^2 + 3}$

47. Graph: $y = x^2$.

Write each number in scientific notation.

48. 615,000

49. 0.0000013

50. MUSICAL INSTRUMENTS The gong shown in the illustration is a percussion instrument used throughout Southeast Asia. The amount of deflection of the horizontal support (in inches) is given by the polynomial

$$0.01875x^4 - 0.15x^3 + 1.2x$$

where x is the distance (in feet) that the gong is hung from one end of the support. Find the deflection if the gong is hung in the middle of the support.

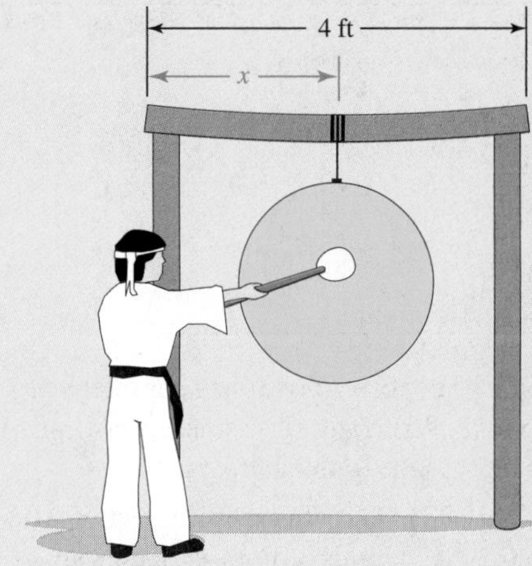

Chapter

5 Factoring and Quadratic Equations

ELEMENTARY **Algebra Now™**

Throughout the chapter, this icon introduces resources on the Elementary AlgebraNow Web site, accessed through http://1pass.thomson.com, that will

- Help you test your knowledge of the material with a pre-test and a post-test
- Provide a personalized learning plan targeting areas you should study

5.1 The Greatest Common Factor; Factoring by Grouping

5.2 Factoring Trinomials of the Form $x^2 + bx + c$

5.3 Factoring Trinomials of the Form $ax^2 + bx + c$

5.4 Factoring Perfect Square Trinomials and the Difference of Two Squares

5.5 Factoring the Sum and Difference of Two Cubes

5.6 A Factoring Strategy

5.7 Solving Quadratic Equations by Factoring

Accent on Teamwork

Key Concept

Chapter Review

Chapter Test

Cumulative Review Exercises

Getty Images

TLE Construction projects come in all shapes and sizes. Whether remodeling a kitchen or erecting a giant skyscraper, algebra and geometry play an important role from start to finish. In this chapter, we explore other applications that involve these branches of mathematics. Using an algebraic process called *factoring,* we will solve applied problems that deal with area, one of the fundamental concepts of geometry.

To learn more about the use of algebra in construction projects, visit *The Learning Equation* on the Internet at http://tle.brookscole.com. (The log-in instructions are in the Preface.) For Chapter 5, the following online lessons are:

- *TLE* Lesson 11: Factoring Trinomials and the Difference of Two Squares
- *TLE* Lesson 12: Solving Quadratic Equations by Factoring

Recall that whole numbers can be factored into products of prime numbers. We will now extend that concept and discuss methods for factoring polynomials.

5.1 The Greatest Common Factor; Factoring by Grouping

- The Greatest Common Factor (GCF)
- Factoring Out the GCF
- Factoring by Grouping

In Chapter 4, we learned how to multiply polynomials. For example, to multiply $3x + 5$ by $4x$, we use the distributive property.

$$4x(3x + 5) = 4x \cdot 3x + 4x \cdot 5$$
$$= 12x^2 + 20x$$

Success Tip

On the game show Jeopardy!, answers are revealed and contestants respond with the appropriate questions. Factoring is similar. Answers to multiplications are given. You are to respond by telling what factors were multiplied.

To factor the polynomial $12x^2 + 20x$, we reverse the previous steps and determine what factors were multiplied to obtain this result. This process is called *factoring the polynomial.*

Multiplication →

$$4x(3x + 5) = 12x^2 + 20x$$

← *Factoring*

When factoring a polynomial, the first step is to determine whether its terms have any common factors.

THE GREATEST COMMON FACTOR (GCF)

To determine whether integers have common factors, it is helpful to write them as products of prime numbers. For example, the prime factorizations of 42 and 90 are given below.

$$42 = 2 \cdot 3 \cdot 7 \qquad 90 = 2 \cdot 3 \cdot 3 \cdot 5$$

The highlighting shows that 42 and 90 have one factor of 2 and one factor of 3 in common. To find their *greatest common factor* (*GCF*), we multiply the common factors: $2 \cdot 3 = 6$. Thus, the GCF of 42 and 90 is 6.

The Greatest Common Factor (GCF)

The **greatest common factor (GCF)** of a list of integers is the largest common factor of those integers.

Recall from arithmetic that the factors of a number divide the number exactly, leaving no remainder. Therefore, the greatest common factor of two or more integers is the *largest natural number that divides each of the integers exactly.*

EXAMPLE 1

Find the GCF of each list of numbers: **a.** 21 and 140, **b.** 24, 60, and 96, and **c.** 9, 10, and 30.

Solution

a. We write each number as a product of primes and determine the factors they have in common.

$$21 = 3 \cdot 7$$
$$140 = 2 \cdot 2 \cdot 5 \cdot 7$$

The Language of Algebra

Recall that the *prime numbers* are: 2, 3, 5, 7, 11, 13, 17, 19, ...

Since the only prime factor common to 21 and 140 is 7, the GCF is 7.

b. To find the GCF of three numbers, we proceed in a similar way.

$$24 = 2 \cdot 2 \cdot 2 \cdot 3$$
$$60 = 2 \cdot 2 \cdot 3 \cdot 5$$
$$96 = 2 \cdot 2 \cdot 2 \cdot 2 \cdot 2 \cdot 3$$

Since 24, 60, and 96 have two factors of 2 and one factor of 3 in common, we have

$$\text{GCF} = 2 \cdot 2 \cdot 3 = 12$$

c. Since there are no prime factors common to all three numbers, the GCF is 1.

$$9 = 3 \cdot 3$$
$$10 = 2 \cdot 5$$
$$30 = 2 \cdot 3 \cdot 5$$

Self Check 1 Find the GCF of each list of numbers: **a.** 24 and 70, **b.** 22, 25, and 98, and **c.** 45, 60, and 75.

To find the greatest common factor of a list of terms, we can use the following approach.

Strategy for Finding the GCF

1. Write each coefficient as a product of prime factors.
2. Identify the numerical and variable factors common to each term.
3. Multiply the common factors identified in Step 2 to obtain the GCF. If there are no common factors, the GCF is 1.

EXAMPLE 2

Find the GCF of each list of terms: **a.** $12x^2$ and $20x$ and **b.** $9a^5b^2$, $15a^4b^2$, and $90a^3b^3$.

Solution

a. ***Step 1:*** We write each coefficient, 12 and 20, as a product of prime factors.

$$12x^2 = 2 \cdot 2 \cdot 3 \cdot x \cdot x$$
$$20x = 2 \cdot 2 \cdot 5 \cdot x$$

Step 2: There are two common factors of 2 and one common factor of x.

Step 3: We multiply the common factors, 2, 2, and x, to obtain the GCF.

$$\text{GCF} = 2 \cdot 2 \cdot x = 4x$$

Success Tip

One way to identify common factors is to circle them:

$$12x^2 = (2) \cdot (2) \cdot 3 \cdot (x) \cdot x$$
$$20x = (2) \cdot (2) \cdot 5 \cdot (x)$$
$$\text{GCF} = 2 \cdot 2 \cdot x = 4x$$

b. ***Step 1:*** The coefficients, 9, 15, and 90, are written as products of primes.

$$9a^5b^2 = 3 \cdot 3 \cdot a \cdot a \cdot a \cdot a \cdot a \cdot b \cdot b$$
$$15a^4b^2 = 3 \cdot 5 \cdot a \cdot a \cdot a \cdot a \cdot b \cdot b$$
$$90a^3b^3 = 2 \cdot 3 \cdot 3 \cdot 5 \cdot a \cdot a \cdot a \cdot b \cdot b \cdot b$$

Success Tip

The exponent on any variable in a GCF is the *smallest* exponent that appears on that variable in all of the terms under consideration.

Step 2: The highlighting identifies one common factor of 3, three common factors of a, and two common factors of b.

Step 3: GCF $= 3 \cdot a \cdot a \cdot a \cdot b \cdot b = 3a^3b^2$

Self Check 2 Find the GCF of each list of terms: **a.** $33c$ and $22c^4$, **b.** $42s^3t^2$, $63s^2t^4$, and $21st^3$.

FACTORING OUT THE GCF

The concept of greatest common factor is used to factor polynomials. For example, to factor $12x^2 + 20x$, we note that there are two terms, $12x^2$ and $20x$. We previously determined that the GCF of $12x^2$ and $20x$ is $4x$. With this in mind, we write each term of $12x^2 + 20x$ as a product of the GCF and one other factor. Then we apply the distributive property: $ab + ac = a(b + c)$.

$12x^2 + 20x = 4x \cdot 3x + 4x \cdot 5$ Write $12x^2$ and write $20x$ as the product of the GCF, $4x$, and one other factor.

$= 4x(3x + 5)$ Write an expression so that the multiplication by $4x$ distributes over the terms $3x$ and 5.

We have found that the factored form of $12x^2 + 20x$ is $4x(3x + 5)$. This process is called **factoring out the greatest common factor.**

EXAMPLE 3

Factor: **a.** $8m - 48$ and **b.** $35a^3b^2 + 14a^2b^3$.

ELEMENTARY Algebra f(x) Now™

Solution

a. The greatest common factor of $8m$ and 48 is 8.

$8m - 48 = 8 \cdot m - 8 \cdot 6$

$= 8(m - 6)$ Factor out the GCF, which is 8.

To check, we multiply: $8(m - 6) = 8 \cdot m - 8 \cdot 6 = 8m - 48$. Since we obtain the original polynomial, $8m - 48$, the factorization is correct.

b. First, find the GCF of $35a^3b^2$ and $14a^2b^3$.

$$\left.\begin{aligned} 35a^3b^2 &= 5 \cdot 7 \cdot a \cdot a \cdot a \cdot b \cdot b \\ 14a^2b^3 &= 2 \cdot 7 \cdot a \cdot a \cdot b \cdot b \cdot b \end{aligned}\right\} \text{The GCF is } 7a^2b^2.$$

Now, we write $35a^3b^2$ and $14a^2b^3$ as the product of the GCF, $7a^2b^2$, and one other factor.

$35a^3b^2 + 14a^2b^3 = 7a^2b^2 \cdot 5a + 7a^2b^2 \cdot 2b$

$= 7a^2b^2(5a + 2b)$ Factor out the GCF, $7a^2b^2$.

We check by multiplying: $7a^2b^2(5a + 2b) = 35a^3b^2 + 14a^2b^3$.

Success Tip

Always verify a factorization by doing the indicated multiplication. The result should be the original polynomial.

Self Check 3 Factor: **a.** $6f - 36$ and **b.** $48s^2t^2 + 84s^3t$.

EXAMPLE 4

Factor: $3x^4 - 5x^3 + x^2$.

ELEMENTARY Algebra f(x) Now™

Solution The polynomial has three terms. We factor out the GCF, which is x^2.

$$3x^4 - 5x^3 + x^2 = x^2(3x^2) - x^2(5x) + x^2(1)$$ The last term has an implied coefficient of 1.

$$= x^2(3x^2 - 5x + 1)$$ Factor out the GCF.

Caution

Sometimes, the GCF of the terms of a polynomial is the same as one of the terms.

We check by multiplying: $x^2(3x^2 - 5x + 1) = 3x^4 - 5x^3 + x^2$.

Self Check 4 Factor: $y^6 - 10y^4 - y^3$.

EXAMPLE 5

Crayons. The amount of wax used to make the crayon shown in the illustration can be found by computing its volume using the formula

$$V = \pi r^2 H + \frac{1}{3}\pi r^2 h$$

Factor the expression on the right-hand side.

h

H

Crayon

r

Solution Each term on the right-hand side of the formula contains a factor of π and r^2.

$$V = \pi r^2 H + \frac{1}{3}\pi r^2 h$$

$$= \pi r^2\left(H + \frac{1}{3}h\right)$$ Factor out πr^2.

The formula can be expressed as $V = \pi r^2\left(H + \frac{1}{3}h\right)$.

It is often useful to factor out a common factor having a negative coefficient.

EXAMPLE 6

Factor -1 out of $-a^3 + 2a^2 - 4$.

Solution First, we write each term of the polynomial as the product of -1 and another factor. Then we factor out the common factor, -1.

Success Tip

The result of Example 6 suggests a quick way to factor out -1. Simply change the sign of each term of $-a^3 + 2a^2 - 4$ and write a $-$ symbol in front of the parentheses.

$$-a^3 + 2a^2 - 4 = (-1)a^3 + (-1)(-2a^2) + (-1)4$$

$$= -1(a^3 - 2a^2 + 4)$$ Factor out -1.

$$= -(a^3 - 2a^2 + 4)$$ The coefficient of 1 need not be written.

We check by multiplying: $-(a^3 - 2a^2 + 4) = -a^3 + 2a^2 - 4$.

Self Check 6 Factor -1 out of $-b^4 - 3b^2 + 2$.

EXAMPLE 7

Factor out the opposite of the GCF in $-20m + 30$.

ELEMENTARY Algebra $f(x)$ Now™

Solution The GCF is 10. To factor out its opposite, we write each term of the polynomial as the product of -10 and another factor. Then we factor out -10.

$$-20m + 30 = (-10)(2m) + (-10)(-3)$$
$$= -10(2m - 3)$$

We check by multiplying: $-10(2m - 3) = -20m + 30$.

Self Check 7 Factor out the opposite of the GCF in $-44c + 55$.

EXAMPLE 8

Factor: $x(x + 4) + 3(x + 4)$.

Solution The polynomial has two terms: $\underbrace{x(x + 4)}_{\text{The first term}} + \underbrace{3(x + 4)}_{\text{The second term}}$.

The GCF of the terms is the binomial $x + 4$, which can be factored out.

$$x(x + 4) + 3(x + 4) = (x + 4)x + (x + 4)3$$ Use the commutative property of multiplication twice to reorder the factors.

$$= (x + 4)(x + 3)$$ Factor out the common factor, which is $(x + 4)$.

Self Check 8 Factor: $2y(y - 1) + 7(y - 1)$.

FACTORING BY GROUPING

To factor the polynomial

$$2x^3 + x^2 + 12x + 6$$

We note that no factor, other than 1, is common to all four terms. However, there is a common factor of x^2 in $2x^3 + x^2$ and a common factor of 6 in $12x + 6$.

$$2x^3 + x^2 = x^2(2x + 1) \quad \text{and} \quad 12x + 6 = 6(2x + 1)$$

Caution

Factoring by grouping can be attempted on any polynomial with four or more terms. However, not every such polynomial can be factored in this way.

We now see that $2x^3 + x^2$ and $12x + 6$ have a common factor of $2x + 1$, which can be factored out.

$$2x^3 + x^2 + 12x + 6 = x^2(2x + 1) + 6(2x + 1)$$ Factor $2x^3 + x^2$ and $12x + 6$.

$$= (2x + 1)(x^2 + 6)$$ Factor out $2x + 1$.

This type of factoring is called **factoring by grouping.**

Factoring by Grouping

1. Group the terms of the polynomial so that the first two terms have a common factor and the last two terms have a common factor.
2. Factor out the common factor from each group.
3. Factor out the resulting common binomial factor. If there is no common binomial factor, regroup the terms of the polynomial and repeat steps 2 and 3.

EXAMPLE 9

Factor: **a.** $2c - 2d + cd - d^2$ and **b.** $x^2 - ax - x + a$.

ELEMENTARY Algebra f(x) Now™

Solution **a.** The first two terms have a common factor of 2 and the last two terms have a common factor of d.

$$2c - 2d + cd - d^2 = 2(c - d) + d(c - d) \quad \text{Factor out 2 from } 2c - 2d \text{ and } d \text{ from } cd - d^2.$$
$$= (c - d)(2 + d) \quad \text{Factor out the common binomial factor, } c - d.$$

We check by multiplying:

$$(c - d)(2 + d) = 2c + cd - 2d - d^2$$
$$= 2c - 2d + cd - d^2 \quad \text{Rearrange the terms to get the original polynomial.}$$

Caution

When factoring polynomials such as the one in Example 9a, don't think that $2(c - d) + d(c - d)$ is in factored form. It is a sum of two terms. To be in factored form, the result must be a product.

b. Since x is a common factor of the first two terms, we can factor it out and proceed as follows.

$$x^2 - ax - x + a = x(x - a) - x + a \quad \text{Factor out } x \text{ from } x^2 - ax.$$

If we factor -1 from $-x + a$, a common binomial factor $(x - a)$ appears, which we can factor out.

$$x^2 - ax - x + a = x(x - a) - 1(x - a)$$
$$= (x - a)(x - 1) \quad \text{Factor out the common factor of } x - a.$$

Check by multiplying.

Self Check 9 Factor: **a.** $7x - 7y + xy - y^2$ and **b.** $b^2 - bc - b + c$.

The next example illustrates that when factoring a polynomial, we should always look for a common factor first.

EXAMPLE 10

Factor: $10k + 10m - 2km - 2m^2$.

ELEMENTARY Algebra f(x) Now™

Solution Since the four terms have a common factor of 2, we factor it out first. Then we use factoring by grouping to factor the polynomial within the parentheses. The first two terms have a common factor of 5. The last two terms have a common factor of $-m$.

$$10k + 10m - 2km - 2m^2 = 2(5k + 5m - km - m^2) \quad \text{Factor out the GCF, 2.}$$
$$= 2[5(k + m) - m(k + m)]$$
$$= 2[(k + m)(5 - m)] \quad \text{Factor out } k + m.$$
$$= 2(k + m)(5 - m)$$

Check by multiplying.

Self Check 10 Factor: $-4t - 4s - 4tz - 4sz$.

Answers to Self Checks **1. a.** 2, **b.** 1, **c.** 15 **2. a.** $11c$, **b.** $21st^2$ **3. a.** $6(f - 6)$, **b.** $12s^2t(4t + 7s)$ **4.** $y^3(y^3 - 10y - 1)$ **6.** $-(b^4 + 3b^2 - 2)$ **7.** $-11(4c - 5)$ **8.** $(y - 1)(2y + 7)$ **9. a.** $(x - y)(7 + y)$ **b.** $(b - c)(b - 1)$ **10.** $-4(t + s)(1 + z)$

5.1 STUDY SET  ELEMENTARY Algebra f(x) Now™

VOCABULARY Fill in the blanks.

1. The letters GCF stand for ________ ________ ________.
2. In the multiplication $2x(x + 7) = 2x^2 + 14x$, the ________ are $2x$ and $x + 7$ and the product is $2x^2 + 14x$.
3. When we write 24 as $2 \cdot 2 \cdot 2 \cdot 3$, we say that 24 has been written as a product of ________.
4. When we write $2x + 4$ as $2(x + 2)$, we say that we have ________ ________ the GCF, 2.
5. To factor $m^3 + 3m^2 + 4m + 12$ by ________, we begin by writing $m^2(m + 3) + 4(m + 3)$.
6. The terms $x(x - 1)$ and $4(x - 1)$ have a common ________ factor, $x - 1$.

CONCEPTS

7. Complete each prime factorization.
   a. $15 = \square \cdot 5$ **b.** $8 = 2 \cdot 2 \cdot \square$
   c. $36 = 2 \cdot 2 \cdot \square \cdot 3$ **d.** $98 = 2 \cdot 7 \cdot \square$
8. Find the GCF of $30x^2$ and $105x^3$.

$$30x^2 = 2 \cdot 3 \cdot 5 \cdot x \cdot x$$
$$105x^3 = 3 \cdot 5 \cdot 7 \cdot x \cdot x \cdot x$$

9. Find the GCF of $12a^2b^2$, $15a^3b$, and $75a^4b^2$.

$$12a^2b^2 = 2 \cdot 2 \cdot 3 \cdot a \cdot a \cdot b \cdot b$$
$$15a^3b = 3 \cdot 5 \cdot a \cdot a \cdot a \cdot b$$
$$75a^4b^2 = 3 \cdot 5 \cdot 5 \cdot a \cdot a \cdot a \cdot a \cdot b \cdot b$$

10. Write a binomial such that the GCF of its terms is $2x^2$.
11. **a.** What property is illustrated here?

$$2x(x - 3) = 2x^2 - 6x$$

   b. Fill in the blank: $2x^2 - 6x = 2x \cdot x - 2x \cdot 3$
   $= \square(x - 3)$
12. Is $3y(3y^2 + 2y - 5)$ the factorization of $9y^3 + 5y^2 - 15y$? Explain.
13. Explain the error in each solution.
   a. Factor out the GCF: $30a^3 - 12a^2 = 6a(5a^2 - 2a)$.
   b. Factor: $6a + 9b + 3 = 3(2a + 3b + 0)$
   $= 3(2a + 3b)$
14. Consider the polynomial $2k - 8 + hk - 4h$.
   a. How many terms does the polynomial have?
   b. Is there a common factor of all the terms?
   c. What is the common factor of the first two terms?
   d. What is the common factor of the last two terms?
15. Is $5(c - d) + r(c - d)$ in factored form? Explain.
16. What is the first step in factoring $8y^2 - 16yz - 6y + 12z$?

NOTATION Complete each factorization.

17. $40m^4 - 8m^3 + 32m = \square(5m^3 - m^2 + \square)$
18. $b^3 - 6b^2 + 2b - 12 = \square(b - 6) + \square(b - 6)$
    $= (\square)(b^2 + 2)$

PRACTICE Find the prime factorization of each number.

19. 12
20. 24
21. 40
22. 62
23. 98
24. 112
25. 225
26. 288

Find the GCF of each list.

27. 18, 24
28. 60, 72
29. m^4, m^3
30. c^2, c^7
31. $20c^2, 12c$
32. $18r, 27r^3$
33. $6m^4n, 12m^3n^2, 9m^3n^3$
34. $15cd^4, 10cd, 40cd^3$

Fill in the blanks.

35. $6x = 3(\quad)$

36. $9y = 3(\quad)$

37. $24y^2 = 8y \cdot \quad$

38. $35h^2 = 7h \cdot \quad$

39. $30t^3 = 15t(\quad)$

40. $54h^3 = 6h^2(\quad)$

41. $32x^3z^4 = 4xz^2(\quad)$

42. $48t^2w^5 = 8tw^2(\quad)$

Factor out the GCF.

43. $3x + 6$

44. $2y - 10$

45. $18m - 9$

46. $24s + 8$

47. $18x + 24$

48. $15s - 35$

49. $d^2 - 7d$

50. $a^2 + 9a$

51. $14c^3 + 63$

52. $33h^4 - 22$

53. $12x^2 - 6x - 24$

54. $27a^2 - 9a + 45$

55. $t^3 + 2t^2$

56. $b^3 - 3b^2$

57. $ab + ac - ad$

58. $rs - rt + ru$

59. $a^3 - a^2$

60. $r^3 + r^2$

61. $24x^2y^3 + 8xy^2$

62. $3x^2y^3 - 9x^4y^3$

63. $12uvw^3 - 18uv^2w^2$

64. $14xyz - 16x^2y^2z$

65. $12r^2 - 3rs + 9r^2s^2$

66. $6a^2 - 12a^3b + 36ab$

67. $\pi R^2 - \pi ab$

68. $\frac{1}{3}\pi R^2h - \frac{1}{3}\pi rh$

69. $3(x + 2) - x(x + 2)$

70. $t(5 - s) + 4(5 - s)$

71. $h^2(14 + r) + 14 + r$

72. $k^2(14 + v) - 7(14 + v)$

Factor out −1 from each polynomial.

73. $-a - b$

74. $-x - 2y$

75. $-2x + 5y$

76. $-3x + 8z$

77. $-3r + 2s - 3$

78. $-6yz + 12xz - 5xy$

Factor each polynomial by factoring out the opposite of the GCF.

79. $-3x^2 - 6x$

80. $-4a^2 + 6a$

81. $-4a^2b^3 + 12a^3b^2$

82. $-25x^4y^3 + 30x^2y^3$

83. $-4a^2b^2c^2 + 14a^2b^2c - 10ab^2c^2$

84. $-10x^4y^3z^2 + 8x^3y^2z - 20x^2y$

Factor by grouping.

85. $2x + 2y + ax + ay$

86. $bx + bz + 5x + 5z$

87. $9p - 9q + mp - mq$

88. $xr + xs + yr + ys$

89. $pm - pn + qm - qn$

90. $2ax + 2bx + 3a + 3b$

91. $2xy - 3y^2 + 2x - 3y$

92. $2ab + 2ac + b + c$

93. $ax + bx - a - b$

94. $2xy + y^2 - 2x - y$

Factor by grouping. Remember to factor out the GCF first.

95. $ax^3 + bx^3 + 2ax^2y + 2bx^2y$

96. $x^3y^2 - 2x^2y^2 + 3xy^2 - 6y^2$

97. $2x^3z - 4x^2z + 32xz - 64z$

98. $4a^2b + 12a^2 - 8ab - 24a$

Factor each polynomial.

99. $x(y + 9) - 11(y + 9)$

100. $2\pi R - 2\pi r$

101. $9mp + 3mq - 3np - nq$

102. $-4abc - 4ac^2 + 2bc + 2c^2$

103. $25x^5y^7z^3 - 45x^3y^2z^6$

104. $(t^3 + 7)x + (t^3 + 7)y$

105. $24m - 12n + 16$

106. $6x^2 - 2x - 15x + 5$

107. $-60P^2 - 80P$

108. $\frac{1}{3}a^2 - \frac{2}{3}a$

APPLICATIONS

109. REARVIEW MIRRORS The dimensions of the three rearview mirrors on an automobile are given in the illustration on the next page. Write an algebraic expression that gives

a. the area of the rearview mirror mounted on the windshield.

b. the total area of the two side mirrors.

c. the total area of all three mirrors. Express the result in factored form.

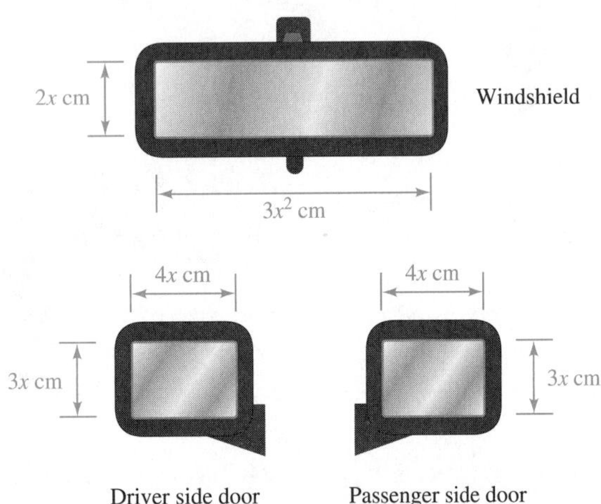

110. COOKING

a. What is the length of a side of the square griddle, in terms of r? What is the area of the cooking surface of the griddle, in terms of r?

b. How many square inches of the cooking surface do the pancakes cover, in terms of r?

c. Find the amount of cooking surface that is not covered by the pancakes. Express the result in factored form.

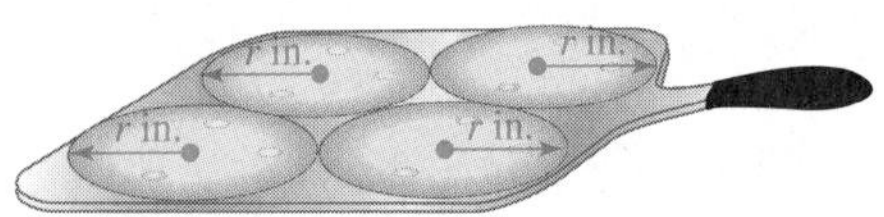

111. AIRCRAFT CARRIERS The rectangular-shaped landing area of $(x^3 + 4x^2 + 5x + 20)$ ft^2 is shaded. The dimensions of the landing area can be found by factoring. What are the length and width of the landing area?

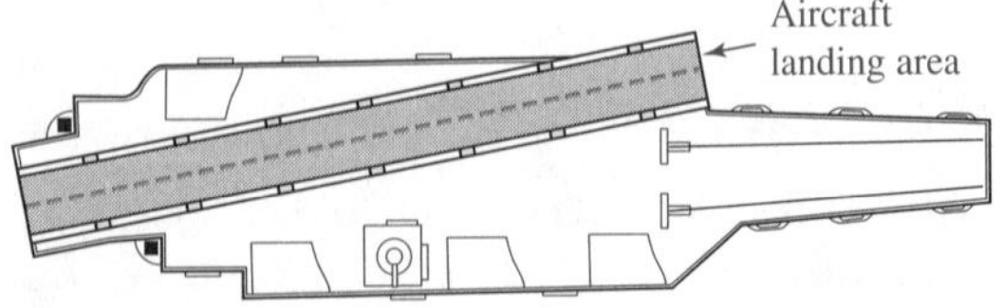

112. INTERIOR DECORATING The expression $\pi rs + \pi Rs$ can be used to find the amount of material needed to make the lamp shade shown. Factor the expression.

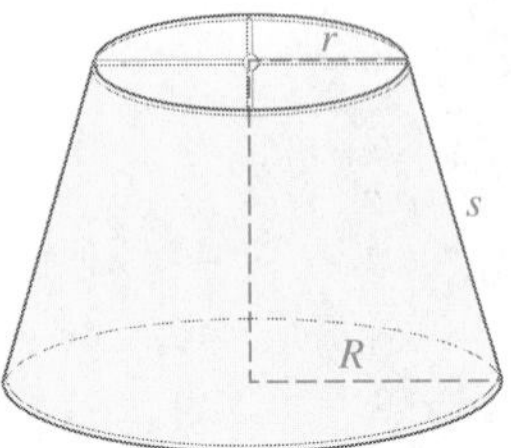

WRITING

113. Explain how to find the GCF of $32a^3$, $16a^2$, and $24a^3$.

114. Explain this diagram.

Multiplication ⟶

$$3x^2(5x^2 - 6x + 4) = 15x^4 - 18x^3 + 12x^2$$

⟵ *Factoring*

115. Explain how factorizations of polynomials are checked.

116. What is a factor? What does it mean to factor?

REVIEW

117. INSURANCE COSTS A college student's good grades earned her a student discount on her car insurance premium. What was the percent of decrease, to the nearest percent, if her annual premium was lowered from \$1,050 to \$925?

118. CALCULATING GRADES A student has test scores of 68%, 75%, and 79% in a government class. What must she score on the last exam to earn a B (80% or better) in the course?

CHALLENGE PROBLEMS

119. Factor: $6x^{4m}y^n + 21x^{3m}y^{2n} - 15x^{2m}y^{3n}$.

120. Factor $ax + ay + bx + by$ by grouping the first two terms and the last two terms. Then rearrange two of the terms of $ax + ay + bx + by$ and factor that polynomial by grouping. Do the results agree?

5.2 Factoring Trinomials of the Form $x^2 + bx + c$

- Factoring Trinomials with a Lead Coefficient of 1
- Factoring Out the GCF
- Prime Polynomials
- The Grouping Method

We have learned how to multiply binomials. For example, to multiply $x + 1$ and $x + 2$, we proceed as follows.

$$(x + 1)(x + 2) = x^2 + 2x + x + 2$$
$$= x^2 + 3x + 2$$

To factor the trinomial $x^2 + 3x + 2$, we will reverse the multiplication process and determine what factors were multiplied to obtain this result. This process is called *factoring the trinomial.* Since the product of two binomials is often a trinomial, many trinomials factor into two binomials.

Multiplication →

$$(x + 1)(x + 2) = x^2 + 3x + 2$$

← *Factoring*

To begin the discussion of trinomial factoring, we consider trinomials of the form $x^2 + bx + c$, such as

$$x^2 + 8x + 15, \quad y^2 - 13y + 12, \quad a^2 + a - 20, \quad \text{and} \quad z^2 - 20z - 21$$

In each case, the **lead coefficient**—the coefficient of the squared variable—is an implied 1.

FACTORING TRINOMIALS WITH A LEAD COEFFICIENT OF 1

To develop a method for factoring trinomials, we will find the product of $x + 6$ and $x + 4$ and make some observations about the result.

$$(x + 6)(x + 4) = x \cdot x + 4x + 6x + 6 \cdot 4 \quad \text{Use the FOIL method to multiply.}$$
$$= x^2 + 10x + 24$$

First term: x^2 Middle term: $10x$ Last term: 24

The Language of Algebra

If a term of a trinomial is a number only, it is called a *constant term.*

$x^2 + 10x + 24$ — Constant term: 24

The result is a trinomial, where

- the first term, x^2, is the product of x and x
- the last term, 24, is the product of 6 and 4
- the coefficient of the middle term, 10, is the sum of 6 and 4

These observations suggest a strategy to use to factor trinomials with a lead coefficient of 1.

EXAMPLE 1 Factor: $x^2 + 8x + 15$.

Solution We assume that $x^2 + 8x + 15$ factors as the product of two binomials, and we represent the binomials using two sets of parentheses. Since the first term of the trinomial is x^2, we enter x and x as the first terms of the binomial factors.

$$x^2 + 8x + 15 = (x \ \square)(x \ \square) \quad \text{Because } x \cdot x \text{ will give } x^2$$

The second terms of the binomials must be two integers whose product is 15 and whose sum is 8. Since the integers must have a positive product and a positive sum, we consider only pairs of positive integer factors of 15. The only such pairs, $1 \cdot 15$ and $3 \cdot 5$, are listed in the table. Then we find the sum of each pair and enter each result in the table.

Positive factors of 15	Sum of the factors of 15
$1 \cdot 15 = 15$	$1 + 15 = 16$
$3 \cdot 5 = 15$	$3 + 5 = 8$

List all of the pairs of positive integers that multiply to give 15.

Add each pair of factors.

Notation

By the commutative property of multiplication, the binomial factors of a trinomial can be written in either order. For Example 1, we can write:

$x^2 + 8x + 15 = (x + 5)(x + 3)$

The second row of the table contains the correct pair of integers 3 and 5, whose product is 15 and whose sum is 8. To complete the factorization, we enter 3 and 5 as the second terms of the binomial factors.

$$x^2 + 8x + 15 = (x + 3)(x + 5)$$

We check by multiplying: $(x + 3)(x + 5) = x^2 + 5x + 3x + 15$

$$= x^2 + 8x + 15$$

Self Check 1 Factor: $y^2 + 7y + 10$.

EXAMPLE 2

Factor: $y^2 - 13y + 12$.

ELEMENTARY Algebra f(x) Now™

Solution Again, we assume that the trinomial factors as the product of two binomials. Since the first term of the trinomial is y^2, the first term of each binomial factor must be y.

$$y^2 - 13y + 12 = (y \ \square)(y \ \square) \quad \text{Because } y \cdot y \text{ will give } y^2$$

The second terms of the binomials must be two integers whose product is 12 and whose sum is -13. Since the integers must have a positive product and a negative sum, we only consider pairs of negative integer factors of 12. The possible pairs are listed in the table.

Negative factors of 12	Sum of the factors of 12
$-1(-12) = 12$	$-1 + (-12) = -13$
$-2(-6) = 12$	$-2 + (-6) = -8$
$-3(-4) = 12$	$-3 + (-4) = -7$

The first row of the table contains the correct pair of integers -1 and -12, whose product is 12 and whose sum is -13. To complete the factorization, we enter -1 and -12 as the second terms of the binomial factors.

$$y^2 - 13y + 12 = (y - 1)(y - 12)$$

We check by multiplying: $(y - 1)(y - 12) = y^2 - y - 12y + 12$
$= y^2 - 13y + 12$

Self Check 2 Factor: $p^2 - 6p + 8$.

EXAMPLE 3

Factor: $a^2 + a - 20$.

Solution Since the first term of the trinomial is a^2, the first term of each binomial factor must be a.

$$a^2 + a - 20 = (a \,\square)(a \,\square) \quad \text{Because } a \cdot a \text{ will give } a^2$$

To determine the second terms of the binomials, we must find two integers whose product is -20 and whose sum is 1. Because the integers must have a negative product, their signs must be different. The possible pairs are listed in the table.

Factors of -20	Sum of the factors of -20
$1(-20) = -20$	$1 + (-20) = -19$
$2(-10) = -20$	$2 + (-10) = -8$
$4(-5) = -20$	$4 + (-5) = -1$
$5(-4) = -20$	$5 + (-4) = 1$
$10(-2) = -20$	$10 + (-2) = 8$
$20(-1) = -20$	$20 + (-1) = 19$

The Language of Algebra

Make sure you understand the following vocabulary: *Many trinomials factor as the product of two binomials.*

Trinomial / Product of two binomials

$a^2 + a - 20 = (a + 5)(a - 4)$

The fourth row of the table contains the correct pair of integers 5 and -4, whose product is -20 and whose sum is 1. To complete the factorization, we enter 5 and -4 as the second terms of the binomial factors.

$$a^2 + a - 20 = (a + 5)(a - 4)$$

Check by multiplying.

Self Check 3 Factor: $m^2 + m - 42$.

EXAMPLE 4

Factor: $z^2 - 4z - 21$.

Solution Since the first term of the trinomial is z^2, the first term of each binomial factor must be z.

$$z^2 - 4z - 21 = (z \,\square)(z \,\square) \quad \text{Because } z \cdot z \text{ will give } z^2$$

To determine the second terms of the binomials, we must find two integers whose product is -21 and whose sum is -4. Because the integers must have a negative product, their signs must be different. The possible pairs are listed in the table.

Factors of -21	Sum of the factors of -21
$1(-21) = -21$	$1 + (-21) = -20$
$3(-7) = -21$	$3 + (-7) = -4$
$7(-3) = -21$	$7 + (-3) = 4$
$21(-1) = -21$	$21 + (-1) = 20$

The second row of the table contains the correct pair of integers 3 and -7, whose product is -21 and whose sum if -4. To complete the factorization, enter 3 and -7 as the second terms of the binomial factors.

$$z^2 - 4z - 21 = (z + 3)(z - 7)$$

Check by multiplying.

Self Check 4 Factor: $q^2 - 2q - 24$.

The following guidelines are helpful when factoring trinomials.

Factoring Trinomials with a Lead Coefficient of 1

To factor a trinomial of the form $x^2 + bx + c$, find two numbers whose product is c and whose sum is b.

1. If c is positive, the numbers have the same sign.
2. If c is negative, the numbers have different signs.

Then write the trinomial as a product of two binomials. You can check by multiplying.

The product of these numbers must be c.

$$x^2 + bx + c = (x \;\square\;)(x \;\square\;)$$

The sum of these numbers must be b.

EXAMPLE 5

Factor: $-h^2 + 2h + 63$.

Solution Because it is easier to factor trinomials that have a positive lead coefficient, we factor out -1. Then we factor the resulting trinomial.

$$\begin{aligned} -h^2 + 2h + 63 &= -1(h^2 - 2h - 63) && \text{Factor out } -1. \\ &= -(h^2 - 2h - 63) && \text{The coefficient of 1 need not be written.} \\ &= -(h + 7)(h - 9) && \text{Factor } h^2 - 2h - 63. \end{aligned}$$

We check by multiplying:

$$\begin{aligned} -(h + 7)(h - 9) &= -(h^2 - 9h + 7h - 63) && \text{Multiply the binomials first.} \\ &= -(h^2 - 2h - 63) && \text{Combine like terms.} \\ &= -h^2 + 2h + 63 && \text{Remove parentheses.} \end{aligned}$$

Self Check 5 Factor: $-x^2 + 11x - 28$.

EXAMPLE 6

Factor: $x^2 - 4xy - 5y^2$.

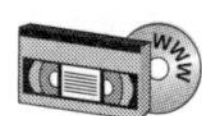

ELEMENTARY Algebra f(x) Now™

Solution Since the first term of the trinomial is x^2, the first term of each binomial factor must be x. Since the third term contains y^2, the last term of each binomial factor must contain y. To complete the factorization, we need to determine the coefficient of each y-term.

$$x^2 - 4xy - 5y^2 = (x \,\square\, y)(x \,\square\, y)$$

The coefficients of y must be two integers whose product is -5 and whose sum is -4. Such a pair is 1 and -5. Instead of writing the first factor as $(x + 1y)$, we write it as $(x + y)$, because $1y = y$.

$$x^2 - 4xy - 5y^2 = (x + y)(x - 5y)$$

Check: $(x + y)(x - 5y) = x^2 - 5xy + xy - 5y^2$
$$= x^2 - 4xy - 5y^2$$

Self Check 6 Factor: $s^2 + 6st - 7t^2$.

FACTORING OUT THE GCF

If the terms of a trinomial have a common factor, it should be factored out first. A trinomial is **factored completely** when no factor can be factored further.

EXAMPLE 7

Factor completely: $2x^4 + 26x^3 + 80x^2$.

ELEMENTARY Algebra f(x) Now™

Solution We begin by factoring out the GCF $2x^2$.

$$2x^4 + 26x^3 + 80x^2 = 2x^2(x^2 + 13x + 40)$$

Next, we factor $x^2 + 13x + 40$. The integers 8 and 5 have a product of 40 and a sum of 13, so the completely factored form of the given trinomial is

$$2x^4 + 26x^3 + 80x^2 = 2x^2(x + 8)(x + 5)$$

Check by multiplying.

Caution

When prime factoring 30, you wouldn't stop here because 6 can be factored:

```
 30
 / \
5   6
```

When factoring polynomials, make sure that no factor can be factored further.

Self Check 7 Factor completely: $4m^5 + 8m^4 - 32m^3$.

EXAMPLE 8

Factor completely: $-13g^2 + 36g + g^3$.

Solution Before factoring the trinomial, we write its terms in descending powers of g.

$$\begin{aligned} -13g^2 + 36g + g^3 &= g^3 - 13g^2 + 36g && \text{Rearrange the terms.} \\ &= g(g^2 - 13g + 36) && \text{Factor out } g. \\ &= g(g - 9)(g - 4) && \text{Factor the trinomial.} \end{aligned}$$

Check by multiplying.

Caution

For multistep factorizations, don't forget to write the GCF in the final factored form.

Self Check 8 Factor completely: $-12t + t^3 + 4t^2$.

PRIME POLYNOMIALS

If a trinomial with integer coefficients cannot be factored using only integers, it is called a **prime trinomial.**

EXAMPLE 9

Factor $x^2 + 2x + 3$, if possible.

ELEMENTARY Algebra f(x) Now™

Solution To factor the trinomial, we must find two integers whose product is 3 and whose sum is 2. The possible factorizations are shown in the table.

Factors of 3	Sum of the factors of 3
$1(3) = 3$	$1 + 3 = 4$
$-1(-3) = 3$	$-1 + (-3) = -4$

The Language of Algebra

When a trinomial is not factorable using only integers, we say it is *prime* and that it does not factor *over* the integers.

Since there are no two integers whose product is 3 and whose sum is 2, $x^2 + 2x + 3$ cannot be factored. It is a prime trinomial.

Self Check 9 Factor $x^2 - 4x + 6$, if possible.

THE GROUPING METHOD

Another way to factor trinomials of the form $x^2 + bx + c$ is to write them as equivalent four-termed polynomials and factor by grouping. To factor $x^2 + 8x + 15$ using this method, we proceed as follows.

1. First, identify b as the coefficient of the x-term, and c as the last term. For trinomials of the form $x^2 + bx + c$, we call c the **key number.**

$$\left.\begin{array}{c} x^2 + bx + c \\ \downarrow \qquad \downarrow \\ x^2 + 8x + 15 \end{array}\right\} b = 8 \text{ and } c = 15$$

2. Now find two integers whose product is the key number, 15, and whose sum is $b = 8$. Since the integers must have a positive product and a positive sum, we consider only positive factors of 15.

Key number = 15

Positive factors of 15	Sum of the factors of 15
$1 \cdot 15 = 15$	$1 + 15 = 16$
$3 \cdot 5 = 15$	$3 + 5 = 8$

The second row of the table contains the correct pair of integers 3 and 5, whose product is 15 and whose sum is 8.

3. Use the integers 3 and 5 as coefficients of two terms to be placed between x^2 and 15.

$$x^2 + 8x + 15 = x^2 + 3x + 5x + 15 \quad \text{Express } 8x \text{ as } 3x + 5x.$$

4. Factor the four-termed polynomial by grouping:

$$x^2 + 3x + 5x + 15 = x(x + 3) + 5(x + 3) \quad \text{Factor } x \text{ out of } x^2 + 3x \text{ and 5 out of } 5x + 15.$$

$$= (x + 3)(x + 5) \quad \text{Factor out } x + 3.$$

Check by multiplying.

The grouping method is an alternative to the method for factoring trinomials discussed earlier in this section. It is especially useful when the constant term, c, has many factors.

Factoring Trinomials of the Form $x^2 + bx + c$ Using Grouping

To factor a trinomial that has a lead coefficient of 1:

1. Identify b and the key number c.
2. Find two numbers whose product is the key number and whose sum is b.
3. Enter the two numbers as coefficients of x in the form shown below. Then factor the polynomial by grouping.

The product of these numbers must be c.

$$x^2 + \square x + \square x + c$$

The sum of these numbers must be b.

4. Check using multiplication.

EXAMPLE 10

Factor: $a^2 + a - 20$.

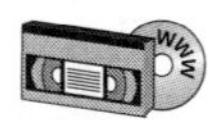

Solution Since $a^2 + a - 20 = a^2 + 1a - 20$, we identify b as 1 and the key number c as -20. We must find two integers whose product is -20 and whose sum is 1. Since the integers must have a negative product, their signs must be different.

Key number $= -20$

Factors of -20	Sum of the factors of -20
$1(-20) = -20$	$1 + (-20) = -19$
$2(-10) = -20$	$2 + (-10) = -8$
$4(-5) = -20$	$4 + (-5) = -1$
$5(-4) = -20$	$5 + (-4) = 1$
$10(-2) = -20$	$10 + (-2) = 8$
$20(-1) = -20$	$20 + (-1) = 19$

The fourth row of the table contains the correct pair of integers 5 and -4, whose product is -20 and whose sum is 1. They serve as the coefficients of $5a$ and $-4a$ that we place between a^2 and -20.

$$a^2 + a - 20 = a^2 + \mathbf{5a} - \mathbf{4a} - 20 \quad \text{Express } a \text{ as } 5a - 4a.$$
$$= a(a + 5) - 4(a + 5) \quad \text{Factor } a \text{ out of } a^2 + 5a \text{ and } -4 \text{ out of } -4a - 20.$$
$$= (a + 5)(a - 4) \quad \text{Factor out } a + 5.$$

Check by multiplying.

Self Check 10 Factor: $m^2 + m - 42$.

EXAMPLE 11

Factor: $2x^3 - 20x^2 + 18x$.

ELEMENTARY Algebra f(x) Now™

Solution First, we factor out $2x$.

$$2x^3 - 20x^2 + 18x = 2x(x^2 - 10x + 9)$$

To factor $x^2 - 10x + 9$ by grouping, we must find two integers whose product is the key number 9 and whose sum is $b = -10$. Such a pair is -9 and -1.

$$x^2 - \mathbf{10x} + 9 = x^2 - \mathbf{9x} - \mathbf{1x} + 9 \quad \text{Express } -10x \text{ as } -9x - 1x.$$
$$= x(x - 9) - 1(x - 9) \quad \text{Factor } x \text{ out of } x^2 - 9x \text{ and } -1 \text{ out of } -1x + 9.$$
$$= (x - 9)(x - 1) \quad \text{Factor out } x - 9.$$

The complete factorization of the original trinomial is

$$2x^3 - 20x^2 + 18x = 2x(x - 9)(x - 1)$$

Check by multiplying.

Self Check 11 Factor: $3m^3 - 27m^2 + 24m$.

EXAMPLE 12

Factor: $x^2 - 4xy - 5y^2$.

Solution We identify b as -4 and the key number c as -5. We must find two integers whose product is -5 and whose sum is -4. Such a pair is -5 and 1. They serve as the coefficients of $-5xy$ and $1xy$ that we place between x^2 and $-5y^2$.

$$x^2 - \mathbf{4xy} - 5y^2 = x^2 - \mathbf{5xy} + \mathbf{1xy} - 5y^2 \quad \text{Express } -4xy \text{ as } -5xy + 1xy.$$
$$= x(x - 5y) + y(x - 5y) \quad \text{Factor } x \text{ out of } x^2 - 5xy \text{ and } y \text{ out of } 1xy - 5y^2.$$
$$= (x - 5y)(x + y) \quad \text{Factor out } x - 5y.$$

Check by multiplying.

Self Check 12 Factor: $q^2 - 2qt - 24t^2$.

Answers to Self Checks **1.** $(y + 2)(y + 5)$ **2.** $(p - 2)(p - 4)$ **3.** $(m + 7)(m - 6)$ **4.** $(q + 4)(q - 6)$ **5.** $-(x - 4)(x - 7)$ **6.** $(s + 7t)(s - t)$ **7.** $4m^3(m + 4)(m - 2)$ **8.** $t(t - 2)(t + 6)$ **9.** prime trinomial **10.** $(m + 7)(m - 6)$ **11.** $3m(m - 8)(m - 1)$ **12.** $(q + 4t)(q - 6t)$

5.2 STUDY SET  ELEMENTARY Algebra $f(x)$ Now™

VOCABULARY Fill in the blanks.

1. A polynomial that has three terms is called a ________. A polynomial that has two terms is called a ________.
2. The statement $x^2 - x - 12 = (x - 4)(x + 3)$ shows that $x^2 - x - 12$ ________ into the product of two binomials.
3. Since $10 = (-5)(-2)$, we say -5 and -2 are ________ of 10.
4. A ________ polynomial cannot be factored by using only integers.
5. The ______ coefficient of the trinomial $x^2 - 3x + 2$ is 1, the ________ of the middle term is -3, and the last ______ is 2.
6. A trinomial is factored ________ when no factor can be factored further.

CONCEPTS Fill in the blanks.

7. Two factorizations of 4 that involve only positive numbers are ____ and ____. Two factorizations of 4 that involve only negative numbers are ____ and ____.
8. **a.** Before attempting to factor a trinomial, be sure that it is written in ________ powers of a variable.

   b. Before attempting to factor a trinomial into two binomials, always factor out any ________ factors first.
9. To factor $x^2 + x - 56$, we must find two integers whose ________ is -56 and whose ______ is 1.
10. Two factors of 18 whose sum is -9 are ____ and ____.
11. $x^2 + 5x + 3$ cannot be factored because we cannot find two integers whose product is ____ and whose sum is ____.
12. Complete the table.

Factors of 8	Sum of the factors of 8
$1(8) = 8$	
$2(4) = 8$	
$-1(-8) = 8$	
$-2(-4) = 8$	

13. If we use the FOIL method to do the multiplication $(x + 5)(x + 4)$, we obtain $x^2 + 9x + 20$.

    a. What step of the FOIL process produced 20?

    b. What steps of the FOIL process produced $9x$?
14. Find two integers whose

    a. product is 10 and whose sum is 7.

    b. product is 8 and whose sum is -6.

    c. product is -6 and whose sum is 1.

    d. product is -9 and whose sum is -8.
15. Given $x^2 + 8x + 15$:

    a. What is the coefficient of the x^2-term?

    b. What is the last term? What is the coefficient of the middle term?

    c. What two integers have a product of 15 and a sum of 8?
16. Complete the factorization table.

Factors of -9	Sum of the factors of -9
$1(-9) = -9$	
$3(-3) = -9$	
$-1(9) = -9$	

17. Consider factoring a trinomial of the form $x^2 + bx + c$.

    a. If c is positive, what can be said about the two integers that should be chosen for the factorization?

    b. If c is negative, what can be said about the two integers that should be chosen for the factorization?
18. What trinomial has the factorization $(x + 8)(x - 2)$?
19. Factor out the GCF: $x(x + 1) + 5(x + 1)$.
20. Factor each polynomial using factoring by grouping.

    a. $x^2 + 2x + 5x + 10$

    b. $y^2 - 4y + 3y - 12$

21. Complete the table.

Key number = 8

Negative factors of 8	Sum of factors of 8
$-1(-8) = 8$	
	$-2 + (-4) = -6$

22. Complete each sentence to explain how to factor $x^2 + 7x + 10$ by grouping.

The product of these numbers must be ▒.

$$x^2 + 7x + 10 = x^2 + \square x + \square x + 10$$

The sum of these numbers must be ▒.

NOTATION

23. To factor a trinomial, a student made a table and circled the correct pair of integers, as shown. Complete the factorization of the trinomial.

$(x \quad)(x \quad)$

Factors	Sum
$1(-6)$	-5
$2(-3)$	-1
$3(-2)$	1
$6(-1)$	5

24. To factor a trinomial by grouping, a student made a table and circled the correct pair of integers, as shown. Enter the correct coefficients.

$x^2 + \quad x + \quad x + 16$

Key number = 16

Factors	Sum
$1 \cdot 16$	17
$2 \cdot 8$	10
$4 \cdot 4$	8

PRACTICE **Complete each factorization.**

25. $x^2 + 3x + 2 = (x \quad 2)(x \quad 1)$

26. $y^2 + 4y + 3 = (y \quad 3)(y \quad 1)$

27. $t^2 - 9t + 14 = (t \quad 7)(t \quad 2)$

28. $c^2 - 9c + 8 = (c \quad 8)(c \quad 1)$

29. $a^2 + 6a - 16 = (a \quad 8)(a \quad 2)$

30. $x^2 - 3x - 40 = (x \quad 8)(x \quad 5)$

Factor each trinomial. If it can't be factored, write "prime."

31. $z^2 + 12z + 11$ **32.** $x^2 + 7x + 10$

33. $m^2 - 5m + 6$ **34.** $n^2 - 7n + 10$

35. $a^2 - 4a - 5$ **36.** $b^2 + 6b - 7$

37. $x^2 + 5x - 24$ **38.** $t^2 - 5t - 50$

39. $a^2 - 10a - 39$ **40.** $r^2 - 9r - 12$

41. $u^2 + 10u + 15$ **42.** $v^2 + 9v + 15$

43. $r^2 - 2r + 4$ **44.** $y^2 - 17y + 72$

45. $r^2 - 9r + 18$ **46.** $u^2 + u - 42$

47. $x^2 + 4xy + 4y^2$ **48.** $m^2 - mn - 12n^2$

49. $a^2 - 4ab - 12b^2$ **50.** $p^2 + pq - 6q^2$

51. $r^2 - 2rs + 4s^2$ **52.** $m^2 + 3mn - 20n^2$

Factor each trinomial. Factor out −1 first.

53. $-x^2 - 7x - 10$ **54.** $-x^2 + 9x - 20$

55. $-t^2 - 15t + 34$ **56.** $-t^2 - t + 30$

57. $-a^2 - 4ab - 3b^2$ **58.** $-a^2 - 6ab - 5b^2$

59. $-x^2 + 6xy + 7y^2$ **60.** $-x^2 - 10xy + 11y^2$

Write each trinomial in descending powers of one variable and factor.

61. $4 - 5x + x^2$ **62.** $y^2 + 5 + 6y$

63. $10y + 9 + y^2$ **64.** $x^2 - 13 - 12x$

65. $r^2 - 2 - r$ **66.** $u^2 - 3 + 2u$

67. $4rx + r^2 + 3x^2$ **68.** $a^2 + 5b^2 + 6ab$

Completely factor each trinomial. Factor out any common monomials first (including −1 if necessary).

69. $2x^2 + 10x + 12$ **70.** $3y^2 - 21y + 18$

71. $-5a^2 + 25a - 30$

72. $-2b^2 + 20b - 18$

73. $z^3 - 29z^2 + 100z$

74. $m^3 - m^2 - 56m$

75. $12xy + 4x^2y - 72y$

76. $48xy + 6xy^2 + 96x$

77. $-r^2 + 14r - 40$

78. $-y^2 - 2y + 99$

79. $-13yz + y^2 - 14z^2$

80. $2x^2 - 12x + 16$

81. $s^2 + 11s - 26$

82. $x^2 + 14x + 45$

83. $a^2 + 10ab + 9b^2$

84. $-r^2 + 14r - 45$

85. $-x^2 + 21x + 22$

86. $-3ab + a^2 + 2b^2$

87. $d^3 - 11d^2 - 26d$

88. $m^2 + 3mn - 10n^2$

APPLICATIONS

89. PETS The cage shown in the illustration is used for transporting dogs. Its volume is $(x^3 + 12x^2 + 27x)$ in.^{3}. The dimensions of the cage can be found by factoring. If the cage is longer than it is tall and taller than it is wide, determine its length, width, and height.

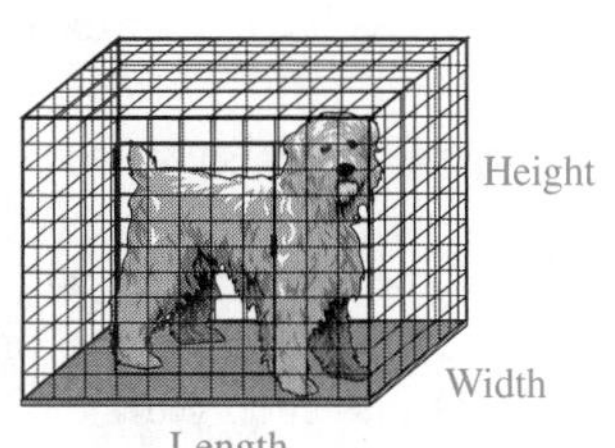

90. PHOTOGRAPHY A picture cube is a clever way to display 6 photographs in a small amount of space. Suppose the surface area of the cube is given by the polynomial $(6s^2 + 12s + 6)$ in.^{2}. Find the expression that represents the length of an edge of the cube.

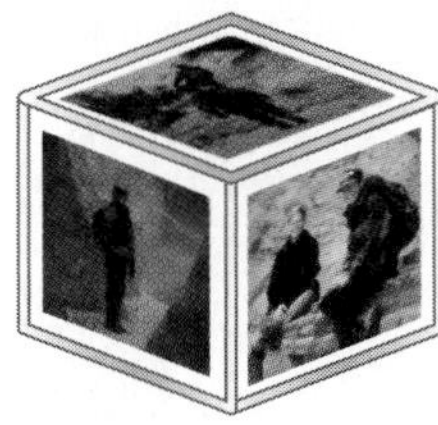

WRITING

91. Explain what it means when we say that a trinomial is the product of two binomials. Give an example.

92. Are $2x^2 - 12x + 16$ and $x^2 - 6x + 8$ factored in the same way? Explain.

93. When factoring $x^2 - 2x - 3$, one student got $(x - 3)(x + 1)$, and another got $(x + 1)(x - 3)$. Are both answers acceptable? Explain.

94. In the partial solution shown below, a student began to factor the trinomial. Write a note to the student explaining his initial mistake.

Factor: $x^2 - 2x - 63$.

$(x - \quad)(x - \quad)$

???

95. Explain the error in the following factorization.

$$x^3 + 8x^2 + 15x = x(x^2 + 8x + 15)$$
$$= (x + 3)(x + 5)$$

96. Explain why the factorization is not complete.

$$2y^2 - 12y + 16 = 2(y^2 - 6y + 8)$$

REVIEW **Simplify each expression. Write each answer without using parentheses or negative exponents.**

97. $\dfrac{x^{12}x^{-7}}{x^3x^4}$

98. $\dfrac{a^4a^{-2}}{a^2a^0}$

99. $(x^{-3}x^{-2})^2$

100. $\left(\dfrac{18a^2b^3c^{-4}}{3a^{-1}b^2c}\right)^{-3}$

CHALLENGE PROBLEMS **Factor completely.**

101. $x^2 - \dfrac{6}{5}x + \dfrac{9}{25}$

102. $x^2 - 0.5x + 0.06$

103. $x^{2m} - 12x^m - 45$

104. $x^2(y + 1) - 3x(y + 1) - 70(y + 1)$

105. Find all positive integer values of c that make $n^2 + 6n + c$ factorable.

106. Find all integer values of b that make $x^2 + bx - 44$ factorable.

5.3 Factoring Trinomials of the Form $ax^2 + bx + c$

• The Trial-and-Check Method • Factoring Out the GCF • The Grouping Method

In this section we will factor trinomials with lead coefficients other than 1, such as

$$2x^2 + 5x + 3, \qquad 6a^2 - 17a + 5, \qquad \text{and} \qquad 4b^2 + 8bc - 45c^2$$

We can use two methods to factor these trinomials. With the first method, we make educated guesses and then check them with multiplication. The correct factorization is determined through a process of elimination. The second method is an extension of factoring by grouping.

THE TRIAL-AND-CHECK METHOD

EXAMPLE 1

Factor: $2x^2 + 5x + 3$.

ELEMENTARY Algebra Now™

Solution We assume that $2x^2 + 5x + 3$ factors as the product of two binomials. Since the first term of the trinomial is $2x^2$, we enter $2x$ and x as the first terms of the binomial factors.

$$(2x \;\square)(x \;\square) \quad \text{Because } 2x \cdot x \text{ will give } 2x^2$$

The second terms of the binomials must be two integers whose product is 3. Since the coefficients of the terms of $2x^2 + 5x + 3$ are positive, we only consider pairs of positive integer factors of 3. Since there is just one such pair, $1 \cdot 3$, we can enter 1 and 3 as the second terms of the binomials, or we can reverse the order and enter 3 and 1.

$$(2x + 1)(x + 3) \qquad \text{or} \qquad (2x + 3)(x + 1)$$

The Language of Algebra

To *interchange* means to put each in the place of the other. We create all of the possible factorizations by *interchanging* the second terms of the binomials.

$(2x + 1)(x + 3)$

$(2x + 3)(x + 1)$

The first possibility is incorrect, because when we find the outer and inner products and combine like terms, we obtain an incorrect middle term of $7x$.

Outer: $6x$

$(2x + 1)(x + 3)$ Multiply and add to find the middle term: $6x + x = 7x$.

Inner: x

The second possibility is correct, because it gives a middle term of $5x$.

Outer: $2x$

$(2x + 3)(x + 1)$ Multiply and add to find the middle term: $2x + 3x = 5x$.

Inner: $3x$

Thus,

$$2x^2 + 5x + 3 = (2x + 3)(x + 1)$$

We check by multiplying: $(2x + 3)(x + 1) = 2x^2 + 2x + 3x + 3$
$= 2x^2 + 5x + 3$

Self Check 1 Factor: $2x^2 + 5x + 2$.

EXAMPLE 2

Factor: $6a^2 - 17a + 5$.

ELEMENTARY Algebra $f(x)$ Now™

Solution Since the first term is $6a^2$, the first terms of the factors must be $6a$ and a or $3a$ and $2a$.

$(6a \ \square)(a \ \square)$ or $(3a \ \square)(2a \ \square)$ Because $6a \cdot a$ or $3a \cdot 2a$ will give $6a^2$

The second terms of the binomials must be two integers whose product is 5. Since the last term of $6a^2 - 17a + 5$ is positive and the coefficient of the middle term is negative, we only consider negative integer factors of the last term. Since there is just one such pair, $-1(-5)$, we can enter -1 and -5, or we can reverse the order and enter -5 and -1 as second terms of the binomials.

$(6a - 1)(a - 5)$ (outer: $-30a$; inner: $-a$) $-30a - a = -31a$.

$(6a - 5)(a - 1)$ (outer: $-6a$; inner: $-5a$) $-6a - 5a = -11a$.

$(3a - 1)(2a - 5)$ (outer: $-15a$; inner: $-2a$) $-15a - 2a = -17a$.

$(3a - 5)(2a - 1)$ (outer: $-3a$; inner: $-10a$) $-3a - 10a = -13a$.

Only the possibility shown in blue gives the correct middle term of $-17a$. Thus,

$$6a^2 - 17a + 5 = (3a - 1)(2a - 5)$$

We check by multiplying: $(3a - 1)(2a - 5) = 6a^2 - 17a + 5$.

Self Check 2 Factor: $6b^2 - 19b + 3$.

EXAMPLE 3

Factor: $3y^2 - 7y - 6$.

Solution Since the first term is $3y^2$, the first terms of the binomial factors must be $3y$ and y.

$(3y \ \square)(y \ \square)$ Because $3y \cdot y$ will give $3y^2$

The second terms of the binomials must be two integers whose product is -6. There are four such pairs: $1(-6)$, $-1(6)$, $2(-3)$, and $-2(3)$. When these pairs are entered, and then reversed, as second terms of the binomials, there are eight possibilities to consider. Four of them can be discarded because they include a binomial whose terms have a common factor. If $3y^2 - 7y - 6$ does not have a common factor, neither can any of its binomial factors.

Success Tip
If the terms of a trinomial do not have a common factor, the terms of each of its binomial factors will not have a common factor.

Success Tip
Reversing the signs within the binomial factors reverses the sign of the middle term. For instance, in Example 3, the factors 2 and -3 give the middle term $-7y$, while -2 and 3 give the middle term $7y$.

Success Tip
In Example 3, all eight possible factorizations were listed. In practice, you will often find the correct factorization without having to examine the entire list of possibilities.

For 1 and −6: $(3y + 1)(y - 6)$ — outer: $-18y$, inner: y; $-18y + y = -17y$ | ~~$(3y - 6)(y + 1)$~~ — $3y - 6$ has a common factor of 3.

For −1 and 6: $(3y - 1)(y + 6)$ — outer: $18y$, inner: $-y$; $18y - y = 17y$ | ~~$(3y + 6)(y - 1)$~~ — $3y + 6$ has a common factor of 3.

For 2 and −3: $(3y + 2)(y - 3)$ — outer: $-9y$, inner: $2y$; $-9y + 2y = -7y$ | ~~$(3y - 3)(y + 2)$~~ — $3y - 3$ has a common factor of 3.

For −2 and 3: $(3y - 2)(y + 3)$ — outer: $9y$, inner: $-2y$; $9y - 2y = 7y$ | ~~$(3y + 3)(y - 2)$~~ — $3y + 3$ has a common factor of 3.

Only the possibility shown in blue gives the correct middle term of $-7y$. Thus,

$$3y^2 - 7y - 6 = (3y + 2)(y - 3)$$

Check by multiplying.

Self Check 3 Factor: $5t^2 - 23t - 10$.

EXAMPLE 4

Factor: $4b^2 + 8bc - 45c^2$.

ELEMENTARY Algebra f(x) Now™

Solution Since the first term is $4b^2$, the first terms of the binomial factors must be $4b$ and b or $2b$ and $2b$. Since the last term contains c^2, the second terms of the binomial factors must contain c.

$$(4b \;\square\; c)(b \;\square\; c) \text{ or } (2b \;\square\; c)(2b \;\square\; c)$$

Because $4b \cdot b$ or $2b \cdot 2b$ gives $4b^2$, and because $c \cdot c$ gives c^2

Success Tip
When using the trial-and-check method to factor $ax^2 + bx + c$, here is one suggestion: if b is relatively small, test the factorizations that use the smaller factors of a and c first.

The coefficients of c must be two integers whose product is -45. Since the coefficient of the last term is negative, the signs of the integers must be different. If we pick factors of $4b$ and b for the first terms, and -1 and 45 for the coefficients of c, the multiplication gives an incorrect middle term of $179bc$.

$(4b - c)(b + 45c)$ — outer: $180bc$, inner: $-bc$; $180bc - bc = 179bc$.

If we pick factors of $4b$ and b for the first terms, and 15 and -3 for the coefficients of c, the multiplication gives an incorrect middle term of $3bc$.

$$(4b + 15c)(b - 3c) \qquad -12bc + 15bc = 3bc.$$

(outer product: $-12bc$; inner product: $15bc$)

If we pick factors of $2b$ and $2b$ for the first terms, and -5 and 9 for the coefficients of c, we have

$$(2b - 5c)(2b + 9c) \qquad 18bc - 10bc = 8bc.$$

(outer product: $18bc$; inner product: $-10bc$)

which gives the correct middle term of $8bc$. Thus,

$$4b^2 + 8bc - 45c^2 = (2b - 5c)(2b + 9c)$$

Check by multiplying.

Self Check 4 Factor: $4x^2 + 4xy - 3y^2$.

Because guesswork is often necessary, it is difficult to give specific rules for factoring trinomials with lead coefficients other than 1. However, the following hints are helpful.

Factoring Trinomials with Lead Coefficients Other Than 1

To factor trinomials with lead coefficients other than 1:

1. Factor out any GCF (including -1 if that is necessary to make $a > 0$ in a trinomial of the form $ax^2 + bx + c$).
2. Write the trinomial as a product of two binomials. The coefficients of the first terms of each binomial factor must be factors of a, and the last terms must be factors of c.

The product of these numbers must be a.

$$ax^2 + bx + c = (\square x \ \square)(\square x \ \square)$$

The product of these numbers must be c.

3. If c is positive, the signs within the binomial factors match the sign of b. If c is negative, the signs within the binomial factors are opposites.
4. Try combinations of first terms and second terms until you find the one that gives the proper middle term. If no combination works, the trinomial is prime.
5. Check by multiplying.

FACTORING OUT THE GCF

EXAMPLE 5

Factor: $2x^2 - 8x^3 + 3x$.

Solution We write the trinomial in descending powers of x,

$$-8x^3 + 2x^2 + 3x$$

and we factor out the opposite of the GCF, which is $-x$.

$$-8x^3 + 2x^2 + 3x = -x(8x^2 - 2x - 3)$$

We now factor $8x^2 - 2x - 3$. Its factorization has the form

$$(8x \;\square)(x \;\square) \text{ or } (2x \;\square)(4x \;\square)$$ Because $8x \cdot x$ or $4x \cdot 2x$ gives $8x^2$

The second terms of the binomials must be two integers whose product is -3. There are two such pairs: $1(-3)$ and $-1(3)$. Since the coefficient of middle term $-2x$ is small, we pick the smaller factors of $8x^2$, which are $2x$ and $4x$, for the first terms and 1 and -3 for the second terms.

$-6x$

$$(2x + 1)(4x - 3)$$ $-6x + 4x = -2x$

$4x$

This factorization gives the correct middle term of $-2x$. Thus,

$$8x^2 - 2x - 3 = (2x + 1)(4x - 3)$$

We can now give the complete factorization of the original trinomial.

$$\begin{aligned} -8x^3 + 2x^2 + 3x &= -x(8x^2 - 2x - 3) \\ &= -x(2x + 1)(4x - 3) \end{aligned}$$

Check by multiplying.

Self Check 5 Factor: $12y - 2y^3 - 2y^2$.

THE GROUPING METHOD

Another way to factor a trinomial of the form $ax^2 + bx + c$ is to write it as an equivalent four-termed polynomial and factor it by grouping. For example, to factor $2x^2 + 5x + 3$, we proceed as follows.

1. Identify the values of a, b, and c.

$$\left.\begin{matrix} ax^2 + bx + c \\ \downarrow \quad\;\; \downarrow \quad\; \downarrow \\ 2x^2 + 5x + 3 \end{matrix}\right\} a = 2, b = 5, \text{ and } c = 3$$

Then, find the product ac, called the **key number:** $ac = 2(3) = 6$.

2. Next, find two numbers whose product is $ac = 6$ and whose sum is $b = 5$. Since the numbers must have a positive product and a positive sum, we consider only positive factors of 6.

Key number = 6

Positive factors of 6	Sum of the factors of 6
$1 \cdot 6 = 6$	$1 + 6 = 7$
$2 \cdot 3 = 6$	$2 + 3 = 5$

The second row of the table contains the correct pair of integers 2 and 3, whose product is 6 and whose sum is 5.

3. Use the factors 2 and 3 as coefficients of two terms to be placed between $2x^2$ and 3.

$$2x^2 + 5x + 3 = 2x^2 + 2x + 3x + 3 \quad \text{Express } 5x \text{ as } 2x + 3x.$$

4. Factor the four-termed polynomial by grouping:

$$2x^2 + 2x + 3x + 3 = 2x(x + 1) + 3(x + 1) \quad \text{Factor } 2x \text{ out of } 2x^2 + 2x \text{ and 3 out of } 3x + 3.$$

$$= (x + 1)(2x + 3) \quad \text{Factor out } x + 1.$$

Check by multiplying.

Factoring by grouping is especially useful when the lead coefficient, a, and the constant term, c, have many factors.

Factoring Trinomials by Grouping

To factor a trinomial by grouping:

1. Factor out any GCF (including -1 if that is necessary to make $a > 0$ in a trinomial of the form $ax^2 + bx + c$).
2. Identify a, b, and c, and find the key number ac.
3. Find two numbers whose product is the key number and whose sum is b.
4. Enter the two numbers as coefficients of x between the first and last terms and factor the polynomial by grouping.

The product of these numbers must be ac.

$$ax^2 + \boxed{}\,x + \boxed{}\,x + c$$

The sum of these numbers must be b.

5. Check by multiplying.

EXAMPLE 6

Factor by grouping: $10x^2 + 13x - 3$.

ELEMENTARY Algebra f(x) Now™

Solution In $10x^2 + 13x - 3$, $a = 10$, $b = 13$, and $c = -3$. The key number is $ac = 10(-3) = -30$. We must find a factorization of -30 in which the sum of the factors is $b = 13$. Since the factors must have a negative product, their signs must be different. The possible factor pairs are listed in the table.

Key number $= -30$

Factors of -30	Sum of the factors of -30
$1(-30) = -30$	$1 + (-30) = -29$
$2(-15) = -30$	$2 + (-15) = -13$
$3(-10) = -30$	$3 + (-10) = -7$
$5(-6) = -30$	$5 + (-6) = -1$
$6(-5) = -30$	$6 + (-5) = 1$
$10(-3) = -30$	$10 + (-3) = 7$
$15(-2) = -30$	$15 + (-2) = 13$
$30(-1) = -30$	$30 + (-1) = 29$

The seventh row contains the correct pair of numbers 15 and -2, whose product is -30 and whose sum is 13. They serve as the coefficients of two terms, $15x$ and $-2x$, that we place between $10x^2$ and -3.

$$10x^2 + 13x - 3 = 10x^2 + 15x - 2x - 3 \quad \text{Express } 13x \text{ as } 15x - 2x.$$

Notation

In Example 6, the middle term, $13x$, may be expressed as $15x - 2x$ or as $-2x + 15x$ when using factoring by grouping. The resulting factorizations will be equivalent.

Finally, we factor by grouping.

$$\begin{aligned} 10x^2 + 15x - 2x - 3 &= 5x(2x + 3) - 1(2x + 3) && \text{Factor out } 5x \text{ from } 10x^2 + 15x. \text{ Factor out } -1 \text{ from } -2x - 3. \\ &= (2x + 3)(5x - 1) && \text{Factor out } 2x + 3. \end{aligned}$$

So $10x^2 + 13x - 3 = (2x + 3)(5x - 1)$. Check by multiplying.

Self Check 6 Factor: $15a^2 + 17a - 4$.

EXAMPLE 7

Factor by grouping: $12x^5 - 17x^4 + 6x^3$.

ELEMENTARY Algebra f(x) Now™

Solution First, we factor out the GCF, which is x^3.

$$12x^5 - 17x^4 + 6x^3 = x^3(12x^2 - 17x + 6)$$

To factor $12x^2 - 17x + 6$, we must find two integers whose product is $12(6) = 72$ and whose sum is -17. Two such numbers are -8 and -9.

$$\begin{aligned} 12x^2 - 17x + 6 &= 12x^2 - 8x - 9x + 6 && \text{Express } -17x \text{ as } -8x - 9x. \\ &= 4x(3x - 2) - 3(3x - 2) && \text{Factor out } 4x \text{ and factor out } -3. \\ &= (3x - 2)(4x - 3) && \text{Factor out } 3x - 2. \end{aligned}$$

The complete factorization of the original trinomial is

$$12x^5 - 17x^4 + 6x^3 = x^3(3x - 2)(4x - 3)$$

Check by multiplying.

Self Check 7 Factor: $21a^4 - 13a^3 + 2a^2$.

Answers to Self Checks **1.** $(2x + 1)(x + 2)$ **2.** $(6b - 1)(b - 3)$ **3.** $(5t + 2)(t - 5)$ **4.** $(2x + 3y)(2x - y)$ **5.** $-2y(y + 3)(y - 2)$ **6.** $(3a + 4)(5a - 1)$ **7.** $a^2(7a - 2)(3a - 1)$

5.3 STUDY SET

ELEMENTARY Algebra Now™

VOCABULARY Fill in the blanks.

1. The trinomial $3x^2 - x - 12$ has a ______ coefficient of 3. The middle term is $-x$ and the last ______ is -12.

2. Given: $5y^2 - 16y + 3 = (5y - 1)(y - 3)$. We say that $5y^2 - 16y + 3$ factors as the product of two __________.

3. The numbers 6 and -2 are two integers whose ________ is -12 and whose ______ is 4.

4. To factor $2m^2 + 11m + 12$ by ________, we write it as $2m^2 + 8m + 3m + 12$.

5. To factor a trinomial by grouping, we begin by finding ac, the key ________.

6. A ________ trinomial cannot be factored using only integers.

CONCEPTS

7. a. Give an example of a pair of positive integer factors of 9.

b. Give an example of a pair of negative integer factors of 16.

c. Give an example of a pair of integer factors of -10.

8. Consider the factorization $(7m - 2)(5m + 1)$.

a. What are the first terms of the binomial factors?

b. What are the second terms of the binomial factors?

9. Consider $(4y - 8)(3y + 2)$.

a. Find the outer product.

b. Find the inner product.

c. Combine the inner and outer products.

10. Check to see whether $(3t - 1)(5t - 6)$ is the correct factorization of $15t^2 - 19t + 6$.

11. If $10x^2 - 27x + 5$ is to be factored, what are the possible first terms of the binomial factors.

12. $(3a - 4)(4a - 3)$ is an incomplete factorization of $12a^2 - 25ab + 12b^2$. What is missing?

13. a. Fill in the blanks. When factoring a trinomial, we write it in __________ powers of the variable. Then we factor out any ______ (including -1 if that is necessary to make the lead coefficient ________).

b. What is the GCF of the terms of $6s^4 + 33s^3 + 36s^2$?

c. Factor out -1 from $-2d^2 + 19d - 8$.

14. Complete each sentence.

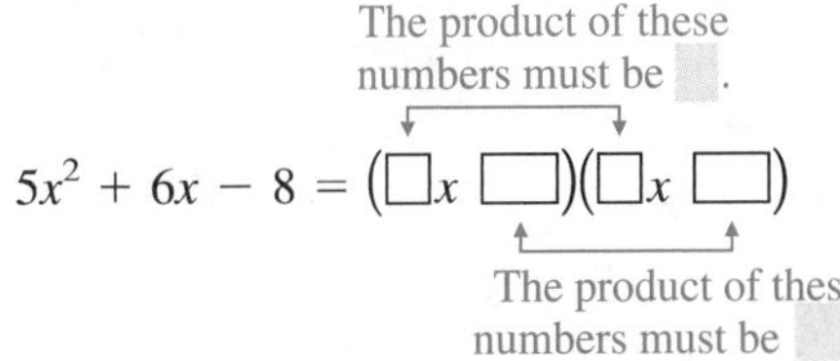

A trinomial has been partially factored. Complete each statement that describes the type of integers we should consider for the blanks.

15. $5y^2 - 13y + 6 = (5y\ \square)(y\ \square)$
Since the last term of the trinomial is ________ and the middle term is ________, the integers must be ________ factors of 6.

16. $5y^2 + 13y + 6 = (5y\ \square)(y\ \square)$
Since the last term of the trinomial is ________ and the middle term is ________, the integers must be ________ factors of 6.

17. $5y^2 - 7y - 6 = (5y\ \square)(y\ \square)$
Since the last term of the trinomial is ________, the signs of the integers will be ________.

18. $5y^2 + 7y - 6 = (5y\ \square)(y\ \square)$
Since the last term of the trinomial is ________, the signs of the integers will be ________.

19. Factor out the GCF: $3x(5x - 2) + 2(5x - 2)$.

20. Factor each polynomial using factoring by grouping.
a. $5x^2 - 10x + 3x - 6$
b. $8h^2 + 32h + h + 4$

21. Complete the key number table.

Key number = 12

Negative factors of 12	Sum of factors of 12
$-1(-12) = 12$	
$-3(-4) = 12$	

22. Complete each sentence to explain how to factor $3x^2 + 16x + 5$ by grouping.

The product of these numbers must be $\square$.

$$3x^2 + 16x + 5 = 3x^2 + \square x + \square x + 5$$

The sum of these numbers must be $\square$.

NOTATION

23. a. Give an example of a trinomial that has a lead coefficient of 1.
b. Give an example of a trinomial that has a lead coefficient that is not 1.

24. Write the terms of the trinomial $40 - t - 4t^2$ in descending powers of the variable.

25. a. Suppose we wish to factor $12b^2 + 20b - 9$ by grouping. Identify a, b, and c.
b. What is the key number, ac?

26. To factor a trinomial by grouping, a student made a table and circled the correct pair of integers, as shown. Enter the correct coefficients.

$6x^2 + \square x + \square x + 6$

Key number = 36

Factors	Sum
$1 \cdot 36$	37
$2 \cdot 18$	20
$3 \cdot 12$	15
$4 \cdot 9$ (circled)	13 (circled)

PRACTICE Complete each factorization.

27. $3a^2 + 13a + 4 = (3a \square 1)(a \square 4)$

28. $2b^2 + 7b + 6 = (2b \square 3)(b \square 2)$

29. $4z^2 - 13z + 3 = (z \square 3)(4z \square 1)$

30. $4t^2 - 4t + 1 = (2t \square 1)(2t \square 1)$

31. $2m^2 + 5m - 12 = (2m \square 3)(m \square 4)$

32. $10u^2 - 13u - 3 = (2u \square 3)(5u \square 1)$

Complete each step of the factorization of the trinomial by grouping.

33. $12t^2 + 17t + 6 = 12t^2 + \square t + \square t + 6$
$= \square(4t + 3) + \square(4t + 3)$
$= (\square)(3t + 2)$

34. $35t^2 - 11t - 6 = 35t^2 + \square t - 21t - 6$
$= 5t(7t + 2) \square 3(7t + 2)$
$= (\square)(5t - 3)$

Factor each trinomial, if possible.

35. $3a^2 + 13a + 4$

36. $2b^2 + 7b + 6$

37. $5x^2 + 11x + 2$

38. $7t^2 + 10t + 3$

39. $4x^2 + 8x + 3$

40. $4z^2 + 13z + 3$

41. $6x^2 + 25x + 21$

42. $6y^2 + 7y + 2$

43. $2x^2 - 3x + 1$

44. $2y^2 - 7y + 3$

45. $4t^2 - 4t + 1$

46. $9x^2 - 32x + 15$

47. $15t^2 - 34t + 8$

48. $7x^2 - 9x + 2$

49. $2x^2 - 3x - 2$

50. $3a^2 - 4a - 4$

51. $12y^2 - y - 1$

52. $8u^2 - 2u - 15$

53. $10y^2 - 3y - 1$

54. $6m^2 + 19m + 3$

55. $12y^2 - 5y - 2$

56. $10x^2 + 21x - 10$

57. $2m^2 + 5m - 10$

58. $10u^2 - 13u - 6$

59. $-13x + 3x^2 - 10$

60. $-14 + 3a^2 - a$

61. $6r^2 + rs - 2s^2$

62. $3m^2 + 5mn + 2n^2$

63. $8n^2 + 91n + 33$

64. $2m^2 + 27m + 70$

65. $4a^2 - 15ab + 9b^2$

66. $12x^2 + 5xy - 3y^2$

67. $130r^2 + 20r - 110$

68. $170h^2 - 210h - 260$

69. $-5t^2 - 13t - 6$

70. $-16y^2 - 10y - 1$

71. $36y^2 - 88y + 32$

72. $70a^2 - 95a + 30$

73. $4x^2 + 8xy + 3y^2$

74. $4b^2 + 15bc - 4c^2$

75. $12y^2 + 12 - 25y$

76. $12t^2 - 1 - 4t$

77. $18x^2 + 31x - 10$

78. $20y^2 - 93y - 35$

79. $-y^3 - 13y^2 - 12y$

80. $-2xy^2 - 8xy + 24x$

81. $3x^2 + 6 + x$

82. $25 + 2u^2 + 3u$

83. $30r^5 + 63r^4 - 30r^3$

84. $6s^5 - 26s^4 - 20s^3$

85. $2a^2 + 3b^2 + 5ab$

86. $11uv + 3u^2 + 6v^2$

87. $pq + 6p^2 - q^2$

88. $-11mn + 12m^2 + 2n^2$

89. $6x^3 - 15x^2 - 9x$

90. $9y^3 + 3y^2 - 6y$

91. $15 + 8a^2 - 26a$

92. $16 - 40a + 25a^2$

93. $16m^3n + 20m^2n^2 + 6mn^3$

94. $-28u^3v^3 + 26u^2v^4 - 6uv^5$

APPLICATIONS

95. FURNITURE The area of a desktop is given by the trinomial $(4x^2 + 20x - 11)$ in.^{2}. Factor it to find the expressions that represent its length and width. Then determine the difference in the length and width of the desktop.

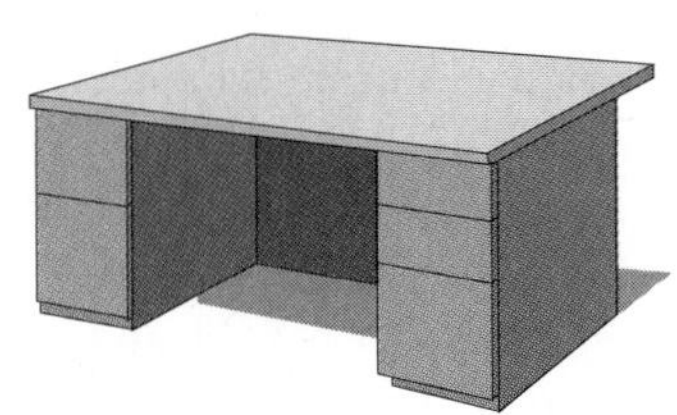

96. STORAGE The volume of an 8-foot-wide portable storage container is given by the trinomial $(72x^2 + 120x - 400)$ ft^3. Its dimensions can be determined by factoring the trinomial. Find the height and the length of the container.

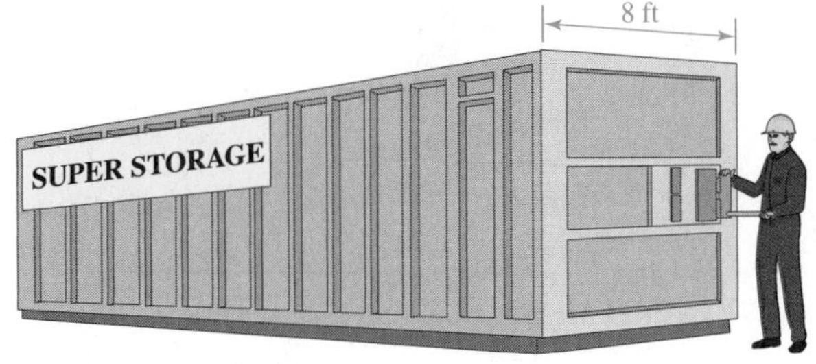

WRITING

97. In the work below, a student began to factor the trinomial. Explain his mistake.

Factor: $3x^2 - 5x + 2$.

$(3x - \quad)(x + \quad)$

???

98. Two students factor $2x^2 + 20x + 42$ and get two different answers:

$(2x + 6)(x + 7)$ and $(x + 3)(2x + 14)$

Do both answers check? Why don't they agree? Is either answer completely correct? Explain.

99. Why is the process of factoring $6x^2 - 5x - 6$ more complicated than the process of factoring $x^2 - 5x - 6$?

100. How can the factorization shown below be checked?

$$6x^2 - 5x - 6 = (3x + 2)(2x - 3)$$

101. Suppose a factorization check of $(3x - 9)(5x + 7)$ gives a middle term $-24x$, but a middle term of $24x$ is actually needed. Explain how to quickly obtain the correct factorization.

102. Suppose we want to factor $2x^2 + 7x - 72$. Explain why $(2x - 1)(x + 72)$ is not a wise choice to try first.

REVIEW **Evaluate each expression.**

103. -7^2

104. $(-7)^2$

105. 7^0

106. 7^{-2}

107. $\dfrac{1}{7^{-2}}$

108. $2 \cdot 7^2$

CHALLENGE PROBLEMS **Factor completely.**

109. $6a^{10} + 5a^5 - 21$

110. $3x^4y^2 - 29x^2y + 56$

111. $8x^2(c^2 + c - 2) - 2x(c^2 + c - 2) - 1(c^2 + c - 2)$

112. Find all integer values of b that make $2x^2 + bx - 5$ factorable.

5.4 Factoring Perfect Square Trinomials and the Difference of Two Squares

- Factoring Perfect Square Trinomials
- Factoring the Difference of Two Squares
- Multistep Factoring

In this section, we will discuss a method that can be used to factor two types of trinomials, called *perfect square trinomials.* We also develop techniques for factoring a type of binomial called the *difference of two squares.*

FACTORING PERFECT SQUARE TRINOMIALS

We have seen that the square of a binomial is a trinomial. We have also seen that the special product rules shown below can be used to quickly find the square of a sum and the square of a difference.

$$(x + y)^2 = x^2 + 2xy + y^2$$

This is the square of the first term of the binomial. This is twice the product of the two terms of the binomial. This is the square of the last term of the binomial.

$$(x - y)^2 = x^2 - 2xy + y^2$$

Notation

In the expression $(y + 3)^2$, the exponent 2 indicates repeated multiplication. It tells how many times $y + 3$ is to be used as a factor:

$$(y + 3)^2 = (y + 3)(y + 3)$$

Trinomials that are squares of a binomial are called **perfect square trinomials.** Some examples of perfect square trinomials are

$y^2 + 6y + 9$ Because it is the square of $(y + 3)$: $(y + 3)^2 = y^2 + 6y + 9$

$t^2 - 14t + 49$ Because it is the square of $(t - 7)$: $(t - 7)^2 = t^2 - 14t + 49$

$4m^2 - 20m + 25$ Because it is the square of $(2m - 5)$: $(2m - 5)^2 = 4m^2 - 20m + 25$

EXAMPLE 1

Determine whether the following trinomials are perfect square trinomials: **a.** $x^2 + 10x + 25$, **b.** $c^2 - 12c - 36$, and **c.** $25y^2 - 30y + 9$.

Solution **a.** To determine whether this is a perfect square trinomial, we note that

$$x^2 + 10x + 25$$

The first term is the square of x. The middle term is twice the product of x and 5: $2 \cdot x \cdot 5 = 10x$. The last term is the square of 5.

Thus, $x^2 + 10x + 25$ is a perfect square trinomial.

b. To determine whether this is a perfect square trinomial, we note that

$$c^2 - 12c - 36$$

The last term, -36, is not the square of a real number.

Since the last term is negative, $c^2 - 12c - 36$ is not a perfect square trinomial.

c. To determine whether this is a perfect square trinomial, we note that

$$25y^2 - 30y + 9$$

The first term is the square of $5y$.

The middle term is twice the product of $5y$ and -3: $2(5y)(-3) = -30y$.

The last term is the square of -3.

Thus, $25y^2 - 30y + 9$ is a perfect square trinomial.

Self Check 1 Determine whether the following are perfect square trinomials: **a.** $y^2 + 4y + 4$, **b.** $b^2 - 6b - 9$, **c.** $4z^2 + 4z + 4$.

Although we can factor perfect square trinomials using methods discussed earlier, we can also factor them by inspecting their terms and applying the special product formulas in reverse.

Factoring Perfect Square Trinomials

$$x^2 + 2xy + y^2 = (x + y)^2$$
$$x^2 - 2xy + y^2 = (x - y)^2$$

EXAMPLE 2

Factor: $N^2 + 20N + 100$.

ELEMENTARY Algebra f(x) Now™

Solution $N^2 + 20N + 100$ is a perfect square trinomial, because:

- The first term N^2 is the square of N.
- The last term 100 is the square of **10**: $10^2 = 100$.
- The middle term is twice the product of N and **10**: $2(N)(10) = 20N$.

To find the factorization, we match the given trinomial to the proper special product formula.

$$\begin{array}{ccl} & x^2 + 2\,x\,y + y^2 = (x + y)^2 \\ & \downarrow \\ N^2 + 20N + 10^2 = & N^2 + 2 \cdot N \cdot 10 + 10^2 = (N + 10)^2 \end{array}$$

Therefore, $N^2 + 20N + 10^2 = (N + 10)^2$. Check by finding $(N + 10)^2$.

Success Tip

Note that the sign of the second term of a perfect square trinomial is the same as the sign of the second term of the squared binomial.

$$x^2 + 2xy + y^2 = (x + y)^2$$
$$x^2 - 2xy + y^2 = (x - y)^2$$

Self Check 2 Factor: $x^2 + 18x + 81$.

EXAMPLE 3

Factor: $9x^2 - 30xy + 25y^2$.

ELEMENTARY Algebra f(x) Now™

Solution $9x^2 - 30xy + 25y^2$ is a perfect square trinomial, because:

- The first term $9x^2$ is the square of $3x$: $(3x)^2 = 9x^2$.
- The last term $25y^2$ is the square of $-5y$: $(-5y)^2 = 25y^2$.
- The middle term is twice the product of $3x$ and $-5y$: $2(3x)(-5y) = -30xy$.

By inspection, we match the given trinomial to the special product formula to find the factorization.

$$9x^2 - 30xy + 25y^2 = (3x)^2 - 2(3x)(5y) + (-5y)^2 \quad 2(3x)(-5y) = -2(3x)(5y).$$
$$= (3x - 5y)^2$$

Therefore, $9x^2 - 30xy + 25y^2 = (3x - 5y)^2$. Check by finding $(3x - 5y)^2$.

Self Check 3 Factor: $16x^2 - 8xy + y^2$.

FACTORING THE DIFFERENCE OF TWO SQUARES

Recall the special product formula for multiplying the sum and difference of two terms:

$$(x + y)(x - y) = x^2 - y^2$$

The Language of Algebra

The expression $x^2 - y^2$ is a *difference of two squares,* whereas $(x - y)^2$ is the *square of a difference.* They are not equivalent because $(x - y)^2 \neq x^2 - y^2$.

The binomial $x^2 - y^2$ is called a **difference of two squares,** because x^2 is the square of x and y^2 is the square of y. If we reverse this formula, we obtain a method for factoring a difference of two squares.

Factoring ⟶

$$x^2 - y^2 = (x + y)(x - y)$$

This pattern is easy to remember if we think of a difference of two squares as the square of a First quantity minus the square of a Last quantity.

Factoring a Difference of Two Squares

To factor the square of a First quantity minus the square of a Last quantity, multiply the First plus the Last by the First minus the Last.

$$F^2 - L^2 = (F + L)(F - L)$$

When factoring a difference of two squares, it is helpful to know the integers that are perfect squares. The number 400, for example, is a perfect square, because $400 = 20^2$. The **perfect integer squares** through 400 are

1, 4, 9, 16, 25, 36, 49, 64, 81, 100, 121, 144, 169, 196, 225, 256, 289, 324, 361, 400

EXAMPLE 4

Factor each of the following, if possible: **a.** $x^2 - 9$, **b.** $1 - b^2$, and **c.** $n^2 - 45$.

Solution **a.** $x^2 - 9$ is the difference of two squares because it can be written as $x^2 - 3^2$. We can match it to the formula for factoring a difference of two squares to find the factorization.

$$\begin{array}{c} F^2 - L^2 = (F + L)(F - L) \\ \downarrow \quad \downarrow \quad\quad \downarrow \quad \downarrow\ \downarrow \quad \downarrow \\ x^2 - 3^2 = (x + 3)(x - 3) \end{array}$$

Therefore, $x^2 - 9 = (x + 3)(x - 3)$.

Check by multiplying: $(x + 3)(x - 3) = x^2 - 9$.

Notation

By the commutative property of multiplication, the factors of a difference of two squares can be written in either order. For example, we can write:

$$x^2 - 9 = (x - 3)(x + 3)$$

b. $1 - b^2$ is the difference of two squares because $1 - b^2 = 1^2 - b^2$. Therefore,

$$1 - b^2 = (1 + b)(1 - b) \quad \text{1 is a perfect integer square: } 1 = 1^2.$$

Check by multiplying.

c. Since 45 is not a perfect integer square, $n^2 - 45$ cannot be factored using integers. It is prime.

Self Check 4 Factor each of the following, if possible: **a.** $c^2 - 4$, **b.** $121 - t^2$, and **c.** $x^2 - 24$.

Terms containing variables such as $25x^2$ are also perfect squares, because they can be written as the square of a quantity:

$$25x^2 = (5x)^2$$

EXAMPLE 5

Factor: $25x^2 - 49$.

ELEMENTARY Algebra *f(x)* Now™

Solution We can write $25x^2 - 49$ in the form $(5x)^2 - 7^2$ and match it to the formula for factoring the difference of two squares:

$$\begin{array}{ccccccc} F^2 & - L^2 & = & (F & + L)(F & - L) \\ \downarrow & \downarrow & & \downarrow & \downarrow \ \downarrow & \downarrow \\ (5x)^2 & - 7^2 & = & (5x & + 7)(5x & - 7) \end{array}$$

Therefore, $25x^2 - 49 = (5x + 7)(5x - 7)$. Check by multiplying.

Self Check 5 Factor: $16y^2 - 9$.

EXAMPLE 6

Factor: $4y^4 - 121z^2$.

ELEMENTARY Algebra *f(x)* Now™

Solution We can write $4y^4 - 121z^2$ in the form $(2y^2)^2 - (11z)^2$ and match it to the formula for factoring the difference of two squares:

$$\begin{array}{ccccccc} F^2 & - & L^2 & = & (F + L) & (F - L) \\ \downarrow & & \downarrow & & \downarrow \quad \downarrow & \downarrow \quad \downarrow \\ (2y^2)^2 & - & (11z)^2 & = & (2y^2 + 11z) & (2y^2 - 11z) \end{array}$$

Therefore, $4y^4 - 121z^2 = (2y^2 + 11z)(2y^2 - 11z)$. Check by multiplying.

Success Tip

Remember that a *difference of two squares* is a binomial. Each term is a square and the terms have different signs. The powers of the variables in the terms must be even.

Self Check 6 Factor: $9m^2 - 64n^4$.

MULTISTEP FACTORING

When factoring a polynomial, always factor out the greatest common factor first.

EXAMPLE 7

Factor: $8x^2 - 8$.

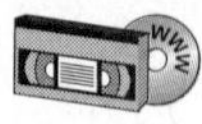

ELEMENTARY Algebra f(x) Now™

Solution We factor out the GCF of 8, and then factor the resulting difference of two squares.

$$8x^2 - 8 = 8(x^2 - 1) \quad \text{The GCF is 8.}$$
$$= 8(x + 1)(x - 1) \quad \text{Think of } x^2 - 1 \text{ as } x^2 - 1^2 \text{ and factor the difference of two squares.}$$

Check by multiplying.

$$8(x + 1)(x - 1) = 8(x^2 - 1) \quad \text{Multiply the binomials first.}$$
$$= 8x^2 - 8 \quad \text{Distribute the multiplication by 8.}$$

Self Check 7 Factor: $2p^2 - 200$.

Sometimes we must factor a difference of two squares more than once to completely factor a polynomial.

EXAMPLE 8

Factor: $x^4 - 16$.

ELEMENTARY Algebra f(x) Now™

Solution

$$x^4 - 16 = (x^2 + 4)(x^2 - 4) \quad \text{Factor the difference of two squares.}$$
$$= (x^2 + 4)(x + 2)(x - 2) \quad \text{Factor another difference of two squares: } x^2 - 4.$$

Caution The binomial $x^2 + 4$ is the **sum of two squares.** In general, after any common factor is removed, a sum of two squares cannot be factored using real numbers.

Self Check 8 Factor: $a^4 - 81$.

Answers to Self Checks **1. a.** yes, **b.** no, **c.** no **2.** $(x + 9)^2$ **3.** $(4x - y)^2$ **4. a.** $(c + 2)(c - 2)$, **b.** $(11 + t)(11 - t)$, **c.** prime **5.** $(4y + 3)(4y - 3)$ **6.** $(3m + 8n^2)(3m - 8n^2)$ **7.** $2(p + 10)(p - 10)$ **8.** $(a^2 + 9)(a + 3)(a - 3)$

5.4 STUDY SET

ELEMENTARY Algebra f(x) Now™

VOCABULARY Fill in the blanks.

1. $x^2 + 6x + 9$ is a ________ square trinomial because it is the square of the binomial $(x + 3)$.
2. The binomial $x^2 - 25$ is called a __________ of two squares and it factors as $(x + 5)(x - 5)$. The binomial $x^2 + 25$ is a ______ of two squares and it does not factor using integers.

CONCEPTS Fill in the blanks.

3. Consider $25x^2 + 30x + 9$.
   - **a.** The first term is the square of ___.
   - **b.** The last term is the square of ___.
   - **c.** The middle term is twice the product of ___ and ___.
4. Consider $49x^2 - 28xy + 4y^2$.
   - **a.** The first term is the square of ___.
   - **b.** The last term is the square of ___.
   - **c.** The middle term is twice the product of ___ and ___.
5. **a.** $x^2 + 2xy + y^2 = (\ \ + \ \ )^2$
   b. $x^2 - 2xy + y^2 = (x \ \ y)^2$
   c. $x^2 - y^2 = (x \ \ y)(\ \ - \ \ )$
6. **a.** $36x^2 = (\ \ )^2$ **b.** $100x^4 = (\ \ )^2$
   c. $4x^2 - 9 = (\ \ )^2 - (\ \ )^2$
7. List the first ten perfect integer squares.

8. Explain why each trinomial is not a perfect square trinomial.
a. $9h^2 - 6h + 7$
b. $j^2 - 8j - 16$
c. $25r^2 + 20r + 16$

9. a. Three incorrect factorizations of $x^2 + 36$ are given below. Use the FOIL method to show why each is wrong.

$(x + 6)(x - 6)$
$(x + 6)(x + 6)$
$(x - 6)(x - 6)$

b. Can $x^2 + 36$ be factored using only integers?

10. $(2b - 7)^2$ is the factorization of a trinomial. What is the trinomial?

11. $(3x + 4y)(3x - 4y)$ is the factorization of a binomial. What is the binomial?

12. $4b(a + 3)(a - 3)$ is the factorization of a binomial. What is the binomial?

NOTATION

13. Give an example of each type of expression.
a. difference of two squares
b. square of a difference
c. perfect square trinomial
d. sum of two squares

14. a. Write $(2s)^2 + 2(2s)(9t) + (9t)^2$ as a perfect square trinomial in simplest form.
b. Write $(6x)^2 - (5y)^2$ as a difference of two squares in simplest form.

15. a. What is the base and what is the exponent of the expression $(x + 3)^2$?
b. What repeated multiplication is represented by $(x - 8)^2$?

16. What is F and what is L?

$$\begin{array}{ccc} F^2 & - & L^2 \\ \downarrow & & \downarrow \\ x^2 & - & 9^2 \end{array}$$

PRACTICE Complete each factorization.

17. $a^2 - 6a + 9 = (a - \square)^2$

18. $t^2 + 2t + 1 = (t \;\square\; 1)^2$

19. $4x^2 + 4x + 1 = (2x \;\square\; 1)^2$

20. $9y^2 - 12y + 4 = (3y - \square)^2$

Factor each polynomial.

21. $x^2 + 6x + 9$

22. $x^2 + 10x + 25$

23. $b^2 + 2b + 1$

24. $m^2 + 12m + 36$

25. $c^2 - 12c + 36$

26. $d^2 - 10d + 25$

27. $y^2 - 8y + 16$

28. $z^2 - 2z + 1$

29. $t^2 + 20t + 100$

30. $r^2 + 24r + 144$

31. $2u^2 - 36u + 162$

32. $3v^2 - 42v + 147$

33. $36x^3 + 12x^2 + x$

34. $4x^4 - 20x^3 + 25x^2$

35. $9 + 4x^2 + 12x$

36. $1 + 4x^2 - 4x$

37. $a^2 + 2ab + b^2$

38. $a^2 - 2ab + b^2$

39. $25m^2 + 70mn + 49n^2$

40. $25x^2 + 20xy + 4y^2$

41. $9x^2y^2 + 30xy + 25$

42. $s^2t^2 - 20st + 100$

43. $t^2 - \frac{2}{3}t + \frac{1}{9}$

44. $p^2 + p + \frac{1}{4}$

45. $s^2 - 1.2s + 0.36$

46. $c^2 + 1.6c + 0.64$

Complete each factorization.

47. $y^2 - 49 = (y + \square)(y - \square)$

48. $p^4 - q^2 = (p^2 + q)(\square - \square)$

49. $t^2 - w^2 = (\square + \square)(t - w)$

50. $49u^2 - 64v^2 = (\square + 8v)(7u - 8v)$

Factor each polynomial, if possible.

51. $x^2 - 16$

52. $x^2 - 25$

53. $4y^2 - 1$

54. $9z^2 - 1$

55. $9x^2 - y^2$

56. $4x^2 - z^2$

57. $16a^2 - 25b^2$

58. $36a^2 - 121b^2$

59. $36 - y^2$

60. $49 - w^2$

61. $a^2 + b^2$

62. $121a^2 + 144b^2$

63. $a^4 - 144b^2$

64. $81y^4 - 100z^2$

65. $t^2z^2 - 64$

66. $900 - B^2C^2$

67. $y^2 - 63$

68. $x^2 - 27$

69. $8x^2 - 32y^2$

70. $2a^2 - 200b^2$

71. $7a^2 - 7$

72. $20x^2 - 5$

73. $-25 + v^2$

74. $-144 + h^2$

75. $6x^4 - 6x^2y^2$

76. $4b^2y - 16c^2y$

77. $x^4 - 81$

78. $y^4 - 625$

79. $a^4 - 16$

80. $b^4 - 256$

81. $c^2 - \frac{1}{16}$

82. $t^2 - \frac{9}{25}$

APPLICATIONS

83. GENETICS The Hardy–Weinberg equation, one of the fundamental concepts in population genetics, is

$$p^2 + 2pq + q^2 = 1$$

where p represents the frequency of a certain dominant gene and q represents the frequency of a certain recessive gene. Factor the left-hand side of the equation.

84. SIGNAL FLAGS The maritime signal flag for the letter X is shown. Find the polynomial that represents the area of the shaded region and express it in factored form.

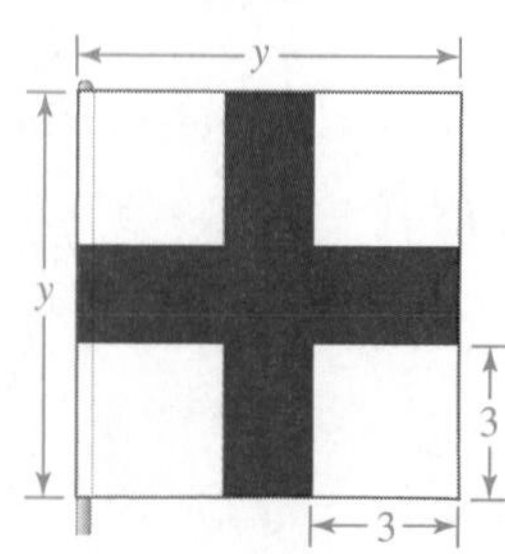

85. PHYSICS The illustration shows a time-sequence picture of a falling apple. Factor the expression, which gives the difference in the distance fallen by the apple during the time interval from t_1 to t_2 seconds.

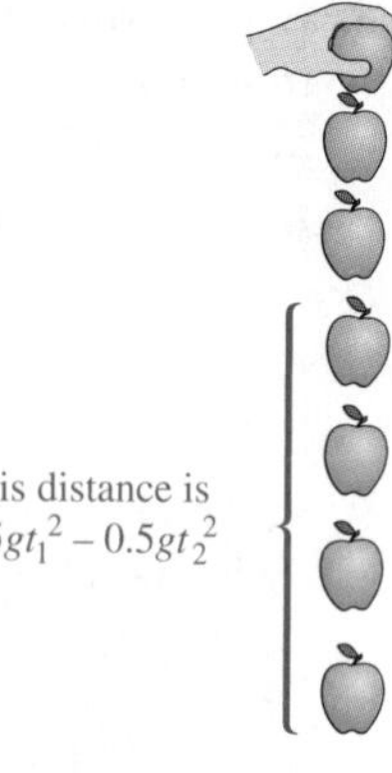

86. DARTS A circular dart board has a series of rings around a solid center, called the bullseye. To find the area of the outer white ring, we can use the formula

$$A = \pi R^2 - \pi r^2$$

Factor the expression on the right-hand side of the equation.

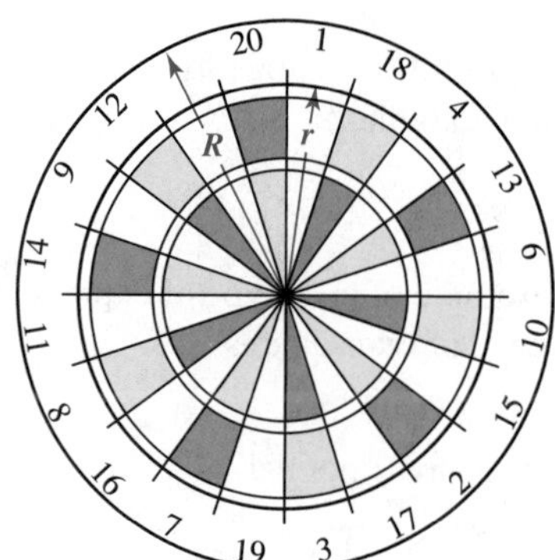

WRITING

87. When asked to factor $x^2 - 25$, one student wrote $(x + 5)(x - 5)$, and another student wrote $(x - 5)(x + 5)$. Are both answers correct? Explain.

88. Explain the error.

$$x^2 - 100 = (x + 50)(x - 50)$$

89. Explain why the following factorization isn't complete.

$$x^4 - 625 = (x^2 + 25)(x^2 - 25)$$

90. Explain why $a^2 + 2a + 1$ is a perfect square trinomial and why $a^2 + 4a + 1$ isn't a perfect square trinomial.

91. Explain the following diagram.

$$x^2 + 2xy + y^2 = (x + y)^2$$
$$x^2 - 2xy + y^2 = (x - y)^2$$

92. Write a comment explaining the error that was made in the following factorization.

Factor: $4x^2 - 16y^2$.
$(2x + 4y)(2x - 4y)$

REVIEW **Perform each division.**

93. $\dfrac{5x^2 + 10y^2 - 15xy}{5xy}$

94. $\dfrac{-30c^2d^2 - 15c^2d - 10cd^2}{-10cd}$

95. $2a - 1\overline{)a - 2 + 6a^2}$

96. $4b + 3\overline{)4b^3 - 5b^2 - 2b + 3}$

CHALLENGE PROBLEMS

97. For what value of c does $80x^2 - c$ factor as $5(4x + 3)(4x - 3)$?

98. Find all values of b so that the trinomial is a perfect square trinomial.

$$0.16x^2 + bxy + 0.25y^2$$

Factor completely.

99. $81x^6 + 36x^3y^2 + 4y^4$

100. $x^{2n} - y^{4n}$

101. $(x + 5)^2 - y^2$

102. $\dfrac{1}{2} - 2a^2$

5.5 Factoring the Sum and Difference of Two Cubes

- Factoring the Sum and Difference of Two Cubes
- Factoring Out the GCF

In this section we will discuss how to factor two types of binomials, called the *sum* and the *difference of two cubes.*

FACTORING THE SUM AND DIFFERENCE OF TWO CUBES

We have seen that the sum of two squares, such as $x^2 + 4$ or $25a^2 + 9b^2$, cannot be factored. However, the sum of two cubes and the difference of two cubes can be factored.

The sum of two cubes

$x^3 + 8$

This is x cubed. This is 2 cubed: $2^3 = 8$.

The difference of two cubes

$a^3 - 64b^3$

This is a cubed. This is $4b$ cubed: $(4b)^3 = 64b^3$.

To find formulas for factoring the sum of two cubes and the difference of two cubes, we need to find the products shown below.

$$(x + y)(x^2 - xy + y^2) = x^3 - x^2y + xy^2 + x^2y - xy^2 + y^3$$
$$= x^3 + y^3 \quad \text{Combine like terms.}$$

$$(x - y)(x^2 + xy + y^2) = x^3 + x^2y + xy^2 - x^2y - xy^2 - y^3 \quad \text{Multiply each term of the trinomial by each term of the binomial.}$$
$$= x^3 - y^3 \quad \text{Combine like terms.}$$

The Language of Algebra

The expression $x^3 + y^3$ is a *sum of two cubes,* whereas, $(x + y)^3$ is the *cube of a sum.* If you expand $(x + y)^3$, you will see that they are not equivalent.

These results justify the formulas for factoring the **sum and difference of two cubes.** They are easier to remember if we think of a sum (or a difference) of two cubes as the cube of a **F**irst quantity plus (or minus) the cube of the **L**ast quantity.

Factoring the Sum and Difference of Two Cubes

To factor the cube of a First quantity plus the cube of a Last quantity, multiply the First plus the Last by the First squared, minus the First times the Last, plus the Last squared.

$$F^3 + L^3 = (F + L)(F^2 - FL + L^2)$$

To factor the cube of a First quantity minus the cube of a Last quantity, multiply the First minus the Last by the First squared, plus the First times the Last, plus the Last squared.

$$F^3 - L^3 = (F - L)(F^2 + FL + L^2)$$

To factor the sum or difference of two cubes, it's helpful to know the cubes of the numbers from 1 to 10:

1, 8, 27, 64, 125, 216, 343, 512, 729, 1,000

EXAMPLE 1

Factor: $x^3 + 8$.

ELEMENTARY Algebra f(x) Now™

Solution $x^3 + 8$ is the sum of two cubes because it can be written as $x^3 + 2^3$. We can match it to the formula for factoring the sum of two cubes to find its factorization.

$$\begin{aligned} F^3 + L^3 &= (F + L)(F^2 - FL + L^2) \\ &\downarrow \\ x^3 + 2^3 &= (x + 2)(x^2 - x2 + 2^2) \\ &= (x + 2)(x^2 - 2x + 4) \quad x^2 - 2x + 4 \text{ does not factor.} \end{aligned}$$

Therefore, $x^3 + 8 = (x + 2)(x^2 - 2x + 4)$. We can check by multiplying.

$$\begin{aligned} (x + 2)(x^2 - 2x + 4) &= x^3 + 2x^2 - 2x^2 - 4x + 4x + 8 \\ &= x^3 + 8 \end{aligned}$$

Caution

In Example 1, a common error is to try to factor $x^2 - 2x + 4$. It is not a perfect square trinomial, because the middle term needs to be $-4x$. Furthermore, it cannot be factored by the methods of Section 5.2. It is prime.

Self Check 1 Factor: $h^3 + 27$.

Expressions containing variables such as $64b^3$ are also perfect cubes, because they can be written as the cube of a quantity:

$$64b^3 = (4b)^3$$

EXAMPLE 2

Factor: $a^3 - 64b^3$.

ELEMENTARY Algebra f(x) Now™

Solution $a^3 - 64b^3$ is the difference of two cubes because it can be written as $a^3 - (4b)^3$. We can match it to the formula for factoring the difference of two cubes to find its factorization.

$$\begin{aligned} F^3 - L^3 &= (F - L)(F^2 + F\,L + L^2) \\ a^3 - (4b)^3 &= (a - 4b)[a^2 + a(4b) + (4b)^2] \\ &= (a - 4b)(a^2 + 4ab + 16b^2) \end{aligned}$$

$a^2 + 4ab + 16b^2$ does not factor.

Therefore, $a^3 - 64b^3 = (a - 4b)(a^2 + 4ab + 16b^2)$. Check by multiplying.

Self Check 2 Factor: $8c^3 - 1$.

You should memorize the formulas for factoring the sum and the difference of two cubes. Note that each has the form

(a binomial)(a trinomial)

and that there is a relationship between the signs that appear in these forms.

same

$$F^3 + L^3 = (F + L)(F^2 - FL + L^2)$$

opposite positive

same

$$F^3 - L^3 = (F - L)(F^2 + FL + L^2)$$

opposite positive

FACTORING OUT THE GCF

If the terms of a binomial have a common factor, the GCF (or opposite of the GCF) should always be factored out first.

EXAMPLE 3 Factor: $-2t^5 + 250t^2$.

ELEMENTARY Algebra f(x) Now™

Solution Each term contains the factor $-2t^2$.

$$\begin{aligned} -2t^5 + 250t^2 &= -2t^2(t^3 - 125) && \text{Factor out } -2t^2. \\ &= -2t^2(t - 5)(t^2 + 5t + 25) && \text{Factor } t^3 - 125. \end{aligned}$$

Therefore, $-2t^5 + 250t^2 = -2t^2(t - 5)(t^2 + 5t + 25)$. Check by multiplying.

Self Check 3 Factor: $4c^3 + 4d^3$.

Answers to Self Checks **1.** $(h + 3)(h^2 - 3h + 9)$ **2.** $(2c - 1)(4c^2 + 2c + 1)$ **3.** $4(c + d)(c^2 - cd + d^2)$

5.5 STUDY SET  ELEMENTARY Algebra$f(x)$Now™

VOCABULARY Fill in the blanks.

1. $x^3 + 27$ is the ______ of two cubes.
2. $a^3 - 125$ is the difference of two _______.
3. The factorization of $x^3 + 8$ is $(x + 2)(x^2 - 2x + 4)$. The first factor is a _________ and the second is a trinomial.
4. To factor $2x^3 - 16$, we begin by factoring out the ______, which is 2.

CONCEPTS Fill in the blanks.

5. a. $F^3 + L^3 = (\quad + \quad)(F^2 - FL + L^2)$
   b. $F^3 - L^3 = (F \quad L)(\quad + FL + \quad)$
6. a. $125 = (\quad)^3$ b. $27m^3 = (\quad)^3$
   c. $8x^3 - 27 = (\quad)^3 - (\quad)^3$
   d. $x^3 + 64y^3 = (\quad)^3 + (\quad)^3$
7. $m^3 + 64$
   This is ___ cubed. This is ___ cubed.
8. $216n^3 - 125$
   This is ___ cubed. This is ___ cubed.
9. List the first six positive integer cubes.
10. $(x - 2)(x^2 + 2x + 4)$ is the factorization of what binomial?
11. The factorization of $y^3 + 27$ is $(y + 3)(y^2 - 3y + 9)$. Is this factored completely, or does $y^2 - 3y + 9$ factor further?
12. Complete each factorization.
    a. $(x - 8)(x^2 + \quad x + \quad)$
    b. $(r - \quad)(r^2 + 9r + \quad)$

NOTATION Give an example of each type of expression.

13. a. sum of two cubes
    b. cube of a sum
14. a. difference of two cubes
    b. cube of a difference

PRACTICE Complete each factorization.

15. $a^3 + 8 = (a + 2)(a^2 - \quad + 4)$
16. $x^3 - 1 = (x - 1)(x^2 + \quad + 1)$
17. $b^3 + 27 = (\quad)(b^2 - 3b + 9)$
18. $z^3 - 125 = (\quad)(z^2 + 5z + 25)$

Factor completely.

19. $y^3 + 1$
20. $x^3 - 8$
21. $a^3 - 27$
22. $b^3 + 125$
23. $8 + x^3$
24. $27 - y^3$
25. $s^3 - t^3$
26. $8u^3 + w^3$
27. $a^3 + 8b^3$
28. $27a^3 - b^3$
29. $64x^3 - 27$
30. $27x^3 + 125$
31. $a^6 - b^3$
32. $a^3 + b^6$
33. $2x^3 + 54$
34. $2x^3 - 2$
35. $-x^3 + 216$
36. $-x^3 - 125$
37. $64m^3x - 8n^3x$
38. $16r^4 + 128rs^3$

APPLICATIONS

39. MAILING BREAKABLES Write an expression that describes the amount of space in the larger box that must be filled with styrofoam chips if the smaller box containing a glass tea cup is to be placed within the larger box for mailing. Then factor the expression.

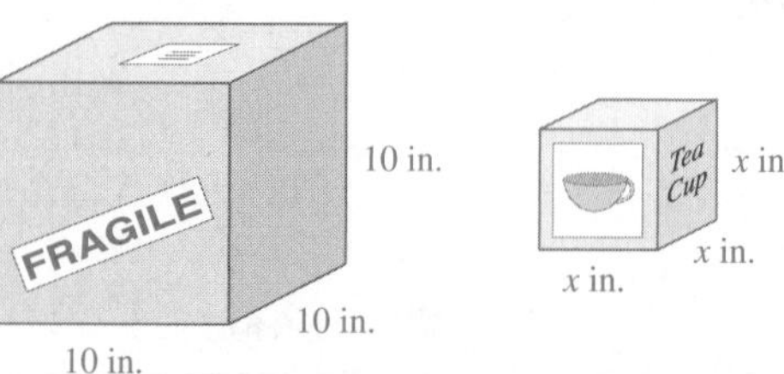

40. MELTING ICE In one hour, the block of ice shown below had melted to the size shown on the right. Write an expression that describes the volume of ice that melted away. Then factor the expression.

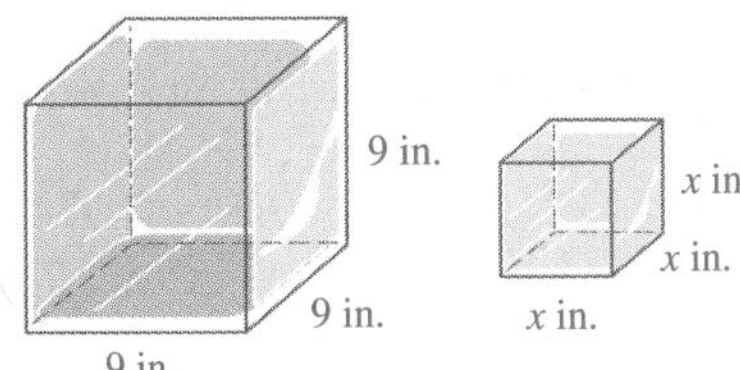

WRITING

41. Explain why $x^3 - 25$ is not a difference of two cubes.

42. Explain this diagram. Then draw a similar diagram for the difference of two cubes.

$$F^3 + L^3 = (F + L)(F^2 - FL + L^2)$$

same (F and F); opposite (+ and −); positive (+ L^2)

REVIEW

43. When expressed as a decimal, is $\frac{7}{9}$ a terminating or a repeating decimal?

44. Solve: $x + 20 = 4x - 1 + 2x$.

45. List the integers.

46. Solve: $2x + 2 = \frac{2}{3}x - 2$.

47. Evaluate $2x^2 + 5x - 3$ for $x = -3$.

48. Solve $T - R = ma$ for R.

CHALLENGE PROBLEMS

49. Consider $x^6 - 1$.
    a. Write the binomial as a difference of two squares. Then factor.
    b. Write the binomial as a difference of two cubes. Then factor.
    c. Why do the factorizations in parts (a) and (b) appear to be different?

50. What binomial multiplied by $(a^2b^2 + 7ab + 49)$ gives a difference of two cubes?

Factor completely using rational numbers.

51. $x^{3m} - y^{3n}$

52. $\frac{125}{8}s^3 + \frac{1}{27}t^3$

5.6 A Factoring Strategy

The factoring methods introduced so far will be used in the remaining chapters to simplify expressions and to solve equations. In such cases, we must determine the factoring method to use ourselves—it will not be specified. This section will give you practice in selecting the appropriate factoring method to use given a randomly chosen polynomial.

The following strategy is helpful when factoring polynomials.

Steps for Factoring a Polynomial

1. Is there a common factor? If so, factor out the GCF.
2. How many terms does the polynomial have?
   If it has *two terms,* look for the following problem types:
   a. The difference of two squares
   b. The sum of two cubes
   c. The difference of two cubes
   If it has *three terms,* look for the following problem types:
   a. A perfect square trinomial
   b. If the trinomial is not a perfect square, use the trial-and-check-method or the grouping method.
   If it has *four or more terms,* try to factor by grouping.
3. Can any factors be factored further? If so, factor them completely.
4. Does the factorization check? Check by multiplying.

EXAMPLE 1

Factor: $2x^4 - 162$.

ELEMENTARY Algebra $f(x)$ Now™

The Language of Algebra

Remember that the instruction to *factor* means to *factor completely.* A polynomial is *factored completely* when no factor can be factored further.

Solution ***Is there a common factor?*** Yes. Factor out the GCF, which is 2.

$$2x^4 - 162 = 2(x^4 - 81)$$

How many terms does it have? The polynomial within the parentheses, $x^4 - 81$, has two terms. It is a difference of two squares.

$$\begin{aligned} 2x^4 - 162 &= 2(x^4 - 81) \\ &= 2(x^2 + 9)(x^2 - 9) \end{aligned}$$

Is it factored completely? No. There is another difference of two squares, $x^2 - 9$.

$$\begin{aligned} 2x^4 - 162 &= 2(x^4 - 81) \\ &= 2(x^2 + 9)(x^2 - 9) \\ &= 2(x^2 + 9)(x + 3)(x - 3) &&\text{$x^2 + 9$ is a sum of two squares and does not factor.} \end{aligned}$$

Therefore, $2x^4 - 162 = 2(x^2 + 9)(x + 3)(x - 3)$.

Does it check? Yes.

$$\begin{aligned} 2(x^2 + 9)(x + 3)(x - 3) &= 2(x^2 + 9)(x^2 - 9) &&\text{Multiply $(x + 3)(x - 3)$ first.} \\ &= 2(x^4 - 81) &&\text{Multiply $(x^2 + 9)(x^2 - 9)$.} \\ &= 2x^4 - 162 \end{aligned}$$

Self Check 1 Factor: $11a^6 - 11a^2$.

EXAMPLE 2

Factor: $-4c^5d^2 - 12c^4d^3 - 9c^3d^4$.

ELEMENTARY Algebra $f(x)$ Now™

Solution ***Is there a common factor?*** Yes. Factor out the opposite of the GCF, $-c^3d^2$, so that the lead coefficient is positive.

$$-4c^5d^2 - 12c^4d^3 - 9c^3d^4 = -c^3d^2(4c^2 + 12cd + 9d^2)$$

How many terms does it have? The polynomial within the parentheses, $4c^2 + 12cd + 9d^2$, has three terms. It is a perfect square trinomial because $4c^2 = (2c)^2$, $9d^2 = (3d)^2$, and $12cd = 2 \cdot 2c \cdot 3d$.

$$\begin{aligned} -4c^5d^2 - 12c^4d^3 - 9c^3d^4 &= -c^3d^2(4c^2 + 12cd + 9d^2) \\ &= -c^3d^2(2c + 3d)^2 \end{aligned}$$

Is it factored completely? Yes. The binomial $2c + 3d$ does not factor further.

Therefore, $-4c^5d^2 - 12c^4d^3 - 9c^3d^4 = -c^3d^2(2c + 3d)^2$.

Does it check? Yes.

$$-c^3d^2(2c + 3d)^2 = -c^3d^2(4c^2 + 12cd + 9d^2) \quad \text{Use a special product formula.}$$
$$= -4c^5d^2 - 12c^4d^3 - 9c^3d^4$$

Self Check 2 Factor: $-32h^4 - 80h^3 - 50h^2$.

EXAMPLE 3

Factor: $y^4 - 3y^3 + y - 3$.

ELEMENTARY Algebra f(x) Now™

Solution ***Is there a common factor?*** No. There is no common factor (other than 1).

How many terms does it have? The polynomial has four terms. Try factoring by grouping.

$$y^4 - 3y^3 + y - 3 = y^3(y - 3) + 1(y - 3) \quad \text{Factor } y^3 \text{ from } y^4 - 3y^3.$$
$$= (y - 3)(y^3 + 1)$$

Is it factored completely? No. We can factor $y^3 + 1$ as a sum of two cubes.

$$y^4 - 3y^3 + y - 3 = y^3(y - 3) + 1(y - 3)$$
$$= (y - 3)(y^3 + 1)$$
$$= (y - 3)(y + 1)(y^2 - y + 1) \quad y^2 - y + 1 \text{ does not factor further.}$$

Therefore, $y^4 - 3y^3 + y - 3 = (y - 3)(y + 1)(y^2 - y + 1)$.

Does it check? Yes.

$$(y - 3)(y + 1)(y^2 - y + 1) = (y - 3)(y^3 + 1) \quad \text{Multiply the last two factors.}$$
$$= y^4 + y - 3y^3 - 3 \quad \text{Use the FOIL method.}$$
$$= y^4 - 3y^3 + y - 3$$

Self Check 3 Factor: $b^4 + b^3 + 8b + 8$

EXAMPLE 4

Factor: $32n - 4n^2 + 4n^3$.

ELEMENTARY Algebra f(x) Now™

Solution ***Is there a common factor?*** Yes. When we write the terms in descending powers of n, we see that the GCF is $4n$.

$$4n^3 - 4n^2 + 32n = 4n(n^2 - n + 8)$$

How many terms does it have? The polynomial within the parentheses has three terms. It is not a perfect square trinomial because the last term, 8, is not a perfect integer square.

To factor the trinomial $n^2 - n + 8$, we must find two integers whose product is 8 and whose sum is -1. As we see in the table, there are no such integers. Thus, $n^2 - n + 8$ is prime.

Negative factors of 8	Sum of the factors of 8
$-1(-8) = 8$	$-1 + (-8) = -9$
$-2(-4) = 8$	$-2 + (-4) = -6$

Is it factored completely? Yes.

Therefore, $4n^3 - 4n^2 + 32n = 4n(n^2 - n + 8)$.

Does it check? Yes.

$$4n(n^2 - n + 8) = 4n^3 - 4n^2 + 32n$$

Self Check 4 Factor: $6m^2 - 54m + 6m^3$.

EXAMPLE 5

Factor: $3y^3 - 4y^2 - 4y$.

ELEMENTARY Algebra $f(x)$ Now™

Solution ***Is there a common factor?*** Yes. The GCF is y.

$$3y^3 - 4y^2 - 4y = y(3y^2 - 4y - 4)$$

How many terms does it have? The polynomial within the parentheses has three terms. It is not a perfect square trinomial because the first term, $3y^2$, is not a perfect square.

If we use grouping to factor $3y^2 - 4y - 4$, the key number is $ac = 3(-4) = -12$. We must find two integers whose product is -12 and whose sum is $b = -4$.

Key number $= -12$

Factors of -12	Sum of the factors of -12
$2(-6) = -12$	$2 + (-6) = -4$

From the table, the correct pair is 2 and -6. These numbers serve as the coefficients of two terms, $2y$ and $-6y$, that we place between $3y^2$ and -4.

$$\begin{aligned} 3y^2 - 4y - 4 &= 3y^2 + 2y - 6y - 4 \\ &= y(3y + 2) - 2(3y + 2) \\ &= (3y + 2)(y - 2) \end{aligned}$$

The trinomial $3y^2 - 4y - 4$ factors as $(3y + 2)(y - 2)$.

Is it factored completely? Yes. Because $3y + 2$ and $y - 2$ do not factor.

Therefore, $3y^3 - 4y^2 - 4y = y(3y + 2)(y - 2)$. Remember to write the GCF y from the first step.

Does it check? Yes.

$$\begin{aligned} y(3y + 2)(y - 2) &= y(3y^2 - 4y - 4) && \text{Multiply the binomials} \\ &= 3y^3 - 4y^2 - 4y \end{aligned}$$

Self Check 5 Factor: $6y^3 + 21y^2 - 12y$.

Answers to Self Checks

1. $11a^2(a^2 + 1)(a + 1)(a - 1)$ **2.** $-2h^2(4h + 5)^2$ **3.** $(b + 1)(b + 2)(b^2 - 2b + 4)$
4. $6m(m^2 + m - 9)$ **5.** $3y(2y - 1)(y + 4)$

5.6 STUDY SET  ELEMENTARY Algebra f(x) Now™

VOCABULARY **Fill in the blanks.**

1. A polynomial is factored ________ when no factor can be factored further.
2. The polynomial $20 + 5x + 12y + 3xy$ has four ______. It can be factored by ________.
3. A polynomial that does not factor using integers is called a ______ polynomial.
4. The binomial $x^2 - 25$ is a ________ of two squares, whereas $x^3 + 125$ is a sum of two ______.
5. $4x^2 - 12x + 9$ is a perfect square ________ because it is the square of $(2x - 3)$.
6. The lead ________ of $x^2 + 3x - 40$ is 1.
7. When factoring a polynomial, always factor out the _____ first.
8. To ______ a factorization, use multiplication.

CONCEPTS **For each of the following polynomials, which factoring method would you use first?**

9. $2x^5y - 4x^3y$
10. $9b^2 + 12y - 5$
11. $x^2 + 18x + 81$
12. $ax + ay - x - y$
13. $x^3 + 27$
14. $y^3 - 64$
15. $m^2 + 3mn + 2n^2$
16. $16 - 25z^2$

NOTATION

17. Find the GCF in the following factorization.

$$8m^5 - 8m^4 - 48m^3 = 8m^3(m + 2)(m - 3)$$

18. Give an example of a product of two binomials.

PRACTICE **Factor each polynomial completely. If a polynomial is not factorable, write "prime."**

19. $2ab^2 + 8ab - 24a$
20. $32 - 2t^4$
21. $-8p^3q^7 - 4p^2q^3$
22. $8m^2n^3 - 24mn^4$
23. $20m^2 + 100m + 125$
24. $3rs + 6r^2 - 18s^2$
25. $x^2 + 7x + 1$
26. $3a^3 + 24b^3$
27. $-2x^5 + 128x^2$
28. $16 - 40z + 25z^2$
29. $a^2c + a^2d^2 + bc + bd^2$
30. $14t^3 - 40t^2 + 6t^4$
31. $-9x^2y^2 + 6xy - 1$
32. $x^2y^2 - 2x^2 - y^2 + 2$
33. $5x^3y^3z^4 + 25x^2y^3z^2 - 35x^3y^2z^5$
34. $2 + 24y + 40y^2$
35. $2c^2 - 5cd - 3d^2$
36. $125p^3 - 64y^3$
37. $8a^2x^3y - 2b^2xy$
38. $a^2 + 8a + 3$
39. $a^2(x - a) - b^2(x - a)$
40. $70p^4q^3 - 35p^4q^2 + 49p^5q^2$
41. $a^2b^2 - 144$
42. $-16x^4y^2z + 24x^5y^3z^4 - 15x^2y^3z^7$
43. $2ac + 4ad + bc + 2bd$
44. $u^2 - 18u + 81$
45. $v^2 - 14v + 49$
46. $28 - 3m - m^2$
47. $39 + 10n - n^2$
48. $81r^4 - 256s^4$
49. $16y^8 - 81z^4$
50. $12x^2 + 14x - 6$
51. $6x^2 - 14x + 8$
52. $12x^2 - 12$
53. $4x^2y^2 + 4xy^2 + y^2$
54. $81r^4s^2 - 24rs^5$
55. $4m^5n + 500m^2n^4$
56. $ae + bf + af + be$
57. $a^2x^2 + b^2y^2 + b^2x^2 + a^2y^2$
58. $6x^2 - x - 16$
59. $4x^2 + 9y^2$
60. $x^4y + 216xy^4$
61. $16a^5 - 54a^2b^3$
62. $25x^2 - 16y^2$

63. $27x - 27y - 27z$

64. $12x^2 + 52x + 35$

65. $xy - ty + xs - ts$

66. $bc + b + cd + d$

67. $35x^8 - 2x^7 - x^6$

68. $x^3 - 25$

69. $5(x - 2) + 10y(x - 2)$

70. $16x^2 - 40x^3 + 25x^4$

71. $49p^2 + 28pq + 4q^2$

72. $x^2y^2 - 6xy - 16$

73. $4t^2 + 36$

74. $m(5 - y) - 5 + y$

75. $p^3 - 2p^2 + 3p - 6$

76. $z^2 + 6yz^2 + 9y^2z^2$

WRITING

77. Which factoring method do you find the most difficult? Why?

78. What four questions make up the factoring strategy for polynomials discussed in this section?

79. What does it mean to factor a polynomial?

80. How is a factorization checked?

REVIEW

81. Graph the real numbers $-3, 0, 2$, and $-\frac{3}{2}$ on a number line.

82. Graph the interval $(-2, 3]$ on a number line.

83. Graph: $y = \dfrac{x}{2} + 1$.

84. Graph: $y < 2 - 3x$.

CHALLENGE PROBLEMS **Factor completely using rational numbers.**

85. $x^4 - 2x^2 - 8$

86. $x(x - y) - y(y - x)$

87. $24 - x^3 + 8x^2 - 3x$

88. $25b^2 + 14b + \dfrac{49}{25}$

89. $x^9 + y^6$

90. $\dfrac{1}{4} - \dfrac{u^2}{81}$

5.7 Solving Quadratic Equations by Factoring

• Quadratic Equations • Solving Quadratic Equations • Applications

The Language of Algebra

Quadratic equations involve the square of a variable, not the fourth power as *quad* might suggest. Why the inconsistency? A closer look at the origin of the word *quadratic* reveals that it comes from the Latin word *quadratus,* meaning square.

In Chapter 2, we solved mixture, investment, and uniform motion problems. To model those situations, we used *linear equations in one variable.*

We will now consider situations that are modeled by *quadratic equations.* Unlike linear equations, quadratic equations contain a variable squared term, such as x^2 or t^2. For example, if a pitcher throws a ball upward with an initial velocity of 63 feet per second (about 45 mph), we can find how long the ball will be in the air by solving the quadratic equation $-16t^2 + 63t + 4 = 0$.

QUADRATIC EQUATIONS

In a linear, or first degree equation, such as $2x + 3 = 8$, the exponent on the variable is 1. A quadratic, or second degree equation, has a term in which the exponent on the variable is 2, and has no other terms of higher degree.

Quadratic Equations A **quadratic equation** is an equation that can be written in the **standard form**

$$ax^2 + bx + c = 0$$

where a, b, and c represent real numbers, and a is not 0.

Some examples of quadratic equations are

$$x^2 - 2x - 63 = 0, \qquad x^2 - 25 = 0, \qquad \text{and} \qquad 2x^2 + 3x = 2$$

The first two equations are in standard form. To write the third equation in standard form, we subtract 2 from both sides to get $2x^2 + 3x - 2 = 0$.

To solve a quadratic equation, we find all values of the variable that make the equation true. Some quadratic equations can be solved using factoring methods in combination with the following property of real numbers.

The Zero-Factor Property

When the product of two real numbers is 0, at least one of them is 0.

If a and b represent real numbers, and

$$\text{if } ab = 0, \text{ then } a = 0 \text{ or } b = 0$$

EXAMPLE 1

Solve: $(4x - 1)(x + 6) = 0$.

ELEMENTARY Algebra f(x) Now™

Solution By the zero-factor property, if the product of $4x - 1$ and $x + 6$ is 0, then $4x - 1$ must be 0, or $x + 6$ must be 0. In symbols, we write this as

$$4x - 1 = 0 \quad \text{or} \quad x + 6 = 0$$

The Language of Algebra

In the zero-factor property, the word *or* means one or the other or both. If the product of two numbers is 0, then one factor is 0, or the other factor is 0, or both factors can be 0.

Now we solve each equation.

$$\begin{array}{rcl|rcl} 4x - 1 &=& 0 \quad \text{or} & x + 6 &=& 0 \\ 4x &=& 1 & x &=& -6 \\ x &=& \frac{1}{4} & & & \end{array}$$

The results must be checked separately to see whether each of them produces a true statement. We substitute $\frac{1}{4}$ and -6 for x in the original equation and simplify.

Check for $\boldsymbol{x = \frac{1}{4}}$

$$\begin{aligned} (4x - 1)(x + 6) &= 0 \\ \left[4\left(\frac{1}{4}\right) - 1\right]\left(\frac{1}{4} + 6\right) &\stackrel{?}{=} 0 \\ (1 - 1)\left(\frac{25}{4}\right) &\stackrel{?}{=} 0 \\ 0\left(\frac{25}{4}\right) &\stackrel{?}{=} 0 \\ 0 &= 0 \end{aligned}$$

Check for $\boldsymbol{x = -6}$

$$\begin{aligned} (4x - 1)(x + 6) &= 0 \\ [4(-6) - 1](-6 + 6) &\stackrel{?}{=} 0 \\ (-24 - 1)(0) &\stackrel{?}{=} 0 \\ -25(0) &\stackrel{?}{=} 0 \\ 0 &= 0 \end{aligned}$$

The resulting true statements indicate that the equation has two solutions: $\frac{1}{4}$ and -6.

Self Check 1 Solve: $(x - 12)(5x + 6) = 0$.

SOLVING QUADRATIC EQUATIONS

In Example 1, the left-hand side of $(4x - 1)(x + 6) = 0$ is in factored form, so we can use the zero-factor property. However, to solve many quadratic equations, we must factor before using the zero-factor property.

EXAMPLE 2

Solve: $x^2 - 2x - 63 = 0$.

Solution We begin by factoring the trinomial on the left-hand side.

$$x^2 - 2x - 63 = 0$$
$$(x + 7)(x - 9) = 0$$
$$x + 7 = 0 \quad \text{or} \quad x - 9 = 0 \qquad \text{Set each factor equal to 0.}$$
$$x = -7 \quad | \quad x = 9 \qquad \text{Solve each equation.}$$

Success Tip

When you see the word *solve* in Example 2, you probably think of steps such as combining like terms, distributing, or doing something to both sides. However, to solve this quadratic equation, we begin by factoring $x^2 - 2x - 63$.

To check the results, we substitute -7 and 9 for x in the original equation and simplify.

Check for x = −7	***Check for x = 9***
$x^2 - 2x - 63 = 0$	$x^2 - 2x - 63 = 0$
$(-7)^2 - 2(-7) - 63 \stackrel{?}{=} 0$	$(9)^2 - 2(9) - 63 \stackrel{?}{=} 0$
$49 - (-14) - 63 \stackrel{?}{=} 0$	$81 - 18 - 63 \stackrel{?}{=} 0$
$63 - 63 \stackrel{?}{=} 0$	$63 - 63 \stackrel{?}{=} 0$
$0 = 0$	$0 = 0$

The solutions of $x^2 - 2x - 63 = 0$ are -7 and 9.

Self Check 2 Solve: $x^2 + 5x + 6 = 0$.

The observations made in Example 2 suggest a strategy for solving quadratic equations by factoring.

The Factoring Method for Solving a Quadratic Equation

1. Write the equation in standard form: $ax^2 + bx + c = 0$.
2. Factor the left-hand side.
3. Use the zero-factor property.
4. Solve each resulting equation.
5. Check the results in the original equation.

EXAMPLE 3

Solve: $x^2 - 25 = 0$.

ELEMENTARY Algebra f(x) Now™

Solution Although $x^2 - 25 = 0$ is missing an x-term, it is a quadratic equation. To solve the equation, we factor the left-hand side and proceed as follows.

Success Tip

In Chapter 9, we discuss methods for solving quadratic equations that cannot be solved by factoring.

$$x^2 - 25 = 0$$
$$(x + 5)(x - 5) = 0 \qquad \text{Factor the difference of two squares.}$$
$$x + 5 = 0 \quad \text{or} \quad x - 5 = 0 \qquad \text{Set each factor equal to 0.}$$
$$x = -5 \quad | \quad x = 5 \qquad \text{Solve each equation.}$$

Check each solution by substituting it into the original equation.

Check for $x = -5$	***Check for $x = 5$***
$x^2 - 25 = 0$	$x^2 - 25 = 0$
$(-5)^2 - 25 \stackrel{?}{=} 0$	$5^2 - 25 \stackrel{?}{=} 0$
$25 - 25 \stackrel{?}{=} 0$	$25 - 25 \stackrel{?}{=} 0$
$0 = 0$	$0 = 0$

The solutions are -5 and 5.

Self Check 3 Solve: $x^2 - 49 = 0$.

EXAMPLE 4

Solve: $6x^2 = 12x$.

ELEMENTARY Algebra f(x) Now™

Solution To write the equation in standard form, we subtract $12x$ from both sides.

$$6x^2 = 12x$$
$$6x^2 - 12x = 12x - 12x$$
$$6x^2 - 12x = 0$$

> **Caution**
>
> In Example 4, a creative, but incorrect, approach is to divide both sides of $6x^2 = 12x$ by $6x$.
>
> $$\frac{6x^2}{6x} = \frac{12x}{6x}$$
>
> You will obtain $x = 2$; however, you will lose the second solution, 0.

To solve this equation, we factor the left-hand side.

$$6x(x - 2) = 0 \quad \text{Factor out } 6x.$$
$$6x = 0 \quad \text{or} \quad x - 2 = 0 \quad \text{Set each factor equal to 0.}$$
$$x = \frac{0}{6} \quad \Big| \quad x = 2 \quad \text{Solve each equation.}$$
$$x = 0$$

The solutions are 0 and 2. Check each solution in the original equation.

Self Check 4 Solve: $5x^2 = 25x$.

EXAMPLE 5

Solve: $2x^2 - 2 = -3x$.

ELEMENTARY Algebra f(x) Now™

Solution We begin by writing the equation in standard form and factoring the left-hand side.

$$2x^2 - 2 = -3x$$
$$2x^2 + 3x - 2 = -3x + 3x \quad \text{To get 0 on the right-hand side, add } 3x \text{ to both sides.}$$
$$2x^2 + 3x - 2 = 0 \quad \text{Combine like terms: } -3x + 3x = 0.$$
$$(2x - 1)(x + 2) = 0 \quad \text{Factor } 2x^2 + 3x - 2.$$
$$2x - 1 = 0 \quad \text{or} \quad x + 2 = 0 \quad \text{Set each factor equal to 0.}$$
$$2x = 1 \quad \Big| \quad x = -2 \quad \text{Solve each equation.}$$
$$x = \frac{1}{2}$$

The solutions are $\frac{1}{2}$ and -2. Check each solution.

Self Check 5 Solve: $3x^2 - 8 = -10x$.

EXAMPLE 6

Solve: $x(18x - 24) = -8$.

Solution We begin by writing the equation in standard form

> **Caution**
>
> To use the zero-factor property, one side of the equation must be 0. In Example 6, it would be incorrect to set each factor equal to -8.
>
> $x = -8$ or $18x - 24 = -8$
>
> If the product of two numbers is -8, one of them does not have to be -8. For example, $2(-4) = -8$.

$$x(18x - 24) = -8$$

$$18x^2 - 24x = -8 \quad \text{Distribute the multiplication by } x.$$

$$18x^2 - 24x + 8 = 0 \quad \text{Add 8 to both sides to make the right-hand side 0.}$$

$$2(9x^2 - 12x + 4) = 0 \quad \text{Factor out 2.}$$

$$2(3x - 2)(3x - 2) = 0 \quad \text{Factor } 9x^2 - 12x + 4.$$

$$2 = 0 \quad \text{or} \quad 3x - 2 = 0 \quad \text{or} \quad 3x - 2 = 0 \quad \text{Set each factor equal to 0.}$$

$$3x = 2 \quad\quad 3x = 2 \quad \text{Solve each equation.}$$

$$x = \frac{2}{3} \quad\quad x = \frac{2}{3}$$

Since 2 is a constant, it cannot equal 0. Therefore, we can discard that possibility. After solving the other equations, we see the two solutions are the same. We call $\frac{2}{3}$ a *repeated solution.* Check by substituting it into the original equation.

Self Check 6 Solve: $x(12x + 36) = -27$.

Some equations involving polynomials with degrees higher than 2 can also be solved using the factoring method.

EXAMPLE 7

Solve: $6x^3 + 12x = 17x^2$.

Solution We can solve this equation as follows:

$$6x^3 + 12x = 17x^2$$

$$6x^3 - 17x^2 + 12x = 0 \quad \text{Subtract } 17x^2 \text{ from both sides to get 0 on the right-hand side.}$$

$$x(6x^2 - 17x + 12) = 0 \quad \text{Factor out } x.$$

$$x(2x - 3)(3x - 4) = 0 \quad \text{Factor } 6x^2 - 17x + 12.$$

$$x = 0 \quad \text{or} \quad 2x - 3 = 0 \quad \text{or} \quad 3x - 4 = 0 \quad \text{Set each factor equal to 0.}$$

$$2x = 3 \quad\quad 3x = 4 \quad \text{Solve each equation.}$$

$$x = \frac{3}{2} \quad\quad x = \frac{4}{3}$$

The solutions are 0, $\frac{3}{2}$, and $\frac{4}{3}$. Check each one.

Self Check 7 Solve: $10x^3 + x^2 = 2x$.

APPLICATIONS

Each of the following problems is modeled by a quadratic equation.

EXAMPLE 8

Softball. A pitcher can throw a fastball underhand at 63 feet per second (about 45 mph). If she throws a ball into the air with that velocity, its height h in feet, t seconds after being released, is given by the formula

$$h = -16t^2 + 63t + 4$$

After the ball is thrown, in how many seconds will it hit the ground?

Solution When the ball hits the ground, its height will be 0 feet. To find the time that it will take for the ball to hit the ground, we set h equal to 0, and solve the quadratic equation for t.

$$h = -16t^2 + 63t + 4$$

$$0 = -16t^2 + 63t + 4 \quad \text{Substitute 0 for the height } h.$$

$$0 = -1(16t^2 - 63t - 4) \quad \text{Factor out } -1.$$

$$0 = -1(16t + 1)(t - 4) \quad \text{Factor } 16t^2 - 63t - 4.$$

$$16t + 1 = 0 \quad \text{or} \quad t - 4 = 0 \quad \text{Set each factor that contains a variable equal to 0.}$$

$$16t = -1 \quad\quad t = 4 \quad \text{Solve each equation.}$$

$$t = -\frac{1}{16}$$

Since time cannot be negative, we discard the solution $t = -\frac{1}{16}$. The second solution, $t = 4$, indicates that the ball hits the ground 4 seconds after being released. Check this answer by substituting 4 for t in $h = -16t^2 + 63t + 4$. You should get $h = 0$.

The Language of Algebra

Consecutive means following one another in uninterrupted order. Elton John holds the record for the most *consecutive* years with a song on the Top 50 music chart: 31 years (from 1970 to 2000).

Consecutive integers are integers that follow one another, such as 15 and 16. They are 1 unit apart. Two consecutive integers can be represented as x and $x + 1$. **Consecutive even integers** are even integers that differ by 2 units, such as 12 and 14. Similarly, **consecutive odd integers** differ by 2 units, such as 9 and 11. We can represent two consecutive even or two consecutive odd integers as x and $x + 2$.

EXAMPLE 9

Women's Tennis. In the 1998 Australian Open, sisters Venus and Serena Williams played against each other for the first time as professionals. Venus was victorious over her younger sister. At that time, their ages were consecutive integers whose product was 272. How old were Venus and Serena when they met in this match?

Analyze the Problem

- Venus is older than Serena.
- Their ages were consecutive integers.
- The product of their ages was 272.
- Find Venus' and Serena's age when they played this match.

Form an Equation Let x = Serena's age when she played in the 1998 Australian Open. Since their ages were consecutive integers, Venus' age was $x + 1$. The word *product* indicates multiplication.

Serena's age	times	Venus' age	was	272.
x	$\cdot$	$(x + 1)$	$=$	272

Solve the Equation

$$x(x + 1) = 272$$

$$x^2 + x = 272 \qquad \text{Distribute the multiplication by } x.$$

$$x^2 + x - 272 = 0 \qquad \text{Subtract 272 from both sides to make the right-hand side 0.}$$

$$(x + 17)(x - 16) = 0 \qquad \text{Factor } x^2 + x - 272. \text{ Two numbers whose product is } -272 \text{ and whose sum is 1 are 17 and } -16.$$

$$x + 17 = 0 \quad \text{or} \quad x - 16 = 0 \qquad \text{Set each factor equal to 0.}$$

$$x = -17 \quad \Big| \quad x = 16 \qquad \text{Solve each equation.}$$

State the Conclusion We discard the solution -17, because an age cannot be negative. Thus, Serena Williams was 16 years old and Venus Williams was $16 + 1 = 17$ years old when they played their first professional match against each other.

Check the Result Since 16 and 17 are consecutive integers, and since $16 \cdot 17 = 272$, the answers check.

EXAMPLE 10

ELEMENTARY Algebra $f(x)$ Now™

X-rays. A rectangular-shaped x-ray film has an area of 80 square inches. The length is 2 inches more than the width. Find its length and width.

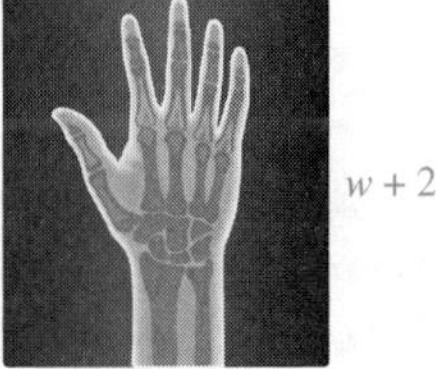

Analyze the Problem

- The area of the film is 80 square inches.
- Its length is 2 inches greater than its width.
- Find its length and width.

Form an Equation Since the width is related to the length, let w = the width of the film. Then $w + 2$ represents the length of the film. To form an equation, we use the formula for the area of a rectangle: $A = lw$.

The length	$\cdot$	the width	equals	the area of the rectangle.
$(w + 2)$	$\cdot$	w	$=$	80

Solve the Equation

$$(w + 2)w = 80$$

$$w^2 + 2w = 80 \qquad \text{Distribute the multiplication by } w.$$

$$w^2 + 2w - 80 = 0 \qquad \text{Subtract 80 from both sides to make the right-hand side 0.}$$

$$(w + 10)(w - 8) = 0 \qquad \text{Factor the trinomial.}$$

$$w + 10 = 0 \quad \text{or} \quad w - 8 = 0 \qquad \text{Set each factor equal to 0.}$$

$$w = -10 \quad \Big| \quad w = 8 \qquad \text{Solve each equation.}$$

State the Conclusion Since the width cannot be negative, we discard the result $w = -10$. Thus, the width of the x-ray film is 8 inches, and the length is $8 + 2 = 10$ inches.

The Language of Algebra

A *theorem* is a mathematical statement that can be proved. The *Pythagorean theorem* is named after *Pythagoras,* a Greek mathematician who lived about 2,500 years ago. He is thought to have been the first to prove the theorem.

Check the Result A rectangle with dimensions of 8 inches by 10 inches does have an area of 80 square inches, and the length is 2 inches more than the width. The answers check.

The next example involves a right triangle. A **right triangle** is a triangle that has a 90° (right) angle. The longest side of a right triangle is the **hypotenuse,** which is the side opposite the right angle. The remaining two sides are the **legs** of the triangle. The **Pythagorean theorem** provides a formula relating the lengths of the three sides of a right triangle.

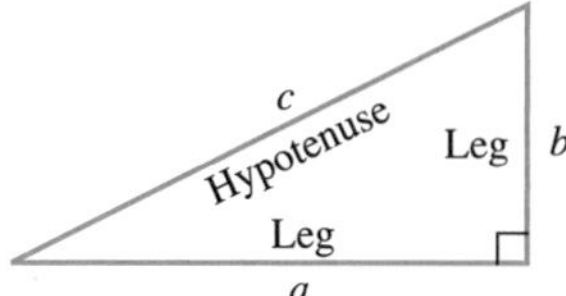

The Pythagorean Theorem

If a and b are the lengths of the legs of a right triangle and c is the length of the hypotenuse, then

$$a^2 + b^2 = c^2$$

EXAMPLE 11

ELEMENTARY Algebra f(x) Now™

Right triangles. The longer leg of a right triangle is 3 units longer than the shorter leg. If the hypotenuse is 6 units longer than the shorter leg, find the lengths of the sides of the triangle.

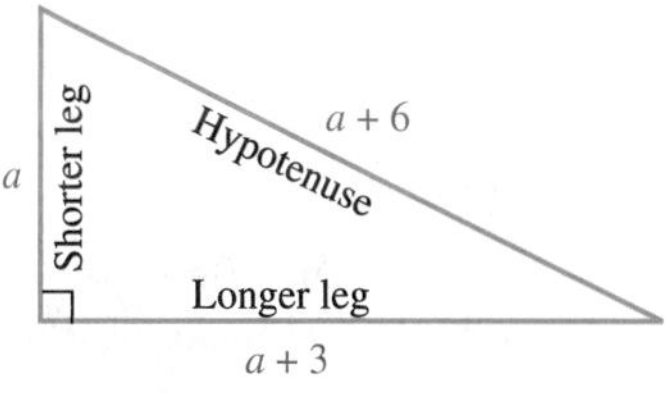

Analyze the Problem We begin by drawing a right triangle and labeling the legs and the hypotenuse.

Form an Equation We let $a =$ the length of the shorter leg. Then the length of the hypotenuse is $a + 6$ and the length of the longer leg is $a + 3$. By the Pythagorean theorem, we have

(The length of the shorter leg)²	plus	(the length of the longer leg)²	equals	(the length of the hypotenuse.)²
a^2	$+$	$(a + 3)^2$	$=$	$(a + 6)^2$

Solve the Equation

$$a^2 + (a + 3)^2 = (a + 6)^2$$

$$a^2 + a^2 + 6a + 9 = a^2 + 12a + 36 \quad \text{Find } (a + 3)^2 \text{ and } (a + 6)^2.$$

$$2a^2 + 6a + 9 = a^2 + 12a + 36 \quad \text{Combine like terms on the left-hand side.}$$

$$a^2 - 6a - 27 = 0 \quad \text{Subtract } a^2\text{, } 12a\text{, and 36 from both sides to make the right-hand side 0.}$$

Now solve the quadratic equation for a.

$$a^2 - 6a - 27 = 0$$

$$(a - 9)(a + 3) = 0 \quad \text{Factor.}$$

$$a - 9 = 0 \quad \text{or} \quad a + 3 = 0 \quad \text{Set each factor equal to 0.}$$

$$a = 9 \quad \Big| \quad a = -3 \quad \text{Solve each equation.}$$

State the Conclusion Since a side cannot have a negative length, we discard the result $a = -3$. Thus, the shorter leg is 9 units long, the hypotenuse is $9 + 6 = 15$ units long, and the longer leg is $9 + 3 = 12$ units long.

Check the Result The longer leg, 12, is 3 units longer than the shorter leg, 9. The hypotenuse, 15, is 6 units longer than the shorter leg, 9, and the lengths satisfy the Pythagorean theorem. So the results check.

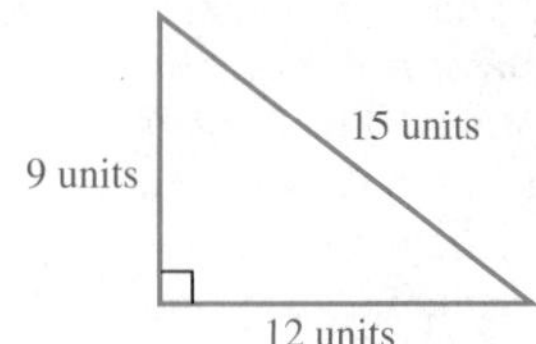

$$9^2 + 12^2 \stackrel{?}{=} 15^2$$
$$81 + 144 \stackrel{?}{=} 225$$
$$225 = 225$$

Answers to Self Checks 1. $12, -\frac{6}{5}$ 2. $-2, -3$ 3. $-7, 7$ 4. $0, 5$ 5. $\frac{2}{3}, -4$ 6. $-\frac{3}{2}$ 7. $0, \frac{2}{5}, -\frac{1}{2}$

5.7 STUDY SET

ELEMENTARY Algebra f(x) Now™

VOCABULARY Fill in the blanks.

1. Any equation that can be written in the form $ax^2 + bx + c = 0$ $(a \neq 0)$ is called a ________ equation.
2. To ______ a quadratic equation, we find all values of the variable that make the equation true.
3. Integers that are 1 unit apart, such as 8 and 9, are called ___________ integers.
4. A ______ triangle is a triangle that has a 90° angle.
5. The longest side of a right triangle is the __________. The remaining two sides are the ______ of the triangle.
6. The ___________ theorem is a formula that relates the lengths of the three sides of a right triangle.

CONCEPTS

7. Which of the following are quadratic equations?
   a. $x^2 + 2x - 10 = 0$ b. $2x - 10 = 0$
   c. $x^2 = 15x$ d. $x^3 + x^2 + 2x = 0$
8. Write each equation in the form $ax^2 + bx + c = 0$.
   a. $x^2 + 2x = 6$ b. $x^2 = 5x$
   c. $3x(x - 8) = -9$ d. $4x^2 = 25$
9. Set $5x + 4$ equal to 0 and solve for x.
10. To solve $3(x + 2)(x - 8) = 0$, which factors should be set equal to 0?
11. What step (or steps) should be performed first before factoring is used to solve each equation?
    a. $x^2 + 7x = -6$
    b. $x(x + 7) = 3$
12. Check to see whether the given number is a solution of the given quadratic equation.
    a. $x^2 - 4x = 0$; $x = 4$
    b. $x^2 + 2x - 4 = 0$; $x = -2$
    c. $4x^2 - x + 3 = 0$; $x = 1$
13. a. Factor: $x^2 + 6x - 16$.
    b. Solve: $x^2 + 6x - 16 = 0$.
14. a. Give an example of two consecutive positive integers.
    b. Fill in the blank. Two consecutive integers can be represented by x and ______.
15. A ball is thrown into the air. Its height h in feet, t seconds after being released, is given by the formula

$$h = -16t^2 + 24t + 6$$

When the ball hits the ground, what is the value of h?

16. In a right triangle, the sum of the ________ of the lengths of the two legs is equal to the square of the length of the ___________.

17. Fill in the blank: If the length of the hypotenuse of a right triangle is c and the lengths of the other two legs are a and b, then $\square = c^2$.

18. a. What kind of triangle is shown here?

$(x + 9)$ ft

x ft

$(x + 1)$ ft

b. What are the lengths of the legs of the triangle?

c. How much longer is the hypotenuse than the shorter leg?

NOTATION Complete each solution.

19.
$$7y^2 + 14y = 0$$
$$\square(y + 2) = 0$$
$$7y = 0 \quad \text{or} \quad \square = 0$$
$$y = \square \quad \bigg| \quad y = -2$$

20.
$$12p^2 - p - 6 = 0$$
$$(\square - 3)(3p + \square) = 0$$
$$\square = 0 \quad \text{or} \quad 3p + 2 = \square$$
$$4p = \square \quad \bigg| \quad 3p = \square$$
$$p = \square \quad \bigg| \quad p = -\frac{2}{3}$$

PRACTICE Solve each equation.

21. $(x - 2)(x + 3) = 0$

22. $(x - 3)(x - 2) = 0$

23. $(2s - 5)(s + 6) = 0$

24. $(3h - 4)(h + 1) = 0$

25. $2(t - 7)(t + 8) = 0$

26. $-(n + 3)(n - 6) = 0$

27. $(x - 1)(x + 2)(x - 3) = 0$

28. $(x + 2)(x + 3)(x - 4) = 0$

29. $x(x - 3) = 0$ **30.** $x(x + 5) = 0$

31. $6x(2x - 5) = 0$ **32.** $5x(5x + 7) = 0$

33. $w^2 - 7w = 0$ **34.** $p^2 + 5p = 0$

35. $3x^2 + 8x = 0$ **36.** $5x^2 - x = 0$

37. $8s^2 - 16s = 0$ **38.** $15s^2 - 20s = 0$

39. $x^2 - 25 = 0$ **40.** $x^2 - 36 = 0$

41. $4x^2 - 1 = 0$ **42.** $9y^2 - 1 = 0$

43. $9y^2 - 4 = 0$ **44.** $16z^2 - 25 = 0$

45. $x^2 = 100$ **46.** $z^2 = 25$

47. $4x^2 = 81$ **48.** $9y^2 = 64$

49. $x^2 - 13x + 12 = 0$ **50.** $x^2 + 7x + 6 = 0$

51. $x^2 - 4x - 21 = 0$ **52.** $x^2 + 2x - 15 = 0$

53. $x^2 - 9x + 8 = 0$ **54.** $x^2 - 14x + 45 = 0$

55. $a^2 + 8a = -15$ **56.** $a^2 - 16a = -64$

57. $4y + 4 = -y^2$ **58.** $-3y + 18 = y^2$

59. $0 = x^2 - 16x + 64$ **60.** $0 = 3h^2 + h - 2$

61. $2x^2 - 5x + 2 = 0$ **62.** $2x^2 + x - 3 = 0$

63. $5x^2 - 6x + 1 = 0$ **64.** $6x^2 - 5x + 1 = 0$

65. $4r^2 + 4r = -1$ **66.** $9m^2 + 6m = -1$

67. $12b^2 + 26b + 12 = 0$ **68.** $25f^2 - 80f + 15 = 0$

69. $-15x^2 + 2 = -7x$ **70.** $-8x^2 - 10x = -3$

71. $x(2x - 3) = 20$ **72.** $x(2x - 3) = 14$

73. $(n + 8)(n - 3) = -30$ **74.** $(2s + 5)(s + 1) = -1$

75. $3b^2 - 30b = 6b - 60$ **76.** $2m^2 - 8m = 2m - 12$

77. $(d + 1)(8d + 1) = 18d$ **78.** $4h(3h + 2) = h + 12$

79. $(x - 2)(x^2 - 8x + 7) = 0$

80. $(x - 1)(x^2 + 5x + 6) = 0$

81. $x^3 + 3x^2 + 2x = 0$ **82.** $x^3 - 7x^2 + 10x = 0$

83. $k^3 - 27k - 6k^2 = 0$ **84.** $j^3 - 22j - 9j^2 = 0$

85. $2x^3 = 2x(x + 2)$ **86.** $x^3 + 7x^2 = x^2 - 9x$

APPLICATIONS

87. OFFICIATING Before a football game, a coin toss is used to determine which team will kick off. The height h (in feet) of a coin above the ground t seconds after being flipped up into the air is given by $h = -16t^2 + 22t + 3$. How long does a team captain have to call heads or tails if it must be done while the coin is in the air (see next page)?

88. DOLPHINS The height h in feet reached by a dolphin t seconds after breaking the surface of the water is given by

$$h = -16t^2 + 32t$$

How long will it take the dolphin to jump out of the water and touch the trainer's hand?

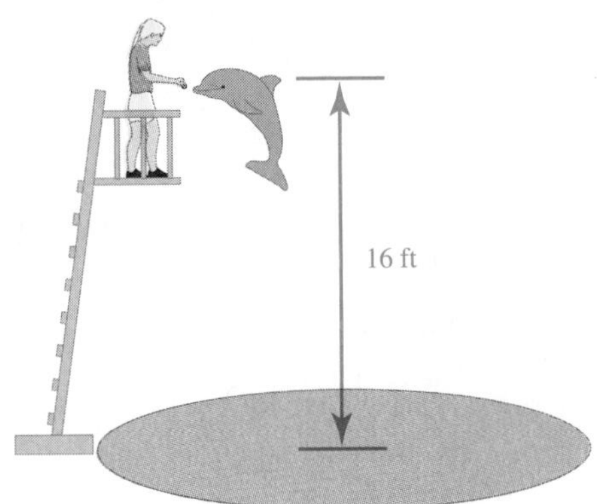

89. EXHIBITION DIVING In Acapulco, Mexico, men diving from a cliff to the water 64 feet below are quite a tourist attraction. A diver's height h above the water t seconds after diving is given by $h = -16t^2 + 64$. How long does a dive last?

90. TIME OF FLIGHT The formula $h = -16t^2 + vt$ gives the height h in feet of an object after t seconds, when it is shot upward into the air with an initial velocity v in feet per second. After how many seconds will the object hit the ground if it is shot with a velocity of 144 feet per second?

91. CHOREOGRAPHY For the finale of a musical, 36 dancers are to assemble in a triangular-shaped series of rows, where each successive row has one more dancer than the previous row. The illustration shows the beginning of such a formation. The relationship between the number of rows r and the number of dancers d is given by

$$d = \frac{1}{2}r(r + 1)$$

Determine the number of rows in the formation. (*Hint:* Multiply both sides of the equation by 2.)

92. CRAFTS The illustration shows how a geometric wall hanging can be created by stretching yarn from peg to peg across a wooden ring. The relationship between the number of pegs p placed evenly around the ring and the number of yarn segments s that criss-cross the ring is given by the formula

$$s = \frac{p(p - 3)}{2}$$

How many pegs are needed if the designer wants 27 segments to criss-cross the ring? (*Hint:* Multiply both sides of the equation by 2.)

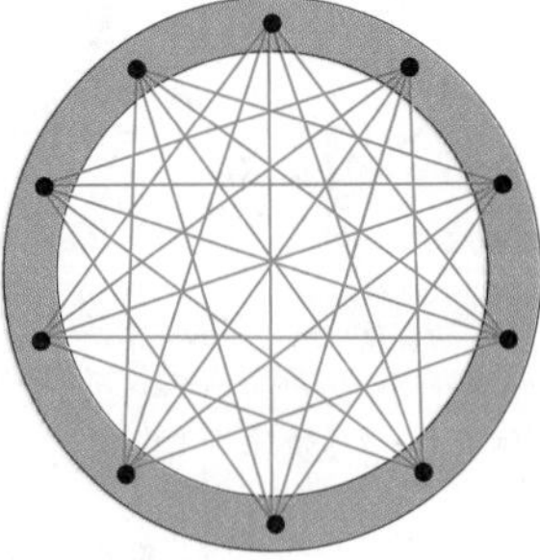

93. CUSTOMER SERVICE At a pharmacy, customers take a ticket to reserve their turn for service. If the product of the ticket number now being served and the next ticket number to be served is 156, what number is now being served?

94. HISTORY Delaware was the first state to enter the Union and Hawaii was the 50th. If we order the positions of entry for the rest of the states, we find that Tennessee entered the Union right after Kentucky, and the product of their order-of-entry numbers is 240. Use the given information to complete these statements:

Kentucky was the ___ th state to enter the Union.

Tennessee was the ___ th state to enter the Union.

95. PLOTTING POINTS The x-coordinate and y-coordinate of a point in quadrant I are consecutive odd integers whose product is 143. Find the coordinates of the point.

96. PRESIDENTS George Washington was born on 2-22-1732 (February 22, 1732). He died in 1799 at the age of 67. The month in which he died and the day of the month on which he died are consecutive even integers whose product is 168. When did Washington die?

97. INSULATION The area of the rectangular slab of foam insulation in the illustration is 36 square meters. Find the dimensions of the slab.

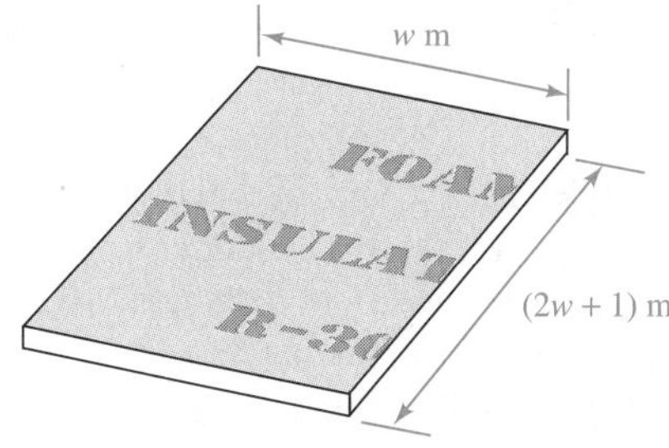

98. SHIPPING PALLETS The length of a rectangular shipping pallet is 2 feet less than 3 times its width. Its area is 21 square feet. Find the dimensions of the pallet.

99. DESIGNING A TENT The length of the base of the triangular sheet of canvas above the door of a tent is 2 feet more than twice its height. The area is 30 square feet. Find the height and the length of the base of the triangle. (*Hint:* Substitute into the formula for the area of a triangle, $A = \frac{1}{2}bh$, then multiply both sides by 2 to clear the equation of the fraction.)

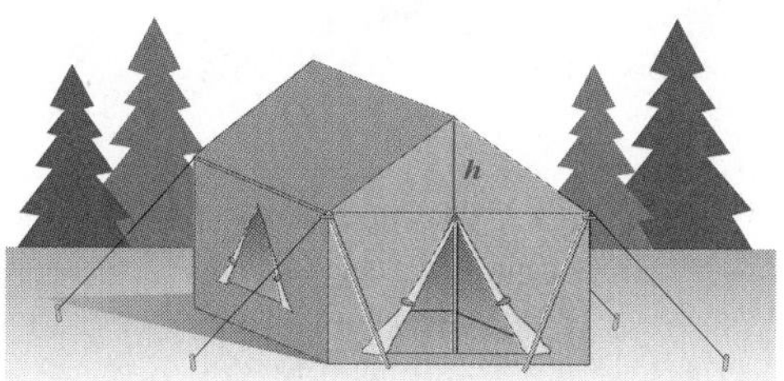

100. TUBING A piece of cardboard in the shape of a parallelogram is twisted to form the tube for a roll of paper towels. The parallelogram has an area of 60 square inches. If its height h is 7 inches more than the length of the base b, what is the length of the base? (*Hint:* The formula for the area of a parallelogram is $A = bh$.)

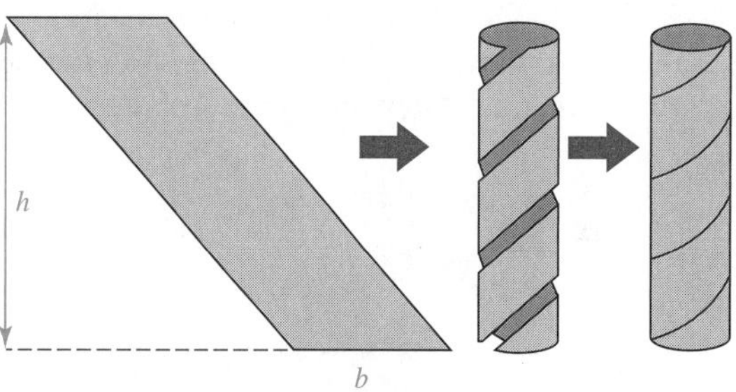

101. WIND DAMAGE A tree was blown over in a wind storm. Find x. Then find the height of the tree when it was standing upright.

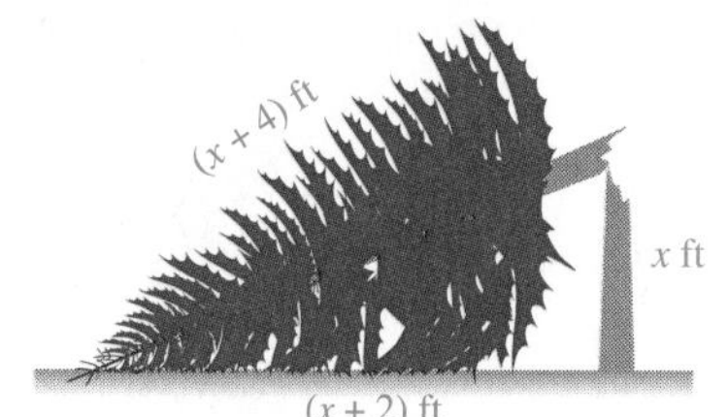

102. BOATING The inclined ramp of the boat launch is 8 meters longer than the rise of the ramp. The run is 7 meters longer than the rise. How long are the three sides of the ramp?

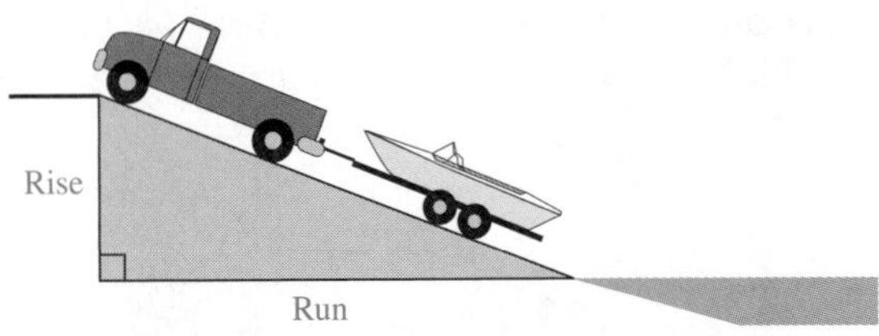

103. CAR REPAIRS To create some space to work under the front end of a car, a mechanic drives it up steel ramps. A ramp is 1 foot longer than the back, and the base is 2 feet longer than the back of the ramp. Find the length of each side of the ramp.

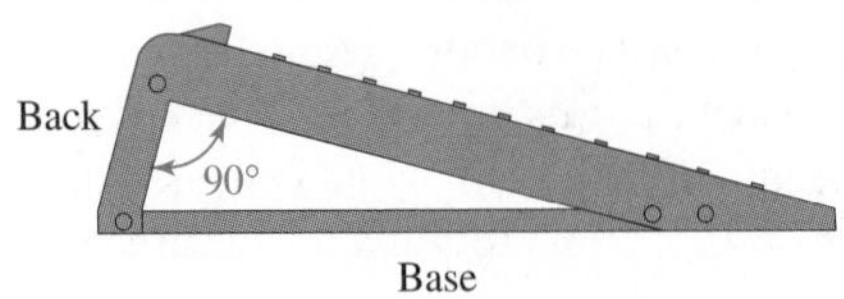

104. GARDENING TOOLS The dimensions (in millimeters) of the teeth of a pruning saw blade are given in the illustration. Find each length.

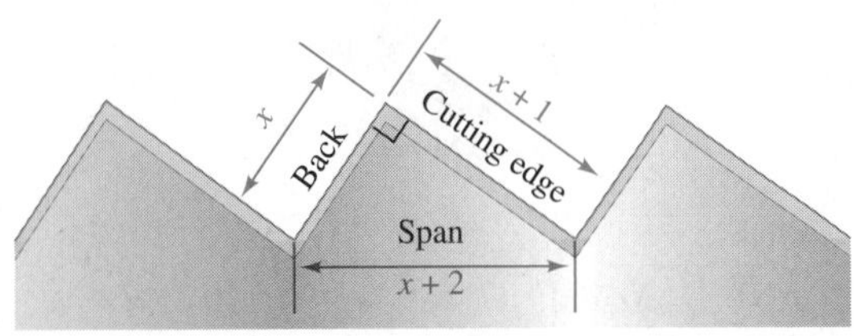

WRITING

105. What is wrong with the logic used to solve $x^2 + x = 6$?

$$x(x + 1) = 6$$

$$x = 6 \quad \text{or} \quad x + 1 = 6$$

$$x = 5$$

So the solutions are 6 and 5.

106. A student solved $x^2 - 5x + 6 = 0$ and obtained two solutions: 2 and 3. Explain the error in his check.

Check: $x^2 - 5x + 6 = 0$

$$2^2 - 5(3) + 6 \stackrel{?}{=} 0$$

$$4 - 15 + 6 \stackrel{?}{=} 0$$

$$-5 \neq 0 \quad \text{False.}$$

2 is not a solution. 3 is not a solution.

107. Suppose that to find the length of the base of a triangle, you write a quadratic equation and solve it to find $b = 6$ or $b = -8$. Explain why one solution should be discarded.

108. What error is apparent in the following illustration?

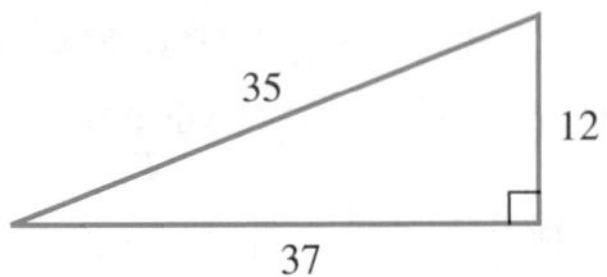

REVIEW

109. EXERCISE A doctor advises a patient to exercise at least 15 minutes but less than 30 minutes per day. Use a compound inequality to express the range of these times t in minutes.

110. SNACKS A bag of peanuts is worth \$0.30 less than a bag of cashews. Equal amounts of peanuts and cashews are used to make 40 bags of a mixture that is worth \$1.05 per bag. How much is a bag of cashews worth?

CHALLENGE PROBLEMS

111. Solve: $x^4 - 625 = 0$.

112. POOL BORDERS The owners of a rectangular swimming pool want to surround the pool with a crushed-stone border of uniform width. They have enough stone to cover 74 square meters. How wide should they make the border?

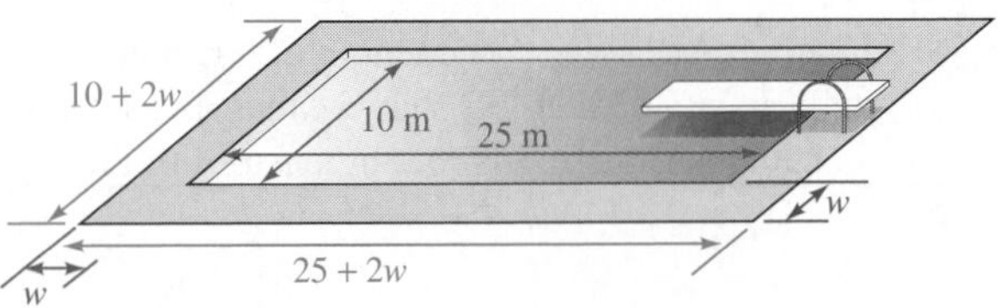

ACCENT ON TEAMWORK

FACTORING MODELS

Overview: In this activity, you will construct geometric models to find factorizations of several trinomials.

Instructions: Form groups of 2 or 3 students.

1. Copy and cut out each of the following figures. On each figure, write its area.

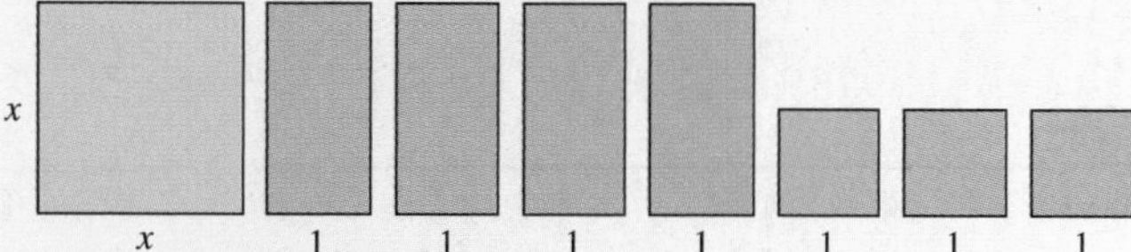

Write a trinomial that represents the *sum* of the areas of the eight figures by combining any like terms: _____ + _____ + _____

2. Now assemble the eight figures to form the large rectangle shown below.

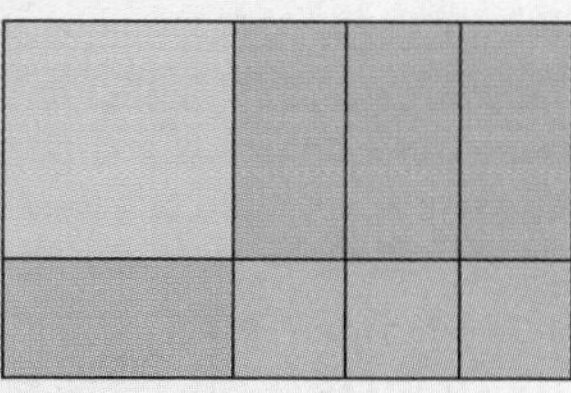

Write an expression that represents the *length* of the rectangle: _____ + _____

Write an expression that represents the *width* of the rectangle: _____ + _____

Express the area of the rectangle as the product of its length and width:
(________)(________)

3. The set of figures used in step 1 and the set of figures used in step 2 are the same. Therefore, the expressions for the areas must be equal. Set your answers from steps 1 and 2 equal to find the factorization of the trinomial $x^2 + 4x + 3$.

______________ = ______________
Answer from step 1 Answer from step 2

4. Make a new model to find the factorization of $x^2 + 5x + 4$. (*Hint:* You will need to make one more 1-by-x figure and one more 1-by-1 figure.)

______________ = ______________

5. Make a new model to find the factorization of $2x^2 + 5x + 2$. (*Hint:* You will need to make one more x-by-x figure.)

______________ = ______________

KEY CONCEPT: FACTORING

Factoring polynomials is the reverse of the process of multiplying polynomials. When we factor a polynomial, we write it as a product of two or more factors.

1. In the following problem, the distributive property is used to multiply a monomial and a binomial.

Find $3(x + 9)$.

$$3(x + 9) = 3 \cdot x + 3 \cdot 9$$
$$= 3x + 27$$

Write this so that it becomes a factoring problem. What would you start with? What would the answer be?

2. In the following problem, we multiply two binomials.

Find $(x + 3)(x + 9)$.

$$(x + 3)(x + 9) = x^2 + 9x + 3x + 27$$
$$= x^2 + 12x + 27$$

Write this so that it becomes a factoring problem. What would you start with? What would the answer be?

A FACTORING STRATEGY

The following flowchart leads you through the correct steps to identify the type(s) of factoring necessary to factor any given polynomial having two or more terms.

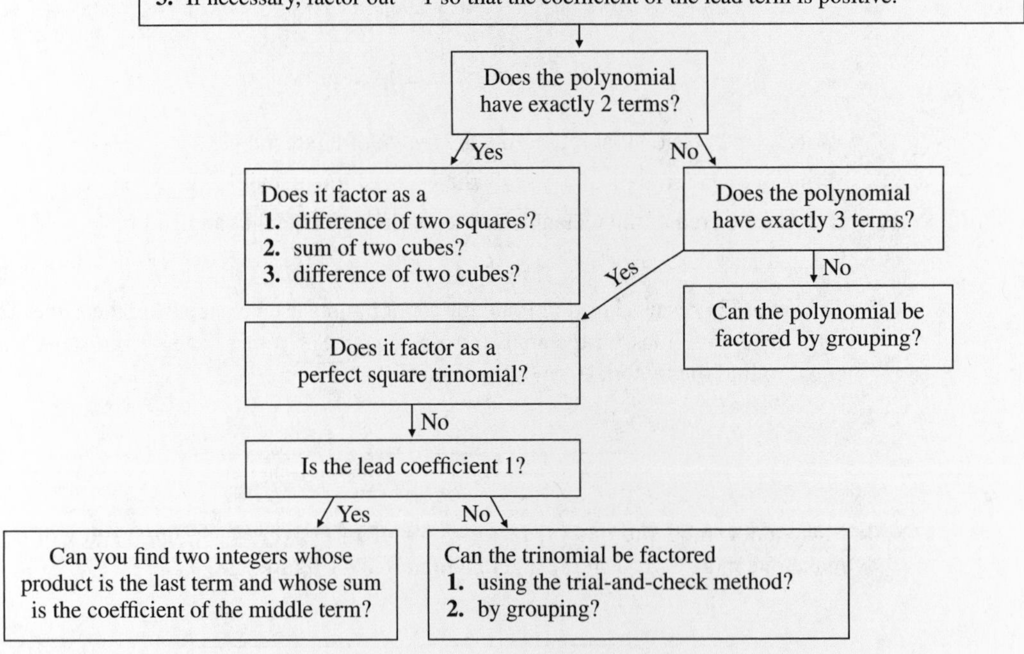

Factor each polynomial completely.

3. $-3a^2 + 21a - 36$

4. $x^2 - 121y^2$

5. $rt + 2r + st + 2s$

6. $v^3 - 8$

7. $6t^2 - 19t + 15$

8. $25y^2 - 20y + 4$

9. $2r^3 - 50r$

10. $46w - 6 + 16w^2$

CHAPTER REVIEW

ELEMENTARY Algebra $f(x)$ Now™

SECTION 5.1 The Greatest Common Factor; Factoring by Grouping

CONCEPTS

A natural number is in *prime-factored form* when it is written as the product of prime numbers.

To find the *greatest common factor* of a list of integers, prime-factor each to identify the common prime factors.

The first step of factoring a polynomial is to see whether the terms of the polynomial have a common factor. If they do, factor out the GCF.

If a polynomial has four or more terms, consider *factoring it by grouping.*

REVIEW EXERCISES

Find the prime factorization of each number.

1. 35

2. 96

Find the GCF of each list.

3. 28 and 35

4. $36a^4$, $54a^3$, and $126a^6$

Factor out the GCF.

5. $3x + 9y$

6. $5ax^2 + 15a$

7. $7s^5 + 14s^3$

8. $\pi ab - \pi ac$

9. $2x^3 + 4x^2 - 8x$

10. $x^2y^2z + xy^3z^2 - xy^2z$

11. $-5ab^2 + 10a^2b - 15ab$

12. $4(x - 2) - x(x - 2)$

Factor out -1.

13. $-a - 7$

14. $-4t^2 + 3t - 1$

Factor.

15. $2c + 2d + ac + ad$

16. $3xy + 9x - 2y - 6$

17. $2a^3 - a + 2a^2 - 1$

18. $4m^2n + 12m^2 - 8mn - 24m$

SECTION 5.2 Factoring Trinomials of the Form $x^2 + bx + c$

To *factor a trinomial* of the form $x^2 + bx + c$ means to write it as the product of two binomials.

To factor $x^2 + bx + c$, find two integers whose product is c and whose sum is b.

19. What is the lead coefficient of $x^2 + 8x - 9$?

20. Complete the table.

Factors of 6	Sum of the Factors of 6
1(6)	
2(3)	
−1(−6)	
−2(−3)	

Before factoring a trinomial, write it in *descending powers* of the variable. Also, factor out -1 if that is necessary to make the lead coefficient positive.

If a trinomial cannot be factored using only integers, it is called a *prime polynomial.*

The *GCF* should always be factored out first. A trinomial is *factored completely* when it is expressed as a product of prime polynomials.

Factor each trinomial, if possible.

21. $x^2 + 2x - 24$

22. $x^2 - 4x - 12$

23. $x^2 - 7x + 10$

24. $t^2 + 10t + 15$

25. $-y^2 + 9y - 20$

26. $10y + 9 + y^2$

27. $c^2 + 3cd - 10d^2$

28. $-3mn + m^2 + 2n^2$

29. Explain how we can check to see if $(x - 4)(x + 5)$ is the factorization of $x^2 + x - 20$.

30. Explain why $x^2 + 7x + 11$ is prime.

Completely factor each trinomial.

31. $5a^5 + 45a^4 - 50a^3$

32. $-4x^2y - 4x^3 + 24xy^2$

SECTION 5.3 Factoring Trinomials of the Form $ax^2 + bx + c$

To factor $ax^2 + bx + c$ using the *trial-and-check* factoring method, we must determine four integers. Use the FOIL method to check your work.

Factors of a

$(\square x \; \square)(\square x \; \square)$

Factors of c

To factor $ax^2 + bx + c$ using the *grouping* method, we write it as

$ax^2 + \square x + \square x + c$

The product of these numbers is ac, and their sum is b.

Factor each trinomial completely, if possible.

33. $2x^2 - 5x - 3$

34. $10y^2 + 21y - 10$

35. $-3x^2 + 14x + 5$

36. $-9p^2 - 6p + 6p^3$

37. $4b^2 - 17bc + 4c^2$

38. $7y^2 + 7y - 18$

39. ENTERTAINING The rectangular-shaped area occupied by a table setting is $(12x^2 - x - 1)$ square inches. Factor the expression to find the binomials that represent the length and width of the table setting.

40. In the following work, a student began to factor $5x^2 - 8x + 3$. Explain his mistake.

$(5x - \quad)(x + \quad)$

SECTION 5.4 Factoring Perfect Square Trinomials and the Difference of Two Squares

Special product formulas are used to factor *perfect square trinomials.*

$$x^2 + 2xy + y^2 = (x + y)^2$$
$$x^2 - 2xy + y^2 = (x - y)^2$$

To factor the *difference of two squares,* use the formula

$$F^2 - L^2 = (F + L)(F - L)$$

Factor each polynomial completely.

41. $x^2 + 10x + 25$

42. $9y^2 + 16 - 24y$

43. $-z^2 + 2z - 1$

44. $25a^2 + 20ab + 4b^2$

Factor each polynomial completely, if possible.

45. $x^2 - 9$

46. $49t^2 - 25y^2$

47. $x^2y^2 - 400$

48. $8at^2 - 32a$

49. $c^4 - 256$

50. $h^2 + 36$

SECTION 5.5 Factoring the Sum and Difference of Two Cubes

To factor the *sum* and *difference of two cubes,* use the formulas

$$F^3 + L^3 = (F + L)(F^2 - FL + L^2)$$
$$F^3 - L^3 = (F - L)(F^2 + FL + L^2)$$

Factor each polynomial completely.

51. $h^3 + 1$

52. $125p^3 + q^3$

53. $x^3 - 27$

54. $16x^5 - 54x^2y^3$

SECTION 5.6 A Factoring Strategy

To factor a random polynomial use the *factoring strategy* discussed in Section 5.6.

Factor each polynomial completely, if possible.

55. $14y^3 + 6y^4 - 40y^2$

56. $s^2t + s^2u^2 + tv + u^2v$

57. $j^4 - 16$

58. $-3j^3 - 24k^3$

59. $12w^2 - 36w + 27$

60. $121p^2 + 36q^2$

61. $2t^3 + 10$

62. $400 - m^2$

63. $x^2 + 64y^2 + 16xy$

64. $18c^3d^2 - 12c^3d - 24c^2d$

SECTION 5.7 Solving Quadratic Equations by Factoring

A *quadratic equation* is an equation of the form

$$ax^2 + bx + c = 0$$

where a, b, and c represent real numbers, and $a \neq 0$.

To use the *factoring method* to solve a quadratic equation:

1. Write the equation in $ax^2 + bx + c = 0$ form.
2. Factor the left-hand side.
3. Use the *zero-factor property:* Set each factor equal to 0.
4. Solve each resulting equation.
5. Check the results in the original equation.

The Pythagorean theorem: If the length of the hypotenuse of a right triangle is c and the lengths of the two legs are a and b, then $c^2 = a^2 + b^2$.

Solve each quadratic equation by factoring.

65. $x^2 + 2x = 0$

66. $x(x - 6) = 0$

67. $x^2 - 9 = 0$

68. $a^2 - 7a + 12 = 0$

69. $2t^2 + 8t + 8 = 0$

70. $2x - x^2 = -24$

71. $5a^2 - 6a + 1 = 0$

72. $2p^3 = 2p(p + 2)$

73. CONSTRUCTION The face of the triangular preformed concrete panel has an area of 45 square meters, and its base is 3 meters longer than twice its height. How long is its base?

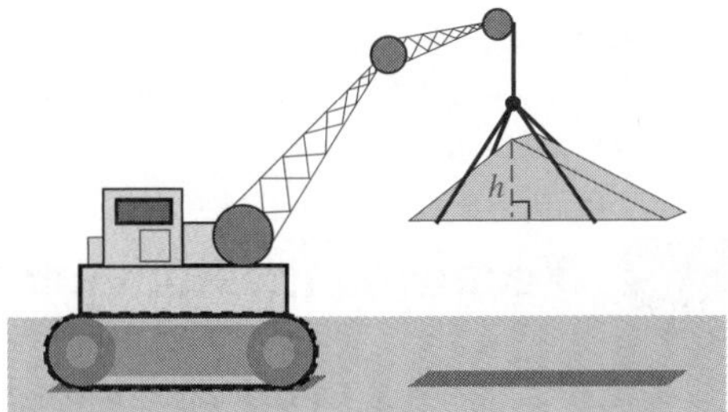

74. ACADEMY AWARDS In 2003, Meryl Streep surpassed Katherine Hepburn as most-nominated actress. The number of times Streep has been nominated and the number of times Hepburn was nominated are consecutive integers whose product is 156. How many times was each nominated?

75. TIGHTROPE WALKERS A circus performer intends to walk up a taut cable to a platform atop a pole, as shown in the illustration. How high above the ground is the platform?

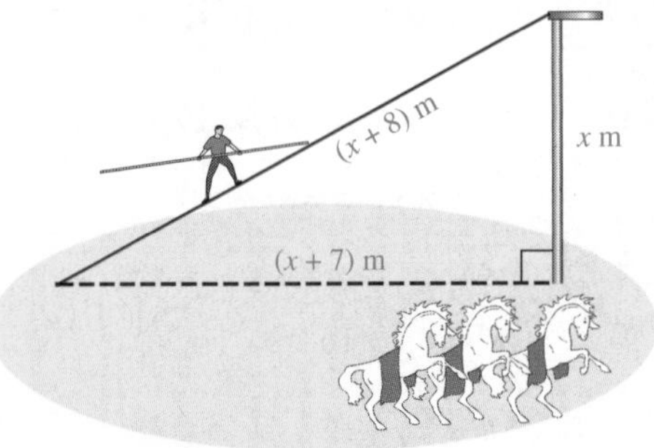

76. BALLOONING A hot-air balloonist accidentally dropped his camera overboard while traveling at a height of 1,600 feet. The height h, in feet, of the camera t seconds after being dropped is given by $h = -16t^2 + 1{,}600$. In how many seconds will the camera hit the ground?

CHAPTER 5 TEST

ELEMENTARY Algebra $f(x)$ Now™

Find the prime factorization of each number.

1. 196
2. 111
3. Find the greatest common factor of $45x^4y^6$ and $30x^3y^8$.

Factor each polynomial completely. If a polynomial cannot be factored, write "prime."

4. $4x + 16$
5. $30a^2b^3 - 20a^3b^2 + 5abc$
6. $q^2 - 81$
7. $x^2 + 9$
8. $16x^4 - 81$
9. $x^2 + 4x + 3$
10. $-x^2 + 9x + 22$
11. $9a - 9b + ax - bx$
12. $2a^2 + 5a - 12$
13. $18x^2 - 60xy + 50y^2$
14. $x^3 + 8$
15. $20m^8 - 15m^6$
16. $3a^3 - 81$
17. LANDSCAPING The combined area of the portions of the square lot that the sprinkler doesn't reach is given by $4r^2 - \pi r^2$, where r is the radius of the circular spray. Factor this expression.

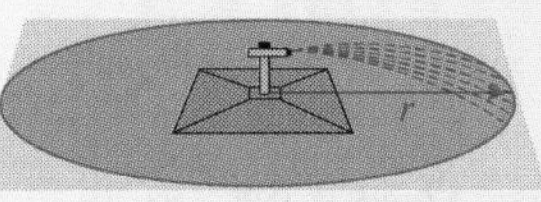

18. CHECKERS The area of a square checkerboard is $25x^2 - 40x + 16$. Find the length of a side.

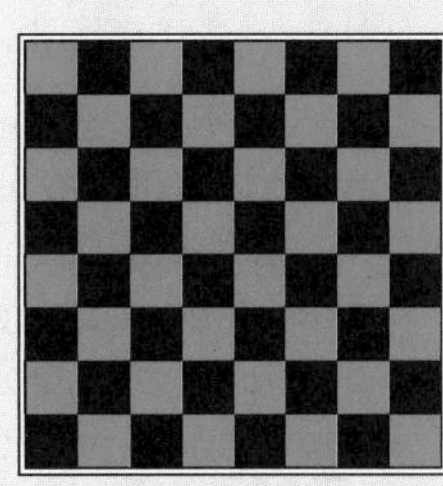

19. Factor $x^2 - 3x - 54$. Show a check of your answer.

Solve each equation.

20. $(x + 3)(x - 2) = 0$
21. $x^2 - 25 = 0$
22. $36x^2 - 6x = 0$
23. $x(x + 6) = -9$
24. $6x^2 + x - 1 = 0$
25. $(a - 2)(a - 5) = 28$
26. $x^3 + 7x^2 = -6x$
27. DRIVING SAFETY Virtually all cars have a "blind spot" where it is difficult for the driver to see a car behind and to the right. The area of the rectangular blind spot shown is 54 square feet. Its length is 3 feet longer than its width. Find its dimensions.

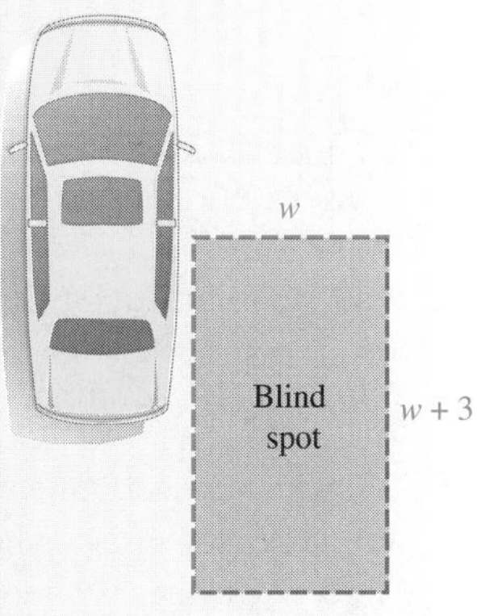

28. What is a quadratic equation? Give an example.
29. Find the length of the hypotenuse of the right triangle shown.

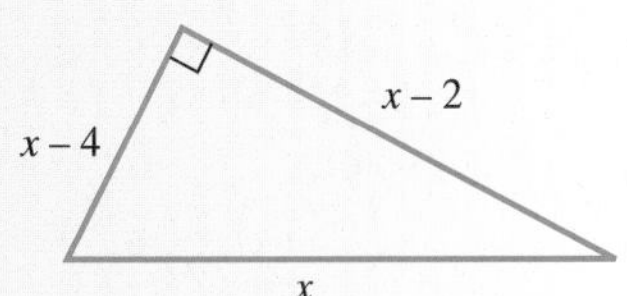

30. If the product of two numbers is 0, what conclusion can be drawn about the numbers?

CHAPTERS 1–5 CUMULATIVE REVIEW EXERCISES

1. HEART RATES Refer to the graph. Determine the difference in the maximum heart beat rate for a 70-year-old as compared to someone half that age.

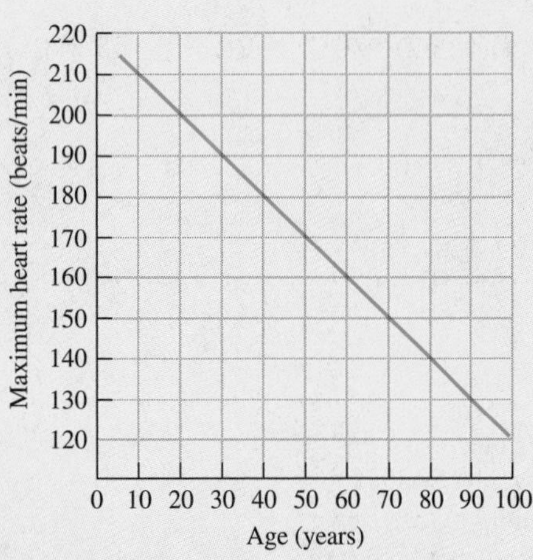

Based on data from *Cardiopulmonary Anatomy and Physiology: Essentials for Respiratory Care*, 2nd ed.

2. Find the prime factorization of 250.

3. Write $\frac{124}{125}$ as a decimal.

4. Tell whether each statement is true or false.
   a. Every integer is a whole number.
   b. Every integer is a rational number.
   c. π is a real number.

5. Find the quotient: $\frac{16}{5} \div \frac{10}{3}$.

6. What is -3 cubed?

Evaluate each expression.

7. $3 + 2[-1 - 4(5)]$

8. $\frac{|-25| - 2(-5)}{9 - 2^4}$

9. Evaluate $\frac{-x - a}{y - b}$ for $x = -2, y = 1, a = 5$, and $b = 2$.

10. Which division is undefined, $\frac{0}{5}$ or $\frac{5}{0}$?

Simplify each expression.

11. $-8y^2 - 5y^2 + 6$

12. $3z + 2(y - z) + y$

Solve each equation.

13. $-(3a + 1) + a = 2$

14. $2 - (4x + 7) = 3 + 2(x + 2)$

15. $\frac{3t - 21}{2} = t - 6$

16. $-\frac{1}{3} - \frac{x}{5} = \frac{3}{2}$

17. Solve $A = P + Prt$ for t.

18. Solve: $-\frac{x}{2} + 4 > 5$. Write the solution set in interval notation and graph it.

19. GEOMETRY TOOLS Find the total distance around the outside edge of the protractor shown below. Round to the nearest tenth of an inch.

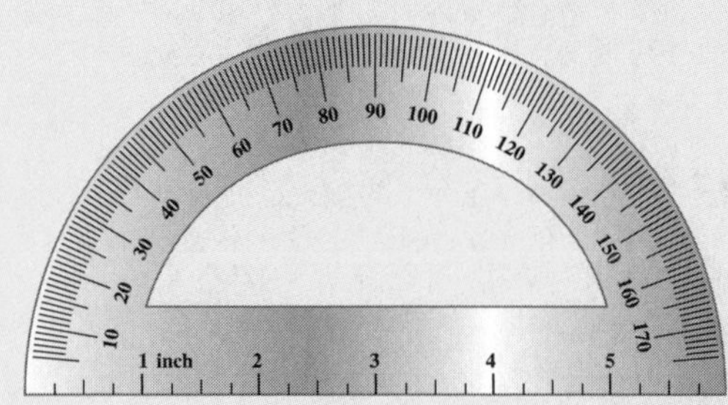

20. Find the distance traveled by a truck traveling for $5\frac{1}{2}$ hours at a rate of 60 miles per hour.

21. What is the formula for simple interest?

22. What is the value of x twenty-dollar bills?

23. PHOTOGRAPHIC CHEMICALS A photographer wishes to mix 6 liters of a 5% acetic acid solution with a 10% solution to get a 7% solution. How many liters of 10% solution must be added?

24. HISTORY George Washington was the first president of the United States. John Adams was the second, and Thomas Jefferson was the third. Grover Cleveland was president two *different* times, as shown on the next page. The sum of the numbers of Cleveland's presidencies is 46. Find these two numbers.

Grover Cleveland

Benjamin Harrison

Grover Cleveland

25. Find the slope and the y-intercept of the graph of $3x - 3y = 6$.
26. Write an equation of the line passing through $(-2, 5)$ and $(-3, -2)$. Answer in slope–intercept form.
27. Graph the line passing through $(-4, 1)$ that has slope $m = -3$.
28. Graph: $8x + 4y \geq -24$.
29. If two lines are parallel, what can be said about their slopes?
30. BEVERAGES The graph shows the annual per-person consumption of coffee and tea in the United States. What was the rate of change in coffee consumption for 1995–2000?

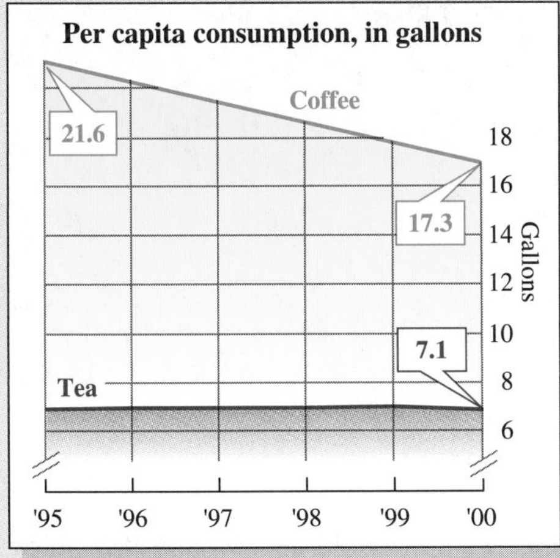

Based on data from Davenport & Co. and the U.S. Department of Agriculture.

31. Write 0.00009011 in scientific notation.
32. Write 1,700,000 in scientific notation.

Simplify each expression. Write each answer without using parentheses or negative exponents.

33. $-y^2(4y^3)$

34. $\dfrac{(x^2y^5)^5}{(x^3y)^2}$

35. $\left(\dfrac{b^5}{b^{-2}}\right)^{-2}$

36. $2x^0$

Perform the operations.

37. $(x^2 - 3x + 8) - (3x^2 + x + 3)$
38. $4b^3(2b^2 - 2b)$
39. $(y - 6)^2$
40. $(3x - 2)(x + 4)$
41. $\dfrac{12a^2b^2 - 8a^2b - 4ab}{4ab}$
42. $x - 3\overline{)2x^2 - 5x - 3}$
43. PLAYPENS
    a. Find the perimeter of the playpen.
    b. Find the area of the floor of the playpen.
    c. Find the volume of the playpen.

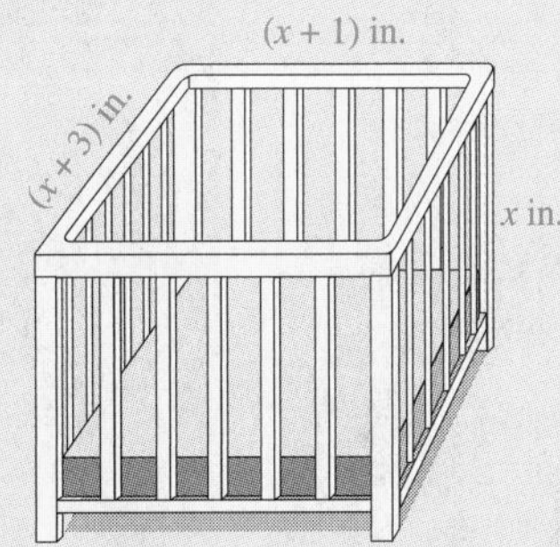

44. What is the degree of the polynomial $7y^3 + 4y^2 + y + 3$?
45. Graph: $y = x^3 + 2$.
46. What is the GCF of $24x^5y^8$ and $54x^6y$?

Factor each polynomial completely.

47. $b^3 - 3b^2$

48. $u^2 - 3 + 2u$

49. $2x^2 - 3x - 2$

50. $9z^2 - 1$

51. $-5a^2 + 25a - 30$

52. $ax + bx + ay + by$

53. $t^3 - 8$

54. $4a^2 - 12a + 9$

Solve each equation.

55. $15s^2 - 20s = 0$

56. $2x^2 - 5x = -2$

Chapter

6 Rational Expressions and Equations

ELEMENTARY **Algebra $f(x)$ Now™**

Throughout the chapter, this icon introduces resources on the Elementary AlgebraNow Web site, accessed through **http://1pass.thomson.com**, that will

- Help you test your knowledge of the material with a pre-test and a post-test
- Provide a personalized learning plan targeting areas you should study

6.1 Simplifying Rational Expressions

6.2 Multiplying and Dividing Rational Expressions

6.3 Addition and Subtraction with Like Denominators; Least Common Denominators

6.4 Addition and Subtraction with Unlike Denominators

6.5 Simplifying Complex Fractions

6.6 Solving Rational Equations

6.7 Problem Solving Using Rational Equations

6.8 Proportions and Similar Triangles

Accent on Teamwork

Key Concept

Chapter Review

Chapter Test

Cumulative Review Exercises

Norbert Schaeffer/CORBIS

In today's fast-paced world, where would we be without technology? Cell phones, computers, CD players, and microwave ovens have become an important part of our daily lives. To design these products, engineers use formulas that involve fractions. Often, these fractions contain variables in their numerators and/or denominators. In this chapter, we will consider such fractions, which are more formally referred to as *rational expressions.*

To learn more about the use of algebra in the field of electronics, visit *The Learning Equation* on the Internet at http://tle.brookscole.com. (The log-in instructions are in the Preface.) For Chapter 6, the online lesson is

- *TLE* Lesson 13: Adding and Subtracting Rational Expressions

In Chapter 1, we discussed methods for adding, subtracting, multiplying, and dividing fractions. We will now apply these skills to a new type of fraction, called a rational expression. The numerators and denominators of rational expressions usually contain variables.

6.1 Simplifying Rational Expressions

- Evaluating Rational Expressions
- Simplifying Rational Expressions
- Factors That Are Opposites

Fractions that are the quotient of two integers are *rational numbers*. Examples are $\frac{1}{2}$ and $\frac{3}{4}$. Fractions such as

$$\frac{3}{2y}, \quad \frac{x}{x+2}, \quad \text{and} \quad \frac{2a^2 - 8a}{a^2 - 6a + 8}$$

that are the quotient of two polynomials are called *rational expressions*.

Rational Expressions

A **rational expression** is an expression of the form $\frac{P}{Q}$, where P and Q are polynomials and Q does not equal 0.

EVALUATING RATIONAL EXPRESSIONS

Rational expressions can have different values depending on the number that is substituted for the variable.

EXAMPLE 1

ELEMENTARY Algebra f(x) Now™

Find the value of $\frac{2x-1}{x^2+1}$ for $x = -3$ and for $x = 0$.

Solution We replace each x in the expression with the given value, evaluate the numerator and denominator separately, and simplify, if possible.

For x = −3

$$\frac{2x-1}{x^2+1} = \frac{2(-3)-1}{(-3)^2+1} = \frac{-6-1}{9+1} = -\frac{7}{10}$$

For x = 0

$$\frac{2x-1}{x^2+1} = \frac{2(0)-1}{0^2+1} = \frac{0-1}{0+1} = -1$$

Self Check 1 Find the value of $\frac{2x-1}{x^2+1}$ for $x = 7$.

Since division by 0 is undefined, we must make sure that the denominator of a rational expression is not equal to 0.

EXAMPLE 2

Find all values of x for which the rational expression is undefined: **a.** $\frac{7x}{x-5}$, **b.** $\frac{3x-2}{x^2-x-6}$, and **c.** $\frac{8}{x^2+1}$.

Solution

a. The numerator of a rational expression can be any value, including 0. A denominator equal to 0 is what makes a rational expression undefined. The denominator of $\frac{7x}{x-5}$ will be 0 if we replace x with 5.

$$\frac{7x}{x-5} = \frac{7(5)}{5-5} = \frac{35}{0}$$

Since $\frac{35}{0}$ is undefined, the expression is undefined for $x = 5$.

b. The expression $\frac{3x-2}{x^2-x-6}$ will be undefined for values of x that make the denominator 0. To find these values, we solve $x^2 - x - 6 = 0$.

The Language of Algebra

Another way that Example 2 could be phrased is: State the *restrictions* on the variable. For $\frac{7x}{x-5}$, we can state the *restriction* by writing $x \neq 5$. For $\frac{3x-2}{x^2-x-6}$, we can write $x \neq 3$ and $x \neq -2$.

$x^2 - x - 6 = 0$ Set the denominator equal to 0.

$(x-3)(x+2) = 0$ Factor the trinomial.

$x - 3 = 0$ or $x + 2 = 0$ Set each factor equal to 0.

$x = 3$ | $x = -2$ Solve each equation.

Since 3 and -2 make the denominator 0, the expression is undefined for $x = 3$ and $x = -2$.

For $x = 3$

$$\frac{3x-2}{x^2-x-6} = \frac{3(3)-2}{3^2-3-6} = \frac{9-2}{9-3-6} = \frac{7}{0}$$

This expression is undefined.

For $x = -2$

$$\frac{3x-2}{x^2-x-6} = \frac{3(-2)-2}{(-2)^2-(-2)-6} = \frac{-6-2}{4+2-6} = \frac{-8}{0}$$

This expression is undefined.

c. No matter what real number is substituted for x, $x^2 + 1$ will not be 0. Thus, no real numbers make $\frac{8}{x^2+1}$ undefined.

Self Check 2 Find any values of x for which the rational expression is undefined: **a.** $\frac{x}{x+9}$, **b.** $\frac{9x+7}{x^2-25}$, and **c.** $\frac{4-x}{x^2+64}$.

SIMPLIFYING RATIONAL EXPRESSIONS

In Section 1.2, we simplified fractions by removing a factor equal to 1. For example, to simplify $\frac{6}{15}$, we factor 6 and 15, and then remove the factor $\frac{3}{3}$.

$$\frac{6}{15} = \frac{2 \cdot 3}{5 \cdot 3} = \frac{2}{5} \cdot \frac{3}{3} = \frac{2}{5} \cdot 1 = \frac{2}{5}$$

To streamline this process, we can replace $\frac{3}{3}$ in $\frac{2 \cdot 3}{5 \cdot 3}$ with the equivalent fraction $\frac{1}{1}$.

$$\frac{6}{15} = \frac{2 \cdot 3}{5 \cdot 3} = \frac{2 \cdot \overset{1}{\cancel{3}}}{5 \cdot \underset{1}{\cancel{3}}} = \frac{2}{5} \quad \text{We are removing } \frac{3}{3} = 1.$$

To **simplify a rational expression** means to write it so that the numerator and denominator have no common factors other than 1.

Simplifying Rational Expressions

1. Factor the numerator and denominator completely to determine their common factors.
2. Remove factors equal to 1 by replacing each pair of factors common to the numerator and denominator with the equivalent fraction $\frac{1}{1}$.

EXAMPLE 3

ELEMENTARY Algebra$f(x)$ Now™

Simplify: $\frac{21x^3}{14x^2}$.

Solution We write the numerator and denominator in factored form and remove a factor equal to 1.

$$\frac{21x^3}{14x^2} = \frac{3 \cdot 7 \cdot x \cdot x \cdot x}{2 \cdot 7 \cdot x \cdot x} \quad \text{Factor the numerator and the denominator.}$$

$$= \frac{3 \cdot \overset{1}{\cancel{7}} \cdot \overset{1}{\cancel{x}} \cdot \overset{1}{\cancel{x}} \cdot x}{2 \cdot \underset{1}{\cancel{7}} \cdot \underset{1}{\cancel{x}} \cdot \underset{1}{\cancel{x}}} \quad \text{Replace } \frac{7}{7} \text{ and each } \frac{x}{x} \text{ with the equivalent fraction } \frac{1}{1}. \text{ This removes the factor } \frac{7 \cdot x \cdot x}{7 \cdot x \cdot x}, \text{ which is equal to 1.}$$

$$= \frac{3x}{2} \quad \text{Do the multiplications in the numerator and in the denominator.}$$

We say that $\frac{21x^3}{14x^2}$ simplifies to $\frac{3x}{2}$.

The Language of Algebra

When a rational expression is simplified, the result is an *equivalent expression.* In Example 3, this means that $\frac{21x^3}{14x^2}$ has the same value as $\frac{3x}{2}$ for all values of x, except those that make either denominator 0.

Self Check 3 Simplify: $\frac{32a^3}{24a}$.

To simplify rational expressions, we often make use of the factoring methods discussed in the preceding chapter.

EXAMPLE 4

ELEMENTARY Algebra$f(x)$ Now™

Simplify: $\frac{30t - 6}{36}$.

Solution

$$\frac{30t - 6}{36} = \frac{6(5t - 1)}{6 \cdot 6} \quad \text{Factor the numerator and denominator.}$$

$$= \frac{\overset{1}{\cancel{6}}(5t - 1)}{\underset{1}{\cancel{6}} \cdot 6} \quad \text{Remove a factor equal to 1 by replacing } \frac{6}{6} \text{ with } \frac{1}{1}.$$

$$= \frac{5t - 1}{6}$$

Self Check 4 Simplify: $\frac{4t - 20}{12}$.

EXAMPLE 5

ELEMENTARY Algebra $f(x)$ Now™

Simplify: $\frac{x^2 + 13x + 12}{x^2 + 12x}$.

Solution

$$\frac{x^2 + 13x + 12}{x^2 + 12x} = \frac{(x + 1)(x + 12)}{x(x + 12)}$$ Factor the numerator. Factor the denominator.

$$= \frac{(x + 1)\overset{1}{\cancel{(x + 12)}}}{x\underset{1}{\cancel{(x + 12)}}}$$ Replace $\frac{x + 12}{x + 12}$ with the equivalent fraction $\frac{1}{1}$. This removes the factor $\frac{x + 12}{x + 12} = 1$.

$$= \frac{x + 1}{x}$$ This rational expression cannot be simplified further.

Caution

Unless otherwise stated, we will now assume that the variables in rational expressions represent real numbers for which the denominator is not zero.

Self Check 5 Simplify: $\frac{x^2 - x - 6}{x^2 - 3x}$.

Caution When simplifying rational expressions, we can only remove factors common to the entire numerator and denominator. It is incorrect to remove terms common to the numerator and denominator.

$$\frac{\overset{1}{\cancel{x}} + 1}{\underset{1}{\cancel{x}}}$$ x is a term of $x + 1$.

$$\frac{a^2 - 3a + \overset{1}{\cancel{2}}}{a + \underset{1}{\cancel{2}}}$$ 2 is a term of $a^2 - 3a + 2$ and a term of $a + 2$.

$$\frac{\overset{1}{\cancel{y^2}} - 36}{\underset{1}{\cancel{y^2}} - y - 7}$$ y^2 is a term of $y^2 - 36$ and a term of $y^2 - y - 7$.

EXAMPLE 6

ELEMENTARY Algebra $f(x)$ Now™

Simplify: **a.** $\frac{3x^2 - 8x - 3}{x^2 - 9}$ and **b.** $\frac{x^2 - 2xy + y^2}{(x - y)^4}$.

Solution

a. $$\frac{3x^2 - 8x - 3}{x^2 - 9} = \frac{(3x + 1)(x - 3)}{(x + 3)(x - 3)}$$ Factor the numerator. Factor the denominator.

$$= \frac{(3x + 1)\overset{1}{\cancel{(x - 3)}}}{(x + 3)\underset{1}{\cancel{(x - 3)}}}$$ Replace $\frac{x - 3}{x - 3}$ with the equivalent fraction $\frac{1}{1}$. This removes the factor $\frac{x - 3}{x - 3} = 1$.

$$= \frac{3x + 1}{x + 3}$$

Notation

For Example 6a, we do not need to write the parentheses in the numerator or denominator of the answer.

b. $$\frac{x^2 - 2xy + y^2}{(x - y)^4} = \frac{(x - y)^2}{(x - y)^4}$$ Factor $x^2 - 2xy + y^2$.

$$= \frac{(x - y)(x - y)}{(x - y)(x - y)(x - y)(x - y)}$$ Write the repeated multiplication indicated by each exponent.

$$= \frac{\overset{1}{\cancel{(x - y)}}\overset{1}{\cancel{(x - y)}}}{\underset{1}{\cancel{(x - y)}}\underset{1}{\cancel{(x - y)}}(x - y)(x - y)}$$ Replace each $\frac{x - y}{x - y}$ with $\frac{1}{1}$.

$$= \frac{1}{(x - y)^2}$$ Use an exponent to write the repeated multiplication in the denominator.

Self Check 6 Simplify: **a.** $\dfrac{4x^2 - 8x - 21}{4x^2 - 49}$ and **b.** $\dfrac{(a + 3b)^5}{a^2 + 6ab + 9b^2}$.

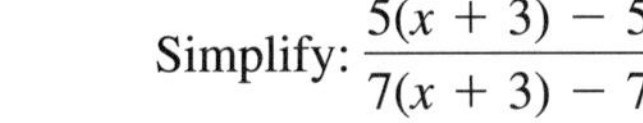

EXAMPLE 7

Simplify: $\dfrac{5(x + 3) - 5}{7(x + 3) - 7}$.

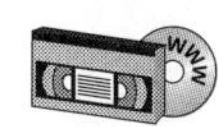

Solution Since $x + 3$ is not a factor of the entire numerator and the entire denominator, it cannot be removed. Instead, we simplify the numerator and denominator separately, factor them, and remove any common factors.

ELEMENTARY Algebra f(x) Now™

$$\frac{5(x + 3) - 5}{7(x + 3) - 7} = \frac{5x + 15 - 5}{7x + 21 - 7} \quad \text{Use the distributive property twice.}$$

$$= \frac{5x + 10}{7x + 14} \quad \text{Combine like terms.}$$

$$= \frac{5(x + 2)}{7(x + 2)} \quad \text{Factor the numerator and the denominator.}$$

$$= \frac{5\overset{1}{\cancel{(x + 2)}}}{7\underset{1}{\cancel{(x + 2)}}} \quad \text{Replace } \frac{x+2}{x+2} \text{ with the equivalent fraction } \frac{1}{1}. \text{ This removes the factor } \frac{x+2}{x+2} = 1.$$

$$= \frac{5}{7}$$

The Language of Algebra

Some rational expressions cannot be simplified. For example, to attempt to simplify the following rational expression, we factor its numerator and denominator:

$$\frac{x^2 + x - 2}{x^2 + x} = \frac{(x + 2)(x - 1)}{x(x + 1)}$$

Since there are no common factors, we say it is in *simplest form* or *lowest terms.*

Self Check 7 Simplify: $\dfrac{4(x - 2) + 4}{3(x - 2) + 3}$.

FACTORS THAT ARE OPPOSITES

If the terms of two polynomials are the same, except that they are opposite in sign, the polynomials are **opposites.** For example, the following pairs of polynomials are opposites.

$2a - 1$ and $1 - 2a$
Compare terms: $2a$ and $-2a$; -1 and 1.

$-3x^2 - x + 5$ and $3x^2 + x - 5$
Compare terms: $-3x^2$, $3x^2$; $-x$ and x; 5 and -5.

Success Tip

When a difference is reversed, the original binomial and the resulting binomial are opposites. Here are some pairs of opposites:

$b - 11$	and	$11 - b$
$x - y$	and	$y - x$
$x^2 - 4$	and	$4 - x^2$

We have seen that the quotient of two real numbers that are opposites is always -1:

$$\frac{2}{-2} = -1 \qquad \frac{-78}{78} = -1 \qquad \frac{3.5}{-3.5} = -1$$

Likewise, the quotient of two binomials that are opposites is always -1.

EXAMPLE 8

Simplify: $\dfrac{2a - 1}{1 - 2a}$.

Solution We can rearrange the terms of the numerator and factor out -1.

$$\frac{2a-1}{1-2a} = \frac{-1+2a}{1-2a}$$ In the numerator, think of $2a - 1$ as $2a + (-1)$. Then change the order of the terms: $2a + (-1) = -1 + 2a$.

$$= \frac{-1(1-2a)}{1-2a}$$ In the numerator, factor out -1.

$$= \frac{-1(\overset{1}{\cancel{1-2a}})}{\underset{1}{\cancel{1-2a}}}$$ Replace $\frac{1-2a}{1-2a}$ with the equivalent fraction $\frac{1}{1}$. This removes the factor $\frac{1-2a}{1-2a}$, which is equal to 1.

$$= \frac{-1}{1}$$

$$= -1$$ Any number divided by 1 is itself.

Self Check 8 Simplify: $\frac{3p-2}{2-3p}$.

In general, we have this fact.

The Quotient of Opposites The quotient of any nonzero polynomial and its opposite is -1.

Caution

It would be incorrect to apply the rule for opposites to a rational expression such as $\frac{x+1}{1+x}$. By the commutative property of addition, this is the quotient of a number and itself. The result is 1, not -1.

$$\frac{x+1}{1+x} = \frac{\overset{1}{\cancel{x+1}}}{\underset{1}{\cancel{x+1}}} = 1$$

For each of the following rational expressions, the numerator and denominator are opposites. Thus, each expression is equal to -1.

$$\frac{x-6}{6-x} = -1 \qquad \frac{2a-9b}{9b-2a} = -1 \qquad \frac{-3x^2-x+5}{3x^2+x-5} = -1$$

This fact can be used to simplify certain rational expressions by removing a factor equal to -1. If a factor of the numerator is the opposite of a factor of the denominator, we can replace them with the equivalent fraction $\frac{-1}{1}$, as shown below.

$$\frac{\overset{-1}{\cancel{x-6}}}{\underset{1}{\cancel{6-x}}}$$ Use slashes / to show that $\frac{x-6}{6-x}$ is replaced with the equivalent fraction $\frac{-1}{1}$.

EXAMPLE 9

Simplify: $\frac{y^2-1}{3-3y}$.

Solution

$$\frac{y^2-1}{3-3y} = \frac{(y+1)(y-1)}{3(1-y)}$$ Factor the numerator. Factor the denominator.

$$= \frac{(y+1)(\overset{-1}{\cancel{y-1}})}{3(\underset{1}{\cancel{1-y}})}$$ Since $y - 1$ and $1 - y$ are opposites, simplify by replacing $\frac{y-1}{1-y}$ with the equivalent fraction $\frac{-1}{1}$.

$$= \frac{-(y+1)}{3}$$

This result may be written in other equivalent forms.

Caution

A − symbol preceding a fraction may be applied to the numerator or to the denominator, but not to both:

$$-\frac{y+1}{3} \neq \frac{-(y+1)}{-3}$$

$$\frac{-(y+1)}{3} = -\frac{y+1}{3}$$ The − symbol in $-(y+1)$ can be written in front of the fraction, and the parentheses can be dropped.

$$\frac{-(y+1)}{3} = \frac{-y-1}{3}$$ The − symbol in $-(y+1)$ represents a factor of -1. Distribute the multiplication by -1 in the numerator.

$$\frac{-(y+1)}{3} = \frac{y+1}{-3}$$ The − symbol in $-(y+1)$ can be applied to the denominator. However, we don't usually use this form.

Self Check 9 Simplify: $\frac{m^2-100}{10m-m^2}$.

Answers to Self Checks **1.** $\frac{13}{50}$ **2. a.** -9, **b.** $-5, 5$, **c.** none **3.** $\frac{4a^2}{3}$ **4.** $\frac{t-5}{3}$ **5.** $\frac{x+2}{x}$ **6. a.** $\frac{2x+3}{2x+7}$, **b.** $(a+3b)^3$ **7.** $\frac{4}{3}$ **8.** -1 **9.** $-\frac{m+10}{m}$

6.1 STUDY SET  ELEMENTARY Algebra f(x) Now™

VOCABULARY Fill in the blanks.

1. A quotient of two polynomials, such as $\frac{x^2+x}{x^2-3x}$, is called a ________ expression.
2. In a rational expression, the polynomial above the fraction bar is called the ________ and the polynomial below the fraction bar is called the ________.
3. Because of the division by 0, the expression $\frac{8}{0}$ is ________.
4. To ________ the rational expression $\frac{x^2-1}{x-3}$ for $x=-2$, we substitute -2 for each x, and simplify.
5. To ________ a rational expression, we remove factors common to the numerator and denominator.
6. The binomials $x-15$ and $15-x$ are called ________, because their terms are the same, except that they are opposite in sign.

CONCEPTS

7. What value of x makes each rational expression undefined?

   a. $\frac{x+2}{x}$ **b.** $\frac{x+2}{x-6}$ **c.** $\frac{x+2}{x+6}$

8. When we simplify $\frac{x^2+5x}{4x+20}$, the result is $\frac{x}{4}$. They have the same value for all real numbers, except $x=-5$. Show that they have the same value for $x=1$.

9. **a.** For the following rational expression, what factor is common to the numerator and denominator?

$$\frac{x^2+2x+1}{x^2+4x+3} = \frac{\overset{1}{\cancel{(x+1)}}(x+1)}{(x+3)\underset{1}{\cancel{(x+1)}}} = \frac{x+1}{x+3}$$

   b. Fill in the blanks: To simplify the rational expression, we replace $\frac{x+1}{x+1}$ with the equivalent fraction ____. This removes the factor $\frac{x+1}{x+1}$, which is equal to ____.

10. Tell whether each pair of polynomials are opposites.

    a. $y+7$ and $y-7$
    b. $b-20$ and $20-b$
    c. x^2-9x and $9x-x^2$
    d. x^2+2x-1 and $-x^2-2x-1$

11. **a.** For the given rational expression, what factor of the numerator and what factor of the denominator are opposites?

$$\frac{x^2-2x}{20-10x} = \frac{x(x-2)}{10(2-x)} = -\frac{x}{10}$$

   b. Fill in the blanks: To simplify the rational expression, we replace $\frac{x-2}{2-x}$ with the equivalent fraction ____. This removes the factor $\frac{x-2}{2-x}$, which is equal to ____.

12. Simplify each expression.

    a. $\frac{x-8}{x-8}$ **b.** $\frac{x-8}{8-x}$

    c. $\frac{x+8}{8+x}$ **d.** $\frac{x-8}{-x+8}$

13. Write each expression in simplest form.

a. $\dfrac{(x + 2)(x - 2)}{(x + 1)(x + 2)}$ **b.** $\dfrac{y(y - 2)}{9(2 - y)}$

c. $\dfrac{(2m + 7)(m - 5)}{(2m + 7)}$ **d.** $\dfrac{x \cdot x}{x \cdot x(x - 30)}$

14. What is the first step in the process of simplifying $\dfrac{3(x + 1) + 2x + 2}{x + 1}$

NOTATION

15. In the following table, the answers to three homework problems are compared to the answers in the back of the book. Are the answers equivalent?

Answer	Book's answer	Equivalent?
$\dfrac{-3}{x + 3}$	$-\dfrac{3}{x + 3}$	
$\dfrac{-x + 4}{6x + 1}$	$\dfrac{-(x - 4)}{6x + 1}$	
$\dfrac{x + 7}{(x - 4)(x + 2)}$	$\dfrac{x + 7}{(x + 2)(x - 4)}$	

16. a. In $\dfrac{(x + 5)\overset{1}{\cancel{(x - 5)}}}{x\underset{1}{\cancel{(x - 5)}}}$, what do the slashes show?

b. In $\dfrac{(x - 3)\overset{-1}{\cancel{(x - 7)}}}{(x + 3)\underset{1}{\cancel{(7 - x)}}}$, what do the slashes show?

PRACTICE Evaluate each expression for $x = 6$.

17. $\dfrac{x - 2}{x - 5}$ **18.** $\dfrac{3x - 2}{x - 2}$

19. $\dfrac{x^2 - 4x - 12}{x^2 + x - 2}$ **20.** $\dfrac{x^2 - 1}{x^3 - 1}$

Evaluate each expression for $y = -3$.

21. $\dfrac{y + 5}{3y - 2}$ **22.** $\dfrac{2y + 9}{y^2 + 25}$

23. $\dfrac{y^3}{3y^2 + 1}$ **24.** $\dfrac{-y}{y^2 - y + 6}$

Find all values of x for which the rational expression is undefined.

25. $\dfrac{x + 5}{8x}$ **26.** $\dfrac{4x - 1}{6x}$

27. $\dfrac{15}{x - 2}$ **28.** $\dfrac{5x}{x + 5}$

29. $\dfrac{15x + 2}{x^2 + 6}$ **30.** $\dfrac{x^2 - 4x}{x^2 + 4}$

31. $\dfrac{x + 1}{2x - 1}$ **32.** $\dfrac{-6x}{3x - 1}$

33. $\dfrac{30x}{x^2 - 36}$ **34.** $\dfrac{2x - 15}{x^2 - 49}$

35. $\dfrac{15}{x^2 + x - 2}$ **36.** $\dfrac{x - 20}{x^2 + 2x - 8}$

Simplify each expression, if possible.

37. $\dfrac{45}{9a}$ **38.** $\dfrac{48}{16y}$

39. $\dfrac{6x^2}{4x^2}$ **40.** $\dfrac{9xy}{6xy}$

41. $\dfrac{2x^2}{x + 2}$ **42.** $\dfrac{5y^2}{y + 5}$

43. $\dfrac{15x^2y}{5xy^2}$ **44.** $\dfrac{12xz}{4xz^2}$

45. $\dfrac{6x + 3}{3y}$ **46.** $\dfrac{4x + 12}{2y}$

47. $\dfrac{x + 3}{3x + 9}$ **48.** $\dfrac{2x - 14}{x - 7}$

49. $\dfrac{a^3 - a^2}{a^4 - a^3}$ **50.** $\dfrac{2c^4 + 2c^3}{4c^5 + 4c^4}$

51. $\dfrac{4x + 16}{5x + 20}$ **52.** $\dfrac{14x - 7}{12x - 6}$

53. $\dfrac{4c + 4d}{d + c}$ **54.** $\dfrac{a + b}{5b + 5a}$

55. $\dfrac{x - 7}{7 - x}$ **56.** $\dfrac{18 - d}{d - 18}$

57. $\dfrac{6x - 30}{5 - x}$ **58.** $\dfrac{6t - 42}{7 - t}$

59. $\dfrac{x^2 - 4}{x^2 - x - 2}$ **60.** $\dfrac{y^2 - 81}{y^2 + 10y + 9}$

61. $\dfrac{x^2 + 3x + 2}{x^2 + x - 2}$

62. $\dfrac{x^2 + x - 6}{x^2 - x - 2}$

63. $\dfrac{2x^2 - 8x}{x^2 - 6x + 8}$

64. $\dfrac{3y^2 - 15y}{y^2 - 3y - 10}$

65. $\dfrac{2 - a}{a^2 - a - 2}$

66. $\dfrac{4 - b}{b^2 - 5b + 4}$

67. $\dfrac{b^2 + 2b + 1}{(b + 1)^3}$

68. $\dfrac{y^2 - 4y + 4}{(y - 2)^3}$

69. $\dfrac{6x^2 - 13x + 6}{3x^2 + x - 2}$

70. $\dfrac{x^2 + 3x + 2}{x^3 + x^2}$

71. $\dfrac{10(c - 3) + 10}{3(c - 3) + 3}$

72. $\dfrac{6(d + 3) - 6}{7(d + 3) - 7}$

73. $\dfrac{6a + 3(a + 2) + 12}{a + 2}$

74. $\dfrac{2y + 4(y - 1) - 2}{y - 1}$

75. $\dfrac{3x^2 - 27}{2x^2 - 5x - 3}$

76. $\dfrac{2x^2 - 8}{3x^2 - 5x - 2}$

77. $\dfrac{-x^2 - 4x + 77}{x^2 - 4x - 21}$

78. $\dfrac{-5x^2 - 2x + 3}{x^2 + 2x - 15}$

79. $\dfrac{x(x - 8) + 16}{16 - x^2}$

80. $\dfrac{x^2 - 3(2x - 3)}{9 - x^2}$

81. $\dfrac{m^2 - 2mn + n^2}{2m^2 - 2n^2}$

82. $\dfrac{c^2 - d^2}{c^2 - 2cd + d^2}$

83. $\dfrac{16a^2 - 1}{4a + 4}$

84. $\dfrac{25m^2 - 1}{5m + 5}$

85. $\dfrac{8u^2 - 2u - 15}{4u^4 + 5u^3}$

86. $\dfrac{6n^2 - 7n + 2}{3n^3 - 2n^2}$

87. $\dfrac{y - xy}{xy - x}$

88. $\dfrac{x^2 + y^2}{x + y}$

89. $\dfrac{6a - 6b + 6c}{9a - 9b + 9c}$

90. $\dfrac{3a - 3b - 6}{2a - 2b - 4}$

91. $\dfrac{15x - 3x^2}{25y - 5xy}$

92. $\dfrac{a + b - c}{c - a - b}$

APPLICATIONS

93. ROOFING The *pitch* of a roof is a measure of how steep the roof is. If pitch $= \frac{\text{rise}}{\text{run}}$, find the pitch of the roof of the cabin shown. Express the result in lowest terms.

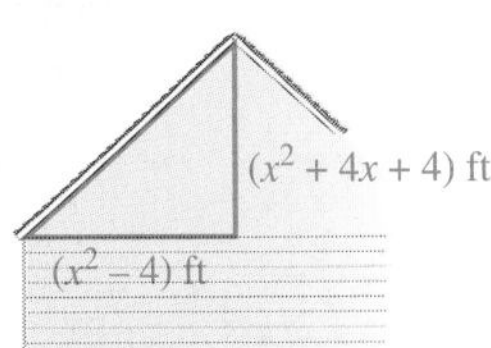

94. GRAPHIC DESIGN A chart of the basic food groups, in the shape of an equilateral triangle, is to be enlarged and distributed to schools for display in their health classes. What is the length of a side of the original design divided by the length of a side of the enlargement? Express the result in lowest terms.

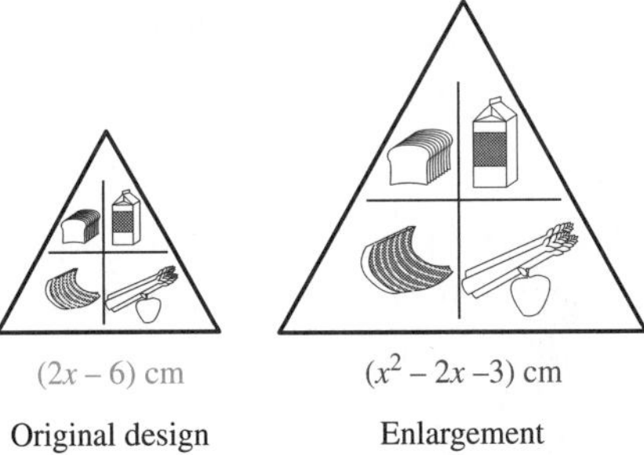

95. ORGAN PIPES The number of vibrations n per second of an organ pipe is given by the formula

$$n = \frac{512}{L}$$

where L is the length of the pipe in feet. How many times per second will a 6-foot pipe vibrate?

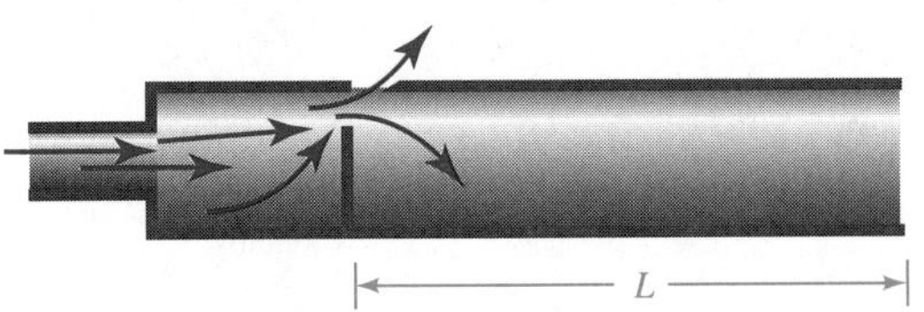

96. RAISING TURKEYS The formula

$$T = \frac{2{,}000m}{m + 1}$$

gives the number T of turkeys on a poultry farm m months after the beginning of the year. How many turkeys will there be on the farm by the end of July?

97. MEDICAL DOSAGES The formula

$$c = \frac{4t}{t^2 + 1}$$

gives the concentration c (in milligrams per liter) of a certain dosage of medication in a patient's blood stream t hours after the medication is administered. Suppose the patient received the medication at noon. Find the concentration of medication in his blood at the following times later that afternoon.

98. MANUFACTURING The average cost c (in dollars) for a company to produce x child car seats is given by the formula

$$c = \frac{50x + 50{,}000}{x}$$

Find the company's average cost to produce a car seat if 1,000 are produced.

WRITING

99. Explain why $\frac{x - 7}{7 - x} = -1$.

100. Explain the difference between a factor and a term. Give several examples.

101. Explain the error in the following work.

$$\frac{x}{x + 2} = \frac{\overset{1}{\cancel{x}}}{\underset{1}{\cancel{x}} + 2} = \frac{1}{3}$$

102. Explain why there are no values for x for which $\frac{x - 7}{x^2 + 49}$ is undefined.

REVIEW

103. State the associative property of addition using the variables a, b, and c.

104. State the distributive property using the variables x, y, and z.

105. If $ab = 0$, what must be true about a or b?

106. What is the product of a number and 1?

CHALLENGE PROBLEMS **Simplify each expression.**

107. $\frac{(x^2 + 2x + 1)(x^2 - 2x + 1)}{(x^2 - 1)^2}$

108. $\frac{2x^2 + 2x - 12}{x^3 + 3x^2 - 4x - 12}$

109. $\frac{x^3 - 27}{x^3 - 9x}$

110. $\frac{b^3 + a^3}{a^2 - ab + b^2}$

6.2 Multiplying and Dividing Rational Expressions

- Multiplying Rational Expressions
- Dividing Rational Expressions
- Unit Conversions

In this section, we will extend the rules for multiplying and dividing fractions to problems involving multiplication and division of rational expressions.

MULTIPLYING RATIONAL EXPRESSIONS

Recall that to multiply fractions, we multiply their numerators and multiply their denominators. For example,

$$\frac{4}{7} \cdot \frac{3}{5} = \frac{4 \cdot 3}{7 \cdot 5}$$ Multiply the numerators and multiply the denominators.

$$= \frac{12}{35}$$ Do the multiplication in the numerator. Do the multiplication in the denominator.

We use the same procedure to multiply rational expressions.

Multiplying Rational Expressions

Let A, B, C, and D represent polynomials, where B and D are not 0,

$$\frac{A}{B} \cdot \frac{C}{D} = \frac{AC}{BD}$$

Then simplify, if possible.

EXAMPLE 1

ELEMENTARY Algebra $f(x)$ Now™

Multiply: **a.** $\frac{x+1}{3} \cdot \frac{x+2}{x}$ and **b.** $\frac{35x^3}{17y} \cdot \frac{y}{5x}$.

Solution

a. $\frac{x+1}{3} \cdot \frac{x+2}{x} = \frac{(x+1)(x+2)}{3x}$ Multiply the numerators and multiply the denominators.

Since the numerator and denominator do not share any common factors, $\frac{(x+1)(x+2)}{3x}$ cannot be simplified. We will leave the numerator in factored form.

Caution

When multiplying rational expressions, always write the result in simplest form by removing any factors common to the numerator and denominator.

b. $\frac{35x^3}{17y} \cdot \frac{y}{5x} = \frac{35x^3 \cdot y}{17y \cdot 5x}$ Multiply the numerators and multiply the denominators.

Since $\frac{35x^3 \cdot y}{17y \cdot 5x}$ is not in simplest form, we simplify it as follows.

$$\frac{35x^3 \cdot y}{17y \cdot 5x} = \frac{5 \cdot 7 \cdot x \cdot x \cdot x \cdot y}{17 \cdot y \cdot 5 \cdot x}$$ Factor $35x^3$.

$$= \frac{\overset{1}{\cancel{5}} \cdot 7 \cdot \overset{1}{\cancel{x}} \cdot x \cdot x \cdot \overset{1}{\cancel{y}}}{17 \cdot \underset{1}{\cancel{y}} \cdot \underset{1}{\cancel{5}} \cdot \underset{1}{\cancel{x}}}$$ Replace $\frac{5}{5}$, $\frac{x}{x}$, and $\frac{y}{y}$ with the equivalent fraction $\frac{1}{1}$. This removes the factor $\frac{5 \cdot x \cdot y}{5 \cdot x \cdot y}$, which is equal to 1.

$$= \frac{7x^2}{17}$$

Self Check 1 Multiply and simplify the result, if possible: **a.** $\frac{y}{y+6} \cdot \frac{12}{y-4}$ and **b.** $\frac{a^4}{8b} \cdot \frac{24b}{11a^3}$.

EXAMPLE 2

ELEMENTARY Algebra $f(x)$ Now™

Multiply: **a.** $\frac{x+3}{2x+4} \cdot \frac{6}{x^2-9}$ and **b.** $\frac{8x^2-8x}{x^2+x-56} \cdot \frac{3x^2-22x+7}{x-x^2}$.

Solution

a. $\frac{x+3}{2x+4} \cdot \frac{6}{x^2-9} = \frac{(x+3)6}{(2x+4)(x^2-9)}$ Multiply the numerators and multiply the denominators.

$$= \frac{(x+3)\cdot 3\cdot 2}{2(x+2)(x+3)(x-3)}$$ Factor 6. Factor out the GCF from $2x+4$. Factor the difference of two squares, x^2-9.

$$= \frac{\overset{1}{\cancel{(x+3)}}\cdot 3\cdot \overset{1}{\cancel{2}}}{\underset{1}{\cancel{2}}(x+2)\underset{1}{\cancel{(x+3)}}(x-3)}$$ Simplify by replacing $\frac{x+3}{x+3}$ and $\frac{2}{2}$ with $\frac{1}{1}$. This removes the factor $\frac{2\cdot(x+3)}{2\cdot(x+3)}$, which is equal to 1.

$$= \frac{3}{(x+2)(x-3)}$$

b. $\frac{8x^2-8x}{x^2+x-56}\cdot\frac{3x^2-22x+7}{x-x^2}$

$$= \frac{(8x^2-8x)(3x^2-22x+7)}{(x^2+x-56)(x-x^2)}$$ Multiply the numerators and multiply the denominators.

$$= \frac{8x(x-1)(3x-1)(x-7)}{(x+8)(x-7)x(1-x)}$$ Factor all four polynomials.

$$= \frac{8\overset{1}{\cancel{x}}\overset{-1}{\cancel{(x-1)}}(3x-1)\overset{1}{\cancel{(x-7)}}}{(x+8)\underset{1}{\cancel{(x-7)}}\underset{1}{\cancel{x}}\underset{1}{\cancel{(1-x)}}}$$ Since $x-1$ and $1-x$ are opposites, simplify by replacing $\frac{x-1}{1-x}$ with $\frac{-1}{1}$.

$$= \frac{-8(3x-1)}{x+8}$$

The result can also be written as $-\frac{8(3x-1)}{x+8}$.

Self Check 2 Multiply and simplify the result, if possible: **a.** $\frac{3n-9}{3n+2}\cdot\frac{9n^2-4}{6}$ and **b.** $\frac{m^2-4m-5}{2m-m^2}\cdot\frac{2m^2-4m}{3m^2-14m-5}$.

EXAMPLE 3

Multiply: **a.** $63x\left(\frac{1}{7x}\right)$ and **b.** $5a\left(\frac{3a-1}{a}\right)$.

Solution **a.** Since any number divided by 1 remains unchanged, we can write any polynomial as a fraction by inserting a denominator of 1.

$$63x\left(\frac{1}{7x}\right) = \frac{63x}{1}\left(\frac{1}{7x}\right)$$ Write $63x$ as a fraction: $63x = \frac{63x}{1}$.

$$= \frac{63x\cdot 1}{1\cdot 7\cdot x}$$ Multiply the numerators and the denominators.

$$= \frac{9\cdot\overset{1}{\cancel{7}}\cdot\overset{1}{\cancel{x}}\cdot 1}{1\cdot\underset{1}{\cancel{7}}\cdot\underset{1}{\cancel{x}}}$$ Write $63x$ in factored form as $9\cdot 7\cdot x$. Then simplify by removing a factor equal to 1: $\frac{7x}{7x}$.

$$= 9$$

b. $5a\left(\frac{3a-1}{a}\right) = \frac{5a}{1}\left(\frac{3a-1}{a}\right)$ Write $5a$ as a fraction: $5a = \frac{5a}{1}$.

$$= \frac{5\overset{1}{\cancel{a}}(3a-1)}{1\cdot\underset{1}{\cancel{a}}}$$ Multiply the numerators and the denominators. Then simplify by removing a factor equal to 1: $\frac{a}{a}$.

$= 5(3a - 1)$

$= 15a - 5$ Distribute the multiplication by 5.

Self Check 3 Multiply and simplify the result, if possible: **a.** $36b\left(\frac{1}{6b}\right)$ and **b.** $4x\left(\frac{x + 3}{x}\right)$.

DIVIDING RATIONAL EXPRESSIONS

Recall that one number is called the **reciprocal** of another if their product is 1. To find the reciprocal of a fraction, we invert its numerator and denominator. We have seen that to divide fractions, we multiply the first fraction by the reciprocal of the second fraction.

$\frac{4}{7} \div \frac{3}{5} = \frac{4}{7} \cdot \frac{5}{3}$ Invert $\frac{3}{5}$ and change the division to a multiplication.

$= \frac{20}{21}$ Multiply the numerators and multiply the denominators.

We use the same procedure to divide rational expressions.

Dividing Rational Expressions

Let A, B, C, and D represent polynomials, where B, C, and D are not 0,

$$\frac{A}{B} \div \frac{C}{D} = \frac{A}{B} \cdot \frac{D}{C} = \frac{AD}{BC}$$

Then simplify, if possible.

EXAMPLE 4

ELEMENTARY Algebra f(x) Now™

Divide: **a.** $\frac{a}{13} \div \frac{17}{26}$ and **b.** $\frac{9x}{35y} \div \frac{15x^2}{14}$.

Solution

a. $\frac{a}{13} \div \frac{17}{26} = \frac{a}{13} \cdot \frac{26}{17}$ Multiply by the reciprocal of $\frac{17}{26}$.

$= \frac{a \cdot 2 \cdot 13}{13 \cdot 17}$ Multiply the numerators and denominators. Then factor 26 as $2 \cdot 13$.

$= \frac{a \cdot 2 \cdot \overset{1}{\cancel{13}}}{\underset{1}{\cancel{13}} \cdot 17}$ Simplify by removing a factor equal to 1.

$= \frac{2a}{17}$

Caution

When dividing rational expressions, always write the result in simplest form (lowest terms), by removing any factors common to the numerator and denominator.

b. $\frac{9x}{35y} \div \frac{15x^2}{14} = \frac{9x}{35y} \cdot \frac{14}{15x^2}$ Multiply by the reciprocal of $\frac{15x^2}{14}$.

$= \frac{3 \cdot 3 \cdot x \cdot 2 \cdot 7}{5 \cdot 7 \cdot y \cdot 3 \cdot 5 \cdot x \cdot x}$ Multiply the numerators and denominators. Then factor 9, 35, 14, and $15x^2$.

$= \frac{3 \cdot \overset{1}{\cancel{3}} \cdot \overset{1}{\cancel{x}} \cdot 2 \cdot \overset{1}{\cancel{7}}}{5 \cdot \underset{1}{\cancel{7}} \cdot y \cdot \underset{1}{\cancel{3}} \cdot 5 \cdot \underset{1}{\cancel{x}} \cdot x}$ Simplify by removing factors equal to 1.

$= \frac{6}{25xy}$

Self Check 4 Divide and simplify the result, if possible: $\frac{8a}{3b} \div \frac{16a^2}{9b^2}$.

EXAMPLE 5

ELEMENTARY Algebra Now™

Divide: $\frac{x^2 + x}{3x - 15} \div \frac{x^2 + 2x + 1}{6x - 30}$.

Solution

$$\frac{x^2 + x}{3x - 15} \div \frac{x^2 + 2x + 1}{6x - 30}$$

$$= \frac{x^2 + x}{3x - 15} \cdot \frac{6x - 30}{x^2 + 2x + 1}$$ Multiply by the reciprocal of $\frac{x^2 + 2x + 1}{6x - 30}$.

$$= \frac{x(x + 1) \cdot 2 \cdot 3(x - 5)}{3(x - 5)(x + 1)(x + 1)}$$ Multiply the numerators and denominators. Then factor the polynomials.

$$= \frac{x\overset{1}{\cancel{(x + 1)}} \cdot 2 \cdot \overset{1}{\cancel{3}}\overset{1}{\cancel{(x - 5)}}}{\underset{1}{\cancel{3}}\underset{1}{\cancel{(x - 5)}}\underset{1}{\cancel{(x + 1)}}(x + 1)}$$ Simplify by removing factors equal to 1.

$$= \frac{2x}{x + 1}$$

The Language of Algebra

To find the reciprocal of $\frac{x^2 + 2x + 1}{6x - 30}$, we invert it. To *invert* means to turn upside down: $\frac{6x - 30}{x^2 + 2x + 1}$. Some amusement park thrill rides have giant loops where the riders become *inverted.*

Self Check 5 Divide and simplify the result, if possible: $\frac{z^2 - 1}{z^2 + 4z + 3} \div \frac{z - 1}{z^2 + 2z - 3}$.

EXAMPLE 6

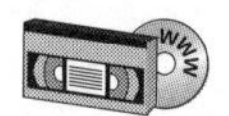

ELEMENTARY Algebra Now™

Divide: $\frac{2x^2 - 3x - 2}{2x + 1} \div (4 - x^2)$.

Solution To divide a rational expression by a polynomial, we write the polynomial as a fraction by inserting a denominator of 1, and then we divide the fractions.

$$\frac{2x^2 - 3x - 2}{2x + 1} \div (4 - x^2)$$

$$= \frac{2x^2 - 3x - 2}{2x + 1} \div \frac{4 - x^2}{1}$$ Write $4 - x^2$ as a fraction with a denominator of 1.

$$= \frac{2x^2 - 3x - 2}{2x + 1} \cdot \frac{1}{4 - x^2}$$ Multiply by the reciprocal of $\frac{4 - x^2}{1}$.

$$= \frac{(2x + 1)(x - 2) \cdot 1}{(2x + 1)(2 + x)(2 - x)}$$ Multiply the numerators and denominators. Then factor $2x^2 - 3x - 2$ and $4 - x^2$.

$$= \frac{\overset{1}{\cancel{(2x + 1)}}\overset{-1}{\cancel{(x - 2)}} \cdot 1}{\underset{1}{\cancel{(2x + 1)}}(2 + x)\underset{1}{\cancel{(2 - x)}}}$$ Since $x - 2$ and $2 - x$ are opposites, simplify by replacing $\frac{x - 2}{2 - x}$ with $\frac{-1}{1}$.

$$= \frac{-1}{2 + x}$$

$$= -\frac{1}{2 + x}$$ Write the $-$ symbol in front of the rational expression.

Self Check 6 Divide and simplify the result, if possible: $(b - a) \div \frac{a^2 - b^2}{a^2 + ab}$.

UNIT CONVERSIONS

Success Tip

Remember that unit conversion factors are equal to 1. Here are some examples:

$\frac{12 \text{ in.}}{1 \text{ ft}} \quad \frac{60 \text{ min}}{1 \text{ hr}} \quad \frac{1 \text{ decade}}{10 \text{ yr}}$

We can use the concepts discussed in this section to make conversions from one unit of measure to another. *Unit conversion factors* play an important role in this process. A **unit conversion factor** is a fraction that has value 1. For example, we can use the fact that 1 square yard = 9 square feet to form two unit conversion factors:

$\frac{1 \text{ yd}^2}{9 \text{ ft}^2} = 1$ Read as "1 square yard per 9 square feet."

$\frac{9 \text{ ft}^2}{1 \text{ yd}^2} = 1$ Read as "9 square feet per 1 square yard."

Since a unit conversion factor is equal to 1, multiplying a measurement by a unit conversion factor does not change the measurement, it only changes the units of measure.

EXAMPLE 7

ELEMENTARY Algebra $f(x)$ Now™

Carpeting. A roll of carpeting is 12 feet wide and 150 feet long. Find the number of square yards of carpeting on the roll.

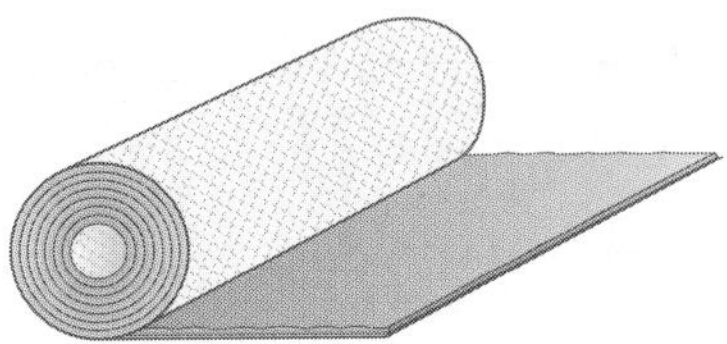

Solution When unrolled, the carpeting forms a rectangular shape with an area of $12 \cdot 150 = 1{,}800$ square feet. We multiply 1,800 ft^2 by a unit conversion factor such that the units of ft^2 are removed and the units of yd^2 are introduced. Then we can remove units common to the numerator and denominator.

$$1{,}800 \text{ ft}^2 = \frac{1{,}800 \text{ ft}^2}{1} \cdot \frac{1 \text{ yd}^2}{9 \text{ ft}^2}$$ Multiply by a unit conversion factor that relates yd^2 to ft^2.

$$= \frac{1{,}800 \cancel{\text{ft}^2}}{1} \cdot \frac{1 \text{ yd}^2}{9 \cancel{\text{ft}^2}}$$ Remove the units of ft^2 that are common to the numerator and denominator.

$$= 200 \text{ yd}^2$$ Divide 1,800 by 9.

There are 200 yd^2 of carpeting on the roll.

EXAMPLE 8

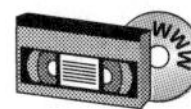

The speed of light. The speed with which light moves through space is about 186,000 miles per second. Express this speed in miles per minute.

Solution The speed of light can be expressed as $\frac{186{,}000 \text{ miles}}{1 \text{ sec}}$. We multiply this fraction by a unit conversion factor such that the units of seconds are removed and the units of minutes are introduced. Since 60 seconds = 1 minute, we will use $\frac{60 \text{ sec}}{1 \text{ min}}$.

$$\frac{186{,}000 \text{ miles}}{1 \text{ sec}} = \frac{186{,}000 \text{ miles}}{1 \text{ sec}} \cdot \frac{60 \text{ sec}}{1 \text{ min}}$$ Multiply by a unit conversion factor that relates seconds to minutes.

$$= \frac{186{,}000 \text{ miles}}{1 \cancel{\text{sec}}} \cdot \frac{60 \cancel{\text{sec}}}{1 \text{ min}}$$ Remove the units of seconds that are common to the numerator and denominator.

$$= \frac{11{,}160{,}000 \text{ miles}}{1 \text{ min}}$$ Multiply 186,000 and 60.

The speed of light is about 11,160,000 miles per minute.

Answers to Self Checks **1. a.** $\frac{12y}{(y+6)(y-4)}$, **b.** $\frac{3a}{11}$ **2. a.** $\frac{(n-3)(3n-2)}{2}$, **b.** $-\frac{2(m+1)}{3m+1}$ **3. a.** 6, **b.** $4x + 12$ **4.** $\frac{3b}{2a}$ **5.** $z - 1$ **6.** $-a$

6.2 STUDY SET  ELEMENTARY Algebra f(x) Now™

VOCABULARY **Fill in the blanks.**

1. The quotient of ________ is -1. For example, $\frac{x-8}{8-x} = -1$.
2. The ________ of $\frac{x^2+6x+1}{10x}$ is $\frac{10x}{x^2+6x+1}$.
3. To find the reciprocal of a rational expression, we ______ its numerator and denominator.
4. A _____ conversion factor is a fraction that is equal to 1, such as $\frac{3 \text{ ft}}{1 \text{ yd}}$.

CONCEPTS **Fill in the blanks.**

5. To multiply rational expressions, multiply their __________ and multiply their ____________. In symbols,

$$\frac{A}{B} \cdot \frac{C}{D} = \boxed{}$$

6. To divide two rational expressions, multiply the first by the _________ of the second. In symbols,

$$\frac{A}{B} \div \frac{C}{D} = \frac{A}{B} \cdot \boxed{}$$

Simplify each fraction.

7. $\frac{(x+7) \cdot 2 \cdot 5}{5(x+1)(x+7)(x-9)}$
8. $\frac{y \cdot y \cdot y(15-y)}{y(y-15)(y+1)}$

Write each polynomial in fractional form.

9. $6n$
10. $3x + 5$

Find the reciprocal of each polynomial.

11. $18x$
12. $\frac{7m+11}{m-2}$
13. $\frac{x^2+1}{x^2+2x+1}$

14. Find the product of the rational expression and its reciprocal.

$$\frac{3}{x+2} \cdot \frac{x+2}{3}$$

15. What units are common to the numerator and denominator?

$$\frac{45 \text{ ft}}{1} \cdot \frac{1 \text{ yd}}{3 \text{ ft}}$$

16. Use the following fact to write two unit conversion factors.

1 square mile = 640 acres

NOTATION

17. The abbreviation for meter is m.
    a. How do we write square meters in symbols?
    b. How do we write cubic meters in symbols?
18. a. What fact is indicated by the unit conversion factor $\frac{1 \text{ day}}{24 \text{ hours}}$?
    b. Fill in the blank: $\frac{1 \text{ day}}{24 \text{ hours}} = \boxed{}$.

PRACTICE **Multiply, and then simplify, if possible.**

19. $\frac{3}{y} \cdot \frac{y}{2}$
20. $\frac{2}{z} \cdot \frac{z}{3}$
21. $\frac{35n}{12} \cdot \frac{16}{7n^2}$
22. $\frac{11m}{21} \cdot \frac{14}{55m^3}$
23. $\frac{2x^2y}{3xy} \cdot \frac{3xy^2}{2}$
24. $\frac{2x^2z}{z} \cdot \frac{5x}{z}$
25. $\frac{10r^2st^3}{6rs^2} \cdot \frac{3r^3t}{2rst}$
26. $\frac{3a^3b}{25cd^3} \cdot \frac{5cd^2}{6ab}$
27. $\frac{z+7}{7} \cdot \frac{z+2}{z}$
28. $\frac{a-3}{a} \cdot \frac{a+3}{5}$

29. $\frac{x+5}{5} \cdot \frac{x}{x+5}$

30. $\frac{a-9}{9} \cdot \frac{8a}{a-9}$

31. $\frac{x-2}{2} \cdot \frac{2x}{2-x}$

32. $\frac{y-3}{y} \cdot \frac{3y}{3-y}$

33. $\frac{5}{m} \cdot m$

34. $p \cdot \frac{10}{p}$

35. $15x\left(\frac{x+1}{15x}\right)$

36. $30t\left(\frac{t-7}{30t}\right)$

37. $12y\left(\frac{y+8}{6y}\right)$

38. $16x\left(\frac{3x+8}{4x}\right)$

39. $(x+8)\frac{x+5}{x+8}$

40. $(y-2)\frac{y+3}{y-2}$

41. $10(h+9)\frac{h-3}{h+9}$

42. $r(r-25)\frac{r+4}{r-25}$

43. $\frac{(x+1)^2}{x+1} \cdot \frac{x+2}{x+1}$

44. $\frac{(y-3)^2}{y-3} \cdot \frac{y-3}{y-3}$

45. $\frac{2x+6}{x+3} \cdot \frac{3}{4x}$

46. $\frac{3y-9}{y-3} \cdot \frac{y}{3y^2}$

47. $\frac{x^2-x}{x} \cdot \frac{3x-6}{3-3x}$

48. $\frac{5z-10}{z+2} \cdot \frac{3}{6-3z}$

49. $\frac{x^2+x-6}{5x} \cdot \frac{5x-10}{x+3}$

50. $\frac{z^2+4z-5}{5z-5} \cdot \frac{5z}{z+5}$

51. $\frac{m^2-2m-3}{2m+4} \cdot \frac{m^2-4}{m^2+3m+2}$

52. $\frac{p^2-p-6}{3p-9} \cdot \frac{p^2-9}{p^2+6p+9}$

53. $\frac{2x^2+17x+21}{x^2+2x-35} \cdot \frac{x^2-25}{2x^2-7x-15}$

54. $\frac{2x^2-9x+9}{x^2-1} \cdot \frac{2x^2-5x+3}{x^2-3x}$

55. $\frac{4x^2-12xy+9y^2}{x^3y^2} \cdot \frac{x^2y}{9y^2-4x^2}$

56. $\frac{ab^4}{16b^2-25a^2} \cdot \frac{25a^2-40ab+16b^2}{a^2b^5}$

57. $\frac{3x^2+5x+2}{x^2-9} \cdot \frac{x-3}{x^2-4} \cdot \frac{x^2+5x+6}{6x+4}$

58. $\frac{x^2-25}{3x+6} \cdot \frac{x^2+x-2}{2x+10} \cdot \frac{6x}{3x^2-18x+15}$

Divide, then simplify, if possible.

59. $\frac{2}{y} \div \frac{4}{3}$

60. $\frac{3}{a} \div \frac{a}{9}$

61. $\frac{3x}{y} \div \frac{2x}{4}$

62. $\frac{3y}{8} \div \frac{2y}{4y}$

63. $\frac{x^2y}{3xy} \div \frac{xy^2}{6y}$

64. $\frac{2xz}{z} \div \frac{4x^2}{z^2}$

65. $24n^2 \div \frac{18n^3}{n-1}$

66. $12m \div \frac{16m^2}{m+4}$

67. $\frac{x+2}{3x} \div \frac{x+2}{2}$

68. $\frac{z-3}{3z} \div \frac{z+3}{z}$

69. $\frac{(z-2)^2}{3z^2} \div \frac{z-2}{6z}$

70. $\frac{(x-3)^4}{x+7} \div \frac{(x-3)^2}{x+7}$

71. $\frac{9a-18}{28} \div \frac{9a^3}{35}$

72. $\frac{3x+6}{40} \div \frac{3x^2}{24}$

73. $\frac{x^2-4}{3x+6} \div \frac{2-x}{x+2}$

74. $\frac{x^2-9}{5x+15} \div \frac{3-x}{x+3}$

75. $\frac{x^2-1}{3x-3} \div (x+1)$

76. $\frac{x^2-16}{x-4} \div (3x+12)$

77. $\frac{x^2-2x-35}{3x^2+27x} \div \frac{x^2+7x+10}{6x^2+12x}$

78. $\frac{x^2-x-6}{2x^2+9x+10} \div \frac{x^2-25}{2x^2+15x+25}$

79. $\frac{36c^2-49d^2}{3d^3} \div \frac{12c+14d}{d^4}$

80. $\frac{25y^2-16z^2}{2yz} \div \frac{10y-8z}{y^2}$

81. $\frac{2d^2+8d-42}{3-d} \div \frac{2d^2+14d}{d^2+5d}$

82. $\dfrac{5x^2 + 13x - 6}{x + 3} \div \dfrac{5x^2 - 17x + 6}{x - 2}$

83. $\dfrac{2r - 3s}{12} \div (4r^2 - 12rs + 9s^2)$

84. $\dfrac{3m + n}{18} \div (9m^2 + 6mn + n^2)$

Complete each unit conversion.

85. $\dfrac{150 \text{ yd}}{1} \cdot \dfrac{3 \text{ ft}}{1 \text{ yd}} = ?$

86. $\dfrac{60 \text{ in.}}{1} \cdot \dfrac{1 \text{ ft}}{12 \text{ in.}} = ?$

87. $\dfrac{30 \text{ meters}}{1 \text{ sec}} \cdot \dfrac{60 \text{ sec}}{1 \text{ min}} = ?$

88. $\dfrac{288 \text{ in.}^2}{1 \text{ year}} \cdot \dfrac{1 \text{ ft}^2}{144 \text{ in.}^2} = ?$

APPLICATIONS

89. INTERNATIONAL ALPHABET The symbols representing the letters A, B, C, D, E, and F in an international code used at sea are printed six to a sheet and then cut into separate cards. If each card is a square, find the area of the large printed sheet shown.

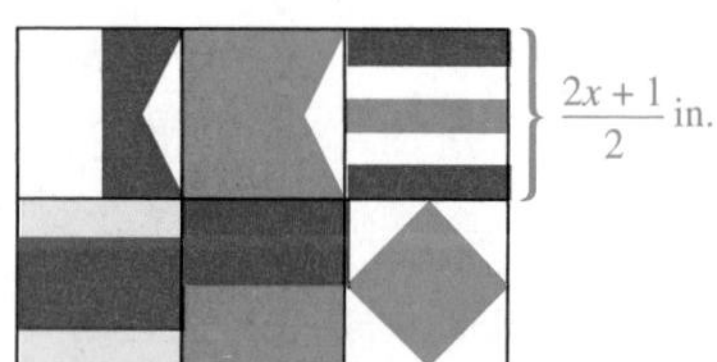

90. PHYSICS EXPERIMENTS The table contains algebraic expressions for the rate an object travels, and the time traveled at that rate in terms of a constant k. Complete the table.

Rate (mph)	Time (hr)	Distance (mi)
$\dfrac{k^2 + k - 6}{k - 3}$	$\dfrac{k^2 - 9}{k^2 - 4}$	

91. TALKING According to the *Sacramento Bee* newspaper, the number of words an average man speaks a day is about 12,000. How many words does an average man speak in 1 year? (*Hint:* 365 days = 1 year.)

92. CLASSROOM SPACE The recommended size of an elementary school classroom in the United States is approximately 900 square feet. Convert this to square yards.

93. NATURAL LIGHT According to the University of Georgia School Design and Planning Laboratory, the basic classroom should have at least 72 square feet of windows for natural light. Convert this to square yards.

94. TRUCKING A cement truck holds 9 cubic yards of concrete. How many cubic feet of concrete does it hold? (*Hint:* 27 cubic feet = 1 cubic yard.)

95. BEARS The maximum speed a grizzly bear can run is about 30 miles per hour. What is its average speed in miles per minute?

96. FUEL ECONOMY Use the information that follows to determine the miles per fluid ounce of gasoline for city and for highway driving for the Chevy Astro Van. (*Hint:* 1 gallon = 128 fluid ounces.)

2003 Chevrolet Astro 2WD

Fuel Economy

Fuel Type	Regular
MPG (city)	**16**
MPG (highway)	**20**

97. TV TRIVIA On the comedy television series *Green Acres* (1965–1971), New York socialites Oliver Wendell Douglas (played by Eddie Albert) and his wife, Lisa Douglas (played by Eva Gabor), move from New York to purchase a 160-acre farm in Hooterville. Convert this to square miles. (*Hint:* 1 square mile = 640 acres.)

98. CAMPING The capacity of backpacks is usually given in cubic inches. Convert a backpack capacity of 5,400 cubic inches to cubic feet. (*Hint:* 1 cubic foot = 1,728 cubic inches.)

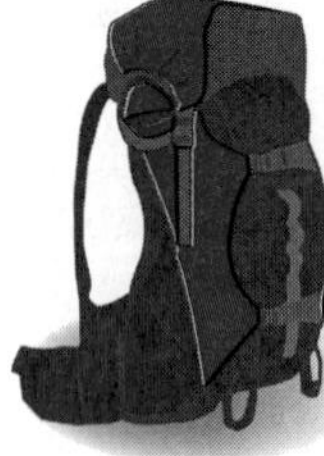

WRITING

99. Explain how to multiply rational expressions.

100. To divide rational expressions, you must first know how to multiply rational expressions. Explain why.

101. Explain how to find the reciprocal of a rational expression.

102. Explain how to write $x^2 + 2x - 1$ as a fraction.

103. Explain why 60 miles per hour and 1 mile per minute are the same speed.

104. Explain why $\frac{1 \text{ ft}}{12 \text{ in.}}$ is equal to 1.

REVIEW

105. HARDWARE An aluminum brace used to support a wooden shelf has a length that is 2 inches less than twice the width of the shelf. The brace is anchored to the wall 8 inches below the shelf, as shown. Find the width of the shelf and the length of the brace.

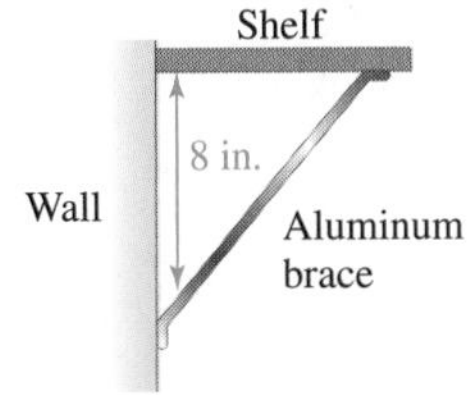

106. FORENSIC MEDICINE The kinetic energy E of a moving object is given by $E = \frac{1}{2}mv^2$, where m is the mass of the object (in kilograms) and v is the object's velocity (in meters per second). Kinetic energy is measured in joules. Examining the damage done to a victim, a police pathologist determines that the energy of a 3-kilogram mass at impact was 54 joules. Find the velocity at impact.

CHALLENGE PROBLEMS **Perform the operations. Simplify, if possible.**

107. $\dfrac{c^3 - 2c^2 + 5c - 10}{c^2 - c - 2} \cdot \dfrac{c^3 + c^2 - 5c - 5}{c^4 - 25}$

108. $\dfrac{x^3 - y^3}{x^3 + y^3} \div \dfrac{x^3 + x^2y + xy^2}{x^2y - xy^2 + y^3}$

109. $\dfrac{y^2}{x + 1} \cdot \dfrac{x^2 + 2x + 1}{x^2 - 1} \div \dfrac{3y}{xy - y}$

110. $\dfrac{x^2 - y^2}{x^4 - x^3} \div \dfrac{x - y}{x^2} \div \dfrac{x^2 + 2xy + y^2}{x + y}$

6.3 Addition and Subtraction with Like Denominators; Least Common Denominators

- Adding and Subtracting Rational Expressions Having Like Denominators
- Finding the Least Common Denominator
- Building Rational Expressions

In this section, we extend the rules for adding and subtracting fractions to problems involving addition and subtraction of rational expressions.

ADDING AND SUBTRACTING RATIONAL EXPRESSIONS HAVING LIKE DENOMINATORS

To add or subtract fractions that have a common denominator, we add or subtract their numerators and write the sum or difference over the common denominator. For example,

$$\frac{3}{7} + \frac{2}{7} = \frac{3 + 2}{7} = \frac{5}{7} \qquad \frac{3}{7} - \frac{2}{7} = \frac{3 - 2}{7} = \frac{1}{7}$$

We use the same procedure to add and subtract rational expressions with like denominators.

Adding and Subtracting Rational Expressions

If $\frac{A}{D}$ and $\frac{B}{D}$ are rational expressions,

$$\frac{A}{D} + \frac{B}{D} = \frac{A + B}{D} \qquad \frac{A}{D} - \frac{B}{D} = \frac{A - B}{D}$$

Then simplify, if possible.

EXAMPLE 1

Add: **a.** $\frac{x}{8} + \frac{3x}{8}$ and **b.** $\frac{4s - 9}{9t} + \frac{7}{9t}$.

ELEMENTARY Algebra f(x) Now™

Solution **a.** The rational expressions have the same denominator, 8. We add their numerators and write the sum over the common denominator.

Caution

When adding or subtracting rational expressions, always write the result in simplest form by removing any factors common to the numerator and denominator.

$$\frac{x}{8} + \frac{3x}{8} = \frac{x + 3x}{8}$$

$$= \frac{4x}{8} \quad \text{Combine like terms in the numerator.}$$

This result can be simplified as follows:

$$= \frac{\overset{1}{\cancel{4}} \cdot x}{\underset{1}{\cancel{4}} \cdot 2} \quad \text{Factor 8 as } 4 \cdot 2\text{. Then simplify by removing a factor equal to 1.}$$

$$= \frac{x}{2}$$

Notation

In Example 1b, the numerator of the result may be written two ways:

Not factored		Factored
$\frac{4s - 2}{9t}$	or	$\frac{2(2s - 1)}{9t}$

Check with your instructor to see which form he or she prefers.

b. $$\frac{4s - 9}{9t} + \frac{7}{9t} = \frac{4s - 9 + 7}{9t} \quad \text{Add the numerators. Write the sum over the common denominator, } 9t.$$

$$= \frac{4s - 2}{9t} \quad \text{Combine like terms in the numerator.}$$

To attempt to simplify the result, we factor the numerator to get $\frac{2(2s - 1)}{9t}$. Since the numerator and denominator do not have any common factors, $\frac{4s - 2}{9t}$ cannot be simplified. Thus,

$$\frac{4s - 9}{9t} + \frac{7}{9t} = \frac{4s - 2}{9t} \quad \text{or} \quad \frac{2(2s - 1)}{9t}$$

Self Check 1 Add and simplify the result, if possible: **a.** $\frac{2x}{15} + \frac{4x}{15}$ and **b.** $\frac{3m - 8}{23n} + \frac{2}{23n}$. ■

EXAMPLE 2

Add: **a.** $\frac{3x + 21}{5x + 10} + \frac{8x + 1}{5x + 10}$ and **b.** $\frac{x^2 + 9x - 7}{2x(x - 6)} + \frac{x^2 - 9x}{(x - 6)2x}$.

ELEMENTARY Algebra f(x) Now™

Solution **a.** $$\frac{3x + 21}{5x + 10} + \frac{8x + 1}{5x + 10} = \frac{3x + 21 + 8x + 1}{5x + 10} \quad \text{Add the numerators. Write the sum over the common denominator, } 5x + 10.$$

$$= \frac{11x + 22}{5x + 10} \quad \text{Combine like terms in the numerator.}$$

$$= \frac{11\overset{1}{\cancel{(x + 2)}}}{5\underset{1}{\cancel{(x + 2)}}} \quad \text{Factor the numerator and the denominator. Then simplify by removing a factor equal to 1.}$$

$$= \frac{11}{5}$$

b. By the commutative property of multiplication, $2x(x - 6) = (x - 6)2x$. Therefore, the denominators are the same. We add the numerators and write the sum over the common denominator.

$$\frac{x^2 + 9x - 7}{2x(x - 6)} + \frac{x^2 - 9x}{(x - 6)2x} = \frac{x^2 + 9x - 7 + x^2 - 9x}{2x(x - 6)}$$

$$= \frac{2x^2 - 7}{2x(x - 6)}$$ Combine like terms in the numerator.

Since the numerator, $2x^2 - 7$, does not factor, $\frac{2x^2 - 7}{2x(x - 6)}$ is in simplest form.

Self Check 2 Add and simplify the result, if possible: **a.** $\frac{m + 3}{3m - 9} + \frac{m - 9}{3m - 9}$ and **b.** $\frac{c^2 - c}{(c - 1)(c + 2)} + \frac{c^2 - 10c}{(c + 2)(c - 1)}$.

EXAMPLE 3

Subtract: $\frac{x + 6}{x^2 + 4x - 5} - \frac{1}{x^2 + 4x - 5}$.

ELEMENTARY Algebra $f(x)$ Now™

Solution Because the rational expressions have the same denominator, we subtract their numerators and keep the common denominator.

$$\frac{x + 6}{x^2 + 4x - 5} - \frac{1}{x^2 + 4x - 5} = \frac{x + 6 - 1}{x^2 + 4x - 5}$$ Subtract the numerators. Write the difference over the common denominator, $x^2 + 4x - 5$.

$$= \frac{x + 5}{x^2 + 4x - 5}$$ Combine like terms in the numerator.

$$= \frac{x + 5}{(x + 5)(x - 1)}$$ Factor the denominator.

$$= \frac{\overset{1}{\cancel{x + 5}}}{\underset{1}{(\cancel{x + 5})}(x - 1)}$$ Simplify by removing a factor equal to 1.

$$= \frac{1}{x - 1}$$

Self Check 3 Subtract and simplify the result, if possible: $\frac{n - 3}{n^2 - 16} - \frac{1}{n^2 - 16}$.

EXAMPLE 4

Subtract: **a.** $\frac{x^2 + 10x}{x + 3} - \frac{4x - 9}{x + 3}$ and **b.** $\frac{x^2}{(x + 7)(x - 8)} - \frac{x^2 + 14x}{(x + 7)(x - 8)}$.

ELEMENTARY Algebra $f(x)$ Now™

Solution **a.** To subtract the numerators, each term of $4x - 9$ must be subtracted from $x^2 + 10x$.

This sign applies to the entire numerator $4x - 9$.

This numerator is written within parentheses to make sure that we subtract both of its terms.

$$\frac{x^2 + 10x}{x + 3} - \frac{4x - 9}{x + 3} = \frac{x^2 + 10x - (4x - 9)}{x + 3}$$ Subtract the numerators. Write the difference over the common denominator.

$$= \frac{x^2 + 10x - 4x + 9}{x + 3}$$ In the numerator, use the distributive property: $-(4x - 9) = -1(4x - 9) = -4x + 9$.

$$= \frac{x^2 + 6x + 9}{x + 3}$$ Combine like terms in the numerator.

$$= \frac{(x + 3)(x + 3)}{x + 3}$$ Factor the numerator.

$$= \frac{\overset{1}{\cancel{(x + 3)}}(x + 3)}{\underset{1}{\cancel{x + 3}}}$$ Simplify by removing a factor equal to 1.

$$= x + 3$$

Notation

A fraction bar groups the terms of the numerator together and the terms of the denominator together. It is helpful to think of an implied pair of parentheses being around each.

$$\frac{4x - 9}{x + 3} = \frac{(4x - 9)}{(x + 3)}$$

b. We subtract the numerators and write the difference over the common denominator.

$$\frac{x^2}{(x + 7)(x - 8)} - \frac{x^2 + 14x}{(x + 7)(x - 8)} = \frac{x^2 - (x^2 + 14x)}{(x + 7)(x - 8)}$$ Write the second numerator within parentheses.

$$= \frac{x^2 - x^2 - 14x}{(x + 7)(x - 8)}$$ Use the distributive property: $-(x^2 + 14x) = -x^2 - 14x$.

$$= \frac{-14x}{(x + 7)(x - 8)}$$ In the numerator, combine like terms.

The result can also be written as $-\frac{14x}{(x + 7)(x - 8)}$.

Success Tip

In Example 4b, $-(x^2 + 14x)$ means $-1(x^2 + 14x)$. The multiplication by -1 changes the signs of x^2 and $14x$:

$$-(x^2 + 14x) = -x^2 - 14x$$

Self Check 4 Subtract and simplify the result, if possible: **a.** $\frac{x^2 + 3x}{x - 1} - \frac{5x - 1}{x - 1}$ and **b.** $\frac{3y^2}{(y + 3)(y - 3)} - \frac{3y^2 + y}{(y + 3)(y - 3)}$.

FINDING THE LEAST COMMON DENOMINATOR

We will now discuss two skills that are needed for adding and subtracting rational expressions that have unlike denominators. To begin, let's consider

$$\frac{11}{8x} + \frac{7}{18x^2}$$

To add these expressions, we must express them as equivalent expressions with a common denominator. The **least common denominator (LCD)** is usually the easiest one to use. The least common denominator of several rational expressions can be found as follows.

Finding the LCD

1. Factor each denominator completely.
2. The LCD is a product that uses each different factor obtained in step 1 the greatest number of times it appears in any one factorization.

EXAMPLE 5

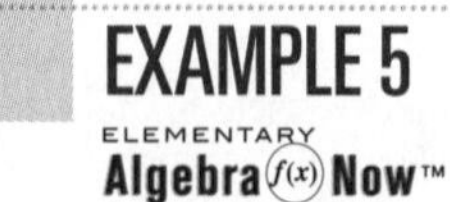

Find the LCD of each pair of rational expressions: **a.** $\frac{11}{8x}$ and $\frac{7}{18x^2}$ and **b.** $\frac{20}{x}$ and $\frac{4x}{x - 9}$.

Solution **a.** We begin by factoring each denominator completely.

$$8x = 2 \cdot 2 \cdot 2 \cdot x \quad \text{Prime factor 8.}$$
$$18x^2 = 2 \cdot 3 \cdot 3 \cdot x \cdot x \quad \text{Prime factor 18.}$$

The factorizations of $8x$ and $18x^2$ contain the factors 2, 3, and x. We form a product using each factor the greatest number of times it appears in any one factorization.

The greatest number of times the factor 2 appears is three times.
The greatest number of times the factor 3 appears is twice.
The greatest number of times the factor x appears is twice.

$$\text{LCD} = 2 \cdot 2 \cdot 2 \cdot 3 \cdot 3 \cdot x \cdot x$$
$$= 72x^2$$

The LCD for $\frac{11}{8x}$ and $\frac{7}{18x^2}$ is $72x^2$.

b. Since the denominators of $\frac{20}{x}$ and $\frac{4x}{x-9}$ are completely factored, the factor x appears once and the factor $x - 9$ appears once. Thus, the LCD is $x(x - 9)$.

Success Tip

For Example 5a, the factorizations can be written:

$$8x = 2^3 \cdot x$$
$$18x^2 = 2 \cdot 3^2 \cdot x^2$$

Note that the highest power of each factor is used to form the LCD.

$$\text{LCD} = 2^3 \cdot 3^2 \cdot x^2 = 72x^2$$

Self Check 5 Find the LCD of each pair of rational expressions: **a.** $\frac{y+7}{6y^3}$ and $\frac{7}{75y}$ and **b.** $\frac{a-3}{a+3}$ and $\frac{21}{a}$.

EXAMPLE 6

Find the LCD of each pair of rational expressions: **a.** $\frac{x}{7x+7}$ and $\frac{x-2}{5x+5}$ and **b.** $\frac{6-x}{x^2+8x+16}$ and $\frac{15x}{x^2-16}$.

Solution **a.** Factor each denominator completely.

$$7x + 7 = 7(x + 1)$$
$$5x + 5 = 5(x + 1)$$

The factorizations of $7x + 7$ and $5x + 5$ contain the factors 7, 5, and $x + 1$. We form a product using each factor the greatest number of times it appears in any one factorization.

The greatest number of times the factor 7 appears is once.
The greatest number of times the factor 5 appears is once.
The greatest number of times the factor $x + 1$ appears is once.

$$\text{LCD} = 7 \cdot 5 \cdot (x + 1) = 35(x + 1)$$

b. Factor each denominator completely.

$$x^2 + 8x + 16 = (x + 4)(x + 4)$$
$$x^2 - 16 = (x + 4)(x - 4)$$

The factorizations of $x^2 + 8x + 16$ and $x^2 - 16$ contain the factors $x + 4$ and $x - 4$.

The greatest number of times the factor $x + 4$ appears is twice.
The greatest number of times the factor $x - 4$ appears is once.

$$\text{LCD} = (x + 4)(x + 4)(x - 4) = (x + 4)^2(x - 4)$$

Notation

Rather than doing a multiplication, it is often better to leave an LCD in factored form. Therefore, in Example 6, we have:

$$\text{LCD} = 35(x + 1)$$
$$\text{LCD} = (x + 4)^2(x - 4)$$

Self Check 6 Find the LCD of each pair of rational expressions: **a.** $\frac{x^3}{x^2 - 6x}$ and $\frac{25x}{2x - 12}$ and **b.** $\frac{m + 1}{m^2 - 9}$ and $\frac{6m^2}{m^2 - 6m + 9}$.

BUILDING RATIONAL EXPRESSIONS

Recall that writing a fraction as an equivalent fraction with a larger denominator is called building the fraction. For example, to write $\frac{3}{5}$ as an equivalent fraction with a denominator of 35, we multiply it by 1 in the form of $\frac{7}{7}$:

$$\frac{3}{5} = \frac{3}{5} \cdot \frac{7}{7} = \frac{21}{35}$$

To add and subtract rational expressions with different denominators, we must write them as equivalent expressions having a common denominator. To do so, we build rational expressions.

Building Rational Expressions To build a rational expression, multiply it by 1 in the form of $\frac{c}{c}$, where c is any nonzero number or expression.

EXAMPLE 7

Write $\frac{7}{15n}$ as an equivalent expression with a denominator of $30n^3$.

ELEMENTARY Algebra f(x) Now™

The Language of Algebra
When we build the rational expression in Example 7, $\frac{7}{15n}$ has the same value as $\frac{14n^2}{30n^3}$ for all values of n, except those that make either denominator 0.

Solution We need to multiply the denominator of $\frac{7}{15n}$ by $2n^2$ to obtain a denominator of $30n^3$. It follows that $\frac{2n^2}{2n^2}$ is the form of 1 that should be used to build an equivalent expression.

$$\frac{7}{15n} = \frac{7}{15n} \cdot \frac{2n^2}{2n^2}$$ Multiply the given rational expression by 1, in the form of $\frac{2n^2}{2n^2}$.

$$= \frac{14n^2}{30n^3}$$ Multiply the numerators. Multiply the denominators.

Self Check 7 Write $\frac{7}{20m^2}$ as an equivalent expression with a denominator of $60m^3$.

EXAMPLE 8

Write $\frac{x + 1}{x^2 + 6x}$ as an equivalent expression with a denominator of $x(x + 6)(x + 2)$.

Solution We factor the denominator to determine what factors are missing.

$$\frac{x + 1}{x^2 + 6x} = \frac{x + 1}{x(x + 6)}$$ Factor out x from $x^2 + 6x$.

It is now apparent that we need to multiply the denominator by $x + 2$ to obtain a denominator of $x(x + 6)(x + 2)$. It follows that $\frac{x + 2}{x + 2}$ is the form of 1 that should be used to build an equivalent expression.

$$\frac{x + 1}{x^2 + 6x} = \frac{x + 1}{x(x + 6)} \cdot \frac{x + 2}{x + 2}$$ Multiply the given rational expression by 1, in the form of $\frac{x + 2}{x + 2}$.

$$= \frac{(x + 1)(x + 2)}{x(x + 6)(x + 2)} \quad \text{Multiply the numerators. Multiply the denominators.}$$

Self Check 8 Write $\frac{x - 3}{x^2 - 4x}$ as an equivalent expression with a denominator of $x(x - 4)(x + 8)$.

Answers to Self Checks **1. a.** $\frac{2x}{5}$, **b.** $\frac{3m - 6}{23n}$ or $\frac{3(m - 2)}{23n}$ **2. a.** $\frac{2}{3}$, **b.** $\frac{2c^2 - 11c}{(c - 1)(c + 2)}$ or $\frac{c(2c - 11)}{(c - 1)(c + 2)}$ **3.** $\frac{1}{n + 4}$ **4. a.** $x - 1$, **b.** $-\frac{y}{(y + 3)(y - 3)}$ **5. a.** $150y^3$, **b.** $a(a + 3)$ **6. a.** $2x(x - 6)$, **b.** $(m + 3)(m - 3)^2$ **7.** $\frac{21m}{60m^3}$ **8.** $\frac{(x - 3)(x + 8)}{x(x - 4)(x + 8)}$

6.3 STUDY SET  Elementary Algebra f(x) Now™

VOCABULARY Fill in the blanks.

1. The rational expressions $\frac{7}{6n}$ and $\frac{n + 1}{6n}$ have a _________ denominator of $6n$.
2. The ______ of the numerators of $\frac{9m}{m + 1}$ and $\frac{6m}{m + 1}$ is $15m$, and the __________ of their numerators is $3m$.
3. The ______ _________ ____________ of $\frac{x - 8}{x + 6}$ and $\frac{6 - 5x}{x}$ is $x(x + 6)$.
4. To ______ a rational expression, we multiply it by a form of 1. For example, $\frac{2}{n^2} \cdot \frac{8}{8} = \frac{16}{8n^2}$.
5. To simplify $5y + 8 - y + 4$, we combine ______ terms.
6. Since $\frac{7x}{x^2 + 1}$ has the same value as $\frac{14x}{2(x^2 + 1)}$ for all values of x, they are called __________ expressions.

CONCEPTS Fill in the blanks.

7. To add or subtract rational expressions that have the same denominator, add or subtract the __________, and write the sum or difference over the common __________.

   In symbols, if $\frac{A}{D}$ and $\frac{B}{D}$ are rational expressions,

   $$\frac{A}{D} + \frac{B}{D} = \frac{A + B}{D} \quad \text{or} \quad \frac{A}{D} - \frac{B}{D} = \frac{A - B}{D}$$

8. When a − symbol precedes an expression within parentheses, such as in $-(15x - 2)$, we drop the parentheses and change the ______ of each term of the expression. For example,

   $$-(15x - 2) = -15x + 2$$

9. Simplify each expression.
   a. $-(6x + 9)$
   b. $x^2 + 3x - 1 + x^2 - 5x$
   c. $7x - 1 - (5x - 6)$
   d. $4x^2 - 2x - (x^2 + x)$
10. The sum of two rational expressions is $\frac{4x + 4}{5(x + 1)}$. Simplify the result.
11. The difference of two rational expressions is $\frac{x^2 - 14x + 49}{x^2 - 49}$. Simplify the result.
12. Fill in the blanks: To find the least common denominator of several rational expressions, _______ each denominator completely. The LCD is a product that uses each different factor the ________ number of times it appears in any one factorization.
13. Factor each denominator completely.
    a. $\frac{17}{40x^2}$
    b. $\frac{x + 25}{2x^2 - 6x}$
    c. $\frac{n^2 + 3n - 4}{n^2 - 64}$
14. Consider the following factorizations.

    $$18x - 36 = 2 \cdot 3 \cdot 3 \cdot (x - 2)$$
    $$3x^2 - 3x - 6 = 3(x - 2)(x + 1)$$

    a. What is the greatest number of times the factor 3 appears in any one factorization?
    b. What is the greatest number of times the factor $x - 2$ appears in any one factorization?

15. Consider $\frac{5}{21a} = \frac{5}{21a} \cdot \frac{a+4}{a+4}$.

a. What rational expression is being built?

b. To build a rational expression, we multiply it by a form of 1. What form of 1 is used here?

16. Fill in the blanks. We need to multiply the denominator of $\frac{x}{x-9}$ by ▭ to obtain a denominator of $3x(x - 9)$. It follows that ▭ is the form of 1 that should be used to build $\frac{x}{x-9}$.

NOTATION

17. Do these rational expressions have the same denominator?

$$\frac{7 - x}{(x + 2)(x - 3)} \qquad \frac{15x + 4}{(x - 3)(x + 2)}$$

18. Write each LCD using exponents.

a. $\text{LCD} = 2 \cdot 3 \cdot 5 \cdot x \cdot x \cdot x$

b. $\text{LCD} = (y + 3)(y + 3)(y - 9)$

PRACTICE **Add or subtract. Simplify the result, if possible.**

19. $\frac{9}{x} + \frac{2}{x}$

20. $\frac{4}{s} + \frac{4}{s}$

21. $\frac{x}{18} + \frac{5}{18}$

22. $\frac{7}{10} + \frac{3y}{10}$

23. $\frac{m-3}{m^3} - \frac{5}{m^3}$

24. $\frac{c+7}{c^4} - \frac{3}{c^4}$

25. $\frac{2x}{y} - \frac{x}{y}$

26. $\frac{16c}{d} - \frac{4c}{d}$

27. $\frac{13t}{99} - \frac{35t}{99}$

28. $\frac{35y}{72} - \frac{44y}{72}$

29. $\frac{x}{9} + \frac{2x}{9}$

30. $\frac{5x}{7} + \frac{9x}{7}$

31. $\frac{50}{r^3 - 25} + \frac{r}{r^3 - 25}$

32. $\frac{75}{h^2 - 27} + \frac{h}{h^2 - 27}$

33. $\frac{6a}{a + 2} - \frac{4a}{a + 2}$

34. $\frac{10b}{b - 6} - \frac{4b}{b - 6}$

35. $\frac{7}{t + 5} - \frac{9}{t + 5}$

36. $\frac{3}{r - 11} - \frac{6}{r - 11}$

37. $\frac{3x - 5}{x - 2} + \frac{6x - 13}{x - 2}$

38. $\frac{8x - 7}{x + 3} + \frac{2x + 37}{x + 3}$

39. $\frac{6x - 5}{3xy} - \frac{3x - 5}{3xy}$

40. $\frac{7x + 7}{5y} - \frac{2x + 7}{5y}$

41. $\frac{3y - 2}{2y + 6} - \frac{2y - 5}{2y + 6}$

42. $\frac{5x + 8}{3x + 15} - \frac{3x - 2}{3x + 15}$

43. $\frac{x + 3}{2y} + \frac{x + 5}{2y}$

44. $\frac{y + 2}{10z} + \frac{y + 4}{10z}$

45. $\frac{6x^2 - 11x}{3x + 2} - \frac{10}{3x + 2}$

46. $\frac{8a^2}{2a + 5} - \frac{4a^2 + 25}{2a + 5}$

47. $\frac{2 - p}{p^2 - p} - \frac{-p + 2}{p^2 - p}$

48. $\frac{7n + 2}{n^2 + 5} - \frac{2 + 7n}{n^2 + 5}$

49. $\frac{3x^2}{x + 1} + \frac{x - 2}{x + 1}$

50. $\frac{10b^2}{3b - 2} - \frac{b^2 + 4}{3b - 2}$

51. $\frac{11w + 1}{3w(w - 9)} - \frac{11w}{3w(w - 9)}$

52. $\frac{y + 2}{y^2(y - 14)} - \frac{y}{y^2(y - 14)}$

53. $\frac{a}{a^2 + 5a + 6} + \frac{3}{a^2 + 5a + 6}$

54. $\frac{b}{b^2 - 4} + \frac{2}{b^2 - 4}$

55. $\frac{2c}{c^2 - d^2} - \frac{2d}{c^2 - d^2}$

56. $\frac{3t}{t^2 - 8t + 7} - \frac{3}{t^2 - 8t + 7}$

57. $\frac{11n - 1}{(n + 4)(n - 2)} - \frac{4n}{(n - 2)(n + 4)}$

58. $\frac{r - 32}{(r - 4)(r - 2)} + \frac{4r}{(r - 2)(r - 4)}$

59. $\frac{1}{t^2 - 2t + 1} - \frac{6 - t}{t^2 - 2t + 1}$

60. $\frac{2}{d^2 - 10d + 25} - \frac{10 - d}{d^2 - 10d + 25}$

Find the LCD of each pair of rational expressions.

61. $\frac{1}{2x}, \frac{9}{6x}$

62. $\frac{4}{9y}, \frac{11}{3y}$

63. $\frac{35}{3a^2b}, \frac{23}{a^2b^3}$

64. $\frac{27}{c^2d}, \frac{17}{2c^2d^3}$

65. $\frac{8}{c}, \frac{8-c}{c+2}$

66. $\frac{d^2-5}{d+9}, \frac{d-3}{d}$

67. $\frac{33}{15a^3}, \frac{9}{10a}$

68. $\frac{m-21}{12m^4}, \frac{m+1}{18m}$

69. $\frac{b-9}{4b+8}, \frac{b}{6}$

70. $\frac{b^2-b}{10b-15}, \frac{11b}{10}$

71. $\frac{-2x}{x^2-1}, \frac{5x}{x+1}$

72. $\frac{7-y^2}{y^2-4}, \frac{y-49}{y+2}$

73. $\frac{3x+1}{3x-1}, \frac{3x}{3x+1}$

74. $\frac{b+1}{b-1}, \frac{b}{b+1}$

75. $\frac{4x-5}{x^2-4x-5}, \frac{3x+1}{x^2-25}$

76. $\frac{44}{s^2-9}, \frac{s+9}{s^2-s-6}$

77. $\frac{5n^2-16}{2n^2+13n+20}, \frac{3n^2}{n^2+8n+16}$

78. $\frac{4y+25}{y^2+10y+25}, \frac{y^2-7}{2y^2+17y+35}$

Build each rational expression into an equivalent expression with the given denominator.

79. $\frac{25}{4}; 20x$

80. $\frac{5}{y}; y^2$

81. $\frac{8}{x}; x^2y$

82. $\frac{7}{y}; xy^2$

83. $\frac{3x}{x+1}; (x+1)^2$

84. $\frac{5y}{y-2}; (y-2)^2$

85. $\frac{2y}{x}; x(x+3)$

86. $\frac{3x}{y}; y(y-9)$

87. $\frac{10}{b-1}; 3(b-1)$

88. $\frac{8}{c+9}; 6(c+9)$

89. $\frac{t+5}{4t+8}; 20(t+2)$

90. $\frac{x+7}{3x-15}; 6(x-5)$

91. $\frac{y+3}{y^2-5y+6}; 4y(y-2)(y-3)$

92. $\frac{3x-4}{x^2+3x+2}; 8x(x+1)(x+2)$

93. $\frac{12-h}{h^2-81}; 3(h+9)(h-9)$

94. $\frac{m^2}{m^2-100}; 9(m+10)(m-10)$

95. $\frac{6t}{t^2+4t+3}; (t+1)(t-2)(t+3)$

96. $\frac{9s}{s^2+6s+5}; (s+1)(s-4)(s+5)$

APPLICATIONS

97. DOING LAUNDRY At a laundromat, it takes 30 minutes to wash a load of clothes and 45 minutes to dry a load. Suppose a customer starts a washer and dryer at the same time. If she unloads, refills, and restarts each appliance when a cycle is complete, in how many minutes will the washing and drying cycles end simultaneously?

98. GEOMETRY

a. What is the difference of the length and width of the rectangle?

b. What is the perimeter?

$\frac{3x+5}{x+2}$ ft

$\frac{5x+11}{x+2}$ ft

WRITING

99. Explain how to add fractions with the same denominator.

100. Explain how to find a least common denominator.

101. Explain the error in the following solution:

$$\frac{2x+3}{x+5} - \frac{x+2}{x+5} = \frac{2x+3-x+2}{x+5} = \frac{x+5}{x+5} = 1$$

102. Explain the error in the following solution:

$$\frac{5x-4}{y} + \frac{x}{y} = \frac{5x-4+x}{y+y} = \frac{6x-4}{2y} = \frac{2(3x-2)}{2y} = \frac{3x-2}{y}$$

103. Explain why the LCD of $\frac{5}{h^2}$ and $\frac{3}{h}$ is h^2 and not h^3.

104. Explain how multiplication by 1 is used to build a rational expression.

REVIEW

105. Give the formula for simple interest.

106. Give the formula that finds the slope of a line.

107. Give the formula for the area of a triangle.

108. Give the slope–intercept form for the equation of a line.

109. Give the formula that is used to solve uniform motion problems.

110. Give the point–slope form for the equation of a line.

CHALLENGE PROBLEMS

111. Perform the operations. Simplify the result, if possible.

$$\frac{3xy}{x-y} - \frac{x(3y-x)}{x-y} - \frac{x(x-y)}{x-y}$$

112. Find the LCD of these rational expressions.

$$\frac{2}{a^3+8}, \quad \frac{a}{a^2-4}, \quad \frac{2a+5}{a^3-8}$$

6.4 Addition and Subtraction with Unlike Denominators

- Adding and Subtracting Rational Expressions With Unlike Denominators
- Denominators That Are Opposites

We have discussed a method for finding the least common denominator (LCD) of two rational expressions. We have also built rational expressions into equivalent expressions having specified denominators. We will now use these skills to add and subtract rational expressions with unlike denominators.

ADDING AND SUBTRACTING RATIONAL EXPRESSIONS WITH UNLIKE DENOMINATORS

The following steps summarize how to add or subtract rational expressions with different denominators.

Adding and Subtracting Rational Expressions With Unlike Denominators

1. Find the LCD.
2. Write each rational expression as an equivalent expression whose denominator is the LCD.
3. Add or subtract the numerators and write the sum or difference over the LCD.
4. Simplify the resulting rational expression, if possible.

EXAMPLE 1

Add: $\frac{9x}{7} + \frac{3x}{5}$.

Solution

Step 1: The LCD is $7 \cdot 5 = 35$.

Step 2: We need to multiply the denominator of $\frac{9x}{7}$ by 5 and we need to multiply the denominator of $\frac{3x}{5}$ by 7 to obtain the LCD, 35. It follows that $\frac{5}{5}$ and $\frac{7}{7}$ are the forms of 1 that should be used to write the equivalent rational expressions.

$$\frac{9x}{7} + \frac{3x}{5} = \frac{9x}{7} \cdot \frac{5}{5} + \frac{3x}{5} \cdot \frac{7}{7}$$ Build the rational expressions so that each has a denominator of 35.

$$= \frac{45x}{35} + \frac{21x}{35}$$ Multiply the numerators. Multiply the denominators.

Step 3: $$= \frac{45x + 21x}{35}$$ Add the numerators. Write the sum over the common denominator.

$$= \frac{66x}{35}$$ Combine like terms in the numerator.

Step 4: Since 66 and 35 have no common factor other than 1, the result cannot be simplified.

Self Check 1 Add: $\frac{y}{2} + \frac{6y}{7}$.

EXAMPLE 2

Subtract: $\frac{13}{18b^2} - \frac{1}{24b}$.

Solution To find the LCD, we form a product that uses each different factor of $18b^2$ and $24b$ the greatest number of times it appears in any one factorization.

$$\left.\begin{aligned} 18b^2 &= 2 \cdot 3 \cdot 3 \cdot b \cdot b \\ 24b &= 2 \cdot 2 \cdot 2 \cdot 3 \cdot b \end{aligned}\right\} \text{LCD} = 2 \cdot 2 \cdot 2 \cdot 3 \cdot 3 \cdot b \cdot b = 72b^2$$

We need to multiply $18b^2$ by 4 to obtain $72b^2$, and $24b$ by $3b$ to obtain $72b^2$. It follows that we should use $\frac{4}{4}$ and $\frac{3b}{3b}$ to build the equivalent rational expressions.

$$\frac{13}{18b^2} - \frac{1}{24b} = \frac{13}{18b^2} \cdot \frac{4}{4} - \frac{1}{24b} \cdot \frac{3b}{3b}$$ Build the rational expressions so that each has a denominator of $72b^2$.

$$= \frac{52}{72b^2} - \frac{3b}{72b^2}$$ Multiply the numerators. Multiply the denominators.

$$= \frac{52 - 3b}{72b^2}$$ Subtract the numerators. Write the difference over the common denominator.

The result cannot be simplified.

Notation

When checking your answers with those in the back of the book, remember that the result of an addition or subtraction of rational expressions can often be presented in several equivalent forms. For instance, the result of Example 2 may also be expressed as

$$\frac{-3b + 52}{72b^2} \quad \text{or} \quad -\frac{3b - 52}{72b^2}$$

Self Check 2 Subtract: $\frac{5}{21z^2} - \frac{3}{28z}$.

EXAMPLE 3

Add: $\dfrac{3}{2x + 18} + \dfrac{27}{x^2 - 81}$.

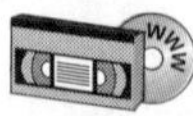

ELEMENTARY Algebra f(x) Now™

Success Tip

In Example 3, we use the distributive property to multiply the numerators of $\frac{3}{2(x+9)}$ and $\frac{x-9}{x-9}$. Note that we don't multiply out the denominators.

$$\frac{3}{2(x+9)} \cdot \frac{x-9}{x-9}$$

The result is $\frac{3x - 27}{2(x+9)(x-9)}$.

Solution After factoring the denominators, we see that the greatest number of times each of the factors 2, $x + 9$, and $x - 9$ appear in any one of the factorizations is once.

$$\left.\begin{aligned} 2x + 18 &= 2(x + 9) \\ x^2 - 81 &= (x + 9)(x - 9) \end{aligned}\right\} \text{LCD} = 2(x + 9)(x - 9)$$

Since we need to multiply $2(x + 9)$ by $x - 9$ to obtain the LCD and $(x + 9)(x - 9)$ by 2 to obtain the LCD, $\frac{x-9}{x-9}$ and $\frac{2}{2}$ are the forms of 1 to use to build the equivalent rational expressions.

$$\frac{3}{2x + 18} + \frac{27}{x^2 - 81} = \frac{3}{2(x + 9)} + \frac{27}{(x + 9)(x - 9)}$$ Write each denominator in factored form.

$$= \frac{3}{2(x + 9)} \cdot \frac{x - 9}{x - 9} + \frac{27}{(x + 9)(x - 9)} \cdot \frac{2}{2}$$ Build the expressions so that each has a denominator of $2(x + 9)(x - 9)$.

$$= \frac{3x - 27}{2(x + 9)(x - 9)} + \frac{54}{2(x + 9)(x - 9)}$$ Multiply the numerators. Multiply the denominators.

Although it is not mandatory, the factors of each denominator are written in the same order.

$$= \frac{3x - 27 + 54}{2(x + 9)(x - 9)}$$ Add the numerators. Write the sum over the common denominator.

$$= \frac{3x + 27}{2(x + 9)(x - 9)}$$ Combine like terms in the numerator.

$$= \frac{3\overset{1}{\cancel{(x + 9)}}}{2\underset{1}{\cancel{(x + 9)}}(x - 9)}$$ Factor the numerator. Then simplify the expression by removing a factor equal to 1.

$$= \frac{3}{2(x - 9)}$$

Self Check 3 Add: $\dfrac{2}{5x + 25} + \dfrac{4}{x^2 - 25}$.

EXAMPLE 4

Subtract: $\dfrac{x}{x - 1} - \dfrac{x - 6}{x - 4}$.

ELEMENTARY Algebra f(x) Now™

Solution The denominators of $\frac{x}{x-1}$ and $\frac{x-6}{x-4}$ are completely factored. The factor $x - 1$ appears once and the factor $x - 4$ appears once. Thus, the LCD $= (x - 1)(x - 4)$.

We need to multiply the first denominator by $x - 4$ to obtain the LCD and the second denominator by $x - 1$ to obtain the LCD. It follows that $\frac{x-4}{x-4}$ and $\frac{x-1}{x-1}$ are the forms of 1 to use to build the equivalent rational expressions.

$$\frac{x}{x-1} - \frac{x-6}{x-4} = \frac{x}{x-1} \cdot \frac{x-4}{x-4} - \frac{x-6}{x-4} \cdot \frac{x-1}{x-1}$$ Build the expressions so that each has a denominator of $(x-1)(x-4)$.

$$= \frac{x^2-4x}{(x-1)(x-4)} - \frac{x^2-7x+6}{(x-4)(x-1)}$$ Multiply the numerators. Multiply the denominators.

By the commutative property of multiplication, these are like denominators.

$$= \frac{x^2-4x-(x^2-7x+6)}{(x-1)(x-4)}$$ Subtract the numerators. Write the difference over the common denominator.

$$= \frac{x^2-4x-x^2+7x-6}{(x-1)(x-4)}$$ Use the distributive property.

$$= \frac{3x-6}{(x-1)(x-4)}$$ Combine like terms in the numerator.

Success Tip

In Example 4, we use the FOIL method to multiply the numerators of $\frac{x-6}{x-4}$ and $\frac{x-1}{x-1}$. Note that we do not multiply out the denominators.

The numerator factors as $3(x-2)$. Since the numerator and denominator have no common factor, the result is in simplest form.

Self Check 4 Subtact: $\frac{x}{x+9} - \frac{x-7}{x+8}$.

EXAMPLE 5

ELEMENTARY Algebra f(x) Now™

Subtract: $\frac{m}{m^2+5m+6} - \frac{2}{m^2+3m+2}$.

Solution Factor each denominator and form the LCD.

$$\left.\begin{aligned} m^2+5m+6 &= (m+2)(m+3) \\ m^2+3m+2 &= (m+2)(m+1) \end{aligned}\right\} \text{LCD} = (m+2)(m+3)(m+1)$$

Examining the factored forms, we see that the first denominator must be multiplied by $m+1$, and the second must be multiplied by $m+3$ to obtain the LCD. To build the expressions, we will use $\frac{m+1}{m+1}$ and $\frac{m+3}{m+3}$.

$$\frac{m}{m^2+5m+6} - \frac{2}{m^2+3m+2}$$

$$= \frac{m}{(m+2)(m+3)} - \frac{2}{(m+2)(m+1)}$$ Write each denominator in factored form.

$$= \frac{m}{(m+2)(m+3)} \cdot \frac{m+1}{m+1} - \frac{2}{(m+2)(m+1)} \cdot \frac{m+3}{m+3}$$ Build each expression.

$$= \frac{m^2+m}{(m+2)(m+3)(m+1)} - \frac{2m+6}{(m+2)(m+1)(m+3)}$$ Multiply numerators. Multiply denominators.

By the commutative property of multiplication, these are like denominators.

$$= \frac{m^2+m-(2m+6)}{(m+2)(m+3)(m+1)}$$ Subtract the numerators. Write the difference over the common denominator. Remember the parentheses.

$$= \frac{m^2 + m - 2m - 6}{(m+2)(m+3)(m+1)}$$ Use the distributive property: $-(2m + 6) = -1(2m + 6) = -2m - 6$.

$$= \frac{m^2 - m - 6}{(m+2)(m+3)(m+1)}$$ Combine like terms in the numerator.

$$= \frac{(m-3)\overset{1}{\cancel{(m+2)}}}{\underset{1}{\cancel{(m+2)}}(m+3)(m+1)}$$ Factor the numerator and simplify the expression by removing a factor equal to 1.

$$= \frac{m-3}{(m+3)(m+1)}$$

Self Check 5 Subtract: $\frac{b}{b^2 - 2b - 8} - \frac{6}{b^2 + b - 20}$.

EXAMPLE 6

Add: $\frac{4b}{a-5} + b$.

ELEMENTARY Algebra f(x) Now™

Solution We can write b as $\frac{b}{1}$. The LCD of $\frac{4b}{a-5}$ and $\frac{b}{1}$ is $1(a - 5)$, or simply $a - 5$. Since we must multiply the denominator of $\frac{b}{1}$ by $a - 5$ to obtain the LCD, we will use $\frac{a-5}{a-5}$ to write an equivalent rational expression.

$$\frac{4b}{a-5} + b = \frac{4b}{a-5} + \frac{b}{1} \cdot \frac{a-5}{a-5}$$ Build $\frac{b}{1}$ so that it has a denominator of $a - 5$.

$$= \frac{4b}{a-5} + \frac{ab - 5b}{a-5}$$ Multiply numerators: $b(a - 5) = ab - 5b$. Multiply denominators: $1(a - 5) = a - 5$.

$$= \frac{4b + ab - 5b}{a-5}$$ Add the numerators. Write the sum over the common denominator.

$$= \frac{ab - b}{a-5}$$ Combine like terms in the numerator: $4b - 5b = -b$.

Although the numerator factors as $b(a - 1)$, the numerator and denominator do not have a common factor. Therefore, the result does not simplify any further.

Self Check 6 Add: $\frac{10y}{n+4} + y$.

DENOMINATORS THAT ARE OPPOSITES

Recall that two polynomials are **opposites** if their terms are the same but they are opposite in sign. For example, $x - 4$ and $4 - x$ are opposites. If we multiply one of these binomials by -1, the result is the other binomial.

$$-1(x - 4) = -x + 4$$
$$= 4 - x$$ Write the expression with 4 first.

$$-1(4 - x) = -4 + x$$
$$= x - 4$$ Write the expression with x first.

These results suggest a general fact.

Multiplying by −1 When a polynomial is multiplied by -1, the result is its opposite.

This fact can be used when adding or subtracting rational expressions whose denominators are opposites.

EXAMPLE 7

Subtract: $\frac{x}{x-7} - \frac{1}{7-x}$.

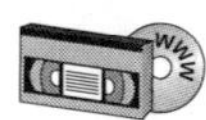

ELEMENTARY **Algebra f(x) Now™**

Success Tip

Either denominator in Example 7 can serve as the LCD. However, it is common to have a result whose denominator is written in descending powers of the variable. Therefore, we chose $x - 7$, as opposed to $7 - x$, as the LCD.

Solution We note that the denominators are opposites. Either can serve as the LCD; we will choose $x - 7$.

We must multiply the denominator of $\frac{1}{7-x}$ by -1 to obtain the LCD. It follows that $\frac{-1}{-1}$ should be the form of 1 that is used to write an equivalent rational expression.

$$\frac{x}{x-7} - \frac{1}{7-x} = \frac{x}{x-7} - \frac{1}{7-x} \cdot \frac{-1}{-1}$$ Build $\frac{1}{7-x}$ so that it has a denominator of $x - 7$.

$$= \frac{x}{x-7} - \frac{-1}{-7+x}$$ Multiply the numerators. Multiply the denominators.

$$= \frac{x}{x-7} - \frac{-1}{x-7}$$ Rewrite the second denominator: $-7 + x = x - 7$.

$$= \frac{x-(-1)}{x-7}$$ Subtract the numerators. Write the difference over the common denominator.

$$= \frac{x+1}{x-7}$$ Simplify the numerator.

The result does not simplify any further.

Self Check 7 Add: $\frac{n}{n-8} + \frac{12}{8-n}$.

Answers to Self Checks **1.** $\frac{19y}{14}$ **2.** $\frac{20-9z}{84z^2}$ **3.** $\frac{2}{5(x-5)}$ **4.** $\frac{6x+63}{(x+9)(x+8)}$ **5.** $\frac{b+3}{(b+2)(b+5)}$ **6.** $\frac{ny+14y}{n+4}$ **7.** $\frac{n-12}{n-8}$

6.4 STUDY SET

ELEMENTARY **Algebra f(x) Now™**

VOCABULARY **Fill in the blanks.**

1. The rational expressions $\frac{x}{x-7}$ and $\frac{1}{x-7}$ have like denominators. The rational expressions $\frac{x+5}{x-7}$ and $\frac{4x}{x+7}$ have ________ denominators.
2. Two polynomials are __________ if their terms are the same, but are opposite in sign.

CONCEPTS

3. Write each denominator in factored form.

   a. $\frac{x+1}{20x^2}$ b. $\frac{3x^2-4}{x^2+4x-12}$

4. The factorizations of the denominators of two rational expressions follow. Find their LCD.

$$12a = 2 \cdot 2 \cdot 3 \cdot a$$
$$18a^2 = 2 \cdot 3 \cdot 3 \cdot a \cdot a$$

5. What is the LCD for $\frac{x-1}{x+6}$ and $\frac{1}{x+3}$?

6. The factorizations of the denominators of two rational expressions follow. Find their LCD.

$$x^2 - 36 = (x+6)(x-6)$$
$$3x - 18 = 3(x-6)$$

7. To build $\frac{x}{x+2}$ so that it has a denominator of $5(x+2)$, we multiply it by 1 in the form of ☐.

8. To build $\frac{x-3}{x-4}$ so that it has a denominator of $(x+5)(x-4)$, we multiply it by 1 in the form of ☐.

9. The LCD for $\frac{1}{9n^2}$ and $\frac{37}{15n^3}$ is

$$\text{LCD} = 3 \cdot 3 \cdot 5 \cdot n \cdot n \cdot n$$

If we want to add these rational expressions, what form of 1 should be used

a. to build $\frac{1}{9n^2}$? **b.** to build $\frac{37}{15n^3}$?

10. The LCD for $\frac{2x+1}{x^2+5x+6}$ and $\frac{3x}{x^2-4}$ is

$$\text{LCD} = (x+2)(x+3)(x-2)$$

If we want to subtract these rational expressions, what form of 1 should be used

a. to build $\frac{2x+1}{x^2+5x+6}$?

b. to build $\frac{3x}{x^2-4}$?

Fill in the blanks. Write each numerator without using parentheses.

11. $\frac{x-3}{x-4} \cdot \frac{8}{8} = \frac{\square}{8(x-4)}$

12. $\frac{x-3}{x-4} \cdot \frac{x+9}{x+2} = \frac{\square}{(x-4)(x+2)}$

13. $\frac{7x}{3x^2(x-5)} + \frac{10}{3x^2(x-5)} = \frac{\square}{3x^2(x-5)}$

14. $\frac{x^2 - 3x + 9 - (x^2 - 5x)}{(x-9)(x+3)} = \frac{\square}{(x-9)(x+3)}$

15. Multiply: $-1(x - 10)$.

16. What is the opposite of $x - 25$?

17. By what must $y - 4$ be multiplied to obtain $4 - y$?

18. To write $\frac{8x}{2-x}$ so that it has a denominator of $x - 2$, we multiply it by 1 in the form of ☐.

19. Write x in fraction form.

20. Fill in the blank. To write $\frac{7}{1}$ so that it has a denominator of $2c + 1$, we multiply it by 1 in the form of ☐.

NOTATION

21. In the following table, a student's answers to three homework problems are compared to the answers in the back of the book. Are the answers equivalent?

Student's answer	Book's answer	Equivalent?
$\frac{m^2+2m}{(m-1)(m-4)}$	$\frac{m^2+2m}{(m-4)(m-1)}$	
$\frac{-5x^2-7}{4x(x+3)}$	$-\frac{5x^2-7}{4x(x+3)}$	
$\frac{-2x}{x-y}$	$\frac{2x}{y-x}$	

22. Simplify each expression.

a. $\frac{6}{6}$ **b.** $\frac{3x}{3x}$

c. $\frac{x+7}{x+7}$ **d.** $\frac{-1}{-1}$

PRACTICE **Perform the operations. Simplify, if possible.**

23. $\frac{x}{3} + \frac{2x}{7}$

24. $\frac{y}{4} + \frac{3y}{5}$

25. $\frac{2y}{9} + \frac{y}{3}$

26. $\frac{4b}{3} - \frac{5b}{12}$

27. $\frac{11}{5x} - \frac{5}{6x}$

28. $\frac{5}{9y} - \frac{1}{4y}$

29. $\frac{7}{m^2} - \frac{2}{m}$

30. $\frac{6}{n^2} + \frac{2}{n}$

31. $\frac{4x}{3} + \frac{2x}{y}$

32. $\frac{7b}{10} + \frac{4b}{t}$

33. $\frac{1}{6c^4} + \frac{8}{9c^2}$

34. $\frac{7}{8b^2} + \frac{5}{6b^3}$

35. $\frac{y}{8} + \frac{y - 3}{16}$

36. $\frac{m}{4} + \frac{m - 4}{8}$

37. $\frac{n}{5} - \frac{n - 2}{15}$

38. $\frac{m}{9} - \frac{m + 1}{27}$

39. $\frac{2 - b}{6} - \frac{b - 7}{21}$

40. $\frac{7 - x}{4} - \frac{x - 3}{14}$

41. $\frac{x - 1}{9x} - \frac{x - 2}{x^3}$

42. $\frac{a - 3}{7a} - \frac{a - 4}{a^3}$

43. $\frac{y + 2}{5y^2} + \frac{y + 4}{15y}$

44. $\frac{x + 3}{x^2} + \frac{x + 5}{2x}$

45. $\frac{x + 5}{xy} - \frac{x - 1}{x^2y}$

46. $\frac{y - 7}{y^2} - \frac{y + 7}{2y}$

47. $\frac{x - 3}{6x} + \frac{x + 4}{8x}$

48. $\frac{3y - 2}{12y} + \frac{3 - y}{18y}$

49. $\frac{a + 2}{b} + \frac{b - 2}{a}$

50. $\frac{x - 1}{x} + \frac{y + 1}{y}$

51. $\frac{x}{x + 1} + \frac{x - 1}{x}$

52. $\frac{t - 2}{t} + \frac{t}{t + 3}$

53. $\frac{1}{5x} + \frac{7x}{x + 5}$

54. $\frac{10h}{h - 3} + \frac{7}{9h}$

55. $\frac{9}{t + 3} + \frac{8}{t + 2}$

56. $\frac{2}{m - 3} + \frac{7}{m - 4}$

57. $\frac{3x}{2x - 1} - \frac{2x}{2x + 3}$

58. $\frac{2y}{5y - 1} - \frac{2y}{3y + 2}$

59. $\frac{4}{a + 2} - \frac{7}{(a + 2)^2}$

60. $\frac{9}{(b - 1)^2} - \frac{2}{b - 1}$

61. $\frac{3m}{m - 2} + \frac{m - 3}{m + 5}$

62. $\frac{2x}{x + 2} + \frac{x + 1}{x - 3}$

63. $\frac{s + 7}{s + 3} - \frac{s - 3}{s + 7}$

64. $\frac{t + 5}{t - 5} - \frac{t - 5}{t + 5}$

65. $\frac{7}{(a + 1)(a + 3)} + \frac{5}{(a + 3)^2}$

66. $\frac{1}{(b - 1)(b - 4)} + \frac{100}{(b - 4)^2}$

67. $\frac{2c - 3}{(2c + 1)(c - 3)} - \frac{3}{c - 3}$

68. $\frac{3y - 7}{(3y + 5)(y - 2)} - \frac{5}{y - 2}$

69. $\frac{b}{b + 1} - \frac{b + 1}{2b + 2}$

70. $\frac{4x + 1}{8x - 12} + \frac{x - 3}{2x - 3}$

71. $\frac{2}{a^2 + 4a + 3} + \frac{1}{a + 3}$

72. $\frac{1}{c + 6} - \frac{-4}{c^2 + 8c + 12}$

73. $\frac{7s}{s^2 + s - 12} - \frac{4}{s + 4}$

74. $\frac{2y}{y^2 - 5y + 6} + \frac{4}{y - 2}$

75. $\frac{x}{x - 2} + \frac{4 + 2x}{x^2 - 4}$

76. $\frac{y}{y + 3} - \frac{2y - 6}{y^2 - 9}$

77. $\dfrac{x+1}{x+2} - \dfrac{x^2+1}{x^2-x-6}$

78. $\dfrac{6w}{w^2-4} - \dfrac{3}{w-2}$

79. $\dfrac{2}{3h-6} + \dfrac{3}{4h+8}$

80. $\dfrac{4}{3d-9} - \dfrac{3}{2d+4}$

81. $\dfrac{8}{y^2-16} - \dfrac{7}{y^2-y-12}$

82. $\dfrac{6}{s^2-9} - \dfrac{5}{s^2-s-6}$

83. $\dfrac{4}{s^2+5s+4} + \dfrac{s}{s^2+2s+1}$

84. $\dfrac{d}{d^2+6d+5} - \dfrac{3}{d^2+5d+4}$

85. $\dfrac{5}{x^2-9x+8} - \dfrac{3}{x^2-6x-16}$

86. $\dfrac{3}{t^2+t-6} + \dfrac{1}{t^2+3t-10}$

87. $\dfrac{x+1}{2x+4} - \dfrac{x^2}{2x^2-8}$

88. $\dfrac{-x}{3x^2-27} + \dfrac{1}{3x+9}$

89. $\dfrac{8}{x} + 6$

90. $\dfrac{2}{y} + 7$

91. $b - \dfrac{3}{a^2}$

92. $c - \dfrac{5}{3b}$

93. $\dfrac{9}{x-4} + x$

94. $\dfrac{9}{m+4} + 9$

95. $\dfrac{x+2}{x+1} - 5$

96. $\dfrac{y+8}{y-8} - 4$

97. $\dfrac{5}{a-4} + \dfrac{7}{4-a}$

98. $\dfrac{4}{b-6} - \dfrac{b}{6-b}$

99. $\dfrac{r+2}{r^2-4} + \dfrac{4}{4-r^2}$

100. $\dfrac{2x+2}{x-2} - \dfrac{2x}{2-x}$

101. $\dfrac{y+3}{y-1} - \dfrac{y+4}{1-y}$

102. $\dfrac{t+1}{t-7} - \dfrac{t+1}{7-t}$

APPLICATIONS

103. Find the total height of the funnel.

104. What is the difference between the diameter of the opening at the top of the funnel and the diameter of its spout?

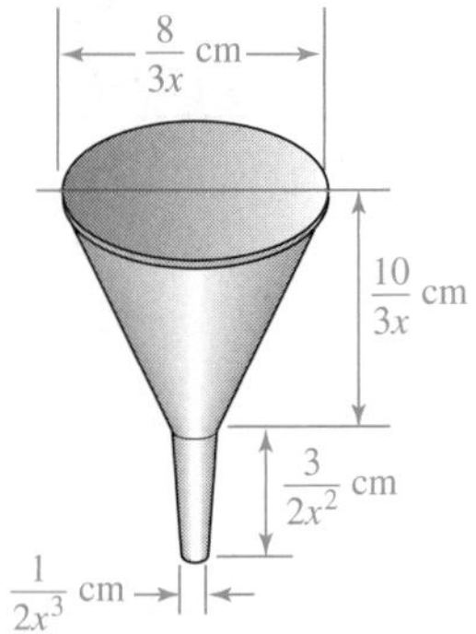

WRITING

105. Explain the error:

$$\frac{3}{x} + \frac{3}{y} = \frac{3+3}{x+y} = \frac{6}{x+y}$$

106. Explain how to add two rational expressions with unlike denominators.

107. When will the LCD of two rational expressions be the product of the denominators of those rational expressions? Give an example.

108. In general, what is the opposite of a polynomial? Explain how to find the opposite of a polynomial. Give examples.

REVIEW

109. Find the slope and y-intercept of the graph of $y = 8x + 2$.

110. Find the slope and y-intercept of the graph of $3x + 4y = -36$.

111. What is the slope of the graph of $y = -2$?

112. Is the graph of the equation $x = 0$ the x-axis or the y-axis?

CHALLENGE PROBLEMS **Perform the operations and simplify the result, if possible.**

113. $\frac{a}{a-1} - \frac{2}{a+2} + \frac{3(a-2)}{a^2+a-2}$

114. $\frac{2x}{x^2-3x+2} + \frac{2x}{x-1} - \frac{x}{x-2}$

115. $\frac{1}{a+1} + \frac{a^2-7a+10}{2a^2-2a-4} \cdot \frac{2a^2-50}{a^2+10a+25}$

116. $1 - \frac{(x-2)^2}{(x+2)^2}$

6.5 Simplifying Complex Fractions

- Simplifying Complex Fractions Using Division
- Simplifying Complex Fractions Using the LCD

A rational expression whose numerator and/or denominator contain rational expressions is called a **complex rational expression** or, more simply, a **complex fraction.** The expression above the main fraction bar of a complex fraction is the numerator, and the expression below the main fraction bar is the denominator. Two examples are:

$$\cfrac{\frac{5x}{3}}{\frac{2x}{9}} \qquad \begin{matrix} \leftarrow \text{Numerator} \rightarrow \\ \leftarrow \text{Main fraction bar} \rightarrow \\ \leftarrow \text{Denominator} \rightarrow \end{matrix} \qquad \cfrac{\frac{1}{2} - \frac{1}{x}}{\frac{x}{3} + \frac{1}{5}}$$

In this section, we will simplify complex fractions.

SIMPLIFYING COMPLEX FRACTIONS USING DIVISION

One method for simplifying complex fractions uses the fact that fractions indicate division.

Simplifying Complex Fractions

1. Write the numerator and the denominator of the complex fraction as single rational expressions.
2. Perform the division by multiplying the numerator of the complex fraction by the reciprocal of the denominator.
3. Simplify the result, if possible.

EXAMPLE 1

ELEMENTARY Algebra f(x) Now™

Simplify: $\cfrac{\frac{5x}{3}}{\frac{2x}{9}}$.

Solution Since the numerator and the denominator of the complex fraction are already single rational expressions, we can perform the division.

$$\cfrac{\frac{5x}{3}}{\frac{2x}{9}} = \frac{5x}{3} \div \frac{2x}{9}$$ Write the division indicated by the main fraction bar using a ÷ symbol.

$= \frac{5x}{3} \cdot \frac{9}{2x}$ To divide rational expressions, multiply the first by the reciprocal of the second.

The Language of Algebra

The second step of this method could also be phrased: Perform the division by *inverting the denominator of the complex fraction and multiplying.*

$= \frac{5x \cdot 9}{3 \cdot 2x}$ Multiply the numerators. Multiply the denominators.

$= \frac{\overset{1}{\cancel{5x}} \cdot \overset{1}{\cancel{3}} \cdot 3}{\underset{1}{\cancel{3}} \cdot 2\underset{1}{\cancel{x}}}$ Factor 9 as $3 \cdot 3$. Then simplify the result by removing factors equal to 1.

$= \frac{15}{2}$

Self Check 1 Simplify: $\dfrac{\frac{7y}{8}}{\frac{21y}{20}}$.

EXAMPLE 2

ELEMENTARY Algebra $f(x)$ Now™

Simplify: $\dfrac{\frac{1}{2} - \frac{1}{x}}{\frac{x}{3} + \frac{1}{5}}$.

Solution We consider the numerator and the denominator of the complex fraction separately. To write the numerator as a single rational expression, we build $\frac{1}{2}$ and $\frac{1}{x}$ to have an LCD of $2x$, and then subtract. To write the denominator as a single rational expression, we build $\frac{x}{3}$ and $\frac{1}{5}$ to have an LCD of 15, and then add.

$$\frac{\frac{1}{2} - \frac{1}{x}}{\frac{x}{3} + \frac{1}{5}} = \frac{\frac{1}{2} \cdot \frac{x}{x} - \frac{1}{x} \cdot \frac{2}{2}}{\frac{x}{3} \cdot \frac{5}{5} + \frac{1}{5} \cdot \frac{3}{3}}$$

The LCD for the numerator is $2x$. The LCD for the denominator is 15.

$$= \frac{\frac{x}{2x} - \frac{2}{2x}}{\frac{5x}{15} + \frac{3}{15}}$$

Multiply the numerators. Multiply the denominators.

$$= \frac{\frac{x - 2}{2x}}{\frac{5x + 3}{15}}$$

Subtract in the numerator and add in the denominator.

Now that the numerator and the denominator of the complex fraction are single rational expressions, we perform the division.

$$\frac{\frac{x - 2}{2x}}{\frac{5x + 3}{15}} = \frac{x - 2}{2x} \div \frac{5x + 3}{15}$$

Write the division indicated by the main fraction bar using a ÷ symbol.

$$= \frac{x-2}{2x} \cdot \frac{15}{5x+3} \quad \text{Multiply by the reciprocal of } \frac{5x+3}{15}.$$

$$= \frac{15(x-2)}{2x(5x+3)} \quad \text{Multiply the numerators. Multiply the denominators.}$$

Since the numerator and denominator have no common factor, the result does not simplify.

Notation

The result after simplifying a complex fraction can often have several equivalent forms. The result for Example 2 could be written:

$$\frac{15x - 30}{2x(5x+3)}$$

Self Check 2 Simplify: $\dfrac{\frac{1}{3} + \frac{1}{x}}{\frac{x}{5} - \frac{1}{2}}$.

EXAMPLE 3

ELEMENTARY Algebra f(x) Now™

Simplify: $\dfrac{\frac{6}{x} + y}{\frac{6}{y} + x}$.

Solution To write $\frac{6}{x} + y$ as a single rational expression, we build y to a fraction with a denominator of x and add. To write $\frac{6}{y} + x$ as a single rational expression, we build x to a fraction with a denominator of y and add.

$$\frac{\frac{6}{x} + y}{\frac{6}{y} + x} = \frac{\frac{6}{x} + \frac{y}{1} \cdot \frac{x}{x}}{\frac{6}{y} + \frac{x}{1} \cdot \frac{y}{y}} \quad \text{The LCD for the numerator is } x. \text{ The LCD for the denominator is } y.$$

$$= \frac{\frac{6}{x} + \frac{xy}{x}}{\frac{6}{y} + \frac{xy}{y}} \quad \text{Multiply the numerators. Multiply the denominators.}$$

$$= \frac{\frac{6+xy}{x}}{\frac{6+xy}{y}} \quad \text{Add in the numerator and denominator.}$$

Now that the numerator and the denominator of the complex fraction are single expressions, we can perform the division.

$$\frac{\frac{6+xy}{x}}{\frac{6+xy}{y}} = \frac{6+xy}{x} \div \frac{6+xy}{y} \quad \text{Write the division using a } \div \text{ symbol.}$$

$$= \frac{6+xy}{x} \cdot \frac{y}{6+xy} \quad \text{Multiply by the reciprocal of } \frac{6+xy}{y}.$$

$$= \frac{y(6+xy)}{x(6+xy)} \quad \text{Multiply the numerators. Multiply the denominators.}$$

Success Tip

Simplifying using division works well when a complex fraction is written, or can be easily written, as a quotient of two single rational expressions.

$$= \frac{y\overset{1}{\cancel{(6 + xy)}}}{x\underset{1}{\cancel{(6 + xy)}}}$$ Simplify the result by removing a factor equal to 1.

$$= \frac{y}{x}$$

Self Check 3 Simplify: $\dfrac{\frac{2}{a} - b}{\frac{2}{b} - a}$.

SIMPLIFYING COMPLEX FRACTIONS USING THE LCD

A second method for simplifying complex fractions uses the concepts of LCD and multiplication by a form of 1.

Simplifying Complex Fractions

1. Find the LCD of all rational expressions in the complex fraction.
2. Multiply the complex fraction by 1 in the form $\frac{\text{LCD}}{\text{LCD}}$.
3. Perform the operations in the numerator and denominator. No fractional expressions should remain within the complex fraction.
4. Simplify the result, if possible.

EXAMPLE 4

ELEMENTARY Algebra f(x) Now™

Simplify: $\dfrac{\frac{1}{2} - \frac{1}{x}}{\frac{x}{3} + \frac{1}{5}}$.

Solution The denominators of the rational expressions that appear in the complex fraction are 2, x, 3, and 5. Thus, their LCD is $2 \cdot x \cdot 3 \cdot 5 = 30x$.

We now multiply the complex fraction by a factor equal to 1, using the LCD: $\frac{30x}{30x} = 1$.

Success Tip

With method 2, each term of the numerator and each term of the denominator of the complex fraction is multiplied by the LCD. Arrows can be helpful in showing this. For Example 4, we can write

$$\frac{\left(\frac{1}{2} - \frac{1}{x}\right)30x}{\left(\frac{x}{3} + \frac{1}{5}\right)30x}$$

$$\frac{\frac{1}{2} - \frac{1}{x}}{\frac{x}{3} + \frac{1}{5}} = \frac{\frac{1}{2} - \frac{1}{x}}{\frac{x}{3} + \frac{1}{5}} \cdot \frac{30x}{30x}$$

$$= \frac{\left(\frac{1}{2} - \frac{1}{x}\right)30x}{\left(\frac{x}{3} + \frac{1}{5}\right)30x}$$ Multiply the numerators. Multiply the denominators.

$$= \frac{\frac{1}{2}(30x) - \frac{1}{x}(30x)}{\frac{x}{3}(30x) + \frac{1}{5}(30x)}$$ In the numerator and the denominator, distribute the multiplication by $30x$.

$$= \frac{15x - 30}{10x^2 + 6x}$$ Perform the multiplications by $30x$.

To attempt to simplify the result, factor the numerator and denominator. Since they do not have a common factor, the result is in simplest form.

$$\frac{15x - 30}{10x^2 + 6x} = \frac{15(x - 2)}{2x(5x + 3)}$$

Success Tip

When simplifying a complex fraction, the same result will be obtained regardless of the method used.

Self Check 4 Use method 2 to simplify: $\dfrac{\frac{1}{4} - \frac{1}{x}}{\frac{x}{5} + \frac{1}{3}}$.

EXAMPLE 5

ELEMENTARY Algebra $f(x)$ Now™

Simplify: $\dfrac{\frac{1}{8} - \frac{1}{y}}{\frac{8 - y}{8}}$.

Solution The denominators of the rational expressions that appear in the complex fraction are 8, y, and 8. Therefore, the LCD is $8y$ and we multiply the complex fraction by a factor equal to 1, using the LCD: $\frac{8y}{8y} = 1$.

$$\frac{\frac{1}{8} - \frac{1}{y}}{\frac{8 - y}{8}} = \frac{\frac{1}{8} - \frac{1}{y}}{\frac{8 - y}{8}} \cdot \frac{8y}{8y}$$

$$= \frac{\left(\frac{1}{8} - \frac{1}{y}\right)8y}{\left(\frac{8 - y}{8}\right)8y}$$ Multiply the numerators. Multiply the denominators.

$$= \frac{\frac{1}{8}(8y) - \frac{1}{y}(8y)}{\left(\frac{8 - y}{8}\right)8y}$$ In the numerator, distribute the multiplication by $8y$.

$$= \frac{y - 8}{(8 - y)y}$$ Perform each multiplication.

$$= \frac{\overset{-1}{\cancel{y - 8}}}{\underset{1}{\cancel{(8 - y)}}y}$$ Since $y - 8$ and $8 - y$ are opposites, replace $\frac{y - 8}{8 - y}$ with $\frac{-1}{1}$.

$$= -\frac{1}{y}$$

The Language of Algebra

After multiplying a complex fraction by $\frac{\text{LCD}}{\text{LCD}}$ and performing the multiplications, the numerator and denominator of the complex fraction will be cleared of fractions.

Self Check 5 Simplify: $\dfrac{\frac{10 - n}{10}}{\frac{1}{10} - \frac{1}{n}}$.

EXAMPLE 6

ELEMENTARY Algebra f(x) Now™

Simplify: $\dfrac{1}{1 + \dfrac{1}{x + 1}}$.

Solution The only rational expression in the complex fraction has the denominator $x + 1$. Therefore, the LCD is $x + 1$. So we multiply the complex fraction by a factor equal to 1, using the LCD: $\frac{x + 1}{x + 1} = 1$.

$$\frac{1}{1 + \dfrac{1}{x + 1}} = \frac{1}{1 + \dfrac{1}{x + 1}} \cdot \frac{x + 1}{x + 1}$$

$$= \frac{1(x + 1)}{\left(1 + \dfrac{1}{x + 1}\right)(x + 1)}$$ Multiply the numerators. Multiply the denominators.

$$= \frac{1(x + 1)}{1(x + 1) + \dfrac{1}{x + 1}(x + 1)}$$ In the denominator, distribute the multiplication by $x + 1$.

$$= \frac{x + 1}{x + 1 + 1}$$ Perform each multiplication.

$$= \frac{x + 1}{x + 2}$$ Combine like terms in the denominator.

The result does not simplify.

Success Tip

Simplifying using the LCD works well when the complex fraction has sums and/or differences in the numerator or denominator.

Self Check 6 Simplify: $\dfrac{2}{\dfrac{1}{x + 2} + 2}$.

Answers to Self Checks **1.** $\frac{5}{6}$ **2.** $\frac{10(x + 3)}{3x(2x - 5)}$ **3.** $\frac{b}{a}$ **4.** $\frac{15(x - 4)}{4x(3x + 5)}$ **5.** $-n$ **6.** $\frac{2(x + 2)}{2x + 5}$

6.5 STUDY SET

ELEMENTARY Algebra f(x) Now™

VOCABULARY **Fill in the blanks.**

1. The expression $\dfrac{\frac{2}{3} - \frac{1}{x}}{\frac{x - 3}{4}}$ is called a ________ rational expression or, more simply, a ________ fraction.

2. In a complex fraction, the ________ is above the main fraction bar and the ________ is below it.

3. To find the ________ of $\frac{x + 8}{x + 7}$, we invert it.

4. The ______ common denominator of all the rational expressions in $\dfrac{\frac{1}{3} - \frac{1}{x}}{\frac{3}{x}}$ is $3x$.

CONCEPTS **Fill in the blanks.**

5. Method 1: To simplify a complex fraction, write its numerator and denominator as ______ rational expressions. Then perform the division by multiplying the numerator of the complex fraction by the ________ of the denominator of the complex fraction.

6. Method 2: To simplify a complex fraction, find the ______ of all rational expressions in the complex fraction. Multiply the complex fraction by ____ in the form $\frac{\text{LCD}}{\text{LCD}}$. Then perform the operations.

7. Consider $\dfrac{\frac{x-3}{4}}{\frac{1}{12}-\frac{x}{6}}$.

a. What is the numerator of the complex fraction?

b. Is the numerator a single rational expression?

c. What is the denominator of the complex fraction?

d. Is the denominator a single rational expression?

8. Fill in the blank.

$$\frac{\frac{12}{y^2}}{\frac{4}{y^3}} = \frac{12}{y^2} \; \square \; \frac{4}{y^3}$$

9. Consider the complex fraction $\dfrac{\frac{1}{y}-\frac{1}{3}}{\frac{5}{6}+\frac{1}{y}}$.

a. What are the denominators of the rational expressions in the complex fraction?

b. What is the LCD of all rational expressions in the complex fraction?

c. To simplify the complex fraction using method 2, it should be multiplied by what form of 1?

10. Fill in the blanks.

$$\frac{\left(\frac{1}{5}-\frac{1}{a}\right)20a}{\left(\frac{a}{4}+\frac{1}{a}\right)20a} = \frac{\frac{1}{5}(\square)-\frac{1}{a}(\square)}{\frac{a}{4}(\square)+\frac{1}{a}(\square)}$$

Find each product.

11. $\frac{x}{12}(24)$

12. $\frac{1}{2}(20x)$

13. $\frac{2}{3}(6y^2)$

14. $\frac{9}{x-8}(x-8)$

NOTATION

15. Write the following expression as a complex fraction.

$$\frac{4x^2}{15} \div \frac{16x}{25}$$

16. Draw arrows to show how the distributive property should be applied in the denominator.

$$\frac{\frac{1}{x-6}(x-6)}{\left(\frac{1}{x-6}+2\right)(x-6)}$$

PRACTICE Simplify each complex fraction.

17. $\dfrac{\frac{x}{2}}{\frac{6}{5}}$

18. $\dfrac{\frac{9}{4}}{\frac{7}{x}}$

19. $\dfrac{\frac{2}{3}}{\frac{3}{4}}$

20. $\dfrac{\frac{3}{5}}{\frac{2}{7}}$

21. $\dfrac{\frac{x}{y}}{\frac{1}{x}}$

22. $\dfrac{\frac{y}{x}}{\frac{x}{xy}}$

23. $\dfrac{\frac{n}{8}}{\frac{1}{n^2}}$

24. $\dfrac{\frac{1}{m}}{\frac{m^3}{15}}$

25. $\dfrac{\frac{1}{x}-3}{\frac{5}{x}+2}$

26. $\dfrac{\frac{1}{y}+3}{\frac{3}{y}-2}$

27. $\dfrac{\frac{2}{3}+1}{\frac{1}{3}+1}$

28. $\dfrac{\frac{3}{5}-2}{\frac{2}{5}-2}$

29. $\dfrac{\frac{4a}{11}}{\frac{6a}{55}}$

30. $\dfrac{\frac{14}{15m}}{\frac{21}{25m}}$

31. $\dfrac{\frac{40x^2}{20x}}{9}$

32. $\dfrac{\frac{18n^2}{6n}}{13}$

33. $\dfrac{\frac{1}{y}-\frac{5}{2}}{\frac{3}{y}}$

34. $\dfrac{\frac{1}{6}-\frac{5}{s}}{\frac{2}{s}}$

35. $\dfrac{\dfrac{d+2}{2}}{\dfrac{d}{3}-\dfrac{d}{4}}$

36. $\dfrac{\dfrac{d^2}{4}+\dfrac{4d}{5}}{\dfrac{d+1}{2}}$

37. $\dfrac{\dfrac{s+15}{16}}{\dfrac{s+15}{8}}$

38. $\dfrac{\dfrac{t-5}{12}}{\dfrac{t-5}{4}}$

39. $\dfrac{\dfrac{1}{2}+\dfrac{3}{4}}{\dfrac{3}{2}+\dfrac{1}{4}}$

40. $\dfrac{\dfrac{2}{3}-\dfrac{5}{2}}{\dfrac{2}{3}-\dfrac{3}{2}}$

41. $\dfrac{\dfrac{x^4}{30}}{\dfrac{7x}{15}}$

42. $\dfrac{\dfrac{5x^2}{24}}{\dfrac{x^5}{56}}$

43. $\dfrac{\dfrac{2}{s}-\dfrac{2}{s^2}}{\dfrac{4}{s^3}+\dfrac{4}{s^2}}$

44. $\dfrac{\dfrac{2}{x^3}-\dfrac{2}{x}}{\dfrac{4}{x}+\dfrac{8}{x^2}}$

45. $\dfrac{-\dfrac{3}{x^3}}{\dfrac{6}{x^5}}$

46. $\dfrac{\dfrac{18}{x^6}}{-\dfrac{2}{x^8}}$

47. $\dfrac{\dfrac{2}{x}+2}{\dfrac{4}{x}+2}$

48. $\dfrac{\dfrac{3}{x}-3}{\dfrac{9}{x}-3}$

49. $\dfrac{\dfrac{2x-8}{15}}{\dfrac{3x-12}{35x}}$

50. $\dfrac{\dfrac{3x-9}{8x}}{\dfrac{5x-15}{32}}$

51. $\dfrac{\dfrac{t-6}{16}}{\dfrac{12-2t}{t}}$

52. $\dfrac{\dfrac{15-30c}{28}}{\dfrac{2c-1}{c}}$

53. $\dfrac{\dfrac{2}{c^2}}{\dfrac{1}{c}+\dfrac{5}{4}}$

54. $\dfrac{\dfrac{7}{s^2}}{\dfrac{1}{s}+\dfrac{10}{3}}$

55. $\dfrac{\dfrac{1}{a^2b}-\dfrac{5}{ab}}{\dfrac{3}{ab}-\dfrac{7}{ab^2}}$

56. $\dfrac{\dfrac{3}{ab^2}+\dfrac{6}{a^2b}}{\dfrac{6}{a}-\dfrac{9}{b^2}}$

57. $\dfrac{\dfrac{3y}{x}-y}{y-\dfrac{y}{x}}$

58. $\dfrac{\dfrac{y}{x}+3y}{y+\dfrac{2y}{x}}$

59. $\dfrac{\dfrac{b^2-81}{18a^2}}{\dfrac{4b-36}{9a}}$

60. $\dfrac{\dfrac{8x-64}{y}}{\dfrac{x^2-64}{y^2}}$

61. $\dfrac{4-\dfrac{1}{8h}}{12+\dfrac{3}{4h}}$

62. $\dfrac{12+\dfrac{1}{3b}}{12-\dfrac{1}{b^2}}$

63. $\dfrac{1}{\dfrac{1}{x}+\dfrac{1}{y}}$

64. $\dfrac{1}{\dfrac{b}{a}-\dfrac{a}{b}}$

65. $\dfrac{\dfrac{1}{x+1}}{1+\dfrac{1}{x+1}}$

66. $\dfrac{\dfrac{1}{x-1}}{1-\dfrac{1}{x-1}}$

67. $\dfrac{\dfrac{x}{x+2}}{\dfrac{x}{x+2}+x}$

68. $\dfrac{\dfrac{2}{x-2}}{\dfrac{2}{x-2}-1}$

69. $\dfrac{\dfrac{5t^2}{9x^2}}{\dfrac{3t}{x^2t}}$

70. $\dfrac{\dfrac{5w^2}{4tz}}{\dfrac{15wt}{z^2}}$

71. $\dfrac{\dfrac{m^2-4}{3m+3}}{\dfrac{2m+4}{m+1}}$

72. $\dfrac{\dfrac{d^2-16}{5d+10}}{\dfrac{2d-8}{25}}$

73. $\dfrac{\dfrac{2}{x}}{\dfrac{2}{y}-\dfrac{4}{x}}$

74. $\dfrac{\dfrac{2y}{3}}{\dfrac{2y}{3}-\dfrac{8}{y}}$

75. $\dfrac{\dfrac{m}{n}+\dfrac{n}{m}}{\dfrac{m}{n}-\dfrac{n}{m}}$

76. $\dfrac{\dfrac{2a}{b}-\dfrac{b}{a}}{\dfrac{2a}{b}+\dfrac{b}{a}}$

77. $\dfrac{3+\dfrac{3}{x-1}}{3-\dfrac{3}{x-1}}$

78. $\dfrac{2-\dfrac{2}{x+1}}{2+\dfrac{2}{x+1}}$

APPLICATIONS

79. GARDENING TOOLS What is the result when the opening of the cutting blades is divided by the opening of the handles? Express the result in simplest form.

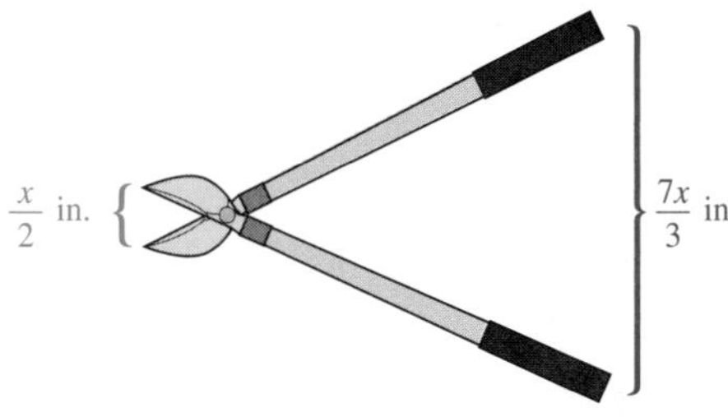

80. EARNED RUN AVERAGE The earned run average (ERA) is a statistic that gives the average number of earned runs a pitcher allows. For a softball pitcher, this is based on a six-inning game. The formula for ERA is

$$\text{ERA} = \frac{\text{earned runs}}{\frac{\text{innings pitched}}{6}}$$

Simplify the complex fraction on the right-hand side of the equation.

81. ELECTRONICS In electronic circuits, resistors oppose the flow of an electric current. To find the total resistance of a parallel combination of two resistors, we can use the formula

$$\text{Total resistance} = \frac{1}{\frac{1}{R_1} + \frac{1}{R_2}}$$

where R_1 is the resistance of the first resistor and R_2 is the resistance of the second. Simplify the complex fraction on the right-hand side of the formula.

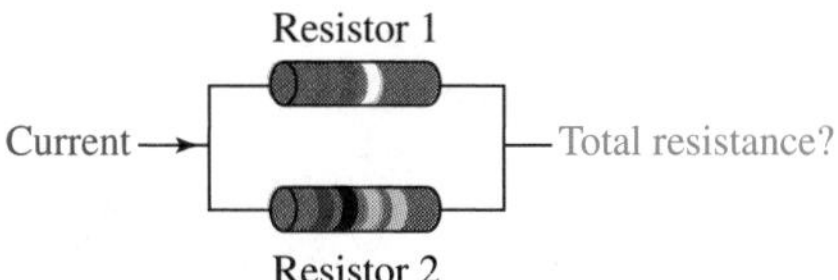

82. DATA ANALYSIS Use the data in the table to find the average measurement for the three-trial experiment.

	Trial 1	Trial 2	Trial 3
Measurement	$\frac{k}{2}$	$\frac{k}{3}$	$\frac{k}{2}$

WRITING

83. What is a complex fraction? Give several examples.

84. Explain how to use method 1 to simplify $\dfrac{1 + \frac{1}{x}}{3 - \frac{1}{x}}$.

85. Explain how to use method 2 to simplify the expression in Problem 84.

86. **a.** List an advantage and a disadvantage of using method 1 to simplify a complex fraction.

b. List an advantage and a disadvantage of using method 2 to simplify a complex fraction.

REVIEW **Simplify each expression. Write each answer without using parentheses or negative exponents.**

87. $(8x)^0$

88. $\dfrac{8x^0}{4}$

89. $\left(\dfrac{3r}{4r^3}\right)^4$

90. $\left(\dfrac{12y^{-3}}{3y^2}\right)^{-2}$

91. $\left(\dfrac{6r^{-2}}{2r^3}\right)^{-2}$

92. $\left(\dfrac{4x^3}{5x^{-3}}\right)^{-2}$

CHALLENGE PROBLEMS **Simplify.**

93. $\dfrac{\frac{h}{h^2 + 3h + 2}}{\frac{4}{h + 2} - \frac{4}{h + 1}}$

94. $\dfrac{\frac{2}{b^2 - 1} - \frac{3}{ab - a}}{\frac{3}{ab - a} - \frac{2}{b^2 - 1}}$

95. $a + \dfrac{a}{1 + \frac{a}{a + 1}}$

96. $\dfrac{y^{-2} + 1}{y^{-2} - 1}$

6.6 Solving Rational Equations

- Solving Rational Equations
- Solving Formulas for a Specified Variable

We have solved equations such as $\frac{2x}{3} = \frac{x}{6} + \frac{3}{2}$ by multiplying both sides of the equation by the LCD. With this approach, the equation that results is equivalent to the original equation but easier to solve because it is cleared of fractions.

Success Tip

Before multiplying both sides of an equation by the LCD, enclose the left-hand and right-hand sides with parentheses:

$$\left(\frac{2x}{3}\right) = \left(\frac{x}{6} + \frac{3}{2}\right)$$

$$\frac{2x}{3} = \frac{x}{6} + \frac{3}{2}$$

$$6\left(\frac{2x}{3}\right) = 6\left(\frac{x}{6} + \frac{3}{2}\right)$$ Multiply both sides of the equation by the LCD of $\frac{2x}{3}$, $\frac{x}{6}$, and $\frac{3}{2}$, which is 6.

$$6\left(\frac{2x}{3}\right) = 6\left(\frac{x}{6}\right) + 6\left(\frac{3}{2}\right)$$ Distribute the multiplication by 6.

$$4x = x + 9$$ Perform each multiplication.

$$3x = 9$$ Subtract x from both sides.

$$x = 3$$ Divide both sides by 3.

Equations such as $\frac{2x}{3} = \frac{x}{6} + \frac{3}{2}$ are called *rational equations.*

Rational Equations A **rational equation** is an equation that contains one or more rational expressions.

Rational equations often have a variable in the denominator. Some examples are:

$$\frac{2}{x} + \frac{1}{2} = \frac{5}{2x}, \qquad y - \frac{12}{y} = 4, \qquad \text{and} \qquad \frac{11x}{x-5} = 6 + \frac{55}{x-5}$$

In this section, we will extend the fraction-clearing strategy to solve rational equations like those listed above.

SOLVING RATIONAL EQUATIONS

To solve a rational equation, we must find all values of the variable that make the equation true. However, before solving a rational equation, we must identify any values of the variable that make an expression in the equation undefined. A number cannot be a solution of the equation if, when substituted for the variable, it makes a denominator equal to 0.

The following steps can be used to solve an equation containing rational expressions.

Solving Rational Equations

1. Determine which numbers cannot be solutions of the equation.
2. Multiply both sides of the equation by the LCD of all rational expressions in the equation. This clears the equation of fractions.
3. Solve the resulting equation.
4. Check all possible solutions in the original equation.

EXAMPLE 1

Solve: $\frac{2}{x} + \frac{1}{2} = \frac{5}{2x}$.

ELEMENTARY Algebra $f(x)$ Now™

Solution If x is 0, the denominators of $\frac{2}{x}$ and $\frac{5}{2x}$ are 0 and the expressions would be undefined. Therefore, 0 cannot be a solution.

Since the denominators are x, 2, and $2x$, we multiply both sides of the equation by the LCD, $2x$, to clear the fractions.

$$\frac{2}{x} + \frac{1}{2} = \frac{5}{2x}$$

$$2x\left(\frac{2}{x} + \frac{1}{2}\right) = 2x\left(\frac{5}{2x}\right) \quad \text{Multiply both sides by } 2x.$$

$$2x\left(\frac{2}{x}\right) + 2x\left(\frac{1}{2}\right) = 2x\left(\frac{5}{2x}\right) \quad \text{Distribute the multiplication by } 2x.$$

$$4 + x = 5 \quad \text{Simplify.}$$

To solve the resulting equation, subtract 4 from both sides.

$$4 + x - 4 = 5 - 4$$
$$x = 1$$

To check, we substitute 1 for x in the original equation.

$$\frac{2}{x} + \frac{1}{2} = \frac{5}{2x}$$

$$\frac{2}{1} + \frac{1}{2} \stackrel{?}{=} \frac{5}{2(1)} \quad \text{Substitute 1 for } x.$$

$$\frac{5}{2} = \frac{5}{2} \quad \text{Evaluate the left-hand side: } \frac{2}{1} + \frac{1}{2} = \frac{4}{2} + \frac{1}{2} = \frac{5}{2}.$$

Since we obtain a true statement, 1 is the solution.

Notation

To streamline multiplication, we can remove factors equal to 1 in the following way:

$$\overset{1}{2x}\left(\frac{2}{\underset{1}{x}}\right)$$

Here, we simplify by replacing $\frac{x}{x}$ with $\frac{1}{1}$.

Success Tip

Don't confuse procedures. To *simplify the expression* $\frac{2}{x} + \frac{1}{2}$, we build each fraction to have the LCD $2x$, add the numerators, and write the sum over the LCD. To *solve the equation* $\frac{2}{x} + \frac{1}{2} = \frac{5}{2x}$, we multiply both sides by the LCD $2x$ to eliminate the denominators.

Self Check 1 Solve: $\frac{1}{3} + \frac{4}{3x} = \frac{5}{x}$.

EXAMPLE 2

Solve: $y - \frac{12}{y} = 4$.

ELEMENTARY Algebra $f(x)$ Now™

Solution If y is 0, the denominator of $\frac{12}{y}$ is 0, and the expression would be undefined. Therefore, 0 cannot be a solution.

Since the only denominator is y, we multiply both sides of the equation by y to clear the fractions.

$$y - \frac{12}{y} = 4$$

$$y\left(y - \frac{12}{y}\right) = y(4) \quad \text{Multiply both sides of the equation by } y.$$

$$y(y) - y\left(\frac{12}{y}\right) = y(4) \quad \text{Distribute the multiplication by } y.$$

$$y^2 - 12 = 4y \quad \text{Simplify.}$$

We can solve the resulting quadratic equation using the factoring method.

$$y^2 - 4y - 12 = 0 \quad \text{Subtract } 4y \text{ from both sides.}$$

$$(y - 6)(y + 2) = 0 \quad \text{Factor the trinomial.}$$

$$y - 6 = 0 \quad \text{or} \quad y + 2 = 0 \quad \text{Set each factor equal to 0.}$$

$$y = 6 \quad \Big| \quad y = -2 \quad \text{Solve each equation.}$$

There are two possible solutions, 6 and -2.

***Check:* $y = 6$**

$$y - \frac{12}{y} = 4$$

$$6 - \frac{12}{6} \stackrel{?}{=} 4$$

$$6 - 2 \stackrel{?}{=} 4$$

$$4 = 4$$

***Check:* $y = -2$**

$$y - \frac{12}{y} = 4$$

$$-2 - \frac{12}{-2} \stackrel{?}{=} 4$$

$$-2 - (-6) \stackrel{?}{=} 4$$

$$4 = 4$$

The solutions of $y - \frac{12}{y} = 4$ are 6 and -2.

Self Check 2 Solve: $x - \frac{24}{x} = -5$.

EXAMPLE 3

ELEMENTARY Algebra $f(x)$ Now™

Solve: $\frac{11x}{x - 5} = 6 + \frac{55}{x - 5}$.

Solution If x is 5, the denominators of $\frac{11x}{x-5}$ and $\frac{55}{x-5}$ are 0, and the expressions are undefined. Therefore, 5 cannot be a solution of the equation.

Since both denominators are $x - 5$, we multiply both sides by the LCD, $x - 5$, to clear the fractions.

Caution

When solving rational equations, each term on both sides must be multiplied by the LCD.

$$\frac{11x}{x - 5} = 6 + \frac{55}{x - 5}$$

$$(x - 5)\left(\frac{11x}{x - 5}\right) = (x - 5)\left(6 + \frac{55}{x - 5}\right) \quad \text{Multiply both sides of the equation by } x - 5.$$

$$(x - 5)\left(\frac{11x}{x - 5}\right) = (x - 5)6 + (x - 5)\left(\frac{55}{x - 5}\right) \quad \text{Distribute the multiplication by } x - 5.$$

$$11x = (x - 5)6 + 55 \quad \text{Simplify: } \frac{x-5}{x-5} = 1.$$

$$11x = 6x - 30 + 55 \quad \text{Distribute the multiplication by 6.}$$

To solve the resulting equation, we proceed as follows:

The Language of Algebra

Extraneous means not a vital part. Mathematicians speak of *extraneous* solutions. Rock groups don't want any *extraneous* sounds (like humming or feedback) coming from their amplifiers. Artists erase any *extraneous* marks on their sketches.

$$\begin{aligned} 11x &= 6x + 25 && \text{Simplify: } -30 + 55 = 25. \\ 5x &= 25 && \text{Subtract } 6x \text{ from both sides.} \\ x &= 5 && \text{Divide both sides by 5.} \end{aligned}$$

We have determined that 5 makes both denominators in the original equation 0. Therefore, 5 is not a solution. Since 5 is the only possible solution, and it must be rejected, it follows that $\frac{11x}{x-5} = 6 + \frac{55}{x-5}$ has no solution.

When solving an equation, a possible solution that does not satisfy the original equation is called an **extraneous solution.** In this example, 5 is an extraneous solution.

Self Check 3 Solve: $\frac{9x}{x-6} = 3 + \frac{54}{x-6}$, if possible.

EXAMPLE 4

Solve: $\frac{x+5}{x+3} + \frac{1}{x^2 + 2x - 3} = 1.$

ELEMENTARY Algebra $f(x)$ Now™

Solution To determine the restrictions on the variable, we factor the second denominator.

$$\frac{x+5}{x+3} + \frac{1}{(x+3)(x-1)} = 1 \qquad \text{If } x \text{ is } -3\text{, the first denominator is 0. If } x \text{ is } -3 \text{ or 1, the second denominator is 0.}$$

We see that -3 and 1 cannot be solutions of the equation, because they make rational expressions in the equation undefined.

Since the denominators are $x + 3$ and $(x + 3)(x - 1)$, we multiply both sides of the equation by the LCD, $(x + 3)(x - 1)$, to clear the fractions.

$$(x+3)(x-1)\left[\frac{x+5}{x+3} + \frac{1}{(x+3)(x-1)}\right] = (x+3)(x-1)[1]$$

$$(x+3)(x-1)\frac{x+5}{x+3} + (x+3)(x-1)\frac{1}{(x+3)(x-1)} = (x+3)(x-1)1 \qquad \text{Distribute the multiplication by } (x+3)(x-1).$$

$$(x-1)(x+5) + 1 = (x+3)(x-1) \qquad \text{Simplify: } \frac{x+3}{x+3} = 1 \text{ and } \frac{x-1}{x-1} = 1.$$

To solve the resulting equation, we multiply the binomials on the left-hand side and the right-hand side, and proceed as follows.

$$\begin{aligned} x^2 + 4x - 5 + 1 &= x^2 + 2x - 3 && \text{Find } (x-1)(x+5) \text{ and } (x+3)(x-1). \\ x^2 + 4x - 4 &= x^2 + 2x - 3 && \text{Combine like terms: } -5 + 1 = -4. \\ 4x - 4 &= 2x - 3 && \text{Subtract } x^2 \text{ from both sides.} \\ 2x - 4 &= -3 && \text{Subtract } 2x \text{ from both sides.} \\ 2x &= 1 && \text{Add 4 to both sides.} \\ x &= \frac{1}{2} && \text{Divide both sides by 2.} \end{aligned}$$

A check will show that $\frac{1}{2}$ is the solution of the original equation.

Self Check 4 Solve: $\frac{1}{x+3} + \frac{1}{x-3} = \frac{5}{x^2-9}$.

SOLVING FORMULAS FOR A SPECIFIED VARIABLE

Recall that a **formula** is an equation that states a known relationship between two or more variables. Many formulas are expressed as rational equations.

EXAMPLE 5

ELEMENTARY Algebra f(x) Now™

Determining a child's dosage. The formula

$$C = \frac{AD}{A+12}$$

is called Young's rule. It is a way to find the approximate child's dose C of a prescribed medication, where A is the age of the child in years and D is the recommended dosage for an adult. Solve the formula for D.

Solution To clear the equation of the fraction, we multiply both sides by $A + 12$.

$$C = \frac{AD}{A+12}$$

$$(A+12)(C) = (A+12)\left(\frac{AD}{A+12}\right)$$

$$(A+12)C = AD \qquad \text{Simplify: } \frac{A+12}{A+12} = 1.$$

$$AC + 12C = AD \qquad \text{Distribute the multiplication by } C.$$

$$\frac{AC+12C}{A} = D \qquad \text{To isolate } D \text{, divide both sides by } A.$$

Solving Young's formula for D, we have $D = \frac{AC + 12C}{A}$.

EXAMPLE 6

Photography. The design of a camera lens uses the formula

$$\frac{1}{f} = \frac{1}{p} + \frac{1}{q}$$

where f is the focal length of the lens, p is the distance from the lens to the object, and q is the distance from the lens to the image. Solve the formula for q.

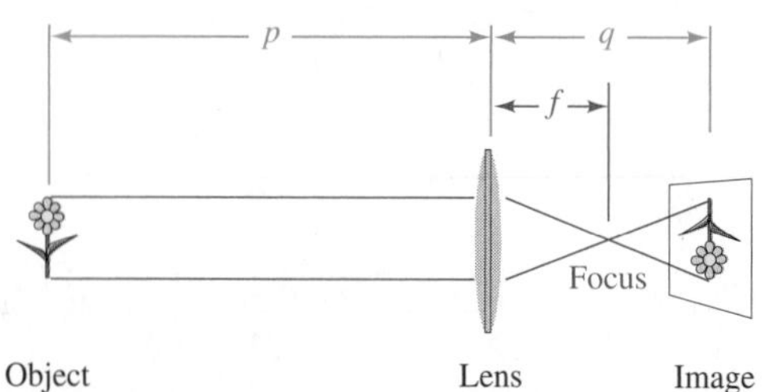

Solution The denominators are f, p, and q. To clear the equation of fractions, we multiply both sides by the LCD, fpq.

$$\frac{1}{f} = \frac{1}{p} + \frac{1}{q}$$

$$fpq\left(\frac{1}{f}\right) = fpq\left(\frac{1}{p} + \frac{1}{q}\right) \quad \text{Multiply both sides of the equation by the LCD.}$$

$$fpq\left(\frac{1}{f}\right) = fpq\left(\frac{1}{p}\right) + fpq\left(\frac{1}{q}\right) \quad \text{Distribute the multiplication by } fpq.$$

$$pq = fq + fp \quad \text{Simplify: } \tfrac{f}{f} = 1, \tfrac{p}{p} = 1, \text{ and } \tfrac{q}{q} = 1.$$

If we subtract fq from both sides, all terms that contain q will be on the left-hand side.

$$pq - fq = fp \quad \text{Subtract } fq \text{ from both sides.}$$

$$q(p - f) = fp \quad \text{Factor out } q \text{ from the terms on the left-hand side.}$$

$$\frac{q(p - f)}{p - f} = \frac{fp}{p - f} \quad \text{To isolate } q\text{, divide both sides by } p - f.$$

$$q = \frac{fp}{p - f} \quad \text{Simplify: } \frac{p - f}{p - f} = 1.$$

Solving the formula for q, we have $q = \dfrac{fp}{p - f}$.

Self Check 6 Solve the formula in Example 6 for p.

Answers to Self Checks **1.** 11 **2.** 3, −8 **3.** no solution **4.** $\frac{5}{2}$ **6.** $p = \dfrac{fq}{q - f}$

6.6 STUDY SET  ELEMENTARY Algebra $f(x)$ Now™

VOCABULARY Fill in the blanks.

1. Equations that contain one or more rational expressions, such as $\frac{x}{x+2} = 4 + \frac{10}{x+2}$, are called ________ equations.

2. To ________ a rational equation we find all the values of the variable that make the equation true.

3. To ________ a rational equation of fractions, multiply both sides by the LCD of all rational expressions in the equation.

4. If x is 3, the denominator of rational expression $\frac{4x}{x-3}$ is 0, and the expression is ________.

5. $x^2 - x + 2 = 0$ is a ________ equation.

6. When solving a rational equation, if we obtain a number that does not satisfy the original equation, the number is called an ________ solution.

CONCEPTS

7. Is 5 a solution of the following equations?

a. $\dfrac{1}{x - 1} = 1 - \dfrac{3}{x - 1}$

b. $\dfrac{x}{x - 5} = 3 + \dfrac{5}{x - 5}$

8. A student was asked to solve a rational equation. The first step of his solution is as follows:

$$12x\left(\frac{5}{x} + \frac{2}{3}\right) = 12x\left(\frac{7}{4x}\right)$$

a. What equation was he asked to solve?

b. What LCD is used to clear the equation of fractions?

9. Consider the rational equation $\frac{x}{x-3} = \frac{1}{x} + \frac{2}{x-3}$.

a. What values of x make a denominator 0?

b. What values of x make a rational expression undefined?

c. What numbers can't be solutions of the equation?

By what should both sides of the equation be multiplied to clear it of fractions?

10. $x + \frac{11}{x} = 3$

11. $\frac{1}{y} = 20 - \frac{5}{y}$

12. $\frac{2x}{x-6} = 4 + \frac{1}{x-6}$

13. $\frac{x}{x+8} = \frac{4}{x-8}$

14. $\frac{x}{5} = \frac{3x}{10} + \frac{7}{2x}$

15. $\frac{x}{x^2-4} = \frac{4}{x-2}$

16. Perform each multiplication.

a. $4x\left(\frac{3}{4x}\right)$

b. $(x+6)(x-2)\left(\frac{3}{x-2}\right)$

17. Fill in the blanks.

$$8x\left(\frac{3}{2x}\right) = 8x\left(\frac{1}{8x}\right) + 8x\left(\frac{5}{4}\right)$$
$$\underline{\quad} = \underline{\quad} + \underline{\quad}$$

18. Factor out a from the terms on the left-hand side of $ab - ca = cb$.

NOTATION

19. Complete the solution.

$$\frac{2}{a} + \frac{1}{2} = \frac{7}{2a}$$
$$\underline{\quad}\left(\frac{2}{a} + \frac{1}{2}\right) = \underline{\quad}\left(\frac{7}{2a}\right)$$
$$\underline{\quad}\left(\frac{2}{a}\right) + \underline{\quad}\left(\frac{1}{2}\right) = \underline{\quad}\left(\frac{7}{2a}\right)$$
$$\underline{\quad} + a = \underline{\quad}$$
$$4 + a - 4 = 7 - 4$$
$$a = \underline{\quad}$$

20. A student solved a rational equation and found 8 to be a possible solution. When she checked 8, she obtained $\frac{3}{0} = \frac{1}{0} + \frac{2}{3}$. What conclusion can be drawn?

21. What operation is indicated when we write fpq?

22. Can $5x\left(\frac{2}{x} + \frac{4}{5}\right)$ be written as $5x \cdot \frac{2}{x} + \frac{4}{5}$? Explain.

PRACTICE **Solve each equation and check the result. If an equation has no solution, so indicate.**

23. $\frac{2}{3} = \frac{1}{2} + \frac{x}{6}$

24. $\frac{7}{4} = \frac{x}{8} + \frac{5}{2}$

25. $\frac{x}{18} = \frac{1}{3} - \frac{x}{2}$

26. $\frac{x}{4} = \frac{1}{2} - \frac{3x}{20}$

27. $\frac{a-1}{7} - \frac{a-2}{14} = \frac{1}{2}$

28. $\frac{3x-1}{6} - \frac{x+3}{2} = \frac{3x+4}{3}$

29. $\frac{3}{x} + 2 = 3$

30. $\frac{2}{x} + 9 = 11$

31. $\frac{x}{x-5} - \frac{5}{x-5} = 3$

32. $\frac{3}{y-2} + 1 = \frac{3}{y-2}$

33. $\frac{a}{4} - \frac{4}{a} = 0$

34. $0 = \frac{t}{3} - \frac{12}{t}$

35. $\frac{2}{y+1} + 5 = \frac{12}{y+1}$

36. $\frac{3}{p+6} - 2 = \frac{7}{p+6}$

37. $-\frac{1}{b} = \frac{5}{b} - 9 - \frac{6}{b}$

38. $\frac{3}{a} = \frac{4}{a} + 8 - \frac{1}{a}$

39. $\frac{1}{t+2} = \frac{t-1}{t+7}$

40. $\frac{d-7}{d-20} = \frac{1}{d+8}$

41. $\frac{5}{n} + \frac{5}{12} = 0$

42. $\frac{2}{m} + \frac{2}{33} = 0$

43. $\frac{1}{8} + \frac{2}{y} = \frac{1}{y} + \frac{1}{10}$

44. $\frac{7}{10} + \frac{4}{c} = \frac{1}{c} + \frac{11}{15}$

45. $\frac{1}{8} + \frac{2}{b} - \frac{1}{12} = 0$

46. $\frac{1}{14} + \frac{2}{n} - \frac{2}{21} = 0$

47. $\frac{4}{5x} = \frac{8}{x-5}$

48. $\frac{1}{6x} = \frac{2}{x-6}$

49. $\frac{1}{4} - \frac{5}{6} = \frac{1}{a}$

50. $\frac{5}{9} - \frac{1}{3} = \frac{1}{b}$

51. $\frac{3}{4h} + \frac{2}{h} = 1$

52. $\frac{5}{3k} + \frac{1}{k} = -2$

53. $\frac{3r}{2} - \frac{3}{r} = \frac{3r}{2} + 3$

54. $\frac{2p}{3} - \frac{1}{p} = \frac{2p - 1}{3}$

55. $\frac{1}{3} + \frac{2}{x - 3} = 1$

56. $\frac{3}{5} + \frac{7}{x + 2} = 2$

57. $\frac{z - 4}{z - 3} = \frac{z + 2}{z + 1}$

58. $\frac{a + 2}{a + 8} = \frac{a - 3}{a - 2}$

59. $\frac{v}{v + 2} + \frac{1}{v - 1} = 1$

60. $\frac{x}{x - 2} = 1 + \frac{1}{x - 3}$

61. $\frac{a^2}{a + 2} - \frac{4}{a + 2} = a$

62. $\frac{z^2}{z + 1} + 2 = \frac{1}{z + 1}$

63. $\frac{5}{x + 4} + \frac{1}{x + 4} = x - 1$

64. $\frac{7}{x - 3} + \frac{1}{x - 3} = x - 5$

65. $\frac{3}{x + 1} - \frac{x - 2}{2} = \frac{x - 2}{x + 1}$

66. $\frac{2}{x - 1} + \frac{x - 2}{3} = \frac{4}{x - 1}$

67. $\frac{b + 2}{b + 3} + 1 = \frac{-7}{b - 5}$

68. $\frac{x - 4}{x - 3} + \frac{x - 2}{x - 3} = x - 3$

69. $\frac{u}{u - 1} + \frac{1}{u} = \frac{u^2 + 1}{u^2 - u}$

70. $\frac{3}{x - 2} + \frac{1}{x} = \frac{2(3x + 2)}{x^2 - 2x}$

71. $\frac{n}{n^2 - 9} + \frac{n + 8}{n + 3} = \frac{n - 8}{n - 3}$

72. $\frac{7}{x - 5} - \frac{3}{x + 5} = \frac{40}{x^2 - 25}$

73. $\frac{x}{x - 1} - \frac{12}{x^2 - x} = \frac{-1}{x - 1}$

74. $y + \frac{2}{3} = \frac{2y - 12}{3y - 9}$

75. $1 - \frac{3}{b} = \frac{-8b}{b^2 + 3b}$

76. $\frac{7}{q^2 - q - 2} + \frac{1}{q + 1} = \frac{3}{q - 2}$

77. $\frac{3}{x - 1} - \frac{1}{x + 9} = \frac{18}{x^2 + 8x - 9}$

78. $\frac{5}{4y + 12} - \frac{3}{4} = \frac{5}{4y + 12} - \frac{y}{4}$

Solve each formula for the indicated variable.

79. $\frac{P}{n} = rt$ for P

80. $\frac{F}{m} = a$ for F

81. $\frac{a}{b} = \frac{c}{d}$ for d

82. $\frac{pc}{s} = \frac{t}{r}$ for c

83. $h = \frac{2A}{b + d}$ for A

84. $T = \frac{3R}{M - n}$ for R

85. $\frac{1}{a} + \frac{1}{b} = 1$ for a

86. $\frac{1}{a} - \frac{1}{b} = 1$ for b

87. $I = \frac{E}{R + r}$ for r

88. $\frac{S}{k + h} = E$ for k

89. $\frac{1}{r} + \frac{1}{s} = \frac{1}{t}$ for r

90. $\frac{5}{x} - \frac{4}{y} = \frac{5}{z}$ for x

91. $F = \frac{L^2}{6d} + \frac{d}{2}$ for L^2

92. $H = \frac{J^3}{cd} - \frac{K^3}{d}$ for J^3

APPLICATIONS

93. MEDICINE Radioactive tracers are used for diagnostic work in nuclear medicine. The **effective half-life** H of a radioactive material in an organism is given by the formula

$$H = \frac{RB}{R + B}$$

where R is the radioactive half-life and B is the biological half-life of the tracer. Solve the formula for R.

94. CHEMISTRY Charles's law describes the relationship between the volume and temperature of a gas that is kept at a constant pressure. It can be expressed as

$$\frac{V_1}{V_2} = \frac{T_1}{T_2}$$

where V_1 and V_2 are variables representing two different volumes, and T_1 and T_2 are variables representing two different temperatures. (Recall that the notation V_1 is read as *V sub one*.) Solve for V_2.

95. ELECTRONICS Most electronic circuits require resistors to make them work properly. Resistors are components that limit current. An important formula about resistors in a circuit is

$$\frac{1}{r} = \frac{1}{r_1} + \frac{1}{r_2}$$

Solve for r.

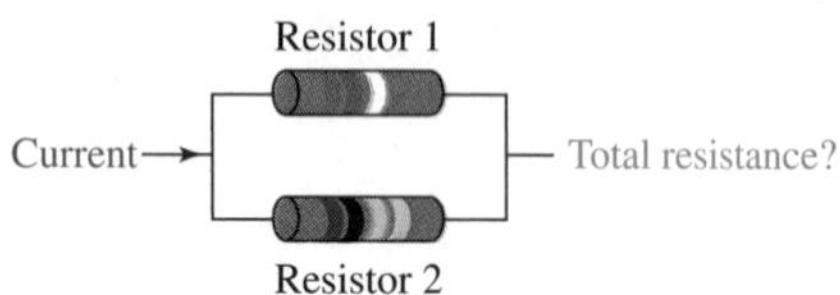

96. MATHEMATICAL FORMULAS To quickly find the sum of a list of fractions, such as

$$\frac{1}{2} + \frac{1}{4} + \frac{1}{8} + \frac{1}{16} + \frac{1}{32} + \frac{1}{64} + \frac{1}{128}$$

mathematicians use the formula

$$S = \frac{a(1 - r^n)}{1 - r}$$

Solve the formula for a.

WRITING

97. Explain how the multiplication property of equality is used to solve rational equations. Give an example.

98. When solving rational equations, how do you know whether a solution is extraneous?

99. What is meant by clearing a rational equation of fractions? Give an example.

100. Explain the difference between the procedure used to simplify

$$\frac{1}{x} + \frac{1}{3}$$

and the procedure used to solve

$$\frac{1}{x} + \frac{1}{3} = \frac{1}{2}$$

REVIEW **Factor completely.**

101. $x^2 + 4x$

102. $x^2 - 16y^2$

103. $2x^2 + x - 3$

104. $6a^2 - 5a - 6$

105. $x^4 - 81$

106. $4x^2 + 10x - 6$

CHALLENGE PROBLEMS

107. Solve: $x^{-2} + 2x^{-1} + 1 = 0$.

108. ENGINES A formula that is used in the design and testing of diesel engines is

$$E = 1 - \frac{T_4 - T_1}{a(T_3 - T_2)}$$

Solve the formula for T_1.

6.7 Problem Solving Using Rational Equations

- Number Problems
- Uniform Motion Problems
- Shared-Work Problems
- Investment Problems

We will now use the five-step problem-solving strategy to solve application problems from a variety of areas, including banking, petroleum engineering, sports, and travel. In each case, we will use a rational equation to model the situation. Then we will apply the skills we have learned to solve the equation. We begin with an example in which we find an unknown number.

NUMBER PROBLEMS

EXAMPLE 1

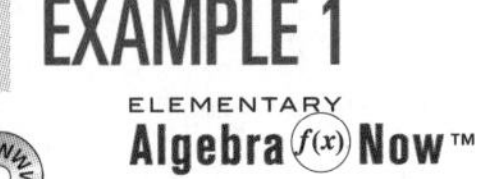

Number problem. If the same number is added to both the numerator and the denominator of the fraction $\frac{3}{5}$, the result is $\frac{4}{5}$. Find the number.

Analyze the Problem

- Begin with the fraction $\frac{3}{5}$.
- Add the same number to the numerator and to the denominator.
- The result is $\frac{4}{5}$.
- Find the number.

Form an Equation Let $n =$ the unknown number. To form an equation, add the unknown number to the numerator and to the denominator of $\frac{3}{5}$. Then set the result equal to $\frac{4}{5}$.

$$\frac{3+n}{5+n} = \frac{4}{5}$$

Solve the Equation To solve this equation, we begin by clearing it of fractions.

$$\frac{3+n}{5+n} = \frac{4}{5}$$

$$5(5+n)\left(\frac{3+n}{5+n}\right) = 5(5+n)\left(\frac{4}{5}\right) \quad \text{Multiply both sides by } 5(5+n)\text{, which is the LCD of the rational expressions appearing in the equation.}$$

$$5(3+n) = (5+n)4 \quad \text{Simplify: } \frac{5+n}{5+n} = 1 \text{ and } \frac{5}{5} = 1.$$

$$15 + 5n = 20 + 4n \quad \text{Distribute the multiplication by 5 and by 4.}$$

$$15 + n = 20 \quad \text{Subtract } 4n \text{ from both sides.}$$

$$n = 5 \quad \text{Subtract 15 from both sides.}$$

State the Conclusion The number is 5.

Check the Result When we add 5 to both the numerator and denominator of $\frac{3}{5}$, we get

$$\frac{3+5}{5+5} = \frac{8}{10} = \frac{4}{5}$$

The result checks.

UNIFORM MOTION PROBLEMS

Recall that we use the distance formula $d = rt$ to solve motion problems. The relationship between distance, rate, and time can be expressed in another way, by solving for t.

$$d = rt \quad \text{Distance = rate} \cdot \text{time.}$$

$$\frac{d}{r} = \frac{rt}{r} \quad \text{Divide both sides by } r.$$

$$\frac{d}{r} = t \quad \text{Simplify: } \frac{r}{r} = 1.$$

$$t = \frac{d}{r}$$

The Language of Algebra

In uniform motion problems, the word *speed* is often used in place of the word *rate.* For example, we can say a car travels at a *rate* of 50 mph or its *speed* is 50 mph.

This alternate form of the distance formula is used to solve the next example.

EXAMPLE 2

Distance runners. A coach can run 10 miles in the same amount of time as his best student-athlete can run 12 miles. If the student runs 1 mile per hour faster than the coach, find the running speeds of the coach and the student.

Analyze the Problem

- The coach runs 10 miles in the same time that the student runs 12 miles.
- The student runs 1 mph faster than the coach.
- Find the speed that each runs.

Form an Equation Since the student's speed is 1 mph faster than the coach's, let $r =$ the speed that the coach runs. Then, $r + 1 =$ the speed that the student runs. The expressions for the rates are entered in the Rate column of the table. The distances run by the coach and by the student are entered in the Distance column of the table.

Using $t = \frac{d}{r}$, we find that the time it takes the coach to run 10 miles, at a rate of r mph, is $\frac{10}{r}$ hours. Similarly, we find that the time it takes the student to run 12 miles, at a rate of $(r + 1)$ mph, is $\frac{12}{r+1}$ hours. These expressions are entered in the Time column of the table.

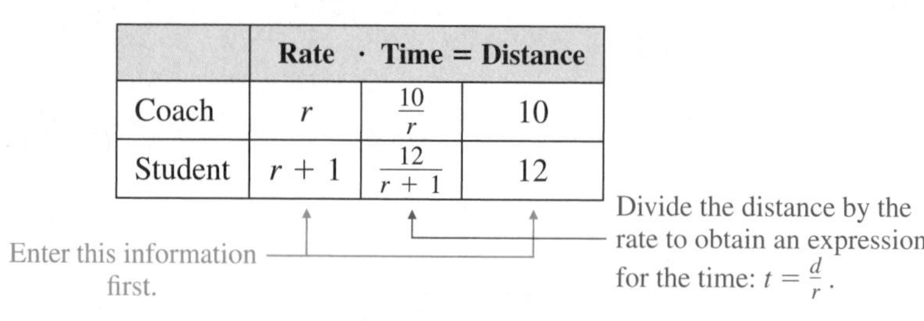

	Rate ·	Time =	Distance
Coach	r	$\frac{10}{r}$	10
Student	$r + 1$	$\frac{12}{r+1}$	12

The time it takes the coach to run 10 miles	equals	the time it takes the student to run 12 miles.
$\frac{10}{r}$	$=$	$\frac{12}{r+1}$

Solve the Equation To solve this rational equation, we begin by clearing it of fractions.

$$\frac{10}{r} = \frac{12}{r+1}$$

$$r(r + 1)\left(\frac{10}{r}\right) = r(r + 1)\left(\frac{12}{r+1}\right) \quad \text{Multiply both sides by the LCD, } r(r+1).$$

$$(r + 1)10 = 12r \quad \text{Simplify.}$$

$$10r + 10 = 12r \quad \text{Distribute the multiplication by 10.}$$
$$10 = 2r \quad \text{Subtract } 10r \text{ from both sides.}$$
$$5 = r \quad \text{Divide both sides by 2.}$$

If $r = 5$, then $r + 1 = 6$.

State the Conclusion The coach's running speed is 5 mph and the student's running speed is 6 mph.

Check the Result The coach will run 10 miles in $\frac{10 \text{ miles}}{5 \text{ mph}} = 2$ hours. The student will run 12 miles in $\frac{12 \text{ miles}}{6 \text{ mph}} = 2$ hours. The times are the same; the results check.

SHARED-WORK PROBLEMS

Problems in which two or more people (or machines) work together to complete a job are called *shared-work problems.* To solve such problems, we must determine the **rate of work** for each person or machine involved. For example, suppose it takes you 4 hours to clean your house. Your rate of work can be expressed as $\frac{1}{4}$ of the job is completed per hour. If someone else takes 5 hours to clean the same house, they complete $\frac{1}{5}$ of the job per hour. In general, a rate of work can be determined in the following way.

Rate of Work If a job can be completed in x hours, the rate of work can be expressed as:

$$\frac{1}{x} \text{ of the job is completed per hour.}$$

If a job is completed in some other unit of time, such as x minutes or x days, then the rate of work is expressed in that unit.

To solve shared-work problems, we must also determine the *amount of work* completed. To do this, we use a formula similar to the distance formula $d = rt$ used for motion problems.

$$\text{Work completed} = \text{rate of work} \cdot \text{time worked} \qquad \text{or} \qquad W = rt$$

EXAMPLE 3

ELEMENTARY Algebra $f(x)$ Now™

Payroll checks. At the end of a pay period, it takes the president of a company 15 minutes to sign all of her employees' payroll checks. What fractional part of the job is accomplished if the president signs checks for 10 minutes?

Solution If all of the checks can be signed in 15 minutes, the president's work rate can be expressed as $\frac{1}{15}$ of the job completed per minute. To find the work that has been completed after 10 minutes, use the work formula.

$$W = rt$$
$$= \frac{1}{15} \cdot 10 \quad \text{Substitute } \frac{1}{15} \text{ for } r\text{, the work rate, and 10 for } t\text{, the time worked.}$$

$$= \frac{10}{15} \quad \text{Multiply.}$$

$$= \frac{2}{3} \quad \text{Simplify the fraction.}$$

In 10 minutes, the president will complete $\frac{2}{3}$ of the job of signing the payroll checks.

Self Check 3 It takes a farmer 8 days to harvest a wheat crop. What part of the job is completed in 6 days?

EXAMPLE 4

ELEMENTARY Algebra $f(x)$ Now™

Filling a tank. An inlet pipe can fill an oil storage tank in 7 days, and a second inlet pipe can fill the same tank in 9 days. If both pipes are used, how long will it take to fill the tank?

Analyze the Problem

- The first pipe can fill the tank in 7 days.
- The second pipe can fill the tank in 9 days.
- How long will it take the two pipes, working together, to fill the tank?

Form an Equation Let $x =$ the number of days it will take to fill the tank if both pipes are used. It is helpful to organize the facts of the problem in a table. Since the pipes will be open for the same amount of time as they fill the tank, enter x as the time worked for each pipe.

The first pipe can fill the tank in 7 days; its rate working alone is $\frac{1}{7}$ of the job per day. The second pipe can fill the tank in 9 days; its rate working alone is $\frac{1}{9}$ of the job per day. To determine the work completed by each pipe, multiply the rate by the time.

	Rate ·	Time =	Work completed
1st pipe	$\frac{1}{7}$	x	$\frac{x}{7}$
2nd pipe	$\frac{1}{9}$	x	$\frac{x}{9}$

Enter this information first. (Rate and Time columns)

Multiply to get each of these entries: $W = rt$.

In shared-work problems, the number 1 represents one whole job completed. So we have

The part of job done by 1st pipe	plus	part of job done by 2nd pipe	equals	1 job completed.
$\frac{x}{7}$	$+$	$\frac{x}{9}$	$=$	1

Solve the Equation

$$\frac{x}{7} + \frac{x}{9} = 1$$

$$63\left(\frac{x}{7} + \frac{x}{9}\right) = 63(1) \quad \text{Clear the equation of fractions by multiplying both sides by the LCD, 63.}$$

$$63\left(\frac{x}{7}\right) + 63\left(\frac{x}{9}\right) = 63$$ Distribute the multiplication by 63.

$$9x + 7x = 63$$ On the left-hand side, do the multiplications.

$$16x = 63$$ Combine like terms.

$$x = \frac{63}{16}$$ Divide both sides by 16.

State the Conclusion If both pipes are used, it will take $\frac{63}{16}$ or $3\frac{15}{16}$ days to fill the tank.

Check the Result In $\frac{63}{16}$ days, the first pipe fills $\frac{1}{7} \cdot \frac{63}{16} = \frac{9}{16}$ of the tank and the second pipe fills $\frac{1}{9} \cdot \frac{63}{16} = \frac{7}{16}$ of the tank. The sum of these efforts, $\frac{9}{16} + \frac{7}{16}$, is $\frac{16}{16}$ or 1 full tank. The result checks.

Example 4 can be solved in a different way by considering *the amount of work done by each pipe in 1 day.* As before, if we let $x =$ the number of days it will take to fill the tank if both inlet pipes are used, then together, in 1 day, they will complete $\frac{1}{x}$ of the job. If we add what the first pipe can do in 1 day to what the second pipe can do in 1 day, the sum is what they can do together in 1 day.

What the first inlet pipe can do in 1 day	plus	what the second inlet pipe can do in 1 day	equals	what they can do together in 1 day.
$\frac{1}{7}$	$+$	$\frac{1}{9}$	$=$	$\frac{1}{x}$

To solve the equation, begin by clearing it of fractions.

$$\frac{1}{7} + \frac{1}{9} = \frac{1}{x}$$

$$63x\left(\frac{1}{7} + \frac{1}{9}\right) = 63x\left(\frac{1}{x}\right)$$ Multiply both sides by the LCD, $63x$.

$$9x + 7x = 63$$ Distribute the multiplication by $63x$ and simplify.

$$16x = 63$$ Combine like terms.

$$x = \frac{63}{16}$$ Divide both sides by 16.

This is the same as the solution obtained in Example 4.

INVESTMENT PROBLEMS

We have used the interest formula $I = Prt$ to solve investment problems. The relationship between interest, principal, rate, and time can be expressed in another way, by solving for P.

$$I = Prt$$ Interest = principal · rate · time.

$$\frac{I}{rt} = \frac{Prt}{rt}$$ Divide both sides by rt.

$$P = \frac{I}{rt}$$ Simplify: $\frac{rt}{rt} = 1$.

This alternate form of the interest formula is used to solve the next example.

EXAMPLE 5

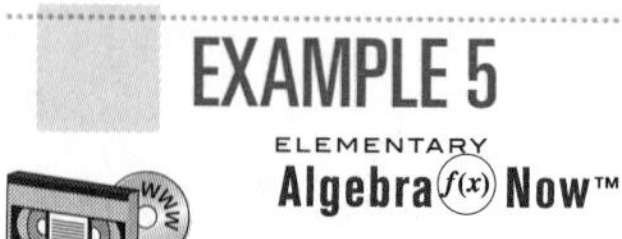

Comparing investments. An amount of money invested for one year in bonds will earn \$120. At a bank, that same amount of money will only earn \$75 interest, because the interest paid by the bank is 3% less than that paid by the bonds. Find the rate of interest paid by each investment.

Analyze the Problem

- The investment in bonds earns \$120 in one year.
- The same amount of money, invested in a bank, earns \$75 in one year.
- The interest rate paid by the bank is 3% less than that paid by the bonds.
- Find the bond's rate of interest and the bank's rate of interest.

Form an Equation Since the interest rate paid by the bank is 3% less than that paid by the bonds, let $r =$ the bond's rate of interest, and $r - 0.03 =$ the bank's interest rate. (Recall that $3\% = 0.03$.)

If an investment earns \$120 interest in 1 year at some rate r, we can use $P = \frac{I}{rt}$ to find that the principal invested was $\frac{120}{r}$ dollars. Similarly, if another investment earns \$75 interest in 1 year at some rate $r - 0.03$, the principal invested was $\frac{75}{r - 0.03}$ dollars. We can organize the facts of the problem in a table.

	Principal ·	Rate ·	Time =	Interest
Bonds	$\frac{120}{r}$	r	1	120
Bank	$\frac{75}{r-0.03}$	$r - 0.03$	1	75

Divide to get each of these entries: $P = \frac{I}{rt}$. Enter this information first.

The amount invested in the bonds	equals	the amount invested in the bank.
$\frac{120}{r}$	$=$	$\frac{75}{r - 0.03}$

Solve the Equation

$$\frac{120}{r} = \frac{75}{r - 0.03}$$

$$r(r - 0.03)\left(\frac{120}{r}\right) = \left(\frac{75}{r - 0.03}\right)r(r - 0.03)$$ Multiply both sides by the LCD, $r(r - 0.03)$.

$$(r - 0.03)120 = 75r$$ Simplify.

$$120r - 3.6 = 75r$$ Distribute the multiplication by 120.

$$45r - 3.6 = 0$$ Subtract $75r$ from both sides.

$$45r = 3.6$$ Add 3.6 to both sides.

$$r = 0.08$$ Divide both sides by 45.

If $r = 0.08$, then $r - 0.03 = 0.05$.

State the Conclusion The bonds pay 0.08, or 8%, interest. The bank's interest rate is 5%.

Check the Result The amount invested at 8% that will earn \$120 interest in 1 year is $\frac{120}{(0.08)1} = \$1{,}500$. The amount invested at 5% that will earn \$75 interest in 1 year is $\frac{75}{(0.05)1} = \$1{,}500$. The amounts invested in the bonds and the bank are the same. The results check.

Answer to Self Check **3.** $\frac{3}{4}$

6.7 STUDY SET

Elementary Algebra $f(x)$ Now™

VOCABULARY Fill in the blanks.

1. In the formula $d = rt$, the variable d stands for the ________ traveled, r is the ______, and t is the ______.
2. In the formula $W = rt$, the variable W stands for the ______ completed, r is the ______, and t is the ______.
3. In the formula $I = Prt$, the variable I stands for the amount of ________ earned, P is the __________, r stands for the annual interest ______, and t is the ______.
4. If a job can be completed in x hours, then the rate of work can be expressed as $\frac{1}{x}$ of the job is completed ______ hour.

CONCEPTS

5. Choose the equation that can be used to solve the following: *If the same number is added to the numerator and the denominator of the fraction $\frac{5}{8}$, the result is $\frac{2}{3}$. Find the number.*

   (i) $\frac{5}{8} + x = \frac{2}{3}$ (ii) $\frac{5 + x}{8} = \frac{2}{3}$

   (iii) $\frac{5 + x}{8 + x} = \frac{2}{3}$ (iv) $\frac{5}{8} = \frac{2 + x}{3 + x}$

6. Choose the equation that can be used to solve the following: *The sum of a number and its reciprocal is $\frac{7}{2}$. Find the number.*

   (i) $x + (-x) = \frac{7}{2}$ (ii) $x + \frac{1}{x} = \frac{7}{2}$

   (iii) $x + \frac{1}{x} = x + \frac{7}{2}$ (iv) $x + \frac{1}{x} = \frac{2}{7}$

7. Solve $d = rt$ for t.
8. Solve $I = Prt$ for P.
9. **a.** Write 9% as a decimal.

   b. Write 0.035 as a percent.

10. It takes a night security officer 45 minutes to check each of the doors in an office building to make sure they are locked. What is the officer's rate of work?
11. If a custodian can sweep a cafeteria floor in 12 minutes, what fractional part of the job does he accomplish in 4 minutes?
12. If the exits at the front of a theater are opened, a full theater can be emptied of all occupants in 6 minutes. How much of the theater is emptied in 1 minute?
13. It takes an elementary school teacher 4 hours to make out the semester report cards. What part of the job does she complete in x hours?
14. Complete the table.

	r	$\cdot\ t$	$= d$
Snowmobile	r		4
4 × 4 truck	$r - 5$		3

15. Complete the table.

	P	$\cdot\ r$	$\cdot\ t$	$= I$
City savings bank		r	1	50
Credit union		$r - 0.02$	1	75

16. Complete the table.

	Rate	· Time	= Work completed
1st printer	$\frac{1}{15}$	x	
2nd printer	$\frac{1}{8}$	x	

NOTATION

17. Write $\frac{55}{9}$ days using a mixed number.

18. How much greater is the rate paid by the mutual funds compared to the rate paid by the certificate of deposit?

	P	$\cdot\ r$	$\cdot\ t$	$= I$
Certificate of deposit	$\frac{600}{r}$	r	1	600
Mutual funds	$\frac{1{,}400}{r + 0.04}$	$r + 0.04$	1	1,400

PRACTICE

19. NUMBER PROBLEM If the same number is added to both the numerator and the denominator of $\frac{2}{5}$, the result is $\frac{2}{3}$. Find the number.

20. NUMBER PROBLEM If the same number is subtracted from both the numerator and the denominator of $\frac{11}{13}$, the result is $\frac{3}{4}$. Find the number.

21. NUMBER PROBLEM If the denominator of $\frac{3}{4}$ is increased by a number and the numerator is doubled, the result is 1. Find the number.

22. NUMBER PROBLEM If a number is added to the numerator of $\frac{7}{8}$ and the same number is subtracted from the denominator, the result is 2. Find the number.

23. NUMBER PROBLEM If a number is added to the numerator of $\frac{3}{4}$ and twice as much is added to the denominator, the result is $\frac{4}{7}$. Find the number.

24. NUMBER PROBLEM If a number is added to the numerator of $\frac{5}{7}$ and twice as much is subtracted from the denominator, the result is 8. Find the number.

25. NUMBER PROBLEM The sum of a number and its reciprocal is $\frac{13}{6}$. Find the numbers.

26. NUMBER PROBLEM The sum of the reciprocals of two consecutive even integers is $\frac{7}{24}$. Find the integers.

APPLICATIONS

27. COOKING If the same number is added to both the numerator and the denominator of the amount of butter used in the following recipe for toffee, the result is the amount of brown sugar to be used. Find the number.

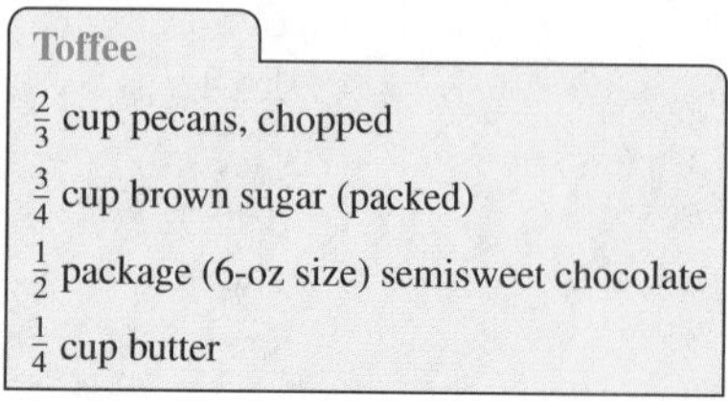

28. TAPE MEASURES If the same number is added to both the numerator and the denominator of the first measurement, the result is the second measurement. Find the number.

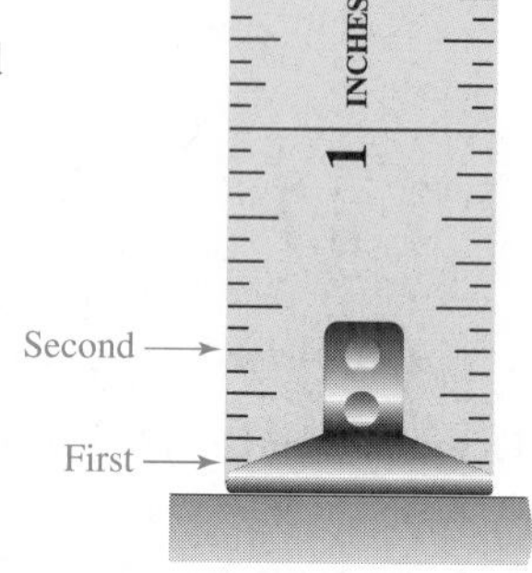

29. FILLING A POOL An inlet pipe can fill an empty swimming pool in 5 hours, and another inlet pipe can fill the pool in 4 hours. How long will it take both pipes to fill the pool?

30. ROOFING HOUSES A homeowner estimates that it will take her 7 days to roof her house. A professional roofer estimates that he could roof the house in 4 days. How long will it take if the homeowner helps the roofer?

31. HOLIDAY DECORATING One crew can put up holiday decorations in the mall in 8 hours. A second crew can put up the decorations in 10 hours. How long will it take if both crews work together to decorate the mall?

32. GROUNDS KEEPING It takes a grounds keeper 45 minutes to prepare a softball field for a game. It takes his assistant 55 minutes to prepare the same field. How long will it take if they work together to prepare the field?

33. FILLING A POOL One inlet pipe can fill an empty pool in 4 hours, and a drain can empty the pool in 8 hours. How long will it take the pipe to fill the pool if the drain is left open?

34. SEWAGE TREATMENT A sludge pool is filled by two inlet pipes. One pipe can fill the pool in 15 days, and the other can fill it in 21 days. However, if no sewage is added, continuous waste removal will empty the pool in 36 days. How long will it take the two inlet pipes to fill an empty sludge pool?

35. GRADING PAPERS On average, it takes a teacher 30 minutes to grade a set of quizzes. It takes her teacher's aide twice as long to do the same grading. How long will it take if they work together to grade a set of quizzes?

36. DOG KENNELS It takes the owner/operator of a dog kennel 6 hours to clean all of the cages. It takes his assistant 2 hours more than that to clean the same cages. How long will it take if they work together?

37. PRINTERS It takes a printer 6 hours to print the class schedules for all of the students enrolled in a community college. A faster printer can print the schedules in 4 hours. How long will it take the two printers working together to print $\frac{3}{4}$ of the class schedules?

38. OFFICE WORK In 5 hours, a secretary can address 100 envelopes. Another secretary can address 100 envelopes in 6 hours. How long would it take the secretaries, working together, to address 300 envelopes. (*Hint:* Think of addressing 300 envelopes as three 100-envelope jobs.)

39. PHYSICAL FITNESS A woman can bicycle 28 miles in the same time as it takes her to walk 8 miles. She can ride 10 mph faster than she can walk. How fast can she walk?

40. COMPARING TRAVEL A plane can fly 300 miles in the same time as it takes a car to go 120 miles. If the car travels 90 mph slower than the plane, find the speed of the plane.

41. PACKAGING FRUIT The diagram shows how apples are processed for market. Although the second conveyor belt is shorter, an apple spends the same amount of time on each belt because the second conveyor moves 1 foot per second slower than the first. Determine the speed of each conveyor belt.

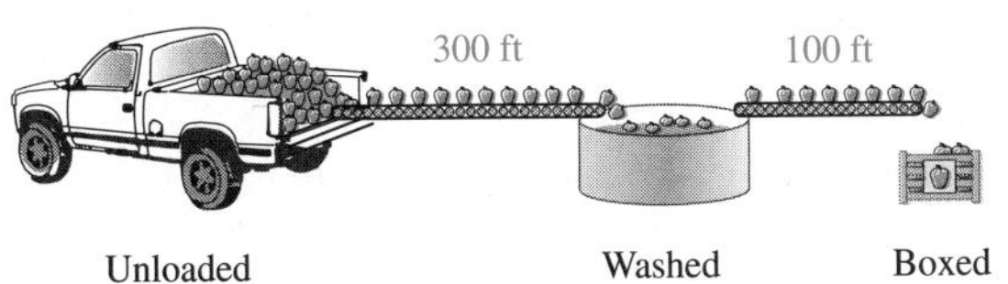

42. BIRDS IN FLIGHT Although flight speed is dependent upon the weather and the wind, in general, a Canada goose can fly about 10 mph faster than a Great Blue heron. In the same time that a Canada goose travels 120 miles, a Great Blue heron travels 80 miles. Find their flying speeds.

43. WIND SPEED When a plane flies downwind, the wind pushes the plane so that its speed is the *sum* of the speed of the plane in still air and the speed of the wind. Traveling upwind, the wind pushes against the plane so that its speed is the *difference* of the speed of the plane in still air and the speed of the wind. Suppose a plane that travels 255 mph in still air can travel 300 miles downwind in the same time as it takes to travel 210 miles upwind. Complete the table and find the speed of the wind.

	Rate ·	Time	= Distance
Downwind	$255 + x$		300
Upwind	$255 - x$		210

44. BOATING A boat that travels 18 mph in still water can travel 22 miles downstream in the same time as it takes to travel 14 miles upstream. Find the speed of the current in the river.

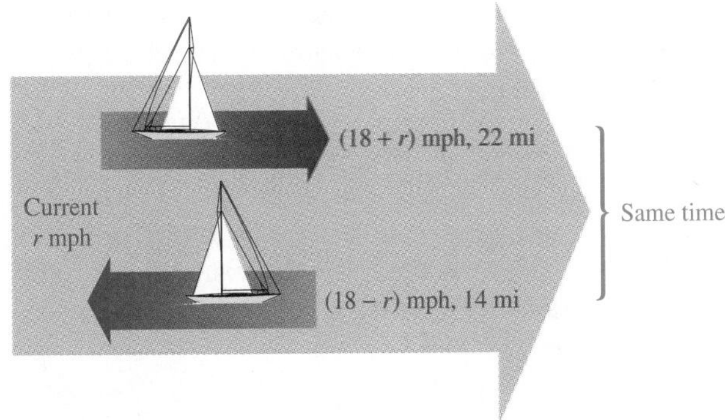

45. COMPARING INVESTMENTS An amount of money invested for 1 year in tax-free bonds will earn \$300. In a certain credit union account, that same amount of money will only earn \$200 interest in a year, because the interest paid is 2% less than that paid by the bonds. Find the rate of interest paid by each investment.

46. COMPARING INVESTMENTS An amount of money invested for 1 year in a savings account will earn \$1,500. That same amount of money, invested in a mini-mall development will earn \$6,500 interest in a year, because the interest paid is 10% more than that paid by the savings account. Find the rate of interest paid by each investment.

47. COMPARING INVESTMENTS Two certificates of deposit (CDs) pay interest at rates that differ by 1%. Money invested for 1 year in the first CD earns \$175 interest. The same principal invested in the second CD earns \$200. Find the two rates of interest.

48. COMPARING INTEREST RATES Two bond funds pay interest at rates that differ by 2%. Money invested for 1 year in the first fund earns \$315 interest. The same amount invested in the second fund earns \$385. Find the lower rate of interest.

WRITING

49. In Example 4, one inlet pipe could fill an oil tank in 7 days, and another could fill the same tank in 9 days. We were asked to find how long it would take if both pipes were used. Explain why each of the following approaches is incorrect.

The time it would take to fill the tank

- is the *sum* of the lengths of time it takes each pipe to fill the tank: 7 days + 9 days = 16 days.
- is the *difference* in the lengths of time it takes each pipe to fill the tank: 9 days − 7 days = 2 days.
- is the *average* of the lengths of time it takes each pipe to fill the tank:

$$\frac{7 \text{ days} + 9 \text{ days}}{2} = \frac{16 \text{ days}}{2} = 8 \text{ days}$$

50. Write a shared-work problem that can be modeled by the equation

$$\frac{x}{3} + \frac{x}{4} = 1$$

REVIEW

51. When expressed as a decimal, is $\frac{7}{9}$ a terminating or a repeating decimal?

52. Solve: $x + 20 = 4x - 1 + 2x$.

53. List the set of integers.

54. Solve: $4x^2 + 8x = 0$.

55. Evaluate $2x^2 + 5x - 3$ for $x = -3$.

56. Solve $T - R = ma$ for R.

CHALLENGE PROBLEMS

57. RIVER TOURS A river boat tour begins by going 60 miles upstream against a 5-mph current. There, the boat turns around and returns with the current. What still-water speed should the captain use to complete the tour in 5 hours?

58. TRAVEL TIME A company president flew 680 miles one way in the corporate jet but returned in a smaller plane that could fly only half as fast. If the total travel time was 6 hours, find the speeds of the planes.

59. SALES A dealer bought some radios for a total of \$1,200. She gave away 6 radios as gifts, sold the rest for \$10 more than she paid for each radio, and broke even. How many radios did she buy?

60. FURNACE REPAIRS A repairman purchased several furnace-blower motors for a total cost of \$210. If his cost per motor had been \$5 less, he could have purchased one additional motor. How many motors did he buy at the regular rate?

6.8 Proportions and Similar Triangles

- Ratios, Rates, and Proportions
- Solving Proportions
- Problem Solving
- Similar Triangles

In this section, we will discuss a problem-solving tool called a *proportion.* A proportion is a type of rational equation that involves two *ratios* or two *rates.*

RATIOS, RATES, AND PROPORTIONS

Ratios enable us to compare numerical quantities. Here are some examples.

- To prepare fuel for a lawnmower, gasoline is mixed with oil in a 50 to 1 ratio.
- In the stock market, winning stocks might outnumber losers by a ratio of 7 to 4.
- Gold is combined with other metals in the ratio of 14 to 10 to make 14-karat jewelry.

Ratios A **ratio** is the quotient of two numbers or the quotient of two quantities that have the same units.

There are three common ways to write a ratio: as a fraction, using the word *to,* or with a colon. For example, the comparison of the number of winning stocks to the number of losing stocks mentioned earlier can be written as

$$\frac{7}{4}, \quad 7 \text{ to } 4, \quad \text{or} \quad 7:4$$

Each of these forms can be read as "the ratio of 7 to 4."

EXAMPLE 1

Translate each phrase into a ratio written in fractional form: **a.** The ratio of 5 to 9 and **b.** 12 ounces to 2 pounds.

Solution

a. The ratio of 5 to 9 is written $\frac{5}{9}$.

b. To write a ratio of two quantities with the same units, we must express 2 pounds in terms of ounces. Since 1 pound = 16 ounces, 2 pounds = 32 ounces. The ratio of 12 ounces to 32 ounces can be simplified so that no units appear in the final form.

$$\frac{12 \text{ ounces}}{32 \text{ ounces}} = \frac{3 \cdot \overset{1}{\cancel{4}}\ \overset{1}{\cancel{\text{ounces}}}}{\underset{1}{\cancel{4}} \cdot 8\ \underset{1}{\cancel{\text{ounces}}}} = \frac{3}{8}$$

Notation

A ratio that is the quotient of two quantities having the same units should be simplified so that no units appear in the final answer.

Self Check 1 Translate each phrase into a ratio written in fractional form: **a.** The ratio of 15 to 2 and **b.** 12 hours to 2 days.

A quotient that compares quantities with different units is called a **rate.** For example, if the 495-mile drive from New Orleans to Dallas takes 9 hours, the average rate of speed is the quotient of the miles driven and the length of time the trip takes.

$$\text{Average rate of speed} = \frac{495 \text{ miles}}{9 \text{ hours}} = \frac{\overset{1}{\cancel{9}} \cdot 55 \text{ miles}}{\underset{1}{\cancel{9}} \cdot 1 \text{ hours}} = \frac{55 \text{ miles}}{1 \text{ hour}}$$

Rates A **rate** is a quotient of two quantities that have different units.

If two ratios or two rates are equal, we say that they are *in proportion.*

Proportion A **proportion** is a mathematical statement that two ratios or two rates are equal.

Some examples of proportions are

$$\frac{1}{2} = \frac{3}{6}, \quad \frac{3 \text{ waiters}}{7 \text{ tables}} = \frac{9 \text{ waiters}}{21 \text{ tables}}, \quad \text{and} \quad \frac{a}{b} = \frac{c}{d}$$

The Language of Algebra

The word *proportion* implies a comparative relationship in size. For a picture to appear realistic, the artist must draw the shapes in the proper *proportion.* Remember the Y2K scare? The massive computer failures predicted by some experts were blown way *out of proportion.*

- The proportion $\frac{1}{2} = \frac{3}{6}$ can be read as "1 is to 2 as 3 is to 6."
- The proportion $\frac{3 \text{ waiters}}{7 \text{ tables}} = \frac{9 \text{ waiters}}{21 \text{ tables}}$ can be read as "3 waiters is to 7 tables as 9 waiters is to 21 tables."
- The proportion $\frac{a}{b} = \frac{c}{d}$ can be read as "a is to b as c is to d."

Each of the four numbers in a proportion is called a **term.** The first and fourth terms are called the **extremes,** and the second and third terms are called the **means.**

First term → a, Second term → b, c ← Third term, d ← Fourth term: $\frac{a}{b} = \frac{c}{d}$

a and d are the extremes. b and c are the means.

For the proportion $\frac{a}{b} = \frac{c}{d}$, we can show that the product of the extremes, ad, is equal to the product of the means, bc, by multiplying both sides of the proportion by bd, and observing that $ad = bc$.

$$\frac{a}{b} = \frac{c}{d}$$

$$bd \cdot \frac{a}{b} = bd \cdot \frac{c}{d} \quad \text{To clear the fractions, multiply both sides by the LCD, } bd.$$

$$ad = bc \quad \text{Simplify: } \frac{b}{b} = 1 \text{ and } \frac{d}{d} = 1.$$

Since $ad = bc$, the product of the extremes equals the product of the means.

The same products ad and bc can be found by multiplying diagonally in the proportion $\frac{a}{b} = \frac{c}{d}$. We call ad and bc **cross products.**

ad bc

$$\frac{a}{b} = \frac{c}{d}$$

The Fundamental Property of Proportions

In a proportion, the product of the extremes is equal to the product of the means.

$$\text{If } \frac{a}{b} = \frac{c}{d}, \text{ then } ad = bc \quad \text{and} \quad \text{if } ad = bc, \text{ then } \frac{a}{b} = \frac{c}{d}.$$

EXAMPLE 2

ELEMENTARY Algebra *f(x)* Now™

Determine whether each equation is a proportion: **a.** $\frac{3}{7} = \frac{9}{21}$ and **b.** $\frac{8}{3} = \frac{13}{5}$.

Solution In each case, we check to see whether the product of the extremes is equal to the product of the means.

a. The product of the extremes is $3 \cdot 21 = 63$. The product of the means is $7 \cdot 9 = 63$. Since the cross products are equal, $\frac{3}{7} = \frac{9}{21}$ is a proportion.

$$3 \cdot 21 = 63 \qquad 7 \cdot 9 = 63$$
$$\frac{3}{7} = \frac{9}{21}$$

b. The product of the extremes is $8 \cdot 5 = 40$. The product of the means is $3 \cdot 13 = 39$. Since the cross products are not equal, the equation is not a proportion: $\frac{8}{3} \neq \frac{13}{5}$.

$$8 \cdot 5 = 40 \qquad 3 \cdot 13 = 39$$
$$\frac{8}{3} = \frac{13}{5}$$

Self Check 2 Determine whether the equation is a proportion: $\frac{6}{13} = \frac{24}{53}$.

SOLVING PROPORTIONS

The fundamental property of proportions provides us with a way to solve proportions.

EXAMPLE 3

Solve: $\frac{12}{18} = \frac{4}{x}$.

Solution To solve for x, we set the cross products equal.

$$\frac{12}{18} = \frac{4}{x}$$
$$12 \cdot x = 18 \cdot 4 \quad \text{In a proportion, the product of the extremes equals the product of the means.}$$
$$12x = 72 \quad \text{Do the multiplications.}$$
$$\frac{12x}{12} = \frac{72}{12} \quad \text{To isolate } x \text{, divide both sides by 12.}$$
$$x = 6$$

Caution

Remember that a cross product is the product of the means or the extremes of a *proportion.* For example, it would be incorrect to try to compute cross products to solve $\frac{12}{18} = \frac{4}{x} + \frac{1}{2}$. It is not a proportion. The right-hand side is not a ratio.

Check: To check the result, we substitute 6 for x in $\frac{12}{18} = \frac{4}{x}$ and find the cross products.

$$12 \cdot 6 = 72 \qquad 18 \cdot 4 = 72$$
$$\frac{12}{18} \stackrel{?}{=} \frac{4}{6}$$

Since the cross products are equal, the solution of $\frac{12}{18} = \frac{4}{x}$ is 6.

Self Check 3 Solve: $\frac{15}{x} = \frac{25}{40}$.

EXAMPLE 4

ELEMENTARY Algebra $f(x)$ Now™

Solve: $\frac{2a+1}{4} = \frac{10}{8}$.

Solution

$$\frac{2a+1}{4} = \frac{10}{8}$$

$$8(2a+1) = 40$$ In a proportion, the product of the extremes equals the product of the means.

$$16a + 8 = 40$$ Distribute the multiplication by 8.

$$16a + 8 - 8 = 40 - 8$$ Subtract 8 from both sides.

$$16a = 32$$ Combine like terms.

$$\frac{16a}{16} = \frac{32}{16}$$ Divide both sides by 16.

$$a = 2$$

The solution is 2.

Success Tip

Since proportions are rational equations, they can also be solved by multiplying both sides by the LCD. For Example 4, an alternate approach is to multiply both sides by 8:

$$8\left(\frac{2a+1}{4}\right) = 8\left(\frac{10}{8}\right)$$

Self Check 4 Solve: $\frac{3x-1}{2} = \frac{12.5}{5}$.

PROBLEM SOLVING

We can use proportions to solve many problems. If we are given a ratio (or rate) comparing two quantities, the words of the problem can be translated to a proportion, and we can solve it to find the unknown.

EXAMPLE 5

ELEMENTARY Algebra $f(x)$ Now™

Grocery shopping. If 6 apples cost \$1.38, how much will 16 apples cost?

Solution

Analyze the Problem We know the cost of 6 apples; we are to find the cost of 16 apples.

Form a Proportion Let c = the cost of 16 apples. If we compare the number of apples to their cost, we know that the two rates are equal.

6 apples is to \$1.38 as 16 apples is to \$c.

6 apples → 6, 16 ← 16 apples

Cost of 6 apples → 1.38, c ← Cost of 16 apples

$$\frac{6}{1.38} = \frac{16}{c}$$

Solve the Proportion

$$6 \cdot c = 1.38(16)$$ In a proportion, the product of the extremes equals the product of the means.

$$6c = 22.08$$ Multiply: 1.38(16) = 22.08.

$$\frac{6c}{6} = \frac{22.08}{6}$$ Divide both sides by 6.

$$c = 3.68$$

State the Conclusion Sixteen apples will cost \$3.68.

Check the Result 16 apples are about 3 times as many as 6 apples, which cost \$1.38. If we multiply \$1.38 by 3, we get an estimate of the cost of 16 apples: $\$1.38 \cdot 3 = \$4.14$. The result, \$3.68, seems reasonable.

Self Check 5 If 9 tickets to a concert cost \$112.50, how much will 15 tickets cost?

Caution When solving problems using proportions, we must make sure that the units of both numerators are the same and the units of both denominators are the same. In Example 5, it would be incorrect to write

$$\text{Cost of 6 apples} \rightarrow \frac{1.38}{6} = \frac{16}{c} \leftarrow \text{16 apples} \quad (\text{6 apples} \rightarrow 6;\ c \leftarrow \text{Cost of 16 apples})$$

EXAMPLE 6

The Language of Algebra

Architects, interior decorators, landscapers, and automotive engineers are a few of the professionals who construct *scale* drawings or *scale* models of the projects they are designing.

Miniatures. A **scale** is a ratio (or rate) that compares the size of a model, drawing, or map to the size of an actual object. The scale indicates that 1 inch on the model carousel is equivalent to 160 inches on the actual carousel. How wide should the model be if the actual carousel is 35 feet wide?

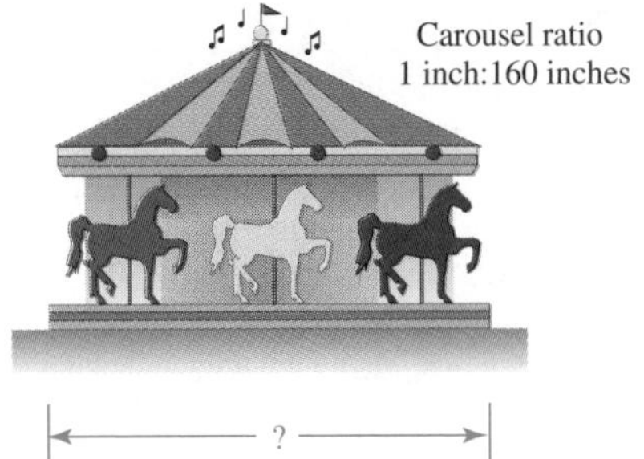

Analyze the Problem We are asked to determine the width of the miniature carousel, if a ratio of 1 inch to 160 inches is used. We would like the width of the model to be given in inches, not feet, so we will express the 35-foot width of the actual carousel as $35 \cdot 12 = 420$ inches.

Form a Proportion Let w = the width of the model. The ratios of the dimensions of the model to the corresponding dimensions of the actual carousel are equal.

1 inch is to 160 inches as w inches is to 420 inches.

$$\begin{array}{l} \text{model} \rightarrow \\ \text{actual} \rightarrow \end{array} \frac{1}{160} = \frac{w}{420} \begin{array}{l} \leftarrow \text{model} \\ \leftarrow \text{actual} \end{array}$$

Solve the Proportion

$420 = 160w$ In a proportion, the product of the extremes is equal to the product of the means.

$\frac{420}{160} = \frac{160w}{160}$ Divide both sides by 160.

$2.625 = w$ Simplify.

State the Conclusion The width of the miniature carousel should be 2.625 in., or $2\frac{5}{8}$ in.

Check the Result A width of $2\frac{5}{8}$ in. is approximately 3 in. When we write the ratio of the model's approximate width to the width of the actual carousel, we get $\frac{3}{420} = \frac{1}{140}$, which is about $\frac{1}{160}$. The answer seems reasonable.

When shopping, *unit prices* can be used to compare costs of different sizes of the same brand to determine the best buy. The **unit price** gives the cost per unit, such as cost per ounce, cost per pound, or cost per sheet. We can find the unit price of an item using a proportion.

EXAMPLE 7

ELEMENTARY Algebra $f(x)$ Now™

Comparison shopping. Which size of toothpaste is the better buy?

$2.19

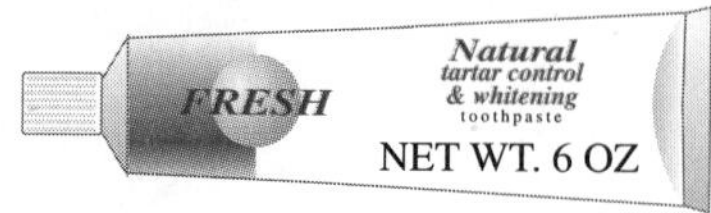

$2.79

Solution To find the unit price for each tube, we let $x =$ the price of 1 ounce of toothpaste. Then we set up and solve the following proportions.

The Language of Algebra

A unit price indicates the cost of 1 unit of an item, such as 1 ounce of bottled water or 1 pound of hamburger. In advanced mathematics, we study unit circles—circles that have a radius of 1 unit.

For the 4-ounce tube:

Ounces → numerator, Price → denominator; Ounce → numerator, Price → denominator

$$\frac{4}{2.19} = \frac{1}{x}$$

$$4x = 2.19$$

$$x = \frac{2.19}{4}$$

$$x \approx 0.55$$ The unit price is approximately \$0.55.

For the 6-ounce tube:

Ounces → numerator, Price → denominator; Ounce → numerator, Price → denominator

$$\frac{6}{2.79} = \frac{1}{x}$$

$$6x = 2.79$$

$$x = \frac{2.79}{6}$$

$$x \approx 0.47$$ The unit price is approximately \$0.47.

The price of 1 ounce of toothpaste from the 4-ounce tube is about 55¢. The price for 1 ounce of toothpaste from the 6-ounce tube is about 47¢. Since the 6-ounce tube has the lower unit price, it is the better buy.

Self Check 7 Which is the better buy: 3 pounds of hamburger for \$6.89 or 5 pounds for \$12.49?

SIMILAR TRIANGLES

If two angles of one triangle have the same measures as two angles of a second triangle, the triangles have the same shape. Triangles with the same shape, but not necessarily the same size, are called **similar triangles.** In the following figure, $\triangle ABC \sim \triangle DEF$. (Read the symbol $\sim$ as "is similar to.")

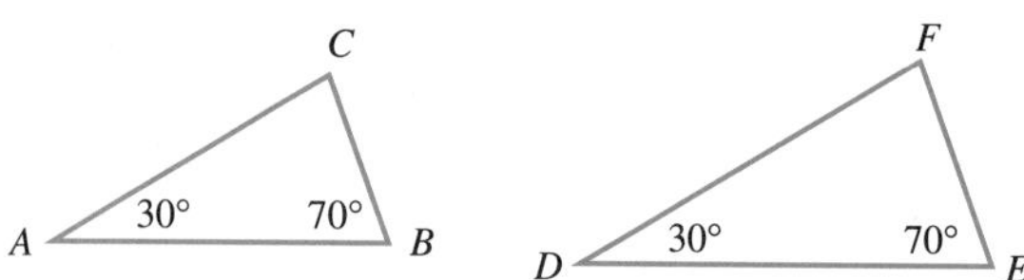

Property of Similar Triangles If two triangles are **similar,** all pairs of corresponding sides are in proportion.

For the similar triangles shown above, the following proportions are true.

$$\frac{AB}{DE} = \frac{BC}{EF}, \quad \frac{BC}{EF} = \frac{CA}{FD}, \quad \text{and} \quad \frac{CA}{FD} = \frac{AB}{DE}$$

Read AB as "the length of segment AB."

EXAMPLE 8

Finding the height of a tree. A tree casts a shadow 18 feet long at the same time as a woman 5 feet tall casts a shadow 1.5 feet long. Find the height of the tree.

Solution **Analyze the Problem** The figure shows the similar triangles determined by the tree and its shadow and the woman and her shadow. Since the triangles are similar, the lengths of their corresponding sides are in proportion. We can use this fact to find the height of the tree.

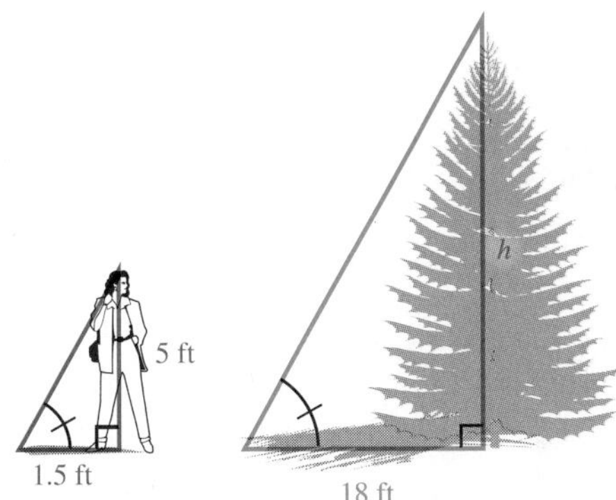

Each triangle has a right angle. Since the sun's rays strike the ground at the same angle, the angles highlighted with a tick mark have the same measure. Therefore, two angles of the smaller triangle have the same measures as two angles of the larger triangle; the triangles are similar.

Form a Proportion If we let $h =$ the height of the tree, we can find h by solving the following proportion.

$$\frac{h}{5} = \frac{18}{1.5} \qquad \frac{\text{Height of the tree}}{\text{Height of the woman}} = \frac{\text{Length of shadow of the tree}}{\text{Length of shadow of the woman}}$$

Solve the Proportion

$1.5h = 5(18)$ In a proportion, the product of the extremes equals the product of the means.

$1.5h = 90$ Multiply.

$\frac{1.5h}{\mathbf{1.5}} = \frac{90}{\mathbf{1.5}}$ Divide both sides by 1.5.

$h = 60$ $\frac{90}{1.5} = 60$.

State the Conclusion The tree is 60 feet tall.

Check the Result $\frac{18}{1.5} = 12$ and $\frac{60}{5} = 12$. The ratios are the same. The result checks.

Self Check 8 Find the height of the tree in Example 8 if the woman is 5 feet 6 inches tall.

Answers to Self Checks **1. a.** $\frac{15}{2}$, **b.** $\frac{1}{4}$ **2.** no **3.** 24 **4.** 2 **5.** \$187.50 **7.** 3 pounds for \$6.89 **8.** 66 ft

6.8 STUDY SET

ELEMENTARY Algebra $f(x)$ Now™

VOCABULARY **Fill in the blanks.**

1. A ______ is the quotient of two numbers or the quotient of two quantities with the same units. A ______ is a quotient of two quantities that have different units.

2. A ________ is a mathematical statement that two ratios or two rates are equal.

3. The ______ of the proportion $\frac{2}{x} = \frac{16}{40}$ are 2, x, 16, and 40.

4. In $\frac{50}{3} = \frac{x}{9}$, the terms 50 and 9 are called the _________ and the terms 3 and x are called the _______ of the proportion.

5. The product of the extremes and the product of the means of a proportion are also known as ______ products.

6. A ______ is a ratio (or rate) that compares the size of a model, drawing, or map to the size of an actual object.

7. Examples of _____ prices are $1.65 per gallon, 17¢ per day, and $50 per foot.

8. Two triangles with the same shape, but not necessarily the same size, are called _______ triangles.

CONCEPTS

9. WEST AFRICA Write the ratio (in fractional form) of the number of red stripes to the number of white stripes on the flag of Liberia.

10. Fill in the blanks: In a proportion, the product of the extremes is ______ to the product of the means.

$$\text{If } \frac{a}{b} = \frac{c}{d}, \text{ then } \_\_\_ = \_\_\_.$$

11. What are the cross products for each proportion?

a. $\frac{6}{5} = \frac{12}{10}$ b. $\frac{15}{2} = \frac{45}{x}$

12. a. Is 45 a solution of $\frac{5}{3} = \frac{75}{x}$?

b. Is -2 a solution of $\frac{a+4}{8} = \frac{3}{16}$?

13. SNACK FOODS In a sample of 25 bags of potato chips, 2 were found to be underweight. Complete the following proportion that could be used to find the number of underweight bags that would be expected in a shipment of 1,000 bags of potato chips.

$$\begin{array}{rcl} \text{number of bags} \rightarrow & \frac{\_\_\_}{\_\_\_} = \frac{\_\_\_}{\_\_\_} & \leftarrow \text{number of bags} \\ \text{number underweight} \rightarrow & & \leftarrow \text{number underweight} \end{array}$$

14. MINIATURES A model of a "high wheeler" bicycle is to be made using a scale of 2 inches to 15 inches. The following proportion was set up to determine the height of the front wheel of the model. Explain the error.

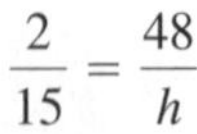

$$\frac{2}{15} = \frac{48}{h}$$

Actual size

15. GROCERY SHOPPING Examine the following pricing stickers. Which item is the better buy?

16. Complete the following proportion that can be used to find the unit price of facial tissue if a box of 85 tissues sells for $1.19.

$$\begin{array}{rcl} \text{price} \rightarrow & \frac{1.19}{85} = \frac{x}{\_\_\_} & \leftarrow \text{price} \\ \text{number of sheets} \rightarrow & & \leftarrow \text{number of sheets} \end{array}$$

17. Two similar triangles are shown. Fill in the blanks to make the proportions true.

$$\frac{AB}{DE} = \frac{\_\_\_}{EF} \qquad \frac{BC}{\_\_\_} = \frac{CA}{FD} \qquad \frac{CA}{FD} = \frac{AB}{\_\_\_}$$

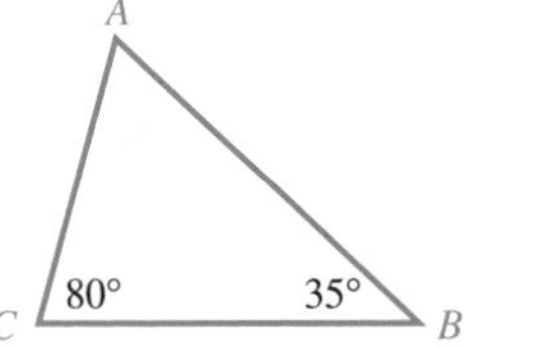

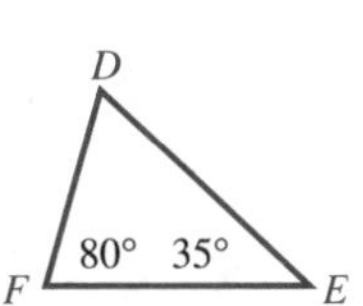

18. The two triangles shown in the following illustration are similar. Complete the proportion.

$$\frac{x}{\_\_\_} = \frac{\_\_\_}{\_\_\_}$$

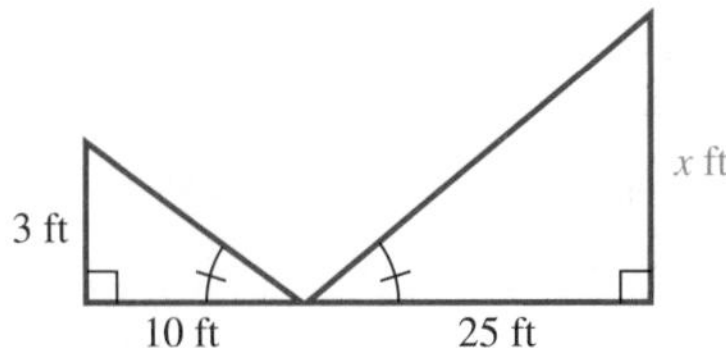

NOTATION **Complete the solution.**

19. Solve for x: $\frac{12}{18} = \frac{x}{24}$.

$$12 \cdot 24 = 18 \cdot x$$
$$288 = 18x$$
$$\frac{288}{18} = \frac{18x}{18}$$
$$16 = x$$

20. Write the ratio of 25 to 4 in two forms.

21. Write each ratio in simplified form.

a. $\frac{12}{15}$ **b.** $\frac{9 \text{ crates}}{7 \text{ crates}}$

22. Fill in the blanks: The proportion $\frac{20}{1.6} = \frac{100}{8}$ can be read: 20 is to 1.6 ____ 100 is ____ 8.

23. Fill in the blank: We read $\triangle XYZ \sim \triangle MNO$ as: triangle XYZ is _______ to triangle MNO.

24. Write the statement *x is approximately 0.75* using symbols.

PRACTICE **Write each ratio as a fraction in simplest form.**

25. 4 boxes to 15 boxes

26. 2 miles to 9 miles

27. 30 days to 24 days

28. 45 people to 30 people

29. 90 minutes to 3 hours

30. 20 inches to 2 feet

31. 13 quarts to 2 gallons

32. 11 dimes to 1 dollar

Tell whether each statement is a proportion.

33. $\frac{9}{7} = \frac{81}{70}$

34. $\frac{5}{2} = \frac{20}{8}$

35. $\frac{7}{3} = \frac{14}{6}$

36. $\frac{13}{19} = \frac{65}{95}$

37. $\frac{9}{19} = \frac{38}{80}$

38. $\frac{40}{29} = \frac{29}{22}$

Solve each proportion.

39. $\frac{2}{3} = \frac{x}{6}$

40. $\frac{3}{6} = \frac{x}{8}$

41. $\frac{5}{10} = \frac{3}{c}$

42. $\frac{7}{14} = \frac{2}{x}$

43. $\frac{6}{x} = \frac{8}{4}$

44. $\frac{4}{x} = \frac{2}{8}$

45. $\frac{x+1}{5} = \frac{3}{15}$

46. $\frac{x-1}{7} = \frac{2}{21}$

47. $\frac{x+7}{-4} = \frac{1}{4}$

48. $\frac{x+3}{12} = \frac{-7}{6}$

49. $\frac{5-x}{17} = \frac{13}{34}$

50. $\frac{4-x}{13} = \frac{11}{26}$

51. $\frac{2x-1}{18} = \frac{9}{54}$

52. $\frac{2x+1}{18} = \frac{14}{3}$

53. $\frac{x-1}{9} = \frac{2x}{3}$

54. $\frac{x+1}{4} = \frac{3x}{8}$

55. $\frac{8x}{3} = \frac{11x+9}{4}$

56. $\frac{3x}{16} = \frac{x+2}{5}$

57. $\frac{2}{3x} = \frac{x}{6}$

58. $\frac{y}{4} = \frac{4}{y}$

59. $\frac{b-5}{3} = \frac{2}{b}$

60. $\frac{2}{c} = \frac{c-3}{2}$

61. $\frac{x-1}{x+1} = \frac{2}{3x}$

62. $\frac{2}{x+6} = \frac{-2x}{5}$

Each pair of triangles is similar. Find the missing side length.

63.

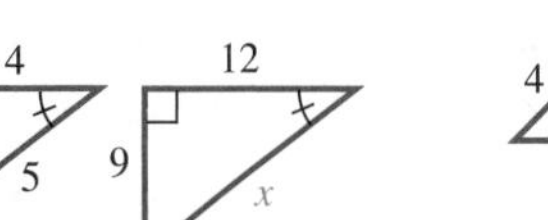

64.

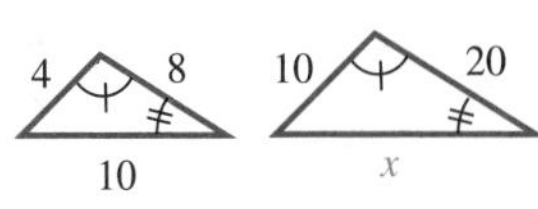

65.

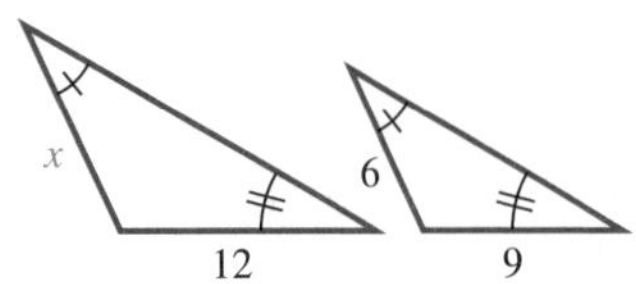

66.

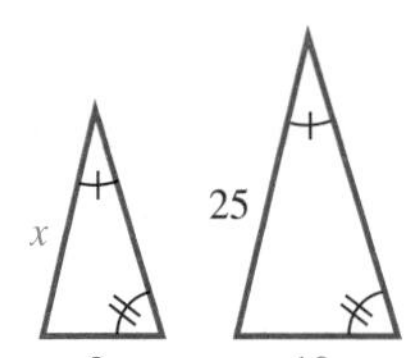

APPLICATIONS

67. GEAR RATIOS Write each ratio in two ways: as a fraction in simplest form and using a colon.

  a. The number of teeth of the larger gear to the number of teeth of the smaller gear

  b. The number of teeth of the smaller gear to the number of teeth of the larger gear

68. FACULTY–STUDENT RATIOS At a college, there are 300 faculty members and 2,850 students. Find the rate of faculty to students. (This is often referred to as the faculty to student ratio, even though the units are different.)

69. SHOPPING FOR CLOTHES If shirts are on sale at two for \$25, how much do five shirts cost?

70. COMPUTING A PAYCHECK Billie earns \$412 for a 40-hour week. If she missed 10 hours of work last week, how much did she get paid?

71. COOKING A recipe for spaghetti sauce requires four 16-ounce bottles of ketchup to make 2 gallons of sauce. How many bottles of ketchup are needed to make 10 gallons of sauce?

72. MIXING PERFUME A perfume is to be mixed in the ratio of 3 drops of pure essence to 7 drops of alcohol. How many drops of pure essence should be mixed with 56 drops of alcohol?

73. CPR A first aid handbook states that when performing cardiopulmonary resuscitation on an adult, the ratio of chest compressions to breaths should be 5 : 2. If 210 compressions were administered to an adult patient, how many breaths should have been given?

74. COOKING A recipe for wild rice soup follows. Find the amounts of chicken broth, rice, and flour needed to make 15 servings.

Wild Rice Soup

A sumptuous side dish with a nutty flavor

3 cups chicken broth	1 cup light cream
$\frac{2}{3}$ cup uncooked rice	2 tablespoons flour
$\frac{1}{4}$ cup sliced onions	$\frac{1}{8}$ teaspoon pepper
$\frac{1}{2}$ cup shredded carrots	Serves: 6

75. NUTRITION The table shows the nutritional facts about a 10-oz chocolate milkshake sold by a fast-food restaurant. Use the information to complete the table for the 16-oz shake. Round to the nearest unit when an answer is not exact.

	Calories	Fat (gm)	Protein (gm)
10-oz chocolate milkshake	355	8	9
16-oz chocolate milkshake			

76. STRUCTURAL ENGINEERING A portion of a bridge is shown. Use the fact that $\frac{AB}{BC}$ is in proportion to $\frac{FE}{ED}$ to find FE.

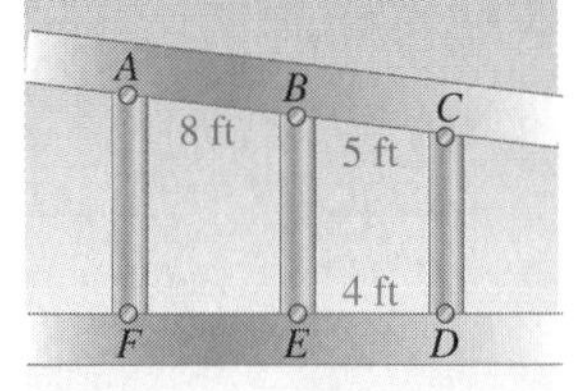

77. QUALITY CONTROL Out of a sample of 500 men's shirts, 17 were rejected because of crooked collars. How many crooked collars would you expect to find in a run of 15,000 shirts?

78. PHOTO ENLARGEMENTS The 3-by-5 photo is to be blown up to the larger size. Find x.

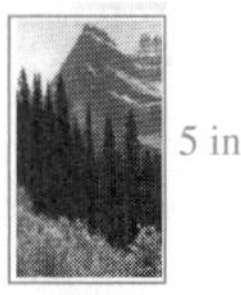

79. MIXING FUEL The instructions on a can of oil intended to be added to lawnmower gasoline are shown on the next page. Are these instructions correct? (*Hint:* There are 128 ounces in 1 gallon.)

Recommended	Gasoline	Oil
50 to 1	6 gal	16 oz

80. DRIVER'S LICENSES Of the 50 states, Alabama has one of the highest ratios of licensed drivers to residents. If the ratio is 800:1,000 and the population of Alabama is 4,500,000, how many residents of that state have a driver's license?

81. CROP DAMAGE To estimate the ground squirrel population on his acreage, a farmer trapped, tagged, and then released a dozen squirrels. Two weeks later, the farmer trapped 35 squirrels and noted that 3 were tagged. Use this information to estimate the number of ground squirrels on his acreage.

82. CONCRETE A 2:3 concrete mix means that for every two parts of sand, three parts of gravel are used. How much sand should be used in a mix composed of 25 cubic feet of gravel?

83. MODEL RAILROADS A model railroad engine is 9 inches long. If the scale is 87 feet to 1 foot, how long is a real engine?

84. MODEL RAILROADS A model railroad caboose is 3.5 inches long. If the scale is 169 feet to 1 foot, how long is a real caboose?

85. BLUEPRINTS The scale for the drawing shown means that a $\frac{1}{4}$-inch length $\left(\frac{1}{4}''\right)$ on the drawing corresponds to an actual size of 1 foot (1′0″). Suppose the length of the kitchen is $2\frac{1}{2}$ inches on the drawing. How long is the actual kitchen?

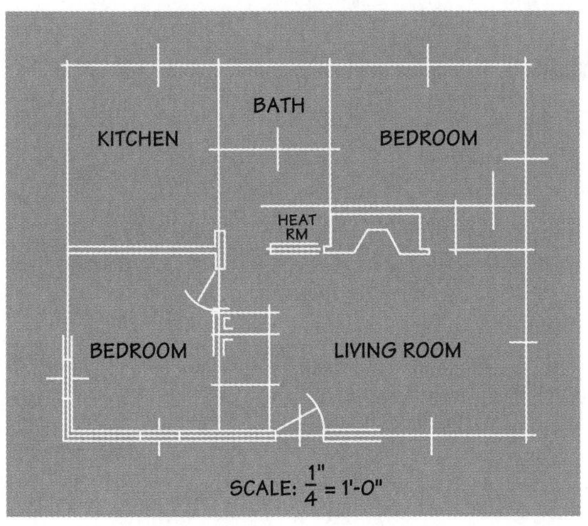

86. THE *TITANIC* A 1:144 scale model of the *Titanic* is to be built. If the ship was 882 feet long, find the length of the model?

For each of the following purchases, determine the better buy.

87. Trumpet lessons: 45 minutes for \$25 or 60 minutes for \$35

88. Memory for a computer: 128 megabytes for \$26 or 512 megabytes for \$110

89. Business cards: 100 for \$9.99 or 150 for \$12.99

90. Dog food: 20 pounds for \$7.49 or 44 pounds for \$14.99

91. Soft drinks: 6-pack for \$1.50 or a case (24 cans) for \$6.25

92. Donuts: A dozen for \$6.24 or a baker's dozen (13) for \$6.65

93. HEIGHT OF A TREE A tree casts a shadow of 26 feet at the same time as a 6-foot man casts a shadow of 4 feet. Find the height of the tree.

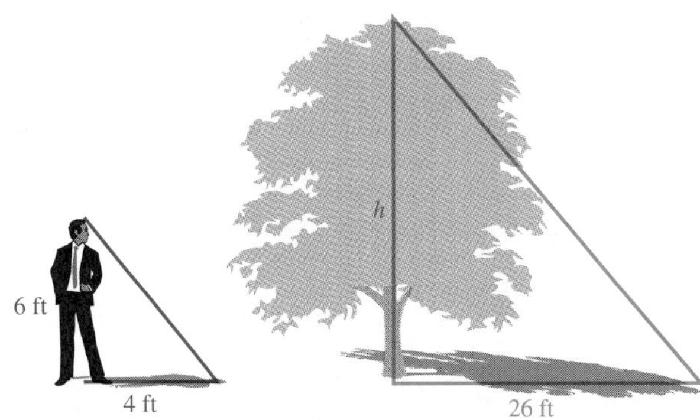

94. HEIGHT OF A BUILDING A man places a mirror on the ground and sees the reflection of the top of a building, as shown. The two triangles in the illustration are similar. Find the height, h, of the building.

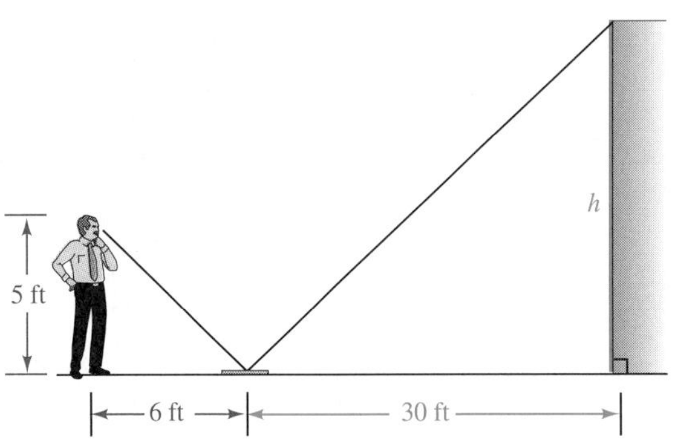

95. SURVEYING To determine the width of a river, a surveyor laid out the similar triangles as shown. Find w.

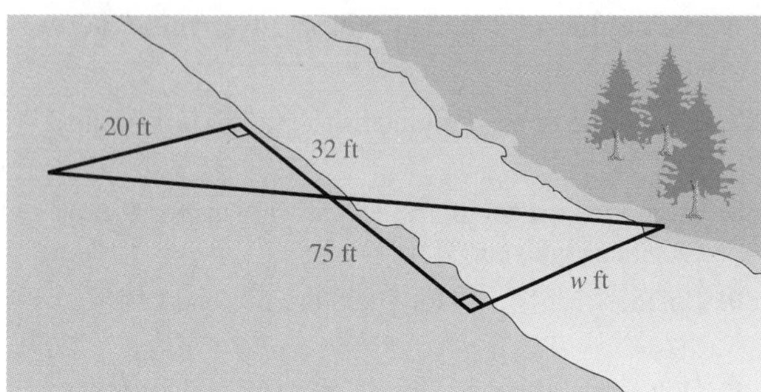

96. FLIGHT PATHS An airplane ascends 100 feet as it flies a horizontal distance of 1,000 feet. How much altitude will it gain as it flies a horizontal distance of 1 mile? (*Hint:* 5,280 feet = 1 mile.)

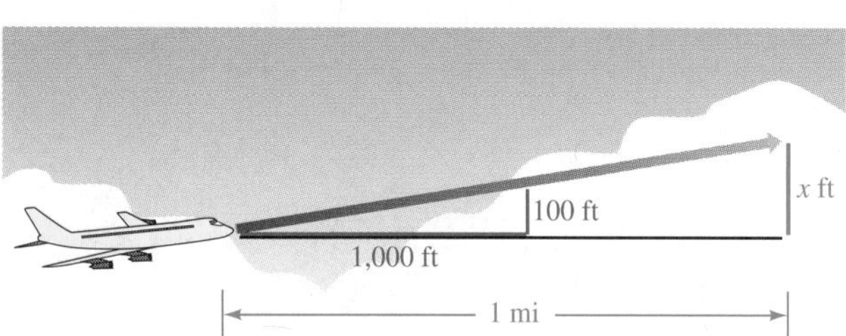

WRITING

97. Explain the difference between a ratio and a proportion.

98. Explain why the concept of cross products cannot be used to solve the equation

$$\frac{x}{3} - \frac{3x}{4} = \frac{1}{12}$$

99. What are similar triangles?

100. What is a unit price? Give an example.

REVIEW

101. Change $\frac{9}{10}$ to a percent.

102. Change $33\frac{1}{3}\%$ to a fraction.

103. Find 30% of 1,600.

104. SHOPPING Maria bought a dress for 25% off the original price of $98. How much did the dress cost?

CHALLENGE PROBLEMS

105. Suppose $\frac{a}{b} = \frac{c}{d}$. Write three other proportions using a, b, c, and d.

106. Verify that $\frac{3}{5} = \frac{12}{20} = \frac{3 + 12}{5 + 20}$. Is the following rule always true? Explain.

$$\frac{a}{b} = \frac{c}{d} = \frac{a + c}{b + d}$$

ACCENT ON TEAMWORK

WHAT IS π?

Overview: In this activity, you will discover an important fact about the ratio of the circumference to the diameter of a circle.

Instructions: Form groups of 2 or 3 students. With a piece of string or a cloth tape measure, find the circumference and the diameter of objects that are circular in shape. You can measure anything that is round: for example, a coin, the top of a can, a tire, or a waste paper basket. Enter your results in a table, as shown below. Convert each measurement to a decimal, and then use a calculator to determine a decimal approximation of the ratio of the circumference C to diameter d.

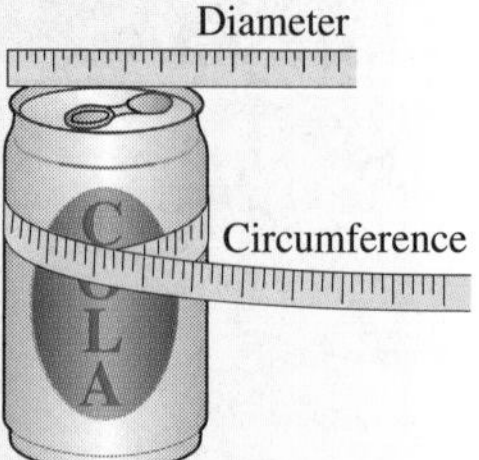

Object	Circumference C	Diameter d	$\frac{C}{d}$ (approx.)
A quarter	$2\frac{15}{16}$ in. = 2.9375 in.	$\frac{15}{16}$ in. = 0.9375 in.	3.13333

Since early history, mathematicians have known that the ratio of the circumference to the diameter of a circle is the same for any size circle, approximately 3. Today, following centuries of study, we know that this ratio is exactly 3.141592653589 . . . .

$$\frac{C}{d} = 3.141592653589\ldots$$

The Greek letter π (pi) is used to represent the ratio of circumference to diameter:

$$\pi = \frac{C}{d}, \qquad \text{where } \pi = 3.141592653589\ldots$$

Are the ratios in your table numerically close to π? Give some reasons why they aren't exactly 3.141592653589 . . . in each case.

USING PROPORTIONS WHEN COOKING

Overview: In this activity, you will use proportions to adjust the ingredients in your favorite recipe so that it would serve everyone in your class.

Instructions: Each student should bring his/her favorite recipe to class. Working in pairs, begin with the first recipe and determine the amount of each ingredient that is needed to make enough of it to serve the exact number of people in the class. For example, if the recipe serves 8 and there are 25 students and an instructor in the class, write and solve proportions to determine the amount of flour, sugar, milk, and so on, to make enough of the recipe to serve 26 people. Then do the same for the second recipe.

When finished, share with the class how you made the calculations, as well as any difficulties that you encountered adjusting the ingredients of the recipes.

KEY CONCEPT: EXPRESSIONS AND EQUATIONS

In this chapter, we have discussed procedures for simplifying rational expressions and procedures for solving rational equations.

SIMPLIFYING RATIONAL EXPRESSIONS

To simplify a rational expression:

1. Factor the numerator and denominator completely to determine all the factors common to both.
2. Remove factors equal to 1 by replacing each pair of factors common to the numerator and denominator with the equivalent fraction $\frac{1}{1}$.

This procedure is also used when multiplying and dividing rational expressions.

1. a. Simplify: $\dfrac{2x^2 - 8x}{x^2 - 6x + 8}$.
   b. What common factor was removed?
2. a. Multiply: $\dfrac{x^2 + 2x + 1}{x} \cdot \dfrac{x^2 - x}{x^2 - 1}$.
   b. What common factors were removed?

BUILDING RATIONAL EXPRESSIONS

To build a rational expression, we multiply it by a form of 1. We use this concept to add and subtract rational expressions having unlike denominators and when simplifying complex fractions.

3. a. Add: $\frac{x}{x+1} + \frac{x-1}{x}$.

b. By what did you multiply the first fraction to rewrite it in terms of the LCD? The second fraction?

4. a. Simplify: $\frac{n - 1 - \frac{2}{n}}{\frac{n}{3}}$.

b. By what did you multiply the numerator and denominator to simplify the complex fraction?

SOLVING RATIONAL EQUATIONS

The multiplication property of equality states that *multiplying both sides of an equation by the same nonzero number does not change its solution.* We use this property when solving rational equations. If we multiply both sides of the equation by the LCD of the rational expressions in the equation, we can clear it of fractions.

5. a. Solve: $\frac{11}{b} + \frac{13}{b} = 12$.

b. By what did you multiply both sides to clear the equation of fractions?

6. a. Solve: $\frac{3}{s+2} - \frac{5}{s^2+s-2} = \frac{1}{s-1}$.

b. By what did you multiply both sides to clear the equation of fractions?

7. a. Solve: $y + \frac{3}{4} = \frac{3y-50}{4y-24}$.

b. By what did you multiply both sides to clear the equation of fractions?

8. a. Solve: $\frac{1}{a} - \frac{1}{b} = 1$ for b.

b. By what did you multiply both sides to clear the equation of fractions?

CHAPTER REVIEW

ELEMENTARY Algebra $f(x)$ Now™

SECTION 6.1 Simplifying Rational Expressions

CONCEPTS

Since division by 0 is undefined, we must make sure that the denominator of a rational expression is not 0.

REVIEW EXERCISES

1. Find the values of x for which the rational expression $\frac{x-1}{x^2-16}$ is undefined.

2. Evaluate $\frac{x^2-1}{x-5}$ for $x = -2$.

Simplify each rational expression, if possible. Assume that no denominators are zero.

3. $\frac{3x^2}{6x^3}$

4. $\frac{5xy^2}{2x^2y^2}$

5. $\frac{x^2}{x^2+x}$

6. $\frac{a^2-4}{a+2}$

To *simplify* a rational expression:

1. Factor the numerator and denominator completely.
2. Remove factors equal to 1 by replacing each pair of factors common to the numerator and denominator with the equivalent fraction $\frac{1}{1}$.

The quotient of any nonzero expression and its opposite is -1.

7. $\dfrac{3p - 2}{2 - 3p}$

8. $\dfrac{8 - x}{x^2 - 5x - 24}$

9. $\dfrac{2x^2 - 16x}{2x^2 - 18x + 16}$

10. $\dfrac{x^2 + x - 2}{x^2 - x - 2}$

11. $\dfrac{4(t + 3) + 8}{3(t + 3) + 6}$

12. Explain the error in the following work: $\dfrac{x + 1}{x} = \dfrac{\overset{1}{\not{x}} + 1}{\underset{1}{\not{x}}} = \dfrac{2}{1} = 2.$

13. DOSAGES Cowling's rule is a formula that can be used to determine the dosage of a prescription medication for children. If C is the proper child's dosage, D is an adult dosage, and A is the child's age in years, then

$$C = \frac{D(A + 1)}{24}$$

Find the daily dosage of an antibiotic for an 11-year-old child if the adult daily dosage is 300 milligrams.

SECTION 6.2 Multiplying and Dividing Rational Expressions

To *multiply* rational expressions, multiply the numerators and multiply the denominators.

$$\frac{A}{B} \cdot \frac{C}{D} = \frac{AC}{BD}$$

Then simplify, if possible.

To divide rational expressions, multiply the first by the reciprocal of the second.

$$\frac{A}{B} \div \frac{C}{D} = \frac{A}{B} \cdot \frac{D}{C} = \frac{AD}{BC}$$

Then simplify, if possible.

Multiply and simplify, if possible.

14. $\dfrac{3xy}{2x} \cdot \dfrac{4x}{2y^2}$

15. $56x\left(\dfrac{12}{7x}\right)$

16. $\dfrac{x^2 - 1}{x^2 + 2x} \cdot \dfrac{x}{x + 1}$

17. $\dfrac{x^2 + x}{3x - 15} \cdot \dfrac{6x - 30}{x^2 + 2x + 1}$

Divide and simplify, if possible.

18. $\dfrac{3x^2}{5x^2y} \div \dfrac{6x}{15xy^2}$

19. $\dfrac{x^2 + 5x}{x^2 + 4x - 5} \div x^2$

20. $\dfrac{x^2 - x - 6}{1 - 2x} \div \dfrac{x^2 - 2x - 3}{2x^2 + x - 1}$

21. Tell whether the given fraction is a unit conversion factor.

a. $\dfrac{1 \text{ ft}}{12 \text{ in.}}$ **b.** $\dfrac{60 \text{ min}}{1 \text{ day}}$ **c.** $\dfrac{2{,}000 \text{ lbs.}}{1 \text{ ton}}$ **d.** $\dfrac{1 \text{ gal}}{4 \text{ qt}}$

To find the *reciprocal* of a rational expression, we invert its numerator and denominator.

A *unit conversion factor* is a fraction that has value 1.

22. TRAFFIC SIGNS Convert the speed limit on the sign from miles per hour to miles per minute.

SECTION 6.3 Addition and Subtraction with Like Denominators; Least Common Denominators

To *add* or *subtract* rational expressions that have the same denominator, add or subtract their numerators and write the sum or difference over the common denominator.

$$\frac{A}{D} + \frac{B}{D} = \frac{A + B}{D}$$

$$\frac{A}{D} - \frac{B}{D} = \frac{A - B}{D}$$

Then simplify, if possible.

Add or subtract and simplify, if possible.

23. $\dfrac{13}{15d} - \dfrac{8}{15d}$

24. $\dfrac{x}{x + y} + \dfrac{y}{x + y}$

25. $\dfrac{3x}{x - 7} - \dfrac{x - 2}{x - 7}$

26. $\dfrac{a}{a^2 - 2a - 8} + \dfrac{2}{a^2 - 2a - 8}$

To find the *LCD,* factor each denominator completely. Form a product using each different factor the greatest number of times it appears in any one factorization.

Find the LCD of each pair of rational expressions.

27. $\dfrac{12}{x}, \dfrac{1}{9}$

28. $\dfrac{1}{2x^3}, \dfrac{5}{8x}$

29. $\dfrac{7}{m}, \dfrac{m + 2}{m - 8}$

30. $\dfrac{x}{5x + 1}, \dfrac{5x}{5x - 1}$

31. $\dfrac{6 - a}{a^2 - 25}, \dfrac{a^2}{a - 5}$

32. $\dfrac{4t + 25}{t^2 + 10t + 25}, \dfrac{t^2 - 7}{2t^2 + 17t + 35}$

To *build* an equivalent rational expression, multiply the given expression by 1 in the form of $\frac{c}{c}$, where c is any nonzero number or expression.

Build each rational expression into an equivalent fraction having the denominator shown in red.

33. $\dfrac{9}{a}$; $7a$

34. $\dfrac{2y + 1}{x - 9}$; $x(x - 9)$

35. $\dfrac{b + 7}{3b - 15}$; $6(b - 5)$

36. $\dfrac{9r}{r^2 + 6r + 5}$; $(r + 1)(r - 4)(r + 5)$

SECTION 6.4 Addition and Subtraction with Unlike Denominators

To *add* or *subtract* rational expressions with unlike denominators:

1. Find the LCD.
2. Write each rational expression as an equivalent expression whose denominator is the LCD.
3. Add or subtract the numerators and write the sum or difference over the LCD.
4. Simplify the resulting rational expression, if possible.

When a polynomial is multiplied by -1, the result is its opposite.

Add or subtract and simplify, if possible.

37. $\dfrac{1}{7} - \dfrac{1}{a}$

38. $\dfrac{x}{x - 1} + \dfrac{1}{x}$

39. $\dfrac{2t + 2}{t^2 + 2t + 1} - \dfrac{1}{t + 1}$

40. $\dfrac{x + 2}{2x} - \dfrac{2 - x}{x^2}$

41. $\dfrac{6}{b - 1} - \dfrac{b}{1 - b}$

42. $\dfrac{8}{c} + 6$

43. $\dfrac{n + 7}{n + 3} - \dfrac{n - 3}{n + 7}$

44. $\dfrac{4}{t + 2} - \dfrac{7}{(t + 2)^2}$

45. $\dfrac{6}{a^2 - 9} - \dfrac{5}{a^2 - a - 6}$

46. $\dfrac{2}{3y - 6} + \dfrac{3}{4y + 8}$

47. Working on a homework assignment, a student added two rational expressions and obtained $\frac{-5n^3 - 7}{3n(n + 6)}$. The answer given in the back of the book was $-\frac{5n^3 + 7}{3n(n + 6)}$. Are the answers equivalent?

48. VIDEO CAMERAS Find the perimeter and the area of the LED screen of the camera.

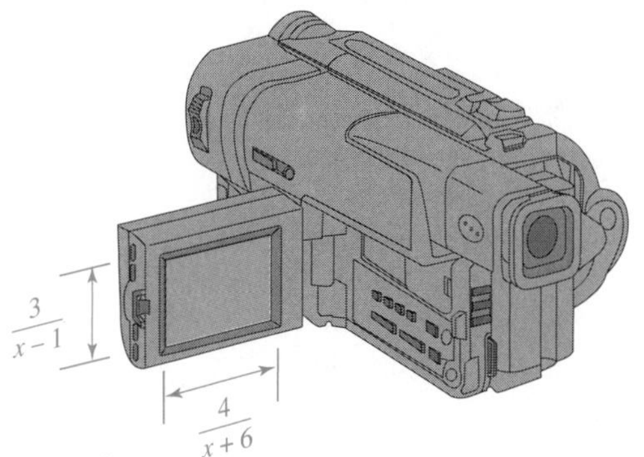

SECTION 6.5 Simplifying Complex Fractions

Complex fractions contain fractions in their numerators and/or their denominators.

Simplify each complex fraction.

49. $\dfrac{\frac{n^4}{30}}{\frac{7n}{15}}$

50. $\dfrac{\frac{r^2 - 81}{18s^2}}{\frac{4r - 36}{9s}}$

To *simplify* a complex fraction:

Method 1
Write the numerator and the denominator as single rational expressions. Perform the indicated division (multiply by the reciprocal).

Method 2
Determine the LCD for all rational expressions in the complex fraction. Multiply the complex fraction by 1 in the form of $\frac{\text{LCD}}{\text{LCD}}$.

51. $\dfrac{\frac{1}{y} + 1}{\frac{1}{y} - 1}$

52. $\dfrac{\frac{7}{a^2}}{\frac{1}{a} + \frac{10}{3}}$

53. $\dfrac{\frac{2}{x - 1} + \frac{x - 1}{x + 1}}{\frac{1}{x^2 - 1}}$

54. $\dfrac{\frac{1}{x^2y} - \frac{5}{xy}}{\frac{3}{xy} - \frac{7}{xy^2}}$

SECTION 6.6 Solving Rational Equations

To *solve* a rational equation:

1. Determine which numbers cannot be solutions.
2. Multiply both sides of the equation by the LCD of all rational expressions in the equation.
3. Solve the resulting equation.
4. Check all possible solutions in the original equation.

A possible solution that does not satisfy the original equation is called an *extraneous* solution.

Solve each equation and check the result. If an equation has no solution, so indicate.

55. $\dfrac{3}{x} = \dfrac{2}{x - 1}$

56. $\dfrac{a}{a - 5} = 3 + \dfrac{5}{a - 5}$

57. $\dfrac{2}{3t} + \dfrac{1}{t} = \dfrac{5}{9}$

58. $a = \dfrac{3a - 50}{4a - 24} - \dfrac{3}{4}$

59. $\dfrac{4}{x + 2} - \dfrac{3}{x + 3} = \dfrac{6}{x^2 + 5x + 6}$

60. $\dfrac{3}{x + 1} - \dfrac{x - 2}{2} = \dfrac{x - 2}{x + 1}$

61. ENGINEERING The efficiency E of a Carnot engine is given by the following formula. Solve it for T_1.

$$E = 1 - \frac{T_2}{T_1}$$

62. Solve for y: $\dfrac{1}{x} = \dfrac{1}{y} + \dfrac{1}{z}$.

SECTION 6.7 Problem Solving Using Rational Equations

To solve a problem follow these steps:

1. Analyze the problem.
2. Form an equation.
3. Solve the equation.
4. State the conclusion.
5. Check the result.

Distance = rate · time

Work completed = rate of work · time

Interest = Principal · rate · time

63. NUMBER PROBLEM If a number is subtracted from the denominator of $\frac{4}{5}$ and twice as much is added to the numerator, the result is 5. Find the number.

64. EXERCISE A jogger can bicycle 30 miles in the same time that it takes her to jog 10 miles. If she can ride 10 mph faster than she can jog, how fast can she jog?

65. HOUSE CLEANING A maid can clean a house in 4 hours. What is her rate of work?

66. HOUSE PAINTING If a homeowner can paint a house in 14 days and a professional painter can paint it in 10 days, how long will it take if they work together?

67. INVESTMENTS In 1 year, a student earned \$100 interest on money she deposited at a savings and loan. She later learned that the money would have earned \$120 if she had deposited it at a credit union, because the credit union paid 1% more interest at the time. Find the rate she received from the savings and loan.

68. WIND SPEED A plane flies 400 miles downwind in the same amount of time as it takes to travel 320 miles upwind. If the plane can fly at 360 mph in still air, find the velocity of the wind.

SECTION 6.8 Proportions and Similar Triangles

A *proportion* is a statement that two ratios or two rates are equal.

In the proportion $\frac{a}{b} = \frac{c}{d}$, a and d are the *extremes,* and b and c are the *means.*

In any proportion, the product of the extremes is equal to the product of the means. That is, the *cross products* are equal.

Determine whether each equation is a proportion.

69. $\frac{4}{7} = \frac{20}{34}$

70. $\frac{5}{7} = \frac{30}{42}$

Solve each proportion.

71. $\frac{3}{x} = \frac{6}{9}$

72. $\frac{x}{3} = \frac{x}{5}$

73. $\frac{x - 2}{5} = \frac{x}{7}$

74. $\frac{2x}{x + 4} = \frac{3}{x - 1}$

75. DENTISTRY The diagram on the right was displayed in a dentist's office. According to the diagram, if the dentist has 340 adult patients, how many will develop gum disease?

3 out of 4 adults will develop gum disease.

The lengths of corresponding sides of *similar triangles* are in proportion.

A *scale* is a ratio (or rate) that compares the size of a model to the size of an actual object.

Unit prices can be used to compare costs of different sizes of the same brand to determine the best buy.

76. A telephone pole casts a shadow 12 feet long at the same time that a man 6 feet tall casts a shadow of 3.6 feet. How tall is the pole?

77. PORCELAIN FIGURINES A model of a flutist, standing and playing at a music stand, was made using a 1/12th scale. If the scale model is 5.5 inches tall, how tall is the flutist?

78. COMPARISON SHOPPING Which is the better buy for recordable compact discs: 150 for \$60 or 250 for \$98?

CHAPTER 6 TEST

ELEMENTARY Algebra Now™

For what numbers is each rational expression undefined?

1. $\dfrac{6x-9}{5x}$

2. $\dfrac{x}{x^2+x-6}$

3. MEMORY The formula $n = \frac{35+5d}{d}$ approximates the number of words n that a certain person can recall d days after memorizing a list of 50 words. How many words will the person remember in 1 week?

4. U.S. SCHOOL CONSTRUCTION The table shows the average school size for projects finished in 2002. Convert the size of a middle school to square yards.

	Square feet
Elementary school	70,174
Middle school	109,512
High school	125,304

(Source: American School & University's 29th Annual Official Education Construction Report)

Simplify each rational expression.

5. $\dfrac{48x^2y}{54xy^2}$

6. $\dfrac{7m-49}{7-m}$

7. $\dfrac{2x^2-x-3}{4x^2-9}$

8. $\dfrac{3(x+2)-3}{6x+5-(3x+2)}$

Find the LCD of each pair of rational expressions.

9. $\dfrac{19}{3c^2d}, \dfrac{6}{c^2d^3}$

10. $\dfrac{4n+25}{n^2-4n-5}, \dfrac{6n}{n^2-25}$

Perform the operations. Simplify, if possible.

11. $\dfrac{12x^2y}{15xy} \cdot \dfrac{25y^2}{16x}$

12. $\dfrac{x^2+3x+2}{3x+9} \cdot \dfrac{x+3}{x^2-4}$

13. $\dfrac{x-x^2}{3x^2+6x} \div \dfrac{3x-3}{3x^3+6x^2}$

14. $\dfrac{a^2-16}{a-4} \div (6a+24)$

15. $\dfrac{3y+7}{2y+3} - \dfrac{3y-6}{2y+3}$

16. $\dfrac{2n}{5m} - \dfrac{n}{2}$

17. $\dfrac{x+1}{x} + \dfrac{x-1}{x+1}$

18. $\dfrac{a+3}{a-1} - \dfrac{a+4}{1-a}$

19. $\dfrac{9}{c-4} + c$

20. $\dfrac{6}{t^2-9} - \dfrac{5}{t^2-t-6}$

Simplify each complex fraction.

21. $\dfrac{\dfrac{3m-9}{8m}}{\dfrac{5m-15}{32}}$

22. $\dfrac{\dfrac{3}{as^2}+\dfrac{6}{a^2s}}{\dfrac{6}{a}-\dfrac{9}{s^2}}$

Solve each equation. If an equation has no solution, so indicate.

23. $\dfrac{1}{3}+\dfrac{4}{3y}=\dfrac{5}{y}$

24. $\dfrac{9n}{n-6}=3+\dfrac{54}{n-6}$

25. $\dfrac{7}{q^2-q-2}+\dfrac{1}{q+1}=\dfrac{3}{q-2}$

26. $\dfrac{2}{3}=\dfrac{2c-12}{3c-9}-c$

27. $\dfrac{y}{y-1}=\dfrac{y-2}{y}$

28. Solve for B: $H=\dfrac{RB}{R+B}$.

29. HEALTH RISKS A medical newsletter states that a "healthy" waist-to-hip ratio for men is 19:20 or less. Does the patient shown in the illustration fall within the "healthy" range?

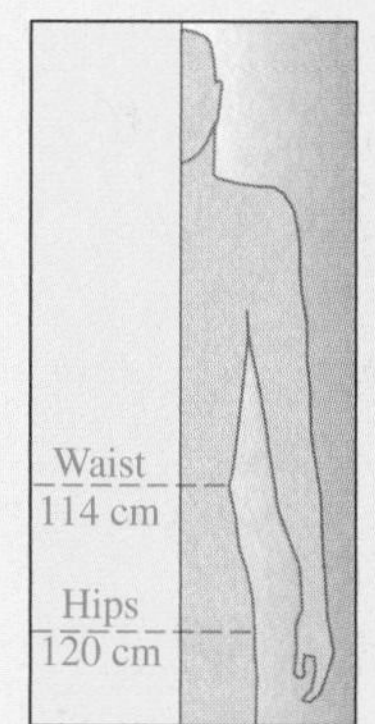

30. CURRENCY EXCHANGE RATES Preparing for a visit to London, a New York resident exchanged 1,500 U.S. dollars for British pounds. (A pound is the basic monetary unit of Great Britain.) If the exchange rate was 5 U.S. dollars for 3 British pounds, how many British pounds did the traveler receive?

31. TV TOWERS A television tower casts a shadow 114 feet long at the same time that a 6-foot-tall television reporter casts a shadow of 4 feet. Find the height of the tower.

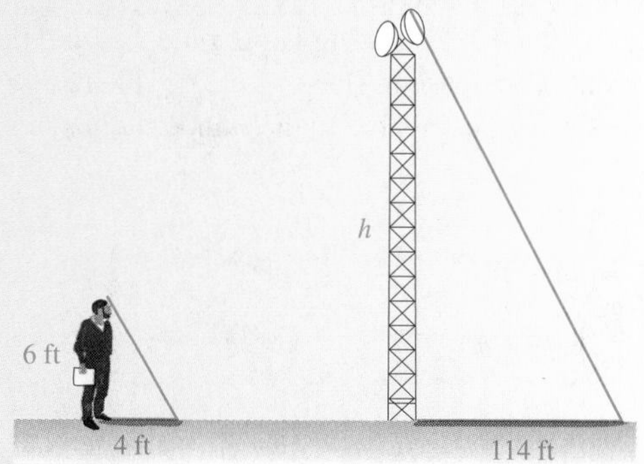

32. COMPARISON SHOPPING Which is the better buy for fabric softener: 80 sheets for \$3.89 or 120 sheets for \$6.19?

33. CLEANING HIGHWAYS One highway worker can pick up all the trash on a strip of highway in 7 hours, and his helper can pick up the trash in 9 hours. How long will it take them if they work together?

34. PHYSICAL FITNESS A man roller-blades at a rate 6 miles per hour faster than he jogs. In the same time it takes him to roller-blade 5 miles he can jog 2 miles. How fast does he jog?

35. Explain the error: $\dfrac{x+5}{5}=\dfrac{x+\overset{1}{\cancel{5}}}{\underset{1}{\cancel{5}}}=x+1$.

36. Explain what it means to clear the following equation of fractions.

$$\frac{u}{u-1}+\frac{1}{u}=\frac{u^2+1}{u^2-u}$$

Why is this a helpful first step in solving the equation?

CHAPTERS 1–6 CUMULATIVE REVIEW EXERCISES

1. Evaluate: $9^2 - 3[45 - 3(6 + 4)]$.

2. GRAND KING SIZE BEDS Because Americans are taller compared to 100 years ago, bed manufacturers are making larger models. Find the percent of increase in sleeping area of the new grand king size bed compared to the standard king size.

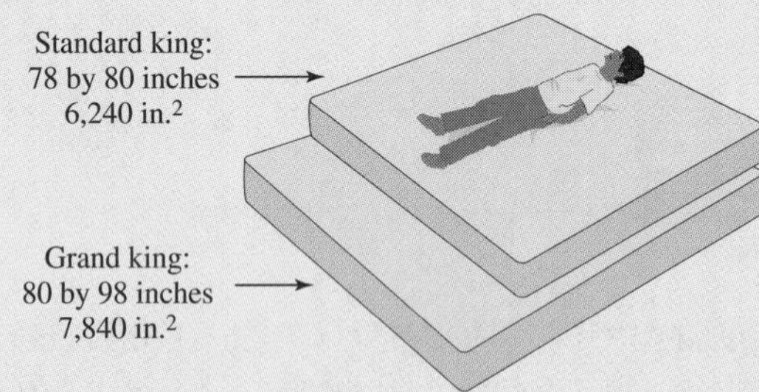

3. Insert the proper symbol, < or >, in the blank to make a true statement.

$$|2 - 4| \;\square\; -(-6)$$

4. Find the average (mean) test score of a student in a history class with scores of 80, 73, 61, 73, and 98.

5. Change 40° C to degrees Fahrenheit.

6. Find the volume of a pyramid that has a square base, measuring 6 feet on a side, and whose height is 20 feet.

7. Tell whether each statement is true or false.
   a. Every integer is a whole number.
   b. 0 is not a rational number.
   c. π is an irrational number.
   d. The set of integers is the set of whole numbers and their opposites.

8. Solve: $2 - 3(x - 5) = 4(x - 1)$.

9. Simplify: $8(c + 7) - 2(c - 3)$.

10. Solve $A - c = 2B + r$ for B.

11. Solve: $7x + 2 \geq 4x - 1$. Write the solution set in interval notation and graph it.

12. Solve: $\frac{4}{5}d = -4$.

13. BLENDING TEA One grade of tea (worth \$3.20 per pound) is to be mixed with another grade (worth \$2 per pound) to make 20 pounds of a mixture that will be worth \$2.72 per pound. How much of each grade of tea must be used?

14. SPEED OF A PLANE Two planes are 6,000 miles apart and their speeds differ by 200 mph. If they travel toward each other and meet in 5 hours, find the speed of the slower plane.

15. Graph: $y = 2x - 3$.

16. Graph: $3x - 2y \leq 6$.

17. What is the slope of a line perpendicular to the line $y = -\frac{7}{8}x - 6$?

18. Find the slope of the line passing through $(-1, 3)$ and $(3, -1)$.

19. Write the equation of a line that has slope 3 and passes through the point $(1, 5)$.

20. CUTTING STEEL The graph shows the amount of wear (in mm) on a cutting blade for a given length of a cut (in m). Find the rate of change in the length of the cutting blade.

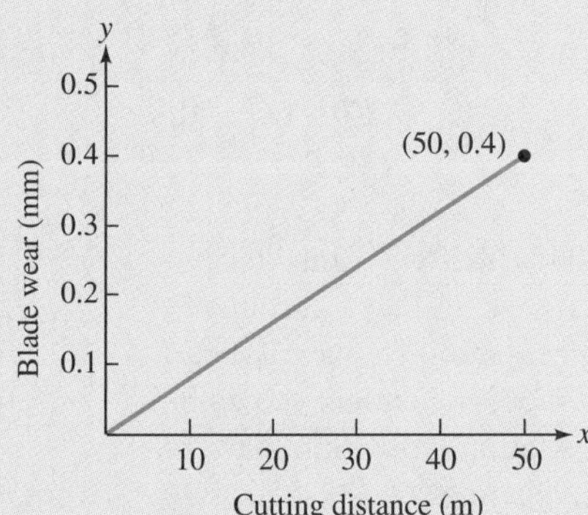

Simplify each expression. Write each answer without using negative exponents.

21. x^4x^3

22. $(x^2x^3)^5$

23. $\left(\frac{y^3y}{2yy^2}\right)^3$

24. $\left(\frac{-2a}{b}\right)^5$

25. $(a^{-2}b^3)^{-4}$

26. $\frac{9b^0b^3}{3b^{-3}b^4}$

27. Write 290,000 in scientific notation.

28. What is the degree of the polynomial $5x^3 - 4x + 16$?

29. CONCENTRIC CIRCLES The area of the ring between the two concentric circles of radius r and R is given by the formula

$$A = \pi(R + r)(R - r)$$

Do the multiplication on the right-hand side of the equation.

30. Graph: $y = -x^3$.

Perform the operations.

31. $(3x^2 - 3x - 2) + (3x^2 + 4x - 3)$

32. $(2x^2y^3)(3x^3y^2)$

33. $(2y - 5)(3y + 7)$

34. $-4x^2z(3x^2 - z)$

35. $\dfrac{6x + 9}{3}$

36. $2x + 3\overline{)2x^3 + 7x^2 + 4x - 3}$

Factor each polynomial completely, if possible.

37. $k^3t - 3k^2t$

38. $2ab + 2ac + 3b + 3c$

39. $2a^2 - 200b^2$

40. $b^3 + 125$

41. $u^2 - 18u + 81$

42. $6x^2 - 63 - 13x$

43. $-r^2 + 2 + r$

44. $u^2 + 10u + 15$

Solve each equation by factoring.

45. $5x^2 + x = 0$

46. $6x^2 - 5x = -1$

47. COOKING The electric griddle shown has a cooking surface of 160 square inches. Find the length and the width of the griddle.

48. For what values of x is the rational expression $\frac{3x^2}{x^2 - 25}$ undefined?

Perform the operations. Simplify, if possible.

49. $\dfrac{x^2 - 16}{x - 4} \div \dfrac{3x + 12}{x}$

50. $\dfrac{4}{x - 3} + \dfrac{5}{3 - x}$

51. $\dfrac{m}{m^2 + 5m + 6} - \dfrac{2}{m^2 + 3m + 2}$

52. Simplify: $\dfrac{2 - \dfrac{2}{x + 1}}{2 + \dfrac{2}{x}}$.

Solve each equation.

53. $\dfrac{7}{5x} - \dfrac{1}{2} = \dfrac{5}{6x} + \dfrac{1}{3}$

54. $\dfrac{u}{u - 1} + \dfrac{1}{u} = \dfrac{u^2 + 1}{u^2 - u}$

55. DRAINING A TANK If one outlet pipe can drain a tank in 24 hours, and another pipe can drain the tank in 36 hours, how long will it take for both pipes to drain the tank?

56. HEIGHT OF A TREE A tree casts a shadow of 29 feet at the same time as a vertical yardstick casts a shadow of 2.5 feet. Find the height of the tree.

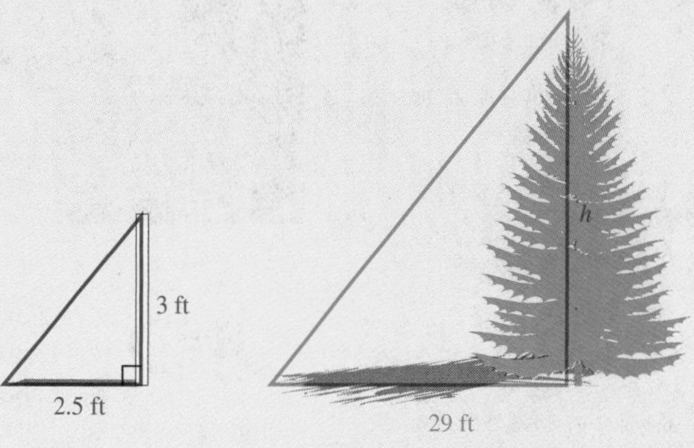

Chapter

7 Solving Systems of Equations and Inequalities

ELEMENTARY **Algebra** *f(x)* **Now**™

Throughout the chapter, this icon introduces resources on the Elementary AlgebraNow Web site, accessed through **http://1pass.thomson.com**, that will

- Help you test your knowledge of the material with a pre-test and a post-test
- Provide a personalized learning plan targeting areas you should study

7.1 Solving Systems of Equations by Graphing

7.2 Solving Systems of Equations by Substitution

7.3 Solving Systems of Equations by Elimination

7.4 Problem Solving Using Systems of Equations

7.5 Solving Systems of Linear Inequalities

Accent on Teamwork

Key Concept

Chapter Review

Chapter Test

Cumulative Review Exercises

Getty Images

Managers make many decisions in the day-to-day operation of a business. For example, a theater owner must determine ticket prices, employees' wages, and show times. Algebra can be helpful in making these decisions. When business applications involve two variable quantities, often we can model them using a pair of equations called a *system of equations.*

To learn more about the use of systems of equations, visit *The Learning Equation* on the Internet at http://tle.brookscole.com. (The log-in instructions are in the Preface.) For Chapter 7, the online lesson is

- *TLE* Lesson 14: Solving Systems of Equations by Graphing

Some problems are more easily solved using a pair of equations in two variables rather than a single equation in one variable. We call a pair of equations with common variables a system of equations.

7.1 Solving Systems of Equations by Graphing

- Systems of Equations and Their Solutions
- The Graphing Method
- Inconsistent Systems and Dependent Equations
- Graphing Calculators

The illustration shows the per capita consumption of chicken and beef in the United States for the years 1985–2002. Plotting both graphs on the same coordinate system makes it easy to compare recent trends. The point of intersection of the graphs indicates that Americans consumed equal amounts of chicken and beef in 1992—about 66 pounds of each, per person.

In this section, we will use a similar graphical approach to solve *systems of equations.*

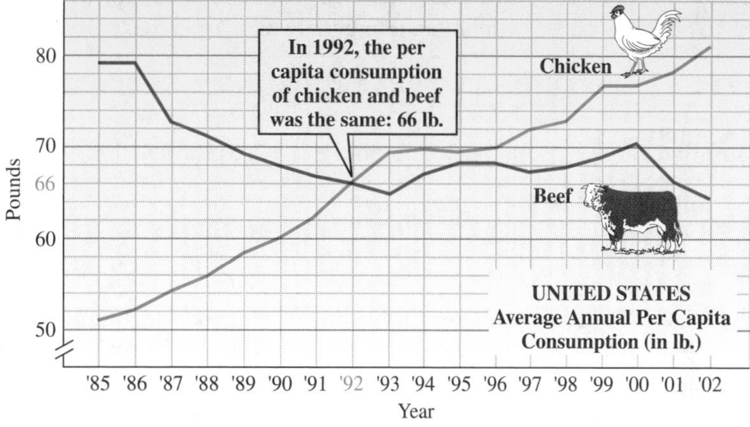

Source: American Meat Institute

SYSTEMS OF EQUATIONS AND THEIR SOLUTIONS

We have previously discussed equations in two variables, such as $x + y = 3$. Because there are infinitely many pairs of numbers whose sum is 3, there are infinitely many pairs (x, y) that satisfy this equation. Some of these pairs are listed in the following table.

The Language of Algebra

We say that (2, 1) *satisfies* $x + y = 3$, because the x-coordinate, 2, and the y-coordinate, 1, make the equation true when substituted for x and y: $2 + 1 = 3$. To *satisfy* means to make content, as in *satisfy* your thirst or a *satisfied* customer.

$$x + y = 3$$

x	y	(x, y)
0	3	(0, 3)
1	2	(1, 2)
2	1	(2, 1)
3	0	(3, 0)

Now consider the equation $x - y = 1$. Because there are infinitely many pairs of numbers whose difference is 1, there are infinitely many pairs (x, y) that satisfy $x - y = 1$. Some of these pairs are listed in the following table.

$x - y = 1$

x	y	(x, y)
0	−1	(0, −1)
1	0	(1, 0)
2	1	(2, 1)
3	2	(3, 2)

The Language of Algebra

A system of equations is two (or more) equations that we consider *simultaneously*—at the same time. Some professional sports teams *simulcast* their games. That is, the announcer's play-by-play description is broadcast on radio and television at the same time.

From the two tables, we see that (2, 1) satisfies both equations.

When two equations are considered simultaneously, we say that they form a **system of equations.** Using a left brace $\{$, we can write the equations from the previous example as a system:

$$\begin{cases} x + y = 3 \\ x - y = 1 \end{cases}$$ Read as "the system of equations $x + y = 3$ and $x - y = 1$."

Because the ordered pair (2, 1) satisfies both of these equations, it is called a **solution of the system.** In general, a system of linear equations can have exactly one solution, no solution, or infinitely many solutions.

EXAMPLE 1

Determine whether (−2, 5) is a solution of each system of equations.

a. $\begin{cases} 3x + 2y = 4 \\ x - y = -7 \end{cases}$ **b.** $\begin{cases} 4y = 18 - x \\ y = 2x \end{cases}$

Solution **a.** Recall that in an ordered pair, the first number is the x-coordinate and the second number is the y-coordinate. To determine whether (−2, 5) is a solution, we substitute −2 for x and 5 for y in each equation.

Check:

$$\begin{aligned} 3x + 2y &= 4 \\ 3(-2) + 2(5) &\stackrel{?}{=} 4 \\ -6 + 10 &\stackrel{?}{=} 4 \\ 4 &= 4 \quad \text{True} \end{aligned}$$

$$\begin{aligned} x - y &= -7 \\ -2 - 5 &\stackrel{?}{=} -7 \\ -7 &= -7 \quad \text{True} \end{aligned}$$

Since (−2, 5) satisfies both equations, it is a solution of the system.

b. Again, we substitute −2 for x and 5 for y in each equation.

Check:

$$\begin{aligned} 4y &= 18 - x \\ 4(5) &\stackrel{?}{=} 18 - (-2) \\ 20 &\stackrel{?}{=} 18 + 2 \\ 20 &= 20 \quad \text{True} \end{aligned}$$

$$\begin{aligned} y &= 2x \\ 5 &\stackrel{?}{=} 2(-2) \\ 5 &= -4 \quad \text{False} \end{aligned}$$

Although (−2, 5) satisfies the first equation, it does not satisfy the second. Thus, (−2, 5) is not a solution of the system.

Self Check 1 Determine whether (4, −1) is a solution of $\begin{cases} x - 2y = 6 \\ y = 3x - 11 \end{cases}$.

THE GRAPHING METHOD

To **solve a system** of equations means to find all of the solutions of the system. One way to solve a system of linear equations is to graph the equations on the same set of axes.

EXAMPLE 2

Solve the following system by graphing: $\begin{cases} 2x + 3y = 2 \\ 3x - 2y = 16 \end{cases}$.

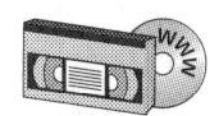

ELEMENTARY Algebra f(x) Now™

Success Tip

Accuracy is crucial when using the graphing method to solve a system. Here are some suggestions for improving your accuracy:

- Use graph paper.
- Use a sharp pencil.
- Use a ruler or straightedge.

Solution We graph both equations on the same coordinate system.

$2x + 3y = 2$

x	y	(x, y)
0	$\frac{2}{3}$	$\left(0, \frac{2}{3}\right)$
1	0	$(1, 0)$
-2	2	$(-2, 2)$

$3x - 2y = 16$

x	y	(x, y)
0	-8	$(0, -8)$
$\frac{16}{3}$	0	$\left(\frac{16}{3}, 0\right)$
2	-5	$(2, -5)$

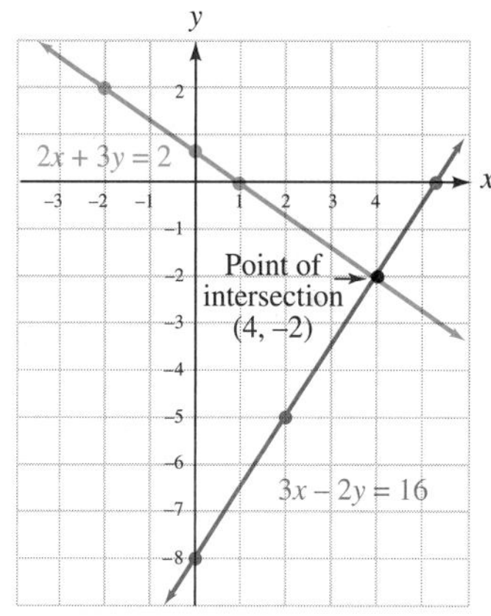

The coordinates of each point on the line graphed in red satisfy $2x + 3y = 2$ and the coordinates of each point on the line graphed in blue satisfy $3x - 2y = 16$. Because the point of intersection is on both graphs, its coordinates satisfy both equations.

It appears that the graphs intersect at the point $(4, -2)$. To verify that it is the solution of the system, we substitute 4 for x and -2 for y in each equation.

Check:

$$\begin{aligned} 2x + 3y &= 2 \\ 2(4) + 3(-2) &\stackrel{?}{=} 2 \\ 8 + (-6) &\stackrel{?}{=} 2 \\ 2 &= 2 \end{aligned} \qquad \begin{aligned} 3x - 2y &= 16 \\ 3(4) - 2(-2) &\stackrel{?}{=} 16 \\ 12 - (-4) &\stackrel{?}{=} 16 \\ 16 &= 16 \end{aligned}$$

Since $(4, -2)$ makes both equations true, it is the solution of the system.

Self Check 2 Solve the following system by graphing: $\begin{cases} 2x - y = -5 \\ x + y = -1 \end{cases}$.

To solve a system of equations in two variables by graphing, follow these steps.

The Graphing Method

1. Carefully graph each equation on the same rectangular coordinate system.
2. If the lines intersect, determine the coordinates of the point of intersection of the graphs. That ordered pair is the solution of the system.
3. Check the proposed solution in each equation of the original system.

INCONSISTENT SYSTEMS AND DEPENDENT EQUATIONS

A system of equations that has at least one solution, like that in Example 2, is called a **consistent system.** A system with no solution is called an **inconsistent system.**

EXAMPLE 3

ELEMENTARY Algebra $f(x)$ Now™

Solve the following system by graphing: $\begin{cases} y = -2x - 6 \\ 4x + 2y = 8 \end{cases}$.

Solution Since $y = -2x - 6$ is written in slope–intercept form, we can graph it by plotting the y-intercept $(0, -6)$ and then drawing a slope of -2. (The rise is -2 and the run is 1.) We graph $4x + 2y = 8$ using the intercept method.

$y = -2x - 6$

$m = -2$ $\quad b = -6$

Slope: $\frac{-2}{1}$

y-intercept: $(0, -6)$

$4x + 2y = 8$

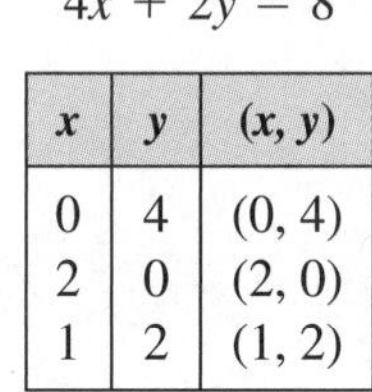

x	y	(x, y)
0	4	(0, 4)
2	0	(2, 0)
1	2	(1, 2)

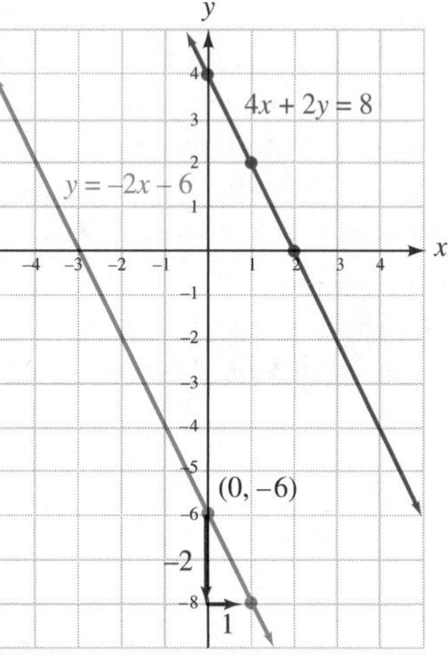

The lines appear to be parallel. We can verify this by writing the second equation in slope–intercept form and observing that the lines have the same slope, -2, and different y-intercepts, $(0, -6)$ and $(0, 4)$.

$y = -2x - 6 \qquad 4x + 2y = 8$

$2y = -4x + 8$ Subtract $4x$ from both sides.

$y = -2x + 4$ Divide both sides by 2.

Different y-intercepts

Same slope

Because the lines are parallel, there is no point of intersection. Such a system has no solution.

Self Check 3 Solve the following system by graphing: $\begin{cases} y = \frac{3}{2}x \\ 3x - 2y = 6 \end{cases}$.

Some systems of equations have infinitely many solutions.

EXAMPLE 4

ELEMENTARY Algebra $f(x)$ Now™

Solve the following system by graphing: $\begin{cases} 4x + 8 = 2y \\ y = 2x + 4 \end{cases}$.

Solution To graph $4x + 8 = 2y$, we use the intercept method and to graph $y = 2x + 4$, we use the slope and y-intercept.

$4x + 8 = 2y$

x	y	(x, y)
0	4	(0, 4)
−2	0	(−2, 0)
−3	−2	(−3, −2)

$y = 2x + 4$

$m = 2 \quad b = 4$

Slope: $\frac{2}{1}$

y-intercept: (0, 4)

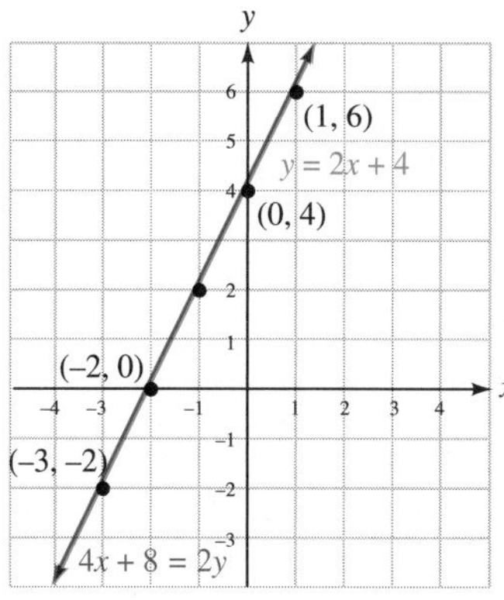

The graphs appear to be identical. We can verify this by writing the first equation in slope–intercept form and observing that it is the same as the second equation.

Since the graphs are the same line, they have infinitely many points in common. For such a system, there are infinitely many solutions.

From the graph, we see that four of the solutions are (−3, −2), (−2, 0), (0, 4), and (1, 6). Checks for two ordered pairs follow.

The Language of Algebra

In Example 4, the graphs of the lines *coincide.* That is, they occupy the same location. To illustrate this concept, think of a clock. At noon and midnight, the hands of the clock *coincide.*

***For* (−3, −2):**

$$4x + 8 = 2y \qquad y = 2x + 4$$
$$4(-3) + 8 \stackrel{?}{=} 2(-2) \qquad -2 \stackrel{?}{=} 2(-3) + 4$$
$$-12 + 8 \stackrel{?}{=} -4 \qquad -2 \stackrel{?}{=} -6 + 4$$
$$-4 = -4 \qquad -2 = -2$$

***For* (0, 4):**

$$4x + 8 = 2y \qquad y = 2x + 4$$
$$4(0) + 8 \stackrel{?}{=} 2(4) \qquad 4 \stackrel{?}{=} 2(0) + 4$$
$$0 + 8 \stackrel{?}{=} 8 \qquad 4 \stackrel{?}{=} 0 + 4$$
$$8 = 8 \qquad 4 = 4$$

Self Check 4 Solve the following system by graphing: $\begin{cases} 6x - 4 = 2y \\ y = 3x - 2 \end{cases}$.

In Examples 2 and 3, the graphs of the equations of the system were different lines. We call equations with different graphs **independent equations.** In Example 4, the equations have the same graph and are called **dependent equations.**

There are three possible outcomes when we solve a system of two linear equations using the graphing method:

The two lines intersect at one point.

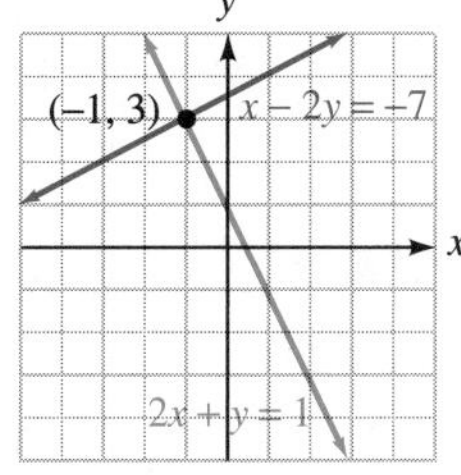

Exactly one solution (the point of intersection)

Consistent system
Independent equations

The two lines are parallel.

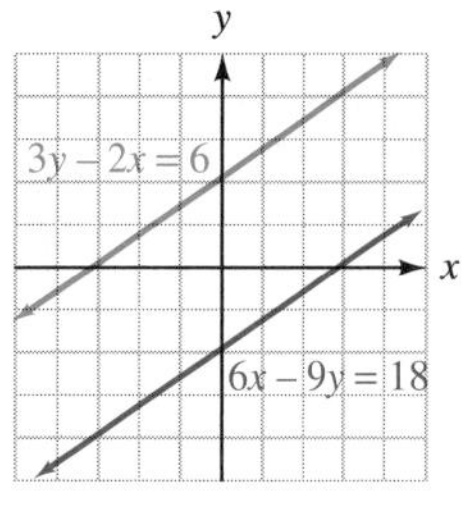

No solution

Inconsistent system
Independent equations

The two lines are identical.

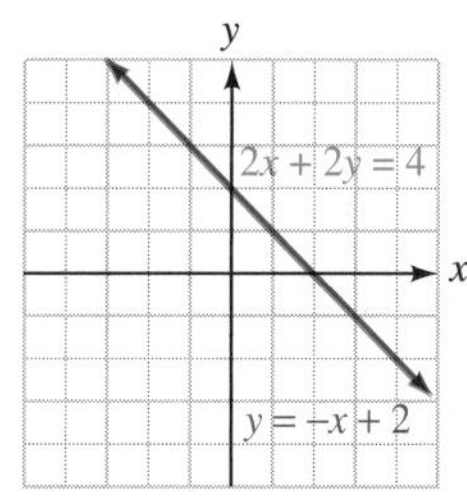

Infinitely many solutions (any point on the line is a solution)

Consistent system
Dependent equations

GRAPHING CALCULATORS

A graphing calculator can be used to solve systems of equations, such as

$$\begin{cases} 2x + y = 12 \\ 2x - y = -2 \end{cases}$$

Before we can enter the equations into the calculator, we must solve them for y.

$$\begin{aligned} 2x + y &= 12 \\ y &= -2x + 12 \end{aligned} \qquad \begin{aligned} 2x - y &= -2 \\ -y &= -2x - 2 \\ y &= 2x + 2 \end{aligned}$$

We enter the resulting equations as Y_1 and Y_2 and graph them on the same axes. If we use the standard window setting, their graphs will look like figure (a).

To find the solution of the system, we can use the INTERSECT feature found on most graphing calculators. With this option, the cursor automatically moves to the point of intersection of the graphs and displays the coordinates of that point. In figure (b), we see that the solution is (2.5, 7).

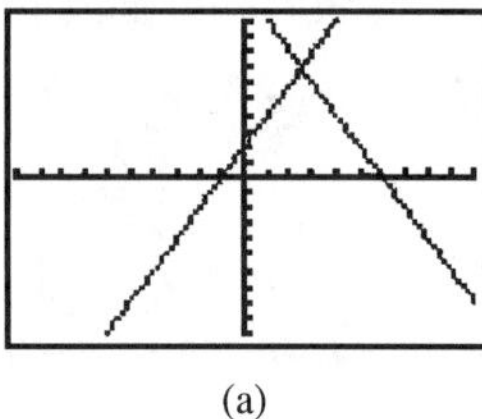

(a)

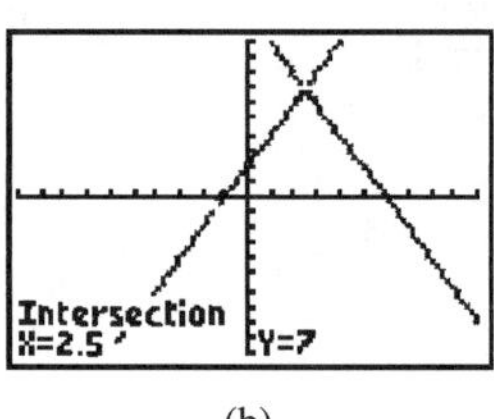

(b)

Answers to Self Checks  **1.** no  **2.** (−2, 1) **3.** no solution **4.** infinitely many solutions

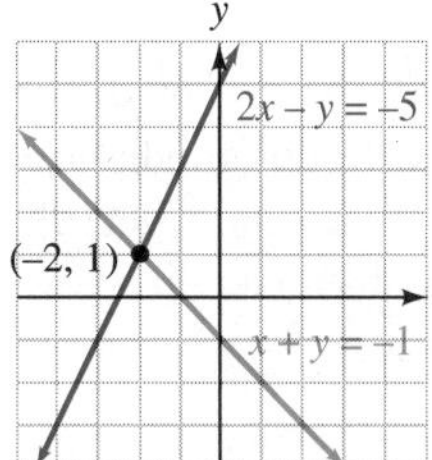

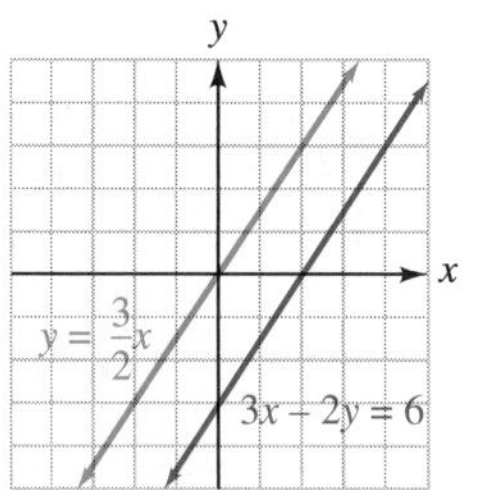

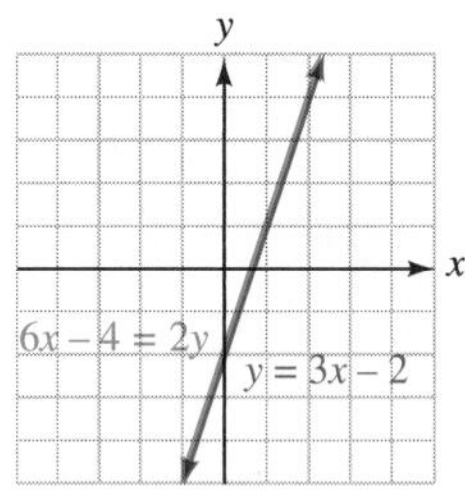

7.1 STUDY SET  ELEMENTARY Algebra f(x) Now™

VOCABULARY **Fill in the blanks.**

1. The pair of equations $\begin{cases} x - y = -1 \\ 2x - y = 1 \end{cases}$ is called a ________ of equations.
2. Because the ordered pair (2, 3) satisfies both equations in Problem 1, it is called a ________ of the system of equations.
3. The x-coordinate of the ordered pair $(-4, 7)$ is -4 and the ____________ is 7.
4. We say that (1, 4) ________ $x + y = 5$, because the x-coordinate, 1, and the y-coordinate, 4, make the equation true when substituted for x and y.
5. The point of ____________ of the lines graphed in part (a) of the following figure is (1, 2).

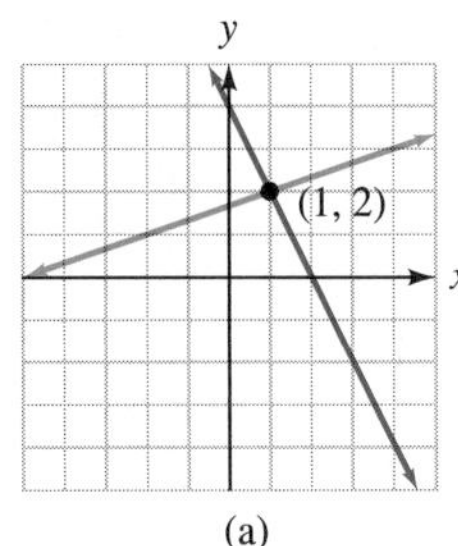

(a)

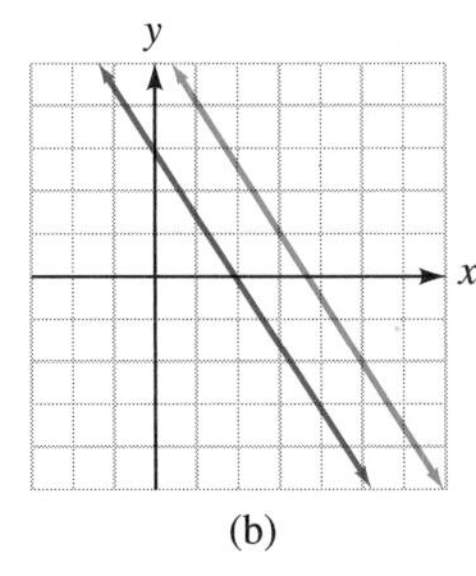

(b)

6. The lines graphed in part (b) of the figure above do not intersect. They are ________ lines.

7. A system of equations that has at least one solution is called a __________ system. A system with no solution is called an ___________ system.

8. We call equations with different graphs ___________ equations. Because __________ equations are different forms of the same equation, they have the same graph.

CONCEPTS

9. The following tables were created to graph the two linear equations in a system. What is the solution of the system?

Equation 1

x	y
0	−5
−5	0
−4	−1
1	−6
2	−7

Equation 2

x	y
0	3
−3	0
−2	1
−4	−1
1	4

Refer to the illustration. Determine whether a true or false statement will result if

10. The coordinates of point A are substituted into the equation for Line 1.

11. The coordinates of point C are substituted into the equation for Line 1.

12. The coordinates of point C are substituted into the equation for Line 2.

13. The coordinates of point B are substituted into the equation for Line 1.

14. To graph $5x - 2y = 10$, we can use the intercept method. Complete the table.

x	y
0	
	0

15. To graph $y = 3x - 2$, we can use the slope and y-intercept. Fill in the blanks.

slope: ___ $= \frac{\quad}{1}$ y-intercept: ___

16. What is the solution of the system graphed on the right? Is it consistent or inconsistent?

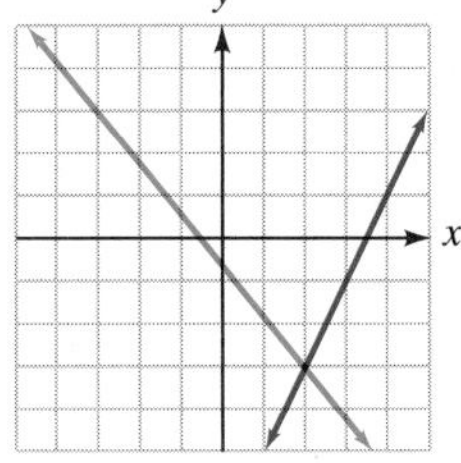

17. How many solutions does the system graphed on the right have? Are the equations dependent or independent?

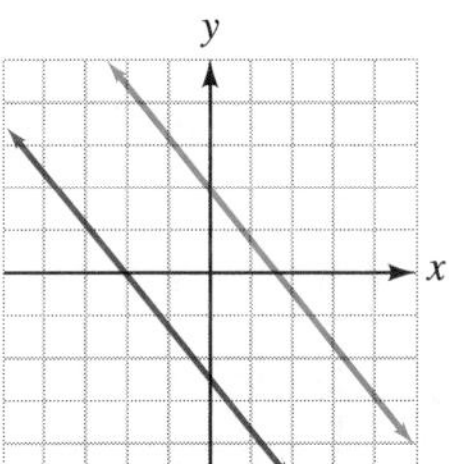

18. How many solutions does the system graphed on the right have? Give three of the solutions. Is the system consistent or inconsistent?

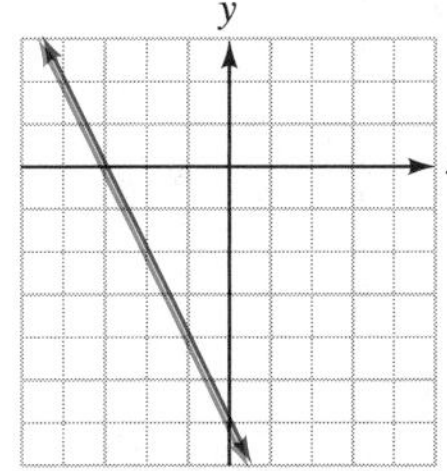

NOTATION

19. The symbol $\stackrel{?}{=}$ is used when checking a solution. What does it mean?

20. Draw the graphs of two linear equations so that the system has one solution, $(-3, -2)$.

PRACTICE **Decide whether the ordered pair is a solution of the given system.**

21. $(1, 1)$, $\begin{cases} x + y = 2 \\ 2x - y = 1 \end{cases}$

22. $(1, 3)$, $\begin{cases} 2x + y = 5 \\ 3x - y = 0 \end{cases}$

23. $(3, -2)$, $\begin{cases} 2x + y = 4 \\ y = 1 - x \end{cases}$

24. $(-2, 4)$, $\begin{cases} 2x + 2y = 4 \\ 3y = 10 - x \end{cases}$

25. $(-2, -4)$, $\begin{cases} 4x + 5y = -23 \\ -3x + 2y = 0 \end{cases}$

26. $(-5, 2)$, $\begin{cases} -2x + 7y = 17 \\ 3x - 4y = -19 \end{cases}$

27. $\left(\frac{1}{2}, 3\right)$, $\begin{cases} 2x + y = 4 \\ 4x - 11 = 3y \end{cases}$

28. $\left(2, \frac{1}{3}\right)$, $\begin{cases} x - 3y = 1 \\ -2x + 6 = -6y \end{cases}$

29. $(2.5, 3.5)$, $\begin{cases} 4x - 3 = 2y \\ 4y + 1 = 6x \end{cases}$

30. $(0.2, 0.3)$, $\begin{cases} 20x + 10y = 7 \\ 20y = 15x + 3 \end{cases}$

Solve each system by graphing. If a system has no solution or infinitely many, so state.

31. $\begin{cases} 2x + 3y = 12 \\ 2x - y = 4 \end{cases}$

32. $\begin{cases} 5x + y = 5 \\ 5x + 3y = 15 \end{cases}$

33. $\begin{cases} x + y = 4 \\ x - y = -6 \end{cases}$

34. $\begin{cases} x + y = 4 \\ x - y = -2 \end{cases}$

35. $\begin{cases} y = 3x + 6 \\ y = -2x - 4 \end{cases}$

36. $\begin{cases} y = x + 3 \\ y = -2x - 3 \end{cases}$

37. $\begin{cases} y = x - 1 \\ 3x - 3y = 3 \end{cases}$

38. $\begin{cases} y = -x + 1 \\ 4x + 4y = 4 \end{cases}$

39. $\begin{cases} y = -\frac{1}{3}x - 4 \\ x + 3y = 6 \end{cases}$

40. $\begin{cases} y = -\frac{1}{2}x - 3 \\ x + 2y = 2 \end{cases}$

41. $\begin{cases} y = -x - 2 \\ y = -3x + 6 \end{cases}$

42. $\begin{cases} y = 2x - 4 \\ y = -5x + 3 \end{cases}$

43. $\begin{cases} -x + 3y = -11 \\ 3x - y = 17 \end{cases}$

44. $\begin{cases} 2x - 3y = -18 \\ 3x + 2y = -1 \end{cases}$

45. $\begin{cases} x + y = 2 \\ y = x \end{cases}$

46. $\begin{cases} x + y = 4 \\ y = x \end{cases}$

47. $\begin{cases} y = \frac{3}{4}x + 3 \\ y = -\frac{x}{4} - 1 \end{cases}$

48. $\begin{cases} y = \frac{2}{3}x + 4 \\ y = -\frac{x}{3} + 7 \end{cases}$

49. $\begin{cases} 2y = 3x + 2 \\ 3x - 2y = 6 \end{cases}$

50. $\begin{cases} 3x - 6y = 18 \\ x = 2y + 3 \end{cases}$

51. $\begin{cases} 4x - 2y = 8 \\ y = 2x - 4 \end{cases}$

52. $\begin{cases} 2y = -6x - 12 \\ 3x + y = -6 \end{cases}$

53. $\begin{cases} x + y = 2 \\ y = x - 4 \end{cases}$

54. $\begin{cases} x + y = 1 \\ y = x + 5 \end{cases}$

55. $\begin{cases} x + 4y = -2 \\ y = -x - 5 \end{cases}$

56. $\begin{cases} 3x + 2y = -8 \\ 2x - 3y = -1 \end{cases}$

57. $\begin{cases} x = 3 \\ 3y = 6 - 2x \end{cases}$

58. $\begin{cases} x = 4 \\ 2y = 12 - 4x \end{cases}$

59. $\begin{cases} y = -3 \\ -x + 2y = -4 \end{cases}$

60. $\begin{cases} y = -4 \\ -2x - y = 8 \end{cases}$

61. $\begin{cases} x + 2y = -4 \\ x - \frac{1}{2}y = 6 \end{cases}$

62. $\begin{cases} \frac{2}{3}x - y = -3 \\ 3x + y = 3 \end{cases}$

Determine the slope and the y-intercept of the graph of each line in the system. Then, use that information to determine the number of solutions of the system.

63. $\begin{cases} y = 6x - 7 \\ y = -2x + 1 \end{cases}$

64. $\begin{cases} y = \frac{1}{2}x + 8 \\ y = 4x - 10 \end{cases}$

65. $\begin{cases} 3x - y = -3 \\ y - 3x = 3 \end{cases}$

66. $\begin{cases} x + 4y = 4 \\ 12y = 12 - 3x \end{cases}$

67. $\begin{cases} y = -x + 6 \\ x + y = 8 \end{cases}$

68. $\begin{cases} 5x + y = 0 \\ y = -5x + 6 \end{cases}$

69. $\begin{cases} 6x + y = 0 \\ 2x + 2y = 0 \end{cases}$

70. $\begin{cases} x + y = 1 \\ 2x - 2y = 5 \end{cases}$

Use a graphing calculator to solve each system, if possible.

71. $\begin{cases} y = 4 - x \\ y = 2 + x \end{cases}$

72. $\begin{cases} 3x - 6y = 4 \\ 2x + y = 1 \end{cases}$

73. $\begin{cases} 6x - 2y = 5 \\ 3x = y + 10 \end{cases}$

74. $\begin{cases} x - 3y = -2 \\ 5x + y = 10 \end{cases}$

APPLICATIONS

75. TRANSPLANTS Refer to the graph. In what year were the number of donors and the number waiting for a liver transplant the same? Estimate the number.

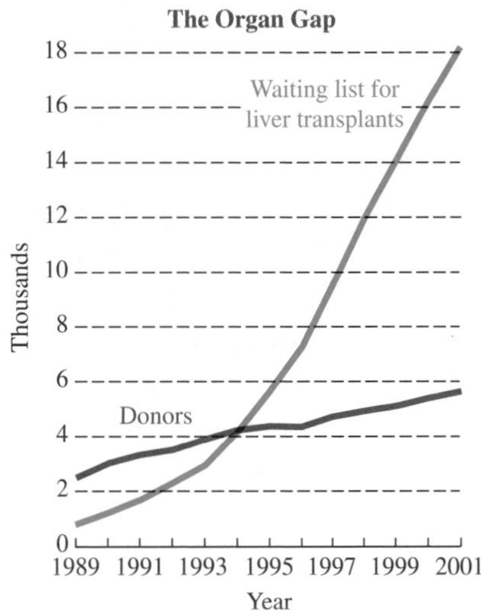

Source: Organ Procurement and Transportation Network

76. COLLEGE BOARDS Many colleges and universities use the SAT test as one indicator of a high school student's readiness to do college level work. Refer to the graph. For any given year, were the average math score and the average verbal score ever the same?

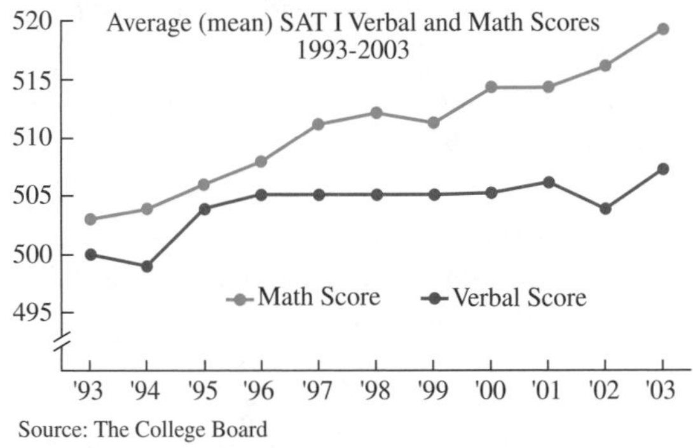

Source: The College Board

77. LATITUDE AND LONGITUDE Refer to the following map.

a. Name three American cities that lie on a latitude line of 30° north.

b. Name three American cities that lie on a longitude line of 90° west.

c. What city lies on both lines?

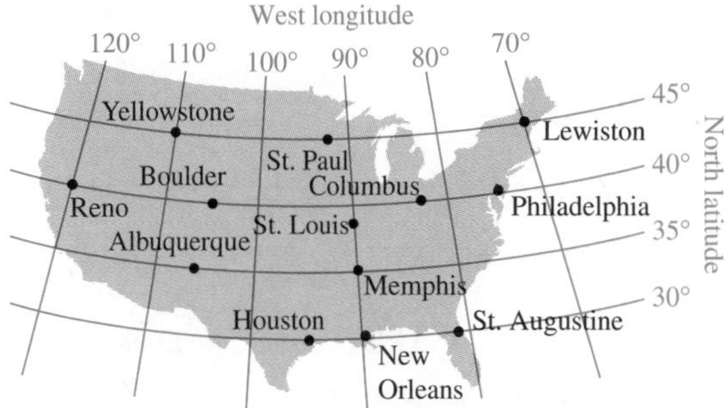

78. ECONOMICS The graph illustrates the law of supply and demand.

a. Complete this sentence: As the price of an item increases, the *supply* of the item ________.

b. Complete this sentence: As the price of an item increases, the *demand* for the item ________.

c. For what price will the supply equal the demand? How many items will be supplied for this price?

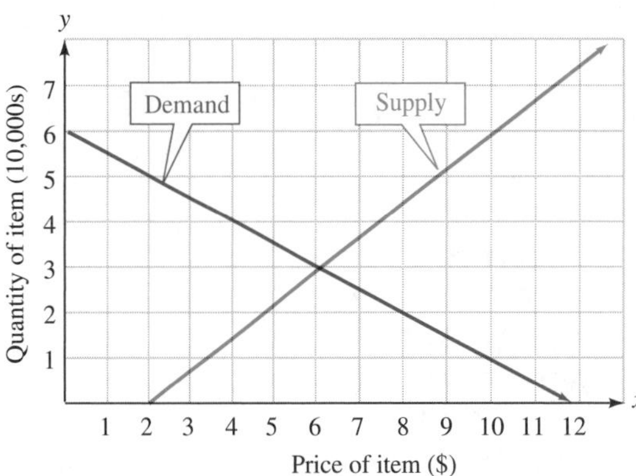

79. DAILY TRACKING POLLS Use the graph on the next page to answer the following.

a. Which political candidate was ahead on October 28 and by how much?

b. On what day did the challenger pull even with the incumbent?

c. If the election was held November 4, who did the poll predict would win, and by how many percentage points?

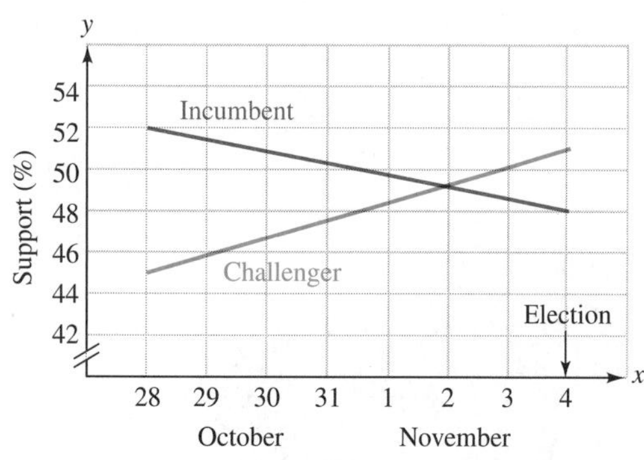

80. AIR TRAFFIC CONTROL The equations describing the paths of two airplanes are $y = -\frac{1}{2}x + 3$ and $3y = 2x + 2$. Graph each equation on the radar screen shown. Is there a possibility of a midair collision? If so, where?

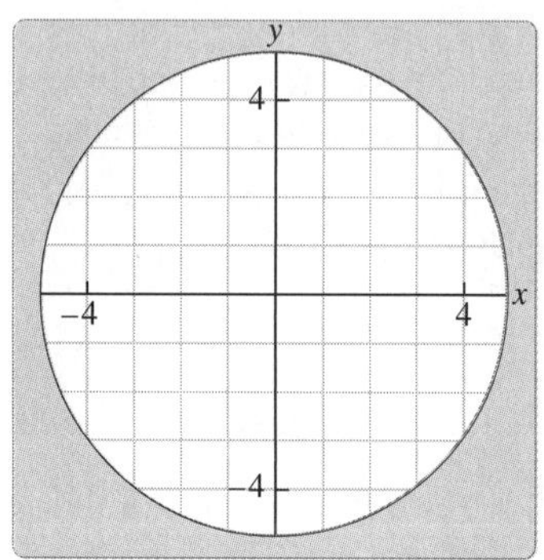

81. TV COVERAGE A television camera is located at $(-2, 0)$ and will follow the launch of a space shuttle, as shown in the graph. (Each unit in the illustration is 1 mile.) As the shuttle rises vertically on a path described by $x = 2$, the farthest the camera can tilt back is a line of sight given by $y = \frac{5}{2}x + 5$. For how many miles of the shuttle's flight will it be in view of the camera?

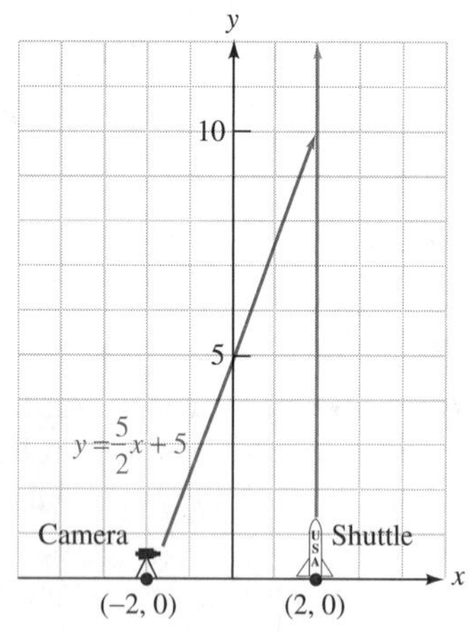

WRITING

82. Explain why it is difficult to determine the solution of the system in the graph.

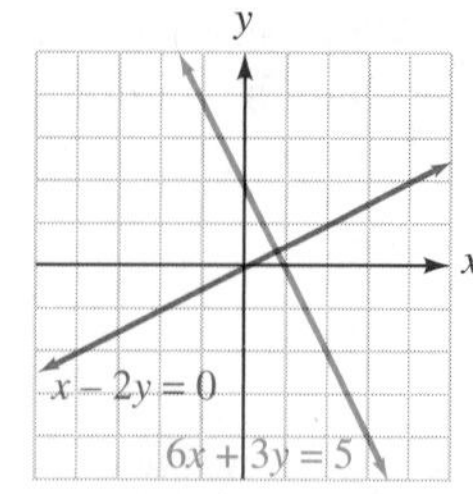

83. Without graphing, how can you tell that the graphs of $y = 2x + 1$ and $y = 2x + 2$ do not intersect?

84. Could a system of two linear equations have exactly two solutions? Explain why or why not.

85. What is an inconsistent system?

86. What are dependent equations?

REVIEW **Perform the operations and simplify, if possible.**

87. $\frac{x + 3}{x^2} + \frac{x + 5}{2x}$

88. $\frac{x^2 - 4}{3x + 6} \div \frac{x - 2}{x + 2}$

89. $\frac{z^2 + 4z - 5}{5z - 5} \cdot \frac{5z}{z + 5}$

90. $\frac{6x - 5}{3xy} - \frac{3x - 5}{3xy}$

CHALLENGE PROBLEMS

91. Can a system of two linear equations in two variables be inconsistent but have dependent equations? Explain.

92. Construct a system of two linear equations that has a solution of $(-2, 6)$.

93. Write a system of two linear equations such that (2, 3) is a solution of the first equation but is not a solution of the second equation.

94. Solve by graphing: $\begin{cases} \frac{1}{3}x - \frac{1}{2}y = \frac{1}{6} \\ \frac{2x}{5} + \frac{y}{2} = \frac{13}{10} \end{cases}$.

7.2 Solving Systems of Equations by Substitution

- The Substitution Method
- Finding a Substitution Equation
- Inconsistent Systems and Dependent Equations

When solving a system of equations by graphing, it is often difficult to determine the coordinates of the intersection point. For example, a solution $\left(\frac{7}{8}, \frac{3}{5}\right)$ is virtually impossible to identify. In this section, we will discuss a second, more precise method for solving systems.

THE SUBSTITUTION METHOD

EXAMPLE 1

ELEMENTARY Algebra f(x) Now™

Solve the system: $\begin{cases} y = 3x - 2 \\ 2x + y = 8 \end{cases}$.

Solution Note that the first equation, $y = 3x - 2$, is solved for y. Because y and $3x - 2$ represent the same value, we can substitute $3x - 2$ for y in the second equation. We call $y = 3x - 2$ the *substitution equation.*

$$\begin{cases} y = 3x - 2 \\ 2x + y = 8 \end{cases}$$

To find the solution of the system, we proceed as follows:

$$\begin{aligned} 2x + y &= 8 && \text{This is the second equation of the system.} \\ 2x + 3x - 2 &= 8 && \text{Substitute } 3x - 2 \text{ for } y. \end{aligned}$$

The resulting equation has only one variable and can be solved for x.

$$\begin{aligned} 2x + 3x - 2 &= 8 \\ 5x - 2 &= 8 && \text{Combine like terms.} \\ 5x &= 10 && \text{Add 2 to both sides.} \\ x &= 2 && \text{Divide both sides by 5.} \end{aligned}$$

Caution

When using the substitution method, a common error is to find the value of one of the variables, say x, and forget to find the value of the other. Remember that a solution of a system is an ordered pair (x, y).

After finding x, we can find y by substituting 2 for x in either equation of the original system. Because the substitution equation, $y = 3x - 2$, is already solved for y, it is easy to substitute into this equation.

$$\begin{aligned} y &= 3x - 2 && \text{Use the substitution equation to find } y. \\ y &= 3(2) - 2 && \text{Substitute 2 for } x. \\ y &= 6 - 2 \\ y &= 4 && \text{This is the } y\text{-value of the solution.} \end{aligned}$$

The ordered pair (2, 4) appears to be the solution of the system. To check, we substitute 2 for x and 4 for y in each equation.

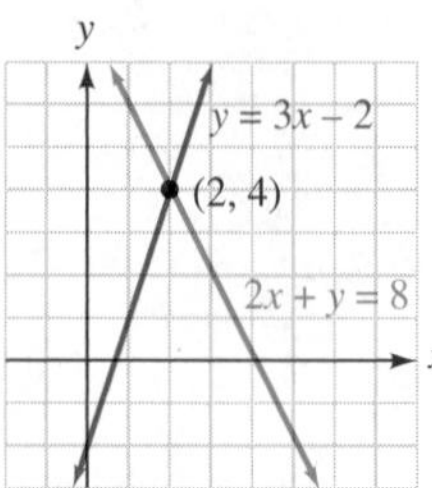

Check:

$$\begin{aligned} y &= 3x - 2 \\ 4 &\stackrel{?}{=} 3(2) - 2 \\ 4 &\stackrel{?}{=} 6 - 2 \\ 4 &= 4 \end{aligned} \qquad \begin{aligned} 2x + y &= 8 \\ 2(2) + 4 &\stackrel{?}{=} 8 \\ 4 + 4 &\stackrel{?}{=} 8 \\ 8 &= 8 \end{aligned}$$

Since (2, 4) satisfies both equations, it is the solution. A graph of the equations of the system shows an intersection point of (2, 4). This illustrates an important fact: *The solution found using the substitution method will be the same as the solution found using the graphing method.*

Self Check 1 Solve the system: $\begin{cases} x + 4y = 7 \\ x = 6y - 3 \end{cases}$.

To solve a system of equations in x and y by the substitution method, follow these steps.

The Substitution Method

1. Solve one of the equations for either x or y. If this is already done, go to step 2.
2. Substitute the expression for x or for y obtained in step 1 into the other equation and solve that equation.
3. Substitute the value of the variable found in step 2 into the substitution equation to find the value of the remaining variable.
4. Check the proposed solution in the equations of the original system. Write the solution as an ordered pair.

EXAMPLE 2

Solve the system: $\begin{cases} 4x + 27 = 7y \\ x = -5y \end{cases}$.

Solution *Step 1:* Note that the second equation, $x = -5y$, is solved for x. Because x and $-5y$ represent the same value, we can substitute $-5y$ for x in the first equation.

$$\begin{cases} 4x + 27 = 7y \\ x = -5y \end{cases}$$

Success Tip

With this method, the basic objective is to use an appropriate substitution to obtain one equation in one variable.

Step 2: When we substitute for x in the first equation, the resulting equation contains only one variable and can be solved for y.

$$\begin{aligned} 4x + 27 &= 7y && \text{This is the first equation of the system.} \\ 4(-5y) + 27 &= 7y && \text{Substitute } -5y \text{ for } x. \\ -20y + 27 &= 7y && \text{Do the multiplication.} \\ 27 &= 27y && \text{Add } 20y \text{ to both sides.} \\ 1 &= y && \text{Divide both sides by 27.} \end{aligned}$$

The Language of Algebra

The phrase *back-substitute* can also be used to describe step 3 of the substitution method. In Example 2, to find x, we *back-substitute* 1 for y in the equation $x = -5y$.

Step 3: To find x, substitute 1 for y in the equation $x = -5y$.

$$\begin{aligned} x &= -5y \\ x &= -5(1) && \text{Substitute 1 for } y. \\ x &= -5 \end{aligned}$$

Step 4: The check below verifies that the solution is $(-5, 1)$.

Check:

$$\begin{aligned} 4x + 27 &= 7y \\ 4(-5) + 27 &\stackrel{?}{=} 7(1) \\ -20 + 27 &\stackrel{?}{=} 7 \\ 7 &= 7 \end{aligned} \qquad \begin{aligned} x &= -5y \\ -5 &\stackrel{?}{=} -5(1) \\ -5 &= -5 \end{aligned}$$

Self Check 2 Solve the system: $\begin{cases} 3x + 40 = 8y \\ x = -4y \end{cases}$.

FINDING A SUBSTITUTION EQUATION

Sometimes neither equation of a system is solved for a variable. In such cases, we can find a substitution equation by solving one of the equations for one of its variables.

EXAMPLE 3

Solve the system: $\begin{cases} 4x + y = 3 \\ 3x + 5y = 15 \end{cases}$.

ELEMENTARY Algebra f(x) Now™

Solution

Step 1: Since the system does not contain an equation solved for x or y, we must choose an equation and solve it for x or y. It is easiest to solve for y in the first equation, because it has a coefficient of 1.

Success Tip

To find a substitution equation, solve one of the equations of the system for one of its variables. If possible, solve for a variable whose coefficient is 1 or -1 to avoid working with fractions.

$$\begin{aligned} 4x + y &= 3 \\ y &= 3 - 4x \end{aligned} \quad \text{Subtract } 4x \text{ from both sides to isolate } y.$$

Because y and $3 - 4x$ represent the same value, we can substitute $3 - 4x$ for y in the second equation of the system.

$$y = 3 - 4x \qquad 3x + 5y = 15$$

Step 2: When we substitute for y in the second equation, the resulting equation contains only one variable and can be solved for x.

Caution

In Example 3, use parentheses when substituting $3 - 4x$ for y so that the multiplication by 5 is distributed over both terms of $3 - 4x$.

$$3x + 5(3 - 4x) = 15$$

$$\begin{aligned} 3x + 5y &= 15 \\ 3x + 5(3 - 4x) &= 15 && \text{Substitute } 3 - 4x \text{ for } y. \\ 3x + 15 - 20x &= 15 && \text{Distribute the multiplication by 5.} \\ 15 - 17x &= 15 && \text{Combine like terms.} \\ -17x &= 0 && \text{Subtract 15 from both sides.} \\ x &= 0 && \text{Divide both sides by } -17. \end{aligned}$$

Step 3: To find y, substitute 0 for x in the equation $y = 3 - 4x$.

$$\begin{aligned} y &= 3 - 4x \\ y &= 3 - 4(0) && \text{Substitute 0 for } x. \\ y &= 3 - 0 \\ y &= 3 \end{aligned}$$

Step 4: The solution appears to be (0, 3). Check it in the original equations.

Check:

$$\begin{aligned} 4x + y &= 3 \\ 4(0) + 3 &\stackrel{?}{=} 3 \\ 0 + 3 &\stackrel{?}{=} 3 \\ 3 &= 3 \end{aligned} \qquad \begin{aligned} 3x + 5y &= 15 \\ 3(0) + 5(3) &\stackrel{?}{=} 15 \\ 0 + 15 &\stackrel{?}{=} 15 \\ 15 &= 15 \end{aligned}$$

Self Check 3 Solve the system: $\begin{cases} 2x - 3y = 10 \\ 3x + y = 15 \end{cases}$.

EXAMPLE 4

Solve the system: $\begin{cases} 3a - 3b = 5 \\ 3 - a = -2b \end{cases}$.

ELEMENTARY Algebra Now™

Solution ***Step 1:*** Since the coefficient of a in the second equation is -1, solve that equation for a.

$$\begin{aligned} 3 - a &= -2b \\ -a &= -2b - 3 \end{aligned} \quad \text{Subtract 3 from both sides.}$$

To obtain a on the left-hand side, multiply both sides of the equation by -1.

$$\begin{aligned} -1(-a) &= -1(-2b - 3) && \text{Multiply both sides by } -1. \\ a &= 2b + 3 && \text{Do the multiplications.} \end{aligned}$$

Because a and $2b + 3$ represent the same value, we can substitute $2b + 3$ for a in the first equation.

$$3a - 3b = 5 \qquad a = 2b + 3$$

Caution

In Example 4, do not substitute $2b + 3$ into the equation from which it came, $3 - a = -2b$. Substitute it into the other equation of the system. This common error leads to an identity.

$$\begin{aligned} 3 - a &= -2b \\ 3 - (2b + 3) &= -2b \\ 3 - 2b - 3 &= -2b \\ 0 &= 0 \end{aligned}$$

Step 2: Substitute $2b + 3$ for a in the first equation and solve for b.

$$\begin{aligned} 3a - 3b &= 5 \\ 3(2b + 3) - 3b &= 5 && \text{Substitute } 2b + 3 \text{ for } a. \\ 6b + 9 - 3b &= 5 && \text{Distribute the multiplication by 3.} \\ 3b + 9 &= 5 && \text{Combine like terms.} \\ 3b &= -4 && \text{Subtract 9 from both sides.} \\ b &= -\frac{4}{3} && \text{Divide both sides by 3.} \end{aligned}$$

Step 3: To find a, substitute $-\frac{4}{3}$ for b in the equation $a = 2b + 3$.

$$\begin{aligned} a &= 2b + 3 \\ a &= 2\left(-\frac{4}{3}\right) + 3 && \text{Substitute } -\tfrac{4}{3} \text{ for } b. \\ a &= -\frac{8}{3} + \frac{9}{3} && \text{Do the multiplication.} \\ a &= \frac{1}{3} && \text{Add.} \end{aligned}$$

Notation

Unless told otherwise, list the values of the variables of a solution in *alphabetical* order. For example, if the equations of a system involve the variables a and b, write the solution in the form (a, b).

Step 4: The solution is $\left(\frac{1}{3}, -\frac{4}{3}\right)$. Check it in the original equations.

Self Check 4 Solve the system: $\begin{cases} 2s - t = 4 \\ 3s - 5t = 2 \end{cases}$.

It is often helpful to clear any equations of fractions and combine any like terms before performing a substitution.

EXAMPLE 5

ELEMENTARY Algebra f(x) Now™

Solve the system: $\begin{cases} \frac{y}{4} = -\frac{x}{2} - \frac{3}{4} \\ 2x - y = -1 + y - x \end{cases}$.

Solution We can clear the first equation of fractions by multiplying both sides by the LCD.

$$\frac{y}{4} = -\frac{x}{2} - \frac{3}{4}$$

$$4\left(\frac{y}{4}\right) = 4\left(-\frac{x}{2} - \frac{3}{4}\right) \quad \text{Multiply both sides by 4.}$$

(1) $$y = -2x - 3 \quad \text{Distribute the multiplication by 4.}$$

We can write the second equation of the system in general form ($Ax + By = C$) by adding x and subtracting y from both sides.

$$2x - y = -1 + y - x$$

$$2x - y + x - y = -1 + y - x + x - y$$

(2) $$3x - 2y = -1 \quad \text{Combine like terms.}$$

Step 1: Equations 1 and 2 form a system, which has the same solution as the original one. To solve this system, we proceed as follows:

(1)
(2)
$$\begin{cases} y = -2x - 3 \\ 3x - 2y = -1 \end{cases}$$

Step 2: To find x, substitute $-2x - 3$ for y in equation 2 and proceed as follows:

$$3x - 2y = -1 \quad \text{This is equation 2.}$$

$$3x - 2(-2x - 3) = -1 \quad \text{Substitute } -2x - 3 \text{ for } y.$$

$$3x + 4x + 6 = -1 \quad \text{Distribute the multiplication by } -2.$$

$$7x + 6 = -1 \quad \text{Combine like terms.}$$

$$7x = -7 \quad \text{Subtract 6 from both sides.}$$

$$x = -1 \quad \text{Divide both sides by 7.}$$

Caution

Always use the original equations when checking a solution. Do not use a substitution equation or an equivalent equation that you found algebraically. If an error was made, a proposed solution that would not satisfy the original system might appear to be correct.

Step 3: To find y, we substitute -1 for x in equation 1.

$$y = -2x - 3 \quad \text{This is equation 1.}$$

$$y = -2(-1) - 3 \quad \text{Substitute } -1 \text{ for } x.$$

$$y = 2 - 3 \quad \text{Do the multiplication.}$$

$$y = -1$$

Step 4: The solution is $(-1, -1)$. Check it in the original system.

Self Check 5 Solve the system: $\begin{cases} \frac{y}{6} = \frac{x}{3} + \frac{1}{2} \\ 2x - y = -3 + y - x \end{cases}$.

INCONSISTENT SYSTEMS AND DEPENDENT EQUATIONS

In the previous section, we solved inconsistent systems and systems of dependent equations graphically. We can also solve these systems using the substitution method.

EXAMPLE 6

ELEMENTARY Algebra $f(x)$ Now™

Solve the system: $\begin{cases} 4y - 12 = x \\ y = \frac{1}{4}x \end{cases}$.

Solution To solve this system, substitute $\frac{1}{4}x$ for y in the first equation and solve for x.

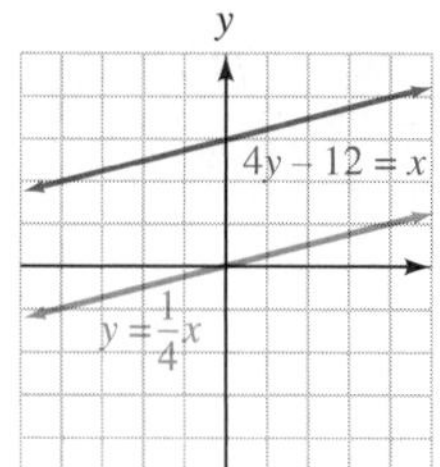

$$4y - 12 = x$$
$$4\left(\frac{1}{4}x\right) - 12 = x \quad \text{Substitute } \frac{1}{4}x \text{ for } y.$$
$$x - 12 = x \quad \text{Do the multiplication.}$$
$$x - 12 - x = x - x \quad \text{Subtract } x \text{ from both sides.}$$
$$-12 = 0$$

Here, the terms involving x drop out, and we get $-12 = 0$. This false statement indicates that the system has no solution, and is inconsistent. The graphs of the equations of the system verify this; they are parallel lines.

Self Check 6 Solve the system: $\begin{cases} x - 4 = y \\ -2y = 4 - 2x \end{cases}$.

EXAMPLE 7

ELEMENTARY Algebra $f(x)$ Now™

Solve the system: $\begin{cases} x = -3y + 6 \\ 2x + 6y = 12 \end{cases}$.

Solution To solve this system, substitute $-3y + 6$ for x in the second equation and solve for y.

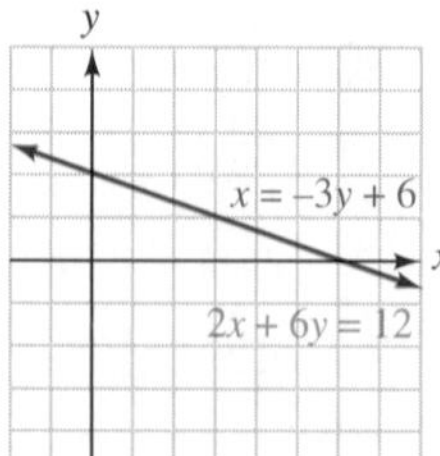

$$2x + 6y = 12$$
$$2(-3y + 6) + 6y = 12 \quad \text{Substitute } -3y + 6 \text{ for } x.$$
$$-6y + 12 + 6y = 12 \quad \text{Distribute the multiplication by 2.}$$
$$12 = 12$$

Here, the terms involving y drop out, and we get $12 = 12$. This true statement indicates that the two equations of the system are equivalent. Therefore, they are dependent equations and the system has infinitely many solutions. The graphs of the equations verify this; they are the same line.

Any ordered pair that satisfies one equation of this system also satisfies the other. To find some solutions, we can substitute some values of x, say 0, 3, and 6, in either equation and solve for y. The results are: (0, 2), (3, 1), and (6, 0).

Self Check 7 Solve the system: $\begin{cases} y = 2 - x \\ 3x + 3y = 6 \end{cases}$.

Answers to Self Checks **1.** (3, 1) **2.** (−8, 2) **3.** (5, 0) **4.** $\left(\frac{18}{7}, \frac{8}{7}\right)$ **5.** (−3, −3) **6.** no solution **7.** infinitely many solutions

7.2 STUDY SET  ELEMENTARY Algebra $f(x)$ Now™

VOCABULARY Fill in the blanks.

1. $\begin{cases} y = x + 3 \\ 3x - y = -1 \end{cases}$ is called a ________ of equations.
2. The ordered pair (1, 4) is a ________ of the system in Problem 1 because it satisfies both equations.
3. When checking a proposed solution of a system of equations, always use the ________ equations.
4. We say that the equation $y = 2x + 4$ is solved for ___.
5. In mathematics, "to ________" means to replace an expression with one that has the same value.
6. In $x - y = 4$, the ________ of x is 1 and the coefficient of y is ____.
7. With the substitution method, the basic objective is to use an appropriate substitution to obtain one equation in ____ variable.
8. To ____ $\frac{x}{2} + \frac{y}{3} = 8$ of fractions, multiply both sides by the LCD, which is 6.
9. When the graphs of the equations of a system are identical lines, the equations are called dependent and the system has ________ many solutions.
10. A system that has no solution is called an ________ system.

CONCEPTS

11. Suppose the substitution method will be used to solve each system. Which equation should be used as the substitution equation?

    a. $\begin{cases} 5x + y = 2 \\ y = -3x \end{cases}$ **b.** $\begin{cases} x = 2y + 1 \\ 7y - 3x = 1 \end{cases}$

12. Multiply both sides of the equation $-x = 4y - 15$ by -1.
13. Suppose the substitution method will be used to solve each system. Find a substitution equation by solving one of the equations for one of the variables.

    a. $\begin{cases} x - 2y = 2 \\ 2x + 3y = 11 \end{cases}$ **b.** $\begin{cases} 2x - 3y = 2 \\ 2x - y = 11 \end{cases}$

14. Is (6, 2) a solution of the system $\begin{cases} x = 3y \\ x - y = 9 \end{cases}$?
15. Suppose $x - 4$ is substituted for y in the equation $x + 3y = 8$. In the following equation, insert parentheses where they are needed.

    $$x + 3x - 4 = 8$$

16. A student uses the substitution method to solve the system $\begin{cases} 4a + 5b = 2 \\ b = 3a - 11 \end{cases}$. She finds that a is 3. What is the easiest way for her to determine the value of b?
17. **a.** What is the LCD of the fractions in $\frac{x}{5} + \frac{2y}{3} = 1$?

    b. Clear the equation of fractions.
18. Write $2x + y = x - 5y + 3$ in the form $Ax + By = C$.
19. Suppose the equation $-2 = 1$ is obtained when a system is solved by the substitution method.

    a. Does the system have a solution?

    b. Which of the following is a possible graph of the system?

i.
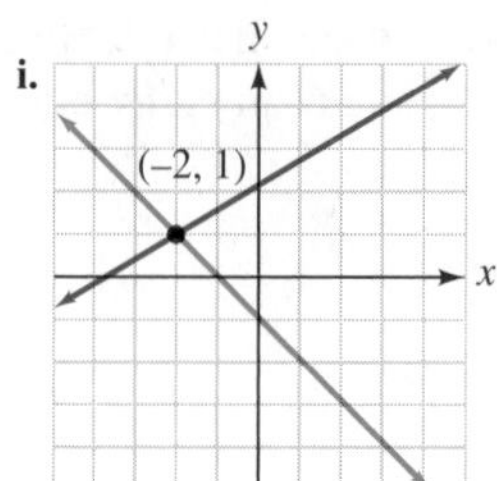

ii.
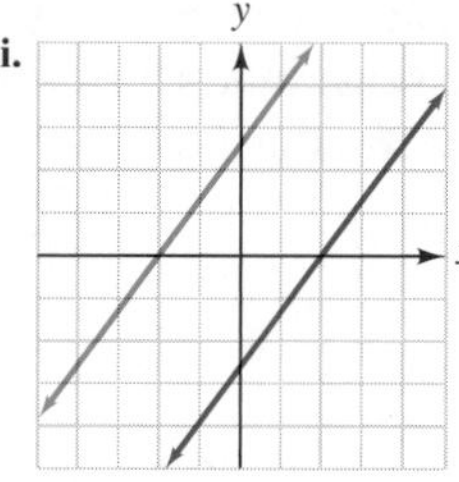

20. Suppose the equation $2 = 2$ is obtained when a system is solved by the substitution method.

a. Does the system have a solution?

b. Which graph is a possible graph of the system?

i.
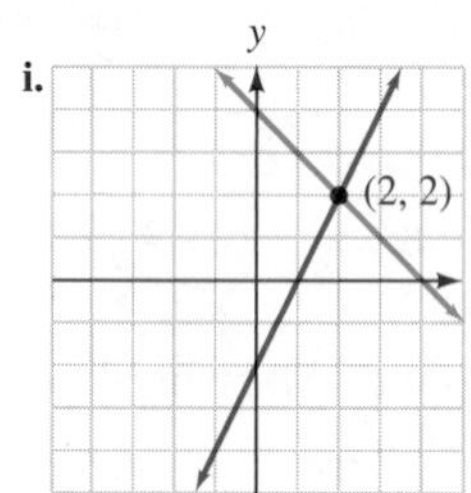

ii.
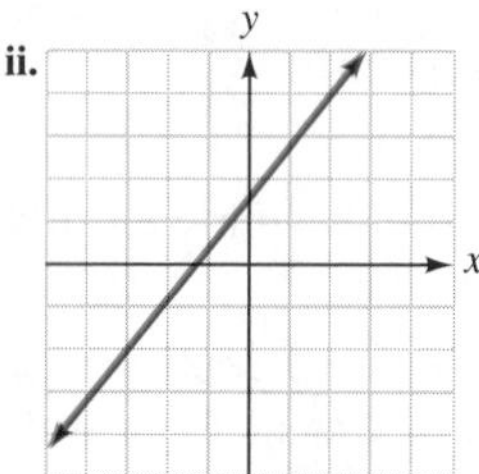

NOTATION **Complete the solution of the system.**

21. Solve: $\begin{cases} y = 3x \\ x - y = 4 \end{cases}$

$$\begin{aligned} x - y &= 4 \quad \text{This is the second equation.} \\ x - (\square) &= 4 \\ -2x &= \square \\ x &= \square \end{aligned}$$

$$\begin{aligned} y &= 3x \quad \text{This is the first equation.} \\ y &= 3(\square) \\ y &= \square \end{aligned}$$

The solution is $(\square, \square)$.

22. The system $\begin{cases} a = 3b + 2 \\ a + 3b = 8 \end{cases}$ was solved and it was found that b is 1 and a is 5. Write the solution as an ordered pair.

PRACTICE **Use the substitution method to solve each system. If a system has no solution or infinitely many, so state.**

23. $\begin{cases} y = 2x \\ x + y = 6 \end{cases}$

24. $\begin{cases} y = 3x \\ x + y = 4 \end{cases}$

25. $\begin{cases} y = 2x - 6 \\ 2x + y = 6 \end{cases}$

26. $\begin{cases} y = 2x - 9 \\ x + 3y = 8 \end{cases}$

27. $\begin{cases} y = 2x + 5 \\ x + 2y = -5 \end{cases}$

28. $\begin{cases} y = -2x \\ 3x + 2y = -1 \end{cases}$

29. $\begin{cases} 3x + y = -4 \\ x = y \end{cases}$

30. $\begin{cases} x + 2y = -6 \\ x = y \end{cases}$

31. $\begin{cases} 2a + 4b = -24 \\ a = 20 - 2b \end{cases}$

32. $\begin{cases} 3a + 6b = -15 \\ a = -2b - 5 \end{cases}$

33. $\begin{cases} 2a - 3b = -13 \\ -b = -2a - 7 \end{cases}$

34. $\begin{cases} a - 3b = -1 \\ -b = -2a - 2 \end{cases}$

35. $\begin{cases} -y = 11 - 3x \\ 2x + 5y = -4 \end{cases}$

36. $\begin{cases} -x = 10 - 3y \\ 2x + 8y = -6 \end{cases}$

37. $\begin{cases} 2x + 3 = -4y \\ x - 6 = -8y \end{cases}$

38. $\begin{cases} 5y + 2 = -4x \\ x + 2y = -2 \end{cases}$

39. $\begin{cases} r + 3s = 9 \\ 3r + 2s = 13 \end{cases}$

40. $\begin{cases} x - 2y = 2 \\ 2x + 3y = 11 \end{cases}$

41. $\begin{cases} 6x - 3y = 5 \\ 2y + x = 0 \end{cases}$

42. $\begin{cases} 5s + 10t = 3 \\ 2s + t = 0 \end{cases}$

43. $\begin{cases} y - 3x = -5 \\ 21x = 7y + 35 \end{cases}$

44. $\begin{cases} 8y = 15 - 4x \\ x + 2y = 4 \end{cases}$

45. $\begin{cases} y = 3x + 6 \\ y = -2x - 4 \end{cases}$

46. $\begin{cases} y = x + 3 \\ y = -2x - 3 \end{cases}$

47. $\begin{cases} x = \frac{1}{3}y - 1 \\ x = y + 1 \end{cases}$

48. $\begin{cases} x = \frac{1}{2}y + 2 \\ x = y + 1 \end{cases}$

49. $\begin{cases} 4x + 5y = 2 \\ 3x - y = 11 \end{cases}$

50. $\begin{cases} 5u + 3v = 5 \\ 4u - v = 4 \end{cases}$

51. $\begin{cases} 3x + 4y = -7 \\ 2y - x = -1 \end{cases}$

52. $\begin{cases} 5x - 2y = -7 \\ 5 - y = -3x \end{cases}$

53. $\begin{cases} 6 - y = 4x \\ 2y = -8x - 20 \end{cases}$

54. $\begin{cases} 9x = 3y + 12 \\ 4 = 3x - y \end{cases}$

55. $\begin{cases} b = \dfrac{2}{3}a \\ 8a - 3b = 3 \end{cases}$

56. $\begin{cases} a = \dfrac{2}{3}b \\ 9a + 4b = 5 \end{cases}$

57. $\begin{cases} 2x + 5y = -2 \\ -\dfrac{x}{2} = y \end{cases}$

58. $\begin{cases} y = -\dfrac{x}{2} \\ 2x - 3y = -7 \end{cases}$

59. $\begin{cases} \dfrac{x}{2} + \dfrac{y}{2} = -1 \\ \dfrac{x}{3} - \dfrac{y}{2} = -4 \end{cases}$

60. $\begin{cases} \dfrac{2a}{3} + \dfrac{b}{5} = 1 \\ \dfrac{a}{3} - \dfrac{2b}{3} = \dfrac{13}{3} \end{cases}$

61. $\begin{cases} 5x = \dfrac{1}{2}y - 1 \\ \dfrac{1}{4}y = 10x - 1 \end{cases}$

62. $\begin{cases} \dfrac{x}{4} + y = \dfrac{1}{4} \\ \dfrac{y}{2} + \dfrac{11}{20} = \dfrac{x}{10} \end{cases}$

63. $\begin{cases} x - \dfrac{4y}{5} = 4 \\ \dfrac{y}{3} = \dfrac{x}{2} - \dfrac{5}{2} \end{cases}$

64. $\begin{cases} 3x - 2y = \dfrac{9}{2} \\ \dfrac{x}{2} - \dfrac{3}{4} = 2y \end{cases}$

65. $\begin{cases} y + x = 2x + 2 \\ 6x - 4y = 21 - y \end{cases}$

66. $\begin{cases} y - x = 3x \\ 2x + 2y = 14 - y \end{cases}$

67. $\begin{cases} x = -3y + 6 \\ 2x + 4y = 6 + x + y \end{cases}$

68. $\begin{cases} 2x - y = x + y \\ -2x + 4y = 6 \end{cases}$

69. $\begin{cases} 4x + 5y + 1 = -12 + 2x \\ x - 3y + 2 = -3 - x \end{cases}$

70. $\begin{cases} 6x + y = -8 + 3x - y \\ 3x - y = 2y + x - 1 \end{cases}$

71. $\begin{cases} 3(x - 1) + 3 = 8 + 2y \\ 2(x + 1) = 8 + y \end{cases}$

72. $\begin{cases} 4(x - 2) = 19 - 5y \\ 3(x - 2) - 2y = -y \end{cases}$

APPLICATIONS

73. DINING When a customer orders the \$5.95 Country Breakfast from the menu, he wants to substitute something from the a la carte menu in place of the hash browns. What items can he pick so that the breakfast doesn't increase in price? Why?

Village Vault Restaurant			
Country Breakfast 2 eggs, 3 pancakes, bacon, sausage, hash browns, coffee			**\$5.95**
A la Carte Menu–Single Servings			
Strawberries	\$1.25	Melon	\$0.95
Croissant	\$1.70	Orange juice	\$1.65
Hash browns	\$0.95	Oatmeal	\$1.95
Muffin	\$1.30	Ham	\$1.80

74. HIGH SCHOOL SPORTS The equations shown model the number of boys and girls taking part in high school sports. In both models, x is the number of years since 1990, and y is the number of participants. If the trends continue, the graphs will intersect. Use the substitution method to predict the year when the number of boys and girls participating in high school sports will be the same.

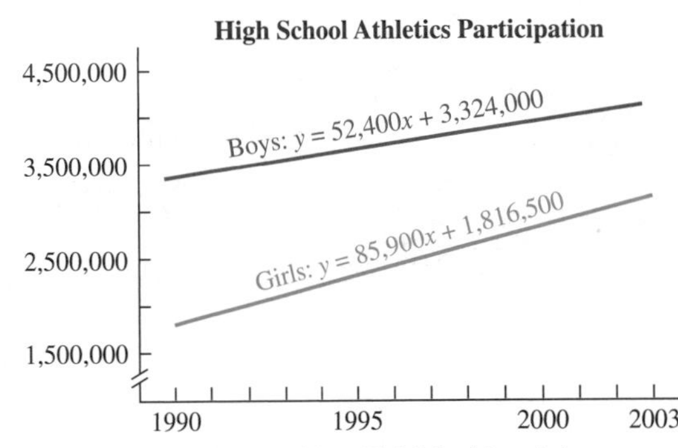

Source: National Federation of State High School Associations

WRITING

75. What concept does this diagram illustrate?

$$\begin{cases} 6x + 5y = 11 \\ y = 3x - 2 \end{cases}$$

76. When using the substitution method, how can you tell whether

a. a system of linear equations has no solution?

b. a system of linear equations has infinitely many solutions?

77. When solving a system, what advantages are there with the substitution method compared to the graphing method?

78. Consider the equation $5x + y = 12$. Explain why it is easier to solve for y than it is for x.

REVIEW **Use a check to decide whether 3 is a solution of the given equation.**

79. $3x - 8 = 1$

80. $9y^2 - 81 = 0$

81. $3(x + 8) + 5x = 2(12 + 4x)$

82. $\frac{7}{x+4} - \frac{1}{2} = \frac{3}{x+4}$

83. $x^3 + 7x^2 = x^2 - 9x$

84. $\frac{11(x - 12)}{2} = 9 - 2x$

CHALLENGE PROBLEMS **Use the substitution method to solve each system.**

85. $\begin{cases} \frac{6x-1}{3} - \frac{5}{3} = \frac{3y+1}{2} \\ \frac{1+5y}{4} + \frac{x+3}{4} = \frac{17}{2} \end{cases}$

86. $\begin{cases} 0.5x + 0.5y = 6 \\ 0.001x - 0.001y = -0.004 \end{cases}$

87. The system $\begin{cases} \frac{1}{2}x = y + 3 \\ x - 2y = 6 \end{cases}$ has infinitely many solutions. Find three of them.

88. Could the substitution method be used to solve the following system?

$$\begin{cases} y = -2 \\ x = 5 \end{cases}$$

How would you solve it?

7.3 Solving Systems of Equations by Elimination

- The Elimination Method
- Using Multiplication to Eliminate a Variable
- Inconsistent Systems and Dependent Equations

In the first step of the substitution method for solving a system of equations, we solve one of the equations for one of the variables. At times, this can be difficult, especially if neither variable has a coefficient of 1 or -1. This is the case for the system

$$\begin{cases} 2x + 5y = 11 \\ 7x - 5y = 16 \end{cases}$$

Solving either equation for x or y involves working with cumbersome fractions. Fortunately, we can solve systems like this one using a simpler algebraic method called the *elimination* or the *addition method.*

THE ELIMINATION METHOD

The elimination method for solving a system is based on the **addition property of equality:** *When equal quantities are added to both sides of an equation, the results are equal.* In symbols, if $A = B$ and $C = D$, then adding the left-hand sides and the right-hand sides of these equations, we have $A + C = B + D$. This procedure is called *adding the equations.*

Add the terms on the left-hand sides.

$$\begin{aligned} A &= B \\ C &= D \\ \hline A + C &= B + D \end{aligned}$$

Add the terms on the right-hand sides.

The Language of Algebra

To *eliminate* means to remove. People allergic to peanuts need to *eliminate* any foods containing peanuts from their diets.

EXAMPLE 1

Solve the system: $\begin{cases} 2x + 5y = 11 \\ 6x - 5y = 13 \end{cases}$.

ELEMENTARY Algebra f(x) Now™

Solution Since $6x - 5y$ and 13 are equal quantities, we can add $6x - 5y$ to the left-hand side and 13 to the right-hand side of the first equation, $2x + 5y = 11$.

$$\begin{aligned} 2x + 5y &= 11 \\ 6x - 5y &= 13 \\ \hline 8x \qquad &= 24 \end{aligned}$$

To add the equations, add the like terms, column by column:

$11 + 13 = 24$

$5y + (-5y) = 0$

$2x + 6x = 8x$

Because the sum of the terms $5y$ and $-5y$ is 0, we say that the variable y has been eliminated. Since the resulting equation has only one variable, we can solve it for x.

$$8x = 24$$

$$x = 3 \quad \text{Divide both sides by 8.}$$

To find the y-value of the solution, substitute 3 for x in either equation of the system.

$2x + 5y = 11$	This is the first equation of the system.
$2(3) + 5y = 11$	Substitute 3 for x.
$6 + 5y = 11$	Multiply.
$5y = 5$	Subtract 6 from both sides.
$y = 1$	Divide both sides by 5.

Now we check the proposed solution (3, 1) in the equations of the original system.

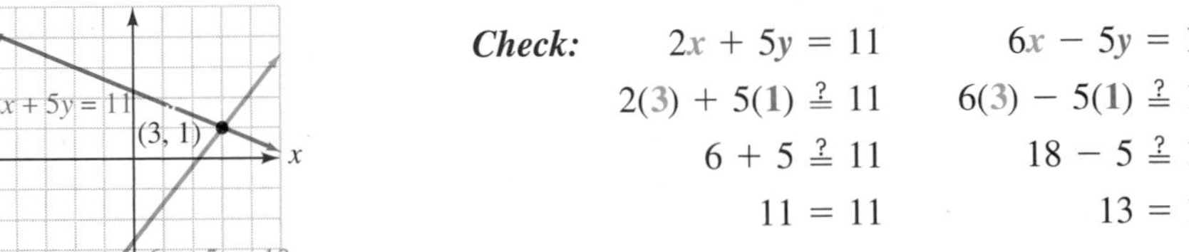

Check:

$2x + 5y = 11$	$6x - 5y = 13$
$2(3) + 5(1) \stackrel{?}{=} 11$	$6(3) - 5(1) \stackrel{?}{=} 13$
$6 + 5 \stackrel{?}{=} 11$	$18 - 5 \stackrel{?}{=} 13$
$11 = 11$	$13 = 13$

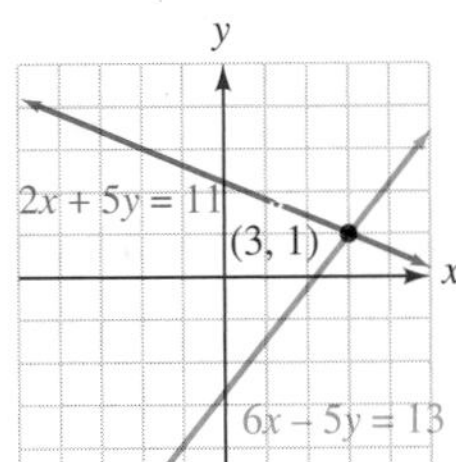

Since (3, 1) satisfies both equations, it is the solution. The graph on the left verifies this.

Self Check 1 Solve the system: $\begin{cases} -4x + 3y = 4 \\ 4x + 5y = 28 \end{cases}$.

To solve a system of equations in x and y by the elimination method, follow these steps.

The Elimination Method

1. Write both equations in general form: $Ax + By = C$.
2. If necessary, multiply one or both of the equations by nonzero quantities to make the coefficients of x (or the coefficients of y) opposites.
3. Add the equations to eliminate the terms involving x (or y).
4. Solve the equation resulting from step 3.
5. Find the value of the other variable by substituting the solution found in step 4 into any equation containing both variables.
6. Check the proposed solution in the equations of the original system. Write the solution as an ordered pair.

USING MULTIPLICATION TO ELIMINATE A VARIABLE

The system in Example 1 was easy to solve using elimination because the coefficients of the terms $5y$ in the first equation and $-5y$ in the second equation were opposites. For many systems, however, we are not able to immediately eliminate a variable by adding. In such cases, we use the multiplication property of equality to create coefficients of x or y that are opposites.

EXAMPLE 2

Solve the system: $\begin{cases} 2x + 7y = -18 \\ 2x + 3y = -10 \end{cases}$.

ELEMENTARY Algebra f(x) Now™

Solution ***Step 1:*** We see that neither the coefficients of x nor the coefficients of y are opposites. Adding these equations as they are does not eliminate a variable.

Step 2: To eliminate x, we can multiply both sides of the second equation by -1. This creates the term $-2x$, whose coefficient is opposite that of the $2x$ term in the first equation.

$$\begin{cases} 2x + 7y = -18 \\ 2x + 3y = -10 \end{cases} \begin{array}{c} \xrightarrow{\text{Unchanged}} \\ \xrightarrow[\text{Multiply by } -1]{} \end{array} \begin{array}{r} 2x + 7y = -18 \\ \mathbf{-1}(2x + 3y) = \mathbf{-1}(-10) \end{array} \begin{array}{c} \xrightarrow{\text{Unchanged}} \\ \xrightarrow[\text{Simplify}]{} \end{array} \begin{cases} 2x + 7y = -18 \\ -2x - 3y = 10 \end{cases}$$

Success Tip

The basic objective of the elimination method is to obtain two equations whose sum will be *one* equation in *one* variable.

Step 3: When the equations are added, x is eliminated.

$$\begin{array}{r} 2x + 7y = -18 \\ \underline{-2x - 3y = \quad 10} \\ 4y = \ -8 \end{array} \quad \text{In the left column: } 2x + (-2x) = 0.$$

Step 4: Solve the resulting equation to find y.

$$\begin{aligned} 4y &= -8 \\ y &= -2 \quad \text{Divide both sides by 4.} \end{aligned}$$

Step 5: To find x, we can substitute -2 for y in either of the equations of the original system, or in $-2x - 3y = 10$. It appears the computations will be simplest if we use $2x + 3y = -10$.

$$\begin{aligned} 2x + 3y &= -10 \\ 2x + 3(-2) &= -10 && \text{Substitute } -2 \text{ for } y. \\ 2x - 6 &= -10 && \text{Multiply.} \\ 2x &= -4 && \text{Add 6 to both sides.} \\ x &= -2 && \text{Divide both sides by 2.} \end{aligned}$$

Step 6: The solution is $(-2, -2)$. Check this result in the original equations.

Self Check 2 Solve the system: $\begin{cases} x + 7y = -24 \\ 3x + 7y = -30 \end{cases}$.

EXAMPLE 3

ELEMENTARY Algebra f(x) Now™

Solve the system: $\begin{cases} 7x + 2y - 14 = 0 \\ 9x = 4y - 28 \end{cases}$.

Solution *Step 1:* To compare coefficients, write each equation in the form $Ax + By = C$. Since each of the original equations will be written in an equivalent form, the resulting system will have the same solution as the original system.

$$\begin{cases} 7x + 2y = 14 & \text{Add 14 to both sides of } 7x + 2y - 14 = 0. \\ 9x - 4y = -28 & \text{Subtract } 4y \text{ from both sides of } 9x = 4y - 28. \end{cases}$$

Step 2: Neither the coefficients of x nor the coefficients of y are opposites. To eliminate y, we can multiply both sides of the first equation by 2. This creates the term $4y$, whose coefficient is opposite that of the $-4y$ term in the second equation.

$$\begin{cases} 7x + 2y = 14 \\ 9x - 4y = -28 \end{cases} \xrightarrow[\text{Unchanged}]{\text{Multiply by 2}} \begin{matrix} 2(7x + 2y) = 2(14) \\ 9x - 4y = -28 \end{matrix} \xrightarrow[\text{Unchanged}]{\text{Simplify}} \begin{cases} 14x + 4y = 28 \\ 9x - 4y = -28 \end{cases}$$

Caution

When using the elimination method, don't forget to multiply both sides of an equation by the appropriate number. For instance, in Example 3:

Multiply both sides by 2

$$2(7x + 2y) = 2(14)$$

Step 3: When the equations are added, y is eliminated.

$$\begin{aligned} 14x + 4y &= 28 \\ 9x - 4y &= -28 \quad \text{In the middle column, } 4y + (-4y) = 0. \\ \hline 23x \quad &= 0 \end{aligned}$$

Step 4: Since the result of the addition is an equation in one variable, we can solve for x.

$$\begin{aligned} 23x &= 0 \\ x &= 0 \quad \text{Divide both sides by 23.} \end{aligned}$$

Step 5: To find y, we can substitute 0 for x in any equation that contains both variables. It appears the computations will be simplest if we use $7x + 2y = 14$.

$$\begin{aligned} 7x + 2y &= 14 \\ 7(0) + 2y &= 14 \quad \text{Substitute 0 for } x. \\ 0 + 2y &= 14 \quad \text{Multiply.} \\ 2y &= 14 \\ y &= 7 \quad \text{Divide both sides by 2.} \end{aligned}$$

Step 6: The solution is (0, 7). Check this result in the original equations.

Self Check 3 Solve the system: $\begin{cases} 3x = 10 - 2y \\ 5x - 6y + 30 = 0 \end{cases}$.

EXAMPLE 4

Solve the system: $\begin{cases} 4a + 7b = -8 \\ 5a + 6b = 1 \end{cases}$.

ELEMENTARY Algebra Now™

Solution *Step 1:* Both equations are written in general form.

Step 2: In this example, we must rewrite both equations to obtain like terms that are opposites. To eliminate a, we can multiply the first equation by 5 to create the term $20a$, and we can multiply the second equation by -4 to create the term $-20a$.

$$\begin{cases} 4a + 7b = -8 \\ 5a + 6b = 1 \end{cases} \xrightarrow[\text{Multiply by } -4]{\text{Multiply by 5}} \begin{matrix} 5(4a + 7b) = 5(-8) \\ -4(5a + 6b) = -4(1) \end{matrix} \xrightarrow[\text{Simplify}]{\text{Simplify}} \begin{cases} 20a + 35b = -40 \\ -20a - 24b = -4 \end{cases}$$

Step 3: When we add the resulting equations, a is eliminated.

$$\begin{aligned} 20a + 35b &= -40 \\ -20a - 24b &= -4 \\ \hline 11b &= -44 \end{aligned}$$ In the left column, $20a + (-20a) = 0$.

Success Tip

In Example 4, we create the term $20a$ from $4a$ and the term $-20a$ from $5a$. Note that the *least common multiple* of 4 and 5 is 20:

4, 8, 12, 16, **20**, 24, 28, . . .
5, 10, 15, **20**, 25, 30, . . .

Step 4: Solve the resulting equation for b.

$11b = -44$

$b = -4$ Divide both sides by 11.

Step 5: To find a, we can substitute -4 for b in any equation that contains both variables. It appears the computations will be simplest if we use $5a + 6b = 1$.

$5a + 6b = 1$

$5a + 6(-4) = 1$ Substitute -4 for b.

$5a - 24 = 1$ Multiply.

$5a = 25$ Add 24 to both sides.

$a = 5$ Divide both sides by 5. This is the a-value of the solution.

Success Tip

With this method, it doesn't matter which variable is eliminated. In Example 4, we could have created terms of $42b$ and $-42b$ to eliminate b. We will get the same solution, $(5, -4)$.

Step 6: Written in (a, b) form, the solution is $(5, -4)$. Check it in the original equations.

Self Check 4 Solve the system: $\begin{cases} 5a + 3b = -7 \\ 3a + 4b = 9 \end{cases}$.

EXAMPLE 5

Solve the system: $\begin{cases} \frac{1}{6}x + \frac{1}{2}y = \frac{1}{3} \\ -\frac{x}{9} + y = \frac{5}{9} \end{cases}$.

Solution ***Step 1:*** To clear the equations of the fractions, multiply both sides of the first equation by 6 and both sides of the second equation by 9.

$$\begin{cases} \frac{1}{6}x + \frac{1}{2}y = \frac{1}{3} \xrightarrow{\text{Multiply by 6}} 6\left(\frac{1}{6}x + \frac{1}{2}y\right) = 6\left(\frac{1}{3}\right) \xrightarrow{\text{Simplify}} \\ -\frac{x}{9} + y = \frac{5}{9} \xrightarrow[\text{Multiply by 9}]{} 9\left(-\frac{x}{9} + y\right) = 9\left(\frac{5}{9}\right) \xrightarrow[\text{Simplify}]{} \end{cases} \begin{cases} x + 3y = 2 \\ -x + 9y = 5 \end{cases}$$

Step 2: The coefficients of x are opposites.

Step 3: The variable x is eliminated when we add the resulting equations.

$$\begin{aligned} x + 3y &= 2 \\ -x + 9y &= 5 \\ \hline 12y &= 7 \end{aligned}$$ In the left column: $x + (-x) = 0$.

Step 4: Now solve to find y.

$12y = 7$

$y = \frac{7}{12}$ Divide both sides by 12.

Step 5: We can find x by substituting $\frac{7}{12}$ for y in any equation containing both variables. However, that computation could be complicated, because $\frac{7}{12}$ is a fraction. Instead, we can begin again with the system that is cleared of fractions, but this time, eliminate y. If we multiply both sides of the first equation by -3, this creates the term $-9y$, whose coefficient is opposite that of the $9y$ term in the second equation.

$$\begin{cases} x + 3y = 2 \\ -x + 9y = 5 \end{cases} \xrightarrow[\text{Unchanged}]{\text{Multiply by } -3} \begin{matrix} -3(x + 3y) = -3(2) \\ -x + 9y = 5 \end{matrix} \xrightarrow[\text{Unchanged}]{\text{Simplify}} \begin{cases} -3x - 9y = -6 \\ -x + 9y = 5 \end{cases}$$

When we add the resulting equations, y is eliminated.

$$\begin{aligned} -3x - 9y &= -6 \\ -x + 9y &= 5 \\ \hline -4x \quad &= -1 \end{aligned}$$

In the middle column: $9y + (-9y) = 0$.

Now we solve the previous equation to find x.

$$-4x = -1$$

$$x = \frac{1}{4} \quad \text{Divide both sides by } -4.$$

Step 6: The solution is $\left(\frac{1}{4}, \frac{7}{12}\right)$. To verify this, check it in the original equations.

Self Check 5 Solve the system: $\begin{cases} -\frac{1}{5}x + y = \frac{8}{5} \\ \frac{x}{8} + \frac{y}{2} = \frac{1}{4} \end{cases}$.

INCONSISTENT SYSTEMS AND DEPENDENT EQUATIONS

We have solved inconsistent systems and systems of dependent equations graphically and by substitution. We can also solve these systems using the elimination method.

EXAMPLE 6

Solve the system: $\begin{cases} 3x - 2y = 2 \\ -3x + 2y = -12 \end{cases}$.

ELEMENTARY Algebra f(x) Now™

Solution If we add the equations as they are, x is eliminated.

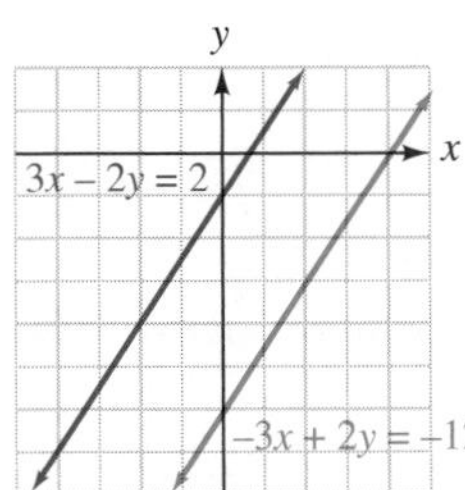

$$\begin{aligned} 3x - 2y &= 2 \\ -3x + 2y &= -12 \\ \hline 0 &= -10 \end{aligned}$$

In the left column: $3x + (-3x) = 0$.
In the middle column: $-2y + 2y = 0$.

In eliminating x, the variable y is eliminated as well. The resulting false statement, $0 = -10$, indicates that the system has no solution and is inconsistent. The graphs of the equations verify this; they are parallel lines.

Self Check 6 Solve the system: $\begin{cases} 2x - 7y = 5 \\ -2x + 7y = 3 \end{cases}$.

EXAMPLE 7

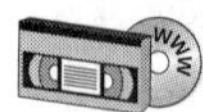
ELEMENTARY Algebra f(x) Now™

Solve the system: $\begin{cases} \dfrac{2x - 5y}{15} = \dfrac{8}{15} \\ -0.2x + 0.5y = -0.8 \end{cases}$.

Solution We can multiply both sides of the first equation by 15 to clear it of fractions and both sides of the second equation by 10 to clear it of decimals.

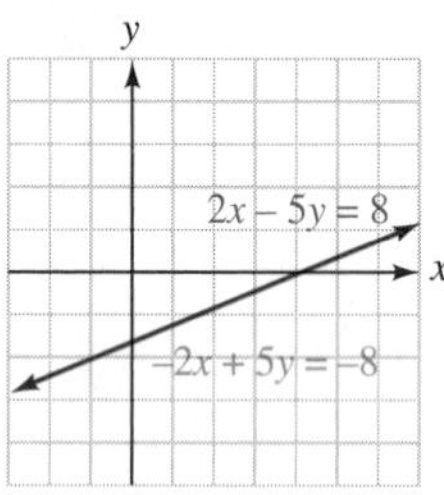

$$\begin{cases} \dfrac{2x - 5y}{15} = \dfrac{8}{15} & \rightarrow & 15\left(\dfrac{2x - 5y}{15}\right) = 15\left(\dfrac{8}{15}\right) & \rightarrow \\ -0.2x + 0.5y = -0.8 & \rightarrow & 10(-0.2x + 0.5y) = 10(-0.8) & \rightarrow \end{cases} \begin{cases} 2x - 5y = 8 \\ -2x + 5y = -8 \end{cases}$$

We add the resulting equations to get

$$\begin{array}{rl} 2x - 5y = & 8 \\ -2x + 5y = & -8 \\ \hline 0 = & 0 \end{array} \quad \begin{array}{l} \text{In the left column: } 2x + (-2x) = 0. \\ \text{In the middle column: } -5y + 5y = 0. \\ \text{In the right column: } 8 + (-8) = 0. \end{array}$$

As in Example 6, both variables are eliminated. However, this time a true statement, $0 = 0$, is obtained. It indicates that the equations are dependent and that the system has infinitely many solutions. The graphs of the equations verify this; they are identical.

To find some solutions, we can substitute some values of x, say -1, 4, and 9, in either equation and solve for y. The results are: $(-1, -2)$, $(4, 0)$, and $(9, 2)$.

Self Check 7 Solve the system: $\begin{cases} \dfrac{3x + y}{6} = \dfrac{1}{3} \\ -0.3x - 0.1y = -0.2 \end{cases}$.

Answers to Self Checks **1.** (2, 4) **2.** (−3, −3) **3.** (0, 5) **4.** (−5, 6) **5.** $\left(-\frac{22}{9}, \frac{10}{9}\right)$ **6.** no solution **7.** infinitely many solutions

7.3 STUDY SET

ELEMENTARY Algebra f(x) Now™

VOCABULARY Fill in the blanks.

1. The coefficients of $3x$ and $-3x$ are ________.
2. The general form of the equation of a line is

   $\square + By = \square$.

3. When the following equations are added, the variable y will be ________.

   $$\begin{array}{r} 5x - 6y = 10 \\ -3x + 6y = 24 \\ \hline \end{array}$$

4. By the ________ property of equality, we know that if we multiply both sides of $2x - 3y = 4$ by 6, we will obtain an equivalent equation.
5. The elimination method for solving a system is based on the ________ property of equality: *When equal quantities are added to both sides of an equation, the results are ________.*
6. The objective of the elimination method is to obtain two equations whose sum will be one equation in ______ variable.

CONCEPTS

7. In the following system, which terms have coefficients that are opposites?

   $$\begin{cases} 3x + 7y = -25 \\ 4x - 7y = 12 \end{cases}$$

8. Add each pair of equations.

a. $\begin{array}{r} 8x + 5y = 11 \\ \underline{-4x + y = -2} \end{array}$ **b.** $\begin{array}{r} 2a + 2b = -6 \\ \underline{3a - 2b = 2} \end{array}$

c. $\begin{array}{r} x - 3y = 15 \\ \underline{-x - y = -14} \end{array}$ **d.** $\begin{array}{r} 5x - y = 7 \\ \underline{-5x + y = -7} \end{array}$

9. a. Fill in the blank:

$$4x + y = 2 \xrightarrow{\text{Multiply by 3}} 3(4x + y) = 3(2) \xrightarrow{\text{Simplify}} \underline{\hspace{2cm}}$$

b. Fill in the blank:

$$x - 3y = 4 \xrightarrow{\text{Multiply by } -2} -2(x - 3y) = -2(4) \xrightarrow{\text{Simplify}} \underline{\hspace{2cm}}$$

10. If the elimination method is used to solve

$$\begin{cases} 3x + 12y = 4 \\ 6x - 4y = 7 \end{cases}$$

a. By what would we multiply the first equation to eliminate x?

b. By what would we multiply the second equation to eliminate y?

11. Suppose the following system is solved using the elimination method and it is found that x is 2. Find the value of y.

$$\begin{cases} 4x + 3y = 11 \\ 3x - 2y = 4 \end{cases}$$

12. What algebraic step should be performed to

a. Clear $\frac{2}{3}x + 4y = -\frac{4}{5}$ of fractions?

b. Clear $0.2x - 0.9y = 6.4$ of decimals?

13. Is $(1, -1)$ a solution of the system $\begin{cases} 2x + 3y = -1 \\ 3x + 5y = -2 \end{cases}$?

14. a. Suppose $0 = 0$ is obtained when a system is solved by the elimination method. Does the system have a solution? Which of the following is a possible graph of the system?

b. Suppose $0 = 2$ is obtained when a system is solved by the elimination method. Does the system have a solution? Which of the following is a possible graph of the system?

i
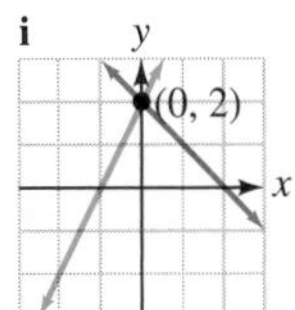

ii
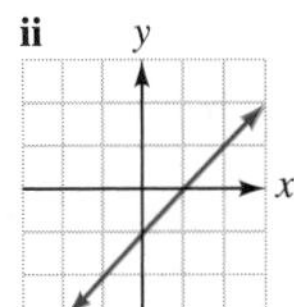

iii
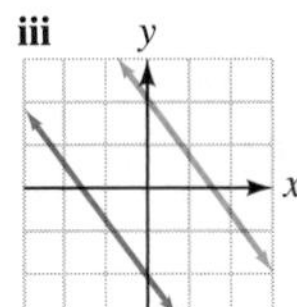

NOTATION **Complete the solution.**

15. Solve: $\begin{cases} x + y = 5 \\ x - y = -3 \end{cases}$.

$$\begin{array}{r} x + y = 5 \\ \underline{x - y = -3} \\ \underline{\hspace{0.8cm}} = 2 \\ x = \underline{\hspace{0.5cm}} \end{array}$$

$x + y = 5$ This is the first equation.

$\underline{\hspace{0.5cm}} + y = 5$

$y = 4$

The solution is $\underline{\hspace{1.5cm}}$.

16. Write each equation of each system in general form: $Ax + By = C$.

a. $\begin{cases} 7x + y + 3 = 0 \\ 8x + 4 = -y \end{cases} \rightarrow \begin{cases} \underline{\hspace{2cm}} \\ \underline{\hspace{2cm}} \end{cases}$

b. $\begin{cases} x = 2y \\ 7 - 5y = 2x \end{cases} \rightarrow \begin{cases} \underline{\hspace{2cm}} \\ \underline{\hspace{2cm}} \end{cases}$

PRACTICE **Use the elimination method to solve each system. If a system has no solution or infinitely many solutions, so indicate.**

17. $\begin{cases} x + y = 5 \\ x - y = 1 \end{cases}$

18. $\begin{cases} x - y = 4 \\ x + y = 8 \end{cases}$

19. $\begin{cases} x + y = 1 \\ x - y = 5 \end{cases}$

20. $\begin{cases} x - y = -5 \\ x + y = 1 \end{cases}$

21. $\begin{cases} x + y = -5 \\ -x + y = -1 \end{cases}$

22. $\begin{cases} -x + y = -3 \\ x + y = 1 \end{cases}$

23. $\begin{cases} 4x + 3y = 24 \\ 4x - 3y = -24 \end{cases}$

24. $\begin{cases} -9x + 5y = -9 \\ -9x - 5y = -9 \end{cases}$

25. $\begin{cases} 2s + t = -2 \\ -2s - 3t = -6 \end{cases}$

26. $\begin{cases} -2x + 4y = 12 \\ 2x + 4y = 28 \end{cases}$

27. $\begin{cases} 5x - 4y = 8 \\ -5x - 4y = 8 \end{cases}$

28. $\begin{cases} 2r + s = -8 \\ -2r + 4s = 28 \end{cases}$

29. $\begin{cases} 4x - 7y = -19 \\ -4x + 7y = 19 \end{cases}$

30. $\begin{cases} x + 20y = -2 \\ -x - 20y = 2 \end{cases}$

31. $\begin{cases} x + 3y = -9 \\ x + 8y = -4 \end{cases}$

32. $\begin{cases} x + 7y = -22 \\ x + 9y = -24 \end{cases}$

33. $\begin{cases} 5c + d = -15 \\ 6c + d = -20 \end{cases}$

34. $\begin{cases} 11c + d = -65 \\ 10c + d = -60 \end{cases}$

35. $\begin{cases} 7x - y = 10 \\ 8x - y = 13 \end{cases}$

36. $\begin{cases} 6x - y = 4 \\ 9x - y = 10 \end{cases}$

37. $\begin{cases} 3x - 5y = -29 \\ 3x - 5y = 15 \end{cases}$

38. $\begin{cases} 2a - 3b = -6 \\ 2a - 3b = 8 \end{cases}$

39. $\begin{cases} 6x - 3y = -7 \\ 9x + y = 6 \end{cases}$

40. $\begin{cases} 9x + 4y = 31 \\ 6x - y = -5 \end{cases}$

41. $\begin{cases} 9x + 4y = 31 \\ 6x - y = -5 \end{cases}$

42. $\begin{cases} 5x - 14y = -12 \\ -x - 6y = -2 \end{cases}$

43. $\begin{cases} 8x + 8y = -16 \\ 3x + y = -4 \end{cases}$

44. $\begin{cases} 4x + 11y = -11 \\ 5x + y = -1 \end{cases}$

45. $\begin{cases} 7x - 50y = -43 \\ x + 3y = 4 \end{cases}$

46. $\begin{cases} x - 2y = -1 \\ 12x + 11y = 23 \end{cases}$

47. $\begin{cases} 8x - 4y = 18 \\ 3x - 2y = 8 \end{cases}$

48. $\begin{cases} 4x + 6y = 5 \\ 8x - 9y = 3 \end{cases}$

49. $\begin{cases} 4x + 3y = 7 \\ 3x - 2y = -16 \end{cases}$

50. $\begin{cases} 3x - 2y = 20 \\ 2x + 7y = 5 \end{cases}$

51. $\begin{cases} 3x + 4y = 12 \\ 4x + 5y = 17 \end{cases}$

52. $\begin{cases} 2x + 11y = -10 \\ 5x + 4y = 22 \end{cases}$

53. $\begin{cases} -3x + 6y = -9 \\ -5x + 4y = -15 \end{cases}$

54. $\begin{cases} -4x + 3y = -13 \\ -6x + 8y = -16 \end{cases}$

55. $\begin{cases} 4a + 7b = 2 \\ 9a - 3b = 1 \end{cases}$

56. $\begin{cases} 5a - 7b = 6 \\ 7a - 6b = 8 \end{cases}$

57. $\begin{cases} 9x = 10y \\ 3x - 2y = 12 \end{cases}$

58. $\begin{cases} 8x = 9y \\ 2x - 3y = -6 \end{cases}$

59. $\begin{cases} 2x + 5y + 13 = 0 \\ 2x + 5 = 3y \end{cases}$

60. $\begin{cases} 3x - 16 = 5y \\ 4x + 5y - 33 = 0 \end{cases}$

61. $\begin{cases} 0 = 4x - 3y \\ 5x = 4y - 2 \end{cases}$

62. $\begin{cases} 6x + 3y = 0 \\ 5y = 2x + 12 \end{cases}$

63. $\begin{cases} 3x - 16 = 5y \\ -3x + 5y - 33 = 0 \end{cases}$

64. $\begin{cases} 2x + 5y - 13 = 0 \\ -2x + 13 = 5y \end{cases}$

65. $\begin{cases} \frac{3}{5}s + \frac{4}{5}t = 1 \\ -\frac{1}{4}s + \frac{3}{8}t = 1 \end{cases}$

66. $\begin{cases} \frac{1}{2}x + \frac{4}{7}y = -1 \\ 5x - \frac{4}{5}y = -10 \end{cases}$

67. $\begin{cases} \frac{1}{2}s - \frac{1}{4}t = 1 \\ \frac{1}{3}s + t = 3 \end{cases}$

68. $\begin{cases} \frac{3}{5}x + y = 1 \\ \frac{4}{5}x - y = -1 \end{cases}$

69. $\begin{cases} -\frac{m}{4} - \frac{n}{3} = \frac{1}{12} \\ \frac{m}{2} - \frac{5n}{4} = \frac{7}{4} \end{cases}$

70. $\begin{cases} -\frac{x}{2} + \frac{y}{3} = 2 \\ \frac{x}{3} + \frac{2y}{3} = \frac{4}{3} \end{cases}$

71. $\begin{cases} \frac{x}{2} - 3y = 7 \\ -x + 6y = -14 \end{cases}$

72. $\begin{cases} -9x + \frac{y}{2} = 7 \\ 18x - y = 0 \end{cases}$

73. $\begin{cases} 2x + y = 10 \\ 0.1x + 0.2y = 1.0 \end{cases}$

74. $\begin{cases} 0.3x + 0.2y = 0 \\ 2x - 3y = -13 \end{cases}$

75. $\begin{cases} 2x - y = 16 \\ 0.03x + 0.02y = 0.03 \end{cases}$

76. $\begin{cases} -5y + 2x = 4 \\ -0.02y + 0.03x = 0.04 \end{cases}$

APPLICATIONS

77. EDUCATION The graph shows educational trends during the years 1980–2001 for persons 25 years or older. The equation $9x + 11y = 352$ approximates the percent y who had less than 12 years of schooling. The equation $5x - 11y = -198$ approximates the percent y who had 4 or more years of college. In each case, x is the number of years since 1980. Use the elimination method to determine in what year the percents were equal.

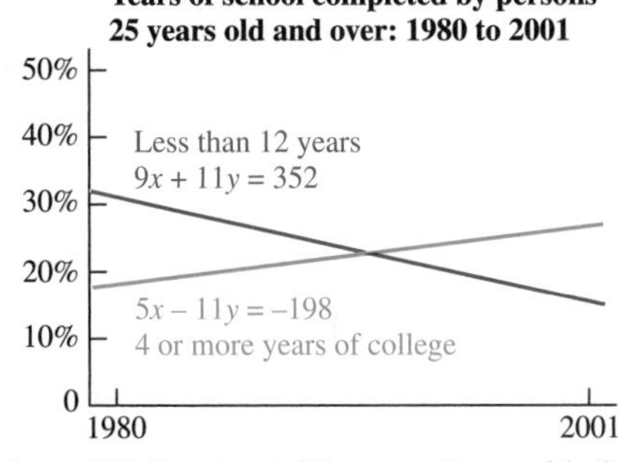

Source: U.S. Department of Commerce, Bureau of the Census

78. EDUCATION Answer Problem 77 by solving the system of equations using the substitution method. (*Hint:* Solve one equation for $11y$, then substitute for $11y$ in the other equation.)

WRITING

79. Why is the method for solving systems that is discussed in this section called the *elimination method?*

80. If the elimination method is to be used to solve this system, what is wrong with the form in which it is written?

$$\begin{cases} 2x - 5y = -3 \\ -2y + 3x = 10 \end{cases}$$

81. Can the system $\begin{cases} 2x + 5y = -13 \\ -2x - 3y = -5 \end{cases}$ be solved more easily using the elimination method or the substitution method? Explain.

82. Explain the error in the following work.

Solve: $\begin{cases} x + y = 1 \\ x - y = 5 \end{cases}$.

$$\begin{aligned} x + y &= 1 \\ +x - y &= 5 \\ \hline 2x \quad &= 6 \\ \frac{2x}{2} &= \frac{6}{2} \\ \boxed{x = 3} \end{aligned}$$

The solution is 3.

REVIEW

83. Translate into symbols: 10 less than x.

84. Solve: $3y + \frac{y + 2}{2} = \frac{2(y + 3)}{3} + 16$.

85. Simplify: $x - x$.

86. Simplify: $3.2m - 4.4 + 2.1m + 16$.

87. Find the area of a triangular-shaped sign with a base of 4 feet and a height of 3.75 feet.

88. Factor: $6x^2 + 7x - 20$.

CHALLENGE PROBLEMS **Use the elimination method to solve each system.**

89. $\begin{cases} \frac{x - 3}{2} + \frac{y + 5}{3} = \frac{11}{6} \\ \frac{x + 3}{3} - \frac{5}{12} = \frac{y + 3}{4} \end{cases}$

90. $\begin{cases} 4(x + 1) = 17 - 3(y - 1) \\ 2(x + 2) + 3(y - 1) = 9 \end{cases}$

7.4 Problem Solving Using Systems of Equations

- Assigning Variables to Two Unknowns
- Geometry Problems
- Number-Value Problems
- The Break Point
- Interest, Uniform Motion, and Mixture Problems

In previous chapters, many applied problems were modeled and solved with an equation in one variable. In this section, the application problems involve two unknowns. It is often easier to solve such problems using a two-variable approach.

ASSIGNING VARIABLES TO TWO UNKNOWNS

The following steps are helpful when solving problems involving two unknown quantities.

Problem-Solving Strategy

1. **Analyze the problem** by reading it carefully to understand the given facts. Often a diagram or table will help you visualize the facts of the problem.
2. Pick different variables to represent two unknown quantities. Translate the words of the problem to **form two equations** involving each of the two variables.
3. **Solve the system** of equations using graphing, substitution, or elimination.
4. **State the conclusion.**
5. **Check the results** in the words of the problem.

EXAMPLE 1

Motion pictures. Each year, Academy Award winners are presented with Oscars. The 13.5-inch statuette has a base on which a gold plated figure stands. The figure itself is 7.5 inches taller than the base. Find the height of the figure and the height of the base.

Analyze the Problem

- The statuette is a total of 13.5 inches tall.
- The figure is 7.5 inches taller than the base.
- Find the height of the figure and the height of the base.

Caution

If two variables are used to represent two unknown quantities, we must form a system of equations to find the unknown.

Form Two Equations Let $x =$ the height of the figure and $y =$ the height of the base. We can translate the words of the problem into two equations, each involving x and y.

The height of the figure	plus	the height of the base	is	13.5 inches.
x	$+$	y	$=$	13.5

The height of the figure	is	the height of the base	plus	7.5 inches.
x	$=$	y	$+$	7.5

The resulting system is $\begin{cases} x + y = 13.5 \\ x = y + 7.5 \end{cases}$.

Solve the System Since the second equation is solved for x, we will use substitution to solve the system.

$$x + y = 13.5 \quad \text{This is the first equation of the system.}$$
$$y + 7.5 + y = 13.5 \quad \text{Substitute } y + 7.5 \text{ for } x.$$
$$2y + 7.5 = 13.5 \quad \text{Combine like terms.}$$

$$2y = 6 \quad \text{Subtract 7.5 from both sides.}$$
$$y = 3 \quad \text{Divide both sides by 2. This is the height of the base.}$$

To find x, substitute 3 for y in the second equation of the system.

$$x = y + 7.5$$
$$x = 3 + 7.5 \quad \text{Substitute for } y.$$
$$x = 10.5 \quad \text{This is the height of the figure.}$$

State the Conclusion The height of the figure is 10.5 inches and the height of the base is 3 inches.

Check the Results The sum of 10.5 inches and 3 inches is 13.5 inches, and the 10.5-inch figure is 7.5 inches taller than the 3-inch base. The results check.

GEOMETRY PROBLEMS

Two angles are said to be **complementary** if the sum of their measures is 90°. Two angles are said to be **supplementary** if the sum of their measures is 180°.

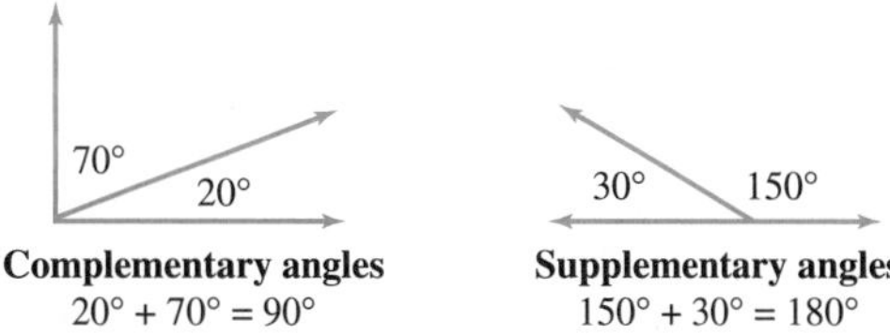

Complementary angles
20° + 70° = 90°

Supplementary angles
150° + 30° = 180°

EXAMPLE 2

Angles. The difference of the measures of two complementary angles is 6°. Find the measure of each angle.

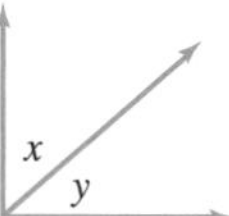

Analyze the Problem

- Since the angles are complementary, the sum of their measures is 90°.
- The word *difference* indicates subtraction. If the measure of the smaller angle is subtracted from the measure of the larger angle, the result will be 6°.
- Find the measure of the larger angle and the measure of the smaller angle.

Form Two Equations Let x = the measure of the larger angle and y = the measure of the smaller angle. We can translate the words of the problem into two equations, each involving x and y.

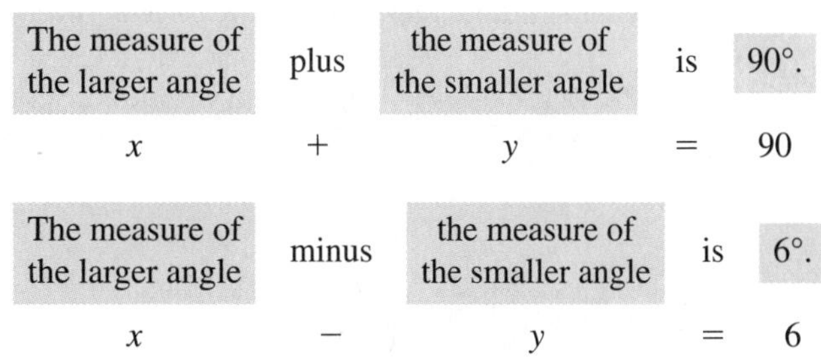

The measure of the larger angle	plus	the measure of the smaller angle	is	90°.
x	$+$	y	$=$	90

The measure of the larger angle	minus	the measure of the smaller angle	is	6°.
x	$-$	y	$=$	6

The resulting system is $\begin{cases} x + y = 90 \\ x - y = 6 \end{cases}$.

Solve the System Since the coefficients of y are opposites, we will use elimination to solve the system.

$$\begin{aligned} x + y &= 90 \\ \underline{x - y} &\underline{= \ \ 6} \\ 2x \qquad &= 96 && \text{Add the equations to eliminate } y. \\ x &= 48 && \text{Divide both sides by 2. This is the measure of the larger angle.} \end{aligned}$$

To find y, substitute 48 for x in the first equation of the system.

$$\begin{aligned} x + y &= 90 \\ 48 + y &= 90 && \text{Substitute 48 for } x. \\ y &= 42 && \text{Subtract 48 from both sides. This is the measure of the smaller angle.} \end{aligned}$$

State the Conclusion The measure of the larger angle is 48° and the measure of the smaller angle is 42°.

Check the Results The sum of 48° and 42° is 90°, and the difference is 6°. The results check.

EXAMPLE 3

History. In 1917, James Montgomery Flagg created the classic *I Want You* poster to help recruiting for World War I. The perimeter of the poster is 114 inches, and its length is 9 inches less than twice its width. Find the length and the width of the poster.

Analyze the Problem

- The perimeter of the rectangular poster is 114 inches.
- The length is 9 inches less than twice the width.
- Find the length and the width of the poster.

Form Two Equations Let l = the length of the poster and w = the width of the poster. The perimeter of a rectangle is the sum of two lengths and two widths, so we have

2	times	the length of the poster	plus	2	times	the width of the poster	is	114 inches.
2	$\cdot$	l	$+$	2	$\cdot$	w	$=$	114

If the length of the poster is 9 inches less than twice the width, we have

The length of the poster	is	2	times	the width of the poster	minus	9 inches.
l	$=$	2	$\cdot$	w	$-$	9

The resulting system is $\begin{cases} 2l + 2w = 114 \\ l = 2w - 9 \end{cases}$.

Solve the System Since the second equation is solved for l, we will use substitution to solve the system.

$$2l + 2w = 114 \quad \text{This is the first equation of the system.}$$
$$2(2w - 9) + 2w = 114 \quad \text{Substitute } 2w - 9 \text{ for } l.$$
$$4w - 18 + 2w = 114 \quad \text{Distribute the multiplication by 2.}$$
$$6w - 18 = 114 \quad \text{Combine like terms.}$$
$$6w = 132 \quad \text{Add 18 to both sides.}$$
$$w = 22 \quad \text{Divide both sides by 6. This is the width of the poster.}$$

To find l, substitute 22 for w in the second equation of the system.

$$l = 2w - 9$$
$$l = 2(22) - 9$$
$$l = 44 - 9$$
$$l = 35 \quad \text{This is the length of the poster.}$$

State the Conclusion The length of the poster is 35 inches and the width is 22 inches.

Check the Results The perimeter is $2(35) + 2(22) = 70 + 44 = 114$ inches, and 35 inches is 9 inches less than twice 22 inches. The results check.

NUMBER-VALUE PROBLEMS

EXAMPLE 4

ELEMENTARY Algebra f(x) Now™

Photography. At a school, two picture packages are available, as shown in the illustration. Find the cost of a class picture and the cost of an individual wallet-size picture.

Solution

Analyze the Problem

- Package 1 contains 1 class picture and 10 wallet-size pictures.
- Package 2 contains 2 class pictures and 15 wallet-size pictures.
- Find the cost of a class picture and the cost of a wallet-size picture.

Form Two Equations Let c = the cost of 1 class picture and w = the cost of 1 wallet-size picture. To write an equation that models the first package, we note that (in dollars) the cost of 1 class picture is c and the cost of 10 wallet-size pictures is $10 \cdot w = 10w$.

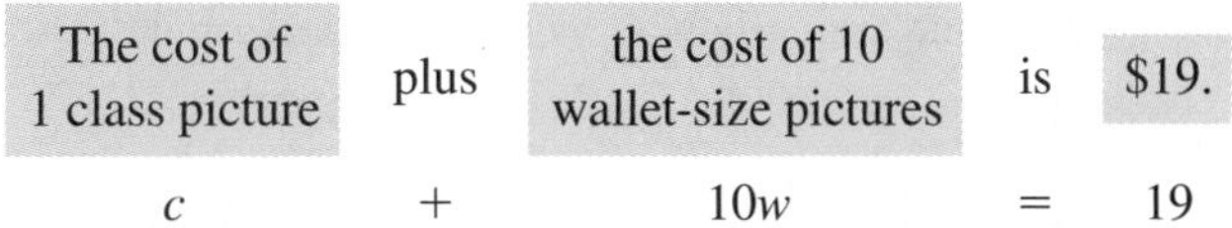

The cost of 1 class picture	plus	the cost of 10 wallet-size pictures	is	$19.
c	$+$	$10w$	$=$	19

To write an equation that models the second package, we note that (in dollars) the cost of 2 class pictures is $2 \cdot c = 2c$, and the cost of 15 wallet-size pictures is $15 \cdot w = 15w$.

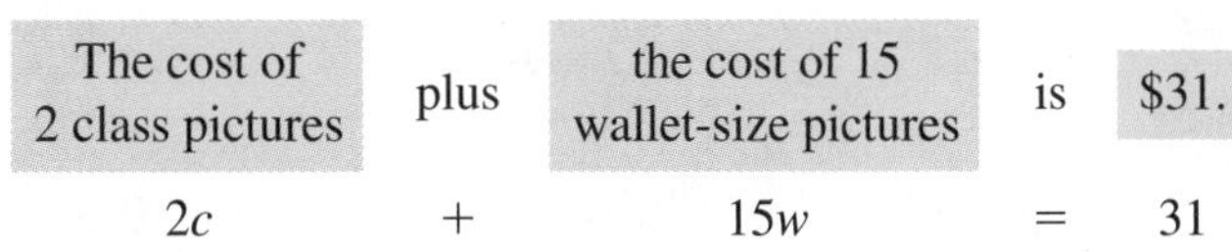

The cost of 2 class pictures	plus	the cost of 15 wallet-size pictures	is	$31.
$2c$	$+$	$15w$	$=$	31

The resulting system is $\begin{cases} c + 10w = 19 \\ 2c + 15w = 31 \end{cases}$.

Solve the System We can use elimination to solve this system. To eliminate c, we proceed as follows.

$$\begin{aligned} -2c - 20w &= -38 && \text{Multiply both sides of } c + 10w = 19 \text{ by } -2. \\ 2c + 15w &= 31 \\ \hline -5w &= -7 && \text{Add the equations to eliminate } c. \\ w &= 1.4 && \text{Divide both sides by } -5. \text{ This is the cost of a wallet-size picture.} \end{aligned}$$

To find c, substitute 1.4 for w in the first equation of the original system.

$$\begin{aligned} c + 10w &= 19 \\ c + 10(1.4) &= 19 && \text{Substitute 1.4 for } w. \\ c + 14 &= 19 && \text{Multiply.} \\ c &= 5 && \text{Subtract 14 from both sides. This is the cost of a class picture.} \end{aligned}$$

State the Conclusion A class picture costs \$5 and a wallet-size picture costs \$1.40.

Check the Results Package 1 has 1 class picture and 10 wallets: \$5 + 10(\$1.40) = \$5 + \$14 = \$19. Package 2 has 2 class pictures and 15 wallets: 2(\$5) + 15(\$1.40) = \$10 + \$21 = \$31. The results check.

THE BREAK POINT

EXAMPLE 5

ELEMENTARY Algebra $f(x)$ Now™

Manufacturing. The setup cost of a machine that mills brass plates is \$750. After setup, it costs \$0.25 to mill each plate. Management is considering the purchase of a larger machine that can produce the same plate at a cost of \$0.20 per plate. If the setup cost of the larger machine is \$1,200, how many plates would the company have to produce to make the purchase worthwhile?

Analyze the Problem We need to find the number of plates (called the **break point**) that will cost equal amounts to produce on either machine.

Form Two Equations We can let c represent the cost of milling p plates. If we call the machine currently being used machine 1, and the new, larger one machine 2, we can form the two equations.

The cost of making p plates on machine 1	is	the setup cost of machine 1	plus	the cost per plate of machine 1	times	the number of plates p to be made.
c	$=$	750	$+$	0.25	$\cdot$	p

The cost of making p plates on machine 2	is	the setup cost of machine 2	plus	the cost per plate of machine 2	times	the number of plates p to be made.
c	$=$	1,200	$+$	0.20	$\cdot$	p

Solve the System Since the costs are equal, we can use the substitution method to solve the system

$$\begin{cases} c = 750 + 0.25p \\ c = 1{,}200 + 0.20p \end{cases}$$

$750 + 0.25p = 1{,}200 + 0.20p$	Substitute $750 + 0.25p$ for c in the second equation.
$0.25p = 450 + 0.20p$	Subtract 750 from both sides.
$0.05p = 450$	Subtract $0.20p$ from both sides.
$p = 9{,}000$	Divide both sides by 0.05.

State the Conclusion If 9,000 plates are milled, the cost will be the same on either machine. If more than 9,000 plates are milled, the cost will be cheaper on the larger machine, because it mills the plates less expensively than the smaller machine.

Check the Result We check the solution by substituting 9,000 for p in each equation of the system and verifying that 3,000 is the value of c in both cases.

If we graph the two equations, we can illustrate the break point.

Machine 1
$c = 750 + 0.25p$

p	c
0	750
1,000	1,000
5,000	2,000

Machine 2
$c = 1{,}200 + 0.20p$

p	c
0	1,200
4,000	2,000
12,000	3,600

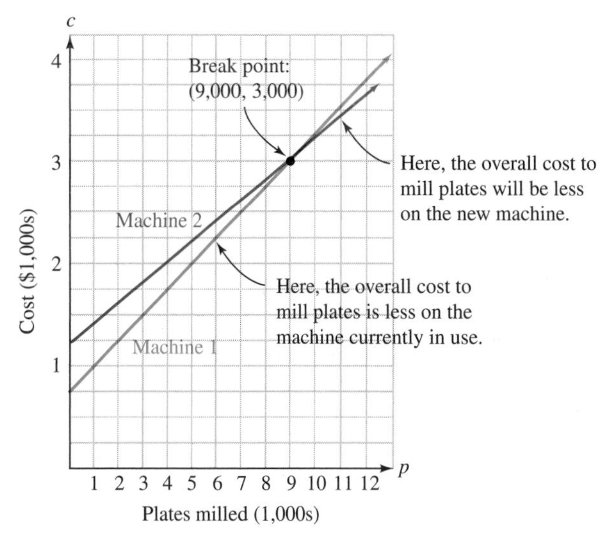

INTEREST, UNIFORM MOTION, AND MIXTURE PROBLEMS

EXAMPLE 6

ELEMENTARY Algebra f(x) Now™

White-collar crime. Investigators discovered that a company secretly moved $150,000 out of the country to avoid paying income tax. Some of the money was invested in a Swiss bank account that paid 8% interest annually. The remainder was deposited in a Cayman Islands account, paying 7% annual interest. The investigation also revealed that the combined interest earned the first year was $11,500. How much money was invested in each account?

Analyze the Problem We are told that an unknown part of the $150,000 was invested at an annual rate of 8% and the rest at 7%. Together, the accounts earned $11,500 in interest.

Caution

In Example 6, it is incorrect to let

~~x = the amount invested in each account~~

This implies that *equal amounts* were invested in the Swiss and Cayman Island accounts. We do not know that.

Form Two Equations Let x = the amount invested in the Swiss account and y = the amount invested in the Cayman Islands account. Because the total investment was \$150,000, we have

The amount invested in the Swiss account	plus	the amount invested in the Cayman Is. account	is	\$150,000.
x	$+$	y	$=$	150,000

We can use the formula $I = Prt$ to determine that x dollars invested for 1 year at 8% earns $x \cdot 0.08 \cdot 1 = 0.08x$ dollars. Similarly, y dollars invested for 1 year at 7% earns $y \cdot 0.07 \cdot 1 = 0.07y$ dollars. If the total combined interest earned was \$11,500, we have

The income on the 8% investment	plus	the income on the 7% investment	is	\$11,500.
$0.08x$	$+$	$0.07y$	$=$	11,500

The resulting system is $\begin{cases} x + y = 150{,}000 \\ 0.08x + 0.07y = 11{,}500 \end{cases}$.

Solve the System To solve the system, clear the second equation of decimals. Then eliminate x.

$$\begin{aligned} -8x - 8y &= -1{,}200{,}000 && \text{Multiply both sides of the first equation by } -8. \\ 8x + 7y &= 1{,}150{,}000 && \text{Multiply both sides of the second equation by 100.} \\ \hline -y &= -50{,}000 \\ y &= 50{,}000 && \text{Multiply both sides by } -1. \end{aligned}$$

To find x, substitute 50,000 for y in the first equation of the original system.

$$\begin{aligned} x + y &= 150{,}000 \\ x + \mathbf{50{,}000} &= 150{,}000 && \text{Substitute.} \\ x &= 100{,}000 && \text{Subtract 50,000 from both sides.} \end{aligned}$$

State the Conclusion \$100,000 was invested in the Swiss bank account, and \$50,000 was invested in the Cayman Islands account.

Check the Results

$$\begin{aligned} \$100{,}000 + \$50{,}000 &= \$150{,}000 && \text{The two investments total \$150,000.} \\ 0.08(\$100{,}000) &= \$8{,}000 && \text{The Swiss bank account earned \$8,000.} \\ 0.07(\$50{,}000) &= \$3{,}500 && \text{The Cayman Islands account earned \$3,500.} \end{aligned}$$

The combined interest is \$8,000 + \$3,500 = \$11,500. The results check.

EXAMPLE 7

Boating. A boat traveled 30 miles downstream in 3 hours and made the return trip in 5 hours. Find the speed of the boat in still water and the speed of the current.

Analyze the Problem Traveling downstream, the speed of the boat will be *faster* than it would be in still water. Traveling upstream, the speed of the boat will be *slower* than it would be in still water.

Form Two Equations Let s = the speed of the boat in still water and c = the speed of the current. Then the speed of the boat going downstream is $s + c$ and the speed of the boat going upstream is $s - c$. We can organize the facts of the problem in a table.

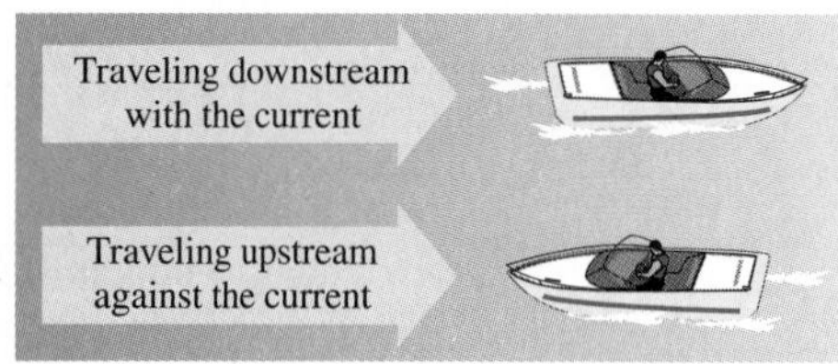

	Rate	· Time	= Distance
Downstream	$s + c$	3	$3(s + c)$
Upstream	$s - c$	5	$5(s - c)$

Enter this information first. (Rate and Time columns)

Set each of these expressions for distance traveled equal to 30.

Since each trip is 30 miles long, the Distance column of the table gives two equations in two variables. To write each equation in general form, use the distributive property.

$$\begin{cases} 3(s + c) = 30 \\ 5(s - c) = 30 \end{cases} \xrightarrow{\text{Distribute}} \begin{cases} 3s + 3c = 30 \\ 5s - 5c = 30 \end{cases}$$

Solve the System To eliminate c, we proceed as follows.

$$\begin{aligned} 15s + 15c &= 150 && \text{Multiply both sides of } 3s + 3c = 30 \text{ by 5.} \\ 15s - 15c &= 90 && \text{Multiply both sides of } 5s - 5c = 30 \text{ by 3.} \\ \hline 30s &= 240 \\ s &= 8 && \text{Divide both sides by 30. This is the speed of the boat in still water.} \end{aligned}$$

To find c, it appears that the computations will be easiest if we use $3s + 3c = 30$.

$$\begin{aligned} 3s + 3c &= 30 \\ 3(8) + 3c &= 30 && \text{Substitute 8 for } s. \\ 24 + 3c &= 30 && \text{Multiply.} \\ 3c &= 6 && \text{Subtract 24 from both sides.} \\ c &= 2 && \text{Divide both sides by 3. This is the speed of the current.} \end{aligned}$$

State the Conclusion The speed of the boat in still water is 8 mph and the speed of the current is 2 mph.

Check the Results With a 2-mph current, the boat's downstream speed will be $8 + 2 = 10$ mph. In 3 hours, it will travel $10 \cdot 3 = 30$ miles. With a 2-mph current, the boat's upstream speed will be $8 - 2 = 6$ mph. In 5 hours, it will cover $6 \cdot 5 = 30$ miles. The results check.

EXAMPLE 8

Medical technology. A laboratory technician has one batch of antiseptic that is 40% alcohol and a second batch that is 60% alcohol. She would like to make 8 liters of solution that is 55% alcohol. How many liters of each batch should she use?

Analyze the Problem Some 60% solution must be added to some 40% solution to make a 55% solution.

Form Two Equations Let x = the number of liters to be used from batch 1 and y = the number of liters to be used from batch 2. We can organize the facts of the problem in a table.

	Amount	· Strength =	Amount of alcohol
Batch 1 (too weak)	x	0.40	$0.40x$
Batch 2 (too strong)	y	0.60	$0.60y$
Mixture	8	0.55	0.55(8)

↑ One equation comes from information in this column.

↑ 40%, 60%, and 55% have been expressed as decimals.

↑ Another equation comes from information in this column.

The information in the table provides two equations.

$$\begin{cases} x + y = 8 \\ 0.40x + 0.60y = 0.55(8) \end{cases}$$

The number of liters of batch 1 plus the number of liters of batch 2 equals the total number of liters in the mixture.

The amount of alcohol in batch 1 plus the amount of alcohol in batch 2 equals the amount of alcohol in the mixture.

Solve the System We can solve this system by elimination. To eliminate x, we proceed as follows.

$$\begin{aligned} -40x - 40y &= -320 \\ 40x + 60y &= 440 \\ \hline 20y &= 120 \\ y &= 6 \end{aligned}$$

Multiply both sides of the first equation by -40.

Multiply both sides of the second equation by 100.

Divide both sides by 20. This is the number of liters of batch 2 needed.

To find x, we substitute 6 for y in the first equation of the original system.

$x + y = 8$

$x + 6 = 8$ Substitute.

$x = 2$ Subtract 6 from both sides. This is the number of liters of batch 1 needed.

State the Conclusion The technician should use 2 liters of the 40% solution and 6 liters of the 60% solution.

Check the Results Note that 2 liters + 6 liters = 8 liters, the required number. Also, the amount of alcohol in the two solutions is equal to the amount of alcohol in the mixture. The results check.

Alcohol in batch 1: $0.40x = 0.40(2) = 0.8$ liters

Alcohol in batch 2: $0.60y = 0.60(6) = 3.6$ liters

Total: 4.4 liters

Alcohol in the mixture: $0.55(8) = 4.4$ liters ■

7.4 STUDY SET

ELEMENTARY Algebra Now™

VOCABULARY Fill in the blanks.

1. A ________ is a letter that stands for a number.
2. An ________ is a statement indicating that two quantities are equal.
3. $\begin{cases} a + b = 20 \\ a = 2b + 4 \end{cases}$ is a ________ of equations.
4. A ________ of a system of equations satisfies both equations.
5. Two angles are said to be ____________ if the sum of their measures is 90°.
6. Two angles are said to be ____________ if the sum of their measures is 180°.

CONCEPTS

7. A length of pipe is to be cut into two pieces. The longer piece is to be 1 foot less than twice the shorter piece. Write two equations that model the situation.

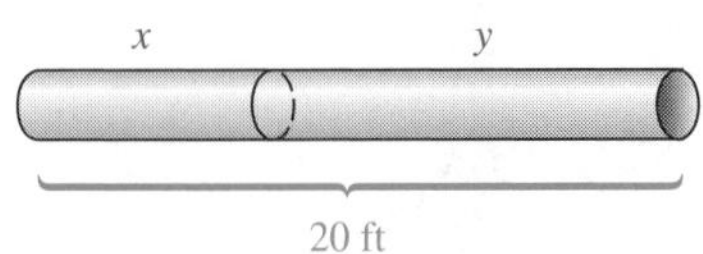

8. Two angles are complementary. The measure of the larger angle is four times the measure of the smaller angle. Write two equations that model the situation.

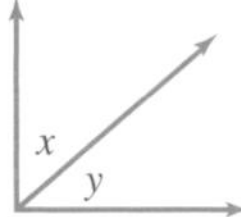

9. Two angles are supplementary. The measure of the smaller angle is 25° less than the measure of the larger angle. Write two equations that model the situation.

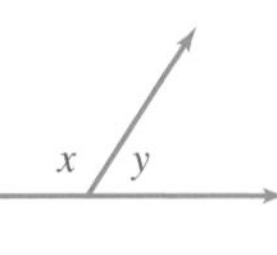

10. The perimeter of a ping pong table is 28 feet. The length is 4 feet more than the width. Write two equations that model the situation.

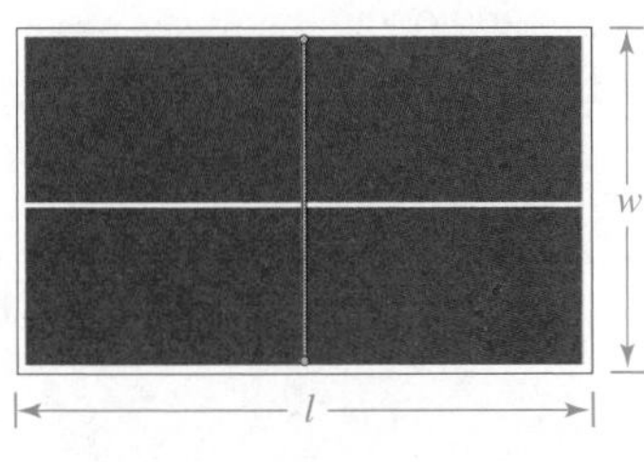

11. Let x = the cost of a chicken taco and y = the cost of a beef taco. Write an equation that models the offer shown in the advertisement.

12. a. Complete the following table.

	Principal ·	Rate ·	Time =	Interest
City Bank	x	5%	1 yr	
USA Savings	y	11%	1 yr	

b. A total of $50,000 was deposited in the two accounts. Use that information to write an equation about the principal.

c. A total of $4,300 was earned by the two accounts. Use that information to write an equation about the interest.

13. In still water, a man can paddle a canoe so that it travels x mph. What will the speed of the canoe be for each situation shown?

a. Downstream

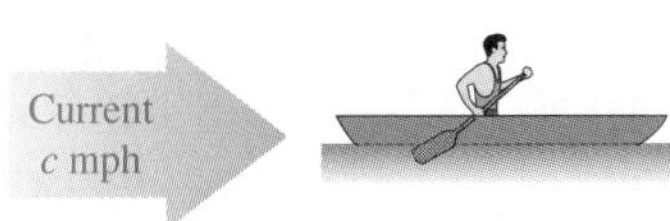

b. Upstream

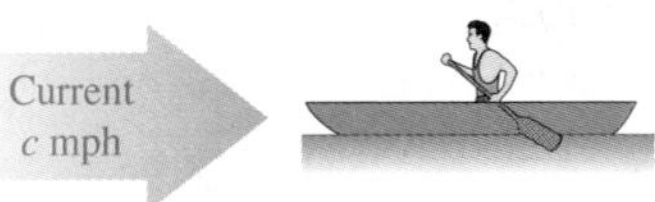

14. Complete the table, which contains information about an airplane flying in windy conditions.

	Rate ·	Time =	Distance
With wind	$x + y$	3	
Against wind	$x - y$	5	

15. a. If the contents of the two test tubes are poured into a third tube, how much solution will the third tube contain?

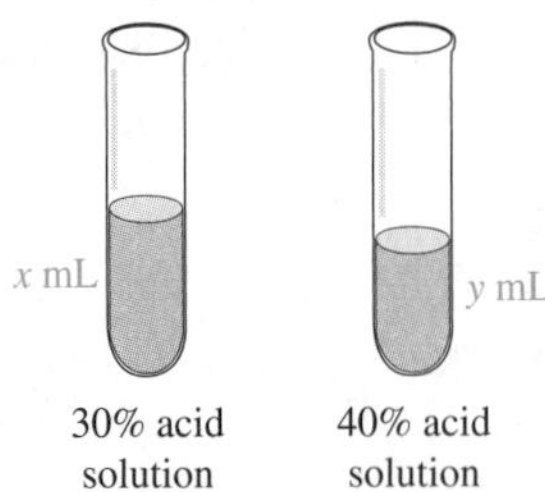

b. Which of the following strengths could the mixture possibly be: 27%, 33%, or 44% acid solution?

16. a. Complete the table, which contains information about mixing two salt solutions to get 12 gallons of a 3% salt solution.

	Amount ·	Strength =	Amount of salt
Weak	x	0.01	
Strong	y	0.06	
Mix			

b. Use the information from the Amount column to write an equation.

c. Use the information from the Amount of salt column to write an equation.

PRACTICE

17. COMPLEMENTARY ANGLES Two angles are complementary. The measure of one angle is 10° more than three times the measure of the other. Find the measure of each angle.

18. SUPPLEMENTARY ANGLES Two angles are supplementary. The measure of one angle is 20° less than 19 times the measure of the other. Find the measure of each angle.

19. SUPPLEMENTARY ANGLES The difference of the measures of two supplementary angles is 80°. Find the measure of each angle.

20. COMPLEMENTARY ANGLES Two angles are complementary. The measure of one angle is 15° more than one-half of the measure of the other. Find the measure of each angle.

APPLICATIONS **Write a system of two equations in two variables to solve each problem.**

21. TREE TRIMMING When fully extended, the arm on a tree service truck is 51 feet long. If the upper part of the arm is 7 feet shorter than the lower part, how long is each part of the arm?

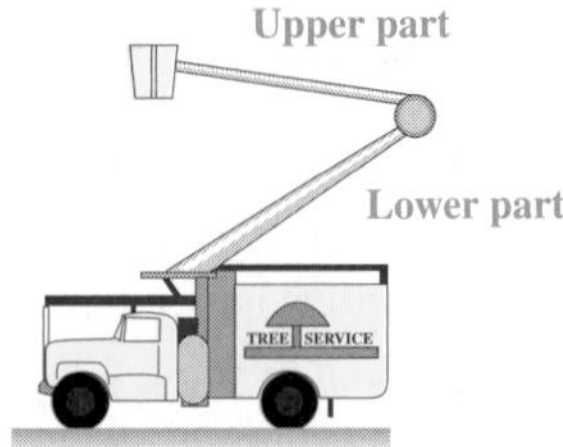

22. ALASKA Most of the 1,422-mile-long Alaskan Highway is actually in Canada. Find the length of the highway that is in Alaska and the length of the highway that is in Canada, if it is known that the difference in the lengths is 1,020 miles.

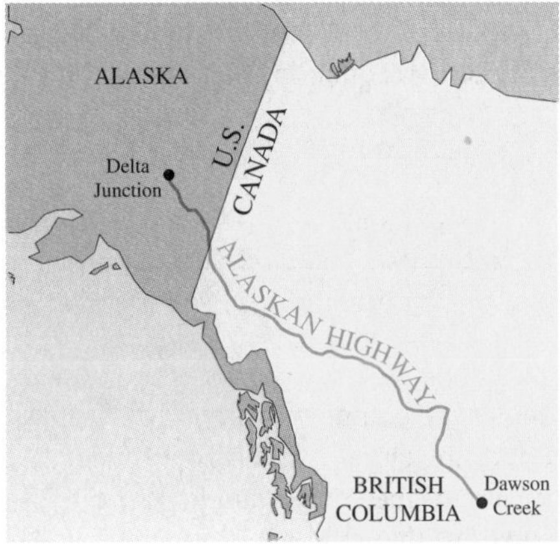

23. EXECUTIVE BRANCH The salaries of the president and vice president of the United States total $592,600 a year. If the president makes $207,400 more than the vice president, find each of their salaries.

24. CAUSES OF DEATH According to the *National Vital Statistics Reports,* in 2001, the number of Americans who died from heart disease was about 7 times the number who died from accidents. If the number of deaths from these two causes totaled approximately 800,000, how many Americans died from each cause in 2001?

25. MARINE CORPS The Marine Corps War Memorial in Arlington, Virginia, portrays the raising of the U.S. flag on Iwo Jima during World War II. Find the measures of the two angles shown if the measure of $\angle 2$ is 15° less than twice the measure of $\angle 1$.

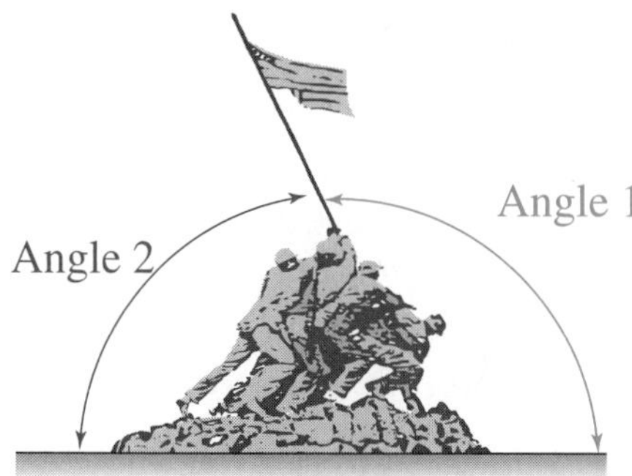

26. PHYSICAL THERAPY To rehabilitate her knee, an athlete does leg extensions. Her goal is to regain a full 90° range of motion in this exercise. Use the information in the illustration to determine her current range of motion in degrees and the number of degrees of improvement she still needs to make.

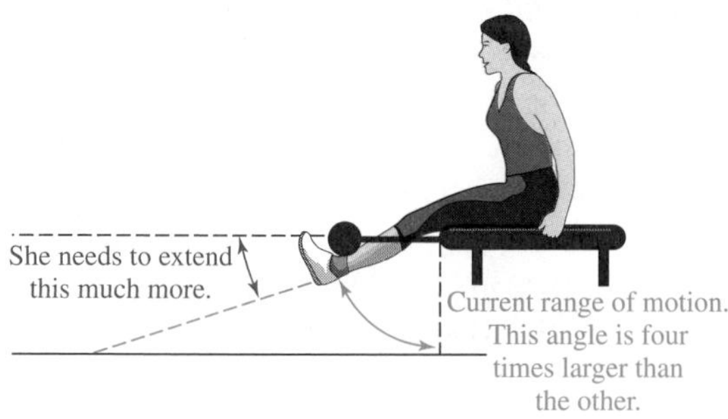

27. THEATER SCREENS At an IMAX theater, the giant rectangular movie screen has a width 26 feet less than its length. If its perimeter is 332 feet, find the length and the width of the screen.

28. ENGLISH ARTISTS In 1770, Thomas Gainsborough painted *The Blue Boy.* The sum of the length and width of the painting is 118 inches. The difference of the length and width is 22 inches. Find the length and width.

29. GEOMETRY A 50-meter path surrounds a rectangular garden. The width of the garden is two-thirds its length. Find the length and width.

30. BALLROOM DANCING A rectangular-shaped dance floor has a perimeter of 200 feet. If the floor were 20 feet wider, its width would equal its length. Find the length and width of the dance floor.

31. BUYING PAINTING SUPPLIES Two partial receipts for paint supplies are shown. (Assume no sales tax was charged.) Find the cost of one gallon of paint and the cost of one paint brush.

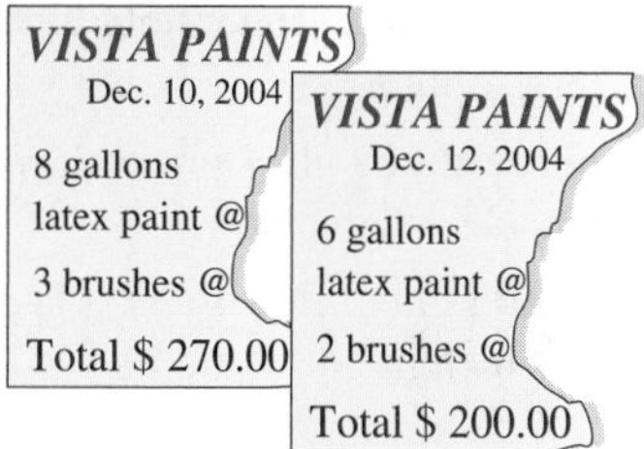

32. WEDDING PICTURES A photographer sells the two wedding picture packages shown. How much does a 10×14 photo cost? How much does an 8×10 photo cost?

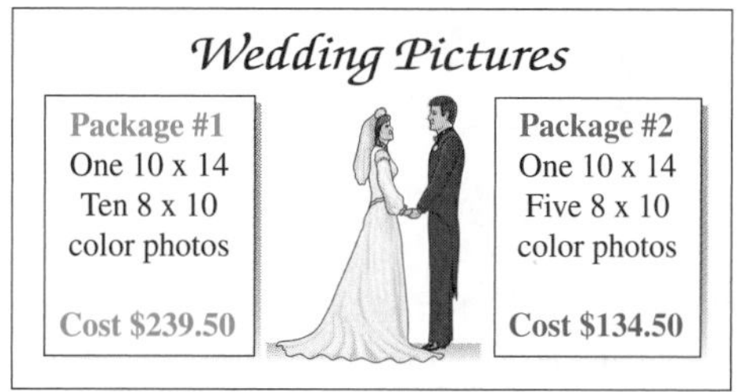

33. SELLING ICE CREAM At a store, ice cream cones cost $0.90 and sundaes cost $1.65. One day the receipts for a total of 148 cones and sundaes were $180.45. How many cones were sold? How many sundaes?

34. BUYING TICKETS The ticket prices for a movie are shown on the next page. Receipts for one showing were $1,440 for an audience of 190 people. How many general admission tickets and how many senior citizen tickets were sold?

35. MAKING TIRES A company has two molds to form tires. One mold has a setup cost of $1,000 and the other has a setup cost of $3,000. The cost to make each tire with the first mold is $15, and the cost to make each tire with the second mold is $10.
   a. Find the break point.
   b. Check your result by graphing both equations on the coordinate system below.
   c. If a production run of 500 tires is planned, determine which mold should be used.

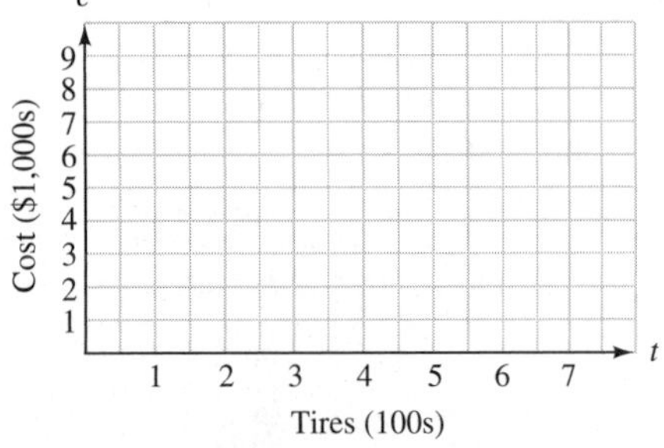

36. CHOOSING A FURNACE A high-efficiency 90+ furnace can be purchased for $2,250 and costs an average of $412 per year to operate in Rockford, Illinois. An 80+ furnace can be purchased for only $1,710, but it costs $466 per year to operate.
   a. Find the break point.
   b. If you intended to live in a Rockford house for 7 years, which furnace would you choose?

37. TELEPHONE RATES A long distance provider offers two plans. Plan 1 has a monthly fee of $10 plus 8¢ per minute. Plan 2 has a monthly fee of $15 plus 4¢ a minute. For what number of minutes will the costs of the plans be the same?

38. SUPPLY AND DEMAND Suppose that the number c of cheesecakes that a bakery will supply a day is given by $c = 60p - 200$, and the number c of cheesecakes that are purchased a day is given by $c = -40p + 1{,}000$, where p is the price (in dollars) of a cheesecake. For what price will supply equal demand?

39. THE GULF STREAM The Gulf Stream is a warm ocean current of the North Atlantic Ocean that flows northward, as shown below. Heading north with the Gulf Stream, a cruise ship traveled 300 miles in 10 hours. Against the current, it took 15 hours to make the return trip. Find the speed of the ship in still water and the speed of the current.

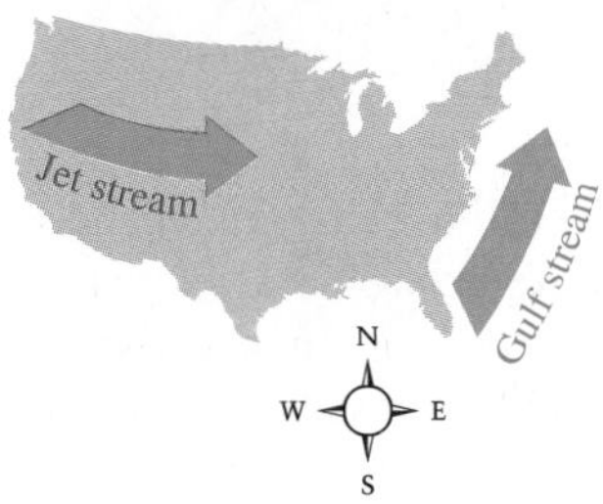

40. THE JET STREAM The jet stream is a strong wind current that flows across the United States, as shown above. Flying with the jet stream, a plane flew 3,000 miles in 5 hours. Against the same wind, the trip took 6 hours. Find the speed of the plane in still air and the speed of the wind current.

41. AVIATION An airplane can fly with the wind a distance of 800 miles in 4 hours. However, the return trip against the wind takes 5 hours. Find the speed of the plane in still air and the speed of the wind.

42. BOATING A boat can travel 24 miles downstream in 2 hours and can make the return trip in 3 hours. Find the speed of the boat in still water and the speed of the current.

43. STUDENT LOANS A college used a $5,000 gift from an alumnus to make two student loans. The first was at 5% annual interest to a nursing student. The second was at 7% to a business major. If the college collected $310 in interest the first year, how much was loaned to each student?

44. FINANCIAL PLANNING In investing \$6,000 of a couple's money, a financial planner put some of it into a savings account paying 6% annual interest. The rest was invested in a riskier mini-mall development plan paying 12% annually. The combined interest earned for the first year was \$540. How much money was invested at each rate?

45. MARINE BIOLOGY A marine biologist wants to set up an aquarium containing 3% salt water. He has two tanks on hand that contain 6% and 2% salt water. How much water from each tank must he use to fill a 16-liter aquarium with a 3% saltwater mixture?

46. COMMEMORATIVE COINS A foundry has been commissioned to make souvenir coins. The coins are to be made from an alloy that is 40% silver. The foundry has on hand two alloys, one with 50% silver content and one with a 25% silver content. How many kilograms of each alloy should be used to make 20 kilograms of the 40% silver alloy?

47. COFFEE SALES A coffee supply store waits until the orders for its special blend reach 100 pounds before making up a batch. Columbian coffee selling for \$8.75 a pound is blended with Brazilian coffee selling for \$3.75 a pound to make a product that sells for \$6.35 a pound. How much of each type of coffee should be used to make the blend that will fill the orders?

	Amount ·	Price =	Total value
Columbian	x	8.75	$8.75x$
Brazilian	y	3.75	$3.75y$
Mixture	100	6.35	100(6.35)

48. MIXING NUTS A merchant wants to mix peanuts with cashews, as shown in the illustration, to get 48 pounds of mixed nuts that will be sold at \$4 per pound. How many pounds of each should the merchant use?

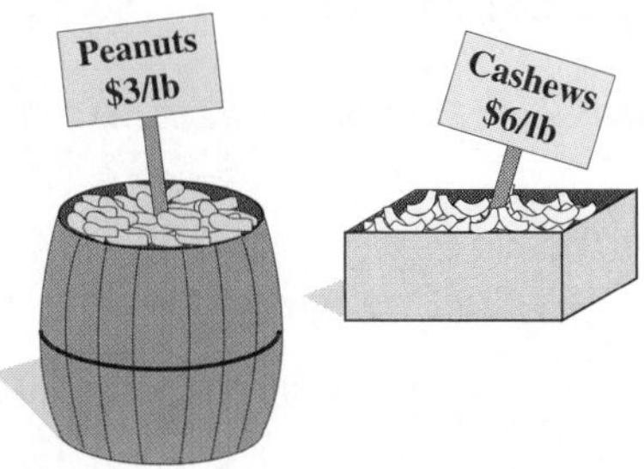

WRITING

49. What is a *break point?* Give an example.

50. A man paid \$89 for two shirts and four pairs of socks. If we let $x =$ the cost of a shirt and $y =$ the cost of a pair of socks, an equation modeling the purchase is $2x + 4y = 89$. Explain why there is not enough information to determine the cost of a shirt or the cost of a pair of socks.

REVIEW **Graph each inequality. Then describe the graph using interval notation.**

51. $x < 4$

52. $x \geq -3$

53. $-1 < x \leq 2$

54. $-2 \leq x \leq 0$

CHALLENGE PROBLEMS

55. On the last scale, how many nails will it take to balance 1 nut?

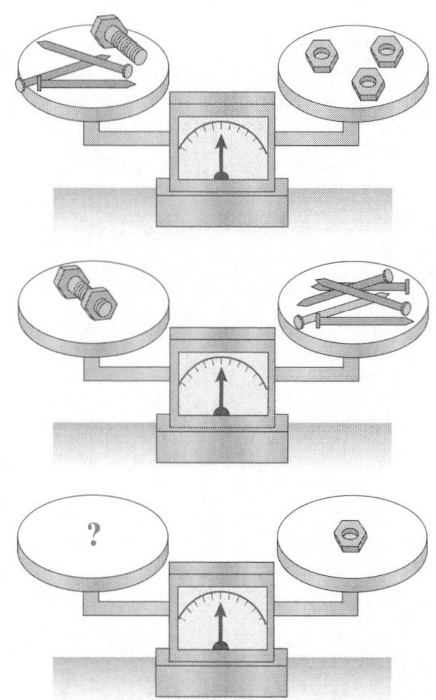

56. FARMING In a pen of goats and chickens, there are 40 heads and 130 feet. How many goats and chickens are in the pen?

7.5 Solving Systems of Linear Inequalities

- Graphing Systems of Linear Inequalities
- An Application

In Section 7.1, we solved systems of linear equations graphically by finding the point of intersection of two lines. Now we consider systems of linear inequalities, such as

$$\begin{cases} x + y \geq -1 \\ x - y \geq 1 \end{cases}$$

To solve systems of linear inequalities, we again find the points of intersection of graphs. In this case, however, we are not looking for an intersection of two lines, but an intersection of two half-planes.

GRAPHING SYSTEMS OF LINEAR INEQUALITIES

A solution of a **system of linear inequalities** is an ordered pair that satisfies each inequality. *To solve a system of linear inequalities* means to find all of its solutions. This can be done by graphing each inequality on the same set of axes and finding the points that are common to every graph in the system.

EXAMPLE 1

Graph the solutions of the system: $\begin{cases} x + y \geq -1 \\ x - y \geq 1 \end{cases}$.

ELEMENTARY Algebra $f(x)$ Now™

Solution To graph $x + y \geq -1$, we begin by graphing the boundary line $x + y = -1$. Since the inequality contains a $\geq$ symbol, the boundary is a solid line. Because the coordinates of the test point (0, 0) satisfy $x + y \geq -1$, we shade (in red) the side of the boundary that contains (0, 0). See part (a) of the figure.

Graph the boundary: The intercept method

$$x + y = -1$$

x	y	(x, y)
0	−1	(0, −1)
−1	0	(−1, 0)

***Shading: Check the test point* (0, 0)**

$$x + y \geq -1$$
$$0 + 0 \stackrel{?}{\geq} -1 \quad \text{Substitute.}$$
$$0 \geq -1$$

(0, 0) is a solution of $x + y \geq -1$.

The Language of Algebra

To solve a system of linear inequalities, we *superimpose* the graphs of the inequalities. That is, we place one graph over the other. Most video camcorders can *superimpose* the date and time over the picture being recorded.

In part (b) of the figure, we superimpose the graph of $x - y \geq 1$ on the graph of $x + y \geq -1$ so that we can determine the points that the graphs have in common. To graph $x - y \geq 1$, we graph the boundary $x - y = 1$ as a solid line. Since the test point (0, 0) does not satisfy $x - y \geq 1$, we shade (in blue) the half-plane that does not contain (0, 0).

Graph the boundary: The intercept method

$$x - y = 1$$

x	y	(x, y)
0	−1	(0, −1)
1	0	(1, 0)

***Shading: Check the test point* (0, 0)**

$$x - y \geq 1$$
$$0 - 0 \stackrel{?}{\geq} 1 \quad \text{Substitute.}$$
$$0 \geq 1$$

(0, 0) is not a solution of $x - y \geq 1$.

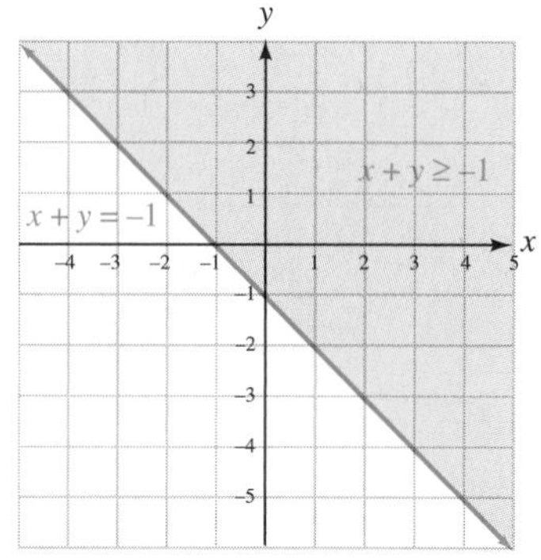

The graph of $x + y \geq -1$ is shaded in red.

(a)

The solutions of the system are shaded in purple. The purple region is the intersection or overlap of the red and blue shaded regions. It includes portions of each boundary.

The graph of $x - y \geq 1$ is shaded in blue. It is drawn over the graph of $x + y \geq -1$.

(b)

Success Tip

Colored pencils are often used to graph systems of inequalities. A standard pencil can also be used. Just draw different patterns of lines instead of shading.

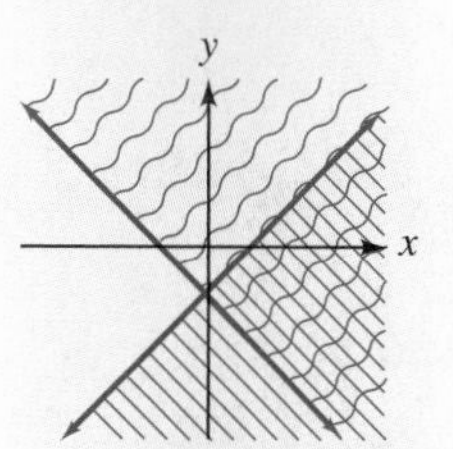

In part (b) of the figure, the area that is shaded twice represents the solutions of the given system. Any point in the doubly shaded region in purple (including the purple portions of each boundary) has coordinates that satisfy both inequalities.

Since there are infinitely many solutions, we cannot check each of them. However, as an informal check, we can select one ordered pair, say (4, 1), that lies in the doubly shaded region and show that its coordinates satisfy both inequalities of the system.

Check:

$$x + y \geq -1 \qquad x - y \geq 1$$
$$4 + 1 \overset{?}{\geq} -1 \qquad 4 - 1 \overset{?}{\geq} 1$$
$$5 \geq -1 \qquad 3 \geq 1$$

The resulting true statements verify that (4, 1) is a solution of the system. If we pick a point that is not in the doubly shaded region, such as (1, 3), (−2, −2), or (0, −4), the coordinates of that point will fail to satisfy one or both of the inequalities.

Self Check 1 Graph the solutions of the system: $\begin{cases} x - y \leq 2 \\ x + y \geq -1 \end{cases}$.

In general, to solve systems of linear inequalities, we will follow these steps.

Solving Systems of Linear Inequalities

1. Graph each inequality on the same rectangular coordinate system.
2. Use shading to highlight the intersection of the graphs (the region where the graphs overlap). The points in this region are the solutions of the system.
3. As an informal check, pick a point from the region and verify that its coordinates satisfy each inequality of the original system.

EXAMPLE 2

ELEMENTARY Algebra f(x) Now™

Graph the solutions of the system: $\begin{cases} y > 3x \\ 2x + y < 4 \end{cases}$.

Solution To graph $y > 3x$, we begin by graphing the boundary line $y = 3x$. Since the inequality contains a $>$ symbol, the boundary is a dashed line. Because the boundary passes through (0, 0), we use (2, 0) as the test point instead. Since (2, 0) does not satisfy $y > 3x$, we shade (in red) the half-plane that does not contain (2, 0). See part (a) of the following figure.

Graph the boundary: Slope and y-intercept

$$y = 3x + 0$$

$m = 3$ $\quad b = 0$

Slope: $\frac{3}{1}$ $\quad$ y-intercept: $(0, 0)$

Shading: Check the test point (2, 0)

$y > 3x$

$\mathbf{0} > 3(\mathbf{2})$ $\quad$ Substitute.

$0 > 6$

Since $0 > 6$ is false, $(2, 0)$ is not a solution of $y > 3x$.

In part (b) of the figure, we superimpose the graph of $2x + y < 4$ on the graph of $y > 3x$ to determine the points that the graphs have in common. To graph $2x + y < 4$, we graph the boundary $2x + y = 4$ as a dashed line. Then we shade (in blue) the half-plane that contains $(0, 0)$, because the coordinates of the test point $(0, 0)$ satisfy $2x + y < 4$.

Graph the boundary: The intercept method

$$2x + y = 4$$

x	y	(x, y)
0	4	(0, 4)
2	0	(2, 0)

Shading: Use the test point (0, 0)

$2x + y < 4$

$2(\mathbf{0}) + \mathbf{0} \overset{?}{<} 4$ $\quad$ Substitute.

$0 < 4$

Since $0 < 4$ is true, $(0, 0)$ is a solution of $2x + y < 4$.

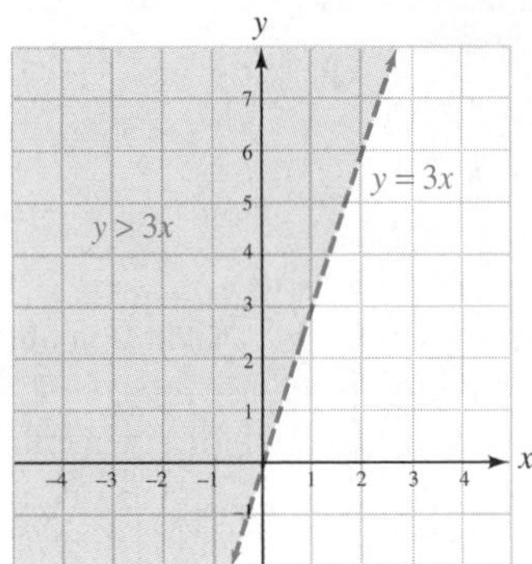

The graph of $y > 3x$ is shaded in red.

(a)

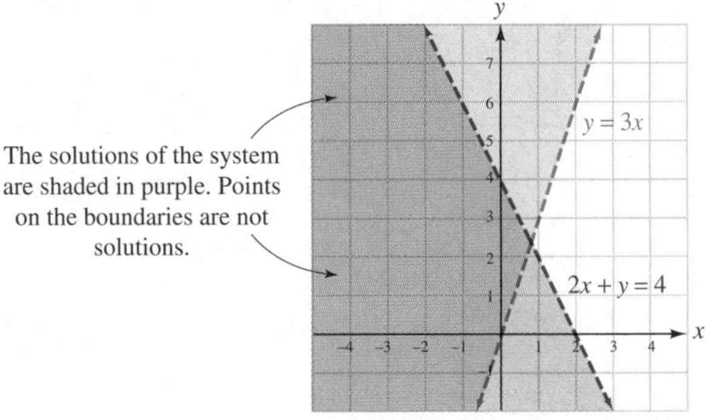

The graph of $2x + y < 4$ is shaded in blue. It is drawn over the graph of $y > 3x$.

(b)

In part (b) of the figure, the area that is shaded twice represents the solutions of the given system. Any point in the doubly shaded region in purple has coordinates that satisfy both inequalities. Pick a point in the region and show that this is true. Note that the region does not include either boundary; points on the boundaries are not solutions of the system.

Self Check 2 Graph the solutions of the system: $\begin{cases} x + 3y < 3 \\ y > \frac{1}{3}x \end{cases}$.

EXAMPLE 3

Graph the solutions of the system: $\begin{cases} x \leq 2 \\ y > 3 \end{cases}$.

ELEMENTARY Algebra f(x) Now™

Solution The boundary of the graph of $x \leq 2$ is the line $x = 2$. Since the inequality contains the symbol $\leq$, we draw the boundary as a solid line. The test point $(0, 0)$ makes $x \leq 2$ true, so we shade the side of the boundary that contains $(0, 0)$. See part (a) of the figure.

Graph the boundary: A table of solutions

$$x = 2$$

x	y	(x, y)
2	0	(2, 0)
2	2	(2, 2)
2	4	(2, 4)

***Shading: Check the test point* (0, 0)**

$$x \leq 2$$
$$0 \leq 2$$

Since $0 \leq 2$ is true, (0, 0) is a solution of $x \leq 2$.

In part (b) of the figure, the graph of $y > 3$ is superimposed over the graph of $x \leq 2$. The boundary of the graph of $y > 3$ is the line $y = 3$. Since the inequality contains the symbol $>$, we draw the boundary as a dashed line. The test point (0, 0) makes $y > 3$ false, so we shade the side of the boundary that does not contain (0, 0).

Graph the boundary: A table of solutions

$$y = 3$$

x	y	(x, y)
0	3	(0, 3)
1	3	(1, 3)
4	3	(4, 3)

***Shading: Check the test point* (0, 0)**

$$y > 3$$
$$0 > 3$$

Since $0 > 3$ is false, (0, 0) is not a solution of $y > 3$.

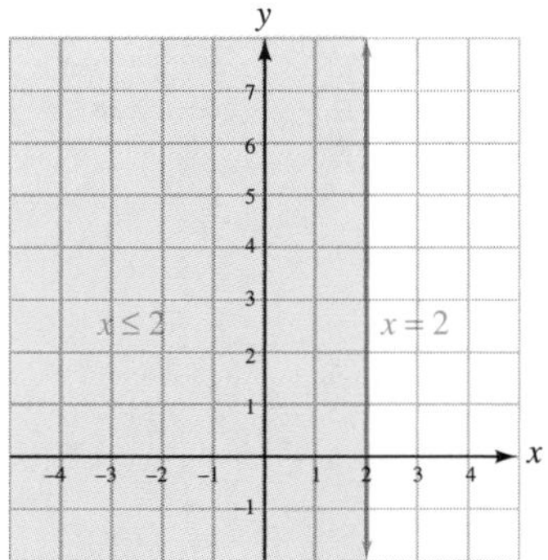

The graph of $x \leq 2$ is shaded in red.

(a)

The solutions of the system are shaded in purple. Points on the purple portion of $x = 2$ are solutions. Points on the dashed boundary line are not.

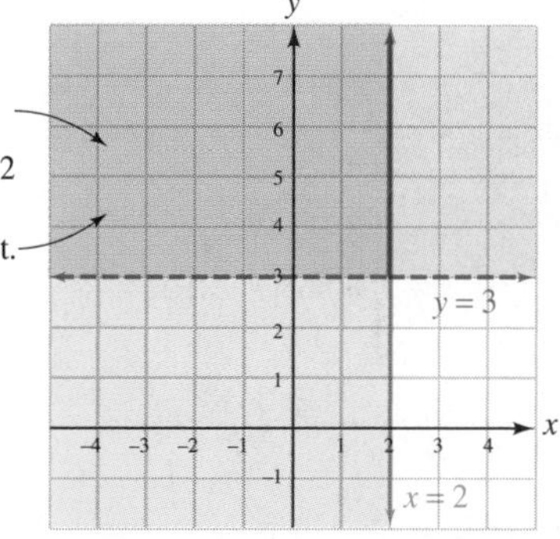

The graph of $y > 3$ is shaded in blue. It is drawn over the graph of $x \leq 2$.

(b)

The area that is shaded twice represents the solutions of the system of inequalities. Any point in the doubly shaded region in purple has coordinates that satisfy both inequalities, including the purple portion of the $x = 2$ boundary. Pick a point in the region and show that this is true.

Self Check 3 Graph the solutions of the system: $\begin{cases} y \leq 1 \\ x > 2 \end{cases}$.

EXAMPLE 4

ELEMENTARY Algebra $f(x)$ Now™

Graph the solutions of the system: $\begin{cases} x \geq 0 \\ y \geq 0 \\ x + 2y \leq 6 \end{cases}$.

Solution This is a system of three linear inequalities. If shading is used to graph them on the same set of axes, it can become difficult to interpret the results. Instead, we can draw directional arrows attached to each boundary line in place of the shading.

- The graph of $x \geq 0$ has the boundary $x = 0$ and includes all points on the y-axis and to the right.
- The graph of $y \geq 0$ has the boundary $y = 0$ and includes all points on the x-axis and above.
- The graph of $x + 2y \leq 6$ has the boundary $x + 2y = 6$. Because the coordinates of the origin satisfy $x + 2y \leq 6$, the graph includes all points on and below the boundary.

The solutions of the system are the points that lie on triangle OPQ and the shaded triangular region that it encloses.

$x + 2y = 6$

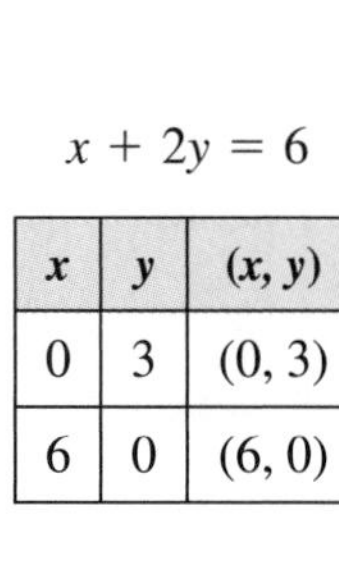

x	y	(x, y)
0	3	(0, 3)
6	0	(6, 0)

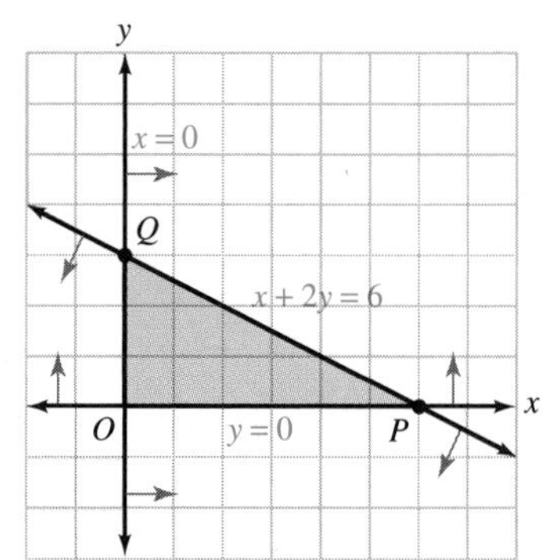

Self Check 4 Graph the solutions of the system: $\begin{cases} x \leq 1 \\ y \leq 2 \\ 2x - y \leq 4 \end{cases}$.

AN APPLICATION

EXAMPLE 5

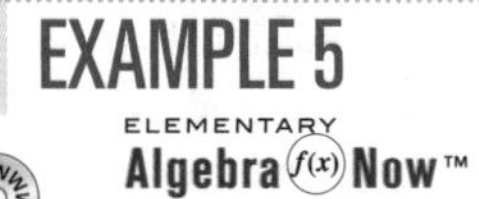

Landscaping. A homeowner budgets from \$300 to \$600 for trees and bushes to landscape his yard. After shopping around, he finds that good trees cost \$150 and mature bushes cost \$75. What combinations of trees and bushes can he afford to buy?

Analyze the Problem

- At least \$300 but not more than \$600 is to be spent for trees and bushes.
- Trees cost \$150 and bushes cost \$75.
- What combination of trees and bushes can he buy?

Form Two Inequalities Let x = the number of trees purchased and y = the number of bushes purchased. We then form the following system of inequalities:

The cost of a tree	times	the number of trees purchased	plus	the cost of a bush	times	the number of bushes purchased	should at least be	\$300.
\$150	·	x	+	\$75	·	y	≥	\$300

The cost of a tree	times	the number of trees purchased	plus	the cost of a bush	times	the number of bushes purchased	should not be more than	\$600.
\$150	$\cdot$	x	$+$	\$75	$\cdot$	y	$\leq$	\$600

Solve the System To solve this system of linear inequalities

$$\begin{cases} 150x + 75y \geq 300 \\ 150x + 75y \leq 600 \end{cases}$$

we use the graphing methods discussed in this section. Neither a negative number of trees nor a negative number of bushes can be purchased, so we restrict the graph to Quadrant I.

State the Conclusion The coordinates of each point highlighted in the graph give a possible combination of the number of trees, x, and the number of bushes, y, that can be purchased. Written as ordered pairs, these possibilities are

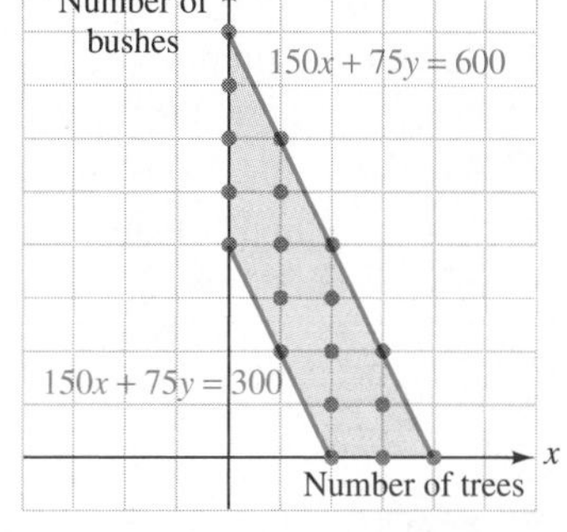

(0, 4), (0, 5), (0, 6), (0, 7), (0, 8),
(1, 2), (1, 3), (1, 4), (1, 5), (1, 6),
(2, 0), (2, 1), (2, 2), (2, 3), (2, 4),
(3, 0), (3, 1), (3, 2), (4, 0)

Check the Result Suppose the homeowner picks the combination of 3 trees and 2 bushes, as represented by (3, 2). Show that this point satisfies both inequalities of the system.

Answers to Self Checks

1.

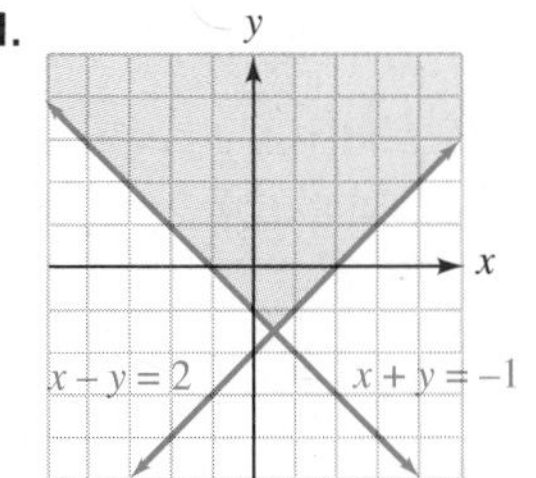

2.

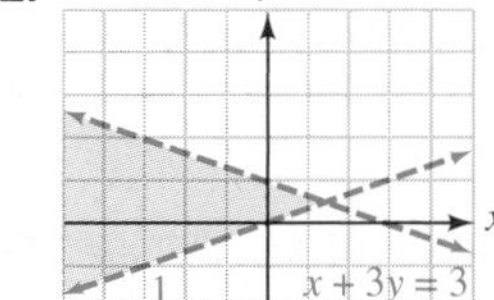

3.

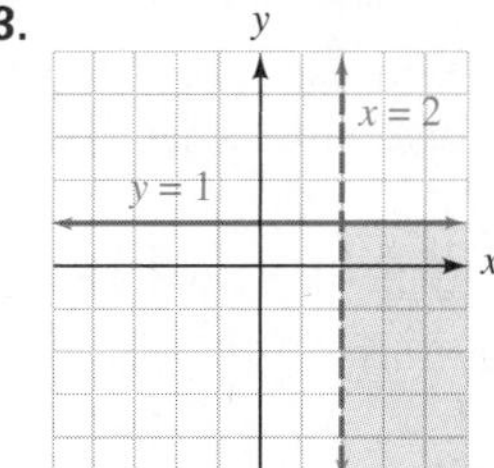

4.

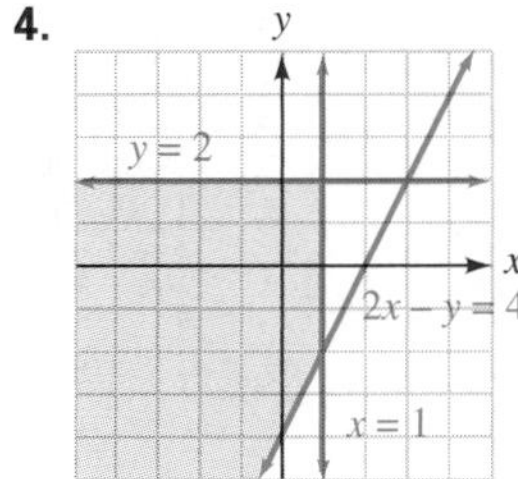

7.5 STUDY SET

ELEMENTARY Algebra $f(x)$ Now™

VOCABULARY Fill in the blanks.

1. $\begin{cases} x + y > 2 \\ x + y < 4 \end{cases}$ is a system of linear ___________.
2. A _________ of a system of linear inequalities is an ordered pair that satisfies each inequality.
3. To _______ a system of linear inequalities means to find all of its solutions.
4. To graph the linear inequality $x + y > 2$, first graph the __________ $x + y = 2$. Then pick the test _______ (0, 0) to determine which half-plane to shade.
5. To find the solutions of a system of two linear inequalities graphically, look for the ____________, or overlap, of the two shaded regions.
6. Any point in the doubly ________ region of the graph of a system of two linear inequalities has coordinates that satisfy both inequalities of the system.

CONCEPTS

7. **a.** What is the equation of the boundary line of the graph of $3x - y < 5$?
   b. Is the boundary a solid or dashed line?
8. **a.** What is the equation of the boundary line of the graph of $y \geq 4x$?
   b. Is the boundary a solid or dashed line?
   c. Why can't (0, 0) be used as a test point to determine what to shade?
9. Find the slope and the y-intercept of the line whose equation is $y = 4x - 3$.
10. Complete the table to find the x- and y-intercepts of the line whose equation is $3x - 2y = 6$.

x	y
0	
	0

11. The boundary of the graph of $2x + y > 4$ is shown.
   a. Does the point (0, 0) make the inequality true?
   b. Should the region above or below the boundary be shaded?

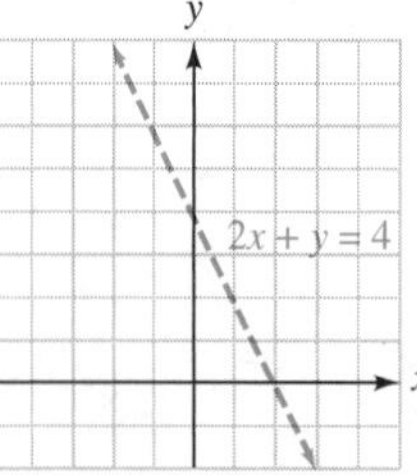

12. Linear inequality 1 is graphed in red and linear inequality 2 is graphed in blue. Decide whether a true or false statement results when the coordinates of the given point are substituted into the given inequality.
   a. A, inequality 1
   b. A, inequality 2
   c. B, inequality 1
   d. B, inequality 2
   e. C, inequality 1
   f. C, inequality 2

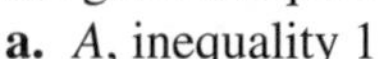

Inequality 1 solutions in red

Inequality 2 solutions in blue

13. The graph of a system of two linear inequalities is shown. Tell whether each point is a solution of the system.
   a. (4, −2)
   b. (1, 3)
   c. the origin

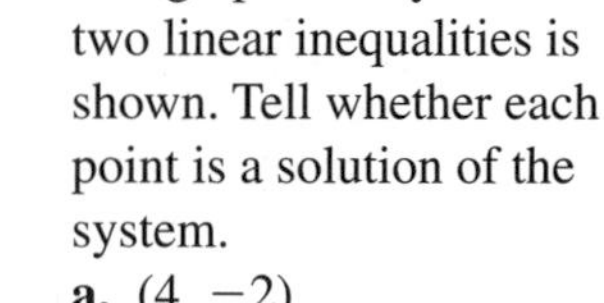

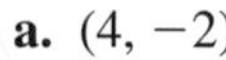

14. Use a check to determine whether each ordered pair is a solution of the system.

$$\begin{cases} x + 2y \geq -1 \\ x - y < 2 \end{cases}$$

   a. (1, 4) **b.** (−2, 0)

15. Match each equation, inequality, or system with the graph of its solution.
   a. $x + y = 2$ **b.** $x + y \geq 2$
   c. $\begin{cases} x + y = 2 \\ x - y = 2 \end{cases}$ **d.** $\begin{cases} x + y \geq 2 \\ x - y \leq 2 \end{cases}$

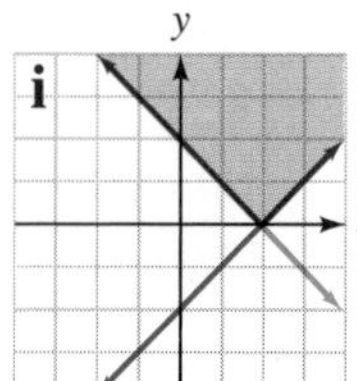

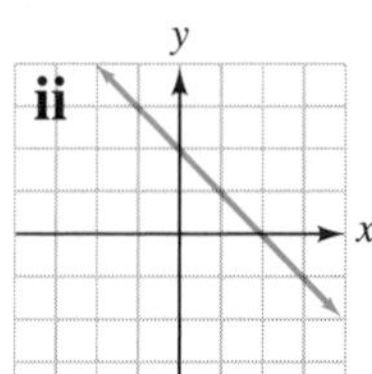

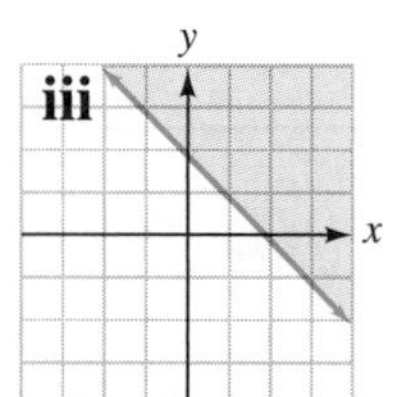

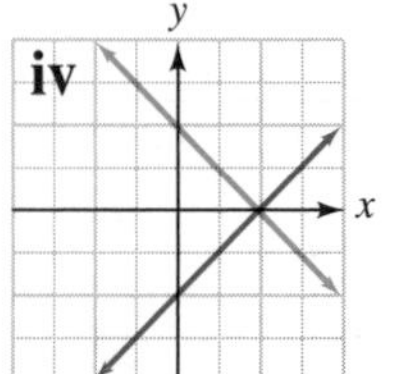

16. Match the system of inequalities with the correct graph.

a. $\begin{cases} x \geq 2 \\ y < 1 \end{cases}$ **b.** $\begin{cases} x > 2 \\ y \leq 1 \end{cases}$

c. $\begin{cases} x \geq 2 \\ y \geq 1 \end{cases}$ **d.** $\begin{cases} x > 2 \\ y > -1 \end{cases}$

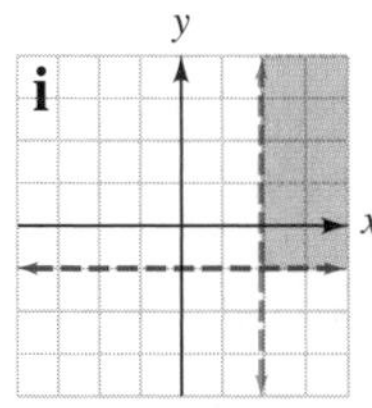

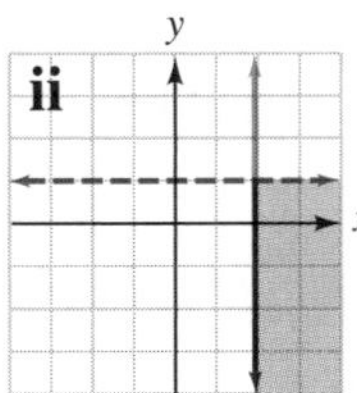

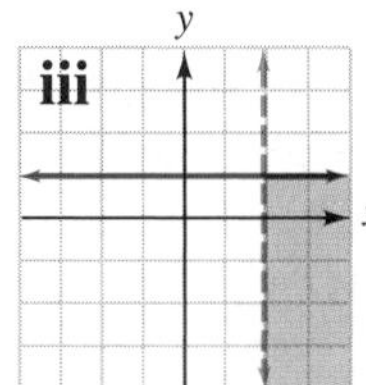

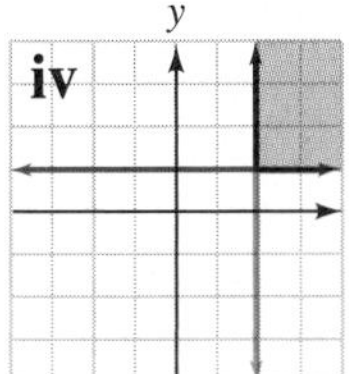

NOTATION

17. Fill in the blank: This graph of the solutions of a system of linear inequalities can be described as the triangle ______ and the triangular region it encloses.

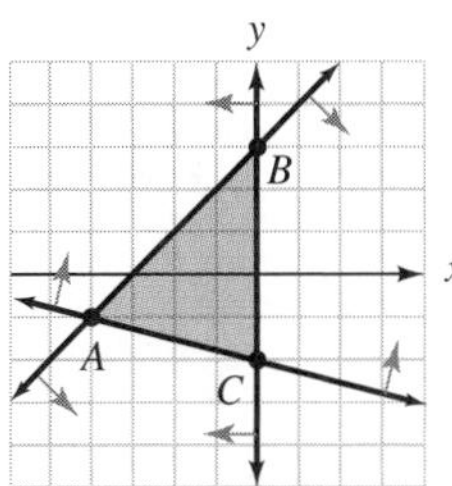

18. Represent each phrase using either $>$, $<$, $\geq$, or $\leq$.

a. is not more than

b. must be at least

c. should not surpass

d. cannot go below

PRACTICE **Graph the solutions of each system.**

19. $\begin{cases} x + 2y \leq 3 \\ 2x - y \geq 1 \end{cases}$

20. $\begin{cases} 2x + y \geq 3 \\ x - 2y \leq -1 \end{cases}$

21. $\begin{cases} x + y < -1 \\ x - y > -1 \end{cases}$

22. $\begin{cases} x + y > 2 \\ x - y < -2 \end{cases}$

23. $\begin{cases} 2x - 3y \leq 0 \\ y \geq x - 1 \end{cases}$

24. $\begin{cases} y > 2x - 4 \\ y \geq -x - 1 \end{cases}$

25. $\begin{cases} x + y < 2 \\ x + y \leq 1 \end{cases}$

26. $\begin{cases} y > -x + 2 \\ y < -x + 4 \end{cases}$

27. $\begin{cases} x \geq 2 \\ y \leq 3 \end{cases}$

28. $\begin{cases} x \geq -1 \\ y > -2 \end{cases}$

29. $\begin{cases} x > 0 \\ y > 0 \end{cases}$

30. $\begin{cases} x \leq 0 \\ y < 0 \end{cases}$

31. $\begin{cases} 3x + 4y \geq -7 \\ 2x - 3y \geq 1 \end{cases}$

32. $\begin{cases} 3x + y \leq 1 \\ 4x - y \geq -8 \end{cases}$

33. $\begin{cases} 2x + y < 7 \\ y > 2 - 2x \end{cases}$

34. $\begin{cases} 2x + y \geq 6 \\ y \leq 4x - 6 \end{cases}$

35. $\begin{cases} 2x - 4y > -6 \\ 3x + y \geq 5 \end{cases}$

36. $\begin{cases} 2x - 3y < 0 \\ 2x + 3y \geq 12 \end{cases}$

37. $\begin{cases} 3x - y + 4 \leq 0 \\ 3y > -2x - 10 \end{cases}$

38. $\begin{cases} 3x + 2y - 12 \geq 0 \\ x < -2 + y \end{cases}$

39. $\begin{cases} y \geq x \\ y \leq \frac{1}{3}x + 1 \end{cases}$

40. $\begin{cases} y > 3x \\ y \leq -x - 1 \end{cases}$

41. $\begin{cases} x + y > 0 \\ y - x < -2 \end{cases}$

42. $\begin{cases} y + 2x \leq 0 \\ y \leq \frac{1}{2}x + 2 \end{cases}$

43. $\begin{cases} x \geq 0 \\ y \geq 0 \\ x + y \leq 3 \end{cases}$

44. $\begin{cases} x - y \leq 6 \\ x + 2y \leq 6 \\ x \geq 0 \end{cases}$

45. $\begin{cases} x - y < 4 \\ y \leq 0 \\ x \geq 0 \end{cases}$

46. $\begin{cases} 2x + y \leq 2 \\ y > x \\ x \geq 0 \end{cases}$

APPLICATIONS

47. BIRDS OF PREY Parts (a) and (b) of the illustration show the individual fields of vision for each eye of an owl. In part (c), shade the area where the fields of vision overlap—that is, the area that is seen by both eyes.

(a)

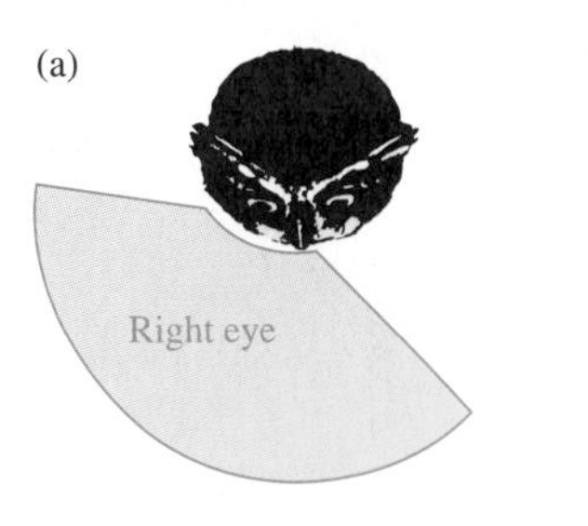

(b)

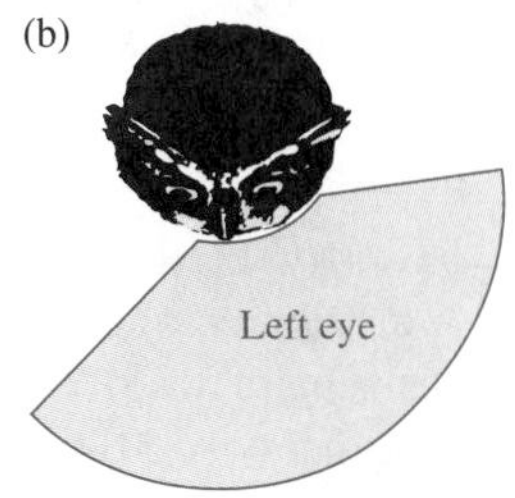

(c)

48. EARTH SCIENCE Shade the area of the earth's surface that is north of the Tropic of Capricorn and south of the Tropic of Cancer.

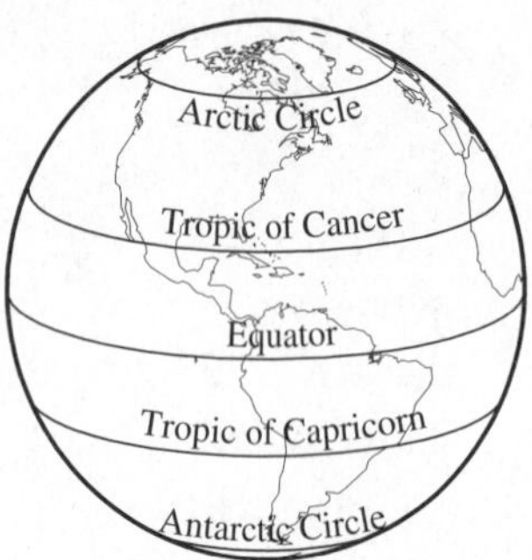

In Problems 49–52, graph each system of inequalities and give two possible solutions.

49. BUYING COMPACT DISCS Melodic Music has compact discs on sale for either \$10 or \$15. If a customer wants to spend at least \$30 but no more than \$60 on CDs, graph a system of inequalities showing the possible combinations of \$10 CDs (x) and \$15 CDs (y) that the customer can buy.

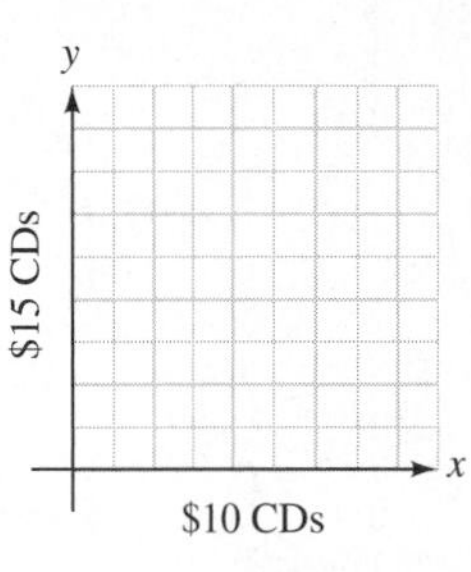

50. BUYING BOATS Dry Boatworks wholesales aluminum boats for \$800 and fiberglass boats for \$600. Northland Marina wants to make a purchase totaling at least \$2,400 but no more than \$4,800. Graph a system of inequalities showing the possible combinations of aluminum boats (x) and fiberglass boats (y) that can be ordered.

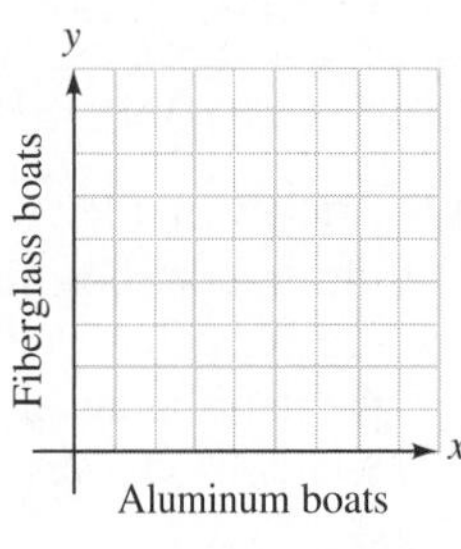

51. BUYING FURNITURE A distributor wholesales desk chairs for \$150 and side chairs for \$100. Best Furniture wants its order to total no more than \$900; Best also wants to order more side chairs than desk chairs. Graph a system of inequalities showing the possible combinations of desk chairs (x) and side chairs (y) that can be ordered.

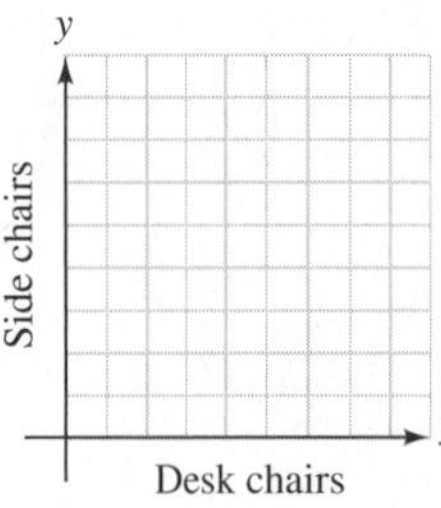

52. ORDERING FURNACE EQUIPMENT J. Bolden Heating Company wants to order no more than \$2,000 worth of electronic air cleaners and humidifiers from a wholesaler that charges \$500 for air cleaners and \$200 for humidifiers. If Bolden wants more humidifiers than air cleaners, graph a system of inequalities showing the possible combinations of air cleaners (x) and humidifiers (y) that can be ordered.

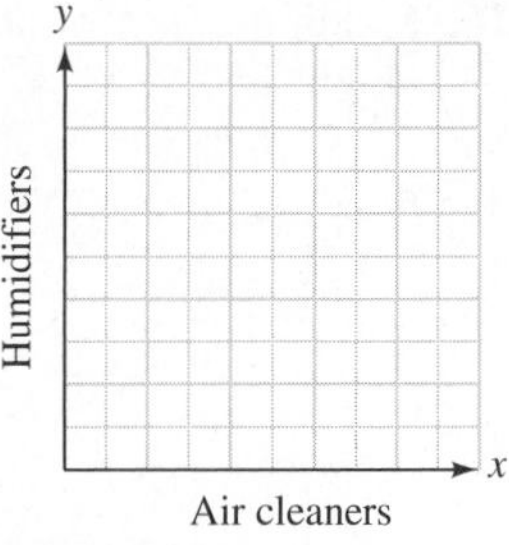

53. PESTICIDES To eradicate a fruit fly infestation, helicopters sprayed an area of a city that can be described by $y \geq -2x + 1$ (within the city limits). Two weeks later, more spraying was ordered over the area described by $y \geq \frac{1}{4}x - 4$ (within the city limits). Show the part of the city that was sprayed twice.

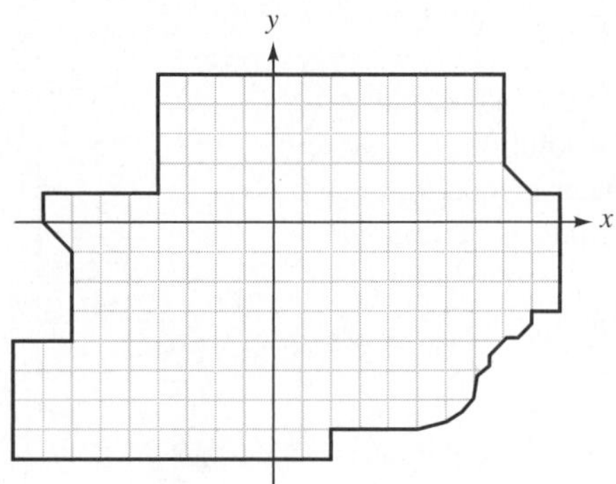

54. REDEVELOPMENT Refer to the following diagram. A government agency has declared an area of a city east of First Street, north of Second Avenue, south of Sixth Avenue, and west of Fifth Street as eligible for federal redevelopment funds. Describe this area of the city mathematically using a system of four inequalities, if the corner of Central Avenue and Main Street is considered the origin.

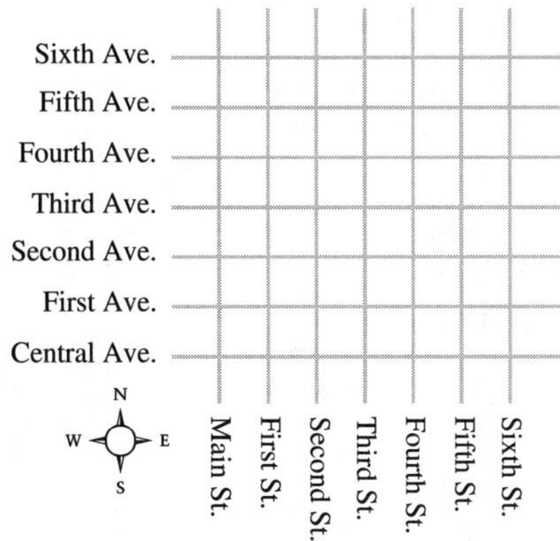

WRITING

55. Explain how to use graphing to solve a system of inequalities.

56. When a solution of a system of linear inequalities is graphed, what does the shading represent?

57. Describe how the graphs of the solutions of these systems are similar and how they differ.

$$\begin{cases} x + y = 4 \\ x - y = 4 \end{cases} \quad \text{and} \quad \begin{cases} x + y \geq 4 \\ x - y \geq 4 \end{cases}$$

58. Explain when a system of inequalities will have no solutions.

REVIEW **Write each number in standard notation.**

59. 2.3×10^2

60. 3.75×10^4

61. 9.76×10^{-4}

62. 7.63×10^{-5}

Write each number in scientific notation.

63. 290,000

64. 1,700,000

65. 0.0000051

66. 0.00073

CHALLENGE PROBLEMS **Graph the solutions of each system.**

67. $\begin{cases} \frac{x}{3} - \frac{y}{2} < -3 \\ \frac{x}{3} + \frac{y}{2} > -1 \end{cases}$

68. $\begin{cases} 3x + y < -2 \\ y > 3(1 - x) \end{cases}$

69. $\begin{cases} 2x + 3y \leq 6 \\ 3x + y \leq 1 \\ x \leq 0 \end{cases}$

70. $\begin{cases} x \geq 0 \\ y \geq 0 \\ 9x + 3y \leq 18 \\ 3x + 6y \leq 18 \end{cases}$

ACCENT ON TEAMWORK

WRITING APPLICATION PROBLEMS

Overview: In Section 7.4, you solved application problems by translating the words of the problem into a system of two equations. In this activity, you will reverse these steps.
Instructions: Form groups of 2 or 3 students. For each type of application, write a problem that could be solved using the given equations. If you need help getting started, refer to the specific problem types in the text. When finished writing the five applications, pick one problem and solve it completely.

A rectangle problem:

$$\begin{cases} 2l + 2w = 320 \\ l = w + 40 \end{cases}$$

A number-value problem:

$$\begin{cases} 5x + 2y = 23 \\ 3x + 7y = 37 \end{cases}$$

An interest problem:

$$\begin{cases} x + y = 75{,}000 \\ 0.03x + 0.05y = 2{,}750 \end{cases}$$

A with–against the wind problem:

$$\begin{cases} 2(x + y) = 600 \\ 3(x - y) = 600 \end{cases}$$

A liquid mixture problem:

$$\begin{cases} x + y = 36 \\ 0.50x + 0.20y = 0.30(36) \end{cases}$$

INTERPRETING GRAPHS OF LINEAR INEQUALITIES

Overview: This activity will strengthen your understanding of shading and boundary lines, which are used when graphing the solutions of systems of linear inequalities.

Instructions: Form groups of 2 or 3 students. Find the solutions of the system of linear inequalities by graphing the inequalities on the given coordinate system.

$$\begin{cases} 3x + 2y < 4 \\ y \le \dfrac{3}{5}x + 2 \end{cases}$$

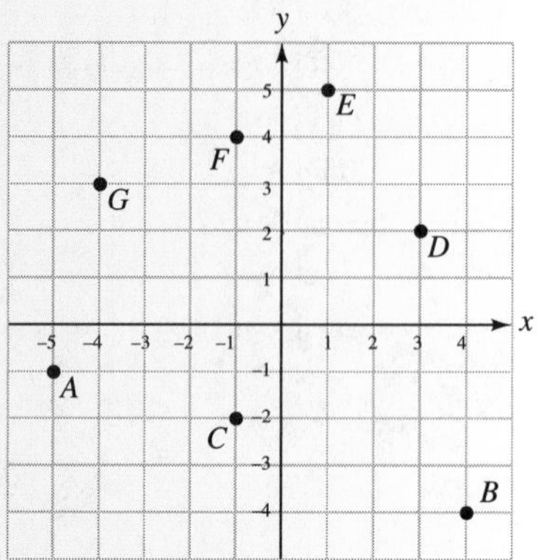

For each point A, B, C, D, E, F, and G on the graph, determine whether its coordinates make the first inequality true or false. Then determine whether each point's coordinates make the second inequality true or false. Write your answers in a table like that shown below.

Point	Coordinates	1st inequality	2nd inequality
A			

KEY CONCEPT: THREE METHODS FOR SOLVING SYSTEMS OF EQUATIONS

Each method that we used to solve systems of linear equations has advantages and disadvantages. Here are some guidelines to follow when deciding whether to use graphing, substitution, or elimination.

Method	Advantages	Disadvantages
Graphing	• You see the solutions. • The graphs allow you to observe trends.	• Inaccurate when the solutions are not integers, or are large numbers off the graph.
Substitution	• Always gives the exact solutions. • Works well if one of the equations is solved for one of the variables, or if it is easy to solve for one of the variables.	• You do not see the solution. • If no variable has a coefficient of 1 or -1, solving for one of the variables often involves fractions.
Elimination	• Always gives the exact solutions. • Works well if no variable has a coefficient of 1 or -1.	• You do not see the solution. • Equations must be written in the form $Ax + By = C$.

SOLVING SYSTEMS OF EQUATIONS BY GRAPHING

A system of linear equations can be solved by graphing both equations and locating the point of intersection of the two lines.

1. FOOD SERVICE The equations in the table give the fees two catering companies charge a Hollywood studio for on-location meal service, where $y =$ the total cost (in dollars) for x meals. Graph the system to find the break point. That is, find the number of meals and the corresponding fee for which the two caterers will charge the same amount.

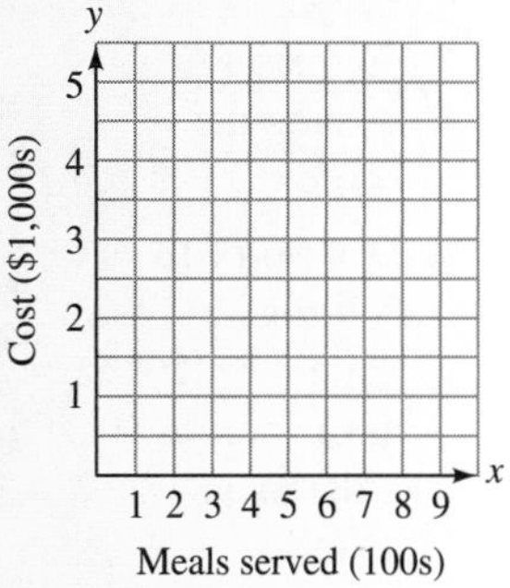

Caterer	Setup fee	Cost per meal	Equation
Sunshine	$1,000	$4	$y = 4x + 1{,}000$
Lucy's	$500	$5	$y = 5x + 500$

SOLVING SYSTEMS OF EQUATIONS BY SUBSTITUTION

When using the substitution method, the objective is to substitute for one of the variables to obtain one equation containing only one unknown.

2. Solve by substitution: $\begin{cases} y = 2x - 9 \\ x + 3y = 8 \end{cases}$

3. Solve by substitution: $\begin{cases} 3x + 4y = -7 \\ 2y - x = -1 \end{cases}$

SOLVING SYSTEMS OF EQUATIONS BY ELIMINATION

When using the elimination method, the objective is to obtain two equations whose sum is an equation containing only one unknown.

4. Solve by elimination: $\begin{cases} x + y = 1 \\ x - y = 5 \end{cases}$

5. Solve by elimination: $\begin{cases} 2x - 3y = -18 \\ 3x + 2y = -1 \end{cases}$

CHAPTER REVIEW

ELEMENTARY Algebra f(x) Now™

SECTION 7.1 Solving Systems of Equations by Graphing

CONCEPTS

A *solution* of a system of equations is an ordered pair that satisfies both equations of the system.

REVIEW EXERCISES

Determine whether the ordered pair is a solution of the system.

1. $(2, -3)$, $\begin{cases} 3x - 2y = 12 \\ 2x + 3y = -5 \end{cases}$

2. $\left(\frac{7}{2}, -\frac{2}{3}\right)$, $\begin{cases} 3y = 2x - 9 \\ 2x + 3y = 5 \end{cases}$

To solve a system *graphically:*

1. Graph each equation on the same rectangular coordinate system.
2. Determine the coordinates of the point of intersection of the graphs. That ordered pair is the solution.
3. Check the solution in each equation of the original system.

A system of equations that has at least one solution is called a *consistent system.* If the graphs are parallel lines, the system has no solution, and it is called an *inconsistent system.*

Equations with different graphs are called *independent equations.* If the graphs are the same line, the system has infinitely many solutions. The equations are called *dependent equations.*

3. COLLEGE ENROLLMENT Estimate the point of intersection of the graphs. Explain its significance.

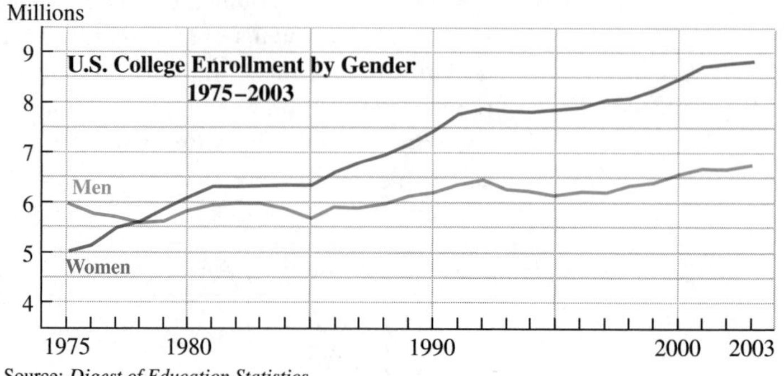

Use the graphing method to solve each system.

4. $\begin{cases} x + y = 7 \\ 2x - y = 5 \end{cases}$

5. $\begin{cases} 2x + y = 5 \\ y = -\dfrac{x}{3} \end{cases}$

6. $\begin{cases} 3x + 6y = 6 \\ x + 2y - 2 = 0 \end{cases}$

7. $\begin{cases} 6x + 3y = 12 \\ y = -2x + 2 \end{cases}$

SECTION 7.2 Solving Systems of Equations by Substitution

To solve a system of equations in x and y by substitution:

1. Solve one of the equations for either x or y.
2. Substitute the expression for x or for y obtained in step 1 into the other equation and solve that equation.

Use the substitution method to solve each system.

8. $\begin{cases} y = 15 - 3x \\ 7y + 3x = 15 \end{cases}$

9. $\begin{cases} x = y \\ 5x - 4y = 3 \end{cases}$

10. $\begin{cases} 6x + 2y = 8 - y + x \\ 3x = 2 - y \end{cases}$

11. $\begin{cases} r = 3s + 7 \\ r = 2s + 5 \end{cases}$

12. $\begin{cases} 9x + 3y - 5 = 0 \\ 3x + y = \dfrac{5}{3} \end{cases}$

13. $\begin{cases} \dfrac{x}{2} + \dfrac{y}{2} = 11 \\ \dfrac{5x}{16} - \dfrac{3y}{16} = \dfrac{15}{8} \end{cases}$

3. Substitute the value of the variable found in step 2 into the substitution equation to find the value of the remaining variable.
4. Check the proposed solution in the equations of the original system.

14. In solving a system using the substitution method, suppose you obtain the result of $8 = 9$.

a. How many solutions does the system have?

b. Describe the graph of the system.

c. What term is used to describe the system?

SECTION 7.3 Solving Systems of Equations by Elimination

To solve a system of equations in x and y using elimination:

1. Write each equation in the form $Ax + By = C$.
2. Multiply one or both equations by nonzero quantities to make the coefficients of x (or y) opposites.
3. Add the equations to eliminate the terms involving x (or y).
4. Solve the equation resulting from step 3.
5. Find the value of the other variable by substituting the value of the variable found in step 4 into any equation containing both variables.
6. Check the solution in the original equations.

Solve each system using the elimination method.

15. $\begin{cases} 2x + y = 1 \\ 5x - y = 20 \end{cases}$

16. $\begin{cases} x + 8y = 7 \\ x - 4y = 1 \end{cases}$

17. $\begin{cases} 5a + b = 2 \\ 3a + 2b = 11 \end{cases}$

18. $\begin{cases} 11x + 3y = 27 \\ 8x + 4y = 36 \end{cases}$

19. $\begin{cases} 9x + 3y = 15 \\ 3x = 5 - y \end{cases}$

20. $\begin{cases} -\dfrac{a}{4} - \dfrac{b}{3} = \dfrac{1}{12} \\ \dfrac{a}{2} - \dfrac{5b}{4} = \dfrac{7}{4} \end{cases}$

21. $\begin{cases} 0.02x + 0.05y = 0 \\ 0.3x - 0.2y = -1.9 \end{cases}$

22. $\begin{cases} -\dfrac{1}{4}x = 1 - \dfrac{2}{3}y \\ 6x - 18y = 5 - 2y \end{cases}$

For each system, tell which method, substitution or elimination, would be easier to use to solve the system and why.

23. $\begin{cases} 6x + 2y = 5 \\ 3x - 3y = -4 \end{cases}$

24. $\begin{cases} x = 5 - 7y \\ 3x - 3y = -4 \end{cases}$

SECTION 7.4 Problem Solving Using Systems of Equations

In this section, we considered ways to solve problems by using *two* variables.

Write a system of two equations in two variables to solve each problem.

25. ELEVATIONS The elevation of Las Vegas, Nevada, is 20 times greater than that of Baltimore, Maryland. The sum of their elevations is 2,100 feet. Find the elevation of each city.

To solve problems involving two unknown quantities:

1. *Analyze* the facts of the problem. Make a table or diagram if necessary.
2. Pick different variables to represent two unknown quantities. *Form* two equations involving the variables.
3. *Solve* the system of equations.
4. *State* the conclusion.
5. *Check* the results.

The *break point* of a linear system is the point of intersection of the graphs.

26. PAINTING EQUIPMENT When fully extended, a ladder is 35 feet in length. If the extension is 7 feet shorter than the base, how long is each part of the ladder?

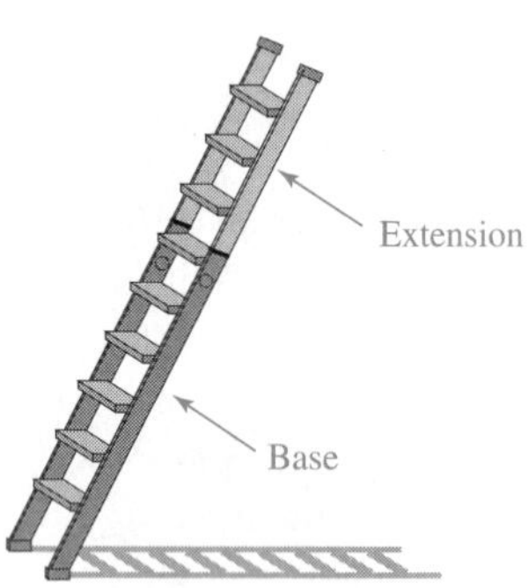

27. CRASH INVESTIGATION In an effort to protect evidence, investigators used 420 yards of yellow "Police Line—Do Not Cross" tape to seal off a large rectangular-shaped area around an airplane crash site. How much area will the investigators have to search if the width of the rectangle is three-fourths of the length?

28. CELEBRITY ENDORSEMENT A company selling a home juicing machine is contemplating hiring either an athlete or an actor to serve as a spokesperson for the product. The terms of each contract would be as follows:

Celebrity	Base pay	Commission per item sold
Athlete	\$30,000	\$5
Actor	\$20,000	\$10

a. For each celebrity, write an equation giving the money y in dollars the celebrity would earn if x juicers were sold.

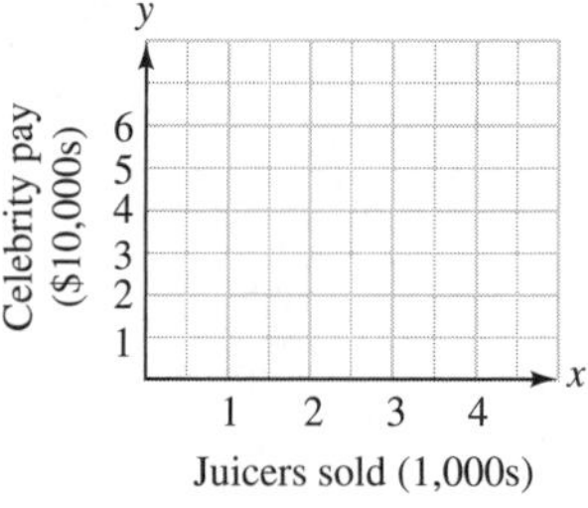

b. For what number of juicers would the athlete and the actor earn the same amount?

c. Graph the equations from part (a). The company expects to sell over 3,000 juicers. Which celebrity would cost the company the least money to serve as a spokesperson?

29. CANDY OUTLET STORE A merchant wants to mix gummy worms worth \$3 per pound and gummy bears worth \$1.50 per pound to make 30 pounds of a mixture worth \$2.10 per pound. How many pounds of each type of candy should he use?

30. BOATING It takes a motorboat 4 hours to travel 56 miles down a river, and 3 hours longer to make the return trip. Find the speed of the current.

31. SHOPPING Packages containing two bottles of contact lens cleaner and three bottles of soaking solution cost \$63.40, and packages containing three bottles of cleaner and two bottles of soaking solution cost \$69.60. Find the cost of a bottle of cleaner and a bottle of soaking solution.

32. INVESTING Carlos invested part of $3,000 in a 10% certificate account and the rest in a 6% passbook account. The total annual interest from both accounts is $270. How much did he invest at 6%?

33. ANTIFREEZE How much of a 40% antifreeze solution must a mechanic mix with a 70% antifreeze solution if he needs 20 gallons of a 50% antifreeze solution?

SECTION 7.5 Solving Systems of Linear Inequalities

A *solution* of a system of linear inequalities is an ordered pair that satisfies each inequality.

To solve a system of linear inequalities:

1. Graph each inequality on the same rectangular coordinate system.
2. Use shading to highlight the intersection of the graphs. The points in this region are the solutions of the system.
3. As an informal check, pick a point from the region and verify that its coordinates satisfy each inequality of the original system.

Solve each system of inequalities.

34. $\begin{cases} 5x + 3y < 15 \\ 3x - y > 3 \end{cases}$

35. $\begin{cases} 3y \le x \\ y > 3x \end{cases}$

36. GIFT SHOPPING A grandmother wants to spend at least $40 but no more than $60 on school clothes for her grandson. If T-shirts sell for $10 and pants sell for $20, write a system of inequalities that describes the possible numbers of T-shirts x and pairs of pants y that she can buy. Graph the system and give two possible solutions.

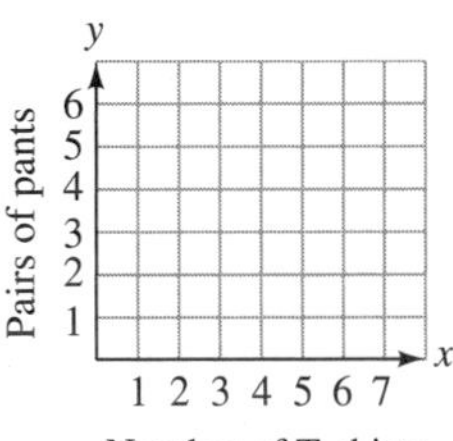

CHAPTER 7 TEST

ELEMENTARY Algebra Now™

Determine whether the ordered pair is a solution of the system.

1. $(5, 3), \begin{cases} 3x + 2y = 21 \\ x + y = 8 \end{cases}$

2. $(-2, -1), \begin{cases} 4x + y = -9 \\ 2x - 3y = -7 \end{cases}$

3. Fill in the blanks.
   a. A ________ of a system of linear equations is an ordered pair that satisfies each equation.
   b. A system of equations that has at least one solution is called a ________ system.
   c. A system of equations that has no solution is called an ________ system.

d. Equations with different graphs are called __________ equations.

e. A system of __________ equations has an infinite number of solutions.

Solve each system by graphing.

4. $\begin{cases} y = 2x - 1 \\ x - 2y = -4 \end{cases}$

5. $\begin{cases} x + y = 5 \\ y = -x \end{cases}$

The following graph shows two different ways in which a salesperson can be paid according to the number of items she sells each month.

6. What is the point of intersection of the graphs? Explain its significance.

7. Which plan is better for the salesperson if she feels that selling 30 items per month is virtually impossible?

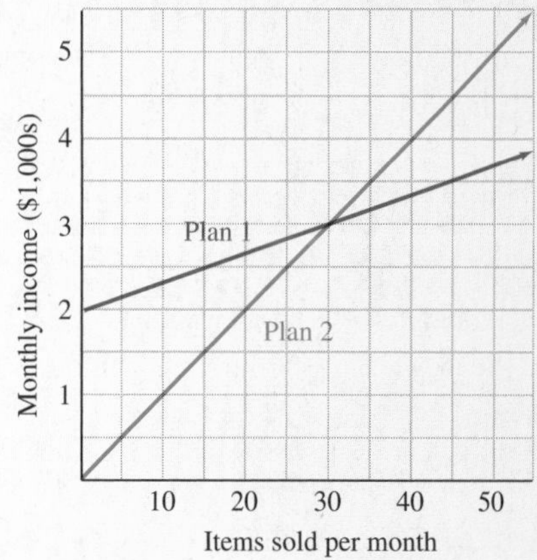

Solve each system by substitution.

8. $\begin{cases} y = x - 1 \\ 2x + y = -7 \end{cases}$

9. $\begin{cases} 3x + 6y = -15 \\ x + 2y = -5 \end{cases}$

10. $\begin{cases} 3a + 4b = -7 \\ 2b - a = -1 \end{cases}$

Solve each system using elimination.

11. $\begin{cases} 3x - y = 2 \\ 2x + y = 8 \end{cases}$

12. $\begin{cases} 4x + 3y = -3 \\ -3x = -4y + 21 \end{cases}$

13. $\begin{cases} 3x - 5y - 16 = 0 \\ \dfrac{x}{2} - \dfrac{5}{6}y = \dfrac{1}{3} \end{cases}$

14. Which method would be most efficient to solve the following system? Explain your answer. **(You do not need to solve the system.)**

$$\begin{cases} 5x - 3y = 5 \\ 3x + 3y = 3 \end{cases}$$

Write a system of two equations in two variables to solve each problem.

15. CHILD CARE On a mother's 22-mile commute to work, she drops her daughter off at a child care facility. The first part of the trip is 6 miles less than the second part. How long is each part of her morning commute?

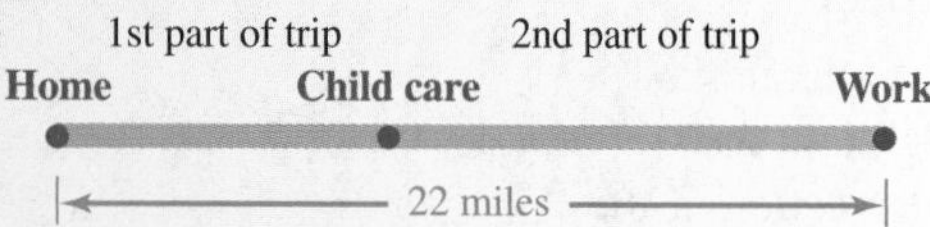

16. VACATIONING It cost a family of 7 a total of $119 for general admission tickets to the San Diego Zoo. How many adult tickets and how many child tickets were purchased?

SAN DIEGO ZOO

General Admission	
ADULT TICKETS	$21
CHILD TICKETS (ages 3-11)	$14

17. FINANCIAL PLANNING A woman invested some money at 8% and some at 9%. The interest for 1 year on the combined investment of $10,000 was $840. How much was invested at 8% and how much was invested at 9%?

18. KAYAKING A man can paddle his kayak x miles per hour in still water. How fast will his kayak travel if he paddles against a river current of c miles per hour?

19. TETHER BALL Translate this fact into an equation in two variables: the angles shown in the illustration are complementary.

20. Solve the system by graphing.

$$\begin{cases} 2x + 3y \le 6 \\ x > 2 \end{cases}$$

21. CLOTHES SHOPPING This system of inequalities describes the number of \$20 shirts, x, and \$40 pairs of pants, y, a person can buy if he or she plans to spend not less than \$80 but not more than \$120. Graph the system. Then give three solutions.

$$\begin{cases} 20x + 40y \ge 80 \\ 20x + 40y \le 120 \end{cases}$$

CHAPTERS 1–7 CUMULATIVE REVIEW EXERCISES

1. CANDY SALES The circle graph shows how \$6,300,000,000 in seasonal candy sales for 2002 was spent. Find the candy sales for Halloween.

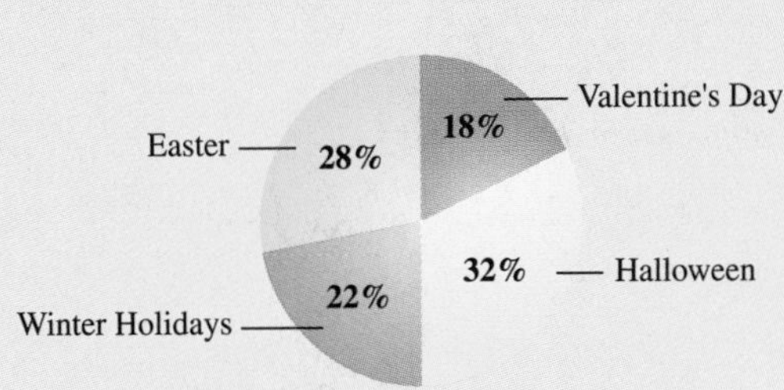

Source: National Confectioners Association

2. List the set of integers.

Evaluate each expression.

3. $3 - 4[-10 - 4(-5)]$

4. $\dfrac{|-45| - 2(-5) + 1^5}{2 \cdot 9 - 2^4}$

5. AIR CONDITIONING Find the volume of air contained in the duct. Round to the nearest tenth of a cubic foot.

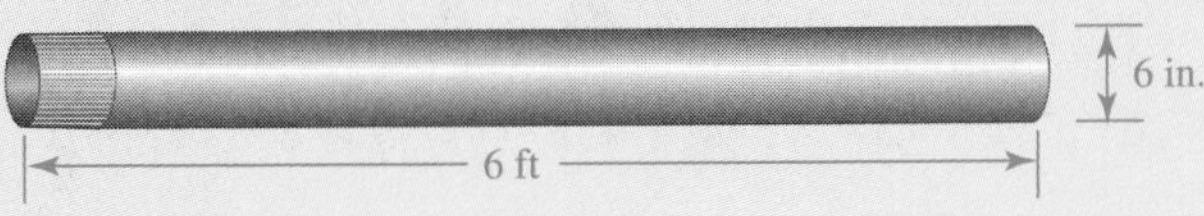

6. Simplify: $3x^2 + 2x^2 - 5x^2$.

Solve each equation.

7. $2 - (4x + 7) = 3 + 2(x + 2)$

8. $\frac{2}{5}y + 3 = 9$

9. Solve $-4x + 6 > 17$ and graph the solution set. Then describe the graph using interval notation.

10. ANGLE OF ELEVATION Find x.

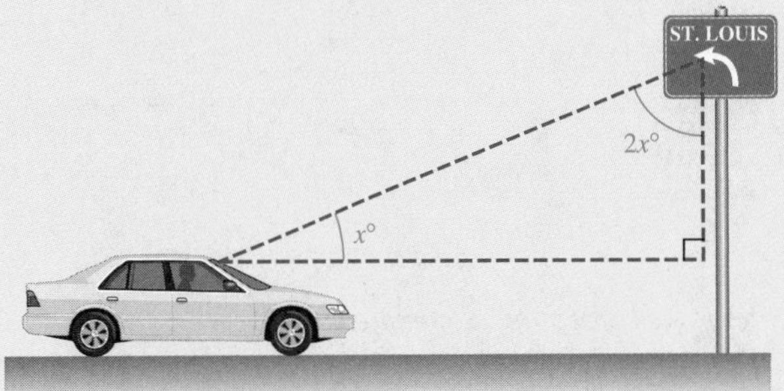

11. STOCK MARKET An investment club invested part of \$45,000 in a high-yield mutual fund that earned 12% annual simple interest. The remainder of the money was invested in Treasury bonds that earned 6.5% simple annual interest. The two investments earned \$4,300 in one year. How much was invested in each account?

12. Give the formula for
 a. the perimeter of a rectangle
 b. the area of a rectangle
 c. the area of a circle
 d. the distance traveled

Graph each equation.

13. $x = -2$ **14.** $y = x^2 + 1$

15. Find the slope and y-intercept of the line. Then write the equation of the line.

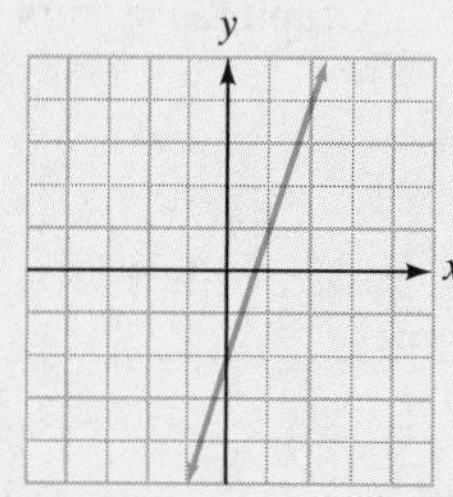

16. Find the slope of the line passing through $(6, -2)$ and $(-3, 2)$.

Simplify each expression. Write each answer without using negative exponents.

17. $(x^5)^2(x^7)^3$ **18.** $\dfrac{16(aa^2)^3}{2a^2a^3}$

19. $\dfrac{2^{-4}}{3^{-1}}$ **20.** $(2x)^0$

Perform the operation(s).

21. $(5x - 8y) - (-2x + 5y)$ **22.** $2x^2(3x^2 + 4x - 7)$

23. $(c + 16)^2$ **24.** $(x + 3)(2x - 3)$

25. $\dfrac{2x - 32}{16x}$ **26.** $3x + 1\overline{)9x^2 + 6x + 1}$

Factor each polynomial completely.

27. $12r^2 - 3rs + 9r^2s^2$ **28.** $u^2 - 18u + 81$

29. $2y^2 - 7y + 3$ **30.** $x^4 - 81$

31. $t^3 - v^3$ **32.** $xy - ty + xs - ts$

Solve each equation.

33. $8s^2 - 16s = 0$ **34.** $x^2 + 2x - 15 = 0$

35. Simplify: $\dfrac{x^2 - 25}{5x + 25}$.

36. Divide: $\dfrac{x^2 - x - 2}{x^2 + x} \div \dfrac{2 - x}{x}$.

37. Subtract: $\dfrac{x + 5}{xy} - \dfrac{x - 1}{x^2y}$.

38. Simplify: $\dfrac{\frac{y}{x} + 3y}{y + \frac{2y}{x}}$.

39. Solve: $\dfrac{3}{x + 1} - \dfrac{x - 2}{2} = \dfrac{x - 2}{x + 1}$

40. The triangles shown are similar. Find a and b.

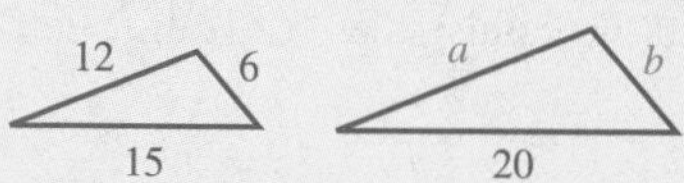

Solve each system by graphing.

41. $\begin{cases} x + 4y = -2 \\ y = -x - 5 \end{cases}$ **42.** $\begin{cases} 2x - 3y < 0 \\ y > x - 1 \end{cases}$

43. Solve $\begin{cases} x - 2y = 2 \\ 2x + 3y = 11 \end{cases}$ by substitution.

44. NUTRITION The table shows per serving nutritional information for egg noodles and rice pilaf. How many servings of each food should be eaten to consume exactly 22 grams of protein and 21 grams of fat?

	Protein (g)	Fat (g)
Egg noodles	5	3
Rice pilaf	4	5

Chapter

8 Radical Expressions and Equations

ELEMENTARY **Algebra** *f(x)* **Now**™

Throughout the chapter, this icon introduces resources on the Elementary AlgebraNow Web site, accessed through **http://1pass.thomson.com**, that will

- Help you test your knowledge of the material with a pre-test and a post-test
- Provide a personalized learning plan targeting areas you should study

Getty/Alan Pappe

8.1 An Introduction to Square Roots

8.2 Simplifying Square Roots

8.3 Adding and Subtracting Radical Expressions

8.4 Multiplying and Dividing Radical Expressions

8.5 Solving Radical Equations

8.6 Higher-Order Roots and Rational Exponents

Accent on Teamwork

Key Concept

Chapter Review

Chapter Test

Cumulative Review Exercises

TLE The Parthenon in Athens, Greece, is one of the best examples of the order and harmony for which ancient Greek architecture is known. Built more than 2,500 years ago, its design involves *golden rectangles* of different sizes. Golden rectangles are considered to have the most pleasing shape to the human eye. The exact dimensions of a golden rectangle involve the irrational number $\sqrt{2}$ (the square root of 2).

To learn more about square roots, visit *The Learning Equation* on the Internet at http://tle.brookscole.com. (The log-in instructions are in the Preface.) For Chapter 8, the online lesson is:

- *TLE* Lesson 15: Using Square Roots

In this chapter, we will simplify radical expressions and solve equations that contain square roots.

8.1 An Introduction to Square Roots

- Finding Square Roots
- Approximating Square Roots
- Square Roots of Negative Numbers
- Square Roots of Variable Expressions
- The Pythagorean Theorem
- The Distance Formula

Addition and subtraction are reverse operations, and so are multiplication and division. In this section, we will discuss another pair of reverse operations: raising a number to a power and finding the root of a number.

FINDING SQUARE ROOTS

When we raise a number to the second power, we are squaring it, or finding its **square.**

- The square of 5 is 25 because $5^2 = 25$.
- The square of -5 is 25, because $(-5)^2 = 25$.

We can reverse the squaring process to find **square roots** of numbers. For example, to find the square roots of 25, we ask ourselves "What number, when squared, is equal to 25?" There are two possible answers.

- 5 is a square root of 25, because $5^2 = 25$.
- -5 is a square root of 25, because $(-5)^2 = 25$.

In general, we have the following definition.

The Definition of Square Root The number b is a **square root** of the number a if $b^2 = a$.

Every positive number has two square roots, one positive and one negative. For example, the two square roots of 9 are 3 and -3, and the two square roots of 144 are 12 and -12. The number 0 is the only real number with exactly one square root. In fact, it is its own square root, because $0^2 = 0$.

A **radical symbol** $\sqrt{}$ represents the **positive** or **principal square root** of a number. When reading this symbol, we usually drop the word *positive* (or *principal*) and simply say *square root.* Since 3 is the positive square root of 9, we can write

$\sqrt{9} = 3$ Read as "the square root of 9 is 3."

The symbol $-\sqrt{}$ is used to represent the **negative square root** of a number. It is the opposite of the principal square root. Since -12 is the negative square root of 144, we can write

$-\sqrt{144} = -12$ Read as "the negative square root of 144 is -12" or "the opposite of the square root of 144 is -12."

Square Root Notation

If a is a positive real number,

1. $\sqrt{a}$ represents the **positive** or **principal square root** of a. It is the positive number we square to get a.
2. $-\sqrt{a}$ represents the **negative square root** of a. It is the opposite of the principal square root of a: $-\sqrt{a} = -1 \cdot \sqrt{a}$.
3. The principal square root of 0 is 0: $\sqrt{0} = 0$.

The number or variable expression under a radical symbol is called the **radicand,** and the radical symbol and radicand are called a **radical.** An algebraic expression containing a radical is called a **radical expression.**

Radical symbol → $\sqrt{16}$ ← Radicand

Radical

EXAMPLE 1

Find each square root: **a.** $\sqrt{16}$, **b.** $\sqrt{1}$, **c.** $\sqrt{0.36}$, **d.** $\sqrt{\frac{4}{9}}$, and **e.** $-\sqrt{225}$.

Solution

a. $\sqrt{16} = 4$ Ask: What positive number, when squared, is 16? The answer is 4.

b. $\sqrt{1} = 1$ Ask: What positive number, when squared, is 1? The answer is 1.

c. $\sqrt{0.36} = 0.6$ Ask: What positive number, when squared, is 0.36? The answer is 0.6.

d. $\sqrt{\frac{4}{9}} = \frac{2}{3}$ Ask: What positive number, when squared, is $\frac{4}{9}$? The answer is $\frac{2}{3}$.

e. $-\sqrt{225}$ is the opposite of the square root of 225. Since $\sqrt{225} = 15$, we have

$-\sqrt{225} = -15$ $-\sqrt{225} = -1 \cdot \sqrt{225} = -1 \cdot 15 = -15$.

The Language of Algebra

We have seen that the symbol $\sqrt{\ }$ is read as "square root." It can also be read as "radical." For example, $\sqrt{16}$ can be read as "*radical* 16."

Self Check 1 Find each square root: **a.** $\sqrt{49}$, **b.** $\sqrt{121}$, **c.** $-\sqrt{0.09}$, and **d.** $\sqrt{\frac{1}{25}}$.

APPROXIMATING SQUARE ROOTS

A number such as 16, 1, 0.36, $\frac{4}{9}$, or 225 that is the square of some rational number is called a **perfect square.** In Example 1, we saw that the square root of a perfect square is a rational number.

If a positive number is not a perfect square, its square root is irrational. For example, $\sqrt{3}$, $\sqrt{29}$, and $\sqrt{83}$ are irrational numbers because 3, 19, and 83 are not perfect squares. Since these square roots are irrational, they cannot be written as the quotient of two integers. Furthermore, their decimal representations are nonterminating and nonrepeating. We can find decimal approximations for them using the square root key $\sqrt{\ }$ on a calculator or from the table of squares roots found in Appendix II.

A table of square roots

n	$\sqrt{n}$
1	1.000
2	1.414
3	1.732
4	2.000
5	2.236

$\sqrt{3} \approx 1.732$ Use a table. Read $\approx$ as "is approximately equal to."

$\sqrt{29} \approx 5.385164807$ Use a calculator.

$\sqrt{83} \approx 9.11$ Use a calculator and round to two decimal places.

Remember that the values found for irrational square roots are only approximations. If we square 9.11, which is an approximation of $\sqrt{83}$, the result is not quite 83.

$$(9.11)^2 = 82.9921$$

Square roots appear in many applied problems.

EXAMPLE 2

ELEMENTARY Algebra $f(x)$ Now™

Pendulums. The **period of a pendulum** is the time required for the pendulum to swing back and forth to complete one cycle. To find the period P (in seconds) of a pendulum L feet long, we can use the formula

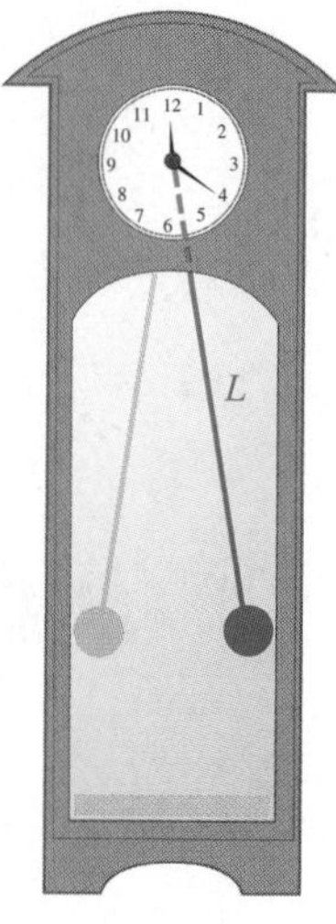

$$P = 1.11\sqrt{L} \quad \text{Read } 1.11\sqrt{L} \text{ as "1.11 times the square root of } L\text{."}$$

Find the period of a clock pendulum that is 5 feet long.

Solution Substitute 5 for L in the formula, use a calculator to find $\sqrt{5}$, and multiply that value by 1.11.

$$P = 1.11\sqrt{L}$$
$$P = 1.11\sqrt{5}$$
$$P \approx 2.482035455$$

The period is approximately 2.5 seconds.

Self Check 2 Find the period of a pendulum that is 3 feet long. Round to the nearest tenth of a second.

SQUARE ROOTS OF NEGATIVE NUMBERS

Not every number has a real-number square root. For example, to find $\sqrt{-4}$, we ask "What number, when squared, is equal to -4?" Since the square of a real number is always positive or 0, no real number squared is -4. Therefore, $\sqrt{-4}$ is not a real number. If we attempt to evaluate $\sqrt{-4}$ using a calculator, an error message will be shown.

Caution

Although they look similar, these radical expressions have very different meanings.

$$-\sqrt{4} = -2$$

$\sqrt{-4}$ is not a real number.

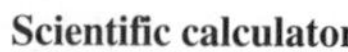

Scientific calculator

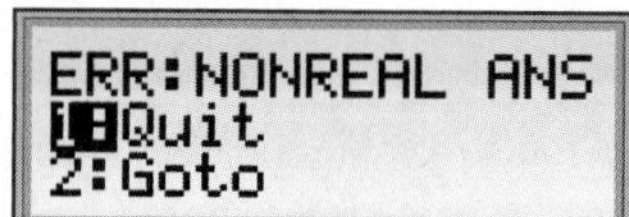

Graphing calculator

Since square roots of negative numbers like $\sqrt{-4}$, $\sqrt{-1}$, and $\sqrt{-39}$ are not real numbers, they do not correspond to any point on the real number line. They are examples of **imaginary numbers,** which are discussed later in the text.

We summarize three important facts about square roots as follows.

Square Roots

1. If a is a perfect square, then $\sqrt{a}$ is rational.
2. If a is a positive number that is not a perfect square, then $\sqrt{a}$ is irrational.
3. If a is a negative number, then $\sqrt{a}$ is not a real number.

EXAMPLE 3 ELEMENTARY Algebra *f(x)* Now™

Classify each square root as rational, irrational, or not a real number: **a.** $\sqrt{55}$, **b.** $\sqrt{-81}$, and **c.** $-\sqrt{400}$.

Solution

a. Since 55 is not a positive perfect square, $\sqrt{55}$ is an irrational number. Rounded to two decimal places, $\sqrt{55} \approx 7.42$.

b. $\sqrt{-81}$ is not a real number because it is the square root of a negative number.

c. Since 400 is a perfect square, $-\sqrt{400}$ is rational: $-\sqrt{400} = -20$.

Self Check 3 Classify each square root as rational, irrational, or not a real number: **a.** $\sqrt{-6}$, **b.** $-\sqrt{37}$, and **c.** $\sqrt{\frac{16}{9}}$.

SQUARE ROOTS OF VARIABLE EXPRESSIONS

We have seen that the square root of a negative number is not a real number. To avoid negative radicands in this chapter, we will assume that *any variables within square root symbols do not represent negative numbers.*

EXAMPLE 4

Find each square root. All variables represent nonnegative numbers. **a.** $\sqrt{x^2}$, **b.** $\sqrt{a^4}$, **c.** $\sqrt{y^6}$, and **d.** $\sqrt{25b^2}$, **e.** $\sqrt{144t^{14}}$.

Solution

a. $\sqrt{x^2} = x$ Ask: What expression, when squared, is x^2? The answer is x.

b. $\sqrt{a^4} = a^2$ Ask: What expression, when squared, is a^4? The answer is a^2.

c. $\sqrt{y^6} = y^3$ Ask: What expression, when squared, is y^6? The answer is y^3.

d. $\sqrt{25b^2} = 5b$ Ask: What expression, when squared, is $25b^2$? The answer is $5b$.

e. $\sqrt{144t^{14}} = 12t^7$ Ask: What expression, when squared, is $144t^{14}$? The answer is $12t^7$.

Self Check 4 Find each square root: **a.** $\sqrt{m^4}$, **b.** $\sqrt{b^2}$, **c.** $\sqrt{t^8}$, and **d.** $\sqrt{81c^2}$.

THE PYTHAGOREAN THEOREM

The Pythagorean theorem is a formula that relates the lengths of the sides of a right triangle.

The Pythagorean Theorem If a and b are the lengths of the legs of a right triangle and c is the length of the hypotenuse,

$$a^2 + b^2 = c^2$$

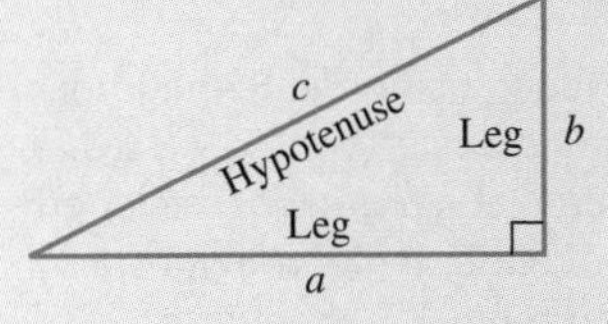

When the lengths of two sides of a right triangle are given, we can use the Pythagorean theorem and the concept of square root to find the length of the third side of the triangle.

EXAMPLE 5

ELEMENTARY Algebra f(x) Now™

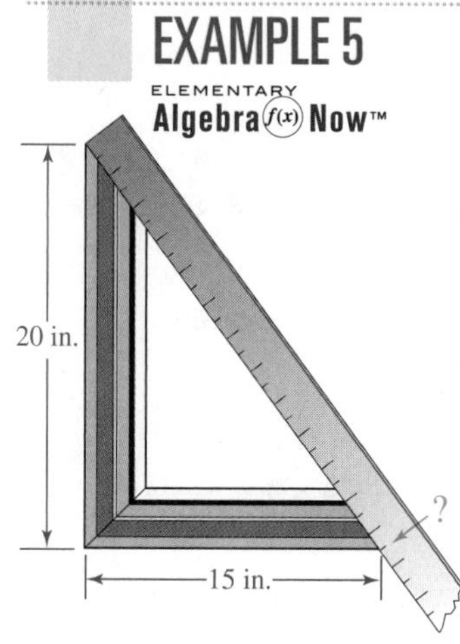

Picture frames. After nailing two pieces of molding together, a frame maker checks her work by making a diagonal measurement. What should she read on the yardstick if the sides of the frame form a right angle?

Analyze the Problem The 15- and 20-inch sides of the frame in the figure are the legs and the diagonal is the hypotenuse of a right triangle. We need to find the length of the hypotenuse.

Form an Equation We can use the Pythagorean theorem to form an equation. We substitute 15 for a, 20 for b, and let c represent the length of the hypotenuse.

Solve the Equation

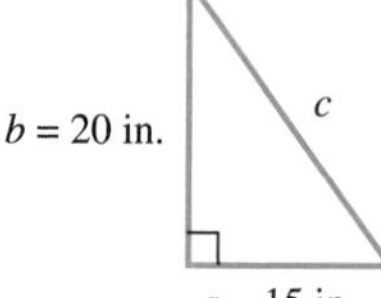

$$a^2 + b^2 = c^2 \quad \text{This is the Pythagorean theorem.}$$
$$15^2 + 20^2 = c^2 \quad \text{Substitute 15 for } a \text{ and 20 for } b.$$
$$225 + 400 = c^2 \quad 15^2 = 225 \text{ and } 20^2 = 400.$$
$$625 = c^2 \quad \text{Add.}$$

To find c, we ask "What number, when squared, is equal to 625?" There are two such numbers: the positive square root of 625 and the negative square root of 625. Since c represents the *length* of the hypotenuse, and it cannot be negative, it follows that c is the positive square root of 625.

$$625 = c^2$$
$$\sqrt{625} = c$$
$$25 = c \quad \sqrt{625} = 25 \text{ because } (25)^2 = 625.$$

State the Conclusion The diagonal measurement should be 25 inches.

Check the Result Sides of length 15, 20, and 25 inches satisfy the Pythagorean theorem.

$$15^2 + 20^2 \stackrel{?}{=} 25^2$$
$$225 + 400 \stackrel{?}{=} 625$$
$$625 = 625$$

EXAMPLE 6

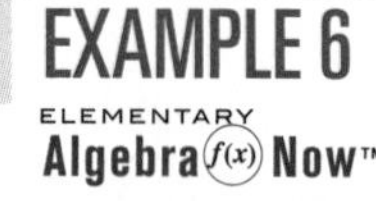

Find the missing side length of the triangle. Give the exact length and an approximation to two decimal places.

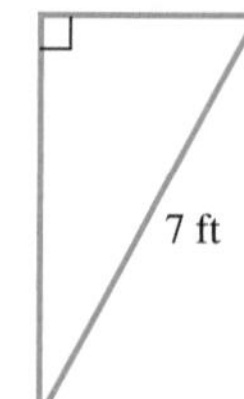

Solution Since the triangle is a right triangle, we can use the Pythagorean theorem to find the missing length. If we substitute 4 for a, 7 for c, and let b represent the unknown length of the other leg, we have

Caution

When using the Pythagorean theorem $a^2 + b^2 = c^2$, we can let a represent the length of either leg of the right triangle. We then let b represent the length of the other leg. The variable c must always represent the length of the hypotenuse.

$$a^2 + b^2 = c^2 \quad \text{This is the Pythagorean theorem.}$$
$$4^2 + b^2 = 7^2 \quad \text{The length of one leg is 4 ft. The length of the hypotenuse is 7 ft.}$$
$$16 + b^2 = 49$$
$$b^2 = 33 \quad \text{To isolate } b^2\text{, subtract 16 from both sides.}$$
$$b = \sqrt{33} \quad \text{If } b^2 = 33\text{, then } b \text{ must be a square root of 33. Because } b \text{ represents a length, it must be the positive square root of 33.}$$
$$b \approx 5.744562647 \quad \text{Use a calculator.}$$

The length of the leg is exactly $\sqrt{33}$ feet, which is approximately 5.74 feet.

Self Check 6 Find the length of the leg of the right triangle. Give the exact length and an approximation to two decimal places.

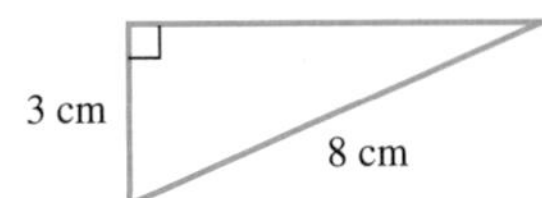

Recall from geometry that an **isosceles triangle** has two sides of equal length. An **isosceles right triangle** is a right triangle with two legs of equal length and angles that measure 45°, 45°, and 90°.

EXAMPLE 7

ELEMENTARY Algebra *f*(*x*) Now™

Roof design. The gable end of a roof is an isosceles right triangle with a span of 48 feet. Find the distance from the eaves to the peak of the roof.

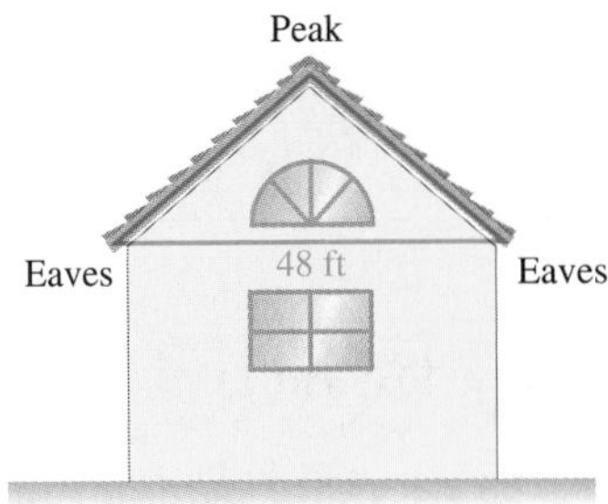

Analyze the Problem The two equal sides of the isosceles triangle form the two legs of a right triangle, and the span of 48 feet is the length of the hypotenuse.

Form an Equation Let x represent the length of each leg of the isosceles right triangle, which is the eaves-to-peak distance. We can use the Pythagorean theorem to form an equation.

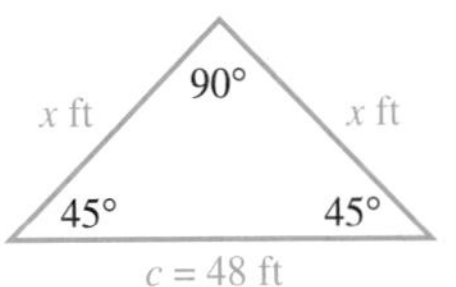

Solve the Equation

$a^2 + b^2 = c^2$	This is the Pythagorean theorem.
$x^2 + x^2 = 48^2$	Substitute x for a and b, and 48 for c.
$2x^2 = 2{,}304$	Combine like terms: $x^2 + x^2 = 2x^2$.
$x^2 = 1{,}152$	To isolate x^2, divide both sides by 2.
$x = \sqrt{1{,}152}$	If $x^2 = 1{,}152$, then x must be a square root of 1,152. Because x represents a length, it must be the positive square root of 1,152.
$x \approx 33.9411255$	Use a calculator.

State the Conclusion The eaves-to-peak distance is exactly $\sqrt{1{,}152}$ feet. Rounded to the nearest foot, the distance is approximately 34 feet.

Check the Result For an eaves-to-peak distance of 34 feet, we have $(34)^2 + (34)^2 = 1{,}156 + 1{,}156 = 2{,}312$, which is approximately $(48)^2 = 2{,}304$. The result seems reasonable.

THE DISTANCE FORMULA

We can use the Pythagorean theorem to develop a formula for finding the distance between two points $P(x_1, y_1)$ and $Q(x_2, y_2)$ on a rectangular coordinate system. The distance d between points P and Q is the length of the hypotenuse of the triangle in the figure. The two legs have lengths $x_2 - x_1$ and $y_2 - y_1$.

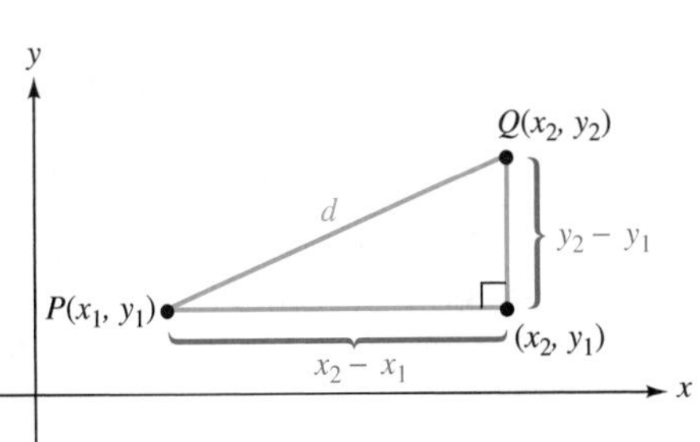

By the Pythagorean theorem, we have

$$d^2 = (x_2 - x_1)^2 + (y_2 - y_1)^2$$

Because d represents the distance between two points, it must be equal to the positive square root of $(x_2 - x_1)^2 + (y_2 - y_1)^2$.

$$d = \sqrt{(x_2 - x_1)^2 + (y_2 - y_1)^2}$$

We call this result the **distance formula.**

The Distance Formula The distance d between the points with coordinates (x_1, y_1) and (x_2, y_2) is given by

$$d = \sqrt{(x_2 - x_1)^2 + (y_2 - y_1)^2}$$

EXAMPLE 8

Find the distance between $(-4, 5)$ and $(3, -1)$.

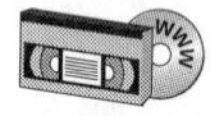

ELEMENTARY Algebra f(x) Now™

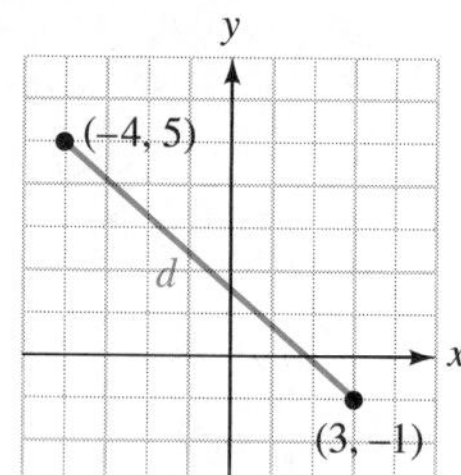

Solution To use the distance formula, it doesn't matter which point we call (x_1, y_1) and which point we call (x_2, y_2). If we choose $(x_1, y_1) = (-4, 5)$ and $(x_2, y_2) = (3, -1)$, we have

$$\begin{aligned} d &= \sqrt{(x_2 - x_1)^2 + (y_2 - y_1)^2} && \text{This is the distance formula.} \\ &= \sqrt{[3 - (-4)]^2 + (-1 - 5)^2} && \text{Substitute 3 for } x_2, -4 \text{ for } x_1, -1 \text{ for } y_2, \text{ and 5 for } y_1. \\ &= \sqrt{7^2 + (-6)^2} && \text{Do the subtractions.} \\ &= \sqrt{49 + 36} && \text{Find the powers.} \\ &= \sqrt{85} \\ &\approx 9.219544457 && \text{Use a calculator.} \end{aligned}$$

The distance between the two points is exactly $\sqrt{85}$ units, or approximately 9.22 units.

Self Check 8 Find the distance between $(-6, 4)$ and $(-1, 2)$.

Answers to Self Checks **1. a.** 7, **b.** 11, **c.** −0.3, **d.** $\frac{1}{5}$ **2.** 1.9 sec **3. a.** not a real number, **b.** irrational, **c.** rational **4. a.** m^2, **b.** b, **c.** t^4, **d.** $9c$ **6.** $\sqrt{55}$ cm ≈ 7.42 cm **8.** $\sqrt{29} \approx 5.39$

8.1 STUDY SET

ELEMENTARY Algebra f(x) Now™

VOCABULARY **Fill in the blanks.**

1. The number b is a ________ root of the number a if $b^2 = a$.
2. The symbol $\sqrt{}$ is called a ________ symbol. It represents the ________ or principal square root of a number.
3. The symbol $-\sqrt{}$ is used to represent the ________ square root of a number.
4. The number or variable expression under a radical symbol is called the ________.
5. A number such as 25, 49, or $\frac{4}{81}$ that is the square of some rational number, is called a ________ square.
6. $\sqrt{-11}$ is not a ________ number.
7. The ________ theorem is a formula that relates the lengths of the sides of a right triangle.

8. The figure illustrates a(n) ________ right triangle.

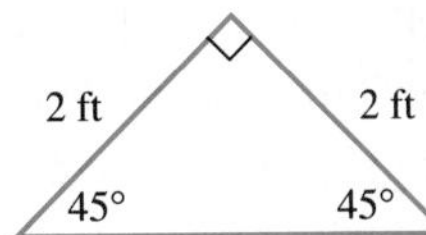

CONCEPTS **Fill in the blanks.**

9. Every positive number has ______ square roots, one positive and one negative.

10. To find the square roots of 36, we ask "What number, when squared, is equal to ___?"

11. The positive square root of 25 is ___ and the negative square root of 25 is ___. In symbols, we write

$$\sqrt{25} = \square \quad \text{and} \quad -\sqrt{25} = \square$$

12. The number 0 is the only real number with exactly one square root: $\sqrt{0} = \square$.

13. 2 is a square root of 4, because $(\square)^2 = 4$ and -2 is a square root of 4, because $(\square)^2 = 4$.

14. The Pythagorean theorem is:

$$a^2 + b^2 = \square$$

15. The distance formula is:

$$d = \sqrt{(x_2 - \square)^2 + (\square - y_1)^2}$$

Classify each square root as rational, irrational, or not a real number.

16. $\sqrt{9}$ **17.** $\sqrt{17}$

18. $\sqrt{-21}$ **19.** $-\sqrt{99}$

Identify the radicand for each radical expression.

20. $\sqrt{64}$ **21.** $-\sqrt{45}$

22. $\sqrt{-8}$ **23.** $\sqrt{16x^2}$

Determine whether each statement is true or false.

24. $-\sqrt{36} = \sqrt{-36}$ **25.** $-\sqrt{43} = -1 \cdot \sqrt{43}$

NOTATION **Complete the solution.**

26. The legs of a right triangle measure 5 and 12 centimeters. Find the length of the hypotenuse.

$$a^2 + b^2 = c^2$$
$$\square^2 + 12^2 = c^2$$
$$25 + \square = c^2$$
$$\square = c^2$$
$$\sqrt{169} = \square$$
$$\square = c$$

PRACTICE **Find each square root without using a calculator.**

27. $\sqrt{49}$ **28.** $\sqrt{25}$

29. $\sqrt{36}$ **30.** $\sqrt{1}$

31. $\sqrt{100}$ **32.** $\sqrt{144}$

33. $\sqrt{400}$ **34.** $\sqrt{900}$

35. $-\sqrt{81}$ **36.** $-\sqrt{36}$

37. $\sqrt{1.21}$ **38.** $\sqrt{1.69}$

39. $\sqrt{169}$ **40.** $\sqrt{196}$

41. $\sqrt{\frac{9}{25}}$ **42.** $\sqrt{\frac{9}{49}}$

43. $-\sqrt{\frac{1}{64}}$ **44.** $-\sqrt{\frac{1}{16}}$

45. $-\sqrt{0.04}$ **46.** $-\sqrt{0.64}$

47. $-\sqrt{289}$ **48.** $-\sqrt{324}$

49. $\sqrt{2{,}500}$ **50.** $\sqrt{625}$

Use a calculator to approximate each square root to three decimal places.

51. $\sqrt{3}$ **52.** $\sqrt{2}$

53. $\sqrt{11}$ **54.** $\sqrt{53}$

55. $\sqrt{95}$ **56.** $\sqrt{99}$

57. $\sqrt{428}$ **58.** $\sqrt{844}$

59. $2\sqrt{3}$ **60.** $3\sqrt{2}$

Find each square root. Assume that all variables represent nonnegative numbers.

61. $\sqrt{m^2}$ **62.** $\sqrt{b^2}$

63. $\sqrt{t^4}$ **64.** $\sqrt{a^6}$

65. $\sqrt{c^{10}}$ **66.** $\sqrt{r^8}$

67. $\sqrt{4y^2}$ **68.** $\sqrt{9b^2}$

69. $\sqrt{64b^4}$ **70.** $\sqrt{100s^{16}}$

Refer to the right triangle.

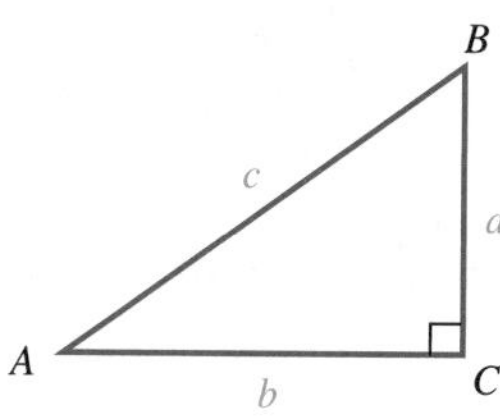

71. Find c if $a = 4$ and $b = 3$.
72. Find c if $a = 5$ and $b = 12$.
73. Find a if $b = 16$ and $c = 34$.
74. Find a if $b = 45$ and $c = 53$.
75. Find b if $c = 125$ and $a = 44$.
76. Find c if $a = 176$ and $b = 57$.

For each isosceles right triangle, find the length of the legs. Give an exact answer and an approximation to two decimal places.

77.

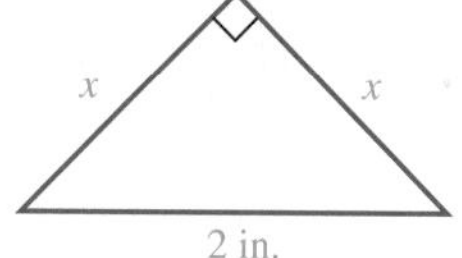

78.

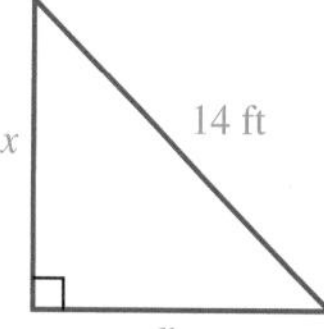

Find the distance between the two points. If an answer contains a radical, give an exact answer and an approximation to two decimal places.

79. (4, 6) and (1, 2)
80. (9, 8) and (3, 0)
81. $(-2, -8)$ and (3, 4)
82. $(-5, -2)$ and (7, 3)
83. $(6, -5)$ and $(0, -4)$
84. $(-2, -5)$ and $(6, -6)$
85. $(9, -1)$ and $(7, -6)$
86. $(5, -6)$ and $(-1, 1)$

APPLICATIONS **If an answer contains a radical, give an exact answer and an approximation to two decimal places.**

87. HIGHWAY SAFETY The formula $s = 4.5\sqrt{d}$ relates the speed s (in mph) of a car and the distance d (in feet) of the skid when a driver hits the brakes on dry pavement. How fast was a car traveling if it leaves a 64-foot-long skid mark?

88. FREE FALL The time t (in seconds) that it takes an object dropped from a distance d (in feet) to fall to the ground is given by the formula $t = 0.25\sqrt{d}$. How long will it take a pair of sunglasses to fall to the ground if a sightseer drops them from the top of a 400-foot-tall observation tower?

89. DRAFTING Among the tools used in drafting are the 30-60-90 and the 45-45-90 triangles shown.
    a. Find the length of the hypotenuse of the 45-45-90 triangle if it is $\sqrt{2}$ times as long as a leg.
    b. Find the length of the side opposite the 60° angle of the 30-60-90 triangle if it is $\sqrt{3}$ times as long as one-half of the hypotenuse.

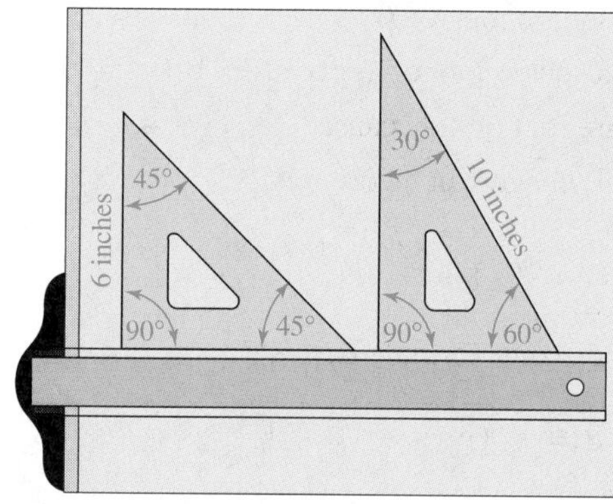

90. GARDENING A rectangular garden has sides of 28 and 45 feet. Find the length of a path that extends from one corner to the opposite corner.

91. BASEBALL A baseball diamond is a square, with each side 90 feet long. How far is it from home plate to second base?

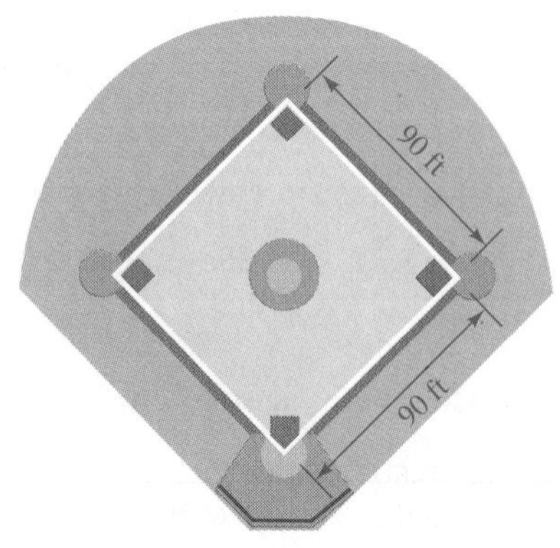

92. TELEVISION The size of a television screen is the diagonal distance from the upper left to the lower right corner. What is the size of the screen shown in the illustration?

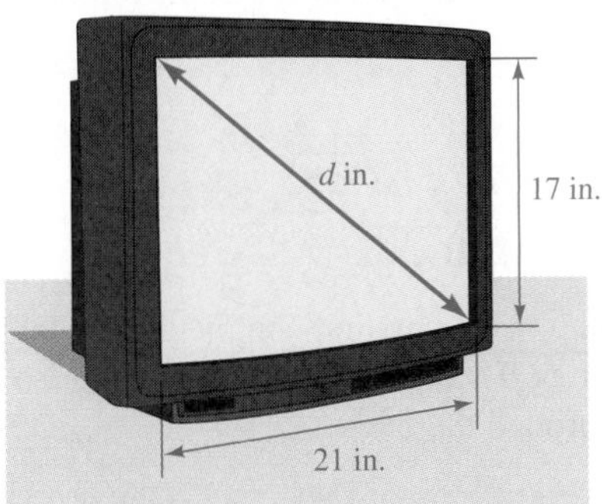

93. QUALITY CONTROL A tool manufacturer measures the distance across each carpenter's square it produces to make sure that the sides meet to form a 90° angle.
   a. Each side of a square is scaled in units of inches. How long is each side?
   b. What diagonal measurement will guarantee that the sides of the square form a 90° angle?

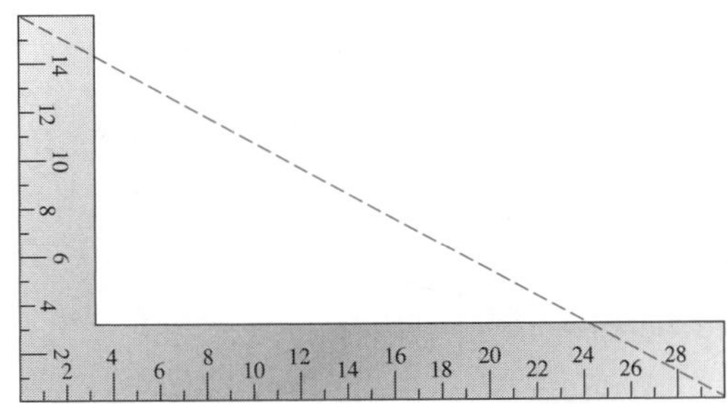

94. WRESTLING The diagonal distance from corner to corner of a square wrestling ring is 24 feet. Find the length of a side of the ring.

95. FOOTBALL On first down and ten, a quarterback tells his tight end to go out 6 yards, cut 45° to the right, and run 6 yards, as shown in the illustration. The tight end follows instructions, catches a pass, and is tackled immediately.
   a. Find x.
   b. Does he gain the necessary 10 yards for a first down?

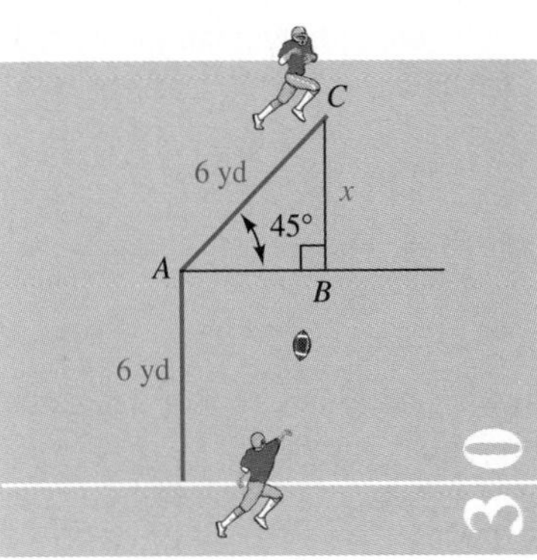

96. DECK DESIGN The plans for a patio deck shown in the illustration call for three redwood support braces directly under the hot tub. Find the length of each support. Round to the nearest tenth of a foot.

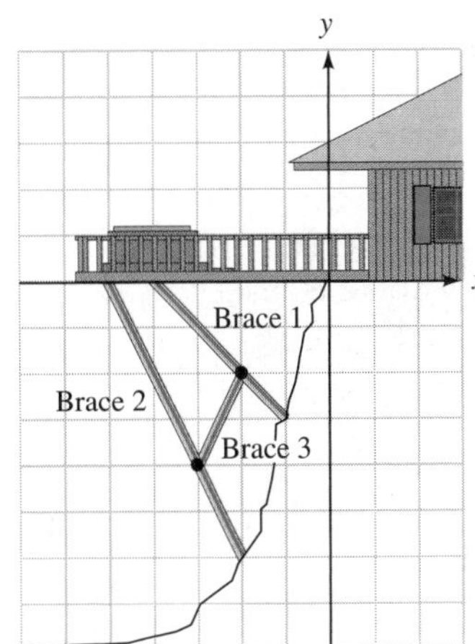

WRITING

97. Is the statement $-\sqrt{9} = \sqrt{-9}$ true or false? Explain your answer.

98. Consider the statement $\sqrt{26} \approx 5.1$. Explain why an $\approx$ symbol is used instead of an $=$ symbol.

99. Suppose you are told that $\sqrt{10} \approx 3.16$. Explain how another key on your calculator (besides the square root key $\sqrt{\ }$) could be used to see whether this is a reasonable approximation.

100. A calculator was used to find $\sqrt{-16}$. Explain the meaning of the message on the display.

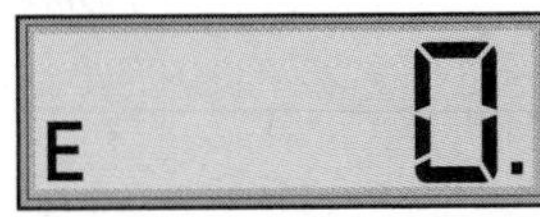

101. Explain why the Pythagorean theorem does not hold for this triangle.

$$3^2 + 4^2 \stackrel{?}{=} 6^2$$
$$9 + 16 \stackrel{?}{=} 36$$
$$25 \neq 36$$

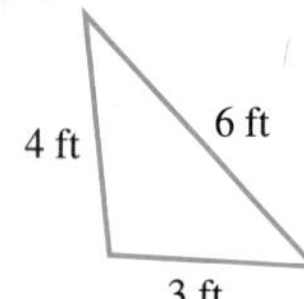

102. Explain why the square root of a negative number cannot be a real number.

REVIEW

103. Add: $(3s^2 - 3s - 2) + (3s^2 + 4s - 3)$.

104. Subtract: $(3c^2 - 2c + 4) - (c^2 - 3c + 7)$.

105. Multiply: $(3x - 2)(x + 4)$.

106. Divide: $x^2 + 13x + 12$ by $x + 1$.

CHALLENGE PROBLEMS

107. SHORTCUTS Instead of walking on the sidewalk, students take a diagonal shortcut across a rectangular vacant lot, as shown in the illustration. How much distance do they save?

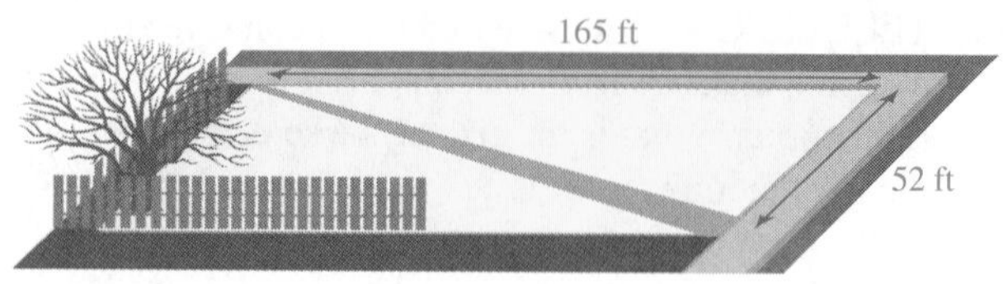

108. Evaluate: $\sqrt{\sqrt{\sqrt{256}}}$.

109. Simplify: $\sqrt{16x^{16n}}$.

110. Graph $y = \sqrt{x}$ by first completing the table of solutions. Then plot the points and draw a smooth line through them.

x	-1	0	1	3	4	5	9	12	16
y									

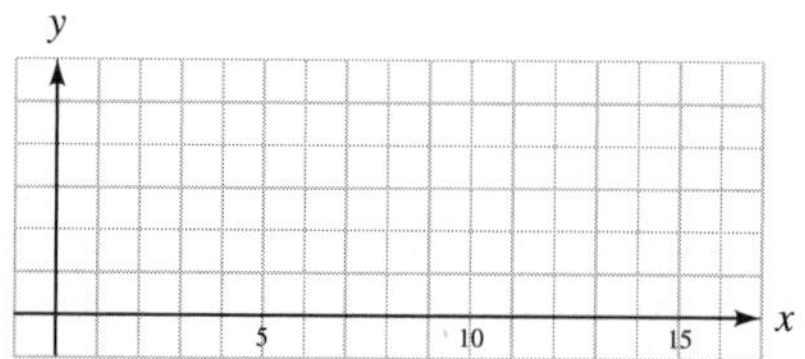

8.2 Simplifying Square Roots

- The Product Rule for Square Roots
- Using the Product Rule to Simplify Square Roots
- The Quotient Rule for Square Roots
- Using the Quotient Rule to Simplify Square Roots

In algebra, it is often helpful to replace an expression with a simpler equivalent expression. This is certainly true when working with square roots. In most cases, square roots should be written in simplified form. We can use the product and quotient rules for square roots to do this.

THE PRODUCT RULE FOR SQUARE ROOTS

To introduce the product rule for square roots, we will find $\sqrt{4 \cdot 25}$ and $\sqrt{4}\sqrt{25}$, and compare the results.

Notation

The product $\sqrt{4}\sqrt{25}$ can also be written as $\sqrt{4} \cdot \sqrt{25}$.

Square root of a product

$$\sqrt{4 \cdot 25} = \sqrt{100}$$
$$= 10$$

Product of square roots

$$\sqrt{4}\sqrt{25} = 2 \cdot 5$$
$$= 10$$

Read as "the square root of 4 times the square root of 25."

In each case, the answer is 10. Thus, $\sqrt{4 \cdot 25} = \sqrt{4}\sqrt{25}$. This result illustrates the *product rule for square roots.*

The Product Rule for Square Roots

The square root of the product of two nonnegative numbers is equal to the product of their square roots.

For any nonnegative real numbers a and b,

$$\sqrt{a \cdot b} = \sqrt{a}\sqrt{b}$$

USING THE PRODUCT RULE TO SIMPLIFY SQUARE ROOTS

A square root is in **simplified form** when each of the following is true.

Simplified Form of a Square Root

1. Except for 1, the radicand has no perfect square factors.
2. No fraction appears in the radicand.
3. No radical appears in the denominator of a fraction.

To simplify square roots, we follow these steps.

Simplifying Square Roots

1. Write the radicand as a product of the greatest perfect square factor and one other factor.
2. Use the product rule for square roots to write the expression as a product of square roots.
3. Find the square root of the perfect square factor.

In step 1 of this method, we factor the radicand using two natural number factors. Since one factor must be a perfect square, it is helpful to memorize these **natural number perfect squares.**

1, 4, 9, 16, 25, 36, 49, 64, 81, 100, 121, 144, 169, 196, 225

EXAMPLE 1

Simplify: **a.** $\sqrt{12}$, **b.** $\sqrt{200}$, and **c.** $\sqrt{35}$.

ELEMENTARY Algebra $f(x)$ Now™

Solution

a. From the previous list, we see that 4 is the greatest perfect square factor of 12. To simplify $\sqrt{12}$, we write 12 as the product of 4 and one other factor. Then we apply the product rule for square roots.

$\sqrt{12} = \sqrt{4 \cdot 3}$ — Write 12 as $12 = 4 \cdot 3$.

(Write the perfect square factor first.)

$= \sqrt{4}\sqrt{3}$ — The square root of a product is equal to the product of the square roots.

$= 2\sqrt{3}$ — Find the square root of the perfect square factor: $\sqrt{4} = 2$. Read as "2 times the square root of 3" or as "2 radical 3."

We say that $2\sqrt{3}$ is the simplified form of $\sqrt{12}$.

Caution

$\sqrt{12} = 2\sqrt{3}$

In simplified form, the radicand should not have any perfect square factors.

b. To simplify $\sqrt{200}$, we note that 200 has several perfect square factors: $200 = 4 \cdot 50$, $200 = 25 \cdot 8$, and $200 = 100 \cdot 2$. Of these, the greatest perfect square factor is 100.

$\sqrt{200} = \sqrt{100 \cdot 2}$ Write 200 as the product of its greatest perfect square factor and one other factor.

$= \sqrt{100}\sqrt{2}$ The square root of a product is equal to the product of the square roots.

$= 10\sqrt{2}$ Find the square root of the perfect square factor: $\sqrt{100} = 10$.

c. For $\sqrt{35}$, the radicand 35 has no perfect square factor other than 1. Therefore, $\sqrt{35}$ cannot be simplified.

Self Check 1 Simplify: **a.** $\sqrt{45}$, **b.** $\sqrt{600}$, and **c.** $\sqrt{33}$.

Prime factorization can be helpful in finding the greatest perfect square factor of a radicand.

EXAMPLE 2

Simplify: $\sqrt{150}$.

ELEMENTARY Algebra f(x) Now™

Solution Since the greatest perfect square factor of 150 is not obvious, we prime-factor 150 and look for pairs of common factors to simplify the square root.

$\sqrt{150} = \sqrt{2 \cdot 3 \cdot 5 \cdot 5}$ Find the prime factorization of 150.

$= \sqrt{2 \cdot 3}\sqrt{5 \cdot 5}$ Use the product rule for square roots. Write $5 \cdot 5$ within one square root because it is a perfect square: $5 \cdot 5 = 25$.

$= \sqrt{6} \cdot 5$ $\sqrt{5 \cdot 5} = \sqrt{25}$, which simplifies to 5.

$= 5\sqrt{6}$ Write the factor of 5 first.

Notation

Common practice is to write a product such as $\sqrt{6} \cdot 5$ with the square root factor last: $5\sqrt{6}$. This way, no misunderstanding can occur about exactly what is under the radical symbol.

Self Check 2 Simplify: $\sqrt{140}$.

Variable expressions can also be perfect squares. For example, x^2, x^4, x^6, and x^8 are perfect squares because

$$x^2 = (x)^2, \quad x^4 = (x^2)^2, \quad x^6 = (x^3)^2, \quad \text{and} \quad x^8 = (x^4)^2$$

Perfect squares like these, which are even powers of a variable, are used to simplify square roots involving variable radicands.

EXAMPLE 3

Simplify: **a.** $\sqrt{x^3}$ and **b.** $\sqrt{a^9}$.

Solution If a radicand contains a variable to an odd power, factor the exponential expression as the product of the greatest possible even power and the variable to the first power.

a. We note that x^2 is the greatest even power of x that is a factor of x^3. To simplify $\sqrt{x^3}$, we write x^3 as the product of x^2 and x and apply the product rule for square roots.

$\sqrt{x^3} = \sqrt{x^2 \cdot x}$ Factor: $x^3 = x^2 \cdot x$.

$= \sqrt{x^2}\sqrt{x}$ The square root of a product is equal to the product of the square roots.

$= x\sqrt{x}$ Find the square root of the perfect square factor: $\sqrt{x^2} = x$.

b. To simplify $\sqrt{a^9}$, we note that a^8 is the greatest even power of a that is a factor of a^9.

$$\sqrt{a^9} = \sqrt{a^8 \cdot a}$$ Factor: $a^9 = a^8 \cdot a$.

$$= \sqrt{a^8}\sqrt{a}$$ The square root of a product is equal to the product of the square roots.

$$= a^4\sqrt{a}$$ Find the square root of the perfect square factor: $\sqrt{a^8} = a^4$.

Self Check 3 Simplify: **a.** $\sqrt{x^5}$ and **b.** $\sqrt{b^{11}}$.

EXAMPLE 4

Simplify: $6\sqrt{27m}$.

ELEMENTARY Algebra f(x) Now™

Solution Since $6\sqrt{27m}$ means $6 \cdot \sqrt{27m}$, we first simplify $\sqrt{27m}$ and then we multiply the result by 6.

Caution

It incorrect to factor $27m$ as $9m \cdot 3$, because $9m$ is not a perfect square.

$$\sqrt{27m} = \sqrt{9m \cdot 3}$$

↑ not a perfect square

$$6\sqrt{27m} = 6\sqrt{9 \cdot 3m}$$ Write $27m$ as a product of its greatest perfect square factor and one other factor: $27m = 9 \cdot 3m$.

$$= 6\sqrt{9}\sqrt{3m}$$ The square root of a product is equal to the product of the square roots.

$$= 6 \cdot 3\sqrt{3m}$$ Find the square root of the perfect square factor: $\sqrt{9} = 3$.

$$= 18\sqrt{3m}$$ Multiply 6 and 3.

Self Check 4 Simplify: $3\sqrt{32y}$.

EXAMPLE 5

Simplify: **a.** $\sqrt{72x^5}$ and **b.** $-3a\sqrt{144a^4b^7}$.

ELEMENTARY Algebra f(x) Now™

Solution **a.** To simplify $\sqrt{72x^5}$, we look for the greatest perfect square factor of $72x^5$. Because

- 36 is the greatest perfect square factor of 72, and
- x^4 is the greatest perfect square factor of x^5,

$36x^4$ is the greatest perfect square factor of $72x^5$.

Caution

$$\sqrt{72x^5} = 6x^2\sqrt{2x}$$

↑

Simplified form: The radicand should not contain any variables with an exponent greater than 1.

$$\sqrt{72x^5} = \sqrt{36x^4 \cdot 2x}$$ Write $72x^5$ as a product of its greatest perfect square factor and one other factor: $72x^5 = 36x^4 \cdot 2x$.

$$= \sqrt{36x^4}\sqrt{2x}$$ The square root of a product is equal to the product of the square roots.

$$= 6x^2\sqrt{2x}$$ Find the square root of the perfect square factor: $\sqrt{36x^4} = 6x^2$.

b. To simplify $-3a\sqrt{144a^4b^7}$, we simplify $\sqrt{144a^4b^7}$ and multiply that result by $-3a$. To simplify $\sqrt{144a^4b^7}$, we look for the greatest perfect square factor of $144a^4b^7$. Because

- 144 is a perfect square,
- a^4 is a perfect square, and
- b^6 is the greatest perfect square factor of b^7,

$144a^4b^6$ is the greatest perfect square factor of $144a^4b^7$.

$$-3a\sqrt{144a^4b^7} = -3a\sqrt{144a^4b^6 \cdot b}$$ Write $144a^4b^7$ as a product of its greatest perfect square factor and one other factor: $144a^4b^7 = 144a^4b^6 \cdot b$.

$$= -3a\sqrt{144a^4b^6}\sqrt{b}$$ The square root of a product is equal to the product of the square roots.

$$= -3a \cdot 12a^2b^3\sqrt{b}$$ $\sqrt{144a^4b^6} = 12a^2b^3$.

$$= -36a^3b^3\sqrt{b}$$ Multiply $-3a$ and $12a^2b^3$.

Self Check 5 Simplify: **a.** $\sqrt{75y^3}$ and **b.** $-5q\sqrt{49p^9q^4}$.

THE QUOTIENT RULE FOR SQUARE ROOTS

To introduce the quotient rule for square roots, we will find $\sqrt{\frac{100}{25}}$ and $\frac{\sqrt{100}}{\sqrt{25}}$, and compare the results.

Square root of a quotient

$$\sqrt{\frac{100}{25}} = \sqrt{4}$$
$$= 2$$

Quotient of square roots

$$\frac{\sqrt{100}}{\sqrt{25}} = \frac{10}{5}$$
$$= 2$$

Read as "the square root of 100 divided by the square root of 25."

Since the answer is 2 in each case, $\sqrt{\frac{100}{25}} = \frac{\sqrt{100}}{\sqrt{25}}$. This result illustrates the *quotient rule for square roots.*

The Quotient Rule for Square Roots

The square root of the quotient of two numbers is equal to the quotient of their square roots.

For any positive real numbers a and b,

$$\sqrt{\frac{a}{b}} = \frac{\sqrt{a}}{\sqrt{b}}$$

USING THE QUOTIENT RULE TO SIMPLIFY SQUARE ROOTS

We use the quotient rule for square roots to simplify square roots in much the same way as we used the product rule.

EXAMPLE 6

Simplify: **a.** $\sqrt{\frac{4}{9}}$, **b.** $\sqrt{\frac{53}{121}}$, and **c.** $\sqrt{\frac{108}{25}}$.

ELEMENTARY Algebra f(x) Now™

Solution In each case, the square root is not in simplified form because the radicand contains a fraction. To write each one in simplified form, we use the quotient rule for square roots.

a. In Section 8.1, we found that $\sqrt{\frac{4}{9}} = \frac{2}{3}$, because $\left(\frac{2}{3}\right)^2 = \frac{4}{9}$. We can also use the quotient rule for square roots to simplify this expression.

$$\sqrt{\frac{4}{9}} = \frac{\sqrt{4}}{\sqrt{9}}$$ The square root of a quotient is equal to the quotient of the square roots.

$$= \frac{2}{3}$$ $\sqrt{4} = 2$ and $\sqrt{9} = 3$.

b. $\sqrt{\dfrac{53}{121}} = \dfrac{\sqrt{53}}{\sqrt{121}}$ — The square root of a quotient is equal to the quotient of the square roots.

$= \dfrac{\sqrt{53}}{11}$ — $\sqrt{121} = 11$.

c. $\sqrt{\dfrac{108}{25}} = \dfrac{\sqrt{108}}{\sqrt{25}}$ — The square root of a quotient is equal to the quotient of the square roots.

$= \dfrac{\sqrt{36 \cdot 3}}{5}$ — To simplify $\sqrt{108}$, factor 108 using its greatest perfect square factor, 36. Write $\sqrt{25}$ as 5.

$= \dfrac{\sqrt{36}\sqrt{3}}{5}$ — In the numerator, use the product rule: The square root of a product is equal to the product of the square roots.

$= \dfrac{6\sqrt{3}}{5}$ — $\sqrt{36} = 6$. This result can also be written as $\dfrac{6}{5}\sqrt{3}$.

Self Check 6 Simplify: **a.** $\sqrt{\dfrac{36}{121}}$, **b.** $\sqrt{\dfrac{29}{64}}$, and **c.** $\sqrt{\dfrac{72}{169}}$.

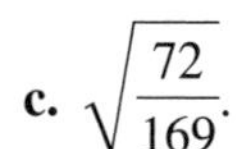

EXAMPLE 7

Simplify: $\sqrt{\dfrac{44x^3}{9xy^2}}$. Assume that the variables represent positive numbers.

ELEMENTARY Algebra f(x) Now™

Solution Before using the quotient rule for square roots, we note that the radicand is a rational expression that can be simplified.

$\sqrt{\dfrac{44x^3}{9xy^2}} = \sqrt{\dfrac{44x^2}{9y^2}}$ — Simplify the radicand by removing a factor of x that is common to the numerator and denominator.

$= \dfrac{\sqrt{44x^2}}{\sqrt{9y^2}}$ — The square root of a quotient is equal to the quotient of the square roots.

$= \dfrac{\sqrt{4x^2}\sqrt{11}}{\sqrt{9y^2}}$ — To simplify $\sqrt{44x^2}$, factor $44x^2$ using its greatest perfect square factor: $\sqrt{4x^2 \cdot 11}$. Then use the product rule: The square root of a product is equal to the product of the square roots.

$= \dfrac{2x\sqrt{11}}{3y}$ — $\sqrt{4x^2} = 2x$ and $\sqrt{9y^2} = 3y$.

Self Check 7 Simplify: $\sqrt{\dfrac{99b^4}{16a^2b^2}}$.

Answers to Self Checks **1. a.** $3\sqrt{5}$, **b.** $10\sqrt{6}$, **c.** cannot be simplified **2.** $2\sqrt{35}$ **3. a.** $x^2\sqrt{x}$, **b.** $b^5\sqrt{b}$ **4.** $12\sqrt{2y}$ **5. a.** $5y\sqrt{3y}$, **b.** $-35p^4q^3\sqrt{p}$ **6. a.** $\frac{6}{11}$, **b.** $\frac{\sqrt{29}}{8}$, **c.** $\frac{6\sqrt{2}}{13}$ **7.** $\frac{3b\sqrt{11}}{4a}$

8.2 STUDY SET  ELEMENTARY Algebra $f(x)$ Now™

VOCABULARY **Fill in the blanks.**

1. The first four natural number perfect ________ are 1, 4, 9, and 16.
2. The ________ of $\sqrt{18x^2}$ is $18x^2$.
3. "To ________ $\sqrt{20}$" means to write it as $2\sqrt{5}$.
4. When we write $24x^2$ as $4x^2 \cdot 6$, we say we have ________ $24x^2$.
5. The word *product* is associated with the operation of ____________ and the word *quotient* with ________.
6. x^4 is an ______ power of the variable x and b^5 is an odd power of the variable b.
7. The expression $\sqrt{4 \cdot 3}$ is a square root of a ________ and $\sqrt{4}\sqrt{3}$ is a product of two _______ roots.
8. The expression $\sqrt{\frac{40}{9}}$ is a square root of a ________ and $\frac{\sqrt{40}}{\sqrt{9}}$ is a quotient of two _______ roots.

CONCEPTS

9. Fill in the blanks.
   a. The square root of the product of two positive numbers is equal to the ________ of their square roots. In symbols, $\sqrt{a \cdot b} = \square$.
   b. The square root of the quotient of two positive numbers is equal to the ________ of their square roots. In symbols,

   $$\sqrt{\frac{a}{b}} = \square, \text{ where } b \neq 0$$

10. Complete each list of perfect squares.
    a. 1, 4, 9, $\square$, 25, 36, $\square$, 64, 81, $\square$, 121, . . .
    b. x^2, $\square$, x^6, $\square$, $\square$, x^{12}, . . .
11. Which natural number perfect square, 1, 4, 9, 16, 25, 36, 49, 64, 81, or 100 is the greatest factor of the given number?
    a. 20 b. 98
    c. 54 d. 48
12. Complete each factorization using a perfect square.
    a. $24 = \square \cdot 6$ b. $63 = \square \cdot 7$
    c. $x^5 = \square \cdot x$ d. $12x^3 = \square \cdot 3x$
13. To simplify $\sqrt{40}$, which of the following expressions should be used?

    $\sqrt{20 \cdot 2}$ $\sqrt{40 \cdot 1}$ $\sqrt{8 \cdot 5}$ $\sqrt{4 \cdot 10}$

14. To simplify $\sqrt{n^5}$, which of the following expressions should be used?

    $\sqrt{n^2 \cdot n^3}$ $\sqrt{n^3 \cdot n^2}$ $\sqrt{n^4 \cdot n}$

15. Write each expression as a product of square roots. Do not simplify.
    a. $\sqrt{81 \cdot 2}$ b. $\sqrt{m^4 \cdot m}$
16. Write each expression as a quotient of square roots. Do not simplify.
    a. $\sqrt{\frac{49}{36}}$ b. $\sqrt{\frac{17}{25}}$
17. Find each square root.
    a. $\sqrt{100}$ b. $\sqrt{t^2}$
    c. $\sqrt{a^4}$ d. $\sqrt{144y^2}$
18. Simplify each expression.
    a. $\sqrt{25}\sqrt{7}$ b. $\sqrt{x^4}\sqrt{x}$
    c. $\frac{\sqrt{64}}{\sqrt{25}}$ d. $\sqrt{\frac{29}{4}}$

NOTATION **Complete the solution to simplify the square root.**

19.
$$\begin{aligned}\sqrt{32} &= \sqrt{\square \cdot 2}\\ &= \sqrt{\square}\sqrt{2}\\ &= \square\sqrt{2}\end{aligned}$$

20. We can read $2\sqrt{6}$ as "2 ______ the square root of 6" or as "2 ________ 6."
21. Simplify each expression.
    a. $3 \cdot 2\sqrt{15}$ b. $5x \cdot 4x\sqrt{14}$
22. Write each expression in better form.
    a. $\sqrt{15} \cdot 9$ b. $\sqrt{3} \cdot (-10)$

PRACTICE **Simplify, if possible. Assume that all variables represent positive numbers.**

23. $\sqrt{20}$ 24. $\sqrt{18}$
25. $\sqrt{50}$ 26. $\sqrt{75}$
27. $\sqrt{27}$ 28. $\sqrt{54}$
29. $\sqrt{98}$ 30. $\sqrt{88}$
31. $\sqrt{48}$ 32. $\sqrt{72}$

33. $\sqrt{44}$
34. $\sqrt{60}$
35. $\sqrt{15}$
36. $\sqrt{21}$
37. $\sqrt{192}$
38. $\sqrt{147}$
39. $\sqrt{250}$
40. $\sqrt{128}$
41. $-\sqrt{700}$
42. $-\sqrt{300}$
43. $\sqrt{162}$
44. $\sqrt{1{,}000}$
45. $3\sqrt{32}$
46. $3\sqrt{52}$
47. $-3\sqrt{72}$
48. $-2\sqrt{28}$
49. $\sqrt{x^5}$
50. $\sqrt{n^3}$
51. $\sqrt{n^9}$
52. $\sqrt{t^7}$
53. $2\sqrt{24}$
54. $2\sqrt{27}$
55. $\sqrt{4k}$
56. $\sqrt{9p}$
57. $\sqrt{12x}$
58. $\sqrt{20y}$
59. $6\sqrt{75t}$
60. $2\sqrt{24s}$
61. $\sqrt{25x^3}$
62. $\sqrt{36y^3}$
63. $\sqrt{10b}$
64. $\sqrt{30s}$
65. $\sqrt{9x^4y}$
66. $\sqrt{16xy^2}$
67. $\sqrt{2c^7d^{11}}$
68. $\sqrt{3x^9y^5}$
69. $-12x\sqrt{16x^2y^3}$
70. $-4x^5y^3\sqrt{36x^3y^3}$
71. $\frac{1}{5}x^2y\sqrt{50x^2y^2}$
72. $\frac{1}{5}x^5y\sqrt{75x^3y^2}$
73. $\sqrt{\frac{25}{9}}$
74. $\sqrt{\frac{36}{49}}$
75. $\sqrt{\frac{81}{64}}$
76. $\sqrt{\frac{121}{144}}$
77. $\sqrt{\frac{6}{121}}$
78. $\sqrt{\frac{15}{64}}$
79. $\sqrt{\frac{26}{25}}$
80. $\sqrt{\frac{17}{169}}$
81. $-\sqrt{\frac{20}{49}}$
82. $-\sqrt{\frac{50}{9}}$
83. $\sqrt{\frac{48}{81}}$
84. $\sqrt{\frac{27}{64}}$
85. $\sqrt{\frac{32}{25}}$
86. $\sqrt{\frac{75}{16}}$
87. $\sqrt{\frac{72x^3}{y^2}}$
88. $\sqrt{\frac{108b^2}{d^4}}$
89. $\sqrt{\frac{125n^5}{64n}}$
90. $\sqrt{\frac{72q^7}{25q^3}}$
91. $\sqrt{\frac{128m^3n^5}{81mn^7}}$
92. $\sqrt{\frac{75p^3q^2}{p^5q^4}}$
93. $\sqrt{\frac{12r^7s^7}{r^5s^2}}$
94. $\sqrt{\frac{m^2n^9}{100mn^3}}$

APPLICATIONS

95. LADDERS Use the Pythagorean theorem to find the exact length of the ladder shown in the illustration. Express the answer in simplified radical form.

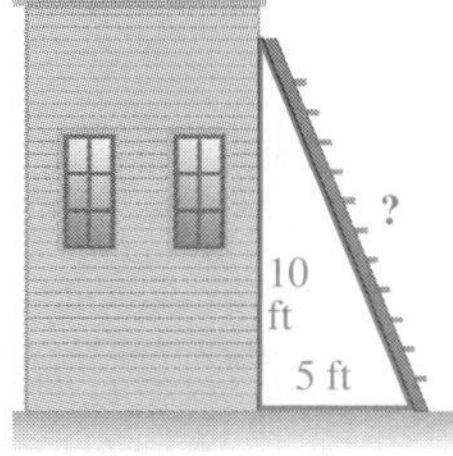

96. DANCING A dance floor is a square with sides of length 12 yards. Use the Pythagorean theorem to find the exact length of the diagonal shown in the illustration. Express the answer in simplified radical form.

97. CROSSWORD PUZZLES A crossword puzzle occupies an area of 28 square inches in a newspaper. Since it is a square, each side of the puzzle is exactly $\sqrt{28}$ inches long.
   a. Express the length in simplified radical form.
   b. Approximate the length to the nearest tenth of an inch.

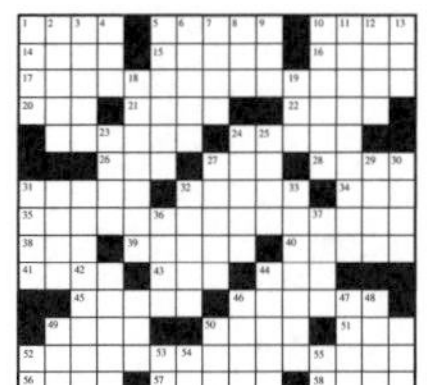

98. AMUSEMENT PARK RIDES The time t (in seconds) that it takes the "Swashbuckler" pirate ship ride to swing from one extreme to the other is given by the formula

$$t = \pi\sqrt{\frac{L}{32}}$$

a. Find t and express it in simplified radical form. Leave π in the answer.

b. Approximate the time to the nearest tenth of a second.

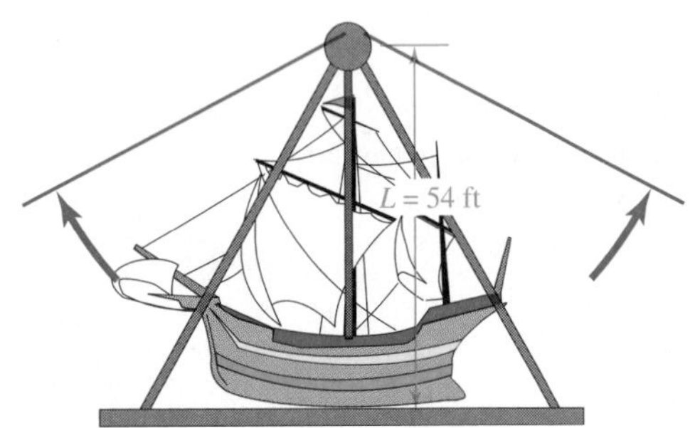

WRITING

99. Explain why each square root is not in simplified form.

a. $\sqrt{8}$ **b.** $\sqrt{\frac{7}{4}}$

100. Explain how to simplify $\sqrt{x^3}$.

101. Explain this property: $\sqrt{a \cdot b} = \sqrt{a}\sqrt{b}$.

102. Explain this property: $\sqrt{\frac{a}{b}} = \frac{\sqrt{a}}{\sqrt{b}}$.

REVIEW Solve each system.

103. $\begin{cases} y = 2x - 6 \\ 2x + y = 6 \end{cases}$

104. $\begin{cases} 2x + y = -2 \\ -2x - 3y = -6 \end{cases}$

105. $\begin{cases} 3x + 4y = -7 \\ 2x - y = -1 \end{cases}$

106. $\begin{cases} 2x + 3y = 8 \\ 3x - 2y = -1 \end{cases}$

CHALLENGE PROBLEMS

107. Show that, in general, $\sqrt{a + b} \neq \sqrt{a} + \sqrt{b}$. That is, the square root of a sum does not equal the sum of the square roots.

108. Find the radicand: $\sqrt{} = 2a^4b\sqrt{7a}$.

8.3 Adding and Subtracting Radical Expressions

- Like Radicals
- Combining Like Radicals
- Simplifying Radicals in a Sum or Difference

We have discussed how to add and subtract like *terms.* In this section, we will discuss how to add and subtract like *radicals.*

LIKE RADICALS

Recall that we use the distributive property to simplify a sum or difference of like terms. For example,

$$3x + 5x = (3 + 5)x \qquad \text{and} \qquad 9y - 6y = (9 - 6)y$$
$$= 8x \qquad\qquad\qquad\qquad = 3y$$

In each case, we say that we have *combined like terms.*

The distributive property can also be used to simplify certain sums and differences of radicals. For example,

$$3\sqrt{2} + 5\sqrt{2} = (3 + 5)\sqrt{2} = 8\sqrt{2} \quad \text{and} \quad 9\sqrt{15} - 6\sqrt{15} = (9 - 6)\sqrt{15} = 3\sqrt{15}$$

The terms $3\sqrt{2}$ and $5\sqrt{2}$ are **like radicals,** and so are $9\sqrt{15}$ and $6\sqrt{15}$. In these two examples, we say that we have *combined like radicals.*

Like Radicals Square root radicals are called **like radicals** when they have the same radicand.

The Language of Algebra
Like radicals are also called **similar radicals.**

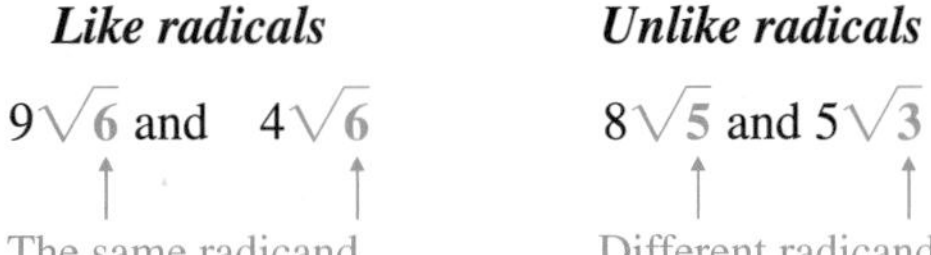

COMBINING LIKE RADICALS

To add or subtract radicals, they must be like radicals. Only like radicals can be combined.

EXAMPLE 1

ELEMENTARY Algebra Now™

Simplify: **a.** $4\sqrt{7} + 6\sqrt{7}$ and **b.** $2\sqrt{m} - 3\sqrt{m}$.

Solution

a. Since $4\sqrt{7}$ and $6\sqrt{7}$ are like radicals, we can add them.

$$4\sqrt{7} + 6\sqrt{7} = 10\sqrt{7} \qquad (4 + 6)\sqrt{7} = 10\sqrt{7}.$$

b. Since $2\sqrt{m}$ and $3\sqrt{m}$ are like radicals, we can subtract them.

$$2\sqrt{m} - 3\sqrt{m} = -\sqrt{m} \qquad (2 - 3)\sqrt{m} = -1\sqrt{m} = -\sqrt{m}.$$

Success Tip
Combining like radicals is similar to combining like terms.

$$4\sqrt{7} + 6\sqrt{7} = 10\sqrt{7}$$

Self Check 1 Simplify: **a.** $20\sqrt{5} + 30\sqrt{5}$ and **b.** $6\sqrt{b} - 4\sqrt{b}$.

EXAMPLE 2

ELEMENTARY Algebra Now™

Simplify, if possible: **a.** $5\sqrt{14x} + \sqrt{14x} - 9\sqrt{14x}$, **b.** $4\sqrt{ab} - 8 + 2\sqrt{ab}$, and **c.** $\sqrt{10} + \sqrt{6}$.

Solution

a. It is helpful to rewrite $\sqrt{14x}$ as $1\sqrt{14x}$.

$$5\sqrt{14x} + \sqrt{14x} - 9\sqrt{14x} = 5\sqrt{14x} + 1\sqrt{14x} - 9\sqrt{14x}$$
$$= -3\sqrt{14x} \qquad (5 + 1 - 9)\sqrt{14x} = -3\sqrt{14x}.$$

Notation
Just as $-1x = -x$,
$-1\sqrt{x} = -\sqrt{x}$.
And just as $1x = x$,
$1\sqrt{x} = \sqrt{x}$.

b. This expression contains two like radicals: $4\sqrt{ab}$ and $2\sqrt{ab}$.

$$4\sqrt{ab} - 8 + 2\sqrt{ab} = 6\sqrt{ab} - 8 \qquad (4 + 2)\sqrt{ab} = 6\sqrt{ab}.$$

c. Because $\sqrt{10}$ and $\sqrt{6}$ are unlike radicals, $\sqrt{10} + \sqrt{6}$ cannot be simplified. A common error is to add the radicands. However, the sum of the square roots is not equal to the square root of the sum.

$$\sqrt{10} + \sqrt{6} \neq \sqrt{16}$$

Self Check 2 Simplify, if possible: **a.** $\sqrt{5y} + 4\sqrt{5y} - 14\sqrt{5y}$, **b.** $1 + 6\sqrt{st} + 9\sqrt{st}$, and **c.** $5\sqrt{29} - \sqrt{11}$.

SIMPLIFYING RADICALS IN A SUM OR DIFFERENCE

If a sum or difference involves radicals that are unlike, make sure each one is written in simplified form. After doing so, like radicals may result that can be combined.

EXAMPLE 3

Simplify: $\sqrt{18} + \sqrt{8}$.

ELEMENTARY Algebra f(x) Now™

Solution We cannot add $\sqrt{18}$ and $\sqrt{8}$ because they are unlike radicals. However, when $\sqrt{18}$ and $\sqrt{8}$ are written in simplified form, we obtain like radicals, which can be added.

$$\sqrt{18} + \sqrt{8} = \sqrt{9 \cdot 2} + \sqrt{4 \cdot 2}$$ Factor 18 and 8 using their greatest perfect square factors.

$$= \sqrt{9}\sqrt{2} + \sqrt{4}\sqrt{2}$$ The square root of a product is equal to the product of the square roots.

$$= 3\sqrt{2} + 2\sqrt{2}$$ $\sqrt{9} = 3$ and $\sqrt{4} = 2$. These are like radicals.

$$= 5\sqrt{2}$$ Think: $(3 + 2)\sqrt{2} = 5\sqrt{2}$.

Caution

We have discussed the product and quotient rules for square roots. There is no sum nor difference rule for square roots:

$$\sqrt{a} + \sqrt{b} \neq \sqrt{a + b}$$
$$\sqrt{a} - \sqrt{b} \neq \sqrt{a - b}$$

Self Check 3 Simplify: $\sqrt{50} + \sqrt{32}$.

EXAMPLE 4

ELEMENTARY Algebra f(x) Now™

Orthopedics. How many feet of cable are used in the following traction setup?

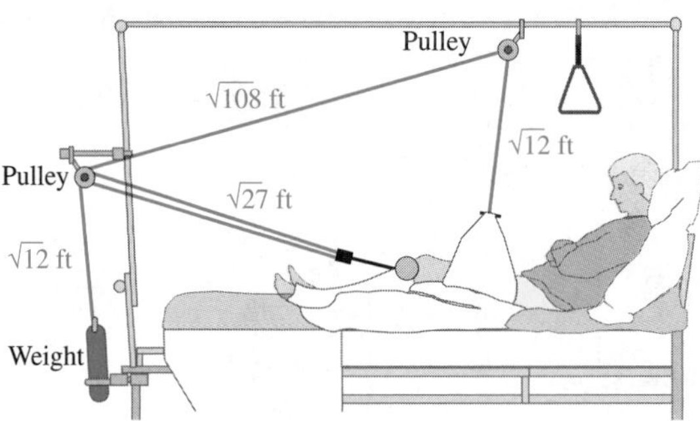

Solution Two lengths of cable are $\sqrt{12}$ feet long, another is $\sqrt{108}$ feet long, and two others are $\sqrt{27}$ feet long. Therefore, the total length of cable is given by

$$2\sqrt{12} + \sqrt{108} + 2\sqrt{27}$$

When we write each radical in simplified form, the results are three like radicals.

$$2\sqrt{12} + \sqrt{108} + 2\sqrt{27}$$

$$= 2\sqrt{4 \cdot 3} + \sqrt{36 \cdot 3} + 2\sqrt{9 \cdot 3}$$ Factor using the greatest perfect squares.

$$= 2\sqrt{4}\sqrt{3} + \sqrt{36}\sqrt{3} + 2\sqrt{9}\sqrt{3}$$ Use the product rule for square roots.

$$= 2 \cdot 2\sqrt{3} + 6\sqrt{3} + 2 \cdot 3\sqrt{3}$$ $\sqrt{4} = 2$, $\sqrt{36} = 6$, and $\sqrt{9} = 3$.

$= 4\sqrt{3} + 6\sqrt{3} + 6\sqrt{3}$ Simplify.

$= 16\sqrt{3}$ Think: $(4 + 6 + 6)\sqrt{3} = 16\sqrt{3}$.

The setup uses $16\sqrt{3}$ feet of cable.

EXAMPLE 5

Simplify: $\sqrt{44x^2} + x\sqrt{99}$.

ELEMENTARY Algebra f(x) Now™

Solution We write each radical in simplified form and then add like radicals.

$\sqrt{44x^2} + x\sqrt{99}$

$= \sqrt{4x^2 \cdot 11} + x\sqrt{9 \cdot 11}$ Factor $44x^2$ and 99.

$= \sqrt{4x^2}\sqrt{11} + x\sqrt{9}\sqrt{11}$ The square root of a product is equal to the product of the square roots:

$= 2x\sqrt{11} + 3x\sqrt{11}$ $\sqrt{4x^2} = 2x$ and $\sqrt{9} = 3$.

$= 5x\sqrt{11}$

Self Check 5 Simplify: $\sqrt{24y^2} + y\sqrt{54}$.

EXAMPLE 6

Simplify: $\sqrt{28x^2y} - 2\sqrt{63y^3}$.

Solution We begin by writing each radical in simplified form.

ELEMENTARY Algebra f(x) Now™

$\sqrt{28x^2y} - 2\sqrt{63y^3}$

$= \sqrt{4x^2 \cdot 7y} - 2\sqrt{9y^2 \cdot 7y}$ Factor $28x^2y$ and $63y^3$.

$= \sqrt{4x^2}\sqrt{7y} - 2\sqrt{9y^2}\sqrt{7y}$ The square root of a product is equal to the product of the square roots.

$= 2x\sqrt{7y} - 2 \cdot 3y\sqrt{7y}$ $\sqrt{4x^2} = 2x$ and $\sqrt{9y^2} = 3y$.

$= 2x\sqrt{7y} - 6y\sqrt{7y}$

Since $2x$ and $6y$ are unlike terms and cannot be subtracted, the expression does not simplify further.

Self Check 6 Simplify: $\sqrt{20mn^2} - \sqrt{80m^3}$.

EXAMPLE 7

Simplify: $\sqrt{27x} + \sqrt{20x}$.

Solution

$\sqrt{27x} + \sqrt{20x} = \sqrt{9 \cdot 3x} + \sqrt{4 \cdot 5x}$ To simplify the radicals, factor $27x$ and $20x$.

$= \sqrt{9}\sqrt{3x} + \sqrt{4}\sqrt{5x}$ The square root of a product is equal to the product of the square roots.

$= 3\sqrt{3x} + 2\sqrt{5x}$ $\sqrt{9} = 3$ and $\sqrt{4} = 2$.

Since the radicands are different, the radicals are unlike and the expression cannot be simplified further.

Self Check 7 Simplify: $\sqrt{75a} + \sqrt{72a}$.

EXAMPLE 8

Simplify: $\sqrt{8x} + \sqrt{3y} - \sqrt{50x} + \sqrt{27y}$.

Solution We simplify the radicals and then combine like radicals, where possible.

$$\sqrt{8x} + \sqrt{3y} - \sqrt{50x} + \sqrt{27y}$$

$$= \sqrt{4 \cdot 2x} + \sqrt{3y} - \sqrt{25 \cdot 2x} + \sqrt{9 \cdot 3y}$$ Factor $8x$, $50x$, and $27y$.

$$= \sqrt{4}\sqrt{2x} + \sqrt{3y} - \sqrt{25}\sqrt{2x} + \sqrt{9}\sqrt{3y}$$ The square root of a product is equal to the product of the square roots.

$$= 2\sqrt{2x} + \sqrt{3y} - 5\sqrt{2x} + 3\sqrt{3y}$$ $\sqrt{4} = 2$, $\sqrt{25} = 5$, and $\sqrt{9} = 3$.

$$= -3\sqrt{2x} + 4\sqrt{3y}$$ Combine like radicals.

Self Check 8 Simplify: $\sqrt{32x} - \sqrt{5y} - \sqrt{200x} + \sqrt{125y}$.

Answers to Self Checks **1. a.** $50\sqrt{5}$, **b.** $2\sqrt{b}$ **2. a.** $-9\sqrt{5y}$, **b.** $15\sqrt{st} + 1$, **c.** cannot be simplified **3.** $9\sqrt{2}$ **5.** $5y\sqrt{6}$ **6.** $2n\sqrt{5m} - 4m\sqrt{5m}$ **7.** $5\sqrt{3a} + 6\sqrt{2a}$ **8.** $-6\sqrt{2x} + 4\sqrt{5y}$

8.3 STUDY SET

ELEMENTARY Algebra f(x) Now™

VOCABULARY **Fill in the blanks.**

1. Square roots such as $\sqrt{2}$ and $5\sqrt{2}$, that have the same radicand, are called like ________.
2. When $\sqrt{8}$ and $\sqrt{18}$ are written in ________ form, the results are the like radicals $2\sqrt{2}$ and $3\sqrt{2}$.
3. Combining like radicals is similar to combining like ________.
4. Combining like radicals refers to the operations of ________ and ________.

CONCEPTS **Decide whether each pair of radicals are like radicals.**

5. $3\sqrt{3}, 4\sqrt{3}$
6. $6\sqrt{10}, 10\sqrt{6}$
7. $7\sqrt{a}, \sqrt{7a}$
8. $-\sqrt{5y}, \sqrt{5y}$
9. Write each radical in simplified form.
   a. $\sqrt{12}$
   b. $\sqrt{4x^3}$
10. Fill in the blanks.
    a. $5\sqrt{6} + 3\sqrt{6} = 8$ ____
    b. $10\sqrt{7y} - 20\sqrt{7y} =$ ____ $\sqrt{7y}$

NOTATION **Complete the solution to find the difference.**

11. $9\sqrt{5} - 3\sqrt{20} = 9\sqrt{5} - 3\sqrt{\;\_\_ \cdot 5}$
    $= 9\sqrt{5} - 3\sqrt{\_\_}\sqrt{5}$
    $= 9\sqrt{5} - 3 \cdot$ ____
    $= 9\sqrt{5} -$ ____
    $=$ ____
12. Write each expression in simpler form.
    a. $-1\sqrt{x}$
    b. $1\sqrt{x}$

PRACTICE **Simplify, if possible. Assume that all variables represent nonnegative numbers.**

13. $5\sqrt{7} + 4\sqrt{7}$
14. $3\sqrt{10} + 4\sqrt{10}$
15. $14\sqrt{21} - 4\sqrt{21}$
16. $7\sqrt{3} - 2\sqrt{3}$
17. $8\sqrt{n} + \sqrt{n}$
18. $6\sqrt{m} + \sqrt{m}$
19. $\sqrt{6} - \sqrt{6}$
20. $3\sqrt{11} - 3\sqrt{11}$

21. $2\sqrt{5} + 2\sqrt{5}$

22. $\sqrt{15} + \sqrt{15}$

23. $-9\sqrt{21} + 6\sqrt{21}$

24. $-2\sqrt{5} + 8\sqrt{5}$

25. $\sqrt{3} + \sqrt{15}$

26. $\sqrt{6} + 8\sqrt{13}$

27. $\sqrt{x} - 4\sqrt{x}$

28. $\sqrt{t} - 9\sqrt{t}$

29. $\sqrt{2} + \sqrt{x}$

30. $\sqrt{3} - 19\sqrt{y}$

31. $2\sqrt{11} + 3\sqrt{11} + 5\sqrt{11}$

32. $2\sqrt{5} + 6\sqrt{5} + 9\sqrt{5}$

33. $4\sqrt{2} + 4\sqrt{2} - 4\sqrt{2}$

34. $9\sqrt{3} - 9\sqrt{3} + 9\sqrt{3}$

35. $5 + 3\sqrt{3} + 3\sqrt{3}$

36. $\sqrt{5} + 2 + 3\sqrt{5}$

37. $-1 + 2\sqrt{r} - 3\sqrt{r}$

38. $-8 - 5\sqrt{c} + 4\sqrt{c}$

39. $\sqrt{12} + \sqrt{27}$

40. $\sqrt{20} + \sqrt{45}$

41. $\sqrt{18} - \sqrt{8}$

42. $\sqrt{32} - \sqrt{18}$

43. $2\sqrt{45} + 2\sqrt{80}$

44. $3\sqrt{80} + 3\sqrt{125}$

45. $2\sqrt{80} - 3\sqrt{125}$

46. $3\sqrt{245} - 2\sqrt{180}$

47. $\sqrt{12} - \sqrt{48}$

48. $\sqrt{48} - \sqrt{75}$

49. $\sqrt{288} - 3\sqrt{200}$

50. $\sqrt{80} - \sqrt{245}$

51. $2\sqrt{28} + 2\sqrt{112}$

52. $4\sqrt{63} + 6\sqrt{112}$

53. $\sqrt{20} + \sqrt{45} + \sqrt{80}$

54. $\sqrt{48} + \sqrt{27} + \sqrt{75}$

55. $\sqrt{200} - \sqrt{75} + \sqrt{48}$

56. $\sqrt{20} + \sqrt{80} - \sqrt{125}$

57. $8\sqrt{6} - 5\sqrt{2} - 3\sqrt{6}$

58. $3\sqrt{2} - 3\sqrt{15} - 4\sqrt{15}$

59. $\sqrt{24} + \sqrt{150} + \sqrt{240}$

60. $\sqrt{28} + \sqrt{63} + \sqrt{18}$

61. $\sqrt{48} - \sqrt{8} + \sqrt{27} - \sqrt{32}$

62. $\sqrt{162} + \sqrt{50} - \sqrt{75} - \sqrt{108}$

63. $\sqrt{72a} - \sqrt{98a}$

64. $\sqrt{25b} - \sqrt{49b}$

65. $\sqrt{2x^2} + \sqrt{8x^2}$

66. $\sqrt{3y^2} - \sqrt{12y^2}$

67. $\sqrt{2d^3} + \sqrt{8d^3}$

68. $\sqrt{3a^3} - \sqrt{12a^3}$

69. $\sqrt{18x^2y} - \sqrt{27x^2y}$

70. $\sqrt{49xy} + \sqrt{xy}$

71. $\sqrt{32x^5} - \sqrt{18x^5}$

72. $\sqrt{27xy^3} - \sqrt{48xy^3}$

73. $3\sqrt{54b^2} + 5\sqrt{24b^2}$

74. $3\sqrt{24x^4y^3} + 2\sqrt{54x^4y^3}$

75. $y\sqrt{490y} - 2\sqrt{360y^3}$

76. $3\sqrt{20x} + 2\sqrt{63y}$

77. $\sqrt{20x^3y} + \sqrt{45x^5y^3} - \sqrt{80x^7y^5}$

78. $x\sqrt{48xy^2} - y\sqrt{27x^3} + \sqrt{75x^3y^2}$

APPLICATIONS

79. ANATOMY Determine the length of the patient's arm if he lets it fall to his side.

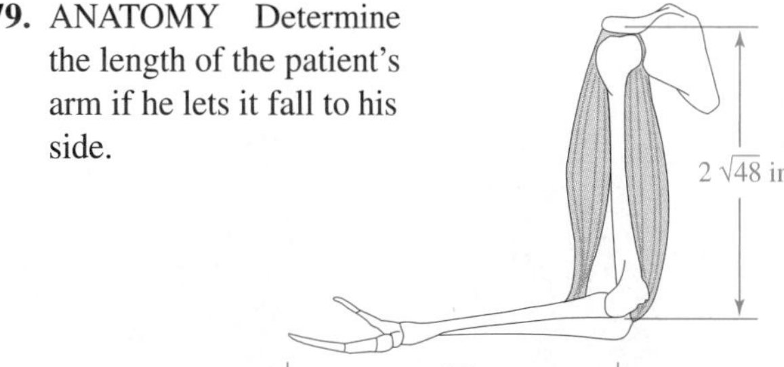

80. PLAYGROUND EQUIPMENT Find the total length of pipe necessary to construct the frame of the swing set.

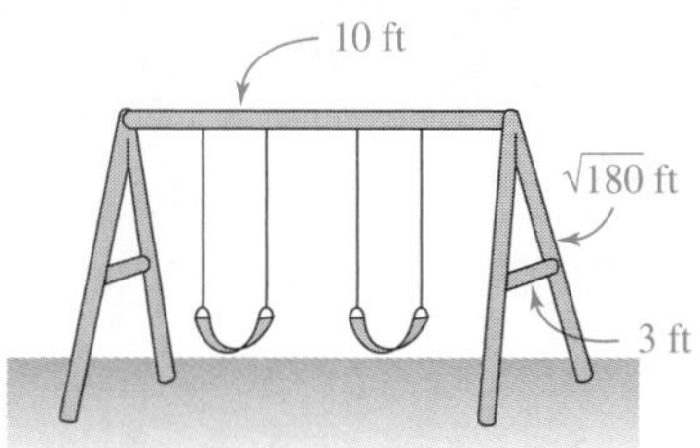

81. HARDWARE Find the difference in the lengths of the "arms" of the door-closing device.

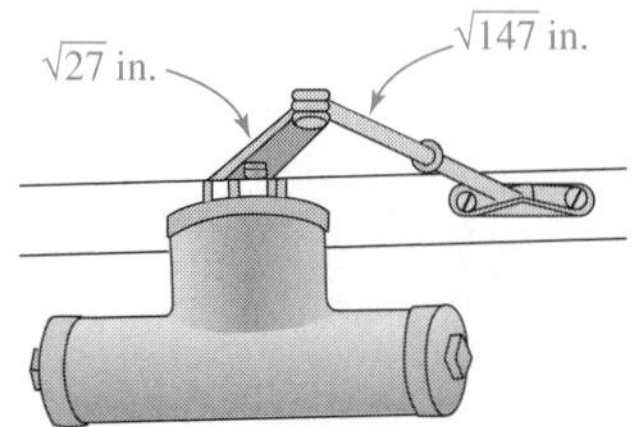

82. BLUEPRINTS What is the length of the motor on the machine?

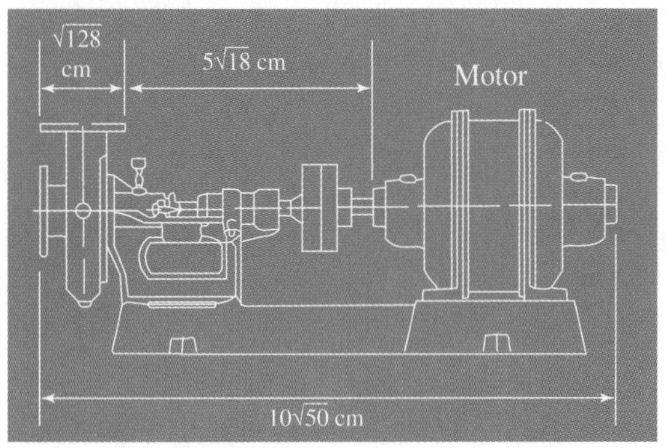

83. CAMPING The length of a center support pole for a tent is given by the formula

$$l = 0.5s\sqrt{3}$$

where s is the length of the side of the tent. Find the total length of the four poles needed for the parents' and children's tents.

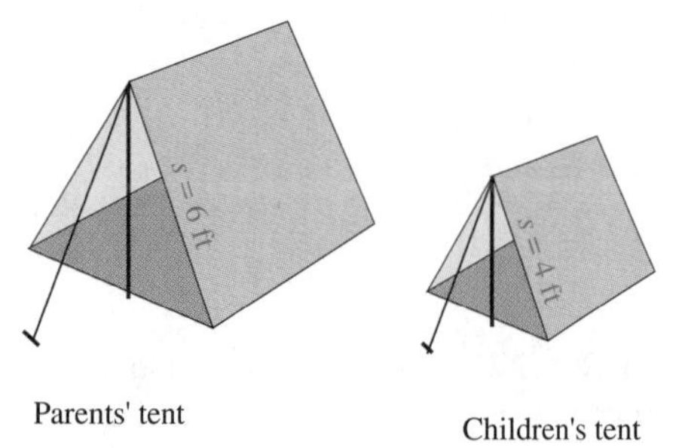

Parents' tent Children's tent

84. FENCING Find the number of feet of fencing needed to enclose the swimming pool complex.

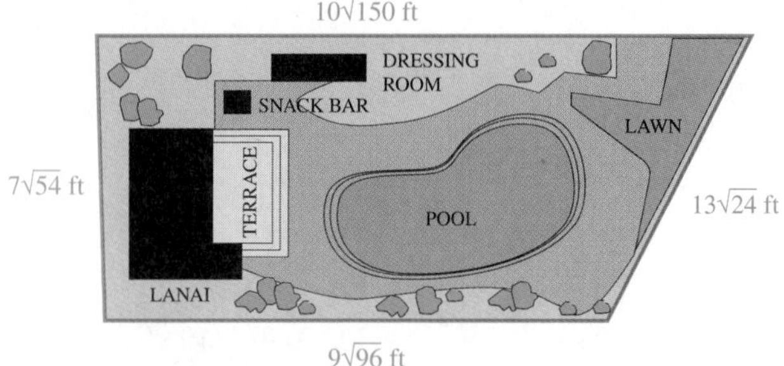

WRITING

85. Explain why $\sqrt{6} + \sqrt{5}$ cannot be simplified further.

86. Is $2\sqrt{5} - \sqrt{5} = 2$? Explain.

87. Are $2\sqrt{3}$ and $3\sqrt{2}$ like radicals? Explain.

88. $\sqrt{2}$ and $\sqrt{50}$ are not like radicals. Can they be added? Explain.

REVIEW **Simplify each expression. Write each answer without using negative exponents.**

89. 3^{-2}

90. $\frac{1}{3^{-2}}$

91. -3^2

92. -3^{-2}

93. x^{-3}

94. $\frac{1}{x^{-3}}$

95. 3^0

96. x^0

CHALLENGE PROBLEMS

97. Fill in the blank: $2y\sqrt{175y} - \sqrt{\quad} = 0$.

98. Simplify: $\frac{1}{10}\sqrt{50x^3} + \frac{1}{8}\sqrt{32x^3}$.

8.4 Multiplying and Dividing Radical Expressions

- Multiplying Square Roots
- Powers of Square Roots
- Multiplying Radical Expressions
- Dividing Radical Expressions
- Rationalizing Denominators

We will now discuss the methods used to multiply and divide square roots.

MULTIPLYING SQUARE ROOTS

We have used the *product rule for square roots* to write square roots in simplified form.

$$\sqrt{a \cdot b} = \sqrt{a} \cdot \sqrt{b}$$

We can also use this rule to multiply square roots.

The Product Rule for Square Roots

The product of the square roots of two nonnegative numbers is equal to the square root of the product of those numbers.

For any nonnegative real numbers a and b,

$$\sqrt{a} \cdot \sqrt{b} = \sqrt{a \cdot b}$$

EXAMPLE 1

ELEMENTARY Algebra f(x) Now™

Multiply and simplify, if possible: **a.** $\sqrt{3} \cdot \sqrt{2}$, **b.** $\sqrt{6x}\left(\sqrt{8x}\right)$, and **c.** $\sqrt{a^3}\sqrt{5a^8}$.

Solution

a. $\sqrt{3} \cdot \sqrt{2} = \sqrt{3 \cdot 2} = \sqrt{6}$ — Multiply the radicands to get 6 and write the square root of that product.

b. $\sqrt{6x}(\sqrt{8x}) = \sqrt{6x \cdot 8x} = \sqrt{48x^2}$ — Multiply the radicands to get $48x^2$ and write the square root of that product.

Since $48x^2$ has a perfect square factor of $16x^2$, we can simplify $\sqrt{48x^2}$.

$$\begin{aligned} \sqrt{48x^2} &= \sqrt{16x^2 \cdot 3} && \text{Factor } 48x^2 \text{ as } 16x^2 \cdot 3. \\ &= \sqrt{16x^2}\sqrt{3} && \text{The square root of a product is the product of the square roots.} \\ &= 4x\sqrt{3} && \sqrt{16x^2} = 4x. \end{aligned}$$

c. $\sqrt{a^3}\sqrt{5a^8} = \sqrt{a^3 \cdot 5a^8} = \sqrt{5a^{11}}$ — Multiply the radicands: $a^3 \cdot 5a^8 = 5a^{8+3} = 5a^{11}$ and write the square root of that product.

Since $5a^{11}$ has a perfect square factor of a^{10}, we can simplify $\sqrt{5a^{11}}$.

$$\begin{aligned} \sqrt{5a^{11}} &= \sqrt{a^{10} \cdot 5a} && \text{Factor } 5a^{11} \text{ as } a^{10} \cdot 5a. \\ &= \sqrt{a^{10}}\sqrt{5a} && \text{The square root of a product is the product of the square roots.} \\ &= a^5\sqrt{5a} && \sqrt{a^{10}} = a^5. \end{aligned}$$

Self Check 1 Multiply and simplify, if possible: **a.** $\sqrt{5} \cdot \sqrt{3}$, **b.** $\sqrt{10y} \cdot \sqrt{2y}$, and **c.** $\sqrt{5b^3}\sqrt{2b^6}$.

EXAMPLE 2

Multiply: $3\sqrt{6} \cdot 4\sqrt{5}$.

Solution The commutative and associative properties of multiplication enable us to change the order of the factors so that we can multiply 3 and 4, and $\sqrt{6}$ and $\sqrt{5}$, separately.

$$\begin{aligned} 3\sqrt{6} \cdot 4\sqrt{5} &= 3 \cdot 4 \cdot \sqrt{6} \cdot \sqrt{5} && 3\sqrt{6} = 3 \cdot \sqrt{6} \text{ and } 4\sqrt{5} = 4 \cdot \sqrt{5}. \\ &= 12\sqrt{6 \cdot 5} && \text{Multiply 3 and 4 and multiply } \sqrt{6} \text{ and } \sqrt{5}. \\ &= 12\sqrt{30} \end{aligned}$$

Self Check 2 Multiply: $8\sqrt{2x} \cdot 4\sqrt{7}$.

POWERS OF SQUARE ROOTS

By definition, when the square root of a positive number is squared, the result is that positive number. This fact about square roots can be stated in the following way.

The Square of a Square Root For any positive real number a, $\left(\sqrt{a}\right)^2 = a$.

EXAMPLE 3

Find: **a.** $\left(\sqrt{5}\right)^2$, **b.** $\left(\sqrt{x+1}\right)^2$, and **c.** $\left(2\sqrt{7}\right)^2$. Assume $x > 0$.

Solution

a. $\left(\sqrt{5}\right)^2 = 5$ The square of the square root of a positive number is that number.

b. $\left(\sqrt{x+1}\right)^2 = x + 1$ If x represents a positive number, then $x + 1$ is positive. The square of the square root of a positive number is that number.

c. We can use the power of a product rule for exponents to evaluate $\left(2\sqrt{7}\right)^2$.

$$\begin{aligned} \left(2\sqrt{7}\right)^2 &= 2^2\left(\sqrt{7}\right)^2 && \text{Raise each factor of the product } 2\sqrt{7} \text{ to the second power.} \\ &= 4 \cdot 7 && \text{Simplify: } 2^2 = 4 \text{ and } \left(\sqrt{7}\right)^2 = 7. \\ &= 28 \end{aligned}$$

Self Check 3 Find: **a.** $\left(\sqrt{11}\right)^2$, **b.** $\left(\sqrt{3y+5}\right)^2$, and **c.** $\left(5\sqrt{2}\right)^2$.

MULTIPLYING RADICAL EXPRESSIONS

To multiply radical expressions containing more than one term, we use the same methods that were used to multiply polynomials with more than one term.

EXAMPLE 4

Multiply and simplify, if possible: $\sqrt{2x}\left(\sqrt{6x^4} + \sqrt{2x}\right)$.

Solution

$$\begin{aligned} \sqrt{2x}\left(\sqrt{6x^4} + \sqrt{2x}\right) &= \sqrt{2x}\sqrt{6x^4} + \sqrt{2x}\sqrt{2x} && \text{Use the distributive property.} \\ &= \sqrt{12x^5} + 2x && \sqrt{2x}\sqrt{2x} = \left(\sqrt{2x}\right)^2 = 2x. \\ &= \sqrt{4x^4 \cdot 3x} + 2x && \text{To simplify } \sqrt{12x^5}\text{, factor } 12x^5. \\ &= \sqrt{4x^4}\sqrt{3x} + 2x && \text{The square root of a product is the product of the square roots.} \\ &= 2x^2\sqrt{3x} + 2x && \sqrt{4x^4} = 2x^2. \end{aligned}$$

Self Check 4 Multiply and simplify, if possible: $\sqrt{6n}\left(\sqrt{3n^6} - \sqrt{6n}\right)$.

EXAMPLE 5

Find the product and simplify: $\left(\sqrt{3x} - 5\right)\left(\sqrt{3x} + 2\right)$.

ELEMENTARY Algebra f(x) Now™

Solution As with polynomials, we can use the FOIL method to find the product.

$$(\sqrt{3x} - 5)(\sqrt{3x} + 2) = \sqrt{3x}\sqrt{3x} + 2\sqrt{3x} - 5\sqrt{3x} - 5 \cdot 2$$ Multiply each term of one polynomial by each term of the other polynomial.

$$= 3x + 2\sqrt{3x} - 5\sqrt{3x} - 10$$ $\sqrt{3x}\sqrt{3x} = (\sqrt{3x})^2 = 3x$.

$$= 3x - 3\sqrt{3x} - 10$$ Combine like radicals.

Self Check 5 Find the product and simplify: $(\sqrt{5a} - 2)(\sqrt{5a} + 3)$.

The special product formulas from Chapter 4 can be used to multiply two-term expressions containing radicals.

EXAMPLE 6

ELEMENTARY Algebra $f(x)$ Now™

Find each product and simplify: **a.** $(\sqrt{6} + \sqrt{y})^2$ and **b.** $(\sqrt{7} - \sqrt{2})(\sqrt{7} + \sqrt{2})$.

Solution **a.** The *square of the sum of two terms* is the square of the first term, plus twice the product of both terms, plus the square of the last term.

$$(\sqrt{6} + \sqrt{y})^2 = (\sqrt{6})^2 + 2 \cdot \sqrt{6}\sqrt{y} + (\sqrt{y})^2$$
$$= 6 + 2\sqrt{6y} + y$$

b. The *product of the sum and difference of two terms* is the square of the first term minus the square of the second term.

$$(\sqrt{7} - \sqrt{2})(\sqrt{7} + \sqrt{2}) = (\sqrt{7})^2 - (\sqrt{2})^2$$
$$= 7 - 2$$
$$= 5$$

Self Check 6 Find each product and simplify: **a.** $(\sqrt{15} - \sqrt{n})^2$ and **b.** $(\sqrt{3} - \sqrt{5})(\sqrt{3} + \sqrt{5})$.

DIVIDING RADICAL EXPRESSIONS

We have used the quotient rule for square roots to simplify square roots of quotients:

$$\sqrt{\frac{a}{b}} = \frac{\sqrt{a}}{\sqrt{b}}, \quad \text{where } b \neq 0$$

We can use this rule to divide square roots.

The Quotient Rule for Square Roots The quotient of the square roots of two numbers is equal to the square root of the quotient of the two numbers.

For any positive real numbers a and b,

$$\frac{\sqrt{a}}{\sqrt{b}} = \sqrt{\frac{a}{b}}$$

EXAMPLE 7

Divide and simplify, if possible: **a.** $\frac{\sqrt{39}}{\sqrt{3}}$, **b.** $\frac{\sqrt{40}}{\sqrt{5}}$, and **c.** $\frac{\sqrt{18x^5}}{\sqrt{2x^3}}$.

ELEMENTARY Algebra f(x) Now™

Solution In each case, the quotient of the square roots is written as the square root of the quotient.

a. $\frac{\sqrt{39}}{\sqrt{3}} = \sqrt{\frac{39}{3}} = \sqrt{13}$ Divide the radicands to get 13 and write the square root of that quotient.

b. $\frac{\sqrt{40}}{\sqrt{5}} = \sqrt{\frac{40}{5}} = \sqrt{8}$ Divide the radicands to get 8 and write the square root of that quotient.

Since 8 has a perfect square factor of 4, we can simplify $\sqrt{8}$.

$$\sqrt{8} = \sqrt{4 \cdot 2} = \sqrt{4}\sqrt{2} = 2\sqrt{2}$$ Factor 8 as $4 \cdot 2$.

c. $\frac{\sqrt{18x^5}}{\sqrt{2x^3}} = \sqrt{\frac{18x^5}{2x^3}} = \sqrt{9x^2} = 3x$ Divide the radicands to get $9x^2$ and write the square root of that quotient.

Self Check 7 Divide and simplify, if possible: **a.** $\frac{\sqrt{42}}{\sqrt{7}}$, **b.** $\frac{\sqrt{90}}{\sqrt{2}}$, and **c.** $\frac{\sqrt{125d^5}}{\sqrt{5d}}$.

RATIONALIZING DENOMINATORS

In the illustration, the length of the diagonal of the square tile is 1 foot. By the Pythagorean theorem, it can be shown that the length of one side of the tile is $\frac{1}{\sqrt{2}}$ feet. This expression is not in simplified form because of the radical in the denominator.

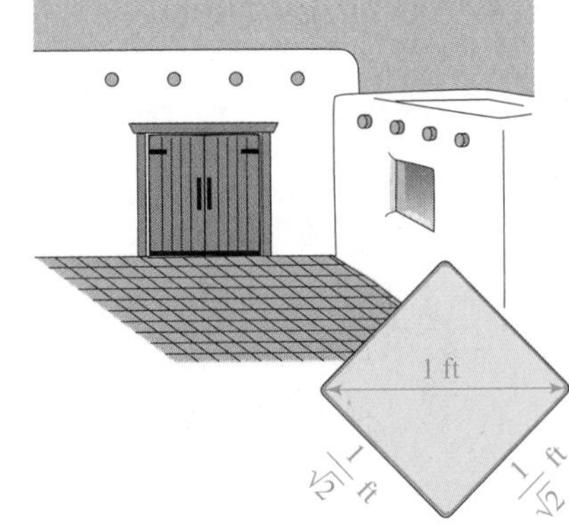

The Language of Algebra
Since $\sqrt{2}$ is an irrational number, the fraction $\frac{1}{\sqrt{2}}$ has an irrational denominator.

It is usually easier to work with radical expressions if their denominators do not contain radicals. We now consider a process in which we change the denominator from a radical that represents an irrational number to a rational number. The process is called **rationalizing the denominator.**

EXAMPLE 8

Rationalize the denominator: $\frac{1}{\sqrt{2}}$.

Solution We want to find a fraction equivalent to $\frac{1}{\sqrt{2}}$ that does not have a radical in its denominator. If we multiply $\frac{1}{\sqrt{2}}$ by $\frac{\sqrt{2}}{\sqrt{2}}$, the denominator becomes $\sqrt{2} \cdot \sqrt{2} = 2$, a rational number.

$$\frac{1}{\sqrt{2}} = \frac{1}{\sqrt{2}} \cdot \frac{\sqrt{2}}{\sqrt{2}}$$ To build an equivalent fraction, multiply by $\frac{\sqrt{2}}{\sqrt{2}} = 1$.

$$= \frac{\sqrt{2}}{2}$$ Multiply the numerators: $1 \cdot \sqrt{2} = \sqrt{2}$.
Multiply the denominators: $\sqrt{2} \cdot \sqrt{2} = (\sqrt{2})^2 = 2$.

Thus, $\frac{1}{\sqrt{2}} = \frac{\sqrt{2}}{2}$. These equivalent fractions represent the same number, but have different forms.

Self Check 8 Rationalize the denominator: $\frac{11}{\sqrt{6}}$.

The procedure for rationalizing square root denominators is as follows.

Rationalizing Denominators

To **rationalize a square root denominator,** multiply the numerator and denominator of the given fraction by the square root that appears in its denominator, or by a square root that makes a perfect square radicand in the denominator.

EXAMPLE 9

ELEMENTARY Algebra *f(x)* Now™

Rationalize each denominator: **a.** $\sqrt{\frac{5}{3}}$ and **b.** $\frac{4}{\sqrt{24x}}$.

Solution **a.** The expression $\sqrt{\frac{5}{3}}$ is not in simplified form, because the radicand is a fraction. To write it in simplified form, we use the quotient rule for square roots and rationalize the denominator.

Caution

Do not attempt to remove a common factor of 3 from the numerator and denominator of $\frac{\sqrt{15}}{3}$. The numerator, $\sqrt{15}$, does not have a factor of 3.

$$\frac{\sqrt{15}}{3} = \frac{\sqrt{3} \cdot \sqrt{5}}{3}$$

$$\sqrt{\frac{5}{3}} = \frac{\sqrt{5}}{\sqrt{3}} \quad \text{The square root of a quotient is equal to the quotient of the square roots.}$$

$$= \frac{\sqrt{5}}{\sqrt{3}} \cdot \frac{\sqrt{3}}{\sqrt{3}} \quad \text{To build an equivalent fraction, multiply by } \frac{\sqrt{3}}{\sqrt{3}} = 1.$$

$$= \frac{\sqrt{15}}{3} \quad \text{Multiply the numerators and multiply the denominators.}$$

b. We could begin by multiplying $\frac{4}{\sqrt{24x}}$ by $\frac{\sqrt{24x}}{\sqrt{24x}}$. However, to work with smaller numbers, it is easier if we simplify $\sqrt{24x}$ first and then rationalize the denominator.

$$\frac{4}{\sqrt{24x}} = \frac{4}{2\sqrt{6x}} \quad \sqrt{24x} = \sqrt{4 \cdot 6x} = \sqrt{4}\sqrt{6x} = 2\sqrt{6x}.$$

$$= \frac{4}{2\sqrt{6x}} \cdot \frac{\sqrt{6x}}{\sqrt{6x}} \quad \text{To build an equivalent fraction, multiply by } \frac{\sqrt{6x}}{\sqrt{6x}} = 1.$$

$$= \frac{4\sqrt{6x}}{2 \cdot 6x} \quad \text{Multiply the numerators and multiply the denominators.}$$

$$= \frac{\overset{1}{\cancel{2}} \cdot \overset{1}{\cancel{2}}\sqrt{6x}}{\underset{1}{\cancel{2}} \cdot \underset{1}{\cancel{2}} \cdot 3x} \quad \text{Factor 4 as } 2 \cdot 2 \text{ and } 6x \text{ as } 2 \cdot 3x. \text{ Simplify the fraction by removing the common factors.}$$

$$= \frac{\sqrt{6x}}{3x}$$

Success Tip

We usually simplify a radical expression before rationalizing the denominator.

Self Check 9 Rationalize each denominator: **a.** $\sqrt{\frac{6}{7}}$ and **b.** $\frac{15}{\sqrt{50y}}$.

In Examples 8 and 9, we rationalized denominators with only one term. We will now discuss a method to rationalize denominators with two terms.

One-termed denominators	*Two-termed denominators*
$\frac{1}{\sqrt{2}}, \frac{\sqrt{5}}{\sqrt{3}}, \frac{4}{\sqrt{24x}}$	$\frac{5}{\sqrt{6}-1}, \frac{3}{\sqrt{7}+2}$

To rationalize the denominator of $\frac{5}{\sqrt{6}-1}$, we multiply the numerator and denominator by $\sqrt{6}+1$, because the product $(\sqrt{6}-1)(\sqrt{6}+1)$ contains no radicals. Radical expressions such as $\sqrt{6}-1$ and $\sqrt{6}+1$ are said to be **conjugates** of each other.

Conjugate Binomials If a and b are real numbers, then

$$a + b \quad \text{and} \quad a - b$$

are called **conjugate binomials.**

EXAMPLE 10

Rationalize the denominator: $\frac{5}{\sqrt{6}-1}$.

ELEMENTARY Algebra $f(x)$ Now™

Solution To rationalize the denominator, we multiply by a form of 1 that uses the conjugate of the denominator.

$$\frac{5}{\sqrt{6}-1} = \frac{5}{\sqrt{6}-1} \cdot \frac{\sqrt{6}+1}{\sqrt{6}+1} \qquad \text{Multiply by } \frac{\sqrt{6}+1}{\sqrt{6}+1} = 1.$$

$$= \frac{5(\sqrt{6}+1)}{(\sqrt{6}-1)(\sqrt{6}+1)} \qquad \text{Multiply the numerators and multiply the denominators.}$$

$$= \frac{5(\sqrt{6}+1)}{6-1} \qquad \text{In the denominator, use a special product formula: } (\sqrt{6}-1)(\sqrt{6}-1) = (\sqrt{6})^2 - 1^2 = 6 - 1.$$

$$= \frac{5(\sqrt{6}+1)}{5}$$

$$= \frac{\overset{1}{\cancel{5}}(\sqrt{6}+1)}{\underset{1}{\cancel{5}}} \qquad \text{Simplify the fraction.}$$

$$= \sqrt{6}+1$$

Self Check 10 Rationalize the denominator: $\frac{3}{\sqrt{7}+2}$.

Answers to Self Checks **1. a.** $\sqrt{15}$, **b.** $2y\sqrt{5}$, **c.** $b^4\sqrt{10b}$ **2.** $32\sqrt{14x}$ **3. a.** 11, **b.** $3y + 5$, **c.** 50 **4.** $3n^3\sqrt{2n} - 6n$ **5.** $5a + \sqrt{5a} - 6$ **6. a.** $15 - 2\sqrt{15n} + n$, **b.** -2 **7. a.** $\sqrt{6}$, **b.** $3\sqrt{5}$, **c.** $5d^2$ **8.** $\frac{11\sqrt{6}}{6}$ **9. a.** $\frac{\sqrt{42}}{7}$, **b.** $\frac{3\sqrt{2y}}{2y}$ **10.** $\sqrt{7} - 2$

8.4 STUDY SET  ELEMENTARY Algebra $f(x)$ Now™

VOCABULARY **Fill in the blanks.**

1. The denominator of $\frac{1}{\sqrt{3}}$ is an ________ number.
2. Two fractions that represent the same number, but have different forms, such as $\frac{1}{\sqrt{2}}$ and $\frac{\sqrt{2}}{2}$, are called ________ fractions.
3. To ________ the denominator of $\frac{4}{\sqrt{5}}$, we multiply it by $\frac{\sqrt{5}}{\sqrt{5}}$.
4. The ________ of $1 + \sqrt{2}$ is $1 - \sqrt{2}$.

CONCEPTS **Fill in the blanks.**

5. $\sqrt{a} \cdot \sqrt{b} = \sqrt{\quad}$
6. $\frac{\sqrt{a}}{\sqrt{b}} = \sqrt{\frac{\quad}{\quad}}$
7. $(\sqrt{a})^2 = \square$
8. $\sqrt{\frac{a}{b}} = \frac{\square}{\square}$
9. $\frac{\sqrt{6}}{\sqrt{6}} = \square$
10. $\frac{1 - \sqrt{15}}{1 - \sqrt{15}} = \square$
11. $7\sqrt{2} \cdot 9\sqrt{6} = 7 \cdot \square \cdot \sqrt{2} \cdot \square$
12. $\sqrt{27x^2} = \sqrt{\quad \cdot 3} = \square$
13. To rationalize the denominator of $\frac{4}{\sqrt{3}}$, we multiply the fraction by $\frac{\sqrt{3}}{\sqrt{3}}$, which is a form of $\square$.
14. To rationalize the denominator of $\frac{1}{5 + \sqrt{6}}$, we multiply the fraction by $\frac{\square}{\square}$.
15. Explain why each expression is not in simplified radical form.

 a. $\sqrt{\frac{3}{4}}$

 b. $\frac{1}{\sqrt{10}}$

Perform each operation, if possible.

16. $\sqrt{2} + \sqrt{3}$
17. $\sqrt{2} \cdot \sqrt{3}$
18. $\sqrt{2} - \sqrt{3}$
19. $\frac{\sqrt{2}}{\sqrt{3}}$

NOTATION **Complete the solution.**

20. $(\sqrt{x} + \sqrt{2})(\sqrt{x} - 3\sqrt{2})$

$= \sqrt{x}\square - \sqrt{x} \cdot 3\sqrt{2} + \sqrt{2}\square - \sqrt{2} \cdot 3\sqrt{2}$

$= x - 3\square + \sqrt{2x} - 3\sqrt{2}\sqrt{2}$

$= x - \square - \square$

PRACTICE **Multiply and simplify, if possible. All variables represent nonnegative numbers.**

21. $\sqrt{3} \cdot \sqrt{5}$
22. $\sqrt{2} \cdot \sqrt{7}$
23. $\sqrt{2x}\sqrt{11}$
24. $\sqrt{10}\sqrt{3a}$
25. $\sqrt{7} \cdot \sqrt{7}$
26. $\sqrt{10} \cdot \sqrt{10}$
27. $\sqrt{5} \cdot \sqrt{10}$
28. $\sqrt{2} \cdot \sqrt{6}$
29. $\sqrt{2}(\sqrt{27})$
30. $\sqrt{15}(\sqrt{5})$
31. $\sqrt{5} \cdot \sqrt{5d}$
32. $\sqrt{15b} \cdot \sqrt{15}$
33. $\sqrt{2}\sqrt{8}$
34. $\sqrt{27}\sqrt{3}$
35. $\sqrt{8}\sqrt{7}$
36. $\sqrt{6}\sqrt{8}$
37. $3\sqrt{2}\sqrt{x}$
38. $4\sqrt{3x}\sqrt{5y}$
39. $\sqrt{x^3}\sqrt{x^5}$
40. $\sqrt{a^7}\sqrt{a^3}$
41. $(5\sqrt{6})(4\sqrt{3})$
42. $(6\sqrt{3})(7\sqrt{2})$
43. $5\sqrt{3} \cdot 2\sqrt{5}$
44. $2\sqrt{5} \cdot 5\sqrt{2}$
45. $3\sqrt{y}(15\sqrt{y})$
46. $4\sqrt{x}(2\sqrt{x})$
47. $\sqrt{6n^4} \cdot \sqrt{10n^5}$
48. $\sqrt{5m^2} \cdot \sqrt{10m^5}$
49. $\sqrt{3s^6} \cdot \sqrt{3s^5}$
50. $\sqrt{15t^8} \cdot \sqrt{5t^3}$
51. $\sqrt{2}(\sqrt{2} + 1)$
52. $\sqrt{5}(\sqrt{5} + 2)$
53. $3\sqrt{3}(\sqrt{27} - \sqrt{2})$
54. $2\sqrt{2}(\sqrt{8} - \sqrt{3})$
55. $\sqrt{x}(\sqrt{3x} - 2)$
56. $\sqrt{y}(\sqrt{y} + 5)$
57. $2\sqrt{x}(\sqrt{9x} + \sqrt{x})$
58. $3\sqrt{z}(\sqrt{4z} - \sqrt{z})$
59. $(\sqrt{2} + 1)(\sqrt{2} - 1)$
60. $(\sqrt{3} - 1)(\sqrt{3} + 1)$
61. $(\sqrt{2} - \sqrt{3})(\sqrt{3} + \sqrt{5})$
62. $(\sqrt{3} + \sqrt{5})(\sqrt{5} - \sqrt{2})$
63. $(\sqrt{2x} + 3)(\sqrt{8x} - 6)$
64. $(\sqrt{5y} - 3)(\sqrt{20y} + 6)$

65. $(2\sqrt{7} - x)(3\sqrt{2} + x)$

66. $(4\sqrt{2} - \sqrt{x})(\sqrt{x} + 2\sqrt{3})$

67. $(\sqrt{3x} + 2)(\sqrt{27x} - 6)$

68. $(\sqrt{6y} - 1)(\sqrt{24y} + 2)$

69. $(\sqrt{6} + \sqrt{3})(\sqrt{6} - \sqrt{3})$

70. $(\sqrt{8} + \sqrt{7})(\sqrt{8} - \sqrt{7})$

71. $(\sqrt{7} - 3)(\sqrt{7} + 3)$

72. $(\sqrt{5} + 4)(\sqrt{5} - 4)$

Find each power. All variables represent nonnegative numbers.

73. $(\sqrt{6})^2$

74. $(\sqrt{11})^2$

75. $(2\sqrt{3})^2$

76. $(3\sqrt{5})^2$

77. $(10\sqrt{x})^2$

78. $(8\sqrt{b})^2$

79. $(\sqrt{y + 2})^2$

80. $(\sqrt{m + 3})^2$

Simplify each expression. All variables represent positive numbers.

81. $\dfrac{\sqrt{12x^3}}{\sqrt{27x}}$

82. $\dfrac{\sqrt{32}}{\sqrt{98x^2}}$

83. $\dfrac{\sqrt{18x}}{\sqrt{25x}}$

84. $\dfrac{\sqrt{27y}}{\sqrt{75y}}$

85. $\dfrac{\sqrt{196x}}{\sqrt{49x^3}}$

86. $\dfrac{\sqrt{50}}{\sqrt{98z^2}}$

Rationalize each denominator and simplify. All variables represent positive numbers.

87. $\dfrac{1}{\sqrt{3}}$

88. $\dfrac{1}{\sqrt{5}}$

89. $\sqrt{\dfrac{13}{7}}$

90. $\sqrt{\dfrac{3}{11}}$

91. $\dfrac{9}{\sqrt{27}}$

92. $\dfrac{4}{\sqrt{20}}$

93. $\dfrac{3}{\sqrt{32}}$

94. $\dfrac{5}{\sqrt{18}}$

95. $\sqrt{\dfrac{12}{5}}$

96. $\sqrt{\dfrac{24}{7}}$

97. $\dfrac{10}{\sqrt{x}}$

98. $\dfrac{12}{\sqrt{y}}$

99. $\dfrac{\sqrt{9y}}{\sqrt{2x}}$

100. $\dfrac{\sqrt{4t}}{\sqrt{3z}}$

101. $\dfrac{3}{\sqrt{3} - 1}$

102. $\dfrac{3}{\sqrt{5} - 2}$

103. $\dfrac{3}{\sqrt{7} + 2}$

104. $\dfrac{5}{\sqrt{8} + 3}$

105. $\dfrac{12}{3 - \sqrt{3}}$

106. $\dfrac{10}{5 - \sqrt{5}}$

107. $\dfrac{5}{\sqrt{3} + \sqrt{2}}$

108. $\dfrac{3}{\sqrt{3} - \sqrt{2}}$

APPLICATIONS

109. AIR HOCKEY Find the area of the playing surface of the air hockey game.

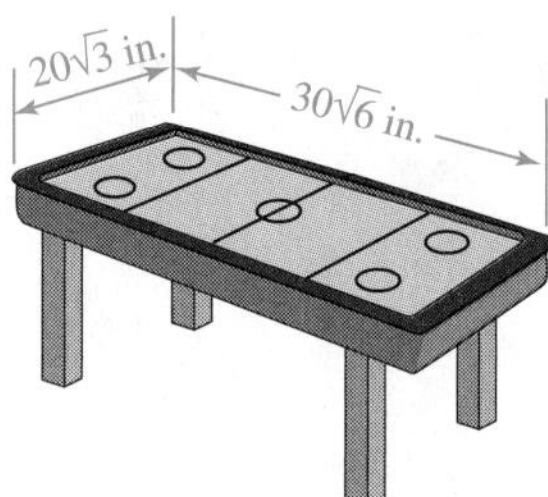

110. PROJECTOR SCREENS To find the length l of a rectangle, we can use the formula

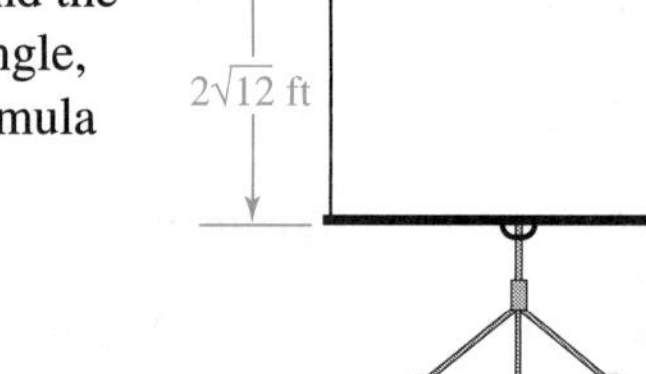

$$l = \frac{A}{w}$$

where A is the area of the rectangle and w is its width. Find the length of the screen if its area is 54 square feet.

111. LAWNMOWERS Use the formula for the area of a circle, $A = \pi r^2$, to find the area of lawn covered by one rotation of the blade. Leave π in your answer.

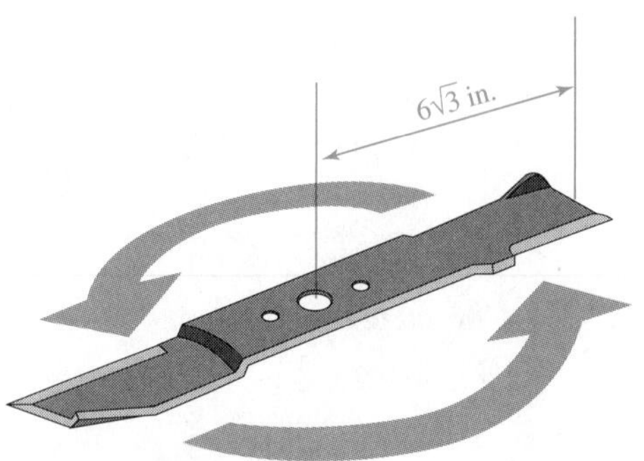

112. AWARDS PLATFORMS Find the total number of cubic feet of concrete needed to construct the Olympic Games awards platforms.

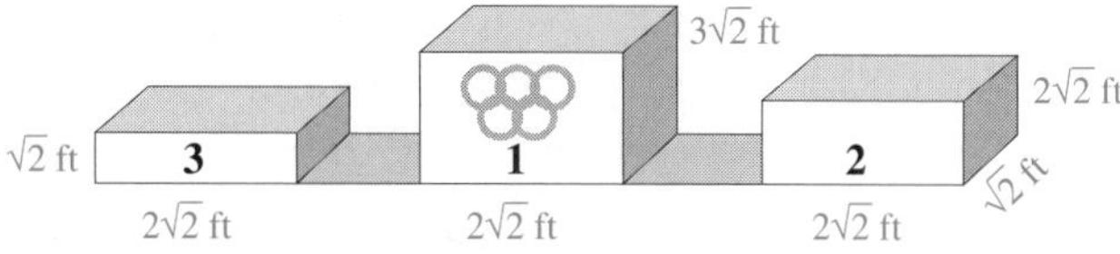

WRITING

113. When rationalizing the denominator of $\frac{5}{\sqrt{6}}$, why must we multiply both the numerator and denominator by $\sqrt{6}$?

114. What does it mean when we say that $\frac{1}{\sqrt{2}}$ and $\frac{\sqrt{2}}{2}$ are equivalent?

115. Explain why multiplying $\frac{9}{\sqrt{5}}$ by $\frac{\sqrt{5}}{\sqrt{5}}$ does not change its value.

116. Is the conjugate of $\sqrt{11} - 8$ the same as the opposite of $\sqrt{11} - 8$? Explain.

REVIEW

117. Is $x = -2$ a solution of $3x - 7 = 5x + 1$?

118. The graph of a line passes through the point (2, 0). Is this the x- or y-intercept?

119. The graph of a straight line rises from left to right. Is the slope of the line positive or negative?

120. Evaluate: $\frac{-4(6 + 2) - 2^4}{|-22 - 4(-5)|}$.

CHALLENGE PROBLEMS

121. Rationalize the denominator. Assume x represents a positive number.

$$\frac{\sqrt{x} - 3}{\sqrt{x} + 3}$$

122. Rationalize the numerator: $\frac{\sqrt{7} + 1}{\sqrt{6} - 4}$.

8.5 Solving Radical Equations

- The Squaring Property of Equality
- Checking Possible Solutions
- Equations Containing One Square Root
- Equations Containing Two Square Roots

When we solve equations containing fractions, we clear them of the fractions by multiplying both sides by the LCD. To solve equations containing radicals, we take a similar approach. The first step is to clear them of the radicals. To do this, we use a new property of equality called the *squaring property.*

THE SQUARING PROPERTY OF EQUALITY

A **radical equation** is an equation that contains a variable in a radicand. Some examples of radical equations are

$$\sqrt{x} = 6, \qquad \sqrt{2x - 1} = 9, \qquad \text{and} \qquad \sqrt{x + 12} = 3\sqrt{x + 4}$$

To solve these equations, we will use the following property.

Squaring Property of Equality If two numbers are equal, their squares are equal.

For any real numbers a and b, if $a = b$, then $a^2 = b^2$.

We can use the squaring property of equality to solve radical equations involving square roots.

EXAMPLE 1

Solve: $\sqrt{x} = 6$.

ELEMENTARY Algebra f(x) Now™

Solution To clear the equation of the square root, we square both sides.

The Language of Algebra

When we square both sides of an equation, we are *raising both sides* to the second power.

$$\sqrt{x} = 6$$

$$\left(\sqrt{x}\right)^2 = (6)^2 \quad \text{Use the squaring property of equality and square both sides.}$$

$$x = 36 \quad \text{Simplify: } \left(\sqrt{x}\right)^2 = x \text{ and } (6)^2 = 36.$$

Checking the result, we have

$$\sqrt{x} = 6$$

$$\sqrt{36} \stackrel{?}{=} 6 \quad \text{Substitute 36 for } x.$$

$$6 = 6 \quad \text{Evaluate the left-hand side.}$$

Since we obtain a true statement, 36 is the solution.

Self Check 1 Solve: $\sqrt{y} = 10$.

CHECKING POSSIBLE SOLUTIONS

If we square both sides of an equation, the resulting equation may not have the same solutions as the original one. For example, consider the equation

$$x = 2$$

The only solution of this equation is 2. However, if we square both sides, we obtain $(x)^2 = (2)^2$, which simplifies to

$$x^2 = 4$$

This new equation has solutions 2 and -2, because $2^2 = 4$ and $(-2)^2 = 4$.

The equations $x = 2$ and $x^2 = 4$ are not equivalent equations because they do not have the same solutions. The solution -2 satisfies $x^2 = 4$ but it does not satisfy $x = 2$. We see that squaring both sides of an equation can produce an equation with solutions that don't satisfy the original one. Therefore, we must check each possible solution in the original equation.

EQUATIONS CONTAINING ONE SQUARE ROOT

To solve radical equations containing square roots, we follow these steps.

Solving Radical Equations

1. Isolate a radical term on one side of the equation.
2. Square both sides of the equation.
3. Solve the resulting equation.
4. Check the possible solutions in the original equation. This step is required.

EXAMPLE 2

Solve: $\sqrt{2x - 1} = 9$.

ELEMENTARY Algebra f(x) Now™

Solution Since the radical is already isolated on one side, we square both sides of the equation to clear the radical.

$$\sqrt{2x-1} = 9$$

$(\sqrt{2x-1})^2 = (9)^2$ Use the squaring property of equality and square both sides.

$2x - 1 = 81$ Simplify: $(\sqrt{2x-1})^2 = 2x - 1$ and $(9)^2 = 81$.

$2x = 82$ To solve the resulting equation, add 1 to both sides.

$x = 41$ Divide both sides by 2.

Check the possible solution in the original equation.

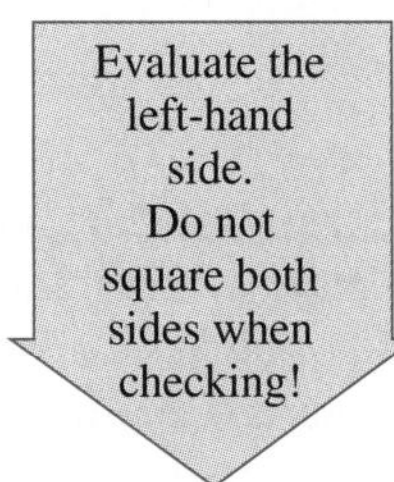

$\sqrt{2x-1} = 9$ This is the original equation.

$\sqrt{2(41)-1} \stackrel{?}{=} 9$ Substitute 41 for x.

$\sqrt{82-1} \stackrel{?}{=} 9$ Do the multiplication within the radical.

$\sqrt{81} \stackrel{?}{=} 9$ Do the subtraction within the radical.

$9 = 9$

The solution is 41.

Self Check 2 Solve: $\sqrt{3x-5} = 2$.

EXAMPLE 3

Solve: $\sqrt{x+2} + 5 = 3$.

ELEMENTARY Algebra f(x) Now™

Solution Since 5 is outside the square root symbol, there are two terms on the left-hand side of the equation. To isolate the radical term, we subtract 5 from both sides.

$\sqrt{x+2} + 5 = 3$

$\sqrt{x+2} = -2$ Subtract 5 from both sides.

We now square both sides to clear the equation of the radical.

$(\sqrt{x+2})^2 = (-2)^2$ Square both sides.

$x + 2 = 4$ Simplify: $(\sqrt{x+2})^2 = x + 2$ and $(-2)^2 = 4$.

$x = 2$ Subtract 2 from both sides.

We check by substituting 2 for x in the original equation.

$\sqrt{x+2} + 5 = 3$ This is the original equation.

$\sqrt{2+2} + 5 \stackrel{?}{=} 3$ Substitute 2 for x.

$\sqrt{4} + 5 \stackrel{?}{=} 3$ Do the addition within the radical.

$2 + 5 \stackrel{?}{=} 3$ Find the square root.

$7 = 3$

Since the result is a false statement, 2 is not a solution of the original equation and must be discarded. The original equation has no solution.

The Language of Algebra

To *isolate the radical* means to use properties of algebra to rewrite the given equation so that the square root term is by itself on one side of the equation.

Success Tip

In Example 3, it is apparent that $\sqrt{x+2} = -2$ has no solution. Since $\sqrt{x+2}$ is the principal square root, and cannot be negative, there is no real number x that could make $\sqrt{x+2}$ equal to a negative number.

Self Check 3 Solve: $\sqrt{x-2} + 5 = 2$.

Example 3 shows that squaring both sides of an equation can lead to possible solutions that do not satisfy the original equation. We call such numbers **extraneous solutions.** In Example 3, 2 is an extraneous solution of $\sqrt{x+2} + 5 = 3$.

EXAMPLE 4

Bridges. The time t in seconds that it takes an object to fall d feet is given by the formula

$$t = \frac{\sqrt{d}}{4}$$

To find the height of the Benjamin Franklin Bridge in Philadelphia, a man stands in the center section and drops a coin into the water. If it takes the coin 3 seconds to hit the water, how high above the water is the bridge?

Solution Since the coin fell for 3 seconds, we substitute 3 for t in the formula and solve for d.

$$t = \frac{\sqrt{d}}{4}$$

$$3 = \frac{\sqrt{d}}{4}$$ Substitute 3 for t.

$$12 = \sqrt{d}$$ To clear the equation of the fraction, multiply both sides by 4.

$$(12)^2 = \left(\sqrt{d}\right)^2$$ To clear the equation of the radical, square both sides.

$$144 = d$$ Simplify: $(12)^2 = 144$ and $\left(\sqrt{d}\right)^2 = d$.

The bridge is 144 feet above the water. Check this result in the original equation.

EXAMPLE 5

Solve: $a = \sqrt{a^2 + 3a - 3}$.

ELEMENTARY Algebra f(x) Now™

Solution Since the radical is isolated on the right-hand side, we square both sides to clear the equation of the radical.

$$a = \sqrt{a^2 + 3a - 3}$$

$$(a)^2 = \left(\sqrt{a^2 + 3a - 3}\right)^2$$ Square both sides.

$$a^2 = a^2 + 3a - 3$$ Simplify.

$$a^2 - a^2 = a^2 + 3a - 3 - a^2$$ Subtract a^2 from both sides.

$$0 = 3a - 3$$ Combine like terms.

$$3 = 3a$$ Add 3 to both sides.

$$1 = a$$ Divide both sides by 3.

We check the possible solution.

$$a = \sqrt{a^2 + 3a - 3}$$

$$1 \stackrel{?}{=} \sqrt{(1)^2 + 3(1) - 3}$$ Substitute 1 for a.

$$1 \stackrel{?}{=} \sqrt{1 + 3 - 3}$$ Evaluate the expression within the radical.

$$1 \stackrel{?}{=} \sqrt{1}$$
$$1 = 1$$

The solution is 1.

Self Check 5 Solve: $b = \sqrt{b^2 - 2b + 10}$.

Sometimes, after clearing an equation of a radical, the result is a quadratic equation.

EXAMPLE 6

Solve: $\sqrt{3 - x} - x = -3$.

ELEMENTARY Algebra $f(x)$ Now™

Solution To isolate the radical, add x to both sides.

$$\sqrt{3 - x} - x = -3$$
$$\sqrt{3 - x} = x - 3 \quad \text{Add } x \text{ to both sides.}$$

We then square both sides to clear the equation of the radical.

$$\left(\sqrt{3 - x}\right)^2 = (x - 3)^2 \quad \text{Square both sides.}$$
$$3 - x = (x - 3)(x - 3) \quad \text{The second power indicates two factors of } x - 3.$$
$$3 - x = x^2 - 6x + 9 \quad \text{Multiply the binomials.}$$

To solve the resulting quadratic equation, we write it in standard form so that the left-hand side is 0.

$$3 = x^2 - 5x + 9 \quad \text{Add } x \text{ to both sides.}$$
$$0 = x^2 - 5x + 6 \quad \text{Subtract 3 from both sides.}$$
$$0 = (x - 3)(x - 2) \quad \text{Factor } x^2 - 5x + 6.$$
$$x - 3 = 0 \quad \text{or} \quad x - 2 = 0 \quad \text{Set each factor equal to 0.}$$
$$x = 3 \quad \Big| \quad x = 2 \quad \text{Solve each equation.}$$

There are two possible solutions to check in the original equation.

Success Tip

Even if you are certain that no algebraic mistakes were made when solving a radical equation, you must still check your solutions. Squaring both sides can introduce extraneous solutions that must be discarded.

***For x* = 3**	***For x* = 2**
$\sqrt{3 - x} - x = -3$	$\sqrt{3 - x} - x = -3$
$\sqrt{3 - 3} - 3 \stackrel{?}{=} -3$	$\sqrt{3 - 2} - 2 \stackrel{?}{=} -3$
$\sqrt{0} - 3 \stackrel{?}{=} -3$	$\sqrt{1} - 2 \stackrel{?}{=} -3$
$0 - 3 \stackrel{?}{=} -3$	$1 - 2 \stackrel{?}{=} -3$
$-3 = -3$	$-1 = -3$

Since a true statement results when 3 is substituted for x, 3 is a solution. Since a false statement results when 2 is substituted for x, 2 is an extraneous solution.

Self Check 6 Solve: $\sqrt{x + 4} - x = -2$.

EQUATIONS CONTAINING TWO SQUARE ROOTS

To solve an equation containing two square roots, we want to have one radical on the left-hand side and one radical on the right-hand side.

EXAMPLE 7

Solve: $\sqrt{x+12} = 3\sqrt{x+4}$.

ELEMENTARY Algebra f(x) Now™

Solution Since each radical is isolated on one side of the equation, we can clear the equation of the radicals by squaring both sides.

$$\sqrt{x+12} = 3\sqrt{x+4}$$

$$\left(\sqrt{x+12}\right)^2 = \left(3\sqrt{x+4}\right)^2 \quad \text{Square both sides.}$$

$$x + 12 = 3^2\left(\sqrt{x+4}\right)^2$$ On the left, simplify: $\left(\sqrt{x+12}\right)^2 = x + 12$. On the right, raise each factor of the product $3\sqrt{x+4}$ to the second power.

$$x + 12 = 9(x+4)$$ To solve the resulting equation, simplify: $3^2 = 9$ and $\left(\sqrt{x+4}\right)^2 = x + 4$.

$$x + 12 = 9x + 36 \quad \text{Distribute the multiplication by 9.}$$

$$12 = 8x + 36 \quad \text{Subtract } x \text{ from both sides.}$$

$$-24 = 8x \quad \text{Subtract 36 from both sides.}$$

$$-3 = x \quad \text{Divide both sides by 8.}$$

Success Tip

Recall that to raise a product to a power, we raise each factor of the product to that power:

$$(xy)^n = x^n y^n$$

Use this rule for exponents to simplify

$$\left(3\sqrt{x+4}\right)^2$$

We check by substituting -3 for x in the original equation.

$$\sqrt{x+12} = 3\sqrt{x+4}$$

$$\sqrt{-3+12} \stackrel{?}{=} 3\sqrt{-3+4}$$

$$\sqrt{9} \stackrel{?}{=} 3\sqrt{1} \quad \text{Evaluate the expressions within the radicals.}$$

$$3 \stackrel{?}{=} 3 \cdot 1 \quad \text{Find each square root.}$$

$$3 = 3$$

The solution is -3.

Self Check 7 Solve: $\sqrt{x-4} = 2\sqrt{x-16}$.

Answers to Self Checks **1.** 100 **2.** 3 **3.** no solution **5.** 5 **6.** 5 **7.** 20

8.5 STUDY SET

ELEMENTARY Algebra f(x) Now™

VOCABULARY Fill in the blanks.

1. A ________ equation is an equation that contains a variable in a radicand.
2. To ________ the radical in $\sqrt{x} + 1 = 6$ means to get $\sqrt{x}$ by itself on one side of the equation.
3. Squaring both sides of an equation can lead to possible solutions that do not satisfy the original equation. We call such numbers ________ solutions.
4. The squaring property of equality states that if two numbers are equal, their ________ are equal.

CONCEPTS Fill in the blanks.

5. The squaring property of equality states that if $a = b$, then $a^2 =$ .
6. To clear the equation $\sqrt{x} = 5$ of the radical, we ________ both sides.
7. Use properties of algebra to isolate each radical term on one side of the equation. Do not solve the equation.

   a. $\sqrt{x-4} - 1 = 2$ **b.** $8 = \sqrt{x} - x$

8. Simplify each expression.

a. $(\sqrt{x})^2$ **b.** $(\sqrt{2x+3})^2$

c. $(4\sqrt{x})^2$ **d.** $(3\sqrt{1-b})^2$

e. $(\sqrt{n^2-2n+6})^2$ **f.** $(x-2)^2$

9. a. Write the quadratic equation in standard form with 0 on the right-hand side.

$$x^2 - 6x + 8 = 15$$

b. Subtract x^2 from both sides of the equation.

$$x^2 = x^2 - 3x + 12$$

10. How would you know, without solving it, that the equation $\sqrt{x-9} = -4$ has no solutions?

NOTATION **Complete each solution.**

11. Solve: $\sqrt{x-3} = 5$.

$$(\sqrt{x-3})^{\square} = 5^{\square}$$
$$\square = 25$$
$$x = \square$$

12. Determine whether -1 is a solution of the equation.

$$2\sqrt{3x+4} = \sqrt{5x+9}$$
$$2\sqrt{3(\square)+4} \stackrel{?}{=} \sqrt{5(\square)+9}$$
$$2\sqrt{\square+4} \stackrel{?}{=} \sqrt{\square+9}$$
$$2\sqrt{1} \stackrel{?}{=} \sqrt{\square}$$
$$2 \cdot \square \stackrel{?}{=} 2$$
$$2 = 2$$

Thus, $\square$ is the solution.

PRACTICE **Solve each equation.**

13. $\sqrt{x} = 3$ **14.** $\sqrt{x} = 5$

15. $\sqrt{y} = 12$ **16.** $\sqrt{y} = 20$

17. $\sqrt{2a} = 4$ **18.** $\sqrt{3a} = 9$

19. $\sqrt{4n} = 6$ **20.** $\sqrt{4n} = 8$

21. $\sqrt{x} = -6$ **22.** $\sqrt{x} = -7$

23. $\sqrt{r} + 4 = 0$ **24.** $\sqrt{r} + 1 = 0$

25. $\sqrt{a} + 2 = 9$ **26.** $\sqrt{a} - 1 = 0$

27. $\sqrt{x} + 1 = 7$ **28.** $\sqrt{x} - 2 = 8$

29. $5\sqrt{x} - 11 = 9$ **30.** $5\sqrt{x} - 6 = 4$

31. $-2 = 2\sqrt{x} - 12$ **32.** $-1 = 2\sqrt{x} - 7$

33. $-\sqrt{x} = -2$ **34.** $-\sqrt{x} = -12$

35. $10 - \sqrt{s} = 7$ **36.** $-4 = 6 - \sqrt{s}$

37. $\sqrt{x+3} = 2$ **38.** $\sqrt{x-2} = 3$

39. $\sqrt{5-T} = 10$ **40.** $\sqrt{3-T} = 2$

41. $\sqrt{6x+19} - 7 = 0$ **42.** $\sqrt{6+2x} = 4$

43. $\sqrt{x-5} - 3 = 4$ **44.** $\sqrt{5x-5} - 5 = 0$

45. $\sqrt{7+2x} + 4 = 1$ **46.** $\sqrt{x+3} + 5 = 4$

47. $x = \sqrt{x^2-2x+16}$ **48.** $x = \sqrt{x^2+3x-21}$

49. $\sqrt{4m^2+6m+6} = -2m$

50. $\sqrt{9t^2+4t+20} = -3t$

51. $x - 3 = \sqrt{x^2-15}$

52. $v - 2 = \sqrt{v^2-16}$

53. $x + 1 = \sqrt{x^2+x+4}$

54. $x - 1 = \sqrt{x^2-4x+9}$

55. $y - 9 = \sqrt{y-3}$ **56.** $m - 9 = \sqrt{m-7}$

57. $\sqrt{15-3t} = t - 5$ **58.** $\sqrt{1-8s} = s + 4$

59. $b = \sqrt{2b-2} + 1$ **60.** $c = \sqrt{5c+1} - 1$

61. $\sqrt{24+10n} - n = 4$ **62.** $\sqrt{7+6y} - y = 2$

63. $6 + \sqrt{m} - m = 0$ **64.** $6 + \sqrt{3m} - m = 0$

65. $\sqrt{3t-9} = \sqrt{t+1}$ **66.** $\sqrt{a-3} = \sqrt{2a-8}$

67. $\sqrt{10-3x} = \sqrt{2x+20}$

68. $\sqrt{1-2x} = \sqrt{x+10}$

69. $\sqrt{3c-8} = \sqrt{c}$ **70.** $\sqrt{2x} = \sqrt{x+8}$

71. $5\sqrt{a} = \sqrt{10a + 15}$

72. $\sqrt{17m - 4} = 4\sqrt{m}$

73. $2\sqrt{3x + 4} = \sqrt{5x + 9}$

74. $\sqrt{3x + 6} = 2\sqrt{2x - 11}$

APPLICATIONS

75. NIAGARA FALLS The time t in seconds that it takes an object to fall d feet is given by the formula

$$t = \frac{\sqrt{d}}{4}$$

The time it took a stuntman to go over the Niagara Falls in a barrel was 3.25 seconds. Find the height of the waterfall.

76. MONUMENTS Gabby Street, a professional baseball player of the 1920s, was known for once catching a ball dropped from the top of the Washington Monument in Washington, D.C. If the ball fell for slightly less than 6 seconds before it was caught, find the approximate height of the monument. (See Problem 75.)

77. PENDULUMS The time t (in seconds) required for a pendulum of length L feet to swing through one back-and-forth cycle, called its period, is given by the formula

$$t = 1.11\sqrt{L}$$

The Foucault pendulum in Chicago's Museum of Science and Industry is used to demonstrate the rotation of the Earth. It completes one cycle in 8.91 seconds. To the nearest tenth of a foot, how long is the pendulum?

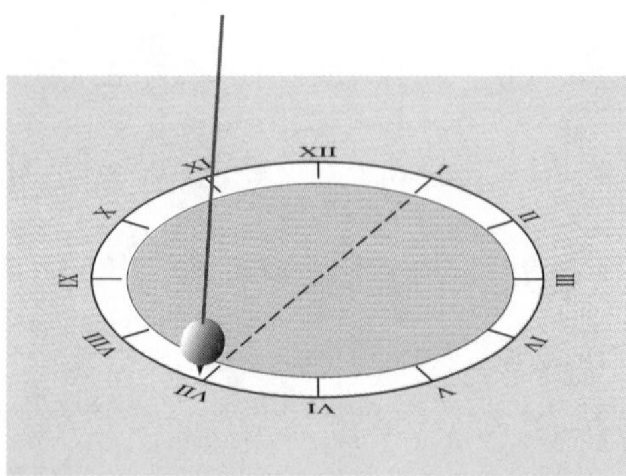

78. POWER USAGE The current I (in amperes), the resistance R (in ohms), and the power P (in watts) are related by the formula

$$I = \sqrt{\frac{P}{R}}$$

Find the power (to the nearest watt) used by a space heater that draws 7 amps when the resistance is 10.2 ohms.

79. ROAD SAFETY The formula $s = k\sqrt{d}$ relates the speed s (in mph) of a car and the distance d (in feet) of the skid when a driver hits the brakes. On wet pavement, $k = 3.24$. How far will a car skid if it is going 55 mph?

80. ROAD SAFETY How far will the car in Problem 79 skid if it is traveling on dry pavement? On dry pavement, $k = 5.34$.

81. HIGHWAY DESIGN A highway curve banked at 8° will accommodate traffic traveling at speed s (in mph) if the radius of the curve is r (feet), according to the equation $s = 1.45\sqrt{r}$. If highway engineers expect traffic to travel at 65 mph, what radius should they specify (to the nearest foot)?

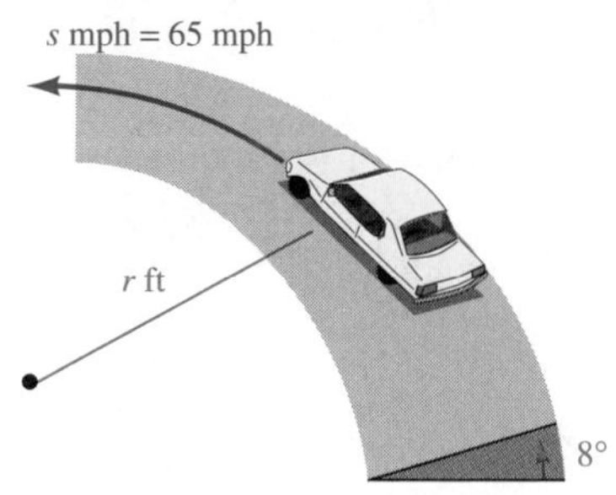

82. GEOMETRY The radius of a cone with volume V and height h is given by the formula

$$r = \sqrt{\frac{3V}{\pi h}}$$

Solve the equation for V.

WRITING

83. Explain why a check is necessary when solving radical equations.

84. Explain why the equation $\sqrt{x} = -2$ has no solution.

85. Explain the error in the following partial solution.

Solve: $\sqrt{1-4y} = y + 2$

$$(\sqrt{1-4y})^2 = (y+2)^2$$
$$1 - 4y = y^2 + 4$$

86. Explain the error in the following partial solution.

Solve: $\sqrt{a+2} - 5 = 4$

$$(\sqrt{a+2} - 5)^2 = 4^2$$

REVIEW **Solve each equation.**

87. $\frac{1}{2} + \frac{x}{5} = \frac{3}{4}$

88. $\frac{1}{3} + \frac{c}{5} = -\frac{3}{2}$

89. $\frac{2}{5}x + 1 = \frac{1}{3} + x$

90. $\frac{2}{3}y + 2 = \frac{1}{5} + y$

CHALLENGE PROBLEMS **Solve each equation.**

91. $\sqrt{\sqrt{x+2}} = 3$

92. $1 + \sqrt{x} = \sqrt{x+3}$

8.6 Higher-Order Roots and Rational Exponents

- Cube Roots
- Higher-Order Roots
- Simplifying Higher-Order Roots
- Rational Exponents

In this section, we will consider *higher-order roots* such as cube roots and fourth roots. We will also extend the definition of exponents to include rational (fractional) exponents.

CUBE ROOTS

When we raise a number to the third power, we are cubing it, or finding its **cube.** We can reverse the cubing process to find **cube roots** of numbers. To find the cube root of 8, we ask "What number, when cubed, is equal to 8?" It follows that 2 is a cube root of 8, because $2^3 = 8$.

In general, we have this definition.

The Definition of Cube Root The number b is a **cube root** of the number a if $b^3 = a$.

All real numbers have one cube root. A positive number has a positive cube root, a negative number has a negative cube root, and the cube root of 0 is 0.

Cube Root Notation The **cube root of *a*** is denoted by $\sqrt[3]{a}$. By definition,

$$\sqrt[3]{a} = b \quad \text{if} \quad b^3 = a$$

Notation

For the square root symbol $\sqrt{}$, the unwritten index is understood to be 2.

$$\sqrt{a} = \sqrt[2]{a}$$

Earlier, we determined that the cube root of 8 is 2. In symbols, we can write: $\sqrt[3]{8} = 2$. The number 3 is called the **index,** 8 is called the **radicand,** and the entire expression is called a **radical.**

Index → $\sqrt[3]{\mathbf{8}}$ ← Radicand

Radical

EXAMPLE 1

Find each cube root: **a.** $\sqrt[3]{27}$, **b.** $\sqrt[3]{-125}$, and **c.** $-\sqrt[3]{64}$.

Solution **a.** $\sqrt[3]{27} = 3$ What number, when cubed, is 27? The answer is $3^3 = 27$.

b. $\sqrt[3]{-125} = -5$ What number, when cubed, is -125? The answer is $(-5)^3 = -125$.

c. $-\sqrt[3]{64}$ is the opposite of the cube root of 64. Since $\sqrt[3]{64} = 4$, we have

$$-\sqrt[3]{64} = -4 \qquad -\sqrt[3]{64} = -1 \cdot \sqrt[3]{64} = -1 \cdot 4 = -4.$$

Self Check 1 Find each cube root: **a.** $\sqrt[3]{1}$, **b.** $\sqrt[3]{-8}$, and **c.** $-\sqrt[3]{216}$.

HIGHER-ORDER ROOTS

Just as there are square roots and cube roots, there are also fourth roots, fifth roots, sixth roots, and so on. In general, we have the following definition.

nth Roots of a

The ***n*th root of *a*** is denoted by $\sqrt[n]{a}$, and

$$\sqrt[n]{a} = b \quad \text{if} \quad b^n = a$$

The number n is called the **index** of the radical. If n is an even natural number, a must be positive or zero, and b must be positive.

EXAMPLE 2

Find each root: **a.** $\sqrt[4]{81}$, **b.** $\sqrt[5]{32}$, **c.** $\sqrt[5]{-1}$, and **d.** $\sqrt[4]{-16}$.

Solution

a. $\sqrt[4]{81} = 3$ What number, raised to the 4th power, is 81? The answer is $3^4 = 81$.

b. $\sqrt[5]{32} = 2$ What number, raised to the 5th power, is 32? The answer is $2^5 = 32$.

c. $\sqrt[5]{-1} = -1$ What number, raised to the 5th power, is -1? The answer is $(-1)^5 = -1$.

d. $\sqrt[4]{-16}$ is not a real number because no real number raised to the fourth power is -16.

Self Check 2 Find each root: **a.** $\sqrt[4]{64}$, **b.** $\sqrt[5]{243}$, **c.** $\sqrt[4]{-1}$, and **d.** $\sqrt[5]{-1{,}024}$.

When n is even and $x \geq 0$, we say that the radical $\sqrt[n]{x}$ represents an **even root.** We can find even roots of expressions that contain variables, provided these variable expressions do not represent negative numbers.

EXAMPLE 3

ELEMENTARY Algebra f(x) Now™

Find each root. Each variable represents a nonnegative number: **a.** $\sqrt[4]{m^4}$, **b.** $\sqrt[4]{16x^{12}}$, and **c.** $\sqrt[6]{x^{12}}$.

Solution

a. $\sqrt[4]{m^4} = m$ What expression, raised to the 4th power, is m^4? The answer is $(m)^4 = m^4$.

b. $\sqrt[4]{16x^{12}} = 2x^3$ What expression, raised to the 4th power, is $16x^{12}$? The answer is $(2x^3)^4 = 16x^{12}$.

c. $\sqrt[6]{x^{12}} = x^2$ What expression, raised to the 6th power, is x^{12}? The answer is $(x^2)^6 = x^{12}$.

Self Check 3 Find each root. Each variable represents a positive number: **a.** $\sqrt[4]{n^8}$, **b.** $\sqrt[4]{81x^{16}}$, and **c.** $\sqrt[6]{m^{18}}$.

When n is odd, we say that the radical $\sqrt[n]{x}$ represents an **odd root.**

EXAMPLE 4

Find each root: **a.** $\sqrt[3]{y^3}$, **b.** $\sqrt[3]{64x^9}$, and **c.** $\sqrt[5]{x^{10}}$.

Solution

a. $\sqrt[3]{y^3} = y$ What expression, when cubed, is y^3? The answer is $(y)^3 = y^3$.

b. $\sqrt[3]{64x^9} = 4x^3$ What expression, when cubed, is $64x^9$? The answer is $(4x^3)^3 = 64x^9$.

c. $\sqrt[5]{x^{10}} = x^2$ What expression, raised to the 5th power, is x^{10}? The answer is $(x^2)^5 = x^{10}$.

Self Check 4 Find each root: **a.** $\sqrt[3]{s^{15}}$, **b.** $\sqrt[3]{1{,}000b^3}$, and **c.** $\sqrt[5]{x^{15}}$.

SIMPLIFYING HIGHER-ORDER ROOTS

The product and quotient rules for square roots can be generalized to apply to higher-order roots.

The Product and Quotient Rules for Radicals

If $\sqrt[n]{a}$ and $\sqrt[n]{b}$ are real numbers,

$$\sqrt[n]{a \cdot b} = \sqrt[n]{a}\sqrt[n]{b} \qquad \sqrt[n]{\frac{a}{b}} = \frac{\sqrt[n]{a}}{\sqrt[n]{b}}, \quad \text{provided } b \neq 0$$

We can use these rules to write higher-order roots in simplified form. To simplify cube roots, it is helpful to be familiar with these natural number **perfect cubes:**

1, 8, 27, 64, 125, 216, 343, 512, 729, 1,000, . . .

To simplify fourth roots, it is helpful to be familiar with these natural number **perfect fourth powers:**

1, 16, 81, 256, 625, . . .

EXAMPLE 5

Simplify: **a.** $\sqrt[3]{54}$, **b.** $\sqrt[4]{405}$, and **c.** $\sqrt[3]{\frac{32}{125}}$.

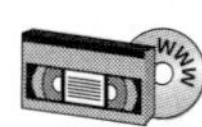

Solution

a. From the list of perfect cubes above, we see that 27 is the greatest perfect cube factor of the radicand 54. To simplify $\sqrt[3]{54}$, we write 54 as the product of 27 and one other factor. Then we apply the product rule for radicals.

$\sqrt[3]{54} = \sqrt[3]{27 \cdot 2}$ Write 54 as a product of its greatest perfect cube factor and one other factor: $54 = 27 \cdot 2$.

$= \sqrt[3]{27}\sqrt[3]{2}$ The cube root of a product is equal to the product of the cube roots.

$= 3\sqrt[3]{2}$ Find the cube root of the perfect cube factor: $\sqrt[3]{27} = 3$.

We say that $3\sqrt[3]{2}$ is the simplified form of $\sqrt[3]{54}$.

b. To determine the greatest perfect fourth power factor of the radicand 405, it is helpful to write it in prime factored form.

$\sqrt[4]{405} = \sqrt[4]{3 \cdot 3 \cdot 3 \cdot 3 \cdot 5}$ Prime-factor 405.

$= \sqrt[4]{3 \cdot 3 \cdot 3 \cdot 3}\sqrt[4]{5}$ Use the product rule for radicals. Write $3 \cdot 3 \cdot 3 \cdot 3$ within one fourth root because it is a perfect fourth power: $3 \cdot 3 \cdot 3 \cdot 3 = 81$.

$= 3\sqrt[4]{5}$ $\sqrt[4]{3 \cdot 3 \cdot 3 \cdot 3} = \sqrt[4]{81}$, which simplifies to 3.

c. We can use the quotient rule for radicals to simplify this expression.

$$\sqrt[3]{\frac{32}{125}} = \frac{\sqrt[3]{32}}{\sqrt[3]{125}}$$ The cube root of the quotient is equal to the quotient of the cube roots.

$$= \frac{\sqrt[3]{8 \cdot 4}}{5}$$ Write 32 as the product of its greatest perfect cube factor and one other factor: $32 = 8 \cdot 4$. In the denominator, $\sqrt[3]{125} = 5$.

$$= \frac{\sqrt[3]{8}\sqrt[3]{4}}{5}$$ The cube root of a product is equal to the product of the cube roots.

$$= \frac{2\sqrt[3]{4}}{5}$$ Find the cube root of the perfect cube factor: $\sqrt[3]{8} = 2$.

Self Check 5 Simplify: **a.** $\sqrt[3]{128}$, **b.** $\sqrt[4]{176}$, and **c.** $\sqrt[3]{\frac{40}{343}}$.

RATIONAL EXPONENTS

It is possible to raise numbers to fractional powers. To give meaning to rational (fractional) exponents, we first consider $\sqrt{7}$. Because $\sqrt{7}$ is the positive number whose square is 7, we have

$$(\sqrt{7})^2 = 7$$

The Language of Algebra

Rational exponents are also called *fractional exponents.*

We now consider the notation $7^{1/2}$. If rational exponents are to follow the same rules as integer exponents, the square of $7^{1/2}$ must be 7, because

$$(7^{1/2})^2 = 7^{1/2 \cdot 2}$$ Keep the base and multiply the exponents.

$$= 7^1$$ Do the multiplication: $\frac{1}{2} \cdot 2 = 1$.

$$= 7$$

Since the square of $7^{1/2}$ and the square of $\sqrt{7}$ are both equal to 7, we define $7^{1/2}$ to be $\sqrt{7}$. Similarly,

$$7^{1/3} = \sqrt[3]{7}, \quad 7^{1/4} = \sqrt[4]{7}, \quad \text{and} \quad 7^{1/5} = \sqrt[5]{7}$$

In general, we have the following definition.

The Definition of $x^{1/n}$

If n represents a positive integer greater than 1 and $\sqrt[n]{x}$ represents a real number,

$$x^{1/n} = \sqrt[n]{x}$$

We can use this definition to simplify exponential expressions that have rational exponents with a numerator of 1.

EXAMPLE 6

Evaluate: **a.** $64^{1/3}$, **b.** $9^{1/2}$, **c.** $(-8)^{1/3}$, and **d.** $16^{1/4}$.

ELEMENTARY Algebra Now™

Solution **a.** To evaluate $64^{1/3}$, we write it in an equivalent radical form. The denominator of the rational exponent is the same as the index of the corresponding radical.

Root

$64^{1/3} = \sqrt[3]{64} = 4$ Because the denominator of the exponent is 3, find the cube root of the base, 64.

b. $9^{1/2} = \sqrt{9} = 3$ Because the denominator of the exponent is 2, find the square root of the base, 9.

c. $(-8)^{1/3} = \sqrt[3]{-8} = -2$ Because the denominator of the exponent is 3, find the cube root of the base, -8.

d. $16^{1/4} = \sqrt[4]{16} = 2$ Because the denominator of the exponent is 4, find the fourth root of the base, 16.

Self Check 6 Evaluate: **a.** $144^{1/2}$, **b.** $(-27)^{1/3}$, and **c.** $81^{1/4}$.

We can extend the definition of $x^{1/n}$ to cover fractional exponents for which the numerator is not 1. For example, because $4^{3/2}$ can be written as $(4^{1/2})^3$, we have

$$4^{3/2} = (4^{1/2})^3 = \left(\sqrt{4}\right)^3 = 2^3 = 8$$

Because $4^{3/2}$ can also be written as $(4^3)^{1/2}$, we have

$$4^{3/2} = (4^3)^{1/2} = 64^{1/2} = \sqrt{64} = 8$$

In general, $x^{m/n}$ can be written as $(x^{1/n})^m$ or as $(x^m)^{1/n}$. Since $(x^{1/n})^m = \left(\sqrt[n]{x}\right)^m$ and $(x^m)^{1/n} = \sqrt[n]{x^m}$, we make the following definition.

The Definition of $x^{m/n}$

If m and n represent positive integers ($n \neq 1$) and $\sqrt[n]{x}$ represents a real number,

$$x^{m/n} = \left(\sqrt[n]{x}\right)^m \quad \text{and} \quad x^{m/n} = \sqrt[n]{x^m}$$

We can use this definition to evaluate exponential expressions that have rational exponents with a numerator that is not 1. To avoid large numbers, we usually find the root of the base first and then calculate the power using the relationship $x^{m/n} = \left(\sqrt[n]{x}\right)^m$.

EXAMPLE 7

Evaluate: **a.** $125^{4/3}$, **b.** $81^{3/4}$, **c.** $-25^{3/2}$, and **d.** $(-27)^{2/3}$.

ELEMENTARY Algebra f(x) Now™

Solution **a.** To evaluate $125^{4/3}$, we write it in an equivalent radical form. The denominator of the rational exponent is the same as the index of the corresponding radical. The numerator of the rational exponent indicates the power to which the radical base is raised.

Power

Root

$125^{4/3} = \left(\sqrt[3]{125}\right)^4 = (5)^4 = 625$ Because the exponent is 4/3, find the cube root of the base, 125, to get 5. Then find the fourth power of 5.

b. $81^{3/4} = \left(\sqrt[4]{81}\right)^3 = (3)^3 = 27$ Because the exponent is 3/4, find the fourth root of the base, 81, to get 3. Then find the third power of 3.

c. For the exponential expression $-25^{3/2}$, the base is 25, not -25.

Caution

We can also evaluate $x^{m/n}$ using $\sqrt[n]{x^m}$, however the resulting radicand is often extremely large. For example,

$$81^{3/4} = \sqrt[4]{81^3}$$
$$= \sqrt[4]{531{,}441}$$
$$= 27$$

$$-25^{3/2} = -(25^{3/2}) = -\left(\sqrt{25}\right)^3 = -(5)^3 = -125$$

Because the exponent is 3/2, find the square root of the base, 25, to get 5. Then find the third power of 5.

d. Because of the parentheses, the base of the exponential expression $(-27)^{2/3}$ is -27.

$$(-27)^{2/3} = \left(\sqrt[3]{-27}\right)^2 = (-3)^2 = 9$$

Because the exponent is 2/3, find the cube root of the base, -27, to get -3. Then find the second power of -3.

Self Check 7 Evaluate: **a.** $100^{3/2}$, **b.** $(-8)^{2/3}$, and **c.** $-32^{4/5}$.

We define negative rational exponents in the following way.

The Definition of $x^{-m/n}$

If $x^{-m/n}$ is a nonzero real number,

$$x^{-m/n} = \frac{1}{x^{m/n}}$$

From the definition, we see that another way to write $x^{-m/n}$ is to write its reciprocal and change the sign of its exponent.

EXAMPLE 8

Evaluate: **a.** $27^{-2/3}$, **b.** $36^{-1/2}$, and **c.** $-625^{-3/4}$.

ELEMENTARY Algebra *f(x)* Now™

Solution **a.** Because the exponent is $-2/3$, we write the reciprocal of $27^{-2/3}$ and change the sign. Then we find the cube root of the base, 27, to get 3. Finally, we find the second power of 3.

Caution

A negative exponent does not indicate a negative number. For example,

$$27^{-2/3} = \frac{1}{9}$$

$$36^{-1/2} = \frac{1}{6}$$

Reciprocal

$$27^{-2/3} = \frac{1}{27^{2/3}} = \frac{1}{\left(\sqrt[3]{27}\right)^2} = \frac{1}{(3)^2} = \frac{1}{9}$$

Root

Power

b. $36^{-1/2} = \frac{1}{36^{1/2}} = \frac{1}{\sqrt{36}} = \frac{1}{6}$

c. In $-625^{-3/4}$, the base is 625.

$$-625^{-3/4} = -\frac{1}{625^{3/4}} = -\frac{1}{\left(\sqrt[4]{625}\right)^3} = -\frac{1}{(5)^3} = -\frac{1}{125}$$

Self Check 8 Evaluate: **a.** $49^{-1/2}$, **b.** $8^{-4/3}$, and **c.** $-16^{-1/4}$.

Answers to Self Checks **1. a.** 1, **b.** -2, **c.** -6 **2. a.** 4, **b.** 3, **c.** not a real number, **d.** -4 **3. a.** n^2, **b.** $3x^4$, **c.** m^3 **4. a.** s^5, **b.** $10b$, **c.** x^3 **5. a.** $4\sqrt[3]{2}$, **b.** $2\sqrt[4]{11}$, **c.** $\frac{2\sqrt[3]{5}}{7}$ **6. a.** 12, **b.** -3, **c.** 3 **7. a.** 1,000, **b.** 4, **c.** -16 **8. a.** $\frac{1}{7}$, **b.** $\frac{1}{16}$, **c.** $-\frac{1}{2}$

8.6 STUDY SET  ELEMENTARY Algebra $f(x)$ Now™

VOCABULARY **Fill in the blanks.**

1. We read $\sqrt[3]{8}$ as the ______ root of 8 and $\sqrt[4]{16}$ as the ______ root of 16.
2. For the radical $\sqrt[3]{27}$, the ______ is 3 and the ______ is 27.
3. "To ______ $\sqrt[3]{54}$" means to write it as $3\sqrt[3]{2}$.
4. In the expression $81^{1/4}$, the ______ is 81 and the ______ is 1/4.
5. The exponential expression $25^{3/2}$ has a ______ (fractional) exponent.
6. The exponential expression $16^{-1/2}$ has a ______ rational exponent.

CONCEPTS **Fill in the blanks.**

7. a. $\sqrt[3]{64} = 4$ because $(\quad)^3 = 64$.
   b. $\sqrt[4]{16x^4} = 2x$ because $(\quad)^4 = 16x^4$.
   c. $\sqrt[5]{-32} = -2$ because $(-2)^5 = \quad$.
8. a. Write the first seven perfect cubes.

   1, , 343

   b. Write the first five perfect fourth powers.

   1, ___, ___, 256, ___
9. a. $\sqrt[n]{a \cdot b} = \quad$ b. $\sqrt[n]{\frac{a}{b}} = \quad$
10. a. $x^{1/n} = \quad$ b. $x^{m/n} = \quad$
    c. $x^{-m/n} = \frac{1}{\quad}$
11. To simplify $\sqrt[3]{24}$, which of the following expressions should be used?

    $\sqrt[3]{12 \cdot 2}$ $\quad$ $\sqrt[3]{4 \cdot 6}$ $\quad$ $\sqrt[3]{8 \cdot 3}$ $\quad$ $\sqrt[3]{24 \cdot 1}$
12. To simplify $\sqrt[4]{48}$, which of the following expressions should be used?

    $\sqrt[4]{24 \cdot 2}$ $\quad$ $\sqrt[4]{16 \cdot 3}$ $\quad$ $\sqrt[4]{8 \cdot 6}$ $\quad$ $\sqrt[4]{48 \cdot 1}$
13. Simplify.
    a. $\sqrt[3]{8}\sqrt[3]{3}$ b. $\sqrt[4]{16}\sqrt[4]{3}$
    c. $\frac{\sqrt[3]{125}}{\sqrt[3]{27}}$ d. $\frac{\sqrt[4]{256}}{\sqrt[4]{81}}$
14. Write each exponential expression in an equivalent form involving a radical. Do not simplify it.
    a. $25^{1/2}$ b. $16^{1/4}$
    c. $(-27)^{2/3}$ d. $49^{3/2}$
15. Write each exponential expression in an equivalent form involving a radical. Do not simplify it.
    a. $36^{-1/2}$ b. $16^{-5/4}$
16. Graph each number on the number line.

$$\left\{\sqrt[3]{27}, 16^{1/2}, \sqrt[5]{-32}, \sqrt[3]{\frac{125}{64}}, 8^{-4/3}\right\}$$

NOTATION

17. Fill in the blanks: The understood index of the radical expression $\sqrt{55}$ is ___. That is, $\sqrt{55} = \sqrt[\quad]{55}$.
18. What is the base of each exponential expression?
    a. $-25^{3/2}$ b. $(-25)^{3/2}$

PRACTICE **Find each root. All variables represent nonnegative numbers.**

19. $\sqrt[3]{8}$
20. $\sqrt[3]{27}$
21. $\sqrt[3]{0}$
22. $\sqrt[3]{1}$
23. $\sqrt[3]{-8}$
24. $\sqrt[3]{-1}$
25. $\sqrt[3]{-64}$
26. $\sqrt[3]{-27}$
27. $\sqrt[3]{\frac{1}{125}}$
28. $\sqrt[3]{\frac{1}{1{,}000}}$
29. $-\sqrt[3]{-1}$
30. $-\sqrt[3]{-27}$
31. $-\sqrt[3]{64}$
32. $-\sqrt[3]{343}$
33. $\sqrt[3]{729}$
34. $\sqrt[3]{512}$
35. $\sqrt[3]{1{,}000}$
36. $\sqrt[3]{125}$
37. $\sqrt[3]{\frac{8}{27}}$
38. $\sqrt[3]{\frac{64}{125}}$
39. $\sqrt[4]{16}$
40. $\sqrt[4]{81}$
41. $\sqrt[4]{256}$
42. $-\sqrt[4]{625}$
43. $-\sqrt[4]{\frac{1}{81}}$
44. $\sqrt[4]{\frac{1}{256}}$
45. $\sqrt[4]{-625}$
46. $\sqrt[4]{-16}$

47. $\sqrt[5]{32}$
48. $\sqrt[5]{-32}$
49. $\sqrt[5]{-1}$
50. $\sqrt[5]{-243}$
51. $\sqrt[6]{64}$
52. $\sqrt[6]{\dfrac{1}{64}}$
53. $\sqrt[3]{m^3}$
54. $\sqrt[3]{n^{12}}$
55. $\sqrt[3]{27a^6}$
56. $\sqrt[3]{-64b^9}$
57. $\sqrt[4]{y^4}$
58. $\sqrt[4]{x^8}$
59. $\sqrt[4]{16b^{12}}$
60. $\sqrt[4]{81h^{16}}$
61. $\sqrt[5]{x^5}$
62. $\sqrt[5]{t^{10}}$
63. $\sqrt[6]{s^{12}}$
64. $\sqrt[6]{d^6}$

Simplify.

65. $\sqrt[3]{24}$
66. $\sqrt[3]{32}$
67. $\sqrt[3]{-128}$
68. $\sqrt[3]{-250}$
69. $\sqrt[3]{72}$
70. $\sqrt[3]{189}$
71. $\sqrt[3]{\dfrac{250}{27}}$
72. $\sqrt[3]{\dfrac{54}{125}}$
73. $\sqrt[4]{162}$
74. $\sqrt[4]{80}$
75. $\sqrt[5]{64}$
76. $\sqrt[5]{160}$

Evaluate.

77. $81^{1/2}$
78. $100^{1/2}$
79. $16^{1/4}$
80. $81^{1/4}$
81. $-144^{1/2}$
82. $-400^{1/2}$
83. $(-125)^{1/3}$
84. $(-8)^{1/3}$
85. $1^{1/2}$
86. $121^{1/2}$
87. $\left(\dfrac{9}{64}\right)^{1/2}$
88. $\left(\dfrac{4}{49}\right)^{1/2}$
89. $8^{1/3}$
90. $27^{1/3}$
91. $16^{3/2}$
92. $4^{5/2}$
93. $27^{2/3}$
94. $8^{4/3}$
95. $8^{-1/3}$
96. $81^{-1/2}$
97. $25^{-3/2}$
98. $9^{-3/2}$
99. $32^{-3/5}$
100. $16^{-3/4}$

APPLICATIONS

101. WINDMILL The power generated by a windmill is related to the speed of the wind by the formula

$$S = \sqrt[3]{\frac{P}{0.02}}$$

where S is the speed of the wind (in mph) and P is the power (in watts). Find the speed of the wind when the windmill is producing 20 watts of power.

102. DEPRECIATION The formula

$$r = 1 - \sqrt[n]{\frac{S}{C}}$$

gives the annual depreciation rate r of an item that had an original cost of C dollars and has a useful life of n years and a salvage value of S dollars. Use the information in the illustration to find (in percent) the annual depreciation rate for the new piece of sound equipment.

OFFICE MEMO

To: Purchasing Dept.
From: Bob Kinsell, Engineering Dept. BK
Re: New sound board

We recommend you purchase the new Sony sound board @ \$81K. This equipment does become obsolete quickly but we figure we can use it for 4 yrs. A college would probably buy it from us then. I bet we could get around \$16K for it.

103. HOLIDAY DECORATING Find the length s of each string of colored lights used to decorate an evergreen tree in the manner shown if $s = (r^2 + h^2)^{1/2}$.

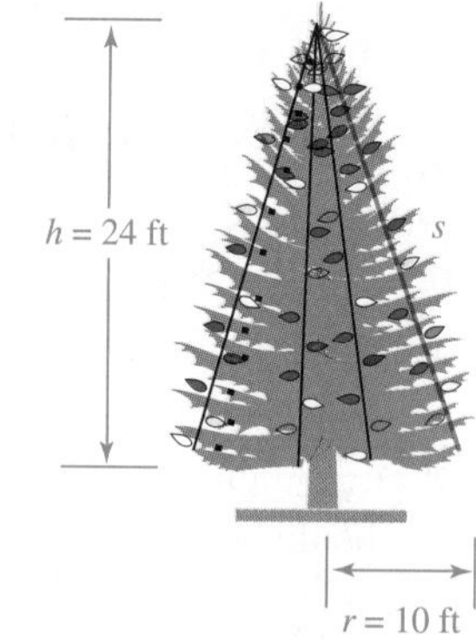

104. VISIBILITY The distance d in miles a person in an airplane can see to the horizon on a clear day is given by the formula $d = 1.22a^{1/2}$, where a is the altitude of the plane in feet. Find d.

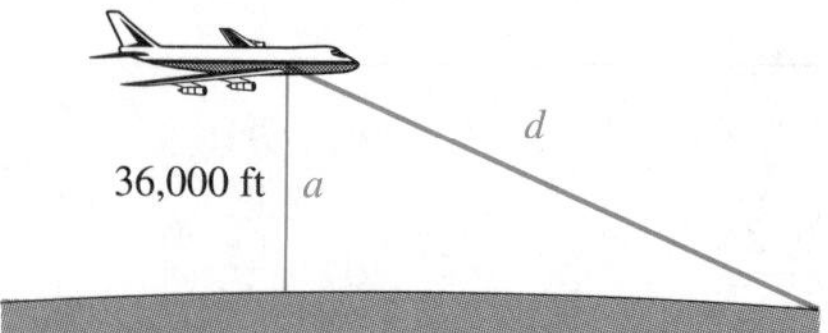

105. SPEAKERS The formula $A = V^{2/3}$ can be used to find the area A of one face of a cube if its volume V is known. Find the amount of floor space occupied by a speaker if it is a cube with a volume of 216 cubic inches.

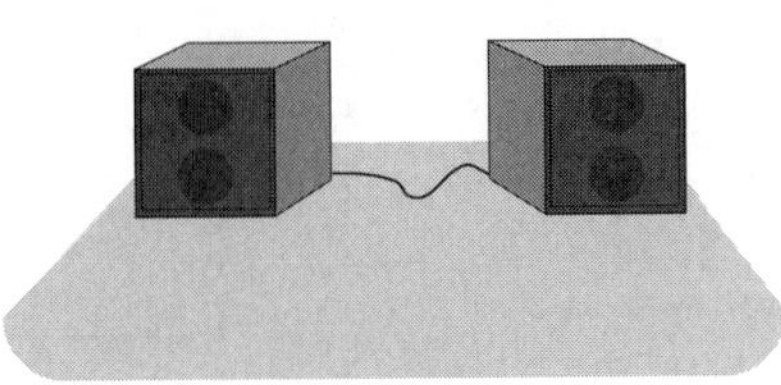

106. MEDICAL TESTS Before a series of X-rays are taken, a patient is injected with a special contrast mixture that highlights obstructions in his blood vessels. The amount of the original dose of contrast material remaining in the patient's bloodstream h hours after it is injected is given by $h^{-3/2}$. How much of the contrast material remains in the patient's bloodstream 4 hours after the injection?

WRITING

107. Explain why a negative number can have a real number for its cube root yet cannot have a real number for its fourth root.

108. What is a rational exponent? Give several examples.

109. Explain this statement: In the expression $16^{3/2}$, the number 3/2 requires that two operations be performed on 16.

110. To evaluate $27^{4/3}$, which expression would be easier to use, $(\sqrt[3]{27})^4$ or $\sqrt[3]{27^4}$? Explain.

REVIEW Graph each equation.

111. $x = 3$

112. $y = -3$

113. $2x - y = -4$

114. $y = 4x - 4$

CHALLENGE PROBLEMS

115. Graph $y = \sqrt[3]{x}$ by first completing the table of solutions. Then plot the points and draw a smooth curve through them.

x	-27	-8	-1	0	1	8	27
y							

116. Evaluate: $\left(\frac{25}{49}\right)^{-1.5}$.

ACCENT ON TEAMWORK

A SPIRAL OF ROOTS

Overview: In this activity, you will gain a better understanding of square roots and get additional practice using the Pythagorean theorem.

Instructions: Form groups of 2 students. You will need an $8\frac{1}{2} \times 11$ inch piece of white, unlined paper, a 3×5 inch index card (to trace a right angle), a ruler, and a pencil.

Draw an isosceles right triangle with legs 1 inch long near the right edge of the paper. Use the Pythagorean theorem to determine the length of the hypotenuse.

Next, construct a second right triangle using the hypotenuse of the first triangle as one leg. Draw a second leg that is 1 inch long. Find the length of the hypotenuse of Triangle 2 using the Pythagorean theorem. Continue creating right triangles, using the previous hypotenuse as one leg and drawing a second leg that is 1 inch long. For each right triangle created, use the Pythagorean theorem to find the length of its hypotenuse. Make a list of the lengths of each hypotenuse, and look for a pattern.

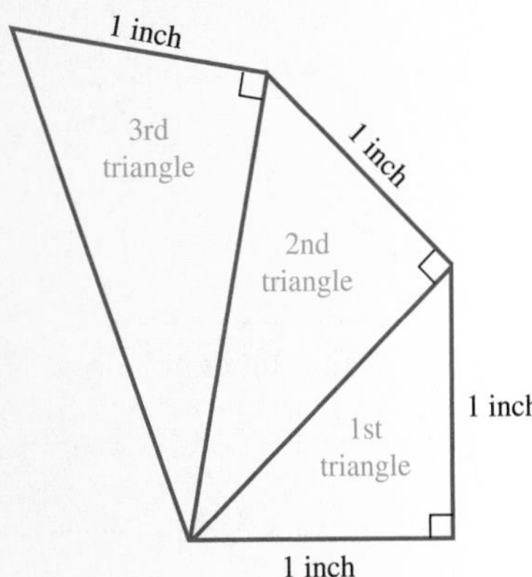

THE PYTHAGOREAN THEOREM

Overview: In this activity, you will use the concept of area to show that *the sum of the squares of the legs of a right triangle is equal to the square of the hypotenuse.*

Instructions: Form groups of 2 or 3 students. You will need three pieces of notebook paper, three pieces of graph paper, scissors, glue, and a pencil.

From the graph paper, cut out the three figures that rest against the sides of the triangle, as shown. Glue the figures on a piece of unlined paper to form a 3-4-5 right triangle. Determine the area of each figure by counting the number of square units and write the area on the figure. Then show that the sum of the areas of the two smaller figures is equal to the area of the largest figure.

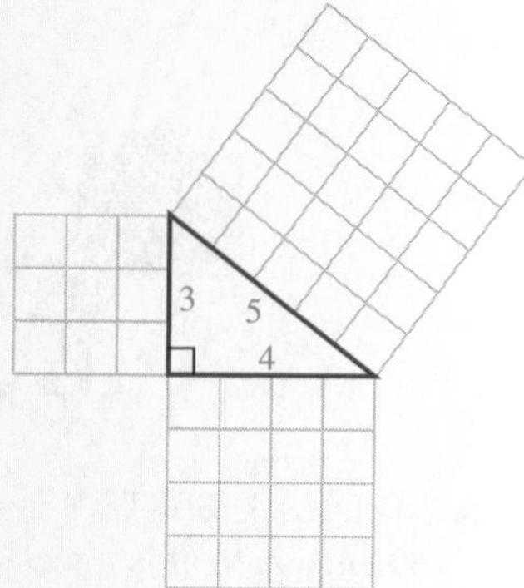

Next, make models of a 6-8-10 and a 5-12-13 right triangle. Show that the sum of the areas of the two smaller figures is equal to the area of the largest figure.

KEY CONCEPT: THE RULES FOR RADICALS

The product and quotient rules for square roots are used in many ways. You need to memorize them.

The Product Rule For any nonnegative real numbers a and b,

$$\sqrt{a \cdot b} = \sqrt{a}\sqrt{b} \quad \text{and} \quad \sqrt{a}\sqrt{b} = \sqrt{a \cdot b}$$

1. Fill in the blanks. The square root of the ________ of two nonnegative numbers is equal to the product of their ________ ________. Also, the product of the ________ ________ of two nonnegative numbers is equal to the square root of the ________ of the two numbers.

2. Use the product rule

a. to simplify the radical expression $\sqrt{24x^2y^3}$.

b. as you add the radical expressions: $\sqrt{8} + \sqrt{18}$.

c. as you subtract the radical expressions: $\sqrt{72a} - \sqrt{98a}$.

d. to multiply the radical expressions: $(2\sqrt{3n})(3\sqrt{5})$.

e. as you rationalize the denominator: $\dfrac{\sqrt{2}}{\sqrt{5}}$.

The Quotient Rule For any positive real numbers a and b,

$$\sqrt{\frac{a}{b}} = \frac{\sqrt{a}}{\sqrt{b}} \quad \text{and} \quad \frac{\sqrt{a}}{\sqrt{b}} = \sqrt{\frac{a}{b}}$$

3. Fill in the blanks: The square root of the ________ of two positive numbers is equal to the quotient of their ________ ________. Also, the quotient of the ________ ________ of two positive numbers is equal to the square root of the ________ of the two numbers.

4. Use the quotient rule

   a. to evaluate $\sqrt{\frac{36}{25}}$.

   b. to simplify $\sqrt{\frac{37}{121}}$.

   c. as you rationalize the denominator: $\sqrt{\frac{5}{3}}$.

   d. to divide the radical expressions: $\frac{\sqrt{18x^5}}{\sqrt{2x^3}}$.

CHAPTER REVIEW

ELEMENTARY Algebra f(x) Now™

SECTION 8.1 An Introduction to Square Roots

CONCEPTS

The number b is a *square root* of a if $b^2 = a$.

The *principal square root* of a positive number a, denoted by $\sqrt{a}$, is the positive square root of a.

The expression within a *radical* symbol $\sqrt{}$ is called the *radicand.*

If a positive number is not a perfect square, its square root is *irrational.* Square roots of negative numbers are called *imaginary numbers.*

REVIEW EXERCISES

1. Fill in the blanks: 6 is a square root of 36 because $(\ \ )^2 = \ \ $.

Find each square root. ***Do not use a calculator.***

2. $\sqrt{25}$
3. $\sqrt{49}$
4. $-\sqrt{144}$
5. $\sqrt{\frac{16}{81}}$
6. $\sqrt{0.64}$
7. $\sqrt{1}$

Use a calculator to approximate each expression to three decimal places.

8. $\sqrt{21}$
9. $-\sqrt{15}$
10. $2\sqrt{7}$

11. Tell whether each number is rational, irrational, or not a real number. $\{\sqrt{-2}, \sqrt{68}, \sqrt{81}, \sqrt{3}\}$

Find each square root. All variables represent nonnegative numbers.

12. $\sqrt{x^2}$
13. $\sqrt{4b^4}$
14. $-\sqrt{y^{12}}$

15. ROAD SIGNS To find the maximum velocity a car can safely travel around a curve without skidding, we can use the formula $v = \sqrt{2.5r}$, where v is the velocity in miles per hour and r is the radius of the curve in feet. How should the road sign be labeled if it is to be posted in front of a curve with a radius of 360 feet?

The Pythagorean theorem: If the length of the hypotenuse of a right triangle is c and the lengths of the two legs are a and b, then

$$a^2 + b^2 = c^2$$

Refer to the right triangle.

16. Find c where $a = 6$ and $b = 8$.

17. Find b where $a = 8$ and $c = 17$.

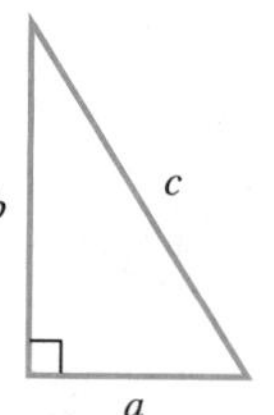

18. THEATER SEATING How much higher is the seat at the top of the incline than the one at the bottom?

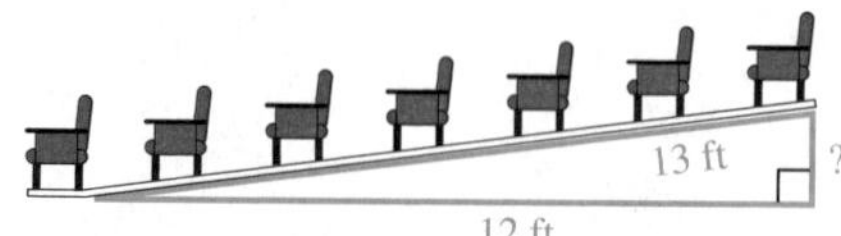

The distance formula:

$$d = \sqrt{(x_2 - x_1)^2 + (y_2 - y_1)^2}$$

Find the distance between the points. If an answer is not exact, round to the nearest hundredth.

19. $(-7, 12), (-4, 8)$

20. $(-15, -3), (-10, -16)$

SECTION 8.2 Simplifying Square Roots

The *product rule* for square roots:

$$\sqrt{a \cdot b} = \sqrt{a}\sqrt{b}$$

The *quotient rule* for square roots:

$$\sqrt{\frac{a}{b}} = \frac{\sqrt{a}}{\sqrt{b}} \quad (b \neq 0)$$

Simplified form of a radical:

1. Except for 1, the radicand has no perfect square factors.
2. No fraction appears in the radicand.
3. No radical appears in the denominator.

Simplify each expression. All variables represent positive numbers.

21. $\sqrt{32}$

22. $\sqrt{x^5}$

23. $\sqrt{80x^2}$

24. $-2\sqrt{63}$

25. $\sqrt{250t^3}$

26. $\sqrt{200x^2y}$

27. $\sqrt{\frac{16}{25}}$

28. $\sqrt{\frac{60}{49}}$

29. $\sqrt{\frac{242x^4}{169x^2}}$

30. FITNESS EQUIPMENT The length of the sit-up board can be found using the Pythagorean theorem. Find its length. Express the answer in simplified radical form. Then express your result as a decimal approximation rounded to the nearest tenth.

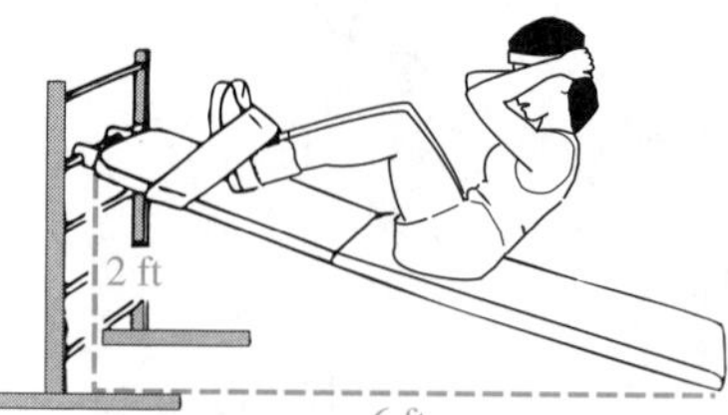

SECTION 8.3 Adding and Subtracting Radical Expressions

Radical expressions can be added or subtracted if they contain like radicals.

Perform the operations. All variables represent nonnegative numbers.

31. $\sqrt{10} + \sqrt{10}$

32. $6\sqrt{x} - \sqrt{x}$

33. $\sqrt{2} + \sqrt{8} - \sqrt{18}$

Square root radicals are called *like radicals* when they have the same radicand.

34. $\sqrt{3} + 4 + \sqrt{27} - 7$ **35.** $5\sqrt{28} - 3\sqrt{63}$ **36.** $3y\sqrt{5xy^3} - y^2\sqrt{20xy}$

37. Explain why we cannot add $3\sqrt{5}$ and $5\sqrt{3}$.

38. GARDENING Find the difference in the lengths of the two wires used to secure the tree.

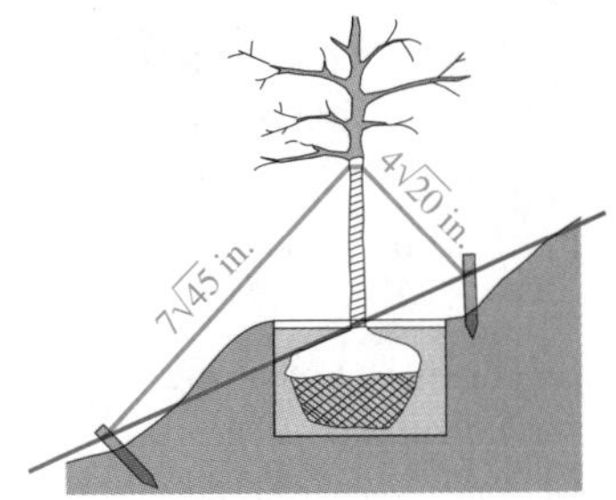

SECTION 8.4 Multiplying and Dividing Radical Expressions

The product of the square roots is equal to the square root of the product.

$$\sqrt{a} \cdot \sqrt{b} = \sqrt{a \cdot b}$$

The quotient of the square roots is equal to the square root of the quotient.

$$\frac{\sqrt{a}}{\sqrt{b}} = \sqrt{\frac{a}{b}} \quad (b \neq 0)$$

Use the FOIL method to multiply two radical expressions, each having two terms.

If the denominator of a fraction is a square root, *rationalize* the denominator by multiplying the numerator and denominator by some appropriate square root.

If a two-term denominator of a fraction contains square roots, multiply the numerator and denominator by the *conjugate* of the denominator.

Perform the operations.

39. $\sqrt{2}\sqrt{3}$

40. $(-5\sqrt{5})^2$

41. $(3\sqrt{3x^3})(4\sqrt{6x^2})$

42. $(\sqrt{15} + 3x)^2$

43. $\sqrt{2}(\sqrt{8} - \sqrt{18})$

44. $(\sqrt{3} + \sqrt{5})(\sqrt{3} - \sqrt{5})$

45. $(\sqrt{x})^2$

46. $(\sqrt{e-1})^2$

47. VACUUM CLEANERS The illustration shows the amount of surface area of a rug suctioned by a vacuum nozzle attachment. Find the area of this section of rug. Express the answer in simplified radical form. Then express your result as a decimal approximation to the nearest tenth.

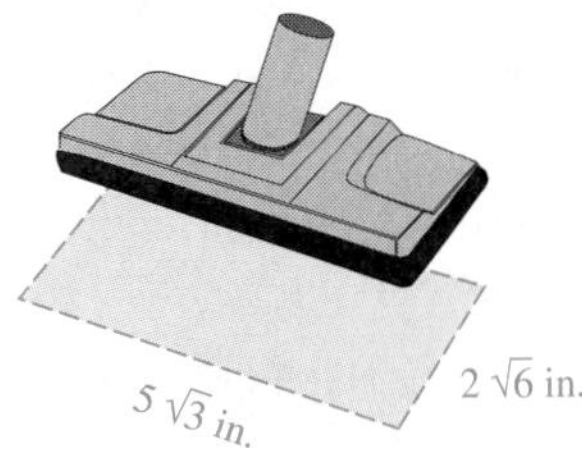

Rationalize each denominator.

48. $\dfrac{1}{\sqrt{7}}$

49. $\sqrt{\dfrac{3}{a}}$

50. $\dfrac{\sqrt{27x^3}}{\sqrt{75x}}$

51. $\dfrac{7}{\sqrt{2} + 1}$

SECTION 8.5 Solving Radical Equations

To solve radical equations containing square roots:

1. Isolate a radical term on one side of the equation.
2. Square both sides of the equation.
3. Solve the resulting equation.
4. Check the possible solutions.

Discard any *extraneous* solutions.

The *squaring property of equality:*

If $a = b$, then $a^2 = b^2$.

Solve each equation and check all solutions.

52. $\sqrt{x} = 9$

53. $\sqrt{2x + 10} = 10$

54. $\sqrt{3x + 4} + 5 = 3$

55. $\sqrt{b + 12} = 3\sqrt{b + 4}$

56. $\sqrt{p^2 - 3} = p + 3$

57. $\sqrt{24 + 10y} - y = 4$

58. FERRIS WHEELS The distance d in feet that an object will fall in t seconds is given by the formula

$$t = \sqrt{\frac{d}{16}}$$

If a person drops their sunglasses from the top of a Ferris wheel and it takes the glasses 2 seconds to hit the ground, how tall is the Ferris wheel?

SECTION 8.6 Higher-Order Roots and Rational Exponents

The cube root of a is denoted by $\sqrt[3]{a}$. By definition, $\sqrt[3]{a} = b$ if $b^3 = a$.

The number b is an nth root of a if $b^n = a$.

In $\sqrt[n]{a}$, the number n is called the *index* of the radical.

When n is even, we say that the radical $\sqrt[n]{x}$ is an *even root.*

When n is odd, $\sqrt[n]{x}$ is an *odd root.*

Real numbers can be raised to fractional powers.

59. Fill in the blanks: $\sqrt[3]{125} = 5$, because ____ $= 125$; 5 is called the ______ root of 125.

Find each root. All variables represent nonnegative numbers.

60. $\sqrt[3]{-27}$

61. $-\sqrt[3]{125}$

62. $\sqrt[4]{81}$

63. $\sqrt[5]{32}$

64. $\sqrt[3]{0}$

65. $\sqrt[3]{-1}$

66. $\sqrt[3]{\frac{1}{64}}$

67. $\sqrt[3]{1}$

68. $\sqrt[3]{x^3}$

69. $\sqrt[3]{27y^6}$

70. $\sqrt[4]{16a^{12}}$

71. $\sqrt[5]{b^{20}}$

Simplify.

72. $\sqrt[3]{54}$

73. $\sqrt[4]{80}$

74. $\sqrt[3]{\frac{56}{125}}$

Evaluate.

75. $49^{1/2}$

76. $(-1{,}000)^{1/3}$

77. $36^{3/2}$

78. $\left(\frac{8}{27}\right)^{2/3}$

79. $4^{-3/2}$

80. $-81^{5/4}$

Rational exponents:

$$x^{1/n} = \sqrt[n]{x}$$
$$x^{m/n} = \left(\sqrt[n]{x}\right)^m = \sqrt[n]{x^m}$$
$$x^{-m/n} = \frac{1}{x^{m/n}}$$

81. DENTISTRY The fractional amount of painkiller remaining in the system of a patient h hours after the original dose was injected into her gums is given by $h^{-3/2}$. How much of the original dose is in the patient's system 16 hours after the injection?

82. Explain why $(-4)^{1/2}$ is not a real number.

CHAPTER 8 TEST

ELEMENTARY Algebra $f(x)$ Now™

Find each square root.

1. $\sqrt{100}$

2. $-\sqrt{\frac{400}{9}}$

3. $\sqrt{0.25}$

4. $\sqrt{1}$

5. ELECTRONICS The current I (in amperes), the resistance R (in ohms), and the power P (in watts) are related by the formula

$$I = \sqrt{\frac{P}{R}}$$

Find the current I drawn by an electrical appliance that uses 980 watts of power when the resistance is 20 ohms.

6. A 26-foot ladder reaches a point on a wall 24 feet above the ground. How far from the wall is the ladder's base?

7. Tell whether each number is rational, irrational, or not a real number.

$$\{\sqrt{19}, \sqrt{-16}, \sqrt{144}\}$$

Simplify each expression. All variables represent positive numbers.

8. $\sqrt{4x^2}$

9. $\sqrt{54x^3}$

10. $\sqrt{\frac{50}{49}}$

11. $\sqrt{\frac{18x^2y^3}{2xy}}$

12. ARCHAEOLOGY A framed grid made up of 1 foot × 1 foot squares was constructed over a dig site. Pieces of pottery were found at points A and B. Find the exact distance between A and B.

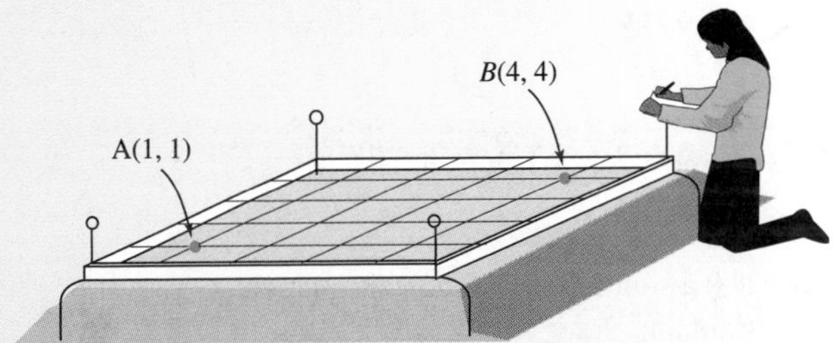

Perform each operation and simplify.

13. $\sqrt{12} + \sqrt{27}$

14. $\left(-2\sqrt{8x}\right)\left(3\sqrt{12x^4}\right)$

15. $\sqrt{3}\left(\sqrt{8} + \sqrt{6}\right)$

16. $\sqrt{8x^3} - x\sqrt{18x}$

17. $\left(\sqrt{2} + \sqrt{3}\right)\left(\sqrt{2} - \sqrt{3}\right)$

18. $\left(2\sqrt{3t}\right)^2$

19. $\left(5\sqrt{x} - 1\right)\left(\sqrt{x} - 4\right)$

20. $\left(\sqrt{x+1}\right)^2$

21. SEWING A corner of fabric is folded over to form a collar and stitched down. From the dimensions given in the following figure, determine the exact number of inches of stitching that must be made. (All measurements are in inches.)

Stitch this flap down.

$\sqrt{40}$

$\sqrt{32}$

$\sqrt{8}$

Rationalize each denominator.

22. $\dfrac{2}{\sqrt{2}}$

23. $\dfrac{\sqrt{3}}{\sqrt{3}-1}$

Solve each equation.

24. $\sqrt{x} = 15$

25. $\sqrt{2-x} - 2 = 6$

26. $\sqrt{3x+9} = 2\sqrt{x+1}$

27. $x - 1 = \sqrt{x-1}$

28. Is 0 a solution of the radical equation $\sqrt{3x+1}$? Explain.

29. Explain why $\sqrt{-9}$ is not a real number.

30. CARPENTRY A carpenter uses a tape measure to see whether the wall he just put up is perfectly square with the floor. Explain what mathematical concept he is applying. If the wall is positioned correctly, what should the measurement on the tape read?

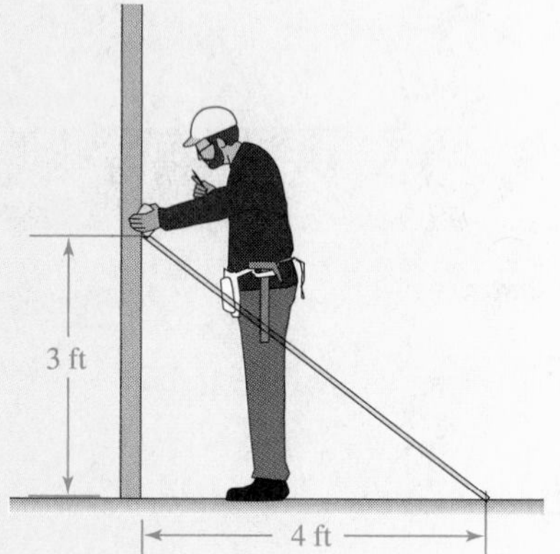

Find each root. All variables represent nonnegative numbers.

31. $\sqrt[3]{125}$

32. $\sqrt[5]{-32}$

33. $\sqrt[3]{\dfrac{1}{64}}$

34. $\sqrt[4]{x^4}$

35. Simplify: $\sqrt[3]{88}$.

Evaluate.

36. $16^{1/2}$

37. $8^{2/3}$

38. $9^{-1/2}$

CHAPTERS 1–8 CUMULATIVE REVIEW EXERCISES

1. Tell whether each statement is true or false.
- **a.** All whole numbers are integers.
- **b.** π is a rational number.
- **c.** A real number is either rational or irrational.

2. Find the value of the expression

$$\frac{-3(3+2)^2 - (-5)}{17 - 3|-4|}$$

3. BACKPACKS Pediatricians advise that children should not carry more than 20% of their own body weight in a backpack. According to this warning, how much weight can a fifth-grade girl who weighs 85 pounds safely carry in her backpack?

4. Simplify: $3p - 6(p + z) + p$.

5. Solve: $2 - (4x + 7) = 3 + 2(x + 2)$.

6. Solve: $3 - 3x \geq 6 + x$. Graph the solution set. Then describe the graph using interval notation.

7. SURFACE AREA The total surface area A of a box with dimensions l, w, and h is given by the formula

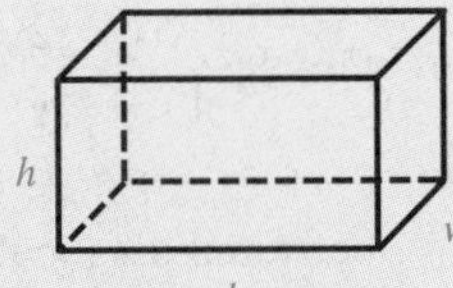

$$A = 2lw + 2wh + 2lh$$

If $A = 202$ square inches, $l = 9$ inches, and $w = 5$ inches, find h.

8. SEARCH AND RESCUE Two search and rescue teams leave base at the same time, looking for a lost boy. The first team, on foot, heads north at 2 mph and the other, on horseback, south at 4 mph. How long will it take them to search a distance of 21 miles between them?

9. BLENDING COFFEE A store sells regular coffee for \$4 a pound and gourmet coffee for \$7 a pound. Using 40 pounds of the gourmet coffee, the owner makes a blend to put on sale for \$5 a pound. How many pounds of regular coffee should he use?

10. Write the equation of the line passing through $(-2, 5)$ and $(4, 8)$.

11. What is the slope of the line defined by each equation?
 a. $y = 3x - 7$ b. $2x + 3y = -10$

Graph each equation.

12. $3x - 4y = 12$ 13. $y = \frac{1}{2}x$

14. $x = 5$ 15. $y = 2x^2 - 3$

16. SHOPPING On the graph, the line approximates the growth in retail sales for U.S. shopping centers during the years 1994–2002. Find the rate of increase in sales by finding the slope of the line.

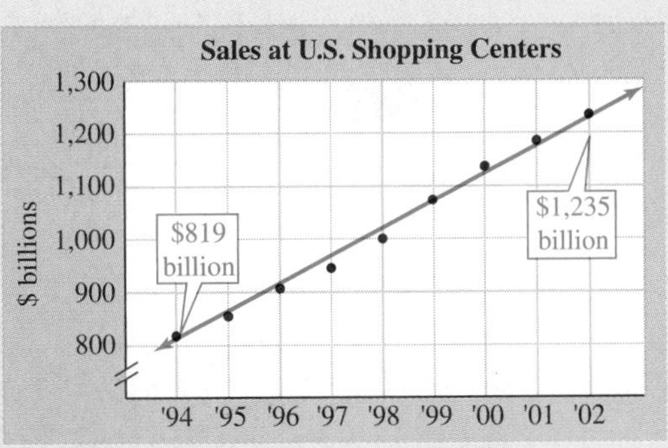

Based on data from International Council of Shopping Centers

17. ASTRONOMY The **parsec,** a unit of distance used in astronomy, is 3×10^{16} meters. The distance to Betelgeuse, a star in the constellation Orion, is 1.6×10^2 parsecs. Use scientific notation to express this distance in meters.

Simplify without using parentheses or negative exponents.

18. $(x^5)^2(x^7)^3$ 19. $\left(\frac{a^3b}{c^4}\right)^5$

20. $4^{-3} \cdot 4^{-2} \cdot 4^5$ 21. $(a^{-2}b^3)^{-4}$

Perform the operations.

22. $(-r^4st^2)(2r^2st)(rst)$ 23. $(-3t + 2s)(2t - 3s)$

24. $(3a^2 - 2a + 4) - (a^2 - 3a + 7)$

25. $2 + x\overline{)3x + 2x^2 - 2}$

Factor completely.

26. $3x^2y - 6xy^2$ 27. $2x^2 + 2xy - 3x - 3y$

28. $a^3 + 8b^3$ 29. $6a^2 - 7a - 20$

Solve each equation.

30. $x^2 + 3x + 2 = 0$ 31. $5x^2 = 10x$

32. $6x^2 - x = 2$ 33. $a^2 - 25 = 0$

34. CHILDREN'S STICKERS A rectangular-shaped sticker has an area of 20 cm^2. The width is 1 cm shorter than the length. Find the length of the sticker.

35. Simplify: $\frac{x^2 + 2x + 1}{x^2 - 1}$.

Perform the operations. Simplify when possible.

36. $\frac{p^2 - p - 6}{3p - 9} \div \frac{p^2 + 6p + 9}{p^2 - 9}$

37. $\frac{x + 2}{x + 5} - \frac{x - 3}{x + 7}$

38. $\frac{3a}{2b} - \frac{2b}{3a}$

39. $\dfrac{\frac{1}{x}+\frac{1}{y}}{\frac{1}{x}-\frac{1}{y}}$

40. Solve: $\dfrac{a+2}{a+3}-1=\dfrac{-1}{a^2+2a-3}$

41. FILLING A POOL An inlet pipe can fill an empty swimming pool in 5 hours, and another inlet pipe can fill the pool in 4 hours. How long will it take both pipes to fill the pool?

42. ONLINE SALES A company found that, on average, it made 9 online sales transactions for every 500 hits on its Internet Web site. If the company's Web site had 360,000 hits in one year, how many sales transactions did it have that year?

Solve each system of equations.

43. $\begin{cases} x = y + 4 \\ 2x + y = 5 \end{cases}$

44. $\begin{cases} \frac{3}{5}s + \frac{4}{5}t = 1 \\ -\frac{1}{4}s + \frac{3}{8}t = 1 \end{cases}$

45. FINANCIAL PLANNING In investing \$6,000 of a couple's money, a financial planner put some of it into a savings account paying 6% annual interest. The rest was invested in a riskier mini-mall development plan paying 12% annually. The combined interest earned for the first year was \$540. How much money was invested at each rate? Use two variables to solve this problem.

46. Graph: $\begin{cases} 3x + 2y \geq 6 \\ x + 3y \leq 6 \end{cases}$.

47. CARGO SPACE How wide a piece of plywood can be stored diagonally in the back of the van?

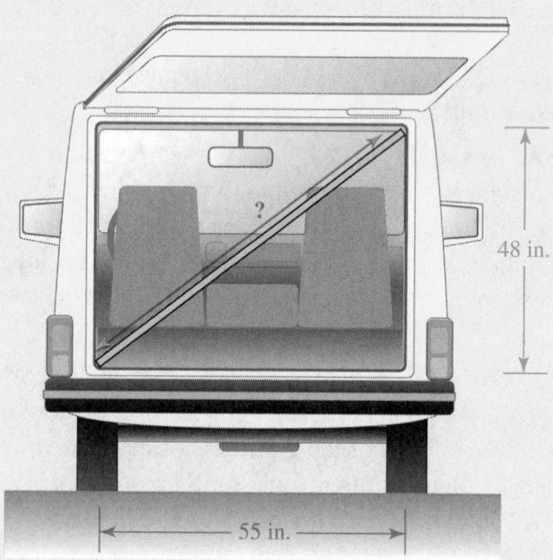

Simplify. All variables represent positive numbers.

48. $-12x\sqrt{16x^2y^3}$

49. $\sqrt{48} - \sqrt{8} + \sqrt{27} - \sqrt{32}$

50. $(\sqrt{y} - 4)(\sqrt{y} - 5)$

51. $\dfrac{4}{\sqrt{20}}$

52. $-\sqrt[3]{-27}$

53. Solve: $\sqrt{6x + 19} - 5 = 2$.

54. Evaluate: $16^{3/2}$.

Chapter

9 Solving Quadratic Equations; Functions

ELEMENTARY **Algebra** $f(x)$ **Now**™

Throughout the chapter, this icon introduces resources on the Elementary AlgebraNow Web site, accessed through **http://1pass.thomson.com**, that will

- Help you test your knowledge of the material with a pre-test and a post-test
- Provide a personalized learning plan targeting areas you should study

9.1 The Square Root Property: Completing the Square

9.2 The Quadratic Formula

9.3 Complex Numbers

9.4 Graphing Quadratic Equations

9.5 An Introduction to Functions

9.6 Variation

Accent on Teamwork

Key Concept

Chapter Review

Chapter Test

Cumulative Review Exercises

Royalty-Free/CORBIS

TLE Graphic designers combine their artistic talents with mathematical skills to create attractive advertisements, photography, and packaging. They perform arithmetic computations with fractions and decimals to determine the proper font size and page margins for a design. They apply algebraic concepts such as graphing and equation solving to design page layouts. And they use formulas to create production schedules and calculate costs.

To learn more about the use of algebra in graphic design, visit *The Learning Equation* on the Internet at http://tle.brookscole.com (The log-in instructions are in the Preface.) For Chapter 9, the online lesson is:

- *TLE* Lesson 16: The Quadratic Formula

We have previously solved quadratic equations by factoring. In this chapter, we will discuss other methods that are used to solve "quadratics."

9.1 The Square Root Property; Completing the Square

- The Square Root Property
- Completing the Square
- Solving Quadratic Equations by Completing the Square
- Making the Lead Coefficient 1

Recall that a **quadratic equation** can be written in the form

$$ax^2 + bx + c = 0$$

where a, b, and c represent real numbers and $a \neq 0$. Some examples are

$$x^2 + 8x - 3 = 0, \qquad 2x^2 - 2 = 4x, \qquad \text{and} \qquad x^2 - 16 = 0$$

We have solved quadratic equations by factoring. To review this method, let's solve $x^2 - 16 = 0$.

$$x^2 - 16 = 0$$
$$(x + 4)(x - 4) = 0 \qquad \text{Factor the difference of two squares.}$$
$$x + 4 = 0 \quad \text{or} \quad x - 4 = 0 \qquad \text{Set each factor equal to 0.}$$
$$x = -4 \quad \Big| \quad x = 4 \qquad \text{Solve each equation.}$$

The solutions are -4 and 4.

Not every quadratic equation can be solved by factoring. In this section, we will discuss a method by which any quadratic equation, factorable or not, can be solved.

THE SQUARE ROOT PROPERTY

We now consider an alternate way to solve the quadratic equation $x^2 - 16 = 0$.

EXAMPLE 1 Solve: $x^2 - 16 = 0$.

ELEMENTARY Algebra f(x) Now™

Solution We begin by isolating x^2 on the left-hand side of the equation.

$$x^2 - 16 = 0$$
$$x^2 = 16 \qquad \text{Add 16 to both sides.}$$

We see that x must be a number whose square is 16. By the definition of square root,

$$x = \sqrt{16} \quad \text{or} \quad x = -\sqrt{16}$$
$$x = 4 \quad \Big| \quad x = -4$$

As with the factor method, we find that the solutions of $x^2 - 16 = 0$ are 4 and -4.

Self Check 1 Use the concept of square root to solve $a^2 - 64 = 0$.

In general, we can use the following property to solve quadratic equations of the form $x^2 = c$.

The Square Root Property For any nonnegative real number c, if $x^2 = c$, then

$$x = \sqrt{c} \quad \text{or} \quad x = -\sqrt{c}$$

EXAMPLE 2

Solve each equation: **a.** $x^2 - 5 = 0$, **b.** $n^2 = 12$, and **c.** $3a^2 = 10$.

ELEMENTARY Algebra f(x) Now™

Solution **a.** We isolate x^2 on the left-hand side and use the square root property.

$$x^2 - 5 = 0$$
$$x^2 = 5 \qquad \text{Add 5 to both sides.}$$
$$x = \sqrt{5} \quad \text{or} \quad x = -\sqrt{5} \qquad \text{Use the square root property.}$$

Check:

For $x = \sqrt{5}$	For $x = -\sqrt{5}$
$x^2 - 5 = 0$	$x^2 - 5 = 0$
$(\sqrt{5})^2 - 5 \stackrel{?}{=} 0$	$(-\sqrt{5})^2 - 5 \stackrel{?}{=} 0$
$5 - 5 \stackrel{?}{=} 0$	$5 - 5 \stackrel{?}{=} 0$
$0 = 0$	$0 = 0$

The solutions are $\sqrt{5}$ and $-\sqrt{5}$. We can use double-sign notation $\pm$ to write the solutions more compactly as

$$\pm\sqrt{5} \qquad \text{Read as “}x\text{ equals positive or negative }\sqrt{5}\text{.”}$$

Caution

When using the square root property to solve an equation, always write the $\pm$ symbol, or you will lose one of the solutions.

b. Since n^2 is already isolated, we simply apply the square root property.

$$n^2 = 12$$
$$n = \pm\sqrt{12} \qquad \text{By the square root property, } n = \sqrt{12} \text{ or } n = -\sqrt{12}.$$
$$n = \pm 2\sqrt{3} \qquad \text{Simplify: } \sqrt{12} = \sqrt{4 \cdot 3} = \sqrt{4}\sqrt{3} = 2\sqrt{3}.$$

Check to verify that $2\sqrt{3}$ and $-2\sqrt{3}$ are solutions.

c. $3a^2 = 10$

$$a^2 = \frac{10}{3} \qquad \text{To isolate } a^2\text{, divide both sides by 3.}$$
$$a = \pm\sqrt{\frac{10}{3}} \qquad \text{By the square root property, } a = \sqrt{\frac{10}{3}} \text{ or } a = -\sqrt{\frac{10}{3}}.$$

To rationalize the denominators of the results, we proceed as follows:

$$a = \pm\sqrt{\frac{10}{3}} = \pm\frac{\sqrt{10}}{\sqrt{3}} \cdot \frac{\sqrt{3}}{\sqrt{3}} = \pm\frac{\sqrt{30}}{3}$$

Verify that $\frac{\sqrt{30}}{3}$ and $-\frac{\sqrt{30}}{3}$ are solutions by checking each in the original equation.

The Language of Algebra

For Example 2c, the exact solutions are $\pm\frac{\sqrt{30}}{3}$. To the nearest hundredth, the approximate solutions are ± 1.83.

Self Check 2 Solve: **a.** $x^2 - 21 = 0$, **b.** $m^2 = 54$, and **c.** $5x^2 = 7$.

We can extend the square root property to solve equations that involve the square of a binomial.

EXAMPLE 3

Solve each equation: **a.** $(x - 3)^2 = 16$, **b.** $(x + 1)^2 = 50$, and **c.** $(2x - 4)^2 = 7$.

ELEMENTARY Algebra f(x) Now™

Solution

Caution

Equations like $x^2 = -16$ and $(x - 3)^2 = -16$ have no real-number solution because no real number squared is negative.

a. We solve this equation by using the square root property.

$$(x - 3)^2 = 16$$

$$x - 3 = \pm\sqrt{16}$$ By the square root property, $x - 3 = \sqrt{16}$ or $x - 3 = -\sqrt{16}$.

$$x - 3 = \pm 4$$ $\sqrt{16} = 4$.

$$x = 3 \pm 4$$ To isolate x, add 3 to both sides. It is standard practice to write the 3 in front of the $\pm$ symbol.

We read 3 ± 4 as "3 plus or minus 4." To find the solutions, we perform the calculation using a plus symbol + and then using a minus symbol −.

$x = 3 + 4$ or $x = 3 - 4$

$x = 7$ | $x = -1$

Check:

For x = 7	***For x = −1***
$(x - 3)^2 = 16$	$(x - 3)^2 = 16$
$(7 - 3)^2 \stackrel{?}{=} 16$	$(-1 - 3)^2 \stackrel{?}{=} 16$
$(-4)^2 \stackrel{?}{=} 16$	$(-4)^2 \stackrel{?}{=} 16$
$16 = 16$	$16 = 16$

The solutions are 7 and −1.

b. $(x + 1)^2 = 50$

$$x + 1 = \pm\sqrt{50}$$ By the square root property, $x + 1 = \sqrt{50}$ or $x + 1 = -\sqrt{50}$.

$$x + 1 = \pm 5\sqrt{2}$$ $\sqrt{50} = \sqrt{25 \cdot 2} = \sqrt{25}\sqrt{2} = 5\sqrt{2}$.

$$x = -1 \pm 5\sqrt{2}$$ To isolate x, subtract 1 from both sides (or add −1 to both sides.)

Check the solutions $-1 + 5\sqrt{2}$ and $-1 - 5\sqrt{2}$.

The Language of Algebra

The $\pm$ symbol is often seen in surveys and polls. Suppose a poll predicts a candidate will receive 48% of the vote, $\pm$4%. That means the candidate's support could be anywhere between $48 - 4 = 44\%$ and $48 + 4 = 52\%$.

c. $(2x - 4)^2 = 7$

$$2x - 4 = \pm\sqrt{7}$$ By the square root property, $2x - 4 = \sqrt{7}$ or $2x - 4 = -\sqrt{7}$.

$$2x = 4 \pm \sqrt{7}$$ Add 4 to both sides.

$$x = \frac{4 \pm \sqrt{7}}{2}$$ Divide both sides by 2.

Check the solutions $\dfrac{4 + \sqrt{7}}{2}$ and $\dfrac{4 - \sqrt{7}}{2}$.

Self Check 3 Solve each equation: **a.** $(x - 2)^2 = 64$, **b.** $(x + 3)^2 = 98$, and **c.** $(3x - 1)^2 = 3$.

COMPLETING THE SQUARE

When the polynomial in a quadratic equation doesn't factor easily, we can solve the equation by *completing the square.* This method is based on the following special products:

$$x^2 + 2bx + b^2 = (x + b)^2 \qquad \text{and} \qquad x^2 - 2bx + b^2 = (x - b)^2$$

In each of these perfect square trinomials, the third term is the square of one-half of the coefficient of x.

- In $x^2 + 2bx + b^2$, the coefficient of x is $2b$. If we find $\frac{1}{2} \cdot 2b$, which is b, and square it, we get the third term, b^2.
- In $x^2 - 2bx + b^2$, the coefficient of x is $-2b$. If we find $\frac{1}{2}(-2b)$, which is $-b$, and square it, we get the third term: $(-b)^2 = b^2$.

We can use these observations to change certain binomials into perfect square trinomials. For example, to change $x^2 + 12x$ into a perfect square trinomial, we find one-half of the coefficient of x, square the result, and add the square to $x^2 + 12x$.

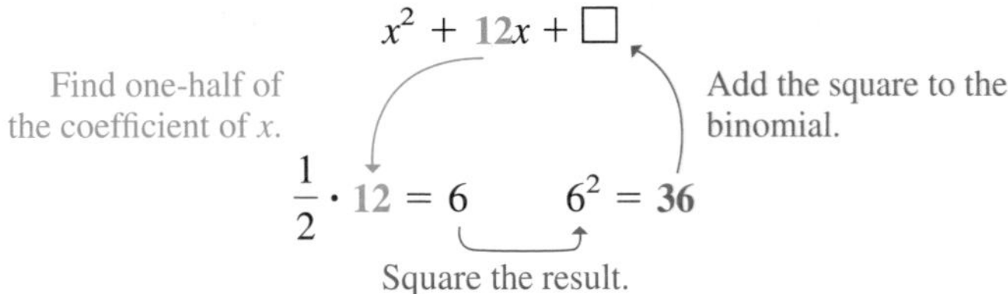

We obtain the perfect square trinomial $x^2 + 12x + 36$ that factors as $(x + 6)^2$. By adding 36 to $x^2 + 12x$, we **completed the square** on $x^2 + 12x$.

Completing the Square

To complete the square on $x^2 + bx$, add the square of one-half of the coefficient of x:

$$x^2 + bx + \left(\frac{1}{2}b\right)^2$$

EXAMPLE 4

ELEMENTARY Algebra *f(x)* Now™

Complete the square and factor the resulting perfect square trinomial: **a.** $x^2 + 6x$ and **b.** $x^2 - 5x$.

Solution

a. The coefficient of x is 6. One-half of 6 is 3, and $3^2 = 9$. If we add 9 to $x^2 + 6x$ it becomes a perfect square trinomial.

$$x^2 + 6x + 9 \qquad \frac{1}{2} \cdot 6 = 3 \text{ and } 3^2 = 9. \text{ Add 9 to the binomial.}$$

This trinomial factors as $(x + 3)^2$.

b. The coefficient of x is -5. One-half of -5 is $-\frac{5}{2}$, and $\left(-\frac{5}{2}\right)^2 = \frac{25}{4}$. If we add $\frac{25}{4}$ to $x^2 - 5x$ it becomes a perfect square trinomial.

$$x^2 - 5x + \frac{25}{4} \qquad \frac{1}{2}(-5) = -\frac{5}{2} \text{ and } \left(-\frac{5}{2}\right)^2 = \frac{25}{4}. \text{ Add } \frac{25}{4} \text{ to the binomial.}$$

This trinomial factors as $\left(x - \frac{5}{2}\right)^2$.

Self Check 4 Complete the square for each expression and factor the resulting perfect square trinomial: **a.** $y^2 - 8y$ and **b.** $y^2 + 3y$.

SOLVING QUADRATIC EQUATIONS BY COMPLETING THE SQUARE

We can use completing the square to solve quadratic equations.

EXAMPLE 5

Solve: $x^2 + 8x - 3 = 0$.

ELEMENTARY Algebra f(x) Now™

Solution To prepare to complete the square, we isolate the variable terms on one side of the equation and the constant term on the other.

The Language of Algebra

In $x^2 + 8x - 3 = 0$, x^2 and $8x$ are called *variable terms* and -3 is called the *constant term.*

$$x^2 + 8x - 3 = 0$$

$$x^2 + 8x = 3 \quad \text{Add 3 to both sides so that the constant term is on the right-hand side.}$$

Now we find one-half of the coefficient of x, square the result, and add the square to both sides. This makes the left-hand side a perfect square trinomial.

$x^2 + 8x + 16 = 3 + 16$ — $\frac{1}{2} \cdot 8 = 4$ and $4^2 = 16$. Add 16 to both sides.

$(x + 4)^2 = 19$ — Factor $x^2 + 8x + 16$ and add.

$x + 4 = \pm\sqrt{19}$ — By the square root property: $x + 4 = \sqrt{19}$ or $x + 4 = -\sqrt{19}$.

$x = -4 \pm \sqrt{19}$ — Subtract 4 from both sides.

The solutions are $-4 + \sqrt{19}$ and $-4 - \sqrt{19}$. We can also approximate each solution.

$$-4 + \sqrt{19} \approx -4 + 4.358898944 \approx 0.36 \qquad -4 - \sqrt{19} \approx -4 - 4.358898944 \approx -8.36$$

Self Check 5 Solve: $x^2 + 10x - 4 = 0$.

EXAMPLE 6

Solve: $x^2 - 7x = 2$.

Solution Since the variable terms are isolated on the left-hand side, we complete the square.

Caution

A common error is to add a constant to one side of an equation to complete the square and forget to add it to the other side.

$$x^2 - 7x = 2$$

$x^2 - 7x + \frac{49}{4} = 2 + \frac{49}{4}$ — $\frac{1}{2}(-7) = -\frac{7}{2}$ and $\left(-\frac{7}{2}\right)^2 = \frac{49}{4}$. Add $\frac{49}{4}$ to both sides.

$\left(x - \frac{7}{2}\right)^2 = \frac{8}{4} + \frac{49}{4}$ — Factor the left-hand side. On the right-hand side, write 2 as $\frac{8}{4}$ so the fractions have the common denominator 4.

$\left(x - \frac{7}{2}\right)^2 = \frac{57}{4}$ — Add the fractions.

$x - \frac{7}{2} = \pm\sqrt{\frac{57}{4}}$ — Use the square root property.

$x - \frac{7}{2} = \pm\frac{\sqrt{57}}{2}$ — $\sqrt{\frac{57}{4}} = \frac{\sqrt{57}}{\sqrt{4}} = \frac{\sqrt{57}}{2}$.

$$x = \frac{7}{2} \pm \frac{\sqrt{57}}{2} \qquad \text{Add } \frac{7}{2} \text{ to both sides.}$$

$$x = \frac{7 \pm \sqrt{57}}{2} \qquad \text{Combine the fractions.}$$

The solutions are $\frac{7 + \sqrt{57}}{2}$ and $\frac{7 - \sqrt{57}}{2}$. If we approximate them to the nearest hundredth, we have

$$\frac{7 + \sqrt{57}}{2} \approx 7.27 \quad \text{and} \quad \frac{7 - \sqrt{57}}{2} \approx -0.27$$

Self Check 6 Solve: $x^2 - 3x - 3 = 0$.

MAKING THE LEAD COEFFICIENT 1

Completing the square can be used to solve any quadratic equation. However, when the coefficient of the squared variable (called the **lead coefficient**) is not 1, we must make it 1 before we can complete the square.

EXAMPLE 7

Solve: $4x^2 + 4x - 3 = 0$.

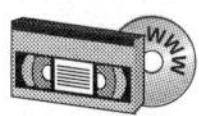

Solution We divide both sides by 4 so that the coefficient of x^2 is 1.

$$4x^2 + 4x - 3 = 0 \qquad \text{The lead coefficient is 4.}$$

$$x^2 + x - \frac{3}{4} = 0 \qquad \text{Divide both sides by 4: } \frac{4x^2}{4} + \frac{4x}{4} - \frac{3}{4} = \frac{0}{4}.$$

$$x^2 + x = \frac{3}{4} \qquad \text{Add } \tfrac{3}{4} \text{ to both sides so that the constant term is on the right-hand side.}$$

$$x^2 + x + \frac{1}{4} = \frac{3}{4} + \frac{1}{4} \qquad \text{Complete the square: } \tfrac{1}{2} \cdot 1 = \tfrac{1}{2} \text{ and } \left(\tfrac{1}{2}\right)^2 = \tfrac{1}{4}. \text{ Add } \tfrac{1}{4} \text{ to both sides.}$$

$$\left(x + \frac{1}{2}\right)^2 = 1 \qquad \text{Factor and add.}$$

$$x + \frac{1}{2} = \pm 1 \qquad \text{Use the square root property.}$$

$$x = -\frac{1}{2} \pm 1 \qquad \text{Subtract } \frac{1}{2} \text{ from both sides.}$$

$$x = -\frac{1}{2} + 1 \quad \text{or} \quad x = -\frac{1}{2} - 1$$

$$x = \frac{1}{2} \quad \Big| \quad x = -\frac{3}{2}$$

The solutions are $\frac{1}{2}$ and $-\frac{3}{2}$. Check each one.

Success Tip

Perhaps you noticed that Example 7 can be solved by factoring:

$(2x - 1)(2x + 3) = 0.$

However, that is not the case for Example 6. These observations illustrate that completing the square can be used to solve any quadratic equation.

Self Check 7 Solve: $2x^2 - 5x - 3 = 0$.

To solve a quadratic equation in x by completing the square, we follow these steps.

Completing the Square to Solve a Quadratic Equation

1. If the coefficient of x^2 is 1, go to step 2. If it is not 1, make it 1 by dividing both sides of the equation by the coefficient of x^2.
2. Get all variable terms on one side of the equation and constants on the other side.
3. Complete the square by finding one-half of the coefficient of x, squaring the result, and adding the square to both sides of the equation.
4. Factor the perfect square trinomial as the square of a binomial.
5. Solve the resulting equation using the square root property.
6. Check your answers in the original equation.

EXAMPLE 8

Solve: $2x^2 - 2 = 4x$.

ELEMENTARY Algebra f(x) Now™

Solution We write the equation in $ax^2 + bx + c = 0$ form and then divide both sides by 2 so that the coefficient of x^2 is 1.

$$2x^2 - 2 = 4x$$

$$2x^2 - 4x - 2 = 0 \quad \text{Subtract } 4x \text{ from both sides.}$$

$$x^2 - 2x - 1 = 0 \quad \text{Divide both sides by 2: } \frac{2x^2}{2} - \frac{4x}{2} - \frac{2}{2} = \frac{0}{2}$$

$$x^2 - 2x = 1 \quad \text{To get the variable terms on one side and the constant term on the other, add 1 to both sides.}$$

$$x^2 - 2x + 1 = 1 + 1 \quad \text{Complete the square: } \tfrac{1}{2}(-2) = -1 \text{ and } (-1)^2 = 1. \text{ Add 1 to both sides.}$$

$$(x - 1)^2 = 2 \quad \text{Factor the perfect square trinomial. Add on the right-hand side.}$$

$$x - 1 = \pm\sqrt{2} \quad \text{Use the square root property.}$$

$$x = 1 \pm \sqrt{2} \quad \text{Add 1 to both sides.}$$

The solutions are $1 \pm \sqrt{2}$. Check each one.

Self Check 8 Solve: $3x^2 + 12 = 18x$.

Answers to Self Checks **1.** 8, −8 **2. a.** $\pm\sqrt{21}$, **b.** $\pm 3\sqrt{6}$, **c.** $\pm\frac{\sqrt{35}}{5}$ **3. a.** 10, −6, **b.** $-3 \pm 7\sqrt{2}$, **c.** $\frac{1}{3} \pm \frac{\sqrt{3}}{3}$ **4. a.** $y^2 - 8y + 16 = (y - 4)^2$, **b.** $y^2 + 3y + \frac{9}{4} = \left(y + \frac{3}{2}\right)^2$ **5.** $-5 \pm \sqrt{29}$ **6.** $\frac{3 \pm \sqrt{21}}{2}$ **7.** $3, -\frac{1}{2}$ **8.** $3 \pm \sqrt{5}$

9.1 STUDY SET

ELEMENTARY Algebra f(x) Now™

VOCABULARY **Fill in the blanks.**

1. $3x^2 + 2x - 1 = 0$ and $x^2 - 15 = 0$ are examples of ________ equations.
2. The ________ root property: If $x^2 = c$, then $x = \sqrt{c}$ or $x = -\sqrt{c}$.
3. In $x^2 - 4x + 1$, the ________ of x is −4.
4. To ________ the square on $x^2 + 6x$, we add the square of one-half of the coefficient of x.
5. The ________ coefficient of $5x^2 - 2x + 7 = 0$ is 5.
6. In $x^2 + 8x + 10 = 0$, the variable terms are x^2 and $8x$, and the ________ term is 10.

CONCEPTS **Fill in the blanks.**

7. If $x^2 = 5$, then $x = \pm$ ____.
8. If $(x - 2)^2 = 7$, then $x - 2 = \pm$ ____.
9. **a.** For $x^2 - 23 = 0$, isolate x^2 on one side of the equation.
   b. For $x^2 - 1 = 4x$, get the variable terms on one side of the equation and the constant term on the other.
10. Solve each equation.
    a. $x - 2 = \pm\sqrt{7}$
    b. $x + 9 = \pm\sqrt{2}$
    c. $4x = 3 \pm \sqrt{2}$
11. What is the result when both sides of $4x^2 + 5x - 16 = 0$ are divided by 4?
12. Factor each perfect square trinomial.
    a. $x^2 + 14x + 49$ **b.** $x^2 - 2x + 1$
13. Find one-half of the given number and square the result.
    a. 6 **b.** -12
    c. 3 **d.** -5
14. Fill in the blanks to complete the square. Then factor the resulting perfect square trinomial.
    a. $x^2 + 8x + \_\_ = (x + \_\_)^2$
    b. $x^2 - 9x + \_\_ = \left(x - \_\_\right)^2$
15. Express each integer as a fraction and add.
    a. $5 + \frac{9}{4}$ **b.** $-2 + \frac{25}{4}$
16. Check to determine whether $2\sqrt{3}$ is a solution of $x^2 - 12 = 0$.

NOTATION

17. Write the statement $x = \sqrt{6}$ or $x = -\sqrt{6}$ using a $\pm$ symbol.
18. Fill in the blanks: $2 \pm \sqrt{3}$ is read as "2 ______ or ______ $\sqrt{3}$."

PRACTICE **Use the square root property to solve each equation.**

19. $x^2 - 36 = 0$
20. $x^2 - 4 = 0$
21. $b^2 - 9 = 0$
22. $a^2 - 25 = 0$
23. $m^2 - 20 = 0$
24. $n^2 - 32 = 0$
25. $x^2 + 81 = 0$
26. $y^2 + 100 = 0$
27. $3m^2 = 27$
28. $4x^2 = 64$
29. $4x^2 = 16$
30. $5x^2 = 125$
31. $x^2 = \frac{9}{16}$
32. $x^2 = \frac{81}{25}$
33. $(x + 1)^2 = 25$
34. $(x - 1)^2 = 49$
35. $(x + 2)^2 = 81$
36. $(x + 3)^2 = 16$
37. $(x - 2)^2 = 8$
38. $(x + 2)^2 = 50$
39. $4(t - 7)^2 - 12 = 0$
40. $2(t - 6)^2 - 22 = 0$

Use the square root property to solve each equation. Approximate the solutions to the nearest hundredth.

41. $x^2 - 14 = 0$
42. $x^2 - 6 = 0$
43. $(y + 5)^2 - 7 = 0$
44. $(y - 15)^2 - 3 = 0$

Complete the square and factor the resulting perfect square trinomial.

45. $x^2 + 2x$
46. $x^2 + 12x$
47. $x^2 - 4x$
48. $x^2 - 14x$
49. $x^2 - 7x$
50. $x^2 + 11x$
51. $b^2 - \frac{2}{3}b$
52. $c^2 - \frac{5}{2}c$

Solve each equation by completing the square.

53. $k^2 - 8k + 12 = 0$
54. $p^2 - 4p + 3 = 0$
55. $x^2 + 6x + 8 = 0$
56. $x^2 + 8x + 12 = 0$
57. $g^2 + 5g - 6 = 0$
58. $s^2 + 5s - 14 = 0$
59. $x^2 + 4x + 1 = 0$
60. $x^2 + 6x + 2 = 0$
61. $x^2 + 8x + 6 = 0$
62. $x^2 + 6x + 4 = 0$

63. $x^2 - 2x - 17 = 0$

64. $x^2 + 10x - 7 = 0$

65. $n^2 - 7n - 5 = 0$

66. $m^2 + 5m - 7 = 0$

67. $x^2 = 4x + 3$

68. $x^2 = 6x - 3$

69. $a^2 - a = 3$

70. $b^2 - 3b = 5$

71. $3x^2 + 9x + 6 = 0$

72. $3d^2 + 24d + 48 = 0$

73. $4x^2 + 4x + 1 = 20$

74. $9x^2 + 12x - 8 = 0$

75. $2x^2 = 3x + 2$

76. $4x^2 = 2 - 7x$

77. $2t^2 + 5t = 4$

78. $4s^2 - 2s = 3$

Solve each equation by completing the square. Approximate the solutions to the nearest hundredth.

79. $x^2 - 2x - 4 = 0$

80. $x^2 - 4x = 2$

81. $n^2 - 5 = 9n$

82. $a^2 - 4 = -7a$

83. $3x^2 - 6x = 1$

84. $2x^2 - 6x = -3$

APPLICATIONS

85. ESCAPE VELOCITY The speed at which a rocket must be fired for it to leave the earth's gravitational attraction is called the *escape velocity.* If the escape velocity v_e, in miles per hour, is given by

$$\frac{v_e^2}{2g} = R$$

where $g = 78{,}545$ and $R = 3{,}960$, find v_e. Round to the nearest mi/hr.

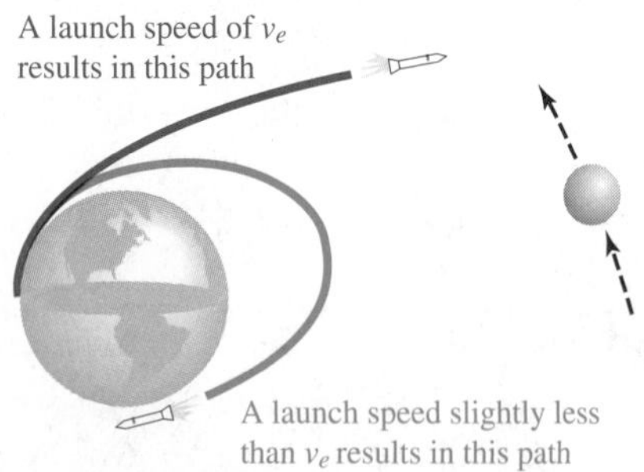

86. CAROUSELS In 1999, the city of Lancaster, Pennsylvania considered installing a classic Dentzel carousel in an abandoned downtown building. After learning that the circular-shaped carousel (like that shown) would occupy 2,376 square feet of floor space and that it was 26 feet high, the proposal was determined to be impractical because of the large remodeling costs. Find the diameter of the carousel to the nearest foot.

87. BICYCLE SAFETY A bicycle training program for children uses a figure-8 course to help them improve their balance and steering. The course is laid out over a paved area covering 800 square feet, as shown. Find its dimensions.

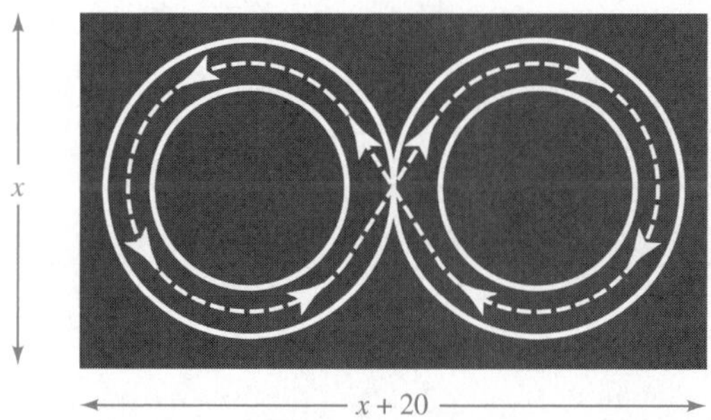

88. BADMINTON A badminton court occupies 880 square feet of the floor space of a gymnasium. If its length is 4 feet more than twice its width, find its dimensions.

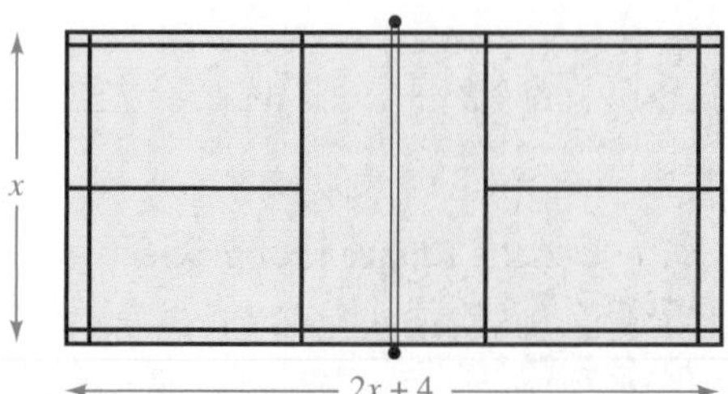

WRITING

89. Give an example of a perfect square trinomial. Why do you think the word "perfect" is used to describe it?

90. Explain the error in the following work.

$$\text{Solve: } x^2 = 28$$
$$x = \pm\sqrt{28}$$
$$x = 2 \pm \sqrt{7}$$

91. Explain why completing the square on $x^2 + 5x$ is more difficult than completing the square on $x^2 + 4x$.

92. In the solution that follows, what error was made?

$$x^2 + 8x = 5$$
$$x^2 + 8x + 16 = 5$$
$$(x + 4)^2 = 5$$
$$x + 4 = \pm\sqrt{5}$$
$$x = -4 \pm \sqrt{5}$$

REVIEW Solve each equation.

93. $\sqrt{5x - 6} = 2$

94. $\sqrt{6x + 1} + 2 = 7$

95. $2\sqrt{x} = \sqrt{5x - 16}$

96. $\sqrt{22y + 86} = y + 9$

CHALLENGE PROBLEMS Solve each equation.

97. $x(x + 3) - \frac{1}{2} = -2$

98. $x[(x - 2) + 3] = 3\left(x - \frac{2}{9}\right)$

9.2 The Quadratic Formula

- The Quadratic Formula
- Quadratic Equations with No Real Solutions
- Applications

We can solve any quadratic equation by completing the square, but the work is sometimes tedious. In this section, we will develop a formula that will enable us to solve quadratic equations with must less effort.

THE QUADRATIC FORMULA

To develop a formula that will produce the solutions of any given quadratic equation, we start with the **general quadratic equation** $ax^2 + bx + c = 0$, with $a > 0$, and solve it for x by completing the square.

$$ax^2 + bx + c = 0$$

$$\frac{ax^2}{a} + \frac{bx}{a} + \frac{c}{a} = \frac{0}{a}$$ Divide both sides by a so that the coefficient of x^2 is 1.

$$x^2 + \frac{b}{a}x + \frac{c}{a} = 0$$ $\frac{ax^2}{a} = x^2$. Write $\frac{bx}{a}$ as $\frac{b}{a}x$.

$$x^2 + \frac{b}{a}x = -\frac{c}{a}$$ Subtract $\frac{c}{a}$ from both sides so that only the terms involving x are on the left-hand side of the equation.

We can complete the square on $x^2 + \frac{b}{a}x$ by adding the square of one-half of the coefficient of x. Since the coefficient of x is $\frac{b}{a}$, we have $\frac{1}{2} \cdot \frac{b}{a} = \frac{b}{2a}$ and $\left(\frac{b}{2a}\right)^2 = \frac{b^2}{4a^2}$.

$$x^2 + \frac{b}{a}x + \frac{b^2}{4a^2} = -\frac{c}{a} + \frac{b^2}{4a^2}$$ To complete the square, add $\frac{b^2}{4a^2}$ to both sides.

$$x^2 + \frac{b}{a}x + \frac{b^2}{4a^2} = -\frac{4ac}{4aa} + \frac{b^2}{4a^2}$$ Multiply $-\frac{c}{a}$ by $\frac{4a}{4a}$. Now the fractions on the right side have the common denominator $4a^2$.

$$\left(x + \frac{b}{2a}\right)^2 = \frac{b^2 - 4ac}{4a^2}$$ On the left-hand side, factor. On the right-hand side, add the fractions.

$$x + \frac{b}{2a} = \pm\sqrt{\frac{b^2 - 4ac}{4a^2}}$$ Use the square root property.

$$x + \frac{b}{2a} = \pm\frac{\sqrt{b^2 - 4ac}}{\sqrt{4a^2}}$$ The square root of a quotient is the quotient of square roots.

$$x + \frac{b}{2a} = \pm\frac{\sqrt{b^2 - 4ac}}{2a}$$ Since $a > 0$, $\sqrt{4a^2} = 2a$.

$$x = -\frac{b}{2a} \pm \frac{\sqrt{b^2 - 4ac}}{2a}$$ To isolate x, subtract $\frac{b}{2a}$ from both sides.

$$x = \frac{-b \pm \sqrt{b^2 - 4ac}}{2a}$$ Combine the fractions.

The Language of Algebra

Your instructor may ask you to *derive* the quadratic formula. That means to solve $ax^2 + bx + c = 0$ for x, using the series of steps shown here, to obtain

$$x = \frac{-b \pm \sqrt{b^2 - 4ac}}{2a}$$

This result is called the **quadratic formula.** To develop this formula, we assumed that a was positive. If a is negative, similar steps are used, and we obtain the same result.

Quadratic Formula The solutions of $ax^2 + bx + c = 0$, with $a \neq 0$, are

$$x = \frac{-b \pm \sqrt{b^2 - 4ac}}{2a}$$

EXAMPLE 1

Solve: $2x^2 + 5x - 3 = 0$.

ELEMENTARY Algebra f(x) Now™

Solution To use the quadratic formula, we need to identify a, b, and c. We do this by comparing the given equation to the general quadratic equation $ax^2 + bx + c = 0$.

$$\begin{array}{c} 2x^2 + 5x - 3 = 0 \\ \uparrow \quad\;\; \uparrow \quad\;\; \uparrow \\ ax^2 + bx + c = 0 \end{array}$$

We see that $a = 2$, $b = 5$, and $c = -3$. To find the solutions, we substitute these values into the formula and evaluate the right-hand side.

$$x = \frac{-b \pm \sqrt{b^2 - 4ac}}{2a}$$ This is the quadratic formula.

$$x = \frac{-5 \pm \sqrt{5^2 - 4(2)(-3)}}{2(2)}$$ Substitute 2 for a, 5 for b, and -3 for c.

$$x = \frac{-5 \pm \sqrt{25 - (-24)}}{4}$$ Evaluate the power and multiply within the radical. Multiply in the denominator.

$$x = \frac{-5 \pm \sqrt{49}}{4}$$ Subtract within the radical.

$$x = \frac{-5 \pm 7}{4}$$ $\sqrt{49} = 7$.

Caution

When writing the quadratic formula, be careful to draw the fraction bar so that it includes the entire numerator. Do not write

$$x = -b \pm \frac{\sqrt{b^2 - 4ac}}{2a}$$

To find the first solution, evaluate the expression using the + symbol. To find the second solution, evaluate the expression using the − symbol.

$$x = \frac{-5 + 7}{4} \quad \text{or} \quad x = \frac{-5 - 7}{4}$$
$$x = \frac{2}{4} \qquad\qquad x = \frac{-12}{4}$$
$$x = \frac{1}{2} \qquad\qquad x = -3$$

Check that $\frac{1}{2}$ and -3 are the solutions.

Self Check 1 Solve: $4x^2 + 7x - 2 = 0$.

EXAMPLE 2

ELEMENTARY Algebra $f(x)$ Now™

Solve $5x^2 + 1 = 5x$ using the quadratic formula. Approximate the solutions to the nearest hundredth.

Solution To identify a, b, and c, we must write the given equation in the form $ax^2 + bx + c = 0$.

$$5x^2 + 1 = 5x$$
$$5x^2 - 5x + 1 = 0 \qquad \text{To get 0 on the right-hand side, subtract } 5x \text{ from both sides.}$$
$$\uparrow \quad \uparrow \quad \uparrow$$
$$ax^2 + bx + c = 0$$

We see that $a = 5$, $b = -5$, and $c = 1$. To find the solutions, substitute these values into the quadratic formula and evaluate the right-hand side.

Success Tip

Perhaps you noticed that Example 1 could be solved by factoring. However, that is not the case for Example 2. These observations illustrate that the quadratic formula can be used to solve any quadratic equation.

$$x = \frac{-b \pm \sqrt{b^2 - 4ac}}{2a}$$
$$x = \frac{-(-5) \pm \sqrt{(-5)^2 - 4(5)(1)}}{2(5)} \qquad \text{Substitute 5 for } a, -5 \text{ for } b, \text{ and 1 for } c.$$
$$x = \frac{5 \pm \sqrt{25 - 20}}{10} \qquad -(-5) = 5. \text{ Evaluate the power and multiply within the radical. Multiply in the denominator.}$$
$$x = \frac{5 \pm \sqrt{5}}{10} \qquad \text{Subtract within the radical.}$$

This fraction does not simplify further. Thus,

$$x = \frac{5 + \sqrt{5}}{10} \quad \text{or} \quad x = \frac{5 - \sqrt{5}}{10}$$

We can use a calculator to approximate the solutions. To the nearest hundredth,

$$\frac{5 + \sqrt{5}}{10} \approx 0.72 \quad \text{or} \quad \frac{5 - \sqrt{5}}{10} \approx 0.28$$

Self Check 2 Solve: $4x^2 + 2 = 7x$. Approximate the solutions to the nearest hundredth.

EXAMPLE 3

Solve: $3x^2 - 4 = -2x$.

ELEMENTARY Algebra f(x) Now™

Solution We begin by writing the given equation in the form $ax^2 + bx + c = 0$.

$$3x^2 - 4 = -2x$$

$$3x^2 + 2x - 4 = 0 \quad \text{To get 0 on the right-hand side, add } 2x \text{ to both sides.}$$

$$\uparrow \quad \uparrow \quad \uparrow$$

$$ax^2 + bx + c = 0$$

We see that $a = 3$, $b = 2$, and $c = -4$. To find the solutions, substitute these values into the quadratic formula and evaluate the right-hand side.

$$x = \frac{-b \pm \sqrt{b^2 - 4ac}}{2a}$$

$$x = \frac{-2 \pm \sqrt{2^2 - 4(3)(-4)}}{2(3)} \quad \text{Substitute 3 for } a\text{, 2 for } b\text{, and } -4 \text{ for } c.$$

$$x = \frac{-2 \pm \sqrt{4 - (-48)}}{6} \quad \text{Evaluate the power and multiply within the radical. Multiply in the denominator.}$$

$$x = \frac{-2 \pm \sqrt{52}}{6} \quad \text{Subtract within the radical.}$$

$$x = \frac{-2 \pm 2\sqrt{13}}{6} \quad \sqrt{52} = \sqrt{4}\sqrt{13} = 2\sqrt{13}.$$

This fraction can be simplified. We begin by factoring the numerator and denominator.

$$x = \frac{2(-1 \pm \sqrt{13})}{2 \cdot 3} \quad \text{Factor out 2 in the numerator. Factor 6 as } 2 \cdot 3.$$

$$x = \frac{\cancel{2}(-1 \pm \sqrt{13})}{\cancel{2} \cdot 3} \quad \text{Remove a factor equal to 1: } \frac{2}{2} = 1.$$

$$x = \frac{-1 \pm \sqrt{13}}{3}$$

Thus, the solutions are

$$\frac{-1 + \sqrt{13}}{3} \quad \text{and} \quad \frac{-1 - \sqrt{13}}{3}$$

Self Check 3 Solve: $2x^2 - 1 = -2x$.

QUADRATIC EQUATIONS WITH NO REAL SOLUTIONS

The next example shows that some quadratic equations have no real-number solutions.

EXAMPLE 4

Solve: $x^2 + 2x + 5 = 0$.

ELEMENTARY Algebra f(x) Now™

Solution To solve by using the quadratic formula, we identify a, b, and c.

$$\begin{array}{l} 1x^2 + 2x + 5 = 0 \\ \uparrow \quad\ \ \uparrow \quad\ \uparrow \\ ax^2 + bx + c = 0 \end{array}$$

Recall that the term x^2 has an understood coefficient 1.

We see that $a = 1, b = 2$, and $c = 5$ and substitute these values into the quadratic formula.

$$x = \frac{-b \pm \sqrt{b^2 - 4ac}}{2a}$$

$$x = \frac{-2 \pm \sqrt{2^2 - 4(1)(5)}}{2(1)} \quad \text{Substitute 1 for } a\text{, 2 for } b\text{, and 5 for } c.$$

$$x = \frac{-2 \pm \sqrt{4 - 20}}{2} \quad \text{Evaluate the power and multiply within the radical. Multiply in the denominator.}$$

$$x = \frac{-2 \pm \sqrt{-16}}{2} \quad \text{Subtract within the radical.}$$

Since $\sqrt{-16}$ is not a real number, there are no real-number solutions.

Self Check 4 Does the equation $2x^2 + x + 1 = 0$ have any real-number solutions?

APPLICATIONS

We have discussed several methods that are used to solve quadratic equations. To determine the most efficient method for a given equation, we can use the following strategy.

Strategy for Solving Quadratic Equations

1. See whether the equation is in a form such that the **square root method** is easily applied.
2. If the square root method can't be used, write the equation in $ax^2 + bx + c = 0$ form.
3. See whether the equation can be solved using the **factoring method.**
4. If you can't factor the quadratic, solve the equation by **completing the square** or by the **quadratic formula.**

EXAMPLE 5

ELEMENTARY Algebra f(x) Now™

Sailing. The height of a triangular sail is 4 feet more than the length of the base. If the sail has an area of 30 square feet, find the length of its base and the height.

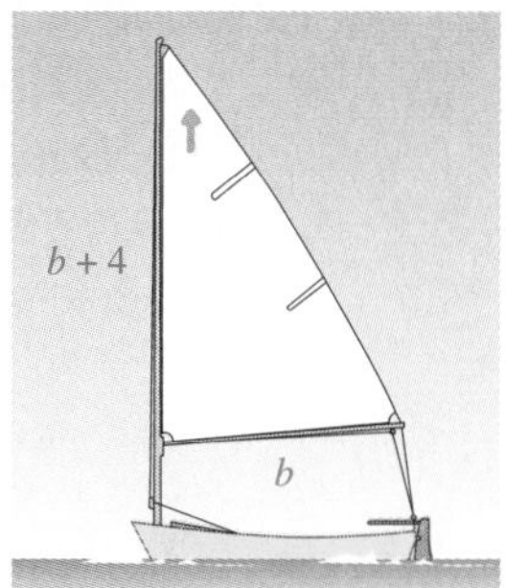

Analyze the Problem

- The height of the sail is 4 feet more than the length of the base.
- The area of the sail is 30 ft^2.
- Find the length of the base and height of the sail.

Form an Equation If we let b = the length of the base of the triangular sail, then $b + 4$ = the height. We can use the formula for the area of a triangle, $A = \frac{1}{2}bh$, to form an equation.

$\frac{1}{2}$	times	the length of the base	times	the height	equals	the area of the triangle.
$\frac{1}{2}$	$\cdot$	b	$\cdot$	$(b + 4)$	$=$	30

Solve the Equation To solve $\frac{1}{2}b(b + 4) = 30$, we write it in quadratic form.

$$\frac{1}{2}b(b + 4) = 30$$

$$b(b + 4) = 60$$ To clear the equation of the fraction, multiply both sides by 2.

$$b^2 + 4b = 60$$ Distribute the multiplication by b.

$$b^2 + 4b - 60 = 0$$ To get 0 on the right-hand side, subtract 60 from both sides.

By inspection, we see that -60 has factors 10 and -6, and that their sum is 4. We can use this pair to solve the quadratic equation by the factoring method.

$$(b + 10)(b - 6) = 0$$ Factor $b^2 + 4b - 60$.

$$b + 10 = 0 \quad \text{or} \quad b - 6 = 0$$ Set each factor equal to 0.

$$\cancel{b = -10} \quad | \quad b = 6$$ Discard the first solution, -10. The length of the base cannot be negative.

State the Conclusion The length of the base of the sail is 6 feet. Since the height is given by $b + 4$, the height of the sail is $6 + 4 = 10$ feet.

Check the Result A height of 10 feet is 4 feet more than the length of the base, which is 6 feet. Also, the area of the triangle is $\frac{1}{2}(6)(10) = 30 \text{ ft}^2$. The results check.

EXAMPLE 6

ELEMENTARY Algebra $f(x)$ Now™

Movie stunts. In a scene for an action movie, a stuntwoman falls from the top of a 95-foot-tall building into a 10-foot-tall airbag directly below her on the ground. The formula $s = 16t^2$ gives the distance s in feet that she falls in t seconds. For how many seconds will she fall before making contact with the airbag?

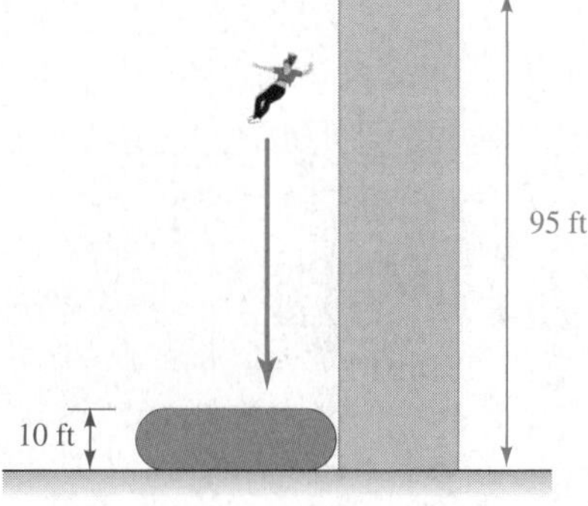

Solution The woman will fall a distance of $95 - 10 = 85$ feet before making contact with the airbag. To find the number of seconds that the fall will last, we substitute 85 for s in the formula and solve for t, the time.

$$s = 16t^2$$

$$\mathbf{85} = 16t^2$$ Substitute 85 for s.

The resulting quadratic equation is easily solved by the square root property.

$$\frac{85}{16} = t^2 \quad \text{To isolate } t^2 \text{, divide both sides by 16.}$$

$$\pm\sqrt{\frac{85}{16}} = t \quad \text{Use the square root property.}$$

$$\pm\frac{\sqrt{85}}{\sqrt{16}} = t \quad \text{The square root of a quotient is the quotient of square roots.}$$

$$\pm\frac{\sqrt{85}}{4} = t \quad \sqrt{16} = 4.$$

The stuntwoman will fall for $\frac{\sqrt{85}}{4}$ seconds (approximately 2.3 seconds) before making contact with the airbag. We discard the other solution, $-\frac{\sqrt{85}}{4}$, because a negative time does not make sense in this example.

EXAMPLE 7

Televisions. A television's screen size is measured diagonally. For the 42-inch plasma television shown in the illustration, the screen's height is 16 inches less than its length. What are the height and length of the screen?

Analyze the Problem A sketch of the screen shows that two adjacent sides and the diagonal form a right triangle. The length of the hypotenuse is 42 inches.

Form an Equation If we let l = the length of the screen, then $l - 16$ is the height of the screen. We can use the Pythagorean theorem to form an equation.

$$a^2 + b^2 = c^2 \quad \text{This is the Pythagorean theorem.}$$

$$l^2 + (l - 16)^2 = 42^2 \quad \text{Substitute } l \text{ for } a,\ l - 16 \text{ for } b, \text{ and 42 for } c.$$

$$l^2 + l^2 - 32l + 256 = 1{,}764 \quad \text{Find } (l - 16)^2 \text{ and } 42^2.$$

$$2l^2 - 32l - 1{,}508 = 0 \quad \text{Subtract 1,764 from both sides.}$$

$$l^2 - 16l - 754 = 0 \quad \text{Divide both sides by 2.}$$

Solve the Equation Because of the large constant term, -754, we will not attempt to solve this quadratic equation by factoring. Instead, we will use the quadratic formula, with $a = 1$, $b = -16$, and $c = -754$.

$$l = \frac{-b \pm \sqrt{b^2 - 4ac}}{2a} \quad \text{In the quadratic formula, replace } x \text{ with } l.$$

$$l = \frac{-(-16) \pm \sqrt{(-16)^2 - 4(1)(-754)}}{2(1)} \quad \text{Substitute 1 for } a,\ -16 \text{ for } b, \text{ and } -754 \text{ for } c.$$

$$l = \frac{16 \pm \sqrt{256 - (-3{,}016)}}{2}$$ Evaluate the power and multiply within the radical. Multiply in the denominator.

$$l = \frac{16 \pm \sqrt{3{,}272}}{2}$$ Subtract within the radical.

Thus,

$$l = \frac{16 + \sqrt{3{,}272}}{2} \quad \text{or} \quad l = \frac{16 - \sqrt{3{,}272}}{2}$$

We can use a calculator to approximate the solutions to the nearest tenth.

$l \approx 36.6$ or $l \approx -20.6$ Discard the negative solution because the length of the screen cannot be negative.

State the Conclusion The length of the television screen is approximately 36.6 inches. Since the height is $l - 16$, the height is approximately $36.6 - 16$ or 20.6 inches.

Check the Result The sum of the squares of the lengths of the sides is $(36.6)^2 + (20.6)^2 = 1{,}763.92$. The square of the length of the hypotenuse is $42^2 = 1{,}764$. Since these are approximately equal, the results seem reasonable.

Answers to Self Checks **1.** $\frac{1}{4}, -2$ **2.** $\frac{7 + \sqrt{17}}{8} \approx 1.39, \frac{7 - \sqrt{17}}{8} \approx 0.36$ **3.** $\frac{-1 \pm \sqrt{3}}{2}$ **4.** no

9.2 STUDY SET  ELEMENTARY Algebra f(x) Now™

VOCABULARY Fill in the blanks.

1. The general ________ equation is $ax^2 + bx + c = 0$, where $a \neq 0$.
2. The formula

   $x =$

   is called the quadratic formula.
3. To ______ a quadratic equation means to find all the values of the variable that make the equation true.
4. $\sqrt{-16}$ is not a _____ number.

CONCEPTS

5. Write each equation in quadratic form.
   a. $x^2 + 2x = -5$ b. $3x^2 = -2x + 1$
6. For each quadratic equation, find a, b, and c.
   a. $x^2 + 5x + 6 = 0$ b. $8x^2 - x = 10$
7. Divide both sides of $2x^2 - 4x + 8 = 0$ by 2, and then find a, b, and c.
8. Evaluate each expression.
   a. $\frac{-2 \pm \sqrt{2^2 - 4(1)(-8)}}{2(1)}$
   b. $\frac{-(-1) \pm \sqrt{(-1)^2 - 4(2)(-4)}}{2(2)}$
9. What common factor do the terms of $10 \pm 15\sqrt{2}$ have?
10. Simplify each expression.
    a. $\frac{-1 \pm \sqrt{45}}{2(7)}$ b. $\frac{6 \pm 2\sqrt{7}}{10}$

11. Simplify: $\dfrac{-(-4) \pm \sqrt{(-4)^2 - 4(2)(-9)}}{2(2)}$

12. A student used the quadratic formula to solve an equation and obtained

$$x = \frac{-3 \pm \sqrt{15}}{2}$$

a. How many solutions does the equation have?

b. What are they?

c. Approximate them to the nearest hundredth.

13. The solutions of a quadratic equation are

$$\frac{-1 \pm \sqrt{5}}{2}$$

Graph them on a number line.

14. Match each quadratic equation with the best method for solving it.

a. $x^2 + 7x - 8 = 0$ **i.** square root method

b. $x^2 + 2x - 31 = 0$ **ii.** factoring method

c. $x^2 = 21$ **iii.** quadratic formula

15. What is the result when both sides of $\frac{1}{2}b(b + 5) = 80$ are multiplied by 2?

16. Write $x^2 + (x + 2)^2 = 25$ in quadratic form and find a, b, and c.

NOTATION Complete the solution.

17. Solve: $x^2 - 5x - 6 = 0$.

$$x = \frac{-b \pm \sqrt{b^2 - 4ac}}{2a}$$

$$x = \frac{-(\square) \pm \sqrt{(-5)^2 - 4(1)(\square)}}{2(\square)}$$

$$x = \frac{5 \pm \sqrt{25 + \square}}{2}$$

$$x = \frac{5 \pm \sqrt{\square}}{2}$$

$$x = \frac{\square \pm 7}{2}$$

$$x = \frac{5 + \square}{2} = \square \quad \text{or} \quad x = \frac{5 - \square}{2} = \square$$

18. What is wrong with this student's work?

Solve: $x^2 + 4x - 5 = 0$.

$$x = -4 \pm \frac{\sqrt{16 - 4(1)(-5)}}{2}$$

19. In reading

$$\frac{-b \pm \sqrt{b^2 - 4ac}}{2a}$$

we say, "the ______________ of b, plus or ________ the ________ root of the quantity b ________ minus 4 ________ a times c, all ________ $2a$."

20. Translate to symbols: the opposite of -3.

PRACTICE Use the quadratic formula to solve each equation.

21. $x^2 - 5x + 6 = 0$

22. $x^2 + 5x + 4 = 0$

23. $x^2 + 7x + 12 = 0$

24. $x^2 - x - 12 = 0$

25. $2x^2 - x - 1 = 0$

26. $2x^2 + 3x - 2 = 0$

27. $6x^2 - 13x = -6$

28. $3x^2 - 5x = 2$

29. $4x^2 + 3x = 1$

30. $4x^2 + 4x = 3$

31. $x^2 + 3x + 1 = 0$

32. $x^2 + 3x - 2 = 0$

33. $3x^2 - x = 3$

34. $5x^2 = 3x + 1$

35. $x^2 + 5 = 2x$

36. $2x^2 + 3x = -3$

37. $x^2 = 1 - 2x$

38. $x^2 = 4 + 2x$

39. $3x^2 = 6x + 2$

40. $3x^2 = -8x - 2$

41. $x^2 + 4x = 3$

42. $x^2 + 6x = 5$

43. $2x^2 + 10x + 11 = 0$

44. $6x^2 - 6x - 1 = 0$

45. $7x^2 - x = -8$

46. $9x^2 - 2x = -4$

47. $-8x - 5 = 2x^2$ **48.** $-11x - 11 = 2x^2$

Use the most convenient method to find all real solutions. If a solution contains a radical, give the exact solution and then approximate it to the nearest hundredth.

49. $(2y - 1)^2 = 25$ **50.** $m^2 + 14m + 49 = 0$

51. $2x^2 + x = 5$ **52.** $2x^2 - x + 2 = 0$

53. $x^2 - 2x - 1 = 0$ **54.** $b^2 = 18$

55. $x^2 - 2x - 35 = 0$ **56.** $x^2 + 5x + 3 = 0$

57. $x^2 + 2x + 7 = 0$ **58.** $3x^2 - x = 1$

59. $4c^2 + 16c = 0$ **60.** $t^2 - 1 = 0$

61. $18 = 3y^2$ **62.** $25x - 50x^2 = 0$

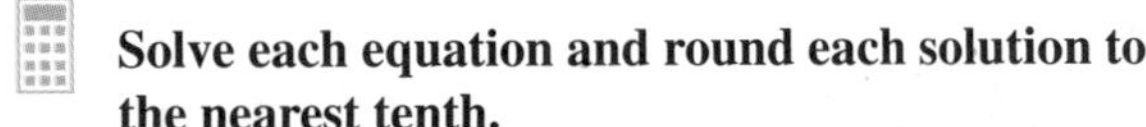

Solve each equation and round each solution to the nearest tenth.

63. $2.4x^2 - 9.5x + 6.2 = 0$

64. $-1.7x^2 + 0.5x + 0.9 = 0$

APPLICATIONS

65. HEIGHT OF A TRIANGLE The triangle shown has an area of 30 square inches. Find its height.

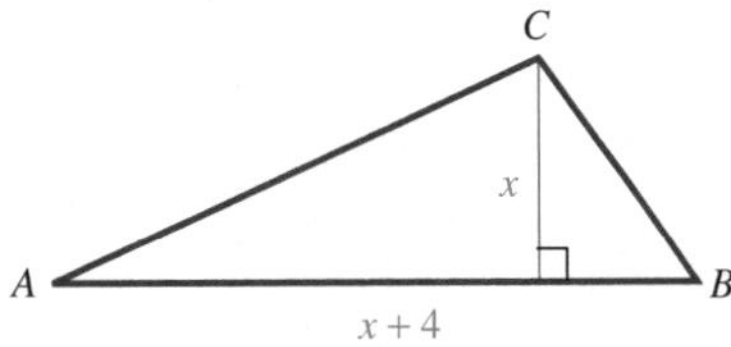

66. NUTRITION A poster that shows the six basic food groups has an area of 90 square inches. The base of the triangular-shaped poster is 3 inches longer than the height. Find the length of the base and the height of the poster.

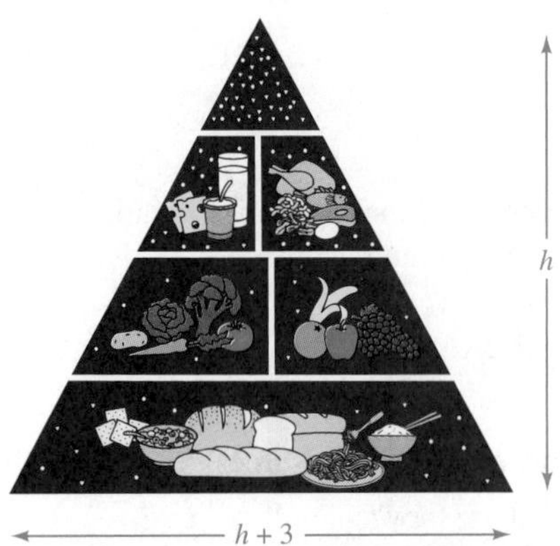

67. FLAGS According to the *Guinness Book of World Records 1998,* the largest flag flown from a flagpole was a Brazilian national flag, a rectangle having an area of 3,102 ft^2. If the flag is 19 feet longer than it is wide, find its width and length.

68. COMICS A comic strip occupies 96 square centimeters of space in a newspaper. The length of the rectangular space is 4 centimeters more than twice its width. Find its dimensions.

69. DAREDEVILS In 1873, Henry Bellini combined a tightrope walk with a leap into the churning Niagara River below, where he was picked up by a boat. If the rope was 200 feet above the water, for how many seconds did he fall before hitting the water? Round to the nearest tenth. (*Hint:* See Example 6.)

70. ROLLER COASTERS Soaring to a height of 420 feet, the *Top Thrill Dragster* at Cedar Point Amusement Park in Sandusky, Ohio, is the tallest and fastest roller coaster in the world. If a rider accidentally dropped a camera as the coaster reached its highest point, how long would it take the camera to hit the ground? Round to the nearest tenth. (*Hint:* See Example 6.)

71. EARTHQUAKES After a powerful earthquake, store owners nailed 50-inch-long boards across a broken display window. Find the height and length of the window, if the height is 10 inches less than the length.

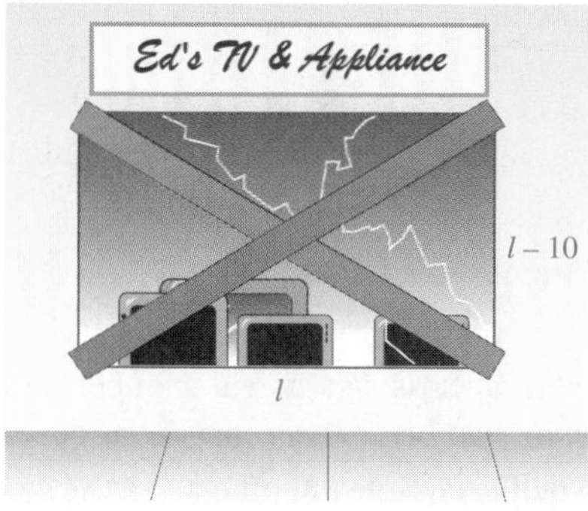

72. THE ABACUS The Chinese abacus shown consists of a frame, parallel wires, and beads that are moved to perform arithmetic computations. The frame is 21 centimeters wider than it is high. Find its height and width.

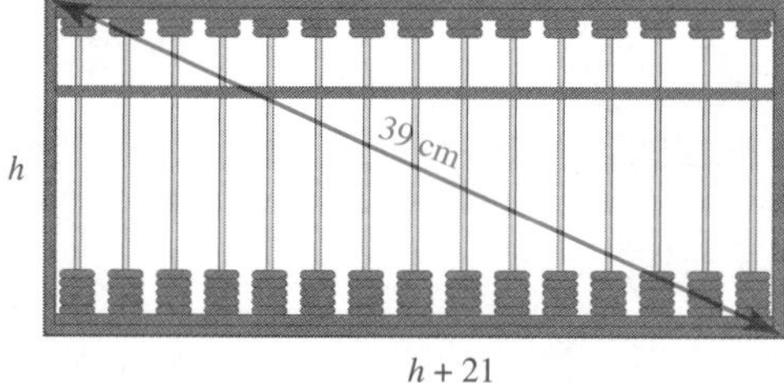

73. INVESTING We can use the formula $A = P(1 + r)^2$ to find the amount $\$A$ that $\$P$ will become when invested at an annual rate of $r\%$ for 2 years. What interest rate is needed to make \$5,000 grow to \$5,724.50 in 2 years?

74. RETAILING When a wholesaler sells n CD players, his revenue R is given by the formula $R = 150n - \frac{1}{2}n^2$. How many players would he have to sell to receive \$11,250? (*Hint:* Multiply both sides of the equation by -2.)

WRITING

75. Do you agree with the following statement? Explain. *The quadratic formula is the easiest method to use to solve quadratic equations.*

76. Rewrite in words:

$$x = \frac{-b \pm \sqrt{b^2 - 4ac}}{2a}$$

77. At times, certain types of solutions to applied problems are discarded. Explain. Give an example of such a situation.

78. In this section, we used diagrams like the following when solving quadratic equations. What is its purpose?

$$\begin{array}{l} 5x^2 + 6x + 7 = 0 \\ \uparrow \quad\;\; \uparrow \quad\; \uparrow \\ ax^2 + bx + c = 0 \end{array}$$

REVIEW **Solve each equation for the indicated variable.**

79. $A = p + prt$; for r

80. $F = \dfrac{GMm}{d^2}$; for M

81. $\dfrac{1}{r} = \dfrac{1}{r_1} + \dfrac{1}{r_2}$; for r

82. $2E = \dfrac{T - t}{9}$; for t

CHALLENGE PROBLEMS

83. DECKING The owner of a pool wants to surround it with a concrete deck of uniform width (shown in gray). If he can afford 368 square feet of decking, how wide can he make the deck?

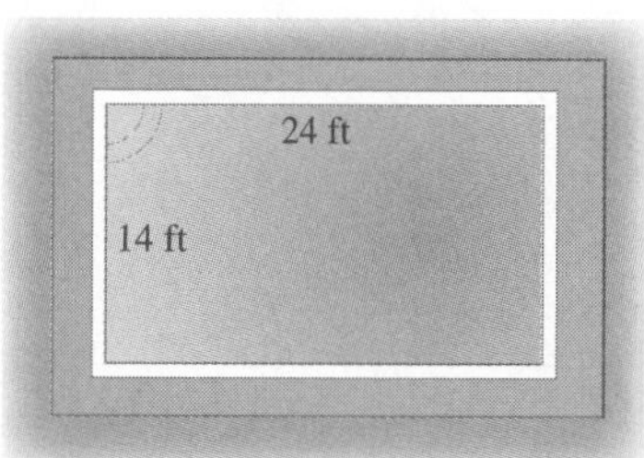

84. METAL FABRICATION A square piece of tin, 12 inches on a side, is to have four equal squares cut from its corners, as shown. If the edges are then to be folded up to make a box with a floor area of 64 square inches, find the depth of the box.

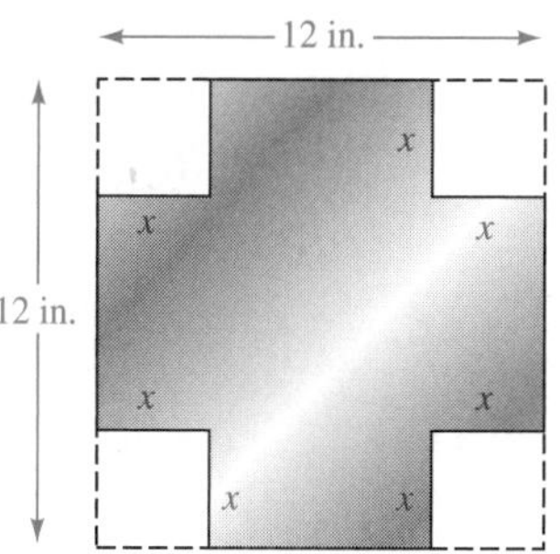

9.3 Complex Numbers

- The Imaginary Number i
- Simplifying Square Roots of Negative Numbers
- Complex Numbers
- Quadratic Equations with Complex Solutions
- Arithmetic of Complex Numbers
- Complex Conjugates

Recall that the square root of a negative number is not a real number. However, an expanded number system, called the *complex number system,* has been devised to give meaning to $\sqrt{-9}$, $\sqrt{-25}$, and the like. To define complex numbers, we use a new type of number that is denoted by the letter i.

THE IMAGINARY NUMBER i

We have seen that some quadratic equations such as $x^2 + x + 1 = 0$ do not have real-number solutions. When we use the quadratic formula to solve it, we encounter the square root of a negative number.

The Language of Algebra

For years, mathematicians thought numbers like $\sqrt{-9}$ and $\sqrt{-25}$ were useless. In the 17th century, René Descartes (1596–1650) called them *imaginary numbers.* Today they have important uses such as describing alternating electric current.

Solve: $x^2 + x + 1 = 0$

$$x = \frac{-b \pm \sqrt{b^2 - 4ac}}{2a}$$

$$x = \frac{-1 \pm \sqrt{1^2 - 4(1)(1)}}{2(1)}$$ Substitute 1 for a, 1 for b, and 1 for c.

$$x = \frac{-1 \pm \sqrt{1 - 4}}{2}$$

$$x = \frac{-1 \pm \sqrt{-3}}{2}$$

$$x = \frac{-1 + \sqrt{-3}}{2} \quad \text{or} \quad x = \frac{-1 - \sqrt{-3}}{2}$$

Since each result contains $\sqrt{-3}$, which is not real, these solutions are not real numbers. To be able to solve quadratic equations such as $x^2 + x + 1 = 0$, we must define the square root of a negative number.

The Number i

The **imaginary number i** is defined as

$$i = \sqrt{-1}$$

From the definition, it follows that $i^2 = -1$.

SIMPLIFYING SQUARE ROOTS OF NEGATIVE NUMBERS

We can use extensions of the product and quotient rules for radicals to write the square root of a negative number as the product of a real number and i.

EXAMPLE 1

Write each expression in terms of i: **a.** $\sqrt{-9}$, **b.** $\sqrt{-5}$, **c.** $-\sqrt{-8}$, and **d.** $\sqrt{-\frac{54}{25}}$.

Solution We write each negative radicand as the product of -1 and a positive number and apply the appropriate rules for radicals. In each case, we replace $\sqrt{-1}$ with i.

Notation

Since it is easy to confuse $\sqrt{5}i$ with $\sqrt{5i}$, we write i first so that it is clear that the i is not within the radical symbol. However, both $i\sqrt{5}$ and $\sqrt{5}i$ are correct.

a. $\sqrt{-9} = \sqrt{-1 \cdot 9} = \sqrt{-1}\sqrt{9} = i \cdot 3 = 3i$

b. $\sqrt{-5} = \sqrt{-1 \cdot 5} = \sqrt{-1}\sqrt{5} = i\sqrt{5}$ or $\sqrt{5}i$

c. $-\sqrt{-8} = -\sqrt{-1 \cdot 4 \cdot 2} = -\sqrt{-1}\sqrt{4}\sqrt{2} = -i \cdot 2 \cdot \sqrt{2} = -2i\sqrt{2}$ or $-2\sqrt{2}i$

d. $\sqrt{-\frac{54}{25}} = \sqrt{-1 \cdot \frac{54}{25}} = \frac{\sqrt{-1 \cdot 54}}{\sqrt{25}} = \frac{\sqrt{-1}\sqrt{9}\sqrt{6}}{\sqrt{25}} = \frac{3i\sqrt{6}}{5}$ or $\frac{3\sqrt{6}}{5}i$

Self Check 1 Write each expression in terms of i: **a.** $\sqrt{-81}$, **b.** $-\sqrt{-11}$, **c.** $\sqrt{-28}$, and **d.** $\sqrt{-\frac{27}{100}}$.

COMPLEX NUMBERS

The imaginary number i is used to define *complex numbers.*

Complex Numbers

A **complex number** is any number that can be written in the form $a + bi$, where a and b are real numbers and $i = \sqrt{-1}$.

Complex numbers of the form $a + bi$, where $b \neq 0$, are also called **imaginary numbers.** Complex numbers of the form bi are also called **pure imaginary numbers.**

For a complex number written in the standard form $a + bi$, we call a the **real part** and b the **imaginary part.**

EXAMPLE 2

Write each number in the form $a + bi$: **a.** 4, **b.** $\sqrt{-16}$, and **c.** $-8 + \sqrt{-45}$.

ELEMENTARY Algebra *f(x)* Now™

Solution

a. $4 = 4 + 0i$ The imaginary part is 0.

b. $\sqrt{-16} = 0 + 4i$ The real part is 0. $\sqrt{-16} = \sqrt{-1}\sqrt{16} = 4i$.

c. $-8 + \sqrt{-45} = -8 + 3i\sqrt{5}$ $\sqrt{-45} = \sqrt{-1}\sqrt{45} = \sqrt{-1}\sqrt{9}\sqrt{5} = 3i\sqrt{5}$.

Self Check 2 Write each number in the form $a + bi$: **a.** -18, **b.** $\sqrt{-36}$, and **c.** $1 + \sqrt{-24}$.

The following illustration shows the relationship between the real numbers, the imaginary numbers, and the complex numbers.

Complex numbers

Real numbers				Imaginary numbers		
-6	$\frac{5}{16}$	-1.75	π	$9 + 7i$	$-2i$	$\frac{1}{4} - \frac{3}{4}i$
48	0	$-\sqrt{10}$	$-\frac{7}{2}$	$0.56i$	$6 + i\sqrt{3}$	

QUADRATIC EQUATIONS WITH COMPLEX SOLUTIONS

We have seen that many quadratic equations do not have real-number solutions. In the complex number system, all quadratic equations have solutions. We can write the solutions in the form $a + bi$.

EXAMPLE 3

Solve: $x^2 + 25 = 0$.

ELEMENTARY Algebra $f(x)$ Now™

Solution We can write the equation as $x^2 = -25$ and use the square root property to solve it.

$$x^2 + 25 = 0$$

$$x^2 = -25 \quad \text{Subtract 25 from both sides.}$$

$$x = \pm\sqrt{-25} \quad \text{Use the square root property.}$$

$$x = \pm 5i \quad \sqrt{-25} = \sqrt{-1}\sqrt{25} = 5i.$$

Notation

It is acceptable to use $a - bi$ as a substitute for the form $a + (-b)i$. In Example 3, we write $0 - 5i$ instead of $0 + (-5)i$.

To express the solutions in the form $a + bi$, we must write 0 for the real part, a. Thus, the solutions are $0 + 5i$ and $0 - 5i$.

Self Check 3 Solve: $x^2 + 100 = 0$. Express the solutions in the form $a + bi$.

EXAMPLE 4

Solve: $4t^2 - 6t + 3 = 0$.

ELEMENTARY Algebra $f(x)$ Now™

Solution We use the quadratic formula to solve the equation.

$$t = \frac{-b \pm \sqrt{b^2 - 4ac}}{2a}$$

$$t = \frac{-(-6) \pm \sqrt{(-6)^2 - 4(4)(3)}}{2(4)} \quad \text{Substitute 4 for } a, -6 \text{ for } b, \text{ and 3 for } c.$$

$$t = \frac{6 \pm \sqrt{36 - 48}}{8}$$

$$t = \frac{6 \pm \sqrt{-12}}{8}$$

$$t = \frac{6 \pm 2i\sqrt{3}}{8} \quad \sqrt{-12} = \sqrt{-1}\sqrt{12} = \sqrt{-1}\sqrt{4}\sqrt{3} = 2i\sqrt{3}.$$

$$t = \frac{\cancel{2}(3 \pm i\sqrt{3})}{\cancel{2} \cdot 4}$$

In the numerator, factor out the common factor 2. In the denominator, factor 8 as $2 \cdot 4$. Then remove a factor equal to 1: $\frac{2}{2} = 1$.

$$t = \frac{3 \pm i\sqrt{3}}{4}$$

Writing each solution as a complex number in the form $a + bi$, we have

$$\frac{3}{4} + \frac{\sqrt{3}}{4}i \quad \text{and} \quad \frac{3}{4} - \frac{\sqrt{3}}{4}i$$

Self Check 4 Solve: $a^2 + 2a + 3 = 0$. Express the solutions in the form $a + bi$.

ARITHMETIC OF COMPLEX NUMBERS

Adding and subtracting complex numbers is similar to adding and subtracting polynomials.

Addition and Subtraction of Complex Numbers

1. To add complex numbers, add their real parts and add their imaginary parts.
2. To subtract complex numbers, add the opposite of the complex number being subtracted.

EXAMPLE 5

Find each sum or difference. **a.** $(5 + 2i) + (1 + 8i)$, **b.** $(-6 - 5i) - (3 - 4i)$, and **c.** $11i + (-2 + 6i)$.

ELEMENTARY Algebra f(x) Now™

Solution

a. $(5 + 2i) + (1 + 8i) = (5 + 1) + (2i + 8i)$ — The sum of the imaginary parts. The sum of the real parts.

$= 6 + 10i$ — $5 + 1 = 6$; $2i + 8i = 10i$.

b. $(-6 - 5i) - (3 - 4i) = (-6 - 5i) + (-3 + 4i)$ — Add the opposite. To find the opposite, change the sign of each term of $3 - 4i$.

$= [-6 + (-3)] + (-5i + 4i)$ — Add the real parts. Add the imaginary parts.

$= -9 - i$ — $-6 + (-3) = -9$; $-5i + 4i = -i$.

c. $11i + (-2 + 6i) = -2 + (11i + 6i)$ — Add the imaginary parts.

$= -2 + 17i$ — $11i + 6i = 17i$.

Success Tip

i is not a variable, but it is helpful to think of it as one when adding, subtracting, and multiplying. For example:

$$4i + 3i = 7i$$
$$8i - 6i = 2i$$
$$i \cdot i = i^2$$

Self Check 5 Find the sum or difference. Write each result in the form $a + bi$:
a. $(3 - 5i) + (-2 + 6i)$, **b.** $(-4 - i) - (-1 - 6i)$, and **c.** $9 + (16 - 4i)$.

Multiplying complex numbers is similar to multiplying polynomials.

EXAMPLE 6

Find each product. **a.** $5i(4 - i)$, **b.** $(2 + 5i)(6 + 4i)$, and **c.** $(3 + 5i)(3 - 5i)$.

ELEMENTARY Algebra f(x) Now™

Solution

a. $5i(4 - i) = 5i \cdot 4 - 5i \cdot i$ — Distribute the multiplication by $5i$.

$= 20i - 5i^2$

$= 20i - 5(-1)$ — $i^2 = -1$.

$= 20i + 5$

$= 5 + 20i$ — Write the result in the form $a + bi$.

b. $(2 + 5i)(6 + 4i) = 12 + 8i + 30i + 20i^2$ — Use the FOIL method.

$= 12 + 38i + 20(-1)$ — $8i + 30i = -38i$. Replace i^2 with -1.

$= 12 + 38i - 20$

$= -8 + 38i$ — $12 - 20 = -8$.

c. $(3 + 5i)(3 - 5i) = 9 - 15i + 15i - 25i^2$ — Use the FOIL method.

$= 9 - 25(-1)$ — $-15i + 15i = 0$. Replace i^2 with -1.

$= 9 + 25$

$= 34$

Written in the form $a + bi$, the product is $36 + 0i$.

Self Check 6 Find the product. Write each result in the form $a + bi$: **a.** $4i(3 - 9i)$, **b.** $(3 - 2i)(5 - 4i)$, and **c.** $(2 + 6i)(2 - 6i)$.

COMPLEX CONJUGATES

In Example 6c, we saw that the product of the imaginary numbers $3 + 5i$ and $3 - 5i$ is the real number 36. We call $3 + 5i$ and $3 - 5i$ *complex conjugates* of each other.

Complex Conjugates The complex numbers $a + bi$ and $a - bi$ are called **complex conjugates.**

For example,

- $7 + 4i$ and $7 - 4i$ are complex conjugates.
- $5 - i$ and $5 + i$ are complex conjugates.
- $-6i$ and $6i$ are complex conjugates, because $-6i = 0 - 6i$ and $6i = 0 + 6i$.

In general, the product of the complex number $a + bi$ and its complex conjugate $a - bi$ is the real number $a^2 + b^2$, as the following work shows:

The Language of Algebra
Recall that the word *conjugate* was used in Chapter 8 when we rationalized the denominators of radical expressions such as $\frac{5}{\sqrt{6} - 1}$.

$$(a + bi)(a - bi) = a^2 - abi + abi - b^2i^2 \quad \text{Use the FOIL method.}$$
$$= a^2 - b^2(-1) \quad -abi + abi = 0. \text{ Replace } i^2 \text{ with } -1.$$
$$= a^2 + b^2$$

We can use this fact when dividing by a complex number.

EXAMPLE 7 Find each quotient: **a.** $\frac{6}{7 - 4i}$ and **b.** $\frac{3 - i}{2 + i}$.

ELEMENTARY Algebra Now™

Solution **a.** We want to build a fraction equivalent to $\frac{6}{7 - 4i}$ that does not have i in the denominator. To make the denominator, $7 - 4i$, a real number, we need to multiply it by its complex conjugate, $7 + 4i$. It follows that $\frac{7 + 4i}{7 + 4i}$ should be the form of 1 that is used to build $\frac{6}{7 - 4i}$.

$$\frac{6}{7 - 4i} = \frac{6}{7 - 4i} \cdot \frac{7 + 4i}{7 + 4i} \quad \text{To build an equivalent fraction, multiply by } \frac{7 + 4i}{7 + 4i} = 1.$$

$$= \frac{42 + 24i}{49 - 16i^2} \quad \text{To multiply the numerators, distribute the multiplication by 6. To multiply the denominators, find } (7 - 4i)(7 + 4i).$$

$$= \frac{42 + 24i}{49 - 16(-1)} \quad \text{Replace } i^2 \text{ with } -1. \text{ The denominator no longer contains } i.$$

$$= \frac{42 + 24i}{49 + 16} \quad \text{Simplify the denominator.}$$

$$= \frac{42 + 24i}{65}$$

$$= \frac{42}{65} + \frac{24}{65}i \quad \text{Write the result in the form } a + bi.$$

b. We can make the denominator of $\frac{3 - i}{2 + i}$ a real number by multiplying it by the complex conjugate of $2 + i$, which is $2 - i$.

$$\frac{3 - i}{2 + i} = \frac{3 - i}{2 + i} \cdot \frac{2 - i}{2 - i}$$ To build an equivalent fraction, multiply by $\frac{2 - i}{2 - i} = 1$.

$$= \frac{6 - 3i - 2i + i^2}{4 - i^2}$$ To multiply the numerators, find $(3 - i)(2 - i)$. To multiply the denominators, find $(2 + i)(2 - i)$.

$$= \frac{6 - 5i + (-1)}{4 - (-1)}$$ Replace i^2 with -1. The denominator no longer contains i.

$$= \frac{5 - 5i}{5}$$ Simplify the numerator and the denominator.

$$= \frac{5(1 - i)}{5}$$ In the numerator, factor out the common factor 5.

$$= 1 - i$$ Simplify the fraction by removing a factor equal to 1: $\frac{5}{5} = 1$.

Caution

A common mistake is to replace i with -1. Remember, $i \neq -1$. By definition $i = \sqrt{-1}$ and $i^2 = -1$.

Self Check 7 Find the quotient. Write each result in the form $a + bi$: **a.** $\frac{5}{4 - i}$ and **b.** $\frac{5 - 3i}{4 + 2i}$.

As the results of Example 7 show, we use the following rule to divide complex numbers.

Division of Complex Numbers To divide complex numbers, multiply the numerator and denominator by the complex conjugate of the denominator.

Answers to Self Checks **1. a.** $9i$, **b.** $-i\sqrt{11}$, **c.** $2i\sqrt{7}$, **d.** $\frac{3i\sqrt{3}}{10}$ **2. a.** $-18 + 0i$, **b.** $0 + 6i$, **c.** $1 + 2i\sqrt{6}$ **3.** $0 \pm 10i$ **4.** $-1 \pm i\sqrt{2}$ **5. a.** $1 + i$, **b.** $-3 + 5i$, **c.** $25 - 4i$ **6. a.** $36 + 12i$, **b.** $7 - 22i$, **c.** $40 + 0i$ **7. a.** $\frac{20}{17} + \frac{5}{17}i$, **b.** $\frac{7}{10} - \frac{11}{10}i$

9.3 STUDY SET  ELEMENTARY Algebra Now™

VOCABULARY **Fill in the blanks.**

1. A ________ number is any number that can be written in the form $a + bi$.
2. For $9 + 2i$, the ______ part is 9 and the __________ part is 2.
3. The complex numbers $6 + 3i$ and $6 - 3i$ are called complex __________.
4. The __________ number i is used to define complex numbers.

CONCEPTS **Fill in the blanks.**

5. **a.** $i =$ ______ **b.** $i^2 =$ ______
6. **a.** To add complex numbers, ______ their real parts and ______ their imaginary parts.
   b. To subtract complex numbers, add the ________ of the complex number being subtracted.
7. $\sqrt{-25} = \sqrt{\quad \cdot 25} = \quad \sqrt{25} = 5$ ___
8. **a.** $5i + 3i =$ ______ **b.** $5i - 3i =$ ______
9. $i \cdot i = i^{\square} =$ ______

10. The product of the complex number $a + bi$ and its complex conjugate $a - bi$ is a ______ number.

11. To simplify $\frac{2 - 3i}{6 - i}$, we multiply it by ______.

12. $\frac{3 \pm \sqrt{-4}}{5} = \frac{3 \pm \square}{5}$

13. Decide whether each statement is true or false.
   a. Every real number is a complex number.
   b. $2 + 7i$ is an imaginary number.
   c. $\sqrt{-16}$ is a real number.
   d. In the complex number system, all quadratic equations have solutions.

14. Complete the diagram.

Complex numbers

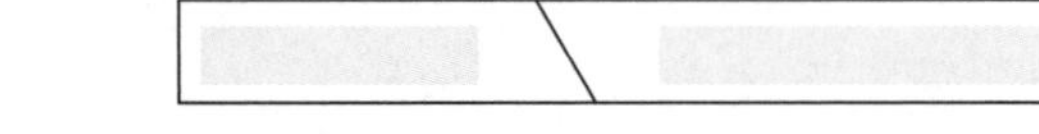

15. Give the complex conjugate of each number.
   a. $2 - 9i$ b. $-8 + i$
   c. $4i$ d. $-11i$

16. Multiply the complex conjugates: $(4 - 5i)(4 + 5i)$.

NOTATION

17. Write each expression so it is clear that i is not within the radical symbol.
   a. $\sqrt{7i}$ b. $2\sqrt{3i}$

18. Write each number in the form $a + bi$.
   a. 12 b. $\frac{3 - 4i}{5}$

PRACTICE Write each expression in terms of *i*.

19. $\sqrt{-9}$
20. $\sqrt{-4}$
21. $\sqrt{-7}$
22. $\sqrt{-11}$
23. $\sqrt{-24}$
24. $\sqrt{-28}$
25. $-\sqrt{-24}$
26. $-\sqrt{-72}$
27. $5\sqrt{-81}$
28. $6\sqrt{-49}$
29. $\sqrt{-\frac{25}{9}}$
30. $\sqrt{-\frac{121}{144}}$

Solve each equation. Write all solutions in the form $a + bi$.

31. $x^2 + 9 = 0$
32. $x^2 + 100 = 0$
33. $x^2 = -36$
34. $x^2 = -49$
35. $x^2 + 8 = 0$
36. $x^2 + 27 = 0$
37. $x^2 = -\frac{16}{9}$
38. $x^2 = -\frac{25}{4}$
39. $x^2 - 3x + 4 = 0$
40. $y^2 + y + 3 = 0$
41. $2x^2 + x + 1 = 0$
42. $2x^2 + 3x + 3 = 0$
43. $x^2 + 2x + 2 = 0$
44. $x^2 - 2x + 6 = 0$
45. $3x^2 + 2x + 1 = 0$
46. $3x^2 - 4x + 2 = 0$
47. $3x^2 - 2x = -3$
48. $5x^2 = 2x - 1$

Perform the operations. Write all answers in the form $a + bi$.

49. $(3 + 4i) + (5 - 6i)$
50. $(5 + 3i) - (6 - 9i)$
51. $(7 - 3i) - (4 + 2i)$
52. $(8 + 3i) + (-7 - 2i)$
53. $(6 - i) + (9 + 3i)$
54. $(5 - 4i) - (3 + 2i)$
55. $(-3 - 8i) - (-3 - 9i)$
56. $(-1 - 8i) - (-1 - 7i)$
57. $(14 - 4i) - 9i$
58. $(20 - 5i) - 17i$
59. $15 + (-3 - 9i)$
60. $-25 + (18 - 9i)$
61. $3(2 - i)$
62. $9(-4 - 4i)$
63. $-4(3 + 4i)$
64. $-7(5 - 3i)$
65. $-5i(5 - 5i)$
66. $2i(7 + 2i)$
67. $(2 + i)(3 - i)$
68. $(4 - i)(2 + i)$
69. $(3 - 2i)(2 + 3i)$
70. $(3 - i)(2 + 3i)$

71. $(4 + i)(3 - i)$

72. $(1 - 5i)(1 - 4i)$

73. $(2 + i)^2$

74. $(3 - 2i)^2$

75. $\dfrac{5}{2 - i}$

76. $\dfrac{26}{3 - 2i}$

77. $\dfrac{3}{5 + i}$

78. $\dfrac{-4}{7 - 2i}$

79. $\dfrac{5i}{6 + 2i}$

80. $\dfrac{-4i}{2 - 6i}$

81. $\dfrac{2 + 3i}{2 - 3i}$

82. $\dfrac{2 - 5i}{2 + 5i}$

83. $\dfrac{3 + 2i}{3 + i}$

84. $\dfrac{3 - 2i}{3 + 2i}$

APPLICATIONS

85. ELECTRICITY In an AC (alternating current) circuit, if two sections are connected in series and have the same current in each section, the voltage V is given by $V = V_1 + V_2$. Find the total voltage in a given circuit if the voltages in the individual sections are $V_1 = 10.31 - 5.97i$ and $V_2 = 8.14 + 3.79i$.

86. ELECTRONICS The impedance Z in an AC (alternating current) circuit is a measure of how much the circuit impedes (hinders) the flow of current through it. The impedance is related to the voltage V and the current I by the formula

$$V = IZ$$

If a circuit has a current of $(0.5 + 2.0i)$ amps and an impedance of $(0.4 - 3.0i)$ ohms, find the voltage.

WRITING

87. What unusual situation discussed at the beginning of this section illustrated the need to define the square root of a negative number?

88. Explain the difference between the opposite of a complex number and its conjugate.

89. What is an imaginary number?

90. In this section, we have seen that $i^2 = -1$. From your previous experience in this course, what is unusual about that fact?

REVIEW

91. Add: $\dfrac{7}{x + 3} + \dfrac{4x}{x + 6}$.

92. Rationalize the denominator: $\dfrac{8}{\sqrt{10}}$.

93. Simplify: $\dfrac{x^2 - 5x + 6}{x - 3}$.

94. Simplify: $\dfrac{\frac{1}{a} + \frac{1}{b}}{\frac{a}{b} - \frac{b}{a}}$.

CHALLENGE PROBLEMS **Perform the operations.**

95. $(2\sqrt{2} + i\sqrt{2})(3\sqrt{2} - i\sqrt{2})$

96. $\dfrac{\sqrt{5} - i\sqrt{3}}{\sqrt{5} + i\sqrt{3}}$

9.4 Graphing Quadratic Equations

- Quadratic Equations in Two Variables
- Finding the Vertex of a Parabola
- Finding the Intercepts of a Parabola
- A Graphing Strategy
- Solving Quadratic Equations Graphically

We have previously graphed the equation $y = x^2$ by constructing a table of solutions, plotting points, and joining them with a smooth curve. The result was a cup-shaped figure called a **parabola.** In this section, we will use our experience with quadratic equations to develop a more comprehensive strategy for graphing parabolas.

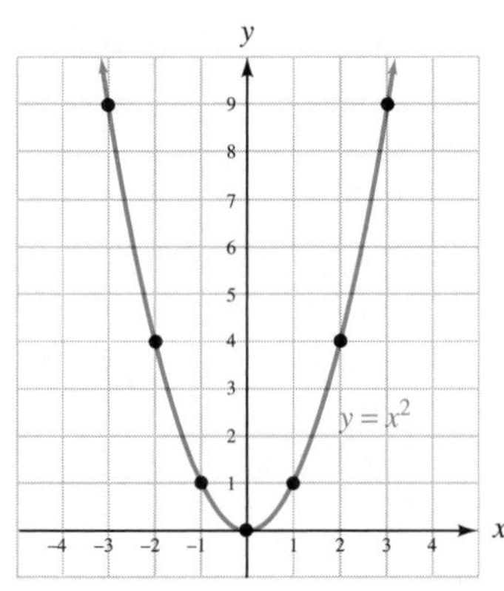

QUADRATIC EQUATIONS IN TWO VARIABLES

Equations that can be written in the form $y = ax^2 + bx + c$, where $a \neq 0$, are called **quadratic equations in two variables.** Some examples are

$$y = x^2 - 3, \qquad y = x^2 - 2x - 3, \qquad \text{and} \qquad y = -2x^2 - 8x - 8$$

To graph quadratic equations like these, we can plot points.

EXAMPLE 1

Graph: $y = x^2 - 3$.

ELEMENTARY Algebra f(x) Now™

Solution We can create a table of solutions by choosing values for x and finding the corresponding values of y. For example, if $x = -3$, we have

$$\begin{aligned} y &= x^2 - 3 \\ &= (-3)^2 - 3 \quad \text{Substitute } -3 \text{ for } x. \\ &= 9 - 3 \\ &= 6 \end{aligned}$$

The Language of Algebra

An axis of symmetry divides a parabola into two matching sides. The sides are said to be mirror images of each other.

The ordered pair $(-3, 6)$ and several others that satisfy the equation are listed in the table. To graph the equation, we plot each point and draw a smooth curve passing through them. The resulting parabola opens upward, and the lowest point on the graph, called the **vertex,** is the point $(0, -3)$. Note that the graph of $y = x^2 - 3$ looks just like the graph of $y = x^2$, except that it is 3 units lower.

$y = x^2 - 3$

x	y	(x, y)
−3	6	(−3, 6)
−2	1	(−2, 1)
−1	−2	(−1, −2)
0	−3	(0, −3)
1	−2	(1, −2)
2	1	(2, 1)
3	6	(3, 6)

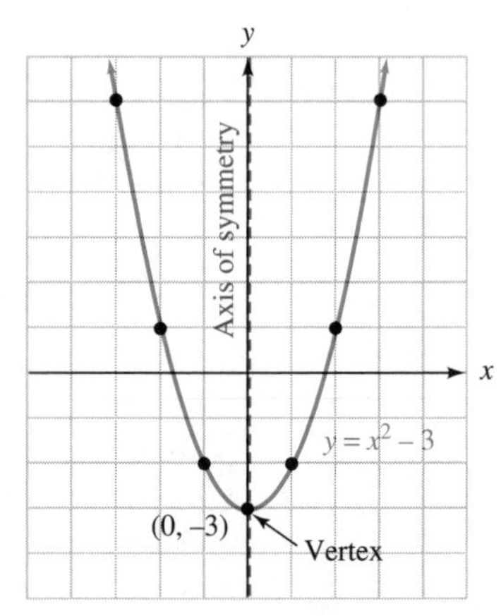

If we fold the graph paper along the y-axis, the two sides of the parabola will match. We call the vertical line through the vertex the **axis of symmetry.**

Self Check 1 Graph: $y = x^2 + 2$.

EXAMPLE 2

Graph: $y = -2x^2 - 4x + 2$.

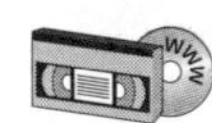

ELEMENTARY Algebra $f(x)$ Now™

Solution We construct a table of solutions, plot points, and draw the graph. The resulting parabola opens downward. The vertex is the highest point on the graph, $(-1, 4)$. The axis of symmetry is the vertical line that passes through the vertex.

$y = -2x^2 - 4x + 2$

x	y	(x, y)
-3	-4	$(-3, -4)$
-2	2	$(-2, 2)$
-1	4	$(-1, 4)$
0	2	$(0, 2)$
1	-4	$(1, -4)$

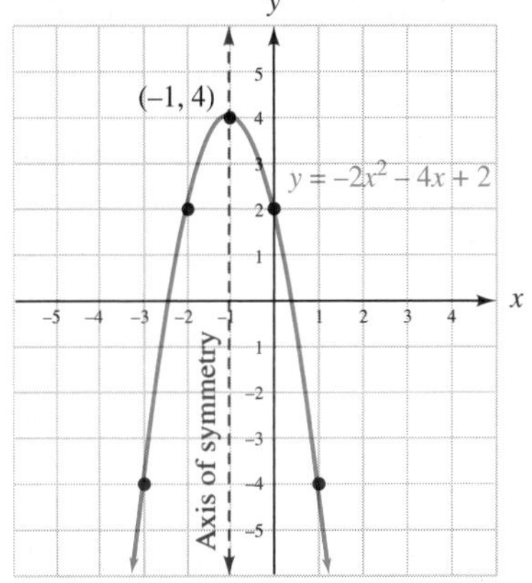

Self Check 2 Graph: $y = -x^2 - 4x - 4$.

In Example 1, the coefficient of the x^2 term in $y = x^2 - 3$ is the positive number 1. In Example 2, the coefficient of the x^2 term in $y = -2x^2 - 4x + 2$ is the negative number -2. These results illustrate the following fact.

Graphs of Quadratic Equations The graph of $y = ax^2 + bx + c$, where $a \neq 0$, is a parabola. It opens upward when $a > 0$ and downward when $a < 0$.

Parabolic shapes can be seen in a wide variety of real-world settings.

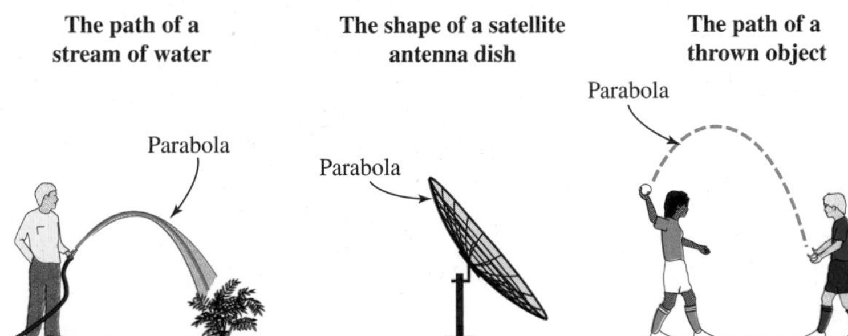

FINDING THE VERTEX OF A PARABOLA

It is usually easier to graph a quadratic equation when we know the coordinates of the vertex of its graph. For a parabola defined by $y = ax^2 + bx + c$, it can be shown that the x-coordinate of the vertex is given by $-\frac{b}{2a}$. This fact enables us to find the y-coordinate of the vertex, as well.

Finding the Vertex of a Parabola

The graph of the quadratic equation $y = ax^2 + bx + c$ is a parabola whose vertex has an x-coordinate of $-\frac{b}{2a}$. To find the y-coordinate of the vertex, substitute $-\frac{b}{2a}$ into the defining equation and find y.

EXAMPLE 3

Find the vertex of the graph of $y = x^2 - 2x - 3$.

ELEMENTARY Algebra f(x) Now™

Solution For $y = x^2 - 2x - 3$, we have $a = 1$, $b = -2$, and $c = -3$. To find the x-coordinate of the vertex, we substitute the values for a and b into the formula $x = -\frac{b}{2a}$.

$$x = -\frac{b}{2a}$$

$$x = -\frac{-2}{2(1)} \quad \text{Substitute 1 for } a \text{ and } -2 \text{ for } b.$$

$$= 1$$

The x-coordinate of the vertex is 1. To find the y-coordinate, we substitute 1 for x:

$$y = x^2 - 2x - 3$$

$$y = 1^2 - 2(1) - 3$$

$$= 1 - 2 - 3$$

$$= -4$$

The vertex of the parabola is the point $(1, -4)$.

Self Check 3 Find the vertex of the graph of $y = x^2 + 6x + 8$.

FINDING THE INTERCEPTS OF A PARABOLA

When graphing quadratic equations, it is often helpful to know the x- and y-intercepts of the parabola. To find the intercepts of a parabola, we use the same steps that we used to find the intercepts of the graphs of linear equations.

Finding Intercepts

To find the y-intercept, substitute 0 for x in the given equation and solve for y.

To find the x-intercepts, substitute 0 for y in the given equation and solve for x.

EXAMPLE 4

Find the y- and x-intercepts of the graph of $y = x^2 - 2x - 3$.

ELEMENTARY Algebra f(x) Now™

Solution To find the y-intercept of the parabola, we let $x = 0$ and solve for y.

$$y = x^2 - 2x - 3$$
$$y = 0^2 - 2(0) - 3 \quad \text{Substitute 0 for } x.$$
$$y = -3$$

The parabola passes through the point $(0, -3)$. We note that the y-coordinate of the y-intercept is the same as the value of the constant term c on the right-hand side of $y = x^2 - 2x - 3$.

To find the x-intercepts of the graph, we set y equal to 0 and solve the resulting quadratic equation.

$$y = x^2 - 2x - 3$$
$$0 = x^2 - 2x - 3 \quad \text{Substitute 0 for } y.$$
$$0 = (x - 3)(x + 1) \quad \text{Factor the trinomial.}$$
$$x - 3 = 0 \quad \text{or} \quad x + 1 = 0 \quad \text{Set each factor equal to 0.}$$
$$x = 3 \quad \Big| \quad x = -1$$

Since there are two solutions, the graph has two x-intercepts: $(3, 0)$ and $(-1, 0)$.

Self Check 4 Find the y- and x-intercepts of the graph of $y = x^2 + 6x + 8$.

A GRAPHING STRATEGY

Success Tip

When graphing, remember that any point on a parabola to the right of the axis of symmetry yields a second point to the left of the axis of symmetry, and vice versa.

Knowing the direction, vertex, axis of symmetry, and intercepts of a parabola is helpful when drawing its graph. For example, to graph $y = x^2 - 2x - 3$, we note that the coefficient of the x^2 term is positive ($a = 1$). Thus, the parabola opens upward. In Examples 3 and 4, we found that the vertex of the graph is $(1, -4)$, the y-intercept is $(0, -3)$, and x-intercepts are $(3, 0)$ and $(-1, 0)$. See part (a) of the figure.

We can locate other points on the parabola using the symmetry of the graph. If the point $(0, -3)$, which is 1 unit to the left of the axis of symmetry, is on the graph, the point $(2, -3)$, which is 1 unit to the right of the axis of symmetry, is also on the graph.

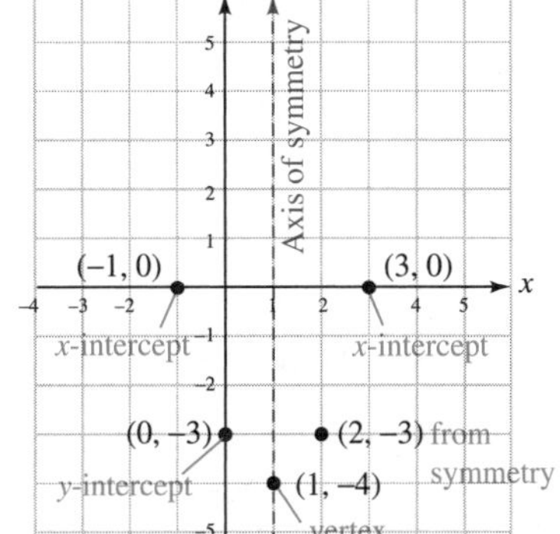

(a)

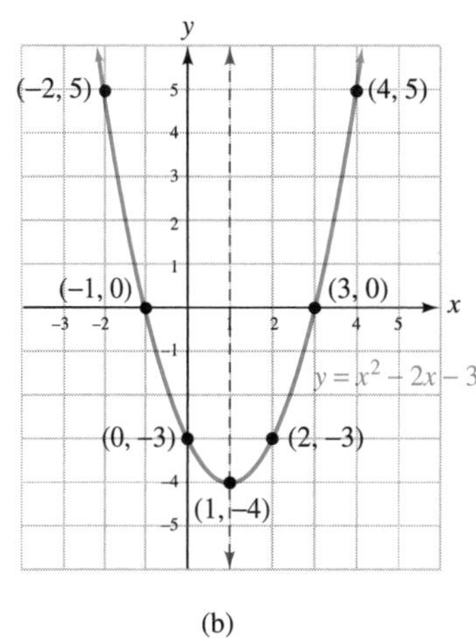

(b)

$y = x^2 - 2x - 3$

x	y	(x, y)
−2	5	(−2, 5)

We can complete the graph by plotting two more points. If $x = -2$, then $y = 5$, and the parabola passes through $(-2, 5)$. Again using symmetry, the parabola must also pass through $(4, 5)$. The completed graph of $y = x^2 - 2x - 3$ is shown in part (b) of the figure.

Much can be determined about the graph of $y = ax^2 + bx + c$ from the coefficients a, b, and c. This information is summarized as follows:

Graphing a Quadratic Equation $y = ax^2 + bx + c$

Determine whether the parabola opens upward or downward by examining a.

The x-coordinate of the vertex of the parabola is $x = -\frac{b}{2a}$.

To find the y-coordinate of the vertex, substitute $-\frac{b}{2a}$ for x into the equation and find y.

The axis of symmetry is the vertical line passing through the vertex.

The y-intercept $(0, y)$ is determined by the value of y when $x = 0$: the y-intercept is $(0, c)$.

The x-intercepts (if any) are determined by the numbers x that make $y = 0$. To find them, solve the quadratic equation $ax^2 + bx + c = 0$.

EXAMPLE 5

Graph: $y = -2x^2 - 8x - 8$.

ELEMENTARY Algebra $f(x)$ Now™

Solution

Step 1 *Determine whether the parabola opens upward or downward.* The equation is in the form $y = ax^2 + bx + c$, with $a = -2$, $b = -8$, and $c = -8$. Since $a < 0$, the parabola opens downward.

Step 2 *Find the vertex and draw the axis of symmetry.* To find the x-coordinate of the vertex, we substitute the values for a and b into the formula $x = -\frac{b}{2a}$.

$$x = -\frac{b}{2a}$$

$$x = -\frac{-8}{2(-2)} \quad \text{Substitute } -2 \text{ for } a \text{ and } -8 \text{ for } b.$$

$$= -2$$

The x-coordinate of the vertex is -2. To find the y-coordinate, we substitute -2 for x in the equation and find y.

$$y = -2x^2 - 8x - 8$$

$$y = -2(-2)^2 - 8(-2) - 8$$

$$= -8 + 16 - 8$$

$$= 0$$

The vertex of the parabola is the point $(-2, 0)$. This point is in blue in the graph.

Step 3 *Find the x- and y-intercepts.* Since $c = -8$, the y-intercept of the parabola is $(0, -8)$. The point $(-4, -8)$, two units to the left of the axis of symmetry, must also be on the graph. We plot both points in black in the graph.

To find the x-intercepts, we set y equal to 0 and solve the resulting quadratic equation.

$$y = -2x^2 - 8x - 8$$

$$0 = -2x^2 - 8x - 8 \quad \text{Set } y = 0.$$

$$0 = x^2 + 4x + 4 \quad \text{Divide both sides by } -2.$$

$$0 = (x + 2)(x + 2) \quad \text{Factor the trinomial.}$$

$$x + 2 = 0 \quad \text{or} \quad x + 2 = 0 \qquad \text{Set each factor equal to 0.}$$
$$x = -2 \quad | \quad x = -2$$

Since the solutions are the same, the graph has only one x-intercept: $(-2, 0)$. This point is the vertex of the parabola and has already been plotted.

Step 4 *Plot another point.* Finally, we find another point on the parabola. If $x = -3$, then $y = -2$. We plot $(-3, -2)$ and use symmetry to determine that $(-1, -2)$ is also on the graph. Both points are in green.

Step 5 Draw a smooth curve through the points, as shown.

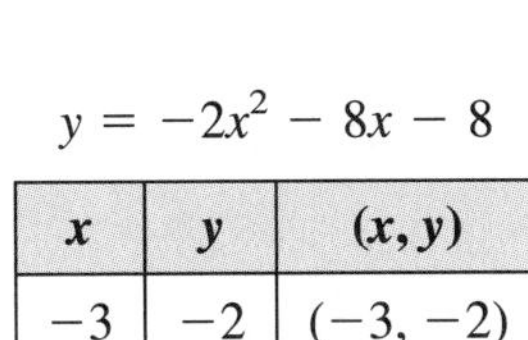

$y = -2x^2 - 8x - 8$

x	y	(x, y)
−3	−2	(−3, −2)

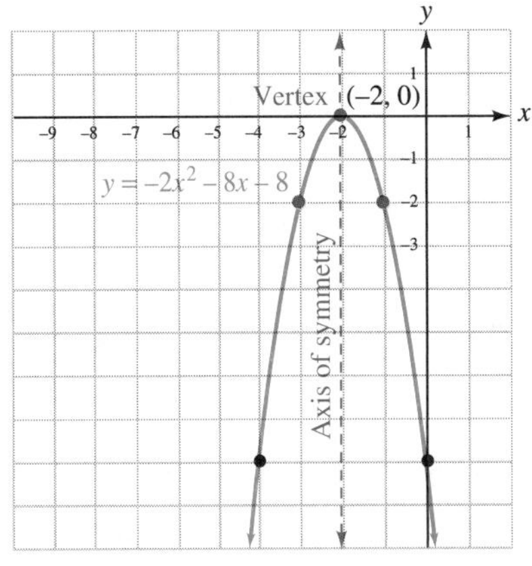

Self Check 5 Use your results from Self Checks 3 and 4 to help graph $y = x^2 + 6x + 8$.

SOLVING QUADRATIC EQUATIONS GRAPHICALLY

The number of x-intercepts of the graph of $y = ax^2 + bx + c$ is the same as the number of real-number solutions of $ax^2 + bx + c = 0$. For example, the graph of $y = x^2 + x - 2$ in part (a) of the following figure has two x-intercepts, and $x^2 + x - 2 = 0$ has two real-number solutions. In part (b), the graph has one x-intercept, and the corresponding equation has one real-number solution. In part (c), the graph does not have an x-intercept, and the corresponding equation does not have any real-number solutions. Note that the solutions of each equation are given by the x-coordinates of the x-intercepts of each respective graph.

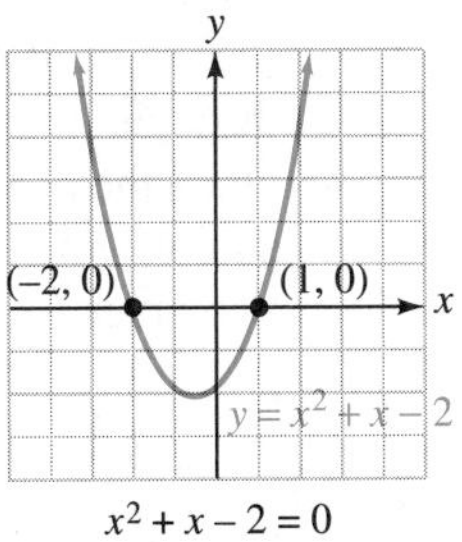

$x^2 + x - 2 = 0$ has two solutions, −2 and 1.

(a)

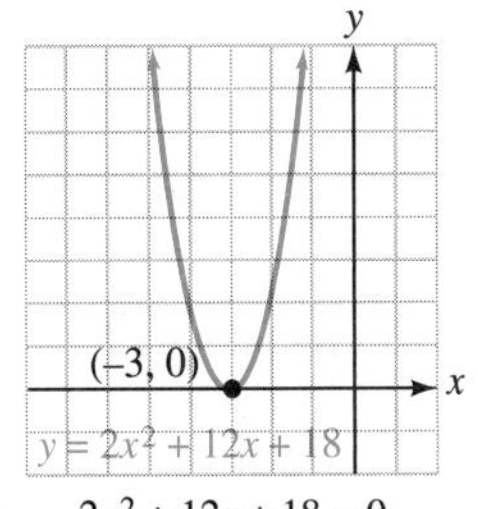

$2x^2 + 12x + 18 = 0$ has one solution, −3.

(b)

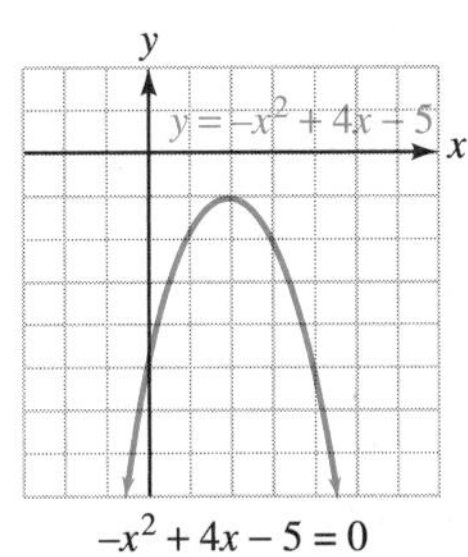

$-x^2 + 4x - 5 = 0$ has no real-number solutions.

(c)

Answers to Self Checks

1.

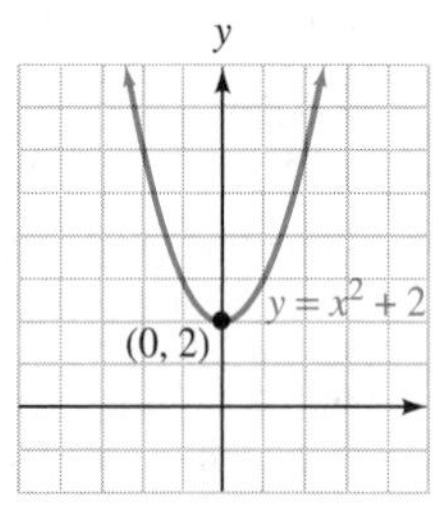

2.

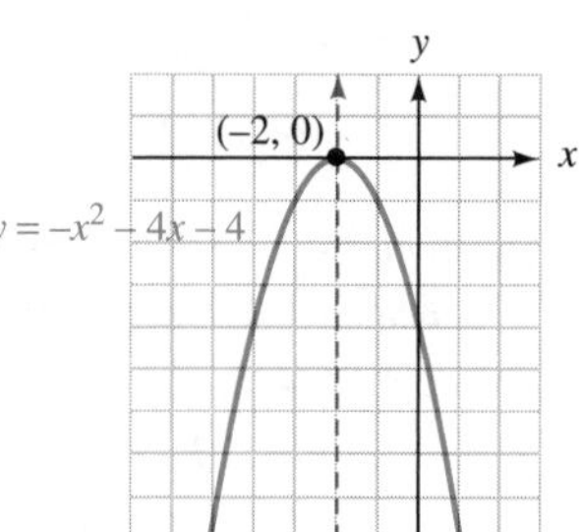

3. $(-3, -1)$

4. y-intercept: $(0, 8)$; x-intercepts: $(-2, 0)$, $(-4, 0)$

5.

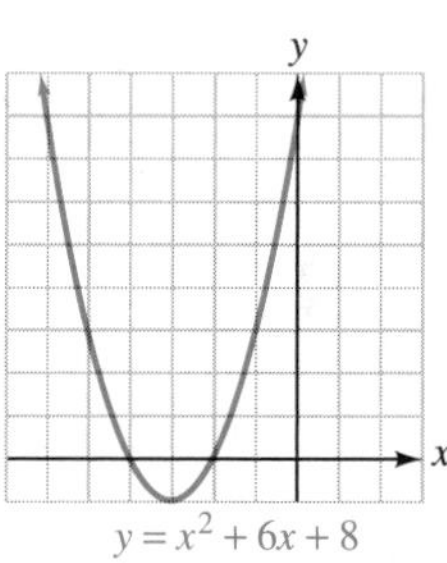

9.4 STUDY SET

ELEMENTARY Algebra $f(x)$ Now™

VOCABULARY Fill in the blanks.

1. $y = 3x^2 + 5x - 1$ is a ________ equation in two variables.
2. The graph of the equation $y = x^2 - 2x + 6$ is a cup-shaped figure called a ________.
3. The lowest point on a parabola that opens upward, and the highest point on a parabola that opens downward, is called the _______ of the parabola.
4. Points where a parabola intersects the x-axis are called the x-________ of the graph.
5. The point where a parabola intersects the y-axis is called the y-________ of the graph.
6. The vertical line that splits the graph of a parabola into two identical parts is called the axis of ________.

CONCEPTS Fill in the blanks.

7. a. The graph of $y = ax^2 + bx + c$ opens downward when a ☐ 0.
   b. The graph of $y = ax^2 + bx + c$ opens upward when $a >$ ☐.
8. The graph of $y = ax^2 + bx + c$ is a parabola whose vertex has an x-coordinate given by ☐.
9. a. To find the y-intercepts of a graph, substitute ☐ for x in the given equation and solve for y.
   b. To find the x-intercepts of a graph, substitute 0 for y in the given equation and solve for ☐.
10. $y = x^2 - 3x - 1$

x	y	(x, y)
3		(3,)

11. a. What do we call the curve shown in the graph?
   b. What are the x-intercepts of the graph?
   c. What is the y-intercept of the graph?
   d. What is the vertex?
   e. Draw the axis of symmetry on the graph.

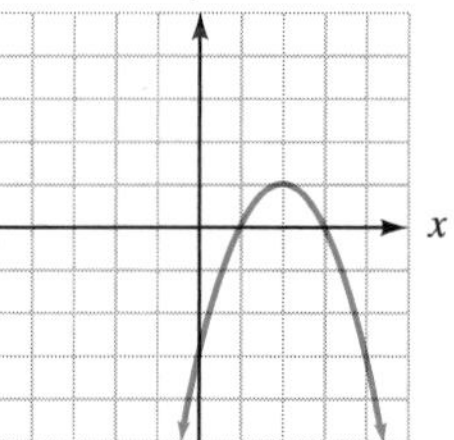

12. The vertex of a parabola is $(1, -3)$, its y-intercept is $(0, -2)$, and it passes through the point $(3, 1)$. Draw the axis of symmetry and use it to help determine two other points on the parabola.

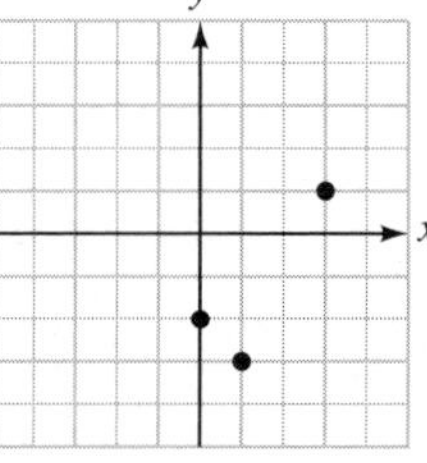

13. For $y = -x^2 + 6x - 7$, the value of $-\frac{b}{2a}$ is 3. Find the y-coordinate of the vertex of the graph of this equation.

14. Draw the graph of a quadratic equation using the given facts about its graph.

- Opens upward
- y-intercept: $(0, -3)$
- Vertex: $(-1, -4)$
- x-intercepts: $(-3, 0)$, $(1, 0)$
-

x	y	(x, y)
2	5	(2, 5)

15. Sketch the graphs of parabolas with zero, one, and two x-intercepts.

16. Sketch the graph of a parabola with equation of the form $y = ax^2 + bx + c$, if possible.

NOTATION

17. Consider the equation $y = 2x^2 + 4x - 8$.
  a. What are a, b, and c?
  b. Find $-\frac{b}{2a}$.

18. Evaluate: $-\frac{-12}{2(-3)}$.

19. Examine the graph of $y = x^2 + 2x - 3$. How many real-number solutions does the equation $x^2 + 2x - 3 = 0$ have? What are they?

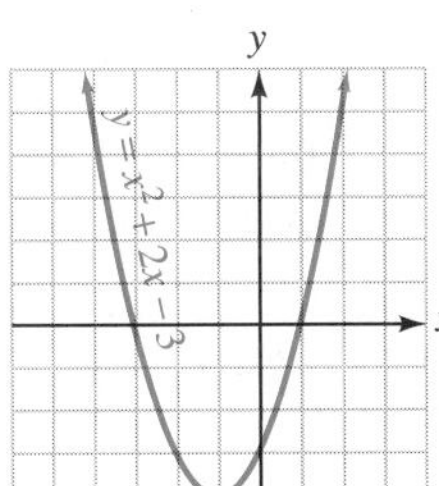

20. Examine the graph of $y = x^2 + 4x + 4$. How many real-number solutions does the equation $x^2 + 4x + 4 = 0$ have? What are they?

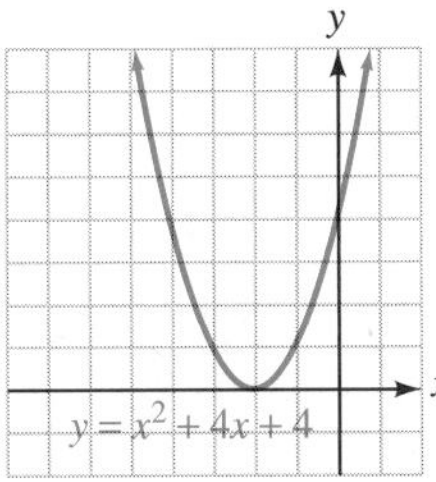

PRACTICE **Decide which direction the graph of each quadratic equation opens.**

21. $y = 2x^2 + 5x - 1$
22. $y = -x^2 - x - 4$
23. $y = -6x^2 - 3x + 5$
24. $y = 5x^2 + 4x - 2$

Find the vertex of the graph of each quadratic equation.

25. $y = 2x^2 - 4x + 1$
26. $y = 2x^2 + 8x - 4$
27. $y = -x^2 + 6x - 8$
28. $y = -x^2 - 2x - 1$

Find the x- and y-intercepts of the graph of the quadratic equation.

29. $y = x^2 - 2x + 1$
30. $y = 2x^2 - 4x$
31. $y = -x^2 - 10x - 21$
32. $y = 3x^2 + 6x - 9$

Graph each quadratic equation. Use the method discussed in Example 5.

33. $y = x^2 + 2x - 3$
34. $y = x^2 + 6x + 5$
35. $y = 2x^2 + 8x + 6$
36. $y = 3x^2 - 12x + 9$
37. $y = -x^2 + 2x + 3$
38. $y = -x^2 + 5x - 4$
39. $y = -x^2 - 2x - 1$
40. $y = -x^2 + 2x - 1$
41. $y = x^2 - 2x$
42. $y = x^2 + x$
43. $y = x^2 + 4x + 4$
44. $y = x^2 - 6x + 9$
45. $y = -x^2 + 2x$
46. $y = -x^2 - 4x$
47. $y = 2x^2 + 3x - 2$
48. $y = 3x^2 - 7x + 2$
49. $y = 4x^2 - 12x + 9$
50. $y = -2x^2 + 4x$

APPLICATIONS

51. BIOLOGY Draw an axis of symmetry over the sketch of the butterfly.

52. CROSSWORD PUZZLES Darken the appropriate squares to the right of the dashed blue line so that the puzzle has symmetry with respect to that line.

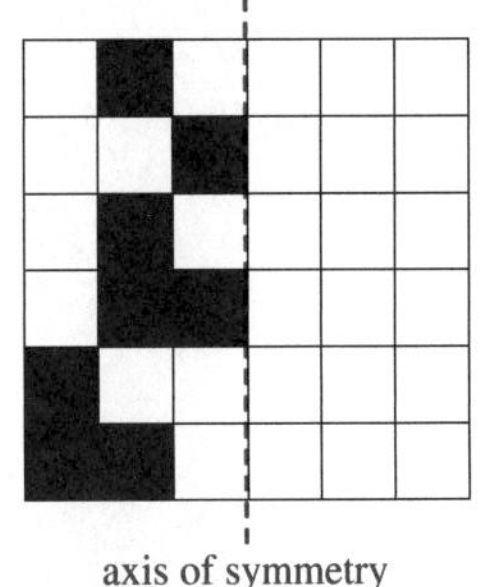

53. COST ANALYSIS A company has found that when it assembles x carburetors in a production run, the manufacturing cost of $\$y$ per carburetor is given by the graph on the next page. What important piece of information does the vertex give?

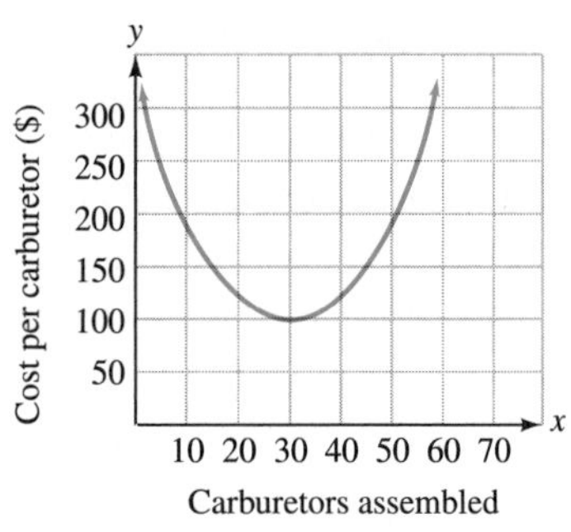

54. HEALTH DEPARTMENT The number of cases of flu seen by doctors at a county health clinic each week during a 10-week period is described by the graph. Write a brief summary report about the flu outbreak. What important piece of information does the vertex give?

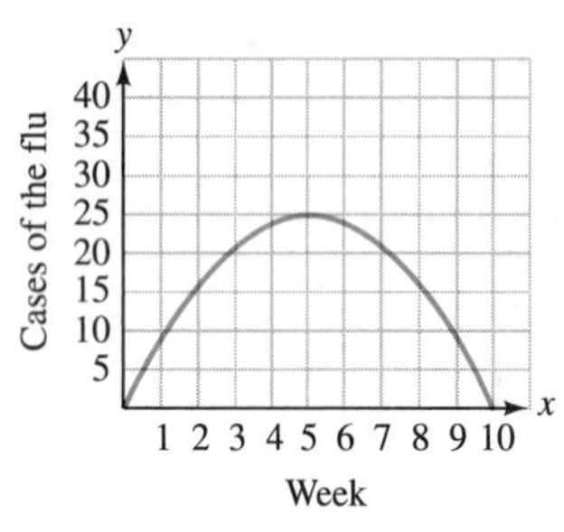

55. TRAMPOLINES The graph shows how far a trampolinist is from the ground (in relation to time) as she bounds into the air and then falls back down to the trampoline.

a. How many feet above the ground is she $\frac{1}{2}$ second after bounding upward?

b. When is she 9 feet above the ground?

c. What is the maximum number of feet above the ground she gets? When does this occur?

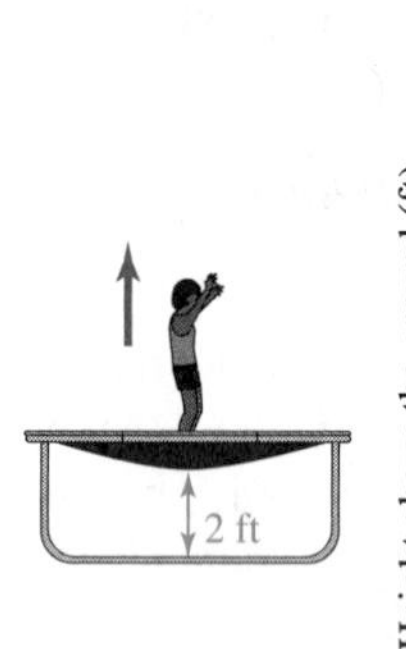

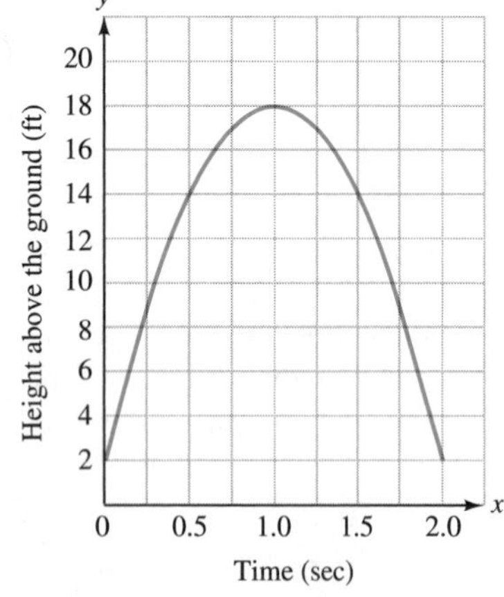

56. PROJECTILE If we disregard air resistance and other outside factors, the path of a projectile, such as a kicked soccer ball, is parabolic. Suppose the path of the soccer ball after it is kicked is given by the quadratic equation $y = -0.5x^2 + 2x$. Plot the points and draw a smooth curve through them to depict the path of the ball.

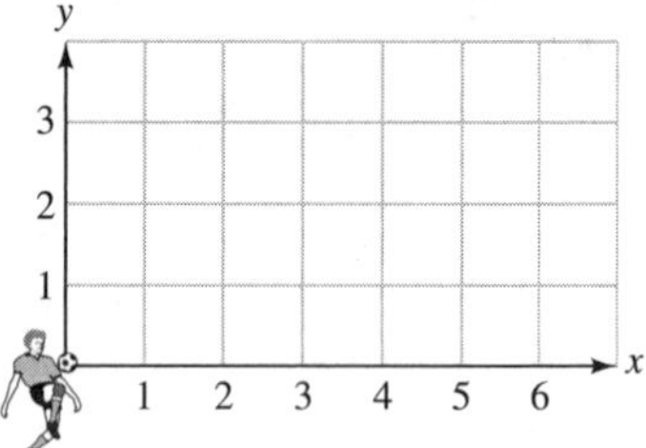

x	0	0.5	1	1.5	2	2.5	3	3.5	4
y									

57. BRIDGES The shapes of the suspension cables in certain types of bridges are parabolic. The suspension cable for the bridge shown in the illustration is described by $y = 0.005x^2$. Finish the mathematical model of the bridge by completing the table of solutions, plotting the points, and drawing a smooth curve through them to represent the cable. Finally, from each plotted point, draw a vertical support cable attached to the roadway.

x	−80	−60	−40	−20	0	20	40	60	80
y									

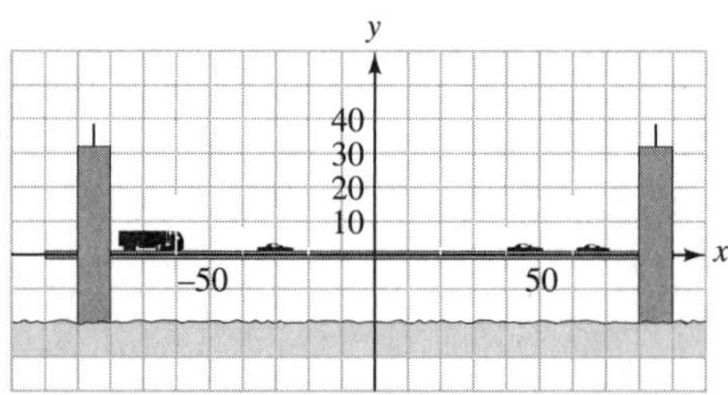

58. TRACK AND FIELD Sketch the parabolic path traveled by the long-jumper's center of gravity from the take-off board to the landing. Let the x-axis represent the ground.

WRITING

59. A mirror is held against the y-axis of the graph of a quadratic equation. What fact about parabolas does this illustrate?

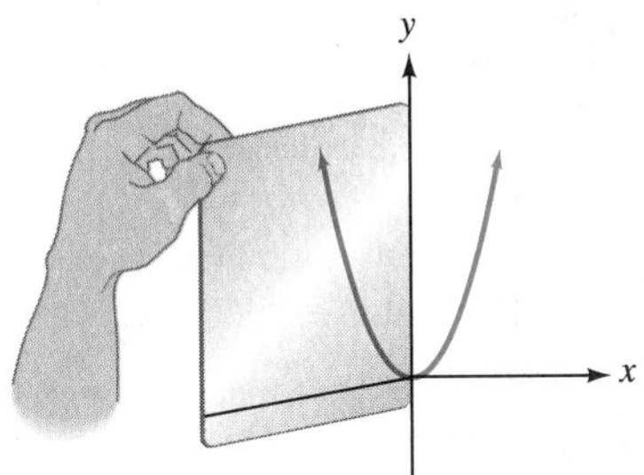

60. Use the example of a stream of water from a drinking fountain to explain the concept of the vertex of a parabola.

61. Explain why the y-intercept of the graph of the quadratic equation $y = ax^2 + bx + c$ is $(0, c)$.

62. **a.** Is it possible for the graph of a parabola not to have an x-intercept? Explain.

b. Is it possible for the graph of a parabola with equation of the form $y = ax^2 + bx + c$ not to have a y-intercept? Explain.

REVIEW Simplify each expression.

63. $\sqrt{8} - \sqrt{50} + \sqrt{72}$

64. $(4\sqrt{x})(-2\sqrt{x})$

65. $3\sqrt{z}(\sqrt{4z} - \sqrt{z})$

66. $\sqrt[3]{27y^3z^6}$

CHALLENGE PROBLEMS

67. Explain how to use the graph of $y = x^2 - x - 2$ to solve the quadratic equation $0 = x^2 - x - 2$.

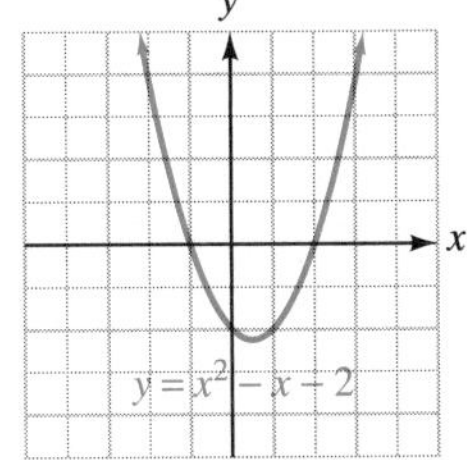

68. What quadratic equation is shown in the graph?

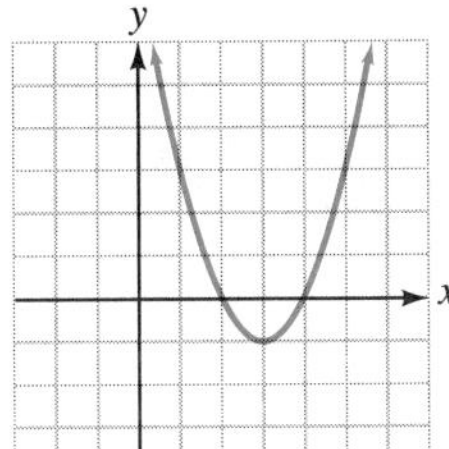

9.5 An Introduction to Functions

- Functions; Domain and Range
- Functions Defined by Equations
- Function Notation
- The Graph of a Function
- The Vertical Line Test
- An Application

The concept of a *function* is one of the most important ideas in all of mathematics. In this section, we will discuss some basic vocabulary and notation associated with functions. To introduce this topic, let's look at a table that we might see on television or printed in a newspaper.

FUNCTIONS; DOMAIN AND RANGE

The following table shows the number of medals won by American athletes at six recent Winter Olympics.

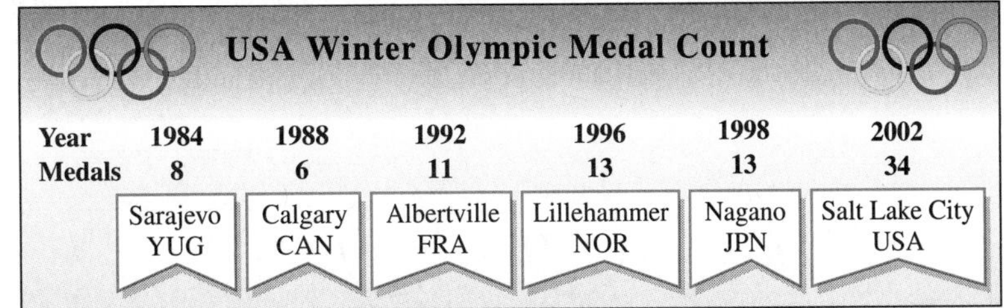

USA Winter Olympic Medal Count

Year	1984	1988	1992	1996	1998	2002
Medals	8	6	11	13	13	34
	Sarajevo YUG	Calgary CAN	Albertville FRA	Lillehammer NOR	Nagano JPN	Salt Lake City USA

For each year listed in the table, there corresponds exactly one medal count. Such a correspondence is an example of a *function.*

Functions A **function** is a rule that assigns to each value of one variable (called the **independent variable**) exactly one value of another variable (called the **dependent variable**).

Domain and Range The set of all possible values that can be used for the independent variable is called the **domain.** The set of all values of the dependent variable is called the **range.**

An **arrow** or **mapping diagram** can be used to show how a function assigns to each member of the domain exactly one member of the range. For the Winter Olympics example, we have

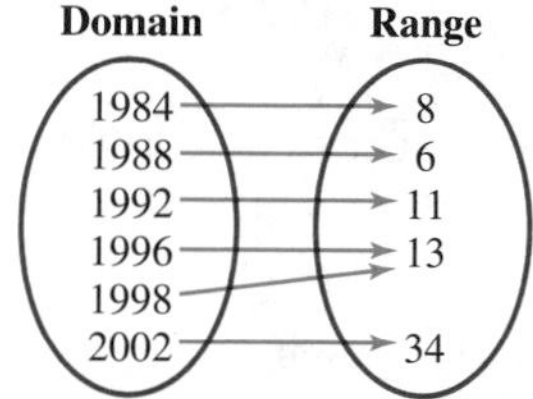

We can restate the definition of a function using the variables x and y.

y* is a Function of *x If to each value of x in the domain there is assigned exactly one value of y in the range, then y is said to be a function of x.

EXAMPLE 1

ELEMENTARY Algebra $f(x)$ Now™

Decide whether the arrow diagram and the tables define y as a function of x.

a.

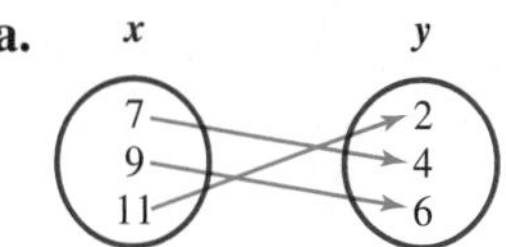

b.

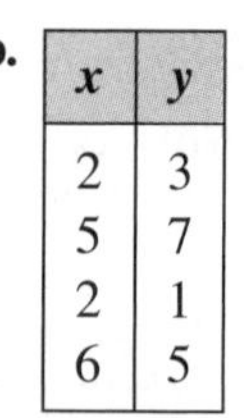

x	y
2	3
5	7
2	1
6	5

c.

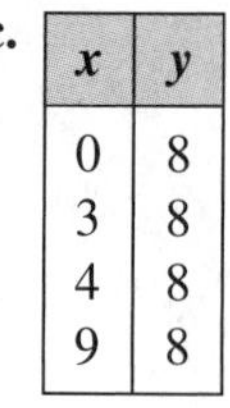

x	y
0	8
3	8
4	8
9	8

Solution

a. The arrow diagram defines a function because to each x-value there is assigned exactly one y-value: $7 \to 4$, $9 \to 6$, and $11 \to 2$.

b. The table does not define a function, because more than one y-value is assigned to the x-value 2. In the first row, 3 is assigned to 2, and in the third row 1 is also assigned to 2.

c. Since to each x-value exactly one y-value is assigned, the table defines a function. It also illustrates an important fact about functions: *To different values of x, the same value of y can be assigned.* In this case, each x-value is assigned the y-value 8.

Self Check 1 Tell whether the arrow diagram and the table define y as a function of x.

a.

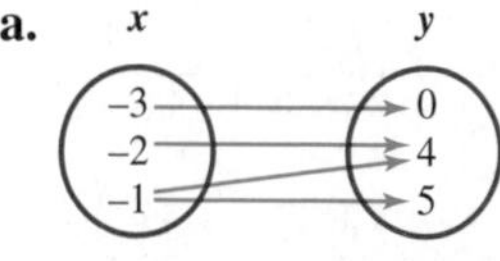

b.

x	y
−6	−6
5	5
8	8

FUNCTIONS DEFINED BY EQUATIONS

A function can be defined by an equation. For example, $y = 2x - 3$ is a rule that assigns to each value of x exactly one value of y. To find the y-value that is assigned to the x-value 4, we substitute 4 for x and evaluate the right-hand side of the equation.

$$
\begin{aligned}
y &= 2x - 3 \\
y &= 2(4) - 3 && \text{Substitute 4 for } x. \\
&= 8 - 3 \\
&= 5
\end{aligned}
$$

The function $y = 2x - 3$ assigns the y-value 5 to an x-value of 4. When making such calculations, the value of x is called an **input** and its corresponding value of y is called an **output.**

FUNCTION NOTATION

A special notation is used to name functions that are defined by equations.

Function Notation The notation $y = f(x)$ denotes that the variable y is a function of x.

Since $y = f(x)$, the equations $y = 2x - 3$ and $f(x) = 2x - 3$ are equivalent. We read $f(x) = 2x - 3$ as "f of x is equal to $2x$ minus 3."

Caution

The symbol $f(x)$ denotes a function. It does not mean $f \cdot x$ (f times x). Read $f(x)$ as f of x.

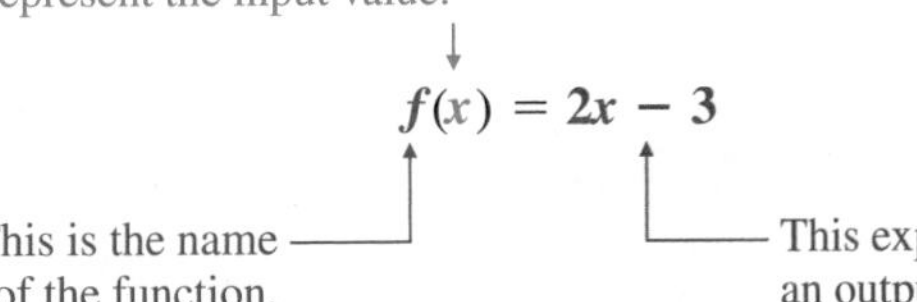

Function notation provides a compact way of denoting the value that is assigned to some number x. For example, if $f(x) = 2x - 3$, the value that is assigned to an x-value 5 is represented by $f(5)$.

$$
\begin{aligned}
f(x) &= 2x - 3 && \text{This is the function.} \\
f(5) &= 2(5) - 3 && \text{Substitute the input of 5 for each } x. \\
&= 10 - 3 && \text{Evaluate the right-hand side.} \\
&= 7 && \text{The output is 7.}
\end{aligned}
$$

Thus, $f(5) = 7$. We read this as "f of 5 is 7." The output 7 is called a **function value.**

EXAMPLE 2

ELEMENTARY Algebra $f(x)$ Now™

For $f(x) = 5x + 7$, find each of the following function values: **a.** $f(2)$, **b.** $f(-4)$, and **c.** $f(0)$.

Solution

a. To find $f(2)$, we substitute the number within the parentheses, 2, for each x in $f(x) = 5x + 7$, and evaluate the right-hand side of the equation.

$$
\begin{aligned}
f(x) &= 5x + 7 \\
f(2) &= 5(2) + 7 && \text{Substitute the input of 2 for each } x.
\end{aligned}
$$

$= 10 + 7$

$= 17$ The output is 17.

Thus, $f(2) = 17$.

The Language of Algebra

Another way to read $f(2) = 17$ is to say "the value of the function is 17 at 2."

b. $f(x) = 5x + 7$

$f(-4) = 5(-4) + 7$ Substitute the input of -4 for each x.

$= -20 + 7$ Evaluate the right-hand side.

$= -13$ The output is -13.

Thus, $f(-4) = -13$.

c. $f(x) = 5x + 7$

$f(0) = 5(0) + 7$ Substitute the input of 0 for each x.

$= 0 + 7$ Evaluate the right-hand side.

$= 7$ The output is 7.

Thus, $f(0) = 7$.

Self Check 2 For $f(x) = -2x + 3$, find each of the following function values: **a.** $f(4)$, **b.** $f(-1)$, and **c.** $f(0)$.

We can think of a function as a machine that takes some input x and turns it into some output $f(x)$, as shown in part (a) of the figure. In part (b), the function machine for $f(x) = x^2 + 2x$ turns the input 4 into the output $4^2 + 2(4) = 24$, and we have $f(4) = 24$.

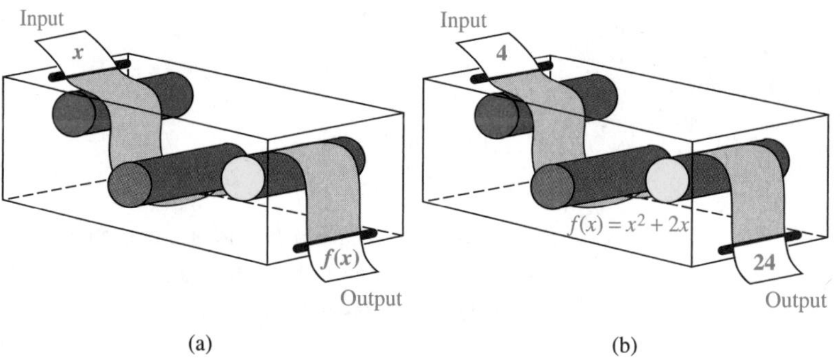

(a) (b)

The letter f used in the notation $y = f(x)$ represents the word *function.* However, other letters can be used to name functions. For example, $y = g(x)$ and $y = h(x)$ also denote functions of x.

EXAMPLE 3

For $g(x) = 3 - 2x$ and $h(x) = x^3 - 1$, find **a.** $g(3)$ and **b.** $h(-2)$.

Solution **a.** To find $g(3)$, we use the function rule $g(x) = 3 - 2x$ and replace x with 3.

$$g(x) = 3 - 2x$$
$$g(3) = 3 - 2(3)$$
$$= 3 - 6$$
$$= -3$$

Thus, $g(3) = -3$.

b. To find $h(-2)$, we use the function rule $h(x) = x^3 - 1$ and replace x with -2.

$$\begin{aligned} h(x) &= x^3 - 1 \\ h(-2) &= (-2)^3 - 1 \\ &= -8 - 1 \\ &= -9 \end{aligned}$$

Thus, $h(-2) = -9$.

Self Check 3 Find $g(0)$ and $h(4)$ for the functions in Example 3.

THE GRAPH OF A FUNCTION

We have seen that a function such as $f(x) = 4x + 1$ assigns to each value of x a single value $f(x)$. The input-output pairs generated by a function can be written in the form $(x, f(x))$. These ordered pairs can be plotted on a rectangular coordinate system to give the graph of the function.

EXAMPLE 4

Graph: $f(x) = 4x + 1$.

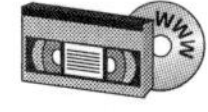

ELEMENTARY Algebra $f(x)$ Now™

Solution We can graph the function by creating a table of function values, plotting the corresponding ordered pairs, and drawing the graph. To make a table, we choose several values for x and find the corresponding values of $f(x)$. If x is -1, we have

$$\begin{aligned} f(x) &= 4x + 1 && \text{This is the function to graph.} \\ f(-1) &= 4(-1) + 1 && \text{Substitute } -1 \text{ for each } x. \\ &= -4 + 1 \\ &= -3 \end{aligned}$$

Thus, $f(-1) = -3$. This means that, when x is -1, $f(x)$ is -3, and it indicates that the ordered pair $(-1, -3)$ lies on the graph of $f(x)$.

Similarly, we find the corresponding values of $f(x)$ for x-values of 0 and 1. Then we plot the resulting ordered pairs and draw a straight line through them to get the graph of $f(x) = 4x + 1$. Since $y = f(x)$, the graph of $f(x) = 4x + 1$ is the same as the graph of the equation $y = 4x + 1$.

$f(x) = 4x + 1$

x	$f(x)$	
−1	−3	→ (−1, −3)
0	1	→ (0, 1)
1	5	→ (1, 5)

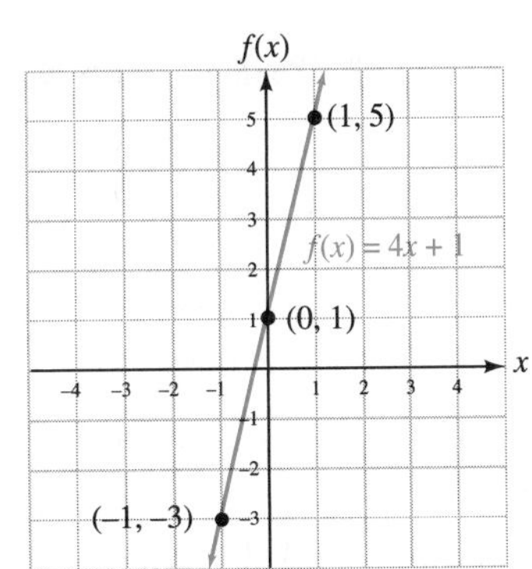

The vertical axis may be labeled y or $f(x)$.

Self Check 4 Graph: $f(x) = -3x - 2$.

We call $f(x) = 4x + 1$ from Example 4 a **linear function** because its graph is a nonvertical straight line. Any linear equation, except those of the form $x = a$, can be written using function notation by writing it in slope–intercept form ($y = mx + b$) and then replacing y with $f(x)$.

EXAMPLE 5

Graph: $f(x) = |x|$.

ELEMENTARY Algebra f(x) Now™

Solution To create a table of function values, we choose values for x and find the corresponding values of $f(x)$. For $x = -4$ and $x = 3$, we have

$$\begin{aligned} f(x) &= |x| \\ f(-4) &= |-4| \\ &= 4 \end{aligned} \qquad \begin{aligned} f(x) &= |x| \\ f(3) &= |3| \\ &= 3 \end{aligned}$$

Thus, $f(-4) = 4$ and $f(3) = 3$.

Similarly, we find the corresponding values of $f(x)$ for several other x-values. When we plot the resulting ordered pairs, we see that they lie in a "V" shape. We join the points to complete the graph as shown. We call $f(x) = |x|$ an **absolute value function.**

$f(x) = |x|$

x	$f(x)$	
−4	4	→ (−4, 4)
−3	3	→ (−3, 3)
−2	2	→ (−2, 2)
−1	1	→ (−1, 1)
0	0	→ (0, 0)
1	1	→ (1, 1)
2	2	→ (2, 2)
3	3	→ (3, 3)
4	4	→ (4, 4)

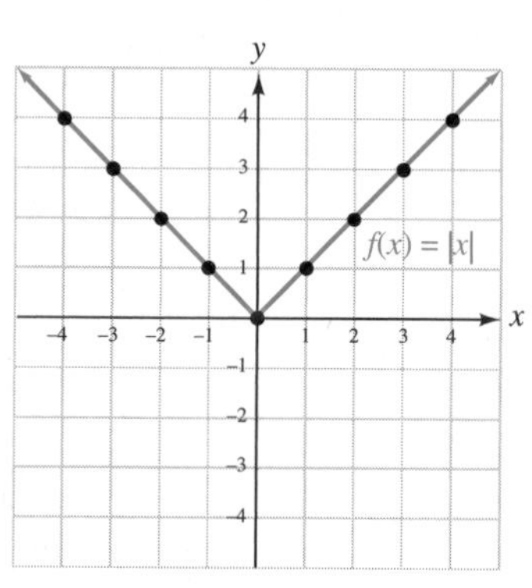

Self Check 5 Graph: $f(x) = |x| + 2$.

THE VERTICAL LINE TEST

If any vertical line intersects a graph more than once, the graph cannot represent a function, because to one value of x there would be assigned more than one value of y.

The Vertical Line Test If a vertical line intersects a graph in more than one point, the graph is not the graph of a function.

The graph shown in red on the next page does not represent a function, because a vertical line intersects the graph at more than one point. The points of intersection indicate that the x-value -1 is assigned two different y-values, 3 and -1.

x	y
-1	3
-1	-1

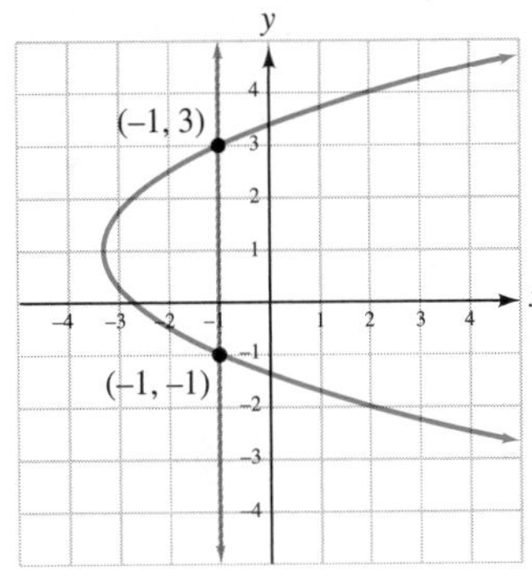

EXAMPLE 6

ELEMENTARY Algebra f(x) Now™

Determine whether each of the following is the graph of a function.

a.

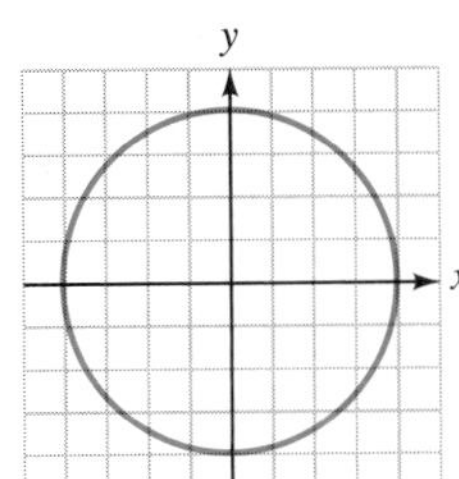

b.

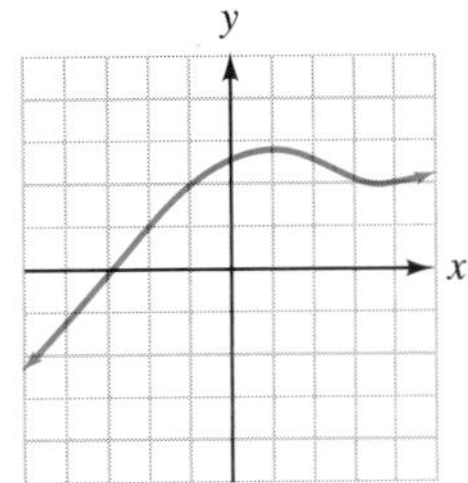

Solution **a.** This is not the graph of a function because the vertical line shown in blue intersects the graph at more than one point. The points of intersection indicate that the x-value 3 is assigned two different y-values, 2.5 and -2.5.

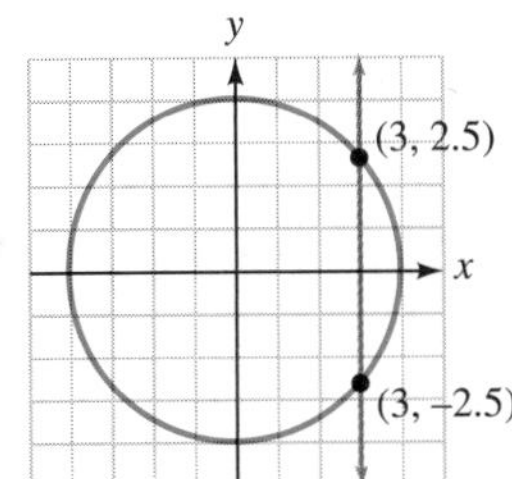

b. This is a graph of a function because no vertical line intersects the graph at more than one point. Several vertical lines are drawn in blue to illustrate this.

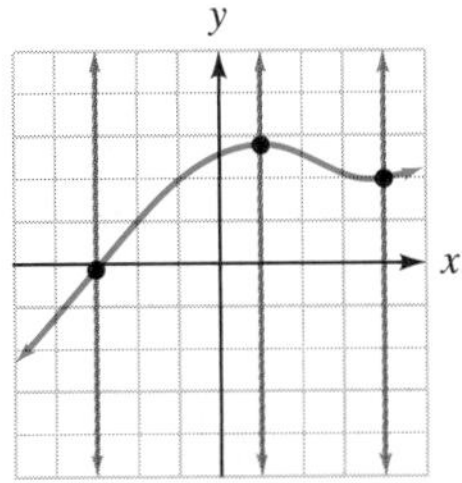

Self Check 6 Determine whether each of the following is the graph of a function.

a.

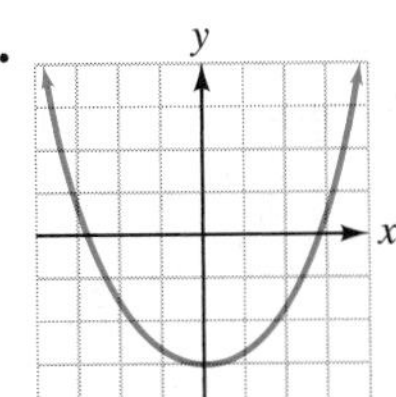

b.

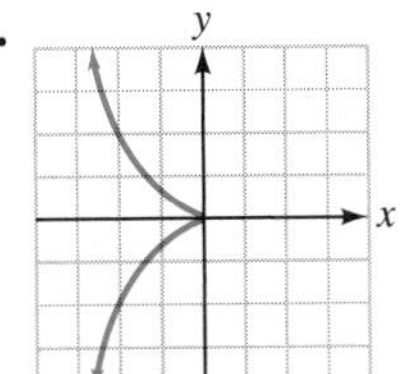

AN APPLICATION

Functions are used to describe certain relationships where one quantity depends upon another. Letters other than f and x are often chosen to more clearly describe these situations.

EXAMPLE 7

ELEMENTARY Algebra $f(x)$ Now™

Party rentals. The function $C(h) = 40 + 5(h - 4)$ gives the cost in dollars to rent an inflatable jumper for h hours. Find the cost of renting the jumper for 10 hours.

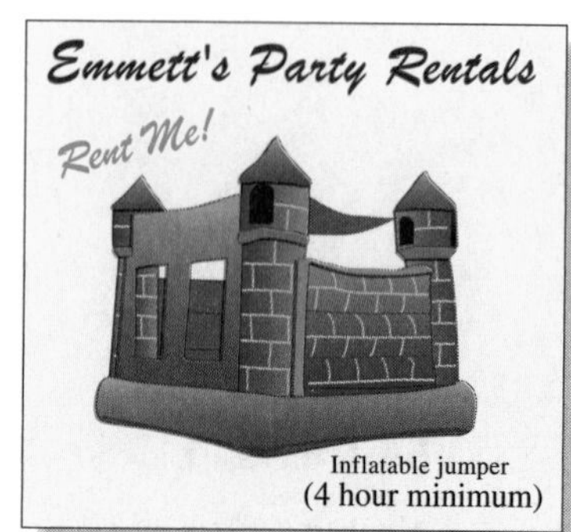

Solution For this application involving hours and cost, the notation $C(h)$ is used. The independent variable is h and the name of the function is C. If the jumper is rented for 10 hours, then h is 10 and we must find $C(10)$.

$$
\begin{aligned}
C(h) &= 40 + 5(h - 4) \\
C(10) &= 40 + 5(10 - 4) && \text{Substitute 10 for each } h. \\
&= 40 + 5(6) \\
&= 40 + 30 \\
&= 70
\end{aligned}
$$

It costs \$70 to rent the jumper for 10 hours.

Self Check 7 Find the cost of renting the jumper for 8 hours.

Answers to Self Checks **1. a.** no, **b.** yes **2. a.** -5, **b.** 5, **c.** 3 **3.** 3, 63

4. $f(x)$, x, $f(x) = -3x - 2$

5. y, x, $y = |x| + 2$

6. a. function, **b.** not a function **7.** \$60

9.5 STUDY SET

ELEMENTARY Algebra $f(x)$ Now™

VOCABULARY **Fill in the blanks.**

1. A ________ is a rule that assigns to each value of one variable (called the independent variable) exactly one value of another variable (called the dependent variable).
2. The set of all possible input values for a function is called the ________, and the set of all output values is called the ________.
3. We can think of a function as a machine that takes some ________ x and turns it into some ________ $f(x)$.
4. If $f(2) = -3$, we call -3 a function ________.
5. The graph of a ________ function is a straight line.
6. The graph of an ________ ________ function is V-shaped.

CONCEPTS

7. FEDERAL MINIMUM WAGE The following table is an example of a function. Use an arrow diagram to show how members of the range are assigned to members of the domain.

Year	1990	1992	1994	1996	1998	2000	2002
Minimum wage ($)	3.80	4.25	4.25	4.75	5.15	5.15	5.15

Source: *Time Almanac 2004*

8. The arrow diagram describes a function.

a. What is the domain of the function?

b. What is the range of the function?

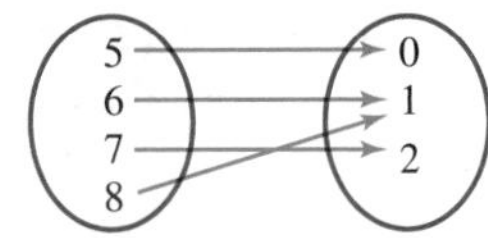

9. For the given input, what value will the function machine output?

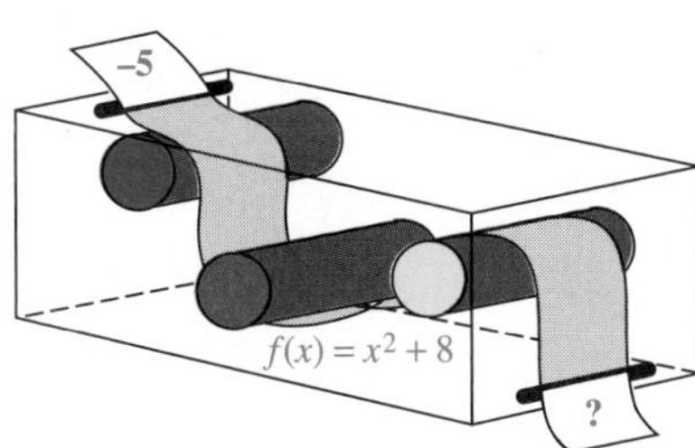

10. For the table of function values, give the corresponding ordered pairs.

$f(x) = x + 1$

x	y	
−3	−2	→
0	1	→
2	3	→

11. Fill in the blank: If a ______ line intersects a graph in more than one point, the graph is not the graph of a function.

12. a. Give the coordinates of the points where the given vertical line intersects the graph.

b. Is this the graph of a function? Explain.

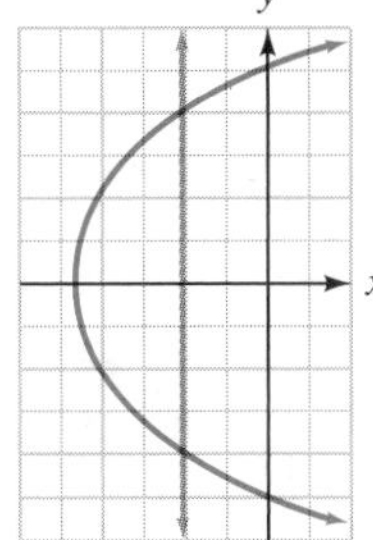

NOTATION Fill in the blanks.

13. We read $f(x) = 5x - 6$ as "f ____ x is $5x$ minus 6."

14. Since $y =$ ____, the equations

$$y = 3x + 2 \text{ and } f(x) = 3x + 2$$

are equivalent.

15. The notation $f(4) = 5$ indicates that when the x-value ____ is input into a function rule, the output is ____. This fact can be shown graphically by plotting the ordered pair (____ , ____).

16. When graphing the function $f(x) = -x + 5$, the vertical axis of the coordinate system can be labeled ____ or ____.

PRACTICE Determine whether each arrow diagram or table defines y as a function of x. If it does not, find ordered pairs that show a value of x that is assigned more than one value of y.

17.

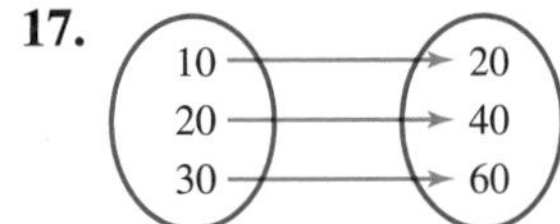

18.

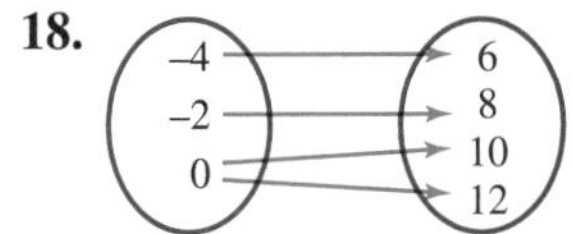

19.

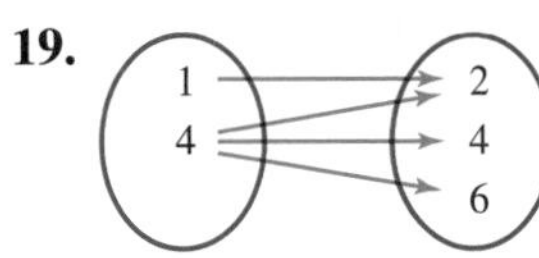

20.

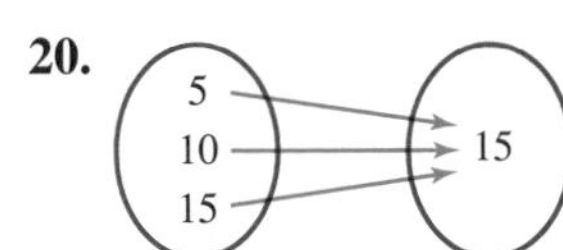

21.

x	y
1	7
2	15
3	23
4	16
5	8

22.

x	y
30	2
30	4
30	6
30	8
30	10

23.

x	y
−4	6
−1	0
0	−3
2	4
−1	2

24.

x	y
1	1
2	2
3	3
4	4

25.

x	y
3	4
3	−4
4	3
4	−3

26.

x	y
−1	1
−3	1
−5	1
−7	1
−9	1

Find each function value.

27. $f(x) = 4x - 1$
a. $f(1)$ **b.** $f(-2)$
c. $f\left(\frac{1}{4}\right)$ **d.** $f(50)$

28. $f(x) = 1 - 5x$
a. $f(0)$ **b.** $f(-75)$
c. $f(0.2)$ **d.** $f\left(-\frac{4}{5}\right)$

29. $f(x) = 2x^2$
a. $f(0.4)$ **b.** $f(-3)$
c. $f(1{,}000)$ **d.** $f\left(\frac{1}{8}\right)$

30. $g(x) = 6 - x^2$
a. $g(30)$ **b.** $g(6)$
c. $g(-1)$ **d.** $g(0.5)$

31. $h(x) = |x - 7|$
a. $h(0)$ **b.** $h(-7)$
c. $h(7)$ **d.** $h(8)$

32. $f(x) = |2 + x|$
a. $f(0)$ **b.** $f(2)$
c. $f(-2)$ **d.** $f(-99)$

33. $g(x) = x^3 - x$
a. $g(1)$ **b.** $g(10)$
c. $g(-3)$ **d.** $g(6)$

34. $g(x) = x^4 + x$
a. $g(1)$ **b.** $g(-2)$
c. $g(0)$ **d.** $g(10)$

35. If $f(x) = 3.4x^2 - 1.2x + 0.5$, find $f(-0.3)$.

36. If $g(x) = x^4 - x^3 + x^2 - x$, find $g(-12)$.

Determine whether each graph is the graph of a function. If it is not, find ordered pairs that show a value of x that is assigned more than one value of y.

37.
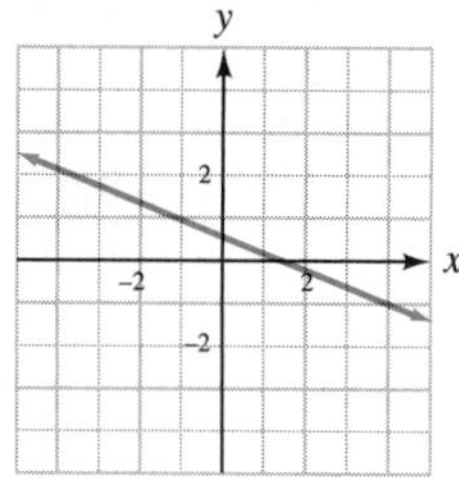

38.
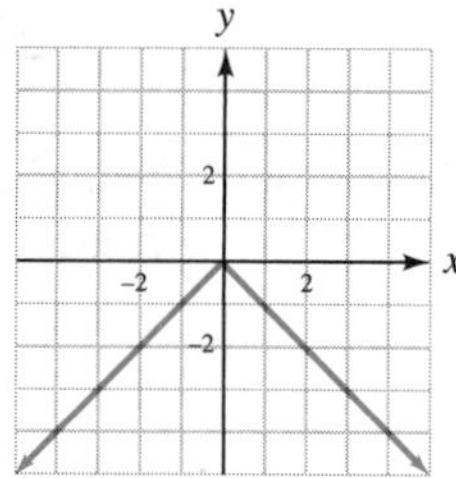

39.
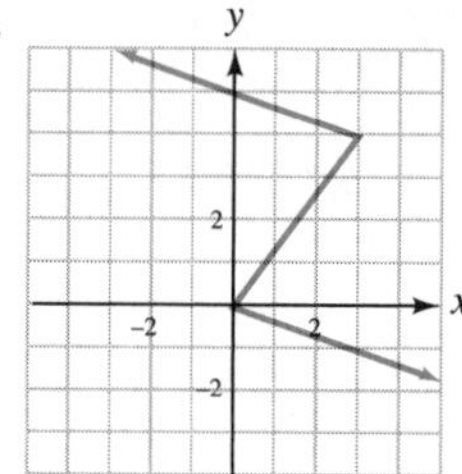

40.
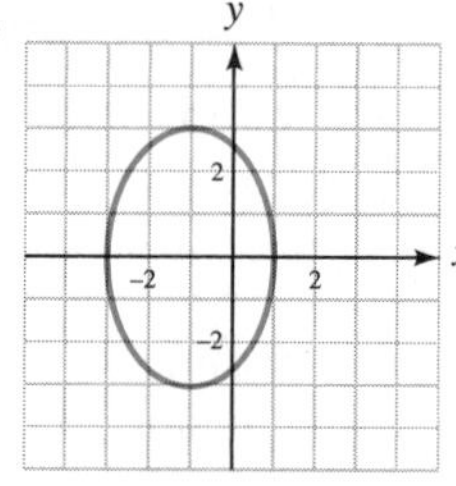

41.
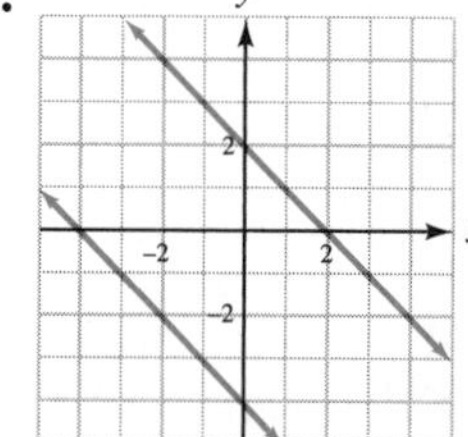

42.
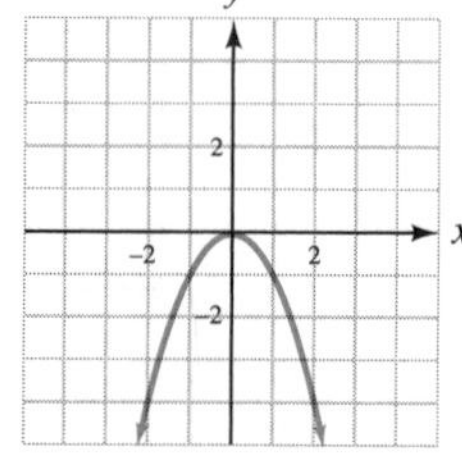

Complete each table of function values and then graph each function.

43. $f(x) = -2 - 3x$

x	$f(x)$
0	
1	
−1	
−2	

44. $h(x) = |1 - x|$

x	$h(x)$
0	
1	
2	
3	
−1	
−2	

Graph each function.

45. $f(x) = \frac{1}{2}x - 2$

46. $f(x) = -\frac{2}{3}x + 3$

47. $s(x) = 2 - x^2$

48. $g(x) = 1 + x^3$

APPLICATIONS

49. REFLECTIONS When a beam of light hits a mirror, it is reflected off the mirror at the same angle that the incoming beam struck the mirror. What type of function could serve as a mathematical model for the path of the light beam shown here?

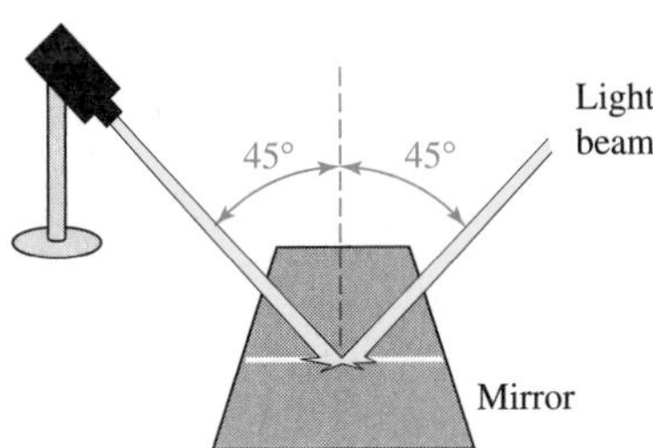

50. VACATIONING The function

$$C(d) = 500 + 100(d - 3)$$

gives the cost in dollars to rent an RV motor home for d days. Find the cost of renting the RV for a vacation that will last 7 days.

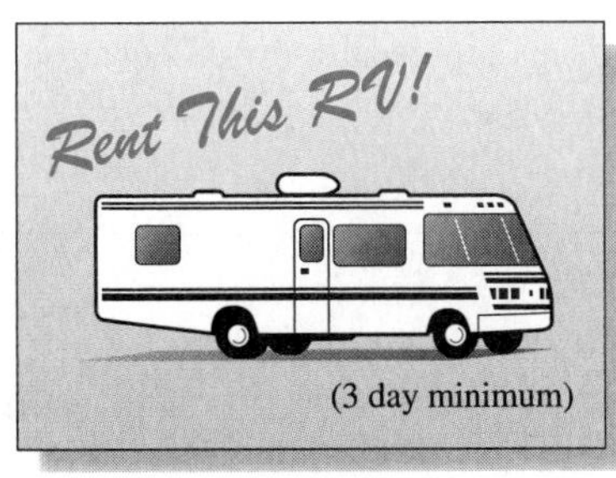

51. LIGHTNING The function $D(t) = \frac{t}{5}$ gives the approximate distance in miles that you are from a lightning strike, where t is the number of seconds between seeing the lightning and hearing the thunder. Find $D(5)$ and explain what it means.

52. STRUCTURAL ENGINEERING The maximum safe load in pounds of the rectangular beam shown in the figure is given by the function $S(t) = \frac{1{,}875t^2}{8}$, where t is the thickness of the beam, in inches. Find the maximum safe load if the beam is 4 inches thick.

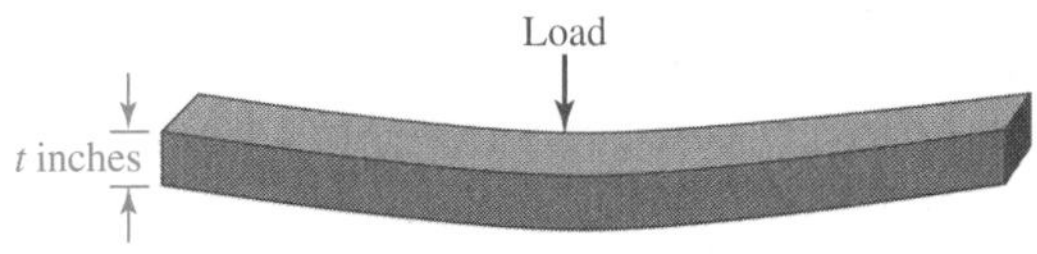

53. LAWN SPRINKLERS The function $A(r) = \pi r^2$ can be used to determine the area that will be watered by a rotating sprinkler that sprays out a stream of water r feet. Find $A(5)$ and $A(20)$. Round to the nearest tenth.

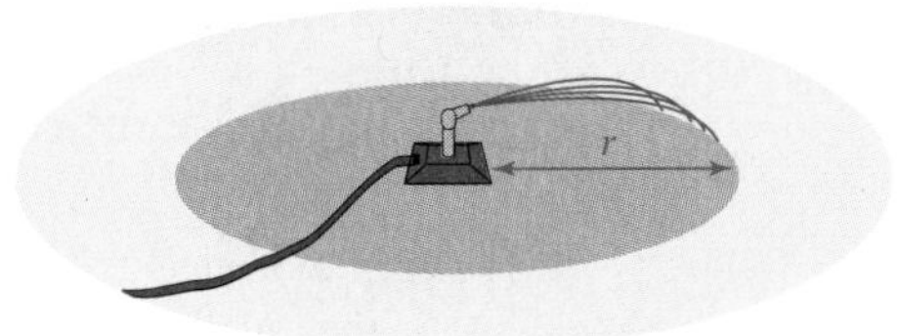

54. PARTS LISTS The function

$$f(r) = 2.30 + 3.25(r + 0.40)$$

approximates the length (in feet) of the belt that joins the two pulleys, where r is the radius (in feet) of the smaller pulley. Find the belt length needed for each pulley in the parts list.

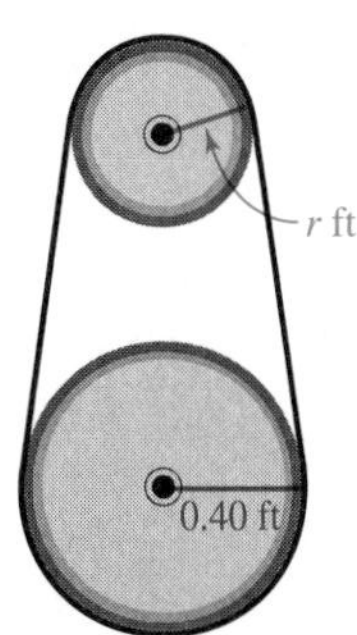

Parts list		
Pulley	***r***	**Belt length**
P-45M	0.32	
P-08D	0.24	

WRITING

55. In the function $y = -5x + 2$, why do you think x is called the *independent* variable and y the *dependent* variable?

56. Explain what a politician meant when she said, "The speed at which the downtown area will be redeveloped is a function of the number of low-interest loans made available to the property owners."

57. A student was asked to determine whether the graph on the right is the graph of a function. What is wrong with the following reasoning?

When I draw a vertical line through the graph, it intersects the graph only once. By the vertical line test, this is the graph of a function.

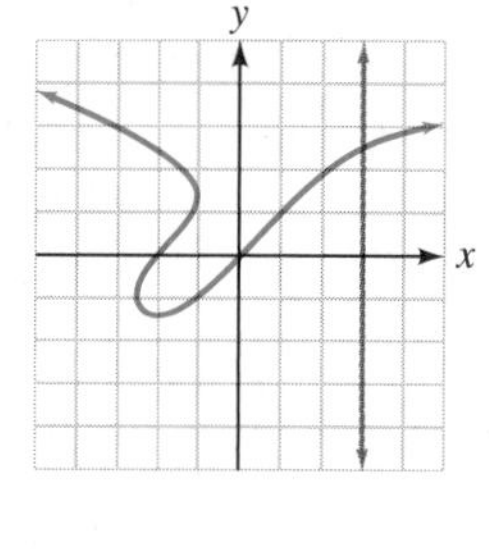

58. In your own words, what is a function?

REVIEW PROBLEMS

59. COFFEE BLENDS A store sells regular coffee for \$4 a pound and gourmet coffee for \$7 a pound. To get rid of 40 pounds of the gourmet coffee, the shopkeeper plans to make a gourmet blend that he will put on sale for \$5 a pound. How many pounds of regular coffee should be used?

60. PHOTOGRAPHIC CHEMICALS A photographer wishes to mix 2 liters of a 5% acetic acid solution with a 10% solution to get a 7% solution. How many liters of 10% solution must be added?

CHALLENGE PROBLEMS

61. Is the graph of $y \geq 3 - x$ a function? Explain.

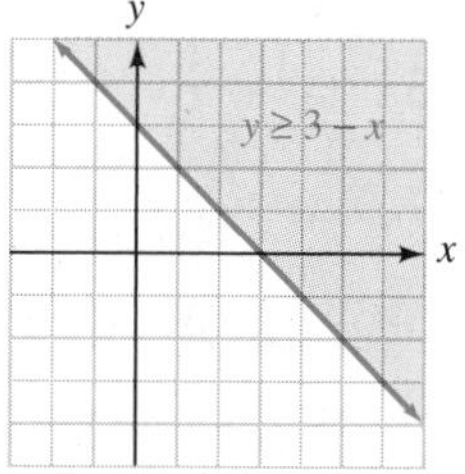

62. If $f(x) = x^2 + x$, find $f(r + 1)$.

9.6 Variation

- Direct Variation
- Inverse Variation
- Joint Variation
- Combined Variation

If the value of one quantity depends on the value of another quantity, we can often describe that relationship using the language of variation:

- The sales tax on an item varies with the price.
- The intensity of light varies with the distance from its source.
- The pressure exerted by water on an object varies with the depth of the object beneath the surface.

In this section, we will discuss two types of variation and see how to represent them algebraically.

DIRECT VARIATION

One type of variation, called **direct variation,** is represented by an equation of the form $y = kx$, where k is a constant (a number). Two variables are said to *vary directly* if one is a constant multiple of the other.

Direct Variation The words *y varies directly with x* mean that

$$y = kx$$

for some constant k, called the **constant of variation.**

Scientists have found that the distance a spring will stretch varies directly with the force applied to it. The more force applied to the spring, the more it will stretch. If d represents the distance stretched and f represents the force applied, this relationship can be expressed by the equation.

$d = kf$ where k is the constant of variation

Suppose that a 150-pound weight stretches a spring 18 inches.

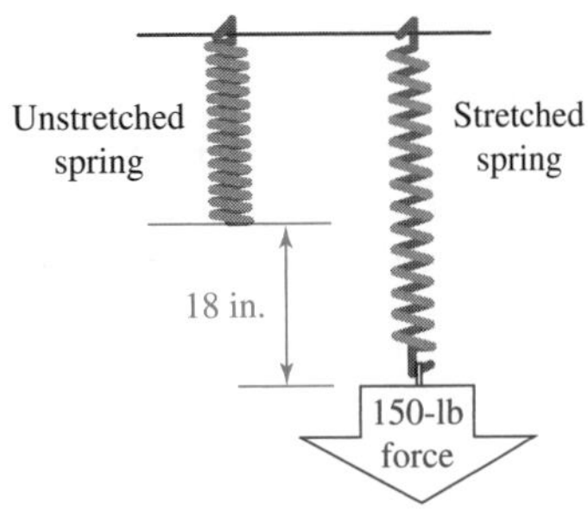

We can find the constant of variation for the spring by substituting 150 for f and 18 for d in the equation $d = kf$ and solving for k:

$$d = kf$$

$$18 = k(150)$$

$$\frac{18}{150} = k \quad \text{Divide both sides by 150 to isolate } k.$$

$$\frac{3}{25} = k \quad \text{Simplify the fraction: } \frac{18}{150} = \frac{\overset{1}{\cancel{6}} \cdot 3}{\underset{1}{\cancel{6}} \cdot 25} = \frac{3}{25}.$$

Therefore, the equation describing the relationship between the distance the spring will stretch and the amount of force applied to it is $d = \frac{3}{25}f$. To find the distance that the same spring will stretch when a 50-pound weight is used, we proceed as follows:

$$d = \frac{3}{25}f \quad \text{This is the equation describing the direct variation.}$$

$$d = \frac{3}{25}(50) \quad \text{Substitute 50 for } f.$$

$$d = 6$$

The spring will stretch 6 inches when the 50-pound weight is used.

The following table shows some other possible values for f and d as determined by the equation $d = \frac{3}{25}f$. When these ordered pairs are graphed and a straight line is drawn through them, it is apparent that as the force f applied to a spring increases, the distance d it stretches increases. Furthermore, the slope of the graph is $\frac{3}{25}$, the constant of variation.

$$d = \frac{3}{25}f$$

f	d
0	0
25	3
50	6
75	9
100	12

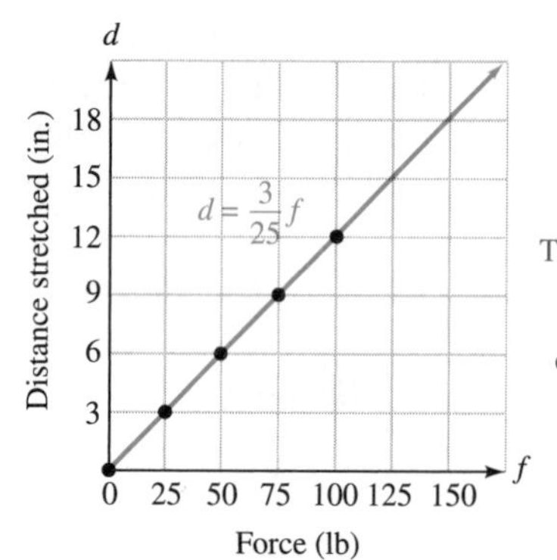

This straight-line graph shows that d varies directly with f.

We can use the following steps to solve variation problems.

Solving Variation Problems

To solve a variation problem:

1. Translate the verbal model into an equation.
2. Substitute the first set of values into the equation from step 1 to determine the value of k.
3. Substitute the value of k into the equation from step 1.
4. Substitute the remaining set of values into the equation from step 3 and solve for the unknown variable.

EXAMPLE 1

Comparing weights. The weight of an object on Earth varies directly with its weight on the moon. If a rock weighed 5 pounds on the moon and 30 pounds on Earth, what would be the weight on Earth of a larger rock weighing 26 pounds on the moon?

Solution

Step 1: We let e represent the weight of the object on Earth and m the weight of the object on the moon. Translating the words *weight on Earth varies directly with weight on the moon,* we get the equation

$$e = km$$

Step 2: To find the constant of variation, k, we substitute 30 for e and 5 for m.

$$e = km$$
$$30 = k(5)$$
$$6 = k \quad \text{Divide both sides by 5.}$$

Step 3: The equation describing the relationship between the weight of an object on Earth and on the moon is

$$e = 6m$$

Step 4: We can find the weight of the larger rock on Earth by substituting 26 for m in the equation from step 3.

$$e = 6m$$
$$e = 6(26)$$
$$e = 156$$

The rock woul/d weigh 156 pounds on Earth.

Self Check 1 The cost of a bus ticket varies directly with the number of miles traveled. If a ticket for a 180-mile trip cost \$45, what would a ticket for a 1,500-mile trip cost?

INVERSE VARIATION

Another type of variation, called **inverse variation,** is represented by an equation of the form $y = \frac{k}{x}$, where k is a constant. Two variables are said to *vary inversely* if one is a constant multiple of the reciprocal of the other.

Inverse Variation The words *y varies inversely with x* mean that

$$y = \frac{k}{x}$$

for some constant k, called the **constant of variation.**

Suppose that the time (in hours) it takes to paint a house varies inversely with the size of the painting crew. As the number of painters increases, the time that it takes to paint the house decreases. If n represents the number of painters and t represents the time it takes to paint the house, this relationship can be expressed by the equation

$$t = \frac{k}{n} \quad \text{where } k \text{ is the constant of variation}$$

If we know that a crew of 8 can paint the house in 12 hours, we can find the constant of variation by substituting 8 for n and 12 for t in the equation $t = \frac{k}{n}$ and solving for k:

$$t = \frac{k}{n}$$

$$12 = \frac{k}{8}$$

$$12 \cdot 8 = k \quad \text{Multiply both sides by 8 to isolate } k.$$

$$96 = k$$

The equation describing the relationship between the size of the painting crew and the time it takes to paint the house is $t = \frac{96}{n}$. We can use this equation to find the time it will take a crew of any size to paint the house. For example, to find the time it would take a four-person crew, we substitute 4 for n in the equation $t = \frac{96}{n}$.

$$t = \frac{96}{n} \quad \text{This is the equation describing the inverse variation.}$$

$$t = \frac{96}{4} \quad \text{Substitute 4 for } n.$$

$$t = 24$$

It would take a four-person crew 24 hours to paint the house.

The following table shows some possible values for n and t as determined by the equation $t = \frac{96}{n}$. When these ordered pairs are graphed and a smooth curve is drawn through them, it is clear that as the number of painters n increases, the time t decreases.

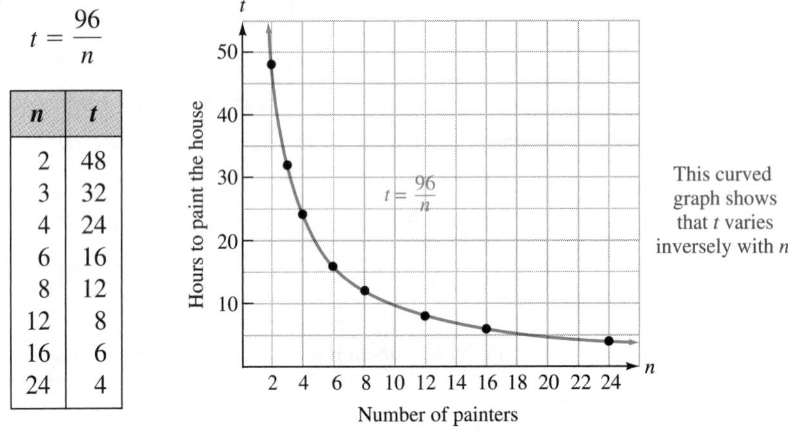

$$t = \frac{96}{n}$$

n	t
2	48
3	32
4	24
6	16
8	12
12	8
16	6
24	4

This curved graph shows that t varies inversely with n.

EXAMPLE 2

Gas law. The volume occupied by a gas varies inversely with the pressure placed on it. That is, the volume decreases as the pressure increases. If a gas occupies a volume of 15 cubic inches when placed under 4 pounds per square inch (psi) of pressure, how much pressure is needed to compress the gas into a volume of 10 cubic inches?

Solution ***Step 1:*** We let V represent the volume occupied by the gas and p represent the pressure. Translating the words *volume occupied by a gas varies inversely with the pressure,* we get the equation

$$V = \frac{k}{p}$$

Step 2: To find the constant of variation, k, we substitute 15 for V and 4 for p.

$$V = \frac{k}{p}$$

$$15 = \frac{k}{4}$$

$$60 = k \quad \text{Multiply both sides by 4.}$$

Step 3: The equation describing the relationship between the volume occupied by the gas and the pressure placed on it is

$$V = \frac{60}{p}$$

Step 4: We can now find the pressure needed to compress the gas into a volume of 10 cubic inches by substituting 10 for V in the equation and solving for p.

$$V = \frac{60}{p}$$

$$10 = \frac{60}{p}$$

$10p = 60$ To clear the equation of the fraction, multiply both sides by p.

$p = 6$ Divide both sides by 10.

It will take 6 psi of pressure to compress the gas into a volume of 10 cubic inches.

Self Check 2 How much pressure is needed to compress the gas in Example 2 into a volume of 8 cubic inches?

JOINT VARIATION

Joint Variation The words *y varies jointly with x and z* mean that

$$y = kxz$$

for some constant k. The constant k is the **constant of variation.**

The area of a rectangle depends on its length l and its width w by the formula

$$A = lw$$

We could say that the area of the rectangle varies jointly with its length and its width. In this example, the constant of variation is $k = 1$.

EXAMPLE 3

ELEMENTARY Algebra $f(x)$ Now™

The area of a triangle varies jointly with the length of its base and its height. If a triangle with an area of 63 square inches has a base of 18 inches and a height of 7 inches, find the area of a triangle with a base of 12 inches and a height of 10 inches.

Solution We let A represent the area of the triangle, b represent the length of the base, and h represent the height. We translate the words *area varies jointly with the length of the base and the height* into the formula

$$A = kbh$$

We are given that $A = 63$ when $b = 18$ and $h = 7$. To find k, we substitute these values into the previous equation and solve for k.

$$A = k\mathbf{bh}$$

$$63 = k(\mathbf{18})(\mathbf{7})$$

$$63 = k(126)$$

$\frac{63}{126} = k$ Divide both sides by 126.

$\frac{1}{2} = k$ Simplify.

Thus, $k = \frac{1}{2}$, and the formula for finding the area is

$$A = \frac{1}{2}bh$$

To find the area of a triangle with a base of 12 inches and a height of 10 inches, we substitute 12 for b and 10 for h in the following formula.

$$A = \frac{1}{2}bh$$
$$A = \frac{1}{2}(12)(10)$$
$$A = 60$$

The area is 60 square inches.

Self Check 3 Find the area of a triangle with a base of 14 inches and a height of 8 inches.

COMBINED VARIATION

Combined variation involves a combination of direct and inverse variation.

EXAMPLE 4

The pressure of a fixed amount of gas varies directly with its temperature and inversely with its volume. A sample of gas at a pressure of 1 atmosphere occupies a volume of 3 cubic meters when its temperature is 273 Kelvin (about 0° Celsius). Find the pressure after the gas is heated to 364 K and compressed to 1 cubic meter.

Solution We let P represent the pressure of the gas, T represent its temperature, and V represent its volume. The words *the pressure varies directly with temperature and inversely with volume* translate into the equation

$$P = \frac{kT}{V}$$

To find k, we substitute 1 for P, 273 for T, and 3 for V.

$$P = \frac{kT}{V}$$
$$1 = \frac{k(273)}{3}$$
$$1 = 91k \qquad \frac{273}{3} = 91.$$
$$\frac{1}{91} = k$$

Since $k = \frac{1}{91}$, the formula is

$$P = \frac{1}{91} \cdot \frac{T}{V} \quad \text{or} \quad P = \frac{T}{91V}$$

To find the pressure under the new conditions, we substitute 364 for T and 1 for V in the previous equation and solve for P.

$$P = \frac{T}{91V}$$

$$P = \frac{364}{91(1)}$$

$$= 4$$

The pressure of the heated and compressed gas is 4 atmospheres.

Self Check 4 Find the pressure after the gas is heated to 373.1 K and compressed into 1 cubic meter.

Answers to Self Checks **1.** \$375 **2.** 7.5 psi **3.** 56 in.^{2} **4.** approximately 4.1 atmospheres

9.6 STUDY SET  ELEMENTARY Algebra f(x) Now™

VOCABULARY **Fill in the blanks. Assume that k is a constant.**

1. The equation $y = kx$ defines ________ variation.
2. The equation $y = \frac{k}{x}$ defines ________ variation.
3. In $y = kx$, the ________ of variation is k.
4. A constant is a ________.
5. The equation $y = kxz$ represents ________ variation.
6. The equation $y = \frac{kx}{z}$ means that y varies ________ with x and ________ with z.

CONCEPTS **Determine whether each graph represents direct variation or inverse variation.**

7.

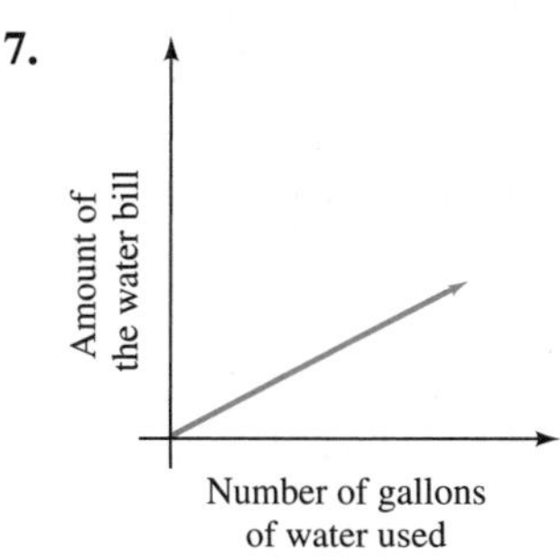

8.

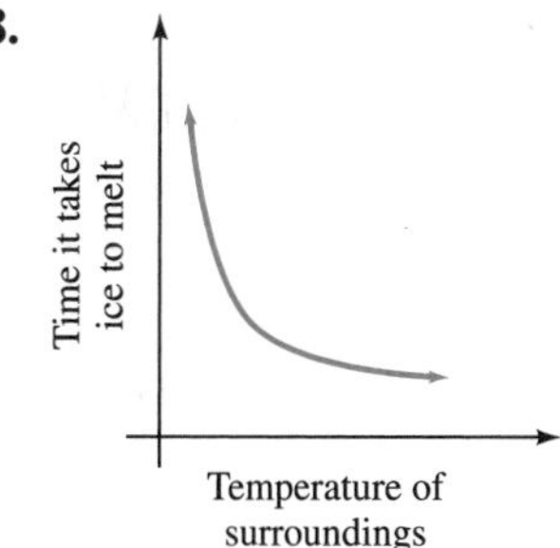

9.

Gallons of gas needed to drive a given distance

mpg rating of a car

10.

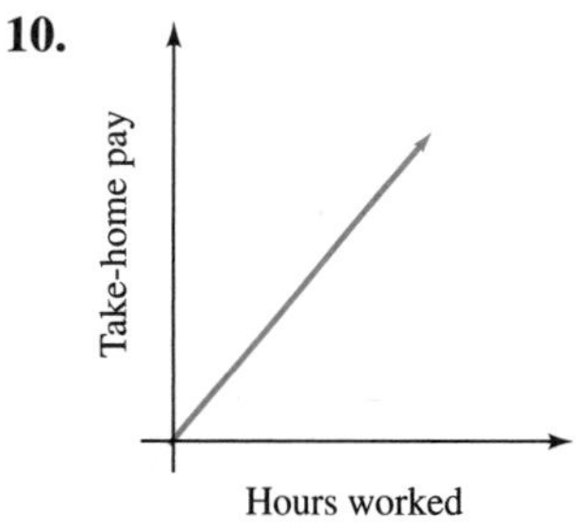

Complete each graph by sketching either a direct variation or an inverse variation.

11.

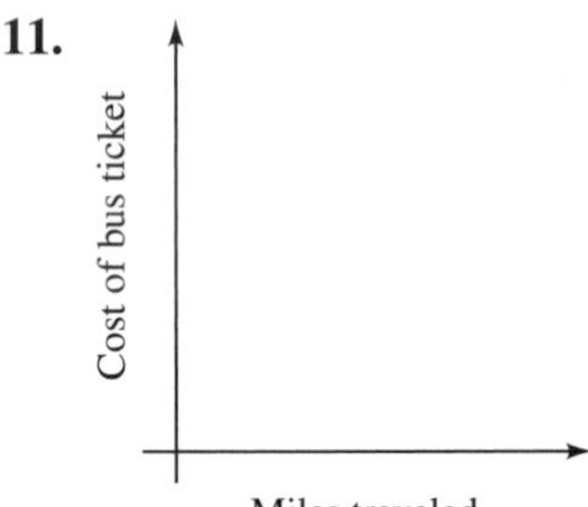

12.

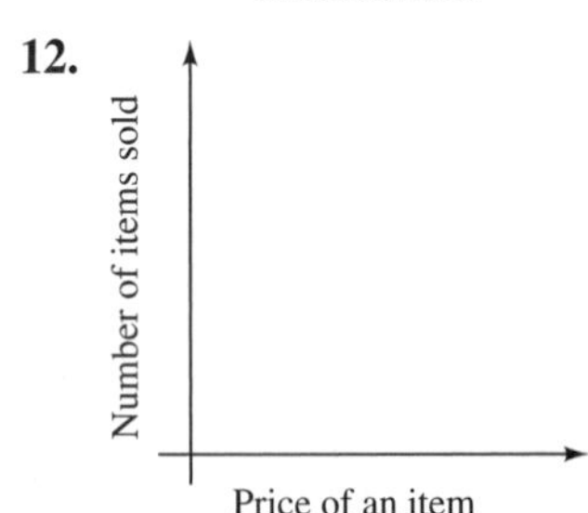

Express each sentence as a formula.

13. The distance d a car can travel while moving at a constant speed varies directly with n, the number of gallons of gasoline it consumes.
14. A farmer's harvest h varies directly with a, the number of acres he plants.
15. For a fixed area, the length l of a rectangle varies inversely with its width w.
16. The value v of a boat varies inversely with its age a.
17. The area A of a circle varies directly with the square of its radius r.
18. The distance s that a body falls varies directly with the square of the time t.
19. The distance d traveled varies jointly with the speed s and the time t.
20. The interest i on a savings account that contains a fixed amount of money varies jointly with the rate r and the time t.
21. The current I varies directly with the voltage V and inversely with the resistance R.
22. The force of gravity F varies directly with the product of the masses m_1 and m_2, and inversely with the square of the distance between them.

NOTATION

23. Decide whether the equation defines direct variation.
    a. $y = kx$ **b.** $y = k + x$
    c. $y = \frac{k}{x}$ **d.** $m = kc$
24. Decide whether each equation defines inverse variation.
    a. $y = kx$ **b.** $y = \frac{k}{x}$
    c. $y = \frac{x}{k}$ **d.** $d = \frac{k}{g}$

PRACTICE **Assume that all variables represent positive numbers.**

25. Assume that y varies directly with x. If $y = 10$ when $x = 2$, find y when $x = 7$.
26. Assume that A varies directly with z. If $A = 30$ when $z = 5$, find A when $z = 9$.
27. Assume that r varies directly with s. If $r = 21$ when $s = 6$, find r when $s = 12$.
28. Assume that d varies directly with t. If $d = 15$ when $t = 3$, find t when $d = 3$.
29. Assume that s varies directly with t^2. If $s = 12$ when $t = 4$, find s when $t = 30$.
30. Assume that y varies directly with x^3. If $y = 16$ when $x = 2$, find y when $x = 3$.
31. Assume that y varies inversely with x. If $y = 8$ when $x = 1$, find y when $x = 8$.
32. Assume that V varies inversely with p. If $V = 30$ when $p = 5$, find V when $p = 6$.
33. Assume that r varies inversely with s. If $r = 40$ when $s = 10$, find r when $s = 15$.
34. Assume that J varies inversely with v. If $J = 90$ when $v = 5$, find J when $v = 45$.
35. Assume that y varies inversely with x^2. If $y = 6$ when $x = 4$, find y when $x = 2$.
36. Assume that i varies inversely with d^2. If $i = 6$ when $d = 3$, find i when $d = 2$.
37. Assume that y varies jointly with r and s. If $y = 4$ when $r = 2$ and $s = 6$, find y when $r = 3$ and $s = 4$.
38. Assume that A varies jointly with x and y. If $A = 18$ when $x = 3$ and $y = 3$, find A when $x = 7$ and $y = 9$.
39. Assume that D varies jointly with p and q. If $D = 20$ when p and q are both 5, find D when p and q are both 10.
40. Assume that z varies jointly with r and the square of s. If $z = 24$ when r and s are 2, find z when $r = 3$ and $s = 4$.

41. Assume that y varies directly with a and inversely with b. If $y = 1$ when $a = 2$ and $b = 10$, find y when $a = 7$ and $b = 14$.

42. Assume that y varies directly with the square of x and inversely with z. If $y = 1$ when $x = 2$ and $z = 10$, find y when $x = 4$ and $z = 5$.

APPLICATIONS

43. COMMUTING DISTANCE The distance that a car can travel without refueling varies directly with the number of gallons of gasoline in the tank. If a car can go 360 miles on a full tank of gas (15 gallons), how far can it go on 7 gallons?

44. COMPUTING FORCES The force of gravity acting on an object varies directly with the mass of the object. The force on a mass of 5 kilograms is 49 newtons. What is the force acting on a mass of 12 kilograms?

45. DOSAGES The recommended dose (in milligrams) of Demerol, a preoperative medication given to children, varies directly with the child's weight in pounds. The proper dosage for a child weighing 30 pounds is 18 milligrams. What would be the correct dosage for a child weighing 45 pounds?

46. MEDICATIONS To fight ear infections in children, doctors often prescribe Ceclor. The recommended dose in milligrams varies directly with the child's body weight in pounds. The correct dosage for a 20-pound child is 124 milligrams. What would be the correct dosage for a 28-pound child?

47. CIDER For the following recipe, the number of inches of stick cinnamon to use varies directly with the number of servings of spiced cider to be made. How many inches of stick cinnamon are needed to make 30 servings?

Hot Spiced Cider

8 cups apple cider or apple juice
$\frac{1}{4}$ to $\frac{1}{2}$ cup packed brown sugar
6 inches stick cinnamon
1 teaspoon whole allspice
1 teaspoon whole cloves
8 thin orange wedges or slices (optional)
8 whole cloves (optional)

Makes 8 servings

48. LUNAR GRAVITY The weight of an object on the moon varies directly with its weight on earth; six pounds on Earth weighs 1 pound on the moon. What would the scale register if the astronaut in the illustration were weighed on the moon?

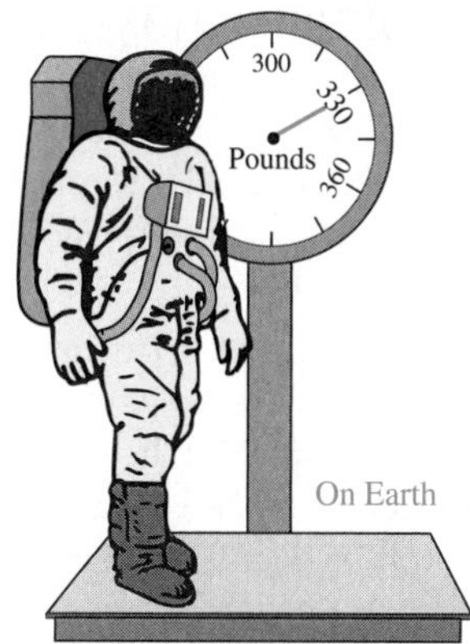

49. COMMUTING TIME The time it takes a car to travel a certain distance varies inversely with its rate of speed. If a certain trip takes 3 hours at 50 miles per hour, how long will the trip take at 60 miles per hour?

50. GEOMETRY For a fixed area, the length of a rectangle is inversely proportional to its width. A rectangle has a width of 12 feet and a length of 20 feet. If its length is increased to 24 feet, find the width that will maintain the same area.

51. ELECTRICITY The current in an electric circuit varies inversely with the resistance. If the current in a circuit is 30 amps when the resistance is 4 ohms, what will the current be for a resistance of 15 ohms?

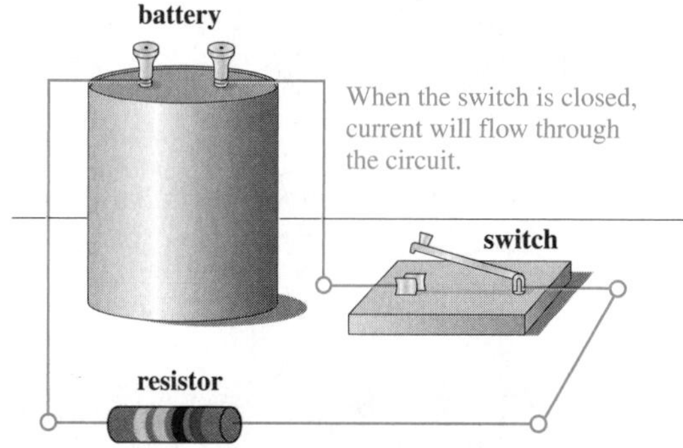

52. FARMING The length of time a given number of bushels of corn will last when feeding cattle varies inversely with the number of animals. If a certain number of bushels will feed 25 cows for 10 days, how long will the feed last for 10 cows?

53. COMPUTING PRESSURES If the temperature of a gas is constant, the volume occupied varies inversely with the pressure. If a gas occupies a volume of 40 cubic meters under a pressure of 8 atmospheres, find the volume when the pressure is changed to 6 atmospheres.

54. COMPUTING DEPRECIATION Assume that the value of a machine varies inversely with its age. If a drill press is worth \$300 when it is 2 years old, find its value when it is 6 years old. How much has the machine depreciated over that 4-year period?

55. COMPUTING INTEREST The interest earned on a fixed amount of money varies jointly with the annual interest rate and the time that the money is left on deposit. If an account earns \$120 at 8% annual interest when left on deposit for 2 years, how much interest would be earned in 3 years at an annual rate of 12%?

56. COST OF A WELL The cost of drilling a water well is jointly proportional to the length and diameter of the steel casing. If a 30-foot well using 4-inch casing costs \$1,200, find the cost of a 35-foot well using 6-inch casing.

57. ELECTRONICS The current in a circuit varies directly with the voltage and inversely with the resistance. If a current of 4 amperes flows when 36 volts is applied to a 9-ohm resistance, find the current when the voltage is 42 volts and the resistance is 11 ohms.

58. BUILDING CONSTRUCTION The deflection of a beam is inversely proportional to its width and the cube of its depth. If the deflection is 2 inches when the width is 4 inches and the depth is 3 inches, find the deflection when the width is 3 inches and the depth is 4 inches.

WRITING

59. Give two examples of quantities that vary directly and two that do not.

60. What is the difference between direct variation and inverse variation?

61. What is a constant of variation?

62. Is there a direct variation or an inverse variation between each pair of quantities? Explain.
    a. The time it takes to type a term paper and the speed at which you type
    b. The time it takes to type a term paper (working at a constant rate) and the length of the term paper

REVIEW Solve each equation.

63. $x^2 - 5x - 6 = 0$

64. $x^2 - 25 = 0$

65. $2(y - 4) = -y^2$

66. $y^3 - y^2 = 0$

67. $5a^3 - 125a = 0$

68. $6t^3 + 35t^2 = 6t$

CHALLENGE PROBLEMS

69. GRAVITY The force with which the Earth attracts an object above the Earth's surface varies inversely with the square of the distance of the object from the center of the Earth. An object 4,000 miles from the center of the Earth is attracted with a force of 80 pounds. Find the force of attraction if the object were 5,000 miles from the center of the Earth.

70. CONSTRUCTION The time it takes to build a highway varies directly with the length of the road and inversely with the number of workers. It takes 100 workers 4 weeks to build 2 miles of highway. How long will it take 80 workers to build 10 miles of highway?

ACCENT ON TEAMWORK

COMPLETING THE SQUARE

Overview: In this activity, you will create a model that will help you visualize the process of completing the square.

Instructions: Form groups of 2 or 3 students. Using construction paper, cut out five pieces like those that make up the model shown. Label each piece with its area. (Recall that the area of a rectangle is the product of its length and width.) Then show that the total area of the model is $x^2 + 4x$ by adding the areas of all the pieces.

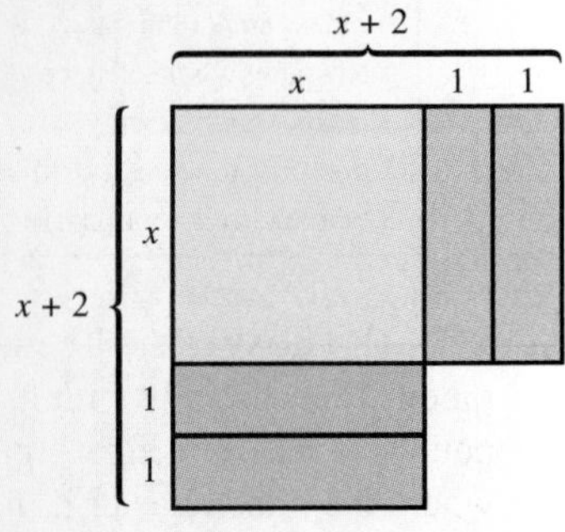

Next, add enough 1×1 squares to fill in the missing corner of the model so that it is itself a large square. Show that the area of the large square is $x^2 + 4x + 4$ by adding the areas of all the pieces. Show that its area could also be expressed as $(x + 2)(x + 2)$.

Finally, write a paragraph that explains how the model illustrates the process of completing the square on $x^2 + 4x$.

If time allows, make another model to complete the square on $x^2 + 8x$.

THE DISCRIMINANT

Overview: In this activity, you will learn how to predict the type of solutions that a quadratic equation has, before solving it.

Instructions: Form groups of 2 or 3 students. The expression $b^2 - 4ac$ is called the **discriminant** of the quadratic equation $ax^2 + bx + c = 0$. It provides us with information about the solutions of a quadratic equation.

- If $b^2 - 4ac > 0$, the equation has two different real-number solutions.
- If $b^2 - 4ac = 0$, the equation has one repeated rational-number solution.
- If $b^2 - 4ac < 0$, the equation has two different imaginary-number solutions.

Find the value of the discriminant for each of the following equations. Then, use the answer to predict the type of solutions that the equation has. Next, solve the equation. Do the solutions match your predictions about them?

1. $x^2 - 2x - 1 = 0$ **2.** $x^2 - 4x + 4 = 0$ **3.** $x^2 + 2x + 2 = 0$

KEY CONCEPT: SOLVING QUADRATIC EQUATIONS

We have studied four methods for solving quadratic equations. Each has its advantages and disadvantages.

Method	Advantages	Disadvantages
Factoring and the zero-factor property	It can be very fast. When each factor is set equal to 0, the resulting equations are usually easy to solve.	Some polynomials may be difficult to factor, others impossible.
Square root property	It is the fastest way to solve equations of the form ax^2 = number or $(ax + b)^2$ = number.	It only applies to equations that are in these forms.
Completing the square	It can be used to solve any quadratic equation. It works well with equations of the form $x^2 + bx + c = 0$, where b is even.	It involves more steps than the other methods. The algebra can be cumbersome if the lead coefficient is not 1.
Quadratic formula	It can be used to solve any quadratic equation.	It involves several computations where sign errors can be made. Often, the result must be simplified.

Solve each quadratic equation using the method listed.

1. $4x^2 - x = 0$; factoring method

2. $a^2 - a - 56 = 0$; factoring method

3. $x^2 = 27$; square root property

4. $(x + 3)^2 = 16$; square root property

5. $x^2 + 4x + 1 = 0$; complete the square

6. $x^2 + 3x + 1 = 0$; quadratic formula

Explain the error in each of the following solutions. Then solve the equation correctly.

7. Solve: $x^2 = 20$

$$x = \sqrt{20}$$
$$x = 2\sqrt{5}$$

8. Solve: $x^2 - 3x = 10$

$$x(x - 3) = 10$$
$$x = 10 \text{ or } x - 3 = 10$$
$$x = 13$$

9. Solve: $x^2 + 6x - 10 = 0$

$$x^2 + 6x = 10$$
$$x^2 + 6x + 9 = 10$$
$$(x + 3)^2 = 10$$
$$x + 3 = \pm\sqrt{10}$$
$$x = -3 \pm \sqrt{10}$$

10. Solve: $x^2 + 5x + 1 = 0$

$a = 1 \quad b = 5 \quad c = 1$

$$x = -5 \pm \frac{\sqrt{5^2 - 4(1)(1)}}{2(1)}$$
$$x = 5 \pm \frac{\sqrt{21}}{2}$$

CHAPTER REVIEW

ELEMENTARY Algebra $f(x)$ Now™

SECTION 9.1 The Square Root Property; Completing the Square

CONCEPTS

We can use the *square root property* to solve $x^2 = c$, where $c > 0$. The two solutions are $x = \sqrt{c}$ or $x = -\sqrt{c}$, which can be written $x = \pm\sqrt{c}$.

REVIEW EXERCISES

Use the square root property to solve each equation.

1. $x^2 = 25$

2. $t^2 = 8$

3. $2x^2 - 1 = 149$

4. $(x - 1)^2 = 25$

5. $4(x - 2)^2 = 9$

6. $(x - 8)^2 = 40$

Use the square root property to solve each equation. Round each solution to the nearest hundredth.

7. $x^2 = 12$

8. $(x - 1)^2 = 55$

To *complete the square* on $x^2 + bx$, add the square of one-half of the coefficient of x:

$$x^2 + bx + \left(\frac{1}{2}b\right)^2$$

To solve a quadratic equation by *completing the square:*

1. If necessary, divide both sides of the equation by the coefficient of x^2 to make its coefficient 1.
2. Get all variable terms on one side of the equation and constants on the other side.
3. Complete the square.
4. Factor the perfect square trinomial.
5. Solve the resulting equation using the square root property.
6. Check your answers in the original equation.

Complete the square to make each expression a perfect square trinomial. Then factor.

9. $x^2 + 4x$

10. $t^2 - 5t$

Solve each quadratic equation by completing the square.

11. $x^2 - 8x + 15 = 0$

12. $x^2 = -5x + 14$

13. $2x^2 + 5x = 3$

14. $2x^2 - 2x - 1 = 0$

15. Solve $x^2 + 4x + 1 = 0$ by completing the square. Round each solution to the nearest hundredth.

16. PLAYGROUND EQUIPMENT A large tractor tire makes a good container for sand. If the circular area that the sandbox covers is 28.3 square feet, what is the radius of the tire? Round to the nearest tenth of a foot.

SECTION 9.2 The Quadratic Formula

The *quadratic formula* gives the solutions of $ax^2 + bx + c = 0$, with $a \neq 0$:

$$x = \frac{-b \pm \sqrt{b^2 - 4ac}}{2a}$$

Write each equation in quadratic form and determine a, b, and c.

17. $x^2 + 2x = -5$

18. $6x^2 = 2x + 1$

Use the quadratic formula to solve each quadratic equation.

19. $x^2 - 2x - 15 = 0$

20. $x^2 - 6x = 7$

21. $6x^2 = 7x + 3$

22. $x^2 - 6x + 7 = 0$

23. Use the quadratic formula to solve $3x^2 + 2x - 2 = 0$. Give the solutions in exact form and then rounded to the nearest hundredth.

24. SECURITY GATES The length of the frame for an iron gate is 14 feet longer than the width. A diagonal crossbrace is 26 feet long. Find the width and length of the gate frame.

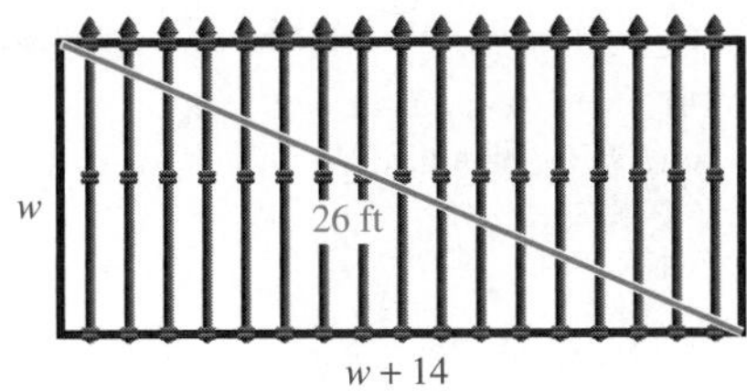

Strategy for solving quadratic equations:

1. Try the square root method.
2. If it doesn't apply, write the equation in $ax^2 + bx + c = 0$ form.
3. Try the factoring method.
4. If it doesn't work, complete the square or use the quadratic formula.

25. MILITARY A pilot releases a bomb from an altitude of 3,000 feet. The bomb's height h in feet above the target t seconds after its release is given by the formula

$$h = 3{,}000 + 40t - 16t^2$$

How long will it be until the bomb hits its target?

Use the most convenient method to find all real solutions of each equation.

26. $x^2 + 6x + 2 = 0$

27. $(y + 3)^2 = 16$

28. $x^2 + 5x = 0$

29. $2x^2 + x = 5$

30. $g^2 - 20 = 0$

31. $a^2 = 4a - 4$

32. $a^2 - 2a + 5 = 0$

SECTION 9.3 Complex Numbers

The *imaginary number i* is defined as

$$i = \sqrt{-1}$$

From the definition, it follows that $i^2 = -1$.

A *complex number* is any number that can be written in the form $a + bi$, where a and b are real numbers and $i = \sqrt{-1}$. We call a the *real part* and b the *imaginary part.*

Complex numbers of the form $a + bi$, where $b \neq 0$, are also called *imaginary numbers.*

Write each expression in terms of i.

33. $\sqrt{-25}$

34. $\sqrt{-18}$

35. $-\sqrt{-49}$

36. $\sqrt{-\frac{9}{64}}$

37. Complete the diagram.

Complex numbers

38. Determine whether each statement is true or false.

a. Every real number is a complex number.

b. $3 - 4i$ is an imaginary number.

c. $\sqrt{-4}$ is a real number.

d. i is a real number.

In the complex number system, all quadratic equations have solutions.

Solve each equation. Write all solutions in the form a + bi.

39. $x^2 = -9$

40. $3x^2 = -16$

41. $2x^2 - 3x + 2 = 0$

42. $x^2 + 2x = -2$

The complex numbers $a + bi$ and $a - bi$ are called *complex conjugates.*

Give the complex conjugate of each number.

43. $3 + 6i$

44. $-1 - 7i$

45. $19i$

46. $-i$

To add complex numbers, add their real parts and add their imaginary parts.

To subtract complex numbers, add the opposite of the complex number being subtracted.

Multiplying complex numbers is similar to multiplying polynomials.

To divide complex numbers, multiply the numerator and denominator by the complex conjugate of the denominator.

Perform the operations. Write all answers in the form a + bi.

47. $(3 + 4i) + (5 - 6i)$

48. $(7 - 3i) - (4 + 2i)$

49. $3i(2 - i)$

50. $(2 + 3i)(3 - i)$

51. $\dfrac{3}{5 + i}$

52. $\dfrac{2 + 3i}{2 - 3i}$

SECTION 9.4 Graphing Quadratic Equations

The *vertex* of a parabola is the lowest (or highest) point on the parabola.

53. Refer to the figure.

a. What are the x-intercepts of the parabola?

b. What is the y-intercept of the parabola?

c. What is the vertex of the parabola?

d. Draw the axis of symmetry of the parabola on the graph.

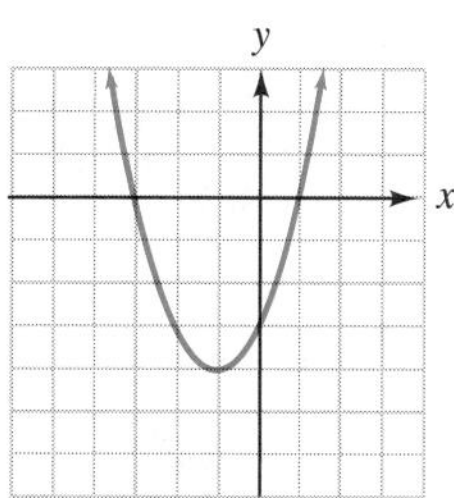

A vertical line through the vertex of a parabola that opens upward or downward is its *axis of symmetry.*

54. What important information can be obtained from the vertex of the parabola in the figure?

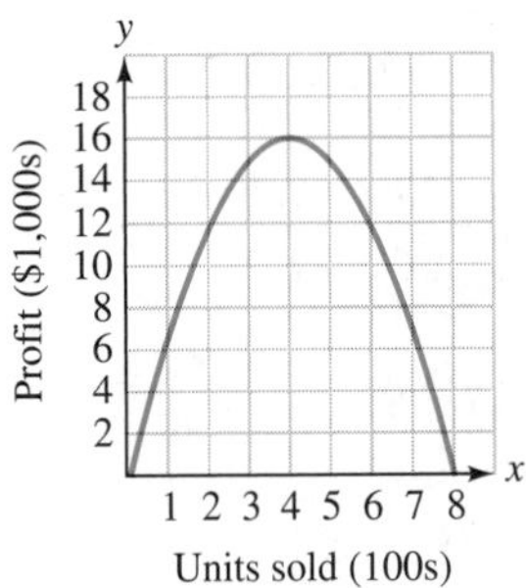

The graph of the quadratic equation $y = ax^2 + bx + c$ is a parabola. It opens upward when $a > 0$ and downward when $a < 0$.

To find the y-intercept, substitute 0 for x in the given equation and solve for y. To find the x-intercepts, substitute 0 for y in the given equation and solve for x.

The x-coordinate of the vertex of the parabola $y = ax^2 + bx + c$ is $x = -\frac{b}{2a}$. To find the y-coordinate of the vertex, substitute $-\frac{b}{2a}$ for x in the equation of the parabola and find y.

The number of x-intercepts of the graph of a quadratic equation $y = ax^2 + bx + c$ is the same as the number of real-number solutions of $ax^2 + bx + c = 0$.

Find the vertex of the graph of each quadratic equation and tell which direction the parabola opens. ***Do not draw the graph.***

55. $y = 2x^2 - 4x + 7$

56. $f(x) = -3x^2 + 18x - 11$

57. Find the x- and y-intercepts of the graph of $y = x^2 + 6x + 5$.

Graph each quadratic equation by finding the vertex, x- and y-intercepts, and axis of symmetry of its graph.

58. $y = x^2 + 2x - 3$

59. $y = -2x^2 + 4x - 2$

60. The graphs of three quadratic equations are shown. Fill in the blanks.

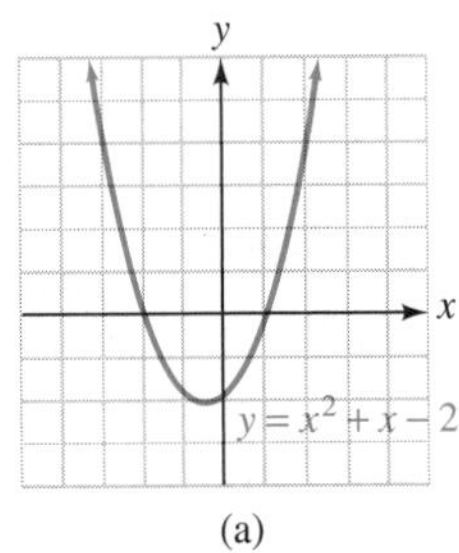

(a)

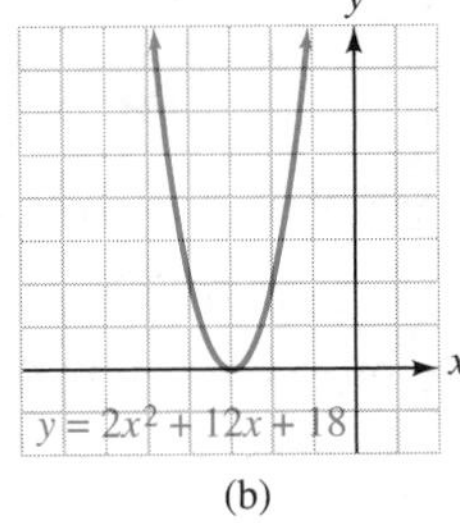

(b)

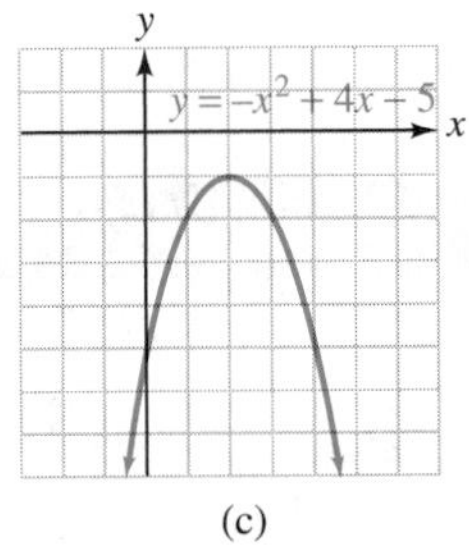

(c)

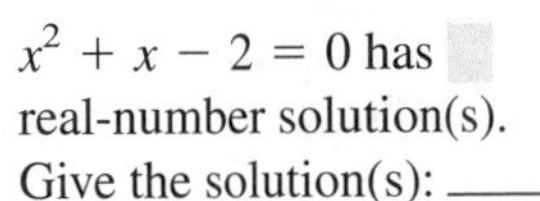

$x^2 + x - 2 = 0$ has ___ real-number solution(s). Give the solution(s): ______

$2x^2 + 12x + 18 = 0$ has ___ real-number solution(s). Give the solution(s): ______

$-x^2 + 4x - 5 = 0$ has ___ real-number solution(s). Give the solution(s): ______

SECTION 9.5 An Introduction to Functions

A *function* is a rule that assigns to each value of one variable (called the *independent variable*) exactly one value of another variable (called the *dependent variable*).

For y to be a function of x, each value of x is assigned exactly one value of y.

Determine whether each arrow diagram or table defines y as a function of x. If it does not, find ordered pairs that show a value of x that is assigned more than one value of y.

61. 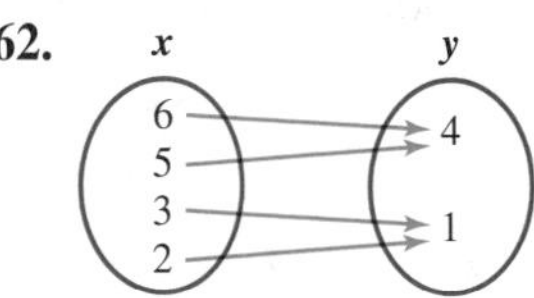

62.

63. 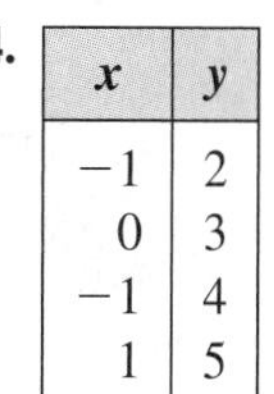

x	y
9	81
7	49
5	25
3	9

64.

x	y
−1	2
0	3
−1	4
1	5

The set of all possible values that can be used for the independent variable is called the *domain*. The set of all values of the dependent variable is called the *range*.

We can think of a function as a machine that takes some input x and turns it into some output $f(x)$, called a *function value*.

65. Label the diagram with the words *domain, range, input,* and *output.* Then find $f(2)$.

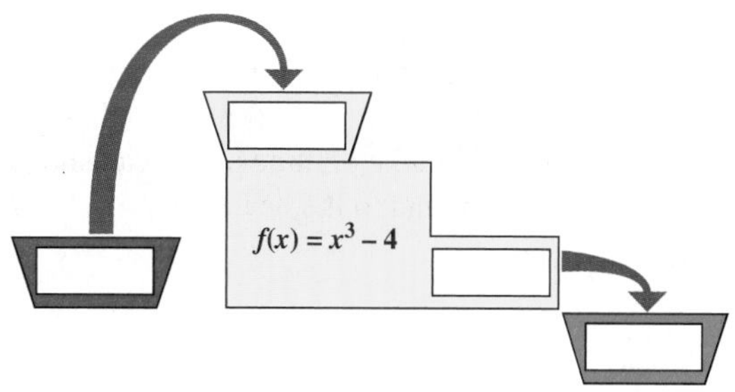

The notation $y = f(x)$ denotes that the variable y is a function of x.

66. Fill in the blank: Since $y =$ ____, the equations $y = 2x - 8$ and $f(x) = 2x - 8$ are equivalent.

For $f(x) = x^2 - 4x$, find each of the following function values.

67. $f(1)$

68. $f(0)$

69. $f(-3)$

70. $f\left(\frac{1}{2}\right)$

For $g(x) = 1 - 6x$, find each of the following function values.

71. $g(1)$

72. $g(-6)$

73. $g(0.5)$

74. $g\left(\frac{3}{2}\right)$

The *vertical line test:* If a vertical line intersects a graph in more than one point, the graph is not the graph of a function.

Determine whether each graph is the graph of a function. If it is not, find ordered pairs that show a value of x that is assigned more than one value of y.

75.

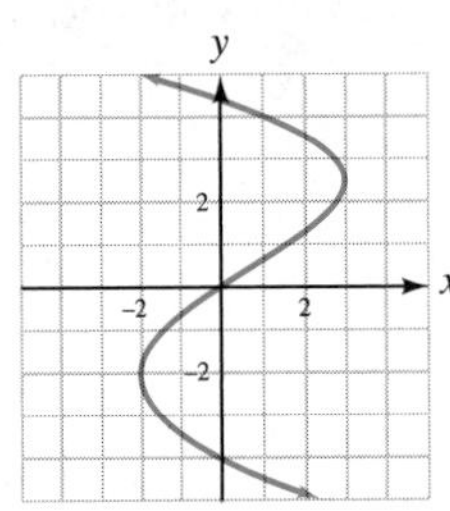

76.

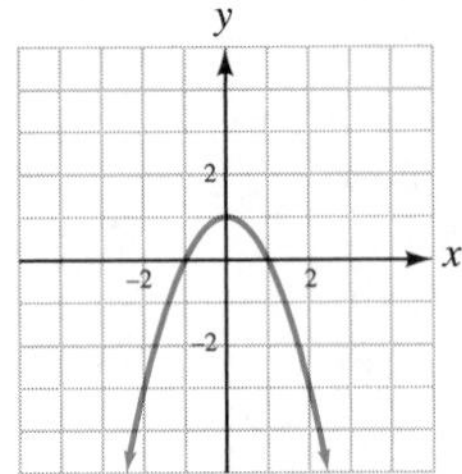

The "input-output" pairs that a function generates can be written in the form $(x, f(x))$. Then these ordered pairs can be plotted on a rectangular coordinate system to give the graph of the function.

77. Complete the table of function values. Then graph the function.

$f(x) = 1 - |x|$

x	$f(x)$
0	
1	
2	
−1	
−2	
−3	

Functions are used to mathematically describe certain relationships where one quantity depends upon another.

78. The function $V(r) = 15.7r^2$ estimates the volume in cubic inches of a can 5 inches tall with a radius of r inches. Find the volume of the can shown in the illustration. Round to the nearest tenth.

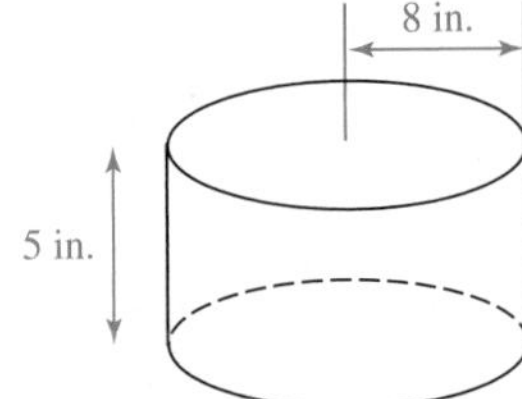

SECTION 9.6 Variation

$y = kx$ represents *direct variation.*

$y = \dfrac{k}{x}$ represents *inverse variation.*

$y = kxz$ represents *joint variation.*

Direct and inverse variation are used together in *combined variation.*

79. PROFIT The profit made by a strawberry farm varies directly with the number of baskets of strawberries sold. If a profit of \$500 is made from the sale of 750 baskets, what is the profit when 1,250 baskets are sold?

80. l varies inversely with w. Find the constant of variation if $l = 30$ when $w = 20$.

81. ELECTRICITY For a fixed voltage, the current in an electrical circuit varies inversely with the resistance in the circuit. If a certain circuit has a current of $2\frac{1}{2}$ amps when the resistance is 150 ohms, find the current in the circuit when the resistance is doubled.

82. Does the graph show direct or inverse variation?

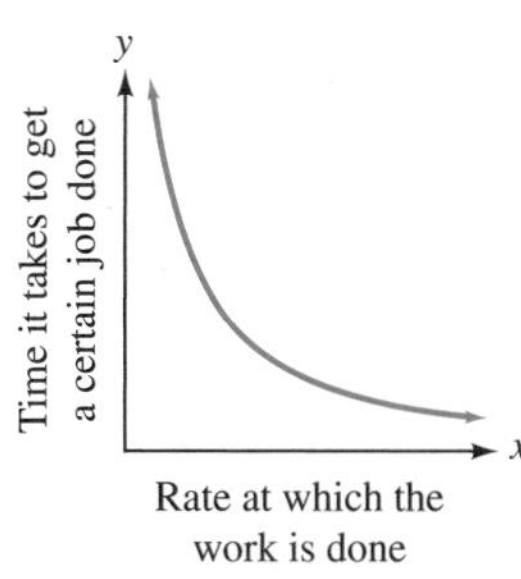

83. R varies jointly with b and c. If $R = 72$ when $b = 4$ and $c = 24$, find R when $b = 6$ and $c = 18$.

84. s varies directly with w and inversely with the square of m. If $s = \frac{7}{4}$ when w and m are 4, find s when $w = 5$ and $m = 7$.

CHAPTER 9 TEST

ELEMENTARY Algebra $f(x)$ Now™

Solve each equation by the square root method.

1. $x^2 = 17$

2. $u^2 = 24$

3. $4y^2 = 25$

4. $(x - 2)^2 = 3$

5. ARCHERY The area of a circular archery target is 5,026.5 cm^2. What is the radius of the target? Round to the nearest centimeter.

6. Find the number required to complete the square on $x^2 - 14x$.

7. Complete the square to solve $a^2 + 2a - 4 = 0$. Give the exact solutions and then round them to the nearest hundredth.

8. Complete the square to solve $2x^2 = 3x + 2$.

Use the quadratic formula to solve each equation.

9. $2x^2 - 5x = 12$

10. $x^2 = 4x - 2$

11. Solve $3x^2 - x - 1 = 0$ using the quadratic formula. Give the exact solutions, and then approximate them to the nearest hundredth.

12. FLAGS According to the *Guinness Book of World Records 1998,* the largest flag in the world is the American "Superflag," which has an area of 128,775 ft^2. If its length is 5 feet less than twice its width, find its width and length.

Write each expression in terms of i.

13. $\sqrt{-100}$

14. $-\sqrt{-18}$

15. What is a complex number? Give an example.

16. Solve: $3x^2 + 2x + 1 = 0$. Write the solutions in the form $a + bi$.

Perform the operations. Write all answers in the form $a + bi$.

17. $(8 + 3i) + (-7 - 2i)$ **18.** $(5 + 3i) - (6 - 9i)$

19. $(2 - 4i)(3 + 2i)$ **20.** $\dfrac{3 - 2i}{3 + 2i}$

21. ADVERTISING When a business runs x advertisements per week on television, the number y of air conditioners it sells is given by the graph. What important information can be obtained from the vertex?

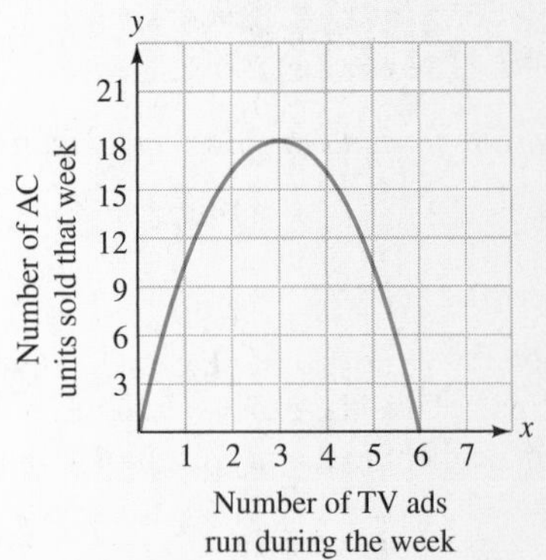

22. Graph the quadratic equation $y = x^2 + 6x + 5$. Find the vertex, the x- and y-intercepts, and the axis of symmetry of the graph.

23. What is the domain of a function? What is the range of a function?

Determine whether the table and the graph define y as a function of x. If not, find ordered pairs that show a value of x that is assigned more than one value of y.

24.

x	y
1	4
2	3
3	2
4	1

25.

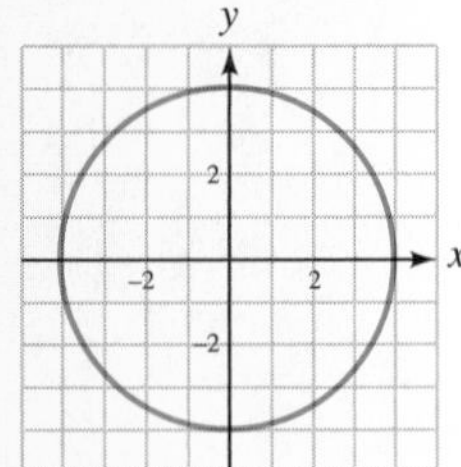

26.

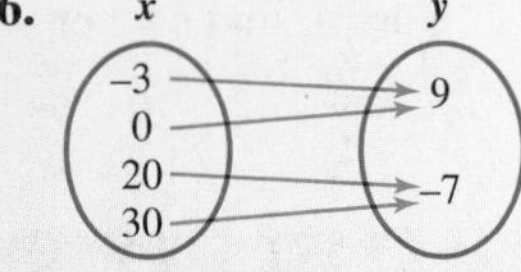

27. If $f(x) = 2x - 7$, find $f(-3)$.

28. If $g(x) = 3.5x^3$ find $g(6)$.

29. TELEPHONE CALLS The function $C(n) = 0.30n + 15$ gives the cost C per month in dollars for making n phone calls. Find $C(45)$ and explain what it means.

30. Graph: $f(x) = |x| - 1$.

31. POGO STICKS The force required to compress a spring varies directly with the change in the length of the spring. If a force of 130 pounds compresses the spring on the pogo stick 6.5 inches, how much force is required to compress the spring 5 inches?

32. Assume that r varies inversely with s. If $r = 40$ when $s = 10$, find r when $s = 15$.

CHAPTERS 1–9 CUMULATIVE REVIEW EXERCISES

1. True or false:
   a. Every rational number can be written as a ratio of two integers.
   b. The set of real numbers corresponds to all points on the number line.
   c. The whole numbers and their opposites form the set of integers.
2. Evaluate: $-4 + 2[-7 - 3(-9)]$.
3. DRIVING SAFETY In cold-weather climates, salt is spread on roads to keep snow and ice from bonding to the pavement. This allows snowplows to remove accumulated snow quickly. According to the graph, when is the accident rate the highest?

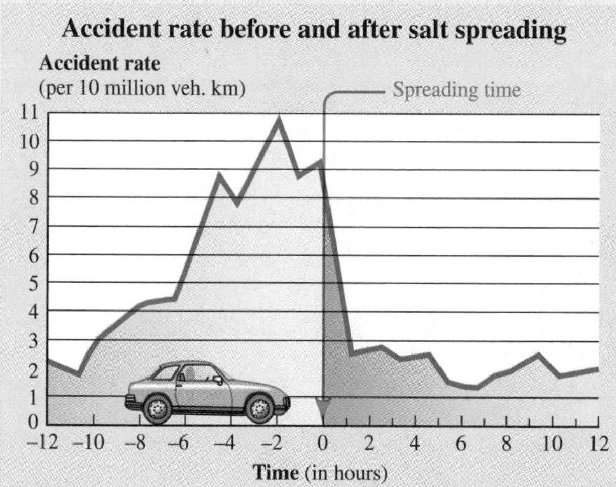

Based on data from the Salt Institute

4. EMPLOYMENT The following newspaper headline appeared in early 2000. How many employees did Xerox have at that time?

Xerox to Cut 5,200 Jobs, or 5.3% of Workforce, on Falling Profits

5. Simplify: $3p - 6(p + z) + p$.
6. Solve: $\frac{5}{6}k = 10$.
7. Solve: $-(3a + 1) + a = 2$.
8. Solve $5x + 7 < 2x + 1$ and graph the solution set. Then use interval notation to describe the solution.
9. ENTREPRENEURS Last year, a women's professional organization made two small-business loans totaling \$28,000 to young women beginning their own businesses. The money was lent at 7% and 10% simple interest rates. If the annual income the organization received from these loans was \$2,560, what was each loan amount?
10. Evaluate $(x - a)^2 + (y - b)^2$ for $x = -2$, $y = 1$, $a = 5$, and $b = -3$.
11. Evaluate: $\left|\frac{4}{5} \cdot 10 - 12\right|$.
12. Find the slope of the line passing through $(-2, -2)$ and $(-12, -8)$.

Graph each equation or inequality.

13. $2y - 2x = 6$
14. $y = -3$
15. $y = -x + 2$
16. $y < 3x$
17. Graph the line passing through $(-2, -1)$ and having slope $\frac{4}{3}$.
18. Graph: $y = x^3 - 2$.
19. Write the equation of the line whose graph has slope $m = -2$ and y-intercept $(0, 1)$.
20. Write the equation of the line whose graph has slope $m = \frac{1}{4}$ and passes through the point $(8, 1)$. Answer in slope–intercept form.
21. What is the slope of the line defined by $4x + 5y = 6$?
22. INSURANCE The graph in red approximates the average annual expenditure on homeowner's insurance in the U.S. for the years 1995–2002. Find the rate of increase over this period of time.

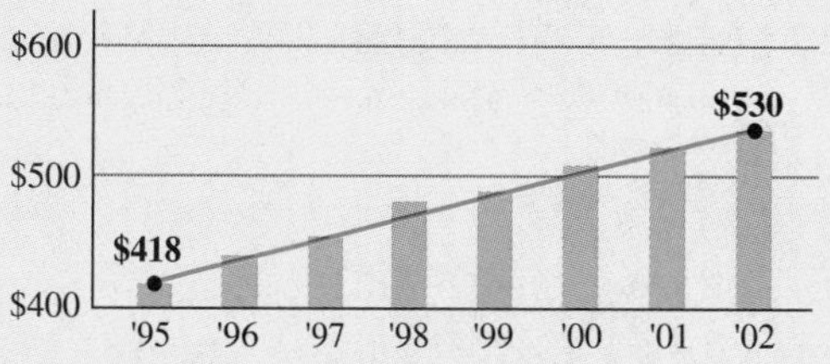

Source: Insurance Information Institute

Simplify each expression. Write each answer without using parentheses or negative exponents.

23. $y^3(y^2y^4)$

24. $\left(\frac{b^2}{3a}\right)^3$

25. $\frac{10a^4a^{-2}}{5a^2a^0}$

26. $\left(\frac{21x^{-2}y^2z^{-2}}{7x^3y^{-1}}\right)^{-2}$

27. FIVE-CARD POKER The odds against being dealt the hand shown are about 2.6×10^6 to 1. Express the odds using standard notation.

28. Write 0.00073 in scientific notation.

Perform the operations.

29. $4(4x^3 + 2x^2 - 3x - 8) - 5(2x^3 - 3x + 8)$

30. $(-2a^3)(3a^2)$

31. $(2b - 1)(3b + 4)$

32. $(2x + 5y)^2$

33. $(3x + y)(2x^2 - 3xy + y^2)$

34. $x - 3\overline{)2x^2 - 3 - 5x}$

Factor each expression completely.

35. $6a^2 - 12a^3b + 36ab$

36. $2x + 2y + ax + ay$

37. $b^3 + 125$

38. $t^4 - 16$

Solve each equation.

39. $3x^2 + 8x = 0$

40. $15x^2 - 2 = 7x$

41. Write a polynomial that represents the perimeter of the rectangle.

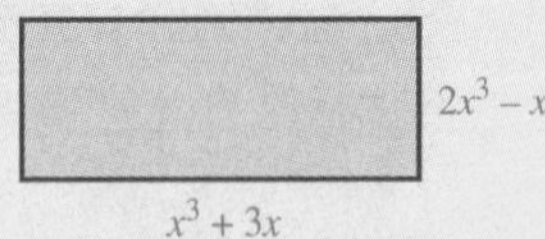

42. HEIGHT OF A TRIANGLE The triangle shown has an area of 22.5 square inches. Find its height.

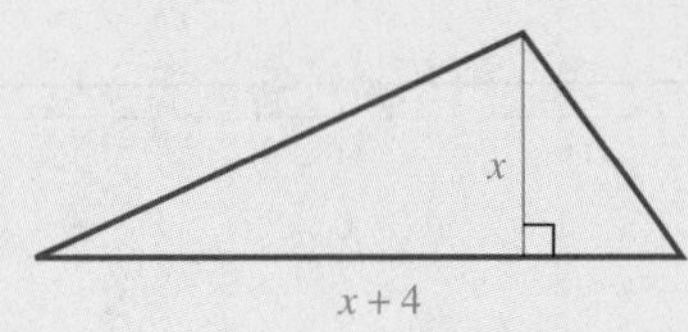

43. For what value is $\frac{x}{x + 8}$ undefined?

44. Simplify: $\frac{3x^2 - 27}{x^2 + 3x - 18}$.

Perform the operations.

45. $\frac{x^2 - x - 6}{2x^2 + 9x + 10} \div \frac{x^2 - 25}{2x^2 + 15x + 25}$

46. $\frac{x + 5}{xy} - \frac{x - 1}{x^2y}$

47. $\frac{x}{x - 2} + \frac{3x}{x^2 - 4}$

48. $\frac{\frac{5}{y} + \frac{4}{y + 1}}{\frac{4}{y} - \frac{5}{y + 1}}$

Solve each equation.

49. $\frac{2p}{3} - \frac{1}{p} = \frac{2p - 1}{3}$

50. $\frac{7}{q^2 - q - 2} + \frac{1}{q + 1} = \frac{3}{q - 2}$

51. Solve the formula $\frac{1}{a} + \frac{1}{b} = 1$ for a.

52. ROOFING A homeowner estimates that it will take him 7 days to roof his house. A professional roofer estimates that he can roof the house in 4 days. How long will it take if the homeowner helps the roofer?

53. LOSING WEIGHT If a person cuts his or her daily calorie intake by 100, it will take 350 days for that person to lose 10 pounds. How long will it take for the person to lose 25 pounds?

54. $\triangle ABC$ and $\triangle DEC$ are similar triangles. Find x.

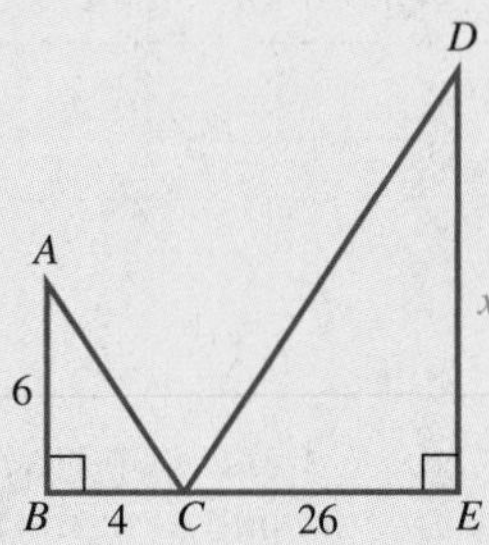

55. Solve using the graphing method.

$$\begin{cases} x + y = 1 \\ y = x + 5 \end{cases}$$

56. Solve using the substitution method.

$$\begin{cases} y = 2x + 5 \\ x + 2y = -5 \end{cases}$$

57. Solve using the addition method.

$$\begin{cases} \frac{3}{5}s + \frac{4}{5}t = 1 \\ -\frac{1}{4}s + \frac{3}{8}t = 1 \end{cases}$$

58. AVIATION With the wind, a plane can fly 3,000 miles in 5 hours. Against the wind, the trip takes 6 hours. Find the airspeed of the plane (the speed in still air). Use two variables to solve this problem.

59. MIXING CANDY How many pounds of each candy must be mixed to obtain 48 pounds of candy that would be worth \$4 per pound? Use two variables to solve this problem.

60. Solve the system of linear inequalities.

$$\begin{cases} 3x + 4y \geq -7 \\ 2x - 3y \geq 1 \end{cases}$$

Simplify each expression. All variables represent positive numbers.

61. $\sqrt{50x^2}$

62. $-\sqrt{100a^6b^4}$

63. $3\sqrt{24} + \sqrt{54}$

64. $(\sqrt{2} + 1)(\sqrt{2} - 3)$

65. $\sqrt{\frac{72x^3}{y^2}}$

66. $\frac{8}{\sqrt{10}}$

67. $\sqrt[3]{\frac{27m^3}{8n^6}}$

68. $\sqrt[4]{16}$

69. $25^{3/2}$

70. $(-8)^{-4/3}$

71. Solve: $\sqrt{6x + 1} + 2 = 7$.

72. Solve $x^2 + 8x + 12 = 0$ by completing the square

73. Solve: $t^2 = 75$.

74. STORAGE CUBES The diagonal distance across the face of each of the stacking cubes is 15 inches. What is the height of the entire storage arrangement? Round to the nearest tenth of an inch.

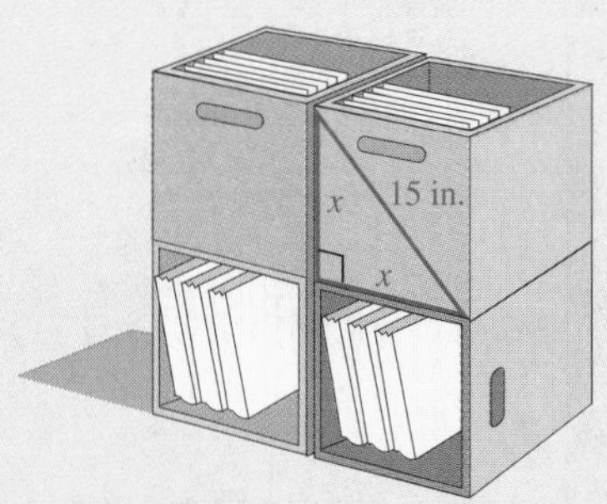

75. Solve $4x^2 - x - 2 = 0$ using the quadratic formula. Give the exact solutions, and then approximate each to the nearest hundredth.

76. QUILTS According to the *Guinness Book of World Records 1998,* the world's largest quilt was made by the Seniors' Association of Saskatchewan, Canada, in 1994. If the length of the rectangular quilt is 11 feet less than twice its width and it has an area of 12,865 ft^2, find its width and length.

Write each expression in terms of i.

77. $\sqrt{-49}$

78. $\sqrt{-54}$

Perform the operations. Express each answer in the form a + bi.

79. $(2 + 3i) - (1 - 2i)$

80. $(7 - 4i) + (9 + 2i)$

81. $(3 - 2i)(4 - 3i)$

82. $\frac{3 - i}{2 + i}$

Solve each equation. Express the solutions in the form $a + bi$.

83. $x^2 + 16 = 0$

84. $x^2 - 4x = -5$

85. Graph the quadratic equation $y = 2x^2 + 8x + 6$. Find the vertex, the x- and y-intercepts, and the axis of symmetry of the graph.

86. POWER OUTPUT The graph shows the power output (in horsepower, hp) of a certain engine for various engine speeds (in revolutions per minute, rpm). For what engine speed does the power output reach a maximum?

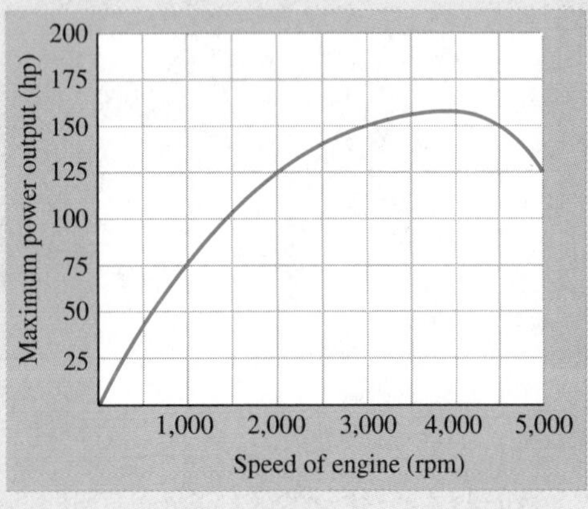

87. If $f(x) = 3x^2 + 3x - 8$, find $f(-1)$.

88. Is the word *domain* associated with the input or the output of a function?

89. Is this the graph of a function?

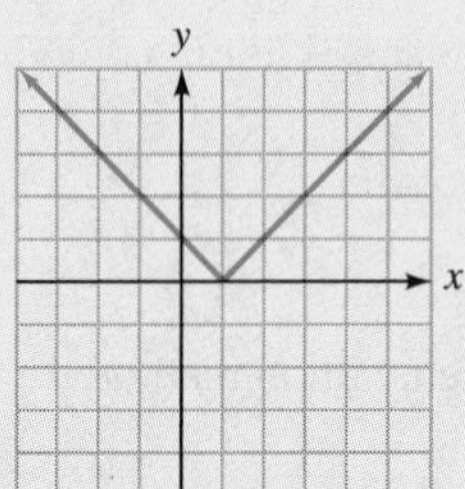

90. GEARS The speed of a gear varies inversely with the number of teeth. If a gear with 10 teeth makes 3 revolutions per second, how many revolutions per second will a gear with 25 teeth make?

Statistics

In this appendix, you will learn about

- The Mean
- The Median
- The Mode

Statistics is a branch of mathematics that deals with the analysis of numerical data. In statistics, three types of averages are commonly used as measures of central tendency of a distribution of numbers: the *mean,* the *median,* and the *mode.*

THE MEAN

We have previously discussed the mean of a distribution.

The Mean The **mean** of several values is the sum of those values divided by the number of values.

$$\text{Mean} = \frac{\text{sum of the values}}{\text{number of values}}$$

EXAMPLE 1

Physiology. As part of a class project, a student measured ten people's reaction time to a visual stimulus. Their reaction times (in seconds) were

0.36, 0.24, 0.23, 0.41, 0.28, 0.25, 0.20, 0.28, 0.39, 0.26

Find the mean reaction time.

Solution To find the mean, we add the values and divide by the number of values.

$$\begin{aligned}\text{Mean} &= \frac{0.36 + 0.24 + 0.23 + 0.41 + 0.28 + 0.25 + 0.20 + 0.28 + 0.39 + 0.26}{10} \\ &= \frac{2.9}{10} \\ &= 0.29\end{aligned}$$

The mean reaction time is 0.29 second.

EXAMPLE 2

Banking. When the mean (average) daily balance of a checking account falls below \$500 in any week, the customer must pay a \$20 service charge. What minimum balance must a customer have on Friday to avoid a service charge?

Security Savings Bank		
Day	**Date**	**Daily balance**
Mon	5/09	\$670.70
Tues	5/10	\$540.19
Wed	5/11	−\$60.39
Thurs	5/12	\$475.65
Fri	5/13	

Analyze the Problem We can find the mean (average) daily balance for the week by adding the daily balances and dividing by 5. If the mean is \$500 or more, there will be no service charge.

Form an Equation We can let $x =$ the minimum balance needed on Friday and translate the words into mathematical symbols.

The sum of the five daily balances | divided by | 5 | is | \$500 .

$$\frac{670.70 + 540.19 + (-60.39) + 475.65 + x}{5} = 500$$

Solve the Equation

$$\frac{670.70 + 540.19 + (-60.39) + 475.65 + x}{5} = 500$$

$$\frac{1{,}626.15 + x}{5} = 500 \quad \text{Simplify the numerator.}$$

$$5\left(\frac{1{,}626.15 + x}{5}\right) = 5(500) \quad \text{Multiply both sides by 5.}$$

$$1{,}626.15 + x = 2{,}500$$

$$x = 873.85 \quad \text{Subtract 1,626.15 from both sides.}$$

State the Conclusion On Friday, the account balance must be at least \$873.85 to avoid a service charge.

Check the Result Check the result by adding the five daily balances and dividing by 5.

THE MEDIAN

The Median The **median** of several values is the middle value. To find the median of several values.

1. Arrange the values in increasing order.
2. If there are an odd number of values, choose the middle value.
3. If there are an even number of values, add the middle two values and divide by 2.

EXAMPLE 3

Finding the median. In Example 1, the following values were the reaction times of ten people to a visual stimulus.

0.36, 0.24, 0.23, 0.41, 0.28, 0.25, 0.20, 0.28, 0.39, 0.26

Find the median of these values.

Solution To find the median, we first arrange the values in increasing order:

0.20, 0.23, 0.24, 0.25, 0.26, 0.28, 0.28, 0.36, 0.39, 0.41

Because there are an even number of values, the median is the sum of the middle two values, 0.26 and 0.28, divided by 2. Thus, the median is

$$\text{Median} = \frac{0.26 + 0.28}{2} = 0.27$$

The median reaction time is 0.27 second.

THE MODE

The Mode The **mode** of several values is the value that occurs most often.

EXAMPLE 4

Finding the mode. Find the mode of the following values.

0.36, 0.24, 0.23, 0.41, 0.28, 0.25, 0.20, 0.28, 0.39, 0.26

Solution Since the value 0.28 occurs most often, it is the mode.

If two different numbers in a distribution tie for occurring most often, there are two modes, and the distribution is called **bimodal.**

Although the mean is probably the most common measure of average, the median and the mode are frequently used. For example, workers' salaries are usually compared to the median (average) salary. To say that the modal (average) shoe size is 10 means that a shoe size of 10 occurs more often than any other shoe size.

APPENDIX I STUDY SET

PRACTICE

In Problems 1–3, use the following distribution of values: 7, 5, 9, 10, 8, 6, 6, 7, 9, 12, 9.

1. Find the mean.
2. Find the median.
3. Find the mode.

In Problems 4–6, use the following distribution of values: 8, 12, 23, 12, 10, 16, 26, 12, 14, 8, 16, 23.

4. Find the median.
5. Find the mode.
6. Find the mean.

7. Find the mean, median, and mode of the following values: 24, 27, 30, 27, 31, 30, and 27.
8. Find the mean, median, and mode of the following golf scores: 85, 87, 88, 82, 85, 91, 88, and 88.

APPLICATIONS

9. FOOTBALL The gains and losses made by a running back on seven plays were −8 yd, 2 yd, −6 yd, 6 yd, 4 yd, −7 yd, and −5 yd. Find his average (mean) yards per carry.
10. SALES If a clerk had the sales shown for one week, find the mean of her daily sales.

Monday	\$1,525
Tuesday	\$ 785
Wednesday	\$1,628
Thursday	\$1,214
Friday	\$ 917
Saturday	\$1,197

11. VIRUSES The table gives the approximate lengths (in centimicrons) of the viruses that cause five common diseases. Find the mean length of the viruses.

Polio	2.5
Influenza	105.1
Pharyngitis	74.9
Chicken pox	137.4
Yellow fever	52.6

12. SALARIES Ten workers in a small business have monthly salaries of \$2,500, \$1,750, \$2,415, \$3,240, \$2,790, \$3,240, \$2,650, \$2,415, \$2,415, and \$2,650. Find the average (mean) salary.
13. JOB TESTING To be accepted into a police training program, a recruit must have an average (mean) score of 85 on a battery of four tests. If a candidate scored 78 on the oral test, 91 on the physical test, and 87 on the psychological test, what is the lowest score she can obtain on the written test and be accepted into the program?
14. GAS MILEAGE Mileage estimates for four cars owned by a small business are shown. If the business buys a fifth car, what must its mileage average be so that the five-car fleet averages 20.8 mpg?

Model	City mileage (mpg)
Chevrolet Lumina	20.3
Jeep Cherokee	14.1
Ford Contour	28.2
Dodge Caravan	16.9

15. SPORT FISHING The weights (in pounds) of the trophy fish caught one week in Catfish Lake were 4, 7, 4, 3, 3, 5, 6, 9, 4, 5, 8, 13, 4, 5, 4, 6, and 9. Find the median and modal averages of the fish caught.
16. SALARIES Find the median and mode of the ten salaries given in Problem 12.

17. FUEL EFFICIENCY The ten most fuel-efficient cars in 1997, based on manufacturer's estimates, are shown. Find the median and mode of the city mileage estimates.

Model	mpg city/hwy
Geo Metro LSi	39/43
Honda Civic HX coupe	35/41
Honda Civic LX sedan	33/38
Mazda Protégé	31/35
Nissan Sentra GXE	30/40
Toyota Paseo	29/37
Saturn SL1	29/40
Dodge Neon Sport Coupe	29/38
Hyundai Accent	29/38
Toyota Tercel DX	28/38

18. FUEL EFFICIENCY Use the data for Problem 17 to find the median and mode of the highway mileage estimates.

WRITING

19. Explain why the mean of two numbers is halfway between the numbers.

20. Can the mean, median, and mode of a distribution be the same number? Explain.

21. Must the mean, median, and mode of a distribution be the same number? Explain.

22. Can the mode of a distribution be greater than the mean? Explain.

II Roots and Powers

n	n^2	$\sqrt{n}$	n^3	$\sqrt[3]{n}$	n	n^2	$\sqrt{n}$	n^3	$\sqrt[3]{n}$
1	1	1.000	1	1.000	51	2,601	7.141	132,651	3.708
2	4	1.414	8	1.260	52	2,704	7.211	140,608	3.733
3	9	1.732	27	1.442	53	2,809	7.280	148,877	3.756
4	16	2.000	64	1.587	54	2,916	7.348	157,464	3.780
5	25	2.236	125	1.710	55	3,025	7.416	166,375	3.803
6	36	2.449	216	1.817	56	3,136	7.483	175,616	3.826
7	49	2.646	343	1.913	57	3,249	7.550	185,193	3.849
8	64	2.828	512	2.000	58	3,364	7.616	195,112	3.871
9	81	3.000	729	2.080	59	3,481	7.681	205,379	3.893
10	100	3.162	1,000	2.154	60	3,600	7.746	216,000	3.915
11	121	3.317	1,331	2.224	61	3,721	7.810	226,981	3.936
12	144	3.464	1,728	2.289	62	3,844	7.874	238,328	3.958
13	169	3.606	2,197	2.351	63	3,969	7.937	250,047	3.979
14	196	3.742	2,744	2.410	64	4,096	8.000	262,144	4.000
15	225	3.873	3,375	2.466	65	4,225	8.062	274,625	4.021
16	256	4.000	4,096	2.520	66	4,356	8.124	287,496	4.041
17	289	4.123	4,913	2.571	67	4,489	8.185	300,763	4.062
18	324	4.243	5,832	2.621	68	4,624	8.246	314,432	4.082
19	361	4.359	6,859	2.668	69	4,761	8.307	328,509	4.102
20	400	4.472	8,000	2.714	70	4,900	8.367	343,000	4.121
21	441	4.583	9,261	2.759	71	5,041	8.426	357,911	4.141
22	484	4.690	10,648	2.802	72	5,184	8.485	373,248	4.160
23	529	4.796	12,167	2.844	73	5,329	8.544	389,017	4.179
24	576	4.899	13,824	2.884	74	5,476	8.602	405,224	4.198
25	625	5.000	15,625	2.924	75	5,625	8.660	421,875	4.217
26	676	5.099	17,576	2.962	76	5,776	8.718	438,976	4.236
27	729	5.196	19,683	3.000	77	5,929	8.775	456,533	4.254
28	784	5.292	21,952	3.037	78	6,084	8.832	474,552	4.273
29	841	5.385	24,389	3.072	79	6,241	8.888	493,039	4.291
30	900	5.477	27,000	3.107	80	6,400	8.944	512,000	4.309
31	961	5.568	29,791	3.141	81	6,561	9.000	531,441	4.327
32	1,024	5.657	32,768	3.175	82	6,724	9.055	551,368	4.344
33	1,089	5.745	35,937	3.208	83	6,889	9.110	571,787	4.362
34	1,156	5.831	39,304	3.240	84	7,056	9.165	592,704	4.380
35	1,225	5.916	42,875	3.271	85	7,225	9.220	614,125	4.397
36	1,296	6.000	46,656	3.302	86	7,396	9.274	636,056	4.414
37	1,369	6.083	50,653	3.332	87	7,569	9.327	658,503	4.431
38	1,444	6.164	54,872	3.362	88	7,744	9.381	681,472	4.448
39	1,521	6.245	59,319	3.391	89	7,921	9.434	704,969	4.465
40	1,600	6.325	64,000	3.420	90	8,100	9.487	729,000	4.481
41	1,681	6.403	68,921	3.448	91	8,281	9.539	753,571	4.498
42	1,764	6.481	74,088	3.476	92	8,464	9.592	778,688	4.514
43	1,849	6.557	79,507	3.503	93	8,649	9.644	804,357	4.531
44	1,936	6.633	85,184	3.530	94	8,836	9.695	830,584	4.547
45	2,025	6.708	91,125	3.557	95	9,025	9.747	857,375	4.563
46	2,116	6.782	97,336	3.583	96	9,216	9.798	884,736	4.579
47	2,209	6.856	103,823	3.609	97	9,409	9.849	912,673	4.595
48	2,304	6.928	110,592	3.634	98	9,604	9.899	941,192	4.610
49	2,401	7.000	117,649	3.659	99	9,801	9.950	970,299	4.626
50	2,500	7.071	125,000	3.684	100	10,000	10.000	1,000,000	4.642

Appendix

III Answers to Selected Exercises

Study Set Section 1.1 (page 6)

1. sum, difference **3.** Variables **5.** equation **7.** horizontal, vertical **9.** equation **11.** algebraic expression **13.** algebraic expression **15.** equation **17. a.** multiplication, subtraction **b.** x **19. a.** addition, subtraction **b.** m **21.**

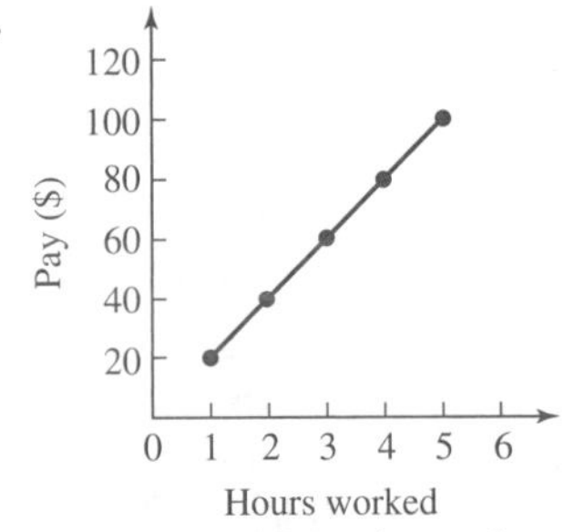

23. They determine that 15-year-old machinery is worth \$35,000. **25.** $5 \cdot 6$, $5(6)$ **27.** $34 \cdot 75$, $34(75)$ **29.** $4x$ **31.** $3rt$ **33.** lw **35.** Prt **37.** $2w$ **39.** xy **41.** $\frac{32}{x}$ **43.** $\frac{90}{30}$ **45.** The product of 8 and 2 is 16. **47.** The difference of 11 and 9 is 2. **49.** The product of 2 and x is 10. **51.** The quotient of 66 and 11 is 6. **53.** $p = 100 - d$ **55.** $7d = h$ **57.** $s = 3c$ **59.** $w = e + 1{,}200$ **61.** $p = r - 600$ **63.** $\frac{l}{4} = m$ **65.** 390, 400, 405 **67.** 1,300, 1,200, 1,100 **69.** $d = \frac{e}{12}$ **71.** 90, 60, 45, 30, 0

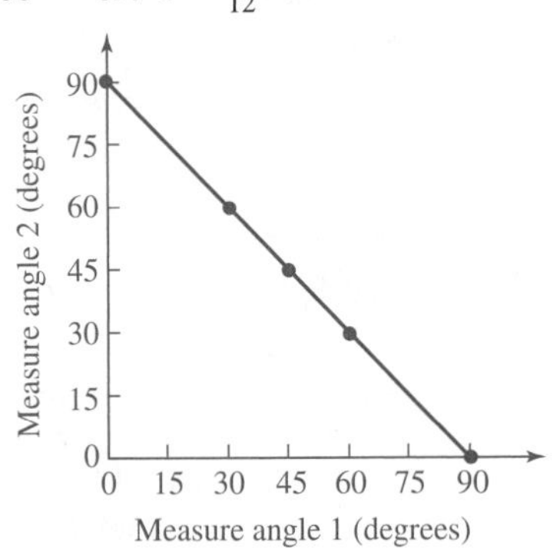

Study Set Section 1.2 (page 18)

1. prime **3.** numerator, denominator **5.** equivalent **7.** least or lowest **9.** 60 **11.** $\frac{4}{12} = \frac{1}{3}$ **13.** $\frac{2}{5}$ **15. a.** 1 **b.** 1 **17. a.** 3 times **b.** 2 times **19. a.** $\frac{5}{16}$ **b.** 1 **c.** $\frac{15}{48}$ **21.** 1, 2, 4, 5, 10, 20 **23.** 1, 2, 4, 7, 14, 28 **25.** $3 \cdot 5 \cdot 5$ **27.** $2 \cdot 2 \cdot 7$ **29.** $3 \cdot 3 \cdot 13$ **31.** $2 \cdot 2 \cdot 5 \cdot 11$ **33.** $\frac{3}{9}$ **35.** $\frac{24}{54}$ **37.** $\frac{35}{5}$ **39.** $\frac{1}{2}$ **41.** $\frac{4}{3}$ **43.** $\frac{3}{4}$ **45.** $\frac{9}{8}$ **47.** lowest terms **49.** $\frac{4}{25}$ **51.** $\frac{3}{10}$ **53.** $\frac{8}{5}$ **55.** $\frac{3}{2}$ **57.** 70 **59.** $10\frac{1}{2}$ **61.** $13\frac{3}{4}$ **63.** $\frac{9}{10}$ **65.** $\frac{5}{8}$ **67.** $\frac{14}{5}$ **69.** 28 **71.** $1\frac{9}{11}$ **73.** $2\frac{1}{2}$ **75.** $\frac{6}{5}$ **77.** $\frac{5}{24}$ **79.** $\frac{19}{15}$ **81.** $\frac{3}{4}$ **83.** $\frac{17}{12}$ **85.** $\frac{22}{35}$ **87.** $\frac{1}{6}$ **89.** $\frac{9}{4}$ **91.** $1\frac{1}{4}$ **93.** $\frac{5}{9}$ **95. a.** $\frac{7}{32}$ in. **b.** $\frac{3}{32}$ in. **97.** $40\frac{1}{2}$ in. **103.** The difference of 7 and 5 is 2. **105.** The quotient of 30 and 15 is 2. **107.** 150, 180

Study Set Section 1.3 (page 28)

1. whole **3.** integers **5.** negative, positive **7.** rational **9.** irrational **11.** real **13.** opposites **15.** $\frac{6}{1}, \frac{-9}{1}, \frac{-7}{8}, \frac{7}{2}, \frac{-3}{10}, \frac{283}{100}$ **17.** 13 and -3 **19. a.** $<$ **b.** $>$ **c.** $>, <$ **d.** $>$ **21.** π in. **23.** square root **25.** is not equal to **27.** Greek **29.** $\frac{-4}{5}, \frac{4}{-5}$ **31.** 0.625 **33.** $0.0\overline{3}$ **35.** 0.42 **37.** $0.\overline{45}$ **39.** $>$ **41.** $>$ **43.** $>$ **45.** $<$ **47.** $<$ **49.** $=$ **51.** $=$ **53.** $>$ **55.** $>$ **57. a.** true **b.** false **c.** false **d.** true **59. a.** $-5 > -6$ **b.** $-25 < 16$ **61.**

$-\frac{35}{8}$ $-\pi$ $-1\frac{1}{2}$ $-0.333...$ $\sqrt{2}$ 3   4.25

-5 -4 -3 -2 -1 0 1 2 3 4 5

63. natural, whole, integers: 9; rational: 9, $\frac{15}{16}$, $3\frac{1}{8}$, 1.765; irrational: 2π, 3π, $\sqrt{89}$; real: all **65.** shell 1; $|-6| > |5|$ **67. a.** ’00: $-$\$81 billion; ’99: $-$\$74 billion; ’02: $-$\$70 billion **b.** ’90; $-$\$40 billion **73.** $\frac{4}{9}$ **75.** $\frac{6}{5}$ **77.** $\frac{13}{30}$

Study Set Section 1.4 (page 37)

1. signed **3.** opposites **5.** identity **7.** $-1 + (-3) = -4$ **9.** -5 **11.** -3 **13. a.** 0 **b.** a **c.** a **d.** $(b + c)$ **15.** positive **17. a.** $1 + (-5)$ **b.** $-80.5 + 15$ **c.** $(20 + 4)$ **19. a.** 0 **b.** 0 **c.** -6 **d.** $-\frac{15}{16}$ **e.** 0 **f.** 0 **21.** $x + y = y + x$ **23.** $8 + 9$

25. −2 **27.** 2 **29.** −8 **31.** 8 **33.** 0 **35.** −77 **37.** 4 **39.** −35 **41.** −8.2 **43.** −20.1 **45.** 0.2 **47.** $-\frac{1}{8}$ **49.** $\frac{5}{12}$ **51.** $-\frac{9}{10}$ **53.** 16 **55.** −15 **57.** −21 **59.** −26 **61.** 0 **63.** 4 **65.** −5 **67.** 1 **69.** −3.2 **71.** 10 **73.** 215 **75.** −112 **77.** 2,150 m **79.** −18, −6, −5, −4 **81. a.** 3,660.66, 1,408.78 **b.** −1,242.86 **83.** $89 million **85.** southward, 132 km **87.** $2.1 million

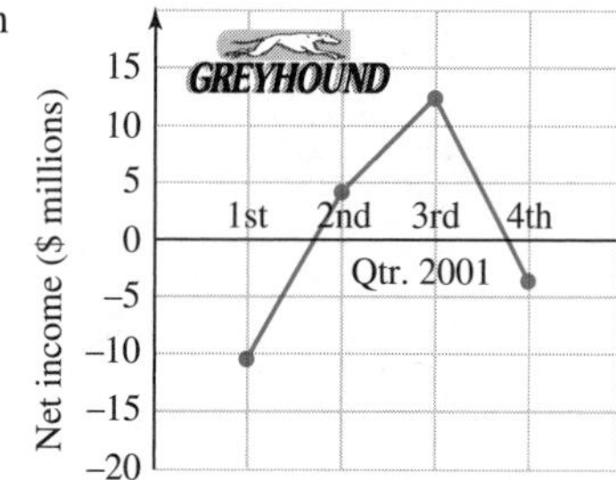

89. true **91.** −9 and 3

Study Set Section 1.5 (page 44)

1. opposites, inverses **3.** range **5. a.** −12 **b.** $\frac{1}{5}$ **c.** −2.71 **d.** 0 **7. a.** number **b.** opposite **9.** + (−9) **11.** no **13.** −10 + (−8) + (−23) + 5 + 34 **15. a.** −500 **b.** y **17.** −3 **19.** −13 **21.** 11 **23.** −21 **25.** −6 **27.** 1 **29.** 12 **31.** 2 **33.** 40 **35.** 0 **37.** 5 **39.** −9 **41.** −88 **43.** 12 **45.** 0 **47.** −4 **49.** −2 **51.** $-\frac{1}{2}$ **53.** $-\frac{5}{16}$ **55.** $-\frac{5}{12}$ **57.** −1.1 **59.** −3.5 **61.** −2.3 **63.** 4.6 **65.** 22 **67.** −25 **69.** −11 **71.** −50 **73.** −7 **75.** −1 **77.** −1 **79.** 0 **81.** 160°F **83.** 1,030 ft **85.** −5.75° **87.** 428 B.C. (−428) **89.** left; −8 **95.** 2 · 3 · 5 **97.** $\frac{21}{56}$ **99.** true

Study Set Section 1.6 (page 54)

1. product, quotient **3.** commutative **5.** undefined **7.** −5 **9.** positive **11.** 1 **13. a.** One of the numbers is 0. **b.** They are reciprocals (multiplicative inverses). **15. a.** NEG **b.** not possible to tell **c.** POS **d.** NEG **17. a.** a **b.** $a(bc)$ **c.** 0 **d.** a **e.** 1 **19. a.** associative property of multiplication **b.** multiplicative inverse **c.** commutative property of multiplication **d.** multiplication property of 1 **21. a.** $-4(-5) = 20$ **b.** $\frac{16}{-8} = -2$ **23.** −16 **25.** 54 **27.** −60 **29.** −24 **31.** −800 **33.** 36 **35.** 2.4 **37.** −0.48 **39.** 0.99 **41.** −15.12 **43.** $-\frac{3}{8}$ **45.** $\frac{1}{12}$ **47.** $-\frac{3}{20}$ **49.** $\frac{15}{16}$ **51.** −520 **53.** 0 **55.** 0 **57.** 60 **59.** 84 **61.** 120 **63.** $\frac{1}{24}$ **65.** −720 **67.** 10 **69.** 3 **71.** −2 **73.** −4 **75.** −1 **77.** 1 **79.** −4 **81.** −20 **83.** −0.005 **85.** 0 **87.** undefined **89.** $-\frac{5}{12}$ **91.** $\frac{15}{4}$ **93.** $1\frac{1}{2}$ **95.** −4.7 **97.** 30.3 **99.** −67 **101.** −6 **103.** −72° **105.** −280 **107.** −193° F **109.** −$43.32 million **111.** −$614,516 **113. a.** 5, −10 **b.** 2.5, −5 **c.** 7.5, −15 **d.** 10, −20 **119.** −5 **121.** $1.08\overline{3}$ **123.** $\frac{3}{7}$

Study Set Section 1.7 (page 65)

1. base, exponent **3.** power **5.** order **7. a.** 54, 34 **b.** 34; multiplication is to be done before addition. **9.** innermost: parentheses; outermost: brackets **11. a.** subtraction, power, addition, multiplication **b.** power, multiplication, subtraction, addition **13. a.** subtraction **b.** division **15. a.** subtraction **b.** power **c.** power **d.** power **17. a.** 3 **b.** x **c.** 1 **d.** 1 **19. a.** −5 **b.** 5 **21.** 3, 9, 18 **23.** 3^4 **25.** 10^2k^3 **27.** $8\pi r^3$ **29.** $6x^2y^3$ **31.** 36 **33.** −256 **35.** −125 **37.** −1,296 **39.** 0.16 **41.** $-\frac{8}{125}$ **43.** −17 **45.** 192 **47.** 38 **49.** 80 **51.** −34 **53.** −28 **55.** 194 **57.** −44 **59.** 0 **61.** −38 **63.** −8 **65.** 12 **67.** 8 **69.** undefined **71.** 201 **73.** 50 **75.** 20 **77.** −396 **79.** 343 **81.** 360 **83.** 12 **85.** −10 **87.** undefined **89.** 59 **91.** 28 **93.** 1,000 **95.** −54 **97.** $\frac{1}{8}$ **99.** −8 **101.** 31 **103.** −39 **105.** −27 **107.** −8 **109.** 11 **111.** $-\frac{8}{9}$ **113.** 1 **115.** 10 **117.** 12 **119.** −1 **121.** 2^2 square units, 3^2 square units, 4^2 square units **123.** $2,106 **125. a.** $11,875 **b.** $95 **127.** 81 in. **133. a.** ii **b.** iii **c.** iv **d.** i

Study Set Section 1.8 (page 75)

1. expressions **3.** constant **5.** evaluate **7.** expression, equation **9. a.** 3 **b.** 11 **c.** −6 **d.** −9 **11. a.** term **b.** factor **c.** factor **d.** term **e.** factor **f.** term **13. a.** 7, 14, 21, $7w$ **b.** 1, 2, 3, $\frac{s}{60}$ **15. a.** x = weight of the car; $2x - 500$ = weight of the van **b.** 3,500 lb **17.** 5, 30; 10, $10d$; 50, $50(x + 5)$ **19.** 5, 25, 45 **21. a.** $8y$ **b.** $2cd$ **c.** $15sx$ **d.** $-9a^3b^2$ **23.** $l + 15$ **25.** $50x$ **27.** $\frac{w}{l}$ **29.** $P + p$ **31.** $k^2 - 2{,}005$ **33.** $J - 500$ **35.** $\frac{1{,}000}{n}$ **37.** $p + 90$ **39.** $35 + h + 300$ **41.** $p - 680$ **43.** $4d - 15$ **45.** $2(200 + t)$ **47.** $|a - 2|$ **49.** 300; $60h$ **51. a.** $3y$ **b.** $\frac{f}{3}$ **53.** $29x$¢ **55.** $\frac{c}{6}$ **57.** $5b$ **59.** $\$5(x + 2)$ **61.** −1, −2, −28 **63.** 41, 11, 2 **65.** 150, −450 **67.** 0, 0, 5 **69.** 20 **71.** −12 **73.** −5 **75.** 156 **77.** $-\frac{1}{5}$ **79.** 17 **81.** 36 **83.** 230 **85.** 50 **87.** 48, 64, 48, 0 **89.** −37° C, −64° C **91.** $1\frac{23}{64}$ in.^{2} **97.** 60 **99.** −225

Key Concept (page 81)

1. f **2.** j **3.** h **4.** a **5.** b **6.** e **7.** c **8.** i **9.** d **10.** g **11.** $C = p + t$ (Answers may vary depending on the variables chosen.) **12.** $b = 2t$ (Answers may vary depending on the variables chosen.) **13.** $x + 4$ = amount of business ($ millions) in the year with the celebrity **14.** 1, 41, 97

Chapter Review (page 82)

1. 1 hr; 100 cars **2.** 100 **3.** 7 P.M. **4.** The difference of 15 and 3 is 12. **5.** The sum of 15 and 3 is 18. **6.** The quotient of 15 and 3 is 5. **7.** The product of 15 and 3 is 45. **8.** 4 · 9; 4(9) **9.** $\frac{9}{3}$ **10.** $8b$ **11.** Prt **12.** equation **13.** expression **14.** 10, 15, 25 **15. a.** 2 · 12, 3 · 8 (answers may vary) **b.** 2 · 2 · 6 (answers may vary) **c.** 1, 2, 3, 4, 6, 8, 12, 24 **16.** $3^3 \cdot 2$ **17.** $7^2 \cdot 3$ **18.** 11 · 7 · 5 **19.** prime **20.** 1 **21.** 0 **22.** $\frac{4}{7}$ **23.** $\frac{4}{3}$ **24.** $\frac{40}{64}$ **25.** $\frac{36}{3}$ **26.** $\frac{7}{64}$ **27.** $\frac{5}{21}$ **28.** $\frac{16}{45}$ **29.** $3\frac{1}{4}$ **30.** $\frac{2}{5}$ **31.** $\frac{5}{22}$ **32.** $\frac{11}{12}$ **33.** $\frac{5}{18}$ **34.** $\frac{17}{96}$ in. **35.** 0 **36.** −$65 billion **37.** −206 ft **38.** < **39.** > **40.** $\frac{7}{10}$ **41.** $\frac{14}{3}$ **42.** 0.004 **43.** $0.7\overline{72}$

44. $-\frac{17}{4}$ −2 0.333... $\frac{7}{8}$ $\sqrt{2}$ π 3.75
−5 −4 −3 −2 −1 0 1 2 3 4 5

45. false **46.** false **47.** true **48.** true **49.** natural: 8; whole: 0, 8; integers: 0, −12, 8; rational: $-\frac{4}{5}$, 99.99, 0, −12, $4\frac{1}{2}$, 0.666 . . . , 8; irrational: $\sqrt{2}$; real: all **50.** > **51.** < **52.** −82 **53.** 12 **54.** −7 **55.** 0 **56.** −11 **57.** −12.3 **58.** $-\frac{3}{16}$ **59.** 11 **60.** commutative property of addition **61.** associative property of addition **62.** addition property of opposites **63.** addition property of 0 **64.** −1 **65.** −10 **66.** 3 **67.** $\frac{9}{16}$ **68.** −4 **69.** −19 **70.** −49 **71.** −15 **72.** 5.7 **73.** −10 **74.** −29 **75.** 65,233 ft **76.** −56 **77.** 54 **78.** 12 **79.** 36 **80.** 6.36 **81.** −2 **82.** $-\frac{2}{15}$ **83.** 0 **84.** associative property of multiplication **85.** commutative property of multiplication **86.** multiplication property of 1 **87.** inverse property of multiplication **88.** 3 **89.** $-\frac{1}{3}$ **90.** −1 **91.** −4 **92.** 3 **93.** $-\frac{6}{5}$ **94.** undefined **95.** −4.5 **96.** high: 2, low: −3 **97.** high: 4, low: −6 **98.** 8^5 **99.** a^4 **100.** $9\pi r^2$ **101.** x^3y^4 **102.** 81 **103.** $-\frac{8}{27}$ **104.** 32 **105.** 50 **106.** 4; power, multiplication, subtraction, addition **107.** 17 **108.** −48 **109.** −9 **110.** 44 **111.** −420 **112.** $-\frac{14}{19}$ **113.** 113 **114.** 0 **115.** $20 **116.** 3 **117.** 1 **118.** 2, −5 **119.** 16, −5, 25 **120.** $\frac{1}{2}$, 1 **121.** 9.6, −1 **122.** $h + 25$ **123.** $s - 15$ **124.** $\frac{1}{2}t$ **125.** $(n + 4)$ in. **126.** $(b - 4)$ in. **127.** $10d$ **128.** $(x - 5)$ years **129.** 30, $10d$ **130.** 0, 19, −16 **131.** 40 **132.** −36

Chapter 1 Test (page 90)

1. $24 **2.** 5 hr **3.** 3, 20, 70 **4.** $2 \cdot 2 \cdot 3 \cdot 3 \cdot 5 = 2^2 \cdot 3^2 \cdot 5$ **5.** $\frac{2}{5}$ **6.** $\frac{3}{2} = 1\frac{1}{2}$ **7.** $\frac{25}{36}$ **8.** $10\frac{1}{15}$ **9.** $3.57 **10.** $0.8\overline{3}$

11. −3.75 −3 $-1\frac{1}{4}$ 0.5 $\sqrt{2}$ $\frac{7}{2}$
−5 −4 −3 −2 −1 0 1 2 3 4 5

12. a. true **b.** false **c.** true **d.** true **14. a.** > **b.** < **c.** < **d.** > **15.** a gain of 0.6 of a rating point **16.** −2 **17.** $\frac{3}{8}$ **18.** −6 **19.** −30 **20.** 2 **21.** −2.44 **22.** 0 **23.** −3 **24.** 0 **25.** 50 **26.** $-\frac{27}{125}$ **27.** 14 **28.** associative property of addition **29. a.** 9^5 **b.** $3x^2z^3$ **30.** 170 **31.** 36 **32.** −12 **33.** −100 **34.** 36 **35.** 4, 17, −59 **36.** $2w + 7$ **37.** $x - 2$ **38.** $25q$ ¢ **40.** 3; 5

Study Set Section 2.1 (page 100)

1. equation **3.** check **5.** equivalent **7.** isolate **9. a.** $x + 5 = 7$ **b.** subtract 5 from both sides **11. a.** $x + 6$ **b.** neither **c.** no **d.** yes **13.** 24 **15.** n **17. a.** c, c **b.** c, c **19. a.** x **b.** y **c.** t **d.** h **21.** 15, 15, 30, 30 **23. a.** is possibly equal to **b.** 27° **25.** yes **27.** no **29.** no **31.** no **33.** no **35.** no **37.** no **39.** yes **41.** yes **43.** 3 **45.** 71 **47.** 9 **49.** 0 **51.** −9 **53.** −3 **55.** −2.3 **57.** −36 **59.** 13 **61.** $\frac{8}{9}$ **63.** $\frac{7}{25}$ **65.** 4 **67.** 41 **69.** 0 **71.** 1 **73.** −6 **75.** 20 **77.** 0.5 **79.** 45 **81.** 0 **83.** −105 **85.** 21 **87.** −2.64 **89.** 1,251,989 **91.** −28 **93.** 65° **95.** 38° **101.** 0 **103.** $45 - x$

Study Set Section 2.2 (page 110)

1. variable **3.** variable, equation, solve **5.** amount, percent, base **7.** $x + 371 + 479 = 1{,}240$ **9.** $x + 11{,}000 = 13{,}500$ **11.** $x + 5 + 8 + 16 = 31$ **13.** ____ is ____ % of ____ ? **15.** $12 = 0.40 \cdot x$ **17. a.** 0.35 **b.** 0.035 **c.** 3.5 **d.** 0.005 **19.** 312 **21.** 26% **23.** 300 **25.** 46.2 **27.** 2.5% **29.** 1,464 **31.** $6x = 330$ **33.** 63 **35.** $322.00 **37.** 27 min **39.** 16 **41.** 5 **43.** 54 **45.** 975 mi. **47.** 54 ft **49.** 135° **51.** 0.48 oz **53.** $684 billion **55.** 78.125% **57.** 19% **59.** 120 **61. a.** 5 g; 25% **b.** 20 g **63.** 1994–1995; about 9.8% **65.** 12% **71.** $\frac{12}{5} = 2\frac{2}{5}$ **73.** no

Study Set Section 2.3 (page 121)

1. simplify **3.** expressions, equations **5.** opposite **7.** coefficient **9. a.** 5, 6, 30 **b.** −8, 2, 4 **11.** They are not like terms. **13.** −1, sign **15. a.** $3a$, $2a$ **b.** 10, 12 **c.** none **d.** $9y^2$, $-8y^2$ **17. a.** 4 + 6, 10 **b.** 30 − 50, −20 **c.** 27 **19. a.** $6(h - 4)$ **b.** $-(z + 16)$ **21.** no, yes, no, no, yes, yes **23.** $63m$ **25.** $-35q$ **27.** $300t$ **29.** $11.2x$ **31.** g **33.** $5x$ **35.** $6y$ **37.** s **39.** $-20r$ **41.** $60c$ **43.** $-96m$ **45.** $5x + 15$ **47.** $36c - 42$ **49.** $24t + 16$ **51.** $0.4x - 1.6$ **53.** $5t + 5$ **55.** $-12x - 20$ **57.** $-78c + 18$ **59.** $-2w + 4$ **61.** $9x + 10$ **63.** $9r - 16$ **65.** $-x + 7$ **67.** $5.6y - 7$ **69.** $40d + 50$ **71.** $-12r - 60$ **73.** $x + y - 5$ **75.** $6x - 21y - 16z$ **77.** $20x$ **79.** 0 **81.** 0 **83.** r **85.** $37y$ **87.** $-s^3$ **89.** 5 **91.** $-10r$ **93.** $3a$ **95.** $-3x$ **97.** x **99.** $\frac{4}{5}t$ **101.** $0.4r$ **103.** $7z - 15$ **105.** $-2x + 5$ **107.** $20d - 66$ **109.** $-3c - 1$ **111.** $s - 12$ **113.** $12c + 34$ **115.** $8x - 9$ **117.** $12x$ **119.** $(4x + 8)$ ft **123.** 0 **125.** 2

Study Set Section 2.4 (page 131)

1. equal **3.** original **5.** identity **7.** subtraction, multiplication **9.** multiplying **11.** $-\frac{5}{4}$ **13.** 30 **15.** 6 **17.** 7, 7, 28, 2, 2 **19. a.** -1 **b.** $\frac{3}{5}$ **21.** 6 **23.** 5 **25.** -7 **27.** -20 **29.** 4 **31.** 2.9 **33.** $\frac{10}{3}$ **35.** -4 **37.** $-\frac{8}{3}$ **39.** 12 **41.** -48 **43.** -12 **45.** 5 **47.** $-\frac{17}{4}$ **49.** 0.04 **51.** -6 **53.** 0 **55.** $\frac{1}{7}$ **57.** $\frac{1}{4}$ **59.** -1 **61.** -41 **63.** 1 **65.** $\frac{9}{2}$ **67.** -7.2 **69.** -82 **71.** 0 **73.** -20 **75.** 3 **77.** 28 **79.** $-\frac{12}{5}$ **81.** $\frac{2}{15}$ **83.** $\frac{27}{5}$ **85.** $\frac{52}{9}$ **87.** $\frac{5}{4}$ **89.** -5 **91.** 80 **93.** 4 **95.** all real numbers **97.** no solution **99.** no solution **101.** all real numbers **103.** 2,991,980 **109.** 0 **111.** $\frac{1}{64}$ **113.** $16x$

Study Set Section 2.5 (page 139)

1. formula **3.** perimeter **5.** radius **7.** circumference **9. a.** $d = rt$ **b.** $r = c + m$ **c.** $p = r - c$ **d.** $I = Prt$ **e.** $C = 2\pi r$ **11.** 11,176,920 mi, 65,280 ft **13. a.** volume **b.** circumference **c.** area **d.** perimeter **15. a.** $(2x + 10)$ cm **b.** $(2x + 6)$ cm^2 **17.** Ax, Ax, By, B, B **19. a.** 3.14 **b.** $98 \cdot \pi$ **c.** the radius of the cylinder; the height of the cylinder **21.** 2.5 mph **23.** $65 million **25.** 3.5% **27.** 4,014°F **29.** $24.55 **31.** about 132 in. **33.** $R = \frac{E}{I}$ **35.** $w = \frac{V}{lh}$ **37.** $r = \frac{C}{2\pi}$ **39.** $h = \frac{3A}{B}$ **41.** $f = \frac{s}{w}$ **43.** $b = P - a - c$ **45.** $r = \frac{T - 2t}{2}$ **47.** $x = \frac{C - By}{A}$ **49.** $m = \frac{2K}{v^2}$ **51.** $c = 3A - a - b$ **53.** $t = T - 18E$ **55.** $r^2 = \frac{s}{4\pi}$ **57.** $v^2 = \frac{2Kg}{w}$ **59.** $r^3 = \frac{3V}{4\pi}$ **61.** $M = 4.2B + 19.8$ **63.** $h = \frac{S - 2\pi r^2}{2\pi r}$ **65.** $y = 9 - 3x$ **67.** $y = \frac{1}{3}x + 3$ **69.** $y = -\frac{3}{4}x - 4$ **71.** $b = \frac{2A}{h} - d$ or $b = \frac{2A - hd}{h}$ **73.** $c = \frac{72 - 8w}{7}$ **75.** 212°F, 0°C **77.** 1,174.6, 956.9 **79.** 36 ft, 48 ft^2 **81.** 50.3 in., 201.1 in.^{2} **83.** 56 in., 144 in.^{2} **85.** 2,450 ft^2 **87.** 27.75 in., 47.8125 in.^{2} **89.** 32 ft^2, 128 ft^3 **91.** 348 ft^3 **93.** 254 in.^{2} **95.** $D = \frac{L - 3.25r - 3.25R}{2}$ **101.** 137.76 **103.** 15%

Study Set Section 2.6 (page 153)

1. perimeter **3.** vertex, base **5. a.** $17, x + 2, 3x$ **b.** 3 ft, 5 ft, 9 ft **7.** 180° **9.** $30,000, 14%, 1 yr **11.** $35t, t, 45t$ **13. a.** $0.06x, 10 - x, 0.03(10 - x), 0.05(10)$ **b.** $0.50(6), 0.25(x), 6 + x, 0.30(6 + x)$ **15.** To multiply a decimal by 100, move the decimal point two places to the right. **17.** Parentheses are needed: $2(2w - 3) + 2w$ **19.** 6,000 **21.** 4 ft, 8 ft **23.** 102 mi, 108 mi, 114 mi, 120 mi **25.** Australia: 12 wk; Japan: 16 wk; Sweden: 10 wk **27.** 250 calories in ice cream, 600 calories in pie **29.** in millions of dollars: $110, $229, $189, $847 **31.** 7 ft, 7 ft, 11 ft **33.** 75 m by 480 m **35.** 20° **37.** 22°, 68° **39.** 17 **41.** 90 **43.** $4,900 **45.** $42,200 at 12%, $22,800 at 6.2% **47.** $7,500 **49.** 2 hr **51.** 4 hr into the flights **53.** 4 hr **55.** $1\frac{1}{3}$ liters **57.** 7.5 oz **59.** 20 gal **61.** 50 lb **63.** 40 lb lemon drops, 60 lb jelly beans **69.** $-50x + 125$ **71.** $3x + 3$ **73.** $16y - 16$

Study Set Section 2.7 (page 167)

1. inequality **3.** solve **5.** interval **7. a.** true **b.** false **c.** true **d.** false **9. a.** $-2 \le 17$ **b.** $x > 32$ **11. a.** same **b.** positive **13. a.**

b. $(8, \infty)$ **c.** all real numbers greater than 8 **15.** three **17. a.** is less than, is greater than **b.** is greater than, or equal to **19.** [, (, ∞, $-\infty$ **21.** 5, 5, 12, 4, 4, 3

23. $(-\infty, 5)$
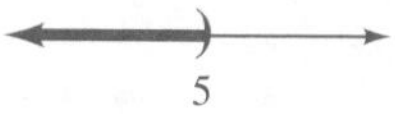

25. $(-3, 1]$

27. $x < -1, (-\infty, -1)$ **29.** $-7 < x \le 2, (-7, 2]$

31. $x > 3, (3, \infty)$
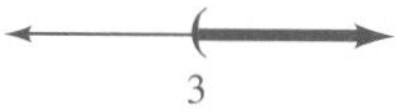

33. $x < -1, (-\infty, -1)$
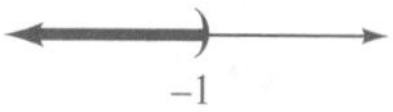

35. $g \ge 10, [10, \infty)$
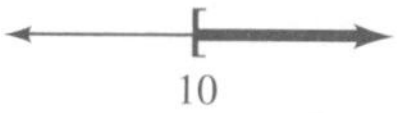

37. $x \ge 3, [3, \infty)$
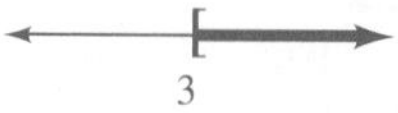

39. $y \le -40, (-\infty, -40]$
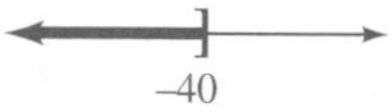

41. $x > \frac{6}{7}, \left(\frac{6}{7}, \infty\right)$
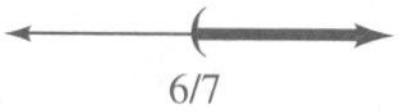

43. $x \le 0.4, (-\infty, 0.4]$

45. $y \ge 20, [20, \infty)$
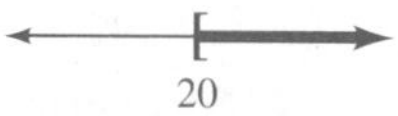

47. $x \ge -24, [-24, \infty)$

49. $n \le 2, (-\infty, 2]$
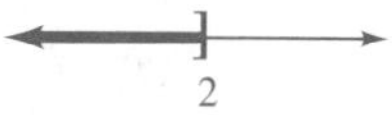

51. $m < 0, (-\infty, 0)$
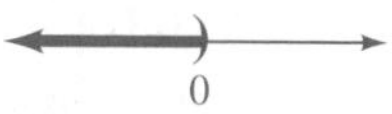

53. $x \ge -10, [-10, \infty)$
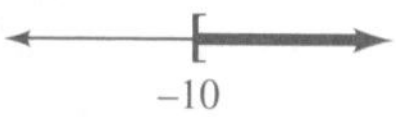

55. $x < -2, (-\infty, -2)$
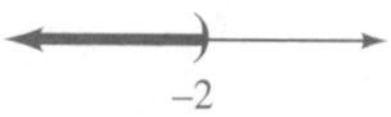

57. $x < -\frac{11}{4}, \left(-\infty, -\frac{11}{4}\right)$
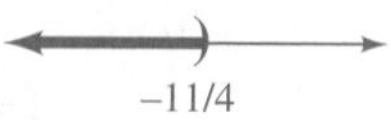

59. $x \le -1, (-\infty, -1]$
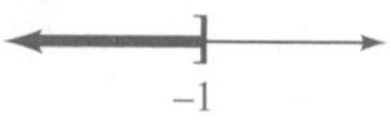

61. $x \ge -13, [-13, \infty)$
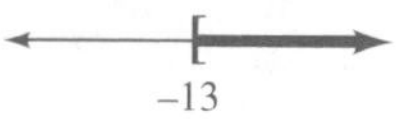

63. $x > 0, (0, \infty)$
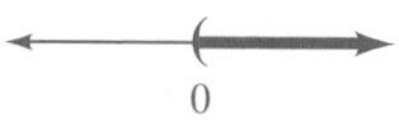

65. $x \le 1.5, (-\infty, 1.5]$

67. $a > 6, (6, \infty)$

69. $x \ge \frac{9}{4}, \left[\frac{9}{4}, \infty\right)$

71. $x \le 20, (-\infty, 20]$

73. $n > \frac{5}{4}, \left(\frac{5}{4}, \infty\right)$

75. $y \le \frac{1}{8}, \left(-\infty, \frac{1}{8}\right]$

77. $x \le \frac{3}{2}, \left(-\infty, \frac{3}{2}\right]$

79. $x \ge 3, [3, \infty)$

81. $7 < x < 10, (7, 10)$

83. $-10 \le x \le 0, [-10, 0]$

85. $-6 \le c \le 10, [-6, 10]$

87. $2 \le x < 3, [2, 3)$

89. $-5 < x < -2, (-5, -2)$

91. $-1 \le x < 2, [-1, 2)$

93. $x \ge 0.03, [0.03, \infty)$

95. 98% or better **97.** 27 mpg or better **99.** 0 ft $< s \le$ 19 ft **101.** $x \ge 35$ ft **103. a.** $0° < a \le 18°$ **b.** $18° \le a \le 50°$ **c.** $30° \le a \le 37°$ **d.** $75° \le a < 90°$ **105. a.** 26 lb $\le w \le$ 31 lb **b.** 12 lb $\le w \le$ 14 lb **c.** 18.5 lb $\le w \le$ 20.5 lb **d.** 11 lb $\le w \le$ 13 lb **109.** -125 **111.** 1, -3, 6

Key Concept (page 171)

1. a. $2x - 8$ **b.** $x = 6$ **2. a.** $y + 5$ **b.** $y = -5$ **3. a.** $\frac{2}{3}a$ **b.** $a = \frac{1}{2}$ **4. a.** $-2x - 10$ **b.** $x \le 5$ **5. a.** 0 **b.** all real numbers **6.** The mistake is on the third line. The student made an equation out of the answer, which is $x - 6$, by writing $0 =$ on the left. Then the student solved that equation.

Chapter Review (page 172)

1. yes **2.** no **3.** no **4.** no **5.** yes **6.** yes **7.** variable, true **8.** 21 **9.** -47 **10.** 13.2 **11.** 107 **12.** 8 **13.** 1 **14.** -96 **15.** 7.8 **16.** 0 **17.** 0 **18.** 160° **19.** 5 **20.** 429 mi **21.** 60° **22.** \$54 billion **23.** \$26.74 **24.** 192.4 **25.** no **26.** 1,567% **27.** $-28w$ **28.** $15r$ **29.** $24x$ **30.** $2.08f$ **31.** $9a$ **32.** r **33.** $5x + 15$ **34.** $-4x - 6 + 2y$ **35.** $-a + 4$ **36.** $3c - 6$ **37.** $20x + 32$ **38.** $-12.6c - 29.4$ **39.** $9p$ **40.** $-7m$ **41.** $4n$ **42.** $-p - 18$ **43.** $0.1k$ **44.** $-8a^3$ **45.** w **46.** $4h - 15$ **47.** $(4x + 4)$ ft **48.** 2 **49.** -30.6 **50.** 30 **51.** -19 **52.** 4 **53.** 1 **54.** $\frac{5}{4}$ **55.** -6 **56.** 6 **57.** $-\frac{22}{75}$ **58.** identity; all real numbers **59.** contradiction; no solution **60.** \$176 **61.** \$11,800 **62.** 3.00 hr **63.** 1,949°F **64.** 168 in. **65.** 1,440 in.^{2} **66.** 76.5 m^2 **67.** 144 in.^{2} **68.** 50.27 cm **69.** 201.06 cm^2 **70.** 4,320 in.^{3} **71.** 9.4 ft^3 **72.** 120 ft^3 **73.** 381.70 in.^{3} **74.** $h = \frac{A}{2\pi r}$ **75.** $G = 3A - 3BC + K$ **76.** $b^2 = c^2 - a^2$ **77.** $y = \frac{3}{4}x + 4$ **78.** 8 ft **79.** 12, 4 **80.** 24.875 in. × 29.875 in. $\left(24\frac{7}{8}\text{ in.} \times 29\frac{7}{8}\text{ in.}\right)$ **81.** 76.5°, 76.5° **82.** \$45x **83.** \$16,000 at 7%, \$11,000 at 9% **84.** 20 **85.** 10 lb of each **86.** $0.12x$ gal

87. $x < 1, (-\infty, 1)$

88. $x \le 12, (-\infty, 12]$

89. $d > \frac{5}{4}, \left(\frac{5}{4}, \infty\right)$

90. $x \ge 3, [3, \infty)$

91. $6 < x < 11, (6, 11)$

92. $-\frac{7}{2} < x \le \frac{3}{2}, \left(-\frac{7}{2}, \frac{3}{2}\right]$

93. 2.40 g $\le w \le$ 2.53 g **94.** The sign length must be 48 inches or less

Chapter 2 Test (page 178)

1. no **2.** 120° **3.** 1,046 **4.** \$76,000 **5.** .878, 1.000 **6.** 3% **7.** $(4x + 6)$ ft **8.** the distributive property **9.** $-20x$ **10.** $224t$ **11.** $-4a + 4$ **12.** $-5.9d^2$ **13.** 0 **14.** -5 **15.** $-\frac{1}{4}$ **16.** 2.5 **17.** $\frac{7}{6}$ **18.** -3 **19.** $r = \frac{A - P}{Pt}$ **20.** \$150 **21.** $-10°$ C **22.** 393 in.^{3} **23.** $\frac{3}{5}$ hr **24.** 10 liters **25.** 68° **26.** \$5,250

27. $x \ge -3, [-3, \infty)$

28. $-3 \le x < 4, [-3, 4)$

29. 1.496 in. $\le w \le$ 1.498 in.

Cumulative Review Exercises Chapters 1–2 (page 180)

1. a. expression **b.** equation **2.** 3, 4, 5 **3.** $2 \cdot 2 \cdot 2 \cdot 5 \cdot 5 = 2^3 \cdot 5^2$ **4.** $\frac{2}{3}$ **5.** $-\frac{2}{9}$ **6.** 6 **7.** $\frac{22}{15} = 1\frac{7}{15}$ **8.** $12\frac{11}{24}$ **9.** 0.9375 **10.** 45 **11. a.** 65 **b.** -12 **12.** the commutative property of multiplication **13.** natural number, whole number, integer, rational number, real number **14.** rational number, real number **15.** rational number, real number **16.** irrational number, real number

17. a. 4^3 **b.** $\pi r^2 h$ **18. a.** -10 **b.** -14 **c.** -64 **d.** 0 **e.** -1 **19. a.** $w + 12$ **b.** $n - 4$ **20.** 4 **21.** 1, -3, 6 **22.** $l = \frac{2{,}000}{d^2}$ (Answers may vary depending on the variables chosen.) **23. a.** 6 ft^2 **b.** 1.2 ft^2 **c.** 20% **24.** 300 **25.** 0 **26.** -2 **27.** 16 **28.** 0 **29.** $-32d$ **30.** $10x - 15y + 5$ **31.** $4x$ **32.** $-8a^2$ **33.** $11t - 50$ **34.** $8t - 20$ **35.** $(x + 3)$ ft **36.** $3x$ ft **37.** 9 **38.** 20 **39.** -0.6 **40.** $\frac{19}{6}$ **41.** -20 **42.** $\frac{5}{4}$ **43.** -2 **44.** no solution, contradiction **45.** 65 m^2 **46.** 376.99 cm^3 **47.** $r^2 = \frac{3V}{\pi h}$ **48.** 37.5 ft-lb **49.** 9.45 lb **50.** 55°, 55° **51.** \$4,000 **52.** 10 oz

53. $x > -2$, $(-2, \infty)$ 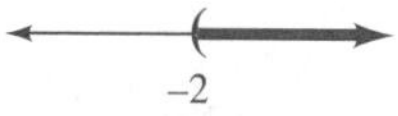

54. $x \le 2$, $(-\infty, 2]$ 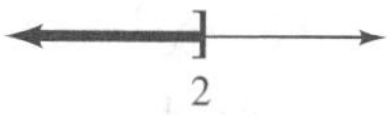

55. $x \ge -1$, $[-1, \infty)$ 

56. $-1 \le x < 2$, $[-1, 2)$ 

Study Set Section 3.1 (page 189)

1. ordered **3.** x-axis, y-axis, origin **5.** rectangular **7. a.** origin, left, up **b.** origin, right, down **9. a.** I and II **b.** II and III **c.** II **d.** IV **11.** 60 beats/min **13.** 140 beats/min **15.** 5 min and 50 min after starting **17.** no difference **19.** about 6 million vehicles **21.** (3, 5) is an ordered pair, 3(5) indicates multiplication, and 5(3 + 5) is an expression containing grouping symbols. **23.** yes **25.** horizontal **27.** (4, 3), (0, 4), $(-5, 0)$, $(-4, -5)$, $(3, -3)$

29.

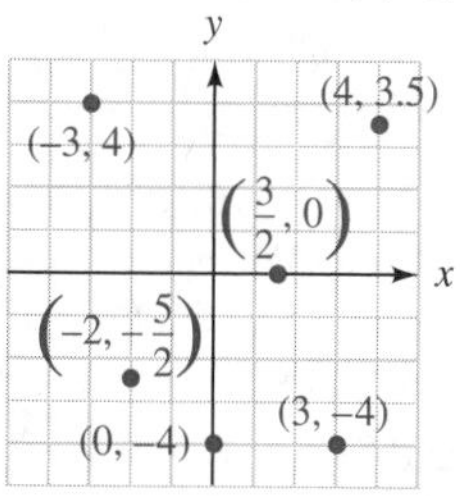

31. rivets: $(-6, 0)$, $(-2, 0)$, (2, 0), (6, 0); welds: $(-4, 3)$, (0, 3), (4, 3); anchors: $(-6, -3)$, $(6, -3)$

33. (E, 4), (F, 3), (G, 2) **35.** Rockford (5, B), Mount Carroll (1, C), Harvard (7, A), intersection (5, E) **37.** (2, 4); 12 sq. units **39. a.** 35 mi **b.** 4 gal **c.** 32.5 mi **41. a.** A 3-year-old car is worth \$7,000. **b.** \$1,000 **c.** 6 yr **43.** 152, 179, 202, 227, 252, 277 **49.** $h = \frac{3(AC + T)}{2}$ **51.** -1

Study Set Section 3.2 (page 202)

1. two **3.** table **5.** linear **7.** infinitely **9. a.** 2 **b.** yes **c.** no **d.** infinitely many **11.** solution, point **13. a.** At least one point is in error. The points should lie on a straight line. Check the computations. **b.** The line is too short. Arrowheads are not drawn. **15. a.** $y = 2x + 1$ **b.** $y = -\frac{5}{3}x - 2$ **c.** $y = \frac{1}{7}x - 3$ or $y = \frac{x}{7} - 3$ **17.** 6, -2, 2, 6 **19.** a, c, a, c **21.** yes **23.** no **25.** no **27.** yes **29.** 11 **31.** 4 **33.** 12, (8, 12), 6, (6, 8) **35.** -13, $(-5, -13)$, -1, $(-1, -1)$

37.

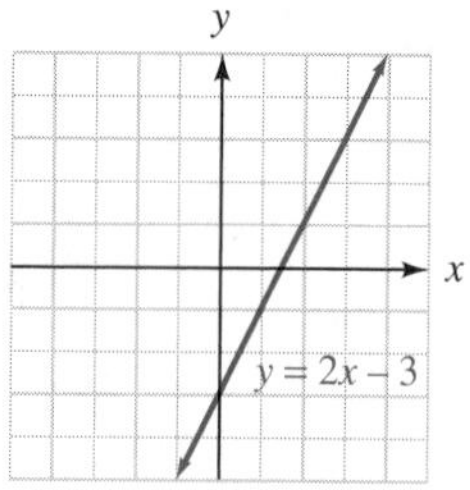

39.

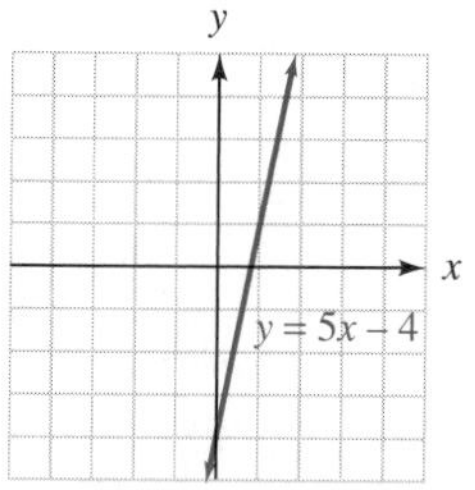

41.

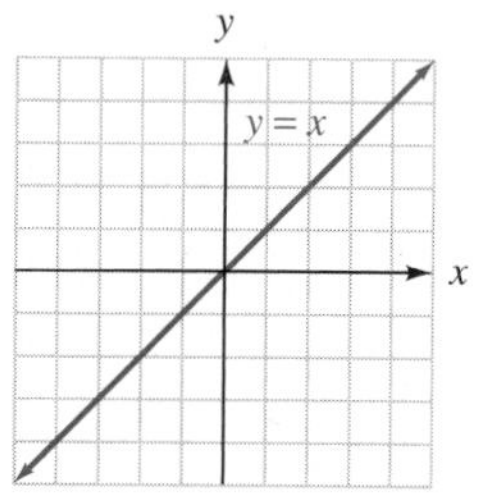

43.

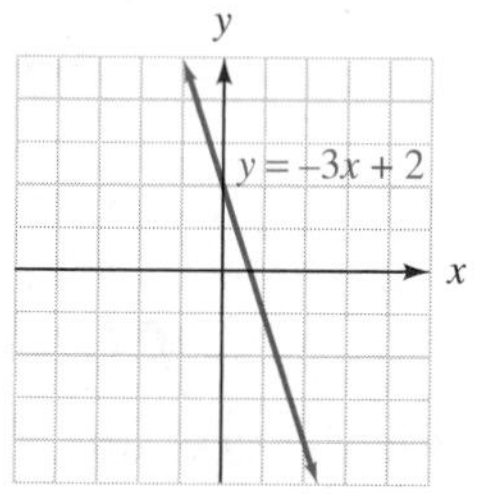

45.

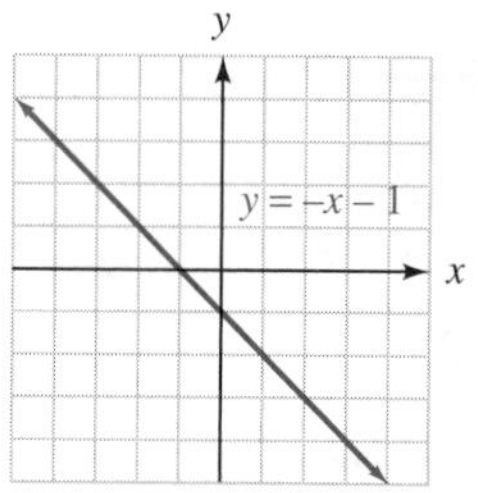

47.

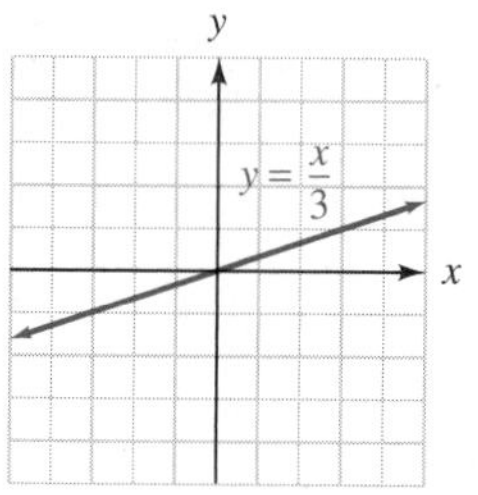

49.

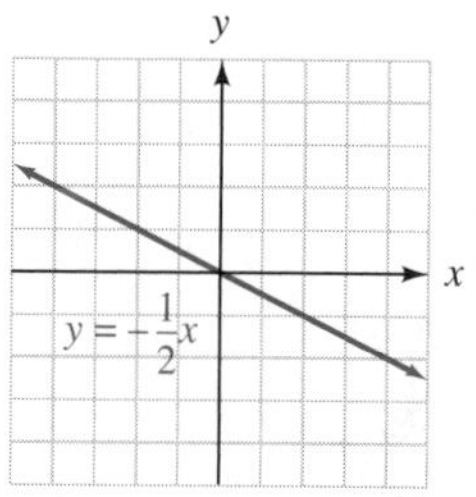

51.

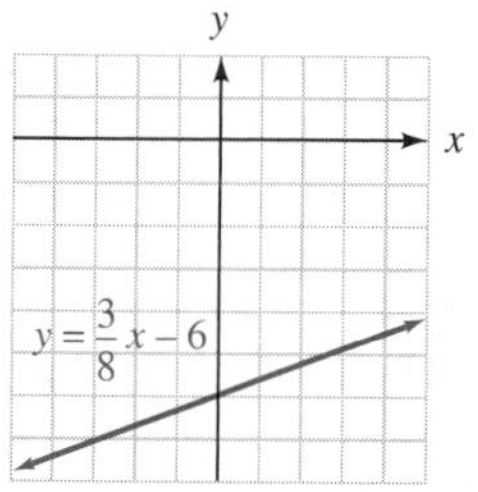

53.

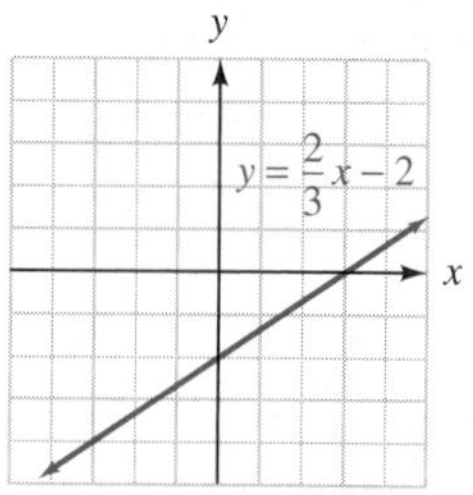

55.

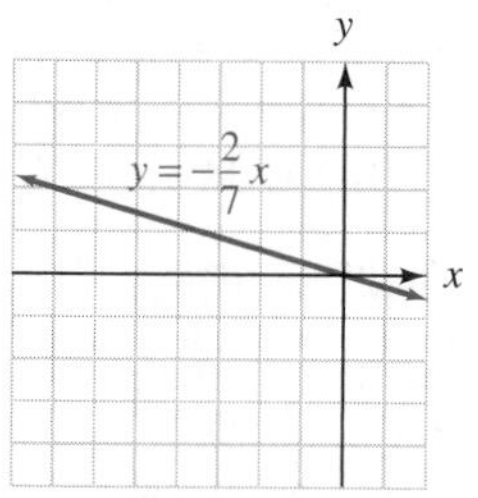

57.

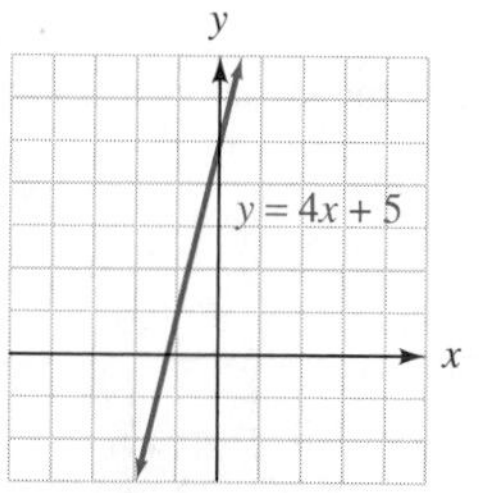

59.

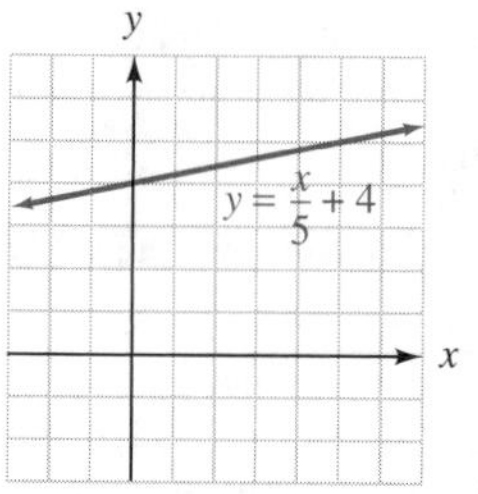

61.

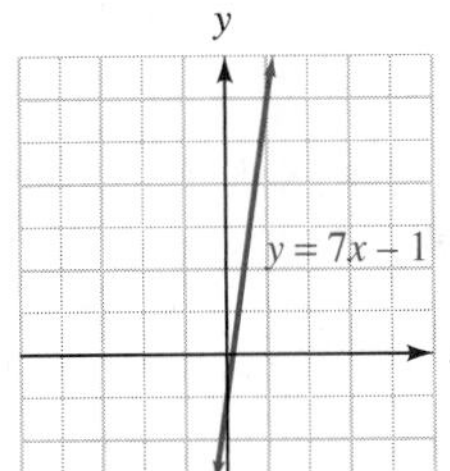

63. -2, $(1, -2)$, 0, $(2, 0)$, 4, $(4, 4)$; 4, $(4, 4)$, 0, $(6, 0)$, -4, $(8, -4)$

65. 3 oz

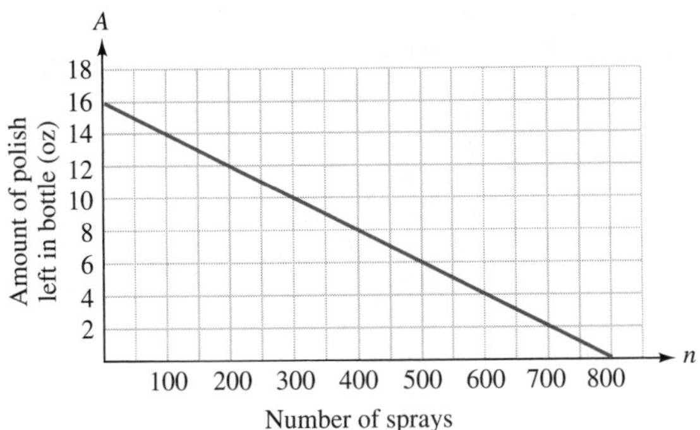

67. about \$70 **69.** about 180 **77.** $5 + 4c$ **79.** 0 **81.** 491

Study Set Section 3.3 (page 214)

1. two **3.** general/standard **5.** intersects/crosses **7.** x-intercept: $(4, 0)$ y-intercept: $(0, 3)$ **9.** x-intercept: $(-5, 0)$ y-intercept: $(0, -4)$ **11.** y-intercept: $(0, 2)$ **13.** y-intercept: $\left(0, \frac{2}{3}\right)$; x-intercept: $\left(-2\frac{1}{2}, 0\right)$ **15. a.** $3x$ **b.** $(0, 3)$ **17.** 2; 1 **19. a.** ii **b.** iv **c.** vi **d.** i **e.** iii **f.** v **21. a.** y-axis **b.** yes **23.** $y = 0$; $x = 0$

25.

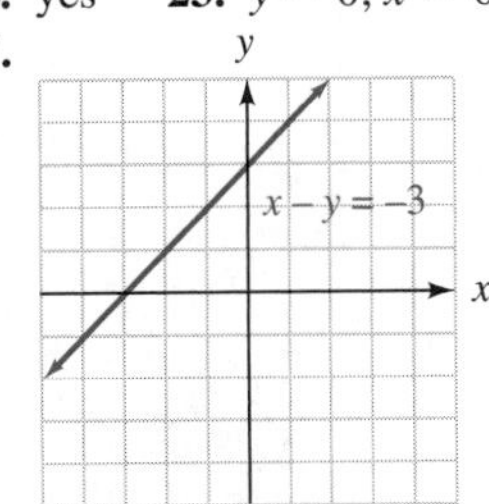

27.

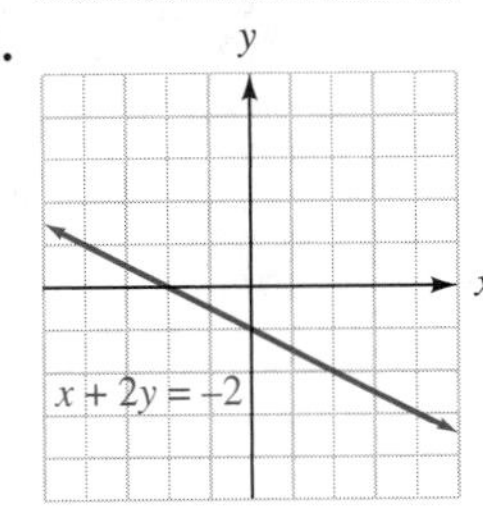

29.

31.

33.

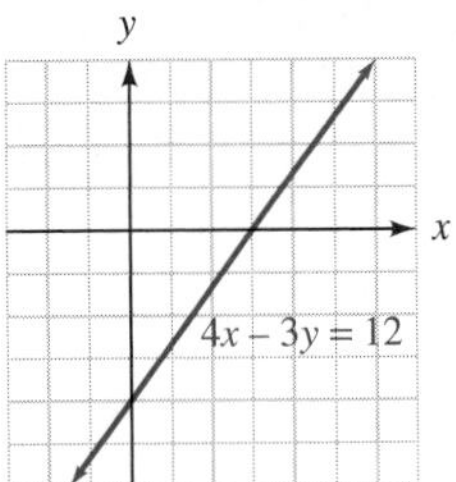

35.

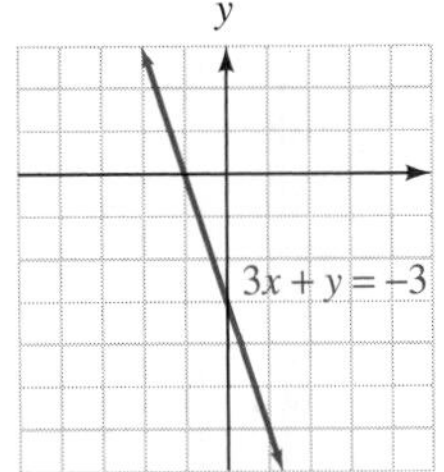

37.

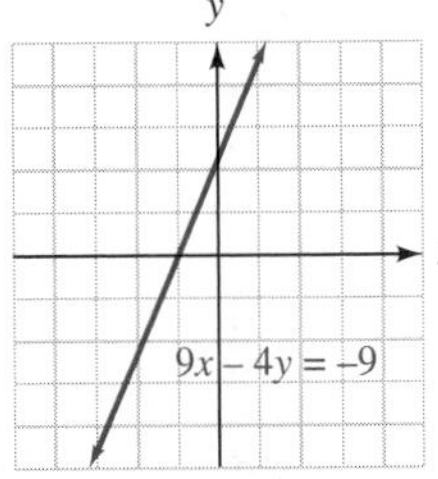

39.

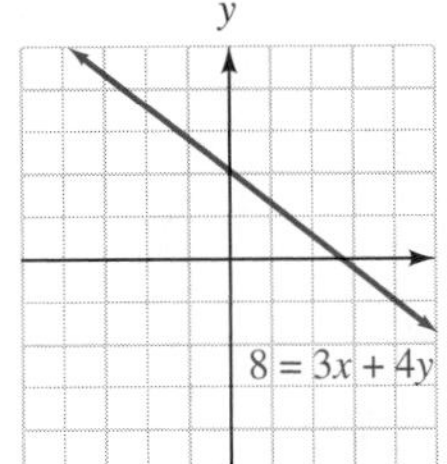

41.

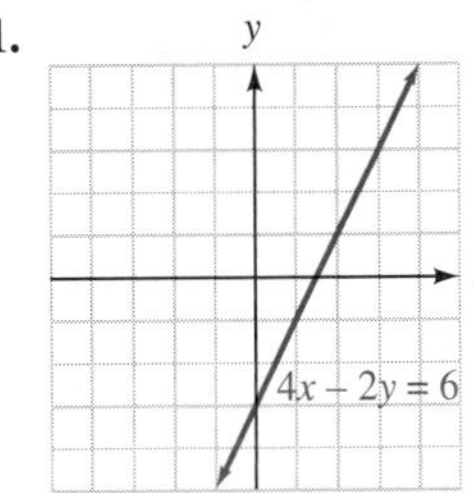

43.

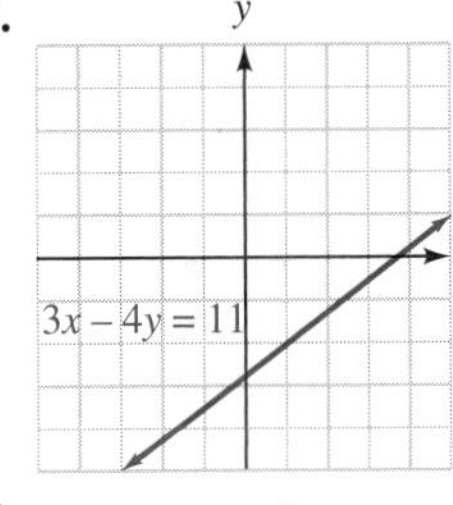

45.

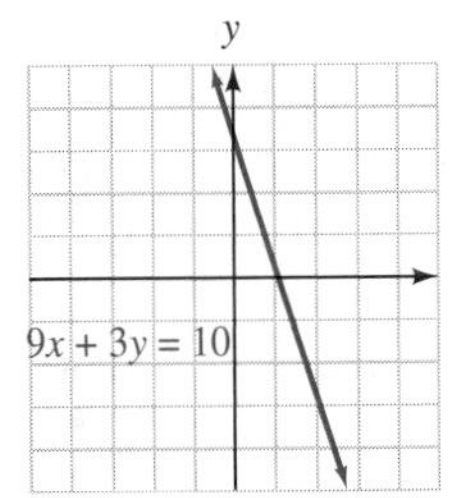

47.

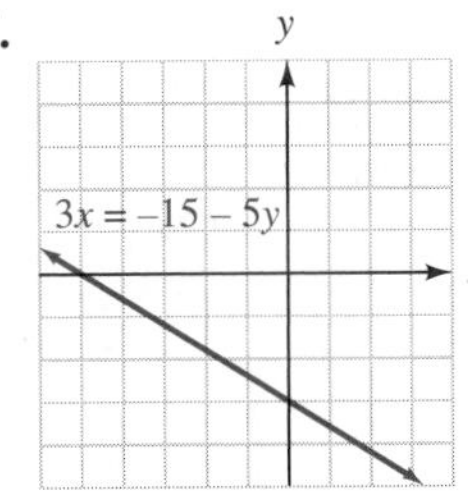

49.

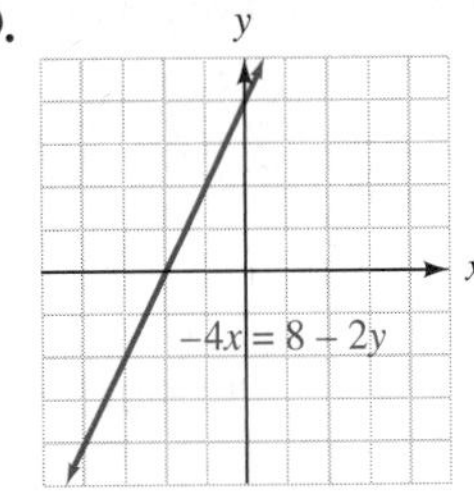

51.

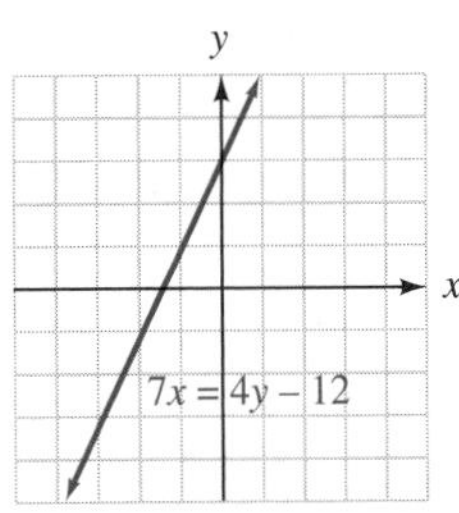

53.

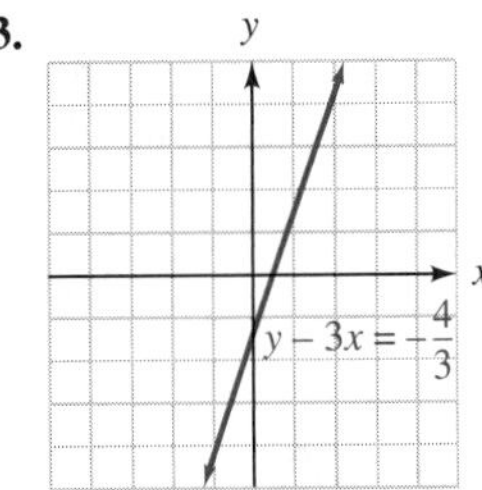

55.

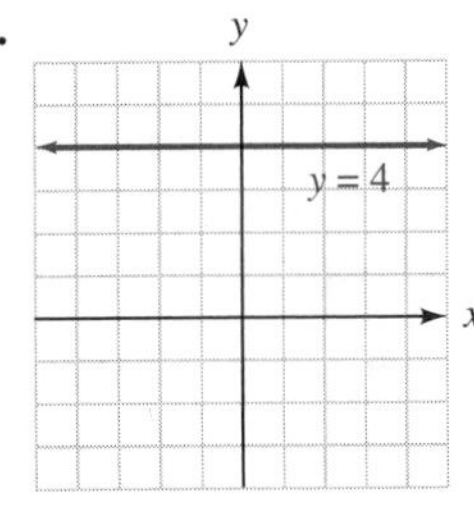

57.

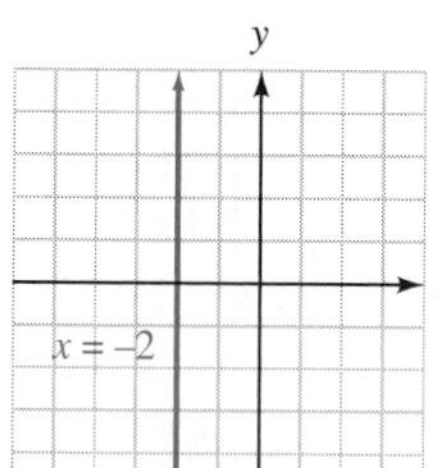

59.

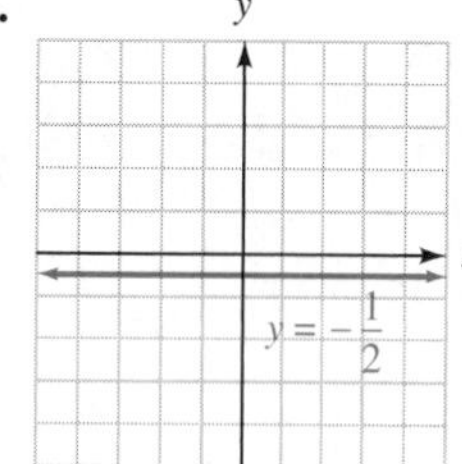

61.

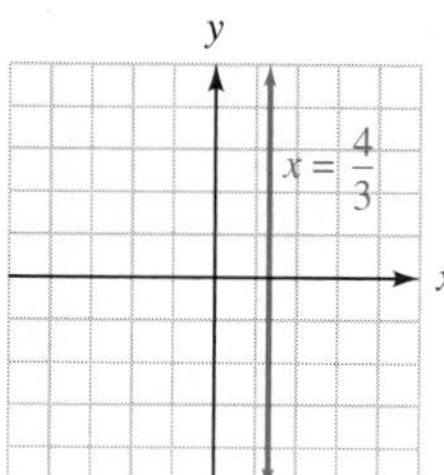

63.

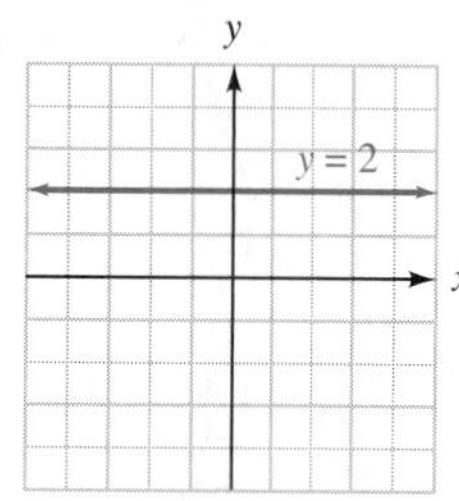

65.

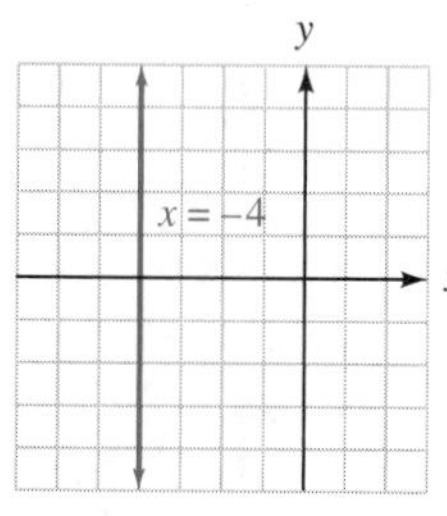

67. a. about −270°C **b.** 0 milliliters

69. a. If only shrubs are purchased, he can buy 200. **b.** If only trees are purchased, he can buy 100. **75.** $\frac{1}{5}$ **77.** $2x - 6$

Study Set Section 3.4 (page 224)

1. ratio **3.** slope **5.** change **7. a.** rises **b.** falls **9.** $\frac{2}{15}$ **11. a.** −3 **b.** 8 **c.** $-\frac{3}{8}$ **13. a.** $\frac{1}{2}$ **b.** $\frac{1}{2}$ **c.** $\frac{1}{2}$ **d.** When finding the slope of a line, any two points on the line give the same result. **15. a.** $m = 0$ **b.** undefined **17. a.** 40% **b.** 15% **19.** $m = \frac{y_2 - y_1}{x_2 - x_1}$ **21.** y^2 means $y \cdot y$ and y_2 means y sub 2. **23.** $\frac{2}{3}$ **25.** $\frac{4}{3}$ **27.** −2 **29.** 0 **31.** $-\frac{1}{5}$ **33.** 1 **35.** −3 **37.** $\frac{5}{4}$ **39.** $-\frac{1}{2}$ **41.** $\frac{3}{5}$ **43.** 0 **45.** undefined **47.** $-\frac{2}{3}$ **49.** −4.75 **51.** $m = \frac{3}{4}$ **53.** $m = 0$ **55.** undefined **57.** 0 **59.** undefined **61.** 0 **63.** $-\frac{2}{5}$ **65.** $\frac{1}{20}$; 5% **67. a.** $\frac{1}{8}$ **b.** $\frac{1}{12}$ **c.** 1: less expensive, steeper; 2: not as steep, more expensive **69.** −875 gal per hr **71.** 300 lb per yr **77.** 40 lb licorice; 20 lb gumdrops

Study Set Section 3.5 (page 237)

1. slope–intercept **3.** parallel **5.** reciprocals **7. a.** no **b.** no **c.** yes **d.** no **e.** yes **f.** yes **9. a.** y, $2x$, 4 **b.** y, $-3x$, + **11.** $y = -\frac{5}{4}x$ **13. a.** parallel **b.** perpendicular **c.** −1 **15. a.** $-\frac{1}{2}$ **b.** 2 **c.** $-\frac{1}{2}$ **d.** Line 1 and Line 2 **17.** $2x$, $-2x$, $5y$, 5, 5, 5, 3, $-\frac{2}{5}$, (0, 3)

19. a. $4x$ **b.** $\frac{4x}{3}$ **c.** x **d.** −2 **21.** a right angle **23.** 4, (0, 2) **25.** −5, (0, −8) **27.** 4, (0, −2) **29.** $\frac{1}{4}$, $\left(0, -\frac{1}{2}\right)$ **31.** $\frac{1}{2}$, (0, 6) **33.** −1, (0, 6) **35.** −1, (0, 8) **37.** $\frac{1}{6}$, (0, −1) **39.** −2, (0, 7) **41.** $-\frac{3}{2}$, (0, 1) **43.** $-\frac{2}{3}$, (0, 2) **45.** $\frac{3}{5}$, (0, −3) **47.** 1, $\left(0, -\frac{11}{6}\right)$ **49.** 1, (0, 0) **51.** −5, (0, 0) **53.** 0, (0, −2) **55.** 0, $\left(0, -\frac{2}{5}\right)$

57. $y = 5x - 3$

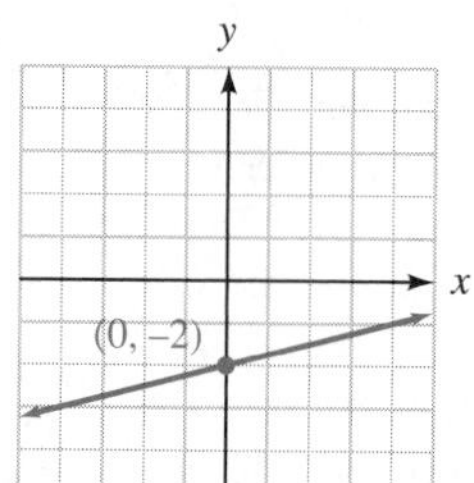

59. $y = \frac{1}{4}x - 2$

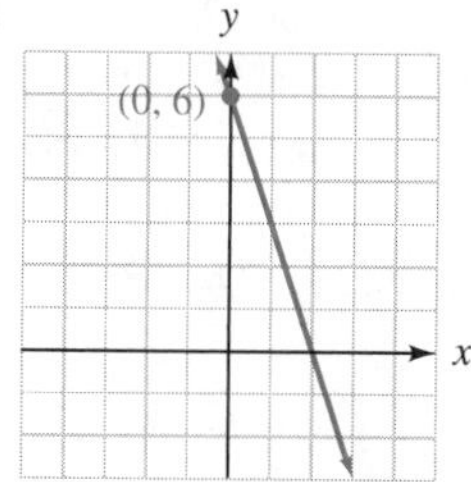

61. $y = -3x + 6$

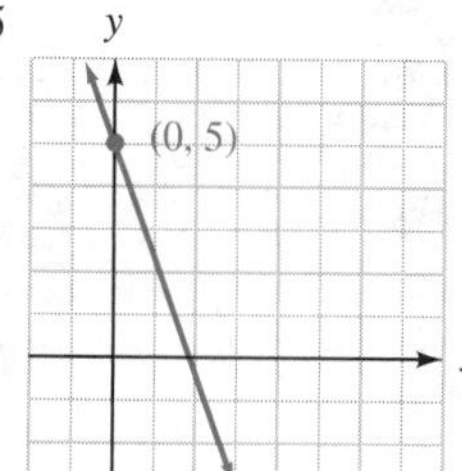

63. $y = -\frac{8}{3}x + 5$

65.

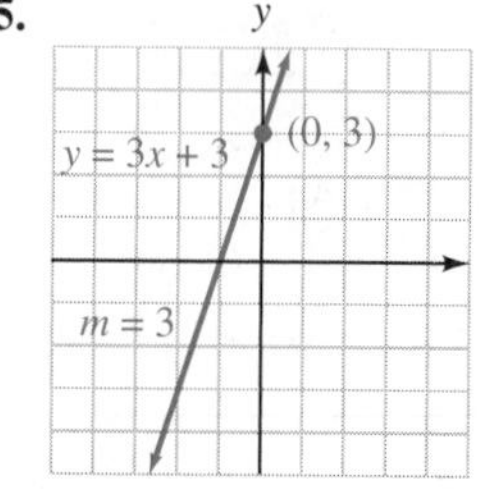

67.

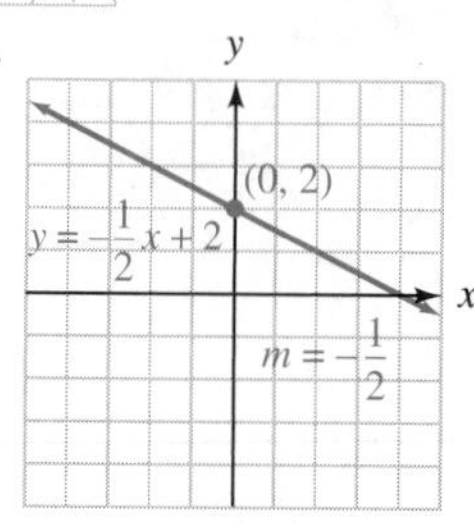

69.
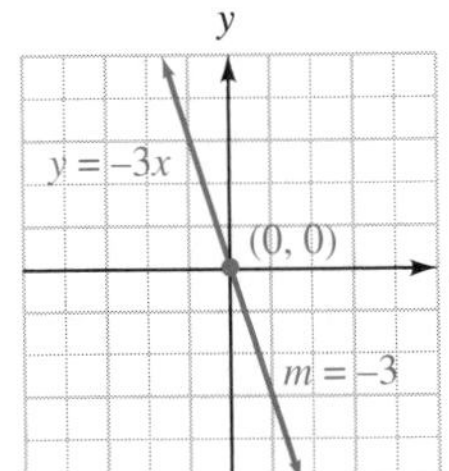

71.
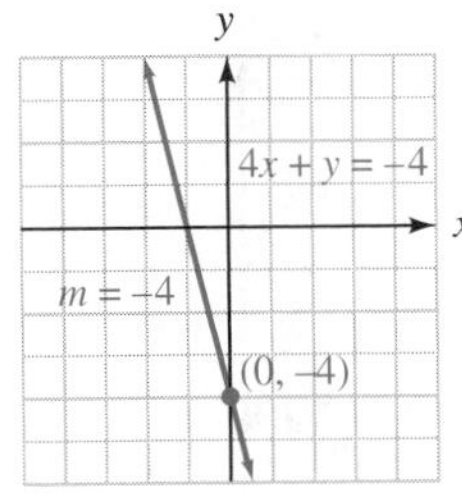

73.
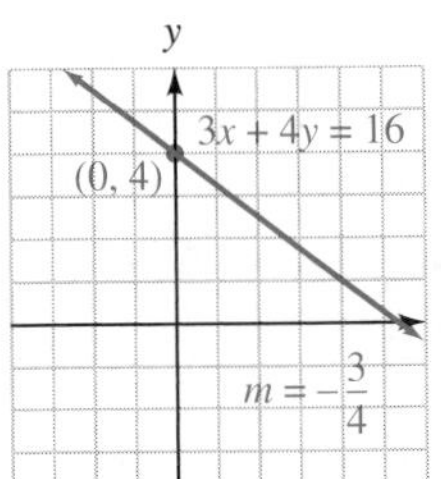

75.
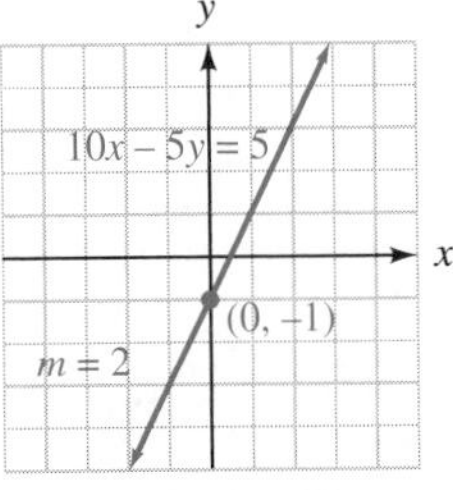

77. parallel **79.** perpendicular **81.** neither **83.** parallel **85.** perpendicular **87.** **a.** $c = 2{,}000h + 5{,}000$ **b.** \$21,000 **89.** $F = 5t - 10$ **91.** $c = -20m + 500$ **93.** **a.** $c = 5x + 20$

c.
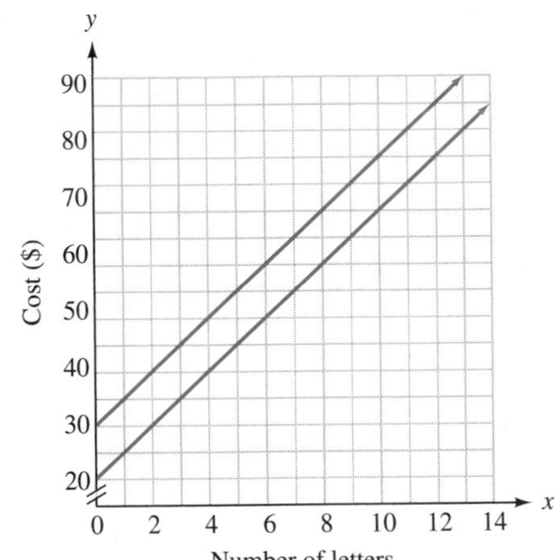

95. $c = 2t + 33.50$ **99.** 42 ft, 45 ft, 48 ft, 51 ft

Study Set Section 3.6 (page 247)

1. point–slope **3.** **a.** point–slope **b.** slope–intercept **5.** **a.** $x + 6$ **b.** $y + 9$ **7.** $-\frac{5}{2}$ **9.** **a.** yes **b.** yes **11.** **a.** $(-2, -3)$ **b.** $\frac{5}{6}$ **c.** $(4, 2)$ **13.** **a.** no **b.** no **c.** yes **d.** yes **15.** (67, 170), (79, 220) **17.** $y - y_1 = m(x - x_1)$ **19.** point–slope, y, slope–intercept **21.** $y - 1 = 3(x - 2)$ **23.** $y + 1 = -\frac{4}{5}(x + 5)$ **25.** $y = \frac{1}{5}x - 1$ **27.** $y = -5x - 37$ **29.** $y = -\frac{4}{3}x + 4$ **31.** $y = -\frac{11}{6}x - \frac{7}{3}$ **33.** $y = -\frac{2}{3}x + 2$ **35.** $y = 8x + 4$ **37.** $y = -3x$ **39.** $y = 2x + 5$ **41.** $y = -\frac{1}{2}x + 1$ **43.** $y = 5$ **45.** $y = \frac{1}{10}x + \frac{1}{2}$ **47.** $x = -8$ **49.** $y = \frac{1}{4}x - \frac{5}{4}$ **51.** $x = 4$ **53.** $y = 5$

55.
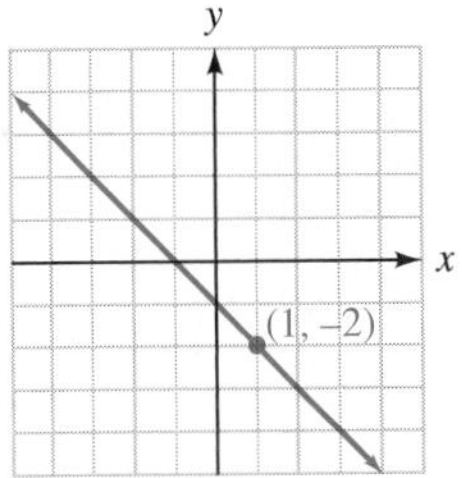

57.
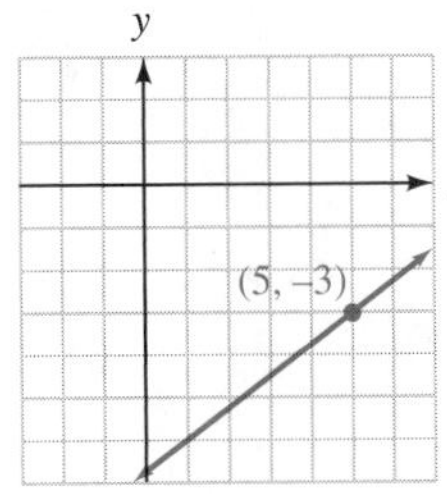

59.
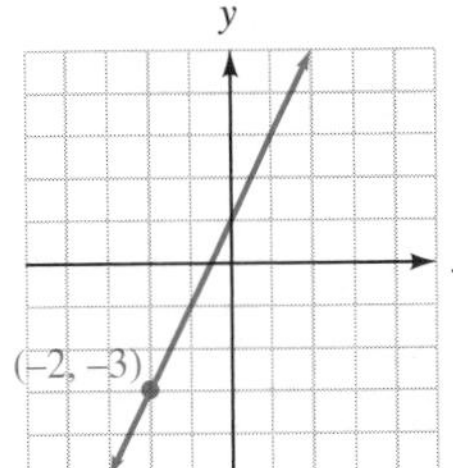

61.
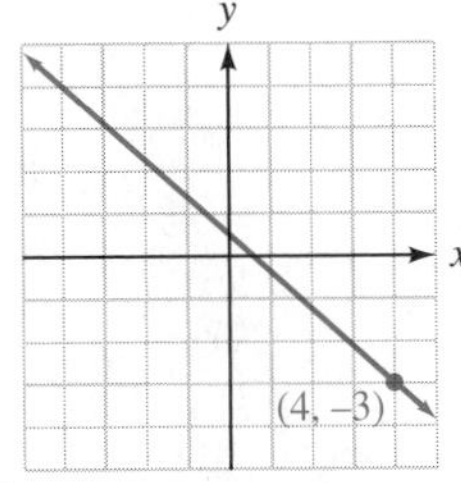

63. $h = 3.9r + 28.9$ **65.** $y = -\frac{2}{5}x + 4$, $y = -7x + 70$, $x = 10$ **67.** **a.** $y = -40m + 920$ **b.** 440 yd^3 **69.** $l = \frac{25}{4}r + \frac{1}{4}$ **71.** **a.** $y = -\frac{1}{4}x + \frac{117}{4}$ or $y = -0.25x + 29.25$ **b.** 19.25 gal **77.** 17 in. by 39 in.

Study Set Section 3.7 (page 258)

1. inequality **3.** solution **5.** boundary **7.** point **9.** **a.** false **b.** false **c.** true **d.** false **11.** **a.** no **b.** yes **c.** no **d.** no **13.** **a.** no **b.** yes **15.** the half-plane opposite that in which the test point lies **17.** **a.** yes **b.** no **c.** no **d.** yes **19.** **a.** horizontal **b.** vertical **21.** **a.** is less than **b.** is greater than **c.** is less than or equal to **d.** is greater than or equal to **23.** $\leq, \geq$ **25.**

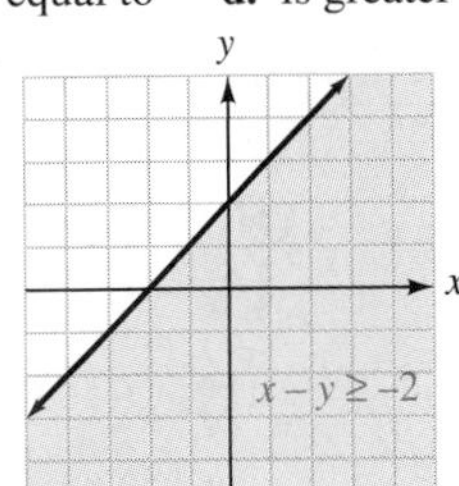

27.
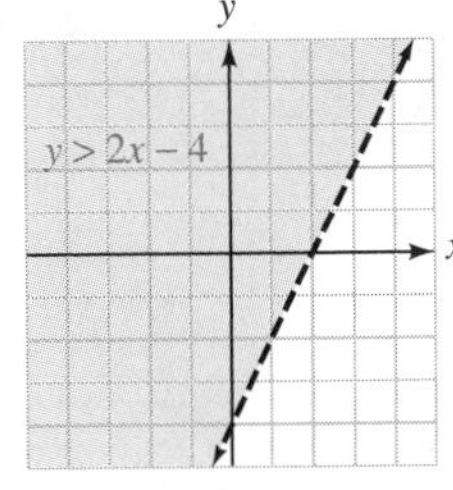

29.
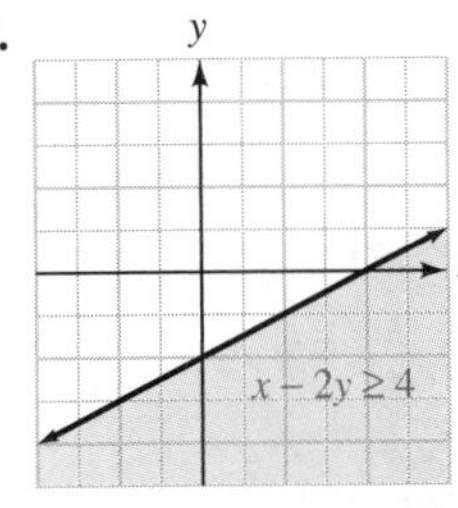

31.
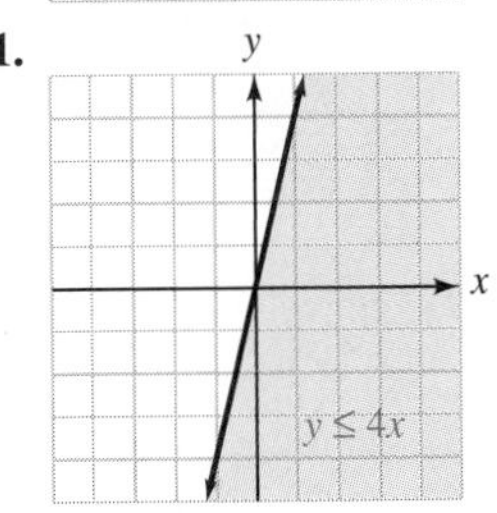

33.
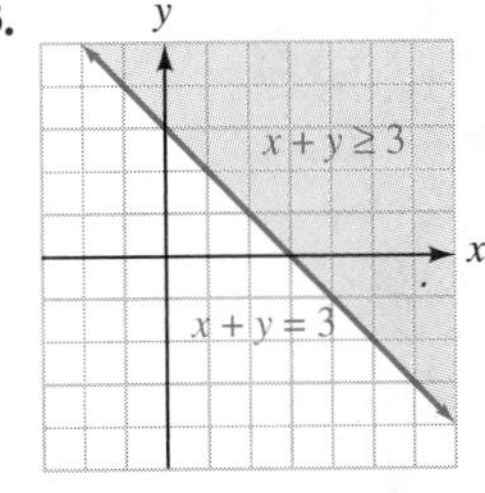

35.

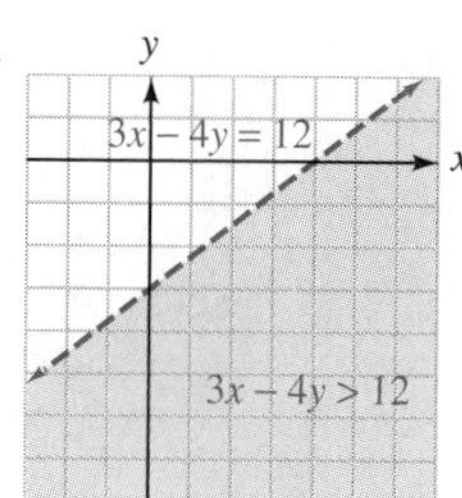

37.

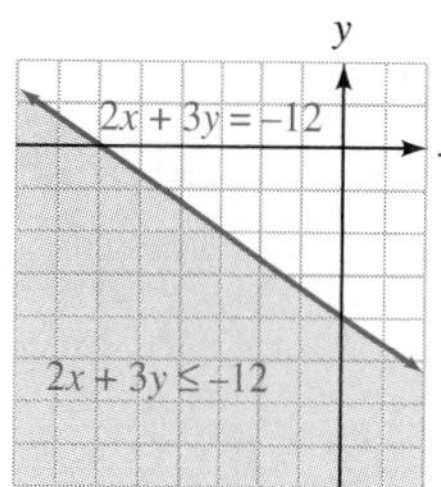

39.

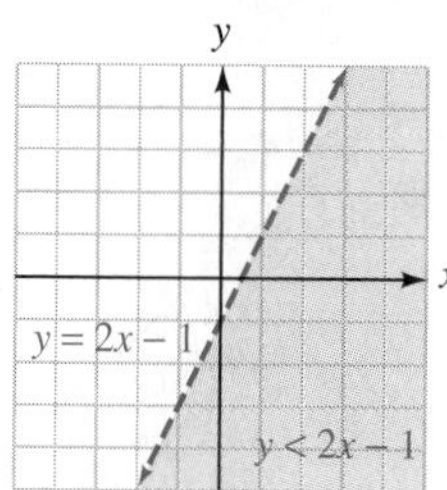

41.

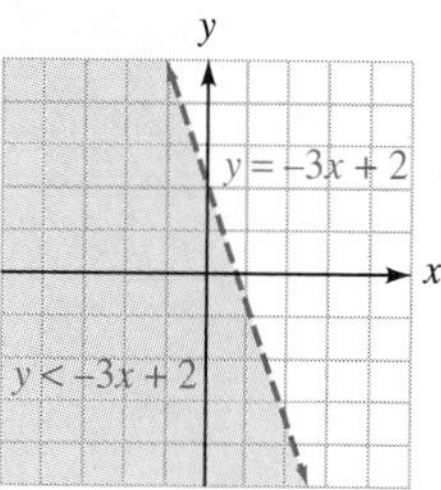

43.

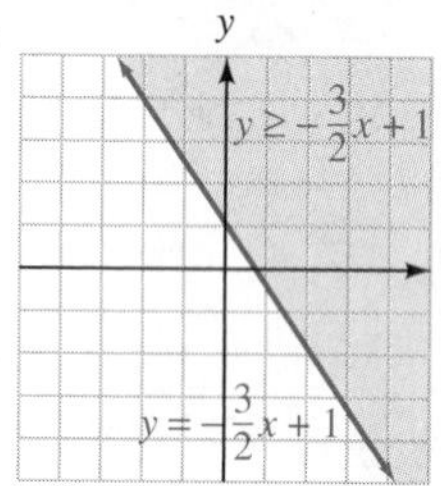

45.

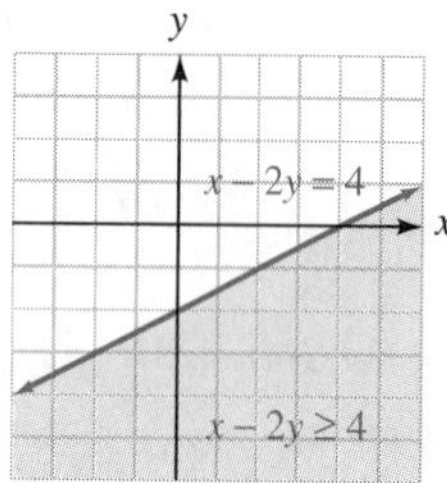

47.

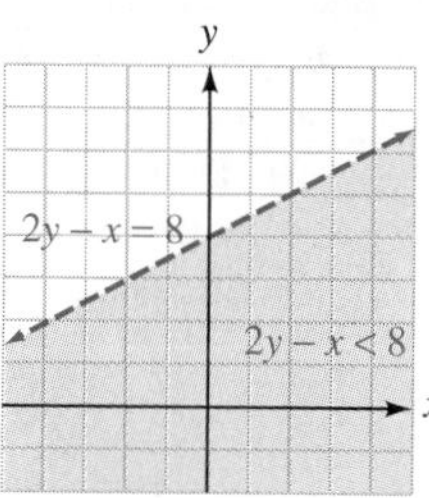

49.

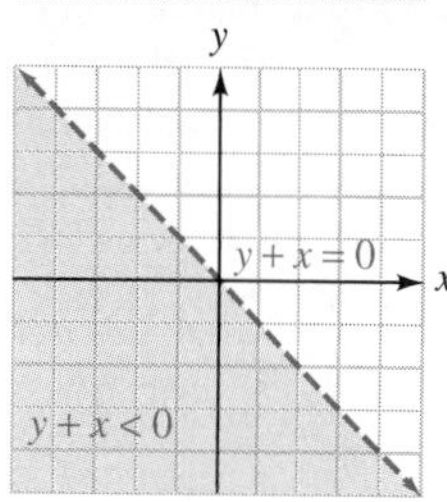

51.

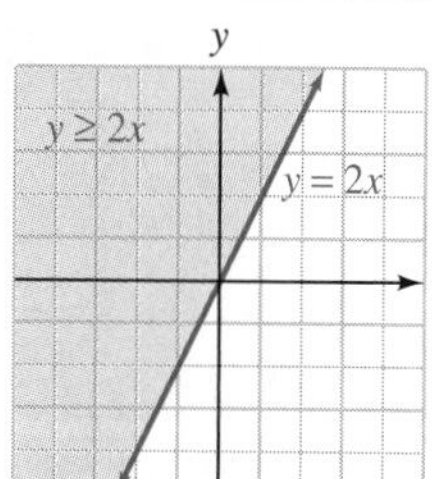

53.

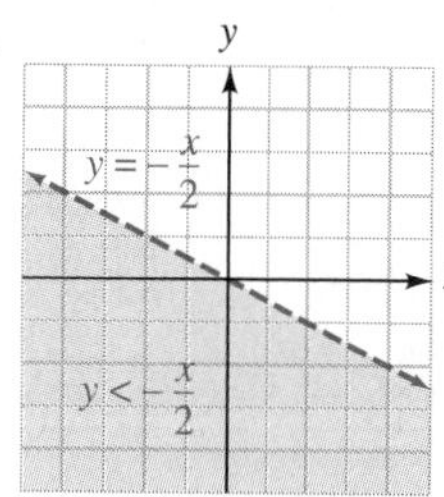

55.

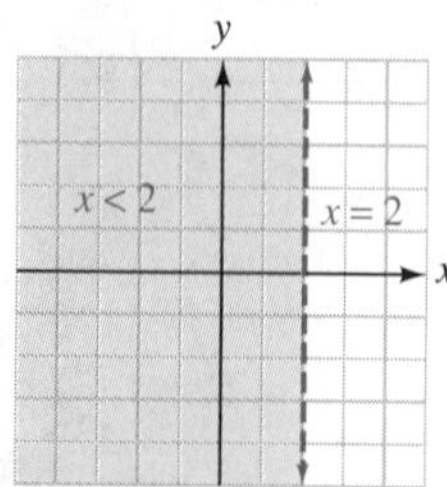

57.

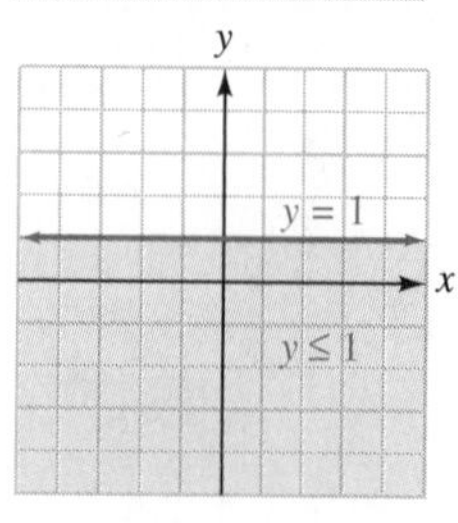

59.

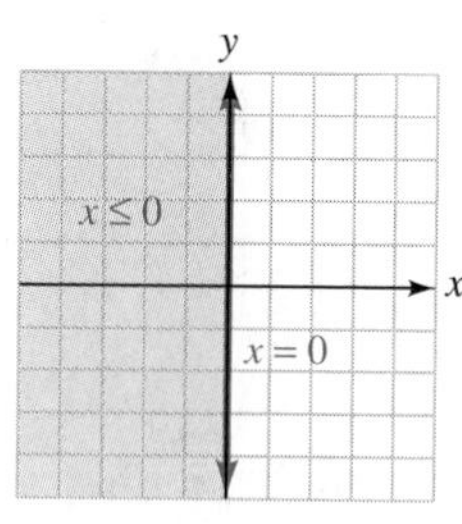

61.

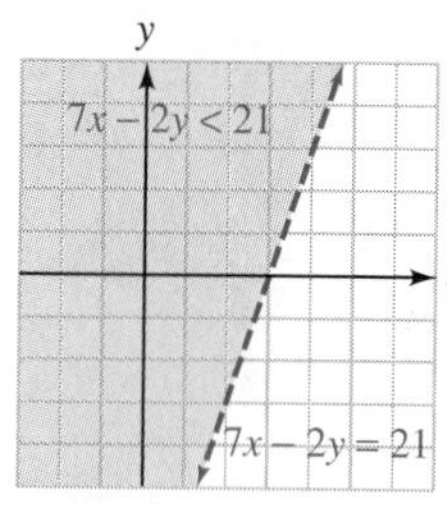

63.

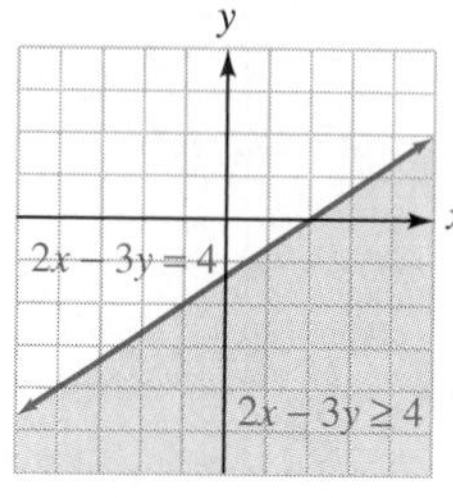

65.

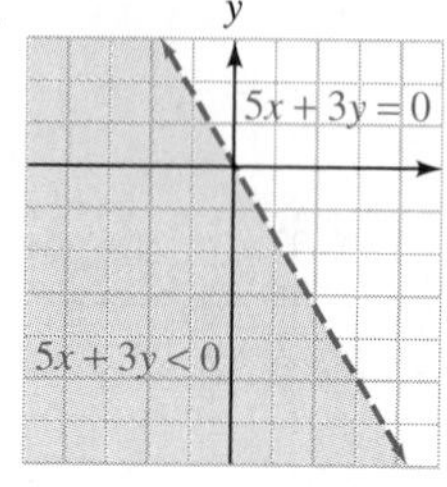

67. ii **69.** (10, 10), (20, 10), (10, 20); answers may vary
71. (50, 50), (30, 40), (40, 40); answers may vary
77. $t = \frac{A - P}{Pr}$ **79.** $15x + 22$

Key Concept (page 263)

1. $y = -3x - 4$ **2.** $y = \frac{1}{5}x + 1$ **3.** 80°F
4. $y = 209x + 2{,}660$; \$8,930 **5.** −2, 4, −3

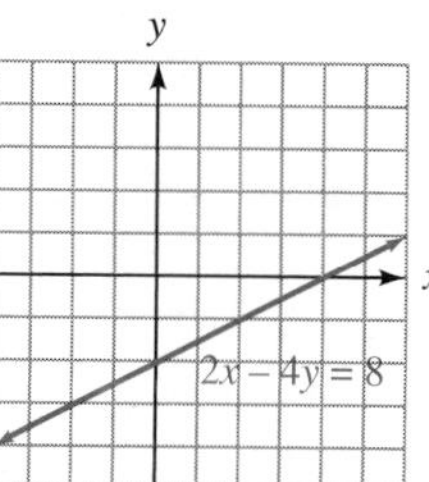

6. a. When new, the press cost \$40,000. **b.** −5,000; the value of the press decreased \$5,000/yr

7. a. (−2, −2) (answers may vary) **b.** −2
c. $y = -2x - 6$ **8.** $x = 1$

Chapter Review (page 264)

1.

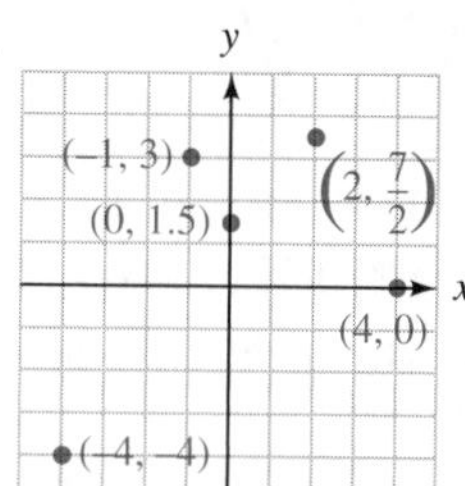

2. (158, 21.5) **3.** quadrant III

4. (0, 0) **5.** (1, 4); 36 square units **6. a.** 2,500; week 2
b. 1,000 **c.** 1st week and 5th week **7.** yes
8. −6, (−2, −6), −8, (−8, 3) **9.** $y = x^2 + 1$ and $y - x^3 = 0$

10.

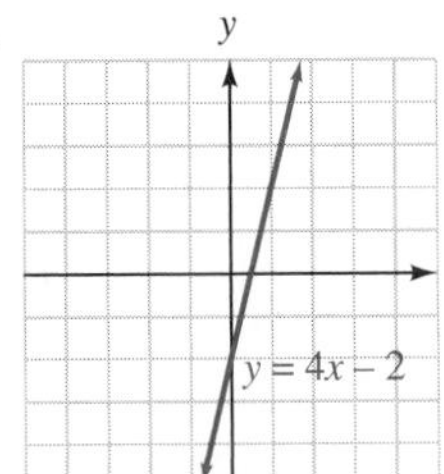

11.

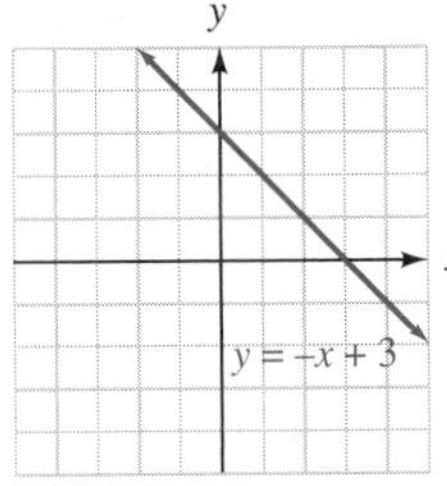

12. a. true **b.** false **13.** about \$195
14. $(-3, 0)$, $(0, 2.5)$ **15.** x-intercept: $(-2, 0)$; y-intercept: $(0, 4)$

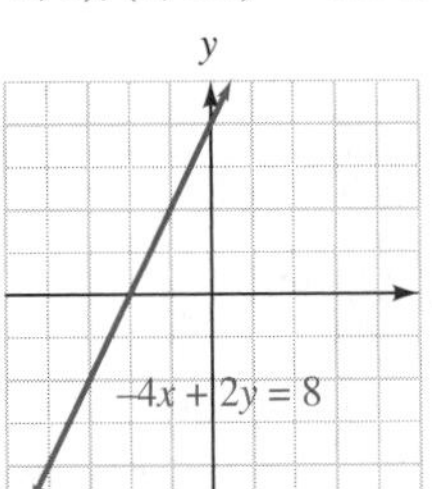

16.

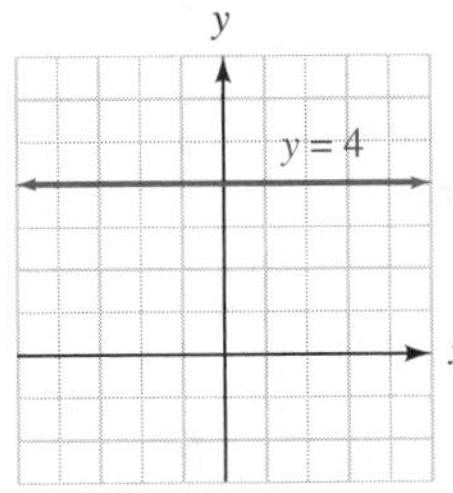

17.

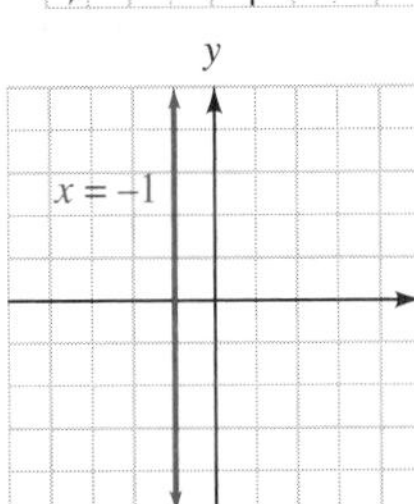

18. (0, 25,000); the equipment was originally valued at \$25,000. (10, 0). In 10 years, the sound equipment had no value.

19. $\frac{1}{4}$ **20.** 0 **21.** -7 **22.** $-\frac{3}{2}$ **23.** $\frac{3}{4}$ **24.** 8.3%
25. a. -1.25 million people per yr **b.** 4.05 million people per yr **26.** $m = \frac{3}{4}$; y-intercept: $(0, -2)$ **27.** $m = -4$; y-intercept: $(0, 0)$ **28.** $m = \frac{1}{8}$; y-intercept $(0, 10)$
29. $m = -\frac{7}{5}$; y-intercept $\left(0, -\frac{21}{5}\right)$ **30.** $y = -6x + 4$
31. $m = 3$; y-intercept: $(0, -5)$.

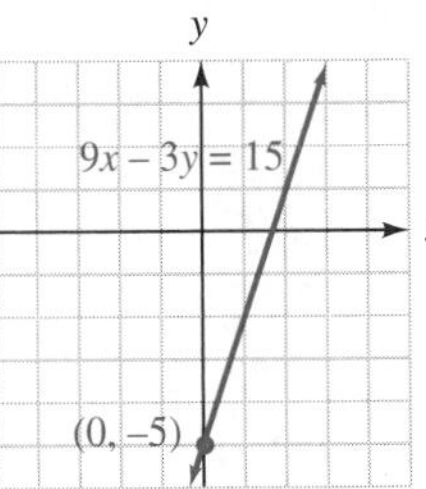

32. a. $c = 300w + 75{,}000$ **b.** 90,600 **33. a.** parallel **b.** perpendicular **34.** $y = 3x + 2$

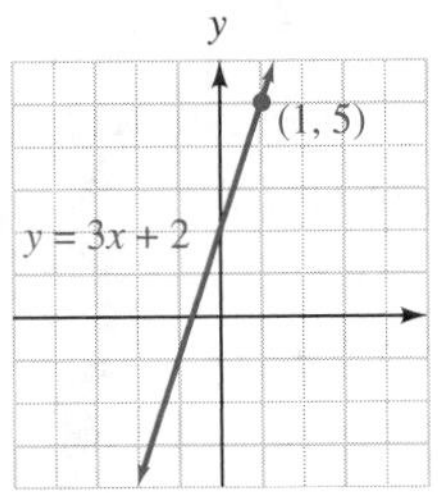

35. $y = -\frac{1}{2}x - 3$

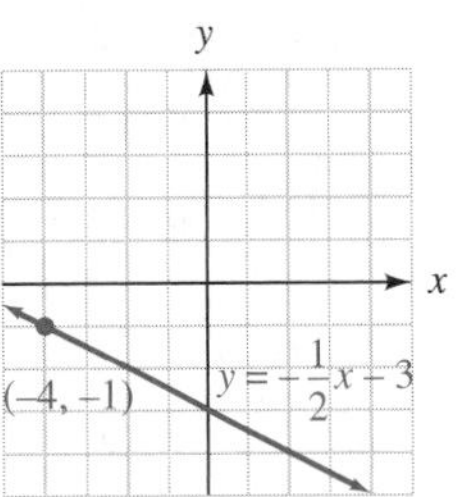

36. $y = \frac{2}{3}x + 5$ **37.** $y = -8$ **38.** $f = -35x + 450$
39. a. yes **b.** yes **c.** yes **d.** no

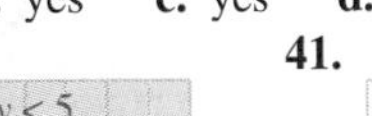

40.

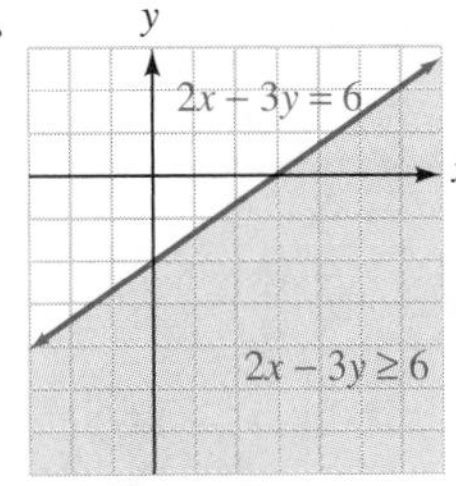

41.

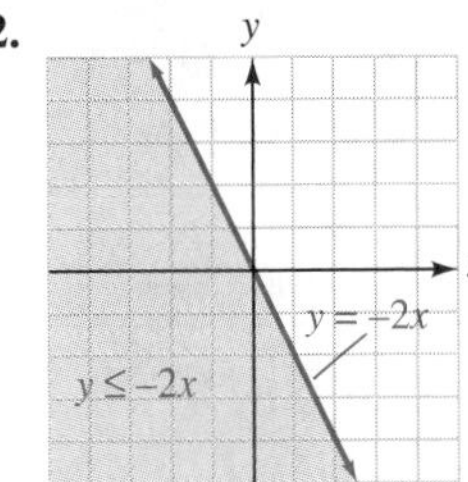

42.

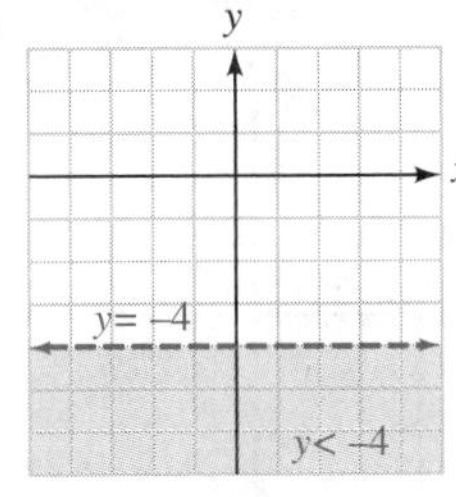

43.

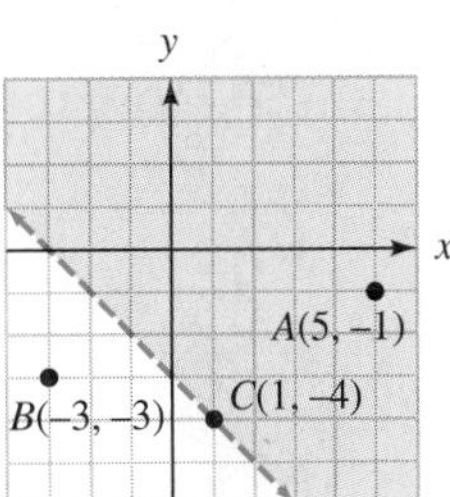

44. a. true **b.** false **c.** false

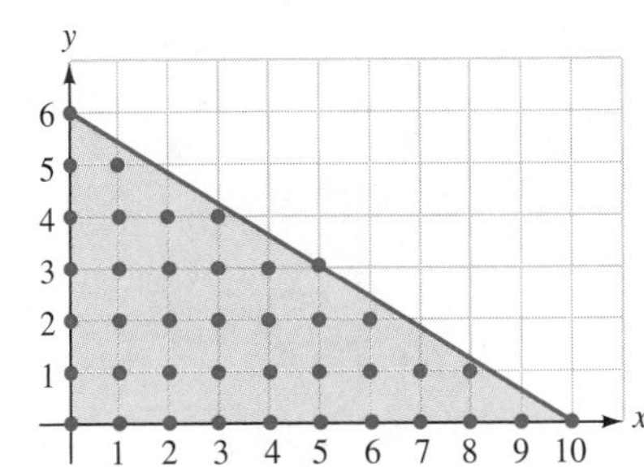

45. (2, 4), (5, 3), (6, 2) answers may vary **46.** An equation contains an = symbol. An inequality contains one of the symbols <, ≤, >, or ≥.

Chapter 3 Test (page 269)

1. 10 **2.** 60 **3.** 1 day before and the 3rd day of the holiday
4. 50 dogs were in the kennel when the holiday began.
5.

II	I
III	IV

6. 1, (2, 1), -6, $(-6, 3)$ **7.** yes

8. a. false **b.** true **9.**

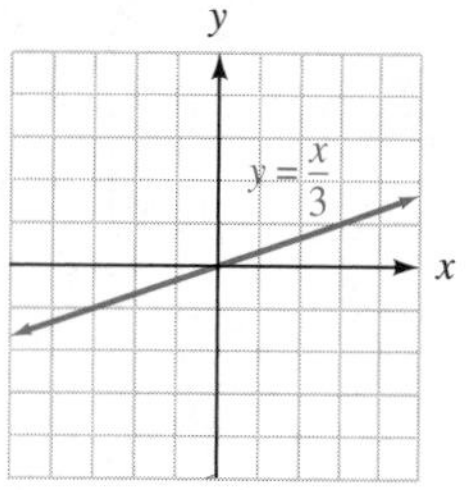

10.

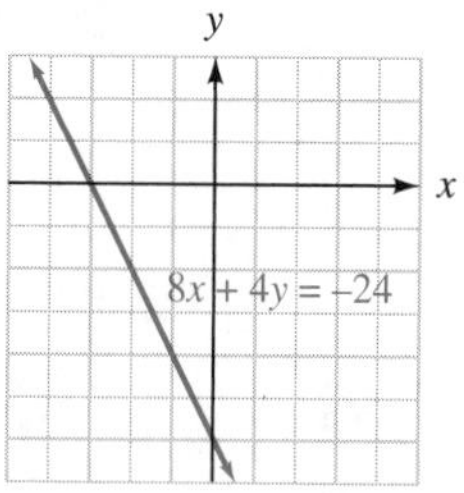

11. x-intercept: (3, 0); y-intercept: (0, −2)

12. $\frac{8}{7}$ **13.** −1 **14.** undefined **15.** $\frac{8}{7}$ **16.** parallel
17. −15 ft per mi **18.** 25 ft per mi
19.

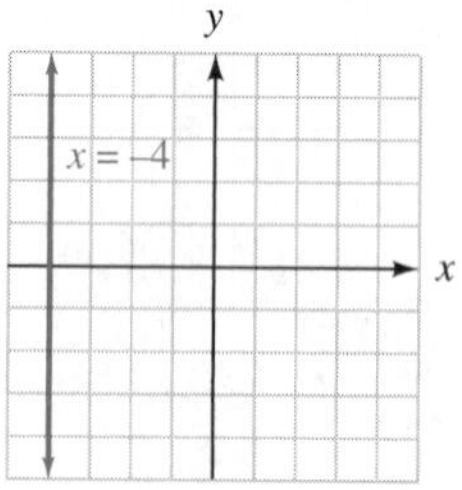

20.

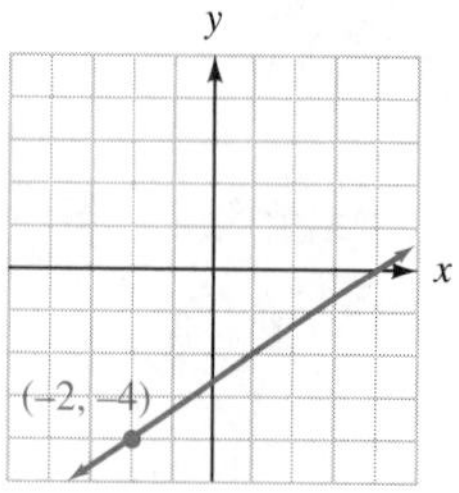

21. $m = -\frac{1}{2}$; (0, 4) **22.** $y = 7x + 19$
23. $v = -1{,}500x + 15{,}000$ **24.** yes **25.** $y = -\frac{1}{5}T + 41$
26.

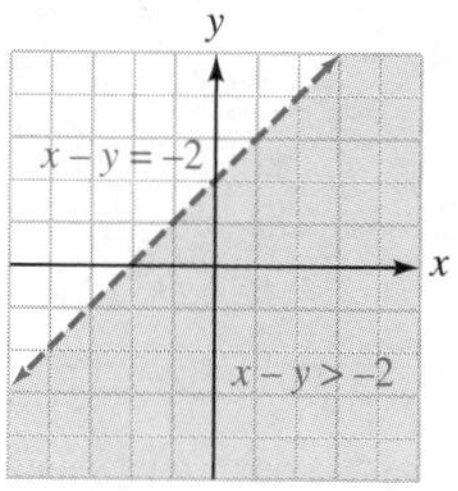

Cumulative Review Exercises Chapters 1–3 (page 271)

1. a. Q3, 2001 **b.** \$3.1 billion in losses **2.** $2^2 \cdot 3^3$
3. 0.004 **4. a.** true **b.** true **c.** true **5.** 1,100 **6.** 2
7. 32 **8.** $500 - x$ **9.** 3, −2 **10. a.** $2x + 8$
b. $2x - 8$ **c.** $-2x - 8$ **d.** $-2x + 8$ **11.** $4a + 10$
12. $-63t$ **13.** $4b^2$ **14.** 0 **15.** 4 **16.** $-160a$
17. $-3y$ **18.** $7x - 12$ **19.** 6 **20.** 2.9 **21.** 9
22. −19 **23.** $\frac{1}{7}$ **24.** 1 **25.** $-\frac{55}{6}$ **26.** no solution, contradiction **27.** 99 **28.** $-\frac{1}{4}$ **29.** 45°
30. $h = \frac{S - 2\pi r^2}{2\pi r}$ **31.** $3\frac{1}{8}$ in., $\frac{39}{64}$ in.^{2} **32.** 10 ft or 11 ft
33. $0.50x$, $0.25(13 - x)$, 0.30(13) **34.** 7.5 hr **35.** 80 lb candy corn, 120 lb gumdrops
36. $x \le 48$,

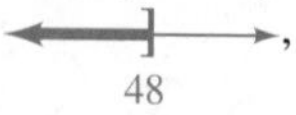

, $(-\infty, 48]$

37. $x > 0$,

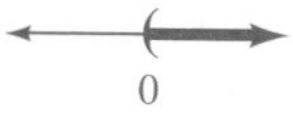

, $(0, \infty)$ **38.** no

39.

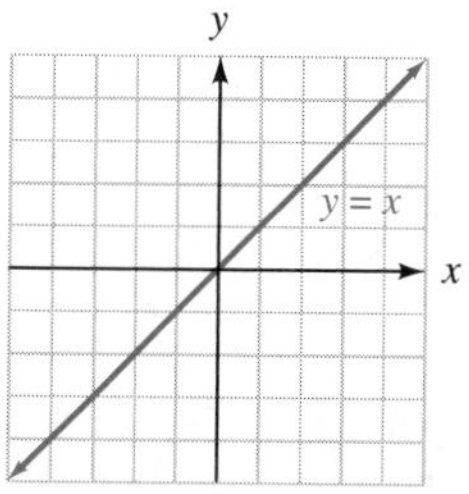

40.

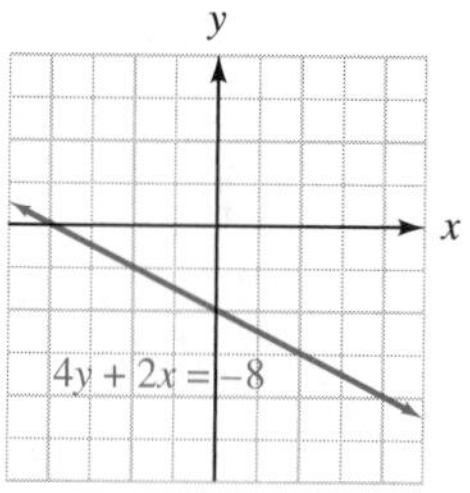

41. 0 **42.** $-\frac{10}{7}$ **43.** $\frac{7}{12}$ **44.** $\frac{2}{3}$, (0, 2)
45. $y = -2x + 1$ **46.** $y + 9 = -\frac{7}{8}(x - 2)$ **47.** yes
48.

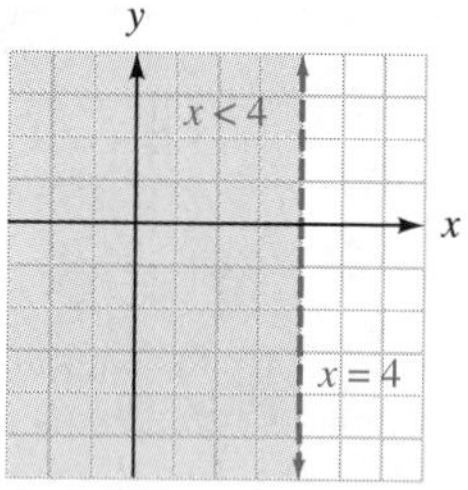

Study Set Section 4.1 (page 281)

1. exponential **3.** factor **5.** power **7. a.** $3x, 3x, 3x, 3x$
b. $(-5y)^3$ **9. a.** $(3x^2)^6$ (answers may vary) **b.** $\left(\frac{3a^3}{b}\right)^2$ (answers may vary) **11. a.** $2x^2$ **b.** 0 **c.** x^4 **d.** a^7
13. a. doesn't simplify **b.** doesn't simplify **c.** x^5 **d.** x
15. a. 16 **b.** −16 **17.** x^6, 18 **19.** base 4, exponent 3
21. base x, exponent 5 **23.** base $-3x$, exponent 2
25. base y, exponent 6 **27.** base $y + 9$, exponent 4
29. base $-3ab$, exponent 7 **31.** $x \cdot x \cdot x \cdot x \cdot x$
33. $\left(\frac{t}{2}\right)\left(\frac{t}{2}\right)\left(\frac{t}{2}\right)$ **35.** $(x - 5)(x - 5)$ **37.** $(4t)^4$ **39.** $-4t^3$
41. $(x - y)^3$ **43.** 12^7 **45.** 2^6 **47.** a^6 **49.** x^7
51. a^9 **53.** $(-7)^5$ **55.** $(8t)^{60}$ **57.** $(n - 1)^3$ **59.** y^9
61. 8^8 **63.** x^{12} **65.** c **67.** $(k - 2)^{14}$ **69.** 3^8
71. y^{15} **73.** m^{500} **75.** a^5b^6 **77.** c^2d^5 **79.** x^3y^3
81. y^4 **83.** c^2d^6 **85.** x^{25} **87.** $243z^{30}$ **89.** $9n^{16}$
91. u^4v^4 **93.** a^9b^6 **95.** $-8r^6s^9$ **97.** $\frac{a^3}{b^3}$ **99.** $\frac{x^{10}}{y^{15}}$
101. $\frac{-32a^5}{b^5}$ **103.** $216k^3$ **105.** $a^{12}b^6$ **107.** a
109. ab^4 **111.** $r^{13}s^3$ **113.** $\frac{y^3}{8}$ **115.** $\frac{27t^{12}}{64}$ **117.** a^{10} mi^2
119. x^9 ft^3 **121. a.** $25x^2$ ft^2 **b.** $9a^2\pi$ ft^2
123. 16 ft, 8 ft, 4 ft, 2 ft **129.** c **131.** d

Study Set Section 4.2 (page 290)

1. base, exponent **3.** reciprocal, 3 **5. a.** 4 − 4, 0, 6, 6, 6, 6, 1 **b.** 1, 1 **7.** 9, 3, 1, $\frac{1}{3}$, $\frac{1}{9}$ **9. a.** x^{m+n} **b.** x^{m-n}
c. x^{mn} **d.** x^ny^n **e.** $\frac{x^n}{y^n}$ **f.** $\frac{1}{x^n}$ **g.** x^n **h.** $\frac{y^n}{x^m}$ **i.** 1
11. once, no, exponents **13.** y^8, −40, 40 **15.** 4, −2, x, −5, $\frac{3}{y}$, −8, 7, −1, −2, −3, a, 0 **17.** 1 **19.** 1 **21.** 2 **23.** 1
25. 1 **27.** $\frac{5}{2}$ **29.** $-15y$ **31.** $\frac{1}{144}$ **33.** $-\frac{1}{4}$ **35.** $\frac{44}{g^6}$

37. $-\frac{1}{1{,}000}$ **39.** $-\frac{1}{64}$ **41.** $\frac{1}{64}$ **43.** $\frac{1}{x^2}$ **45.** $-\frac{1}{b^5}$ **47.** 36 **49.** -8 **51.** $\frac{8}{7}$ **53.** 125 **55.** $\frac{3}{16}$ **57.** $\frac{b^2}{a^5}$ **59.** r^{20} **61.** $-p^{10}$ **63.** $\frac{1}{h^7}$ **65.** $8s$ **67.** h^7 **69.** $\frac{1}{16y^4}$ **71.** $\frac{1}{a^3b^6}$ **73.** 8 **75.** $\frac{9}{y^8}$ **77.** $\frac{1}{y}$ **79.** $\frac{1}{r^6}$ **81.** $\frac{4t^2}{s^5}$ **83.** $\frac{125}{d^6}$ **85.** $-2a^4$ **87.** $\frac{1}{x^9}$ **89.** t^{10} **91.** y^5 **93.** 2 **95.** $\frac{1}{a^2b^4}$ **97.** $\frac{1}{x^6y^3}$ **99.** $\frac{1}{x^3}$ **101.** $-\frac{y^{10}}{32x^{15}}$ **103.** a^{14} **105.** $\frac{256x^{28}}{81}$ **107.** $\frac{y^{14}}{9z^{10}}$ **109.** $\frac{9}{4g^2}$ **111.** 2 **113.** $\frac{1}{2}$ **115.** $\frac{8}{9}$ **117.** $10^2, 10^1, 10^0, 10^{-1}, 10^{-2}, 10^{-3}, 10^{-4}$ **119.** about \$4,605 **121.** It gives the initial number of bacteria b. **125.** 13.5 yr **127.** $y = \frac{3}{4}x - 5$

Study Set Section 4.3 (page 297)

1. scientific **3.** notation **5.** powers **7. a.** right **b.** left **9. a.** 10^{-7} **b.** 10^9 **11. a.** positive **b.** negative **13. a.** 7.7 **b.** 5.0 **c.** 8 **15. a.** $(5.1 \times 1.5)(10^9 \times 10^{22})$ **b.** $\frac{8.8}{2.2} \times \frac{10^{30}}{10^{19}}$ **17.** 1, 10, integer **19.** 230 **21.** 812,000 **23.** 0.00115 **25.** 0.000976 **27.** 6,001,000 **29.** 2.718 **31.** 0.06789 **33.** 0.00002 **35.** 9,000,000,000 **37.** 2.3×10^4 **39.** 1.7×10^6 **41.** 6.2×10^{-2} **43.** 5.1×10^{-6} **45.** 5.0×10^9 **47.** 3.0×10^{-7} **49.** 9.09×10^8 **51.** 3.45×10^{-2} **53.** 9.0×10^0 **55.** 1.718×10^{18} **57.** 1.23×10^{-14} **59.** 714,000 **61.** 0.004032 **63.** 30,000 **65.** 0.0043 **67.** 0.0308 **69.** 200,000 **71.** $9.038030748 \times 10^{15}$ **73.** $1.734152992 \times 10^{-12}$ **75.** 2.57×10^{13} mi **77.** 63,800,000 mi^2 **79.** 6.22×10^{-3} mi **81.** g, x, u, v, i, m, r **83.** 1.7×10^{-18} g **85.** 3.09936×10^{16} ft **87.** 2.08×10^{11} dollars **89. a.** 1.7×10^6; 1,700,000 **b.** 1986: 2.05×10^6, 2,050,000 **95.** 5 **97.** $c = 30t + 45$

Study Set Section 4.4 (page 306)

1. polynomial **3.** term **5.** degree **7.** x, decreasing or descending **9.** nonlinear **11. a.** yes **b.** no **c.** no **d.** yes **e.** yes **f.** yes **13.** no **15.** Not enough ordered pairs were found—the correct graph is a parabola. **17.** $-2, -2, 4, -2, -8, -14, -15$ **19.** $y = x^2 - 1$, $y = x^3 - 1$ **21.** binomial **23.** trinomial **25.** monomial **27.** binomial **29.** trinomial **31.** none of these **33.** trinomial **35.** 4th **37.** 2nd **39.** 1st **41.** 4th **43.** 12th **45.** 0th **47. a.** -44 **b.** 37 **49. a.** 9 **b.** 8 **51. a.** -22 **b.** -406 **53. a.** 28 **b.** 4 **55. a.** 2 **b.** 0 **57.** 9, 4, 1, 0, 1, 4, 9

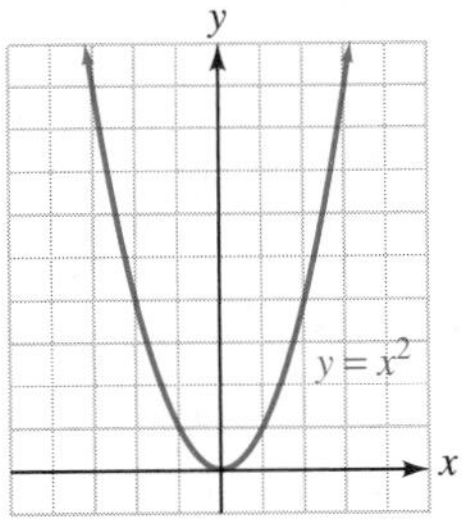

59.

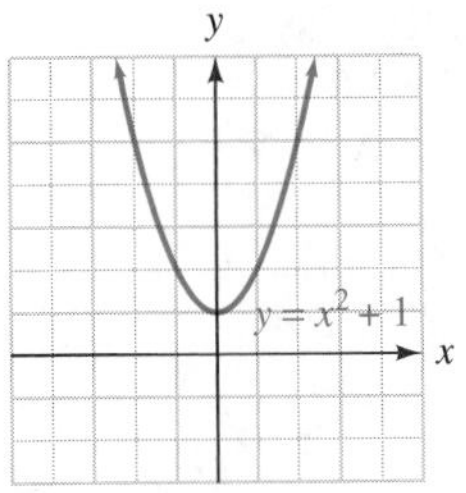

61.

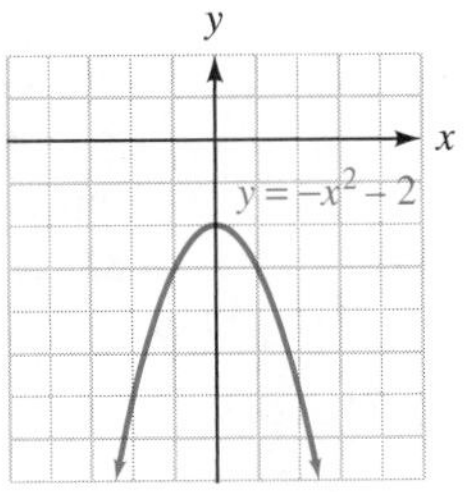

63.

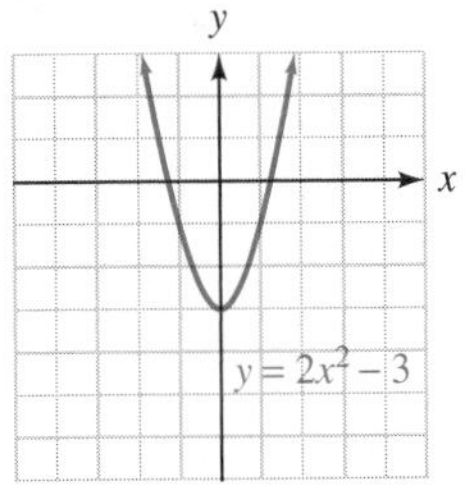

65.

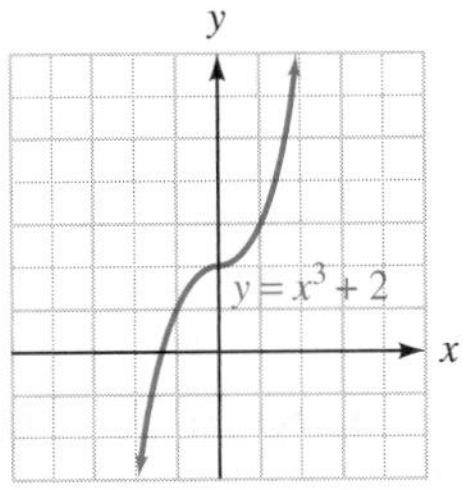

67.

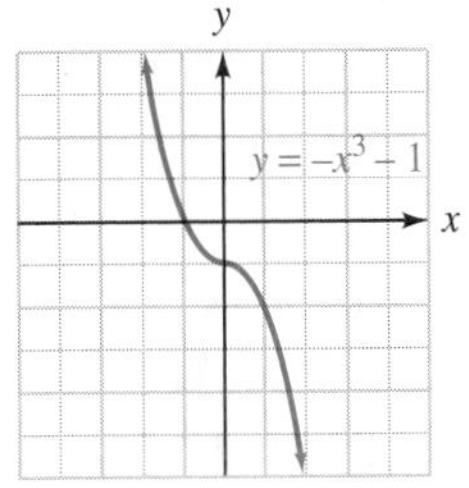

69. 91 **71.** 63 ft

73.

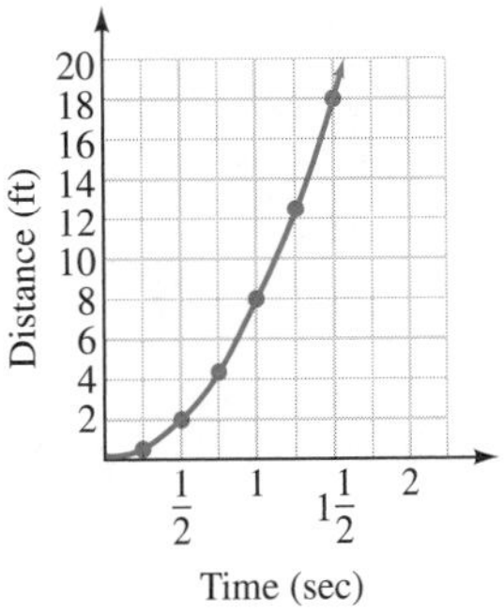

75. a. It costs 8¢ to make a 2-in. bolt. **b.** 12¢ **c.** a 4-in. bolt

79. $y \geq -3$; $[-3, \infty)$

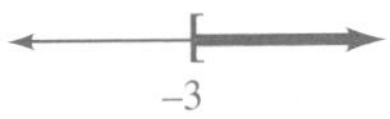

81. x^{18} **83.** y^9

Study Set Section 4.5 (page 314)

1. polynomials **3.** Like **5.** descending **7.** combine **9.** $-4x^2, +$ **11. a.** $5x^2$ **b.** $14m^3$ **c.** $8a^3b - ab$ **d.** $6cd + 4c^2d$ **13. a.** $-5x^2 + 8x - 23$ **b.** $5y^4 - 3y^2 + 7$ **15.** $-, +, -, 2x^2, 2$ **17.** true **19.** $12t^2$ **21.** $-48u^3$ **23.** $-0.1x$ **25.** $2st$ **27.** $6r$ **29.** $-ab$ **31.** $x^2 + x$ **33.** $13x^3$ **35.** $-4x^3y + x^2y + 5$ **37.** $\frac{7}{12}c^2 - \frac{1}{2}cd + d^2$ **39.** $7x + 4$ **41.** $13a^2 + a$ **43.** $7x - 7y$ **45.** $3x - 4y$ **47.** $6x^2 + x - 5$ **49.** $7b + 4$ **51.** $3x + 1$ **53.** $-5h^3 + 5h^2 + 30$ **55.** $-1.94x^2 + 3.4x + 0.01$ **57.** $\frac{5}{4}r^4 + \frac{7}{3}r^2 - 2$ **59.** $5x^2 + x + 11$ **61.** $5x^2 + 6x - 8$ **63.** $-x^3 + 6x^2 + x + 14$ **65.** $-x^3 + 4x^2 + 5x + 6$ **67.** $7x^3 - 2x^2 - 2x + 15$ **69.** $6x - 2$

71. $-5x^2 - 8x - 4$ **73.** $4y^3 - 12y^2 + 8y + 8$
75. $4s^2 - 5s + 7$ **77.** $3y^5 - 6y^4 + 1.2$
79. $t^3 + 3t^2 + 6t - 5$ **81.** $-3x^2 + 5x - 7$
83. a. $(x^2 - 8x + 12)$ ft **b.** $(x^2 + 2x - 8)$ ft
85. $(11x - 12)$ ft **87. a.** $(6x + 5)$ ft **b.** $(4x^2 + 26)$ ft
89. a. $(22t + 20)$ ft **b.** 108 ft **97.** 180°
99.

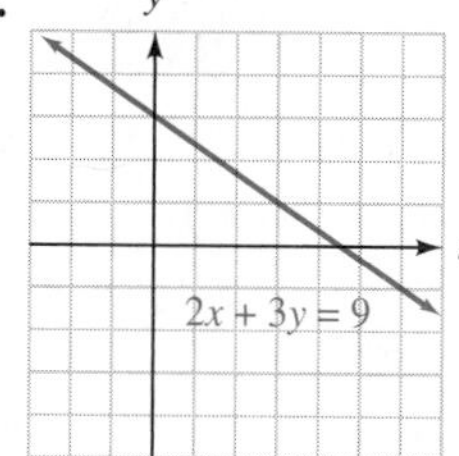

Study Set Section 4.6 (page 323)

1. monomials **3.** binomials **5.** like **7. a.** coefficients, variable **b.** each, each **c.** any, third **d.** length
9. F, L, I, O, $6x^2$, $8x$, $15x$, 20 **11. a.** $2x + 4$ **b.** -12
c. $x^2 + 4x - 32$ **13.** $7x, 7x, 7x, 21x^3$ **15.** $15m^2$
17. $12x^5$ **19.** $-24b^6$ **21.** $6x^5y^5$ **23.** $-2a^{11}$ **25.** $6c^6$
27. $\frac{1}{16}a^{10}$ **29.** $-3x^4y^5z^8$ **31.** $3x + 12$ **33.** $-4t^2 - 28$
35. $3x^2 - 6x$ **37.** $-6x^4 + 2x^3$ **39.** $6x^{14} - 72x^{13}$
41. $\frac{5}{8}t^8 + 5t^4$ **43.** $0.12p^9 - 1.8p^7$ **45.** $-12x^4z + 4x^2z^2$
47. $6x^4 + 8x^3 - 14x^2$ **49.** $12a^3 + 9a^2 - 12a$
51. $18a^6 - 12a^5$ **53.** $y^2 + 2y - 15$ **55.** $2t^2 + 5t - 12$
57. $6y^2 - y - 35$ **59.** $4x^2 - 4x - 15$
61. $a^2 + 2ab + b^2$ **63.** $12a^2 - 5ab - 2b^2$
65. $t^4 + t^2 - 12$ **67.** $-6t^2 + 13st - 6s^2$
69. $8a^2 + \frac{17}{9}ar - \frac{5}{12}r^2$ **71.** $8x^2 - 12x - 8$
73. $3a^3 - 3ab^2$ **75.** $x^3 - x + 6$
77. $4t^3 + 11t^2 + 18t + 9$ **79.** $2x^3 + 7x^2 - 16x - 35$
81. $-3x^3 + 25x^2y - 56xy^2 + 16y^3$
83. $r^4 - 5r^3 + 2r^2 - 7r - 15$ **85.** $x^3 - 3x + 2$
87. $12x^3 + 17x^2 - 6x - 8$
89. $6a^4 + 5a^3 + 5a^2 + 10a + 4$ **91.** $(6x^2 + x - 1)$ cm^2
93. $(9x^2 + 6x + 1)$ ft^2 **95.** $(4x^2 - 6x + 2)$ cm^2
97. $(6.28x^2 - 6.28)$ in.^{2} **99.** $(35x^2 + 43x + 12)$ cm^2
101. $(2x^3 - 4x^2 - 6x)$ in.^{3} **107.** 1 **109.** $-\frac{2}{3}$
111. (0, 2)

Study Set Section 4.7 (page 330)

1. term, term **3.** sum **5.** difference **7. a.** $x^{m \cdot n}$
b. x^ny^n **9. a.** 2 **b.** 2 **11.** second, square
13. x, 4, 4, $8x$ **15.** s, $-$, 5, 25 **17.** $x^2 + 2x + 1$
19. $r^2 + 4r + 4$ **21.** $m^2 - 12m + 36$
23. $f^2 - 16f + 64$ **25.** $d^2 - 49$ **27.** $n^2 - 36$
29. $16x^2 + 40x + 25$ **31.** $49m^2 - 28m + 4$
33. $y^4 + 18y^2 + 81$ **35.** $4v^6 - 32v^3 + 64$
37. $16f^2 - 0.16$ **39.** $9n^2 - 1$ **41.** $1 - 6y + 9y^2$
43. $x^2 - 4xy + 4y^2$ **45.** $4a^2 - 12ab + 9b^2$
47. $s^2 + \frac{3}{2}s + \frac{9}{16}$ **49.** $a^2 + 2ab + b^2$ **51.** $r^2 - 2rs + s^2$
53. $36b^2 - \frac{1}{4}$ **55.** $r^2 + 20rs + 100s^2$
57. $36 - 24d^3 + 4d^6$ **59.** $-64x^2 - 48x - 9$
61. $-25 + 36g^2$ **63.** $12x^3 + 36x^2 + 27x$
65. $-80d^3 + 40d^2 - 5d$ **67.** $4d^5 - 4dg^6$
69. $x^3 + 12x^2 + 48x + 64$ **71.** $n^3 - 18n^2 + 108n - 216$
73. $8g^3 - 36g^2 + 54g - 27$
75. $n^4 - 8n^3 + 24n^2 - 32n + 16$ **77.** $3t^2 + 12t - 9$
79. $2x^2 + xy - y^2$ **81.** $13x^2 - 8x + 5$ **83.** $36m + 36$
85. $\frac{\pi}{4}D^2 - \frac{\pi}{4}d^2$ **87.** $(x^2 + 12x + 36)$ in.^{2}
89. $(36x^2 + 36x + 6)$ ft^2 **95.** $3^3 \cdot 7$ **97.** $\frac{5}{6}$ **99.** $\frac{21}{40}$

Study Set Section 4.8 (page 338)

1. numerator, denominator **3.** polynomial **5.** descending
7. placeholder **9. a.** x^{m-n} **b.** $\frac{1}{x^n}$
11. a. $7x^3 + 5x^2 - 3x - 9$ **b.** $6x^4 - x^3 + 2x^2 + 9x$
13. $-2x$ **15.** It is correct.
17. $7x^2, x^3, 7x^2, 7, 2 - 2, 4x^3, 1$
19. a. $5x^4 + 0x^3 + 2x^2 + 0x - 1$
b. $-3x^5 + 0x^4 - 2x^3 + 0x^2 + 4x - 6$ **21.** $\frac{4}{3}$ **23.** x^3
25. $5m^5$ **27.** $\frac{4h^2}{3}$ **29.** $-\frac{1}{5d^4}$ **31.** $\frac{r^2}{s}$ **33.** $\frac{2x^2}{y}$ **35.** $\frac{4r}{y^2}$
37. $-\frac{13}{3rs}$ **39.** $2x + 3$ **41.** $2x^5 - 8x^2$ **43.** $\frac{h^2}{4} + \frac{2}{h}$
45. $-2w^2 - \frac{1}{w^4}$ **47.** $3s^5 - 6s^2 + 4s$
49. $c^3 + 3c^2 - 2c - \frac{5}{c}$ **51.** $\frac{1}{5y} - \frac{2}{5x}$ **53.** $3a - 2b$
55. $3x^2y - 2x - \frac{1}{y}$ **57.** $5x - 6y + 1$ **59.** $x + 6$
61. $y + 12$ **63.** $3a - 2$ **65.** $b + 3$ **67.** $2x + 1$
69. $x - 7$ **71.** $3x + 2$ **73.** $x^2 + 2x - 1$
75. $2x^2 + 2x + 1$ **77.** $x^2 + x + 1$ **79.** $x + 1$
81. $2x - 3$ **83.** $x^2 - x + 1$ **85.** $a^2 - 3a + 10 + \frac{-30}{a + 3}$
87. $x + 1 + \frac{-1}{2x + 3}$ **89.** $2x + 2 + \frac{-3}{2x + 1}$
91. $x^2 + 2x - 1 + \frac{6}{2x + 3}$ **93.** $2x^2 + x + 1 + \frac{2}{3x - 1}$
95. $(2x^2 - x + 3)$ in. **97. a.** $t = \frac{d}{r}$ **b.** $3x^2$ **c.** $x + 4$
99. $(3x - 2)$ ft **101.** $4x^2 + 3x + 7$ **107.** $y = -\frac{11}{6}x - \frac{7}{3}$
109. -80

Key Concept (page 343)

1. a. x, descending, x **b.** 4 **c.** 3, 2, 1, 0 **d.** 3
e. 1, -2, 6, -8 **2. a.** binomial **b.** none of these
c. trinomial **d.** monomial **3.** $7x^3$ **4.** m^{10} **5.** $-2a^2b$
6. $-42y^{13}$ **7.** $\frac{2c^2}{d}$ **8.** $25f^6$ **9.** combine **10.** signs
11. each, each **12.** term **13.** long
14. $14x^3 - x^2 - 10x + 4$ **15.** $12s^3t + 10s^2t^2 - 18st^3$
16. $2x^2 - 13x - 24$ **17.** $4x^4 + 12x^2 + 9$
18. $16h^{10} - 64t^2$ **19.** $y^3 + 4y^2 - 3y - 18$
20. $3x^4 + 9x^5 - 6x^3$ **21.** $x^2 + 2x + 3$

Chapter Review (page 344)

1. a. $-3 \cdot x \cdot x \cdot x \cdot x$ **b.** $(\frac{1}{2}pq)(\frac{1}{2}pq)(\frac{1}{2}pq)$ **2. a.** base x, exponent 6 **b.** base $2x$, exponent 6 **3.** 125 **4.** 64
5. -64 **6.** 4 **7.** 7^{12} **8.** m^2n^2 **9.** y^{21} **10.** $81x^4$

11. b^{12} **12.** $-y^2z^5$ **13.** $256s^3$ **14.** $4x^4y^2$ **15.** x^{15}
16. $\frac{x^2}{y^2}$ **17.** $(m - 25)^{12}$ **18.** $125yz^4$ **19.** $64x^{12}$ in.^{3}
20. y^4 m^2 **21.** 1 **22.** 1 **23.** 9 **24.** $\frac{1}{1,000}$ **25.** $\frac{4}{3}$
26. $-\frac{1}{25}$ **27.** $\frac{1}{x^5}$ **28.** $-\frac{6}{y}$ **29.** $\frac{8}{49}$ **30.** x^{14} **31.** $-\frac{27}{r^9}$
32. $\frac{1}{16z^2}$ **33.** 7.28×10^2 **34.** 9.37×10^{15}
35. 1.36×10^{-2} **36.** 9.42×10^{-3} **37.** 1.8×10^{-4}
38. 7.53×10^5 **39.** 726,000 **40.** 0.0000000391
41. 2.68 **42.** 57.6 **43.** 0.03 **44.** 160
45. 6,310,000,000; 6.31×10^9 **46.** $1.0 \times 10^5 = 100{,}000$
47. a. 4 **b.** $3x^3$ **c.** 3, −1, 1, 10 **d.** 10 **48. a.** 7th, monomial **b.** 3rd, monomial **c.** 2nd, binomial **d.** 5th, trinomial **e.** 6th, binomial **f.** 4th, none of these
49. 3, −13 **50.** 8 in.
51.

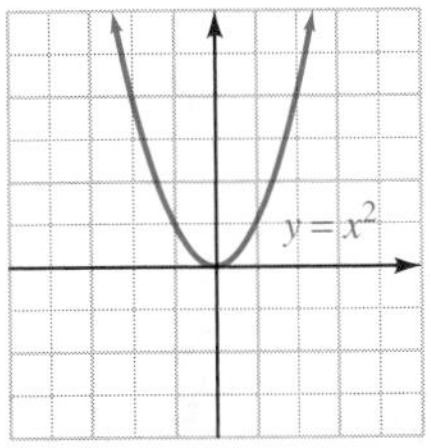

52.

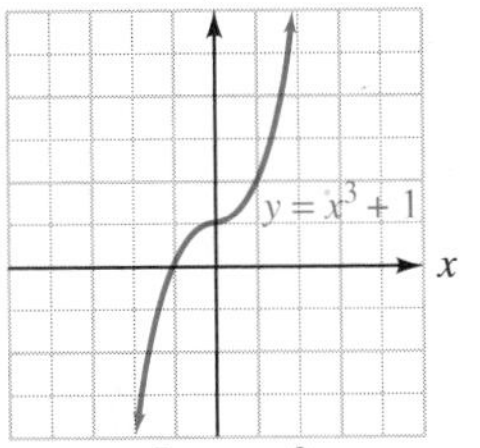

53. $13y^3$ **54.** $-4a^3b + a^2b + 5$ **55.** $\frac{7}{12}x^2 - \frac{3}{4}xy + y^2$
56. $-c^5 - 5c^4 + 12$ **57.** $25r^6 + 9r^3 + 5r$
58. $4a^2 + 4a - 6$ **59.** $4r^3s - 7r^2s^2 - 7rs^3 - 2s^4$
60. $5x^2 + 19x + 3$ **61.** $-z^3 + 2z^2 + 5z - 17$
62. $4x^2 + 2x + 8$ **63.** $8x^3 - 7x^2 + 19x$
64. $(x^2 + x + 3)$ in. **65.** $10x^3$ **66.** $-6x^{10}z^5$ **67.** $120b^{11}$
68. $2h^{14} + 8h^{11}$ **69.** $x^2y^3 - x^3y^2$ **70.** $9n^4 - 15n^3 + 6n^2$
71. $6x^6 + 12x^5$ **72.** $a^6b^4 - a^5b^5 + a^3b^6 - 7a^3b^2$
73. $x^2 + 5x + 6$ **74.** $2x^2 - x - 1$ **75.** $6a^2 - 6$
76. $6a^2 - 6$ **77.** $2a^2 - ab - b^2$ **78.** $6n^8 - 13n^6 + 5n^4$
79. $8a^3 - 27$ **80.** $56x^4 + 15x^3 - 21x^2 - 3x + 2$
81. $8x^3 + 1$ **82.** $(6x + 10)$ in.; $(2x^2 + 11x - 6)$ in.^{2}; $(6x^3 + 33x^2 - 18x)$ in.^{3} **83.** $x^2 + 6x + 9$ **84.** $4x^2 - 0.81$
85. $a^2 - 6a + 9$ **86.** $x^2 + 8x + 16$ **87.** $4y^2 - 4y + 1$
88. $y^4 - 1$ **89.** $36r^4 + 120r^2s + 100s^2$
90. $-64a^2 + 48a - 9$ **91.** $80r^4s - 80s^5$
92. $36b^3 - 96b^2 + 64b$ **93.** $t^2 - \frac{3}{2}t + \frac{9}{16}$
94. $m^3 + 6m^2 + 12m + 8$ **95.** $24c^2 - 10c + 37$
96. $(x^2 - 4)$ in.^{2} **97.** $2n^3$ **98.** $-\frac{2x}{3y^2}$ **99.** $\frac{a^3}{6} - \frac{4}{a^4}$
100. $3a + 4b - 5$ **101.** $x - 5$ **102.** $2x + 1$
103. $3x^2 + 2x + 1 + \frac{2}{2x - 1}$ **104.** $3x^2 - x - 4$
105. $(y + 3)(3y + 2) = 3y^2 + 11y + 6$ **106.** $x + 5$

Chapter 4 Test (page 349)

1. $2x^3y^4$ **2.** −36 **3.** y^6 **4.** $32x^{21}$ **5.** 3 **6.** $\frac{2}{y^3}$
7. $\frac{1}{125}$ **8.** $(x + 1)^9$ **9.** y^3 **10.** $\frac{64a^3}{b^3}$
11. $1{,}000y^{12}$ in.^{3} **12.** 6.25×10^{18} **13.** 0.000093
14. 9,200 **15.** x^4, 1, 4, $8x^2$, 8, 2, −12, −12, 0, 4
16. binomial **17.** 5th degree **18.**

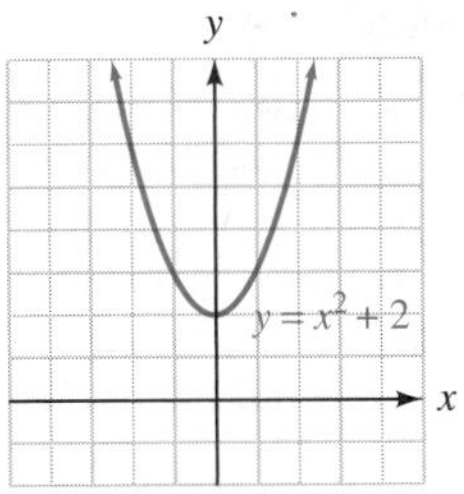

19. 0 ft; The rock hits the canyon floor 18 seconds after being dropped. **20.** $-4a^3b + a^2b + 5$ **21.** $5a^2 - 5a + 3$
22. $6b^3c - 2bc - 12$ **23.** $-3y^3 + 18y^2 - 28y + 35$
24. $-4x + 8y$ **25.** $-4x^5y$ **26.** $3y^4 - 6y^3 + 9y^2$
27. $6x^2 - 7x - 20$ **28.** $2x^3 - 7x^2 + 14x - 12$
29. $1 - 100c^2$ **30.** $49b^6 - 42b^3 + 9$ **31.** $2x^2 + xy - y^2$
32. $\frac{a}{4b} - \frac{b}{2a}$ **33.** $x - 2$ **34.** $3x^2 + 2x + 1 + \frac{2}{2x - 1}$
35. $(x - 5)$ ft

Cumulative Review Exercises Chapters 1–4 (page 351)

1. a. 1993 **b.** 1986 **c.** 1996–1997 **2.** $\frac{5}{8}$ **3.** $\frac{22}{35}$
4. irrational **5.** $250 - x$ **6.** 18, 16 **7.** $2^2 \cdot 5^2$
8.

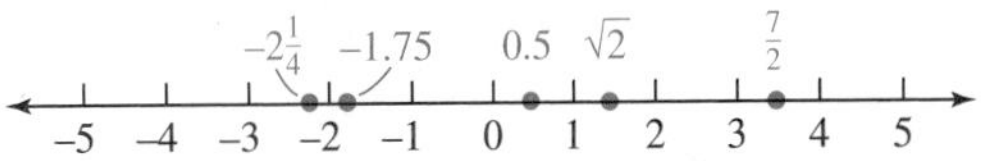

9. $0.\overline{6}$ **10.** associative property of multiplication **11.** $10d$ cents **12.** r **13.** $18x$ **14.** $3d - 11$ **15.** $-78c + 18$
16. 0 **17.** 1 **18.** 4 **19.** −2 **20.** 13 **21.** 41
22. $\frac{10}{9}$ **23.** −24
24. $x < -14$, $(-\infty, -14)$

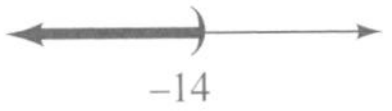

25. $h = \frac{2A}{b + B}$ **26.** 20 lb of \$1.90 candy; 10 lb of \$2.20 candy
27.

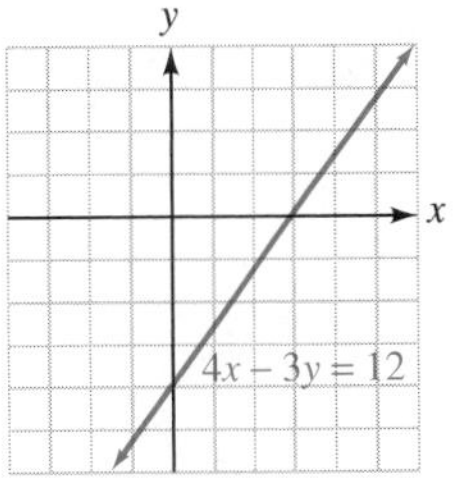

28.

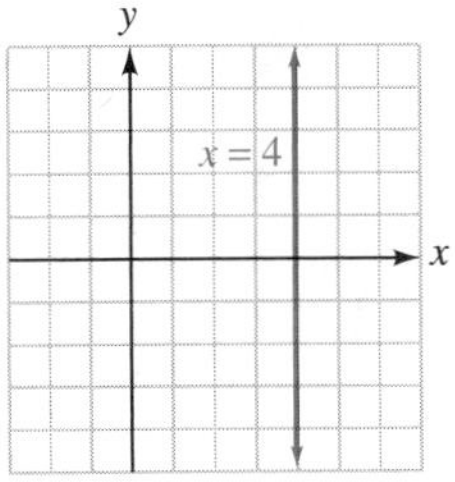

29. $\frac{1}{2}$ **30.** 0 **31.** $\frac{2}{3}$ **32.** $y = \frac{2}{3}x + 5$
33. $3x - 4y = -22$ **34.** $y = 4$ **35.** perpendicular
36. $-6x^4 - 17x^2 - 68x + 11$ **37.** $9x^4y^8$ **38.** $\frac{1}{16y^4}$
39. x^{14} **40.** $a^2b^7c^6$ **41.** a^7 **42.** $\frac{64t^{12}}{27}$ **43.** $7c^2 + 7c$
44. $12x^3 + 36x^2 + 27x$ **45.** $6t^2 + 7st - 3s^2$ **46.** $2x + 1$
47.

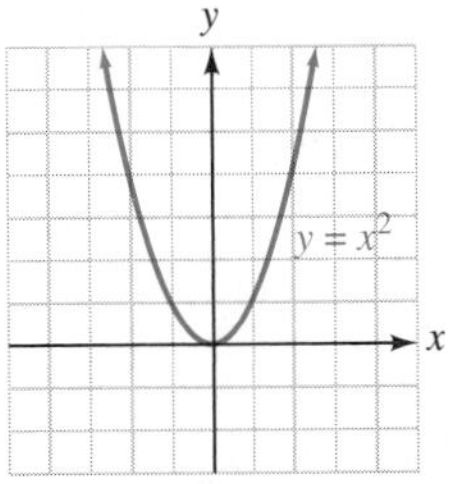

48. 6.15×10^5
49. 1.3×10^{-6} **50.** 1.5 in.

Study Set Section 5.1. (page 360)

1. greatest common factor **3.** primes **5.** grouping **7. a.** 3 **b.** 2 **c.** 3 **d.** 7 **9.** $3a^2b$ **11. a.** the distributive property **b.** $2x$ **13. a.** The GCF is $6a^2$, not $6a$. **b.** The 0 in the first line should be 1. **15.** no **17.** $8m$, 4 **19.** $2^2 \cdot 3$ **21.** $2^3 \cdot 5$ **23.** $2 \cdot 7^2$ **25.** $3^2 \cdot 5^2$ **27.** 6 **29.** m^3 **31.** $4c$ **33.** $3m^3n$ **35.** $2x$ **37.** $3y$ **39.** $2t^2$ **41.** $8x^2z^2$ **43.** $3(x + 2)$ **45.** $9(2m - 1)$ **47.** $6(3x + 4)$ **49.** $d(d - 7)$ **51.** $7(2c^3 + 9)$ **53.** $6(2x^2 - x - 4)$ **55.** $t^2(t + 2)$ **57.** $a(b + c - d)$ **59.** $a^2(a - 1)$ **61.** $8xy^2(3xy + 1)$ **63.** $6uvw^2(2w - 3v)$ **65.** $3r(4r - s + 3rs^2)$ **67.** $\pi(R^2 - ab)$ **69.** $(x + 2)(3 - x)$ **71.** $(14 + r)(h^2 + 1)$ **73.** $-(a + b)$ **75.** $-(2x - 5y)$ **77.** $-(3r - 2s + 3)$ **79.** $-3x(x + 2)$ **81.** $-4a^2b^2(b - 3a)$ **83.** $-2ab^2c(2ac - 7a + 5c)$ **85.** $(x + y)(2 + a)$ **87.** $(p - q)(9 + m)$ **89.** $(m - n)(p + q)$ **91.** $(2x - 3y)(y + 1)$ **93.** $(a + b)(x - 1)$ **95.** $x^2(a + b)(x + 2y)$ **97.** $2z(x - 2)(x^2 + 16)$ **99.** $(y + 9)(x - 11)$ **101.** $(3p + q)(3m - n)$ **103.** $5x^3y^2z^3(5x^2y^5 - 9z^3)$ **105.** $4(6m - 3n + 4)$ **107.** $-20P(3P + 4)$ **109. a.** $6x^3$ cm^2 **b.** $24x^2$ cm^2 **c.** $6x^2(x + 4)$ cm^2 **111.** $(x^2 + 5)$ ft; $(x + 4)$ ft **117.** 12%

Study Set Section 5.2 (page 371)

1. trinomial, binomial **3.** factors **5.** lead, coefficient, term **7.** $4 \cdot 1, 2 \cdot 2, -4(-1), -2(-2)$ **9.** product, sum **11.** 3, 5 **13. a.** Last: $5 \cdot 4$ **b.** Outer: $x \cdot 4$ and Inner: $5 \cdot x$ **15. a.** 1 **b.** 15, 8 **c.** 5, 3 **17. a.** They are both positive or both negative. **b.** One will be positive, the other negative. **19.** $(x + 1)(x + 5)$ **21.** $-1 + (-8) = -9, -2(-4) = 8$ **23.** $+ 3, - 2$ **25.** $+, +$ **27.** $-, -$ **29.** $+, -$ **31.** $(z + 11)(z + 1)$ **33.** $(m - 3)(m - 2)$ **35.** $(a - 5)(a + 1)$ **37.** $(x + 8)(x - 3)$ **39.** $(a - 13)(a + 3)$ **41.** prime **43.** prime **45.** $(r - 3)(r - 6)$ **47.** $(x + 2y)(x + 2y)$ **49.** $(a - 6b)(a + 2b)$ **51.** prime **53.** $-(x + 5)(x + 2)$ **55.** $-(t + 17)(t - 2)$ **57.** $-(a + 3b)(a + b)$ **59.** $-(x - 7y)(x + y)$ **61.** $(x - 4)(x - 1)$ **63.** $(y + 9)(y + 1)$ **65.** $(r - 2)(r + 1)$ **67.** $(r + 3x)(r + x)$ **69.** $2(x + 3)(x + 2)$ **71.** $-5(a - 3)(a - 2)$ **73.** $z(z - 4)(z - 25)$ **75.** $4y(x + 6)(x - 3)$ **77.** $-(r - 10)(r - 4)$ **79.** $(y - 14z)(y + z)$ **81.** $(s + 13)(s - 2)$ **83.** $(a + 9b)(a + b)$ **85.** $-(x - 22)(x + 1)$ **87.** $d(d - 13)(d + 2)$ **89.** $(x + 9)$ in., x in., $(x + 3)$ in. **97.** $\frac{1}{x^2}$ **99.** $\frac{1}{x^{10}}$

Study Set Section 5.3 (page 381)

1. lead, term **3.** product, sum **5.** number **7. a.** $1 \cdot 9 = 9$ **b.** $-2(-8) = 16$ **c.** $-2(5) = -10$ **9. a.** $8y$ **b.** $-24y$ **c.** $-16y$ **11.** $10x$ and x, $5x$ and $2x$ **13. a.** descending, GCF, positive **b.** $3s^2$ **c.** $-(2d^2 - 19d + 8)$ **15.** positive, negative, negative **17.** negative, different **19.** $(5x - 2)(3x + 2)$ **21.** $-1 + (-12) = -13, -2(-6) = 12, -2 + (-6) = -8, -3 + (-4) = -7$ **23. a.** $x^2 + 6x + 1$ **b.** $3x^2 + 6x + 1$ **25. a.** 12, 20, −9 **b.** −108 **27.** $+, +$ **29.** $-, -$ **31.** $-, +$ **33.** 9, 8, $3t$, 2, $4t + 3$ **35.** $(3a + 1)(a + 4)$ **37.** $(5x + 1)(x + 2)$ **39.** $(2x + 3)(2x + 1)$ **41.** $(6x + 7)(x + 3)$ **43.** $(2x - 1)(x - 1)$ **45.** $(2t - 1)(2t - 1)$ **47.** $(15t - 4)(t - 2)$ **49.** $(2x + 1)(x - 2)$ **51.** $(4y + 1)(3y - 1)$ **53.** $(5y + 1)(2y - 1)$ **55.** $(3y - 2)(4y + 1)$ **57.** prime **59.** $(3x + 2)(x - 5)$ **61.** $(3r + 2s)(2r - s)$ **63.** $(8n + 3)(n + 11)$ **65.** $(4a - 3b)(a - 3b)$ **67.** $10(13r - 11)(r + 1)$ **69.** $-(5t + 3)(t + 2)$ **71.** $4(9y - 4)(y - 2)$ **73.** $(2x + 3y)(2x + y)$ **75.** $(4y - 3)(3y - 4)$ **77.** $(18x - 5)(x + 2)$ **79.** $-y(y + 12)(y + 1)$ **81.** prime **83.** $3r^3(5r - 2)(2r + 5)$ **85.** $(2a + 3b)(a + b)$ **87.** $(3p - q)(2p + q)$ **89.** $3x(2x + 1)(x - 3)$ **91.** $(2a - 5)(4a - 3)$ **93.** $2mn(4m + 3n)(2m + n)$ **95.** $(2x + 11)$ in., $(2x - 1)$ in.; 12 in. **103.** −49 **105.** 1 **107.** 49

Study Set Section 5.4 (page 388)

1. perfect **3. a.** $5x$ **b.** 3 **c.** $5x$, 3 **5. a.** x, y **b.** $-$ **c.** $+$, x, y **7.** 1, 4, 9, 16, 25, 36, 49, 64, 81, 100 **9. a.** $x^2 - 36$, $x^2 + 12x + 36$, $x^2 - 12x + 36$ **b.** no **11.** $9x^2 - 16y^2$ **13. a.** $x^2 - 9$ **b.** $(x - 9)^2$ **c.** $x^2 + 2xy + y^2$ **d.** $x^2 + 25$ **15. a.** $x + 3$; 2 **b.** $(x - 8)(x - 8)$ **17.** 3 **19.** + **21.** $(x + 3)^2$ **23.** $(b + 1)^2$ **25.** $(c - 6)^2$ **27.** $(y - 4)^2$ **29.** $(t + 10)^2$ **31.** $2(u - 9)^2$ **33.** $x(6x + 1)^2$ **35.** $(2x + 3)^2$ **37.** $(a + b)^2$ **39.** $(5m + 7n)^2$ **41.** $(3xy + 5)^2$ **43.** $\left(t - \frac{1}{3}\right)^2$ **45.** $(s - 0.6)^2$ **47.** 7, 7 **49.** t, w **51.** $(x + 4)(x - 4)$ **53.** $(2y + 1)(2y - 1)$ **55.** $(3x + y)(3x - y)$ **57.** $(4a + 5b)(4a - 5b)$ **59.** $(6 + y)(6 - y)$ **61.** prime **63.** $(a^2 + 12b)(a^2 - 12b)$ **65.** $(tz + 8)(tz - 8)$ **67.** prime **69.** $8(x + 2y)(x - 2y)$ **71.** $7(a + 1)(a - 1)$ **73.** $(v + 5)(v - 5)$ **75.** $6x^2(x + y)(x - y)$ **77.** $(x^2 + 9)(x + 3)(x - 3)$ **79.** $(a^2 + 4)(a + 2)(a - 2)$ **81.** $\left(c + \frac{1}{4}\right)\left(c - \frac{1}{4}\right)$ **83.** $(p + q)^2$ **85.** $0.5g(t_1 + t_2)(t_1 - t_2)$ **93.** $\frac{x}{y} + \frac{2y}{x} - 3$ **95.** $3a + 2$

Study Set Section 5.5 (page 394)

1. sum **3.** binomial **5. a.** F, L **b.** $-$, F^2, L^2 **7.** m, 4 **9.** 1, 8, 27, 64, 125, 216 **11.** It is factored completely. **13. a.** $x^3 + 8$ **b.** $(x + 8)^3$ **15.** $2a$ **17.** $b + 3$ **19.** $(y + 1)(y^2 - y + 1)$ **21.** $(a - 3)(a^2 + 3a + 9)$ **23.** $(2 + x)(4 - 2x + x^2)$ **25.** $(s - t)(s^2 + st + t^2)$ **27.** $(a + 2b)(a^2 - 2ab + 4b^2)$ **29.** $(4x - 3)(16x^2 + 12x + 9)$ **31.** $(a^2 - b)(a^4 + a^2b + b^2)$ **33.** $2(x + 3)(x^2 - 3x + 9)$ **35.** $-(x - 6)(x^2 + 6x + 36)$ **37.** $8x(2m - n)(4m^2 + 2mn + n^2)$ **39.** $(1{,}000 - x^3)$ in.^{3}; $(10 - x)(100 + 10x + x^2)$ **43.** repeating **45.** $\{\ldots, -4, -3, -2, -1, 0, 1, 2, 3, 4, \ldots\}$ **47.** 0

Study Set Section 5.6 (page 399)

1. completely **3.** prime **5.** trinomial **7.** GCF **9.** factor out the GCF **11.** perfect square trinomial **13.** sum of two cubes **15.** trinomial factoring **17.** $8m^3$ **19.** $2a(b + 6)(b - 2)$ **21.** $-4p^2q^3(2pq^4 + 1)$ **23.** $5(2m + 5)^2$ **25.** prime **27.** $-2x^2(x - 4)(x^2 + 4x + 16)$ **29.** $(c + d^2)(a^2 + b)$ **31.** $-(3xy - 1)^2$ **33.** $5x^2y^2z^2(xyz^2 + 5y - 7xz^3)$ **35.** $(2c + d)(c - 3d)$ **37.** $2xy(2ax + b)(2ax - b)$ **39.** $(x - a)(a + b)(a - b)$ **41.** $(ab + 12)(ab - 12)$ **43.** $(c + 2d)(2a + b)$ **45.** $(v - 7)^2$ **47.** $-(n + 3)(n - 13)$ **49.** $(4y^4 + 9z^2)(2y^2 + 3z)(2y^2 - 3z)$ **51.** $2(3x - 4)(x - 1)$ **53.** $y^2(2x + 1)^2$ **55.** $4m^2n(m + 5n)(m^2 - 5mn + 25n^2)$ **57.** $(x^2 + y^2)(a^2 + b^2)$ **59.** prime **61.** $2a^2(2a - 3b)(4a^2 + 6ab + 9b^2)$ **63.** $27(x - y - z)$ **65.** $(x - t)(y + s)$ **67.** $x^6(7x + 1)(5x - 1)$ **69.** $5(x - 2)(1 + 2y)$ **71.** $(7p + 2q)^2$ **73.** $4(t^2 + 9)$ **75.** $(p - 2)(p^2 + 3)$

81. 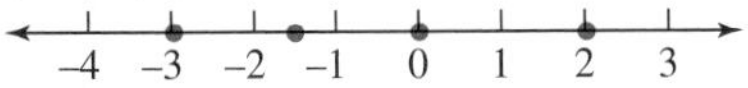

83. 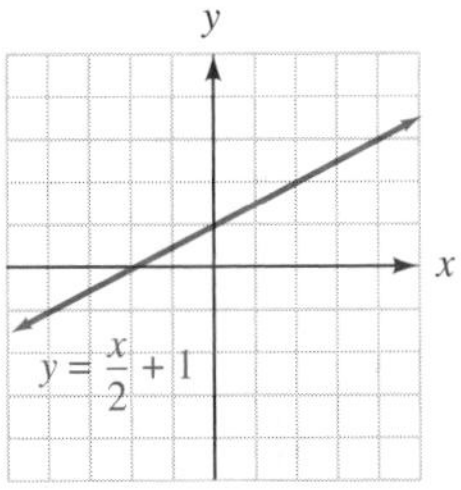

Study Set Section 5.7 (page 408)

1. quadratic **3.** consecutive **5.** hypotenuse, legs **7. a.** yes **b.** no **c.** yes **d.** no **9.** $-\frac{4}{5}$ **11. a.** Add 6 to both sides. **b.** Distribute the multiplication by x and subtract 3 from both sides. **13. a.** $(x - 2)(x + 8)$ **b.** $2, -8$ **15.** 0 **17.** $a^2 + b^2$ **19.** $7y, y + 2, 0$ **21.** $2, -3$ **23.** $\frac{5}{2}, -6$ **25.** $7, -8$ **27.** $1, -2, 3$ **29.** $0, 3$ **31.** $0, \frac{5}{2}$ **33.** $0, 7$ **35.** $0, -\frac{8}{3}$ **37.** $0, 2$ **39.** $-5, 5$ **41.** $-\frac{1}{2}, \frac{1}{2}$ **43.** $-\frac{2}{3}, \frac{2}{3}$ **45.** $-10, 10$ **47.** $-\frac{9}{2}, \frac{9}{2}$ **49.** $12, 1$ **51.** $-3, 7$ **53.** $8, 1$ **55.** $-3, -5$ **57.** $-2, -2$ **59.** $8, 8$ **61.** $\frac{1}{2}, 2$ **63.** $\frac{1}{5}, 1$ **65.** $-\frac{1}{2}, -\frac{1}{2}$ **67.** $-\frac{2}{3}, -\frac{3}{2}$ **69.** $\frac{2}{3}, -\frac{1}{5}$ **71.** $-\frac{5}{2}, 4$ **73.** $-3, -2$ **75.** $2, 10$ **77.** $\frac{1}{8}, 1$ **79.** $2, 7, 1$ **81.** $0, -1, -2$ **83.** $0, 9, -3$ **85.** $0, -1, 2$ **87.** $\frac{3}{2} = 1.5$ sec **89.** 2 sec **91.** 8 **93.** 12 **95.** (11, 13) **97.** 4 m by 9 m **99.** $h = 5$ ft, $b = 12$ ft **101.** 6, 16 ft **103.** 3 ft, 4 ft, 5 ft **109.** $15 \text{ min} \le t < 30 \text{ min}$

Key Concept (page 414)

1. Factor $3x + 27$; $3(x + 9)$ **2.** Factor $x^2 + 12x + 27$; $(x + 3)(x + 9)$ **3.** $-3(a - 3)(a - 4)$ **4.** $(x + 11y)(x - 11y)$ **5.** $(r + s)(t + 2)$ **6.** $(v - 2)(v^2 + 2v + 4)$ **7.** $(3t - 5)(2t - 3)$ **8.** $(5y - 2)^2$ **9.** $2r(r + 5)(r - 5)$ **10.** $2(8w - 1)(w + 3)$

Chapter Review (page 415)

1. $5 \cdot 7$ **2.** $2^5 \cdot 3$ **3.** 7 **4.** $18a^3$ **5.** $3(x + 3y)$ **6.** $5a(x^2 + 3)$ **7.** $7s^3(s^2 + 2)$ **8.** $\pi a(b - c)$ **9.** $2x(x^2 + 2x - 4)$ **10.** $xy^2z(x + yz - 1)$ **11.** $-5ab(b - 2a + 3)$ **12.** $(x - 2)(4 - x)$ **13.** $-(a + 7)$ **14.** $-(4t^2 - 3t + 1)$ **15.** $(c + d)(2 + a)$ **16.** $(y + 3)(3x - 2)$ **17.** $(2a^2 - 1)(a + 1)$ **18.** $4m(n + 3)(m - 2)$ **19.** 1 **20.** $7, 5, -7, -5$ **21.** $(x + 6)(x - 4)$ **22.** $(x - 6)(x + 2)$ **23.** $(x - 5)(x - 2)$ **24.** prime **25.** $-(y - 5)(y - 4)$ **26.** $(y + 9)(y + 1)$ **27.** $(c + 5d)(c - 2d)$ **28.** $(m - 2n)(m - n)$ **29.** Multiply to see if $(x - 4)(x + 5) = x^2 + x - 20$. **30.** There are no two integers whose product is 11 and whose sum is 7. **31.** $5a^3(a + 10)(a - 1)$ **32.** $-4x(x + 3y)(x - 2y)$ **33.** $(2x + 1)(x - 3)$ **34.** $(2y + 5)(5y - 2)$ **35.** $-(3x + 1)(x - 5)$ **36.** $3p(2p + 1)(p - 2)$ **37.** $(4b - c)(b - 4c)$ **38.** prime **39.** $(4x + 1)$ in., $(3x - 1)$ in. **40.** The signs of the second terms must be negative. **41.** $(x + 5)^2$ **42.** $(3y - 4)^2$ **43.** $-(z - 1)^2$ **44.** $(5a + 2b)^2$ **45.** $(x + 3)(x - 3)$ **46.** $(7t + 5y)(7t - 5y)$ **47.** $(xy + 20)(xy - 20)$ **48.** $8a(t + 2)(t - 2)$ **49.** $(c^2 + 16)(c + 4)(c - 4)$ **50.** prime **51.** $(h + 1)(h^2 - h + 1)$ **52.** $(5p + q)(25p^2 - 5pq + q^2)$ **53.** $(x - 3)(x^2 + 3x + 9)$ **54.** $2x^2(2x - 3y)(4x^2 + 6xy + 9y^2)$ **55.** $2y^2(3y - 5)(y + 4)$ **56.** $(t + u^2)(s^2 + v)$ **57.** $(j^2 + 4)(j + 2)(j - 2)$ **58.** $-3(j + 2k)(j^2 - 2jk + 4k^2)$ **59.** $3(2w - 3)^2$ **60.** prime **61.** $2(t^3 + 5)$ **62.** $(20 + m)(20 - m)$ **63.** $(x + 8y)^2$ **64.** $6c^2d(3cd - 2c - 4)$ **65.** $0, -2$ **66.** $0, 6$ **67.** $-3, 3$ **68.** $3, 4$ **69.** $-2, -2$ **70.** $6, -4$ **71.** $\frac{1}{5}, 1$ **72.** $0, -1, 2$ **73.** 15 m **74.** Streep: 13; Hepburn: 12 **75.** 5 m **76.** 10 sec

Chapter 5 Test (page 419)

1. $2^2 \cdot 7^2$ **2.** $3 \cdot 37$ **3.** $15x^3y^6$ **4.** $4(x + 4)$ **5.** $5ab(6ab^2 - 4a^2b + c)$ **6.** $(q + 9)(q - 9)$ **7.** prime **8.** $(4x^2 + 9)(2x + 3)(2x - 3)$ **9.** $(x + 3)(x + 1)$ **10.** $-(x - 11)(x + 2)$ **11.** $(a - b)(9 + x)$ **12.** $(2a - 3)(a + 4)$ **13.** $2(3x - 5y)^2$ **14.** $(x + 2)(x^2 - 2x + 4)$ **15.** $5m^6(4m^2 - 3)$ **16.** $3(a - 3)(a^2 + 3a + 9)$ **17.** $r^2(4 - \pi)$ **18.** $(5x - 4)$ **19.** $(x - 9)(x + 6)$; Multiply the binomials: $(x - 9)(x + 6) = x^2 + 6x - 9x - 54 = x^2 - 3x - 54$. **20.** $-3, 2$ **21.** $-5, 5$ **22.** $0, \frac{1}{6}$ **23.** $-3, -3$ **24.** $\frac{1}{3}, -\frac{1}{2}$ **25.** $9, -2$ **26.** $0, -1, -6$ **27.** 6 ft by 9 ft **28.** A quadratic equation is an equation that can be written in the form $ax^2 + bx + c = 0$; $x^2 - 2x + 1 = 0$. (Answers may vary.) **29.** 10 **30.** At least one of them is 0.

Cumulative Review Exercises Chapters 1–5 (page 420)

1. about 35 beats/min difference **2.** $2 \cdot 5^3$ **3.** 0.992 **4. a.** false **b.** true **c.** true **5.** $\frac{24}{25}$ **6.** -27 **7.** -39

8. -5 **9.** 3 **10.** $\frac{5}{0}$ **11.** $-13y^2 + 6$ **12.** $3y + z$ **13.** $-\frac{3}{2}$ **14.** -2 **15.** 9 **16.** $-\frac{55}{6}$ **17.** $t = \frac{A - P}{Pr}$ **18.** $(-\infty, -2), x < -2$

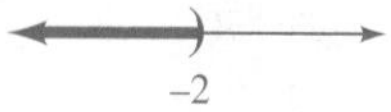

19. 15.4 in. **20.** 330 mi **21.** $I = Prt$ **22.** $\$20x$ **23.** 4 L **24.** 22nd president, 24th president **25.** 1; $(0, -2)$ **26.** $y = 7x + 19$

27.

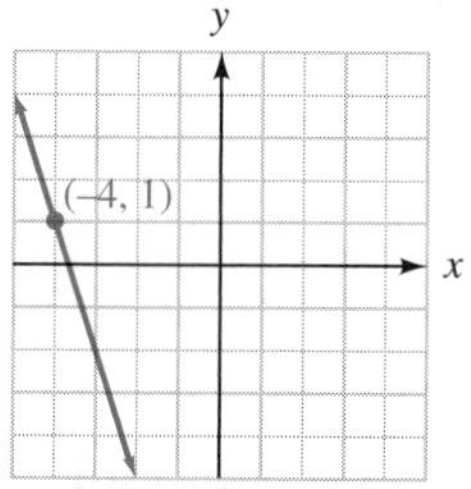

28.

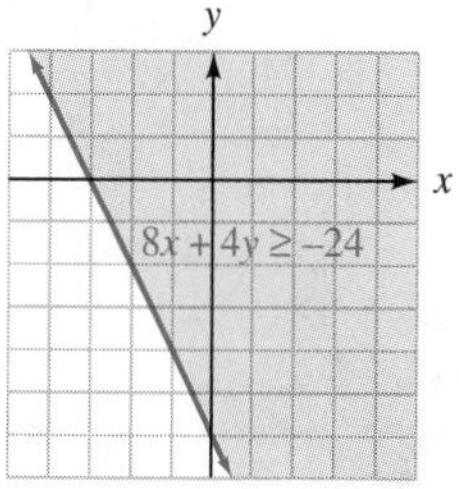

29. They are the same. **30.** a decrease of 0.86 gal/year **31.** 9.011×10^{-5} **32.** 1.7×10^6 **33.** $-4y^5$ **34.** x^4y^{23} **35.** $\frac{1}{b^{14}}$ **36.** 2 **37.** $-2x^2 - 4x + 5$ **38.** $8b^5 - 8b^4$ **39.** $y^2 - 12y + 36$ **40.** $3x^2 + 10x - 8$ **41.** $3ab - 2a - 1$ **42.** $2x + 1$ **43. a.** $(4x + 8)$ in. **b.** $(x^2 + 4x + 3)$ in.^{2} **c.** $(x^3 + 4x^2 + 3x)$ in.^{3} **44.** 3

45.

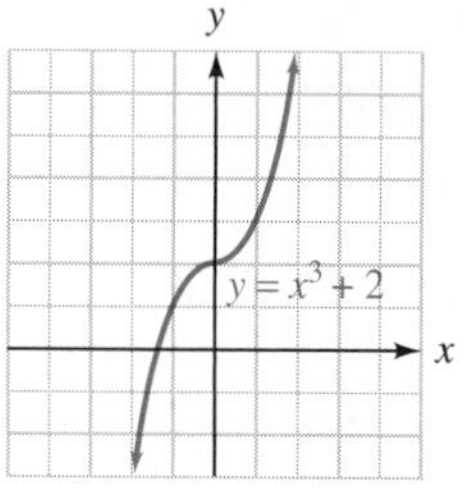

46. $6x^5y$ **47.** $b^2(b - 3)$ **48.** $(u + 3)(u - 1)$ **49.** $(2x + 1)(x - 2)$ **50.** $(3z + 1)(3z - 1)$ **51.** $-5(a - 3)(a - 2)$ **52.** $(x + y)(a + b)$ **53.** $(t - 2)(t^2 + 2t + 4)$ **54.** $(2a - 3)^2$ **55.** $0, \frac{4}{3}$ **56.** $\frac{1}{2}, 2$

Study Set Section 6.1 (page 429)

1. rational **3.** undefined **5.** simplify **7. a.** 0 **b.** 6 **c.** -6 **9. a.** $x + 1$ **b.** $\frac{1}{1}, 1$ **11. a.** $x - 2, 2 - x$ **b.** $\frac{-1}{1}, -1$ **13. a.** $\frac{x - 2}{x + 1}$ **b.** $-\frac{y}{9}$ **c.** $m - 5$ **d.** $\frac{1}{x - 30}$ **15.** yes, yes, yes **17.** 4 **19.** 0 **21.** $-\frac{2}{11}$ **23.** $-\frac{27}{28}$ **25.** 0 **27.** 2 **29.** none **31.** $\frac{1}{2}$ **33.** $-6, 6$ **35.** $-2, 1$ **37.** $\frac{5}{a}$ **39.** $\frac{3}{2}$ **41.** does not simplify **43.** $\frac{3x}{y}$ **45.** $\frac{2x + 1}{y}$ **47.** $\frac{1}{3}$ **49.** $\frac{1}{a}$ **51.** $\frac{4}{5}$ **53.** 4 **55.** -1 **57.** -6 **59.** $\frac{x + 2}{x + 1}$ **61.** $\frac{x + 1}{x - 1}$ **63.** $\frac{2x}{x - 2}$ **65.** $-\frac{1}{a + 1}$ **67.** $\frac{1}{b + 1}$ **69.** $\frac{2x - 3}{x + 1}$ **71.** $\frac{10}{3}$ **73.** 9 **75.** $\frac{3(x + 3)}{2x + 1}$ **77.** $-\frac{x + 11}{x + 3}$ **79.** $\frac{4 - x}{4 + x}$ or $-\frac{x - 4}{x + 4}$ **81.** $\frac{m - n}{2(m + n)}$ **83.** does not simplify **85.** $\frac{2u - 3}{u^3}$ **87.** does not simplify **89.** $\frac{2}{3}$ **91.** $\frac{3x}{5y}$ **93.** $\frac{x + 2}{x - 2}$ **95.** $85\frac{1}{3}$ **97.** 2, 1.6, and 1.2 milligrams per liter **103.** $(a + b) + c = a + (b + c)$ **105.** One of them is zero.

Study Set Section 6.2 (page 438)

1. opposites **3.** invert **5.** numerators, denominators, $\frac{AC}{BD}$ **7.** $\frac{2}{(x + 1)(x - 9)}$ **9.** $\frac{6n}{1}$ **11.** $\frac{1}{18x}$ **13.** $\frac{x^2 + 2x + 1}{x^2 + 1}$ **15.** ft **17. a.** m^2 **b.** m^3 **19.** $\frac{3}{2}$ **21.** $\frac{20}{3n}$ **23.** x^2y^2 **25.** $\frac{5r^3t^3}{2s^2}$ **27.** $\frac{(z + 7)(z + 2)}{7z}$ **29.** $\frac{x}{5}$ **31.** $-x$ **33.** 5 **35.** $x + 1$ **37.** $2y + 16$ **39.** $x + 5$ **41.** $10h - 30$ **43.** $x + 2$ **45.** $\frac{3}{2x}$ **47.** $-(x - 2)$ **49.** $\frac{(x - 2)^2}{x}$ **51.** $\frac{(m - 2)(m - 3)}{2(m + 2)}$ **53.** $\frac{x + 5}{x - 5}$ **55.** $-\frac{2x - 3y}{xy(3y + 2x)}$ **57.** $\frac{x + 1}{2(x - 2)}$ **59.** $\frac{3}{2y}$ **61.** $\frac{6}{y}$ **63.** $\frac{2}{y}$ **65.** $\frac{4(n - 1)}{3n}$ **67.** $\frac{2}{3x}$ **69.** $\frac{2(z - 2)}{z}$ **71.** $\frac{5(a - 2)}{4a^3}$ **73.** $-\frac{x + 2}{3}$ **75.** $\frac{1}{3}$ **77.** $\frac{2(x - 7)}{x + 9}$ **79.** $\frac{d(6c - 7d)}{6}$ **81.** $-(d + 5)$ **83.** $\frac{1}{12(2r - 3s)}$ **85.** 450 ft **87.** 1,800 meters per min **89.** $\frac{12x^2 + 12x + 3}{2}$ in.^{2} **91.** 4,380,000 **93.** 8 yd^2 **95.** $\frac{1}{2}$ mile per minute **97.** $\frac{1}{4}$ mi^2 **105.** $w = 6$ in., $l = 10$ in.

Study Set Section 6.3 (page 447)

1. common **3.** least common denominator **5.** like **7.** numerators, denominator, $A + B, A - B$ **9. a.** $-6x - 9$ **b.** $2x^2 - 2x - 1$ **c.** $2x + 5$ **d.** $3x^2 - 3x$ **11.** $\frac{x - 7}{x + 7}$ **13. a.** $2 \cdot 2 \cdot 2 \cdot 5 \cdot x \cdot x$ **b.** $2x(x - 3)$ **c.** $(n + 8)(n - 8)$ **15. a.** $\frac{5}{21a}$ **b.** $\frac{a + 4}{a + 4}$ **17.** yes **19.** $\frac{11}{x}$ **21.** $\frac{x + 5}{18}$ **23.** $\frac{m - 8}{m^3}$ **25.** $\frac{x}{y}$ **27.** $-\frac{2t}{9}$ **29.** $\frac{x}{3}$ **31.** $\frac{r + 50}{r^3 - 25}$ **33.** $\frac{2a}{a + 2}$ **35.** $-\frac{2}{t + 5}$ **37.** 9 **39.** $\frac{1}{y}$ **41.** $\frac{1}{2}$ **43.** $\frac{x + 4}{y}$ **45.** $2x - 5$ **47.** 0 **49.** $3x - 2$ **51.** $\frac{1}{3w(w - 9)}$ **53.** $\frac{1}{a + 2}$ **55.** $\frac{2}{c + d}$ **57.** $\frac{7n - 1}{(n + 4)(n - 2)}$ **59.** $\frac{t - 5}{t^2 - 2t + 1}$ **61.** $6x$ **63.** $3a^2b^3$ **65.** $c(c + 2)$ **67.** $30a^3$ **69.** $12(b + 2)$ **71.** $(x + 1)(x - 1)$ **73.** $(3x + 1)(3x - 1)$ **75.** $(x + 1)(x + 5)(x - 5)$ **77.** $(2n + 5)(n + 4)^2$ **79.** $\frac{125x}{20x}$ **81.** $\frac{8xy}{x^2y}$ **83.** $\frac{3x^2 + 3x}{(x + 1)^2}$ **85.** $\frac{2xy + 6y}{x(x + 3)}$ **87.** $\frac{30}{3(b - 1)}$ **89.** $\frac{5t + 25}{20(t + 2)}$ **91.** $\frac{4y^2 + 12y}{4y(y - 2)(y - 3)}$ **93.** $\frac{36 - 3h}{3(h + 9)(h - 9)}$ **95.** $\frac{6t^2 - 12t}{(t + 1)(t - 2)(t + 3)}$ **97.** in 90 minutes **105.** $I = Prt$ **107.** $A = \frac{1}{2}bh$ **109.** $d = rt$

Study Set Section 6.4 (page 455)

1. unlike **3. a.** $2 \cdot 2 \cdot 5 \cdot x \cdot x$ **b.** $(x - 2)(x + 6)$ **5.** $(x + 6)(x + 3)$ **7.** $\frac{5}{5}$ **9. a.** $\frac{5n}{5n}$ **b.** $\frac{3}{3}$ **11.** $8x - 24$ **13.** $7x + 10$ **15.** $-x + 10 = 10 - x$ **17.** -1 **19.** $\frac{x}{1}$ **21.** yes, no, yes **23.** $\frac{13x}{21}$ **25.** $\frac{5y}{9}$ **27.** $\frac{41}{30x}$ **29.** $\frac{7 - 2m}{m^2}$ **31.** $\frac{4xy + 6x}{3y}$ **33.** $\frac{16c^2 + 3}{18c^4}$ **35.** $\frac{3y - 3}{16}$ **37.** $\frac{2n + 2}{15}$ **39.** $\frac{-9b + 28}{42}$ **41.** $\frac{x^3 - x^2 - 9x + 18}{9x^3}$ **43.** $\frac{y^2 + 7y + 6}{15y^2}$ **45.** $\frac{x^2 + 4x + 1}{x^2y}$ **47.** $\frac{7}{24}$ **49.** $\frac{a^2 + 2a + b^2 - 2b}{ab}$

51. $\frac{2x^2 - 1}{x(x + 1)}$ **53.** $\frac{35x^2 + x + 5}{5x(x + 5)}$ **55.** $\frac{17t + 42}{(t + 3)(t + 2)}$
57. $\frac{2x^2 + 11x}{(2x - 1)(2x + 3)}$ **59.** $\frac{4a + 1}{(a + 2)^2}$ **61.** $\frac{4m^2 + 10m + 6}{(m - 2)(m + 5)}$
63. $\frac{14s + 58}{(s + 3)(s + 7)}$ **65.** $\frac{12a + 26}{(a + 1)(a + 3)^2}$ **67.** $-\frac{4c + 6}{(2c + 1)(c - 3)}$
69. $\frac{b - 1}{2(b + 1)}$ **71.** $\frac{1}{a + 1}$ **73.** $\frac{3}{s - 3}$ **75.** $\frac{x + 2}{x - 2}$
77. $-\frac{2}{x - 3}$ **79.** $\frac{17h - 2}{12(h - 2)(h + 2)}$ **81.** $\frac{1}{(y + 3)(y + 4)}$
83. $\frac{s^2 + 8s + 4}{(s + 4)(s + 1)(s + 1)}$ **85.** $\frac{2x + 13}{(x - 8)(x - 1)(x + 2)}$
87. $-\frac{1}{2(x - 2)}$ **89.** $\frac{6x + 8}{x}$ **91.** $\frac{a^2b - 3}{a^2}$ **93.** $\frac{x^2 - 4x + 9}{x - 4}$
95. $-\frac{4x + 3}{x + 1}$ **97.** $-\frac{2}{a - 4}$ **99.** $\frac{1}{r + 2}$ **101.** $\frac{2y + 7}{y - 1}$
103. $\frac{20x + 9}{6x^2}$ cm **109.** 8; (0, 2) **111.** 0

Study Set Section 6.5 (page 464)

1. complex, complex **3.** reciprocal **5.** single, reciprocal
7. a. $\frac{x - 3}{4}$ **b.** yes **c.** $\frac{1}{12} - \frac{x}{6}$ **d.** no
9. a. y, 3, 6, and y **b.** $6y$ **c.** $\frac{6y}{6y}$ **11.** $2x$ **13.** $4y^2$
15. $\frac{4x^2}{15}\Big/\frac{16x}{25}$ **17.** $\frac{5x}{12}$ **19.** $\frac{8}{9}$ **21.** $\frac{x^2}{y}$ **23.** $\frac{n^3}{8}$ **25.** $\frac{1 - 3x}{5 + 2x}$
27. $\frac{5}{4}$ **29.** $\frac{10}{3}$ **31.** $18x$ **33.** $\frac{2 - 5y}{6}$ **35.** $\frac{6d + 12}{d}$
37. $\frac{1}{2}$ **39.** $\frac{5}{7}$ **41.** $\frac{x^3}{14}$ **43.** $\frac{s^2 - s}{2 + 2s}$ **45.** $-\frac{x^2}{2}$ **47.** $\frac{1 + x}{2 + x}$
49. $\frac{14x}{9}$ **51.** $-\frac{t}{32}$ **53.** $\frac{8}{4c + 5c^2}$ **55.** $\frac{b - 5ab}{3ab - 7a}$
57. $\frac{3 - x}{x - 1}$ **59.** $\frac{b + 9}{8a}$ **61.** $\frac{32h - 1}{96h + 6}$ **63.** $\frac{xy}{y + x}$
65. $\frac{1}{x + 2}$ **67.** $\frac{1}{x + 3}$ **69.** $\frac{5t^2}{27}$ **71.** $\frac{m - 2}{6}$ **73.** $\frac{y}{x - 2y}$
75. $\frac{m^2 + n^2}{m^2 - n^2}$ **77.** $\frac{x}{x - 2}$ **79.** $\frac{3}{14}$ **81.** $\frac{R_1R_2}{R_2 + R_1}$
87. 1 **89.** $\frac{81}{256r^8}$ **91.** $\frac{r^{10}}{9}$

Study Set Section 6.6 (page 473)

1. rational **3.** clear **5.** quadratic **7. a.** yes **b.** no
9. a. 3, 0 **b.** 3, 0 **c.** 3, 0 **11.** y **13.** $(x + 8)(x - 8)$
15. $(x + 2)(x - 2)$ **17.** 12, 1, $10x$
19. $2a$, $2a$, $2a$, $2a$, $2a$, 4, 7, 3 **21.** multiplication; $fpq = f \cdot p \cdot q$ **23.** 1 **25.** $\frac{3}{5}$ **27.** 7 **29.** 3 **31.** no solution; 5 is extraneous **33.** -4, 4 **35.** 1 **37.** no solution; 0 is extraneous **39.** -3, 3 **41.** -12 **43.** -40 **45.** -48 **47.** $-\frac{5}{9}$ **49.** $-\frac{12}{7}$ **51.** $\frac{11}{4}$ **53.** -1 **55.** 6 **57.** 1 **59.** 4 **61.** no solution; -2 is extraneous
63. 2, -5 **65.** -4, 3 **67.** -2, 1 **69.** 2 **71.** 0
73. 3, -4 **75.** 1, -9 **77.** -5 **79.** $P = nrt$
81. $d = \frac{bc}{a}$ **83.** $A = \frac{h(b + d)}{2}$ **85.** $a = \frac{b}{b - 1}$
87. $r = \frac{E - IR}{I}$ **89.** $r = \frac{st}{s - t}$ **91.** $L^2 = 6dF - 3d^2$
93. $R = \frac{HB}{B - H}$ **95.** $r = \frac{r_1r_2}{r_2 + r_1}$ **101.** $x(x + 4)$
103. $(2x + 3)(x - 1)$ **105.** $(x^2 + 9)(x + 3)(x - 3)$

Study Set Section 6.7 (page 483)

1. distance, rate, time **3.** interest, principal, rate, time
5. iii **7.** $t = \frac{d}{r}$ **9. a.** 0.09 **b.** 3.5% **11.** $\frac{1}{3}$ **13.** $\frac{x}{4}$
15. $\frac{50}{r}, \frac{75}{r - 0.02}$ **17.** $6\frac{1}{9}$ days **19.** 4 **21.** 2 **23.** 5
25. $\frac{2}{3}, \frac{3}{2}$ **27.** 8 **29.** $2\frac{2}{9}$ hr **31.** $4\frac{4}{9}$ hr **33.** 8 hr
35. 20 min **37.** $1\frac{4}{5}$ hr = 1.8 hr **39.** 4 mph **41.** 1st: $1\frac{1}{2}$ ft per sec; 2nd: $\frac{1}{2}$ ft per sec **43.** $\frac{300}{255 + x}, \frac{210}{255 - x}$, 45 mph
45. Credit union: 4%; bonds: 6% **47.** 7% and 8%
51. repeating **53.** $\{\ldots, -4, -3, -2, -1, 0, 1, 2, 3, 4, \ldots\}$
55. 0

Study Set Section 6.8 (page 493)

1. ratio, rate **3.** terms **5.** cross **7.** unit **9.** $\frac{6}{5}$
11. a. 60, 60 **b.** $15x$, 90 **13.** 25, 2, 1,000, x **15.** 24 twelve-oz bottles **17.** BC, EF, DE **19.** x, 288, 18, 18, 16
21. a. $\frac{4}{5}$ **b.** $\frac{9}{7}$ **23.** similar **25.** $\frac{4}{15}$ **27.** $\frac{5}{4}$ **29.** $\frac{1}{2}$
31. $\frac{13}{8}$ **33.** no **35.** yes **37.** no **39.** 4 **41.** 6
43. 3 **45.** 0 **47.** -8 **49.** $-\frac{3}{2}$ **51.** 2 **53.** $-\frac{1}{5}$
55. -27 **57.** 2, -2 **59.** 6, -1 **61.** $-\frac{1}{3}$, 2 **63.** 15
65. 8 **67. a.** $\frac{3}{2}$, 3:2 **b.** $\frac{2}{3}$, 2:3 **69.** \$62.50 **71.** 20
73. 84 **75.** 568, 13, 14 **77.** 510 **79.** not exactly, but close **81.** 140 **83.** 65 ft, 3 in. **85.** 10 ft **87.** 45 minutes for \$25 **89.** 150 for \$12.99 **91.** 6-pack for \$1.50
93. 39 ft **95.** $46\frac{7}{8}$ ft **101.** 90% **103.** 480

Key Concept (page 499)

1. a. $\frac{2x}{x - 2}$ **b.** $x - 4$ **2. a.** $x + 1$ **b.** $x, x - 1, x + 1$
3. a. $\frac{2x^2 - 1}{x(x + 1)}$ **b.** $\frac{x}{x}, \frac{x + 1}{x + 1}$ **4. a.** $\frac{3(n^2 - n - 2)}{n^2}$ **b.** $3n$
5. a. 2 **b.** b **6. a.** 5 **b.** $(s + 2)(s - 1)$ **7. a.** 2, 4
b. $4(y - 6)$ **8. a.** $b = \frac{a}{1 - a}$ **b.** ab

Chapter Review (page 500)

1. 4, -4 **2.** $-\frac{3}{7}$ **3.** $\frac{1}{2x}$ **4.** $\frac{5}{2x}$ **5.** $\frac{x}{x + 1}$ **6.** $a - 2$
7. -1 **8.** $-\frac{1}{x + 3}$ **9.** $\frac{x}{x - 1}$ **10.** does not simplify
11. $\frac{4}{3}$ **12.** x is not a common factor of the numerator and the denominator; x is a term of the numerator. **13.** 150 mg
14. $\frac{3x}{y}$ **15.** 96 **16.** $\frac{x - 1}{x + 2}$ **17.** $\frac{2x}{x + 1}$ **18.** $\frac{3y}{2}$
19. $\frac{1}{x(x - 1)}$ **20.** $-x - 2$ **21. a.** yes **b.** no **c.** yes
d. yes **22.** $\frac{1}{3}$ mile per minute **23.** $\frac{1}{3d}$ **24.** 1
25. $\frac{2x + 2}{x - 7}$ **26.** $\frac{1}{a - 4}$ **27.** $9x$ **28.** $8x^3$ **29.** $m(m - 8)$
30. $(5x + 1)(5x - 1)$ **31.** $(a + 5)(a - 5)$
32. $(2t + 7)(t + 5)^2$ **33.** $\frac{63}{7a}$ **34.** $\frac{2xy + x}{x(x - 9)}$ **35.** $\frac{2b + 14}{6(b - 5)}$
36. $\frac{9r^2 - 36r}{(r + 1)(r - 4)(r + 5)}$ **37.** $\frac{a - 7}{7a}$ **38.** $\frac{x^2 + x - 1}{x(x - 1)}$
39. $\frac{1}{t + 1}$ **40.** $\frac{x^2 + 4x - 4}{2x^2}$ **41.** $\frac{b + 6}{b - 1}$ **42.** $\frac{6c + 8}{c}$
43. $\frac{14n + 58}{(n + 3)(n + 7)}$ **44.** $\frac{4t + 1}{(t + 2)^2}$ **45.** $\frac{1}{(a + 3)(a + 2)}$
46. $\frac{17y - 2}{12(y - 2)(y + 2)}$ **47.** yes **48.** $\frac{14x + 28}{(x + 6)(x - 1)}$ units, $\frac{12}{(x + 6)(x - 1)}$ square units **49.** $\frac{n^3}{14}$ **50.** $\frac{r + 9}{8s}$ **51.** $\frac{1 + y}{1 - y}$
52. $\frac{21}{3a + 10a^2}$ **53.** $x^2 + 3$ **54.** $\frac{y - 5xy}{3xy - 7x}$ **55.** 3 **56.** no solution; 5 is extraneous **57.** 3 **58.** 2, 4 **59.** 0

60. −4, 3 **61.** $T_1 = \frac{T_2}{1 - E}$ **62.** $y = \frac{xz}{z - x}$ **63.** 3
64. 5 mph **65.** $\frac{1}{4}$ of the job per hour **66.** $5\frac{5}{6}$ days
67. 5% **68.** 40 mph **69.** no **70.** yes **71.** $\frac{9}{2}$ **72.** 0
73. 7 **74.** 4, $-\frac{3}{2}$ **75.** 255 **76.** 20 ft **77.** 5 ft 6 in.
78. 250 for $98

Chapter 6 Test (page 506)

1. 0 **2.** −3, 2 **3.** 10 words **4.** 12,168 yd^2 **5.** $\frac{8x}{9y}$
6. −7 **7.** $\frac{x + 1}{2x + 3}$ **8.** 1 **9.** $3c^2d^3$
10. $(n + 1)(n + 5)(n - 5)$ **11.** $\frac{5y^2}{4}$ **12.** $\frac{x + 1}{3(x - 2)}$
13. $-\frac{x^2}{3}$ **14.** $\frac{1}{6}$ **15.** $\frac{13}{2y + 3}$ **16.** $\frac{4n - 5mn}{10m}$
17. $\frac{2x^2 + x + 1}{x(x + 1)}$ **18.** $\frac{2a + 7}{a - 1}$ **19.** $\frac{c^2 - 4c + 9}{c - 4}$
20. $\frac{1}{(t + 3)(t + 2)}$ **21.** $\frac{12}{5m}$ **22.** $\frac{a + 2s}{2as^2 - 3a^2}$ **23.** 11
24. no solution, 6 is extraneous **25.** 1 **26.** 1, 2 **27.** $\frac{2}{3}$
28. $B = \frac{HR}{R - H}$ **29.** yes **30.** 900 **31.** 171 ft
32. 80 sheets for $3.89 **33.** $3\frac{15}{16}$ hr **34.** 4 mph **35.** 5 is not a common factor of the numerator and denominator.
36. We multiply both sides of the equation by the LCD of the rational expressions appearing in the equation. The resulting equation is easier to solve.

Cumulative Review Exercises Chapters 1–6 (page 508)

1. 36 **2.** about 26% **3.** < **4.** 77 **5.** 104°F
6. 240 ft^3 **7. a.** false **b.** false **c.** true **d.** true
8. 3 **9.** $6c + 62$ **10.** $B = \frac{A - c - r}{2}$
11. $x \geq -1$, $[-1, \infty)$ 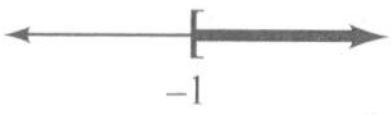 **12.** −5
13. 12 lb of the $3.20 tea and 8 lb of the $2 tea **14.** 500 mph
15.
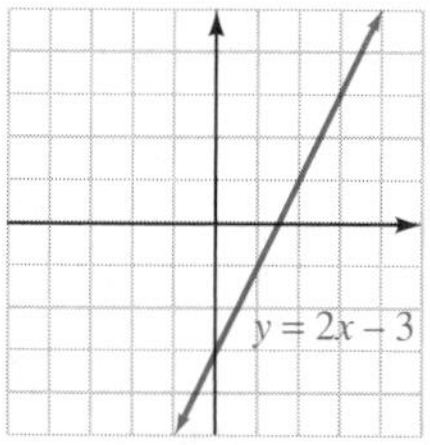

16.
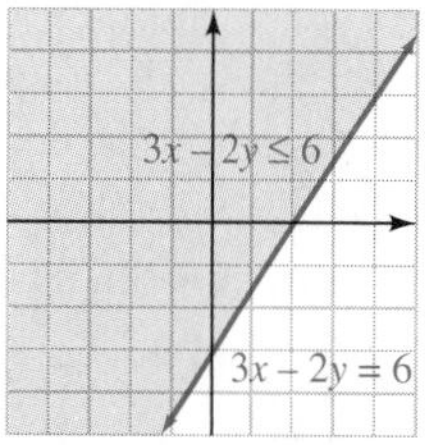

17. $\frac{8}{7}$ **18.** −1 **19.** $y = 3x + 2$ **20.** 0.008 mm/m
21. x^7 **22.** x^{25} **23.** $\frac{y^3}{8}$ **24.** $-\frac{32a^5}{b^5}$ **25.** $\frac{a^8}{b^{12}}$ **26.** $3b^2$
27. 2.9×10^5 **28.** 3 **29.** $A = \pi R^2 - \pi r^2$
30.
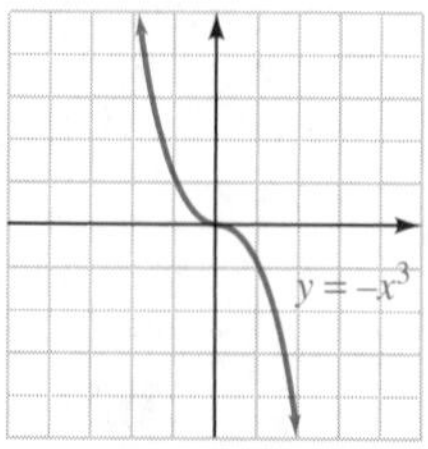

31. $6x^2 + x - 5$ **32.** $6x^5y^5$
33. $6y^2 - y - 35$
34. $-12x^4z + 4x^2z^2$
35. $2x + 3$ **36.** $x^2 + 2x - 1$
37. $k^2t(k - 3)$
38. $(b + c)(2a + 3)$
39. $2(a + 10b)(a - 10b)$
40. $(b + 5)(b^2 - 5b + 25)$
41. $(u - 9)^2$ **42.** $(2x - 9)(3x + 7)$ **43.** $-(r - 2)(r + 1)$
44. prime **45.** 0, $-\frac{1}{5}$ **46.** $\frac{1}{3}, \frac{1}{2}$ **47.** 16 in., 10 in.
48. 5, −5 **49.** $\frac{x}{3}$ **50.** $-\frac{1}{x - 3}$ **51.** $\frac{m - 3}{(m + 3)(m + 1)}$
52. $\frac{x^2}{(x + 1)^2}$ **53.** $\frac{17}{25}$ **54.** 2 **55.** $14\frac{2}{5}$ hr **56.** 34.8 ft

Study Set Section 7.1 (page 516)

1. system **3.** y-coordinate **5.** intersection
7. consistent, inconsistent **9.** (−4, −1) **11.** true
13. false **15.** 3, 3, (0, −2) **17.** no solution; independent
19. is possibly equal to **21.** yes **23.** yes **25.** no
27. no **29.** yes
31.
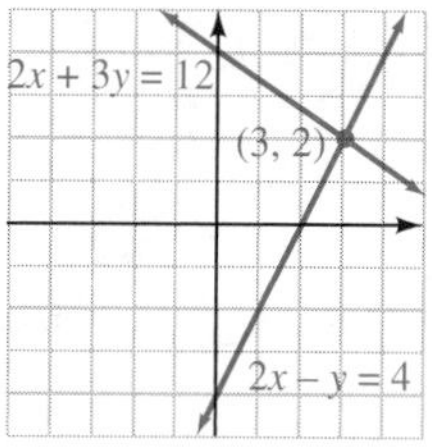

33.
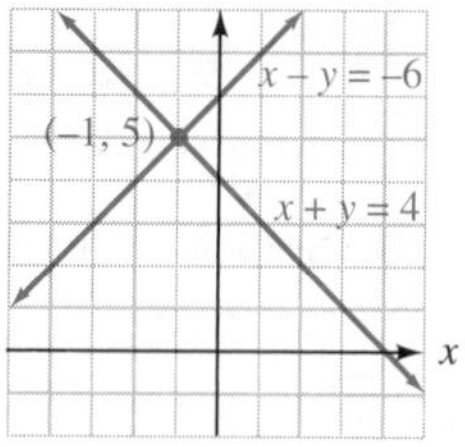

35.
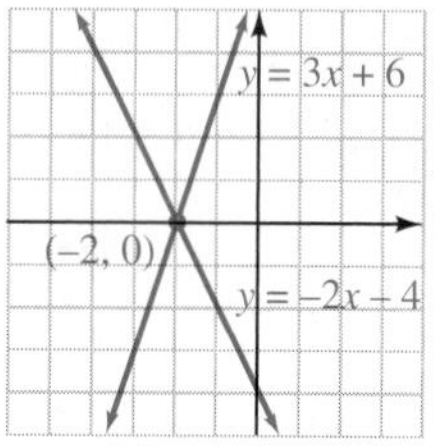

37.
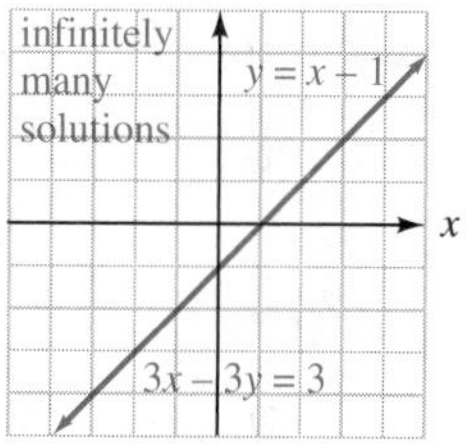

39.
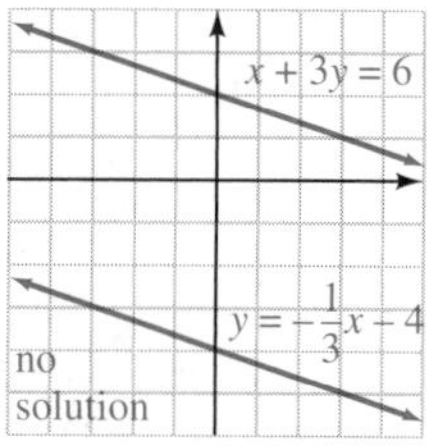

41.
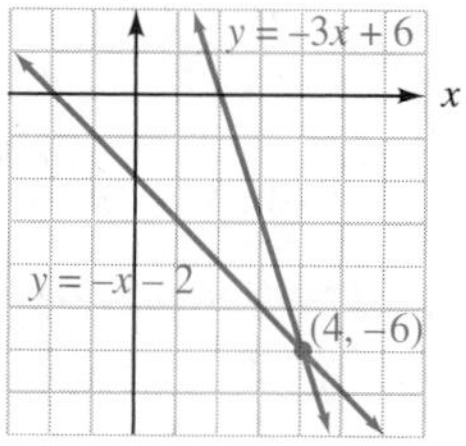

43.
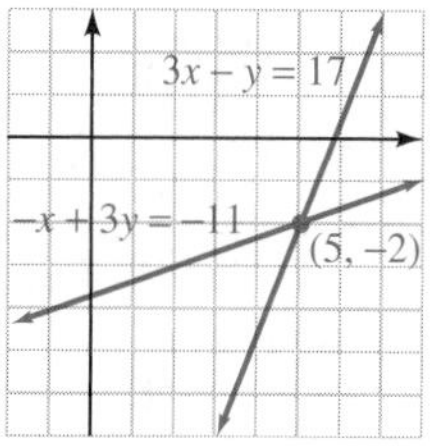

45.
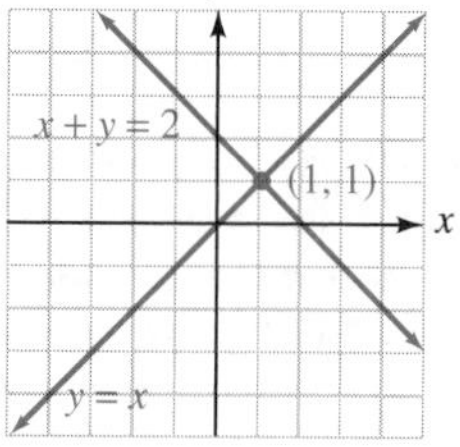

47.
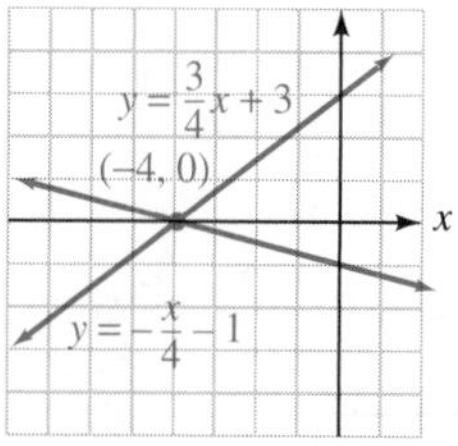

49.
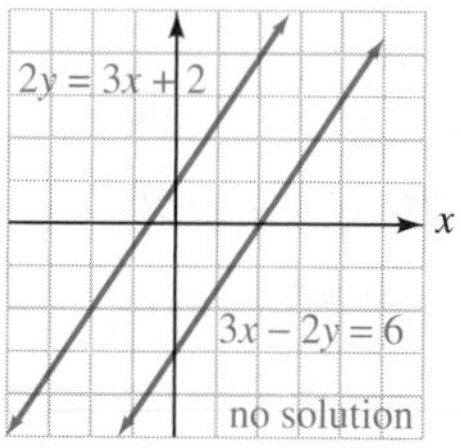

51. 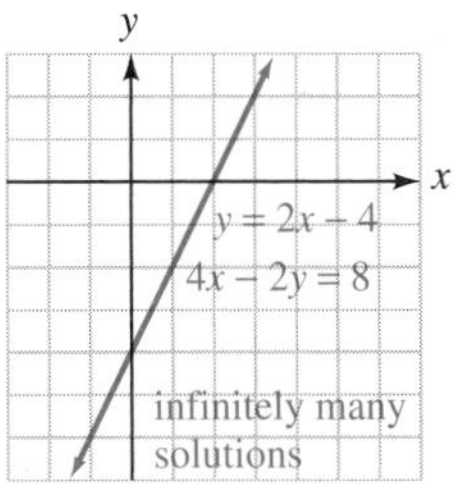

53. 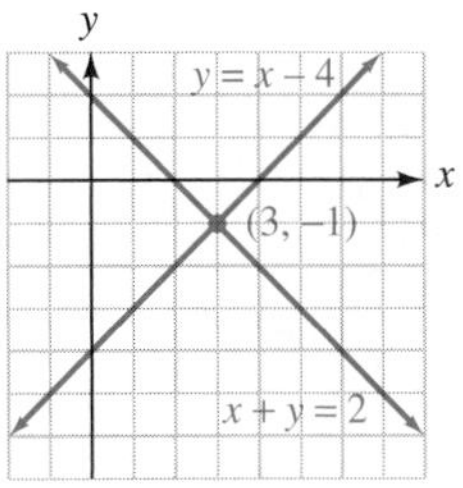

55. 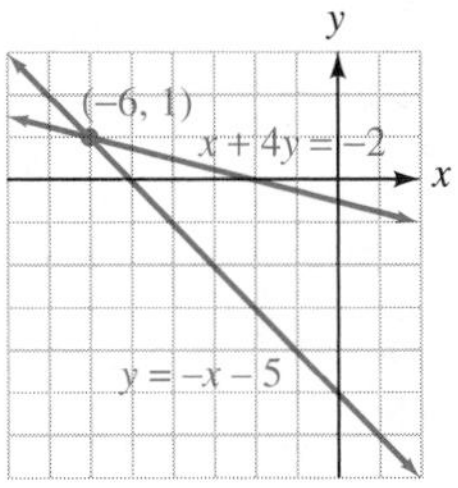

57. 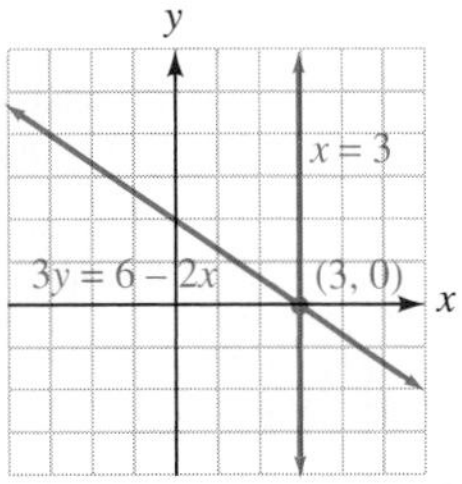

59. 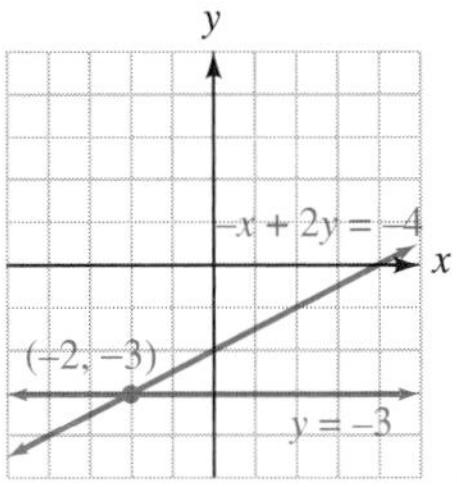

61. 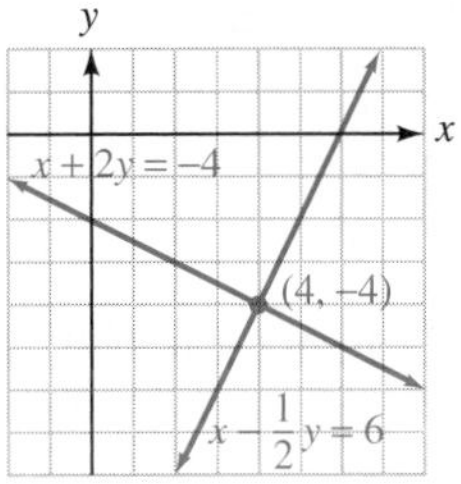

63. 1 solution **65.** same line; infinitely many solutions **67.** no solution **69.** 1 solution **71.** (1, 3) **73.** no solution **75.** 1994; 4,100 **77. a.** Houston, New Orleans, St. Augustine **b.** St. Louis, Memphis, New Orleans **c.** New Orleans **79. a.** the incumbent; 7% **b.** November 2 **c.** the challenger; 3 **81.** 10 mi **87.** $\frac{x^2 + 7x + 6}{2x^2}$ **89.** z

Study Set Section 7.2 (page 527)

1. system **3.** original **5.** substitute **7.** one **9.** infinitely **11. a.** $y = -3x$ **b.** $x = 2y + 1$ **13. a.** $x = 2 + 2y$ **b.** $y = -11 + 2x$ **15.** $x + 3(x - 4) = 8$ **17. a.** 15 **b.** $3x + 10y = 15$ **19. a.** no **b.** ii **21.** $3x$, 4, −2, −2, −6, −2, −6 **23.** (2, 4) **25.** (3, 0) **27.** (−3, −1) **29.** (−1, −1) **31.** no solution **33.** (−2, 3) **35.** (3, −2) **37.** $\left(-4, \frac{5}{4}\right)$ **39.** (3, 2) **41.** $\left(\frac{2}{3}, -\frac{1}{3}\right)$ **43.** infinitely many solutions **45.** (−2, 0) **47.** (−2, −3) **49.** (3, −2) **51.** (−1, −1) **53.** no solution **55.** $\left(\frac{1}{2}, \frac{1}{3}\right)$ **57.** (4, −2) **59.** (−6, 4) **61.** $\left(\frac{1}{5}, 4\right)$ **63.** $\left(10, \frac{15}{2}\right)$ **65.** (9, 11) **67.** infinitely many solutions **69.** (−4, −1) **71.** (4, 2) **73.** melon **79.** yes **81.** yes **83.** no

Study Set Section 7.3 (page 536)

1. opposites **3.** eliminated **5.** addition, equal **7.** $7y$ and $-7y$ **9. a.** $12x + 3y = 6$ **b.** $-2x + 6y = -8$ **11.** 1 **13.** yes **15.** $2x$, 1, 1, (1, 4) **17.** (3, 2) **19.** (3, −2) **21.** (−2, −3) **23.** (0, 8) **25.** (−3, 4) **27.** (0, −2) **29.** infinitely many solutions **31.** (−12, 1) **33.** (−5, 10) **35.** (3, 11) **37.** no solution **39.** $\left(\frac{1}{3}, 3\right)$ **41.** $\left(\frac{1}{3}, 7\right)$ **43.** (−1, −1) **45.** (1, 1) **47.** $\left(1, -\frac{5}{2}\right)$ **49.** (−2, 5) **51.** (8, −3) **53.** (3, 0) **55.** $\left(\frac{13}{75}, \frac{14}{75}\right)$ **57.** (10, 9) **59.** (−4, −1) **61.** (6, 8) **63.** no solution **65.** (−1, 2) **67.** (3, 2) **69.** (1, −1) **71.** infinitely many solutions **73.** $\left(\frac{10}{3}, \frac{10}{3}\right)$ **75.** (5, −6) **77.** 1991 **83.** $x - 10$ **85.** 0 **87.** 7.5 ft^2

Study Set Section 7.4 (page 549)

1. variable **3.** system **5.** complementary **7.** $x + y = 20$, $y = 2x - 1$ **9.** $x + y = 180$, $y = x - 25$ **11.** $5x + 2y = 10$ **13. a.** $(x + c)$ mph **b.** $(x - c)$ mph **15. a.** $(x + y)$ mL **b.** 33% **17.** 20°, 70° **19.** 50°, 130° **21.** 22 ft, 29 ft **23.** president: $400,000; vice president: $192,600 **25.** 65°, 115° **27.** 96 ft, 70 ft **29.** 15 m, 10 m **31.** $30, $10 **33.** 85, 63 **35. a.** 400 tires

b. 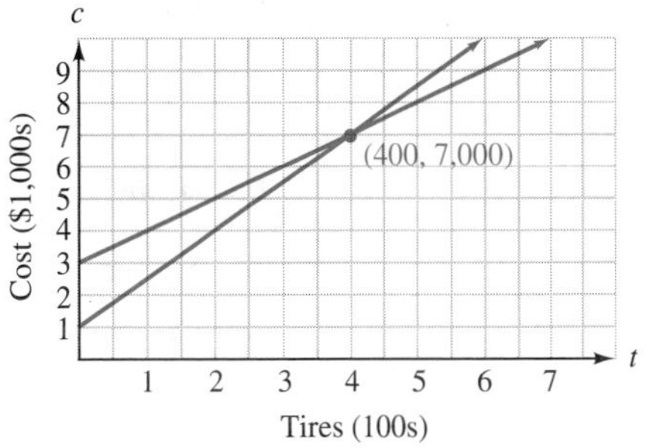

c. the second mold **37.** 125 min **39.** 25 mph, 5 mph **41.** 180 mph, 20 mph **43.** nursing: $2,000; business: $3,000 **45.** 4 L 6% salt water, 12 L 2% salt water **47.** 52 lb $8.75, 48 lb $3.75 **51.** $(-\infty, 4)$ 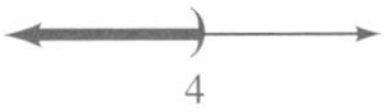

53. $(-1, 2]$ 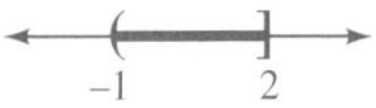

Study Set Section 7.5 (page 560)

1. inequalities **3.** solve **5.** intersection **7. a.** $3x - y = 5$ **b.** dashed **9.** slope: $4 = \frac{4}{1}$; y-intercept: (0, −3) **11. a.** no **b.** above **13. a.** yes **b.** no **c.** no **15. a.** ii **b.** iii **c.** iv **d.** i **17.** *ABC*

19. 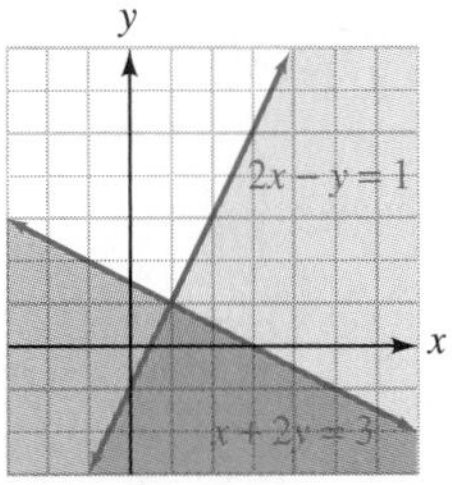

21. 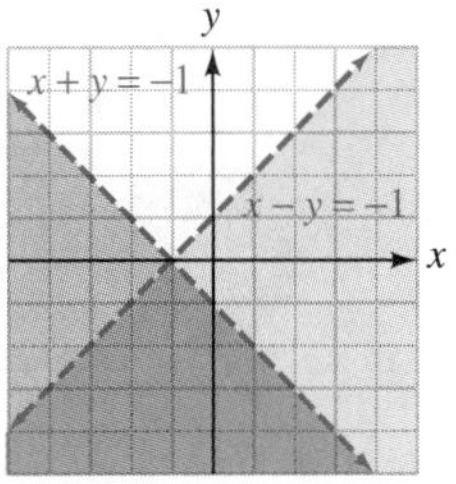

23. 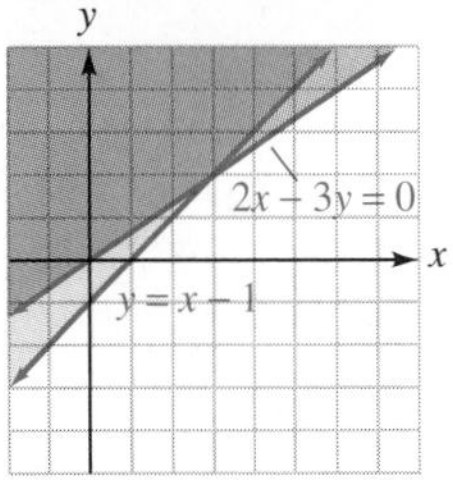

25. 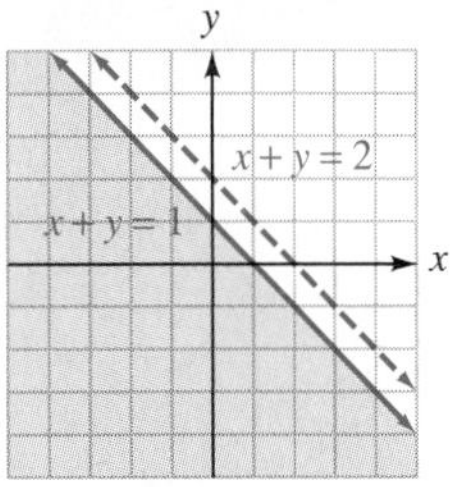

27. 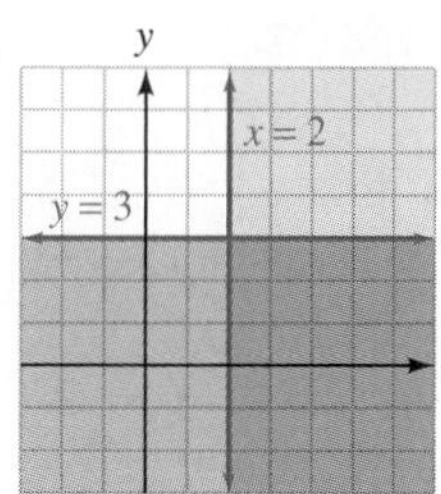

29. 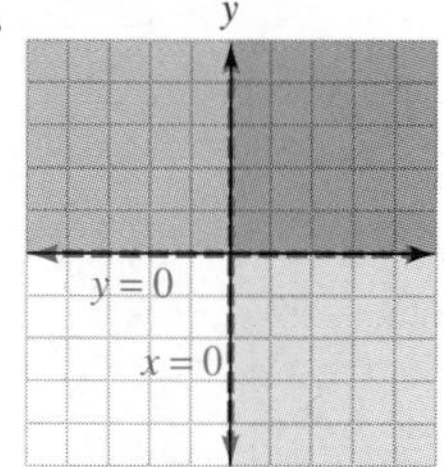

31. 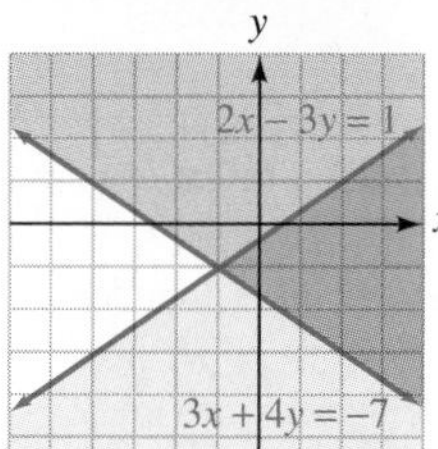

33. 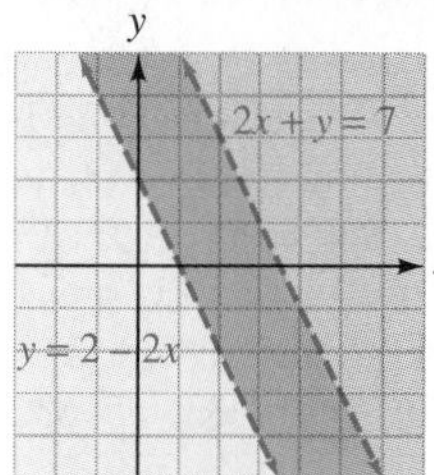

35. 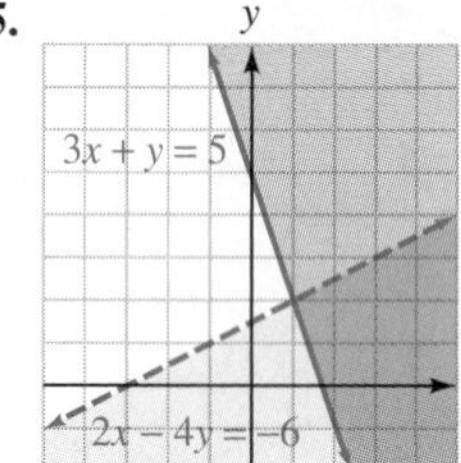

37. 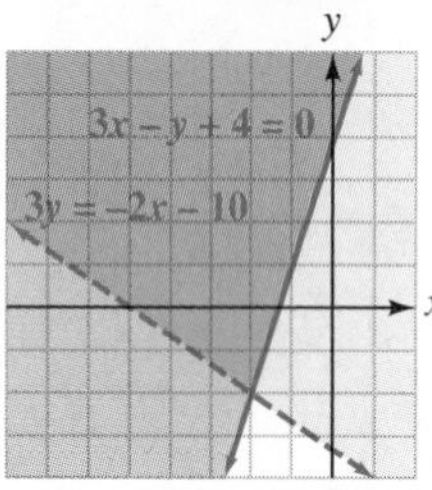

39. 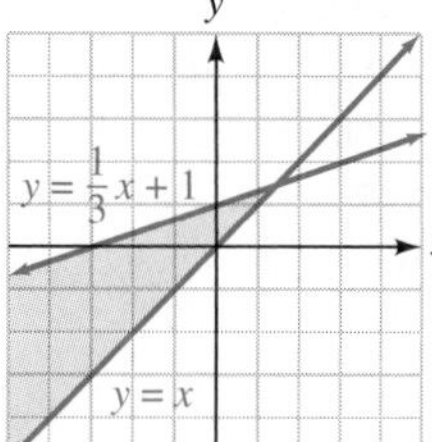

41. 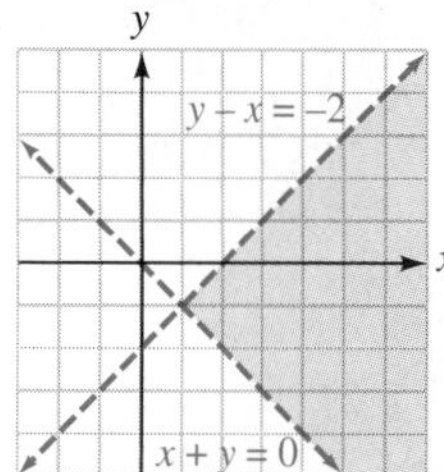

43. 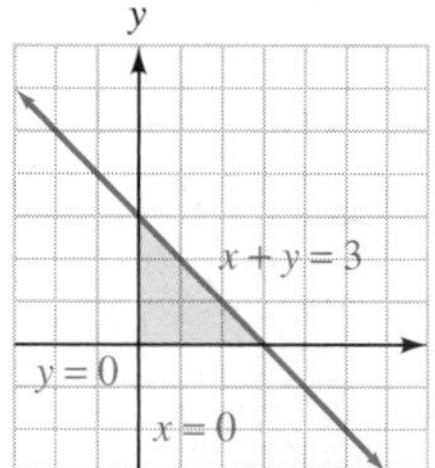

45. 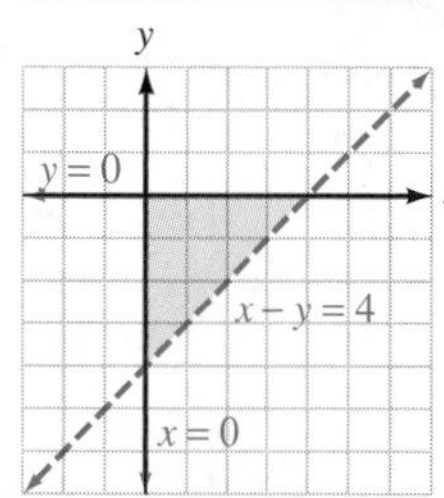

47. (c)
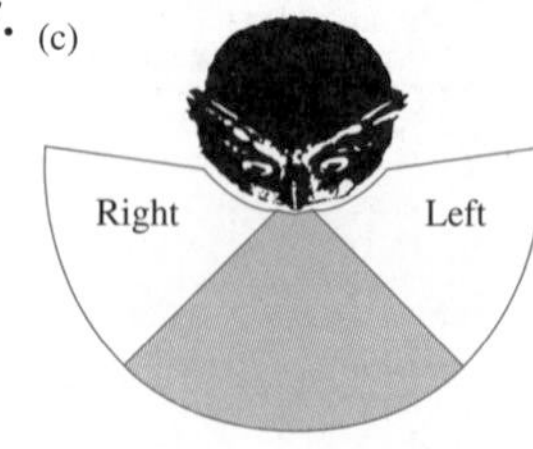

49. 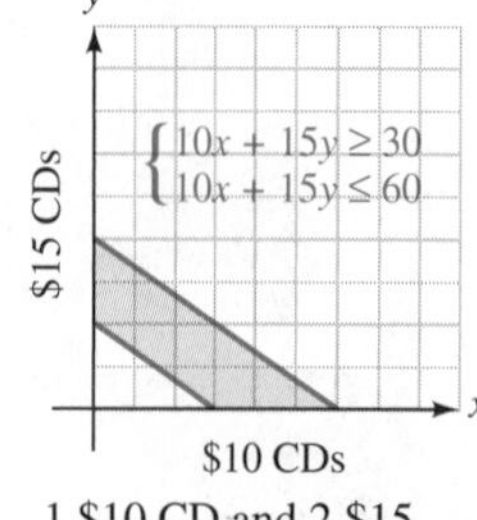

1 $10 CD and 2 $15 CDs; 4 $10 CDs and 1 $15 CD

51. 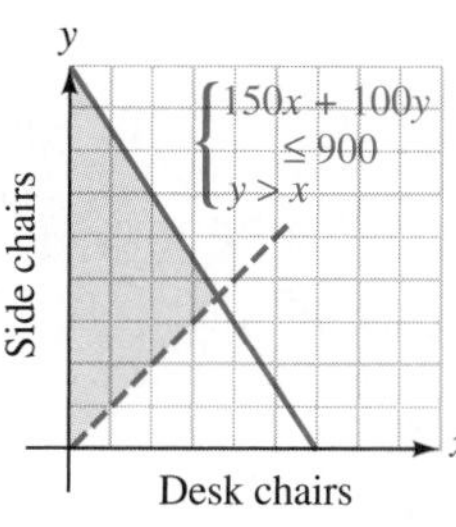

2 desk chairs and 4 side chairs; 1 desk chair and 5 side chairs

53. 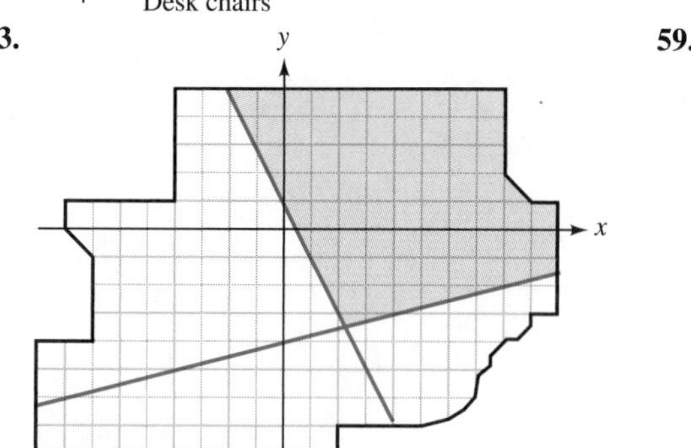

59. 230

61. 0.000976 63. 2.9×10^5 65. 5.1×10^{-6}

Key Concept (page 564)

1. 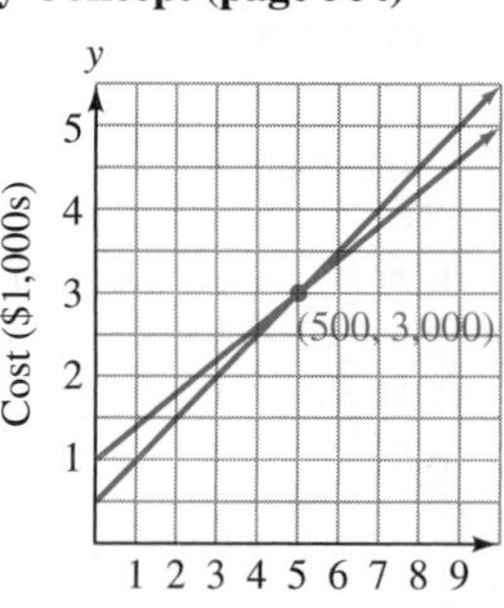

(500, 3,000); 500 meals, $3,000 2. (5, 1) 3. (−1, −1) 4. (3, −2) 5. (−3, 4)

Chapter Review (page 565)

1. yes 2. yes 3. (1978, 5,600,000); In 1978, the same number of men as women were enrolled in college, about 5.6 million of each.

4. 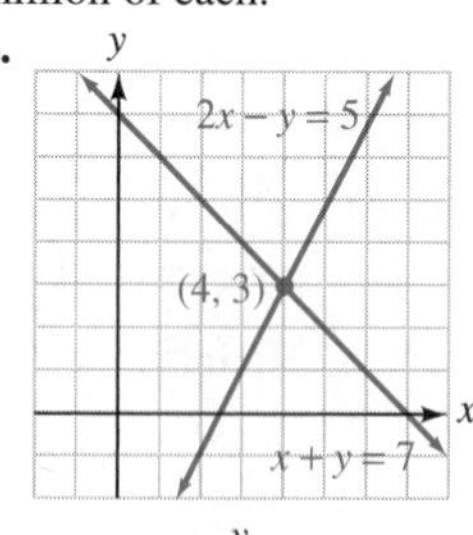

5. 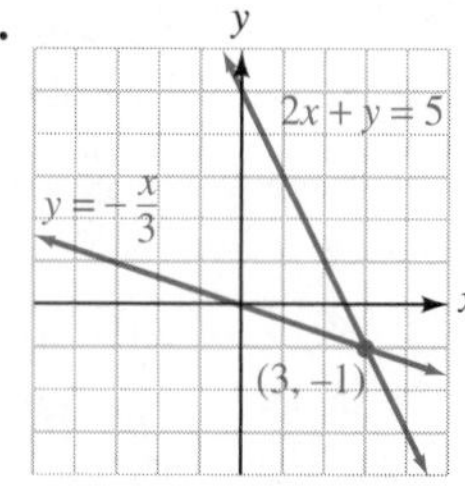

6. 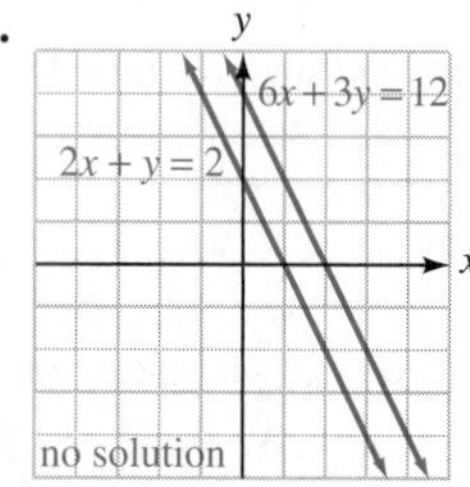

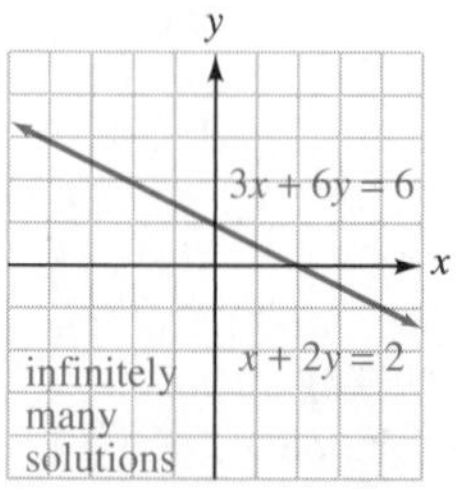

7.
y
6x + 3y = 12
2x + y = 2
x
no solution

8. (5, 0) **9.** (3, 3) **10.** $\left(-\frac{1}{2}, \frac{7}{2}\right)$ **11.** (1, −2)
12. infinitely many solutions **13.** (12, 10) **14. a.** no solution **b.** two parallel lines **c.** inconsistent system
15. (3, −5) **16.** $\left(3, \frac{1}{2}\right)$ **17.** (−1, 7) **18.** (0, 9)
19. infinitely many solutions **20.** (1, −1) **21.** (−5, 2)
22. no solution **23.** Elimination; no variables have a coefficient of 1 or −1 **24.** Substitution; equation 1 is solved for x. **25.** Las Vegas: 2,000 ft; Baltimore: 100 ft
26. base: 21 ft; extension: 14 ft **27.** 10,800 yd^2
28. a. $y = 5x + 30{,}000$, $y = 10x + 20{,}000$ **b.** 2,000
c. the athlete

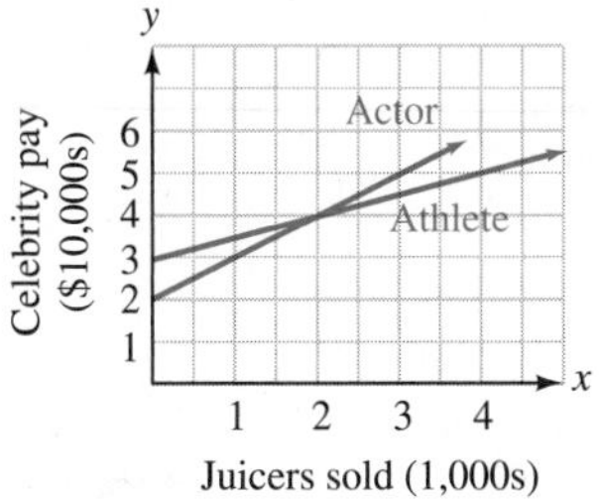

29. 12 lb worms, 18 lb bears **30.** 3 mph
31. \$16.40, \$10.20 **32.** \$750 **33.** $13\frac{1}{3}$ gal 40%, $6\frac{2}{3}$ gal 70%

34.

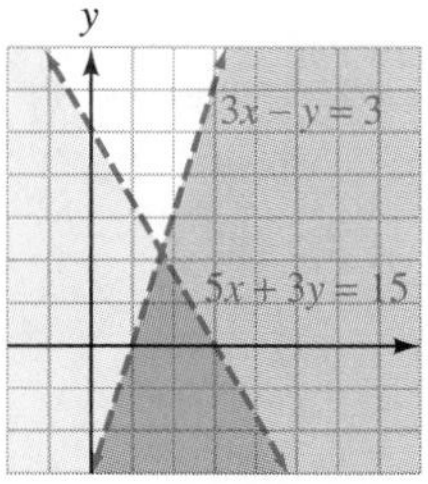

35.

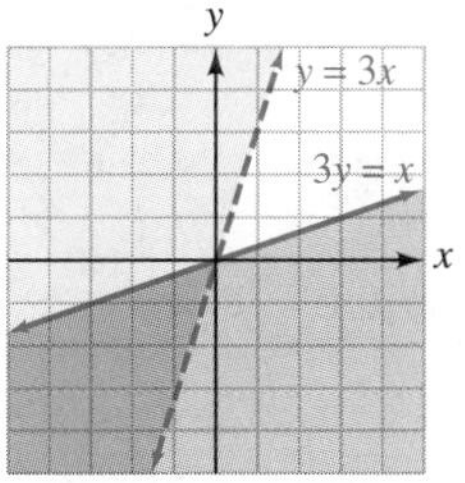

36.

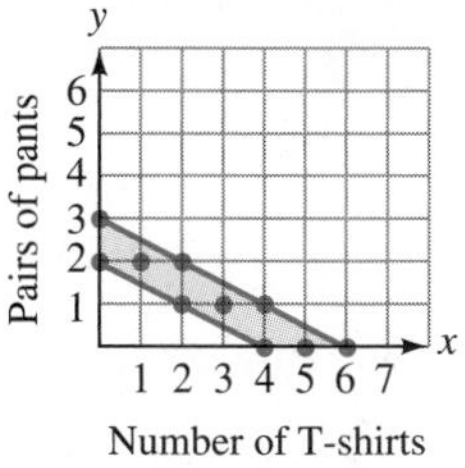

$10x + 20y \geq 40$, $10x + 20y \leq 60$; (3, 1): 3 shirts and 1 pair of pants; (1, 2): 1 shirt and 2 pairs of pants (answers may vary)

Chapter 7 Test (page 569)

1. yes **2.** no **3. a.** solution **b.** consistent **c.** inconsistent **d.** independent **e.** dependent
4. (2, 3)

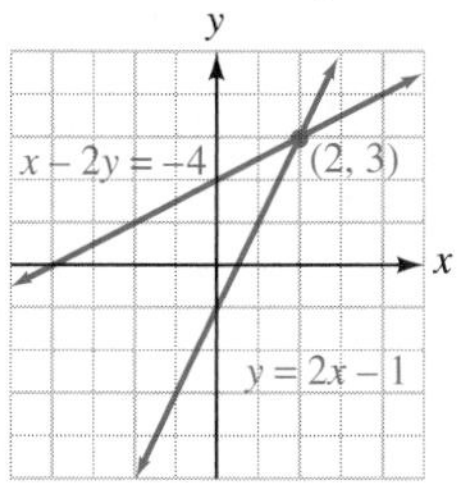

5.

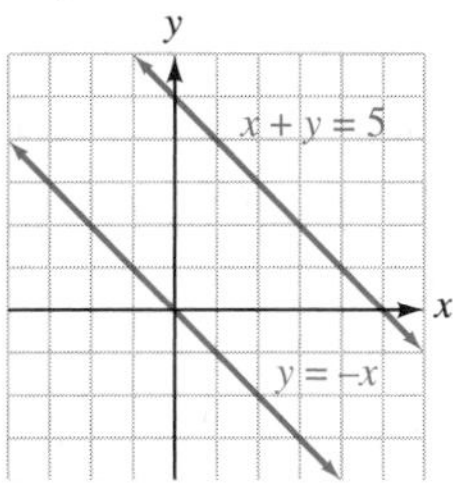

no solution

6. (30, 3,000); if 30 items are sold, the salesperson gets paid the same by both plans, \$3,000. **7.** Plan 1 **8.** (−2, −3)
9. infinitely many solutions **10.** (−1, −1) **11.** (2, 4)
12. (−3, 3) **13.** no solution **14.** Elimination; the terms involving y can be eliminated easily. **15.** 8 mi, 14 mi
16. 3 adults' tickets; 4 children's tickets **17.** \$6,000, \$4,000
18. $(x - c)$ mph **19.** $x + y = 90$
20.

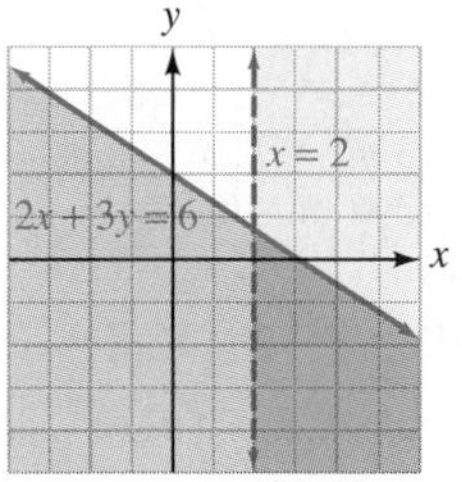

21.

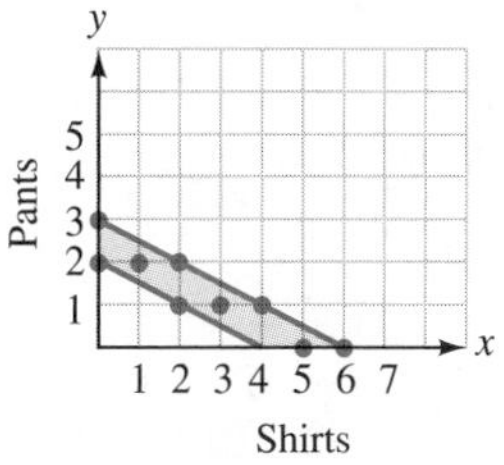

(1, 2), (2, 2), (3, 1) (answers may vary)

Cumulative Review Exercises Chapters 1–7 (page 571)

1. \$2,016,000,000 **2.** {. . . , −3, −2, −1, 0, 1, 2, 3, . . .}
3. −37 **4.** 28 **5.** 1.2 ft^3 **6.** 0 **7.** −2 **8.** 15
9. $x < -\frac{11}{4}$; $\left(-\infty, -\frac{11}{4}\right)$

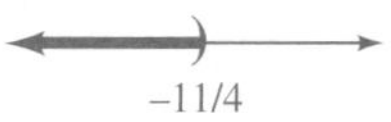

10. 30
11. mutual fund: \$25,000; bonds: \$20,000
12. a. $P = 2l + 2w$ **b.** $A = lw$ **c.** $A = \pi r^2$ **d.** $d = rt$
13.

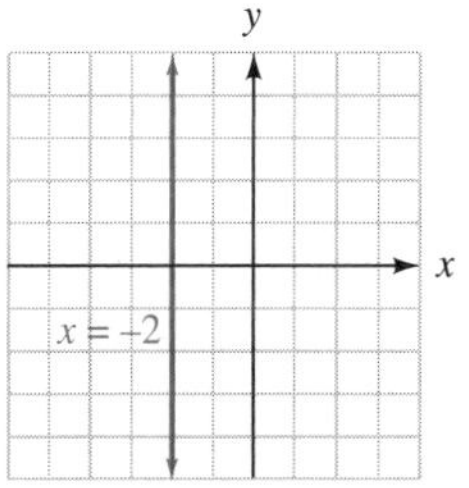

14.

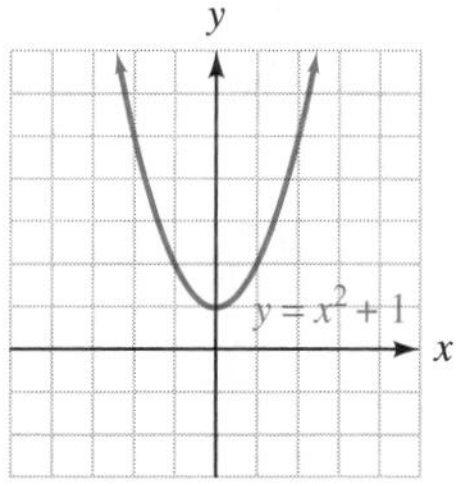

15. $m = 3$, (0, −2); $y = 3x - 2$ **16.** $-\frac{4}{9}$ **17.** x^{31}
18. $8a^4$ **19.** $\frac{3}{16}$ **20.** 1 **21.** $7x - 13y$
22. $6x^4 + 8x^3 - 14x^2$ **23.** $c^2 + 32c + 256$
24. $2x^2 + 3x - 9$ **25.** $\frac{1}{8} - \frac{2}{x}$ **26.** $3x + 1$
27. $3r(4r - s + 3rs^2)$ **28.** $(u - 9)^2$ **29.** $(2y - 1)(y - 3)$
30. $(x^2 + 9)(x + 3)(x - 3)$ **31.** $(t - v)(t^2 + tv + v^2)$
32. $(x - t)(y + s)$ **33.** 0, 2 **34.** 3, −5 **35.** $\frac{x - 5}{5}$
36. −1 **37.** $\frac{x^2 + 4x + 1}{x^2 y}$ **38.** $\frac{3x + 1}{x + 2}$ **39.** −4, 3 **40.** 16, 8
41.

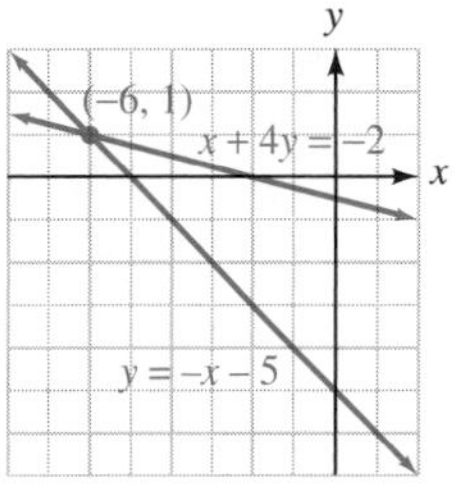

42.

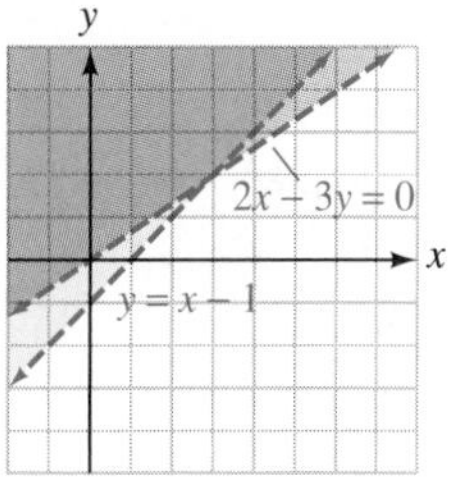

43. (4, 1) **44.** noodles: 2 servings, rice: 3 servings

Study Set Section 8.1 (page 580)

1. square **3.** negative **5.** perfect **7.** Pythagorean
9. two **11.** 5, −5, 5, −5 **13.** 2, −2 **15.** x_1, y_2
17. irrational **19.** irrational **21.** 45 **23.** $16x^2$
25. true **27.** 7 **29.** 6 **31.** 10 **33.** 20 **35.** −9

37. 1.1 **39.** 13 **41.** $\frac{3}{5}$ **43.** $-\frac{1}{8}$ **45.** -0.2 **47.** -17 **49.** 50 **51.** 1.732 **53.** 3.317 **55.** 9.747 **57.** 20.688 **59.** 3.464 **61.** m **63.** t^2 **65.** c^5 **67.** $2y$ **69.** $8b^2$ **71.** 5 **73.** 30 **75.** 117 **77.** $\sqrt{2}$ in. $\approx$ 1.41 in. **79.** 5 **81.** 13 **83.** $\sqrt{37} \approx 6.08$ **85.** $\sqrt{29} \approx 5.39$ **87.** 36 mph **89. a.** $6\sqrt{2}$ in. $\approx$ 8.49 in. **b.** $5\sqrt{3}$ in. $\approx$ 8.66 in. **91.** $\sqrt{16{,}200}$ ft $\approx$ 127.28 ft **93. a.** 16 in., 30 in. **b.** 34 in. **95. a.** $\sqrt{18}$ yd $\approx$ 4.24 yd **b.** yes: 6 yd + 4.24 yd > 10 yd **103.** $6s^2 + s - 5$ **105.** $3x^2 + 10x - 8$

Study Set Section 8.2 (page 590)

1. squares **3.** simplify **5.** multiplication, division **7.** product, square **9. a.** product, $\sqrt{a}\sqrt{b}$ **b.** quotient, $\frac{\sqrt{a}}{\sqrt{b}}$ **11. a.** 4 **b.** 49 **c.** 9 **d.** 16 **13.** $\sqrt{4 \cdot 10}$ **15. a.** $\sqrt{81}\sqrt{2}$ **b.** $\sqrt{m^4}\sqrt{m}$ **17. a.** 10 **b.** t **c.** a^2 **d.** $12y$ **19.** 16, 16, 4 **21. a.** $6\sqrt{15}$ **b.** $20x^2\sqrt{14}$ **23.** $2\sqrt{5}$ **25.** $5\sqrt{2}$ **27.** $3\sqrt{3}$ **29.** $7\sqrt{2}$ **31.** $4\sqrt{3}$ **33.** $2\sqrt{11}$ **35.** $\sqrt{15}$ **37.** $8\sqrt{3}$ **39.** $5\sqrt{10}$ **41.** $-10\sqrt{7}$ **43.** $9\sqrt{2}$ **45.** $12\sqrt{2}$ **47.** $-18\sqrt{2}$ **49.** $x^2\sqrt{x}$ **51.** $n^4\sqrt{n}$ **53.** $4\sqrt{6}$ **55.** $2\sqrt{k}$ **57.** $2\sqrt{3x}$ **59.** $30\sqrt{3t}$ **61.** $5x\sqrt{x}$ **63.** in simplified form **65.** $3x^2\sqrt{y}$ **67.** $c^3d^5\sqrt{2cd}$ **69.** $-48x^2y\sqrt{y}$ **71.** $x^3y^2\sqrt{2}$ **73.** $\frac{5}{3}$ **75.** $\frac{9}{8}$ **77.** $\frac{\sqrt{6}}{11}$ **79.** $\frac{\sqrt{26}}{5}$ **81.** $-\frac{2\sqrt{5}}{7}$ **83.** $\frac{4\sqrt{3}}{9}$ **85.** $\frac{4\sqrt{2}}{5}$ **87.** $\frac{6x\sqrt{2x}}{y}$ **89.** $\frac{5n^2\sqrt{5}}{8}$ **91.** $\frac{8m\sqrt{2}}{9n}$ **93.** $2rs^2\sqrt{3s}$ **95.** $\sqrt{125}$ ft $= 5\sqrt{5}$ ft **97. a.** $2\sqrt{7}$ in. **b.** 5.3 in. **103.** (3, 0) **105.** $(-1, -1)$

Study Set Section 8.3 (page 596)

1. radicals **3.** terms **5.** yes **7.** no **9. a.** $2\sqrt{3}$ **b.** $2x\sqrt{x}$ **11.** 4, 4, $2\sqrt{5}$, $6\sqrt{5}$, $3\sqrt{5}$ **13.** $9\sqrt{7}$ **15.** $10\sqrt{21}$ **17.** $9\sqrt{n}$ **19.** 0 **21.** $4\sqrt{5}$ **23.** $-3\sqrt{21}$ **25.** $\sqrt{3} + \sqrt{15}$ **27.** $-3\sqrt{x}$ **29.** $\sqrt{2} + \sqrt{x}$ **31.** $10\sqrt{11}$ **33.** $4\sqrt{2}$ **35.** $5 + 6\sqrt{3}$ **37.** $-1 - \sqrt{r}$ **39.** $5\sqrt{3}$ **41.** $\sqrt{2}$ **43.** $14\sqrt{5}$ **45.** $-7\sqrt{5}$ **47.** $-2\sqrt{3}$ **49.** $-18\sqrt{2}$ **51.** $12\sqrt{7}$ **53.** $9\sqrt{5}$ **55.** $10\sqrt{2} - \sqrt{3}$ **57.** $5\sqrt{6} - 5\sqrt{2}$ **59.** $7\sqrt{6} + 4\sqrt{15}$ **61.** $7\sqrt{3} - 6\sqrt{2}$ **63.** $-\sqrt{2a}$ **65.** $3x\sqrt{2}$ **67.** $3d\sqrt{2d}$ **69.** $3x\sqrt{2y} - 3x\sqrt{3y}$ **71.** $x^2\sqrt{2x}$ **73.** $19b\sqrt{6}$ **75.** $-5y\sqrt{10y}$ **77.** $2x\sqrt{5xy} + 3x^2y\sqrt{5xy} - 4x^3y^2\sqrt{5xy}$ **79.** $18\sqrt{3}$ in. **81.** $4\sqrt{3}$ in. **83.** $10\sqrt{3}$ ft **89.** $\frac{1}{9}$ **91.** -9 **93.** $\frac{1}{x^3}$ **95.** 1

Study Set Section 8.4 (page 605)

1. irrational **3.** rationalize **5.** $a \cdot b$ **7.** a **9.** 1 **11.** 9, $\sqrt{6}$ **13.** 1 **15. a.** The radicand is a fraction. **b.** There is a radical in the denominator. **17.** $\sqrt{6}$ **19.** $\frac{\sqrt{6}}{3}$ **21.** $\sqrt{15}$ **23.** $\sqrt{22x}$ **25.** 7 **27.** $5\sqrt{2}$ **29.** $3\sqrt{6}$ **31.** $5\sqrt{d}$ **33.** 4 **35.** $2\sqrt{14}$ **37.** $3\sqrt{2x}$ **39.** x^4 **41.** $60\sqrt{2}$ **43.** $10\sqrt{15}$ **45.** $45y$ **47.** $2n^4\sqrt{15n}$ **49.** $3s^5\sqrt{s}$ **51.** $2 + \sqrt{2}$ **53.** $27 - 3\sqrt{6}$ **55.** $x\sqrt{3} - 2\sqrt{x}$ **57.** $8x$ **59.** 1 **61.** $\sqrt{6} + \sqrt{10} - 3 - \sqrt{15}$ **63.** $4x - 18$ **65.** $6\sqrt{14} + 2x\sqrt{7} - 3x\sqrt{2} - x^2$ **67.** $9x - 12$ **69.** 3 **71.** -2 **73.** 6 **75.** 12 **77.** $100x$ **79.** $y + 2$ **81.** $\frac{2x}{3}$ **83.** $\frac{3\sqrt{2}}{5}$ **85.** $\frac{2}{x}$ **87.** $\frac{\sqrt{3}}{3}$ **89.** $\frac{\sqrt{91}}{7}$ **91.** $\sqrt{3}$ **93.** $\frac{3\sqrt{2}}{8}$ **95.** $\frac{2\sqrt{15}}{5}$ **97.** $\frac{10\sqrt{x}}{x}$ **99.** $\frac{3\sqrt{2xy}}{2x}$ **101.** $\frac{3\sqrt{3} + 3}{2}$ **103.** $\sqrt{7} - 2$ **105.** $6 + 2\sqrt{3}$ **107.** $5\sqrt{3} - 5\sqrt{2}$ **109.** $1{,}800\sqrt{2}$ in.^{2} **111.** 108π in.^{2} **117.** no **119.** positive

Study Set Section 8.5 (page 612)

1. radical **3.** extraneous **5.** b^2 **7. a.** $\sqrt{x - 4} = 3$ **b.** $x + 8 = \sqrt{x}$ **9. a.** $x^2 - 6x - 7 = 0$ **b.** $0 = -3x + 12$ **11.** 2, 2, $x - 3$, 28 **13.** 9 **15.** 144 **17.** 8 **19.** 9 **21.** no solution **23.** no solution **25.** 49 **27.** 36 **29.** 16 **31.** 25 **33.** 4 **35.** 9 **37.** 1 **39.** -95 **41.** 5 **43.** 54 **45.** no solution **47.** 8 **49.** -1 **51.** 4 **53.** 3 **55.** 12 **57.** 5 **59.** 1, 3 **61.** -2, 4 **63.** 9 **65.** 5 **67.** -2 **69.** 4 **71.** 1 **73.** -1 **75.** 169 ft **77.** 64.4 ft **79.** about 288 ft **81.** 2,010 ft **87.** $\frac{5}{4}$ **89.** $\frac{10}{9}$

Study Set Section 8.6 (page 621)

1. cube, fourth **3.** simplify **5.** rational **7. a.** 4 **b.** $2x$ **c.** -32 **9. a.** $\sqrt[n]{a}\sqrt[n]{b}$ **b.** $\frac{\sqrt[n]{a}}{\sqrt[n]{b}}$ **11.** $\sqrt[3]{8 \cdot 3}$ **13. a.** $2\sqrt[3]{3}$ **b.** $2\sqrt[4]{3}$ **c.** $\frac{5}{3}$ **d.** $\frac{4}{3}$ **15. a.** $\frac{1}{\sqrt{36}}$ **b.** $\frac{1}{(\sqrt[4]{16})^5}$ **17.** 2, 2 **19.** 2 **21.** 0 **23.** -2 **25.** -4 **27.** $\frac{1}{5}$ **29.** 1 **31.** -4 **33.** 9 **35.** 10 **37.** $\frac{2}{3}$ **39.** 2 **41.** 4 **43.** $-\frac{1}{3}$ **45.** not a real number **47.** 2 **49.** -1 **51.** 2 **53.** m **55.** $3a^2$ **57.** y **59.** $2b^3$ **61.** x **63.** s^2 **65.** $2\sqrt[3]{3}$ **67.** $-4\sqrt[3]{2}$ **69.** $2\sqrt[3]{9}$ **71.** $\frac{5\sqrt[3]{2}}{3}$ **73.** $3\sqrt[4]{2}$ **75.** $2\sqrt[5]{2}$ **77.** 9 **79.** 2 **81.** -12 **83.** -5 **85.** 1 **87.** $\frac{3}{8}$ **89.** 2 **91.** 64 **93.** 9 **95.** $\frac{1}{2}$ **97.** $\frac{1}{125}$ **99.** $\frac{1}{8}$ **101.** 10 mph **103.** 26 ft **105.** 36 in.^{2} **111.** **113.**

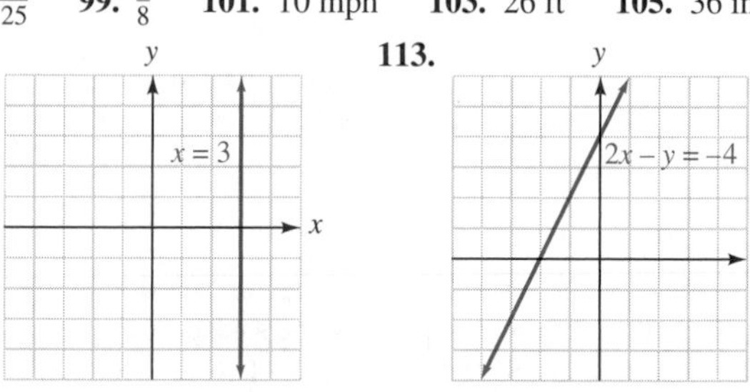

Key Concept (page 624)

1. product, square roots, square roots, product
2. a. $2xy\sqrt{6y}$ **b.** $5\sqrt{2}$ **c.** $-\sqrt{2a}$ **d.** $6\sqrt{15n}$
e. $\frac{\sqrt{10}}{5}$ **3.** quotient, square roots, square roots, quotient
4. a. $\frac{6}{5}$ **b.** $\frac{\sqrt{37}}{11}$ **c.** $\frac{\sqrt{15}}{3}$ **d.** $3x$

Chapter Review (page 625)

1. 6, 36 **2.** 5 **3.** 7 **4.** -12 **5.** $\frac{4}{9}$ **6.** 0.8 **7.** 1 **8.** 4.583 **9.** -3.873 **10.** 5.292 **11.** not a real number; irrational; rational; irrational **12.** x **13.** $2b^2$ **14.** $-y^6$ **15.** 30 mph **16.** 10 **17.** 15 **18.** 5 ft **19.** 5 **20.** $\sqrt{194} \approx 13.93$ **21.** $4\sqrt{2}$ **22.** $x^2\sqrt{x}$ **23.** $4x\sqrt{5}$ **24.** $-6\sqrt{7}$ **25.** $5t\sqrt{10t}$ **26.** $10x\sqrt{2y}$ **27.** $\frac{4}{5}$ **28.** $\frac{2\sqrt{15}}{7}$ **29.** $\frac{11x\sqrt{2}}{13}$ **30.** $2\sqrt{10}$ ft; 6.3 ft **31.** $2\sqrt{10}$ **32.** $5\sqrt{x}$ **33.** 0 **34.** $-3 + 4\sqrt{3}$ **35.** $\sqrt{7}$ **36.** $y^2\sqrt{5xy}$ **37.** The radicands are different. **38.** $13\sqrt{5}$ in. **39.** $\sqrt{6}$ **40.** 125 **41.** $36x^2\sqrt{2x}$ **42.** $15 + 6x\sqrt{15} + 9x^2$ **43.** -2 **44.** -2 **45.** x **46.** $e - 1$ **47.** $30\sqrt{2}$ in.^{2}; 42.4 in.^{2} **48.** $\frac{\sqrt{7}}{7}$ **49.** $\frac{\sqrt{3a}}{a}$ **50.** $\frac{3x}{5}$ **51.** $7\sqrt{2} - 7$ **52.** 81 **53.** 45 **54.** no solution **55.** -3 **56.** -2 **57.** $-2, 4$ **58.** 64 ft **59.** 5^3, cube **60.** -3 **61.** -5 **62.** 3 **63.** 2 **64.** 0 **65.** -1 **66.** $\frac{1}{4}$ **67.** 1 **68.** x **69.** $3y^2$ **70.** $2a^3$ **71.** b^4 **72.** $3\sqrt[3]{2}$ **73.** $2\sqrt[4]{5}$ **74.** $\frac{2\sqrt[3]{7}}{5}$ **75.** 7 **76.** -10 **77.** 216 **78.** $\frac{4}{9}$ **79.** $\frac{1}{8}$ **80.** -243 **81.** $\frac{1}{64}$ of the original dose **82.** There is no real number that, when squared, gives -4.

Chapter 8 Test (page 629)

1. 10 **2.** $-\frac{20}{3}$ **3.** 0.5 **4.** 1 **5.** 7 amps **6.** 10 ft **7.** irrational, not a real number, rational **8.** $2x$ **9.** $3x\sqrt{6x}$ **10.** $\frac{5\sqrt{2}}{7}$ **11.** $3y\sqrt{x}$ **12.** $3\sqrt{2}$ ft **13.** $5\sqrt{3}$ **14.** $-24x^2\sqrt{6x}$ **15.** $2\sqrt{6} + 3\sqrt{2}$ **16.** $-x\sqrt{2x}$ **17.** -1 **18.** $12t$ **19.** $5x - 21\sqrt{x} + 4$ **20.** $x + 1$ **21.** $(6\sqrt{2} + 2\sqrt{10})$ in. **22.** $\sqrt{2}$ **23.** $\frac{3 + \sqrt{3}}{2}$ **24.** 225 **25.** -62 **26.** 5 **27.** 2, 1 **28.** No; when 0 is substituted for x, the result is not true. **29.** There is no real number that, when squared, gives -9. **30.** the Pythagorean theorem; 5 ft **31.** 5 **32.** -2 **33.** $\frac{1}{4}$ **34.** x **35.** $2\sqrt[3]{11}$ **36.** 4 **37.** 4 **38.** $\frac{1}{3}$

Cumulative Review Exercises Chapters 1–8 (page 630)

1. a. true **b.** false **c.** true **2.** -14 **3.** 17 lb **4.** $-2p - 6z$ **5.** -2 **6.** $x \le -\frac{3}{4}$, $\left(-\infty, -\frac{3}{4}\right]$,

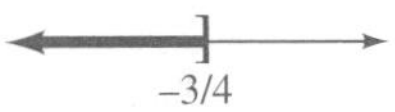

7. 4 in. **8.** 3.5 hr **9.** 80 **10.** $x - 2y = -12$ **11. a.** 3 **b.** $-\frac{2}{3}$

12.

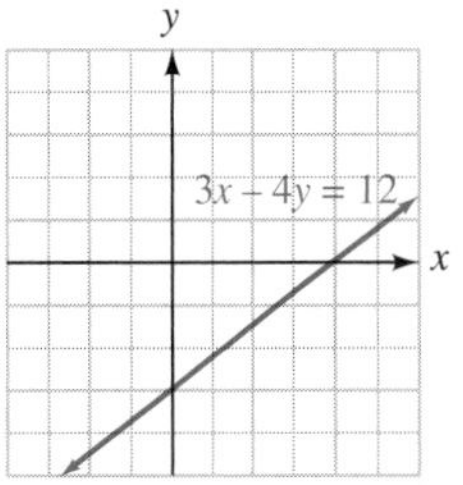

13.

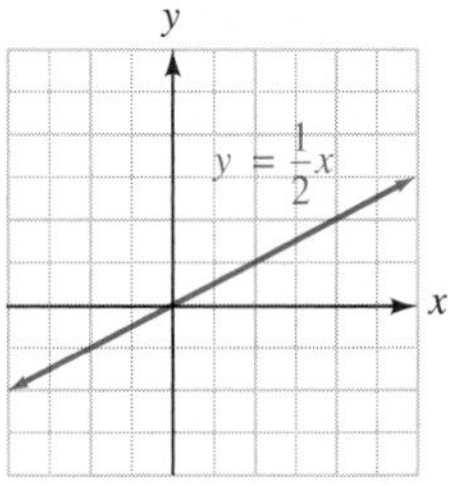

14.

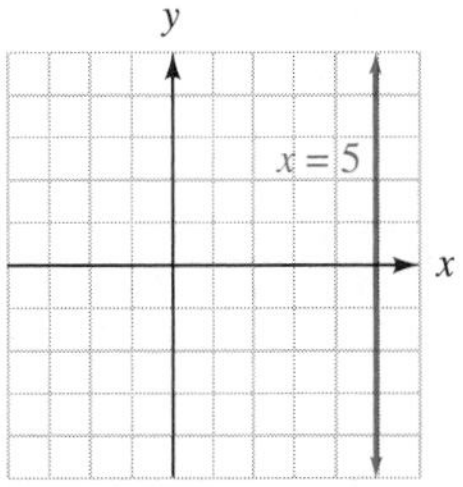

15.

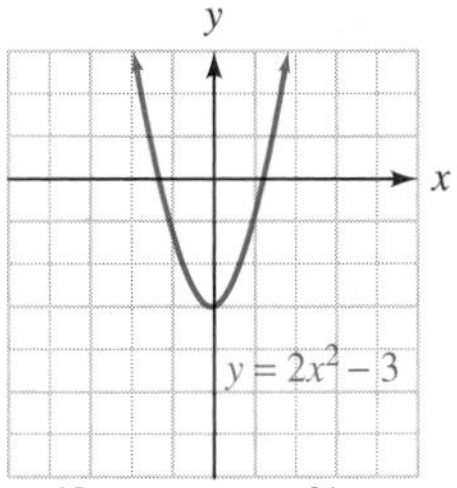

16. \$52 billion per yr **17.** 4.8×10^{18} m **18.** x^{31} **19.** $\frac{a^{15}b^5}{c^{20}}$ **20.** 1 **21.** $\frac{a^8}{b^{12}}$ **22.** $-2r^7s^3t^4$ **23.** $-6t^2 + 13st - 6s^2$ **24.** $2a^2 + a - 3$ **25.** $2x - 1$ **26.** $3xy(x - 2y)$ **27.** $(x + y)(2x - 3)$ **28.** $(a + 2b)(a^2 - 2ab + 4b^2)$ **29.** $(3a + 4)(2a - 5)$ **30.** $-1, -2$ **31.** 0, 2 **32.** $\frac{2}{3}, -\frac{1}{2}$ **33.** $-5, 5$ **34.** 5 cm **35.** $\frac{x + 1}{x - 1}$ **36.** $\frac{(p + 2)(p - 3)}{3(p + 3)}$ **37.** $\frac{7x + 29}{(x + 5)(x + 7)}$ **38.** $\frac{9a^2 - 4b^2}{6ab}$ **39.** $\frac{y + x}{y - x}$ **40.** 2 **41.** $2\frac{2}{9}$ hr **42.** 6,480 **43.** $(3, -1)$ **44.** $(-1, 2)$ **45.** 6%: \$3,000; 12%: \$3,000
46.

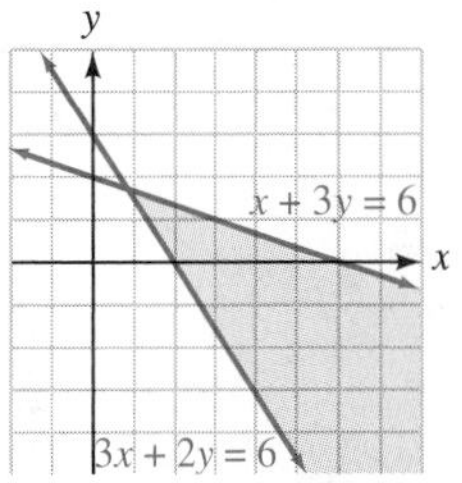

47. 73 in. **48.** $-48x^2y\sqrt{y}$ **49.** $7\sqrt{3} - 6\sqrt{2}$ **50.** $y - 9\sqrt{y} + 20$ **51.** $\frac{2\sqrt{5}}{5}$ **52.** 3 **53.** 5 **54.** 64

Study Set Section 9.1 (page 640)

1. quadratic **3.** coefficient **5.** lead **7.** $\sqrt{5}$ **9. a.** $x^2 = 23$ **b.** $x^2 - 4x = 1$ **11.** $x^2 + \frac{5}{4}x - 4 = 0$ **13. a.** 9 **b.** 36 **c.** $\frac{9}{4}$ **d.** $\frac{25}{4}$ **15. a.** $\frac{29}{4}$ **b.** $\frac{17}{4}$ **17.** $x = \pm\sqrt{6}$ **19.** ± 6 **21.** ± 3 **23.** $\pm 2\sqrt{5}$ **25.** no real solution **27.** ± 3 **29.** ± 2 **31.** $\pm\frac{3}{4}$ **33.** $-6, 4$ **35.** $7, -11$ **37.** $2 \pm 2\sqrt{2}$ **39.** $7 \pm \sqrt{3}$ **41.** ± 3.74 **43.** $-2.35, -7.65$ **45.** $x^2 + 2x + 1 = (x + 1)^2$ **47.** $x^2 - 4x + 4 = (x - 2)^2$ **49.** $x^2 - 7x + \frac{49}{4} = \left(x - \frac{7}{2}\right)^2$ **51.** $b^2 - \frac{2}{3}b + \frac{1}{9} = \left(b - \frac{1}{3}\right)^2$ **53.** 2, 6 **55.** $-2, -4$ **57.** $1, -6$ **59.** $-2 \pm \sqrt{3}$ **61.** $-4 \pm \sqrt{10}$ **63.** $1 \pm 3\sqrt{2}$ **65.** $\frac{7}{2} \pm \frac{\sqrt{69}}{2}$ **67.** $2 \pm \sqrt{7}$ **69.** $\frac{1}{2} \pm \frac{\sqrt{13}}{2}$ **71.** $-1, -2$ **73.** $-\frac{1}{2} \pm \sqrt{5}$ **75.** $2, -\frac{1}{2}$ **77.** $-\frac{5}{4} \pm \frac{\sqrt{57}}{4}$ **79.** $3.24, -1.24$ **81.** $9.52, -0.52$ **83.** $2.15, -0.15$ **85.** 24,941 mi/hr **87.** 20 ft by 40 ft **93.** 2 **95.** 16

Study Set Section 9.2 (page 650)

1. quadratic **3.** solve **5. a.** $x^2 + 2x + 5 = 0$ **b.** $3x^2 + 2x - 1 = 0$ **7.** 1, −2, 4 **9.** 5 **11.** $\frac{2 \pm \sqrt{22}}{2}$
13.

$\frac{-1-\sqrt{5}}{2}$ $\frac{-1+\sqrt{5}}{2}$
−3 −2 −1 0 1 2 3

15. $b(b + 5) = 160$ **17.** −5, −6, 1, 24, 49, 5, 7, 6, 7, −1 **19.** opposite (negative), minus, square, squared, times, over **21.** 2, 3 **23.** −3, −4 **25.** $1, -\frac{1}{2}$ **27.** $\frac{3}{2}, \frac{2}{3}$ **29.** $\frac{1}{4}, -1$ **31.** $\frac{-3 \pm \sqrt{5}}{2}$ **33.** $\frac{1 \pm \sqrt{37}}{6}$ **35.** no real solutions **37.** $-1 \pm \sqrt{2}$ **39.** $\frac{3 \pm \sqrt{15}}{3}$ **41.** $-2 \pm \sqrt{7}$ **43.** $\frac{-5 \pm \sqrt{3}}{2}$ **45.** no real solutions **47.** $\frac{-4 \pm \sqrt{6}}{2}$ **49.** −2, 3 **51.** $\frac{-1 \pm \sqrt{41}}{4}$; −1.85, 1.35 **53.** $1 \pm \sqrt{2}$; −0.41; 2.41 **55.** −5, 7 **57.** no real solutions **59.** −4, 0 **61.** $\pm\sqrt{6}$; ±2.45 **63.** 0.8, 3.1 **65.** 6 in. **67.** 47 ft by 66 ft **69.** 3.5 sec **71.** 30 in., 40 in. **73.** 7% **79.** $r = \frac{A - p}{pt}$ **81.** $r = \frac{r_1 r_2}{r_2 + r_1}$

Study Set Section 9.3 (page 659)

1. complex **3.** conjugates **5. a.** $\sqrt{-1}$ **b.** −1 **7.** $-1, \sqrt{-1}, i$ **9.** 2, −1 **11.** $\frac{6+i}{6+i}$ **13. a.** true **b.** true **c.** false **d.** true **15. a.** $2 + 9i$ **b.** $-8 - i$ **c.** $0 - 4i$ **d.** $0 + 11i$ **17. a.** $i\sqrt{7}$ **b.** $2i\sqrt{3}$ **19.** $3i$ **21.** $i\sqrt{7}$ or $\sqrt{7}i$ **23.** $2i\sqrt{6}$ or $2\sqrt{6}i$ **25.** $-2i\sqrt{6}$ or $-2\sqrt{6}i$ **27.** $45i$ **29.** $\frac{5}{3}i$ **31.** $0 \pm 3i$ **33.** $0 \pm 6i$ **35.** $0 \pm 2i\sqrt{2}$ **37.** $0 \pm \frac{4}{3}i$ **39.** $\frac{3}{2} \pm \frac{\sqrt{7}}{2}i$ **41.** $-\frac{1}{4} \pm \frac{\sqrt{7}}{4}i$ **43.** $-1 \pm i$ **45.** $-\frac{1}{3} \pm \frac{\sqrt{2}}{3}i$ **47.** $\frac{1}{3} \pm \frac{2\sqrt{2}}{3}i$ **49.** $8 - 2i$ **51.** $3 - 5i$ **53.** $15 + 2i$ **55.** $0 + i$ **57.** $14 - 13i$ **59.** $12 - 9i$ **61.** $6 - 3i$ **63.** $-12 - 16i$ **65.** $-25 - 25i$ **67.** $7 + i$ **69.** $12 + 5i$ **71.** $13 - i$ **73.** $3 + 4i$ **75.** $2 + i$ **77.** $\frac{15}{26} - \frac{3}{26}i$ **79.** $\frac{1}{4} + \frac{3}{4}i$ **81.** $-\frac{5}{13} + \frac{12}{13}i$ **83.** $\frac{11}{10} + \frac{3}{10}i$ **85.** $18.45 - 2.18i$ **91.** $\frac{4x^2 + 19x + 42}{(x + 3)(x + 6)}$ **93.** $x - 2$

Study Set Section 9.4 (page 668)

1. quadratic **3.** vertex **5.** intercept **7. a.** < **b.** 0 **9. a.** 0 **b.** x **11. a.** parabola **b.** (1, 0), (3, 0) **c.** (0, −3) **d.** (2, 1) **13.** 2 **17. a.** 2, 4, −8 **b.** −1 **19.** two; −3, 1 **21.** upward **23.** downward **25.** (1, −1) **27.** (3, 1) **29.** (1, 0); (0, 1) **31.** (−3, 0), (−7, 0); (0, −21) **33.**

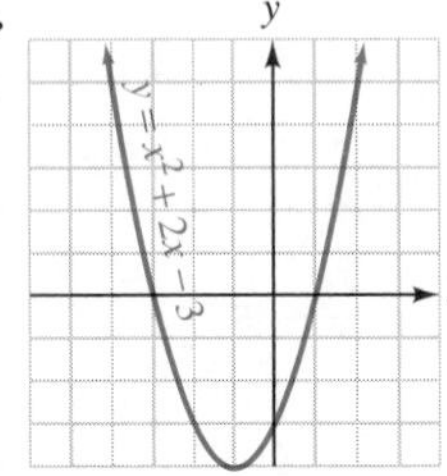

35.

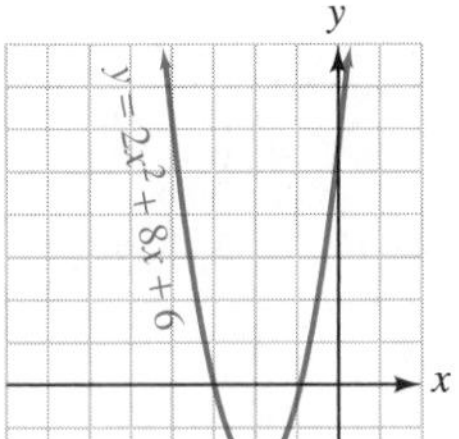

37.

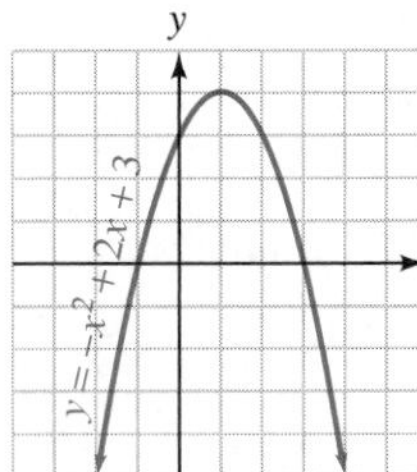

39.

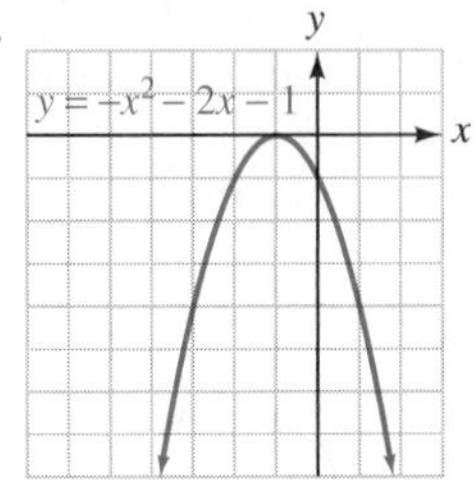

41.

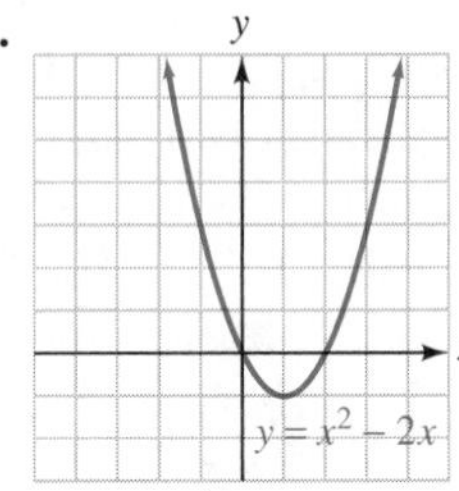

43.

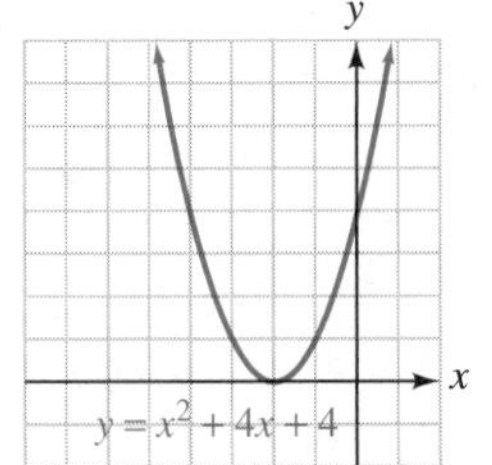

45.

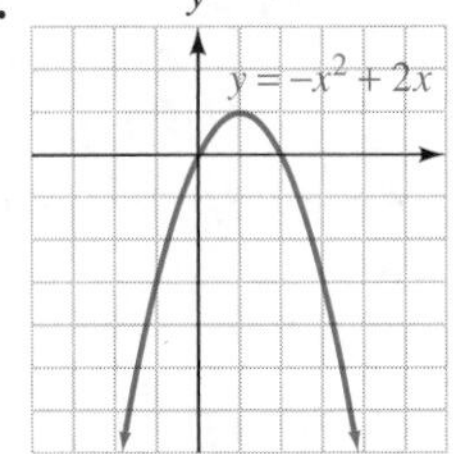

47.

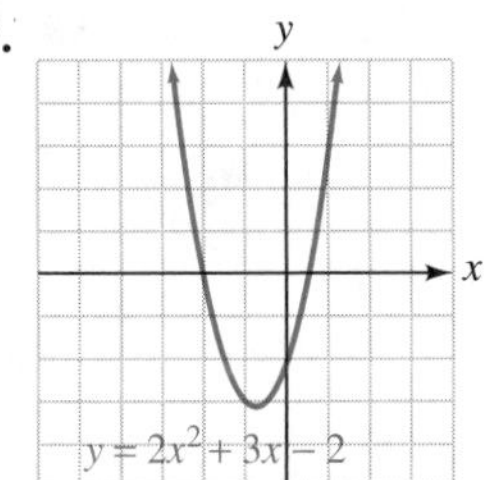

49.

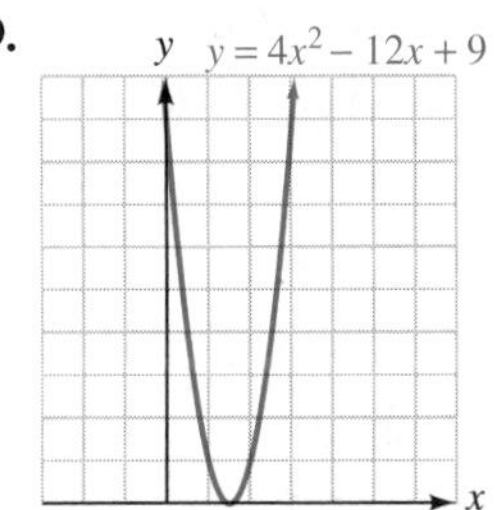

53. The cost to manufacture a carburetor is lowest ($100) for a production run of 30 units. **55. a.** 14 ft **b.** 0.25 sec and 1.75 sec **c.** 18 ft; 1.0 sec **57.** 32, 18, 8, 2, 0, 2, 8, 18, 32 **63.** $3\sqrt{2}$ **65.** $3z$

Study Set Section 9.5 (page 678)

1. function **3.** input, output **5.** linear
7.

Domain	Range
1990	3.80
1992	4.25
1994	4.25
1996	4.75
1998	5.15
2000	5.15
2002	5.15

9. 33 **11.** vertical

13. of **15.** 4, 5, 4, 5 **17.** yes **19.** no; (4, 2), (4, 4), (4, 6) **21.** yes **23.** no; (−1, 0), (−1, 2) **25.** no; (3, 4), (3, −4) or (4, 3), (4, −3) **27. a.** 3 **b.** −9 **c.** 0 **d.** 199 **29. a.** 0.32 **b.** 18 **c.** 2,000,000 **d.** $\frac{1}{32}$ **31. a.** 7 **b.** 14 **c.** 0 **d.** 1 **33. a.** 0 **b.** 990 **c.** −24 **d.** 210 **35.** 1.166 **37.** yes **39.** no; (3, 4), (3, −1)

(answers may vary) **41.** no; (0, 2), (0, −4) (answers may vary)
43. −2, −5, 1, 4 **45.**

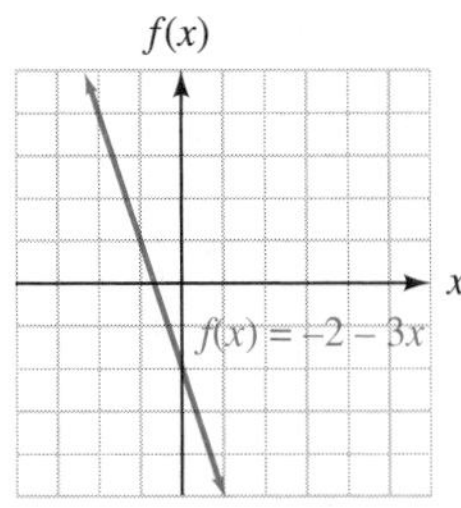

47.

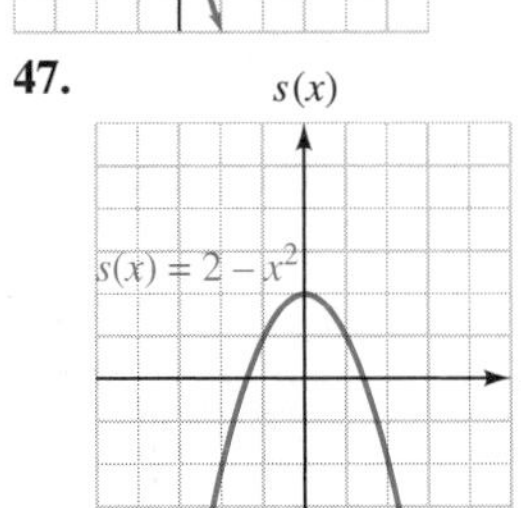

49. $f(x) = |x|$
51. $D(5) = 1$; If it is 5 seconds between seeing the lightning and hearing the thunder, the lightning strike was 1 mile away.
53. 78.5 ft^2, 1,256.6 ft^2
59. 80 lb

Study Set Section 9.6 (page 689)

1. direct **3.** constant **5.** joint **7.** direct **9.** inverse
11.

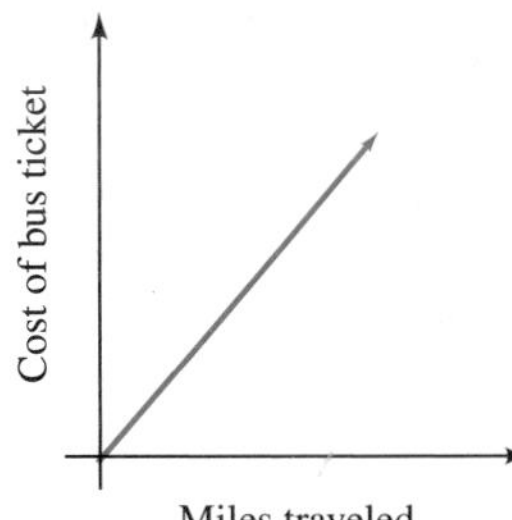

13. $d = kn$ **15.** $l = \frac{k}{w}$

17. $A = kr^2$ **19.** $d = kst$ **21.** $I = \frac{kV}{R}$ **23. a.** yes **b.** no **c.** no **d.** yes **25.** 35 **27.** 42 **29.** 675 **31.** 1 **33.** $\frac{80}{3}$ **35.** 24 **37.** 4 **39.** 80 **41.** $\frac{5}{2}$ **43.** 168 mi **45.** 27 mg **47.** $22\frac{1}{2}$ **49.** $2\frac{1}{2}$ hr **51.** 8 amps **53.** $53\frac{1}{3}$ m^3 **55.** \$270 **57.** $3\frac{9}{11}$ amp **63.** −1, 6 **65.** 2, −4 **67.** 0, 5, −5

Key Concept (page 693)

1. $0, \frac{1}{4}$ **2.** 8, −7 **3.** $\pm 3\sqrt{3}$ **4.** 1, −7 **5.** $-2 \pm \sqrt{3}$ **6.** $\frac{-3 \pm \sqrt{5}}{2}$ **7.** A ± symbol should be written in front of $\sqrt{20}$ and $2\sqrt{5}$; $\pm 2\sqrt{5}$ **8.** Write $x^2 - 3x = 10$ as $x^2 - 3x - 10 = 0$; 5, −2 **9.** 9 wasn't added to both sides; $-3 \pm \sqrt{19}$ **10.** The fraction bar should be extended; $x = \frac{5 \pm \sqrt{21}}{2}$

Chapter Review (page 694)

1. ±5 **2.** $\pm 2\sqrt{2}$ **3.** $\pm 5\sqrt{3}$ **4.** −4, 6 **5.** $\frac{7}{2}, \frac{1}{2}$ **6.** $8 \pm 2\sqrt{10}$ **7.** ±3.46 **8.** −6.42, 8.42 **9.** $x^2 + 4x + 4 = (x + 2)^2$ **10.** $t^2 - 5t + \frac{25}{4} = \left(t - \frac{5}{2}\right)^2$ **11.** 3, 5 **12.** 2, −7 **13.** $\frac{1}{2}, -3$ **14.** $\frac{1 \pm \sqrt{3}}{2}$ **15.** −0.27, −3.73 **16.** 3.0 ft **17.** $x^2 + 2x + 5 = 0$; 1, 2, 5 **18.** $6x^2 - 2x - 1 = 0$; 6, −2, −1 **19.** 5, −3 **20.** 7, −1 **21.** $\frac{3}{2}, -\frac{1}{3}$ **22.** $3 \pm \sqrt{2}$ **23.** $\frac{-1 \pm \sqrt{7}}{3}$; −1.22, 0.55 **24.** 10 ft, 24 ft **25.** 15 sec **26.** $-3 \pm \sqrt{7}$ **27.** 1, −7 **28.** 0, −5 **29.** $\frac{-1 \pm \sqrt{41}}{4}$ **30.** $\pm 2\sqrt{5}$ **31.** 2, 2 **32.** no real solutions **33.** $5i$ **34.** $3i\sqrt{2}$ **35.** $-7i$ **36.** $\frac{3}{8}i$ **37.** Real numbers, Imaginary numbers **38. a.** true **b.** true **c.** false **d.** false **39.** $0 \pm 3i$ **40.** $0 \pm \frac{4\sqrt{3}}{3}i$ **41.** $\frac{3}{4} \pm \frac{\sqrt{7}}{4}i$ **42.** $-1 \pm i$ **43.** $3 - 6i$ **44.** $-1 + 7i$ **45.** $0 - 19i$ **46.** $0 + i$ **47.** $8 - 2i$ **48.** $3 - 5i$ **49.** $3 + 6i$ **50.** $9 + 7i$ **51.** $\frac{15}{26} - \frac{3}{26}i$ **52.** $-\frac{5}{13} + \frac{12}{13}i$ **53. a.** (−3, 0), (1, 0) **b.** (0, −3) **c.** (−1, −4) **54.** The maximum profit of \$16,000 is obtained from the sale of 400 units. **55.** (1, 5); upward **56.** (3, 16); downward **57.** (−5, 0), (−1, 0); (0, 5)
58. **59.**

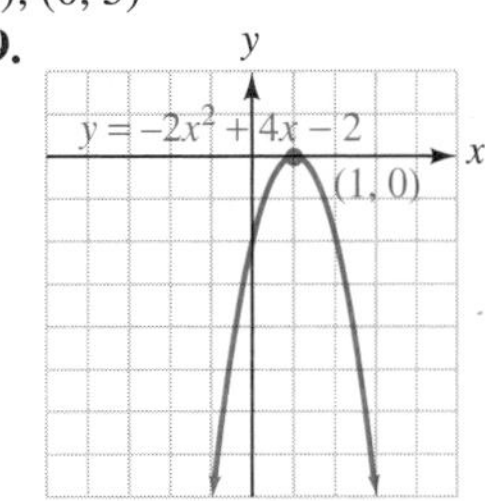

60. a. 2, −2, 1 **b.** 1, −3 **c.** no, none **61.** yes **62.** yes **63.** yes **64.** no; (−1, 2), (−1, 4) **65.** 4, Domain, Input, Output, Range **66.** $f(x)$ **67.** −3 **68.** 0 **69.** 21 **70.** $-\frac{7}{4}$ **71.** −5 **72.** 37 **73.** −2 **74.** −8 **75.** no; (1, 0.5), (1, 4), (1, −4.5) **76.** yes **77.** 1, 0, −1, 0, −1, −2

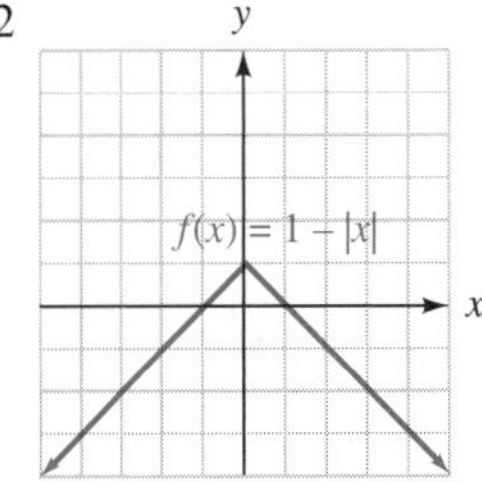

78. 1,004.8 $in.^3$ **79.** \$833.33 **80.** 600 **81.** 1.25 amps **82.** inverse variation **83.** 81 **84.** $\frac{5}{7}$

Chapter 9 Test (page 701)

1. $\pm\sqrt{17}$ **2.** $\pm 2\sqrt{6}$ **3.** $\pm\frac{5}{2}$ **4.** $2 \pm \sqrt{3}$ **5.** 40 cm **6.** 49 **7.** $-1 \pm \sqrt{5}$; −3.24, 1.24 **8.** $2, -\frac{1}{2}$ **9.** $-\frac{3}{2}, 4$ **10.** $2 \pm \sqrt{2}$ **11.** $\frac{1 \pm \sqrt{13}}{6}$; −0.43, 0.77 **12.** 255 ft, 505 ft **13.** $10i$ **14.** $-3i\sqrt{2}$ **15.** A complex number is any number that can be written in the form $a + bi$, where a and b are real numbers and $i = \sqrt{-1}$; $2 + 8i$ **16.** $-\frac{1}{3} \pm \frac{\sqrt{2}}{3}i$ **17.** $1 + i$ **18.** $-1 + 12i$ **19.** $14 - 8i$ **20.** $\frac{5}{13} - \frac{12}{13}i$ **21.** The most air conditioners sold in a week (18) occurred when 3 ads were run.

22.

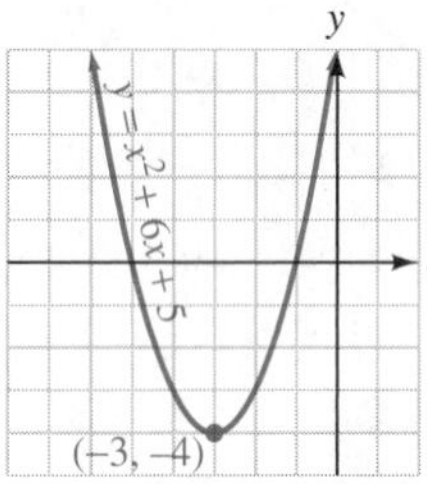

23. The set of all possible values that can be used for the independent variable (inputs). The set of all values of the dependent variable (outputs). **24.** yes **25.** no; (2, 3.5), (2, −3.5) **26.** yes **27.** −13 **28.** 756

29. $C(45) = 28.50$; It costs \$28.50 to make 45 calls.

30.

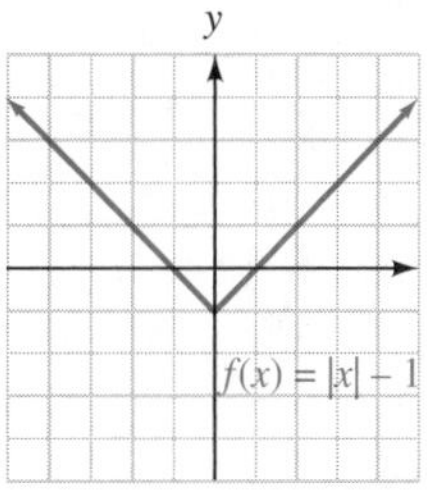

31. 100 lb **32.** $\frac{80}{3}$

Cumulative Review Exercises Chapters 1–9 (page 703)

1. a. true **b.** true **c.** true **2.** 36 **3.** 2 hours before salt is spread **4.** about 98,113 **5.** $-2p - 6z$ **6.** 12 **7.** $-\frac{3}{2}$ **8.** $x < -2$; $(-\infty, -2)$

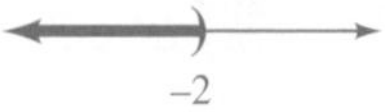

9. \$8,000 at 7%, \$20,000 at 10% **10.** 65 **11.** 4 **12.** $\frac{3}{5}$

13.

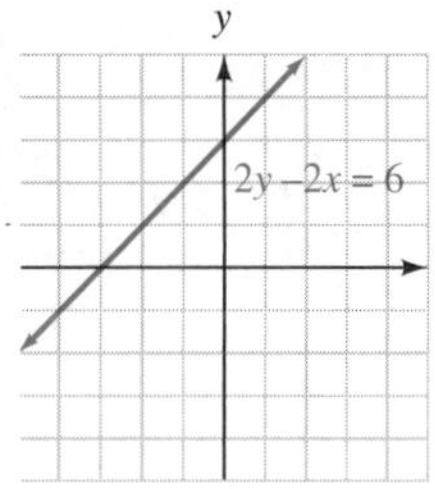

14.

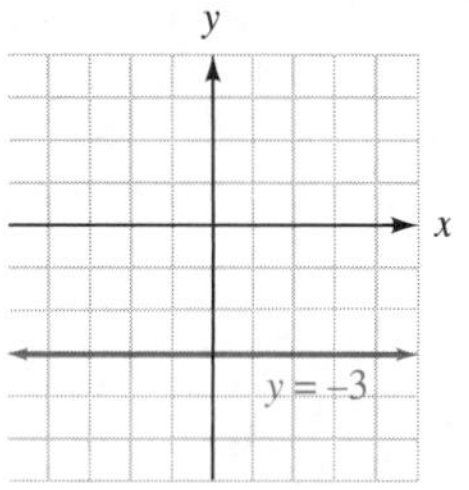

15.

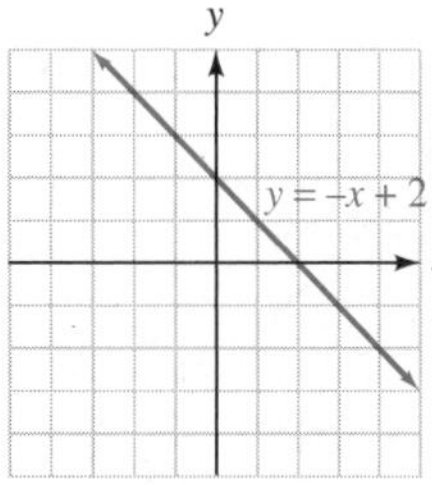

16.

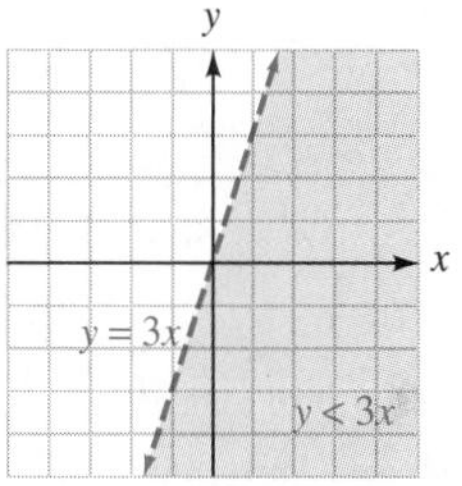

17.

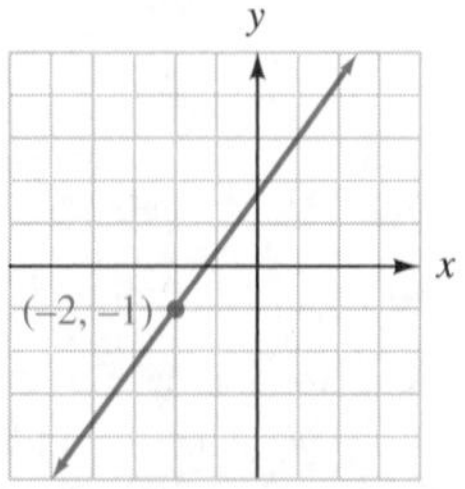

18.

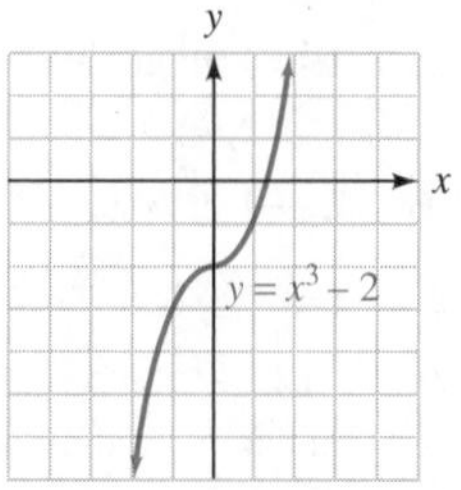

19. $y = -2x + 1$ **20.** $y = \frac{1}{4}x - 1$ **21.** $-\frac{4}{5}$

22. an increase of \$16 per yr **23.** y^9 **24.** $\frac{b^6}{27a^3}$ **25.** 2 **26.** $\frac{x^{10}z^4}{9y^6}$ **27.** 2,600,000 to 1 **28.** 7.3×10^{-4}

29. $6x^3 + 8x^2 + 3x - 72$ **30.** $-6a^5$ **31.** $6b^2 + 5b - 4$ **32.** $4x^2 + 20xy + 25y^2$ **33.** $6x^3 - 7x^2y + y^3$ **34.** $2x + 1$ **35.** $6a(a - 2a^2b + 6b)$ **36.** $(x + y)(2 + a)$ **37.** $(b + 5)(b^2 - 5b + 25)$ **38.** $(t^2 + 4)(t + 2)(t - 2)$ **39.** $0, -\frac{8}{3}$ **40.** $\frac{2}{3}, -\frac{1}{5}$ **41.** $6x^3 + 4x$ **42.** 5 in. **43.** −8 **44.** $\frac{3(x + 3)}{x + 6}$ **45.** $\frac{x - 3}{x - 5}$ **46.** $\frac{x^2 + 4x + 1}{x^2y}$ **47.** $\frac{x^2 + 5x}{x^2 - 4}$ **48.** $\frac{9y + 5}{4 - y}$ **49.** 3 **50.** 1 **51.** $a = \frac{b}{b - 1}$ **52.** $2\frac{6}{11}$ days **53.** 875 days **54.** 39

55.

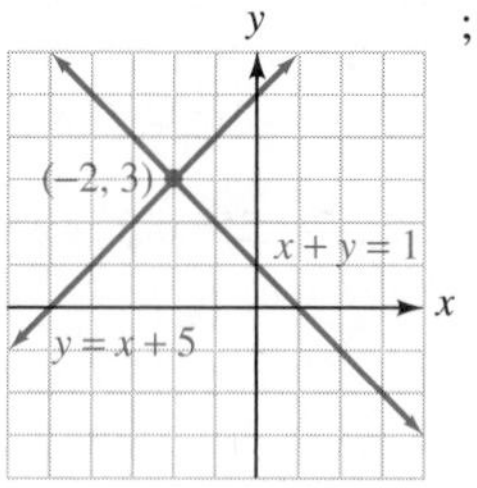

; (−2, 3) **56.** (−3, −1)

57. (−1, 2) **58.** 550 mph **59.** 32 lb of hard candy, 16 lb soft candy **60.**

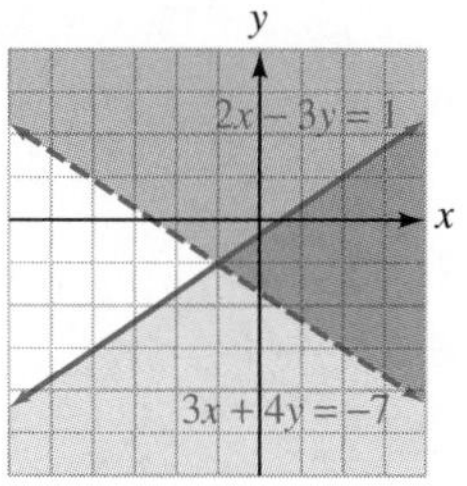

61. $5x\sqrt{2}$

62. $-10a^3b^2$ **63.** $9\sqrt{6}$ **64.** $-1 - 2\sqrt{2}$ **65.** $\frac{6x\sqrt{2x}}{y}$ **66.** $\frac{4\sqrt{10}}{5}$ **67.** $\frac{3m}{2n^2}$ **68.** 2 **69.** 125 **70.** $\frac{1}{16}$ **71.** 4 **72.** −2, −6 **73.** $\pm 5\sqrt{3}$ **74.** 21.2 in. **75.** $\frac{1 \pm \sqrt{33}}{8}$; −0.59, 0.84 **76.** 83 ft × 155 ft **77.** $7i$ **78.** $3i\sqrt{6}$ **79.** $1 + 5i$ **80.** $16 - 2i$ **81.** $6 - 17i$ **82.** $1 - i$ **83.** $0 \pm 4i$ **84.** $2 \pm i$ **85.**

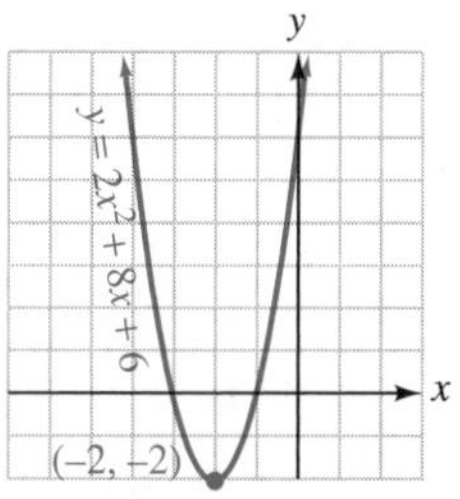

86. 4,000 rpm **87.** −8 **88.** input **89.** yes **90.** 1.2

Appendix I (page A-4)

1. 8 **3.** 9 **5.** 12 **7.** 28, 27, 27 **9.** −2 yd **11.** 74.5 centimicrons **13.** 84 **15.** 5 lb, 4 lb **17.** median: 29.5, mode: 29

Index

A

Absolute value, 27
Absolute value function, 676
Absolute zero, 216
Adams, John, 420
Adding
  complex numbers, 657
  fractions, 15
  polynomials, 311
  rational expressions, 441, 450
  real numbers, 32–34
Addition property
  of equality, 94, 530
  of inequalities, 161
  of opposites, 36
  of zero, 36
Additive inverses, 36
Algebraic expression(s), 5, 69
  evaluating, 73
Angle(s), 99
  complementary, 156, 541
  right, 99
  straight, 99
  supplementary, 156, 541
Area, 136
Arithmetic mean, 64
Arrow diagram, 672
Associative property
  of addition, 35
  of multiplication, 50
Axis(es)
  horizontal, 2
  vertical, 2
  x-, 185
  y-, 185
Axis of symmetry, 663

B

Bar graph(s), 2, 184
Base of an exponent, 58, 275
Base angles of an isosceles triangle, 147
Bimodal, A-4
Binomial(s), 301
  conjugate, 604
  multiplying the sum and difference of two terms, 328
  squaring a, 327
Blue Boy, 551
Boundary line, 253
Braces, 22
Bracket, 161
Break point, 544
Building fractions, 12
Building rational expressions, 446

C

Cartesian coordinate system, 184
Center of a circle, 137
Circle(s), 137
  center of, 137
  circumference of, 137
  diameter of, 137
  radius of, 137
Circle graph(s), 108, 184
Circumference, 137
Cleveland, Grover, 420, 421
Closed circle, 161
Coefficient, 69, 300
Combined variation, 688
Combining like terms, 120, 310
Commutative property
  of addition, 35
  of multiplication. 49
Complementary angles, 156, 541
Completing the square, 637, 640
Complex conjugates, 658
Complex fractions, 459
  simplifying, 459, 462
Complex numbers, 655, 657
  adding, 657
  dividing, 658
  imaginary part, 655
  multiplying, 657
  real part, 655
  subtracting, 657
Complex number system, 654
Complex rational expressions, 459
Composite numbers, 10
Compound inequalities, 165
Compound interest, 134
Conjugate binomials, 604
Conjugates, 604
Consecutive even integers, 405
Consecutive integers, 155, 405
Consecutive odd integers, 405
Consistent system, 514
Constant of variation, 682, 685, 687
Constant term, 69, 301
Contradictions, 129
Coordinate plane, 185
Coordinates
  x-, 185
  y-, 185
Cross products, 488
Cube, 615
Cube roots, 615
Cube root notation, 615

D

Decimal(s)
  repeating, 23
  terminating, 23
Degree
  of a polynomial, 301
  of a term, 301
Denominator(s), 10
  rationalizing, 602, 603
Dependent equations, 515
Dependent variable, 672
Descartes, René, 184
Diameter, 137
Difference
  of two cubes, 391
  of two squares, 386
Direct variation, 682
Discriminant, 693
Distance formula, 580
Distance problems, 135
Distributive property, 116, 117
Dividing
  a polynomial by a monomial, 334
  a polynomial by a polynomial, 335
  complex numbers, 658
  fractions, 12
  rational expressions, 435
  real numbers, 51–52
Division property(ies), 53
  of equality, 98
  of inequalities, 163
Domain of a function, 672

E

Elements of a set, 22
Equation(s), 4, 93
  contradiction, 129
  dependent, 515
  equivalent, 94
  general quadratic, 643
  identity, 129
  independent, 515

in one variable, 195
in two variables, 195
left-hand side, 93
linear, 198
linear in one variable, 124
of horizontal lines, 211
of vertical lines, 211
quadratic, 400, 634
radical, 607
rational, 468
right-hand side, 93
solutions of, 93
solving quadratic, 402
solving radical, 608
standard form of, 400
system of, 512
Equivalent equations, 94
Equivalent fractions, 12
Evaluating
algebraic expressions, 73
polynomials, 302
rational expressions, 423
Even root, 616
Exponent(s), 58, 275
base of, 58, 275
natural-number, 275
negative, 286
power rule for, 278
product rule for, 276
quotient rule for, 277
rational, 618
rules of, 288
zero, 285
Exponential expression(s), 59, 275
Exponential notation, 58
Expression(s), 5
algebraic, 69
exponential, 59, 275
rational, 422
Extraneous solution(s), 471, 610
Extrapolation, 188
Extremes of a proportion, 488

F

Factor, 9
Factoring
by grouping, 358
out the GCF, 356
perfect square trinomials, 385
the difference of two cubes, 392
the difference of two squares, 386
the sum of two cubes, 392
trinomials, 374
Flagg, James Montgomery, 542
FOIL method, 320
Formula(s), 5, 133, 472
distance, 580
quadratic, 644
Fraction(s)
adding, 15
building, 12
complex, 459
denominator of, 10
dividing, 12
equivalent, 12
improper, 18
least common denominator (LCD), 16
lowest common denominator (LCD), 16
lowest terms of a, 13
multiplying, 11
negative, 23
numerator of, 10
reciprocal of, 12
simplest form of a, 13
simplifying, 13, 15
special forms, 11
subtracting, 15
Fraction bar, 10
Function(s), 672
absolute value, 676
domain of, 672
linear, 676
range of, 672
Function notation, 673
Function value, 673

G

Gainsborough, Thomas, 551
General quadratic equation, 643
Generation time, 292
Grade of an incline, 226
Graph(s)
of a number, 27
bar, 2
circle, 108
line, 2
of a function, 675
of an equation, 198
of quadratic equations, 663
Graphing
an inequality, 160
compound inequalities, 165
horizontal lines, 210
linear equations, 198
linear inequalities, 253, 257
points, 185
polynomials, 303
quadratic equations, 666
vertical lines, 210
Graphing calculators, 212
Greatest common factor (GCF), 354
Grouping symbols, 62

H

Half-life, 475
Half-planes, 253
Harrison, Benjamin, 421
Horizontal axis, 2
Hypotenuse, 407

I

Identity, 129
Identity element for addition, 36
Ihm Al-Jabr wa'l muqabalah, 2
Imaginary number(s), 576, 654
Improper fractions, 18
Inconsistent system, 514
Independent equations, 515
Independent variable, 672
Index, 615
Inequality(ies), 159
solution of, 160
system of linear, 554
Inequality symbol(s), 26, 159
Input, 673
Integers, 22
consecutive, 155, 405
consecutive even, 405
consecutive odd, 405
Intercept method of graphing a line, 207
Intercepts, 664
x-, 206
y-, 206
Interest, 134
compound, 134
simple, 134
Interest problems, 134
Interpolation, 188
Interval notation, 160
Intervals, 160
Inverse variation, 685
Investment problems, 149
Irrational numbers, 24
Isosceles right triangle, 579
Isosceles triangle(s), 147, 579
base angles of, 147
vertex angle of, 147

J

Jefferson, Thomas, 420
Joint variation, 687

K

Key number, 368, 378

L

Lead coefficient, 301, 363, 639
Lead term of a polynomial, 301
Least common denominator (LCD), 16, 444
finding the, 441
Legs of a right triangle, 407
Like radicals, 593
Like terms, 119, 310
Linear equations, 198
in one variable, 124
Linear function, 676
Linear inequality, 252
solution of, 252
Line graph(s), 2, 184
Lowest common denominator (LCD), 16
Lowest terms of a fraction, 13

M

Mapping diagram, 672
Markup, 133
Mathematical model, 2
Mean(s), A-2
arithmetic, 64
of a proportion, 488
Median, A-3
Members of a set, 22
Mixed number, 17
Mixture problems, 151
Mode, A-4
Monomial, 301
Motion problems, 150
Multiplication property
of equality, 97
of inequalities, 163
of one, 51
of zero, 50
Multiplicative identity, 12
Multiplicative inverse(s), 51, 125
Multiplying
binomials, 320
by -1, 455

complex numbers, 657
fractions, 11
monomials, 318
polynomials, 321
polynomials by binomials, 319
rational expressions, 433
real numbers, 47–49

N

Natural-number exponents, 275
Natural number perfect squares, 585
Natural numbers, 22
Negative exponents, 286
Negative fractions, 23
Negative inequality symbol, 161
Negative numbers, 22
Negative reciprocals, 235
Negative sign, 22
Negative square root, 574
Nonlinear equations, 305
Notation
cube root, 615
exponential, 58
function, 673
interval, 160
scientific, 294
square root, 575
subscript, 219
*n*th roots of *a*, 616
Number(s)
complex, 655
composite, 10
imaginary, 576, 654
irrational, 24
mixed, 17
natural, 22
negative, 22
positive, 22
prime, 10
rational, 23, 24
real, 25
whole, 9, 22
Number line, 22
Number-value problems, 148
Numerator, 10

O

Odd root, 616
Open circle, 160
Opposite(s), 22, 36, 427, 454
of an opposite, 41
of a sum. 118
of opposites, 428
Ordered pair, 185
Order of operations, 61
Origin, 185
Output, 673

P

Parabola, 304, 661
Parallel lines, 234
Parenthesis(es), 35, 160
Pendulums, 576
Percent, 107
of decrease, 109
of increase, 109
Percent sentence, 107
Perfect cubes, 617
Perfect fourth powers, 617
Perfect square(s), 575
natural number, 585
square trinomials, 384
Perimeter, 136
Period of a pendulum, 576
Perpendicular lines, 235
Pi, 24
Pie chart, 108
Pitch, 223
Plotting points, 185
Point–slope form, 242
Polynomial(s), 300
adding, 311
dividing, 334–338
evaluating, 302
in *x*, 301
multiplying, 318–321
subtracting, 312
Positive numbers, 22
Positive sign, 22
Positive square root, 574
Power rule for exponents, 278
Powers of a product and a quotient, 279
Present value, 289
Prime-factored form, 10
Prime factorization, 10
Prime numbers, 10
Prime trinomial, 368
Principal square root, 574
Problem-solving strategy, 540
Problem solving, 103
Product, 3
Product rule
for exponents, 276
for radicals, 617
Product rule for square roots, 585, 599
Profit, 133
Property(ies)
addition of equality, 94, 530
addition of inequalities, 161
addition of opposites, 36
addition of zero, 36
associative of addition, 35
associative of multiplication, 50
commutative of addition, 35
commutative of multiplication, 49
distributive, 116, 117
division of equality, 98
division of inequalities, 163
multiplication of 1, 13, 51
multiplication of equality, 97
multiplication of inequalities, 163
multiplication of zero, 50
of addition, 35
of division, 53
opposite of a sum, 118
product for radicals, 617
quotient for radicals, 617
square root, 635
squaring of equality, 607
subtraction of equality, 96
subtraction of inequalities, 161
zero-factor, 401
Proportion(s), 487
extremes of, 488
fundamental property of, 4
means of, 488
solving, 489
term of, 488
Pure imaginary number, 655
Pythagorean theorem, 407, 577

Q

Quadrants, 185
Quadratic equation(s), 400, 634
in two variables, 662
graphing, 666
solving, 402
Quadratic formula, 644
Quotient, 3
Quotient rule
for exponents, 277
for radicals, 617
for square roots, 588, 601

R

Radical, 575, 615
Radical equations, 607
solving, 608
Radical expressions, 575
Radical symbol, 574
Radicand, 575, 615
Radius, 137
Range, 43
of a function, 672
Rate(s), 487
of change, 223
of work, 479
Ratio(s), 217, 487
Rational equations, 468
solving, 468
Rational exponents, 618
Rational expressions, 423
adding, 441, 450
building, 446
complex, 459
dividing, 435
multiplying, 433
subtracting, 441, 45
Rationalizing denom
Rational numbers, 2
Real numbers, 25
subtracting, 4
adding, 32-
dividing,
multipl
Reciproc
Rectan
Rect
Rer
R
ation, 294
elements of, 22
members of, 22

Set-builder notation, 23
Signed numbers, 32
Similar triangles, 492
Simple interest, 134
Simplest form of a fraction, 13
Simplified form of a square root, 585
Simplifying
  complex fractions, 459, 462
  fractions, 13, 15
  products, 115
  rational expressions, 423, 425
  square roots, 585
Slope–intercept form, 231
Slope of a line, 217, 219
Slopes
  of horizontal lines, 222
  of parallel lines, 234
  of perpendicular lines, 235
  undefined for vertical lines, 222
Slope triangle, 218
…s, 471
…93
…nators, 602, 603
Squa…
Squaring…
Squaring p…
Standard form…
Straight angles, …
Strategy
  for finding the GCF…
  for factoring, 395
Subscript notation, 219

Subtracting
  complex numbers, 657
  fractions, 15
  polynomials, 312
  rational expressions, 441, 450
  real numbers, 41, 42
Subtraction property
  of inequalities, 161
Sum, 3
  of two cubes, 391
Supplementary angles, 156, 541
Symbols
  for division, 4
  for multiplication, 4
  grouping, 62
  inequality, 26, 159
  negative inequality, 161
  radical, 574
System of equations, 512
  consistent, 514
  inconsistent, 514
  solving by elimination, 531
  solving by graphing, 513
  solving by substitution, 522
System of linear inequalities, 554
  solving, 555

T

Table of solutions (values), 196
Term(s)
  like, 119
  unlike, 119
Terminating decimals, 23
Term of an expression, 69
…erms, 300
…st point, 255
…eorems
  …ythagorean, 407, 577
  …sson 1: The real numbers, 1
  …son 2: Order of operations, 1
  …on 3: Translating written phrases, 92
  …n 4: Solving equations part 1, 92
  … 5: Solving equations part 2, 92
  … 6: Equations containing two
    …bles, 183
  …: Rate of change and the slope of a
    …3
  … The product rule for exponents,
  …e quotient rule for exponents,
  …gative exponents, 274
  …toring trinomials and the
    … two squares, 353
  …ng quadratic equations by
  Lesson 13: Adding and subtracting rational
    expressions, 422
  Lesson 14: Solving systems of equations by
    graphing, 510
  Lesson 15: Using square roots, 573
  Lesson 16: The quadratic formula, 633
Trial-and-check method for factoring
  trinomials, 374
Triangle(s)
  isosceles, 147
  similar, 492
Trinomial(s), 301
  perfect square, 384
  prime, 368

U

Unit conversion factor, 437
Unit price, 491
Unknown, 93
Unlike terms, 119

V

Variable(s), 4, 93
  dependent, 672
  independent, 672
Variation
  combined, 688
  direct, 682
  inverse, 685
  joint, 687
Verbal model, 3
Vertex, 662
  of a parabola, 664
Vertex angle, 147
Vertical axis, 2
Vertical line test, 676
Viewing window, 212
Volume, 137

W

Washington, George, 420
Whole numbers, 9, 22

X

x-axis, 185
x-coordinate, 185
x-intercept, 206

Y

y-axis, 185
y-coordinate, 185
y-intercept, 206

Z

Zero-factor property, 401
Zero exponents, 285